Signals and Systems

S. Palani

Signals and Systems

Second Edition

S. Palani
Professor (Retired)
National Institute of Technology
Tiruchirappalli, India

ISBN 978-3-030-75744-1 ISBN 978-3-030-75742-7 (eBook)
https://doi.org/10.1007/978-3-030-75742-7

Jointly published with ANE Books Pvt. Ltd.
The print edition is not for sale in South Asia (India, Pakistan, Sri Lanka, Bangladesh, Nepal and Bhutan) and Africa. Customers from South Asia and Africa can please order the print book from: ANE Books Pvt. Ltd.
ISBN of the Co-Publisher's edition: 978-9-383-65627-1

This Springer imprint is published by the registered company Springer Nature Switzerland AG
The registered company address is: Gewerbestrasse 11, 6330 Cham, Switzerland

Preface to Second Edition

I have ventured to bring out the second edition of the book *Signals and Systems* in its new form due to the success and wide patronage extended to the previous edition and reprints by the members of the teaching faculty and student community. The present edition as in the previous edition covers the undergraduate syllabus in *Signals and Systems* for the B.E. degree courses. A thorough revision of all the chapters in the previous edition has been undertaken. Few errors noticed in the previous edition have been removed and appropriate corrections have been made. Signal representation is a vital topic to understand the importance of the theoretical concepts in *Signals and Systems*. A large number of numerical problems have been included in Chap. 1 which describes signal representation (both continuous and discrete time signals). Similarly, the classification of systems is well explained in Chap. 2 with graphical illustration wherever possible. More number of numerical problems have been added in Chap. 4 which describes Fourier Series Analysis. Further, the properties of FS are well explained and applied in solving many FS problems by cutting short lengthy procedures. Similarly, in Chap. 6, explanation is provided for the Fourier Transform method of Analysis and for the properties of FT which are frequently used to solve numerical problems in an easier way. However, in FS and FT, conventional methods of solving the numerical problems are also retained. In Chap. 8, numerical problems using LT properties have been solved. I hope that the readers of this book would appreciate the above attempts. Since every theoretical concept is explained by a variety of numerical examples which are presented in a graded manner, the book is voluminous. I take this opportunity to thank Ane Books Pvt. Ltd and the publisher for taking up this difficult job.

Pudukkottai, India S. Palani

Preface to First Edition

The book SIGNALS AND SYSTEMS presents a comprehensive treatment of signals and linear systems for the undergraduate level study. It is a rich subject with diverse applications such as signal processing, control systems and communication systems. It provides an integrated treatment of continuous-time and discrete-time forms of signals and systems. These two forms are treated side by side. Even though continuous-time and discrete-time theory have many mathematical properties common between them, the physical processes that are modelled by continuous-time systems are very much different from the discrete-time systems counterpart.

I have written this book with the material I have collected during my long experience of teaching signals and systems to the undergraduate level students in national level reputed institutions. The book in the present form is written to meet the requirements of undergraduate syllabus of Indian Universities in general and Anna University in particular for B.E./B.Tech. degree courses. The organization of the chapters is as follows.

Chapter 1 deals with the representation of signals and systems. It motivates the reader as to what signals and systems are and how they are related to other areas such as communication systems, control systems and digital signal processing. In this chapter, various terminologies related to signals and systems are defined. Further, mathematical description, representation and classifications of signals and systems are explained.

Chapter 2 presents a detailed descriptions of system classifications. Under broader category, systems are classified as continuous-time and discrete-time systems. Each of them is further classified as linear and non-linear, time invariant and time varying, static and dynamic, causal and non-causal, stable and unstable and invertible and non-invertible. Systems are identified accordingly.

A comprehensive treatment of time domain analysis of continuous-time and discrete-time systems are given in Chapter 3. It develops convolution from the representation of an input signals as a superposition of impulses. To find the convolution of two time signals, both analytical as well as graphical methods are explained.

Chapter 4 deals with the Fourier representation of continuous-time signals. Continuous time periodic signals are represented by trigonometric Fourier series, polar Fourier series and exponential Fourier series.

In Chapter 5, discrete-time signals is represented by exponential Fourier series and their properties are derived. The Fourier spectra of discrete-time signal is also determined in this chapter.

It is not possible to find Fourier series representation of non-periodic signals. In Chapter 6, Fourier transform is introduced which can represent periodic as well as non-periodic signals. In this chapter the Fourier transform for continuous-time signal is explained.

In Chapter 7, the representation of discrete-time signal using discrete time Fourier transform is explained. Further, discrete Fourier transform and Fast Fourier Transform algorithm are also explained here. The Laplace transform is a very powerful tool in the analysis of continuous time signals and systems.

In Chapter 8, the Laplace transform method is explained and its properties derived. The use of Laplace transform to solve differential. equation is described. Finally different forms of structure realization of continuous-time systems are discussed.

Chapter 9 is devoted to the z-transform and its application to discrete time signals and systems. The properties of z-transform and techniques for inversion are introduced in this chapter. The use of z-transform for solving difference equation is explained. Different forms of structure realization of discrete-time system is also explained in this chapter.

In Chapter 10, the sampling theorem is explained. The necessary condition to avoid aliasing is also explained here.

The notable features of this book includes the following:

1. The syllabus content of signals and systems for undergraduate level of most of the Indian Universities in general and Anna University in particular has been covered.
2. The organization of the chapter are sequential in nature.
3. Large number of numerical examples have been worked out.
4. Chapter objectives and summary are given in each chapter.
5. For the students to practice, short and long questions with answers are given at the end of each chapter.

I take this opportunity to thank Shri. Sunil, Managing Director Ane Books India, for coming forward to publish the book. I would like to express my sincere thanks to Shri. R. Krishnamoorthi, sales manager Ane Books India who took the initiatives to publish the book in a short span of time. I would like to express my sincere thanks to Mr. V. Ashok who has done a wonderful job to key the voluminous book like this in a very short time and beautifully too. My sincere thanks are also due to my colleague

Mr. N. Sathurappan who gave some useful suggestions. I would also like to thank my wife Dr. S. Manimegalai, M.B.B.S., M.D., who was the source of inspiration while preparing this book.

S. Palani

Contents

About the Author

Dr. S. Palani obtained his B.E. degree in Electrical Engineering in the year 1966 from the University of Madras, M.Tech. in Control Systems Engineering from Indian Institute of Technology Kharagpur in 1968, and Ph.D. in Control Systems Engineering from the University of Madras in 1982. He has a wide teaching experience of over four decades. He started his teaching career in the year 1968 at the erstwhile Regional Engineering College (now National Institute of Technology), Tiruchirapalli in the department of EEE and occupied various positions. As Professor and Head, he took the initiative to start the Instrumentation and Control Engineering Department. After a meritorious service of over three decades in REC, Tiruchirapalli, he joined Sudharsan Engineering College, Pudukkottai as the founder Principal. He established various departments with massive infrastructure. Since 2006, he is the Dean and Professor of the ECE department in the same college.

He has published more than a hundred research papers in reputed national and international journals and conferences and has won many cash awards. Under his guidance, seventeen research scholars were awarded Ph.D. He has carried out several research projects worth about several lakhs rupees funded by the Government of India and AICTE. As the theme leader of the Indo - UK, REC Project on energy, he has visited many Universities and industries in the United Kingdom. He is the author of the books titled Control Systems Engineering, Signals and Systems, Digital Signal Processing, Linear System Analysis, and Automatic Control Systems. His areas of research include the design of Controllers for Dynamic systems, Digital Signal Processing, and Image Processing. Dr. S. Palani is the reviewer of technical papers of reputed International Journals. He has chaired/organized many International conferences and Workshops.

Chapter 1
Representation of Signals

Learning Objectives

- To define various terminologies related to signals and systems.
- To classify signals and systems.
- To give mathematical description and representation of signals and systems.
- To perform basic operations on CT and DT signals.
- To classify CT and DT signals as periodic and non-periodic, odd and even and power and energy signals.

1.1 Introduction

The concepts of signals and systems play a very important role in many areas of science and technology. These concepts are very extensively applied in the field of circuit analysis and design, long distance communication, power system generation and distribution, electron devices, electrical machines, biomedical engineering, aeronautics, process control and speech and image processing to mention a few. **Signals represent some independent variables that contain some information about the behavior of some natural phenomenon.** Voltages and currents in electrical and electronic circuits, electromagnetic radio waves, human speech and sounds produced by animals are some of the examples of signals. **When these signals are operated on some objects, they give out signals in the same or modified form. These objects are called systems.** A system is, therefore, defined as the interconnection of objects with a definite relationship between objects and attributes. Signals appearing at various stages of the system are attributes. R, L, C components, spring, dash-pots, mass,

S. Palani, *Signals and Systems*,
https://doi.org/10.1007/978-3-030-75742-7_1

etc. are the objects. The electrical and electronic circuits comprising of R, L, C components and amplifiers, the transmitter and receiver in a communication system, the petrol and diesel engines in an automobile, chemical plants, nuclear reactor, human beings, animals, a government establishment, *etc.* are all examples of systems.

1.2 Terminologies Related to Signals and Systems

Before we give mathematical descriptions and representations of various terminologies related to signals and systems, the following terminologies which are very frequently used are defined as follows:

1.2.1 Signal

A signal is defined as a physical phenomenon that carries some information or data. The signals are usually functions of independent variable time. There are some cases where the signals are not functions of time. The electrical charge distributed in a body is a signal which is a function of space and not time.

1.2.2 System

A system is defined as the set of interconnected objects with a definite relationship between objects and attributes. The inter-connected components provide desired function.

Objects are parts or components of a system. For example, switches, springs, masses, dash-pots, *etc.* in mechanical systems and inductors, capacitors and resistors in an electrical system are the objects. The displacement of mass, spring and dash-pot and the current flow and the voltage across the inductor, capacitor and resistor are the attributes. There is a definite relationship between the objects and attributes. The voltages across R, L, C series components can be expressed as $v_R = iR$, $V_L = L\frac{di}{dt}$ and $V_C = \frac{1}{C}\int idt$. If this series circuit is excited by the voltage source $e_i(t)$, the $e_i(t)$ is the input attribute or the input signal. If the voltage across any of the objects R, L and C is taken, then such an attribute is called the output signal. The block diagram representation of input and output (voltage across the resistor) signals and the system is shown in Fig. 1.1.

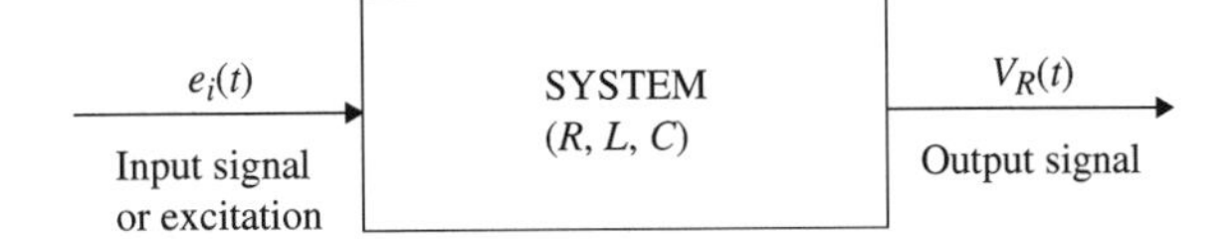

Fig. 1.1 Block diagram representation of signals and systems

1.3 Continuous and Discrete Time Signals

Signals are broadly classified as follows:

1. Continuous Time signal (CT signal).
2. Discrete Time signal (DT signal).

The signal that is specified for every value of time t is called continuous time signal and is denoted by $x(t)$. On the other hand, the signal that is specified at the discrete value of time is called discrete time signal. The discrete time signal is represented as a sequence of numbers and is denoted by $x[n]$ where n is an integer. Here time t is divided into n discrete time intervals. The Continuous Time signal (CT) and Discrete Time signal (DT) are represented in Figs. 1.2 and 1.3, respectively.

It is to be noted that in continuous time signal representation the independent variable t which has unit as sec. is put in the parenthesis $(\cdot)$ and in discrete time signal the independent variable n which is an integer is put inside the square parenthesis $[\cdot]$. Accordingly, the dependent variables of the continuous time signal/system are

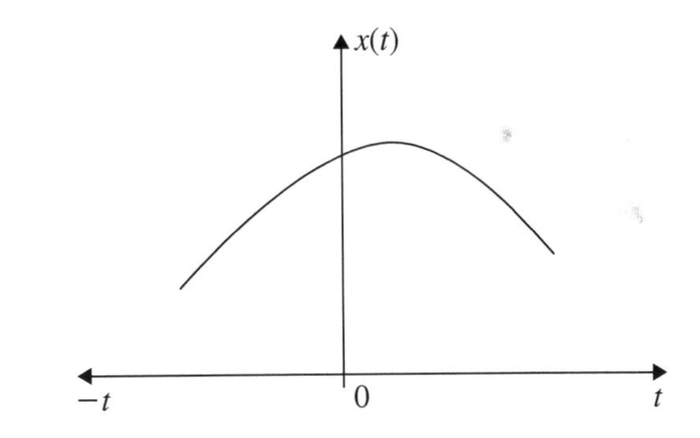

Fig. 1.2 CT signal

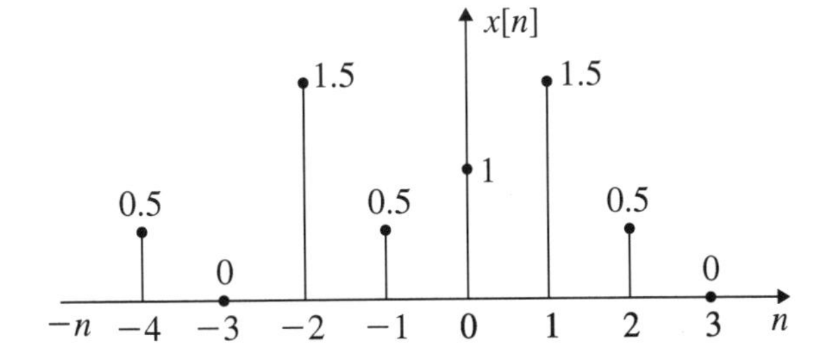

Fig. 1.3 DT signal

denoted as $x(t)$, $g(t)$, $u(t)$, etc. Similarly the dependent variables of discrete time signals/systems are denoted as $x[n]$, $g[n]$, $u[n]$, *etc.*

A discrete time signal $x[n]$ is represented by the following two methods:

1.

$$x[n] = \begin{cases} \left(\frac{1}{a}\right)^n & n \geq 0 \\ 0 & n < 0 \end{cases} \tag{1.1}$$

Substituting various values of n where $n \geq 0$ in Eq. (1.1) the sequence for $x[n]$ which is denoted by $x\{n\}$ is written as follows:

$$x[n] = \left\{1, \frac{1}{a}, \frac{1}{a^2}, \ldots, \frac{1}{a^n}\right\}$$

2. The sequence is also represented as given below.

$$x[n] = \{3,\ 2,\ \underset{\uparrow}{5},\ 4,\ 6,\ 8,\ 2\}$$

The arrow indicates the value of $x[n]$ at $n = 0$ which is 5 in this case. The numbers to the left of the arrow indicate to the negative sequence $n = -1, -2$, *etc.* The numbers to the right of the arrow correspond to $n = 1, 2, 3, 4$, *etc.* Thus, for the above sequence, $x[-1] = 2; x[-2] = 3; x[0] = 5; x[1] = 4; x[2] = 6; x[3] = 8$ and $x[4] = 2$. If no arrow is marked for a sequence, the sequence starts from the first term in the extreme left. Consider the sequence

$$x[n] = \{5,\ 3,\ 4,\ 2\}.$$

Here, $x[0] = 5; x[1] = 3; x[2] = 4$ and $x[3] = 2$. There is no negative sequence here.

■ Example 1.1

Graphically represent the following sequence:

$$x[n] = \{1,\ 0,\ -1,\ 1\}$$

Solution: The graphical representation $x[n] = \{1,\ 0,\ -1,\ 1\}$ is shown in Fig. 1.4.

■ Example 1.2

Graphically represent the following sequence:

$$x[n] = \{-2,\ 1,\ 0,\ \underset{\uparrow}{1},\ 2,\ 0,\ 1\}$$

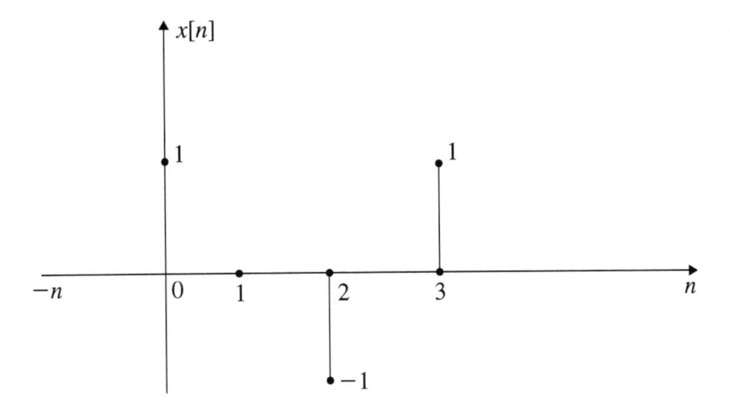

Fig. 1.4 Graphical representation of $x[n]$

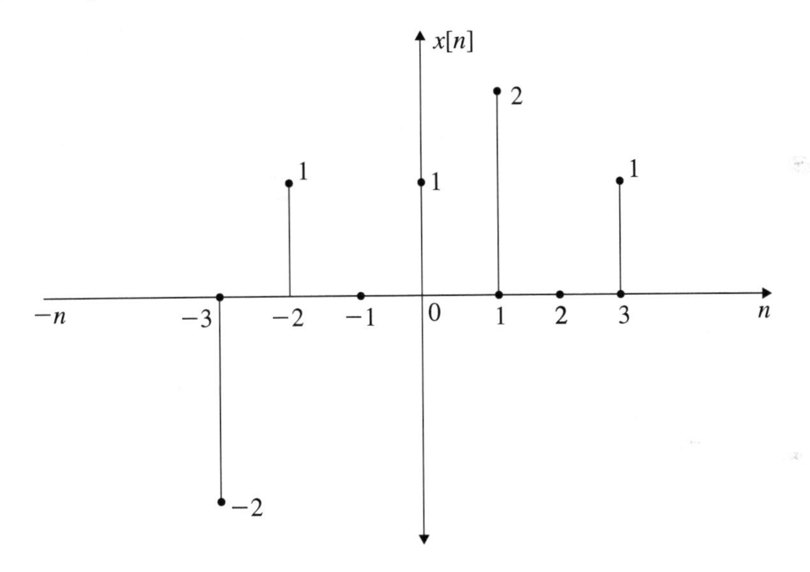

Fig. 1.5 Graphical representation of $x[n]$

Solution: The sequence

$$x[n] = \{-2,\ 1,\ 0,\ \underset{\uparrow}{1},\ 2,\ 0,\ 1\}$$

is represented in Fig. 1.5.

1.4 Basic Continuous Time Signals

Basic signals play a very important role in signals and systems analysis. The following are the basic continuous time signals which serve as a basis to represent other signals. The basic continuous time signals are as follows:

1. Unit impulse function.
2. Unit step function.
3. Unit ramp function.
4. Unit parabolic function.
5. Unit rectangular pulse (or Gate) function.
6. Unit area triangular function.
7. Unit signum function.
8. Unit Sinc function.
9. Sinusoidal signal.
10. Real exponential signal.
11. Complex exponential signal.

The mathematical description and graphical representation of the above signals are discussed below. Similar to continuous time signals, basic discrete time signals are also available. The descriptions of these signals will immediately follow this.

1.4.1 Unit Impulse Function

The unit impulse function is also known as **Dirac delta** function which is represented in Fig. 1.6. The unit impulse function is denoted as $\delta(t)$ and its mathematical description is given below.

$$\delta(t) = \begin{cases} 0 & t \neq 0 \\ 1 & t = 0 \end{cases} \tag{1.2}$$

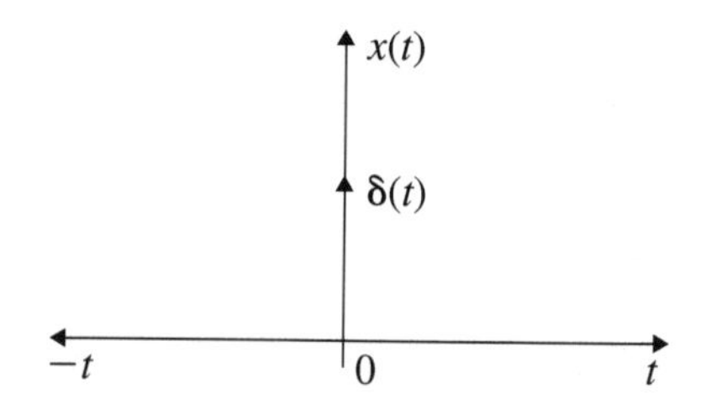

Fig. 1.6 Unit impulse function

1.4.1.1 Importance of Impulse Function

1. By applying impulse signal to a system, one can get the impulse response of the system. From impulse response, it is possible to get the transfer function of the system.
2. For a linear time invariant system, if the area under the impulse response curve is finite, then the system is said to be stable.
3. From the impulse response of the system, one can easily get the step response and ramp response by integrating it once and twice, respectively.
4. Impulse signal is easy to generate and apply to any system.

1.4.1.2 Some Properties of Impulse Function

1. $\delta(at) = \frac{1}{a}\delta(t)$
2. $\delta(-t) = \delta(t)$
3. $x(t)\delta(t) = x(0)\delta(t)$
4. $x(t)\delta(t - t_0) = x(t_0)\delta(t - t_0)$
5. $\int_{-\infty}^{\infty} \delta(t)dt = 1$
6. $t\delta(t) = 0$
7. $t\frac{d\delta(t)}{dt} = -\delta(t)$
8. $x(t) * \delta(t - t_0) = x(t - t_0)$

1.4.2 Unit Step Function

The unit step function is shown in Fig. 1.7. The function is defined as follows:

$$u(t) = \begin{cases} 1 & t \geq 0 \\ 0 & t < 0 \end{cases} \tag{1.3}$$

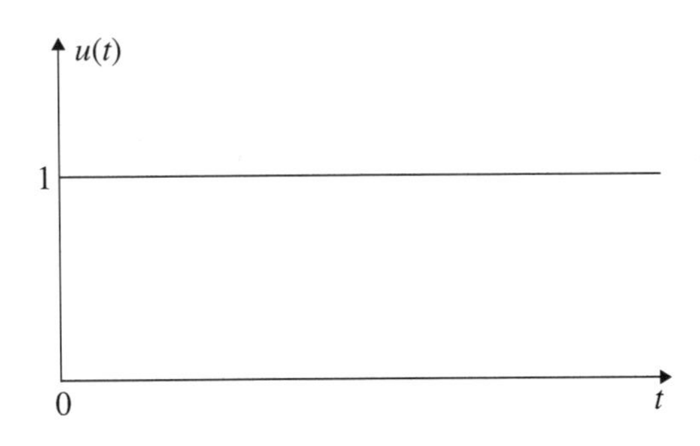

Fig. 1.7 Unit step function

The step function is denoted by $u(t)$. Any causal signal which begins at $t = 0$ (which has a value of zero for $t < 0$) is multiplied by the signal by $u(t)$. For example, **a causal exponentially decaying signal e^{-at} ($t \geq 0$) is represented as $x(t) = e^{-at}u(t)$. Similarly e^{-at} ($t < 0$) is represented as $x(t) = e^{-at}u(-t)$.**

1.4.2.1 Importance of Step Function

1. Step function is easy to generate and apply to the system.
2. By differentiating the step response, the impulse response can be obtained. By integrating the step response, the ramp response can be obtained.
3. Step signal is considered as a white noise which is drastic. If the system response is satisfactory for a step signal, it is likely to give a satisfactory response to other types of signals.
4. Application of step signal is equivalent to the application of numerous sinusoidal signals with a wide range of frequencies.

1.4.3 Unit Ramp Function

The unit ramp function is represented in Fig. 1.8. It is defined by the following mathematical equation:

$$r(t) = \begin{cases} t & t \geq 0 \\ 0 & t < 0 \end{cases} \tag{1.4}$$

For a causal signal ($t \geq 0$), the ramp function can also be expressed as

$$r(t) = t\,u(t) \tag{1.5}$$

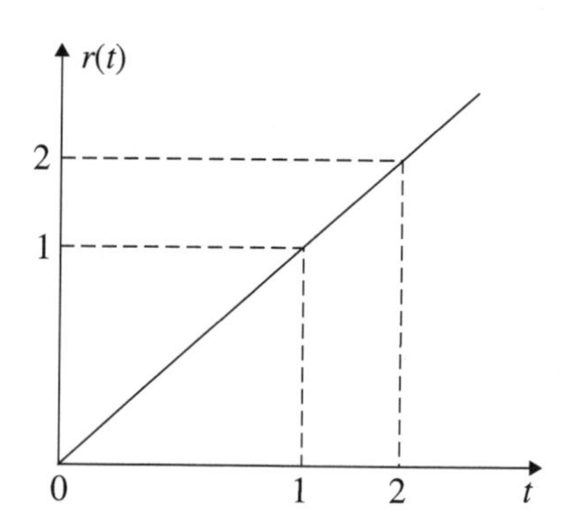

Fig. 1.8 Unit ramp function

1.4.3.1 Relationships between Impulse, Step and Ramp Signals

1. Integrating the unit step signal $u(t)$, we get

$$\boxed{\int u(t)dt = \int dt = t} \tag{1.6}$$

By integrating the unit step function, unit ramp function is obtained. In the reverse process, by differentiating a ramp function, a step function is obtained.

2. The continuous time unit step function is the running integral of the unit impulse function which is expressed as

$$u(t) = \int_{-\infty}^{t} \delta(\tau)d\tau$$

$$\boxed{\frac{du(t)}{dt} = \delta(t)} \tag{1.7}$$

3. By differentiating the ramp function twice, the impulse function is obtained.

$$\begin{aligned} r(t) &= t \\ \frac{dr(t)}{dt} &= 1 = u(t) \end{aligned} \tag{1.8}$$

$$\boxed{\frac{d^2r(t)}{dt^2} = \frac{du(t)}{dt} = \delta(t)} \tag{1.9}$$

Thus, the impulse function is obtained by differentiating the ramp function twice. By the reverse process, by integrating the impulse function twice, the ramp function is obtained which is mathematically expressed as follows:

$$\boxed{r(t) = \iint \delta(t)\, dt} \tag{1.10}$$

The relationships between the impulse, step and ramp signals are represented below.

$$\delta(t) \overset{\text{integrate}}{\longrightarrow} u(t) \overset{\text{integrate}}{\longrightarrow} r(t)$$

$$r(t) \overset{\text{differentiate}}{\longrightarrow} u(t) \overset{\text{differentiate}}{\longrightarrow} \delta(t)$$

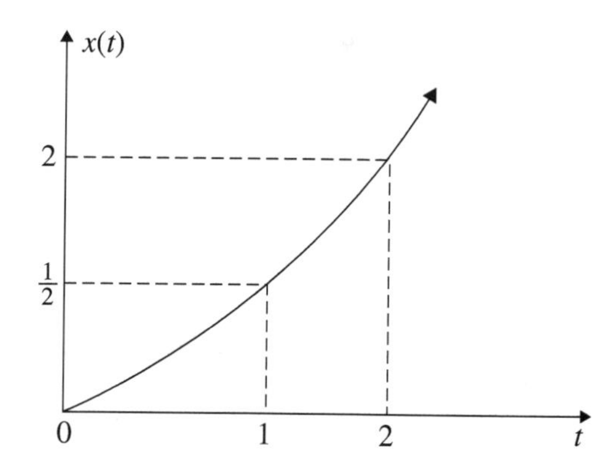

Fig. 1.9 Unit parabolic function

1.4.4 Unit Parabolic Function

The unit parabolic function $x(t)$ is represented in Fig. 1.9. The mathematical expression is given below.

$$\boxed{x(t) = \frac{1}{2}t^2} \qquad t \geq 0 \tag{1.11}$$

If the parabolic function is differentiated, unit ramp function is obtained. Thus,

$$\frac{dx(t)}{dt} = t \qquad t \geq 0.$$

Step, ramp and parabolic functions are called singularity functions.

1.4.5 Unit Rectangular Pulse (or Gate) Function

The unit area rectangular pulse which is also called gate function is represented in Fig. 1.10. Mathematically it is described as follows:

$$x(t) = \begin{cases} 1 & \text{for } |t| \leq \frac{T}{2} \\ 0 & \text{otherwise} \end{cases} \tag{1.12}$$

The above equation is also written in the following form:

$$x(t) = 1 \qquad -\frac{T}{2} \leq t \leq \frac{T}{2}$$

The function is written as $x(t) = \text{rect}(\frac{t}{T})$.

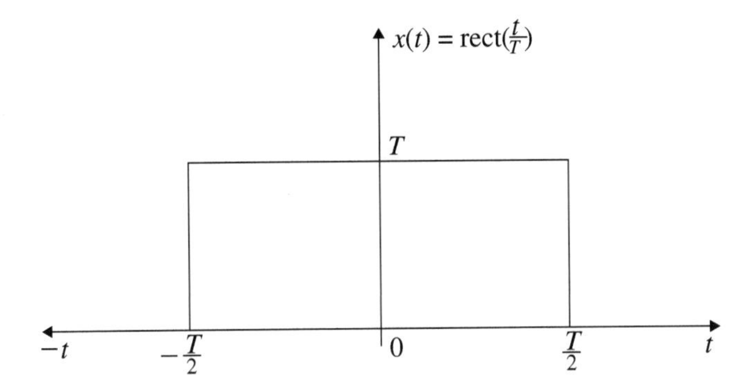

Fig. 1.10 Unit area rectangular pulse (or gate) function

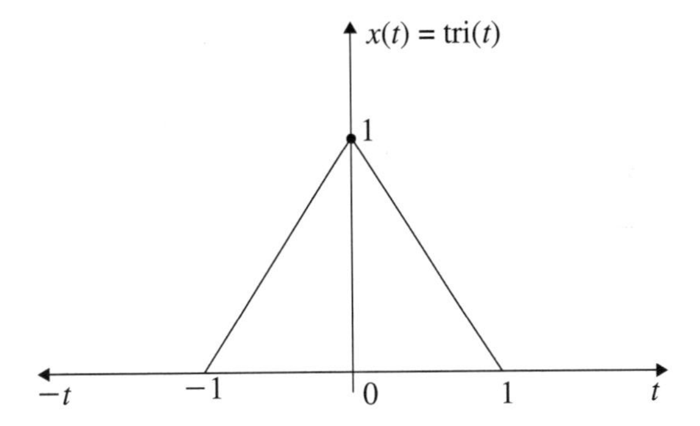

Fig. 1.11 Unit area triangular function

1.4.6 *Unit Area Triangular Function*

The unit area triangular function is represented in Fig. 1.11. It is symbolically written as $x(t) = \text{tri}(t)$. It is defined as

$$\text{tri}(t) = \begin{cases} [1 - |t|] & |t| \leq 1 \\ 0 & |t| > 1 \end{cases} \tag{1.13}$$

The above equation can be written in the following form also:

$$\begin{aligned} \text{tri}(t) &= [1 + t] \qquad -1 \leq t \leq 0 \\ &= [1 - t] \qquad 0 \leq t \leq 1 \end{aligned}$$

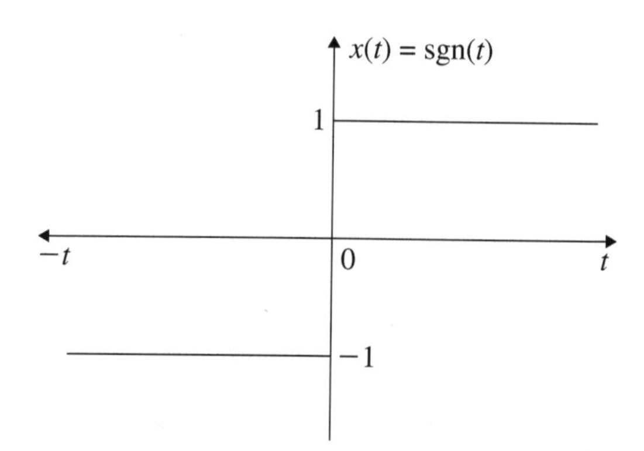

Fig. 1.12 Representation of unit signum function

1.4.7 Unit Signum Function

The signum function is written in the abbreviated form as sgn(t). It represents the characteristics of an ideal relay. This is shown in Fig. 1.12. It is defined by the following equations:

$$\operatorname{sgn}(t) = \begin{cases} 1 & t > 0 \\ 0 & t = 0 \\ -1 & t < 0 \end{cases} \tag{1.14}$$

1.4.8 Unit Sinc Function

The unit sinc function is represented in Fig. 1.13. It is defined as

$$\operatorname{sinc}(t) = \frac{\sin \pi t}{\pi t} \qquad -\infty < t < \infty. \tag{1.15}$$

1.4.9 Sinusoidal Signal

The sinusoidal signal is represented in Fig. 1.14. It is defined as

$$x(t) = A \sin(\omega t - \phi) \tag{1.16}$$

where $A =$ Peak amplitude, $\omega =$radian frequency and $\phi =$phase shift.

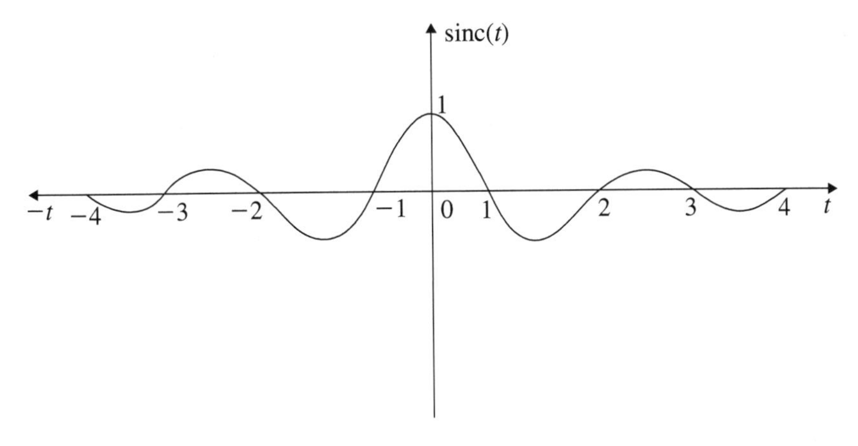

Fig. 1.13 Representation of unit sinc function

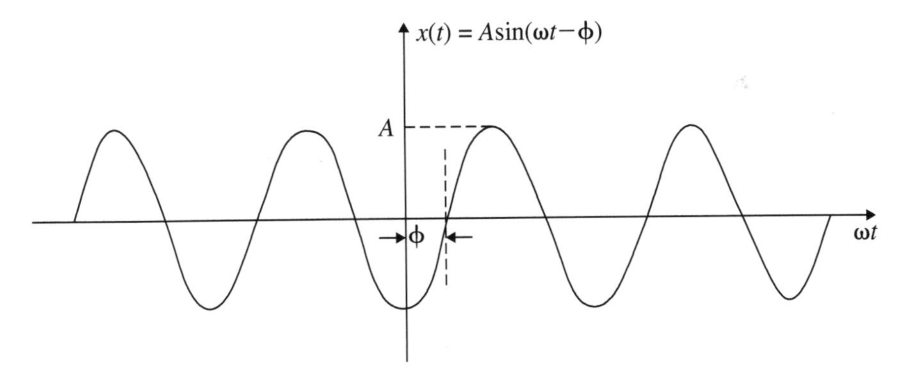

Fig. 1.14 Representation of sinusoidal signal

1.4.10 Real Exponential Signal

Let

$$x(t) = e^{st} \tag{1.17}$$

where $s = \sigma + j\omega$ is a complex number. The signal $x(t)$ in Eq. (1.17) is called general complex exponential. Equation (1.17) is written in the following form:

$$\begin{aligned} x(t) &= e^{(\sigma+j\omega)t} \\ &= e^{\sigma t} e^{j\omega t} \\ &= e^{\sigma t}(\cos\omega t + j\sin\omega t) \end{aligned} \tag{1.18}$$

If $\omega = 0$,

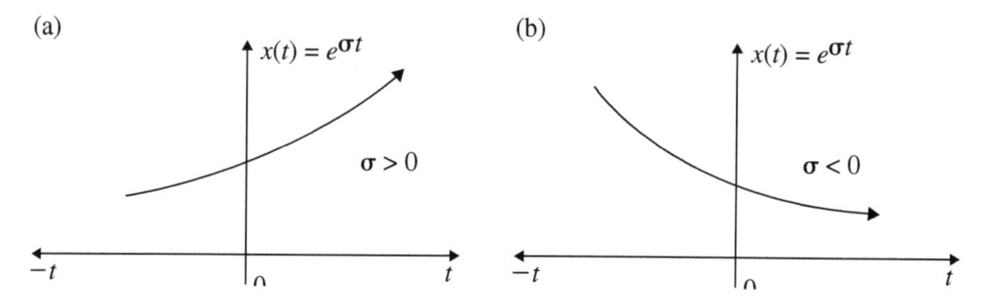

Fig. 1.15 Representation of real exponential signals. **a** Growing exponential; **b** Decaying exponential

$$\boxed{x(t) = e^{\sigma t}} \tag{1.19}$$

Equation (1.19) is real exponential. The plot of $x(t)$ with respect to t for $\sigma > 0$ and $\sigma < 0$ is shown in Fig. 1.15a and b, respectively. For $\sigma > 0$, the signal is exponentially growing and for $\sigma < 0$, it is exponentially decaying.

1.4.11 Complex Exponential Signal

The signal $x(t)$ in Eq. (1.18) is the general complex exponential which has real part as $e^{\sigma t} \cos \omega t$ and the imaginary part $e^{\sigma t} \sin \omega t$. For $\sigma = 0$, the signal $x(t)$ is a sinusoid. For $\sigma > 0$, $x(t)$ is a sinusoid which is exponentially building and is shown in Fig. 1.16a. For $\sigma < 0$, the signal $x(t) = e^{-\sigma t}(\cos \omega t + j \sin \omega t)$ is exponentially decaying and is shown in Fig. 1.16b.

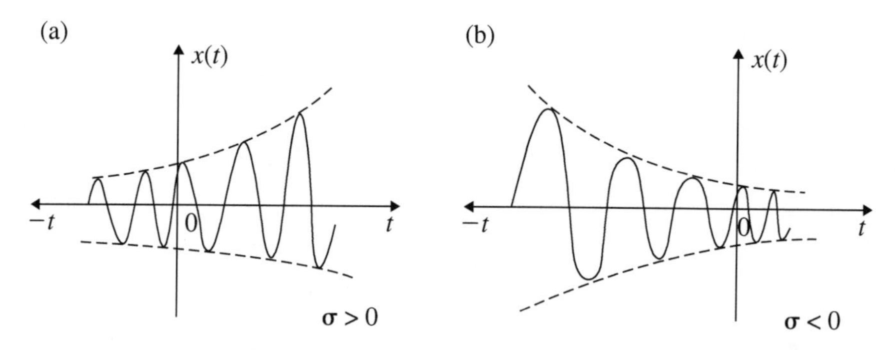

Fig. 1.16 Complex exponential signals. **a** Exponentially growing ($\sigma > 0$); **b** Exponentially decaying ($\sigma < 0$)

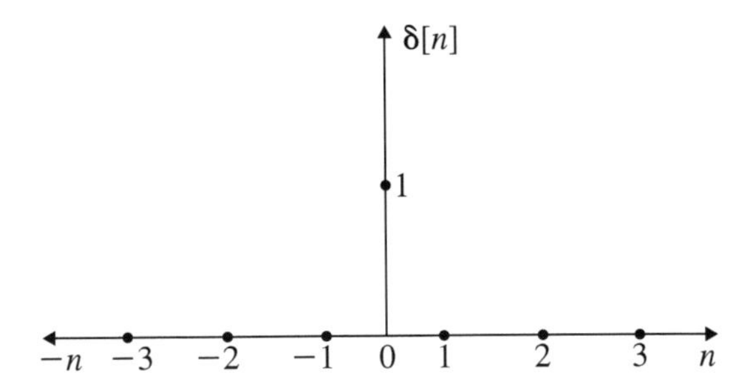

Fig. 1.17 Basic unit impulse sequence

1.5 Basic Discrete Time Signals

Similar to continuous time signals, basic discrete signals are available. However, these signals are represented at discrete intervals of time "*n*" where *n* is an integer. Representation of basic discrete time signals is discussed below.

1.5.1 The Unit Impulse Sequence

The basic impulse sequence is shown in Fig. 1.17. The unit impulse sequence also called sample is defined as

$$\delta[n] = \begin{cases} 1 & n = 0 \\ 0 & n \neq 0 \end{cases} \tag{1.20}$$

$\delta[n]$ is also called Kronicker delta function.

1.5.2 The Basic Unit Step Sequence

The basic unit step sequence is represented in Fig. 1.18. It is denoted by $u(n)$. It is defined as

$$u[n] = \begin{cases} 1 & n \geq 0 \\ 0 & n < 0 \end{cases} \tag{1.21}$$

Any discrete sequences $x[n]$ for $n \geq 0$ is expressed as $x[n]u[n]$. For $n < 0$, it is expressed as $x[n]u[-n]$. It is be noted that at $n = 0$, the value of $u[n] = 1$.

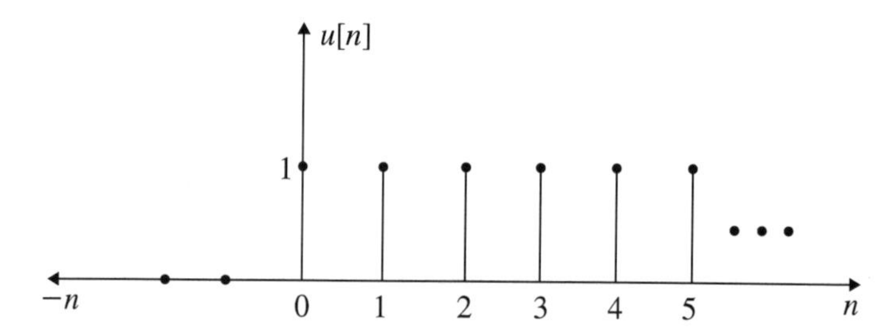

Fig. 1.18 Basic unit step sequence

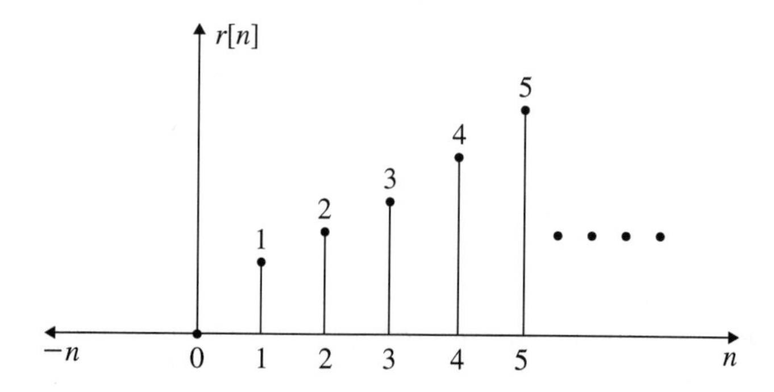

Fig. 1.19 Basic unit ramp sequence

1.5.3 The Basic Unit Ramp Sequence

The basic unit ramp sequence which is denoted by $r[n]$ is represented in Fig. 1.19. It is defined as

$$r[n] = \begin{cases} n & n \geq 0 \\ 0 & n < 0 \end{cases} \tag{1.22}$$

1.5.4 Unit Rectangular Sequence

The discrete time unit rectangular sequence is shown in Fig. 1.20a. It is defined as

$$\text{rect}[n] = \begin{cases} 1 & |n| \leq N \\ 0 & |n| > N \end{cases} \tag{1.23}$$

The above equation can also be expressed as

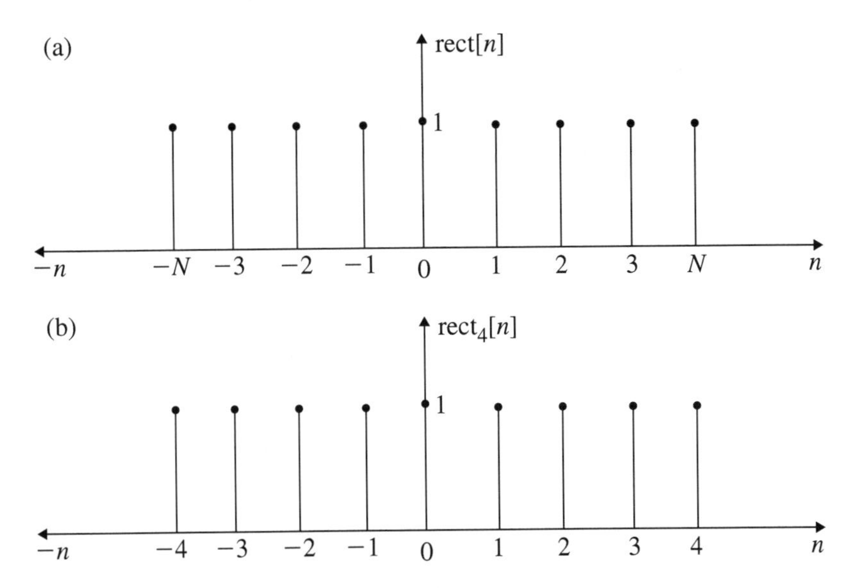

Fig. 1.20 Unit rectangular sequence

$$\text{rect}[n] = \begin{cases} 1 & -N \leq n \leq N \\ 0 & \text{otherwise} \end{cases}$$

N indicates the width of the rectangular sequence on both sides of $-n$ and $+n$. For example, the notation $\text{rect}_4[n]$ indicates four samples for $0 < n \leq 4$ and four samples for $-4 \leq n < 0$ and one sample at $n = 0$. Thus, there will be nine samples for $\text{rect}_4[n]$. This is represented in Fig. 1.20b.

1.5.5 Sinusoidal Sequence

The discrete time sinusoidal signal is defined by the following mathematical expression:

$$x[n] = Ae^{-\alpha n} \sin(\omega_0 n + \phi) \tag{1.24}$$

where A and α are real numbers and ϕ is the phase shift. Depending on the value of α, the sinusoidal sequence is divided into the following categories:

- A purely sinusoidal sequence ($\alpha = 0$).
- Decaying sinusoidal sequence ($\alpha > 0$).
- Growing sinusoidal sequence ($\alpha < 0$).

The above sinusoidal sequences are illustrated in Fig. 1.21a–c, respectively.

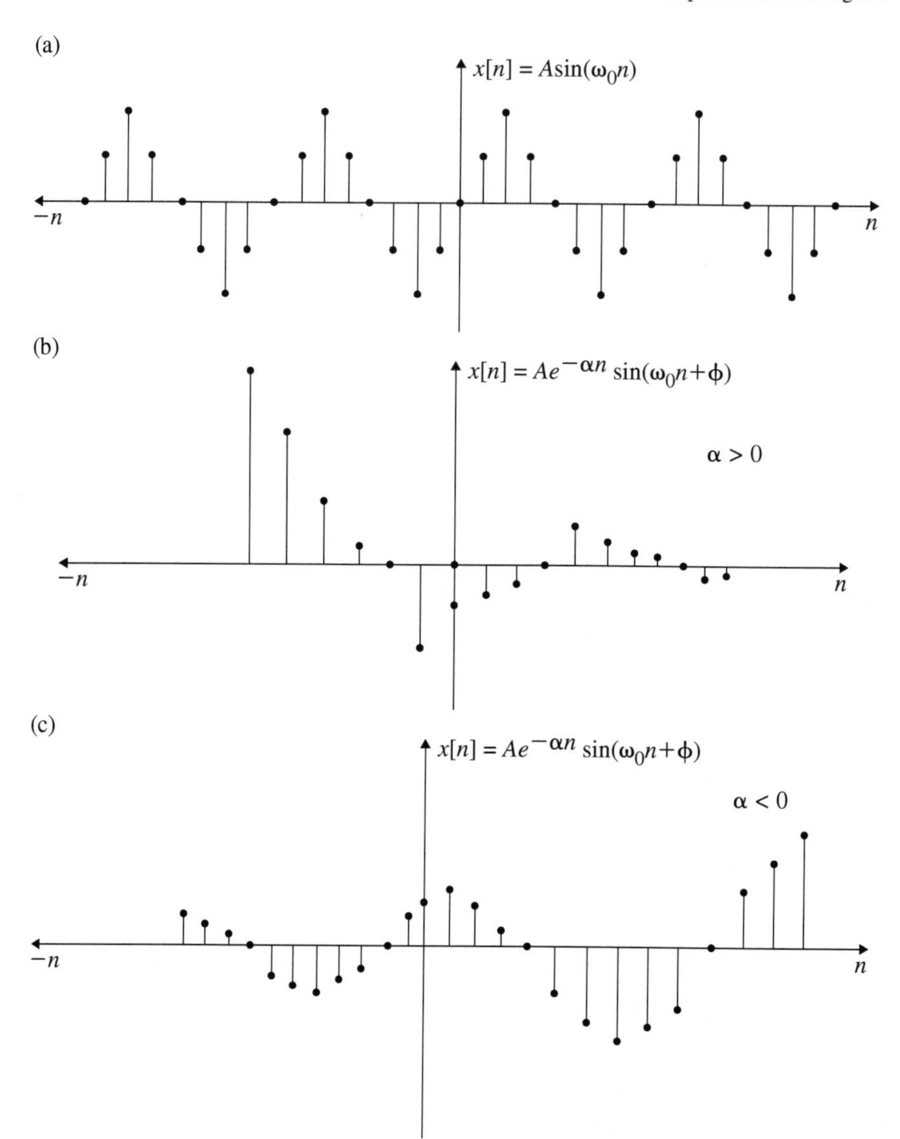

Fig. 1.21 Discrete time sinusoidal signal. **a** Purely sinusoidal; **b** Decaying sinusoidal; **c** Growing sinusoidal

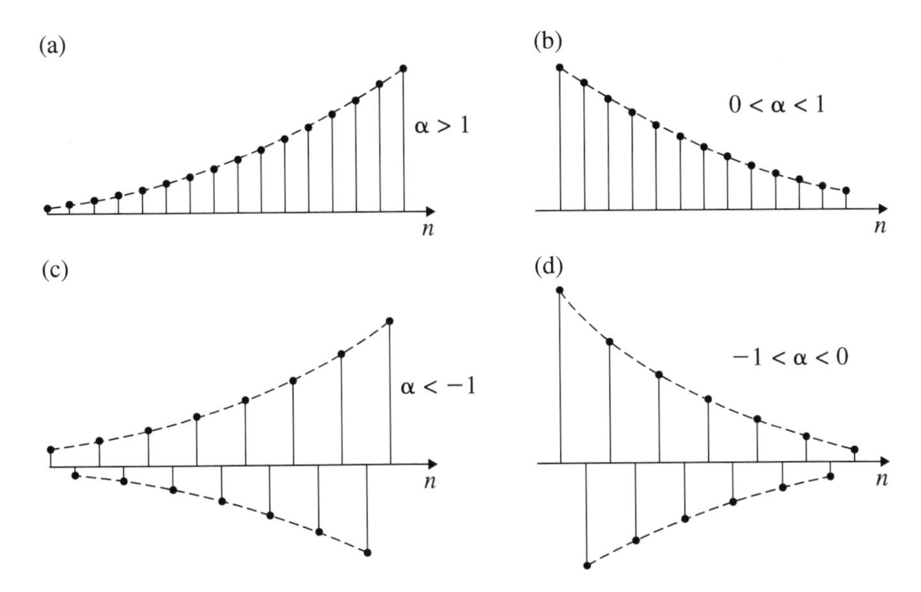

Fig. 1.22 Discrete time real exponential sequences. **a** $\alpha > 1$; **b** $0 < \alpha < 1$; **c** $\alpha < -1$; **d** $-1 < \alpha < 0$

1.5.6 Discrete Time Real Exponential Sequence

The general complex exponential sequence is defined as

$$x[n] = A\alpha^n \tag{1.25}$$

where A and α are in general complex numbers.

In Eq. (1.25) if A and α are real, the sequence is called real exponential. These sequences for various values of α are shown in Fig. 1.22. Depending on the value of α, the sequence is classified as:

1. Exponentially growing signal ($\alpha > 1$, Fig. 1.22a).
2. Exponentially decaying signal ($0 < \alpha < 1$, Fig. 1.22b).
3. Exponentially growing for alternate value of n ($\alpha < -1$, Fig. 1.22c).
4. Exponentially decaying for alternate value of n ($-1 < \alpha < 0$, Fig. 1.22d).

1.6 Basic Operations on Continuous Time Signals

The basic operations performed on continuous time signals are given below:

1. Addition of CT signals.
2. Multiplications of CT signals.

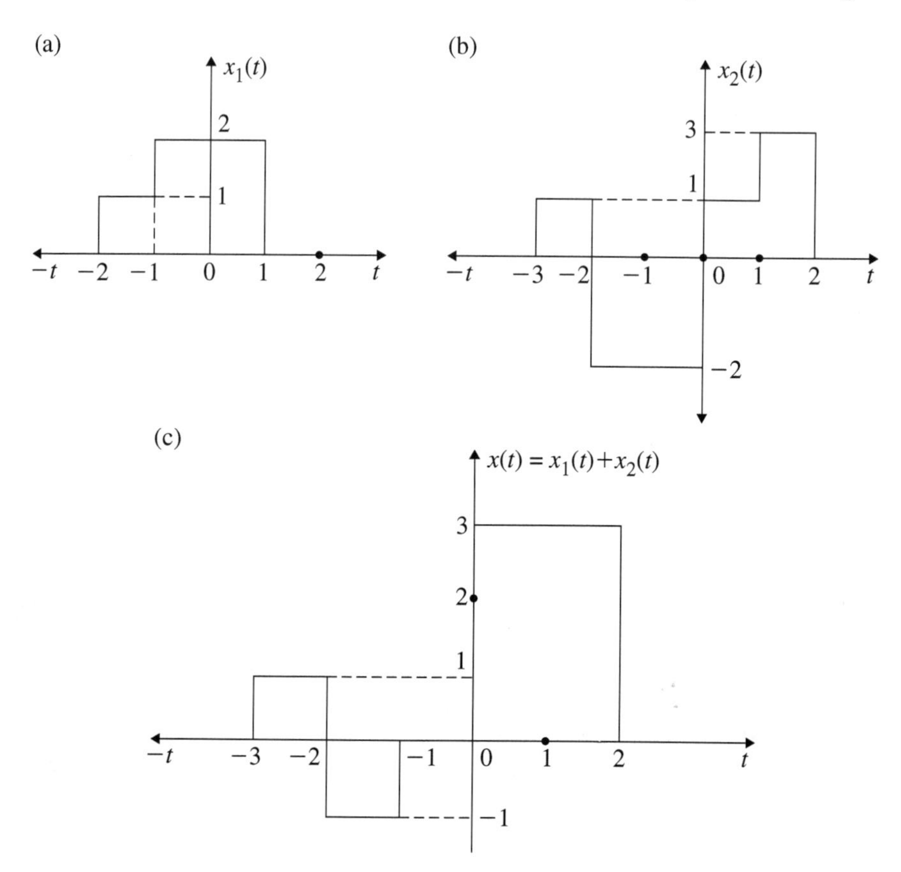

Fig. 1.23 Additions of two CT signals

3. Amplitude scaling of CT signals.
4. Time scaling of CT signals.
5. Time shifting of CT signals.
6. Reflection or folding of CT signals.
7. Inverted CT signal.

1.6.1 Addition of CT Signals

Consider the signals $x_1(t)$ and $x_2(t)$ which are shown in Fig. 1.23a, b. The amplitude of these two signals at each instant of time is added to get their sum. The following table is prepared.

From Table 1.1, $x(t) = x_1(t) + x_2(t)$ is plotted and is shown in Fig. 1.23c.

Table 1.1 Addition of two CT signals

t	-3	-2	-1	0	1	2
$x_1(t)$	0	1	2	2	0	0
$x_2(t)$	1	-2	-2	1	3	0
$x(t) = x_1(t) + x_2(t)$	1	-1	0	3	3	0

Table 1.2 Multiplication of two CT signals

t	-3	-2	-1	0	1	2
$x_1(t)$	0	1	2	2	0	0
$x_2(t)$	1	-2	-2	1	3	0
$x(t) = x_1(t) \times x_2(t)$	0	-2	-4	2	0	0

1.6.2 Multiplications of CT Signals

Consider the two signals $x_1(t)$ and $x_2(t)$ shown in Fig. 1.23a and b, respectively. These signals $x_1(t)$ and $x_2(t)$ are multiplied to get $x(t)$

$$x(t) = x_1(t) \times x_2(t)$$

The functions $x_1(t)$ and $x_2(t)$ at different time intervals are determined from figure and multiplied. Table 1.2 is prepared to get $x(t)$ at different time intervals. Table 1.2 is transformed to plot $x(t) = x_1(t) \times x_2(t)$ which is shown in Fig. 1.24.

1.6.3 Amplitude Scaling of CT Signals

Consider the signals $x(t)$ sketched and shown in Fig. 1.25a. This signal when multiplied by a factor A is expressed as $Ax(t)$. At any time t, the amplitude of $x(t)$ is multiplied by A. This type of signal transformation is called amplitude scaling. The signal $3x(t)$ is shown in Fig. 1.25b. At any instant t, $x(t)$ is multiplied by a factor 3.

Consider the signal $\frac{x(t)}{2}$. At any time t, the amplitude of $x(t)$ shown in Fig. 1.25a is divided by the factor 2. The above transformation is plotted in Fig. 1.25c.

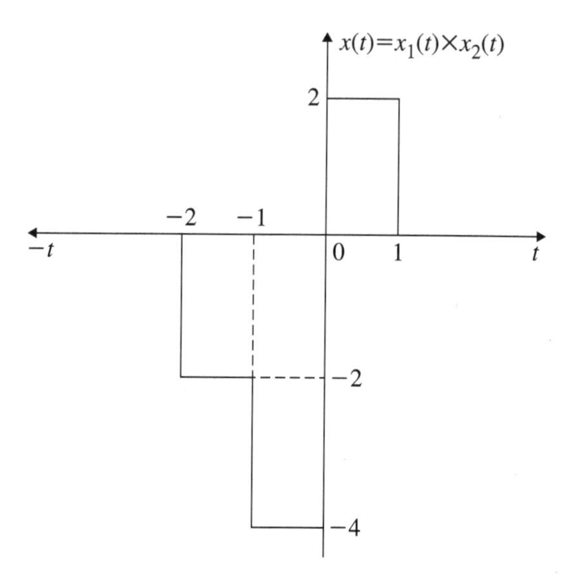

Fig. 1.24 Multiplications of two CT signals

1.6.4 Time Scaling of CT Signals

The compression or expansion of a signal in time is known as time scaling. Consider the signal $x(t)$ shown in Fig. 1.26a. The signal is time compressed and shown in Fig. 1.26b as $x(4t)$. For any given magnitude of $x(t)$, the time is divided by the factor 4. The time expanded signal $x(\frac{t}{4})$ is shown in Fig. 1.26c. Here, for any given magnitude of $x(t)$, the time is multiplied by the factor 4. In general, for any given amplitude of $x(t)$, $x(at)$ is time compressed by a factor a and $x(\frac{t}{a})$ is time expanded by a factor a.

1.6.5 Time Shifting of CT Signals

Consider the signal $x(t) = u(t)$, the unit step function. The step function is shown in Fig. 1.27a as $u(t)$. The transformation $t = t - t_0$ where t_0 is any arbitrary constant amounts to shifting $u(t)$ to the right by t_0 unit if t_0 is positive and is denoted as $u(t - t_0)$. If t_0 is negative, the function is shifted to the left by t_0 unit and is denoted as $u(t + t_0)$. The right shifted $u(t - t_0)$ is shown in Fig. 1.27b and left shifted $u(t + t_0)$ is shown in Fig. 1.27c. The signal $u(-t)$ is shown in Fig. 1.27d and is obtained by folding $u(t)$ as shown in Fig. 1.27a. $u(-t) = 1$ for $t < 0$. If we fold across the vertical axis, the signal to the right of the vertical axis is transformed to its left and *vice versa*. That is why it is called **folded signal**. The signal $u(-t - t_0)$ is obtained by shifting

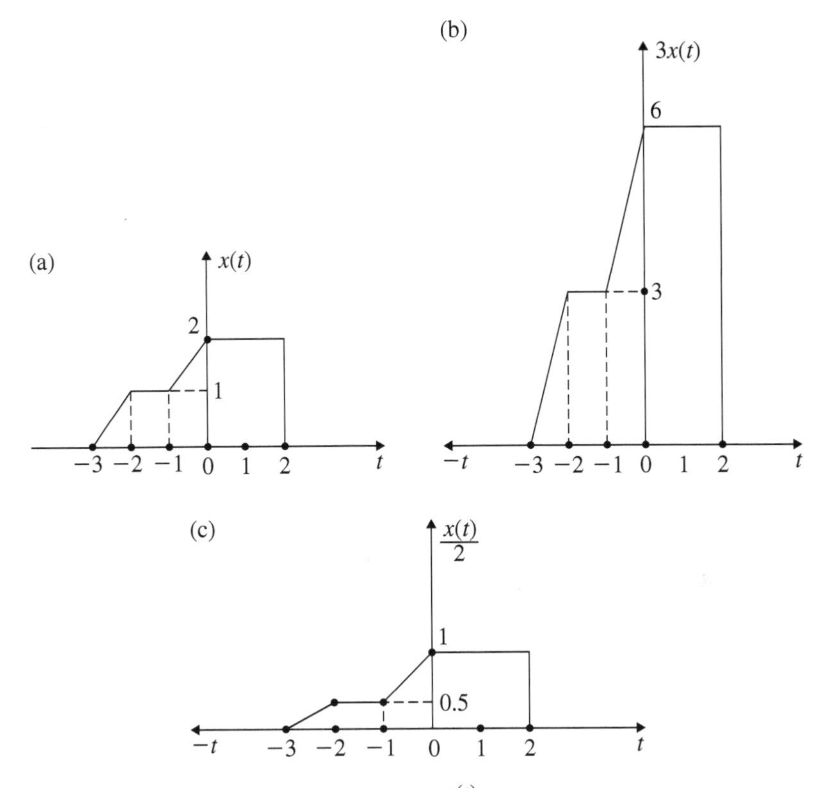

Fig. 1.25 Amplitude scaling. **a** $x(t)$; **b** $3x(t)$ and **c** $\frac{x(t)}{2}$

the signal $u(-t)$ to the left by t_0 unit as shown in Fig. 1.27e. The signal $u(-t+t_0)$ is obtained by shifting the signal $u(-t)$ to the right by t_0 unit and is shown in Fig. 1.27f.

Summary of Shifting of CT signal

1. **It $x(t)$ is given, then $x(t+t_0)$ is plotted by shifting $x(t)$ to the left by t_0.**
2. **It $x(t)$ is given, then $x(t-t_0)$ is plotted by shifting $x(t)$ to the right by t_0.**
3. **It $x(-t)$ is given, then $x(-t-t_0)$ is plotted by shifting $x(-t)$ to the left by t_0.**
4. **It $x(-t)$ is given, then $x(-t+t_0)$ is plotted by shifting $x(-t)$ to the right by t_0.**
5. **In general for $x(t+t_0)$ and $x(-t-t_0)$ the time shift is made to the left of $x(t)$ and $x(-t)$, respectively, by t_0. For $x(t-t_0)$ and $x(-t+t_0)$ the time shift is made to the right of $x(t)$ and $x(-t)$, respectively, by t_0.**

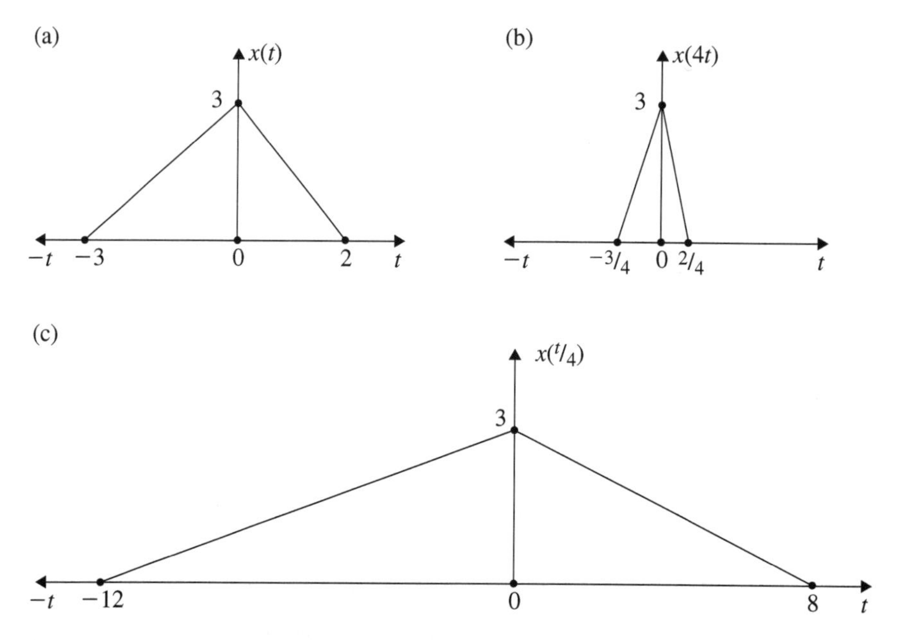

Fig. 1.26 Time scaling of CT signals

1.6.6 Signal Reflection or Folding

Consider the signal $x(t)$ shown in Fig. 1.28a. The signal $x(-t)$ is obtained by putting a mirror along the vertical axis. The signal to the right of the vertical axis gets reflected to the left and *vice versa*. Alternatively, if we make a folding across the vertical axis, the signal in the right of the vertical axis is printed in the left and *vice versa*. The signal so obtained is $x(-t)$.

1.6.7 Inverted CT Signal

Consider the CT signal $x(t)$ shown in Fig. 1.29a. The inverted signal $-x(t)$ is obtained by inverting its amplitude. By this the signal above the horizontal axis (time axis) comes below the axis and *vice versa*. Alternatively, if a mirror is put along the horizontal axis, the signal above the axis gets reflected below the axis and *vice versa*.

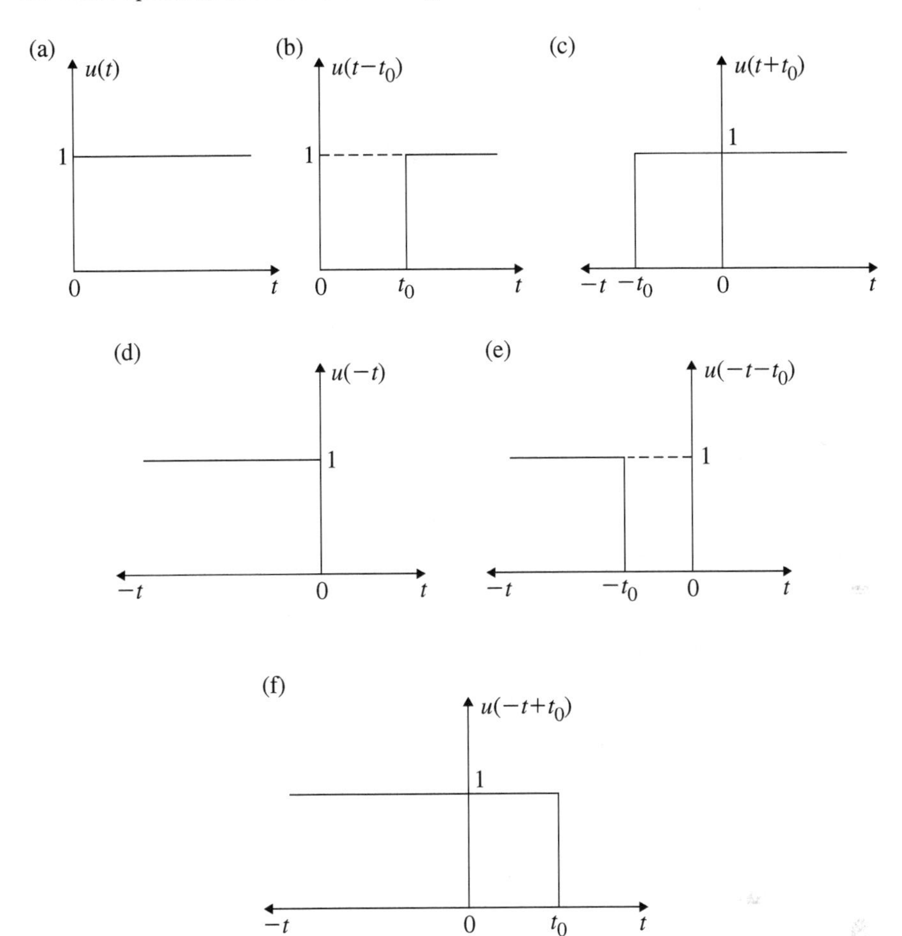

Fig. 1.27 Representation of time shifting CT signals

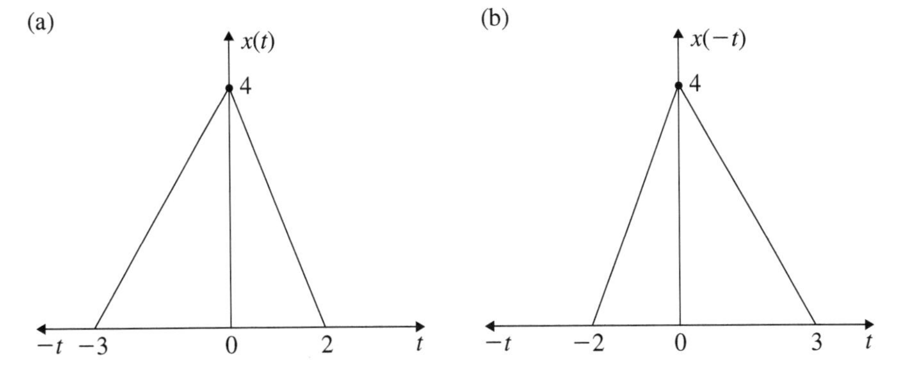

Fig. 1.28 CT signal reflection or folding

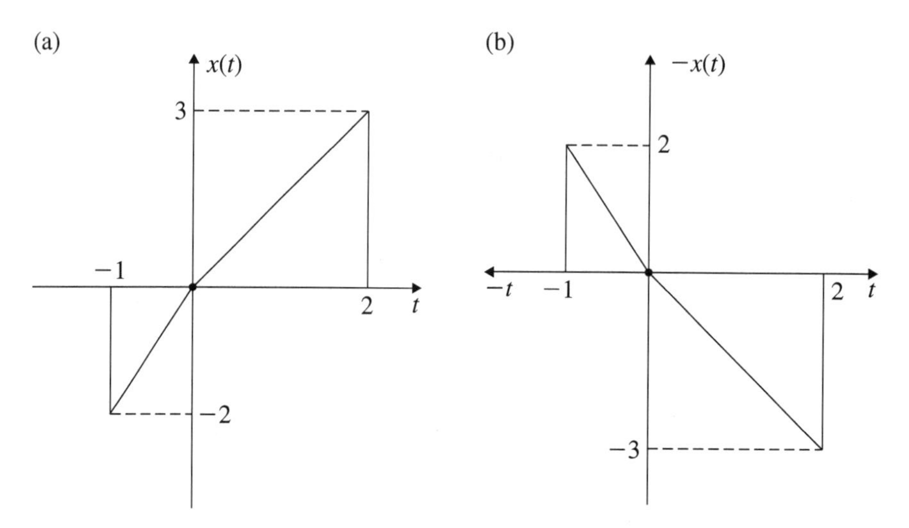

Fig. 1.29 Inverted CT signal

1.6.8 Multiple Transformation

The transformation namely amplitude scaling, time reversal, time shifting, time scaling, *etc.* when applied simultaneously, the sequence of operation is important. If not followed correctly, it would give erroneous results.

Consider the following signal:

$$y(t) = Ax\left(\frac{-t - t_0}{a}\right)$$

The following sequence of transformation is followed:

1. $y(t)$ is written in the following form:

$$y(t) = Ax\left(-\frac{t}{a} - \frac{t_0}{a}\right)$$

2. Plot $x(t)$.
3. Plot $Ax(t)$ using amplitude scaling.
4. Plot $Ax(-t)$ using time reversal.
5. Plot $Ax(-t - \frac{t_0}{a})$ by shifting $Ax(-t)$ to the left by $\frac{t_0}{a}$ (time shifting).
6. Plot $Ax(-\frac{t}{a} - \frac{t_0}{a})$ by time expansion.

The following examples illustrate the above sequence of operation.

■ Example 1.3

Consider the signal $y(t) = 5x(-3t + 1)$ where $x(t)$ is shown in Fig. 1.30a. Plot $y(t)$ and $-y(t)$.

Solution:

$$y(t) = 5x(-3t + 1)$$

1. The given signal $x(t)$ is represented in Fig. 1.30a.
2. The signal $x(t)$ is amplitude scaled and plotted in Fig. 1.30b.
3. $5x(-t)$ is obtained by folding $5x(t)$ in Fig. 1.30b and is plotted in Fig. 1.30c.
4. $5x(-t)$ is time shifted by one unit to the right and $5x(-t + 1)$ is obtained and shown in Fig. 1.30d.
5. $5x(-t + 1)$ is time compressed by a factor 3 and $5x(-3t + 1)$ is obtained. This is shown in Fig. 1.30e.
6. $5x(-3t + 1)$ amplitude inverted to get $-5x(-3t + 1)$. This is shown in Fig. 1.30f.

■ Example 1.4

Consider the signal

$$x(t) = \text{rect}(t)$$

Plot $y(t) = 5\text{rect}(\frac{t-3}{4})$.

Solution:

$$x(t) = 5\text{rect}\frac{(t-3)}{4}$$

1. $x(t)$ can be written as $x(t) = 5\text{rect}\left(\frac{t}{3} - \frac{3}{4}\right)$. The plot of $\text{rect}(t)$ is shown in Fig. 1.31a.
2. The time delayed ($t_0 = 3/4$) signal is right shifted by 3/4 and with its amplitude multiplied by 5 is shown in Fig. 1.31b.
3. The time shifted signal represented in step 2 is to be time expanded by a factor
4. This is shown in Fig. 1.31c as $y(t) = 5\text{rect}\frac{(t-3)}{4}$.

■ Example 1.5

For the signal shown in Fig. 1.32a, sketch

$$y(t) = -3x\left(-\frac{2}{3}t + 1.5\right)$$

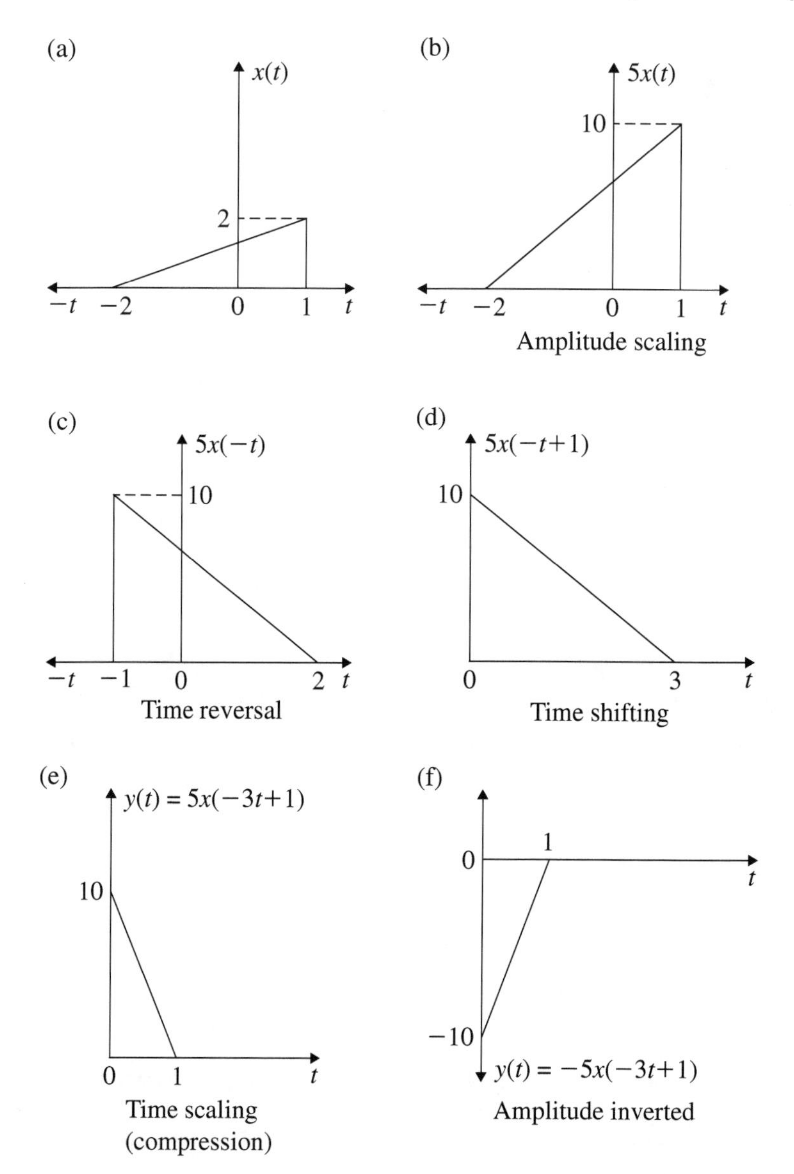

Fig. 1.30 Basic operations on CT signal

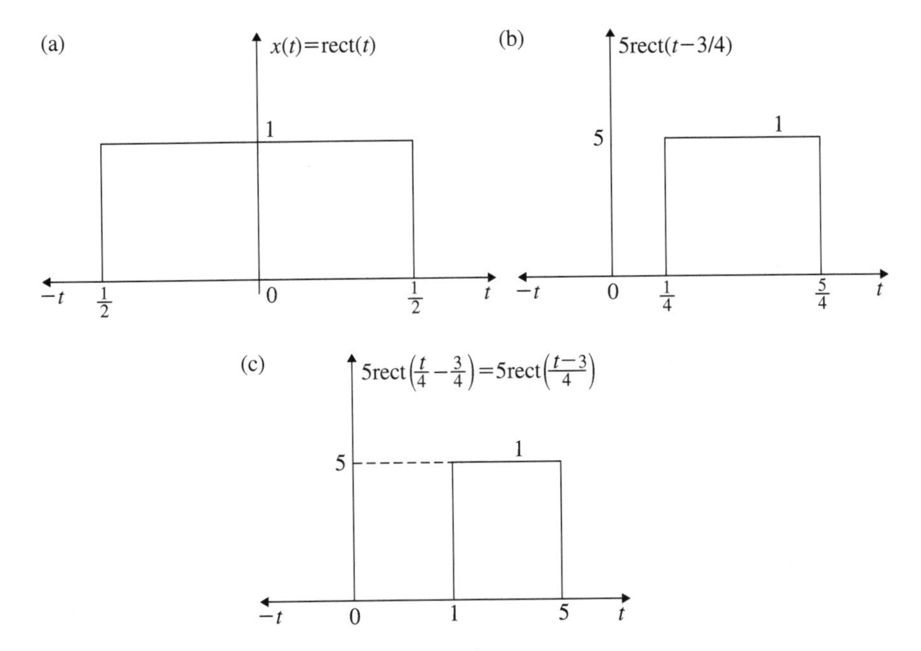

Fig. 1.31 Sketch of $5\text{rect}\frac{(t-3)}{4}$

Solution:

1. $x(t)$ is sketched as shown in Fig. 1.32a.
2. By time reversal $x(-t)$ is obtained and sketched as shown in Fig. 1.32b.
3. By amplitude scaling and inversion $-3x(t)$ is obtained and is shown in Fig. 1.32c.
4. The signal obtained in step 3 is right shifted by $t = 1.5$ and $-3x(-t + 1.5)$ is shown in Fig. 1.32e.
5. By time scaling expanded by 3/2, we get $-3x(-(2/3)t + 1.5)$ which is shown in Fig. 1.32f.

■ Example 1.6

For the signal $x(t)$ shown in Fig. 1.33 give mathematical equation in terms of step signals.

Solution: The signal $x(t)$ shown in Fig. 1.33 is in the form of stair case in the time interval $-3 \leq t \leq 3$. The mathematical expression in terms of step signals can be derived as explained below:

1. For the time interval $-3 \leq t < \infty$, the step signal is generated as $u(t + 3)$.
2. For the time interval $-2 \leq t < \infty$, the step signal is generated as $u(t + 2)$.
3. For the time interval $-1 \leq t < \infty$, the step signal is generated as $2u(t + 1)$.

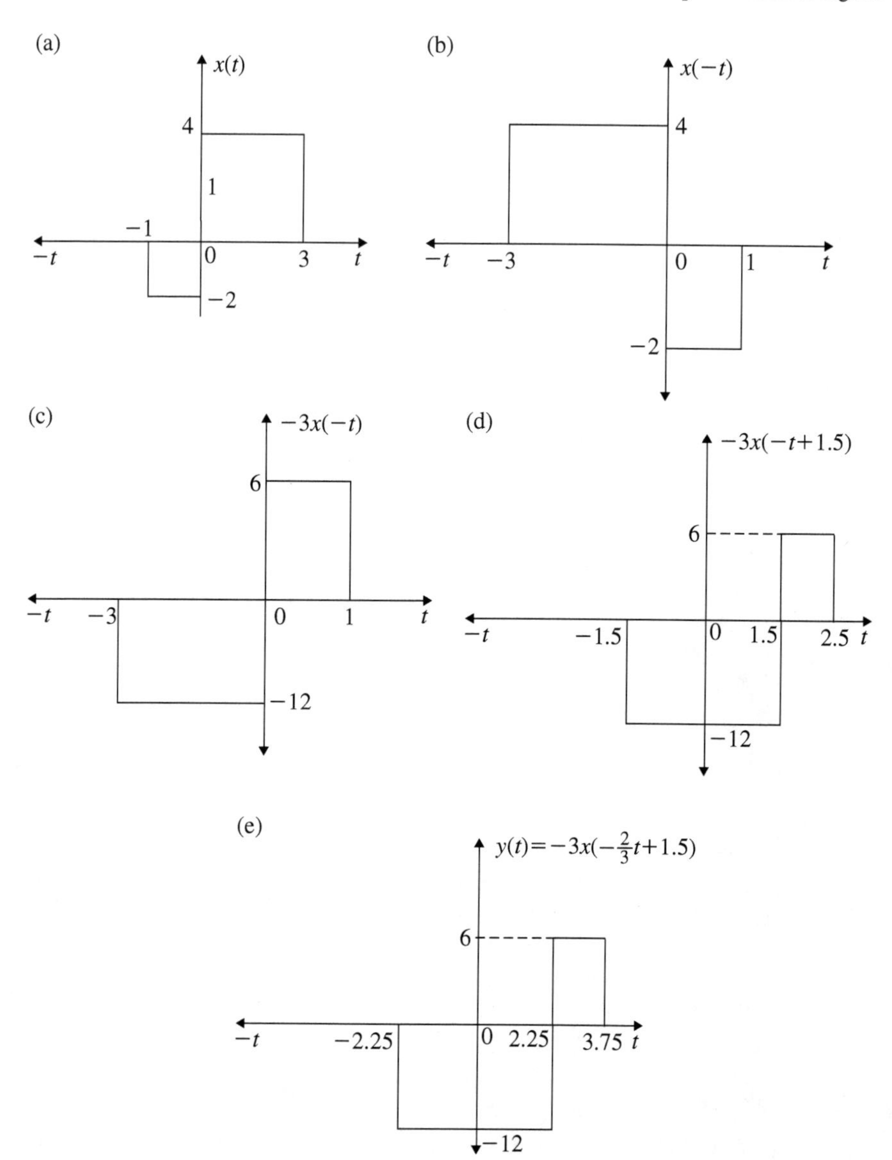

Fig. 1.32 Sketch of $y(t) = -3x\left(-\frac{2}{3}t + 1.5\right)$

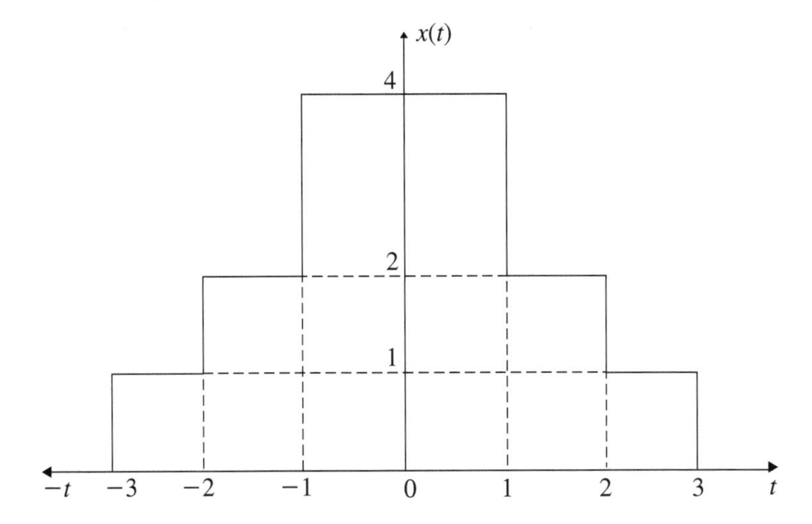

Fig. 1.33 The signal $x(t)$ for Example 1.6

If the above three step signals are added, we will get the stair-case signal $-3 \leq t \leq 1$. However, these step signals are extended to $t \to \infty$ also and hence, they are to be canceled by negative going steps at $t = 1, t = 2$ and $t = 3$ as $-2u(t-1), -u(t-2)$ and $u(t-3)$, respectively. Thus, $x(t)$ is obtained by adding these step signals.

$$\boxed{x(t) = u(t+3) + u(t+2) + 2u(t+1) - 2u(t-1) - u(t-2) - u(t-3)}$$

■ Example 1.7

For a signal $x(t)$ shown in Fig. 1.34a, sketch

(a) $x(3t+2)$

(b) $x\left(\frac{-t}{2} - 1\right)$

(Anna University, June, 2007)

Solution: **To plot $x(3t+2)$**

1. $x(t)$ is represented in Fig. 1.34a. $x(t)$ is moved to the left by $t = 2$ and is shown in Fig. 1.34b.
2. By time compression by a factor 3, from Fig. 1.34b, $x(3t+2)$ is obtained and is shown in Fig. 1.34c.

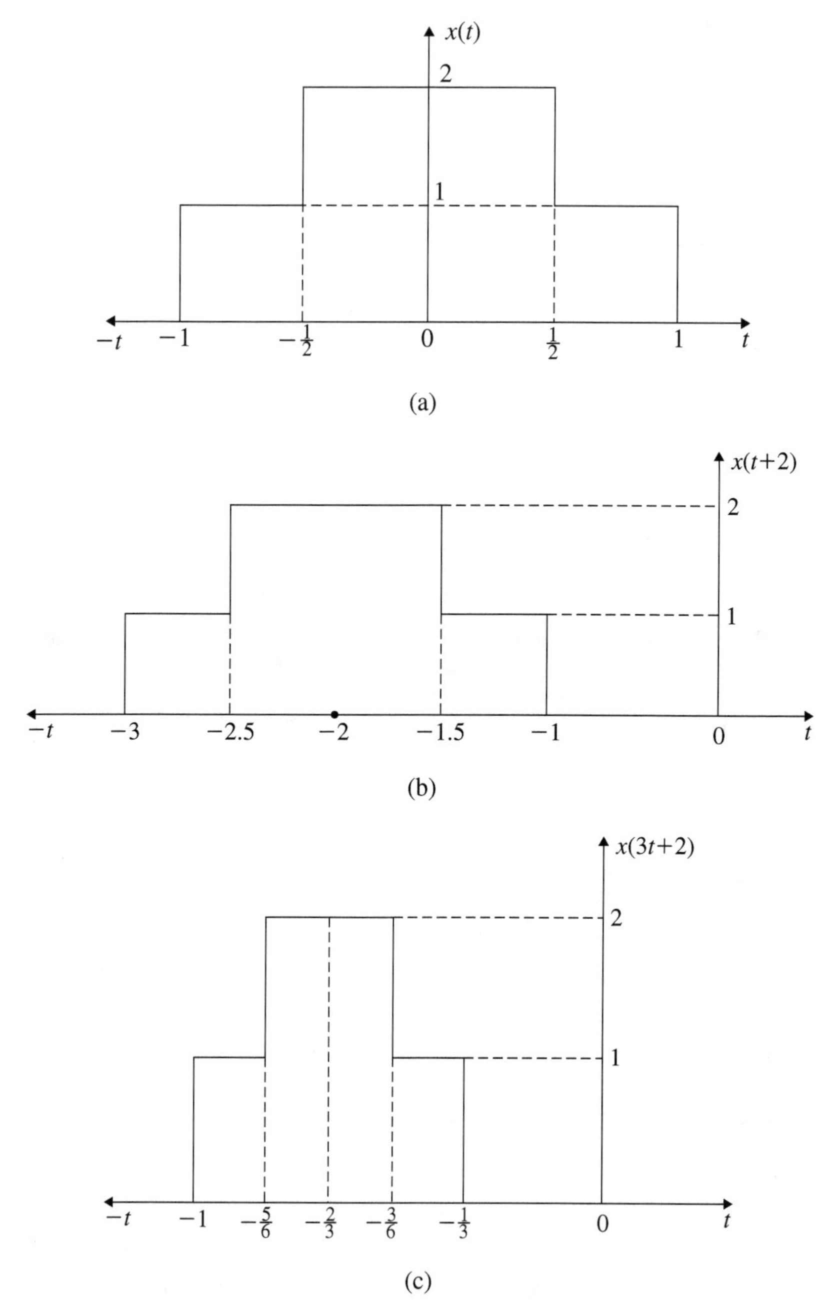

Fig. 1.34 **a** Plot of $x(t)$. **b** Time shifted $x(t)$. **c** Time compressed $x(t)$. **d** Folded $x(t)$. **e** Time shifted $x(-t)$. **f** Time expansion of $x(-t-1)$ to get $x(-\frac{t}{2}-1)$

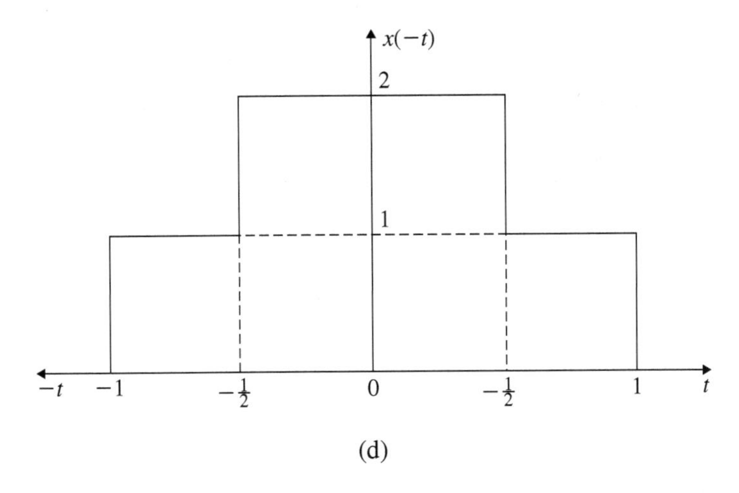

(d)

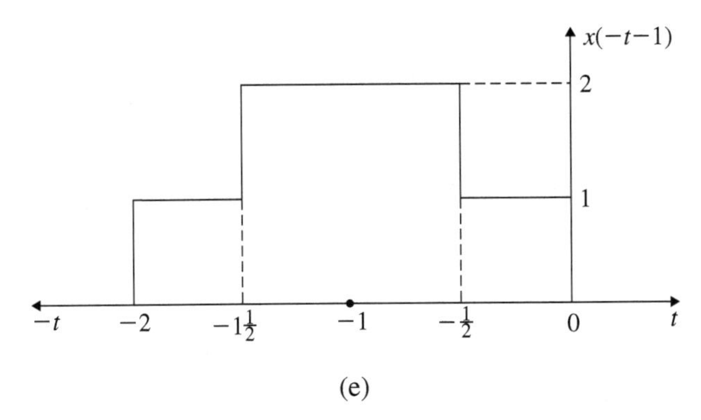

(e)

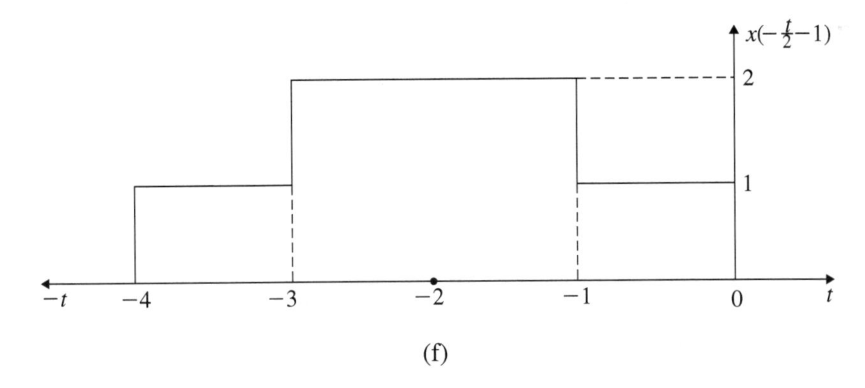

(f)

Fig. 1.34 (continued)

Solution: **To plot** $\boldsymbol{x(-(\frac{t}{2})-1)}$

1. By folding $x(t)$ represented in Fig. 1.34a, $x(-t)$ is obtained and is shown in Fig. 1.34d.
2. $x(-t-1)$ is obtained by shifting $x(-t)$ by $t=1$ to the left. $x(-t-1)$ is sketched as shown in Fig. 1.34e.
3. By time expansion, the time of the signal $x(-t-1)$ is multiplied by the factor 2, and $x(-\frac{t}{2}-1)$ is obtained. This is shown in Fig. 1.34f.

■ Example 1.8

The rectangular signal $x(t)=\text{rect}(t/2)$ is shown in Fig. 1.35a. Sketch the following signals:

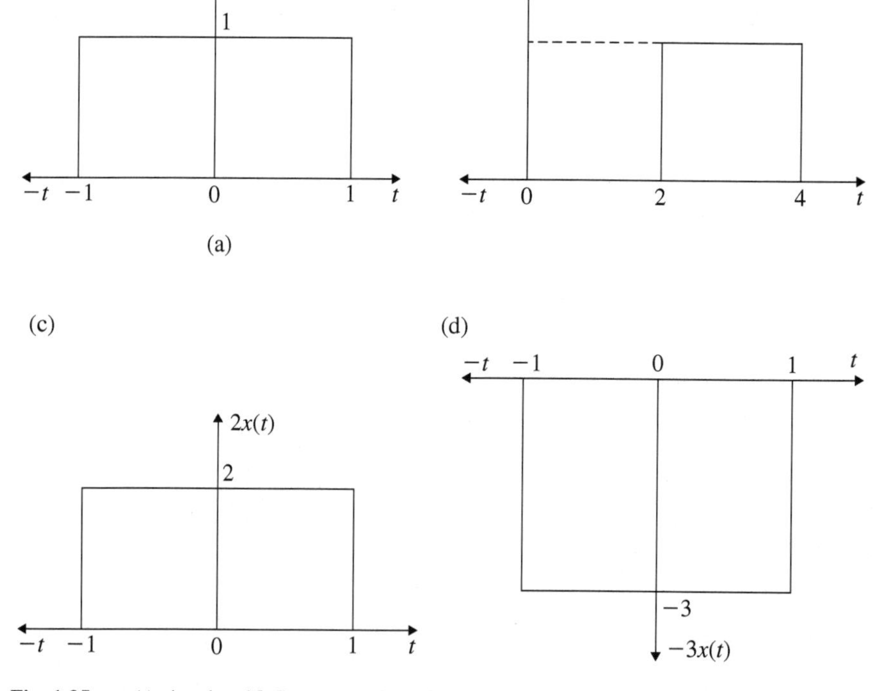

Fig. 1.35 **a** $x(t)$ signal and **b** Representation of $x(t-3)$. **c** Representation of $2x(t)$ and **d** Representation of $-3x(t)$. **e** Representation of $x(t-2)$ and **f** Representation of $3x(t)$. **g** Representation of $x(t-2)+3x(t)$

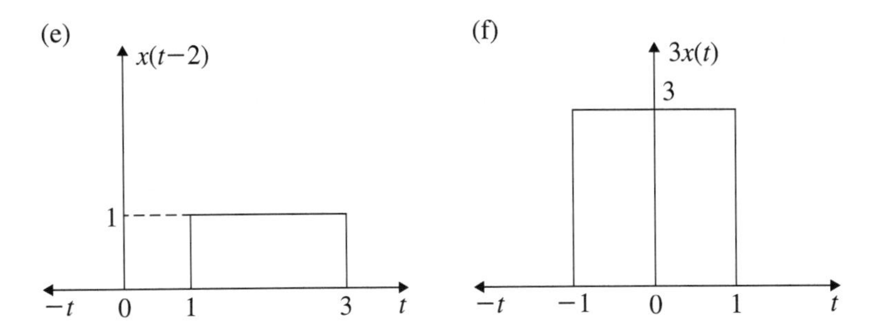

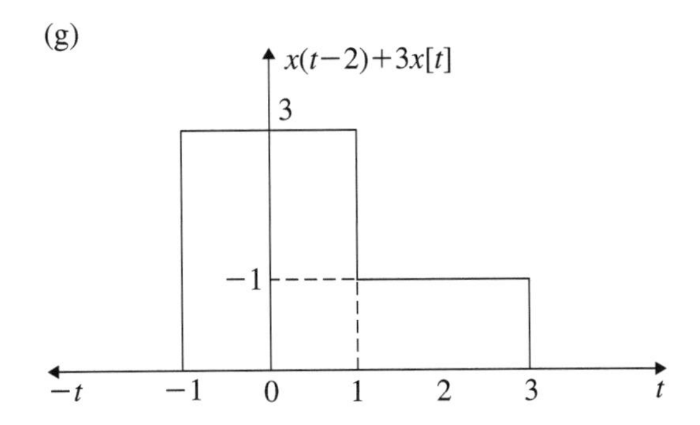

Fig. 1.35 (continued)

(a) $x(t-3)$
(b) $2x(t)$
(c) $-3x(t)$
(d) $x(t-2)+3x(t)$

Solution:

(a) **To represent the signal $x(t-3)$**
$x(t-3)$ is obtained by time shifting $x(t)$ by 3 unit of time toward right. This is shown in Fig. 1.35b.

(b) **To represent the signal $2x(t)$**
This is an amplitude scaled signal. The amplitude of $x(t)$ is multiplied by the factor 2 and is shown in Fig. 1.35c.

(c) **To represent the signal $-3x(t)$**
The signal $x(t)$ is amplitude inverted and multiplied by a factor 3. This is shown in Fig. 1.35d.

(d) **To represent the signal $x(t-2)+3x(t)$**
The time delayed $x(t-2)$ is obtained by shifting $x(t)$ to the right by a factor 2. This is represented in Fig. 1.35e. The signal $x(t)$ is amplitude multiplied by a factor 3 and $3x(t)$ is obtained. This is shown in Fig. 1.35f. By adding the signals shown in Fig. 1.35e and in Fig. 1.35f, $x(t-2)+3x(t)$ is obtained and is represented in Fig. 1.35g.

■ Example 1.9

Consider the triangular wave form $x(t)$ shown in Figure 1.36(a). Sketch the following wave forms:

(a) $x(2t+3)$

(b) $x\left(\dfrac{t+3}{2}\right)$

(c) $x\left(\dfrac{t}{2}-3\right)$

(d) $x(-2t+3)$

(e) $x(-2t-3)$

Solution:

(a) **To sketch $x(2t+3)$**
Figure 1.36a shows $x(t)=\text{tri}(t)$. By time shifting by $t=3$ toward left, $x(t+3)$ is obtained and this is sketched in Fig. 1.36b. $x(t+3)$ is time compressed by a factor of 2 to get $x(2t+3)$. This is sketched in Fig. 1.36c.

(b) **To sketch $x\left(\frac{t+3}{2}\right)$**
The signal $x\left(\frac{t+3}{2}\right)$ is written as $x\left(\frac{t}{2}+1.5\right)$. The signal $x(t)$ is time shifted to the left by 1.5 unit to get $x(t+1.5)$. This is sketched in Fig. 1.36d. $x(t+1.5)$ is time expanded by a factor 2 to get $x\left(\frac{t}{2}+1.5\right)$ which is nothing but $x\left(\frac{t+3}{2}\right)$. This is sketched in Fig. 1.36e.

(c) **To sketch $x\left(\frac{t}{2}-3\right)$**
$x(t-3)$ is obtained from $x(t)$ by time shifting the signal $x(t)$ to the right by 3 unit and is shown in Fig. 1.36f. By time expansion of $x(t-3)$ by a factor 2, $x\left(\frac{t}{2}-3\right)$ is obtained and sketched as shown in Fig. 1.36g.

(d) **To sketch the signal $x(-2t+3)$**
Signal $x(-t)$ is obtained by folding $x(t)$ and it is shown in Fig. 1.36h. $x(-t)$ is time shifted to the right by 3 unit to get $x(-t+3)$. This is shown in Fig. 1.36i. The signal $x(-t+3)$ is time compressed by a factor 2 to get $x(-2t+3)$. This is sketched in Fig. 1.36j.

(e) **To sketch the signal $x(-2t-3)$**
$x(-t)$ is shown in Fig. 1.36h. From Fig. 1.36h, $x(-t)$ is time shifted toward left by 3 units to get $x(-t-3)$. This is shown in Fig. 1.36k. $x(t-3)$ is time compressed by a factor 2 to get $x(-2t-3)$. This is sketched in Fig. 1.36l.

■ Example 1.10

A continuous time signal $x(t)$ is shown in Figure 1.37(a). Sketch and label carefully each of the following signals:

(a) $x(t-1)$

(b) $x(2-t)$

(c) $x(t)\left[\delta\left(t+\frac{3}{2}\right)-\delta\left(t-\frac{3}{2}\right)\right]$

(d) $x(2t+1)$

(*Anna University, April, 2008*)

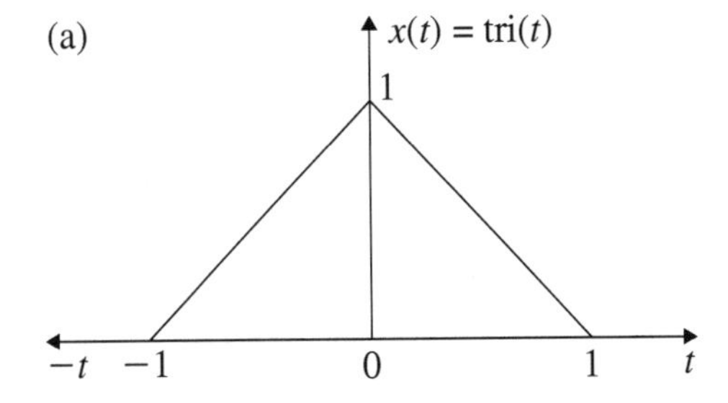

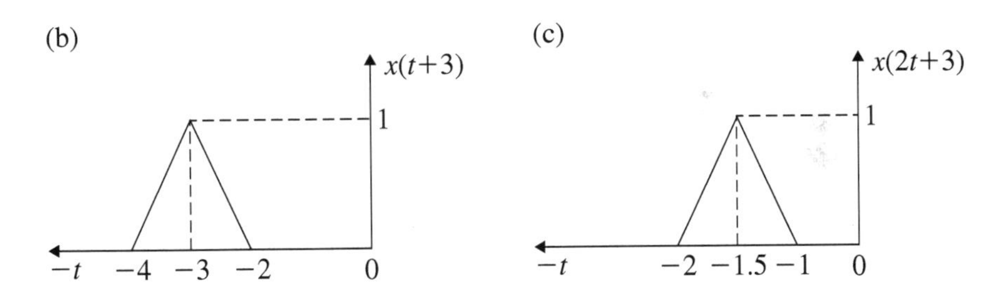

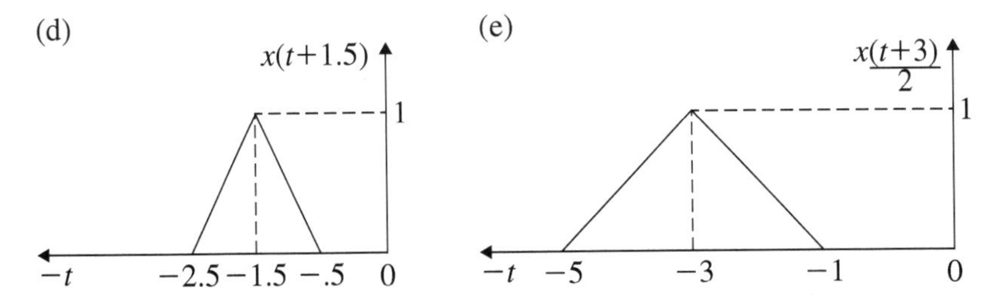

Fig. 1.36 **a** $x(t)=\text{tri}(t)$. **b** $x(t+3)$; **c** $x(2t+3)$. **d** $x(t+1.5)$; **e** $x\left(\frac{t+3}{2}\right)$. **f** $x(t-3)$; **g** $x\left(\frac{t}{2}-3\right)$. **h** $x(-t)$; *i* $x(-t+3)$; *j* $x(-2t+3)$. **k** $x(-t-3)$; **l** $x(-2t-3)$

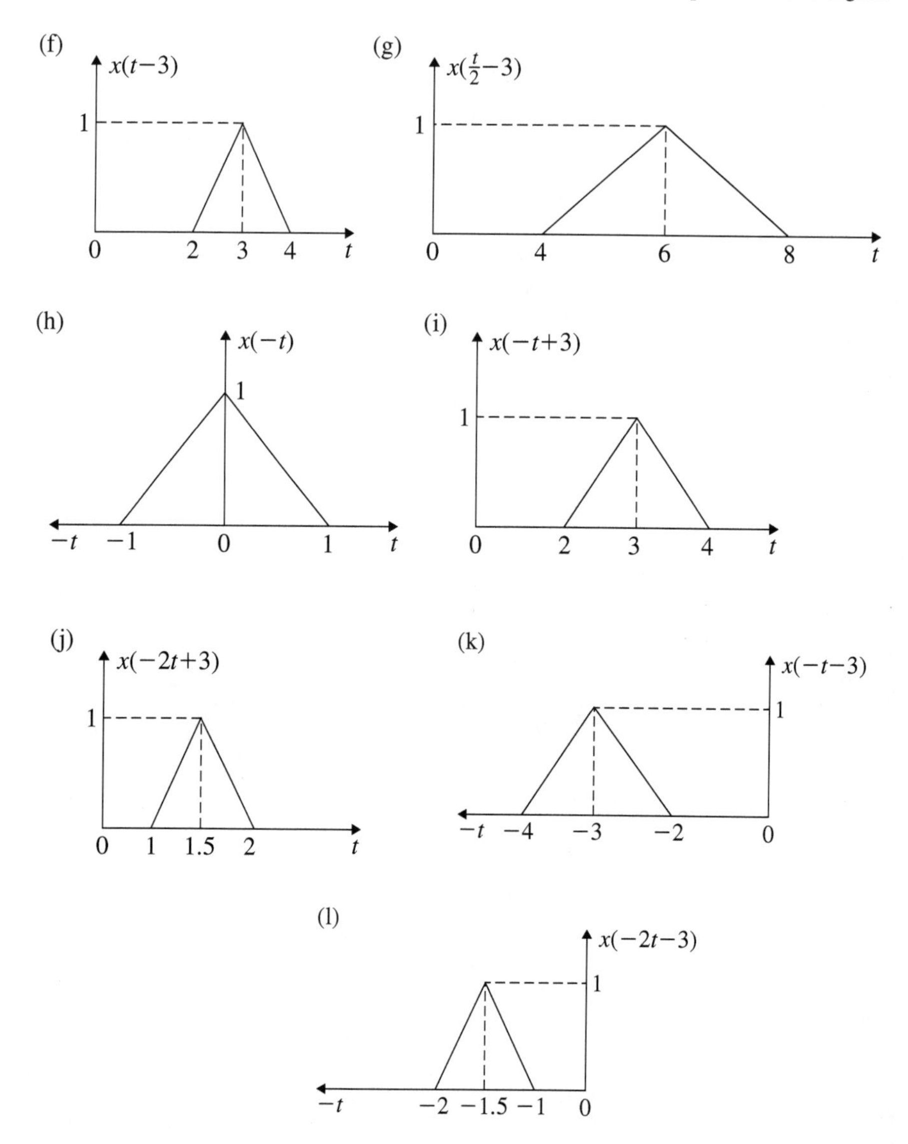

Fig. 1.36 (continued)

Solution:

(a) **To sketch $x(t-1)$**
$x(t-1)$ is the time delayed signal of $x(t)$ by one unit. $x(t)$ is shifted to the right by $t = 1$ and it is sketched as shown in Fig. 1.37b.

(b) **To sketch $x(2-t)$**
The folded signal of $x(t)$ is $x(-t)$ and is shown in Fig. 1.37c. $x(-t)$ is right shifted by 2 unit to get $x(2-t)$ and is shown in Fig. 1.37d.

(c) **To sketch $x(t)[\delta(t+\frac{3}{2})-\delta(t-\frac{3}{2})]$**
$\delta(t+\frac{3}{2})$ and $\delta(t-\frac{3}{2})$ are shown in Fig. 1.37e, which occur as unit impulses at $t=-\frac{3}{2}$ and $t=\frac{3}{2}$, respectively. At $t=-\frac{3}{2}$, $x(t)=-\frac{1}{2}$ and $\delta(t+\frac{3}{2})=1$. Using the property of impulse $x(t)\delta(t-t_0)=x(t_0)\delta(t-t_0)$, we get $x(t)\delta(t+\frac{3}{2})=-\frac{1}{2}$. Similarly at $t=\frac{3}{2}$, $x(t)=\frac{1}{2}$ and $-\delta(t-\frac{3}{2})=-1$. Hence, $x(t)\delta(t-\frac{3}{2})=-\frac{1}{2}$. This is sketched as shown in Fig. 1.37f.

(d) **To sketch $x(2t+1)$**
From Fig. 1.37a, $x(t+1)$ is derived by shifting $x(t)$ to the left by $t=1$. This is shown in Fig. 1.37g. By time compression of $x(t+1)$ by a factor 2, $x(2t+1)$ is obtained and sketched as shown in Fig. 1.37h.

■ Example 1.11

Sketch the signal $x(t)=[u(t)-u(t-a)]$ where $a>0$.

Solution:

(1) The unit step signal $u(t)$ is shown in Fig. 1.38a.
(2) The unit step signal with a time delay a and amplitude inverted is shown in Fig. 1.38b.
(3) If the above two step signals are added, a pulse signal is obtained and is sketched as shown in Fig. 1.38c which gives $u(t)-u(t-a)$. The above signal is defined as

$$x(t)=1 \qquad 0\le t\le a$$

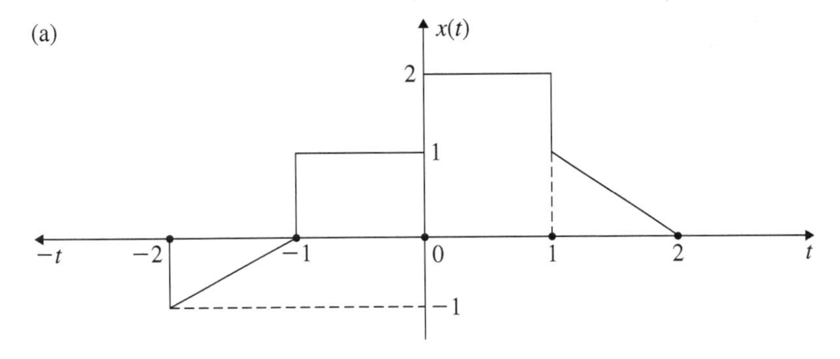

Fig. 1.37 **a** $x(t)$ plot. **b** $x(t-1)$ plot

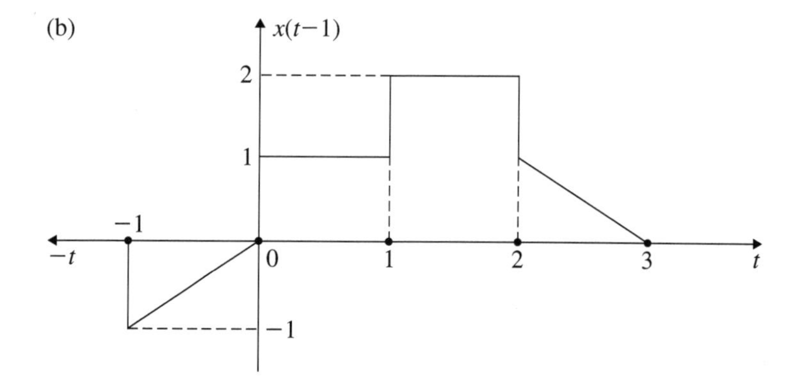

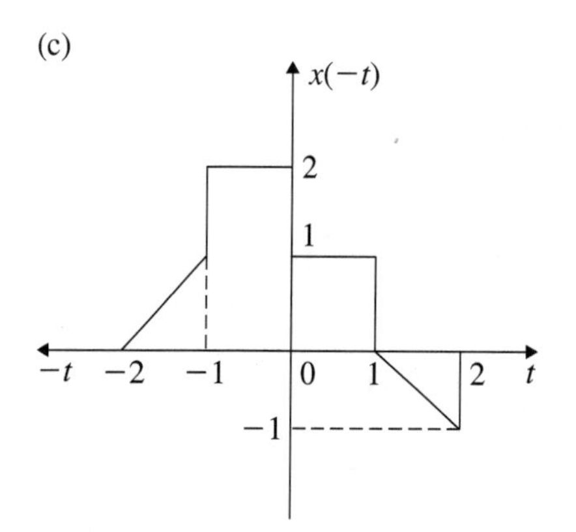

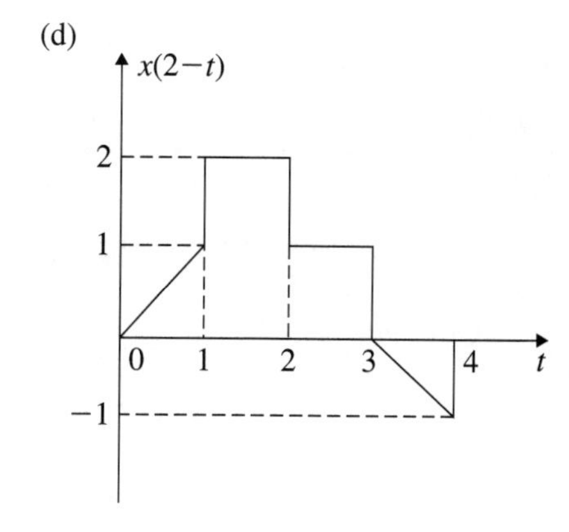

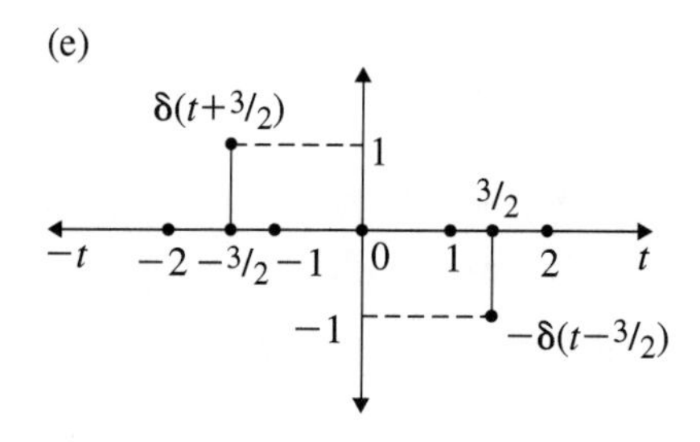

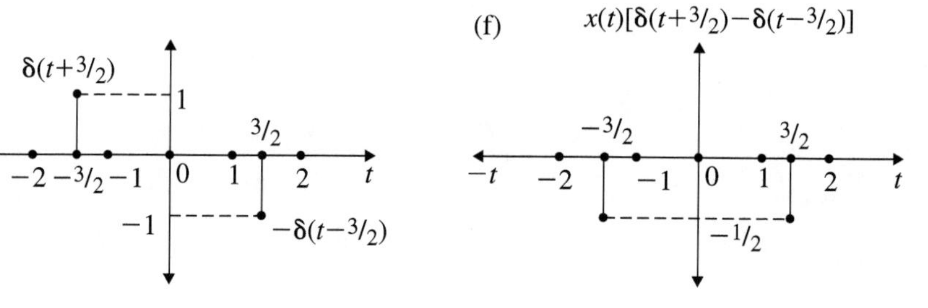

Fig. 1.37 (continued)

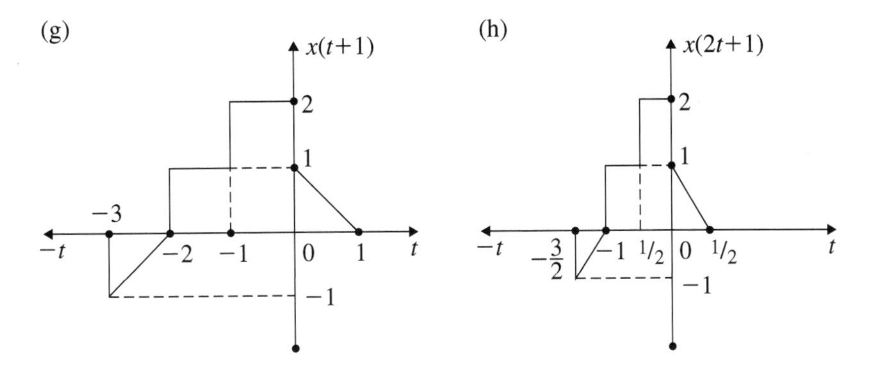

Fig. 1.37 (continued)

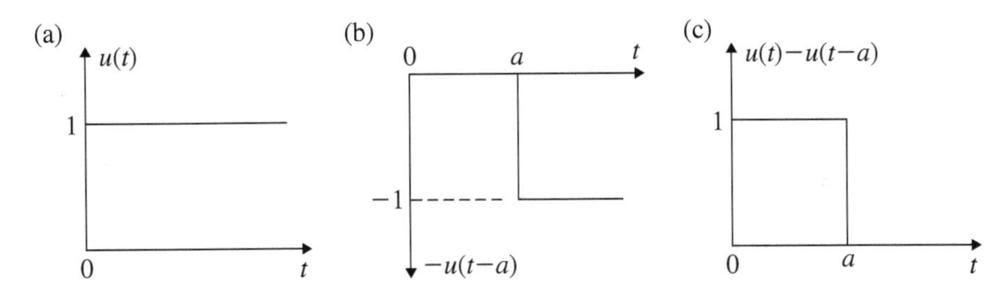

Fig. 1.38 Pulse signal from two step signals

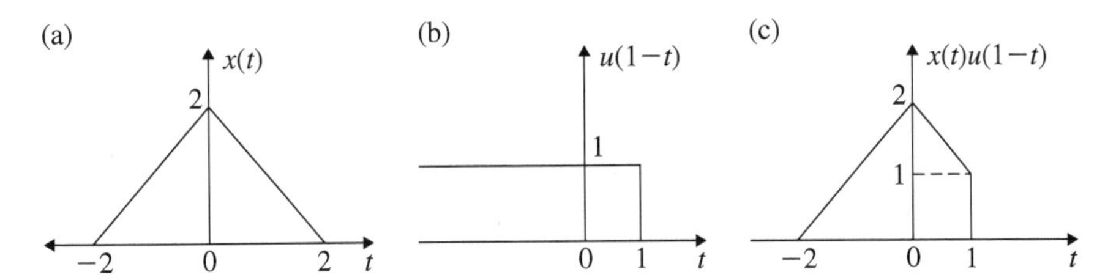

Fig. 1.39 Product of triangular and time delayed step signals

■ Example 1.12

Consider the signal $x(t)$ shown in Fig. 1.39a. Sketch the signal $x(t)u(1-t)$.

Solution:

1. The signal $x(t)$ is shown in Fig. 1.39a. The signal $u(1-t)$ is shown in Fig. 1.39b.
2. The signal $x(t)$ is multiplied by the factor 1 for the intervals $-2 \le t \le 0$ and $0 \le t \le 1$. During these time intervals, the slope of the straight lines of the triangles are $+1$ and -1, respectively. Hence, $x(t)$ is retained as it is. At $t = 1$, $x(t) = 1$ and $u(1-t) = 1$. Hence, $x(t)u(1-t) = 1$
3. For $t > 1, u(1-t) = 0$ and hence, $x(t)u(1-t) = 0$. This is sketched in Fig. 1.39c.

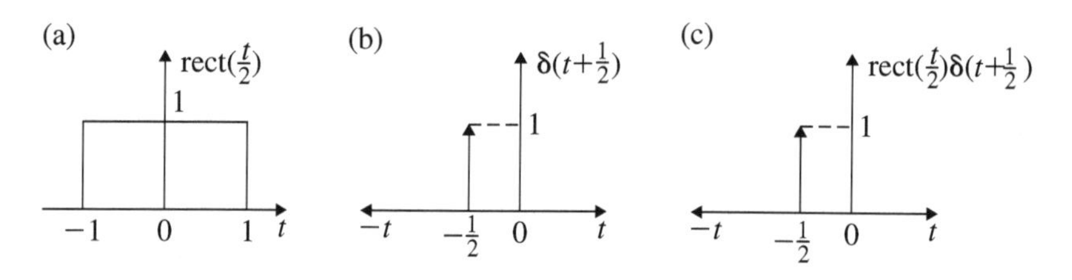

Fig. 1.40 Product of rectangular and time advanced impulse

■ Example 1.13

Consider the signal $\text{rect}(\frac{t}{2})$. Sketch the signal $\text{rect}(\frac{t}{2})\ \delta(t + \frac{1}{2})$.

Solution:

1. The rectangular pulse $\text{rect}(\frac{t}{2})$ is shown in Fig. 1.40a.
2. The time advanced impulse $\delta(t + \frac{1}{2})$ is defined as follows:

$$\delta\left(t + \frac{1}{2}\right) = 1 \quad \text{if } t = -\frac{1}{2}$$
$$= 0 \quad \text{otherwise}$$

   This is sketched in Fig. 1.40b.
3. At $t = -\frac{1}{2}$, the magnitude of $\text{rect}(t) = 1$. Hence, using the property $x(t)$ $\delta(t + t_0) = x(t_0)$, we sketch $x(t)\delta(t + t_0)$ as an impulse at $t = -\frac{1}{2}$ which is shown in Fig. 1.40c.

■ Example 1.14

$$x(t) = 10e^{-3t+4}$$

Determine $x(t + 2)$, $x(-t + 2)$ and $x(\frac{t}{4} - 5)$.

Solution:

1. For $t = t + 2$,

$$x(t) = 10e^{-3t+4}$$

$$x(t + 2) = 10e^{-3(t+2)+4}$$

$$\boxed{x(t + 2) = 10e^{-3t-2}}$$

2. For $t = -t + 2$,

$$x(-t + 2) = 10e^{-3(-t+2)+4}$$

$$\boxed{x(-t+2) = 10e^{3t-2}}$$

3. For $t = (\frac{t}{4} - 5)$,

$$x\left(\frac{t}{4} - 5\right) = 10e^{-3(\frac{t}{4}-5)+4}$$

$$\boxed{x\left(\frac{t}{4} - 5\right) = 10e^{-\frac{3}{4}t+19}}$$

■ Example 1.15

Decompose the signal $x(t)$ shown in Fig. 1.41a in terms of basic signals such as delta, step and ramp.

(Anna University, December, 2007)

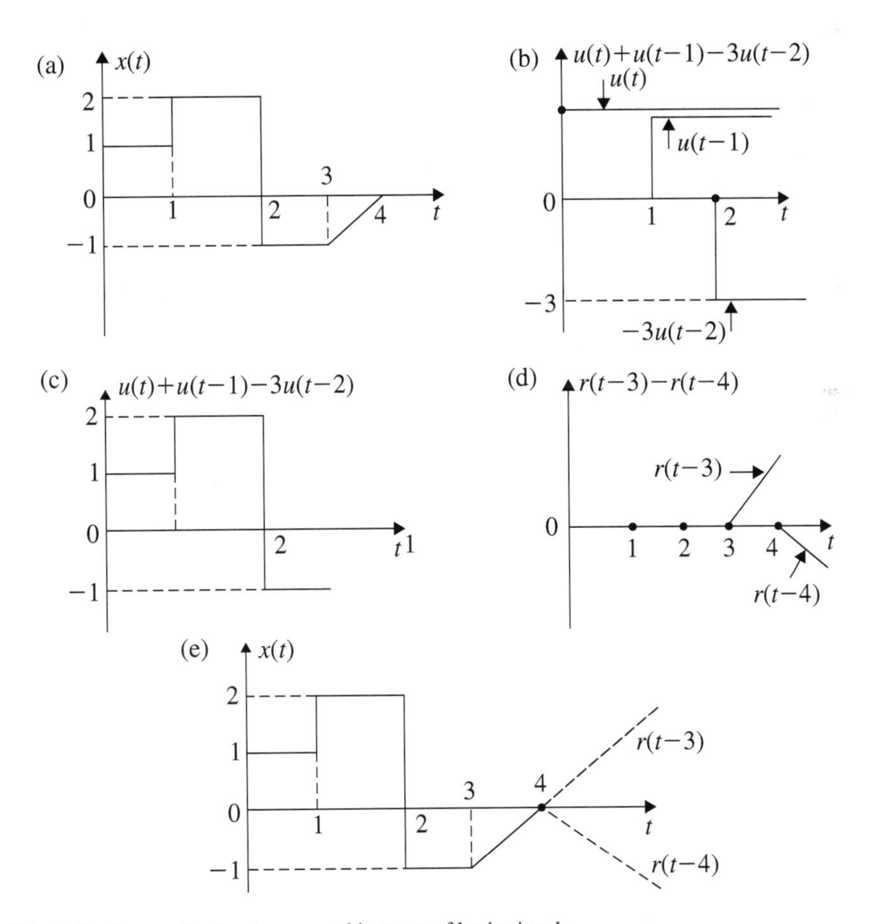

Fig. 1.41 Composite signal expressed in terms of basic signals

Solution:

1. The given signal $x(t)$ is shown in Fig. 1.41a.
2. The signals $u(t) + u(t-1) - 3u(t-2)$ are shown in Fig. 1.41b and their sum is shown in Fig. 1.41c.
3. The signals $r(t-3)$ and $r(t-4)$ are shown in Fig. 1.41d.
4. Signals in Fig. 1.41c, d are summed up and they are shown in Fig. 1.41e which is nothing but $x(t)$. Hence,

$$\boxed{x(t) = u(t) - u(t-1) - 3u(t-2) + r(t-3) - r(t-4)}$$

5. For the time $3 \leq t < 4$, the ramp signal is to be generated with a +ve slope of 1. The equation of this straight line signal is

$$r(t) = t + c$$

At $t = 3$, $r(t) = -1$

$$\begin{aligned}
-1 &= 3 + c \\
c &= -4 \\
r(t) &= (t-4) \qquad 3 \leq t < 4 \\
&= (t-4)[u(t-3) - u(t-4)]
\end{aligned}$$

$$\boxed{x(t) = u(t) + u(t-1) - 3u(t-2) + (t-4)[u(t-3) - u(t-4)]}$$

■ Example 1.16

Sketch the signals

$$\begin{aligned}
&\text{(a)} \qquad x(t) = -4\text{sgn}\, 3t \\
&\text{(b)} \qquad x(t) = 5\text{sinc}\, 10t
\end{aligned}$$

Solution:

(a) $\boldsymbol{x(t) = -4\text{sgn}3t}$
The signal sgn t is shown in Fig. 1.42a. The signum function is inverted and multiplied by a factor 4. The time compression by a factor 3 does not apply in this case as the signal remains constant for $-\infty < t < \infty$. The signal is shown in Fig. 1.42b.

(b) $\boldsymbol{x(t) = 5\text{sinc}10t}$
The signal sinc t is sketched in Fig. 1.42c. The sinc function amplitude is multi-

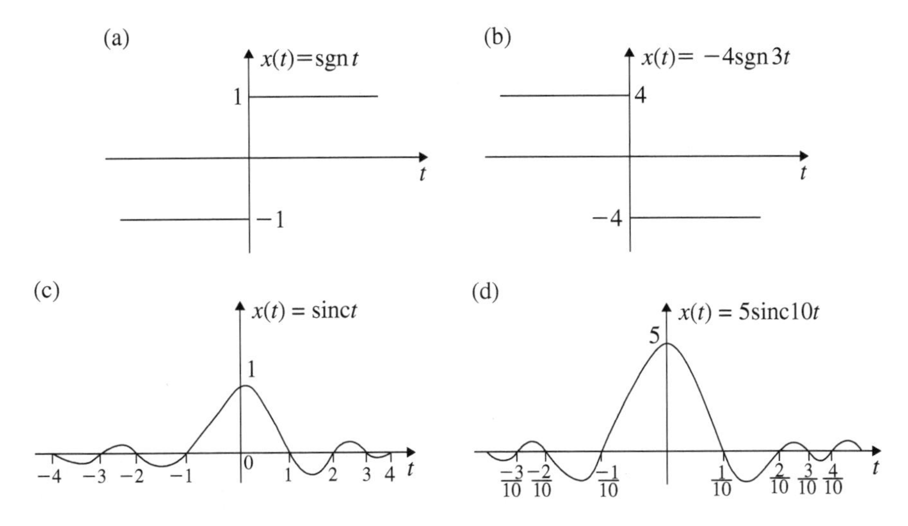

Fig. 1.42 Representation of signum and sinc functions

plied by the factor 5 and the time is compressed by the factor 10. $x(t) = 5 \sin 10t$ is represented in Fig. 1.42d.

■ Example 1.17

Consider the signal shown in Figure1.43(a) and sketch the following signals.

1. $x(t)u(t-1)$
2. $x(t)u(t+1)$
3. $x(t)u(-t-1)$
4. $x(t)u(-t+1)$
5. $x(t-1)u(t-1)$
6. $x(2t+1)u(t+1)$
7. $x(-2t-1)u(-t-1)$
8. $x(-2t+1)u(-t+1)$
9. $x(t)[u(t+1) - u(t-2)]$

Solution:

1.

$$u(t-1) = \begin{cases} 1 & t \geq 1 \\ 0 & t < 1 \end{cases}$$

$x(t)$ is sketched in Fig. 1.43a. It will be multiplied by 1 for $t \geq 1$ and by 0 for $t < 1$. $x(t)u(t-1)$ is shown in Fig. 1.43b.

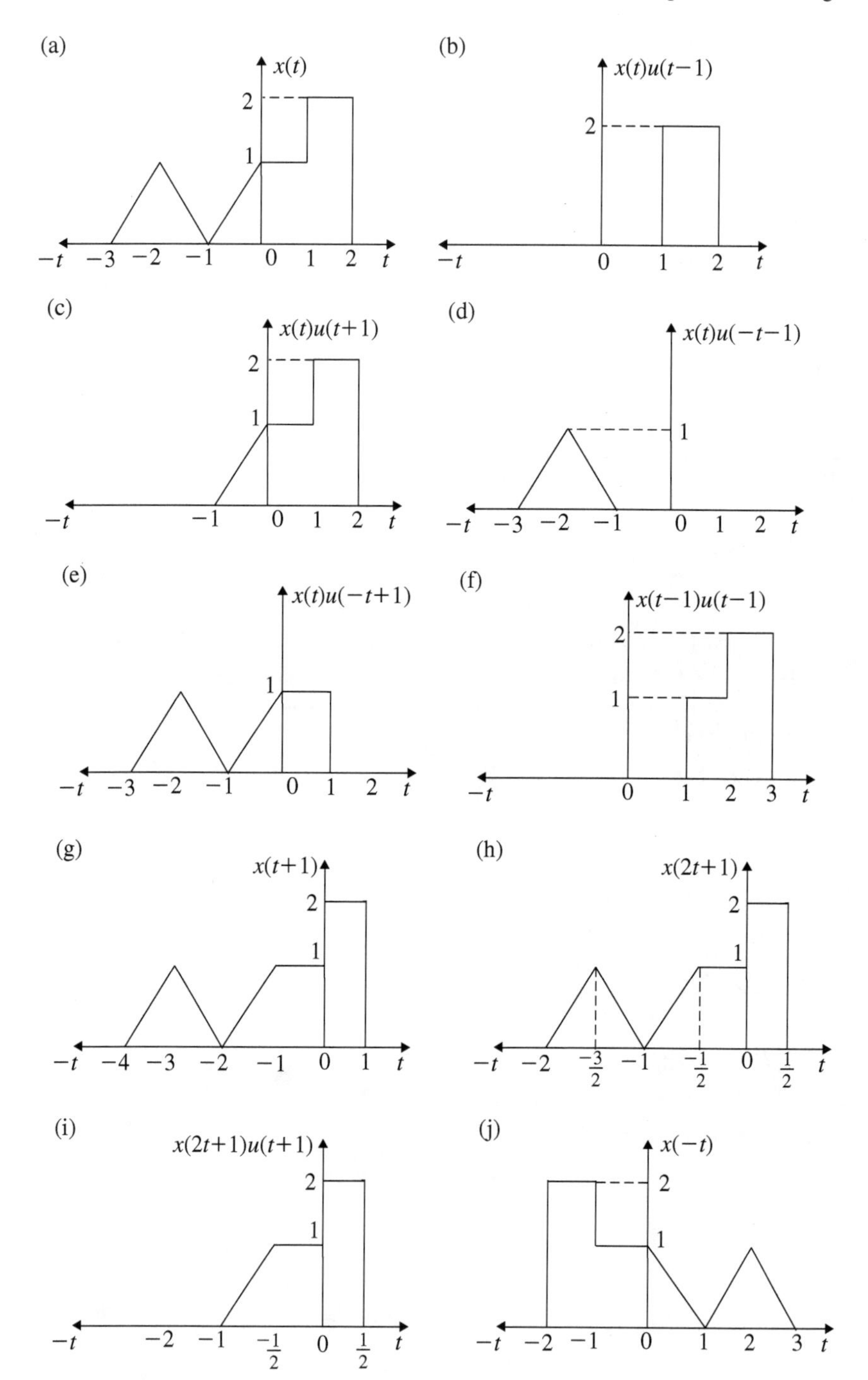

Fig. 1.43 Sketching of signal $x(t)$ for Example 1.17

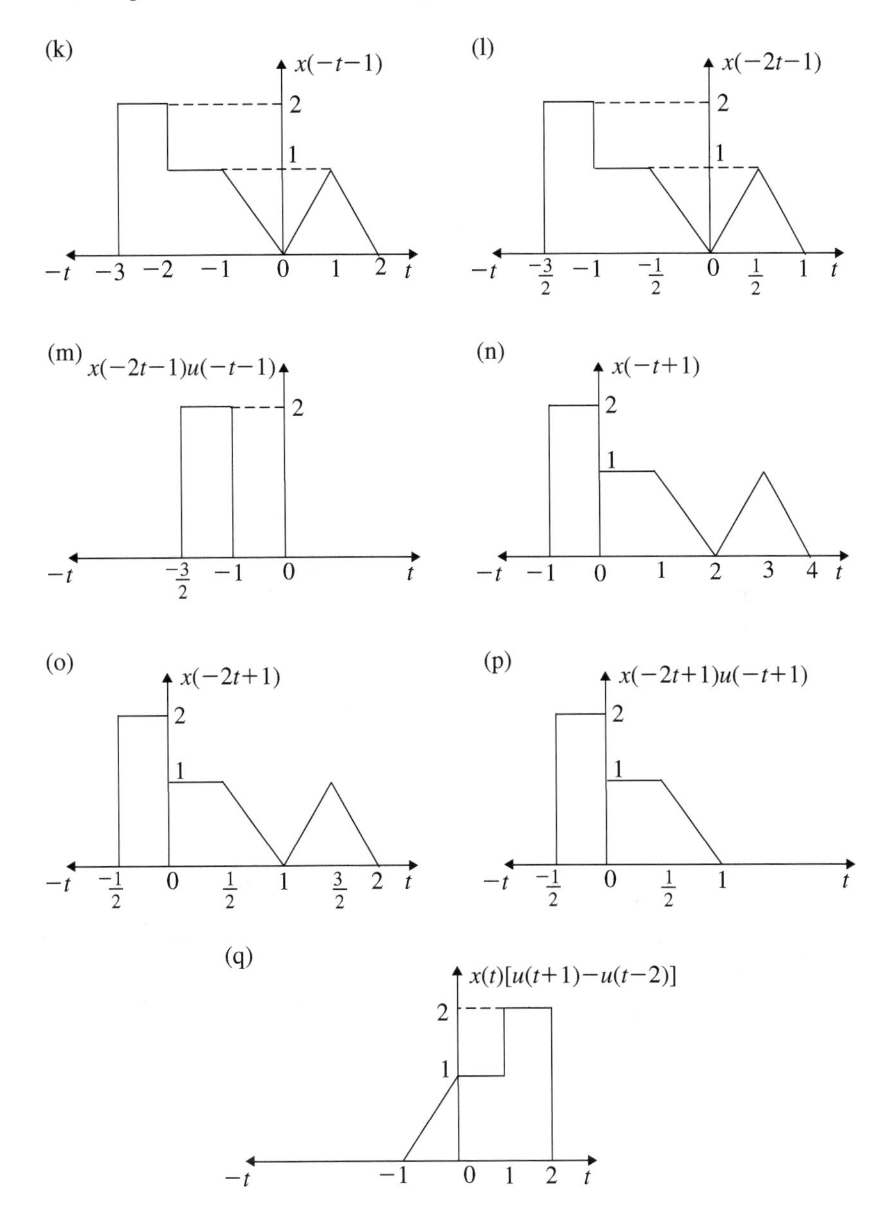

Fig. 1.43 (continued)

2.

$$u(t+1) = \begin{cases} 1 & t \geq -1 \\ 0 & t < -1 \end{cases}$$

$x(t)$ is multiplied by 1 for $t \geq -1$ and by 0 for $t < -1$. $x(t)u(t+1)$ is shown in Fig. 1.43c.

3.

$$u(-t-1) = \begin{cases} 1 & t \leq -1 \\ 0 & t > -1 \end{cases}$$

$x(t)$ is multiplied by 1 for $t \leq -1$ and by 0 for $t > -1$. $x(t)u(-t-1)$ is shown in Fig. 1.43d.

4.

$$u(-t+1) = \begin{cases} 1 & t \leq 1 \\ 0 & t > 1 \end{cases}$$

$x(t)$ is multiplied by 1 for $t \leq 1$ and by 0 for $t > 1$. $x(t)u(-t+1)$ is shown in Fig. 1.43e.

5.

$$u(t-1) = \begin{cases} 1 & t \geq 1 \\ 0 & t < 1 \end{cases}$$

$x(t)$ is right shifted by 1 to get $x(t-1)$. This is shown in Fig. 1.43e. $x(t-1)$ for $t > 1$ is identified and plotted as $x(t-1)u(t-1)$ and shown in Fig. 1.43f.

6.

$$u(t+1) = \begin{cases} 1 & t \geq -1 \\ 0 & t < -1 \end{cases}$$

$x(t)$ is left shifted by 1 and $x(t+1)$ is plotted as shown in Fig. 1.43g. $x(t+1)$ is time compressed by 2 and is shown in Fig. 1.43h. $x(2t+1)$ for $t \geq -1$ is multiplied by 1 and for $t < -1$ by 0 and $x(2t+1)u(t+1)$ is plotted as shown in Fig. 1.43i.

7.

$$u(-t-1) = \begin{cases} 1 & t \leq -1 \\ 0 & t > -1 \end{cases}$$

$x(-t)$ is sketched by reflection of $x(t)$ and is shown in Fig. 1.43j. It is left shifted by 1 to get $x(-t-1)$ and is shown in Fig. 1.43k. $x(-t-1)$ is time compressed by 2 and is shown as $x(-2t-1)$ in Fig. 1.43l. For $t \leq -1, x(-2t-1)$ is multiplied by 1 and for $t > -1$ by 0 and $x(-2t-1)u(-t-1)$ is obtained and plotted as shown in Fig. 1.43m.

8.

$$u(-t+1) = \begin{cases} 1 & t \leq 1 \\ 0 & t > 1 \end{cases}$$

$x(-t)$ is right shifted by 1 and plotted as $x(-t+1)$ as shown in Fig. 1.43n. $x(-t+1)$ is time compressed by 2 and $x(-2t+1)$ is plotted as shown in Fig. 1.43o. For $t < 1, x(-2t+1)$ is multiplied by 1 and for $t \geq 1$, it is multiplied by zero. Thus, $x(-2t+1)u(-t+1)$ is obtained and sketched as shown in Fig. 1.43p.

9.

$$[u(t+1) - u(t-2)] = \begin{cases} 1 & -1 \leq t < 2 \\ 0 & \text{otherwise} \end{cases}$$

$x(t)$ is multiplied by 1 for the time interval $-1 \leq t < 2$ and by 0 elsewhere. Thus, $x(t)[u(t+1) - u(t-2)]$ is obtained and sketched as shown in Fig. 1.43q.

■ Example 1.18

(I) Consider the CT signals shown in Fig. 1.44a–c. For each of these figures give the mathematical description. Derive the first derivative of these equations in terms of singularity equations. Sketch the waveform of the derivative signals.

(II) Consider the mathematical description of a certain signal which is given as

$$x(t) = (2t-6)[u(t+1) - u(t-4)]$$

Sketch the signal $x(t)$.

Solution:

(I) (a)

$$x(t) = u(t) - u(t-a)$$

The above equation represents a rectangular pulse of width a and height 1 and is represented in Fig. 1.44a. The derivative of the above equation is obtained as

$$\begin{aligned} \frac{dx(t)}{dt} &= \frac{du(t)}{dt} - \frac{du(t-a)}{dt} \\ &= \delta(t) - \delta(t-a) \end{aligned}$$

The impulses at $t = 0$ and $t = a$ are shown in Fig. 1.44d. From Fig. 1.44a, it is evident that the derivative of a pulse can be obtained just by observation. For $x(t)$ being constant, the derivative will be zero. At $t = 0$, there is an up going pulse of strength 1 and at $t = a$, there is a down going pulse of strength -1.

(b)

$$x(t) = \begin{cases} t & 0 \leq t \leq a \\ 0 & \text{otherwise} \end{cases}$$

The above equation represents a triangle with unity slope with a base of a and is sketched as shown in Fig. 1.44b. Further, the above equation can be written as follows:

$$x(t) = t[u(t) - u(t-a)]$$

Differentiating the two time functions using $u - v$ method, we get

$$\begin{aligned} \frac{dx(t)}{dt} &= t\left[\frac{du(t)}{dt} - \frac{du(t-a)}{dt}\right] + [u(t) - u(t-a)] \\ &= t\delta(t) - t\delta(t-a) + [u(t) - u(t-a)] \\ &= 0 - a\delta(t-a) + u(t) - u(t-a) \end{aligned}$$

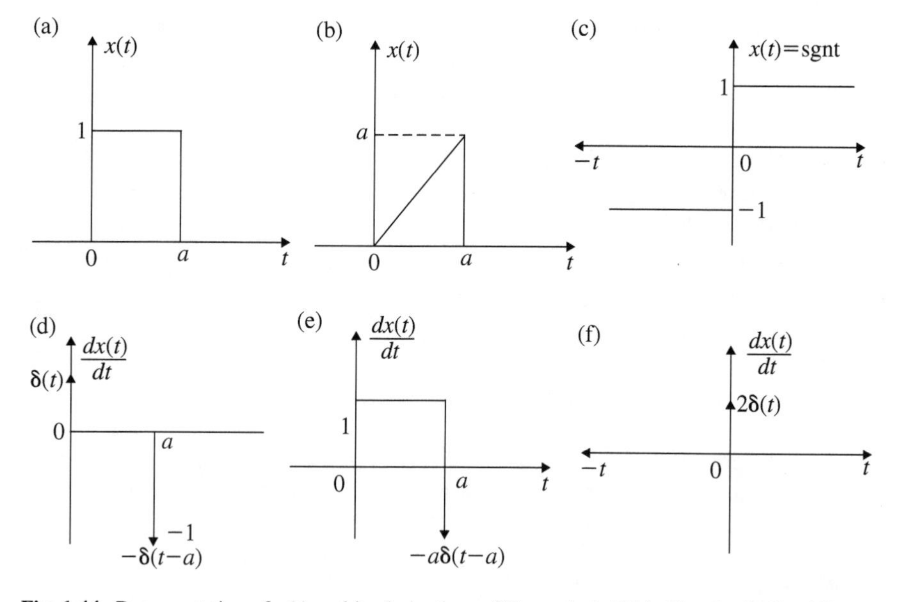

Fig. 1.44 Representation of $x(t)$ and its derivatives of Example 1.18(a). Sketch of $x(t)$ of Example 1.18(b)

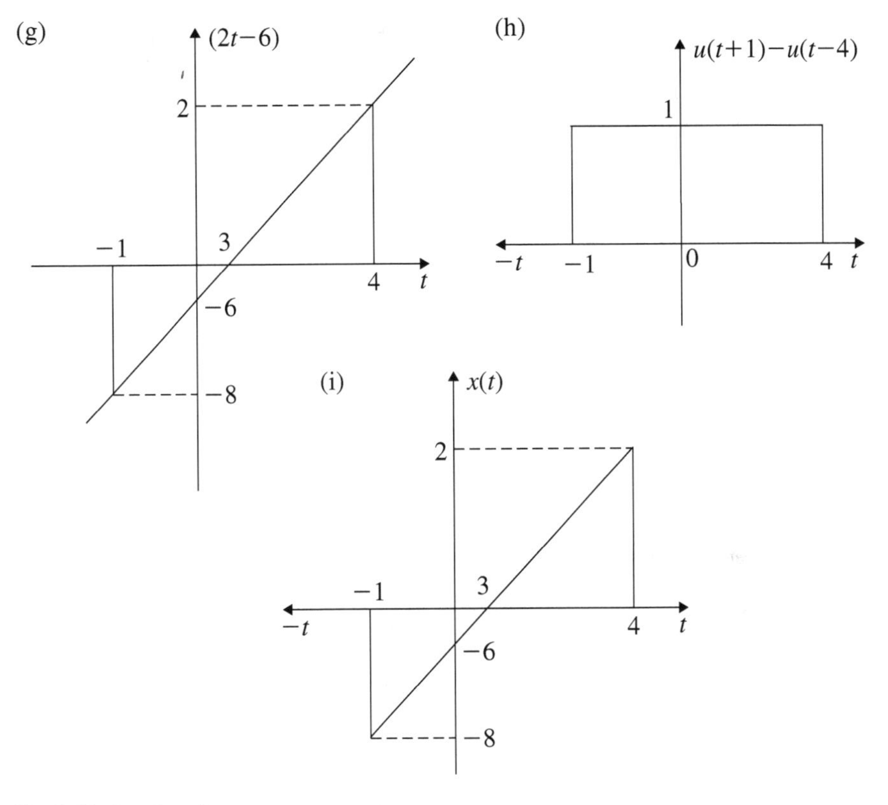

Fig. 1.44 (continued)

Thus, the derivative of the triangle is obtained as the sum of a rectangular pulse of width a and an impulse of strength $-a$ occurring at $t = a$. This is represented in Fig. 1.44e. This can be also obtained just by observation. For the straight line of the triangle, the slope is one, and correspondingly, the rectangular pulse in the derivative of $x(t)$ will have the height which is nothing but the slope (which is 1 here) and the width of the rectangular pulse is "a" which is the base of the triangle. The triangle has a negative going impulse at $t = a$. The rectangular pulse together with the impulse is shown in Fig. 1.44e.

(c)

$$\begin{aligned} x(t) &= \text{sgn}\, t \\ &= u(t) - u(-t) \\ \frac{dx(t)}{dt} &= \delta(t) - \delta(-t) \\ &= \delta(t) + \delta(t) \\ &= 2\delta(t) \end{aligned}$$

$x(t)$ is shown in Fig. 1.44c and its derivative in Fig. 1.44f. By observation of Fig. 1.44c, its derivation can be obtained. Since signal is constant at all time t except $t = 0$, its derivative is zero. At $t = 0$, there is an up going impulse from -1 to 1. This impulse is represented as $2\delta(t)$.

(II) 1. $(2t - 6)$ is a straight line of slope $+2$ and is passing through $t = 3$ with zero magnitude. This is sketched as shown in Fig. 1.44g.

2.

$$[u(t+1) - u(t-4)] = \begin{cases} 1 & -1 \leq t < 4 \\ 0 & \text{otherwise} \end{cases}$$

The signal of the equation is sketched in Fig. 1.44h.

3. $(2t - 1)$ will be multiplied by 1 for $-1 \leq t < 4$ and by zero elsewhere. The plot of $x(t)$ is shown in Fig. 1.44i.

■ Example 1.19

Sketch the signal $x(t)$ which has the following mathematical description.

$$x(t) = \begin{cases} \frac{3}{2}t + 3 & -2 \leq t \leq 0 \\ 3e^{-3t} & 0 \leq t < 4 \\ 0 & \text{otherwise} \end{cases}$$

Also, sketch the signals $x(t/4)$ and $x(5t)$ and give their mathematical descriptions.

Solution:

1. $x(t) = \frac{3}{2}t + 3$ for $-2 \leq t > 0$ is a straight line with a slope 3/2 and $x(t) = 3e^{-3t}$ for $0 \leq t < 4$ is an exponential decay. These two are combined and shown as $x(t)$ in Fig. 1.45a.
2. $x(t/4)$ is the time expanded signal. In the signal description of $x(t/4)$, time t in $x(t)$ is replaced by $t/4$ and the time axis t is expanded by 4. Thus, the following mathematical description is written for $x(t/4)$

$$x(t/4) = \begin{cases} \frac{3}{8}t + 3 & -2 \leq \frac{t}{4} < 0 \\ & \text{or } -8 \leq t > 0 \end{cases}$$

$$= \begin{cases} 3e^{-\frac{3}{4}t} & 0 \leq \frac{t}{4} < 4 \\ & \text{or } 0 \leq t < 16 \\ 0 & \text{otherwise} \end{cases}$$

$x(t/4)$ with the above mathematical description is sketched as shown in Fig. 1.45b.

3. $x(5t)$ is the time compressed signal. To sketch $x(5t)$, time t in $x(t)$ is replaced by $5t$ and the time axis is compressed by a factor 5. Thus, the following mathematical description is given for $x(5t)$

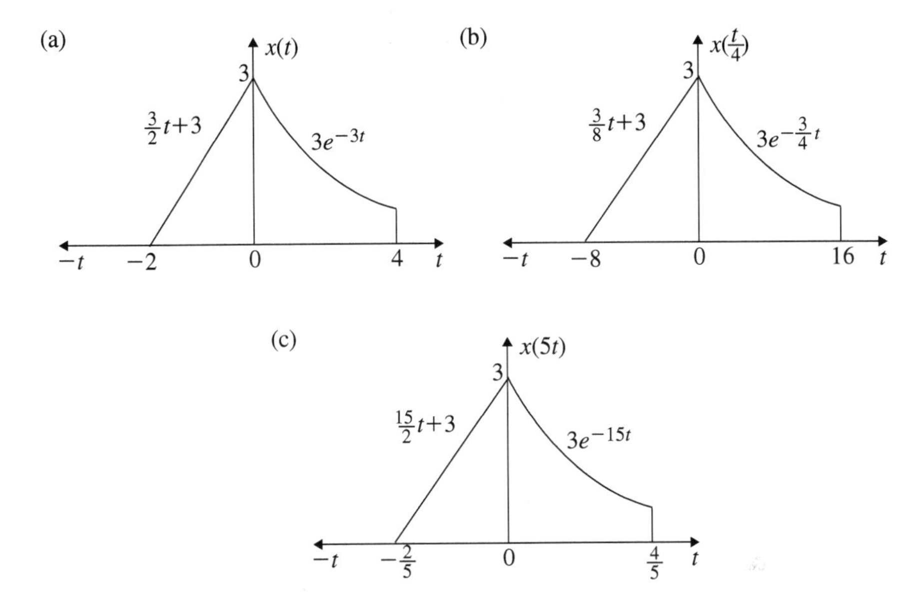

Fig. 1.45 Graphical representation $x(t)$, $x(t/4)$ and $x(5t)$ of Example 1.19

$$x(5t) = \begin{cases} \frac{15}{2}t + 3 & -2 \leq 5t < 0 \\ & \text{or } \frac{-2}{5} \leq t < 0 \end{cases}$$

$$= \begin{cases} 3e^{-15t} & 0 \leq 5t < 4 \\ & \text{or } 0 \leq t < \frac{4}{5} \end{cases}$$

With the above mathematical description $x(5t)$ is sketched and is shown in Fig. 1.45c.

■ Example 1.20

The signal $x(t)$ is given by

$$x(t) = \begin{cases} (t+1) & -1 \leq t < 0 \\ 1 & 0 \leq t < 2 \\ 0 & \text{otherwise} \end{cases}$$

Find the mathematical and graphical representation for $x(-t)$, $x(3-t)$ and $x(t/2)$.

(*Anna University, 2011*)

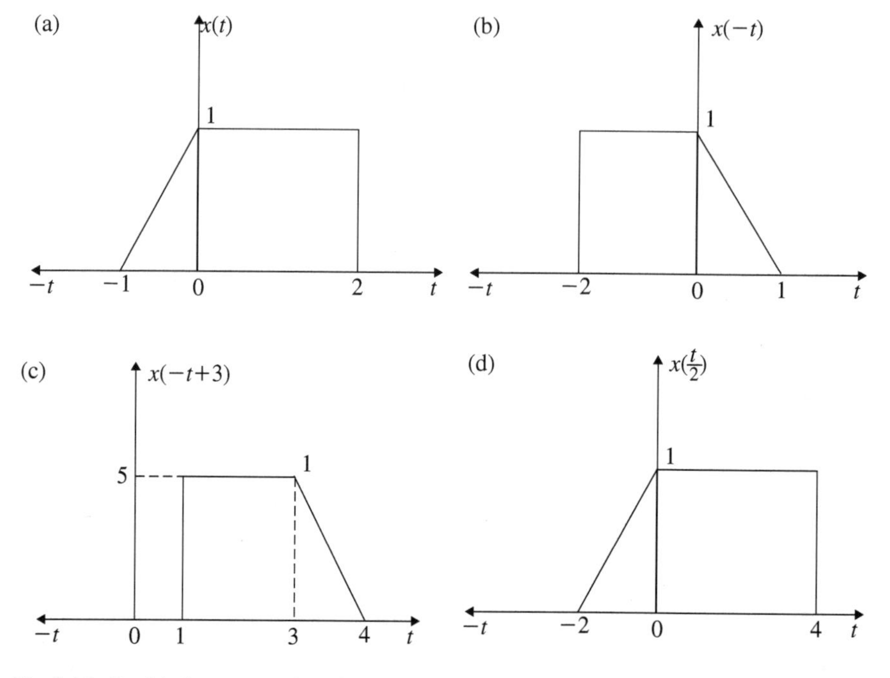

Fig. 1.46 Graphical representation of $x(t)$, of Example 1.20

Solution:

1. The graphical representation of $x(t)$ is shown in Fig. 1.46a. $x(t)$ can be mathematically represented by a single equation for all t as

$$\begin{aligned} x(t) &= (t+1)[u(t+1)-u(t)] + [u(t)-u(t-2)] \\ &= (t+1)u(t+1) - tu(t) - u(t-2) \end{aligned}$$

2. $x(-t)$ is obtained from $x(t)$ by signal reflection and is shown in Fig. 1.46b. $x(-t)$ is mathematically represented as

$$x(-t) = \begin{cases} 1 & -2 \le t < 0 \\ (1-t) & 0 \le t < 1 \\ 0 & \text{otherwise} \end{cases}$$

The above equations can also be written as a single equation for all time t as

$$\begin{aligned} x(t) &= [u(t+2)-u(t)] + (1-t)[u(t)-u(t-1)] \\ &= [u(t+2) - tu(t) + (t-1)u(t-1)] \end{aligned}$$

3. $x(-t)$ is right shifted by 3 sec. to get $x(3-t)$ and is graphically represented in Fig. 1.46c. The following mathematical equation is written for $x(3-t)$.

$$x(3-t) = \begin{cases} 1 & 1 \le t < 3 \\ 4-t & 3 \le t < 4 \\ 0 & \text{otherwise} \end{cases}$$

The following single equation for $x(3-t)$ for all time t is written.

$$\begin{aligned} x(3-t) &= [u(t-1) - u(t-3)] + (4-t)[u(t-3) - u(t-4)] \\ &= u(t-1) - (t-3)u(t-3) + (t-4)u(t-4) \end{aligned}$$

4. $x(t/2)$ is obtained by time expansion of $x(t)$ and is shown in Fig. 1.46d. The following mathematical equations are written for $x(t/2)$.

$$x\left(\frac{t}{2}\right) = \begin{cases} \frac{1}{2}t+1 & -2 \le t < 0 \\ 1 & 0 \le t < 4 \\ 0 & \text{otherwise} \end{cases}$$

The above equations can be written as a single equation as

$$\begin{aligned} x\left(\frac{t}{2}\right) &= \left(\frac{1}{2}t+1\right)[u(t+2) - u(t)] + [u(t) - u(t-4)] \\ &= \left(\frac{1}{2}t+1\right)u(t+2) - \frac{1}{2}tu(t) - u(t-4) \end{aligned}$$

■ Example 1.21

Carefully sketch the following signals. Express the signals in terms of singularity functions.

1. $x(t) = 5\text{rect}\left(\frac{t}{4}\right)$
2. $x(t) = \text{rect}\frac{(t+2)}{4}$
3. $x(t) = \text{rect}(-4t+5)$
4. $x(t) = \begin{cases} 1 & 0 \le t < 3 \\ 0 & 3 \le t < 5 \\ 1 & 5 \le t < 8 \\ 0 & \text{otherwise} \end{cases}$

   Express $x(t)$as products as well as sum of step functions.
5. $x(t) = 2\text{rect}\frac{(t-5)}{2}$

6. $x(t) = \text{rect}\dfrac{(t-5)}{5} + \text{rect}\dfrac{(t-6)}{2}$

7. $x(t) = r(t) + r(t-1) + 2r(t-2) - 2r(t-3) - 3u(t-6)$

8. $x(t) = r(t+3) - r(t) + r(t-3)$

9. $x(t) = u(t)u(5-t)$

10. $x(t) = u(t) + \delta(t-3)$

11. $x(t) = u(t)\delta(t-3)$

12. $x(t) = \begin{cases} r(t+5) & -5 \le t < -3 \\ 2 & -3 \le t < 2 \\ 4 & 2 \le t \le 4 \\ -r(t-4) & 4 \le t < 5 \\ 0 & \text{otherwise} \end{cases}$

13. $x(t) = u\,(t+4)\,r(-t+2)$

14. $x(t) = \text{rect}\left(\dfrac{t}{4}\right)\text{tri}(t+1.5)$

15. $x(t) = \text{rect}\left(\dfrac{t}{2}\right)\text{tri}\left(\dfrac{t}{4}\right)$

16. $x(t) = r(2t+1)\text{rect}\dfrac{(t-2)}{4}$

Solution:

1.

$$x(t) = 5\text{rect}\left(\frac{t}{4}\right)$$

rect(t) is plotted as shown in Fig. 1.47a.

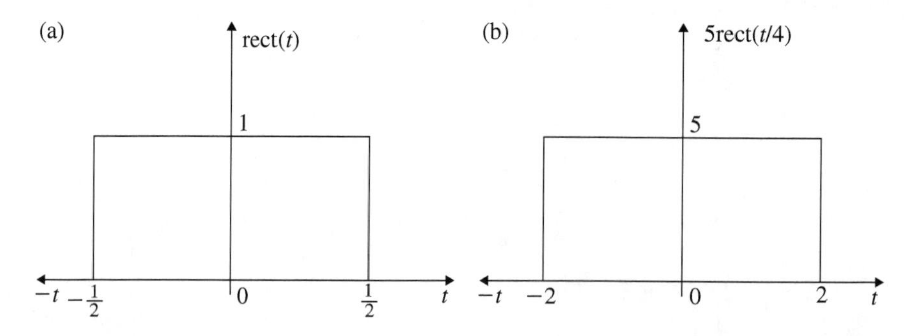

Fig. 1.47 Rectangular CT signal

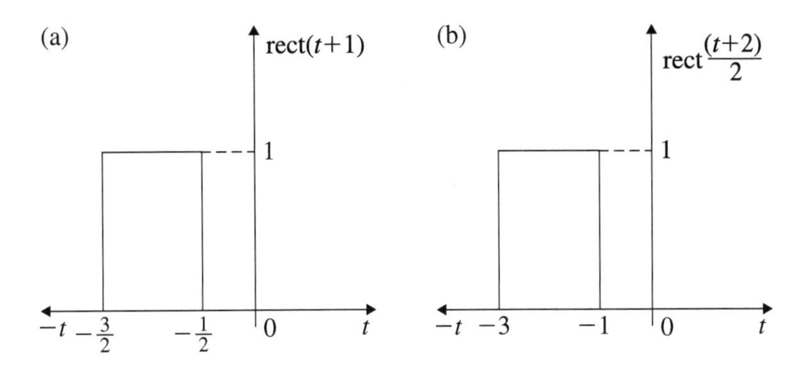

Fig. 1.48 Time shifted rectangular CT signal

Its amplitude is multiplied by 5 and time is expanded by 4 to get $x(t) = 5\text{rect}\left(\frac{t}{4}\right)$ which is shown in Fig. 1.47b. The mathematical equation for the above signal is

$$x(t) = 5[u(t+2) - u(t-2)]$$

or

$$x(t) = 5u(t+2)u(-t+2)$$

2.

$$\text{rect}\frac{(t+2)}{2} = \text{rect}\left(\frac{t}{2}+1\right)$$

$x(t)$ is left shifted by 1 sec and plotted as $x(t+1)$ which is shown in Fig. 1.48a. $\text{rect}(t+1)$ is time expanded by a factor 2 to get $\text{rect}(\frac{t}{2}+1)$ which is nothing but$\text{rect}\frac{(t+2)}{2}$. This is plotted in Fig. 1.48b. The mathematical equation for this signal is written as

$$x(t) = u(t+3) - u(t+1)$$

or

$$x(t) = u(t+3)u(-t-1)$$

3. The signal $x(-t)$ is sketched in Fig. 1.49a. $x(-t)$ is right shifted by 5 sec to get $x(-t+5)$ and is sketched in Fig. 1.49b. $x(-t+5)$ is time compressed by a factor 4 to get $x(-4t+5)$ which is sketched as shown in Fig. 1.49c. The equation for $x(-4t+5)$ can be written as

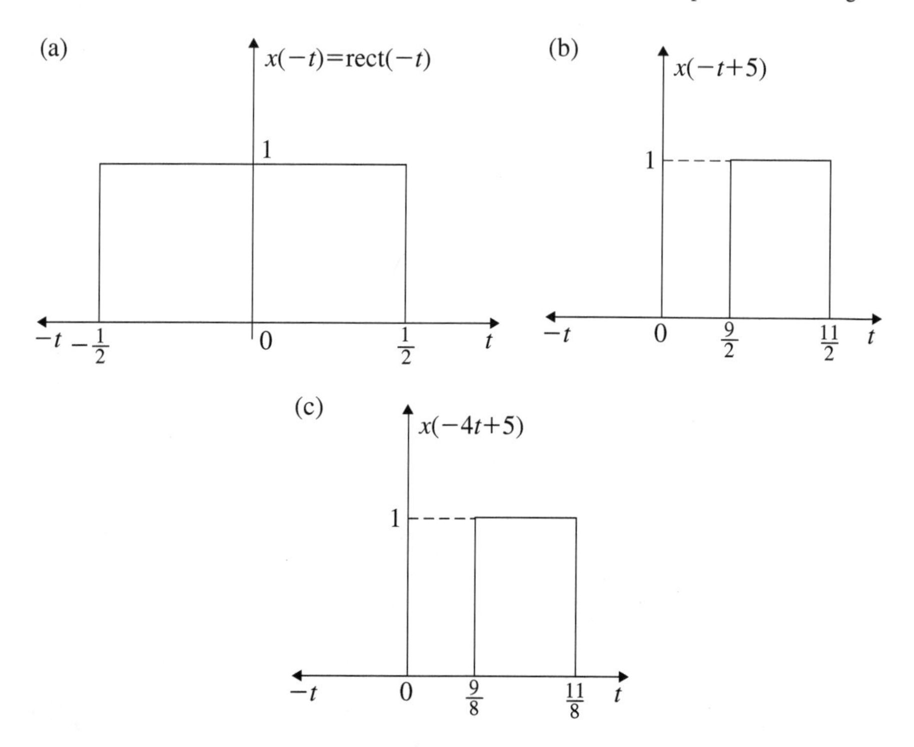

Fig. 1.49 $x(t) = \text{rect}(-4t + 5)$ graphical representation

$$\begin{aligned} x(t) &= u\left(t - \frac{9}{8}\right) - u\left(t - \frac{11}{8}\right) \\ &= u\left(t - \frac{9}{8}\right) u\left(-t + \frac{11}{8}\right) \end{aligned}$$

4.

$$x(t) = \begin{cases} 1 & 0 \le t < 3 \\ 0 & 3 \le t < 5 \\ 1 & 5 \le t < 8 \\ 0 & \text{otherwise} \end{cases}$$

The signal $x(t)$ is sketched as shown in Fig. 1.50. $x(t)$ can be expressed in the mathematical forms as

$$x(t) = u(t) - u(t - 3) + u(t - 5) - u(t - 8)$$

or

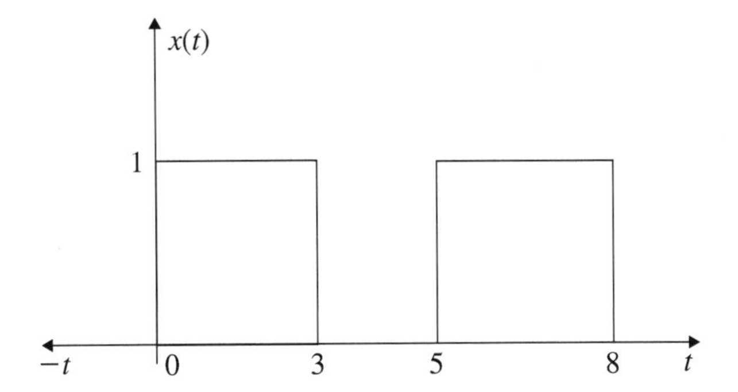

Fig. 1.50 Graphical representation of Example 1.21.4

$$x(t) = u(t)u(3-t) + u(t-5)u(8-t)$$

5.

$$\begin{aligned} x(t) &= 2\text{rect}\frac{(t-5)}{2} \\ &= 2\text{rect}\left(\frac{t}{2} - \frac{5}{2}\right) \end{aligned}$$

rect$(t - \frac{5}{2})$ is sketched by right shifting 2rect(t) by $\frac{5}{2}$ as shown in Fig. 1.51a and 2rect$(t - \frac{5}{2})$ is obtained. This signal is time expanded by a factor of 2 to get $x(t) = 2\text{rect}\frac{(t-5)}{2}$ and is sketched as shown in Fig. 1.51b.
The mathematical expression for $x(t)$ is written as

$$x(t) = 2[u(t-4) - u(t-6)]$$

or

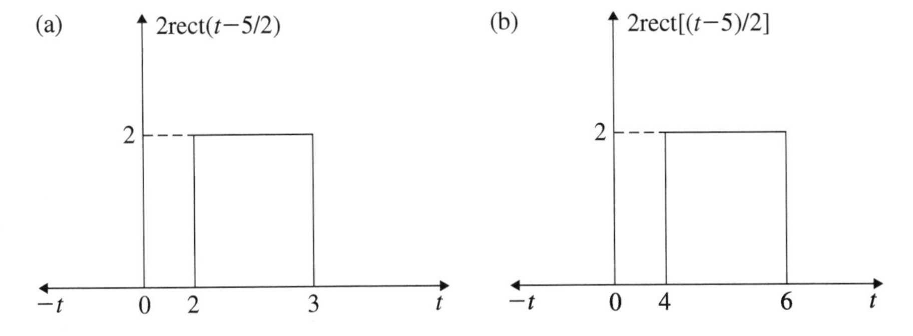

Fig. 1.51 Graphical representation of Example 1.21.5

$$x(t) = 2u(t-4)u(6-t)$$

6.

$$x(t) = \text{rect}\frac{(t-5)}{(5)} + \text{rect}\frac{(t-6)}{2}$$

rect$(t-1)$ is plotted by shifting rect(t) by 1 to the right and is shown in Fig. 1.52a.

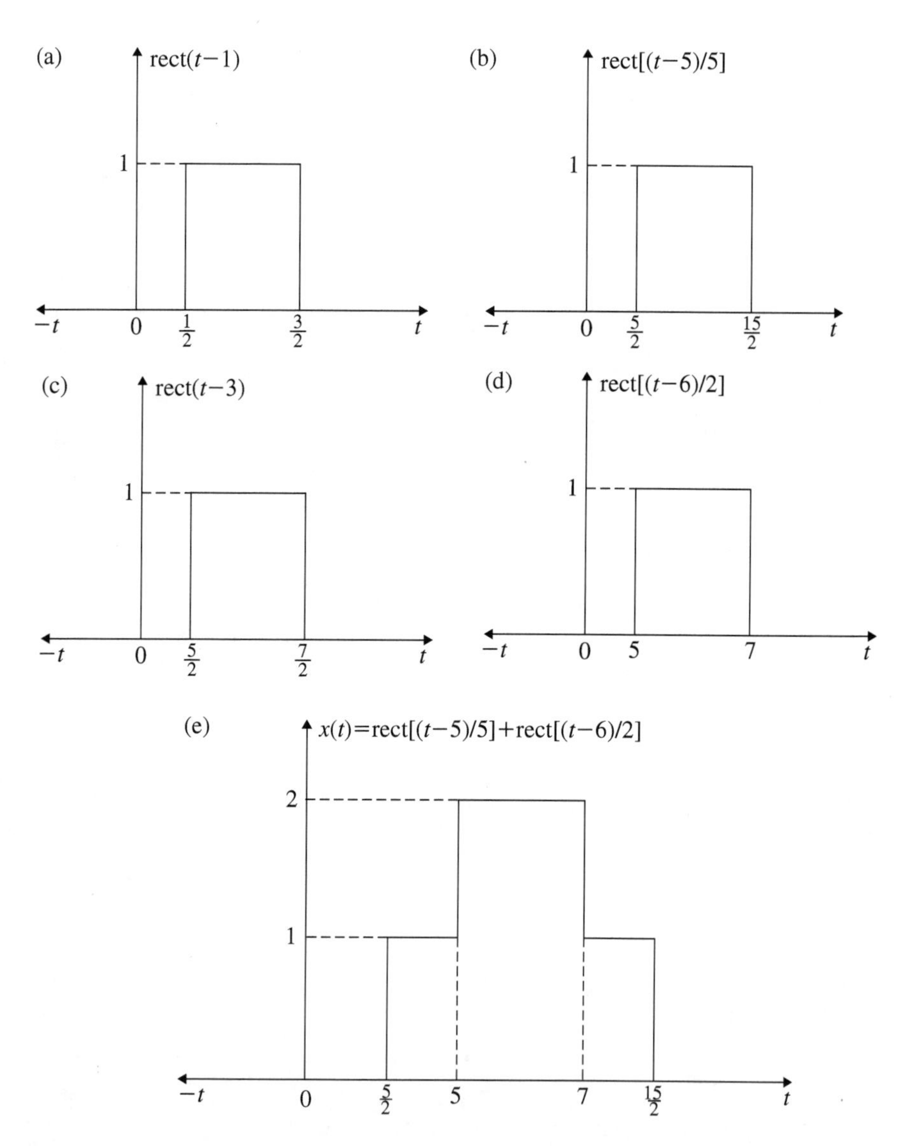

Fig. 1.52 Graphical representation of Example 1.21.6

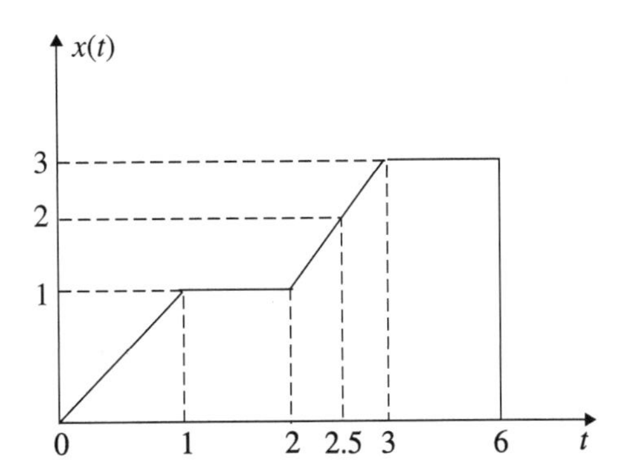

Fig. 1.53 Graphical representation of Example 1.21.7

$$\text{rect}\frac{(t-5)}{5} = \text{rect}\left(\frac{t}{5} - 1\right)$$

Now $\text{rect}(t-1)$ is time expanded by 5 and $\text{rect}\frac{(t-5)}{5}$ is plotted as shown in Fig. 1.52b.

$$\text{rect}\frac{(t-6)}{2} = \text{rect}\left(\frac{t}{2} - 3\right)$$

$\text{rect}(t-3)$ is plotted by right shifting $\text{rect}(t)$ by 3. $\text{rect}(t-3)$ (Fig. 1.52c) is time expanded by 2 to get $\text{rect}\frac{(t-3)}{2}$ which is shown in Fig. 1.52(d). Figure 1.52b, d is added to get Fig. 1.52e which represents

$$x(t) = \text{rect}\frac{(t-5)}{5} + \text{rect}\frac{(t-6)}{2}$$

From Fig. 1.52e, the following equation is derived for $x(t)$.

$$\boxed{x(t) = u\left(t - \frac{5}{2}\right) + u(t-5) - u(t-7) - u\left(t - \frac{15}{2}\right)}$$

7.

$$x(t) = r(t) - r(t-1) + 2r(t-2) - 2r(t-3) - 3u(t-6)$$

For $0 \leq t < 1$, the ramp signal with slope one is drawn. ramp$(t-1)$ cancels $r(t)$ and hence, $x(t)$ remains constant at 1. At $t = 2$, the ramp $(t-2)$ starts with slope 2 and exits upto $t = 3$. At $t = 3$ a negative going ramp with slope -2 makes $x(t)$ flat with amplitude 3. At $t = 6$, a negative step function with amplitude -3 makes $x(t) = 0$. Thus, $x(t)$ is sketched as shown in Fig. 1.53. $x(t)$ can be expressed in terms of the singularity step and ramp functions as follows.

For $0 \leq t < 1$,

$$x(t) = t$$

and for $2 \leq t > 3$,

$$x(t) = 2t - 3$$

$$x(t) = \begin{cases} t & 0 \leq t < 1 \\ 1 & 1 \leq t < 2 \\ (2t-3) & 2 \leq t < 3 \\ 3 & 3 \leq t < 6 \end{cases}$$

The above equation can be written as a single equation for all time t.

$$\begin{aligned} x(t) = {} & t[u(t) - u(t-1)] + [u(t-1) - u(t-2)] \\ & + (2t-3)[u(t-2) - u(t-3)] + 3[u(t-3) - u(t-6)] \end{aligned}$$

$$\boxed{\begin{aligned} x(t) = {} & tu(t) - (t-1)u(t-1) + 2(t-2)u(t-2) \\ & -2(t-3)u(t-3) - 3u(t-6) \end{aligned}}$$

The above equation can be checked for any time t. For example for $t = 2.5$,

$$\begin{aligned} x(t) &= 2.5 - (2.5 - 1) + 2(2.5 - 2) \\ &= 2.5 - 1.5 + 1 \\ &= 2 \end{aligned}$$

8.

$$x(t) = r(t+3) - r(t) - 2r(t-3)$$

The above equation can be written as follows.

$$x_1 = t + c$$

For $t = -3$,

$$\begin{aligned} x_1(t) &= 0 \\ 0 &= -3 + c \\ \text{or} \quad c &= 3 \\ x_1(t) &= t + 3 \\ x_3(t) &= -2t + c \end{aligned}$$

For $t = 3$,

$$x_3(t) = 3$$
$$3 = -6 + c$$
or $$c = 9$$
$$x_3(t) = -2t + 9$$

$$x(t) = \begin{cases} t+3 & -3 \le t < 0 \\ 3 & 0 \le t < 3 \\ -2t+9 & t \ge 3 \\ 0 & \text{otherwise} \end{cases}$$

$$x(t) = x_1(t) + x_2(t) + x_3(t)$$

The above equation is represented in Fig. 1.54.

$$x(t) = (t+3)[u(t+3) - u(t)] + 3[u(t) - u(t-3)] + (9-2t)u(t-3)$$

$$\boxed{x(t) = (t+3)u(t+3) - tu(t) - 2(t-3)u(t-3)}$$

9.

$$x(t) = u(t)u(5-t)$$

$u(t)$ and $u(5-t)$ are sketched in Fig. 1.55a and b, respectively. The product of these two signals exits when they overlap. As seen from these figures the overlapping occurs for $0 \le t < 5$ and it forms a pulse as shown in Fig. 1.55c. From Fig. 1.55c, the following equation is written for $x(t)$.

$$\boxed{x(t) = u(t) - u(t-5)}$$

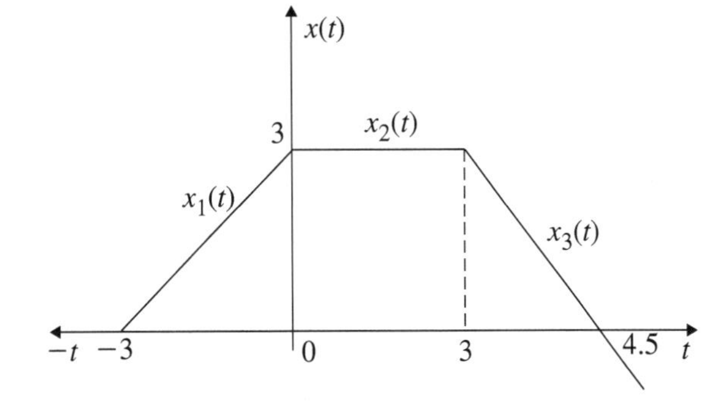

Fig. 1.54 Graphical representation of Example 1.21.8

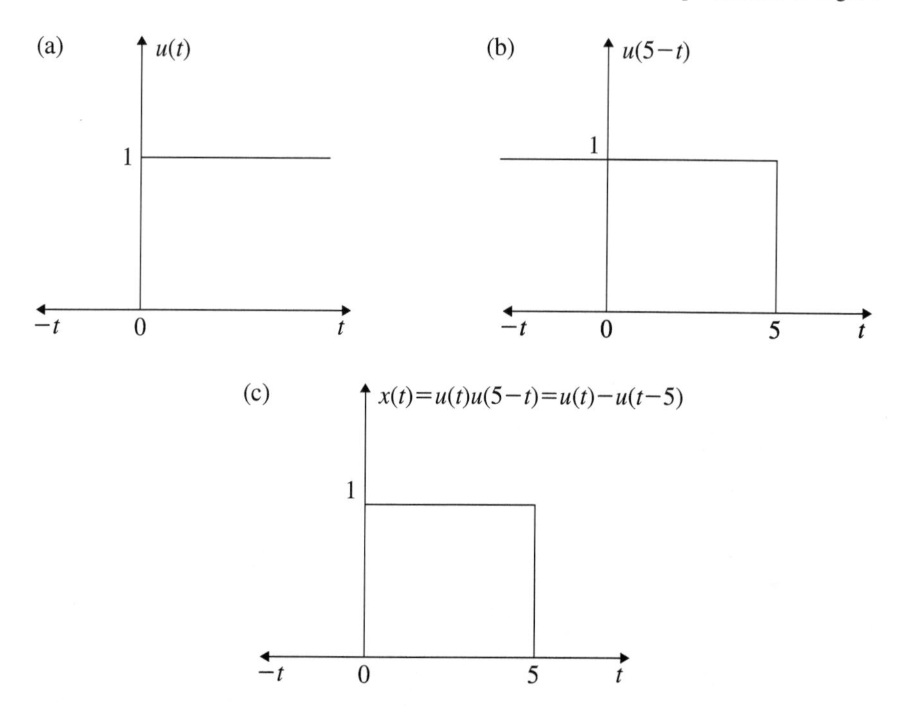

Fig. 1.55 Graphical representation of Example 1.21.9

Fig. 1.56 Graphical representation of Example 1.21.10

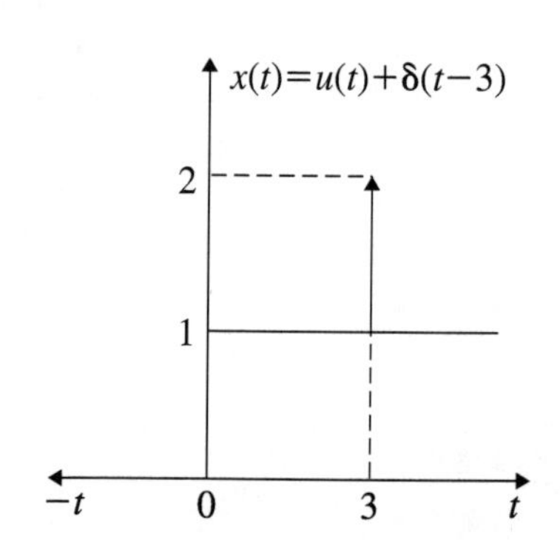

10.

$$x(t) = u(t) + \delta(t - 3)$$

The above signal is the summation of a step and impulse at $t = 3$. The sketch of $x(t)$ is shown in Fig. 1.56.

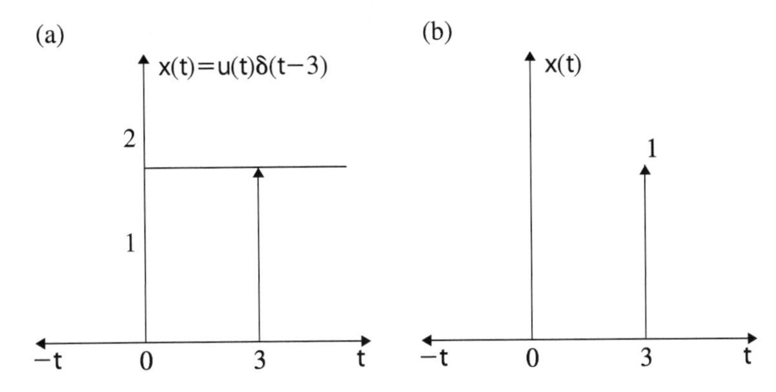

Fig. 1.57 Graphical representation of Example 1.21.11

11.

$$
\begin{aligned}
x(t) &= u(t)\delta(t-3) \\
&= u(3)\delta(t-3)
\end{aligned}
$$

$u(t)$ and $\delta(t-3)$ are represented in Fig. 1.57a. $x(t) = u(t)\delta(t-3)$ is nothing but $u(t)$ at $t = 3$. At $t \neq 3$, $x(t) = 0$. This is represented in Fig. 1.57b.

12.

$$
x(t) = \begin{cases} r(t+5) & -5 \leq t < -3 \\ 2 & -3 \leq t < 2 \\ 4 & 2 \leq t \leq 4 \\ -r(t-4) & 4 \leq t < 5 \\ 0 & \text{otherwise} \end{cases}
$$

The above equations are sketched as $x(t)$ in Fig. 1.58. For $-5 \leq t < -3, x(t)$ is a straight line with a slope one and constant $c = 5$. In other words, $x(t) = (t+5)$ for $-5 \leq t < -3$. Similarly, for $4 \leq t < 5$, it has a negative slope of 4 and constant $c = 20$. Here,

$$
x(t) = -4t + 20 \quad 4 \leq t < 5
$$

The following single equation in terms of step and ramp singularity functions is written for the signal $x(t)$

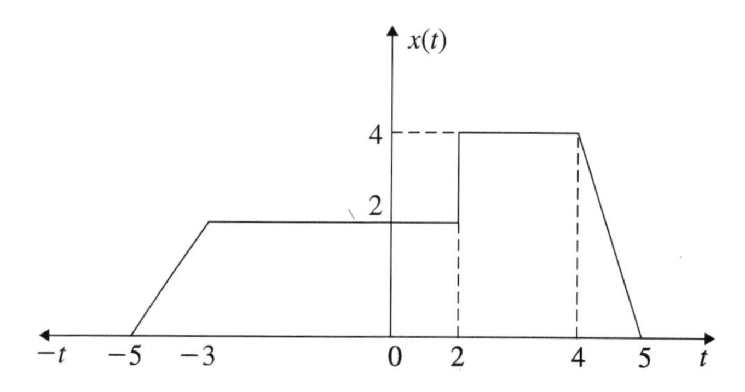

Fig. 1.58 Graphical representation of Example 1.21.12

$$\begin{aligned} x(t) &= (t+5)[u(t+5)-u(t+3)] + 2[u(t+3)-u(t-2)] \\ &\quad +4[u(t-2)-u(t-4)] + (-4t+20)[u(t-4)-u(t-5)] \\ &= (t+5)u(t+5) - [(t+5-2)u(t+3)] + 2u(t-2) \\ &\quad +[(-4t+20-4)u(t-4)] + 4(t-5)u(t-5) \end{aligned}$$

$$\boxed{\begin{aligned} x(t) = (t+5)u(t+5) - (t+3)u(t+3) + 2u(t-2) \\ -4(t-4)u(t-4) + 4(t-5)u(t-5) \end{aligned}}$$

The validity of the above equation may be checked for any time t. For example, for $t = 3$,

$$\begin{aligned} x(t)|_{t=3} &= 8 - 6 + 2 \\ &= 4 \end{aligned}$$

13.

$$x(t) = u(t+4)\mathrm{ramp}(-t+2)$$

The signal ramp(t) is plotted in Fig. 1.59a as $r(t)$. By folding $r(t)$, $r(-t)$ is plotted in Fig. 1.59b. $r(-t)$ is right shifted by 2 to get $r(-t+2)$ which is sketched in Fig. 1.59c. The step signal $u(t)$ is left shifted by 4 to get $u(t+4)$. This is shown in Fig. 1.59d. When $r(-t+2)$ and $u(t+4)$ are multiplied to get $x(t)$, the signal $x(t) = 0$ for $t \leq -4$ and also for $t > 2$. This is shown in Fig. 1.59f.
The equation for $x_1(t)$ in Fig. 1.59f is

$$\begin{aligned} x_1(t) &= -mt + c \\ &= -t + c \end{aligned}$$

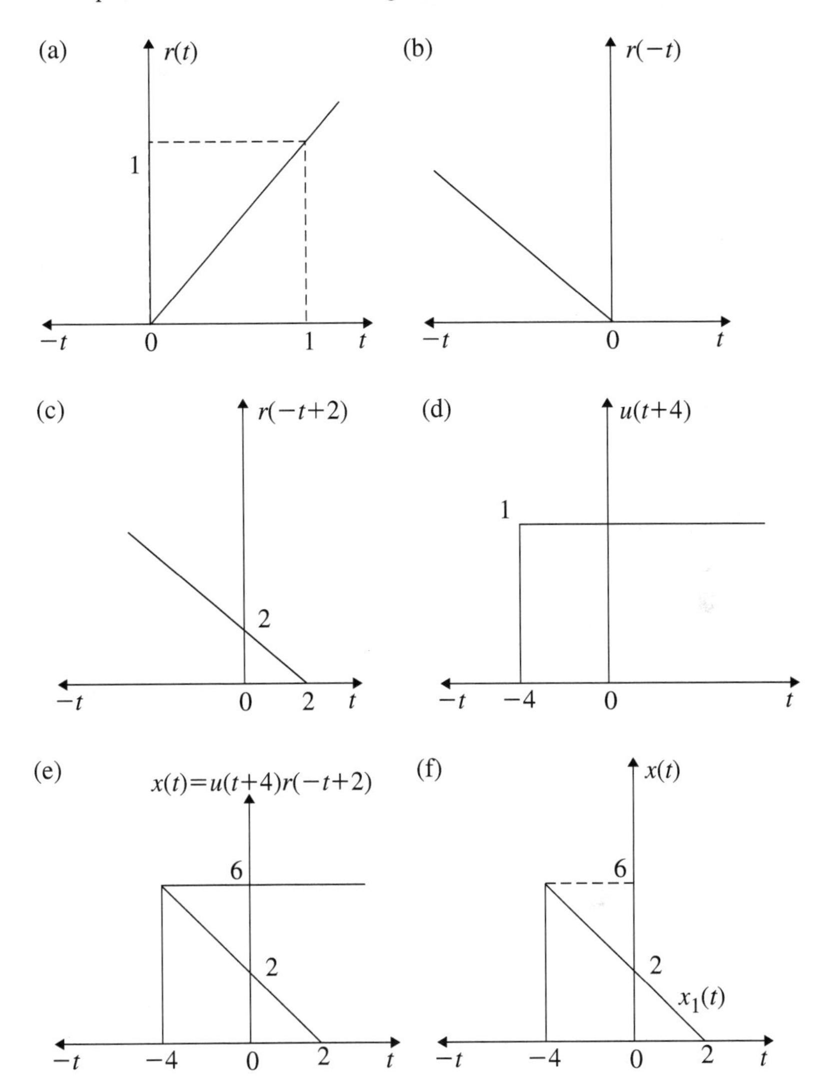

Fig. 1.59 Graphical representation of Example 1.21.13

For $t = 2$,

$$\begin{aligned} x_1(t) &= 0 \\ 0 &= -2 + c \\ c &= 2 \\ x_1(t) &= (2 - t) \end{aligned}$$

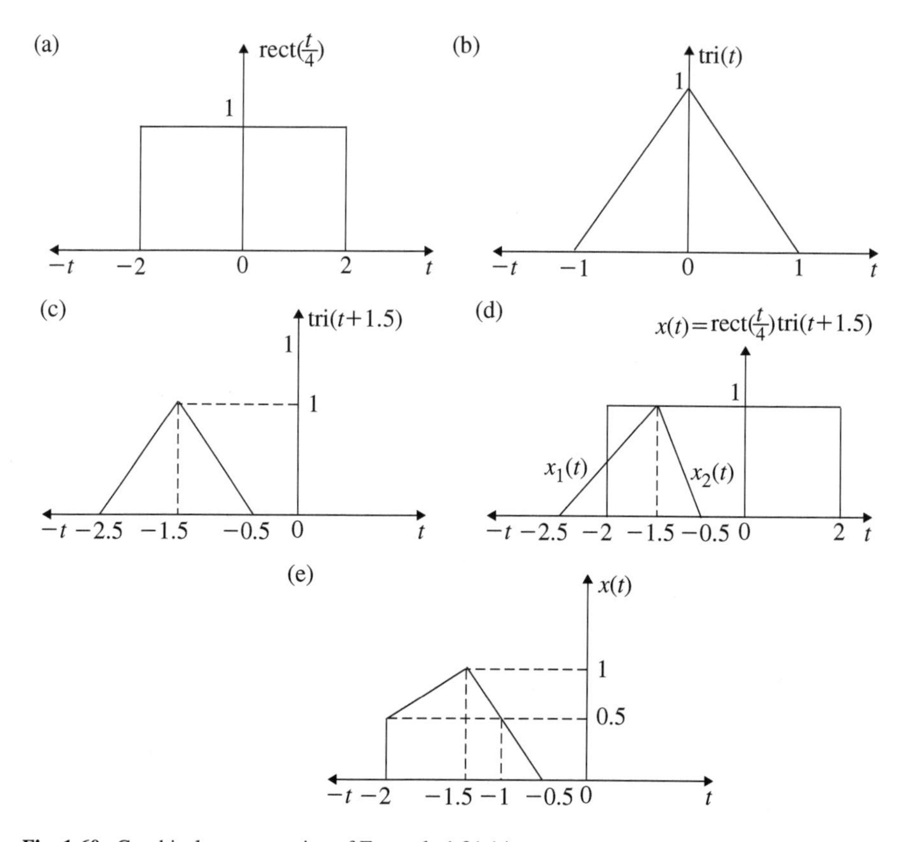

Fig. 1.60 Graphical representation of Example 1.21.14

For Fig. 1.59f, the following equation is written

$$x(t) = (2 - t)[u(t + 4) - u(t - 2)]$$

$$\boxed{x(t) = -t[u(t + 4) - u(t - 2)] + 2[u(t + 4) - u(t - 2)]}$$

14.

$$x(t) = \text{rect}\left(\frac{t}{4}\right)\text{tri}(t + 1.5)$$

The signal rect $\left(\frac{t}{4}\right)$ is shown in Fig. 1.60a. The signal tri(t) is shown in Fig. 1.60b. tri(t) is left shifted by 1.5 to get tri$(t + 1.5)$ which is shown in Fig. 1.60c. The product of rect$(\frac{t}{4})$ and tri$(t + 1.5)$ is shown in Fig. 1.60d. $x(t)$ is obtained by multiplying the above two signals when they overlap. This is shown in Fig. 1.60e.

From Fig. 1.60d, the equations for x_1 and x_2 (which are straight lines) are obtained as follows:

$$x_1(t) = mt + c$$

where $m = 1$. For $t = -1.5$

$$\begin{aligned} x_1(t) &= 1 \\ 1 &= -1.5 + c \\ c &= 2.5 \\ x_1(t) &= t + 2.5 \\ x_2(t) &= -t + c \end{aligned}$$

For $t = -1.5$

$$\begin{aligned} x_2(t) &= 1 \\ 1 &= 1.5 + c \\ c &= -0.5 \\ x_2(t) &= -(t + 0.5) \end{aligned}$$

From Fig. 1.60e, the following equation is written

$$x(t) = (t + 2.5)[u(t + 2) - u(t + 1.5)] - (t + 0.5)[u(t + 1.5) - u(t + 0.5)]$$

$$\boxed{x(t) = [(t + 2.5)u(t + 2) - (2t + 3)u(t + 1.5) + (t + 0.5)u(t + 0.5)]}$$

15.

$$x(t) = \text{rect}\left(\frac{t}{2}\right) \text{tri}\left(\frac{t}{4}\right)$$

rect $\left(\frac{t}{4}\right)$ is sketched as shown in Fig. 1.61a. tri$\left(\frac{t}{4}\right)$ is sketched as shown in Fig. 1.61b. Combined rect $\left(\frac{t}{2}\right)$ tri $\left(\frac{t}{4}\right)$ is shown in Fig. 1.61c. The overlapping of these two signals takes place during the period $-1 \leq t \leq 1$. In this period tri$\left(\frac{t}{2}\right)$ is multiplied by 1 which gives $x(t)$. This is shown in Fig. 1.61d. $x(t)$ contains two straight lines $x_1(t)$ and $x_2(t)$ and they have slopes of $\frac{1}{4}$ and $-\frac{1}{4}$, respectively. These equations are written as follows:

$$x_1(t) = \frac{1}{4}t + c$$

For $t = 0$,

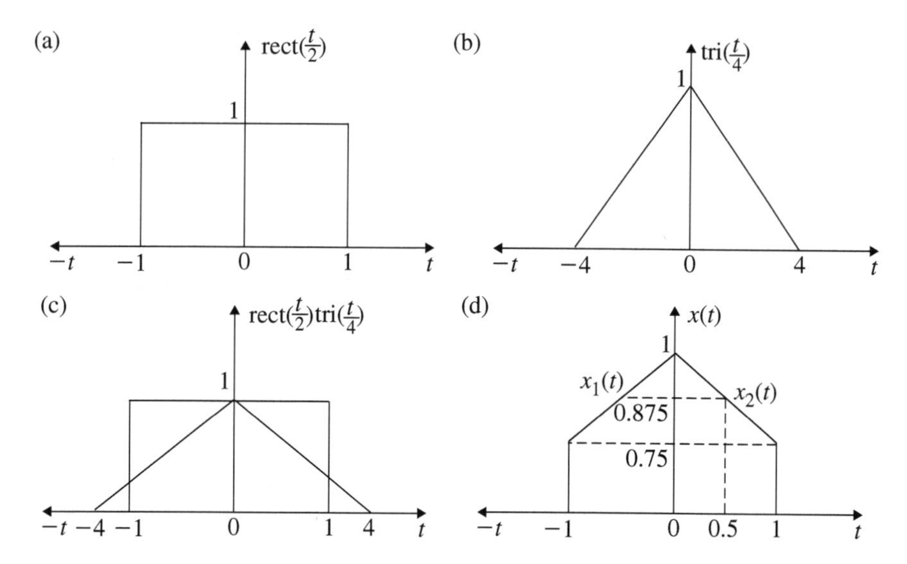

Fig. 1.61 Graphical representation of Example 1.21.15

$$x_1(t) = 1$$
$$1 = 0 + c$$
$$x_1(t) = \frac{1}{4}t + 1$$
$$x_2(t) = -\frac{1}{4}t + c$$

For $t = 0$,

$$x_2(t) = 1$$
$$1 = 0 + c$$

$$x_2(t) = -\frac{1}{4}t + 1$$
$$x(t) = x_1(t)[u(t+1) - u(t)] + x_2(t)[u(t) - u(t-1)]$$
$$= (0.25t + 1)[u(t+1) - u(t)] + (1 - 0.25t)[u(t) - u(t-1)]$$

$$\boxed{x(t) = 0.25(t+4)u(t+1) - 0.5tu(t) + 0.25(t-4)u(t-1)}$$

The above equation can be checked as follows. For $t = 0.5$

$$x_2(t) = -\frac{1}{4} \times 0.5 + 1$$
$$= 0.875$$
$$x(t)|_{t=0.5} = 0.25 \times 4.5 - 0.5 \times 0.5$$
$$= 0.875$$

16.

$$x(t) = \text{ramp}(2t + 1)\text{rect}\frac{(t-2)}{4}$$
$$\text{rect}\frac{(t-2)}{4} = \text{rect}\left(\frac{t}{4} - \frac{1}{2}\right)$$

The signal rect $(t - (1/2))$ is shown in Fig. 1.62a. It is time expanded by 4 and rect $\frac{(t-2)}{4}$ is shown in Fig. 1.62b. The signal ramp(t) is represented in Fig. 1.62c. The ramp $r(t)$ is left shifted by 1 to get $r(t+1)$ which is shown in Fig. 1.62d. The signal $r(t+1)$ is time compressed by 2 and $r(2t+1)$ is shown in Fig. 1.62e. The combined signal $r(2t+1)\text{rect}\frac{(t-2)}{4}$ is shown in Fig. 1.62f. These two signals overlap in the time interval $0 \leq t < 4$ and their product gives $x(t)$ which is represented in Fig. 1.62g. From Fig. 1.62g, the following equation is written

$$x(t) = (2t + 1)\,[u(t) - u(t-4)]$$

$$\boxed{x(t) = 2t[u(t) - u(t-4)] + [u(t) - u(t-4)]}$$

■ Example 1.22

Consider the signal $y(t) = 2x(-4t - 5)$ where $y(t)$ is described by

$$y(t) = \begin{cases} 2t & 0 \leq t < 1 \\ 2 & 1 \leq t < 3 \\ 0 & \text{otherwise} \end{cases}$$

Find the original signal $x(t)$ and plot with respect to t. Give the mathematical description for $x(t)$.

Solution:

1.

$$y(t) = 2x(-4t - 5)$$

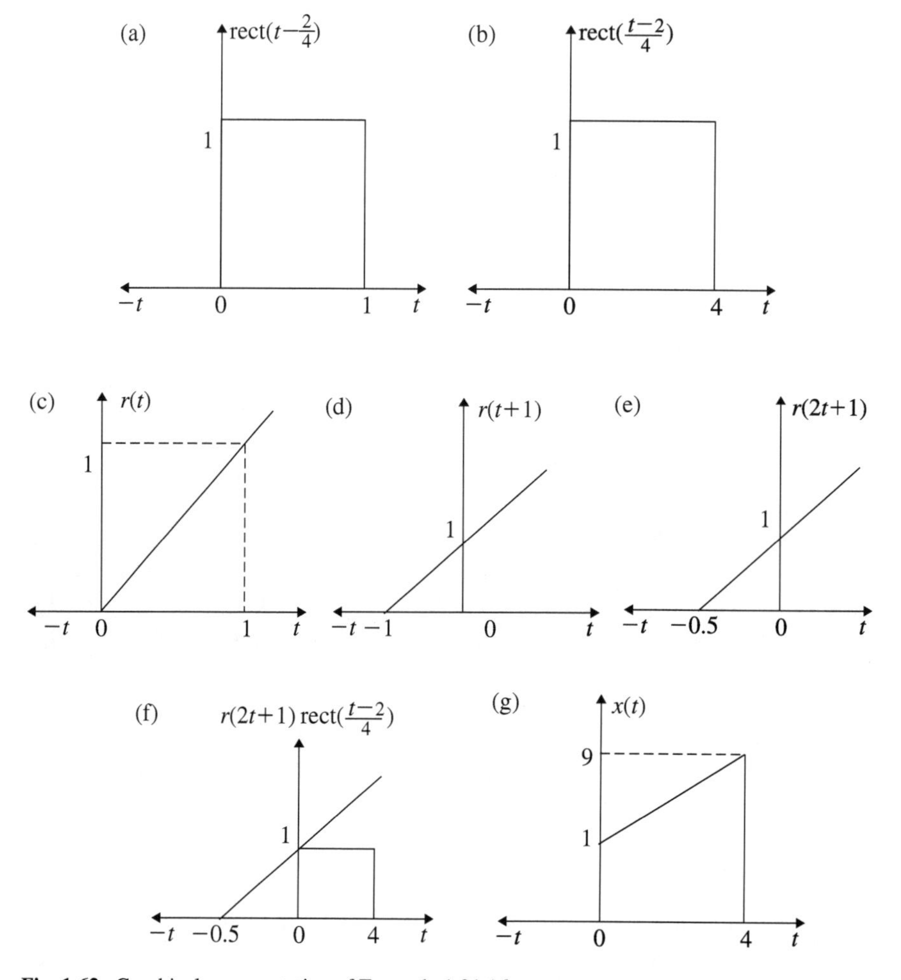

Fig. 1.62 Graphical representation of Example 1.21.16

$y(t)$ as per the mathematical description given is plotted as shown in Fig. 1.63a. The following operations on $x(t)$ are performed in the sequential order to get $y(t)$.

(a) Signal reflection.
(b) Time shifting.
(c) Time compression and amplitude scaling.

2. To get $x(t)$ from $y(t)$, the reverse process is done in the following sequence.

(a) Amplitude scaling and time expansion.
(b) Time shifting.
(c) Signal reflection.

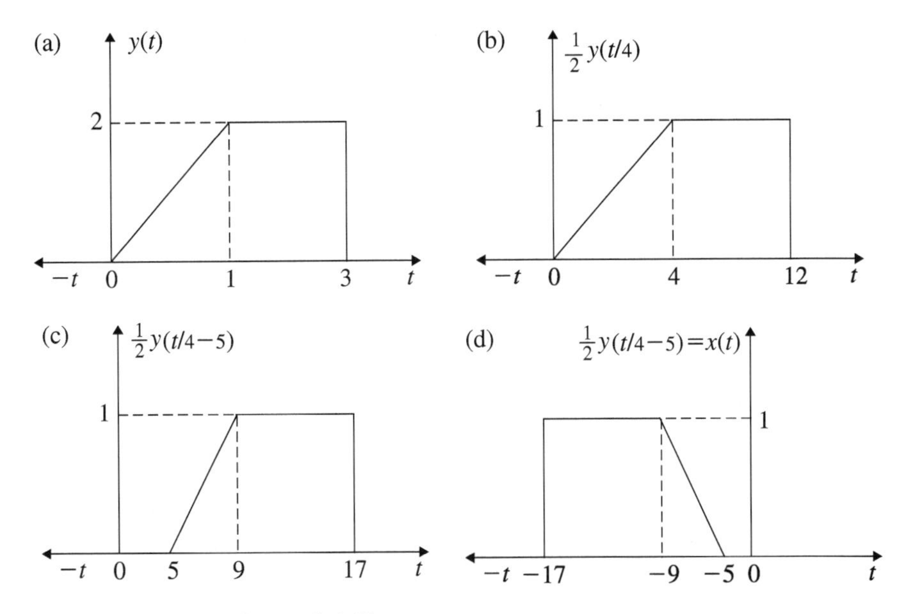

Fig. 1.63 Plot of $y(t)$ of Example 1.22

3. **Amplitude scaling and time expansion**: The amplitude of $y(t)$ is truncated by 2 and time expanded by 4. The plot of $\frac{1}{2}y\left(\frac{t}{4}\right)$ is shown in Fig. 1.63b.
4. $x(-4t-5)$ is left shifted by 5 sec. Therefore, $\frac{1}{2}y\left(\frac{t}{4}\right)$ is to be right shifted by 5 sec. Thus, the plot of $\frac{1}{2}y(\frac{t}{4}-5)$ is obtained as given in Fig. 1.63c.
5. Finally time reflection is done to get $\frac{1}{2}y(\frac{-t}{4}-5)$, which gives $x(t)$. This is shown in Fig. 1.63d.
6. The mathematical description of $x(t)$ is given below. For $-9 \leq t \leq -5$, the straight line has a slope of $-1/4$ with a constant $-5/4$. Thus, we write

$$x(t) = \begin{cases} -\frac{1}{4}t - \frac{5}{4} & -9 \leq t < -5 \\ 1 & -17 \leq t < -9 \\ 0 & \text{otherwise} \end{cases}$$

■ Example 1.23

Consider the signal

$$y(t) = \frac{1}{4}x(-3t+4)$$

shown in Fig. 1.64a. Find the original signal and give its mathematical description by a single equation for all time t.

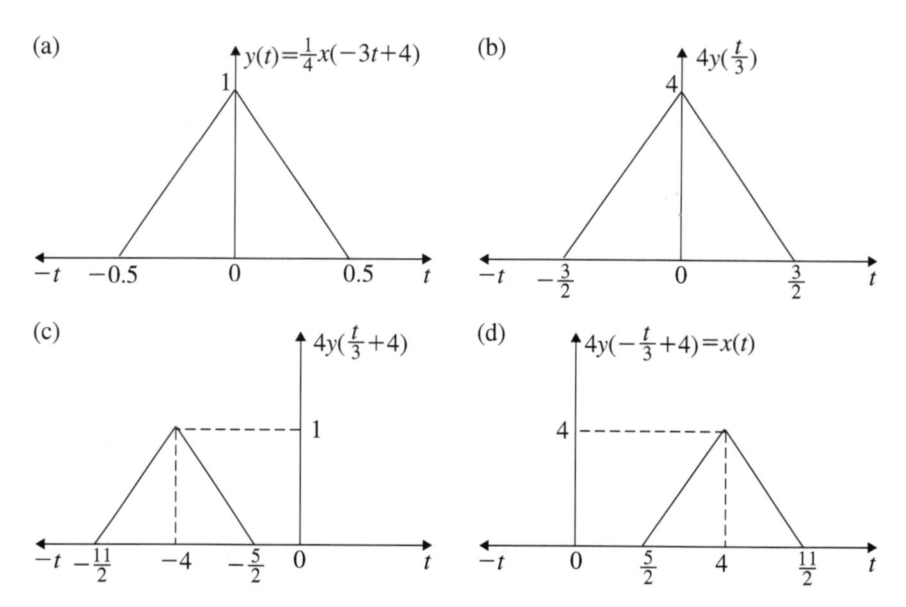

Fig. 1.64 Sketch of signals of Example 1.23

Solution:

1. The multiple transformation of the signal $x(t)$ to get $y(t)$ takes place in the following sequence.

   (a) Time reversal.
   (b) Time shifting.
   (c) Amplitude and time scaling.

From the transformed signal $y(t)$, the original signal $x(t)$ is obtained in the reverse order.

(a) Amplitude and time scaling.
(b) Time shifting.
(c) Time reversal.

2. The signal $y(t) = \frac{1}{4}x(-3t + 4)$ is shown in Fig. 1.64a. Here, the amplitude of $x(t)$ is truncated by 4 and time is compressed by 3. In the transformation the amplitude of $y(t)$ is amplified by a factor of 4 and time should be expanded. The first transformation of $y(t)$ becomes $4y(t/3)$ and is shown in Fig. 1.64b.
3. $y(t)$ is right shifted by 4 sec. It should be left shifted by 4 sec to get $4y(\frac{t}{3} + 4)$ which is sketched in Fig. 1.64c.
4. Finally to get $x(t)$, $4y(\frac{t}{3} + 4)$ is time reversed as $4y(-t/3 + 4)$. This is sketched in Fig. 1.64d.
5. To get the mathematical description of $x(t)$, the following equations are written

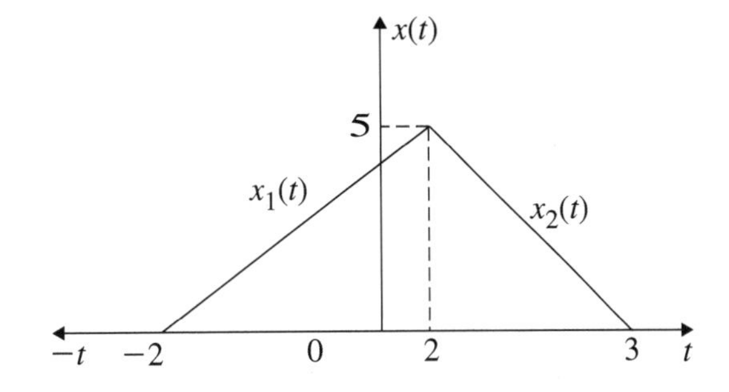

Fig. 1.65 Signal $x(t)$ of Example 1.24

$$x(t) = \begin{cases} \frac{8}{3}t - \frac{20}{3} & \frac{5}{2} \le t < 4 \\ \frac{-8}{3}t + \frac{44}{3} & 4 \le t < \frac{11}{2} \\ 0 & \text{otherwise} \end{cases}$$

6. As a single equation for all time t, the following equation is written

$$x(t) = \left[\frac{8}{3}t - \frac{20}{3}\right]\left[u\left(t - \frac{5}{2}\right) - u(t-4)\right] + \left[-\frac{8}{3}t + \frac{44}{3}\right]\left[u(t-4) - u\left(t - \frac{11}{2}\right)\right]$$

$$\boxed{x(t) = \frac{1}{3}[(8t-20)u(t-2.5) + (-16t+64)u(t-4) + (8t-44)u(t-5.5)]}$$

■ Example 1.24

Consider the signal shown in Fig. 1.65.

(a) Give mathematical description for $x(t)$.
(b) Express $x(t)$ by a single expression for all t.

Solution:

(a) $x(t)$ can be decomposed as $x_1(t)$ and $x_2(t)$ where

$$\begin{aligned} x_1(t) &= m_1 t + c_1 && -2 \le t > 2 \\ x_2(t) &= -m_2 t + c_2 && 2 \le t > 3 \\ x(t) &= 0 && \text{otherwise} \\ x(t) &= x_1(t) + x_2(t) \end{aligned}$$

From Fig. 1.65, the slope $m_1 = \frac{5}{4}$

$$x_1(t) = \frac{5}{4}t + c_1$$

For $t = -2, x_1(t) = 0$

$$0 = \frac{-5}{2} + c_1$$
$$c_1 = \frac{5}{2} = 2.5$$

From Fig. 1.65, the slope $m_2 = 5$

$$x_2 = -5t + c_2$$

For $t = 3, x_2(t) = 0$

$$0 = -5 \times 3 + c_2$$
$$c_2 = 15$$
$$x_2(t) = -5t + 15$$

(b) $x_1(t)$ can be written by a single equation for all t as follows:

$$x_1(t) = \left[\frac{5}{4}t + 2.5\right][u(t+2) - u(t-2)]$$

$x_2(t)$ can be written by a single equation for all t as

$$x_2 = [-5t + 15][u(t-2) - u(t-3)]$$
$$x(t) = x_1(t) + x_2(t)$$
$$= \frac{5}{4}tu(t+2) + 2.5u(t+2) - \frac{5}{4}tu(t-2) - 2.5u(t-2) - 5tu(t-2)$$
$$+15u(t-2) + 5t(t-3) - 15u(t-3)$$

$$\boxed{x(t) = \left(\frac{5}{4}t + 2.5\right)u(t+2) + \left(\frac{-25}{4}t + 12.5\right)u(t-2) + (5t - 15)u(t-3)}$$

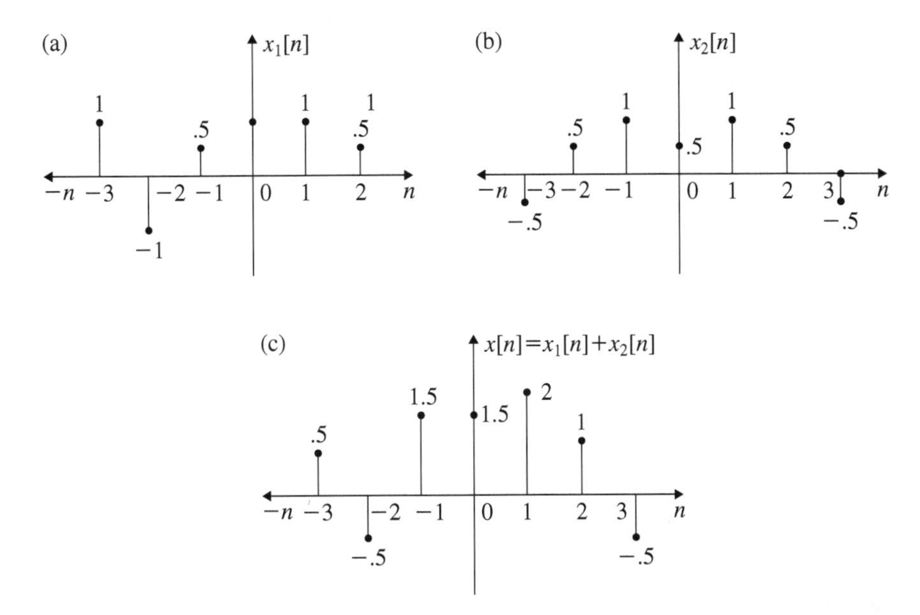

Fig. 1.66 Addition of DT signals

1.7 Basic Operations on Discrete Time Signals

The basic operations that are applied to continuous time signals are also applicable to discrete time signals. The time t in CT signal is replaced by n in DT signals. The basic operations as applied to DT signals are explained below.

1.7.1 Addition of Discrete Time Sequence

Addition of discrete time sequence is done by adding the signals at every instant of time. Consider the signals $x_1[n]$ and $x_2[n]$ shown in Fig. 1.66a and b, respectively. The addition of these signals at every n is done and represented as $y[n] = x_1[n] + x_2[n]$. This is shown in Fig. 1.66c.

1.7.2 Multiplication of DT Signals

The multiplication of two DT signals $x_1[n]$ and $x_2[n]$ is obtained by multiplying the signal values at each instant of time n. Consider the signal $x_1[n]$ and $x_2[n]$ represented in Fig. 1.67a and b. At each instant of time n, the samples of $x_1[n]$ and $x_2[n]$ are multiplied and represented as shown in Fig. 1.67c.

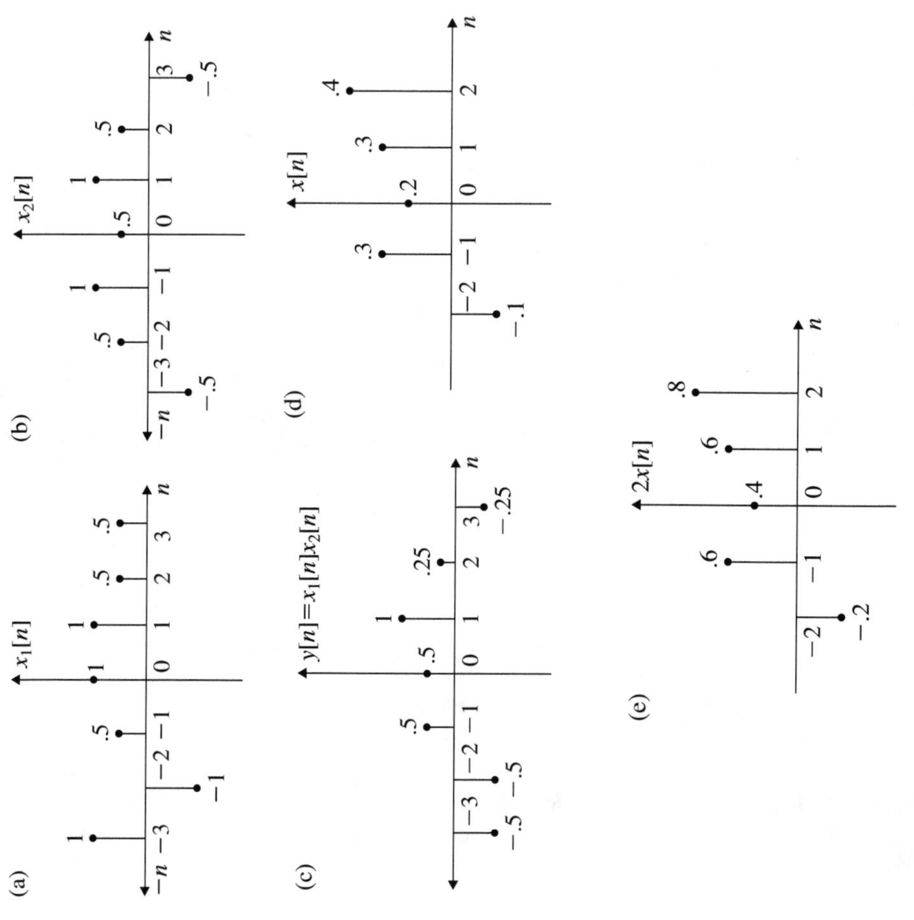

Fig. 1.67 Multiplications of two DT signals. Amplitude scaling of DT signals

1.7.3 Amplitude Scaling of DT Signal

Let $x[n]$ be a discrete time signal. The signal $Ax[n]$ is represented by multiplying the amplitude of the sequence by A at each instant of time n. Consider the signal $x[n]$ shown in Fig. 1.67d. The signal $2x[n]$ is represented and shown in Fig. 1.67e.

1.7.4 Time Scaling of DT Signal

The time compression or expansion of a DT signal in time is known as time scaling. Consider the signal $x[n]$ shown in Fig. 1.68a. The time compressed signal $x[2n]$ and time expanded signal $x[\frac{n}{2}]$ are shown in Fig. 1.68b and c, respectively. One should note that while doing compression and expansion of DT signal, **only for integer value of n the samples exist. For non-integer value of n, the samples do not exist**.

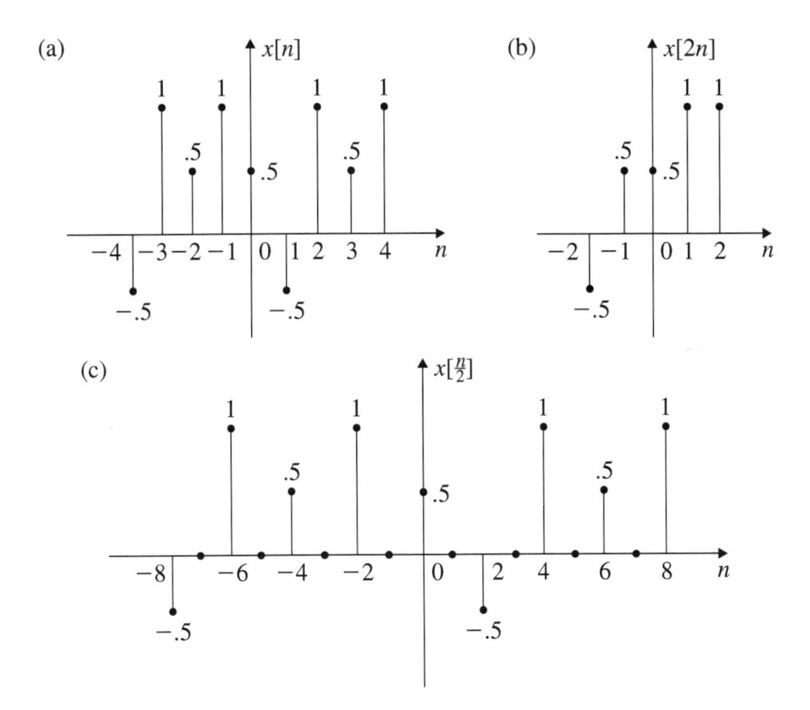

Fig. 1.68 Time scaling of DT signal

Time Compression

Let

$$y[n] = x[2n]$$
$$y[-2] = x[-4] = -0.5$$

$$y[-1] = x[-2] = 0.5$$
$$y[0] = x[0] = 0.5$$
$$y[1] = x[2] = 1$$
$$y[2] = x[4] = 1.$$

The plot of $x[2n]$ is shown in Fig. 1.68b.

Time Expansion

Let

$$y[n] = x\left[\frac{n}{2}\right]$$
$$y[-8] = x[-4] = -0.5$$
$$y[-6] = x[-3] = 1$$
$$y[-4] = x[-2] = 0.5$$
$$y[-2] = x[-1] = 1$$
$$y[0] = x[0] = 0.5$$
$$y[2] = x[1] = -0.5$$
$$y[4] = x[2] = 1$$
$$y[6] = x[3] = 0.5$$
$$y[8] = x[4] = 1.$$

The plot of $x[\frac{n}{2}]$ is shown in Fig. 1.68c.

1.7.5 Time Shifting of DT Signal

As in the case of CT signal, time shifting property is applied to DT signal also. Let $x[n]$ be the DT signal. Let n_0 be the time by which $x[n]$ is time shifted. Since n is an integer, n_0 is also an integer. The following points are applicable while DT signal is time shifted.

- For the DT signals $x[-n - n_0]$ and $x[n + n_0]$, the signals $x[-n]$ and $x[n]$ are to be left shifted by n_0.

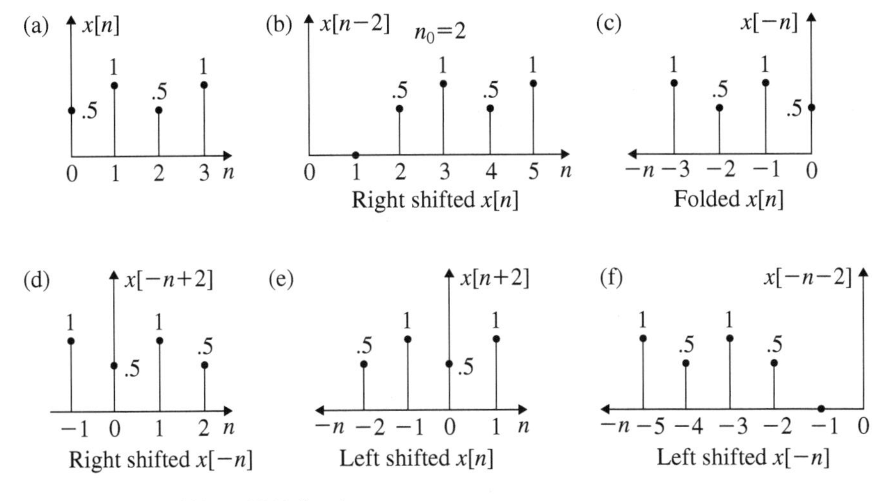

Fig. 1.69 Time shifting of DT signal

- For the DT signals $x[n - n_0]$ and $x[-n + n_0]$, the signals $x[n]$ and $x[-n]$ are to be right shifted by n_0.

Figure 1.69 shows time shifting of DT signal.

In Fig. 1.69a, the sequence $x[n]$ is shown. The sequence $x[n - 2]$ which is right shifted by 2 samples is shown in Fig. 1.69b. $x[-n]$ which is the folded signal is shown in Fig. 1.69c. $x[-n + 2]$ which is left shifted of $x[-n]$ is shown in Fig. 1.69d. $x[n + 2]$ which is right shifted of $x[n]$ is shown in Fig. 1.69e. $x[-n - 2]$ which is left shifted of $x[-n]$ is shown in Fig. 1.69f.

1.7.6 Multiple Transformation

The transformations namely amplitude scaling, time reversal, time shifting, time scaling, *etc.* are applied to represent DT sequence. The sequence of operation of these transformations is important and followed as described below.

Consider the following DT signal:

$$y[n] = Ax\left[-\frac{n}{a} + n_0\right]$$

1. Plot $x[n]$ sequence and obtain $Ax[n]$ by amplitude scaling.
2. Using time reversal (folding), plot $Ax[-n]$.
3. Using time shifting, plot $Ax[-n + n_0]$ where $n_0 > 0$. The time shift is to be right of $x[-n]$ by n_0 samples.
4. Using time scaling, plot $Ax[-\frac{n}{a} + n_0]$ where a is in integer. In the above case, keeping amplitude constant, time is expanded by a.

The following examples illustrate the above operations.

■ Example 1.25

Let $x[n]$ and $y[n]$ be as given in Fig. 1.70a and b, respectively. Plot

(a) $x[2n]$

(b) $x[3n-1]$

(c) $x[n-2]+y[n-2]$

(d) $y[1-n]$

(Anna University, December, 2006)

Solution: The DT signals $x[n]$ and $y[n]$ are plotted as shown in Fig. 1.70a and b, respectively.

(a) **To plot $x[2n]$**
Here, the DT sequence is time compressed by a factor 2. Hence, the samples only with even numbers are divided by a factor 2 and the corresponding amplitudes marked and shown in Fig. 1.70c. When odd values of n are divided by the factor 2, it becomes a fraction and they are skipped.

(b) **To plot $x[3n-1]$**
The plot of $x[n-1]$ is obtained by right shifting of $x[n]$ by $n_0 = 1$. This is shown in Fig. 1.70d. When $x[n-1]$ is time compressed by a factor 3, $x[3n-1]$ is obtained. Only integers which are divisible by 3 in the sequence $x[n-1]$ are to be taken to plot $x[3n-1]$. Thus, samples for $n=0$ and $n=3$ will be plotted as shown in Fig. 1.70e.

(c) **To plot $x[n-2]+y[n-2]$**
The sequence $x[n-2]$ is obtained by right shifting of $x[n]$ by 2 and is shown in Fig. 1.70f. Similarly, the sequence $y[n-2]$ is obtained by right shifting of $y[n]$ by 2 and is shown in Fig. 1.70g. The sequence $x[n-2]+y[n-2]$ is obtained by summing up the sequences in Fig. 1.70f, g for all n and is shown in Fig. 1.70h.

(d) **To plot $y[1-n]$**
The sequence $y[-n]$ is obtained by folding $y[n]$ and is shown in Fig. 1.70i. $y[-n]$ is right shifted by 1 sample to get the sequence $y[1-n]$. This is shown in Fig. 1.70j.

■ Example 1.26

Consider the sequence shown in Fig. 1.71a. Express the sequence in terms of step function.

Solution: The unit step sequence $u[n]$ is shown in Fig. 1.71b. The unit negative step sequence with a time delay of $n_0 = 4$ is shown in Fig. 1.71b. It is evident from

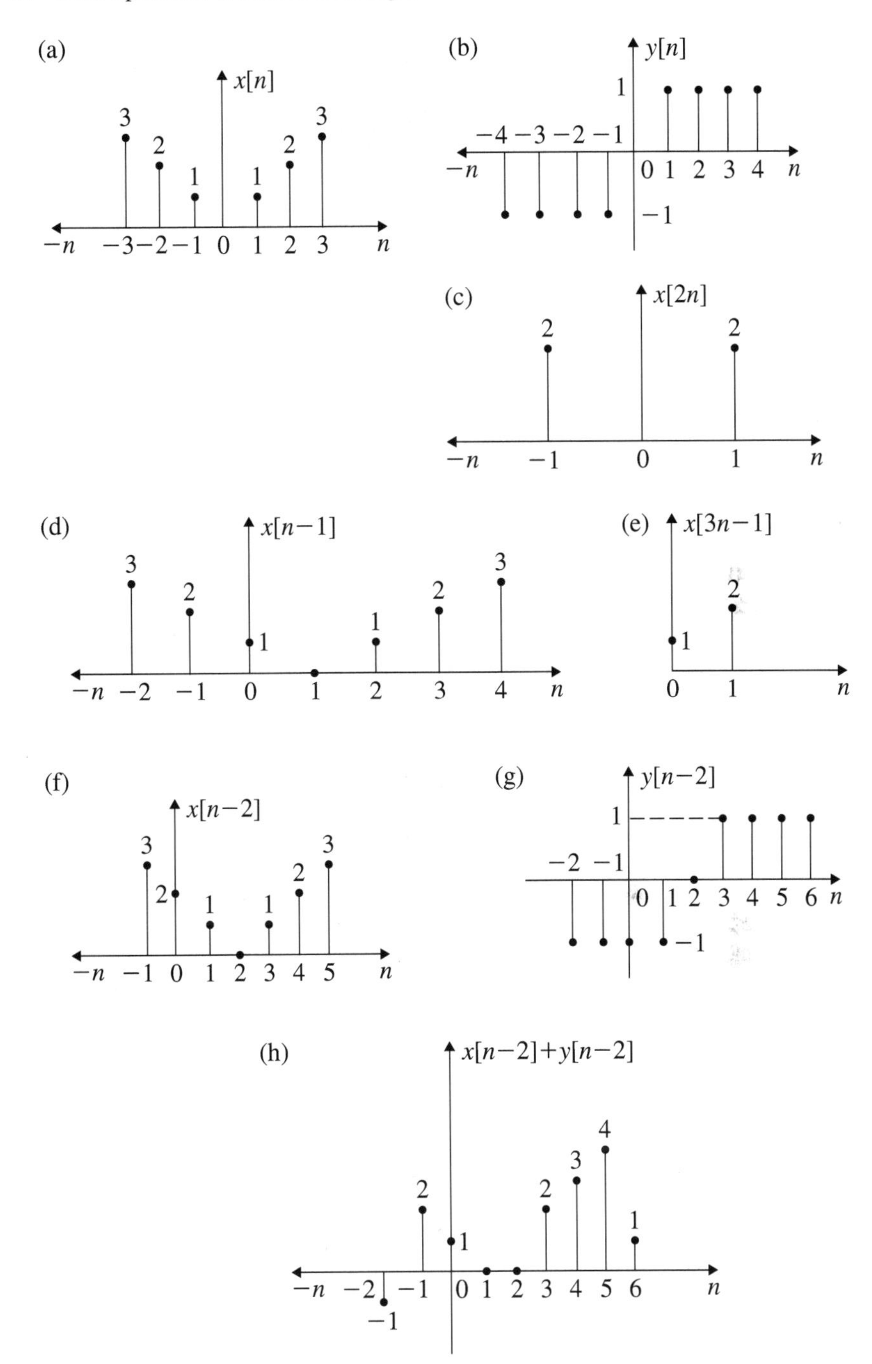

Fig. 1.70 Two discrete sequences

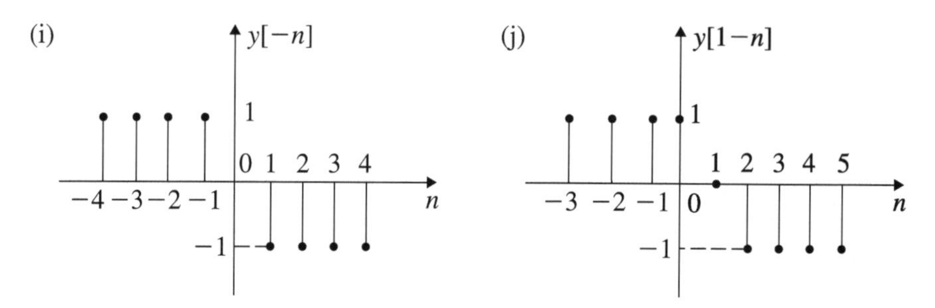

Fig. 1.70 (continued)

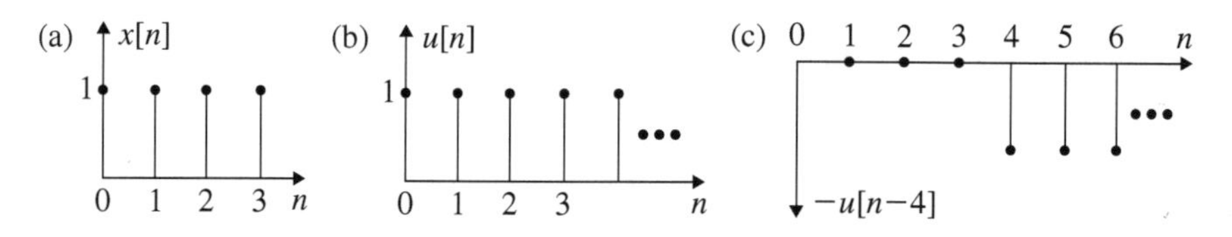

Fig. 1.71 Sequences expressed in terms of step sequences

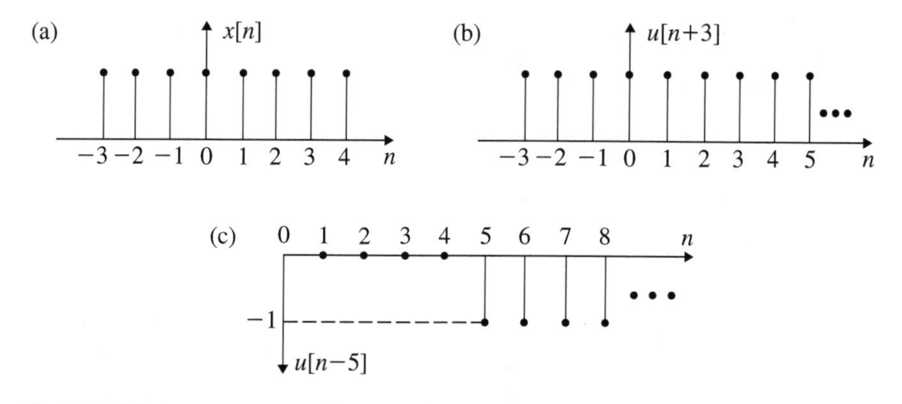

Fig. 1.72 DT sequences expressed in terms of step sequences

Fig. 1.71 that $\{u[n] - u[n-4]\}$ gives the required $x[n]$ sequence which is represented in Fig. 1.71a. Thus, $x[n] = \{u[n] - u[n-4]\}$.

■ Example 1.27

Consider the sequence shown in Fig. 1.72a. Express the sequence in terms of step function.

Solution:

1. Figure 1.72a represents the sequence $x[n]$ in the interval $-3 \le n \le 4$.
2. Consider $u[n+3]$ which is represented in Fig. 1.72b. The sequence interval is $-3 \le n < \infty$.

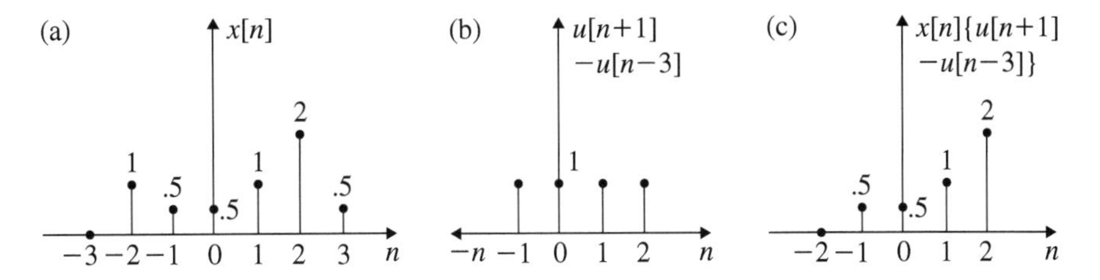

Fig. 1.73 Multiplication of DT sequences

3. Consider the step sequence with a time delay of $n_0 = 5$ and inverted. This can be written as $-u[n-5]$ for the interval $5 \leq n < \infty$. This is represented in Fig. 1.72c.
4. Now consider the sum of the sequences $u[n+3]$ and $-u[n-5]$. This is nothing but $x[n]$. Thus,

$$\boxed{x[n] = u[n+3] - u[n-5]}$$

■ Example 1.28

A discrete time sequence $x[n]$ is shown in Fig. 1.73a. Find

$$x[n]\{u[n+1] - u[n-3]\}$$

Solution:

1. $x[n]$ sequence is represented in Fig. 1.73a.
2. $\{u[n+1] - u[n-3]\}$ sequence is nothing but the time delayed unit step sequence with $n_0 = 3$ being subtracted from the time advanced unit step sequence with $n_0 = 1$. This sequence is represented in Fig. 1.73b.
3. Multiplying sample wise of Fig. 1.73a, b, the required sequence $x[n]\{u[n+1] - u[n-3]\}$ is obtained and represented in Fig. 1.73c.

■ Example 1.29

Sketch $x[n] = a^n$ where $-2 \leq n \leq 2$ for the two cases shown below:

$$(1) \qquad a = \left(-\frac{1}{4}\right)$$

$$(2) \qquad a = -4$$

(Anna University, May, 2007)

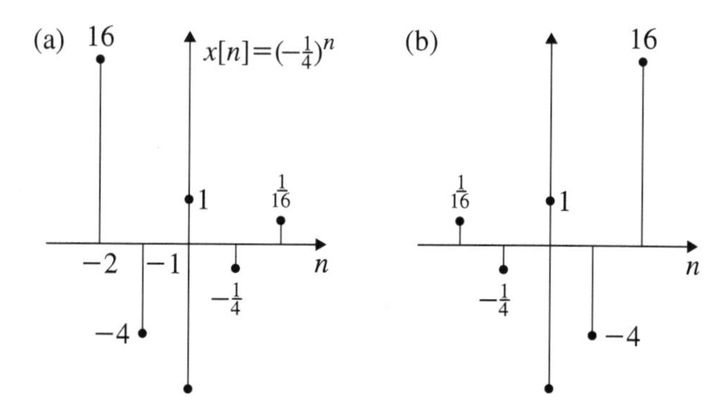

Fig. 1.74 DT sequences of Example 1.29

Solution:
For $x[n] = (-\frac{1}{4})^n$ and $x[n] = (-4)^n$ where $-2 \leq n \leq 2$, $x[n]$ is found and tabulated below:

n	-2	-1	0	1	2
$x[n] = (-\frac{1}{4})^n$	16	-4	1	$-\frac{1}{4}$	$\frac{1}{16}$
$x[n] = (-4)^n$	$\frac{1}{16}$	$-\frac{1}{4}$	1	-4	16

The samples of $x[n]$ are plotted and shown in Fig. 1.74. $x[n] = (-\frac{1}{4})^n$ is represented in Fig. 1.74a and $x[n] = (-4)^n$ is represented in Fig. 1.74b.

■ Example 1.30

Given

$$x[n] = \{2, 3, 4, 1, \underset{\uparrow}{6}, 7, 5, 2, 4\}$$

sketch the following signals.

1. $x[n]$
2. $x[n]u[n]$
3. $x[n]u[-n]$
4. $x[n-2]u[n]$
5. $x[n-2]u[n-2]$
6. $x[n-2]u[n+2]$
7. $x[n]u[-n-3]$
8. $x[2n]$
9. $x[2n+3]u[n+3]$

10. $x[-n]u[-n]$
11. $x[\frac{n}{2}-1]u[n+4]$
12. $x[n]u[3-n]$
13. $x[n](u[n+2]-u[n-3])$

Solution:

1. The given sequence $x[n]$ is plotted as shown in Fig. 1.75a.
2.

$$u[n] = \begin{cases} 1 & n \geq 0 \text{ for all } n \\ 0 & n < 0 \end{cases}$$

$x[n]$ is multiplied by 1 for all n when $n \geq 0$ and by 0 when $n < 0$. $x[n]u[n]$ is sketched in Fig. 1.75b.

3.

$$u[-n] = \begin{cases} 1 & n \leq 0 \text{ for all } n \\ 0 & n > 0 \end{cases}$$

$x[n]$ is multiplied by 1 for all n when $n \leq 0$ and by 0 when $n > 0$. $x[n]u[-n]$ is sketched in Fig. 1.75c.

4.

$$u[n] = \begin{cases} 1 & n \geq 0 \text{ for all } n \\ 0 & n < 0 \end{cases}$$

$x[n]$ is right shifted by 2 samples to get $x[n-2]$. For $n \geq 0, x[n-2]$ is multiplied by 1 and for $n < 0$, it is multiplied by 0. $x[n-2]u[n]$ is sketched in Fig. 1.75d.

5.

$$u[n-2] = \begin{cases} 1 & n \geq 2 \text{ for all } n \\ 0 & n < 2 \end{cases}$$

$x[n-2]$ is right shifted by 2 samples to get $x[n-2]$. For $n \geq 2$, $x[n-2]$ is multiplied by 1 and for $n < 2$, it is multiplied by 0. $x[n-2]u[n-2]$ is sketched in Fig. 1.75e.

6.

$$u[n+2] = \begin{cases} 1 & n \geq -2 \text{ for all } n \\ 0 & n < -2 \end{cases}$$

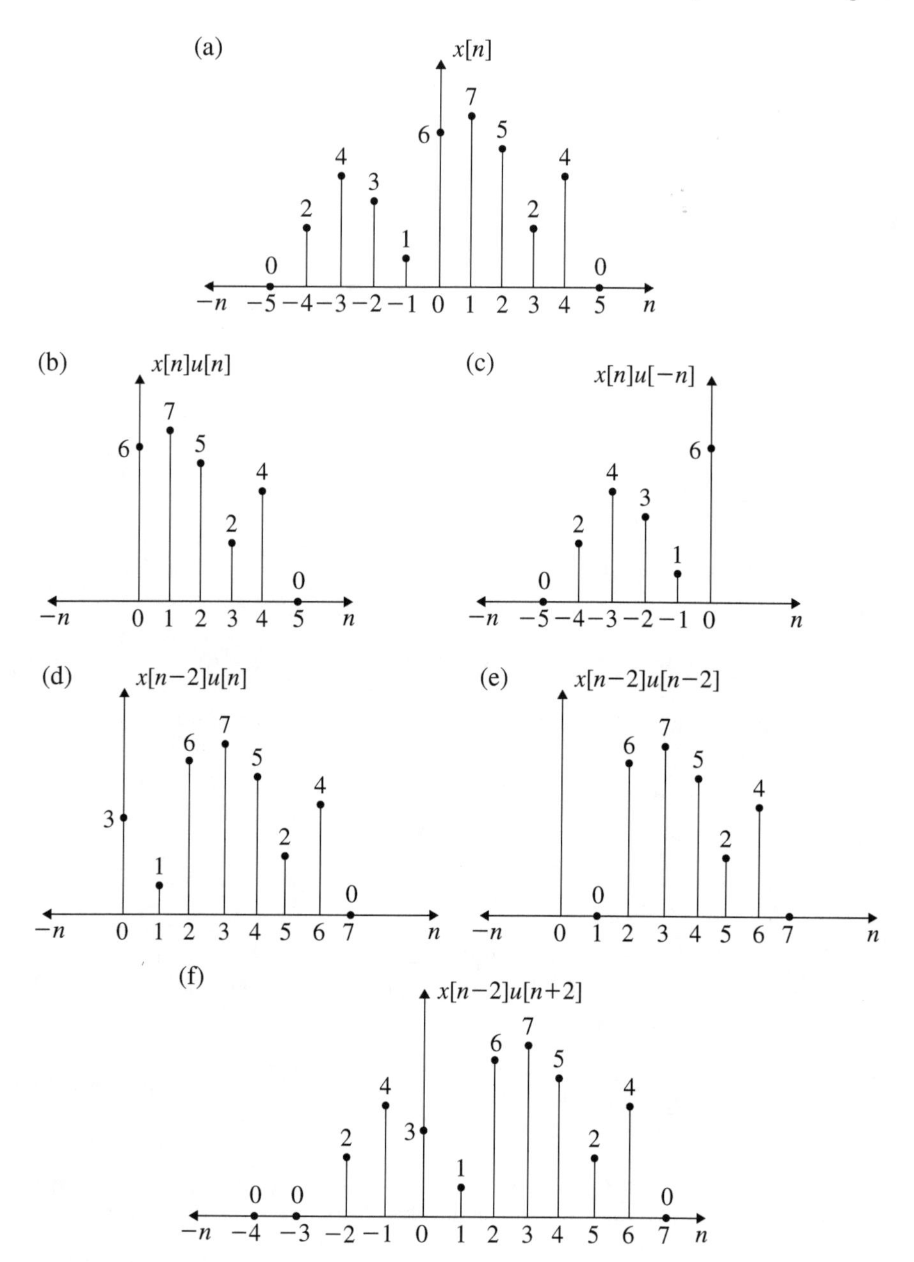

Fig. 1.75 DT signal $x[n]$ when scaled and time shifted

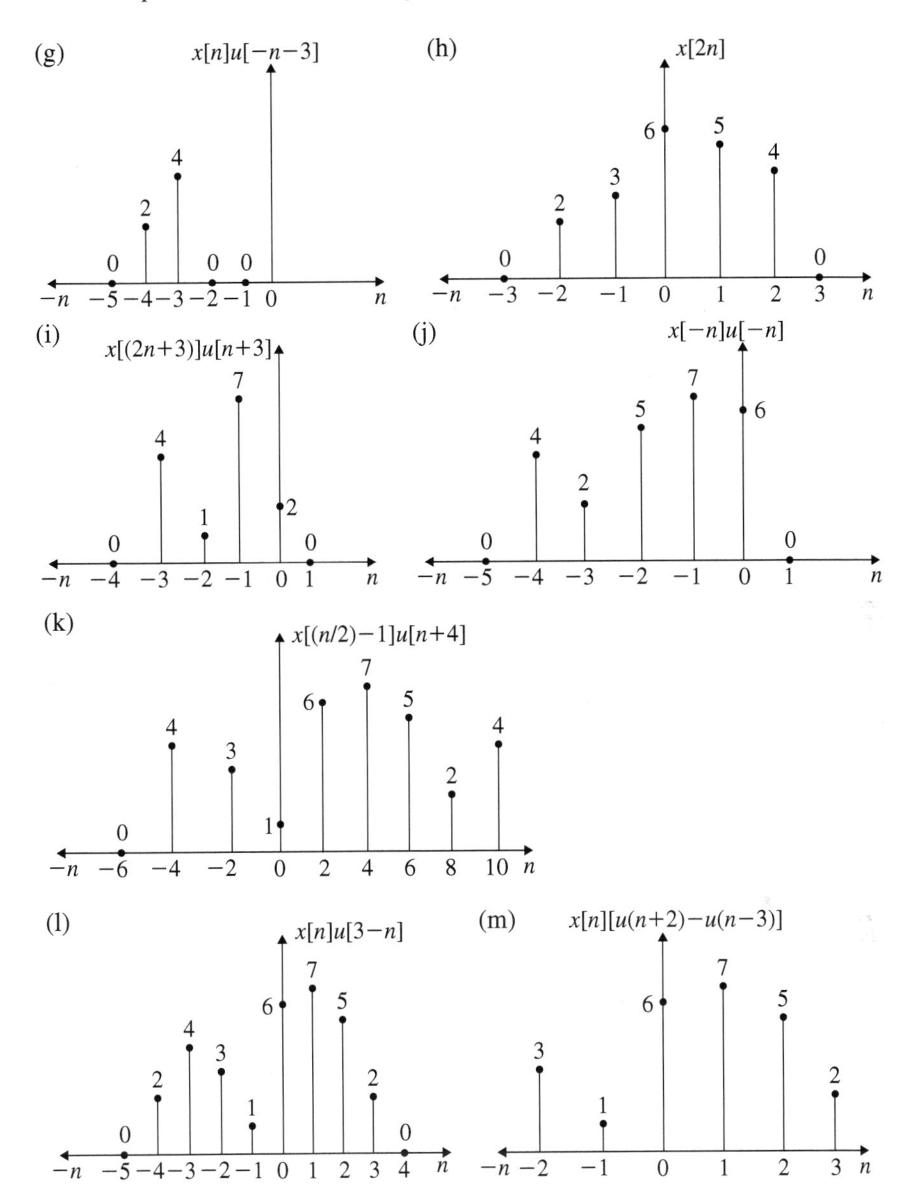

Fig. 1.75 (continued)

$x[n]$ is right shifted by 2 samples to get $x[n-2]$. For $n \geq -2$, for all n, $x[n-2]$ is multiplied by 1 and elsewhere it is multiplied by 0. The sketch of $x[n-2]u[n+2]$ is shown in Fig. 1.75f.

7.

$$u[-n-3] = \begin{cases} 1 & n \leq -3 \text{ for all } n \\ 0 & n > -3 \end{cases}$$

$x[n]$ is multiplied by 1 for all n and $n \leq -3$. For $n > -3$, $x[n]$ is multiplied by 0. The sketch of $x[n]u[-n-3]$ is shown in Fig. 1.75g.

8. $x[n]$ is time compressed by 2 and $x[2n]$ is obtained. $x[2n]$ is sketched as shown in Fig. 1.75h.

9.

$$u[n+3] = \begin{cases} 1 & n \geq -3 \text{ for all } n \\ 0 & n < -3 \end{cases}$$

$x[n]$ is left shifted by 3 samples to get $x[n+3]$. $x[n+3]$ is time compressed by 2 to get $x[2n+3]$. $x[2n+3]$ is multiplied by 1 for $n \geq -3$ for all n and elsewhere it is multiplied by 0. The sketch of $x[2n+3]u[n+3]$ is shown in Fig. 1.75i.

10.

$$u[-n] = \begin{cases} 1 & n \leq 0 \text{ for all } n \\ 0 & n > 0 \end{cases}$$

$x[-n]$ is sketched by reflecting $x[n]$. For $n \leq 0$, $x[-n]$ is multiplied by 1 for all n. For $n > 0$, $x[-n]$ is multiplied by 0. The plot of $x[-n]u[-n]$ is shown in Fig. 1.75j.

11.

$$u[n+4] = \begin{cases} 1 & n \geq -4 \text{ for all } n \\ 0 & n < -4 \end{cases}$$

$x[n]$ is right shifted by one sample to get $x[n-1]$. It is time expanded by multiplying every n by 2 and thus $x[\frac{n}{2}-1]$ is obtained. $x[\frac{n}{2}-1]$ is multiplied by 1 for $n \geq -4$ for all n. Elsewhere $x[\frac{n}{2}-1]$ is zero. The plot for $x[\frac{n}{2}-1]u[n+4]$ is shown in Fig. 1.75k.

12.

$$u[3-n] = \begin{cases} 1 & n \leq 3 \text{ for all } n \\ 0 & n > 3 \end{cases}$$

$x[n]$ is multiplied by 1 for $n \leq 3$ for all n and elsewhere by zero. The sketch of $x[n]u[3-n]$ is shown in Fig. 1.75l.

13.

$$u[n+2] - u[n-3] = \begin{cases} 1 & -2 \leq n \leq 3 \text{ for all } n \\ 0 & \text{otherwise} \end{cases}$$

$x[n]$ is multiplied by 1 for $-2 \leq n \leq 3$ for all n and elsewhere by zero. The plot of $x[n][u(n+2) - u(n-3)]$ is shown in Fig. 1.75m.

■ Example 1.31

Carefully sketch the following discrete time functions.

1. (a) $x[n] = \text{rect}_4[n]$
   (b) $x[n] = \text{rect}_4[n+2]$
2. (a) $x[n] = \text{ramp}[-n]$
   (b) $x[n] = \text{ramp}[\frac{n}{2}]$
3. (a) $x[n] = \delta[3n]$
   (b) $x[n] = \delta[\frac{2}{3}n]$
4. $x(n) = u[3n-2]$
5. $x(n) = \text{tri}_4[n]$
6. $x(n) = \text{ramp}[n+2] - 2\text{ramp}[n] - \text{ramp}[n-2]$
7. $x(n) = 4\sin\left[\frac{\pi}{3}n\right]\text{rect}_4[n]$
8. $x(n) = 2\sin\left[\frac{\pi}{2}n\right]\text{rect}_3[n-2]$
9. $x(n) = n(u[n] - u[n-5])$
10. $x(n) = (n-2)(u[n-2] - u[n-5])$
11. $x(n) = (8-n)(u[n-6] - u[n-10])$

Solution:

1.

$$x[n] = \text{rect}_4[n]$$

Suffix four in $\text{rect}_4[n]$ indicates the width of the rectangular pulse on either side of n. $x[n] = \text{rect}_4[n]$ is sketched as shown in Fig. 1.76a. $\text{rect}_4[n]$ is left shifted by two samples to get $x[n] = \text{rect}_4[n+2]$ which is sketched in Fig. 1.76b.

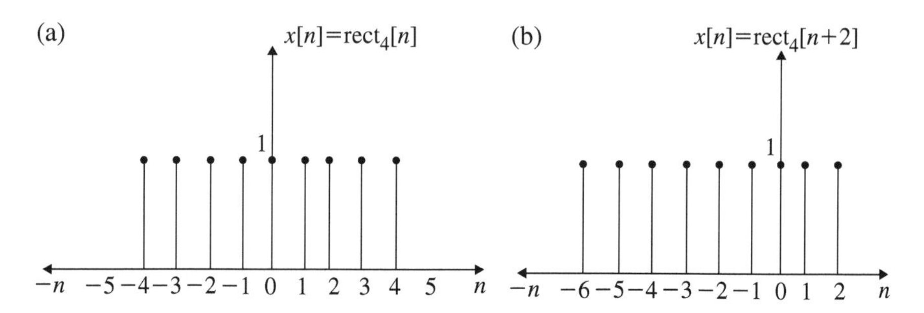

Fig. 1.76 Graphical representation of **a** $x[n] = \text{rect}_4[n]$ and **b** $x[n] = \text{rect}_4[n + 2]$

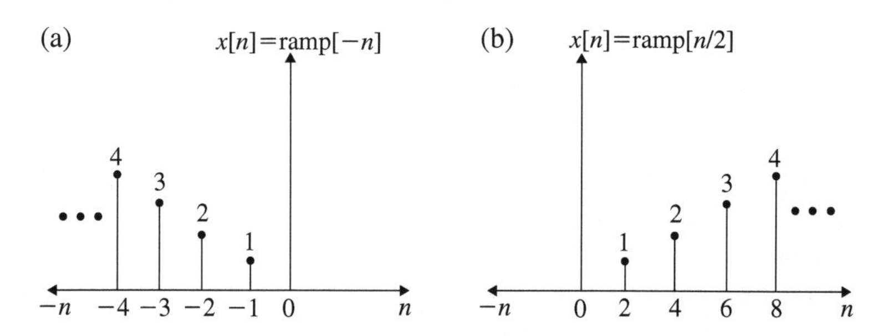

Fig. 1.77 Graphical representation of **a** $x[n] = \text{ramp}[-n]$ and **b** $x[n] = \text{ramp}[\frac{n}{2}]$

2. (a)

$$\text{ramp}[n] = \begin{cases} n & 0 \leq n < \infty \\ 0 & n < 0 \end{cases}$$

$$\text{ramp}[-n] = \begin{cases} -n & -\infty \leq n \leq 0 \\ 0 & n > 0 \end{cases}$$

$x[n] = \text{ramp}[-n]$ is represented in Fig. 1.77a.

(b) $x[n] = \text{ramp}[n]$ is time expanded and is sketched as $x[n] = \text{ramp}[\frac{n}{2}]$ as shown in Fig. 1.77b.

3. (a)

$$\delta[3n] = \begin{cases} 1 & n = 0 \\ 0 & n \neq 0 \end{cases}$$

$\delta[an] = 1$ for all integer values of a at $n = 0$. For $n \neq 0$, $\delta[an] = 0$. This is represented in Fig. 1.78a.

(b) In $\delta[\frac{2}{3}n]$, $\frac{2}{3}$ is not an integer. Hence, for all n, $\delta[\frac{2}{3}n] = 0$ as represented in Fig. 1.78b.

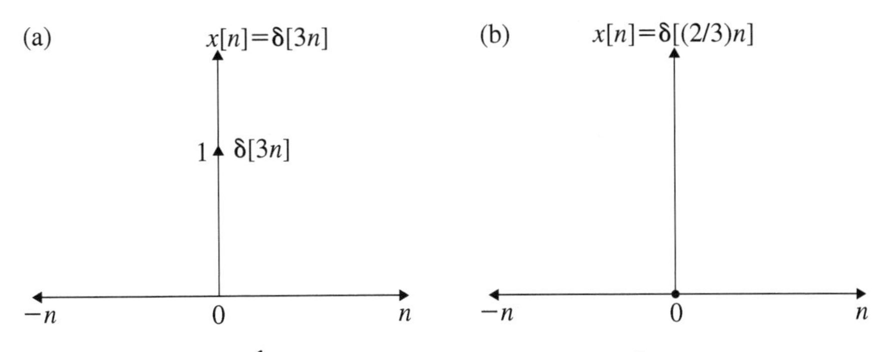

Fig. 1.78 Graphical representation of **a** $x[n] = \delta[3n]$ and **b** $x[n] = \delta[\frac{2}{3}n]$

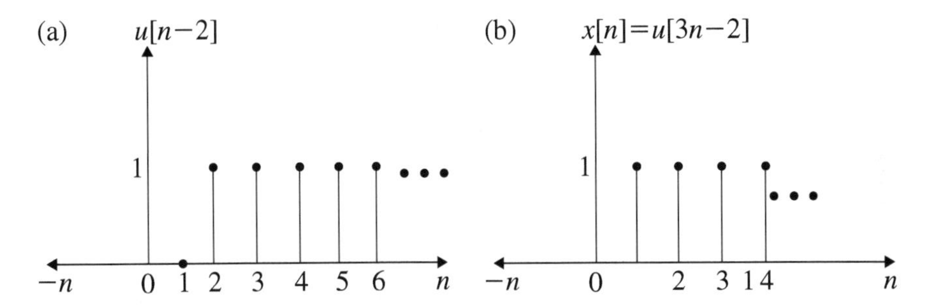

Fig. 1.79 Graphical representation of **a** $u[n-2]$ and **b** $x[n] = u[3n-2]$

4. The step sequence $u[n]$ is right shifted by two samples to get $u[n-2]$. This is represented in Fig. 1.79a. $u[n-2]$ is time compressed by 3 to get $x[n] = u[3n-2]$ and this shown in Fig. 1.79b.
5. (Fig. 1.80)
6.

$$x[n] = \text{ramp}[n+2] - 2\text{ramp}[n] + \text{ramp}[n-2]$$

The components of $x[n]$ namely ramp$[n+2]$, -2ramp$[n]$ and ramp$[n-2]$ are sketched as shown in Fig. 1.81a–c, respectively. The sum of these components at n is represented as $x[n]$ in Fig. 1.81d.

7.

$$x[n] = 4\sin\left[\frac{\pi}{3}n\right]\text{rect}_4[n]$$

Rectangular pulse of width four on both sides is shown in Fig. 1.82a. Its amplitude is four. $\sin\frac{\pi}{3}n$ is shown in Fig. 1.82b. The sinusoid has a radian frequency of $\Omega_0 = \frac{\pi}{3}$ radians. The periodicity of the sinusoid is given by

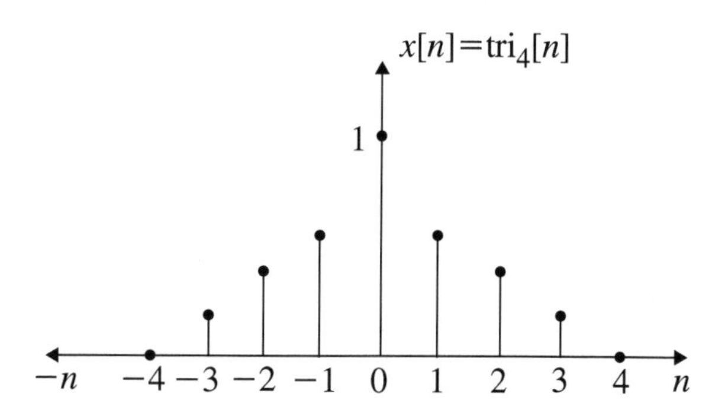

Fig. 1.80 Graphical representation of $x[n] = \text{tri}_4[n]$

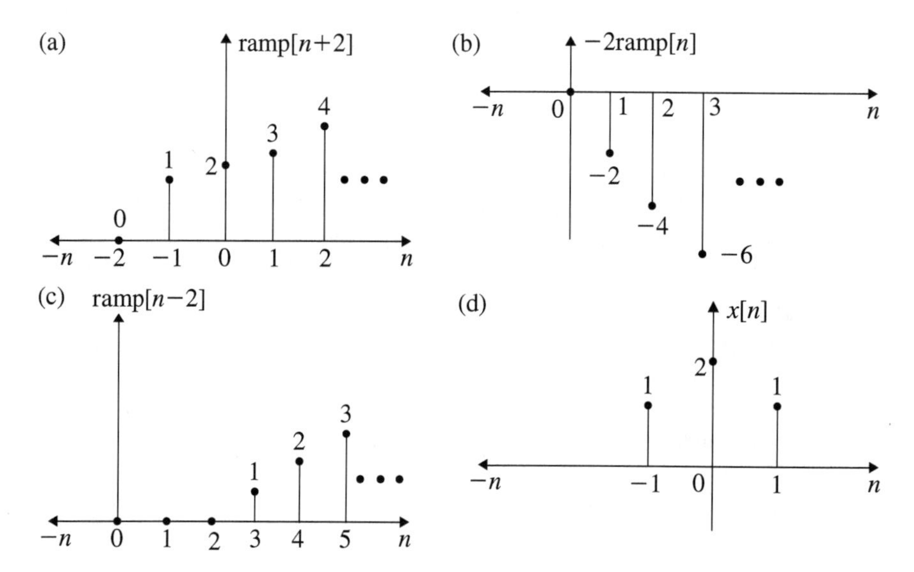

Fig. 1.81 Graphical representation of $x[n]$ of Example 1.31.6

$$\begin{aligned}N_0 &= \frac{2\pi}{\Omega_0} \\ &= \frac{2\pi}{\pi} \times 3 \\ &= 6 \text{ samples per cycle}\end{aligned}$$

Thus, each sample is separated by $\frac{\pi}{3}$ radians or 60°. The product of 4 $\sin[\frac{\pi}{3}n]\text{rect}_4[n]$ is obtained by multiplying the sample strength of $4\sin[(\pi/3)n]$ with the corresponding sample of $\text{rect}_4[n]$ both of which have the same n. Thus, $x[n]$ is obtained and is shown in Fig. 1.82c. $x[n]$ is limited to the width of ± 4 which is the width of $\text{rect}_4[n]$. In Fig. 1.82b, c, the dotted line graph represents the sinusoid of the CT signal corresponding to $\sin[\frac{\pi n}{3}]$.

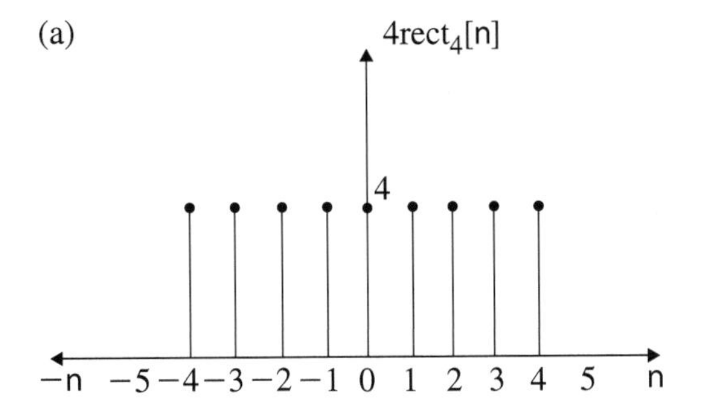

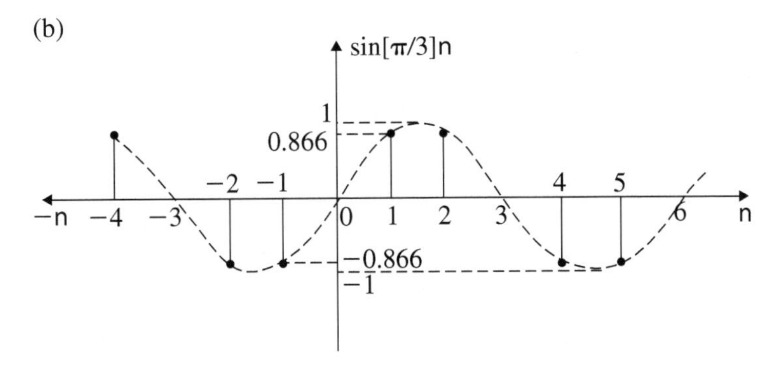

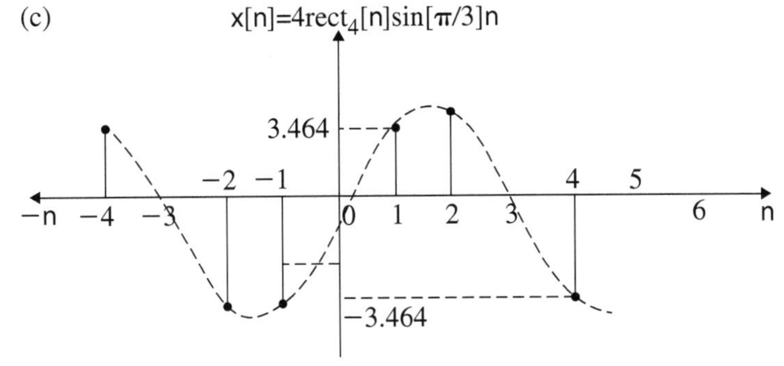

Fig. 1.82 Graphical representation of $x[n] = 4\sin[\frac{\pi}{3}n]\mathrm{rect}_4[n]$

8.

$$x[n] = 2\sin\left[\frac{\pi}{2}n\right]\mathrm{rect}_3[n-2]$$

$\mathrm{rect}_3[n-2]$ is sketched with right shifted $\mathrm{rect}_3[n]$ by two samples with its amplitude 2. This is represented in Fig. 1.83a. $\sin[\frac{\pi}{2}n]$ has a radian frequency $\Omega_0 = \frac{\pi}{2}$. The number of samples in one cycle (2π radians) is

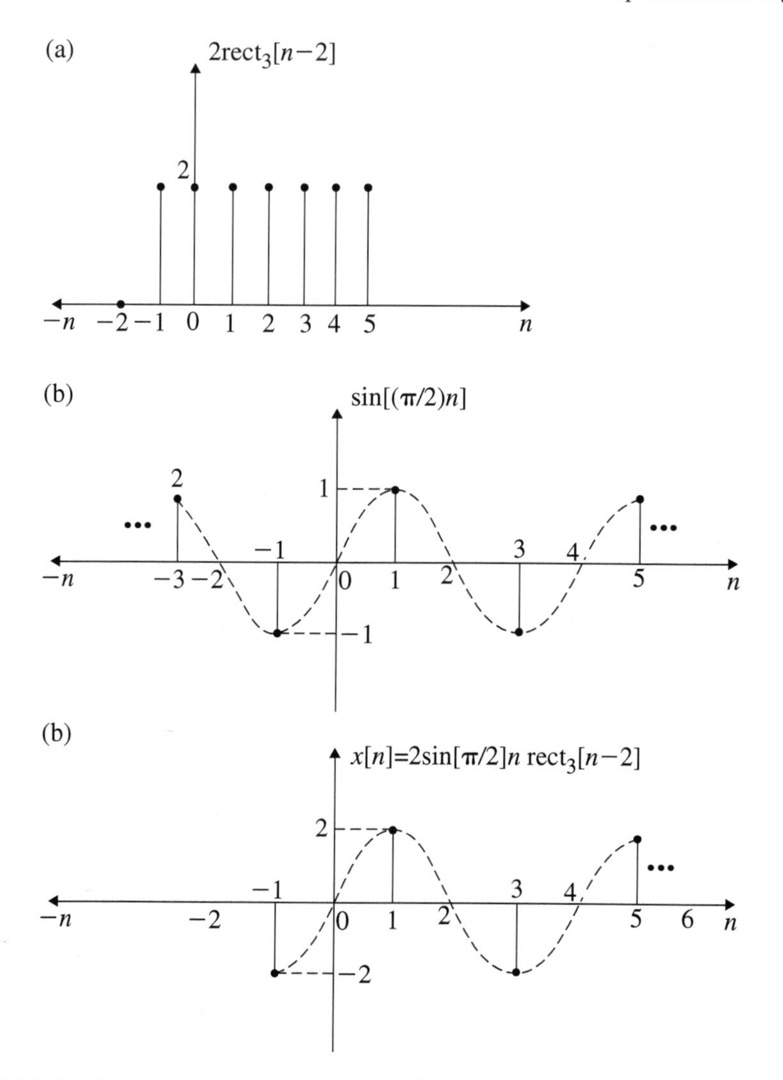

Fig. 1.83 Graphical representation of $x[n] = 2\sin[\frac{\pi}{2}n]\mathrm{rect}_3[n-2]$

$$\begin{aligned} N_0 &= \frac{2\pi}{\Omega_0} \\ &= \frac{2\pi}{\pi} \times 2 \\ &= 4 \text{ samples/cycle} \end{aligned}$$

For odd values of n, peak occurs and for even value of n, the pulse strength is zero. The periodic $\sin \frac{n\pi}{2}$ DT signal is represented in Fig. 1.83b. $x[n]$ is obtained by multiplying a sample at $n = n_1$ in $2\mathrm{rect}_3[n-2]$ with a corresponding sample

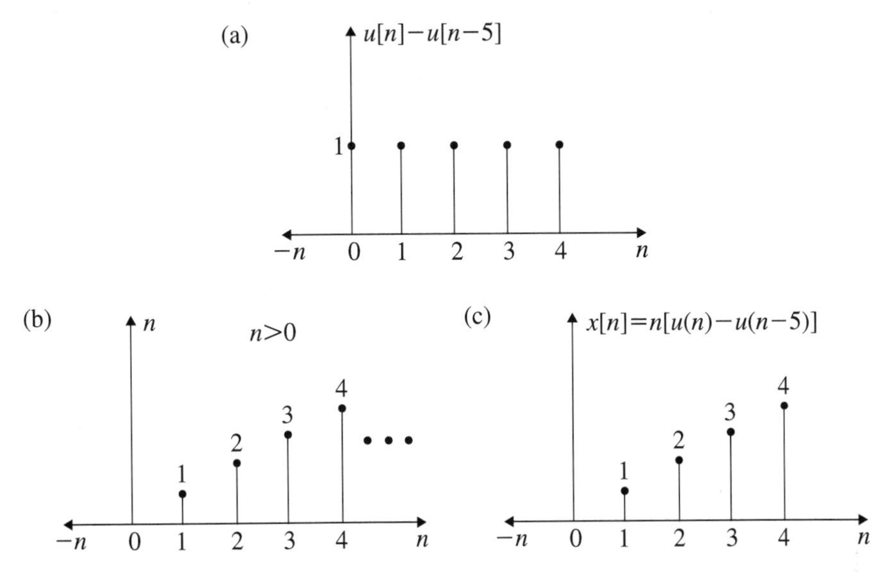

Fig. 1.84 Graphical representation of signal of Example 1.31.9

at $n = n_1$ in $\sin[\frac{\pi}{2}n]$ signal. $x[n]$ exists only in the sample interval $-1 \leq n \leq 5$. For the values of $n, x[n] = 0$. $x[n]$ is sketched as shown in Fig. 1.83c.

9.

$$x[n] = n(u[n] - u[n-5])$$

$u[n]$ is a unit step sequence and $u[n-5]$ is the time shifted (right shift) by 5 samples. $u[n] - u[n-5]$ exists only during $0 \leq n \leq 4$. For $n \geq 5$, the negative going $-u[n-5]$ will cancel out with the positive going pulses of $u[n]$. Figure 1.84a represents $u[n] - u[n-5]$. The ramp signal n is represented in Fig. 1.84b. The product of these two signals is represented in Fig. 1.84c. $x[n]$ exists only during the sample interval $0 \leq n \leq 4$. For any other value $x[n] = 0$.

10.

$$x[n] = [n-2](u[n-2] - u[n-5])$$

$$u[n-2] - u[n-5] = \begin{cases} 1 & 2 \leq n < 5 \\ 0 & \text{otherwise} \end{cases}$$

The above function is represented in Fig. 1.85a. $[n-2]$ is a ramp function that is right shifted by two samples from the origin. This is represented in Fig. 1.85b. Each sample of Fig. 1.85a is multiplied by the corresponding sample occurring at the same instant in Fig. 1.85b and the product is obtained as $x[n]$ which is represented in Fig. 1.85c.

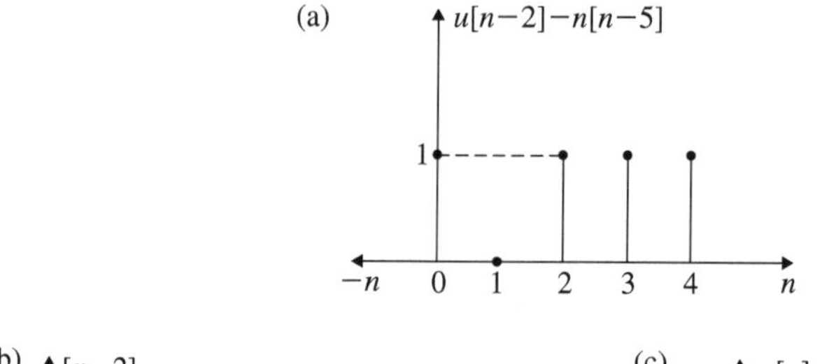

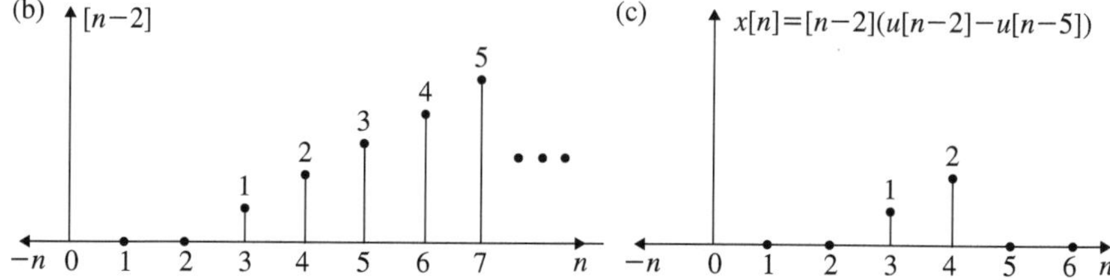

Fig. 1.85 Graphical representation of $x[n] = [n-2](u[n-2] - u[n-5])$

11.

$$x[n] = [8-n](u[n-6] - u[n-10])$$

$$u[n-6] - u[n-10] = \begin{cases} 1 & 6 \leq n < 10 \\ 0 & \text{otherwise} \end{cases}$$

The above function is represented in Fig. 1.86a. $[-n]$ is the folded version of $[n]$ and is shown in Fig. 1.86b. $[-n]$ is right shifted by eight samples to get $[8-n]$ which is represented in Fig. 1.86c. $[8-n]$ and $u[n-6] - u[n-10]$ overlap each other for $n = 6$ and $n = 7$ only. For $n = 6$, the product of these two functions is 2. For $n = 7$, the product is 1. Thus, $x[n]$ is plotted as shown in Fig. 1.86d.

■ Example 1.32

Sketch the following signals. Find the radian and cyclic frequencies and the period in each case.

1. $x[n] = \sin(\frac{\pi}{8}n)$
2. $x[n] = \cos(\frac{\pi}{12}n)$
3. $x[n] = \cos(\frac{\pi}{12}n - \frac{\pi}{3})$
4. $x[n] = \cos(\frac{\pi}{12}n + \frac{\pi}{3})$

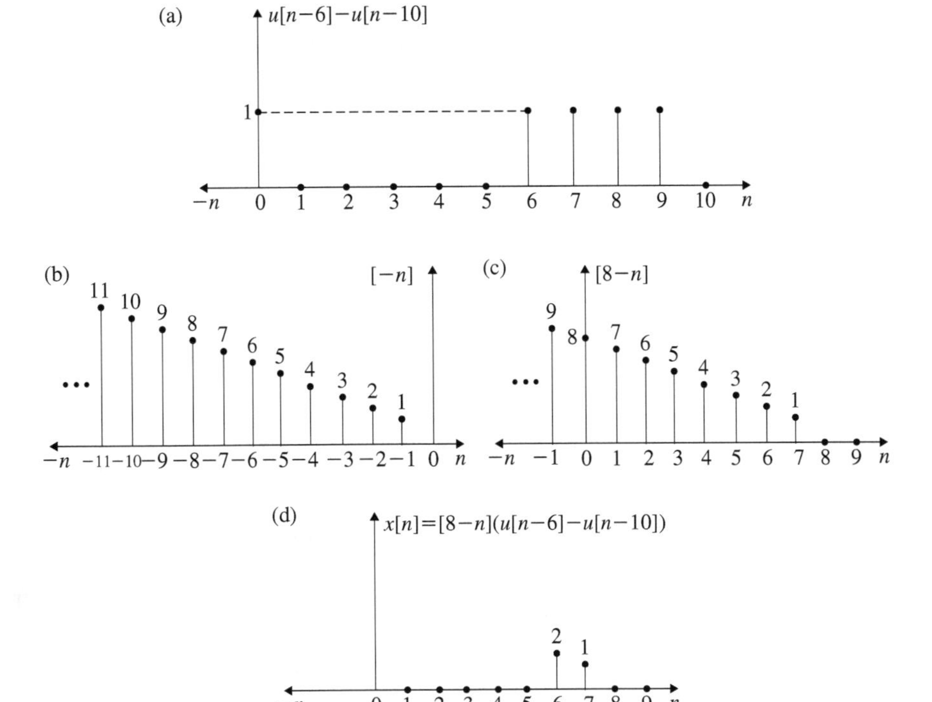

Fig. 1.86 Graphical representation of $x[n] = [8 - n](u[n - 6] - u[n - 10])$

Solution:

1.

$$x[n] = \sin \frac{\pi}{8} n$$

The radian frequency is given by

$$\Omega_0 = \frac{\pi}{8} \text{ radian per sample}$$

The number of samples per cycle is given by

$$\begin{aligned} N_0 &= \frac{2\pi}{\Omega_0} \\ &= \frac{2\pi}{\pi} \times 8 \\ &= 16 \text{ samples per cycle} \end{aligned}$$

The cyclic frequency is given by

$$f_0 = \frac{1}{N_0}$$
$$= \frac{1}{16} \text{ cycles per sample}$$

The plot of $x[n]$ is shown in Fig. 1.87a for two cycles.

2.

$$x[n] = \cos\left(\frac{\pi}{12}n\right)$$

The radian frequency is given by

$$\Omega_0 = \frac{\pi}{12} \text{radians/samples}$$

The cosine wave repeats itself for every 2π radians. The number of samples per cycle is given by

$$N_0 = \frac{2\pi}{\Omega_0}$$
$$= \frac{2\pi}{\pi} \times 12$$
$$= 24 \text{ samples per cycle.}$$

The cyclic frequency is given by

$$f_0 = \frac{1}{N_0}$$
$$= \frac{1}{24} \text{ cycles per sample}$$

For $n = 0$,

$$\cos\frac{\pi}{12}n = 1$$

and is maximum. For $n = 12$,

$$\cos\frac{\pi}{12} \times 12 = -1$$

For $n = \pm 6$,

$$\cos\frac{\pi}{12} \times 6 = 0$$

For $n = -12$,

$$\cos\frac{\pi}{12} \times 12 = -1$$

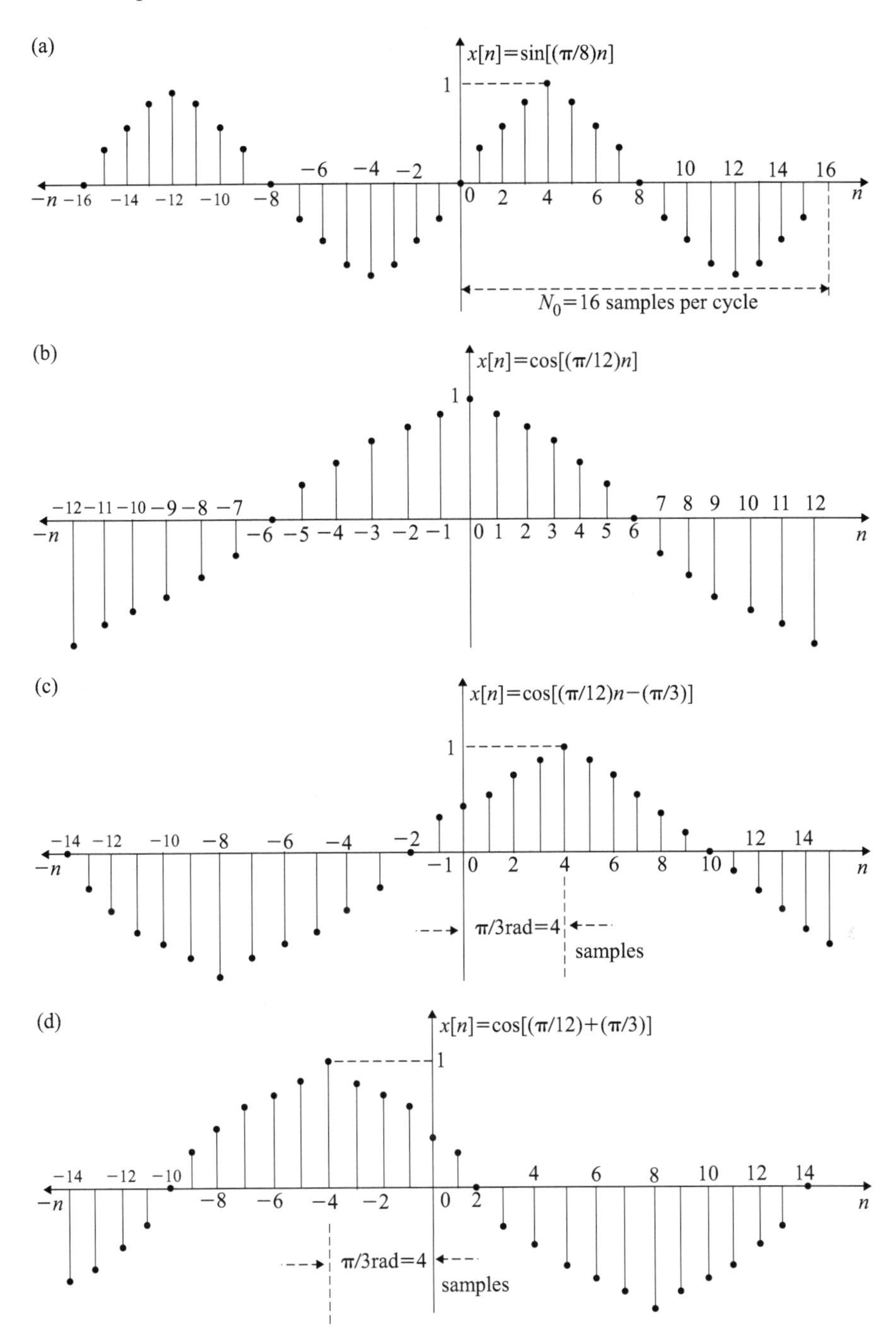

Fig. 1.87 Graphical representation of $x[n]$ of Example 1.32

The complete wave of $\cos(\frac{\pi}{12}n)$ is sketched in Fig. 1.87b for one cycle.

3.

$$x[n] = \cos\left(\frac{\pi}{12}n - \frac{\pi}{3}\right)$$

This cosine wave is the same as the one given in Example 1.32.2 but it lags behind by $\frac{\pi}{3}$ radians. Since 2π corresponds to 24 samples, $\frac{\pi}{3}$ corresponds to

$$\frac{24}{2\pi} \times \frac{\pi}{3} = 4 \text{ samples}$$

The original $\cos(\frac{\pi}{12}n)$ wave is right shifted by 4 samples and is sketched as shown in Fig. 1.87c.

4.

$$x[n] = \cos\left(\frac{\pi}{12}n + \frac{\pi}{3}\right)$$

Here, $\cos(\frac{\pi}{12}n + \frac{\pi}{3})$ leads $\cos(\frac{\pi}{12}n)$ by $\frac{\pi}{3}$ radians. $\frac{\pi}{3}$, as stated in the previous problem, corresponds to 4 samples. Here, $\cos(\frac{\pi}{12}n)$ is left shifted by 4 samples to get $\cos(\frac{\pi}{12}n + \frac{\pi}{3})$ and this is sketched in Fig. 1.87d.

■ Example 1.33

Given

$$x[n] = \{1,\ 2,\ \underset{\uparrow}{3},\ -4,\ 6\}$$

Plot the signal $x[-n-1]$.

(*Anna University, May, 2007*)

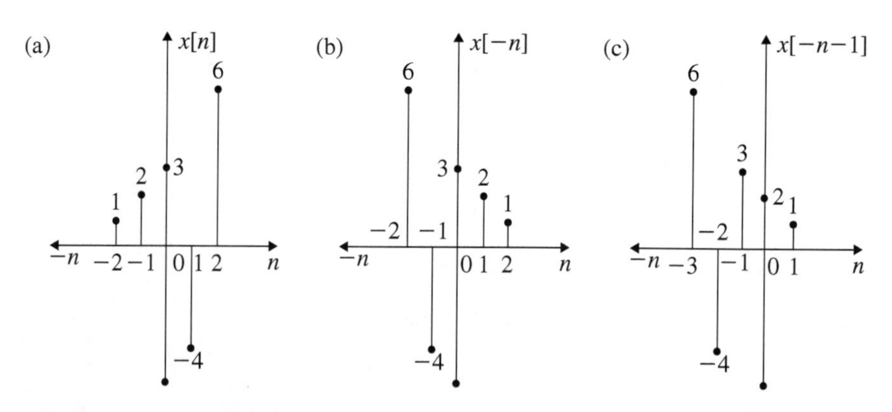

Fig. 1.88 DT sequences of Example 1.33

Solution:

1. The sequence $x[n]$ is represented in Fig. 1.88a.
2. By folding $x[n]$, $x[-n]$ is obtained and represented in Fig. 1.88b.
3. $x[-n]$ is shifted to the left by one sample and $x[-n-1]$ is obtained. This is represented in Fig. 1.88c.

1.8 Classification of Signals

Signals which are classified in the broad category of continuous and discrete time signals are further classified as follows.

1. Deterministic and non-deterministic (random) signals.
2. Periodic and non-periodic (aperiodic) signals.
3. Odd and even signals.
4. Power and energy signals.

1.8.1 Deterministic and Non-deterministic Continuous Signals

Deterministic signals are signals which are characterized mathematically. The amplitude of such signals at any time interval t can be determined at all time t. Consider the signals described by the following equations:

$$x(t) = A$$

$$x(t) = A \sin \omega t$$

The above signals represent a step signal and a sinusoidal signal, respectively, and they are shown in Fig. 1.89a, b. At any instant of time t, the amplitude of the step signal which is deterministic can be easily determined. On the other hand, consider the sinusoidal signal polluted with noise shown in Fig. 1.89b. The magnitude of such a signal cannot be easily determined since the noise variation is random.

1.8.2 Periodic and Non-periodic Continuous Signals

Consider the continuous time signal described by the following equation:

$$x(t + nT_0) = x(t) \text{ for all } t \tag{1.26}$$

where n is any integer value. **A continuous time signal $x(t)$ is said to be periodic with period T_0 if it repeats itself in a minimum positive interval. The minimum**

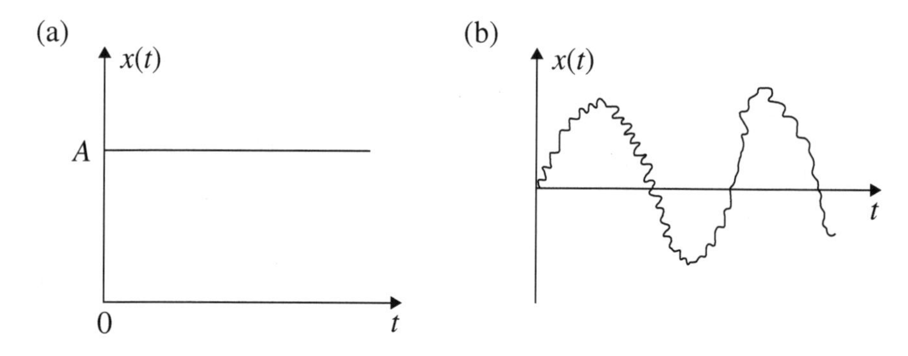

Fig. 1.89 Continuous. **a** Deterministic signal; **b** Random signal

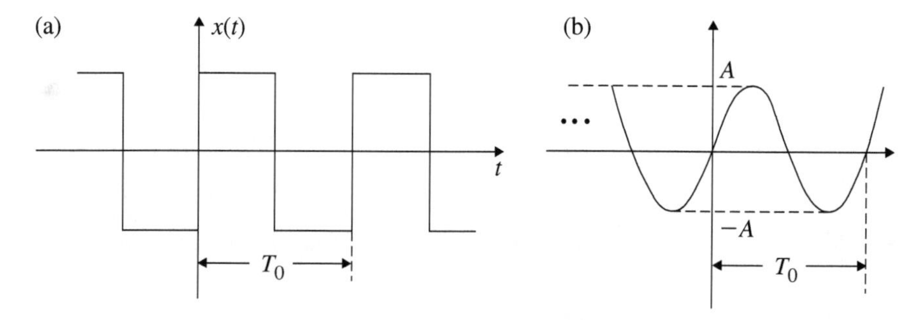

Fig. 1.90 Examples of periodic signals. **a** Rectangular wave; **b** Sine wave

positive interval over which a function repeats is called fundamental period T_0. The fundamental frequency f is expressed as

$$f_0 = \frac{1}{T_0} \tag{1.27}$$

where f_0 is expressed in cycles per sec. The fundamental radian frequency is expressed as

$$\begin{aligned} \omega_0 &= 2\pi f_0 \\ &= \frac{2\pi}{T_0} \end{aligned} \tag{1.28}$$

Here ω is expressed in rad./sec. The periodic rectangular wave and sine wave are shown in Fig. 1.90a and b, respectively.

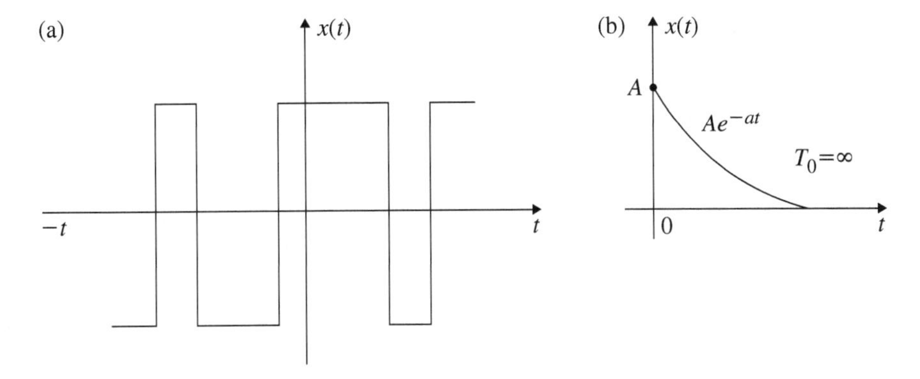

Fig. 1.91 Non-periodic signals. **a** Rectangular; **b** Exponential decay

Any continuous time signal which is not periodic is said to be non-periodic or aperiodic signal. Figure 1.91a represents a non-periodic rectangular wave and Fig. 1.91b represents an exponential decay. The non-periodic signal does not repeat itself with respect to time.

1.8.3 Fundamental Period of Two Periodic Signals

Consider the periodic signal of two periodic functions with two different fundamental periods as given below.

$$x(t) = A_1 \sin\left(2\pi \frac{t}{T_1}\right) + A_2 \sin\left(2\pi \frac{t}{T_2}\right) \tag{1.29}$$

where T_1 and T_2 are the fundamental periods of two sine waves. The fundamental period of the composite signal $x(t)$ is given by the shortest time by which these signals have an integer number. If each of these two signals repeat exactly an integer number of times in some minimum time interval, then they will repeat exactly an integer number of times again in the next time interval. This is calculated as the Least Common Multiple (LCM) of the two fundamental periods. Thus, the fundamental period of a periodic signal, which is composed of more than one periodic signal, is obtained by taking the least common multiple of the fundamental periods of all the signals. The fundamental frequency of the sum of the signals is the greatest common divisor of the two frequencies. **It is to be remembered that if any of the composite signal is non-periodic, then the overall function is also non-periodic.**

Instead of sum of two functions, if a signal is a product of two functions, the method of finding the fundamental period remains the same. Consider the following composite signal:

$$x(t) = A \sin\left(2\pi \frac{t}{T_1}\right) \sin\left(2\pi \frac{t}{T_2}\right) \tag{1.30}$$

The fundamental period of the two sine functions are T_1 and T_2. The fundamental period of $x(t)$ is calculated as the least common multiple of T_1 and T_2. The sum of the product of two or more periodic signals is periodic **iff** (if and only if) their ratio of their fundamental periods is rational. The following steps are followed to determine this:

1. Determine the fundamental period of the individual signal in the sum or product.
2. Find the ratio of the fundamental period of the first signal with the fundamental period of every other signal.
3. If these ratios are rational, then the sum or the product of the composite signal is periodic.
4. The fundamental period of the composite signal is determined by taking the least common multiple of the fundamental period. Alternatively, the greatest common divisor of the fundamental frequency of each signal gives the fundamental frequency of the composite signal.

For example if T_1, T_2 and T_3 are the fundamental periods of three signals which are the sums of the composite signal, then the ratios $\frac{T_1}{T_2}$ and $\frac{T_1}{T_3}$ should be an integer multiple or rational. $\frac{T_1}{T_2} = \frac{5}{3}$ is an integer or rational number. On the other hand, $\frac{T_1}{T_2} = \frac{5}{3.17}$ is not an integer number and it is not rational.

Sinusoidal and complex exponentials are examples of continuous time periodic signals. Consider the following sinusoidal signal.

$$x(t) = A \sin(\omega_0 t + \theta) \tag{1.31}$$

$$\begin{aligned} x(t + T_0) &= A \sin(\omega_0 (t + T_0) + \theta) \\ &= A \sin(\omega_0 t + \omega_0 T_0 + \theta) \end{aligned} \tag{1.32}$$

A sine function repeats itself when its total argument is increased or decreased by any integer multiple of 2π radians. Thus, in Eq. (1.32), if we put $\omega_0 T_0 = 2\pi$,

$$x(t + T_0) = A \sin(\omega_0 t + \theta) = x(t)$$

In other words, the fundamental period of a sine function is

$$\boxed{T_0 = \frac{2\pi}{\omega_0}} \tag{1.33}$$

Now consider the complex exponential.

$$x(t) = e^{j\omega_0 t}$$
$$x(t+T_0) = e^{j\omega_0(t+T_0)} \tag{1.34}$$
$$= e^{j\omega_0 t} e^{j\omega_0 T_0} \tag{1.35}$$

If we put $e^{j\omega_0 T_0} = 1$, Eq. (1.35) becomes

$$x(t+T_0) = e^{j\omega_0 t} = x(t)$$

Thus, the condition for the complex exponential to be periodic is that

$$e^{j\omega_0 T_0} = 1$$
$$\text{or} \quad \omega_0 T_0 = 2\pi \qquad [e^{j2\pi} = \cos 2\pi + j \sin 2\pi = 1]$$

$$\boxed{T_0 = \frac{2\pi}{\omega_0}} \tag{1.36}$$

■ Example 1.34

Test the periodicity of the following signals:

(a) $x(t) = 3\cos\left(5t + \frac{\pi}{6}\right)$
(b) $x(t) = e^{j10t}$
(c) $x(t) = \tan(5t + \theta)$
(d) $x(t) = 1$

(Anna University, May, 2006)

Solution:

(a) $\boldsymbol{x(t) = 3\cos\left(5t + \frac{\pi}{6}\right)}$

$$\omega_0 = 5\,\text{rad./sec.}$$

Using Eq. (1.33), we get

$$\boxed{T_0 = \frac{2\pi}{\omega_0} = \frac{2\pi}{5}\ \text{sec.}}$$

The given signal is periodic with the fundamental period $T_0 = \frac{2\pi}{5}$ sec.

(b) $\boldsymbol{x(t) = e^{j10t}}$

$$\omega_0 = 10\,\text{rad./sec.}$$

Using Eq. (1.36), we get

$$\begin{aligned} T_0 &= \frac{2\pi}{\omega_0} \\ &= \frac{2\pi}{10} = 0.2\pi \text{ sec.} \end{aligned}$$

The given signal is periodic with the fundamental period

$$\boxed{T_0 = 0.2\pi \text{ sec.}}$$

(c) $\boldsymbol{x(t) = \tan(5t + \theta)}$

$$\begin{aligned} x(t + T_0) &= \tan(5(t + T_0) + \theta) \\ &= \tan(5t + 5T_0 + \theta) \end{aligned}$$

The tangent function repeats itself for every π rad. of its total argument. Thus, if $5T_0 = \pi$,

$$\begin{aligned} x(t + T_0) &= \tan(5t + \theta) \\ &= x(t) \end{aligned}$$

Hence,

$$\boxed{T_0 = \frac{\pi}{5} \text{ sec.}}$$

(d) $x(t)$ is a d.c. signal and it does not repeat itself. Hence, it is not periodic.

■ Example 1.35

If $x_1(t)$ and $x_2(t)$ are periodic signals of period T_1 and T_2, show that the sum $x(t) = x_1(t) + x_2(t)$ is a periodic signal if $T_1/T_2 = n/m$ which is a rational number.

Solution: For the signals $x_1(t)$ and $x_2(t)$ to be periodic, the following equations hold good.

$$\begin{aligned} x_1(t) &= x_1(t + mT_1) \\ x_2(t) &= x_2(t + nT_2) \end{aligned}$$

Now,

$$\begin{aligned} x(t) &= x_1(t) + x_2(t) \\ x(t+T) &= x_1(t+T) + x_2(t+T) \\ &= x_1(t+mT_1) + x_2(t+nT_2) \end{aligned}$$

From the above equations, we get

$$\begin{aligned} T &= mT_1 = nT_2 \\ \frac{T_1}{T_2} &= \frac{n}{m} = \text{a rational number.} \end{aligned}$$

■ Example 1.36

If $x_1(t)$ and $x_2(t)$ are the periodic signals with fundamental periods T_1 and T_2, respectively, show that the product $x(t) = x_1(t)x_2(t)$ will be periodic if $\frac{T_1}{T_2}$ is a rational number.

Solution: For the periodic signals $x_1(t)$ and $x_2(t)$, the following equations are written:

$$\begin{aligned} x_1(t) &= x_1(t+T_1) = x_1(t+mT_1) \\ x_2(t) &= x_2(t+T_2) = x_2(t+nT_2) \\ x(t) &= x_1(t+mT_1)x_2(t+nT_2) \end{aligned}$$

Also

$$x(t+T) = x_1(t+T)x_2(t+T)$$

From the above two equations, we get

$$\begin{aligned} T &= mT_1 = nT_2 \\ \frac{T_1}{T_2} &= \frac{n}{m} = \text{a rational number.} \end{aligned}$$

■ Example 1.37

Test whether the following signals are periodic. If periodic determine the fundamental period and frequency.

(a) $x(t) = e^{j(\pi t - 2)}$

(b) $x(t) = \cos^2 t$

(c) $x(t) = E_v \cos 4\pi t$

(d) $x(t) = e^{(j\pi - 2)t}$

Solution:

(a) $\boldsymbol{x(t) = e^{j(\pi t-2)}}$

$$\begin{aligned} x(t) &= e^{j(\pi t-2)} \\ &= e^{-j2}e^{j\pi t} \end{aligned}$$

The signal is a complex exponential with e^{-j2} being a constant. Comparing this with standard complex exponential, we get

$$\begin{aligned} e^{j\pi t} &= e^{j\omega_0 t} \\ \omega_0 &= \pi \\ T_0 &= \frac{2\pi}{\omega_0} = \frac{2\pi}{\pi} \\ T_0 &= 2 \text{ sec.} \\ f_0 &= \frac{1}{T_0} = \frac{1}{2} \\ f_0 &= 0.5 \text{ Hertz.} \end{aligned}$$

The signal is a periodic one with fundamental period $T_0 = 2$ sec. and fundamental frequency $f_0 = 0.5$ Hertz.

(b) $\boldsymbol{x(t) = \cos^2 t}$

$$\begin{aligned} \cos^2 t &= \frac{1}{2}[1 + \cos 2t] \\ &= \frac{1}{2} + \frac{1}{2}\cos 2t \\ &= x_1(t) + x_2(t) \end{aligned}$$

where

$$x_1(t) = \frac{1}{2} \text{ which is a d.c. signal}$$

and

$$x_2(t) = \frac{1}{2}\cos 2t$$

For $x_1(t)$, the fundamental radian frequency

$$\begin{aligned} \omega_0 &= 2 \\ T_0 &= \frac{2\pi}{\omega_0} = \frac{2\pi}{2} = \pi \text{ sec.} \end{aligned}$$

The fundamental frequency $f_0 = \frac{1}{T_0} = \frac{1}{\pi}$ Hz.

(c) $\boldsymbol{x(t) = E_v \cos 4\pi t}$
The even function of $x(t)$ is

$$
\begin{aligned}
E_v x(t) &= \frac{1}{2}[x(t) + x(-t)] \\
&= \frac{1}{2}[\cos 4\pi t + \cos(-4\pi t)] \\
&= \cos 4\pi t \\
\omega_0 &= 4\pi \\
T_0 &= \frac{2\pi}{\omega_0} = \frac{2\pi}{4\pi} = 0.5 \text{ sec.} \\
f_0 &= \frac{1}{T_0} = \frac{1}{0.5} = 2 \text{ Hz}
\end{aligned}
$$

(d) $\boldsymbol{x(t) = e^{(j\pi-2)t}}$

$$
\begin{aligned}
x(t) &= e^{(j\pi-2)t} \\
&= e^{-2t} e^{j\pi t}
\end{aligned}
$$

The function $e^{j\pi t}$ is periodic with fundamental period 2 sec. as seen in problem (a). However, the function e^{-2t} is non-periodic and becomes zero at $t \to \infty$. Hence, the composite signal $x(t)$ is aperiodic.

■ Example 1.38

Consider the following continuous time signal:

$$x(t) = 2\cos 3\pi t + 7\cos 9t$$

Find the periodicity of the signal.

(*Anna University, May, 2005*)

Solution:

$$x(t) = x_1(t) + x_2(t)$$

where

$$
\begin{aligned}
x_1(t) &= 2\cos 3\pi t \\
x_2(t) &= 7\cos 9t
\end{aligned}
$$

If T_1 is the fundamental period of $x_1(t)$,

$$\begin{aligned}
\omega_1 &= 3\pi \\
T_1 &= \frac{2\pi}{\omega_1} = \frac{2\pi}{3\pi} = \frac{2}{3} \text{ (rational)} \\
x_2(t) &= 7\cos 9t \\
\omega_2 &= 9 \\
T_2 &= \frac{2\pi}{\omega_2} = \frac{2\pi}{9} \text{ (not rational)} \\
\frac{T_1}{T_2} &= \frac{2}{3}\frac{9}{2\pi} = \frac{3}{\pi} \text{ (not rational)}
\end{aligned}$$

The signal $x(t)$ is not periodic.

■ Example 1.39

Find the fundamental period and frequency of the following signals:

(a) $x(t) = 5\sin 24\pi t + 7\sin 36\pi t$

(b) $x(t) = 5\cos \pi t \sin 3\pi t$

Solution:

(a) **Method 1:**

$$\begin{aligned}
x(t) &= 5\sin 24\pi t + 7\sin 36\pi t \\
&= x_1(t) + x_2(t)
\end{aligned}$$

where

$$\begin{aligned}
x_1(t) &= 5\sin 24\pi t \\
x_2(t) &= 7\sin 36\pi t
\end{aligned}$$

Let T_1 and T_2 be the fundamental periods of $x_1(t)$ and $x_2(t)$, respectively.

$$\begin{aligned}
\omega_1 &= 24\pi \\
T_1 &= \frac{2\pi}{\omega_1} = \frac{2\pi}{24\pi} = \frac{1}{12} \text{ (rational)} \\
\omega_2 &= 36\pi \\
T_2 &= \frac{2\pi}{\omega_2} = \frac{2\pi}{36\pi} = \frac{1}{18} \text{ (rational)} \\
\frac{T_1}{T_2} &= \frac{1}{12} \times 18 = \frac{3}{2} \text{ (rational)}
\end{aligned}$$

The composite signal is a periodic signal. Since T_1 and T_2 are rational, $x(t)$ is periodic. The fundamental period is obtained as follows. From the ratio of $\frac{T_1}{T_2}$,

$$2T_1 = 3T_2 = T_0$$
$$T_0 = \frac{2}{12} = \frac{1}{6} \text{ sec.}$$

or

$$T_0 = \frac{3}{18} = \frac{1}{6} \text{ sec.}$$
$$f_0 = \frac{1}{T_0} = 6 \text{ Hz.}$$

$$\boxed{\begin{aligned} T_0 &= \frac{1}{6} \text{ sec.} \\ f_0 &= 6 \text{ Hz.} \end{aligned}}$$

Method 2:
In this method, the Least Common Multiple (LCM) for T_1 and T_2 is obtained which gives T_0. In case, T_1 and T_2 are fractions, they are made integers by multiplying by a least number. For T_1 and T_2 thus obtained, LCM is found. T_0 is obtained by dividing by the same number which was chosen to make T_1 and T_2 as integers. In the above example,

(1)

$$T_1 = \frac{1}{12} \quad \text{and} \quad T_2 = \frac{1}{18}$$

By multiplying T_1 and T_2 by 36, $T_1 = 3$ and $T_2 = 2$.
(2) The LCM for the new T_1 and T_2 is easily obtained as 6.
(3) T_0 is obtained by dividing LCM by 36.

$$T_0 = \frac{\text{LCM}}{36} = \frac{6}{36} = \frac{1}{6} \text{ sec.}$$

$$\boxed{\begin{aligned} T_0 &= \frac{1}{6} \text{ sec.} \\ f_0 &= 6 \text{ Hz.} \end{aligned}}$$

(b)

$$x(t) = 5 \cos \pi t \sin 3\pi t$$
$$= x_1(t)x_2(t)$$

where

$$x_1(t) = 5\cos \pi t$$
$$x_2(t) = \sin 3\pi t$$

The product of two functions is expressed as the sum of the two functions using the following formula.

$$\begin{aligned}\sin(A+B) - \sin(A-B) &= [\sin A \cos B + \cos A \sin B - \sin A \cos B + \cos A \sin B] \\ &= 2\cos A \sin B \\ \cos A \sin B &= \frac{1}{2}[\sin(A+B) - \sin(A-B)]\end{aligned}$$

The given function can, therefore, be written as

$$5\cos \pi t \sin 3\pi t = \frac{5}{2}(\sin 4\pi t - \sin 2\pi t)$$

Let

$$\begin{aligned}x_1(t) &= \frac{5}{2}\sin 4\pi t \\ \omega_{01} &= 4\pi \\ T_{01} &= \frac{2\pi}{\omega_{01}} \\ &= \frac{1}{2} \\ x_2(t) &= \frac{5}{2}\sin 2\pi t \\ \omega_{02} &= 2\pi \\ T_{02} &= \frac{2\pi}{\omega_{02}} \\ &= 1 \\ \frac{T_{01}}{T_{02}} &= \frac{1}{2} \\ T_0 &= 2T_{01} = T_{02}\end{aligned}$$

$$\boxed{\begin{aligned}T_0 &= 1\ \text{sec.} \\ f_0 &= \frac{1}{T_0} = 1\ \text{Hz}\end{aligned}}$$

■ Example 1.40

Find whether the following signal is periodic. If periodic, determine the fundamental period and frequency. Also determine the fundamental period of each function in the composite signal in the time of the fundamental period.

$$x(t) = \sin(2\pi t - \pi) - 5\cos\left(3\pi t + \frac{\pi}{4}\right) - 8\cos\left(5\pi t - \frac{\pi}{8}\right)$$

Solution:

$$x(t) = x_1(t) + x_2(t) + x_3(t)$$

where

$$\begin{aligned} x_1(t) &= \sin(2\pi t - \pi) \\ x_2(t) &= -5\cos\left(3\pi t + \frac{\pi}{4}\right) \\ x_3(t) &= -8\cos\left(5\pi t + \frac{\pi}{8}\right) \end{aligned}$$

Let T_1, T_2 and T_3 be the fundamental periods of $x_1(t)$, $x_2(t)$ and $x_3(t)$, respectively.

$$\begin{aligned} \omega_1 &= 2\pi \\ T_1 &= \frac{2\pi}{\omega_1} \\ &= \frac{2\pi}{2\pi} = 1 \text{ sec. (rational)} \\ \omega_2 &= 3\pi \\ T_2 &= \frac{2\pi}{\omega_2} \\ &= \frac{2\pi}{3\pi} = \frac{2}{3} \text{ sec. (rational)} \\ \omega_3 &= 5\pi \\ T_3 &= \frac{2\pi}{\omega_3} \\ &= \frac{2\pi}{5\pi} = \frac{2}{5} \text{ sec. (rational)} \\ \frac{T_1}{T_2} &= \frac{1 \times 3}{2} \\ &= \frac{3}{2} \text{ sec. (rational)} \\ \frac{T_1}{T_3} &= \frac{1 \times 5}{2} \end{aligned}$$

$$= \frac{5}{2} \text{ sec. (rational)}$$

Hence, the composite signal $x(t)$ is periodic. The fundamental periods are obtained by taking LCM of T_1, T_2 and T_3 as explained below.

(1)

$$T_1 = 1; \quad T_2 = \frac{2}{3}; \quad T_3 = \frac{2}{5}$$

Multiply by 15 to make them integers. The new periods are obtained as $T_1 = 15$, $T_2 = 10$ and $T_3 = 6$.

(2) The LCM is obtained as

$$\begin{array}{r|lll} 5 & 15, & 10, & 6 \\ \hline 3 & 3, & 2, & 6 \\ \hline 2 & 1, & 2, & 2 \\ \hline & 1, & 1, & 1 \end{array}$$

The LCM $= 5 \times 3 \times 2 = 30$.

(3)

$$T_0 = \frac{\text{LCM}}{15} = \frac{30}{15} = 2 \text{ sec.}$$

$$\boxed{\begin{aligned} T_0 &= 2 \text{ sec.} \\ f_0 &= \frac{1}{T_0} = 0.5 \text{ Hz.} \end{aligned}}$$

The fundamental period of $x_1(t)$ during $T_0 = 2$ sec. is

$$T_{01} = \frac{T_0}{T_1} = \frac{2}{1} = 2$$

The fundamental period of $x_2(t)$ during $T_0 = 2$ sec. is

$$T_{02} = \frac{T_0}{T_2} = \frac{2}{2} \times 3 = 3$$

The fundamental period of $x_3(t)$ during $T_0 = 2$ sec. is

$$T_{03} = \frac{T_0}{T_3} = \frac{2}{2} \times 5 = 5$$

■ Example 1.41

Determine whether the following signals are periodic. If periodic, find the period

1. $x_1(t) = \sin 4\pi t$
2. $x_2(t) = \sin 23t$
3. $x_3(t) = \sin 4\pi t + \sin 23t$

Solution:

1.

$$\begin{aligned} x_1(t) &= \sin 4\pi t \\ \omega_{01} &= 4\pi \\ T_{01} &= \frac{2\pi}{\omega_{01}} \\ &= \frac{2\pi}{4\pi} \\ &= 0.5\,\text{sec.} \end{aligned}$$

The signal is periodic and the period $T_0 = 0.5\,\text{sec.}$

2.

$$\begin{aligned} x_2(t) &= \sin 23t \\ \omega_{02} &= 23 \\ T_{02} &= \frac{2\pi}{\omega_{02}} \\ &= \frac{2\pi}{23}\ \text{sec.} \end{aligned}$$

The signal is periodic. The fundamental period $T_0 = \frac{2\pi}{23}\,\text{sec.}$

3.

$$\begin{aligned} x_3(t) &= \sin 4\pi t + \sin 23t \\ &= x_1(t) + x_2(t) \\ T_{01} &= \frac{1}{2} \\ T_{02} &= \frac{2\pi}{23} \\ \frac{T_{01}}{T_{02}} &= \frac{1}{2} \times \frac{23}{2\pi} \\ &= \frac{23}{4\pi}\ \text{(irrational)} \end{aligned}$$

Since $\frac{T_{01}}{T_{02}}$ is irrational, the signal $x_3(t)$ is not periodic.

■ Example 1.42

For the following signals

1. Sketch the signals.
2. Determine analytically which are periodic (if periodic) and give the period.

   (a) $x(t) = 4\cos 5\pi t$
   (b) $x(t) = 4\cos(5\pi t - 0.25\pi)$
   (c) $x(t) = 4u(t) + 2\sin(3t)$
   (d) $x(t) = u(t) - 0.5$

Solution:

1. (a)

$$x(t) = 4\cos 5\pi t$$

This is a cosine wave with $\omega_0 = 5\pi$ and $T_0 = \frac{2\pi}{\omega_0} = 0.4$ sec. At $t = 0$, the maximum value of 4 is reached, it becomes zero at $t = 0.1\,\text{sec}(\frac{1}{4}T_0)$, reaches negative maximum at $t = 0.2\,\text{sec}(\frac{1}{2}T_0)$, becomes zero at $t = 0.3sec(3/4T_0)$ and reaches maximum at $t = 0.4\,\text{sec}(T_0)$ and thus completes one cycle. The same wave is repeated for negative time. The signal is sketched as shown in Fig. 1.92a.

(b)

$$\begin{aligned} x(t) &= 4\cos(5\pi t - 0.25\pi) \\ &= 4\cos(5\pi t - 45°) \end{aligned}$$

The signal $4\cos(5\pi t - 45°)$ lags behind the signal $4\cos(5\pi t)$ by 0.2π radians or 45°. This is sketched as shown in Fig. 1.92a.

(c)

$$x(t) = 4u(t) + 2\sin 3t$$

$$4u(t) = \begin{cases} 4 + 2\sin 3t & 0 < t < \infty \\ 2\sin 3t & t < 0 \end{cases}$$

$$x(t) = 2\sin 3t$$

This is a sinusoidal signal of maximum amplitude 2 with $\omega_0 = 3$. The fundamental period $T_0 = \frac{2\pi}{\omega_0} = \frac{2}{3}\pi$

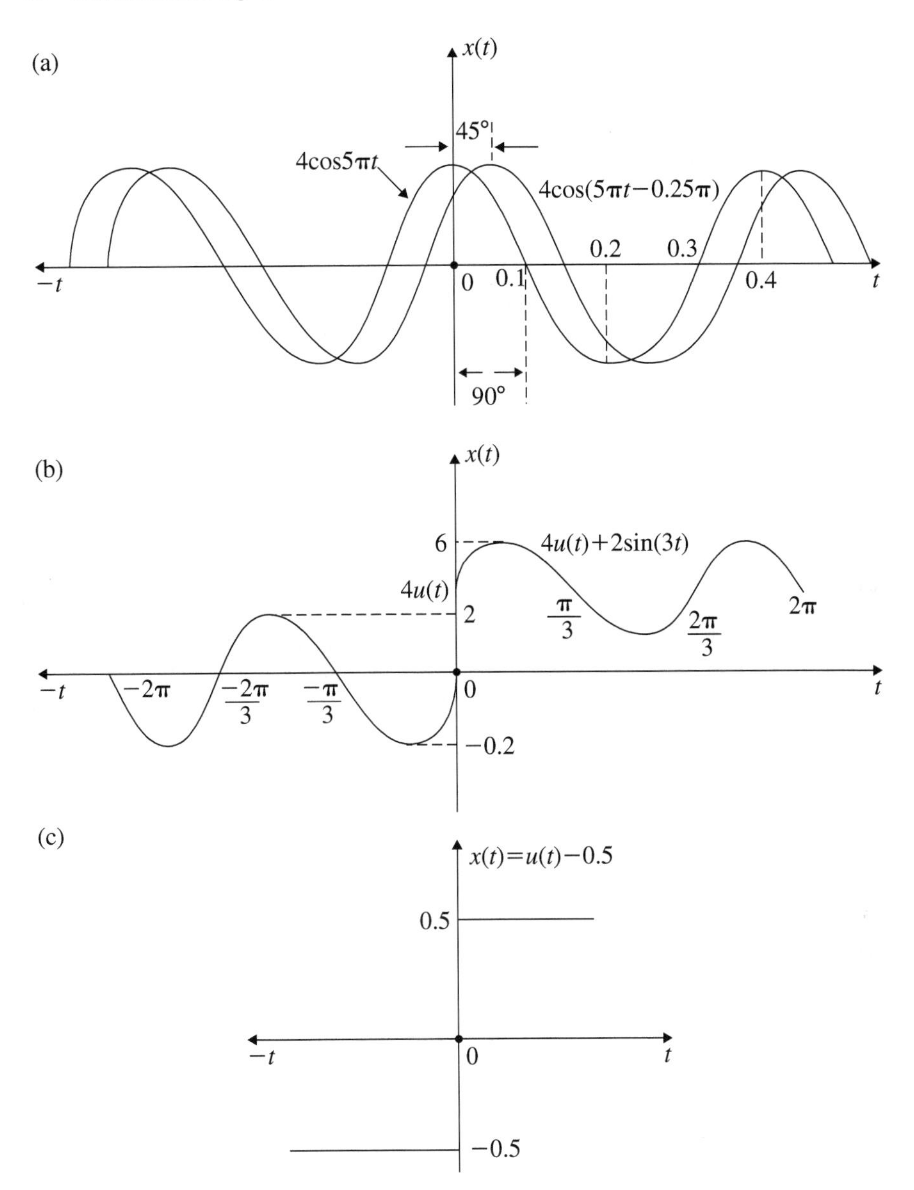

Fig. 1.92 Representation of signals of Example 1.42

$$x(t) = \begin{cases} 4u(t) + 2\sin(3t) & 0 \le t < \infty \\ 2\sin(3t) & -\infty \le t < 0 \end{cases}$$

$2\sin(3t)$ is superimposed with 4 for $t > 0$. For $t < 0$, $2\sin 3t$ is sketched as shown in Fig. 1.92b.

(d)

$$x(t) = u(t) - 0.5$$

$$x(t) = \begin{cases} 0.5 & 0 \le t < \infty \\ -0.5 & -\infty \le t < 0 \end{cases}$$

The signal is sketched as shown in Fig. 1.92c.

2. (a)

$$x(t) = 4\cos 5\pi t$$

Let T_0 be the periodicity of the signal.

$$\begin{aligned} x(t+T_0) &= 4\cos 5\pi(t+T_0) \\ &= 4\cos(5\pi t + 5\pi T_0) \end{aligned}$$

If $5\pi T_0 = 2\pi$

$$\begin{aligned} x(t+T_0) &= 4\cos(5\pi t + 2\pi) \\ &= 4\cos(5\pi t) \\ &= x(t) \end{aligned}$$

$x(t)$ is periodic with period $T_0 = \frac{2}{5} = 0.4$ sec.

$$\boxed{T_0 = 0.4 \text{ sec.}}$$

(b)

$$\begin{aligned} x(t) &= 4\cos(5\pi t - 0.25\pi) \\ x(t+T_0) &= 4\cos(5\pi(t+T_0) - 0.25\pi) \\ &= 4\cos(5\pi t - 0.25\pi + 5\pi T_0) \\ &= 4\cos(5\pi t - 0.25\pi) \end{aligned}$$

if $5\pi T_0 = 2\pi$

$$x(t+T_0) = x(t)$$

The signal $x(t)$ is periodic with period $T_0 = 0.4\,\text{sec}$.

$$\boxed{T_0 = 0.4\text{ sec.}}$$

(c)

$$x(t) = 4u(t) + 2\sin 3t$$

$2\sin 3t$ is a periodic signal for $t > 0$ with period $T_0 = \frac{2\pi}{\omega_0}$ where $\omega_0 = 3$. For $t < 0$,

$$x(t) = 2\sin 3t$$

This is also periodic with period $T_0 = \frac{2\pi}{3}$ sec. However, at $t = 0$, $x(t)$ has discontinuity and it does not recur at any other time. Hence, the signal $x(t)$ is non-periodic. This can be analytically proved as follows:

$$x(t + T_0) = 4u(t + T_0) + 2\sin 3(t + T_0)$$

For $T_0 = \frac{2}{3}\pi$

$$\begin{aligned} x(t + T_0) &= 4u\left(t + \frac{2}{3}\pi\right) + 2\sin 3t \\ &\neq x(t) \end{aligned}$$

$$\boxed{\text{The Signal is Non-periodic.}}$$

(d)

$$\begin{aligned} x(t) &= u(t) - 0.5 \\ x(t + T_0) &= u(t + T_0) - 0.5 \\ &\neq x(t) \end{aligned}$$

for any T_0

$$\boxed{\text{The Signal is Non-periodic.}}$$

1.8.4 *Odd and Even Functions of Continuous Time Signals*

One of the properties of signals is their symmetry when the time is reversed. They are classified as even and odd signals. A continuous time signal $x(t)$ is said to be an

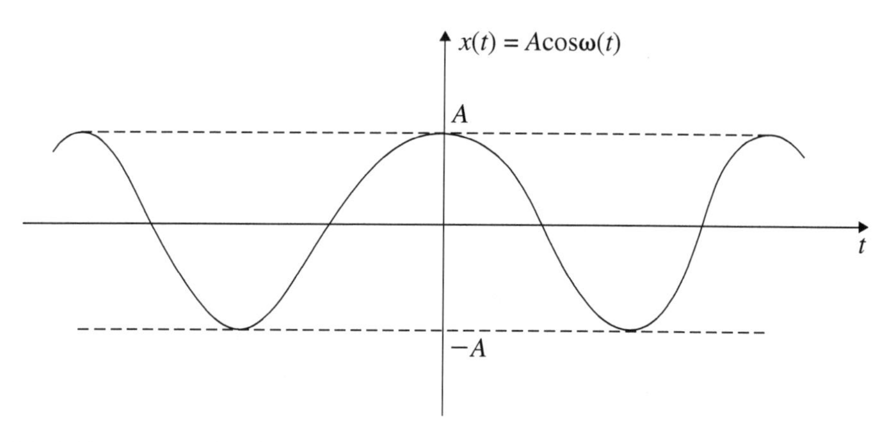

Fig. 1.93 Representation of an even (symmetric) function

even signal if it satisfies the following condition:

$$x(-t) = x(t) \text{ for all } t \tag{1.37}$$

It is identical under folding about the origin. A signal $x(t)$ is said to be an odd signal if it satisfies the condition

$$x(-t) = -x(t) \text{ for all } t \tag{1.38}$$

An odd signal must necessarily be zero at $t = 0$. While even signals are symmetric about the vertical axis odd signals are anti-symmetric (asymmetric) about the time origin. Consider the following signal:

$$\begin{aligned} x(t) &= A\cos\omega t \\ x(-t) &= A\cos(-\omega t) \\ &= A\cos\omega t \\ &= x(t) \end{aligned}$$

The above even signal is shown in Fig. 1.93. Consider the following signal:

$$\begin{aligned} x(t) &= A\sin\omega t \\ x(-t) &= A\sin(-\omega t) \\ &= -A\sin\omega t \\ &= -x(t) \end{aligned}$$

The above odd signal is shown in Fig. 1.94. The odd function is zero at $t = 0$ as seen in Fig. 1.94.

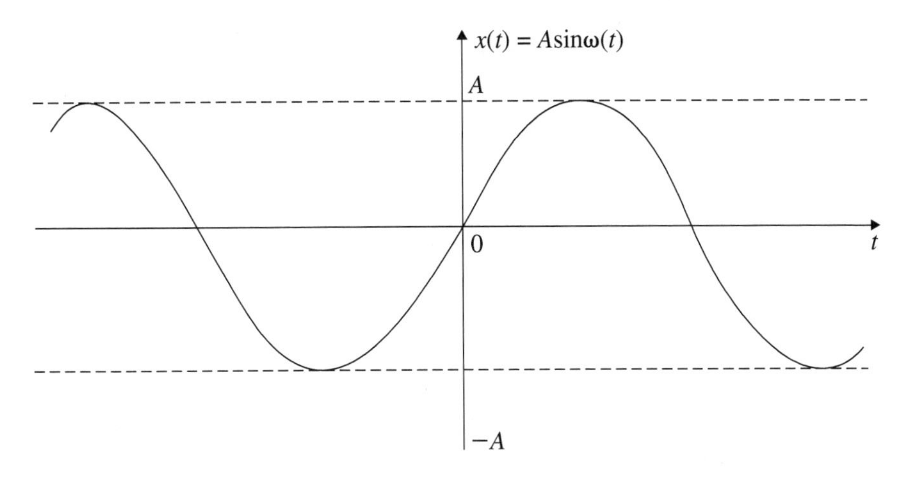

Fig. 1.94 Representing of an odd (anti-symmetric) function

1.8.4.1 Even and Odd Components of a Signal

A continuous time signal $x(t)$ can be expressed in terms of odd and even components. Let $x_e(t)$ and $x_0(t)$ represent the even and odd components of $x(t)$. We may write $x(t)$ as

$$x(t) = x_e(t) + x_0(t) \tag{1.39}$$

Putting $t = -t$ in Eq. (1.39), we get

$$x(-t) = x_e(-t) + x_0(-t) \tag{1.40}$$

For an even function $x_e(-t) = x_e(t)$ and for an odd function $x_0(-t) = -x_0(t)$. Equation (1.40) is written as

$$x(-t) = x_e(t) - x_0(t) \tag{1.41}$$

Adding Eqs. (1.39) and (1.41) the following equation is obtained:

$$\boxed{x_e(t) = \frac{1}{2}[x(t) + x(-t)]} \tag{1.42}$$

Subtracting Eq. (1.41) from Eq. (1.39), we get

$$\boxed{x_0(t) = \frac{1}{2}[x(t) - x(-t)]} \tag{1.43}$$

■ Example 1.43

Show that the even function has its odd part zero.

Solution: From Eq. (1.42) the even function of $x(t)$ can be written as

$$x_e(t) = \frac{1}{2}[x(t) + x(-t)]$$

For an even function $x(-t) = x(t)$. Hence, the above equation can be written as

$$x_e(t) = \frac{1}{2}[x(t) + x(t)] = x(t)$$

From Eq. (1.43) the odd function of $x(t)$ can be written as

$$\begin{aligned} x_0(t) &= \frac{1}{2}[x(t) - x(-t)] \\ &= \frac{1}{2}[x(t) - x(t)] \\ &= 0 \end{aligned}$$

Thus, it is proved that for an even function the odd part is zero.

■ Example 1.44

Show that the odd function has its even part zero.

Solution: Let $x(t)$ be an odd function. For an odd function, $x(-t) = -x(t)$. The even function of $x(t)$ can be written as

$$\begin{aligned} x_e(t) &= \frac{1}{2}[x(t) + x(-t)] \\ &= \frac{1}{2}[x(t) - x(t)] \\ &= 0 \\ x_0(t) &= \frac{1}{2}[x(t) - x(-t)] \\ &= \frac{1}{2}[x(t) + x(t)] \\ &= x(t) \end{aligned}$$

Thus, for an odd function $x(t)$, the even part of $x(t) = 0$.

■ Example 1.45

Show that the product of two even signals is an even signal.

Solution: Let $x_1(t)$ and $x_2(t)$ be the two even signals. Let $x(t)$ be the product of these two signals.

$$x(t) = x_1(t)x_2(t)$$

For an even function, $x(-t) = x(t)$ and $x_1(-t) = x_1(t)$ and $x_2(-t) = x_2(t)$. The above equation is written as follows. Substituting $t = -t$, we get

$$\begin{aligned} x(-t) &= x_1(-t)x_2(-t) \\ &= x_1(t)x_2(t) = x(t) \end{aligned}$$

Thus, $x(t) = x(-t)$ which is even.

■ Example 1.46

Show that the product of two odd signals is an even signal.

Solution: Let $x_1(t)$ and $x_2(t)$ be two odd signals.

For the odd signals, $x_1(-t) = -x_1(t)$ and $x_2(t) = -x_2(t)$. Let $x(t)$ be the product of $x_1(t)$ and $x_2(t)$.

$$x(t) = x_1(t)x_2(t)$$

Putting $t = -t$ in the above equation, we get

$$\begin{aligned} x(-t) &= x_1(-t)x_2(-t) \\ &= x_1(t)x_2(t) \\ &= x(t) \end{aligned}$$

Thus, it is proved that $x(t) = x(-t)$. The product of two odd signals is an even signal.

■ Example 1.47

Prove that the product of an odd and an even signal is an odd signal.

Solution: Let $x_1(t)$ be an odd signal and $x_2(t)$ be an even signal.

Then $x_1(-t) = -x_1(t)$ and $x_2(-t) = x_2(t)$. Let $x(t)$ be the product of $x_1(t)$ and $x_2(t)$.

$$x(t) = x_1(t)x_2(t)$$

Putting $t = -t$ in the above equation, we get

$$\begin{aligned} x(-t) &= x_1(-t)x_2(-t) \\ &= -x_1(t)x_2(t) \\ &= -x(t) \end{aligned}$$

Thus, $x(t) = -x(-t)$ which is odd. The product of an odd and an even signal is an odd signal.

■ Example 1.48

Show that the sum of the two even functions is an even function and the sum of the two odd functions is an odd function.

Solution: Let $x(t)$ be expressed as the sum of two functions $x_1(t)$ and $x_2(t)$.

$$x(t) = x_1(t) + x_2(t)$$

Substituting $t = -t$ in the above equation, we get

$$x(-t) = x_1(-t) + x_2(-t) \tag{a}$$

If $x_1(t)$ and $x_2(t)$ are even functions, the above equation is written as

$$\begin{aligned} x(-t) &= x_1(t) + x_2(t) \\ &= x(t) \end{aligned}$$

This shows that $x(t)$ which is the sum of two even functions is an even function. If $x_1(t)$ and $x_2(t)$ are odd functions, equation (a) can be written as

$$\begin{aligned} x(-t) &= x_1(-t) + x_2(-t) \\ &= -(x_1(t) + x_2(t)) \\ &= -x(t) \end{aligned}$$

Thus, $x(t)$ which is the sum of two odd functions is an odd function.

■ Example 1.49

Find whether the following signals are odd or even. Find the odd and even components.

(a) $x(t) = t^2 - 5t + 10$

(b) $x(t) = t^4 + 4t^2 + 6$

(c) $x(t) = t^3 + 3t$

(d) $x(t) = 10\sin\left(10\pi t + \frac{\pi}{4}\right)$

(e) $x(t) = e^{j10t}$

Solution:

(a) $\boldsymbol{x(t) = t^2 - 5t + 10}$
Put $t = -t$

$$\begin{aligned} x(-t) &= t^2 + 5t + 10 \\ &\neq x(t) \\ &\neq -x(t) \end{aligned}$$

The function is neither even nor odd.

$$\begin{aligned} x_e(t) &= \frac{1}{2}[x(t) + x(-t)] \\ &= \frac{1}{2}[t^2 - 5t + 10 + t^2 + 5t + 10] \end{aligned}$$

$$\boxed{x_e(t) = (t^2 + 10)}$$

$$\begin{aligned} x_0(t) &= \frac{1}{2}[x(t) - x(-t)] \\ &= \frac{1}{2}[t^2 - 5t + 10 - t^2 - 5t - 10] \end{aligned}$$

$$\boxed{x_0(t) = -5t}$$

(b) $\boldsymbol{x(t) = t^4 + 4t^2 + 6}$
Put $t = -t$

$$\begin{aligned} x(-t) &= t^4 + 4t^2 + 6 = x(t) \\ x(t) &= x(-t) \end{aligned}$$

The function is even. The odd part should be zero which can be verified as

$$\begin{aligned} x_0(t) &= \frac{1}{2}[x(t) - x(-t)] \\ &= \frac{1}{2}[t^4 + 4t^2 + 6 - t^4 - 4t^2 - 6] \\ &= 0 \end{aligned}$$

$$\boxed{x_e(t) = x(t) = t^4 + 4t^2 + 6}$$

(c) $\boldsymbol{x(t) = t^3 + 3t}$
Put $t = -t$

$$x(-t) = -(t^3 + 3t) = -x(t)$$

The function is odd. The even component is zero.

$$\boxed{\begin{aligned} x_0(t) &= t^3 + 3t \\ x_e(t) &= 0 \end{aligned}}$$

(d) $\boldsymbol{x(t) = 10\sin\left(10\pi t + \frac{\pi}{4}\right)}$
Put $t = -t$

$$\begin{aligned}
x(-t) &= 10\sin\left(-10\pi t + \frac{\pi}{4}\right) \\
&= -10\sin\left(10\pi t - \frac{\pi}{4}\right) \\
&= -10\left[\sin 10\pi t \cos\frac{\pi}{4} - \cos 10\pi t \sin\frac{\pi}{4}\right] \\
&= \frac{-10}{\sqrt{2}}[\sin 10\pi t - \cos 10\pi t] \\
&\neq x(t) \\
&\neq -x(t)
\end{aligned}$$

The above signal is neither even nor odd.

$$\begin{aligned}
x(t) &= 10\left[\sin 10\pi t \cos\frac{\pi}{4} + \cos 10\pi t \sin\frac{\pi}{4}\right] \\
&= \frac{10}{\sqrt{2}}[\sin 10\pi t + \cos 10\pi t] \\
x_e(t) &= \frac{1}{2}[x(t) + x(-t)] \\
&= \frac{10}{2\sqrt{2}}[\sin 10\pi t + \cos 10\pi t - \sin 10\pi t + \cos 10\pi t]
\end{aligned}$$

$$\boxed{x_e(t) = \frac{10}{\sqrt{2}}\cos 10\pi t}$$

$$x_0(t) = \frac{1}{2}[x(t) - x(-t)]$$
$$= \frac{10}{2\sqrt{2}}[\sin 10\pi t + \cos 10\pi t + \sin 10\pi t - \cos 10\pi t]$$

$$\boxed{x_0(t) = \frac{10}{\sqrt{2}} \sin 10\pi t}$$

(e) $\boldsymbol{x(t) = e^{j10t}}$

$$x(-t) = e^{-j10t}$$
$$x(t) \neq x(-t)$$
$$x(t) \neq -x(-t)$$

The signal is neither odd nor even.

$$x_e(t) = \frac{1}{2}[x(t) - x(-t)]$$
$$= \frac{1}{2}[e^{j10t} + e^{-j10t}]$$

$$\boxed{x_e(t) = \cos 10t}$$

$$x_0(t) = \frac{1}{2}[x(t) - x(-t)]$$
$$= \frac{1}{2}[e^{j10t} - e^{-j10t}]$$

$$\boxed{x_0(t) = j \sin 10t}$$

Note: In all the above cases, $x_0(t)$ passes through the origin at $t = 0$.

■ Example 1.50

Sketch the even and odd components of a step signal shown in Fig. 1.95a.

Solution:

The step function is shown in Fig. 1.95a. $x(-t)$ is shown in Fig. 1.95b. In Fig. 1.95c, the sum of $x(t)$ and $x(-t)$ is represented. The even function $x_e(t) = \frac{1}{2}[x(t) + x(-t)]$

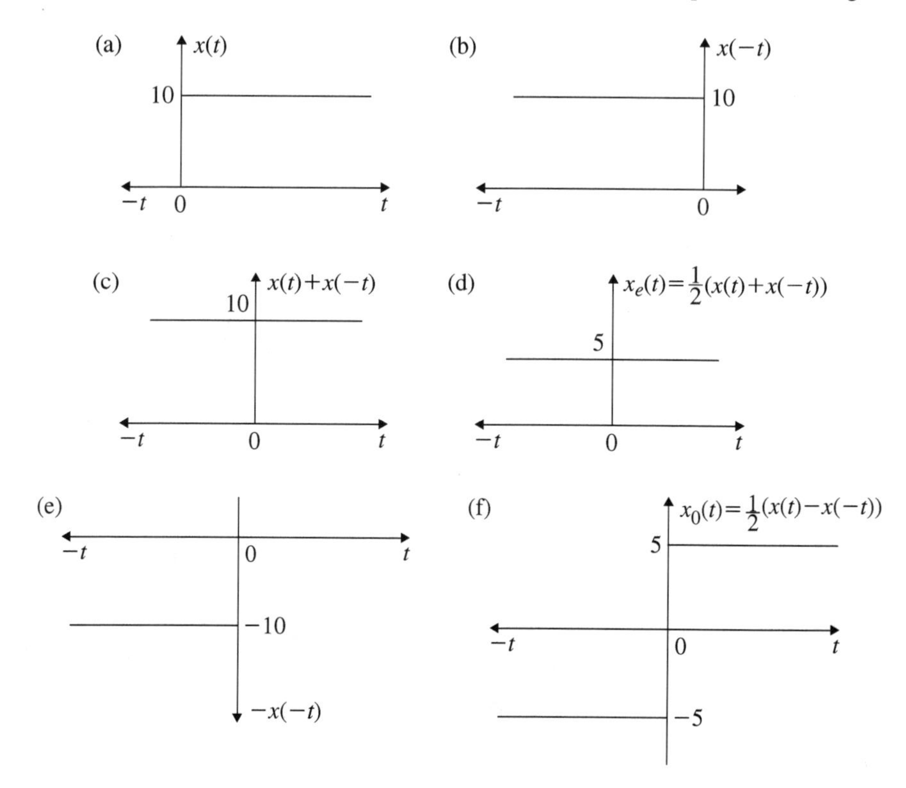

Fig. 1.95 Even and odd components of a step function

is shown in Fig. 1.95d. In Fig. 1.95e, $-x(-t)$ is represented. The odd function $x_0(t) = \frac{1}{2}[x(t) - x(-t)]$ is represented in Fig. 1.95f.

■ Example 1.51

Sketch the even and odd components of the pulse signal shown in Fig. 1.96a.

Solution:

$x(t)$ is shown in Fig. 1.96a. In Fig. 1.96b, $x(-t)$ is represented. The sum of $x(t) + x(-t)$ is shown in Fig. 1.96c. The even component of $x(t)$ which is $x_e(t) = \frac{1}{2}[x(t) + x(-t)]$ is shown in Fig. 1.96d. In Fig. 1.96e, $-x(-t)$ is shown. The odd component of $x(t)$ which is $x_0(t) = \frac{1}{2}[x(t) - x(-t)]$ is represented in Fig. 1.96f.

■ Example 1.52

Sketch the even and odd components of the triangular wave shown in Fig. 1.97a.

Solution:

Figure 1.97a represents the $x(t)$ which is a triangular wave. $x(-t)$ is represented in Fig. 1.97b. $x(t) + x(-t)$ is represented in Fig. 1.97c. From this figure, the even

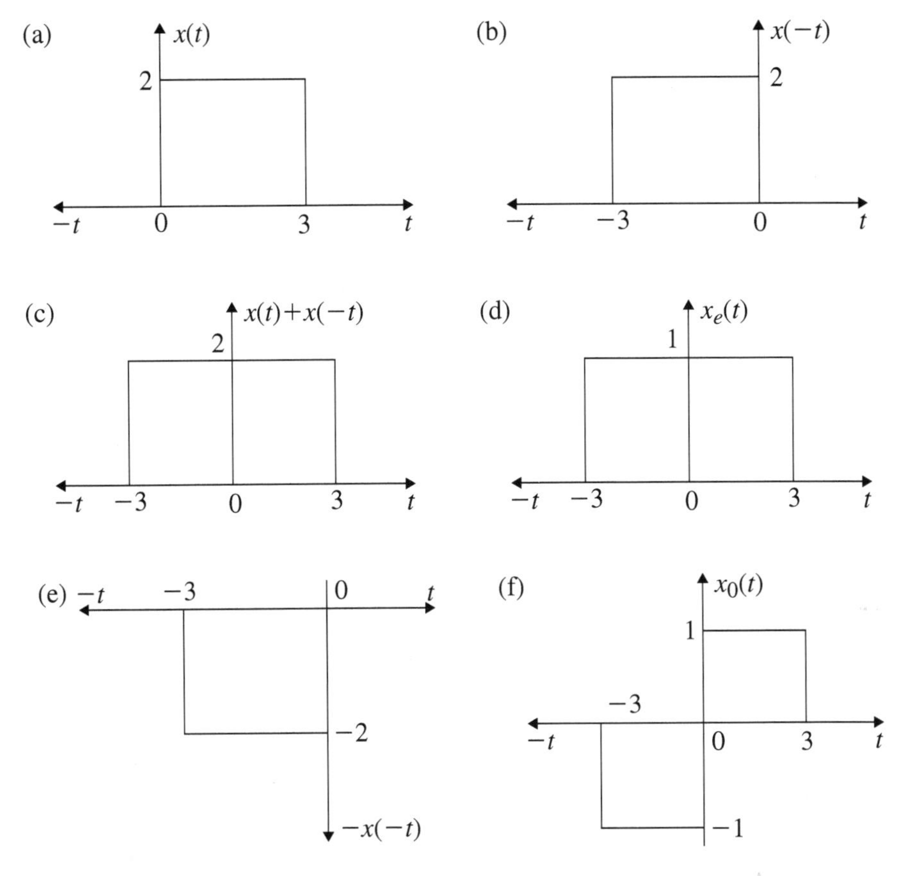

Fig. 1.96 Even and odd components of a pulse signal

component is obtained by dividing the amplitude by 2 and $x_e(t)$ is shown in Fig. 1.97d. In Fig. 1.97e, $-x(-t)$ is represented which is obtained by inverting Fig. 1.97b. Adding Fig. 1.97a, e, $[x(t) - x(-t)]$ is obtained and represented in Fig. 1.97f. By dividing the amplitude of Fig. 1.97f by 2, $x_0(t)$ which is $\frac{1}{2}[x(t) - x(-t)]$ is obtained and sketched as shown in Fig. 1.97g.

■ Example 1.53

Sketch the even and odd components of exponential signal $x(t) = 10e^{-2t}$.

Method (a):

Solution:

$x(t) = 10e^{-2t}$ is sketched and shown in Fig. 1.98a. Figure 1.98a is time reversed to get $x(-t)$ and is sketched in Fig. 1.98b. The sum of $x(t)$ and $x(-t)$ is sketched as

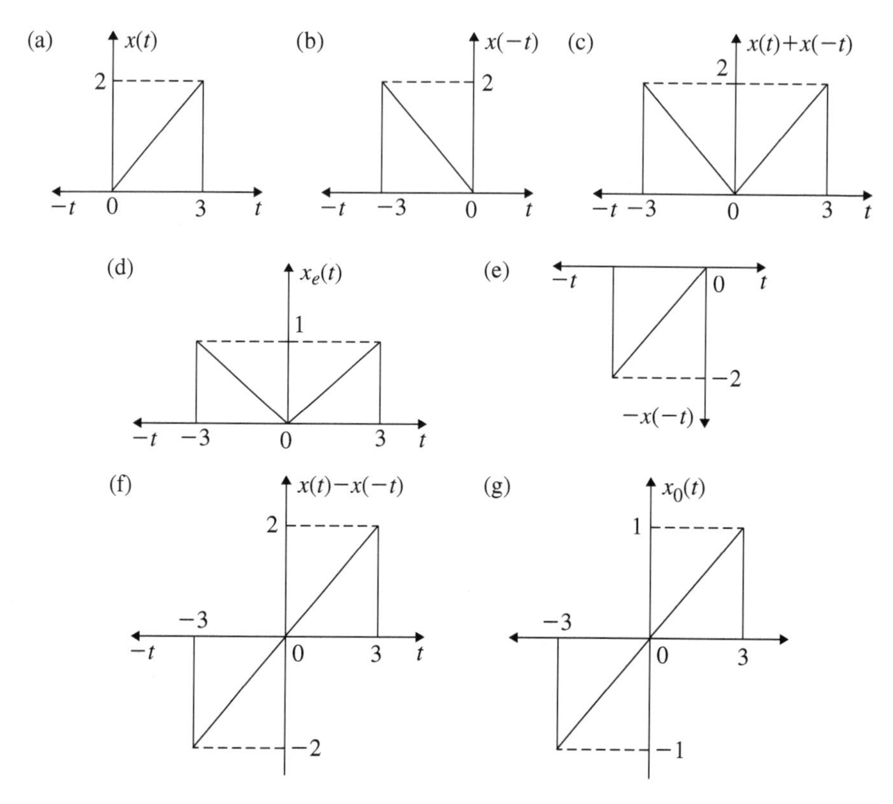

Fig. 1.97 Even and odd components of a triangular wave

shown in Fig. 1.98c. The amplitude of Fig. 1.98c is reduced by a factor 2. This gives $x_e(t) = \frac{1}{2}[x(t) + x(-t)]$ and is shown in Fig. 1.98d. Figure 1.98a is inverted and time reversed to get $-x(-t)$ which is sketched in Fig. 1.98e. The sum of Fig. 1.98a, e gives $[x(t) - x(-t)]$ and this is sketched and shown in Fig. 1.98f. The amplitude of Fig. 1.98f is reduced by a factor 2 which gives odd signal $x_0(t) = \frac{1}{2}[x(t) - x(-t)]$. This is shown in Fig. 1.98g.

■ Example 1.54

Sketch the even and odd parts of the signal shown in Figs. 1.99a and 1.100a.

(*Anna University, May, 2009*)

Solution:

$x(t)$ is graphically represented in Fig. 1.99a. By time folding of Fig. 1.99a, $x(-t)$ is obtained and is shown in Fig. 1.99b. These figures are graphically added to get $x(t) + x(-t)$ and represented in Fig. 1.99c. To get the even signal of $x(t)$, the amplitude of the signal is divided by a factor 2 and is represented in Fig. 1.99d. The signal $x(t)$ is time folded and inverted to get $-x(-t)$. This is represented in Fig. 1.99e. Figure 1.99a,

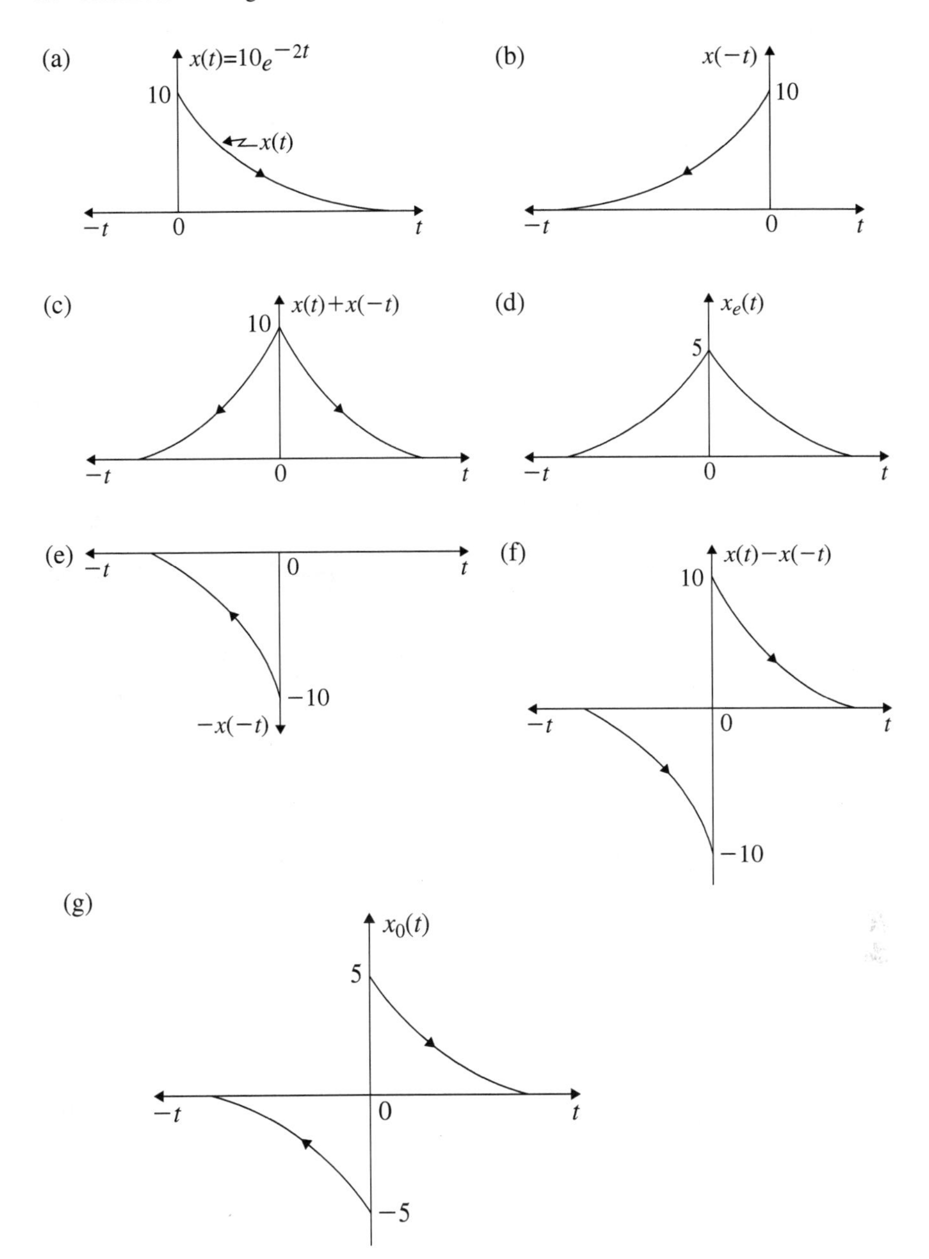

Fig. 1.98 Representation of even and odd function of exponential decay

e is graphically added to get $x(t) - x(-t)$ which is represented in Fig. 1.99f. The amplitude of the signal in Fig. 1.99f is divided by a factor 2 which gives $x_0(t)$ of $x(t)$. This is represented in Fig. 1.99g.

Note the even component $x_e(t)$ in Fig. 1.99d. It is symmetrical with respect to the vertical axis and when time folded identical mirror images are obtained. Similarly, the odd component $x_0(t)$ represented in Fig. 1.99g passes through the origin at $t = 0$ and it is also anti-symmetry.

Consider Fig. 1.100a where $x(t)$ is represented. By folding $x(t)$, we get $x(-t)$ and is shown in Fig. 1.100b. $x(-t)$ when inverted, we get $-x(-t)$ and is represented in Fig. 1.100c.

$$x_0(-t) = \frac{1}{2}(x_1(t) - x(-t))$$

This is obtained by combining Fig. 1.100a and c after dividing the amplitude by a factor 2. $x_0(t)$ is shown in Fig. 1.100d. The even component is expressed as

$$x_e(t) = \frac{1}{2}(x(t) + x(-t))$$

By combining Fig. 1.100a, b and by dividing the amplitude by a factor 2, the even component is obtained. The even component is represented in Fig. 1.100e.

■ Example 1.55

Find the even and odd component of the following signal:

$$x(t) = \cos t + \sin t + \cos t \sin t$$

(Anna University, May, 2007)

Solution:

$$x(t) = \cos t + \sin t + \cos t \sin t$$

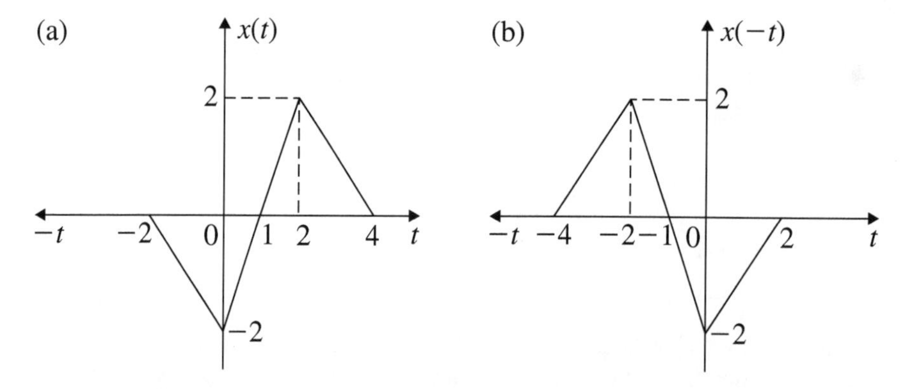

Fig. 1.99 Representation of even and odd signals of Example 1.54(a)

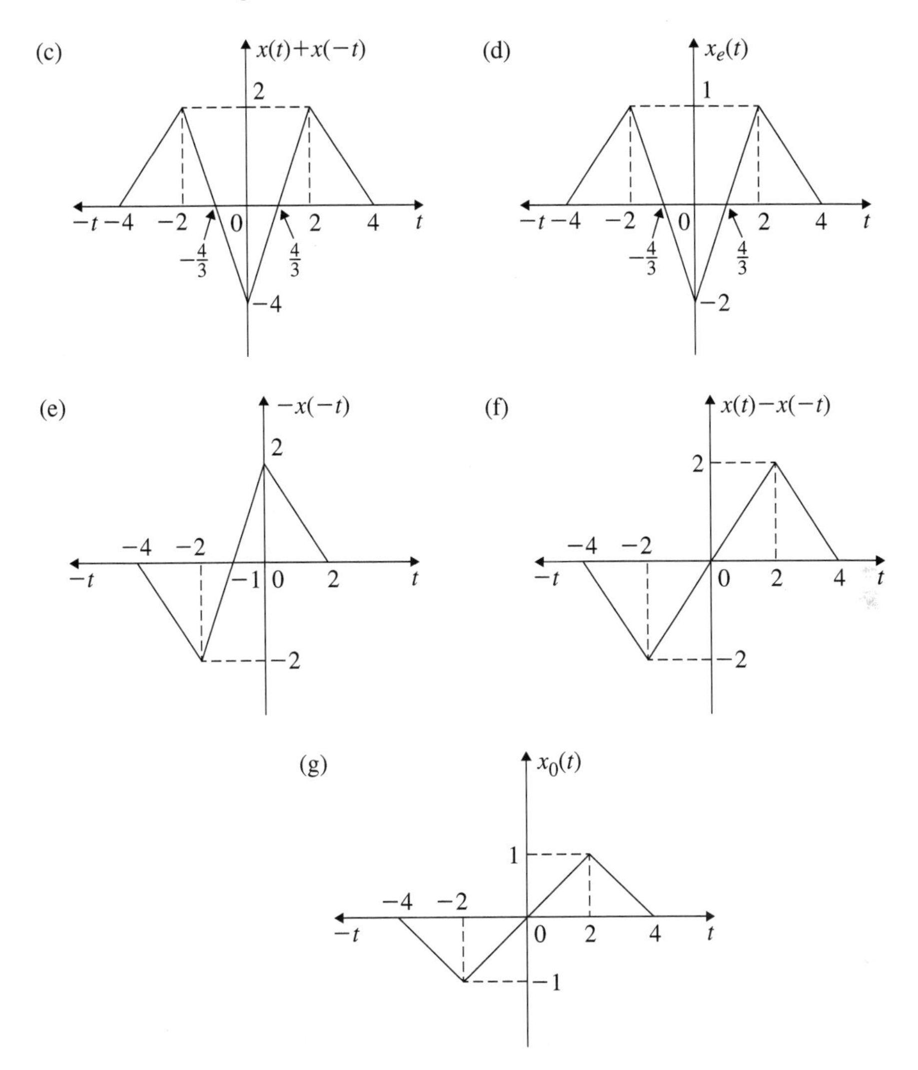

Fig. 1.99 (continued)

Put $t = -t$

$$\begin{aligned} x(-t) &= \cos(-t) + \sin(-t) + \cos(-t)\sin(-t) \\ &= \cos t - \sin t - \cos t \sin t \end{aligned}$$

$$\begin{aligned} x_e(t) &= \frac{1}{2}[x(t) + x(-t)] \\ &= \frac{1}{2}[\cos t + \sin t + \cos t \sin t + \cos t - \sin t - \cos t \sin t] \end{aligned}$$

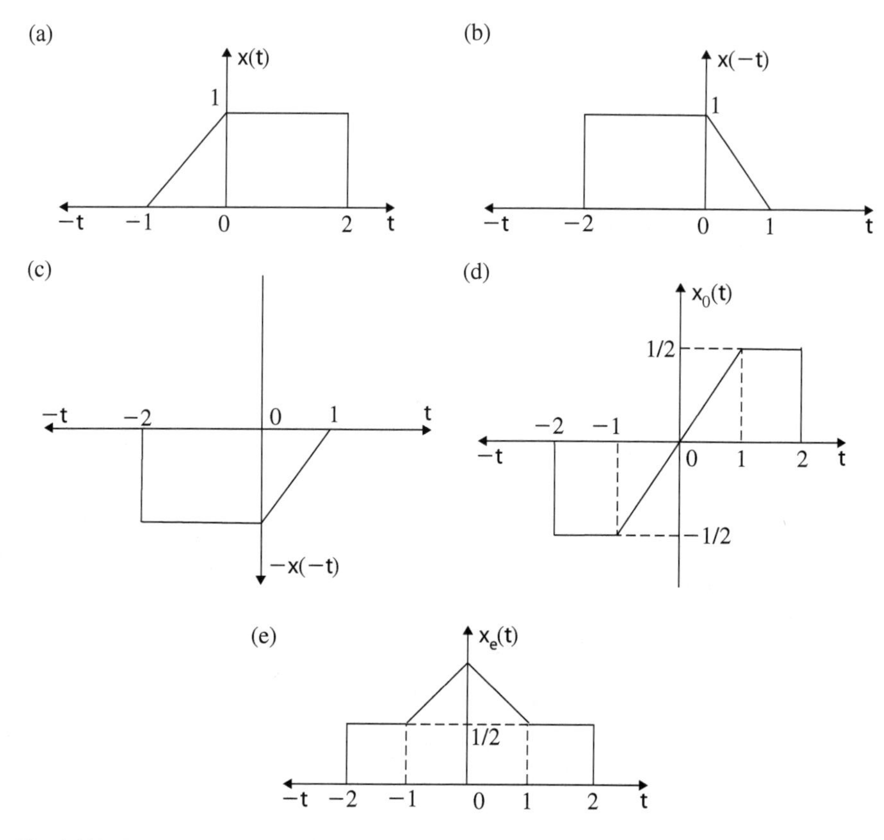

Fig. 1.100 Representation of odd and even components of Example 1.54(b)

$$\boxed{x_e(t) = \cos t}$$

The odd component of $x(t)$ is obtained as follows:

$$\begin{aligned} x_0(t) &= \frac{1}{2}[x(t) - x(-t)] \\ &= \frac{1}{2}[\cos t + \sin t + \cos t \sin t - \cos t + \sin t + \cos t \sin t] \end{aligned}$$

$$\boxed{x_0(t) = \sin t[1 + \cos t]}$$

■ Example 1.56

Find the odd and ever components of the following signals and sketch the same.

1. $x(t) = \sin \omega_0 t$
2. $x(t) = \sin \omega_0 t u(t)$
3. $x(t) = \cos \omega_0 t$
4. $x(t) = \cos \omega_0 t u(t)$

Solution:

1.

$$x(t) = \sin \omega_0 t$$
$$x(-t) = -\sin \omega_0 t$$
$$x_0(t) = \frac{1}{2}[x(t) - x(-t)]$$

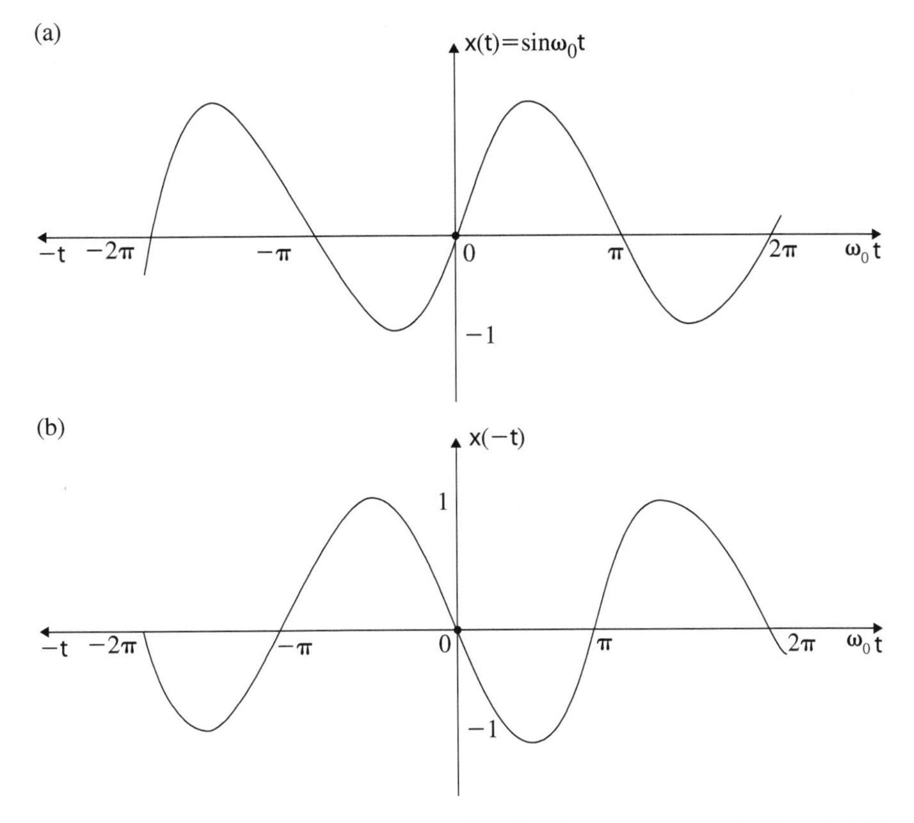

Fig. 1.101 Representation of $x(t) = \sin \omega_0 t$ and its odd and even components for Example 1.56. Representation of $x(t) = \sin \omega_0 t u(t)$ and its odd and even components for Example 1.56. Representation of $x(t) = \cos \omega_0 t$. Representation of $x_0(t) = \frac{1}{2} \cos \omega_0 t[u(t) - u(-t)]$

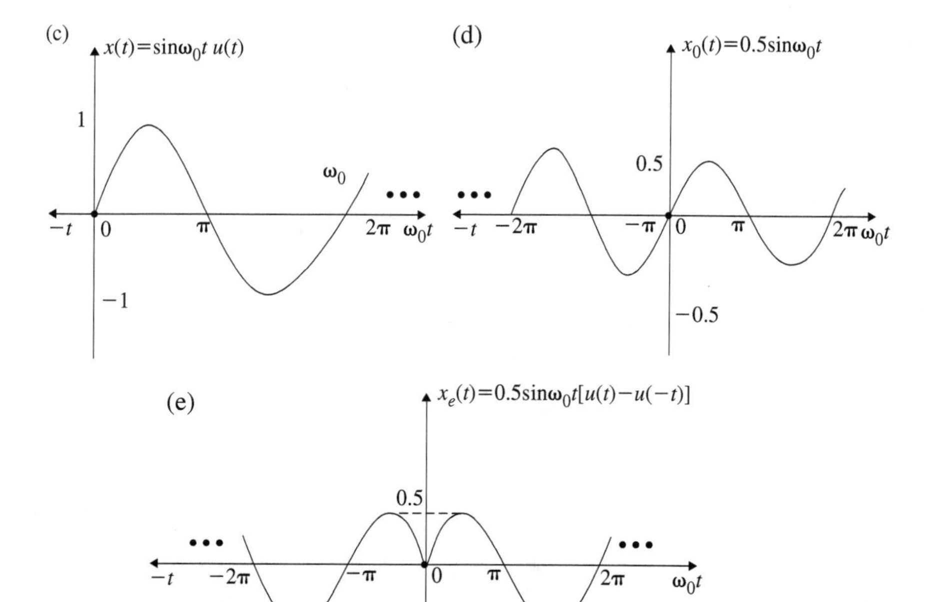

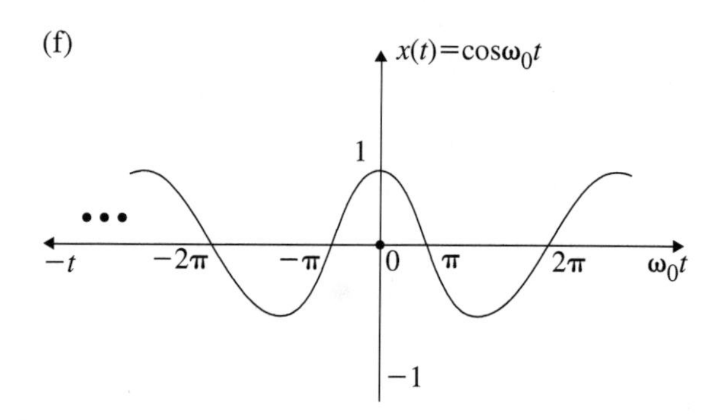

(f)

$x(t)=\cos\omega_0 t$

1

$-t$ -2π $-\pi$ 0 π 2π $\omega_0 t$

-1

Fig. 1.101 (continued)

$$
\begin{aligned}
&= \frac{1}{2}[\sin\omega_0 t + \sin\omega_0 t] \\
&= \sin\omega_0 t \\
&= x(t)
\end{aligned}
$$

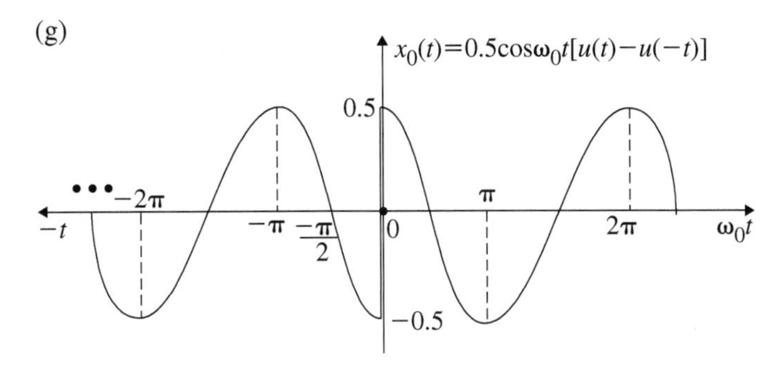

Fig. 1.101 (continued)

$$x_e(t) = \frac{1}{2}[x(t) + x(-t)]$$
$$= \frac{1}{2}[\sin \omega_0 t - \sin \omega_0 t]$$
$$= 0$$

$x(t)$ and $x(-t)$ are sketched as shown in Fig. 1.101a and b, respectively. From Fig. 1.101b, we easily get $-x(-t)$ by inverting $x(-t)$. By adding $\frac{1}{2}$ of $x(t)$ and $-x(-t)$ we will get $x(t)$ itself. Similarly, by adding $\frac{1}{2}$ of $x(t)$ and $x(-t)$ point by point, it becomes zero.

2.

$$x(t) = \sin \omega_0 t u(t)$$
$$x(-t) = -\sin \omega_0 t u(-t)$$
$$-x(-t) = \sin \omega_0 t u(-t)$$
$$x_0(t) = \frac{1}{2}[x(t) - x(-t)]$$
$$= \frac{1}{2}[\sin \omega_0 t u(t) + \sin \omega_0 t u(-t)]$$

$$x_0(t) = \frac{1}{2} \sin \omega_0 t[u(t) + u(-t)]$$
$$= \frac{1}{2} \sin \omega_0 t$$
$$x_e(t) = \frac{1}{2}[x(t) + x(-t)]$$
$$= \frac{1}{2}[\sin \omega_0 t u(t) - \sin \omega_0 t u(-t)]$$
$$= \frac{1}{2} \sin \omega_0 t[u(t) - u(-t)]$$

The signals $x(t)$, $x_0(t)$ and $x_e(t)$ are sketched as shown in Fig. 1.101c–e, respectively.

3.

$$\begin{aligned} x(t) &= \cos\omega_0 t \\ x(-t) &= \cos\omega_0 t \\ x_0(t) &= \frac{1}{2}[x(t) - x(-t)] \\ &= \frac{1}{2}[\cos\omega_0 t - \cos\omega_0 t] = 0 \end{aligned}$$

$$\begin{aligned} x_e(t) &= \frac{1}{2}[x(t) + x(-t)] \\ &= \frac{1}{2}[\cos\omega_0 t + \cos\omega_0 t] \\ &= \cos\omega_0 t \\ &= x(t) \end{aligned}$$

The signal $x(t)$ is represented in Fig. 1.101f.

4.

$$\begin{aligned} x(t) &= \cos\omega_0 t u(t) \\ x(-t) &= \cos\omega_0 t u(-t) \\ -x(-t) &= -\cos\omega_0 t u(-t) \\ x_0(t) &= \frac{1}{2}[x(t) - x(-t)] \\ &= \frac{1}{2}\cos\omega_0 t[u(t) - u(-t)] \\ x_e(t) &= \frac{1}{2}[x(t) + x(-t)] \end{aligned}$$

$$\begin{aligned} x_e(t) &= \frac{1}{2}\cos\omega_0 t[u(t) + u(-t)] \\ &= \frac{1}{2}\cos\omega_0 t \\ &= \frac{1}{2}x(t) \end{aligned}$$

The odd component is represented in Fig. 1.101g. The even component is nothing but $\frac{1}{2}x(t)$ represented in Fig. 1.101f with maximum amplitude being reduced to $1/2$.

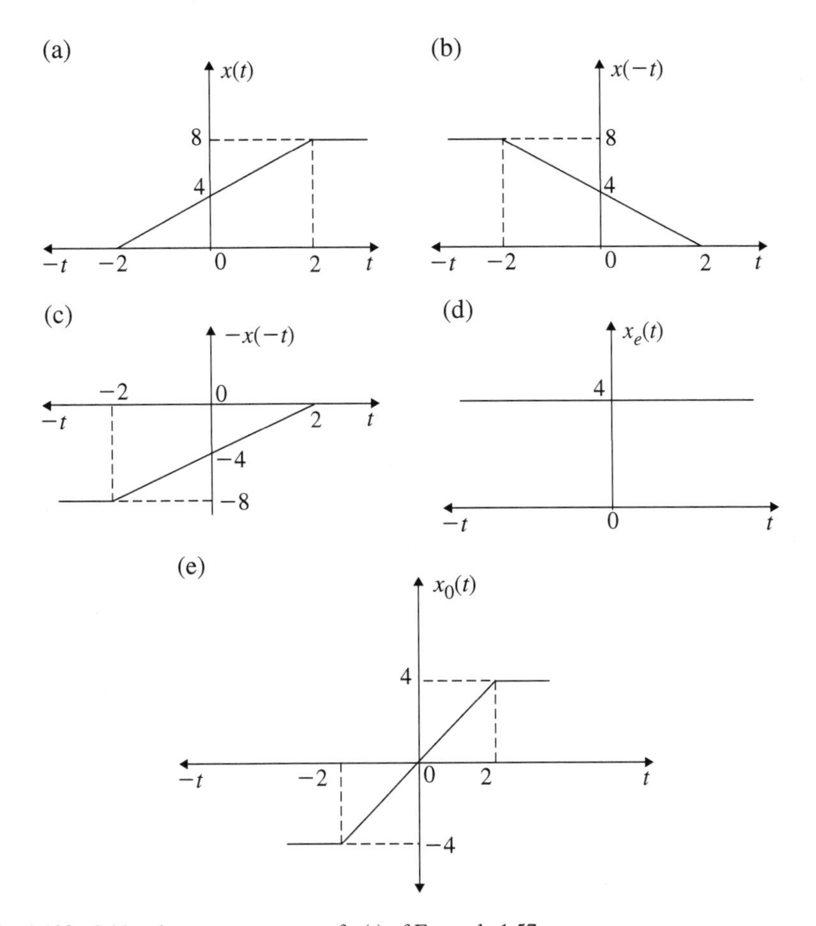

Fig. 1.102 Odd and even components of $x(t)$ of Example 1.57

■ Example 1.57

A certain CT signal is described by the following mathematical equations.

$$x(t) = \begin{cases} (2t+4) & -2 \leq t < 2 \\ 8 & t \geq 2 \\ 0 & t \leq -2 \end{cases}$$

Sketch the signal $x(t)$. Sketch the odd and even components of $x(t)$ and give the mathematical description of these components.

Solution:

$$x(t) = \begin{cases} 2t+4 & -2 \le t \le 2 \\ 8 & t \ge 2 \\ 0 & t \le -2 \end{cases}$$

To satisfy the above mathematical equations, $x(t)$ is sketched as shown in Fig. 1.102a. $x(-t)$ is sketched by signal reflection (folding) and is shown in Fig. 1.102b. By signal inversion of $x(-t)$, $-x(-t)$ is obtained and is shown in Fig. 1.102c. The odd and even components of $x(t)$ are obtained as given below.

$$x_e(t) = \frac{1}{2}[x(t) + x(-t)]$$

For $-\infty \le t \le -2$,

$$x(t) = 0$$
$$x(-t) = 8$$

Hence,

$$x_e(t) = \frac{1}{2}[0 + 8]$$
$$= 4$$

For $-2 \le t \le 2$,

$$x(t) = 2t + 4$$
$$x(-t) = -2t + 4$$
$$x_e(t) = \frac{1}{2}[2t + 4 - 2t + 4]$$
$$= 4$$

For $t \ge 2$,

$$x(t) = 8$$
$$x(-t) = 0$$
$$x_e(t) = \frac{1}{2}[8 + 0]$$
$$= 4$$

Thus,

$$x_e(t) = 4$$

for all t. The even component of $x(t)$ is sketched and shown in Fig. 1.102d.

$$x_0(t) = \frac{1}{2}[x(t) - x(-t)]$$

For $t \leq -2$,

$$\begin{aligned} x(t) &= 0 \\ -x(t) &= -8 \\ x_0(t) &= \frac{1}{2}[0 - 8] \\ &= -4 \end{aligned}$$

For $-2 \leq t \leq 2$,

$$\begin{aligned} x(t) &= 2t + 4 \\ -x(t) &= 2t - 4 \\ x_0(t) &= \frac{1}{2}[2t + 4 + 2t - 4] \\ &= 2t \end{aligned}$$

This is the equation of a straight line with slope 2 and passing through the origin. For $t \geq 2$,

$$\begin{aligned} x(t) &= 8 \\ -x(-t) &= 0 \\ x_0(t) &= \frac{1}{2}[8 + 0] \\ &= 4 \end{aligned}$$

The odd component of $x(t)$ is sketched as shown in Fig. 1.102e.

1.8.5 Energy and Power of Continuous Time Signals

Consider the electric circuit shown in Fig. 1.103 in which a resistor R is connected across the voltage source $v(t)$. The current flowing through the resistor is $i(t)$. The instantaneous power consumed by the resistor is given by

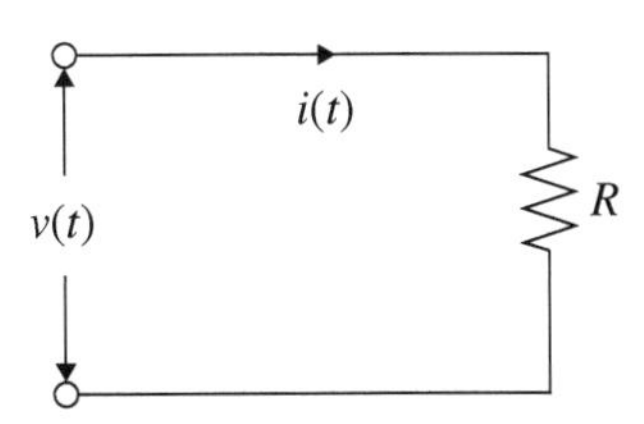

Fig. 1.103 Electric circuit with a resistor

$$P = i^2(t)R$$
$$= \frac{v^2(t)}{R} \tag{1.44}$$

If we assume $R = 1$ ohm, the power is expressed as normalized power which is given by

$$P = v^2(t) \tag{1.45}$$

The average power consumption by the circuit over the time $t_1 \leq t \leq t_2$ is given by the following equation:

$$P = \frac{1}{(t_2 - t_1)} \int_{t_1}^{t_2} v^2(t)\, dt \tag{1.46}$$

The average energy consumption which is the product of power and time is given as

$$E = \int_{t_1}^{t_2} P\, dt = \int_{t_1}^{t_2} v^2(t)\, dt \tag{1.47}$$

Similar to voltages and currents, many other physical variables such as force, temperature, pressure, charge, *etc.* are available for other types of systems. As a convention, similar terminologies for power and energy of continuous signal $x(t)$ and discrete signal $x[n]$ are defined and used. However, the "power" and "energy" defined here are not related to physical power and energy. Thus, if $x(t)$ represents a continuous time signal, then the average power over an infinite time interval T is defined as

$$\boxed{P = \underset{T\to\infty}{Lt} \frac{1}{2T} \int_{-T}^{T} |x(t)|^2\, dt} \tag{1.48}$$

The expression for the total energy is expressed as

$$\boxed{E = \underset{T\to\infty}{Lt} \int_{-T}^{T} |x(t)|^2\, dt} \tag{1.49}$$

If the energy signal does not converge, such signals have infinite energy. On the other hand, if E converges, then the signal has finite energy. From Eqs. (1.48) and (1.49), the following inferences are drawn and given in Table 1.3.

Signals may be neither energy nor power signals. But they cannot be both an energy signal and a power signal. If it is one, it cannot be the other.

Table 1.3 Properties of power and energy signal

Energy signal	Power signal
1. The total energy is obtained using $E = \underset{T\to\infty}{Lt} \int_{-T}^{T} \lvert x(t)\rvert^2\, dt$	1. The average power is obtained using $P = \underset{T\to\infty}{Lt} \frac{1}{2T} \int_{-T}^{T} \lvert x(t)\rvert^2\, dt$
2. For the energy signal $0 < E < \infty$, the average power $P = 0$	2. For the power signal $0 < P < \infty$, the energy E should be ∞
3. Non-periodic signals are energy signals	3. Periodic signals are power signals. However, all power signals need not be periodic
4. Energy signals are not time limited	4. Power signals exist over infinite time

■ Example 1.58

Find the power, RMS value and energy of the following signals:

$$\text{(a)} \qquad x(t) = A\, u(t)$$
$$\text{(b)} \qquad x(t) = e^{-3t} u(t)$$

Solution:

(a) $\boldsymbol{x(t) = A\, u(t)}$

$$P = \underset{T\to\infty}{Lt} \frac{1}{2T} \int_{-T}^{T} A^2 dt$$

For $x(t) = A\, u(t)$, the signal starts at $t = 0$.

$$P = \underset{T\to\infty}{Lt} \frac{1}{2T} A^2 \int_{0}^{T} dt = \frac{A^2}{2T} \left[t\right]_0^T$$
$$= \underset{T\to\infty}{Lt}\, A^2 \frac{T}{2T} = \frac{A^2}{2}$$

$$\boxed{P = \frac{A^2}{2} \text{ watts}}$$

RMS value of power is

$$P_{\text{RMS}} = \sqrt{P} = \frac{A}{\sqrt{2}}$$

$$\boxed{P_{\text{RMS}} = \frac{A}{\sqrt{2}}}$$

Since power is finite, energy E is infinite.

(b) $\boldsymbol{x(t) = e^{-3t}u(t)}$

For this signal, t varies from 0 to ∞.

$$\begin{aligned}
E &= \underset{T\to\infty}{Lt} \int_0^T (e^{-3t})^2 dt \\
&= \underset{T\to\infty}{Lt} \int_0^T e^{-6t} dt \\
&= \underset{T\to\infty}{Lt} \frac{(-1)}{6} \left[e^{-6t}\right]_0^T \\
&= \frac{1}{6} \underset{T\to\infty}{Lt} \left[1 - e^{-6T}\right]
\end{aligned}$$

$$\boxed{E = \frac{1}{6} \text{ Joules}}$$

Since E is finite, power $P = 0$.

■ Example 1.59

Find the power and energy of the following signals:

$$\begin{aligned}
&\text{(a)} \qquad x(t) = A\cos(\omega_0 t + \phi) \\
&\text{(b)} \qquad x(t) = A\sin(\omega_0 t + \phi)
\end{aligned}$$

Solution:

(a) $\boldsymbol{x(t) = A\cos(\omega_0 t + \phi)}$

Since the signal is periodic, it is necessarily a power signal and its energy $E = \infty$. The power of the signal is determined as follows:

$$P = \underset{T\to\infty}{Lt} \frac{1}{2T} \int_{-T}^{T} A^2 \cos^2(\omega_0 t + \phi) dt$$

But,

$$\cos^2(\omega_0 t + \phi) = \frac{1 + \cos 2(\omega_0 t + \phi)}{2}$$

$$P = \underset{T\to\infty}{Lt} \frac{A^2}{4T} \int_{-T}^{T} [1 + \cos 2(\omega_0 t + \phi)]dt$$

Now consider the integral

$$\int_{-T}^{T} \cos 2(\omega_0 t + \phi)dt$$

$$\begin{aligned}
&= \frac{1}{2\omega_0}[\sin 2(\omega_0 t + \phi)]_{-T}^{T} \\
&= \frac{1}{2\omega_0}[\sin 2(\omega_0 T + \phi) - \sin 2(-\omega_0 T + \phi)] \\
&= \frac{1}{2\omega_0}[\sin 2\phi - \sin 2\phi] \qquad [\because \quad \omega_0 T = 2\pi] \\
&= 0
\end{aligned}$$

$$\begin{aligned}
P &= \frac{A^2}{4} \underset{T\to\infty}{Lt} \frac{1}{T}[t]_{-T}^{T} \\
&= \frac{A^2}{4} \underset{T\to\infty}{Lt} \frac{1}{T} 2T
\end{aligned}$$

$$\boxed{P = \frac{A^2}{2}}$$

(b) $\boldsymbol{x(t) = A \sin(\omega_0 t + \phi)}$

$$\begin{aligned}
P &= \underset{T\to\infty}{Lt} \frac{1}{2T} \int_{-T}^{T} A^2 \sin^2(\omega_0 t + \phi)dt \\
&= \underset{T\to\infty}{Lt} \frac{A^2}{2T} \int_{-T}^{T} \frac{[1 - \cos 2(\omega_0 t + \phi)]}{2} dt \\
&= \underset{T\to\infty}{Lt} \frac{A^2}{4T} \left[\int_{-T}^{T} dt - \int_{-T}^{T} \cos 2(\omega_0 t + \phi) \right] dt
\end{aligned}$$

Since $\int_{-T}^{T} \cos 2(\omega_0 t + \phi)dt = 0$,

$$P = \underset{T\to\infty}{Lt} \frac{A^2}{4T}\left[t\right]_{-T}^{T}$$

$$\boxed{P = \frac{A^2}{2}}$$

Since P is finite, $E = \infty$.

■ Example 1.60

Find the power and energy of the following signals:

$$x(t) = 5\cos(10t + \phi) + 10\sin(5t + \phi)$$

Solution:

$$\begin{aligned} x(t) &= 5\cos(10t + \phi) + 10\sin(5t + \phi) \\ &= x_1(t) + x_2(t) \end{aligned}$$

where

$$\begin{aligned} x_{1(t)} &= 5\cos(10t + \phi) \\ x_2(t) &= 10\sin(5t + \phi) \end{aligned}$$

Let P_1 and P_2 be the powers of $x_1(t)$ and $x_2(t)$, respectively.

$$\begin{aligned} P_1 &= \frac{A^2}{2} = \frac{25}{2} = 12.5 \\ P_2 &= \frac{A^2}{2} = \frac{100}{2} = 50 \end{aligned}$$

The average power

$$\begin{aligned} P &= P_1 + P_2 \\ &= 12.5 + 50 \end{aligned}$$

$$\boxed{P = 62.5 \text{ watts}}$$

Since the power is finite, energy $E = \infty$.

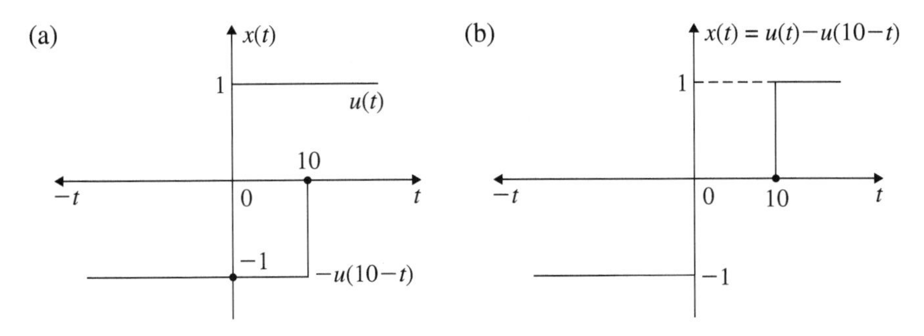

Fig. 1.104 Representation of $x(t) = u(t) - u(10 - t)$

■ Example 1.61

Find the power and energy of the following signal:

$$x(t) = 5t \qquad -10 < t < 10$$

Solution:
Energy of the signal E is

$$E = \int_{-10}^{10} (5t)^2 dt = 25 \left[\frac{t^3}{3} \right]_{-10}^{10}$$
$$= \frac{25}{3} \times 2000$$

$$\boxed{E = \frac{50000}{3} \text{ Joules}}$$

Power of the signal P is zero.

■ Example 1.62

Find the energy and power of the following signal:

$$x(t) = u(t) - u(10 - t)$$

Solution:
The signal $u(t)$ and $-u(10 - t)$ are represented in Fig. 1.104a. In Fig. 1.104b, $x(t) = u(t) - u(10 - t)$ is sketched. From Fig. 1.104b, the following equation for power is written:

$$P = \underset{T\to\infty}{Lt} \frac{1}{2T} \left[\int_{-T}^{0} (-1)^2 dt + \int_{10}^{T} (1)^2 dt \right]$$

$$= \underset{T\to\infty}{Lt} \frac{1}{2T} \left\{ [t]_{-T}^{0} + [t]_{10}^{T} \right\}$$

$$= \frac{1}{2} \underset{T\to\infty}{Lt} \frac{1}{T} [T + T - 10]$$

$$= \frac{1}{2} \underset{T\to\infty}{Lt} \left[2 - \frac{10}{T} \right] = 1$$

$$\boxed{P = 1 \text{ watt}}$$

If the power is finite, the energy $E = \infty$.

Example 1.63

Determine the power and RMS value of the following signal.

$$x(t) = e^{jat} \cos \omega_0 t$$

(*Anna University, 2007*)

Solution:

$$P = \underset{T\to\infty}{Lt} \frac{1}{2T} \int_{-T}^{T} |e^{jat} \cos \omega_0 t|^2 dt$$

$$e^{jat} = \cos at + j \sin at$$

$$|e^{jat}| = \sqrt{\cos^2 at + \sin^2 at} = 1$$

$$P = \underset{T\to\infty}{Lt} \frac{1}{2T} \int_{-T}^{T} \cos^2 \omega_0 t dt$$

$$= \underset{T\to\infty}{Lt} \frac{1}{4T} \int_{-T}^{T} (1 + \cos 2\omega_0 t) dt$$

Since $\int_{-T}^{T} \cos 2\omega_0 t dt = 0$, using Eq. (1.33),

$$P = \underset{T\to\infty}{Lt} \frac{1}{4T} \int_{-T}^{T} dt = \frac{1}{4T} 2T$$

$$\boxed{P = 0.5 \text{ watt}}$$

RMS value of power is

$$\boxed{P_{\text{RMS}} = \frac{1}{\sqrt{2}} = 0.707}$$

■ Example 1.64

Find the power and energy of the following signals:

(a) $x(t) = 10e^{j2\pi t}u(t)$

(b) $x(t) = e^{j(2t+\pi/4)}$

(Anna University, April, 2007)

Solution:

(a) $\boldsymbol{x(t) = 10e^{j2\pi t}u(t)}$

$$P = \underset{T\to\infty}{Lt}\frac{1}{2T}\int_0^T |10e^{j2\pi t}|^2 dt \qquad [x(t) = 0 \text{ for } t < 0]$$

$$= \frac{100}{2}\underset{T\to\infty}{Lt}\frac{1}{T}\int_0^T dt \qquad |e^{j2\pi t}| = 1$$

$$= 50\,\frac{1}{T}[T] = 50$$

$$\boxed{P = 50 \text{ watts}}$$

Since power is finite, $E = \infty$.

(b) $\boldsymbol{x(t) = e^{j(2t+\pi/4)}}$

$$|x(t)| = \left|e^{j(2t+\pi/4)}\right| = 1$$

$$P = \underset{T\to\infty}{Lt}\frac{1}{2T}\int_{-T}^T dt = \frac{1}{2T}2T = 1$$

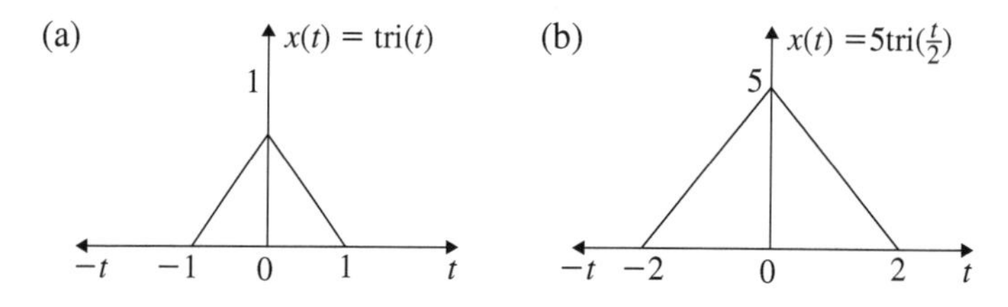

Fig. 1.105 Representation of triangular signals of Example 1.65

$$\boxed{P = 1}$$

Since power is finite, $E = \infty$.

■ Example 1.65

Find the energy of the following signal:

$$x(t) = 5 \operatorname{tri}\left(\frac{t}{2}\right)$$

Solution:

The triangular signal $x(t) = \operatorname{tri}(t)$ is shown in Fig. 1.105a. By amplitude multiplication and time expansion, $x(t) = 5 \operatorname{tri}\left(\frac{t}{2}\right)$ is obtained and shown in Fig. 1.105b. For Fig. 1.105b, the following equation is written:

$$x(t) = \frac{5}{2}t + c \qquad -2 \leq t \leq 0$$

c is obtained as 5.

$$x(t) = -\frac{5}{2}t + c \qquad 0 \leq t \leq 2$$

c is obtained as 5.

Let E_1 be the energy for the time interval $-2 \leq t \leq 0$ and E_2 energy for the time interval $0 \leq t \leq 2$.

$$\begin{aligned}
E_1 &= \int_{-2}^{0} \left(\frac{5}{2}t + 5\right)^2 dt \\
&= \left[\frac{25}{12}t^3 + 25t + \frac{25}{2}t^2\right]_{-2}^{0} \\
&= \frac{50}{3}
\end{aligned}$$

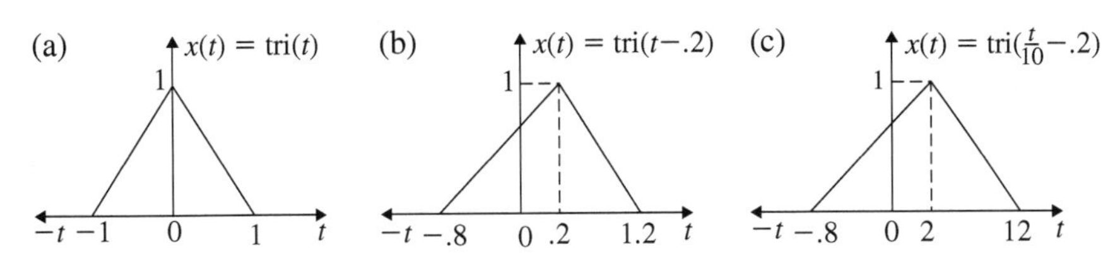

Fig. 1.106 Representation of $x(t) = \text{tri}(\frac{t}{10} - 0.2)$

$$E_2 = \int_0^2 \left(-\frac{5}{2}t + 5\right)^2 dt$$

$$= \int_0^2 \left(\frac{25}{4}t^2 + 25t - 25t\right) dt$$

$$= \left[\frac{25}{4}\frac{t^3}{3} + 25t - \frac{25}{2}t^2\right]_0^2$$

$$= \frac{50}{3}$$

$$E = E_1 + E_2 = \frac{50}{3} + \frac{50}{3}$$

$$\boxed{E = \frac{100}{3} \text{ Joules}}$$

Since energy is finite, the average power $P = 0$.

■ Example 1.66

Find the energy of the following signal:

$$x(t) = \text{tri}\left(\frac{t-2}{10}\right)$$

Solution:

$$x(t) = \text{tri}\left(\frac{t-2}{10}\right)$$
$$= \text{tri}(0.1t - .2)$$

Figure 1.106a shows $x(t) = \text{tri}(t)$. The time shifted signal $x(t) = \text{tri}(t - 0.2)$ is shown in Fig. 1.106b. The time shift is 0.2 toward right. By time elongation by factor

10, $x(t) = \text{tri}(\frac{t}{10} - .2)$ is obtained and is shown in Fig. 1.106c. For Fig. 1.106c, the following equations are written:

$$x(t) = \frac{1}{10}t + c \qquad -8 \leq t \leq 2$$

For $t = 2, x(t) = 1$

$$\begin{aligned} 1 &= \frac{2}{10} + c \\ c &= 0.8 \\ x(t) &= 0.1t + 0.8 \\ x(t) &= -\frac{1}{10}t + c \qquad 2 \leq t \leq 12 \end{aligned}$$

For $t = 2, x(t) = 1$

$$\begin{aligned} 1 &= \frac{-2}{10} + c \\ c &= 1.2 \\ x(t) &= -0.1t + 1.2. \end{aligned}$$

Energy of the signal is given as

$$\begin{aligned} E &= \int_{-8}^{2} (0.1t + 0.8)^2 dt + \int_{2}^{12} (-0.1t + 1.2)^2 dt \\ &= E_1 + E_2 \end{aligned}$$

where

$$E_1 = \int_{-8}^{2} (0.1t + 0.8)^2 dt$$

and

$$E_2 = \int_{2}^{12} (-0.1t + 1.2)^2 dt$$

$$\begin{aligned} E_1 &= \frac{1}{100}\int_{-8}^{2} (t + 8)^2 dt \\ &= \frac{1}{100}\int_{-8}^{2} (t^2 + 16t + 64) dt \end{aligned}$$

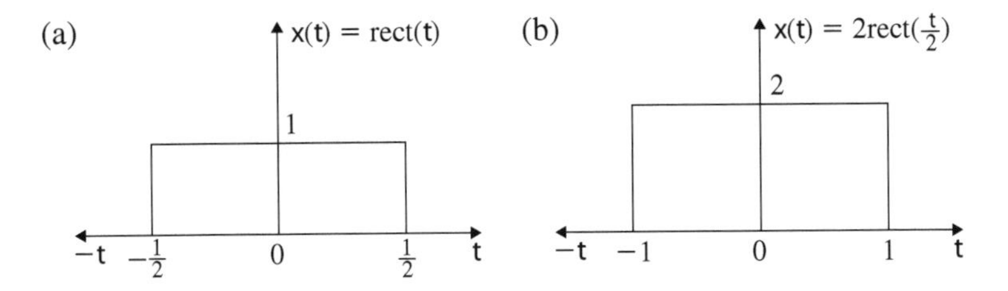

Fig. 1.107 Representation of rectangular function

$$= \frac{1}{100}\left[\frac{t^3}{3} + 8t^2 + 64t\right]_{-8}^{2}$$

$$= \frac{10}{3}$$

$$E_2 = \int_2^{12} \frac{1}{100}(12 - t)^2 dt$$

$$= \frac{1}{100}\int_2^{12} (t^2 - 24t + 144)dt$$

$$= \frac{1}{100}\left[\frac{t^3}{3} - 12t^2 + 144t\right]_2^{12}$$

$$= \frac{10}{3}$$

$$E = E_1 + E_2 = \frac{10}{3} + \frac{10}{3} = \frac{20}{3}$$

$$\boxed{E = \frac{20}{3} \text{ Joules}}$$

Since the energy is finite, the average power is zero.

■ Example 1.67

Find the energy of the following signal:

$$x(t) = 2\,\mathrm{rect}\left(\frac{t}{2}\right).$$

Solution: The rectangular or unit gate function is represented in Fig. 1.107a. It is defined as

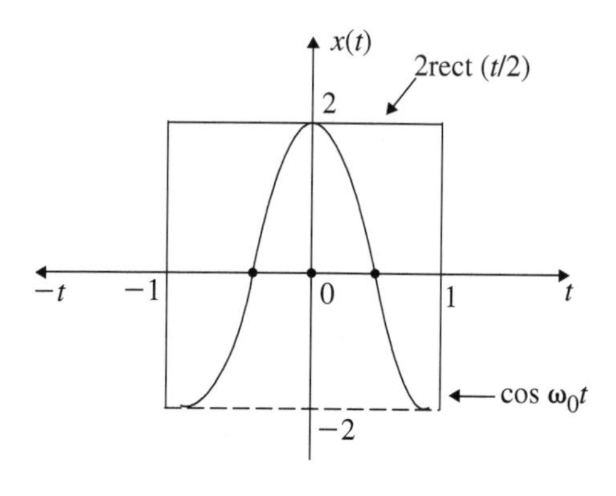

Fig. 1.108 Representation of $x(t) = 2\,\text{rect}(\frac{t}{2})\cos\omega_0 t$

$$\begin{aligned} x(t) &= 1 \quad -\frac{1}{2} \le t \le \frac{1}{2} \\ &= 0 \quad \text{otherwise} \end{aligned}$$

The rectangular signal with amplitude scaling and time elongation is shown in Fig. 1.107b. From Fig. 1.107b, the following equation for energy is written:

$$E = \int_{-1}^{1} (2)^2 dt = 4\big[t\big]_{-1}^{1} = 8$$

$$\boxed{E = 8\,\text{Joules}}$$

Since the energy is finite, the average power = 0.

■ Example 1.68

Find the energy of the following signal:

$$x(t) = 2\,\text{rect}\left(\frac{t}{2}\right)\cos\omega_0 t$$

Solution:

$$x(t) = 2\,\text{rect}\left(\frac{t}{2}\right)\cos\omega_0 t \quad -1 \le t \le 1$$

The above function is represented in Fig. 1.108. Here $T_0 = 2\,\text{sec}$. Hence, $\omega_0 = (2\pi/T_0) = \pi$ rad/sec.

$$\begin{aligned} E &= \int_{-1}^{1} (2\cos\omega_0 t)^2 dt \\ &= 4\int_{-1}^{1}\left(\frac{1}{2} + \frac{1}{2}\cos 2\omega_0 t\right) dt \end{aligned}$$

Since

$$\int_{-1}^{1} \cos 2\omega_0 t dt = 0 \qquad \text{(see Example 1.43)}$$

$$\textit{therefore,} \qquad E = 2\int_{-1}^{1} dt = 2\big[t\big]_{-1}^{1} = 4$$

$$\boxed{E = 4 \text{ Joules}}$$

Since the energy is finite, the average power $P = 0$.

■ Example 1.69

A trapezoidal pulse $x(t)$ is defined by

$$x(t) = \begin{cases} (5-t) & 4 \le t \le 5 \\ 1 & -4 \le t \le 4 \\ (t+5) & -5 \le t \le -4 \end{cases}$$

(a) Determine total energy of $x(t)$.
(b) Sketch $x(2t-3)$.
(c) If $y(t) = \frac{dx(t)}{dt}$, determine the total energy of $y(t)$.

(Anna University, December, 2007)

Solution:

(a) **To determine the total energy of $x(t)$.**
The given trapezoid pulse $x(t)$ is represented in Fig. 1.109a. The total energy of the signal is determined as described below:

$$\begin{aligned} E &= \int_{-5}^{-4} (t+5)^2 dt + \int_{-4}^{4} (1)^2 dt + \int_{4}^{5} (5-t)^2 dt \\ &= \int_{-5}^{-4} (t^2+10t+25) dt + \int_{-4}^{4} dt + \int_{4}^{5} (t^2-10t+25) dt \\ &= \left[\frac{t^3}{3} + 5t^2 + 25t\right]_{-5}^{-4} + \Big[t\Big]_{-4}^{4} + \left[\frac{t^3}{3} - 5t^2 + 25t\right]_{4}^{5} \\ &= -\frac{64}{3} + 80 - 100 + \frac{125}{3} - 125 + 125 + 8 + \frac{125}{3} - 125 \\ &\quad +125 - \frac{64}{3} + 80 - 100 \\ &= \frac{1}{3} + 8 + \frac{1}{3} \end{aligned}$$

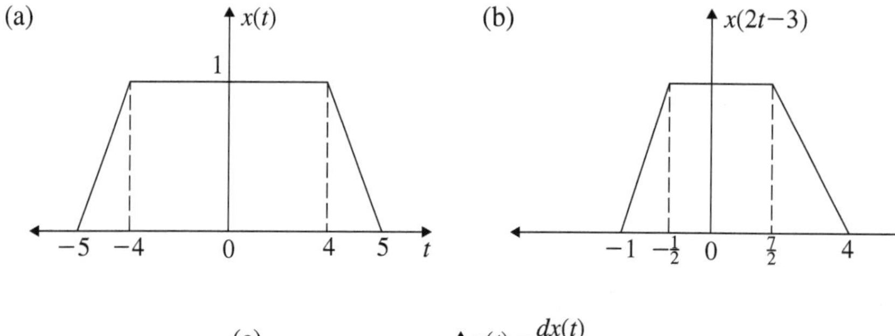

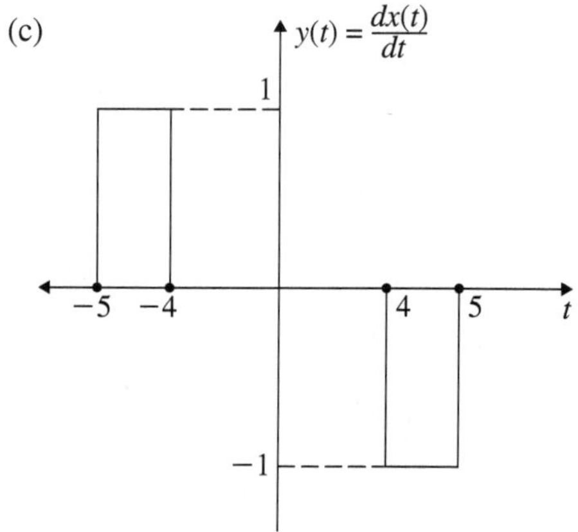

Fig. 1.109 Example 1.69

$$\boxed{E = \frac{26}{3} \text{ Joules}}$$

(b) **To sketch $x(2t-3)$**
$x(t)$ in Fig. 1.109a is right shifted by $t_0 = 3$ and time compressed by a factor 2. $x(2t-3)$ is shown in Fig. 1.109b.

(c) **To determine the total energy for $y(t) = \frac{dx}{dt}$.**

$$x(t) = 5 + t \qquad -5 \le t \le -4$$

$$y(t) = \frac{dx(t)}{dt} = 1 \qquad -5 \le t \le -4$$

$$x(t) = 1 \qquad -4 \le t \le 4$$

$$y(t) = \frac{dx(t)}{dt} = 0 \qquad -4 \le t \le 4$$

$$x(t) = 5 - t \qquad 4 \le t \le 5$$

$$y(t) = \frac{dx(t)}{dt} = -1 \qquad 4 \le t \le 5$$

The sketch of the above equations is shown in Fig. 1.109c. From this figure, the total energy is calculated as follows.

$$E = \int_{-5}^{-4} (1)^2 dt + \int_{4}^{5} (-1)^2 dt$$
$$= [t]_{-5}^{-4} + [t]_{4}^{5} = 1 + 1$$

$$\boxed{E = 2 \text{ Joules}}$$

■ Example 1.70

Consider the following CT signal.

$$x(t) = 2\delta(t+5) - 2\delta(t-6)$$

Calculate the energy of the signal

$$y(t) = \int_{-\infty}^{t} x(\tau) d\tau$$

Solution:

$$y(t) = \int_{-\infty}^{t} x(\tau) d\tau$$
$$= \int_{-\infty}^{t} 2\delta(\tau+5) d\tau - \int_{-\infty}^{t} 2\delta(\tau-6) d\tau$$
$$= 2u(t+5) - 2u(t-6)$$

$y(t)$ is represented in Fig. 1.110. The energy of the signal $y(t)$ is calculated as

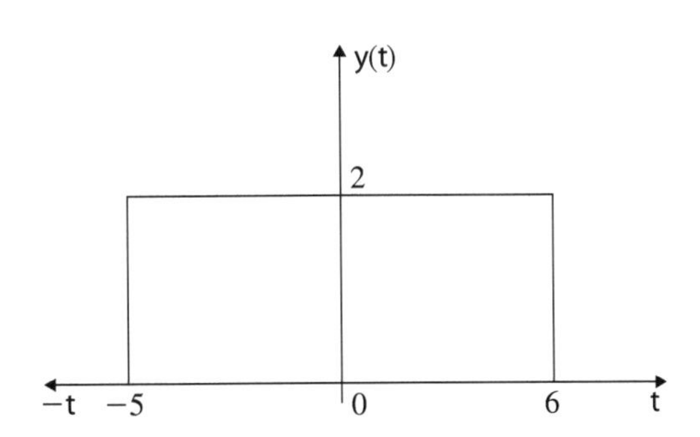

Fig. 1.110 Representation of signal of Example 1.70

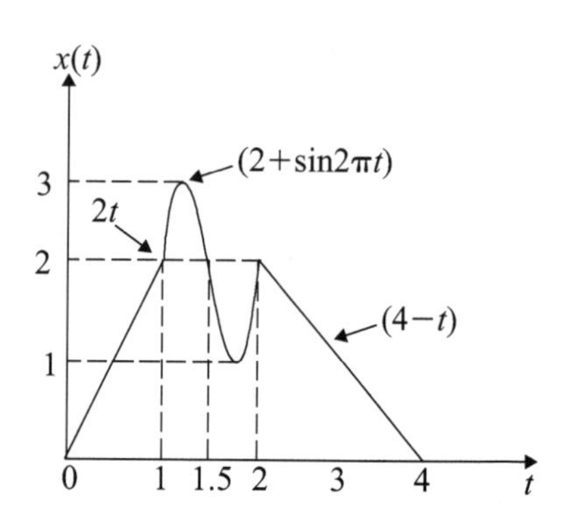

Fig. 1.111 Sketch of $x(t)$ of Example 1.71

$$
\begin{aligned}
E &= \int_{-5}^{6} |y(t)|^2 dt \\
&= \int_{-5}^{6} (2)^2 dt \\
&= 4\Big[t\Big]_{-5}^{6}
\end{aligned}
$$

$$\boxed{E = 44}$$

■ Example 1.71

A certain signal is described by the following mathematical equations (Fig. 1.111).

$$
x(t) = \begin{cases} 2t & 0 \le t < 1 \\ 2 + \sin 2\pi t & 1 \le t < 2 \\ 4 - t & 2 \le t < 4 \\ 0 & \text{otherwise} \end{cases}
$$

Sketch the signal $x(t)$. What is the energy of the signal?

Solution: $x(t)$ is split up as

$$x(t) = x_1(t) + x_2(t) + x_3(t)$$

where

$$\begin{aligned} x_1(t) &= 2t & 0 \le t < 1 \\ x_2(t) &= 2 + \sin 2\pi t & 1 \le t < 2 \\ x_3(t) &= 4 - t & 2 \le t < 4 \end{aligned}$$

E_1, E_2 and E_3 are the corresponding energies for $x_1(t), x_2(t)$ and $x_3(t)$, respectively

$$\begin{aligned} E_1 &= \int_0^1 |x_1(t)|^2 dt \\ &= \int_0^1 4t^2 dt \\ &= \frac{4}{3}\left[t^3\right]_0^1 \\ &= \frac{4}{3} \end{aligned}$$

$$\begin{aligned} E_2 &= \int_1^2 |x_2(t)|^2 dt \\ &= \int_1^2 (2 + \sin 2\pi t)^2 dt \\ &= \int_1^2 (4 + \sin^2 2\pi t + 4 \sin 2\pi t) dt \\ &= \int_1^2 (4 + 0.5(1 - \cos 4\pi t) + 4 \sin 2\pi t) dt \\ &= \int_1^2 \left(\frac{9}{2} - \frac{1}{2}\cos 4\pi t + 4 \sin 2\pi t\right) dt \\ &= \left[\frac{9}{2}t - \frac{1}{2}\frac{\sin 4\pi t}{4\pi} - \frac{4}{2\pi}\cos 2\pi t\right]_1^2 \\ &= \left[9 - 0 - \frac{2}{\pi}\right] - \left[\frac{9}{2} - 0 - \frac{2}{\pi}\right] \\ &= \frac{9}{2} \end{aligned}$$

$$\begin{aligned} E_3 &= \int_2^4 (4 - t)^2 dt \\ &= \int_2^4 (t^2 - 8t + 16) dt \end{aligned}$$

$$= \left[\frac{t^3}{3} - 4t^2 + 16t\right]_2^4$$

$$= \left[\frac{64}{3} - 64 + 64\right] - \left[\frac{8}{3} - 16 + 32\right]$$

$$= \frac{64}{3} - \frac{8}{3} - 16$$

$$= \frac{8}{3}$$

$$E = E_1 + E_2 + E_3$$

$$= \frac{4}{3} + \frac{9}{2} + \frac{8}{3}$$

$$\boxed{E = 8.5}$$

1.9 Classification of Discrete Time Signals

Like continuous time signals, discrete time signals are also classified as

1. Periodic and non-periodic signals.
2. Odd and even signals.
3. Power and energy signals.

They are discussed below with suitable examples.

1.9.1 Periodic and Non-Periodic DT Signals

A discrete time signal (sequence) $x[n]$ is said to be periodic with period N which is a positive integer if

$$x[n+N] = x[n] \quad \text{for all } n \tag{1.50}$$

Consider the DT sequence shown in Fig. 1.112. The signal gets repeated for every N. For Fig. 1.112, the following equation is written:

$$x[n+mN] = x[n] \quad \text{for all } n \tag{1.51}$$

where m is any integer. The smallest positive integer N in Eq. (1.51) is called the fundamental period N_0. Any sequence which is not periodic is said to be non-periodic or aperiodic.

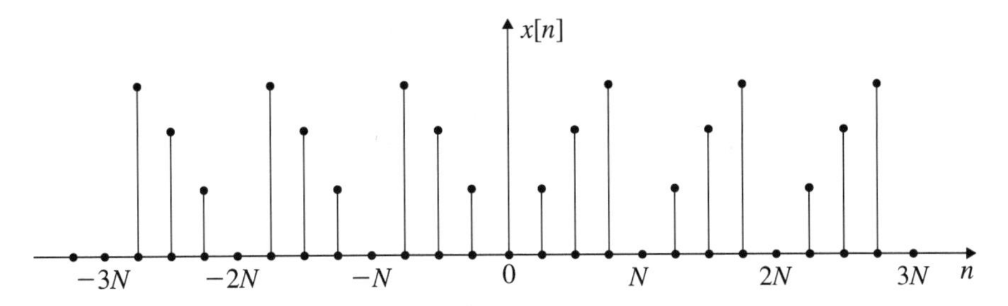

Fig. 1.112 Periodic sequence

■ Example 1.72

Show that complex exponential sequence $x[n] = e^{j\omega_0 n}$ is periodic and find the fundamental period.

Solution:

$$\begin{aligned} x[n] &= e^{j\omega_0 n} \\ x[n+N] &= e^{j\omega_0 (n+N)} \\ &= e^{j\omega_0 n} e^{j\omega_0 N} \end{aligned}$$

$$\begin{aligned} &= e^{j\omega_0 n} \quad \text{if } e^{j\omega_0 N} = 1 \\ \omega_0 N &= m2\pi \quad \text{where } m \text{ is any integer.} \end{aligned}$$

$$\boxed{N = m\frac{2\pi}{\omega_0}}$$

or

$$\frac{\omega_0}{2\pi} = \frac{m}{N} = \text{rational number.}$$

Thus, $e^{j\omega_0 n}$ is periodic if $\frac{m}{N}$ is rational. For $m = 1$, $N = N_0$. The corresponding frequency $F_0 = \frac{1}{N_0}$ is the fundamental frequency. F_0 is expressed in cycles and not Hz. Similarly ω_0 is expressed in radians and not in radians per second.

■ Example 1.73

Consider the following DT Signal.

$$x[n] = \sin(\omega_0 n + \phi)$$

Under what condition, the above signal is periodic.

Solution:

$$
\begin{aligned}
x[n] &= \sin(\omega_0 n + \phi) \\
x[n+N] &= \sin(\omega_0 (n+N) + \phi) \\
&= \sin(\omega_0 n + \omega_0 N + \phi) \\
&= \sin(\omega_0 n + \phi) \quad \text{if } \omega_0 N = 2\pi m \\
&= x[n]
\end{aligned}
$$

$$
\boxed{\frac{\omega_0}{2\pi} = \frac{m}{N} = \text{rational}}
$$

■ Example 1.74

If $x_1[n]$ and $x_2[n]$ are periodic, then show that the sum of the composite signal $x[n] = x_1[n] + x_2[n]$ is also periodic with the Least Common Multiple (LCM) of the fundamental period of individual signal.

Solution: Let N_1 and N_2 be the fundamental periods of $x_1[n]$ and $x_2[n]$, respectively. Since both $x_1[n]$ and $x_2[n]$ are periodic,

$$
\begin{aligned}
x_1[n] &= x_1[n + mN_1] \\
x_2[n] &= x_2[n + kN_2] \\
x[n] &= x_1[n] + x_2[n] \\
&= x_1[n + mN_1] + x_2[n + kN_2]
\end{aligned}
$$

For $x[n]$ to be periodic with period N,

$$
\begin{aligned}
x[n+N] &= x_1[n+N] + x_2[n+N] \\
x[n] &= x[n+N] \\
x_1[n + mN_1] + x_2[n + kN_2] &= x_1[n+N] + x_2[n+N]
\end{aligned}
$$

The above equation is satisfied if

$$
mN_1 = kN_2 = N
$$

m and k which are integers are chosen to satisfy the above equation. It implies that N is the LCM of N_1 and N_2.

On similar line, it can be proved that if $x_1[n]$ and $x_2[n]$ are periodic signals with fundamental period N_1 and N_2, respectively, then $x[n] = x_1[n]x_2[n]$ is periodic if

$$
mN_1 = kN_2 = N
$$

■ Example 1.75

Find whether the following signals are periodic. If periodic, determine the fundamental period.

$$\text{(a)} \quad x[n] = e^{j\pi n}$$
$$\text{(b)} \quad x[n] = \cos\left[\frac{n}{8} - \pi\right]$$
$$\text{(c)} \quad x[n] = \sin^2 \frac{\pi}{4} n$$

Solution:

(a) $\boldsymbol{x[n] = e^{j\pi n}}$

$$\omega_0 = \pi$$
$$N = \frac{2\pi}{\omega_0} m$$

$$\boxed{N = \frac{2\pi}{\pi} = 2} \quad \text{if } m = 1$$

$x[n]$ is periodic with fundamental period 2.

(b) $\boldsymbol{x[n] = \cos\left[\frac{n}{8} - \pi\right]}$

$$\omega_0 = \frac{1}{8}$$
$$N = \frac{2\pi}{\omega_0} m = 16\pi m$$

For any integer value of m, N is not integer. Hence, $x[n]$ is not periodic.

$$\boxed{x[n] \text{ is not periodic}}$$

(c) $\boldsymbol{x[n] = \sin^2 \frac{\pi}{4} n}$

$$\begin{aligned}
x[n] &= \sin^2 \frac{\pi}{4} n \\
&= \frac{1}{2} - \frac{1}{2} \cos \frac{2\pi}{4} n \\
&= x_1[n] + x_2[n] \\
x_1[n] &= \frac{1}{2} = \frac{1}{2}(1)^n \text{ is periodic with } N_1 = 1 \\
x_2[n] &= -\frac{1}{2} \cos \frac{\pi}{2} n \\
\omega_0 &= \frac{\pi}{2} \\
N_2 &= \frac{2\pi}{\omega_0} m = 4m = 4 \qquad \text{for } m = 1 \\
\frac{N_1}{N_2} &= \frac{1}{4} \\
\text{or} \quad 4N_1 &= N_2 = N
\end{aligned}$$

$$\boxed{N = 4}$$

■ Example 1.76

Find the periodicity of the following DT signal

$$x[n] = \sin \frac{2\pi}{3} n + \cos \frac{\pi}{2} n$$

(*Anna University, December, 2007*)

Solution:

$$\begin{aligned}
x[n] &= \sin \frac{2\pi}{3} n + \cos \frac{\pi}{2} n \\
&= x_1[n] + x_2[n] \\
x_1[n] &= \sin \frac{2}{3} \pi n \\
\omega_1 &= \frac{2}{3} \pi \\
N_1 &= \frac{2\pi}{\omega_1} m_1 = \frac{2\pi}{2\pi} 3m_1 = 3 \quad \text{for } m_1 = 1 \\
x_2[n] &= \cos \frac{\pi}{2} n \\
\omega_2 &= \frac{\pi}{2} \\
N_2 &= \frac{2\pi}{\omega_2} m_2 = \frac{2\pi}{\pi} 2m_2 = 4 \quad \text{for } m_2 = 1
\end{aligned}$$

$$\frac{N_1}{N_2} = \frac{3}{4} \text{ or } 4N_1 = 3N_2 = N$$

$$\boxed{N = 12}$$

■ Example 1.77

Determine whether the following signal is periodic. If periodic, find its fundamental period.

$$x[n] = \cos\left(\frac{n\pi}{2}\right)\cos\left(\frac{n\pi}{4}\right)$$

(*Anna University, December, 2006*)

Solution: **Method 1:**

$$\begin{aligned}
x[n] &= \cos\left(\frac{n\pi}{2}\right)\cos\left(\frac{n\pi}{4}\right) \\
&= x_1[n]x_2[n] \\
x_1[n] &= \cos\frac{n\pi}{2} \\
\omega_1 &= \frac{\pi}{2} \\
N_1 &= \frac{2\pi}{\omega_1}m_1 = \frac{2\pi}{\pi}2m_1 = 4 \quad \text{for } m_1 = 1 \\
x_2[n] &= \cos\frac{n\pi}{4} \\
\omega_2 &= \frac{\pi}{4} \\
N_2 &= \frac{2\pi}{\omega_2}m_2 = \frac{2\pi}{\pi}4m_2 = 8 \quad \text{for } m_2 = 1 \\
\frac{N_1}{N_2} &= \frac{4}{8} = \frac{1}{2} \quad \text{or} \\
2N_1 &= N_2 = N
\end{aligned}$$

$$\boxed{N = 8}$$

The signal is periodic and the fundamental period $N = 8$.

Method 2:

$$x[n] = \cos\left(\frac{n\pi}{2}\right)\cos\left(\frac{n\pi}{4}\right)$$

Using the following formula, we get

$$\begin{aligned}
\cos(A+B)+\cos(A-B) &= \cos A\cos B - \sin A\sin B + \cos A\cos B + \sin A\sin B \\
&= 2\cos A\sin B \\
\cos A\cos B &= \frac{1}{2}\cos(A+B) + \frac{1}{2}\cos(A-B)
\end{aligned}$$

Substituting $A = n\pi/2$ and $B = n\pi/2$, we get

$$\begin{aligned}
x[n] &= \frac{1}{2}\cos\pi n\left(\frac{1}{2}+\frac{1}{4}\right) + \frac{1}{2}\cos\pi n\left(\frac{1}{2}-\frac{1}{4}\right) \\
&= \frac{1}{2}\cos\frac{3}{4}\pi n + \frac{1}{2}\cos\frac{1}{4}\pi n
\end{aligned}$$

Choosing

$$x_1[n] = \frac{1}{2}\cos\frac{3}{4}\pi n$$

and

$$x_2[n] = \frac{1}{2}\cos\frac{1}{4}\pi n$$

we get

$$\begin{aligned}
\omega_1 &= \frac{3}{4}\pi \\
N_1 &= \frac{2\pi}{\omega_1}m_1 = \frac{2\pi 4}{3\pi}m_1 = 8 \quad \text{for } m_1 = 3 \\
\omega_2 &= \frac{1}{4}\pi \\
N_2 &= \frac{2\pi}{\omega_2}m_2 = \frac{2\pi 4}{\pi}m_2 = 8 \quad \text{for } m_2 = 1 \\
\frac{N_1}{N_2} &= \frac{8}{8} = 1 \\
N_1 &= N_2 = N = 8
\end{aligned}$$

$$\boxed{N = 8\,\text{sec.}}$$

■ Example 1.78

Test whether the following signals are periodic or not and if periodic, calculate the fundamental period.

(a) $x[n] = \cos\left(\frac{\pi}{2}n\right) + \sin\left(\frac{\pi}{8}n\right) + 3\cos\left(\frac{\pi}{4}n + \frac{\pi}{3}\right)$

(b) $x[n] = e^{j\frac{2\pi}{3}n} + e^{j\frac{3\pi}{4}n}$

(Anna University, December, 2007)

Solution:

(a)

$$\begin{aligned}
x[n] &= \cos\left(\frac{\pi}{2}n\right) + \sin\left(\frac{\pi}{8}n\right) + 3\cos\left(\frac{\pi}{4}n + \frac{\pi}{3}\right) \\
&= x_1[n] + x_2[n] + x_3[n] \\
x_1[n] &= \cos\frac{\pi}{2}n \\
\omega_1 &= \frac{\pi}{2}; \quad N_1 = \frac{2\pi}{\omega_1} = \frac{2\pi\, 2}{\pi} \quad \text{for } m_1 = 1 \\
N_1 &= 4 \\
x_2[n] &= \sin\left(\frac{\pi}{8}n\right) \\
\omega_2 &= \frac{\pi}{8}; \quad N_2 = \frac{2\pi}{\omega_2}m_2 = \frac{2\pi\, 8}{\pi} \quad \text{for } m_2 = 1 \\
N_2 &= 16 \\
x_3[n] &= 3\cos\left(\frac{\pi}{4}n + \frac{\pi}{3}\right) \\
\omega_3 &= \frac{\pi}{4}; \quad N_3 = \frac{2\pi}{\omega_3}m_3 = \frac{2\pi\, 4}{\pi} \quad \text{for } m_3 = 1 \\
N_3 &= 8
\end{aligned}$$

To find the LCM of N_1, N_2 and N_3.

$$\begin{array}{r|l}
4 & 4,\ 8,\ 16 \\ \hline
2 & 1,\ 2,\ 4 \\ \hline
 & 1,\ 1,\ 2
\end{array}$$

$$\text{LCM} = 4 \times 2 \times 2 = 16$$

$$\boxed{N = 16}$$

The signal is periodic.

(b)

$$
\begin{aligned}
x[n] &= e^{j\frac{2\pi}{3}n} + e^{j\frac{3\pi}{4}n} \\
&= x_1[n] + x_2[n] \\
x_1[n] &= e^{j\frac{2\pi}{3}n} \\
\omega_1 &= \frac{2\pi}{3};\ N_1 = \frac{2\pi}{\omega_1}m_1 = \frac{2\pi}{2\pi}3 \quad \text{for } m_1 = 1 \\
N_1 &= 3 \\
x_2[n] &= e^{j\frac{3\pi}{4}n} \\
\omega_2 &= \frac{3\pi}{4};\ N_2 = \frac{2\pi}{\omega_2}m_2 = \frac{2\pi}{3\pi}4m_2
\end{aligned}
$$

$$
\begin{aligned}
N_2 &= 8 \quad \text{for } m_2 = 3 \\
\frac{N_1}{N_2} &= \frac{3}{8} \\
8N_1 &= 3N_2 = N = 24
\end{aligned}
$$

$$
\boxed{N = 24}
$$

The signal is periodic with fundamental period $N = 24$.

1.9.2 Odd and Even DT Signals

Similar to continuous time signals, DT signals are also classified as odd and even signals. The relationships are analogous to CT signals.

A discrete time signal $x[n]$ is said to be an even signal if

$$
x[-n] = x[n] \tag{1.52}
$$

A discrete time signal $x[n]$ is said to be an odd signal if

$$
x[-n] = -x[n] \tag{1.53}
$$

The signal $x[n]$ can be expressed as the sum of odd and even signals as

$$
x[n] = x_e[n] + x_0[n] \tag{1.54}
$$

The even and odd components of $x[n]$ can be expressed as

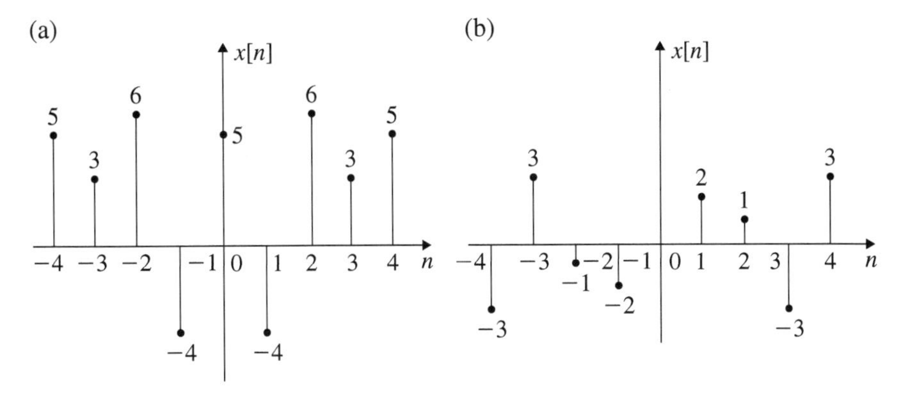

Fig. 1.113 **a** Even function and **b** Odd function

$$x_e[n] = \frac{1}{2}\left[x[n] + x[-n]\right] \tag{1.55}$$

$$x_0[n] = \frac{1}{2}\left[x[n] - x[-n]\right] \tag{1.56}$$

It is to be noted that

- An even function has an odd part which is zero.
- An odd function has an even part which is zero.
- The product of two even signals or of two odd signals is an even signal.
- The product of an odd and an even signal is an odd signal.
- At $n = 0$, the odd signal is zero.

The even and odd signals are represented in Fig. 1.113a and b, respectively.

■ Example 1.79

Determine whether the following functions are odd or even:

$$\text{(a)} \qquad x[n] = \sin 2\pi n$$

$$\text{(b)} \qquad x[n] = \cos 2\pi n$$

Solution:

(a) $\boldsymbol{x[n] = \sin 2\pi n}$

$$\begin{aligned} x[-n] &= \sin(-2\pi n) = -\sin 2\pi n \\ &= -x[n] \end{aligned}$$

$$\boxed{\text{This is an odd signal.}}$$

(b) $\boldsymbol{x[n] = \cos 2\pi n}$

$$\begin{aligned} x[-n] &= \cos(-2\pi n) = \cos 2\pi n \\ &= x[n] \end{aligned}$$

$$\boxed{\text{This is an even signal.}}$$

■ Example 1.80

Find the even and odd components of DT signal given below. Verify the same by graphical method.

$$x[n] = \{-2,\ 1,\ \underset{\uparrow}{3},\ -5,\ 4\}$$

Solution: $x[-n]$ is obtained by folding $x[n]$. Thus,

$$x[-n] = \{4,\ -5,\ \underset{\uparrow}{3},\ 1,\ -2\}$$

$$-x[-n] = \{-4,\ 5,\ \underset{\uparrow}{-3},\ -1,\ 2\}$$

$$\begin{aligned} x_e[n] &= \frac{1}{2}\left[x[n] + x[-n]\right] \\ &= \frac{1}{2}\left[\{-2,\ 1,\ \underset{\uparrow}{3},\ -5,\ 4\} + \{4,\ -5,\ 3,\ \underset{\uparrow}{1},\ -2\}\right] \\ &= \frac{1}{2}\left[(-2+4),\ (1-5),\ \underset{\uparrow}{(3+3)},\ (-5+1),\ (4-2)\right] \end{aligned}$$

$$\boxed{x_e[n] = \{1,\ -2,\ \underset{\uparrow}{3},\ -2,\ 1\}}$$

$$\begin{aligned} x_0[n] &= \frac{1}{2}\left[x[n] - x[-n]\right] \\ &= \frac{1}{2}\left[\{-2,\ 1,\ \underset{\uparrow}{3},\ -5,\ 4\} + \{-4,\ 5,\ -3,\ \underset{\uparrow}{-1},\ 2\}\right] \\ &= \frac{1}{2}\left[(-2-4),(1+5),\underset{\uparrow}{(3-3)},(-5-1),(4+2)\right] \end{aligned}$$

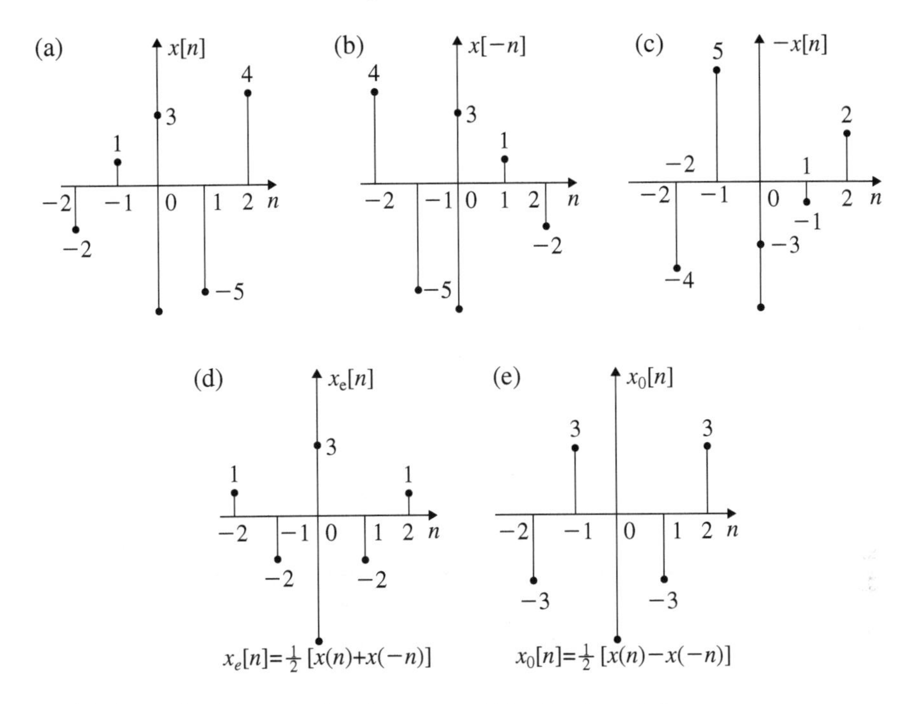

Fig. 1.114 Graphical determination of even and odd function from $x[n]$

$$\boxed{x_0[n] = \{-3,\ 3,\ \underset{\uparrow}{0},\ -3,\ 3\}}$$

Odd and even components by graphical method.
Solution:

1. $x[n]$ is represented in Fig. 1.114a.
2. $x[-n]$ is obtained by folding $x[n]$ which is represented in Fig. 1.114b.
3. $-x[n]$ is obtained by inverting $x[-n]$ of Fig. 1.114b. This is represented in Fig. 1.114c.
4. $x_e[n] = \frac{1}{2}\,[x[n] + x[-n]]$. Figure 1.114a, b sample wise are added and their amplitudes are divided by the factor 2. This gives $x_e[n]$ and is represented in Fig. 1.114d.
5. $x_0[n] = \frac{1}{2}\,[x[n] - x[-n]]$. Figure 1.114a, c sample wise are added and their amplitudes are divided by a factor 2 to get $x_0[n]$. This is represented in Fig. 1.114e.

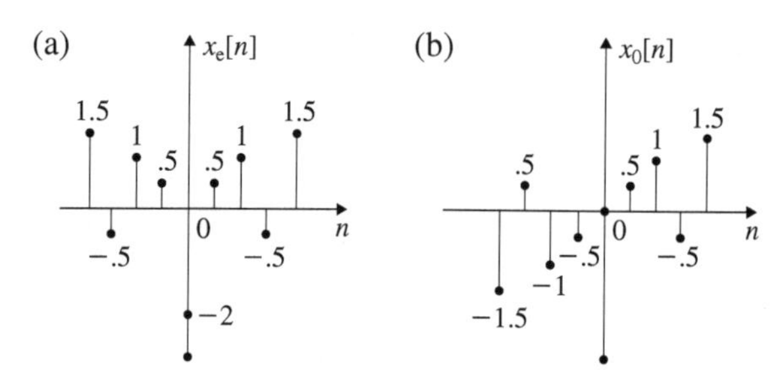

Fig. 1.115 **a** Even function and **b** Odd function

■ Example 1.81

Find the even and odd components of the following DT signal and sketch the same.

$$x[n] = \{-2, 1, 2, -1, 3\}$$

(Anna University, December, 2007)

Solution:

$$x[n] = \{-2, 1, 2, -1, 3\}$$

$$x[-n] = \{3, -1, 2, 1, \underset{\uparrow}{-2}\}$$

$$x_e[n] = \frac{1}{2}\{x[n] + x[-n]\}$$

$$= \frac{1}{2}[\{\underset{\uparrow}{-2}, 1, 2, -1, 3\} + \{3, -1, 2, 1, \underset{\uparrow}{-2}\}]$$

$$= \{1.5, -.5, 1, .5, \underset{\uparrow}{-2}, .5, 1, -.5, 1.5\}$$

$$x_0[n] = \frac{1}{2}[x[n] - x[-n]]$$

$$= \frac{1}{2}[\{\underset{\uparrow}{-2}, 1, 2, -1, 3\} - \{3, -1, 2, 1, \underset{\uparrow}{-2}\}]$$

$$x_0[n] = \{-1.5, .5, -1, -.5, \underset{\uparrow}{0}, .5, 1, -.5, 1.5\}$$

Even and odd components of $x[n]$ are represented in Fig. 1.115a and b, respectively.

■ Example 1.82

Given $x[n]$ and $y[n]$

$$\begin{aligned} x[-1] &= 2 \\ x[n] &= 1 \qquad 0 \le n \le 5 \\ x[6] &= 0.5 \\ &= 0 \qquad \text{otherwise} \\ y[n] &= 2u[n] \end{aligned}$$

Plot

1. $x[n/2]$
2. $x[n]y[n/2]$
3. Even part of $x[n]$
4. $x[n] + y[n/2]\delta[n-1]$

(*Anna University, December, 2011*)

Solution:

1. For the given equation, $x[n]$ is represented as shown in Fig. 1.116a. $x[n/2]$ is obtained by time expansion and is shown in Fig. 1.116b.
2.
$$y[n] = 2u[n]$$
$$y\left[\frac{n}{2}\right] = 2u\left[\frac{n}{2}\right]$$
The step sequence is time expanded by 2 and amplitude multiplied by 2. $y[\frac{n}{2}]$ is sketched as shown in Fig. 1.116c.
3. The even part of a function is given by
$$x_{ev}[n] = \frac{1}{2}[x[n] + x[-n]]$$
$x[n]$ shown in Fig. 1.116a and $x[-n]$ shown in Fig. 1.116f are added and divided by a factor 2 to get $x_{ev}[n]$ and are shown in Fig. 1.116g.
4. $y[n/2]$ is shown in Fig. 1.116c. Now
$$y\left[\frac{n}{2}\right]\delta[n-1] = y\left[\frac{n}{2}\right]\Big|_{n=1}$$

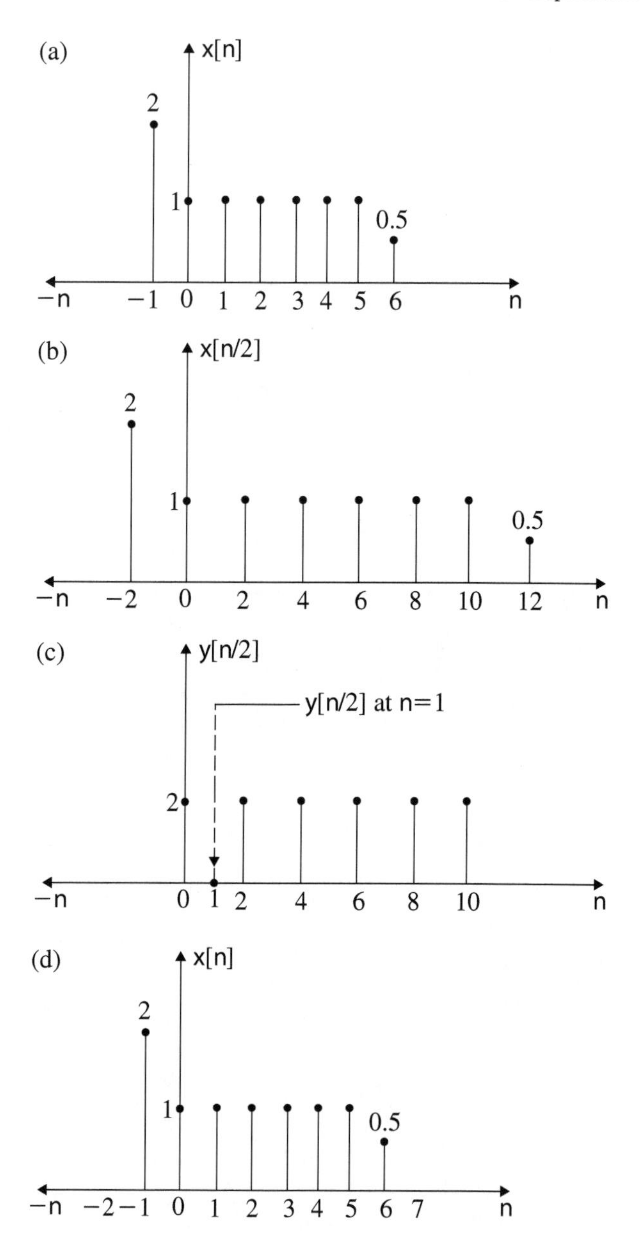

Fig. 1.116 Plot of $x[n]$ of Example 1.82. Figure of Example 1.82

(e)

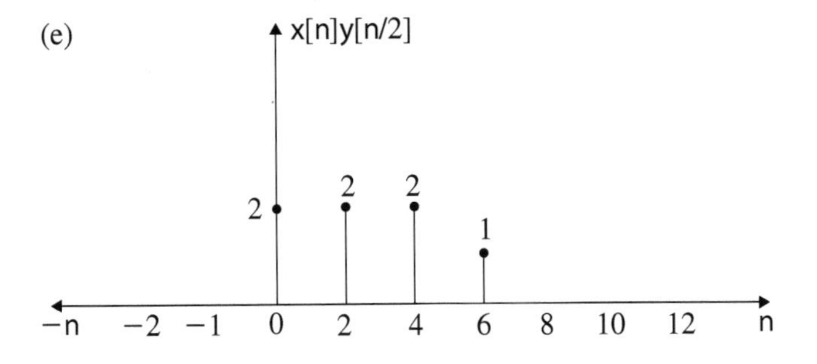

(f)

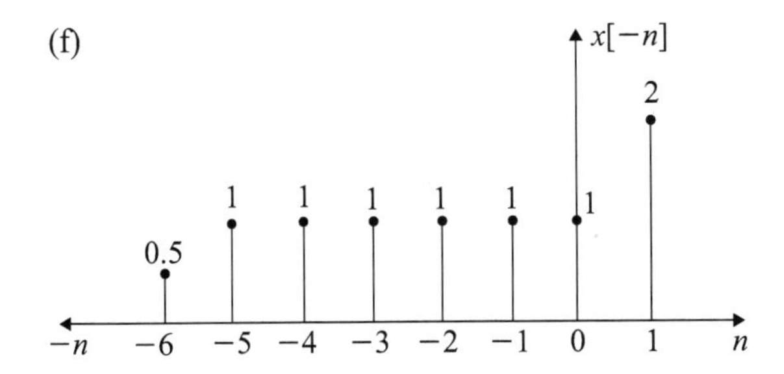

(g)

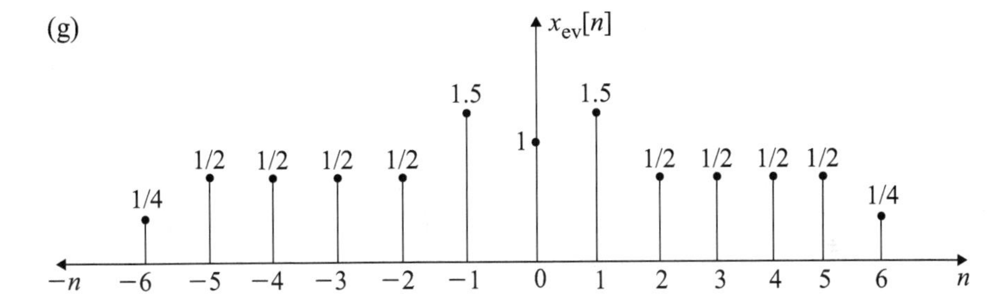

Fig. 1.116 (continued)

where $n = 1$

$$x[n] + y\left[\frac{n}{2}\right]\delta[n-1] = x[n]$$

$x[n]$ is shown in Fig. 1.116a.

1.9.3 Energy and Power of DT Signals

For a discrete time signal $x[n]$, the total energy is defined as

$$E = \sum_{n=-\infty}^{\infty} |x[n]|^2 \tag{1.57}$$

The average power is defined as

$$P = \underset{N\to\infty}{Lt} \frac{1}{(2N+1)} \sum_{n=-N}^{N} |x[n]|^2 \tag{1.58}$$

From the definitions of energy and power, the following inferences are derived:

1. $x[n]$ is an energy sequence iff $0 < E < \infty$. For finite energy signal, the average power $P = 0$.
2. $x[n]$ is a power sequence iff $0 < P < \infty$. For a sequence with average power P being finite, the total energy $E = \infty$.
3. Periodic signal is a power signal and *vice versa* is not true. Here the energy of the signal per period is finite.
4. Signals which do not satisfy the definitions of total energy and average power are neither termed as power signal nor energy signal. The following summation formulae are very often used while evaluating the average power and total energy of DT sequence.

1.

$$\sum_{n=0}^{N-1} a^n = \frac{(1-a^n)}{(1-a)} \qquad a \neq 1 \tag{1.59}$$
$$= N \qquad a = 1$$

2.

$$\sum_{n=0}^{\infty} a^n = \frac{1}{(1-a)} \qquad a < 1 \tag{1.60}$$

3.

$$\sum_{n=m}^{\infty} a^n = \frac{a^m}{(1-a)} \qquad a < 1 \tag{1.61}$$

4.

$$\sum_{n=0}^{\infty} na^n = \frac{a}{(1-a)^2} \qquad a < 1 \tag{1.62}$$

■ Example 1.83

Determine whether the following signals are energy signals or power signals:

$$
\begin{aligned}
&\text{(a)} \quad x[n] = A\delta[n]\\
&\text{(b)} \quad x[n] = u[n]\\
&\text{(c)} \quad x[n] = \text{ramp}\, n\\
&\text{(d)} \quad x[n] = A\\
&\text{(e)} \quad x[n] = 2e^{j(\pi n+\theta)}\\
&\text{(f)} \quad x[n] = \cos\frac{\pi}{2}n
\end{aligned}
$$

Solution:

(a) $\boldsymbol{x[n] = A\delta[n]}$

$$
\begin{aligned}
x[n] &= A\delta[n]\\
&= A \qquad n = 0\\
&= 0 \qquad n \neq 0
\end{aligned}
$$

$$\text{Energy } E = \sum_{n=0}^{0}(A)^2$$

$$\boxed{E = A^2}$$

For unit impulse, $A = 1$ and $E = 1$.

(b) $\boldsymbol{x[n] = u[n];\ n \geq 0}$

$$
\begin{aligned}
P &= \underset{N\to\infty}{Lt} \frac{1}{(2N+1)} \sum_{n=0}^{N} |x(n)|^2\\
&= \underset{N\to\infty}{Lt} \frac{1}{(2N+1)} \sum_{n=0}^{N} 1
\end{aligned}
$$

But $\sum_{n=0}^{N} 1 = (N+1)$

$$
\begin{aligned}
P &= \underset{N\to\infty}{Lt} \frac{(N+1)}{(2N+1)}\\
&= \underset{N\to\infty}{Lt} \frac{N(1+\frac{1}{N})}{N(2+\frac{1}{N})} = \frac{1}{2}
\end{aligned}
$$

$$\boxed{\begin{array}{l} P = \dfrac{1}{2} \\ E = \infty \end{array}}$$

(c) $\boldsymbol{x[n] = \mathbf{ramp}\, n;\ n \geq 0}$

$$P = \underset{N\to\infty}{Lt} \frac{1}{(2N+1)} \sum_{n=0}^{N} |x[n]|^2$$

$$P = \underset{N\to\infty}{Lt} \frac{1}{(2N+1)} \sum_{n=0}^{N} n^2$$

But $\sum_{n=0}^{N} n^2 = \frac{N(N+1)(2N+1)}{6}$

$$P = \underset{N\to\infty}{Lt} \frac{N(N+1)(2N+1)}{(2N+1)6}$$

$$\boxed{P = \infty}$$

$$E = \underset{N\to\infty}{Lt} \sum_{n=0}^{N} n^2$$

$$= \underset{N\to\infty}{Lt} \frac{N(N+1)(2N+1)}{6} = \infty$$

$$\boxed{E = \infty}$$

The signal $x[n] = n$ is neither power signal nor energy signal.

(d) $\boldsymbol{x[n] = A}$

$$P = \underset{N\to\infty}{Lt} \frac{1}{(2N+1)} \sum_{n=-\infty}^{\infty} A^2$$

$$= \underset{N\to\infty}{Lt} \frac{A^2}{(2N+1)}(2N+1) \qquad \left[\sum_{n=-\infty}^{\infty} 1 = (2N+1)\right]$$

$$\boxed{\begin{array}{l} P = A^2 \\ E = \infty \end{array}}$$

(e) $\boldsymbol{x[n] = 2e^{j(\pi n+\theta)}}$

$$P = \underset{N\to\infty}{Lt} \frac{1}{(2N+1)} \sum_{-N}^{N} |2e^{j(n\pi+\theta)}|^2$$

$$P = \underset{N\to\infty}{Lt} \frac{1}{2N+1} 4 \sum_{-N}^{N} |e^{j(n\pi+\theta)}|^2$$

But $|e^{j(n\pi+\theta)}| = 1$ and $\sum_{-N}^{N} 1 = (2N+1)$

$$P = \underset{N\to\infty}{Lt} 4\frac{(2N+1)}{(2N+1)} = 4$$

$$\boxed{\begin{array}{c} P = 4 \\ E = \infty \end{array}}$$

(f) $\boldsymbol{x[n] = \cos \frac{\pi}{2} n}$

$$P = \frac{1}{(2N+1)} \sum_{-N}^{N} \cos^2 \frac{\pi}{2} n$$

Since $\sum_{-N}^{N} \cos \pi n = 0$,

$$\begin{aligned} P &= \underset{N\to\infty}{Lt} \frac{1}{(2N+1)} \sum_{-N}^{N} \frac{(1+\cos \pi n)}{2} \\ &= \frac{1}{2} \underset{N\to\infty}{Lt} \frac{(2N+1)}{(2N+1)} \\ &= \frac{1}{2} \end{aligned}$$

$$\boxed{\begin{array}{c} P = \frac{1}{2} \\ E = \infty \end{array}}$$

■ Example 1.84

Determine the energy of the signal shown in Fig. 1.117 whose

$$x[n] = \left(\frac{1}{3}\right)^n u[n]$$

(*Anna University, December, 2007*)

Fig. 1.117 $x[n] = \left(\frac{1}{3}\right)^2 u[n]$

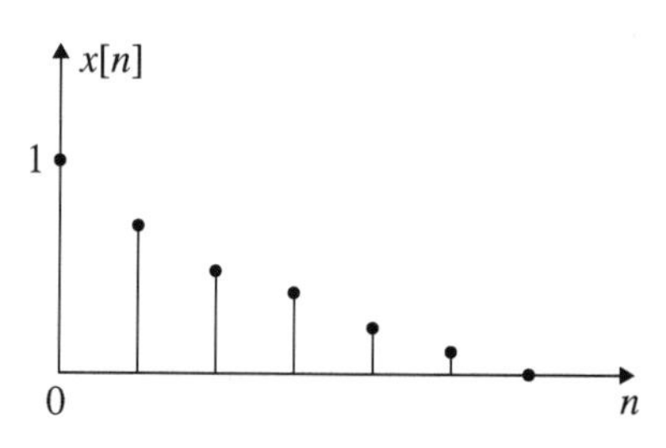

Solution:

$$\begin{aligned}
E &= \underset{N\to\infty}{Lt} \sum_{n=0}^{N} \left(\frac{1}{3}\right)^{2n} \\
&= \underset{N\to\infty}{Lt} \sum_{n=0}^{N} \left(\frac{1}{9}\right)^{n} \\
&= 1 + \frac{1}{9} + \left(\frac{1}{9}\right)^2 + \ldots \\
&= \frac{1}{1-\frac{1}{9}}
\end{aligned}$$

$$\boxed{\begin{aligned} E &= \frac{9}{8} \\ P &= 0 \end{aligned}}$$

■ Example 1.85

Find the energy of the following sequence:

$$x[n] = n \qquad 0 \le n \le 4$$

Solution:

$$\begin{aligned}
x[n] &= n \\
&= \{0,\ 1,\ 2,\ 3,\ 4\} \\
E &= \sum_{n=0}^{4} n^2 \\
&= 0 + 1 + 4 + 9 + 16
\end{aligned}$$

$$\boxed{E = 30}$$

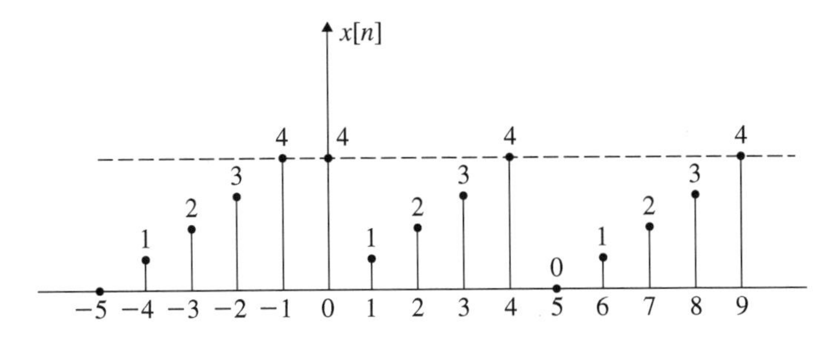

Fig. 1.118 $x[n]$ of Example 1.86

■ Example 1.86

Determine the average power and the energy per period of the sequence shown in Fig. 1.118.

Solution: The fundamental period N of the signal is 5. Hence, the average power per period is

$$\begin{aligned} P &= \frac{1}{5}\sum_{n=0}^{4} n^2 \\ &= \frac{1}{5}[0+1+4+9+16] \end{aligned}$$

$$\boxed{P = 6}$$

Average energy per period is

$$\begin{aligned} E &= \sum_{n=0}^{4} n^2 \\ &= [0+1+4+9+16] \end{aligned}$$

$$\boxed{E = 30}$$

$$\text{Total Energy} = \infty$$

■ Example 1.87

Find the energy and power of the following signal:

$$x[n] = a^n u[n]$$

for the following cases:

$$\text{(a)} \quad |a| < 1$$
$$\text{(b)} \quad |a| = 1$$
$$\text{(c)} \quad |a| > 1$$

Solution:

(a) $\boldsymbol{x[n] = a^n u[n]}$ **where** $\boldsymbol{|a| < 1}$ **and** $\boldsymbol{n \geq 0}$

$$E = \sum_{n=0}^{\infty}(a^n)^2$$
$$= 1 + a^2 + a^4 + \ldots$$

$$\boxed{E = \frac{1}{1 - |a|^2}}$$

$$P = 0$$

(b) $\boldsymbol{x[n] = a^n u[n]}$ **where** $\boldsymbol{|a| = 1}$

$$E = \underset{N\to\infty}{Lt} \sum_{0}^{N} 1^n = \underset{N\to\infty}{Lt} (N + 1)$$

$$\boxed{E = \infty}$$

$$P = \underset{N\to\infty}{Lt} \frac{1}{(2N + 1)} \sum_{0}^{N}(1)^n$$

$$P == \underset{N\to\infty}{Lt} \frac{(N + 1)}{(2N + 1)}$$
$$= \underset{N\to\infty}{Lt} \frac{N(1 + \frac{1}{N})}{N(2 + \frac{1}{N})}$$

$$\boxed{P = \frac{1}{2}}$$

(c) $\boldsymbol{x[n] = a^n u[n]}$ **where** $\boldsymbol{a > 1}$

$$E = \underset{N\to\infty}{Lt} \sum_{0}^{N} a^n$$
$$= 1 + a + a^2 + \cdots + a^N$$

$$\boxed{E = \infty}$$

$$P = \underset{N\to\infty}{Lt} \frac{1}{(2N+1)} \sum_{n=0}^{N} a^n$$
$$= \underset{N\to\infty}{Lt} \frac{1}{(N+1)} \frac{(1-a^{N+1})}{(1-a)}$$

$$\boxed{P = \infty}$$

The signal is neither energy nor power signal.

■ Example 1.88

Find the energy of the following signal:

$$x[n] = nu[n] - 2nu[n-4] + nu[n-8]$$

Solution:

$$x[n] = nu[n] - 2nu[n-4] + nu[n-8]$$
$$= x_1[n] + x_2[n] + x_3[n]$$

$x_1[n]$, $x_2[n]$ and $x_3[n]$ are shown in Fig. 1.119a–c, respectively. Figure 1.119d represents $x[n]$. From Fig. 1.119d, the energy of the signal $x[n]$ is obtained as

$$E = 1^2 + 2^2 + 3^2 + 4^2 + 3^2 + 2^2 + 1^2$$

$$\boxed{E = 44}$$

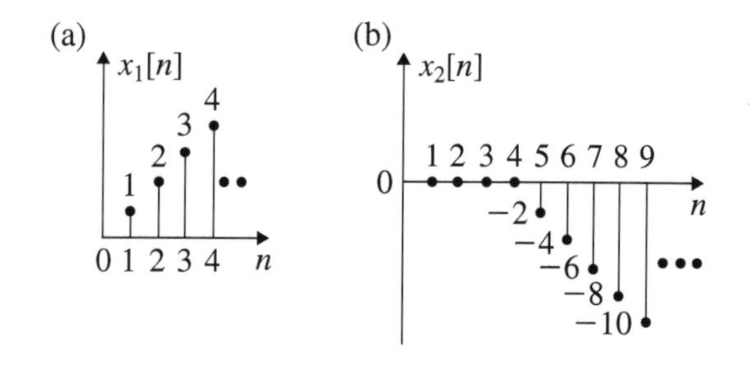

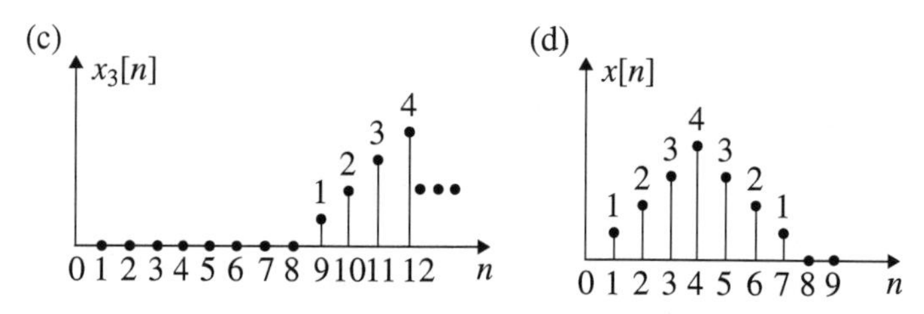

Fig. 1.119 DT energy signal of Example 1.88

■ Example 1.89

Determine the value of power and energy of each of the following signals:

$$\text{(a)} \qquad x[n] = e^{j(\frac{\pi n}{2}+\frac{\pi}{8})}$$

$$\text{(b)} \qquad x[n] = \left(\frac{1}{2}\right)^n u[n]$$

Solution: *(Anna University, April, 2008)*

(a) $\boldsymbol{x[n] = e^{j(\frac{\pi n}{2}+\frac{\pi}{8})}}$

$$P = \underset{N\to\infty}{Lt} \frac{1}{2N+1}\sum_{-N}^{N} |e^{j(\frac{\pi n}{2}+\frac{\pi}{8})}|^2$$

$$= \underset{N\to\infty}{Lt} \frac{1}{(2N+1)}\sum_{-N}^{N} 1$$

$$P = \frac{(2N+1)}{(2N+1)} = 1$$

$$\boxed{P = 1 \text{ and } E = \infty}$$

(b) $\boldsymbol{x[n] = \left[\frac{1}{2}\right]^n u[n]}$

$$E = \underset{N\to\infty}{Lt} \sum_{0}^{N} \left(\frac{1}{2}\right)^{2n}$$

$$= \underset{N\to\infty}{Lt} \sum_{0}^{N} \left(\frac{1}{4}\right)^{n}$$

$$= \frac{1}{1-\frac{1}{4}} = \frac{4}{3}$$

$$\boxed{E = \frac{4}{3} \text{ and } P = 0}$$

■ Example 1.90

Find the energy of the following DT signal

$$x[n] = \left(\frac{1}{2}\right)^n \qquad n \geq 0$$
$$= 3^n \qquad n < 0$$

(*Anna University, April, 2005*)

Solution:

$$E = \left[\sum_{-\infty}^{-1} (3)^{2n} + \sum_{0}^{\infty} \left(\frac{1}{2}\right)^{2n}\right]$$

$$= \left[\sum_{-\infty}^{-1} (9)^{n} + \sum_{0}^{\infty} \left(\frac{1}{4}\right)^{n}\right]$$

$$= \left[\sum_{1}^{\infty} (9)^{-n} + \frac{1}{(1-\frac{1}{4})}\right]$$

$$= \left[\sum_{1}^{\infty} \left(\frac{1}{9}\right)^{n} + \frac{4}{3}\right]$$

$$= \left[\frac{1}{9} + \frac{1}{9^2} + \frac{1}{9^3} + \cdots\right] + \frac{4}{3}$$

$$= \frac{1}{9}\left[1 + \frac{1}{9} + \frac{1}{9^2} + \frac{1}{9^3} + \cdots\right] + \frac{4}{3}$$

$$= \frac{1}{9}\frac{1}{[1-\frac{1}{9}]} + \frac{4}{3}$$

$$= \frac{1}{8} + \frac{4}{3}$$

$$\boxed{E = \frac{35}{24}}$$

Summary

1. Signals are broadly classified as Continuous Time (CT) and Discrete Time (DT) signals. They are further classified as deterministic and stochastic, periodic and non-periodic, odd and even and energy and power signals.
2. Basic CT and DT signals include impulse, step, ramp, parabolic, rectangular pulse, triangular pulse, signum function, sinc function, sinusoid, real and complex exponentials.
3. Basic operations on CT and DT signals include addition, multiplication, amplitude scaling, time scaling , time shifting, reflection or folding and amplitude inverted signals.
4. In time shifting of CT signal, for $x(t+t_0)$ and $x(-t-t_0)$, the time shift is made to the left of $x(t)$ and $x(-t)$, respectively, by t_0. For $x(t-t_0)$ and $x(-t+t_0)$, the time shift is made to the right of the $x(t)$ and $x(-t)$, respectively, by t_0. Similar operation holds good for DT signals $x[n+n_0]$, $x[-n-n_0]$, $x[n-n_0]$ and $x[-n+n_0]$ when shifted by n_0.
5. To plot CT and DT signals, the operation performed is in the following sequence. The signal is folded (if necessary), time shifted, time scaled, amplitude scaled and inverted.
6. Signals are classified as even signals and odd signals. Even signals are **symmetric** about the vertical axis whereas odd signals are **anti-symmetric** about the time origin. Odd signals pass through the origin. The product of two even signals or two odd signals is an even signal. The product of an even and an odd signal is an odd signal.
7. A CT signal which repeats itself for every T seconds or a DT signal for every N sequence is called a periodic signal. If the signal is not periodic, it is called an aperiodic or non-periodic signal. The necessary condition for the composite of two or more signals to be periodic is that the individual signal should be periodic.
8. A signal is an energy signal iff the total energy of the signal satisfies the condition $0 < E < \infty$. A signal is called a power signal iff the average power of the signal satisfies the condition $0 < P < \infty$. If the energy of a signal

is finite, the average power is zero. If the power of the signal is finite, the signal has infinite energy. All periodic signals are power signals. However, all power signals need not be periodic. Signals which are deterministic and non-periodic are usually energy signals. Some signals are neither energy signal nor power signal.

Exercises

I. Short Answer Type Questions

1. **How are signals classified?**
   Signals are generally classified as CT and DT signal. They are further classified as deterministic and non-deterministic, odd and even, periodic and non-periodic and power and energy signals.
2. **What are odd and even signals?**
   A continuous CT signal is said to be an even signal if it satisfies the condition $x(-t) = x(t)$ for all t. It is said to be an odd signal if $x(-t) = -x(t)$ for all t. For a DT signal, if $x[-n] = x[n]$ condition is satisfied, it is an even sequence (signal). If $x[-n] = -x[n]$, the sequence is called odd sequence.
3. **How even and odd components of a signal are mathematically expressed for CT and DT signals?**

$$x_e(t) = \frac{1}{2}[x(t) + x(-t)]$$
$$x_0(t) = \frac{1}{2}[x(t) - x(-t)]$$
$$x_e[n] = \frac{1}{2}\{x[n] + x[-n]\}$$
$$x_0[n] = \frac{1}{2}\{x[n] - x[-n]\}$$

4. **What are periodic and non-periodic signals?**
   A continuous time signal is said to be a periodic signal if it repeats itself for every T sec. It satisfies the condition $x(t) = x(t + T)$ for all t. A discrete time signal is said to be a period signal if it satisfies the condition $x[n] = x[n + N]$ for all n. A signal which is not periodic is said to be non-periodic.
5. **What is the fundamental period of a periodic signal? What is fundamental frequency?**
   A CT signal is said to be periodic if it satisfies the condition $x(t) = x(t + T)$. If this condition is satisfied for $T = T_0$, it is also satisfied for $T = 2T_0, 3T_0, \ldots$. The smallest value of T that satisfies the above condition is called fundamental period. The fundamental frequency $f_0 = \frac{1}{T_0}$ Hz. It is also expressed as $\omega_0 = \frac{2\pi}{T_0}$ rad/sec.

6. **What are power and energy signals?**
   For a CT signal, the total energy is defined as

$$E = \underset{T\to\infty}{Lt} \int_{-T}^{T} |x(t)|^2 dt$$

   and the average power is defined as

$$P = \underset{T\to\infty}{Lt} \frac{1}{2T} \int_{-T}^{T} |x(t)|^2 dt$$

   The square root of P is called Root Mean Square (RMS) value of $x(t)$. For a DT signal $x[n]$, the total energy is defined as

$$E = \sum_{n=-\infty}^{\infty} x^2[n]$$

   The average power is defined as

$$P = \underset{T\to\infty}{Lt} \frac{1}{2N+1} \sum_{n=-N}^{N} x^2[n]$$

7. **Determine whether the signal $x[n] = \cos[0.1\pi n]$ is periodic.**
   The signal $x[n]$ is periodic with fundamental period $N_0 = 20$.
8. **Find whether the signal $x[n] = 5\cos[6\pi n]$ is periodic.**
   The signal is periodic with fundamental period $N_0 = 1$.
9. **What is the condition that the signal $x(t) = e^{at}u(t)$ to be energy signal?**
   For the signal $x(t) = e^{at}u(t)$ to be energy signal, $a < 0$.
10. **Is the unit step signal an energy signal?**
    The unit step has an average power $P = \frac{1}{2}$. It is a power signal.
11. **Determine the power and RMS value of the signal $x(t) = e^{jat}\cos\omega_0 t$.**
    Average power $P = \frac{1}{2}$ and RMS power $P_{\text{RMS}} = \frac{1}{\sqrt{2}}$.
12. **What is the periodicity of $x(t) = e^{j100\pi t + 30^\circ}$?**
    The periodicity of the signal $x(t)$ is $T = \frac{1}{50}$sec.
13. **Find the average power of the signal.**

$$x[n] = u[n] - u[n-N]$$

    The average power $P = 1$.
14. **Find the total energy of**

$$x[n] = \{1, \underset{\uparrow}{1}, 1\}$$

The total energy $E = 3$.

15. **If the discrete time signal $x[n] = \{0, 0, 0, 3, 2, 1, -1, -7, 6\}$, then find $y[n] = x[2n - 3]$.**

$$y[n] = \{0, 0, 0, 3, 1, -7\}$$

16. **What is the energy of the signal $x[n] = u[n] - u[n - 6]$?**

$$E = 6$$

17. **Find the equivalence of the following functions (a) $\delta(at)$; (b) $\delta(-t)$; (c) $t\delta(t)$; (d) $\sin t\delta(t)$; (e) $\cos\delta(t)$ and (f) $x(t)\delta(t - t_0)$.**

$$\begin{aligned}
&(a) \quad \delta(at) = \frac{1}{a}\delta(t)\\
&(b) \quad \delta(-t) = \delta(t)\\
&(c) \quad t\delta(t) = 0\\
&(d) \quad \sin t\delta(t) = 0\\
&(e) \quad \cos t\delta(t) = \delta(t)\\
&(f) \quad x(t)\delta(t - t_0) = x(t_0)
\end{aligned}$$

18. **How do you represent an exponential e^{-at} for $t \geq 0$ and $t < 0$?**
The everlasting exponential e^{-at} is expressed as $e^{-at}u(t)$ for $t \geq 0$ and $e^{-at}u(-t)$ for $t < 0$.
19. **Find the value of $\frac{t^2+5}{t^2+6}\delta(t - 2)$.**

$$\frac{(t^2 + 5)}{(t^2 + 6)}\delta(t - 2) = 0.9\,\delta(t - 2)$$

20. **Find the odd and even components of e^{j2t}.**

$$x_e(t) = \cos 2t$$

$$x_0(t) = \sin 2t$$

II. Long Answer Type Questions

1. A triangular pulse signal $x(t)$ is shown in Fig. 1.120a. Sketch the following signals. (a) $x(4t)$; (b) $x(4t + 3)$; (c) $x(-3t + 2)$; (d) $x(\frac{t}{3} + 2)$; (e) $x(3t - 2)$ and (f) $x(4t + 3) + x(2t)$.
2. Sketch the following CT functions. (a) $x(t) = 8u(5 - t)$; (b) $x(t) = 3\delta(t + 2)$; (c) $x(t) = \text{ramp}(t + 1)$; (d) $x(t) = 5\text{rect}\frac{(t+1)}{4}$; (e) $x(t) = -\text{tri}\frac{t-1}{4}$; (f) $x(t) =$

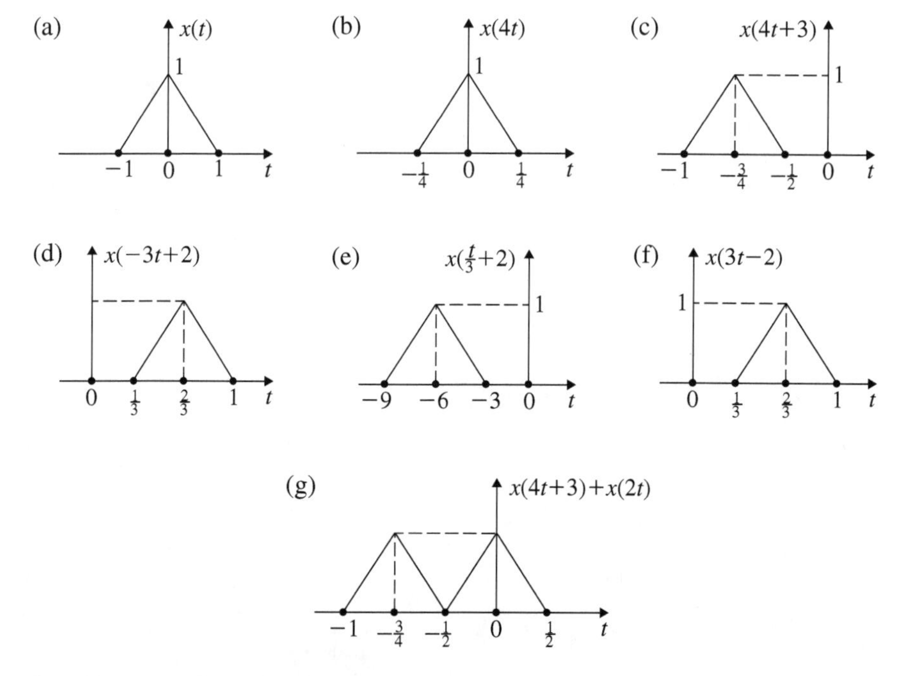

Fig. 1.120 Operations of CT signals

$u(t) - u(t-5)$; (g) $x(t) = u(t) - u(t+5)$; (h) $x(t) = -\text{ramp}(t)u(t-3)$; (i) $x(t) = u(t)(t + \frac{1}{3})\text{ramp}(\frac{1}{3} - t)$ and (j) $x(t) = \text{rect}(t+2) - \text{rect}(t-2)$.

3. Determine whether each of the following CT signals are periodic. If periodic, determine the fundamental period (Fig. 1.121).

(a) $x(t) = e^{j2t}$

(b) $x(t) = e^{(-2+j3)t}$

(c) $x(t) = \sin\left(60\pi t + \frac{\pi}{4}\right)$

(d) $x(t) = \cos\left(60\pi t - \frac{\pi}{4}\right) - \sin 20\pi t$

(e) $x(t) = \sin\left(8\pi t + \frac{\pi}{3}\right) + 5\cos\left(\frac{\pi t}{3} + \frac{\pi}{2}\right) + 6\cos\left(7\pi t - \frac{\pi}{2}\right)$

(f) $x(t) = 30\sin\left(8\pi t + \frac{\pi}{3}\right)\cos\left(2\pi t + \frac{\pi}{2}\right)\sin\left(5\pi t - \frac{\pi}{2}\right)$

(a) Periodic with period $T_0 = \pi$ sec. (b) Not periodic. (c) Periodic. $T_0 = \frac{1}{30}$ sec. (d) Periodic $T_0 = \frac{1}{10}$ sec. (e) Periodic $T_0 = 6$ sec. (f) Periodic $T_0 = 2$ sec.

4. Sketch the even and odd parts of the following signals shown in Fig. 1.122a, b.

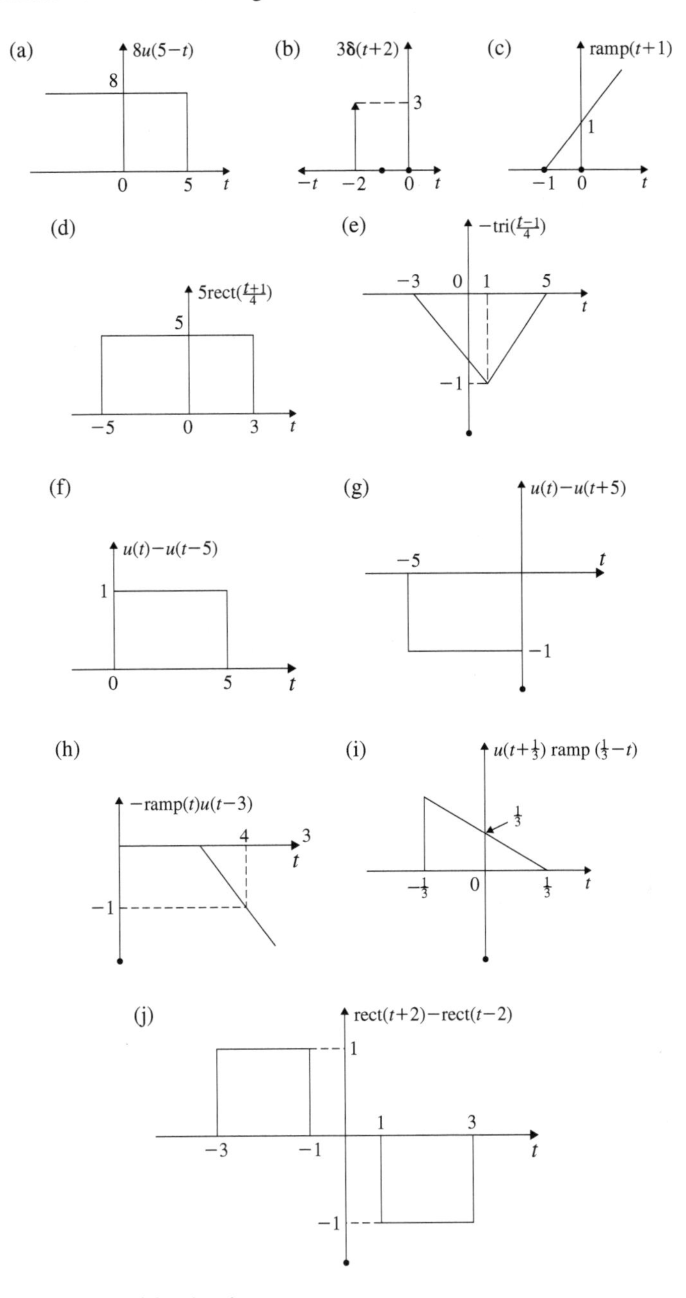

Fig. 1.121 Operations of CT signals

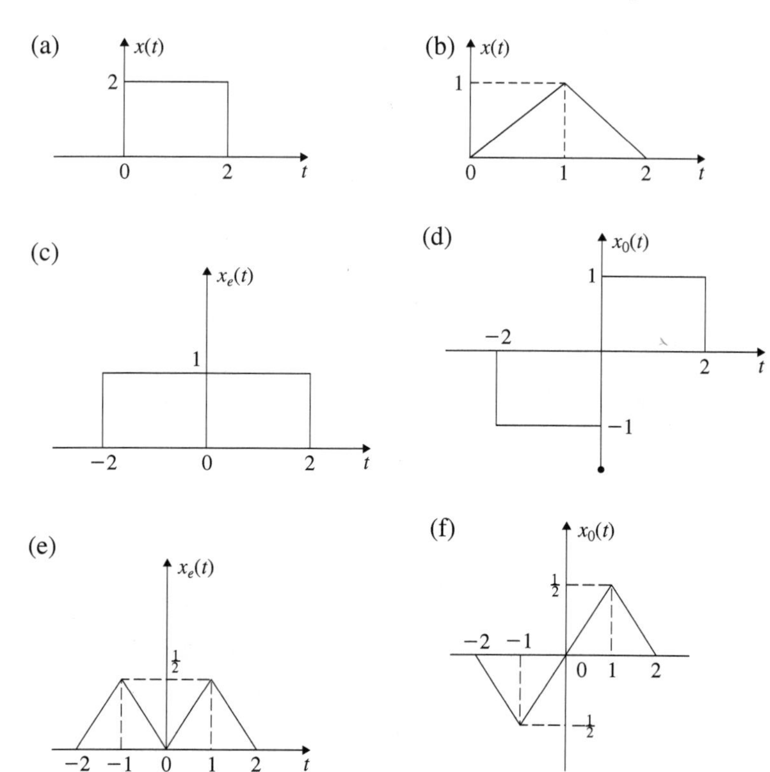

Fig. 1.122 Even and odd signals of CT signals

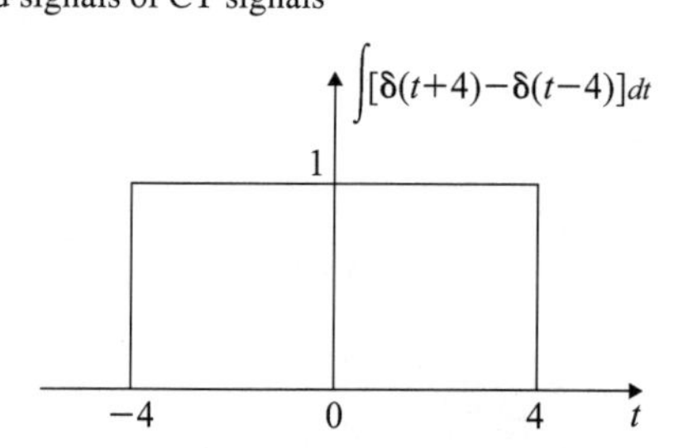

Fig. 1.123 Representation of $x(t) = \int[\delta(t+4) - \delta(t-4)]dt$

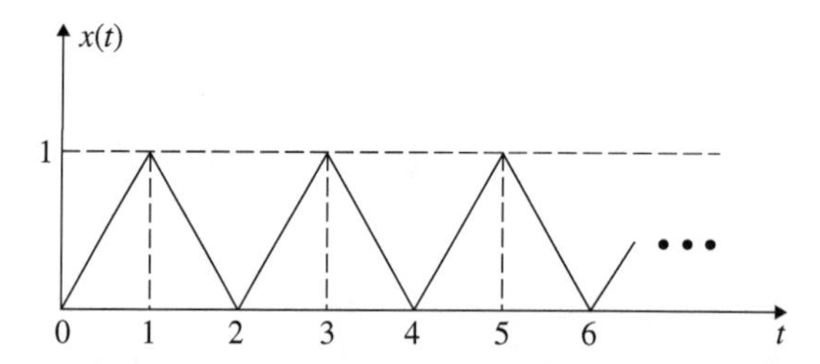

Fig. 1.124 Triangular wave

5. Consider the CT signal $x(t) = \delta(t+4) - \delta(t-4)$. Sketch $\int x(t)dt$ and find the energy of the signal (Fig. 1.123).
   Energy $E = 8$.
6. Find the energy of the following CT signal. (a) $x(t) = \text{tri}3t$; (b) $x(t) = 2\text{tri}(\frac{t}{2})$; (c) $x(t) = \text{rect}10t$; (d) $2\,\text{rect}(\frac{t}{10})$; (e) $\sin(2\pi t)$.
   (a) $E = \frac{2}{9}$; (b) $E = \frac{16}{3}$; (c) $E = \frac{1}{5}$; (d) $E = 80$ and (e) $E = \frac{1}{2}$.
7. What is the average power of the triangular wave shown in Fig. 1.124.
   Average power $P = \frac{1}{3}$ watts.
8. For the following DT signal, find even and odd components

$$x[n] = \{1, -3, 2, 5, 4\}$$

$$x_e[n] = \{2, 2.5, 1, -1.5, \underset{\uparrow}{1}, -1.5, 1, 2.5, 2\}$$

$$x_0[n] = \{-2, -2.5, -1, 1.5, \underset{\uparrow}{0}, -1.5, 1, 2.5, 2\}$$

9. Find whether the following signals are periodic. If periodic, find the fundamental period. (a) $x[n] = \cos(\frac{n}{8} - \pi)$; (b) $x[n] = \cos(\frac{\pi}{8} + \frac{\pi}{2}) + \cos(\frac{\pi}{6} - \frac{\pi}{2})$; (c) $x[n] = \cos(\frac{5\pi N}{12} + \frac{\pi}{2}) + \sin\frac{10\pi n}{8}$ and (d)$x[n] = e^{j3n} + e^{j4\pi n}$.
   (a) Not periodic. (b) Periodic with fundamental period $N_0 = 48$ samples/sec. (c) Periodic with fundamental period $N_0 = 24$ samples/sec. (d) Non-periodic.
10. Given $x[n]$ and $y[n]$

$$\begin{aligned} x[-1] &= 2 \\ x[n] &= 1 \qquad 1 \le n \le 5 \\ x[6] &= \frac{1}{2} \\ &= 0 \qquad \text{for other } n \end{aligned}$$

Plot (a) $x[\frac{n}{2}]$ and (b) $E_v x[n]$. *(Anna University, 2007)*.

(a)

$$x[n] = \{\underset{\uparrow}{2}, 1, 1, 1, 1, 1, 1, 0.5\}$$

$$x\left[\frac{n}{2}\right] = \{\underset{\uparrow}{2}, 0, 1, 0, 1, 0, 1, 0, 1, 0, 1, 0, 1, 0.5\}$$

(b)

$$x_e[n] = \{0.25, 0.5, 0.5, 0.5, 0.5, 0.5, \underset{\uparrow}{1}, 0.5, 0.5, 0.5, 0.5, 0.5, 0.25\}$$

11. Find whether the following signal is periodic. If periodic, find the fundamental period.

$$x[n] = \cos\left(2\pi n + \frac{\pi}{2}\right) + \sin\left(5\pi n - \frac{\pi}{4}\right) + \sin\left(8\pi n + \frac{\pi}{2}\right)$$

The signal is periodic. Their fundamental period $N_0 = 2$ samples/sec.

12. Determine whether the signal $x(t) = 3\cos t + 4\cos(t/3)$ is periodic. If periodic, find the period.
The fundamental period $T = 6\pi$ is irrational and hence, the signal is not periodic.

13. Find the odd and even components of $x[n] = \delta[n]$.

$$x_0[n] = 0; \qquad x_e[n] = \delta[n]$$

14. Find the odd and even components of $x(t) = u(t)$.

$$x_e(t) = \frac{1}{2}; \qquad x_0(t) = \frac{1}{2}\text{sgnt}$$

15. Evaluate $x(t) = \cos\left(\frac{\pi}{8}t\right)\delta(t-4)$.

$$x(t) = 0$$

Chapter 2
Continuous and Discrete Time Systems

Learning Objectives

Under broader category, systems are classified as continuous and discrete time systems. The objectives of the chapters are to further classify them as

- Linear and non-linear systems.
- Time invariant and time varying systems.
- Static and dynamic systems.
- Causal and non-causal systems.
- Stable and unstable systems.
- Invertible and non-invertible systems.
- To define the above properties of the system.
- To illustrate these properties with numerical examples.

2.1 Introduction

A system is an interconnection of objects with a definite relationship with the objects and attributes. Consider a simple R, L, C series electric circuit. The components (objects) R, L and C when connected together form the system. The current flow in the series circuit and the voltages across the elements R, L and C are the attributes. If i is the current flowing in the circuit, the voltage across the resistor R is iR. Thus, the object R and the attribute i have a definite relationship between them. The voltages across any of these objects R, L and C can be taken as the output. Thus, the system when excited by a signal, processes and produces signals as outputs in the same form or in a modified form. Electrical motors, communication systems, automotive vehicles, human body, government, stock markets etc. are examples of systems. The block diagram representation of a system is shown in Fig. 2.1.

S. Palani, *Signals and Systems*,
https://doi.org/10.1007/978-3-030-75742-7_2

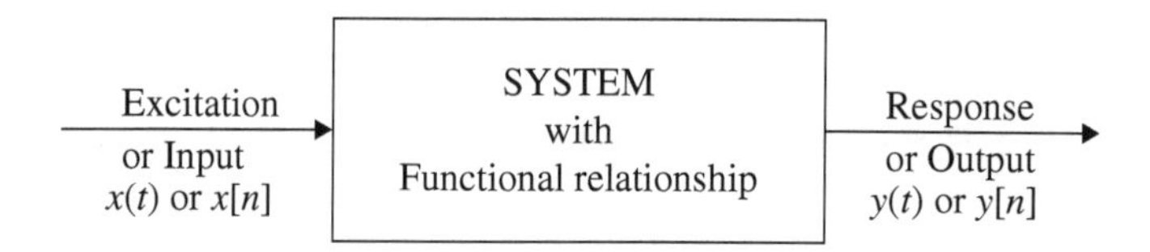

Fig. 2.1 Block diagram representation of system

In Fig. 2.1 the system is excited by the input signal $x(t)$ or $x[n]$. It is being processed by the functional relationship of the system and the response is obtained as $y(t)$ or $y[n]$. The functional relationship includes differential equation or difference equation or the system transfer function which is $H(s)$ for CT system and $H(z)$ for DT system.

2.2 Linear Time Invariant Continuous (LTIC) Time System

The block diagram of a continuous time system is shown in Fig. 2.2a. $x(t)$ is the input signal which is continuous. The system with the functional relationship $H(s)$ produces the output $y(t)$ which is also continuous. The system dynamics or the functional relationship is written in the form of differential equation connecting $x(t)$ and $y(t)$. If the Laplace transform of $x(t)$ and $y(t)$ are $X(s)$ and $Y(s)$ respectively, the system functional relationship is written as

$$\frac{Y(s)}{X(s)} = H(s) \tag{2.1}$$

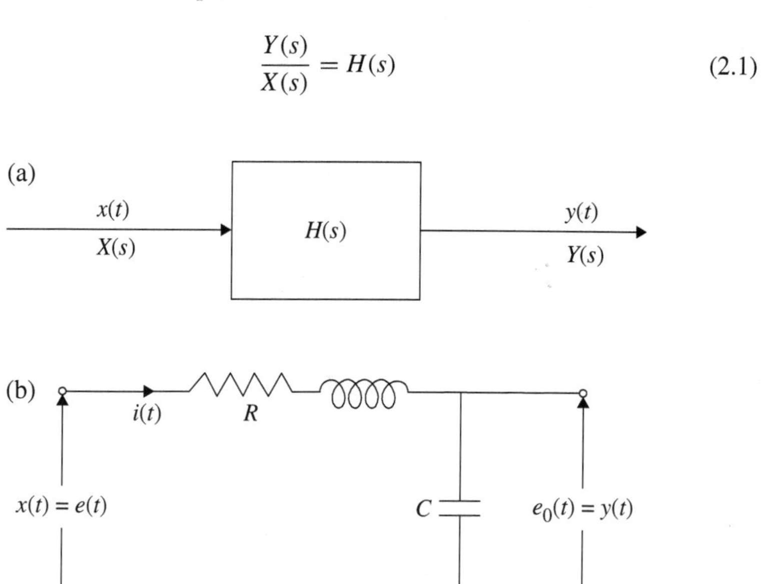

Fig. 2.2 **a** Block diagram of CT system. **b** *R-L-C* series electric circuit

$H(s)$ is called system function or system transfer function.

Consider the electric network shown in Fig. 2.2b. The following dynamic equation is written for Fig. 2.2b:

$$e(t) = Ri(t) + L\frac{di(t)}{dt} + \frac{1}{C}\int i(t)dt \tag{2.2}$$

$$e_0(t) = \frac{1}{C}\int i(t)dt \tag{2.3}$$

In the continuous time system shown in Fig. 2.2b. $e(t)$ is represented by $x(t)$ and $e_0(t)$ is represented by $y(t)$. The system dynamic equations are given in Equations (2.2) and (2.3).

2.3 Linear Time Invariant Discrete (LTID) Time System

Consider the discrete time system represented in block diagram as shown in Fig. 2.3. Here $H[z]$ represents the functional relationship of $x[n]$ and $y[n]$. The input and output sequences $x[n]$ and $y[n]$ occur at only discrete interval of time n where n is an integer. In DT system, the input and output are related by the difference equation which is given below:

$$y[n-2] + a_1 y[n-1] + a_2 y[n] = bx[n] \tag{2.4}$$

In Equation (2.4), $y[n]$ is the output sequence. $y[n-1]$ and $y[n-2]$ are the delayed output at $n = 1$ and $n = 2$ respectively.

2.4 Properties (Classification) of Continuous Time System

The continuous time system possesses the following properties and it is classified accordingly.

1. Linear and non-linear systems.
2. Time invariant and time varying systems.

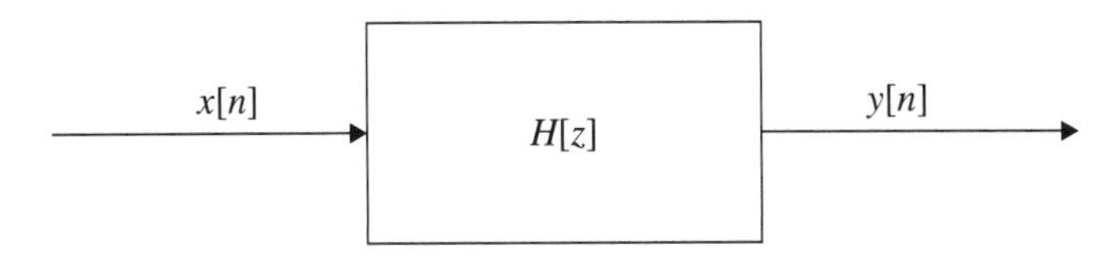

Fig. 2.3 Block diagram of DT system

3. Causal and non-causal systems.
4. Static and dynamic systems (Systems without and with memory).
5. Stable and unstable systems.
6. Invertible and non-invertible systems.

The above properties of LTIC time system are defined, described and illustrated with examples below.

2.4.1 Linear and Non-linear Systems

For a linear system if an input $x_1(t)$ produces an output $y_1(t)$ and another input $x_2(t)$ when applied separately produces an output $y_2(t)$, then when both inputs $x(t) = [x_1(t) + x_2(t)]$ are applied to the system simultaneously will produce an output $y(t) = y_1(t) + y_2(t)$. Thus

$$\begin{aligned} x_1(t) &= y_1(t) \\ x_2(t) &= y_2(t) \\ [x_1(t) + x_2(t)] &= [y_1(t) + y_2(t)] \end{aligned} \tag{2.5}$$

Equation (2.5) obeys the **Additivity** property of superposition theorem. Further, the linear system should also satisfy the **homogeneity** or **scaling** property of superposition theorem. According to this property, if

$$\begin{aligned} a_1x_1(t) &= a_1y_1(t) \\ a_2x_2(t) &= a_2y_2(t) \end{aligned}$$

then,

$$[a_1x_1(t) + a_2x_2(t)] = [a_1y_1(t) + a_2y_2(t)] \tag{2.6}$$

Thus, for a continuous system to be linear, the weighted sum of several inputs produces the weighted sum of outputs. In other words, it should satisfy the homogeneity and additivity properties of superposition theorem. If the above conditions are not satisfied the system is said to be non-linear.

Further it is necessary that for zero input, the output should also be zero for the system to be linear.

Step By Step Procedure to Test Linearity

1. Let

$$y_1(t) = f(x_1(t))$$
$$y_2(t) = f(x_2(t))$$

Find the weighted sum of the output

$$y_3(t) = a_1 y_1(t) + a_2 y_2(t)$$
$$y_3(t) = a_1 f(x_1(t)) + a_2 f(x_2(t))$$

where a_1 and a_2 are called the weights.

2. For the linear combination of input $[a_1 x_1(t) + a_2 x_2(t)]$ find the output for the weighted sum of the input.

$$y_4(t) = f[a_1 x_1(t) + a_2 x_2(t)]$$

3. If

$$y_3(t) = y_4(t)$$

the system is linear. Otherwise the system is non-linear. The following examples, illustrate the method of testing the linearity of continuous time systems.

4. If the output is not zero for zero input, the system will be non-linear.

■ Example 2.1

Consider the following input-output equation of a certain system.

$$y(t) = [2x(t)]^2$$

Determine whether the system is linear or non-linear.

Solution:

$$\begin{aligned} y(t) &= [2x(t)]^2 \\ &= 4x^2(t) \\ y_1(t) &= 4x_1^2(t) \\ y_2(t) &= 4x_2^2(t) \end{aligned}$$

The weighted sum of the output is,

$$\begin{aligned} y_3(t) &= a_1 y_1(t) + a_2 y_2(t) \\ &= 4a_1 x_1^2(t) + 4a_2 x_2^2(t) \end{aligned}$$

The output due to the weighted sum of the input $[a_1 x_1 + a_2 x_2]$ is,

$$\begin{aligned} y_4(t) &= 4[a_1x_1(t) + a_2x_2(t)]^2 \\ &= 4[a_1^2x_1^2(t) + a_2^2x_2^2(t) + 2a_1a_2x_1(t)x_2(t)] \\ y_3(t) &\neq y_4(t) \end{aligned}$$

Hence, the system is non-linear.

■ Example 2.2

Consider the following systems. Determine whether each of them is linear.

$$\begin{aligned} &\text{(a)} \quad y(t) = 5x(t)\sin 10t \\ &\text{(b)} \quad y(t) = 3x(t) + 5 \\ &\text{(c)} \quad y(t) = t^2x(t+1) \\ &\text{(d)} \quad y(t) = E_v x(t) \\ &\text{(e)} \quad y(t) = x(t^2) \\ &\text{(f)} \quad y(t) = \int_{-\infty}^{t} 10x(\tau)d\tau \\ &\text{(g)} \quad y(t) = e^{-2x(t)} \\ &\text{(h)} \quad y(t) = x(t-7) - x(5-t) \end{aligned}$$

Solution:

(a) $\boldsymbol{y(t) = 5x(t)\sin 10t}$

$$\begin{aligned} y_1(t) &= 5x_1(t)\sin 10t \\ y_2(t) &= 5x_2(t)\sin 10t \end{aligned}$$

The weighted sum of the output is,

$$y_3(t) = a_1y_1(t) + a_2y_2(t) = 5\sin 10t(a_1x_1(t) + a_2x_2(t))$$

The output due to the weighted sum of the input is,

$$\begin{aligned} y_4(t) &= 5\sin 10t(a_1x_1(t) + a_2x_2(t)) \\ y_3(t) &= y_4(t) \end{aligned}$$

The system is Linear.

(b) $\boldsymbol{y(t) = 3x(t) + 5}$

$$\begin{aligned} y_1(t) &= 3x_1(t) + 5 \\ y_2(t) &= 3x_2(t) + 5 \\ y_3(t) &= a_1y_1(t) + a_2y_2(t) \\ &= 3(a_1x_1(t) + a_2x_2(t)) + 5(a_1 + a_2) \\ y_4(t) &= 3(a_1x_1(t) + a_2x_2(t)) + 5 \\ y_3(t) &\neq y_4(t) \end{aligned}$$

Further, if $x(t) = 0$, $y(t) = 5$ and not zero.

The system is Non-linear.

(c) $\boldsymbol{y(t) = t^2x(t + 1)}$

$$\begin{aligned} y_1(t) &= t^2x_1(t + 1) \\ y_2(t) &= t^2x_2(t + 1) \\ y_3(t) &= a_1y_1(t) + a_2y_2(t) \\ &= t^2[a_1x_1(t + 1) + a_2x_2(t + 1)] \\ y_4(t) &= t^2[a_1x_1(t + 1) + a_2x_2(t + 1)] \\ y_3(t) &= y_4(t) \end{aligned}$$

The system is Linear.

(d) $\boldsymbol{y(t) = E_v x(t)}$

$$\begin{aligned} y(t) &= \frac{1}{2}[x(t) + x(-t)] \\ y_1(t) &= \frac{1}{2}[x_1(t) + x_1(-t)] \\ y_2(t) &= \frac{1}{2}[x_2(t) + x_2(-t)] \end{aligned}$$

The weighted sum of the output is,

$$\begin{aligned} y_3(t) &= a_1y_1(t) + a_2y_2(t) \\ &= \frac{1}{2}[a_1x_1(t) + a_2x_2(t) + a_1x_1(-t) + a_2x_2(-t)] \end{aligned}$$

The output due to the weighted sum of the input is,

$$y_4(t) = \frac{1}{2}[a_1(x_1(t) + x_1(-t)) + a_2(x_2(t) + x_2(-t))]$$
$$= \frac{1}{2}[a_1x_1(t) + a_2x_2(t) + a_1x_1(-t) + a_2x_2(-t)]$$
$$y_3(t) = y_4(t)$$

The system is Linear.

(e) $\boldsymbol{y(t) = x(t^2)}$

$$y_1(t) = x_1(t^2)$$
$$y_2(t) = x_2(t^2)$$

The weighted sum of the output is,

$$y_3(t) = a_1y_1(t) + a_2y_2(t)$$
$$= a_1x_1(t^2) + a_2x_2(t^2)$$

The output due to the weighted sum of the input is,

$$y_4(t) = a_1x_1(t^2) + a_2x_2(t^2)$$
$$y_3(t) = y_4(t)$$

The system is Linear.

(f) $\boldsymbol{y(t) = 10 \int_{-\infty}^{t} x(\tau)d\tau}$

$$y_1(t) = 10 \int_{-\infty}^{t} x_1(\tau)d\tau$$
$$y_2(t) = 10 \int_{-\infty}^{t} x_2(\tau)d\tau$$

The weighted sum of the output is,

$$y_3(t) = a_1y_1(t) + a_2y_2(t)$$
$$= 10\left[a_1 \int_{-\infty}^{t} x_1(\tau)d\tau + a_2 \int_{-\infty}^{t} x_2(\tau)d\tau\right]$$

The output due to the weighted sum of the input is,

$$y_4(t) = 10\left[\int_{-\infty}^{t} \{a_1x_1(\tau) + a_2x_2(\tau)\}\, d\tau\right]$$
$$= 10\left[\left\{\int_{-\infty}^{t} a_1x_1(\tau)d\tau + \int_{-\infty}^{t} a_2x_2(\tau)d\tau\right\}\right]$$
$$y_3(t) = y_4(t)$$

The system is Linear.

(g) $\boldsymbol{y(t) = e^{-2x(t)}}$

For $x(t) = 0, y(t) = 1$ and not zero. Hence the system is non-linear. Also

$$y_1(t) = e^{-2x_1(t)}$$
$$y_2(t) = e^{-2x_2(t)}$$
$$y_3(t) = a_1y_1(t) + a_2y_2(t) = a_1e^{-2x_1(t)} + a_2e^{-2x_2(t)}$$
$$y_4(t) = e^{-2(a_1x_1(t)+a_2x_2(t))} = e^{-2a_1x_1(t)}e^{-2a_2x_2(t)}$$
$$y_3(t) \neq y_4(t)$$

The system is Non-linear.

(h) $\boldsymbol{y(t) = x(t-7) - x(5-t)}$

$$y_1(t) = x_1(t-7) - x_1(5-t)$$
$$y_2(t) = x_2(t-7) - x_2(5-t)$$

The weighted sum of the output is,

$$y_3(t) = a_1y_1(t) + a_2y_2(t)$$
$$= a_1[x_1(t-7) - x_1(5-t)] + a_2[x_2(t-7) - x_2(5-t)]$$

The output due to the weighted sum of the input is,

$$y_4(t) = a_1[x_1(t-7) - x_1(5-t)] + a_2[x_2(t-7) - x_2(5-t)]$$
$$y_3(t) = y_4(t)$$

The system is Linear.

Linearity Test for the System Described by Differential Equation

Step 1. Write down the system differential equation with responses $y_1(t)$ and $y_2(t)$ for the inputs $x_1(t)$ and $x_2(t)$ respectively.
Step 2. Multiply the $y_1(t)$ response equation with a_1 and $y_2(t)$ response equation with a_2 and add them.
Step 3. Write down the differential equation for the sum of the inputs $x(t) = a_1x_1(t) + a_2x_2(t)$.
Step 4. If $y(t) = a_1y_1(t) + a_2y_2(t)$ obtained in Steps 2 and 3 are same, the given differential equation is linear. Otherwise the differential equation is non-linear.

The following examples illustrate the above method.

Example 2.3

Determine whether the following differential equations are linear or non-linear:

$$\begin{aligned}
&\text{(a)} \qquad \frac{dy(t)}{dt} + 10y(t) = x(t)\\
&\text{(b)} \qquad \frac{dy(t)}{dt} + 10\sin y(t) = 2x(t)\\
&\text{(c)} \qquad y(t)\frac{dy(t)}{dt} + 10y(t) = 2x(t)\\
&\text{(d)} \qquad \frac{dy(t)}{dt} + 5y(t) = x(t)\frac{dx(t)}{dt}\\
&\text{(e)} \qquad \frac{dy(t)}{dt} + 5y(t) = x^2(t)\\
&\text{(f)} \qquad \frac{5dy(t)}{dt} + 7y(t) + 15 = x(t)
\end{aligned}$$

Solution:

(a)

$$\frac{dy(t)}{dt} + 10y(t) = x(t)$$

Let y_1 be the output response due to the input x_1 and y_2 be the response due to the input x_2. Thus we may write the output responses due to the weights a_1 and a_2 as

$$\begin{aligned}
y_1 &\rightarrow x_1\\
a_1y_1 &\rightarrow a_1x_1\\
y_2 &\rightarrow x_2\\
a_2y_2 &\rightarrow a_2x_2
\end{aligned}$$

The output response due to the weight a_1 is

$$\frac{d(a_1 y_1)}{dt} + 10a_1 y_1 = a_1 x_1$$

The output response due to the weight a_2 is

$$\frac{d(a_2 y_2)}{dt} + 10a_2 y_2 = a_2 x_2$$

The weighted sum of the response due to each input signal is,

$$\frac{d}{dt}[a_1 y_1(t)] + 10a_1 y_1(t) = a_1 x_1(t)$$
$$\frac{d}{dt}[a_2 y_2(t)] + 10a_2 y_2(t) = a_2 x_2(t)$$

Adding the above two equations we get the weighted sum of the output as

$$\frac{d}{dt}[a_1 y_1(t) + a_2 y_2(t)] + 10[a_1 y_1(t) + a_2 y_2(t)] = [a_1 x_1(t) + a_2 x_2(t)] \quad (a)$$

The response of the system due to weighted sum of input is given as,

$$a_1 \frac{dy_1(t)}{dt} + a_2 \frac{dy_2(t)}{dt} + 10[a_1 y_1(t) + a_2 y_2(t)] = [a_1 x_1(t) + a_2 x_2(t)]$$

$$\frac{d}{dt}[a_1 y_1(t) + a_2 y_2(t)] + 10[a_1 y_1(t) + a_2 y_2(t)] = [a_1 x_1(t) + a_2 x_2(t)] \quad (b)$$

Equations (a) and (b) are same. Also Eq. (a) or (b) is identical to the system equation

$$\frac{dy(t)}{dt} + 10y(t) = x(t)$$

with the input

$$x(t) = a_1 x_1(t) + a_2 x_2(t)$$

and the output

$$y(t) = a_1 y_1(t) + a_2 y_2(t)$$

Therefore, when the input $[a_1 x_1(t) + a_2 x_2(t)]$ is applied, the system output response is $[a_1 y_1(t) + a_2 y_2(t)]$. Hence the system is linear. Also, when the input is zero, the system response is obtained from

$$\frac{dy(t)}{dt} + 10y(t) = 0$$

The above equation, when solved with zero initial conditions, $y(t) = 0$. For zero input, the output is also zero.

The system is Linear.

(b) $\frac{dy(t)}{dt} + 10 \sin y(t) = 2x(t)$

$$\frac{dy(t)}{dt} + 10 \sin y(t) = 2x(t)$$

The weighted sum of responses due to $a_1x_1(t)$ and $a_2x_2(t)$ are,

$$\frac{d}{dt}[a_1y_1(t)] + 10 \sin a_1y(t) = 2a_1x_1(t)$$
$$\frac{d}{dt}[a_2y_2(t)] + 10 \sin a_2y_2(t) = 2a_2x_2(t)$$

The weighted sum of the responses is obtained by adding the above two equations.

$$\frac{d}{dt}[a_1y_1(t) + a_2y_2(t)] + 10 \sin a_1y_1(t) + 10 \sin a_2y_2(t) = 2[a_1x_1(t) + a_2x_2(t)] \quad (a)$$

The output response due to weighted sum of inputs $x(t) = a_1x_1(t) + a_2x_2(t)$ is,

$$a_1\frac{d}{dt}y_1(t) + a_2\frac{d}{dt}y_2(t) + 10a_1 \sin y_1(t) + 10a_2 \sin y_2(t) = 2[a_1x_1(t) + a_2x_2(t)]$$

$$\frac{d}{dt}[a_1y_1(t) + a_2y_2(t) + 10[a_1 \sin y_1(t) + a_2 \sin y_2(t)] = 2[a_1x_1(t) + a_2x_2(t)] \quad (b)$$

Equations (a) and (b) are not the same. Hence, it is not linear.

The system is Non-linear.

(c) $y(t)\frac{dy(t)}{dt} + 10y(t) = 2x(t)$

$$y(t)\frac{dy(t)}{dt} + 10y(t) = 2x(t)$$

The weighted output responses due to inputs $a_1x_1(t)$ and $a_2x_2(t)$ are,

$$a_1y_1(t)\frac{d}{dt}[a_1y_1(t)] + 10a_1y_1(t) = 2a_1x_1(t)$$

$$a_2y_2(t)\frac{d}{dt}[a_2y_2(t)] + 10a_2y_2(t) = 2a_2x_2(t)$$

The sum of the weighted response is due to $x(t) = a_1x_1(t) + a_2x_2(t)$ is obtained by adding the above two equations.

$$a_1^2y_1(t)\frac{d}{dt}[y_1(t)] + a_2^2y_2(t)\frac{d}{dt}[y_2(t)] + 10[a_1y_1(t) + a_2y_2(t)]=2[a_1x_1(t) + a_2x_2(t)] \quad (a)$$

The response due to weighted sum of inputs $x(t) = a_1x_1(t) + a_2x_2(t)$ is,

$$a_1y_1(t)\frac{d}{dt}y_1(t) + 10a_1y_1(t) + a_2y_2(t)\frac{d}{dt}y_2(t) + 10a_2y_2(t) = 2[a_1x_1(t) + a_2x_2(t)]$$

$$a_1y_1(t)\frac{d}{dt}y_1(t) + a_2y_2(t)\frac{d}{dt}y_2(t) + 10[a_1y_1(t) + a_2y_2(t)] = 2[a_1x_1(t) + a_2x_2(t)] \quad (b)$$

Equations (a) and (b) are not equal. Hence, the system is not linear.

The system is Non-linear.

(d) $\frac{dy(t)}{dt} + 5y(t) = x(t)\frac{dx(t)}{dt}$.
The output response of the system due to the weight a_1 is

$$\frac{d(a_1y_1)}{dt} + 5a_1y_1 = a_1x_1\frac{dx_1}{dt}$$

The output response of the system due to the weight a_2 is

$$\frac{d(a_2y_2)}{dt} + 5a_2y_2 = a_2x_2\frac{dx_2}{dt}$$

The weighted sum of the responses due to the above weight is obtained adding the above two equations and is given below as,

$$\frac{d}{dt}(a_1y_1 + a_2y_2) + 5(a_1y_1 + a_2y_2) = a_1x_1\frac{dx_1}{dt} + a_2x_2\frac{dx_2}{dt} \quad (a)$$

The output of the system due to the input with weight a_1 is

$$a_1\frac{dy_1}{dt} + 5a_1y_1 = a_1x_1\frac{dx_1}{dt}$$

The output of the system due to the input with weight a_2 is

$$a_2\frac{dy_2}{dt} + 5a_2y_2 = a_2x_2\frac{dx_2}{dt}$$

The output due to the sum of the weighted input is obtained by adding the above two equations

$$\frac{d}{dt}(a_1y_1 + a_2y_2) + 5(a_1y_1 + a_2y_2) = a_1x_1\frac{dx_1}{dt} + a_2x_2\frac{dx_2}{dt} \qquad (b)$$

Equation (a) = Equation (b)

The output response due to the weights a_1 and a_2 is the same when the input is given the same weights. Further for $x(t) = 0, y(t) = 0$ which can be obtained by solving the equation

$$\frac{dy(t)}{dt} + 5y(t) = 0$$

with zero initial conditions.

The system is Linear.

It is to be noted that when the system is described by the differential equation, the system linearity is decided by the differential equation describing the dynamics of the system and it is independent of the input.

(e) $\boldsymbol{\frac{dy(t)}{dt} + 5y(t) = x^2(t)}$.

The output response of the system due to the weights a_1 and a_2 are written as

$$\frac{d(a_1y_1)}{dt} + 5a_1y_1 = a_1x_1^2 \quad \text{and} \quad \frac{d(a_2y_2)}{dt} + 5a_2y_2 = a_2x_2^2$$

Adding the above two equations we get,

$$\frac{d}{dx}(a_1y_1 + a_2y_2) + 5(a_1y_1 + a_2y_2) = a_1x_1^2 + a_2x_2^2 \qquad (a)$$

The outputs of the system due to the input with weights a_1 and a_2 are given by

$$a_1\frac{dy_1}{dt} + 5a_1y_1 = a_1x_1^2 \quad \text{and} \quad a_2\frac{dy_2}{dt} + 5a_2y_2 = a_2x_2^2$$

The output due to the sum of the weighted inputs is obtained by adding the above two equations

$$\frac{d}{dx}(a_1y_1 + a_2y_2) + 5(a_1y_1 + a_2y_2) = a_1x_1^2 + a_2x_2^2 \qquad (b)$$

Equation (a) = Equations (b)

Further for $x(t) = 0, y(t) = 0$ which can be obtained by solving the equation

$$\frac{dy(t)}{dt} + 5y(t) = 0$$

with zero initial conditions.

The system is Linear.

(f) $\frac{5dy}{dt} + 7y + 15 = x$.
Let y_1 be the response due to x_1 and y_2 be the response due to x_2

$$y_1 \rightarrow x_1 \quad \text{and} \quad y_2 \rightarrow x_2$$

The output response y_1 due to the weight a_1 is

$$5\frac{d(a_1y_1)}{dx} + 7a_1y_1 + 15 = a_1x_1$$

The output response y_2 due to the weight a_2 is

$$5\frac{d(a_2y_2)}{dx} + 7a_2y_2 + 15 = a_2x_2$$

Adding the above two equations we get

$$5\frac{d(a_1y_1 + a_2y_2)}{dx} + 7(a_1y_1 + a_2y_2) + 15 = (a_1x_1 + a_2y_2) \qquad (a)$$

Let a_1 weight be given to the input x_1. The response y_1 due to this weight is obtained from

$$5a_1\frac{dy_1}{dx} + 7a_1y_1 + 15a_1 = a_1x_1$$

Let a_2 weight be given to the input x_2. The response y_2 due to this weight is obtained from

$$5a_2\frac{dy_2}{dx} + 7a_2y_2 + 15a_2 = a_2x_2$$

The output response equation due to the weighted sum of the inputs is obtained by adding the above two equations.

$$5\frac{d}{dx}(a_1y_1 + a_2y_2) + 7(a_1y_1 + a_2y_2) + 15(a_1 + a_2) = (a_1x_1 + a_2y_2) \qquad (b)$$

Equation (a) is not equal to Eq. (b). Thus the weighted sum of the output is not equal to the output due to the weighted sum of the input. Equation (b) is not the same as the original equation.

The system is Non-linear.

2.4.2 Time Invariant and Time Varying Systems

A continuous time system is said to be time invariant if the parameters of the system do not change with time. The characteristics of such system are fixed over a time. The input-output of a certain continuous time system is shown in (2.4) a and b respectively. If the input is delayed by t_0 seconds, the characteristic of the output response remains the same but delayed by t_0 seconds. This is illustrated in Fig. 2.4c and d respectively. This property is also illustrated in Fig. 2.4e and f in block dia-

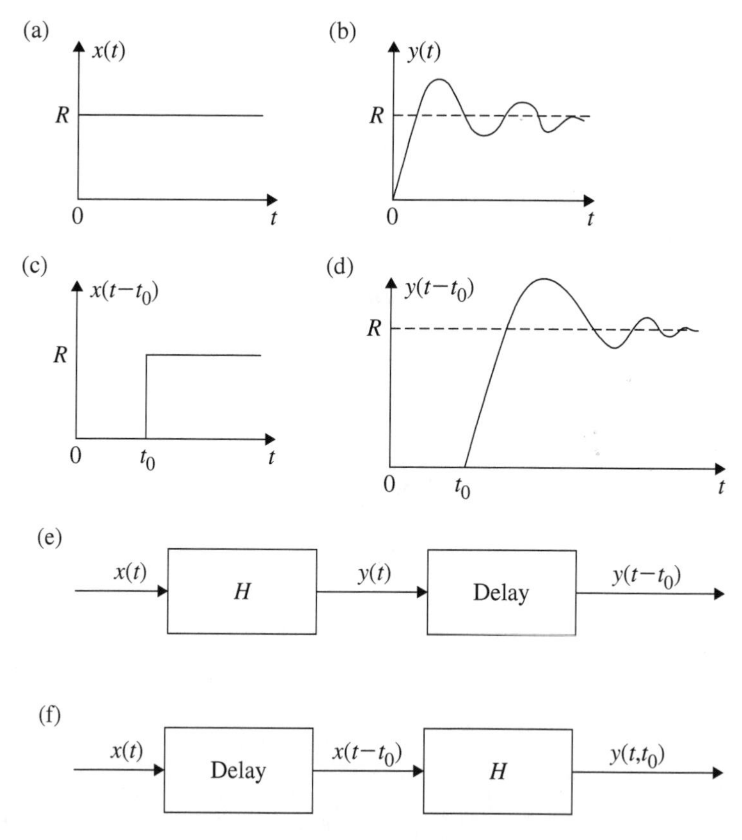

Fig. 2.4 Time invariancy property

gram form. In Fig. 2.4e the output $y(t)$ of the system H is delayed by t_0 seconds to get $y(t - t_0)$ as the delayed output. The delayed output $y(t - t_0)$ of system H can also be obtained by delaying the input $x(t)$ as $x(t - t_0)$. This is illustrated in Fig. 2.4f. This time delay of the system commutes only if the system is time invariant. The above property will not apply if the system is time varying which can be easily proved. Thus, to identify the time invariant system, the following steps are followed:

Step 1. For the delayed input $x(t - t_0)$ obtain the output $y(t, t_0)$.
Step 2. Obtain the expression for the delayed output $y(t - t_0)$ by substituting $t = (t - t_0)$.
Step 3. If $y(t, t_0) = y(t - t_0)$, then the system is time invariant. Otherwise it is a time varying system.

The following examples illustrate the method of identifying time invariancy.

■ Example 2.4

Check whether the following systems are time invariant or not:

(a) $y(t) = tx(t)$

(b) $y(t) = \cos x(t)$

(c) $y(t) = x(t) \cos x(t)$

(d) $y(t) = e^{-2x(t)}$

(e) $\frac{d^2}{dt} y(t) + 2\frac{d}{dt} y(t) + 5y(t) = x(t)$

(f) $\frac{d^2}{dt} y(t) + 2t\frac{d}{dt} y(t) + 5y(t) = x(t)$

(g) $y(t) = \left[\frac{dx(t)}{dt}\right]^2$

(h) $y(t) = at^2 x(t) + btx(t - 2)$

(*Anna University, 2013*)

Solution:

(a) $\boldsymbol{y(t) = tx(t)}$

1. For the delayed input $x(t - t_0)$, the output $y(t, t_0)$ is obtained as

$$y(t, t_0) = tx(t - t_0)$$

2. The delayed output $y(t - t_0)$ is obtained by substituting $t = t - t_0$ in the given equation

$$y(t - t_0) = (t - t_0)x(t - t_0)$$

3. $y(t-t_0) \neq y(t,t_0)$
4.

The system is Time Varying.

(b) $\boldsymbol{y(t) = \cos x(t)}$

1. $y(t,t_0) = \cos x(t-t_0)$ [For Delayed input]
2. $y(t-t_0) = \cos x(t-t_0)$ [Delayed output]
3. $y(t-t_0) = y(t,t_0)$
4.

The system is Time Invariant.

(c) $\boldsymbol{y(t) = x(t)\cos x(t)}$

1. $y(t,t_0) = x(t-t_0)\cos x(t-t_0)$ [For Delayed input]
2. $y(t-t_0) = x(t-t_0)\cos x(t-t_0)$ [Delayed output]
3. $y(t-t_0) = y(t,t_0)$
4.

The system is Time Invariant.

(d) $\boldsymbol{y(t) = e^{-2x(t)}}$

1. The output due to delayed input is,

$$y(t,t_0) = e^{-2x(t-t_0)}$$

2. The delayed output is obtained by putting $t = t - t_0$

$$y(t-t_0) = e^{-2x(t-t_0)}$$

3. $y(t-t_0) = y(t,t_0)$
4.

The system is Time Invariant.

(e) $\boldsymbol{\frac{d^2}{dt}y(t) + 2\frac{d}{dt}y(t) + 5y(t) = x(t)}$

The coefficients of the given differential equation are 1, 2 and 5 and they are constants. They do not vary with time. Hence

The system is Time Invariant.

(f) $\frac{d^2}{dt^2}y(t) + 2t\frac{d}{dt}y(t) + 5y(t) = x(t)$

The coefficient of $\frac{dy(t)}{dt}$ is $2t$ and it varies with respect to time. Hence

$$\boxed{\text{The system is Time Varying.}}$$

(g) $y(t) = \left[\frac{d}{dt}x(t)\right]^2$

1. For the delayed input $x(t - t_0)$ the output is obtained as

$$y(t, t_0) = \left[\frac{d}{dt}x(t - t_0)\right]^2$$

2. The delayed output is obtained by putting $t = t - t_0$ in the given equation

$$y(t - t_0) = \left[\frac{d}{dt}x(t - t_0)\right]^2$$

3. $y(t - t_0) = y(t, t_0)$
4.

$$\boxed{\text{The system is Time Invariant.}}$$

(h) $y(t) = at^2x(t) + btx(t - 2)$

The output $y(t, t_0)$ due to the delayed input $x(t - t_0)$ is

$$y(t, t_0) = at^2x(t - t_0) + btx(t - t_0 - 2) \qquad (a)$$

The delayed output $y(t - t_0)$ is obtained by substituting $t = t - t_0$.

$$y(t - t_0) = a(t - t_0)^2x(t - t_0) + b(t - t_0)x(t - t_0 - 2) \qquad (b)$$

From equations (a) and (b) we see

$$y(t, t_0) \neq y(t - t_0)$$

$$\boxed{\text{The system is Time Varying.}}$$

2.4.3 Static and Dynamic Systems (Memoryless and System with Memory)

Consider the R-C series electrical circuit shown in Fig. 2.4a. The charge in the capacitor is determined by the current that has flown through it. By this mechanism the capacitor remembers about some thing about its past. Similarly consider the mechanical system in Fig. 2.4b. The stored energy in the mechanical spring depends on the past history of the applied force. The present response of these systems which have energy storing elements **depends not only on the present excitation but also on the past excitation** which are remembered by these elements. Such systems are called dynamic systems or systems with memory.

Consider the electrical network of Fig. 2.4a in which only a resistor is connected. The current flowing through the resistor depends on the present value of the excitation. The response does not depend on the excitation at any other time. Such systems which have no energy storing elements are called static systems or systems without memory.

A dynamic system is, therefore, defined as a system in which the output signal at any specified time depends on the values of the input signals at the specific time at other time also.

A static system is defined as a system in which the output signal at any specified time depends on the present value of the input signal alone. Static system is also called as instantaneous system

Consider the input $x(t)$ and output $y(t)$ at $t = 0$ as represented in Fig. 2.4(c). If the output at any instant of time depends upon the input which occurs at the same instant of time without any deviation, $(t_0 = 0)$ the input is called the present input. If the output at any instant t_0 depends only on the value of the input $x(t)$ for $t < t_0$ the input with respect to the output is called as past input. On the other hand if $t > t_0$ then the input is called future input. Thus we have

$$\begin{array}{|l|}\hline t = t_0, \text{ Present input.} \\ t < t_0, \text{ Past input.} \\ t > t_0, \text{ Future input.} \\ \hline\end{array}$$

The following examples illustrate the method of identifying static and dynamic systems.

Example 2.5

Determine whether the following systems are static or dynamic:

$$\begin{aligned} &\text{(a)} \quad y(t) = x(t+1) + 5 \\ &\text{(b)} \quad y(t) = x(t^2) \\ &\text{(c)} \quad y(t) = x(t) \sin 2t \end{aligned}$$

(d) $y(t) = x(t-3) + x(3-t)$

(e) $y(t) = x\left(\frac{t}{4}\right)$

(f) $y(t) = \int_{-\infty}^{t} x(\tau)d\tau$

(g) $\frac{dy(t)}{dt} + 5y(t) = 2x(t)$

(h) $y(t) = 2x(t) + 3$

(i) $y(t) = e^{-2x(t)}$

Solution:

(a) $\boldsymbol{y(t) = x(t+1) + 5}$

$$y(0) = x(1) + 5$$

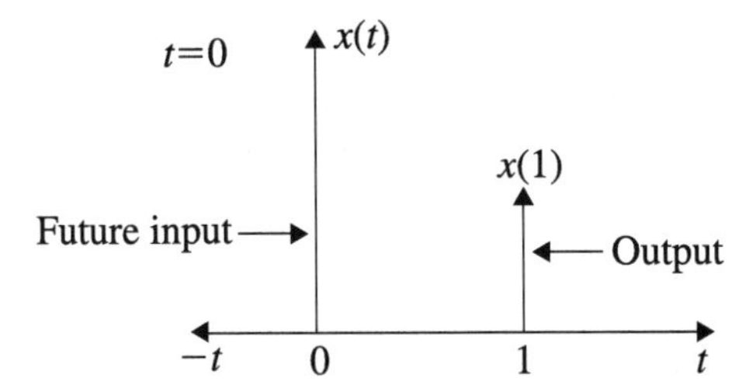

The system response depends on the future input $x(t+1)$ where $t > t_0$. Hence

The system is Dynamic.

(b) $\boldsymbol{y(t) = x(t^2)}$

For $t = 1$,

$$y(1) = x(1) \quad [t = t_0 \text{ Present input}]$$

For $t = 2$,

$$y(2) = x(4)$$

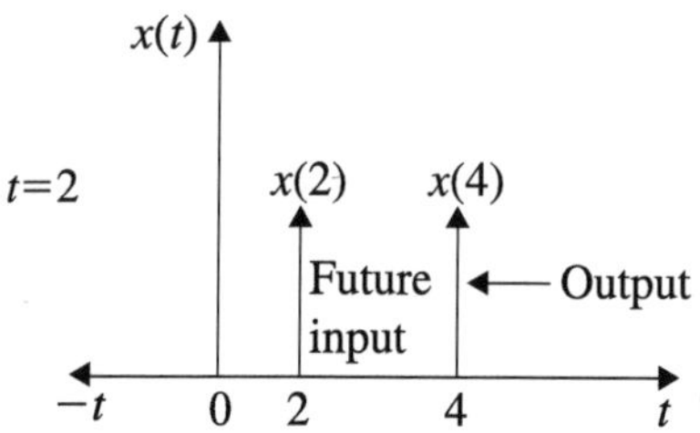

The response depends on the present and future inputs. The output $x(4)$ depends upon the future input $x(2)$. Hence

The system is Dynamic.

(c) $\boldsymbol{y(t) = x(t) \sin 2t}$

The system response depends on the present value of the input $x(t)$. Due to $\sin 2t$, only its magnitude varies from -1 to $+1$. Hence, the output depends upon the present input since $y(1) = x(1) \sin 2$.

The system is Static.

(d) $\boldsymbol{y(t) = x(t-3) + x(3-t)}$

For $t = 0$,

$$y(0) = x(-3) + x(3)$$

Consider the output

$$y(0) = x(-3) + x(3)$$

The input-output are represented below.

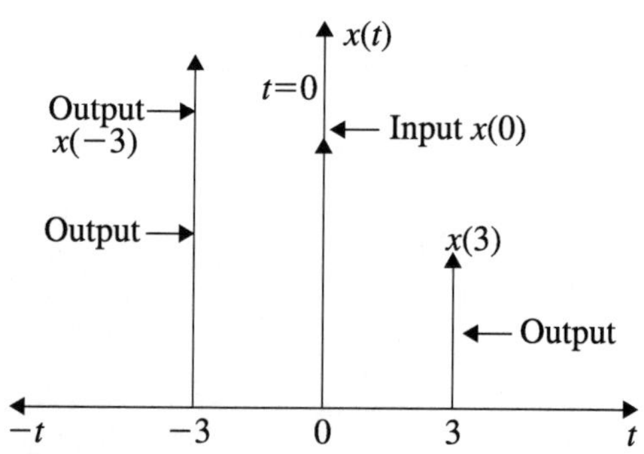

The output $x(-3)$ depends upon the past input. The output $x(3)$ depends on Future input. The system response depends on past and future values of input. Hence

The system is Dynamic.

(e) $\boldsymbol{y(t) = x\left(\frac{t}{4}\right)}$

$$\text{For } t = 0, \quad y(0) = x(0)$$
$$\text{For } t = 1, \quad y(1) = x\left(\frac{1}{4}\right)$$
$$\text{For } t = -1, \quad y(-1) = x\left(-\frac{1}{4}\right)$$

Consider the output $y(0) = x(0)$ for $t = 0$. The output depends upon the present input. Consider the output at $t = 1$

$$y(1) = x\left(\frac{1}{4}\right)$$

The input-output are shown below.

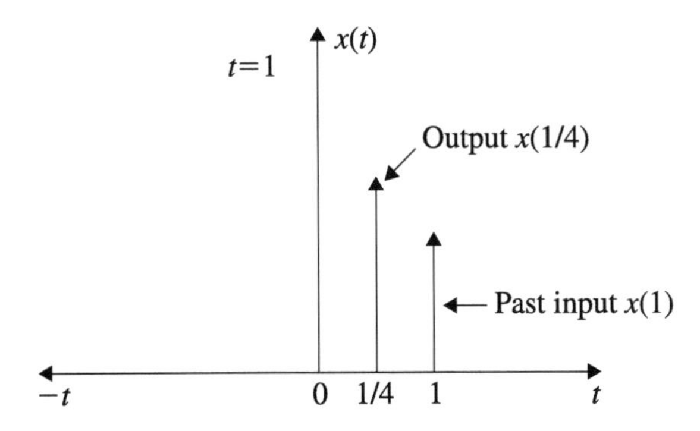

The output $x(1/4)$ depends upon the past input $x(1)$. Now consider the output at $t = -1$

$$y(-1) = x\left(-\frac{1}{4}\right)$$

The input-output are represented below.

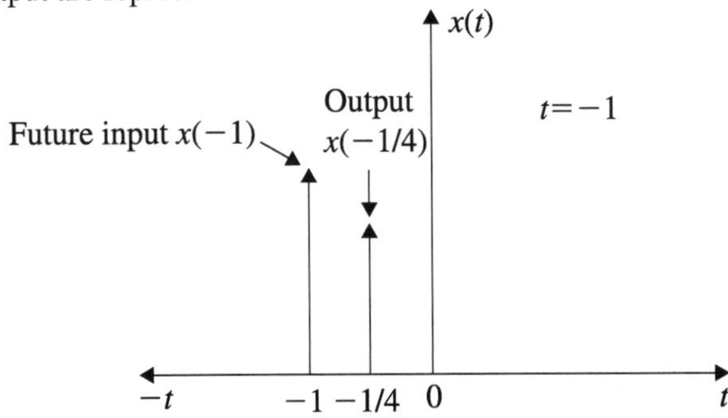

The output $x(-1/4)$ depends upon future input $x(-1)$. The system response depends on present, future and past values of input. Hence

The system is Dynamic.

(f) $\boldsymbol{y(t) = \int_{-\infty}^{t} x(\tau)d\tau}$
By integrating the input, the output is retained and stored in a memory from time t to the infinite past. Hence

The system is Dynamic.

(g) $\boldsymbol{\frac{dy(t)}{dt} + 5y(t) = 2x(t)}$
The input-output is described by a first order differential equation. It requires an energy storing element which remembers the past history of the input applied. Hence

The system is Dynamic.

(h) $\boldsymbol{y(t) = 2x(t) + 3}$
The output always depends on the present input. Hence

The system is Static.

(i) $\boldsymbol{y(t) = e^{-2x(t)}}$
The output always depends on the present input only. Hence

The system is Static.

2.4.4 Causal and Non-causal Systems

Consider a continuous time system excited by the signal $x(t)$. **If the response (output) depends on the present and past values of the input $x(t)$, the system is said to be causal**. In a causal signal, the output cannot start before the input is applied. Hence, the causal system is also called **non-anticipative system**. On the other hand, if the system acts on the knowledge of future input, before it is being applied such systems are called **anticipative or non-causal systems**. Real time systems are all causal systems.

Consider the system described by the following input-output equation

$$y(t) = x(t-3) + x(t+3) \tag{2.7}$$

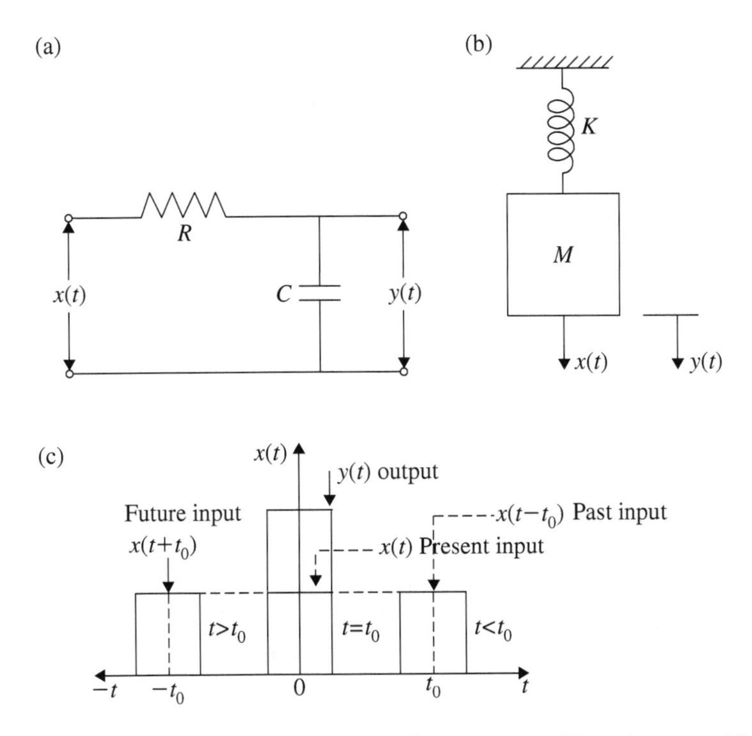

Fig. 2.5 **a**, **b** Dynamic systems. **c** Representation of present, past and future inputs graphically

For the input shown in Fig. 2.5a, the output $y(t)$ is sketched and shown in Fig. 2.5b. The output $y(t)$ at time t is given by the sum of the input values at $(t - 3)$ which is 3 second before and at $(t + 3)$ which is 3 second after. This is illustrated in Fig. 2.5b. Here the system responds to the future input $x(t + 3)$ and it is non-causal system and cannot be realizable in real time. The following examples illustrate the method of identifying causal and non-causal systems.

■ Example 2.6

Consider the continuous time systems described below by their input-output equations. Identify whether they are causal or non-causal.

$$\text{(a)} \quad y(t) = x\left(\frac{t}{4}\right)$$

$$\text{(b)} \quad y(t) = x(t)\sin(1 + t)$$

$$\text{(c)} \quad y(t) = x(t^2)$$

$$\text{(d)} \quad y(t) = x(\sqrt{t})$$

$$\text{(e)} \quad y(t) = x(t + 1)$$

$$\text{(f)} \quad y(t) = x(t - 1)$$

(g) $$y(t) = \frac{d}{dt}x(t)$$

(h) $$y(t) = \int_{t-4}^{t+4} x(\tau)d\tau$$

Solution:

(a) $\boldsymbol{y(t) = x\left(\frac{t}{4}\right)}$

$$\text{For } t = 0, \quad y(0) = x(0)$$
$$\text{For } t = -4, \quad y(-4) = x(-1)$$
$$\text{For } t = 1, \quad y(1) = x\left(\frac{1}{4}\right)$$

In the above three cases, the input-output represented as given below. The output depends upon the present, future and past inputs. Since the output depends on future value of input which is evident from $y(-4) = x(-1)$.

The system is Non-causal.

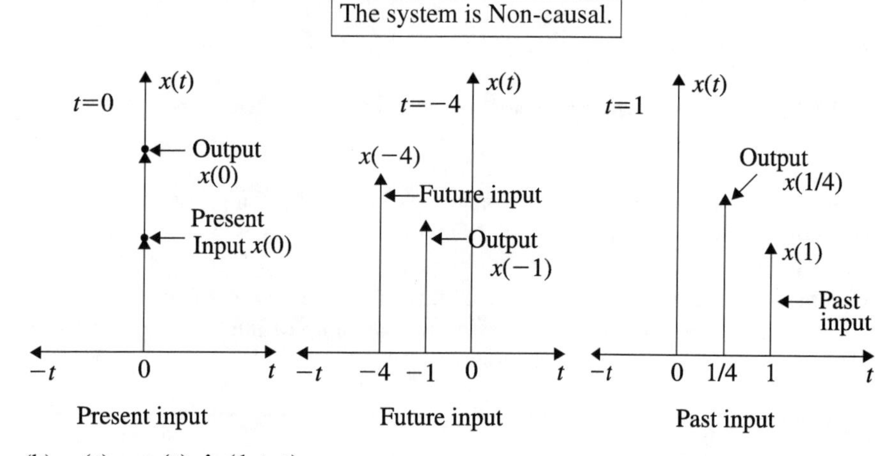

(b) $\boldsymbol{y(t) = x(t)\sin(1+t)}$

$$y(0) = x(0)\sin(1)$$
$$y(1) = x(1)\sin(2)$$
$$y(-1) = x(-1)\sin(0)$$

Thus at all time, the output depends on the present input only. Hence

The system is Causal.

(c) $\boldsymbol{y(t) = x(t^2)}$

$$\text{For } t = 0, \quad y(0) = x(0)$$
$$\text{For } t = 1, \quad y(1) = x(1)$$
$$\text{For } t = 2, \quad y(2) = x(4)$$

The system output depends on the present input as seen from $y(0) = x(0)$ and $y(1) = x(1)$.

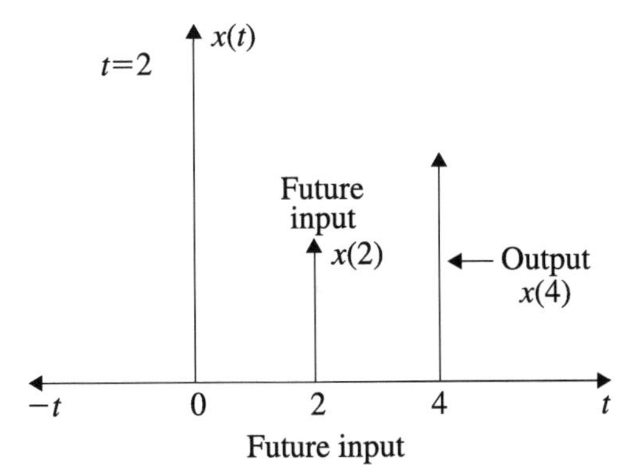

The system output $y(t)$ at $t = 2$, which is $y(2) = x(4)$ depends on the future input $x(t)$. Hence

The system is Non-causal.

(d) $\boldsymbol{y(t) = x(\sqrt{t})}$
At $t = 0.64$

$$y(.64) = x(.8)$$

The output depends on the future input. Hence

The system is Non-causal.

(e) $\boldsymbol{y(t) = x(t+1)}$
For $t = 0$,

$$y(0) = x(1)$$

The system output depends on the future input.

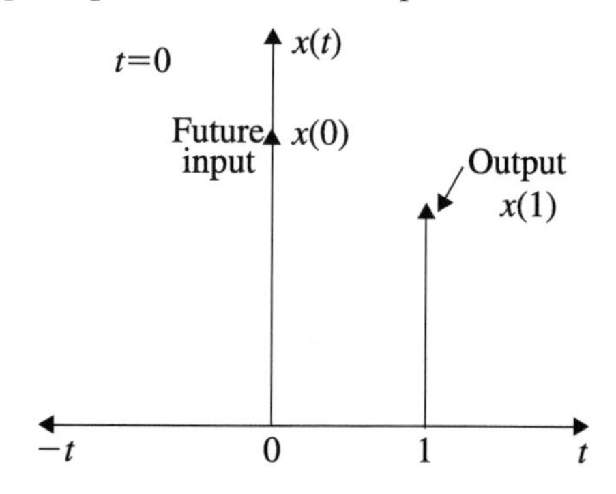

Hence

The system is Non-causal.

(f) $\boldsymbol{y(t) = x(t-1)}$

$$\begin{aligned} &\text{For } t = 0, \quad y(0) = x(-1) \\ &\text{For } t = 1, \quad y(1) = x(0) \\ &\text{For } t = 2, \quad y(2) = x(1) \end{aligned}$$

The output depends on the past values of the input.

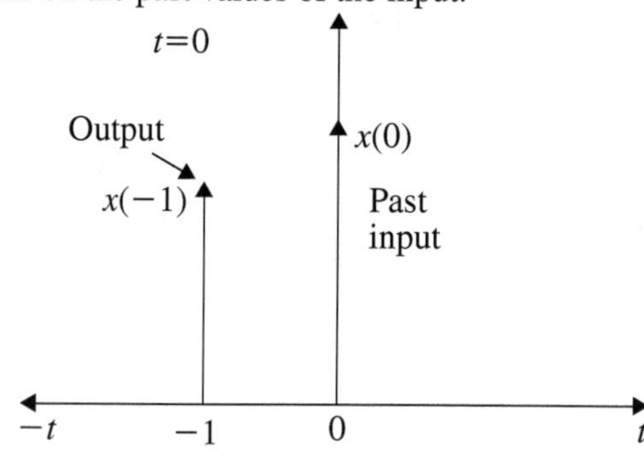

Hence

The system is Causal.

(g) $y(t) = \frac{d}{dt}x(t)$

$$y(0) = \frac{d}{dt}x(0)$$
$$y(1) = \frac{d}{dt}x(1)$$

The output depends on the present input. Hence

The system is Causal.

(h) $y(t) = \int_{t-4}^{t+4} x(\tau)d\tau$

$$y(t) = \Big[x(\tau)\Big]_{t-4}^{t+4}$$
$$= x(t+4) - x(t-4)$$

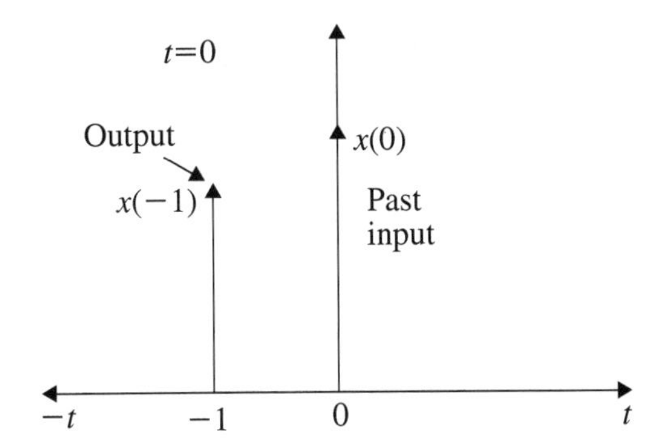

For $t = 0$,

$$y(0) = x(4) - x(-4)$$

The output $y(0)$ depends on future input $x(4)$. Hence

The system is Non-causal.

2.4.5 *Stable and Unstable Systems*

Consider a cone which is resting on its base as shown in Fig. 2.7a. The cone at this position when given a small disturbance, will stay in the same position with a small displacement which is the new equilibrium state. Now this position of the cone is said to be in stable state. On the other hand, consider the cone resting on its tip. When the cone is given a small displacement (say an impulse) the contact of the tip with

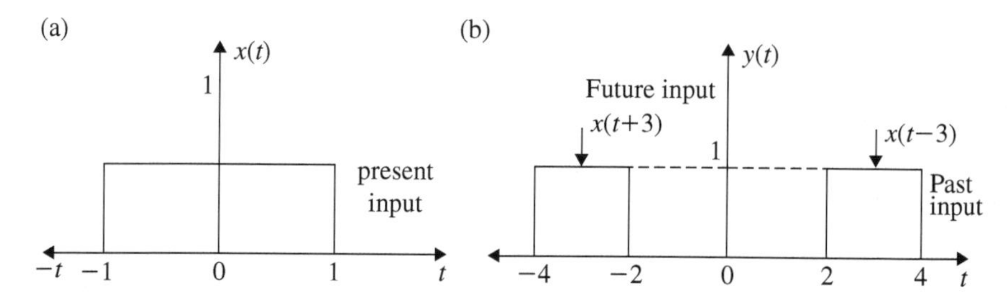

Fig. 2.6 A non-causal system

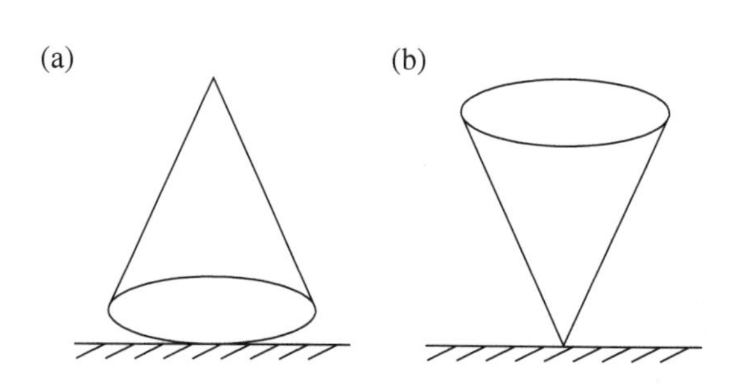

Fig. 2.7 Stable and unstable systems

the resting surface is lost and it rolls over the surface. The output position (resting on the tip) is never reached. This state of the cone is said to be unstable.

Consider a linear time invariant continuous time system which is excited by an impulse as shown in block diagram of Fig. 2.6a. The output response of the system is shown in Fig. 2.6b and c. In Fig. 2.6b the area under the impulse response curve is finite. It can be mathematically proved, that such systems whose area of the impulse response curve is finite, are said to be stable. On the other hand, consider Fig. 2.6c. The area under this impulse curve is infinite. Systems, which possess such an impulse curve are said to be unstable.

A linear time invariant continuous time system is said to be Bounded Input Bounded Output (BIBO) stable, if for any bounded input, it produces bounded output. This also implies that for BIBO stability, the area under the impulse response (output) curve should be finite.

The BIBO stability concept is mathematically expressed as follows. Let the input-output of a linear time invariant system be expressed as,

$$y(t) = f[x(t)] \qquad \text{for all } t \tag{2.8}$$

If $|x(t)|$ is bounded, $|y(t)|$ should also be bounded for the system to be stable.

$$|y(t)| \leq M_y < \infty \qquad \text{for all } t \tag{2.9}$$

$$|x(t)| \leq M_x < \infty \qquad \text{for all } t \tag{2.10}$$

where $|M_x|$ and $|M_y|$ represent positive values. It can be easily established that the necessary and sufficient condition for the LTIC time system to be stable is,

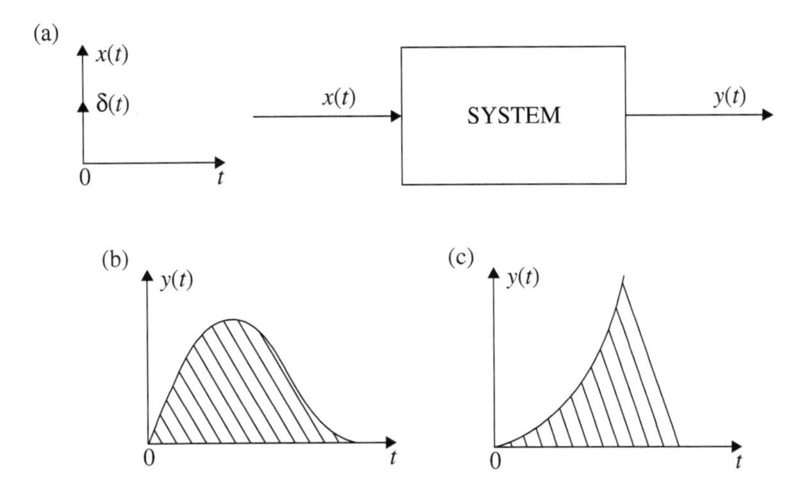

Fig. 2.8 Impulse response of stable and unstable systems

$$\boxed{y(t) = \int_{-\infty}^{\infty} |x(t)| dt < \infty} \tag{2.11}$$

The following examples illustrate the method of finding the stability of LTIC time system.

■ Example 2.7

Determine whether the systems described by the following equations: are BIBO stable.

(a) $y(t) = tx(t)$
(b) $y(t) = e^{-2|t|}$
(c) $y(t) = x(t)\sin t$
(d) $y(t) = te^{2t}u(t)$
(e) $y(t) = e^{4t}u(t-3)$
(f) $y(t) = e^{-2t}\sin 2t\, u(t)$

Solution:

(a) $\boldsymbol{y(t) = tx(t)}$
If $x(t)$ is bounded, $y(t)$ varies with respect to time and becomes unbounded. Hence

$$\boxed{\text{The system is BIBO Unstable.}}$$

(b) $\boldsymbol{y(t) = e^{-2|t|}}$
Here

$$
\begin{aligned}
x(t) &= e^{-2t} && 0 \leq t < \infty \\
&= e^{2t} && -\infty < t < 0 \\
y(t) &= \int_{-\infty}^{\infty} x(t)dt \\
&= \int_{-\infty}^{0} e^{2t}dt + \int_{0}^{\infty} e^{-2t}dt \\
&= \left[\frac{1}{2}e^{2t}\right]_{-\infty}^{0} - \left[\frac{1}{2}e^{-2t}\right]_{0}^{\infty} \\
&= \frac{1}{2}[1+1] = 1 < \infty
\end{aligned}
$$

The output is bounded and the system is stable.

The system is BIBO Stable.

(c) $\boldsymbol{y(t) = x(t)\sin t}$
It $x(t)$ is bounded, $y(t)$ is also bounded because $\sin t$ will take a maximum value of $+1$ and -1. Hence, $y(t)$ is bounded.

The system is BIBO Stable.

(d) $\boldsymbol{y(t) = te^{2t}u(t)}$
Here the output varies linearly as t and also exponentially increasing due to e^{2t}. Hence, $|y(t)| = \infty$ and the system is BIBO unstable. Mathematically this can be proved as follows. For a causal system, $|y(t)|$ can be written as

$$
|y(t)| = \int_{0}^{\infty} te^{2t}dt
$$

The following integration formula is used to evaluate the above integral.

$$
\int_{0}^{\infty} te^{at}dt = \frac{1}{a^2}\left[e^{at}\{at-1\}\right]_{0}^{\infty}
$$

$$
\begin{aligned}
|y(t)| &= \frac{1}{4}\left[e^{2t}\{2t-1\}\right]_{0}^{\infty} \\
&= \frac{1}{4}[e^{\infty}\{2\infty-1\}+1] \\
&= \infty
\end{aligned}
$$

The system is BIBO Unstable.

(e) $\boldsymbol{y(t) = e^{4t}u(t-3)}$

The output response is exponentially increasing as t increases with a time delay of $t = 3$. Hence, the system is unstable. This is mathematically proved as follows:

$$
\begin{aligned}
|y(t)| &= \int_{-\infty}^{\infty} |x(t)|dt \\
&= \int_{3}^{\infty} e^{4t}dt \\
&= \frac{1}{4}\Big[e^{4t}\Big]_3^{\infty} \\
&= \infty - \frac{1}{4}e^{12} \\
&= \infty
\end{aligned}
$$

The system is BIBO Unstable.

(f) $\boldsymbol{y(t) = e^{-2t}\sin 2t\, u(t)}$

The output response is a function of exponential decay and a sinusoid. The sinusoid will have a maximum value of $+1$ and -1. As t increases, $y(t)$ will exponentially decrease and the output is bounded. The result can be mathematically obtained as follows. For a causal signal $u(t)$

$$
|y(t)| = \int_{0}^{\infty} e^{-2t}\sin 2t\, dt
$$

Using the formula,

$$
\int_{0}^{\infty} e^{at}\sin bt\, dt = \frac{\left[e^{at}\{a\sin bt - b\cos at\}\right]_0^{\infty}}{a^2 + b^2}
$$

we get,

$$
\begin{aligned}
|y(t)| &= \frac{2}{2^2 + 2^2}\Big[e^{-2t}\{\sin 2t - \cos 2t\}\Big]_0^{\infty} \\
&= \frac{1}{4} < \infty
\end{aligned}
$$

The system is BIBO Stable.

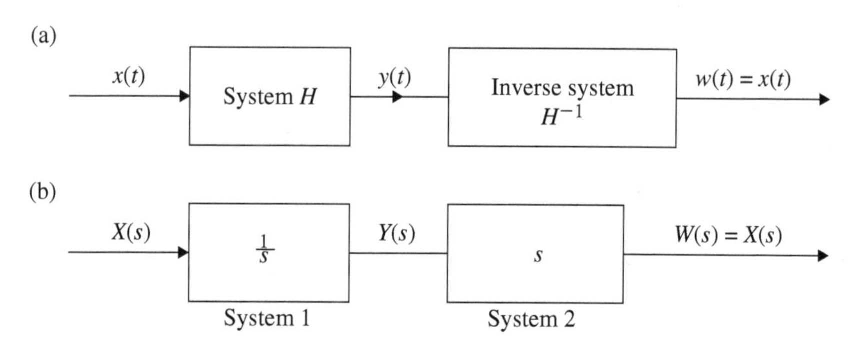

Fig. 2.9 Representation of inverse system

2.4.6 *Invertibility and Inverse System*

Consider the system H which is excited with $x(t)$. The system produces the output $y(t)$. This signal is applied as the input to the inverse system H^{-1} which produces the output $x(t)$. The block diagram representation of the system and the inverse system is shown in Fig. 2.9a. Form Fig. 2.9a, the inverse system is defined as follows.

A system is said to be invertible if the distinct inputs give distinct output.

Consider the system shown in Fig. 2.9b. The input-output relationship of system 1 is described as,

$$\frac{d}{dt}y(t) = x(t)$$

Consider system 2, the input-output of this system is described by

$$\frac{d}{dt}y(t) = x(t)$$

When these two systems are cascaded, the output response of the interconnected system is same as the excitation of the system itself. **The system which makes this possible is called inverse system. Here unique excitation produces unique response**.

■ Example 2.8

Consider the systems described by the equations given below:

(a) The impulse $h(t)$ is given as

$$h(t) = \delta(t) - 3e^{-3t}u(t) + 4e^{-4t}u(t)$$

(b)

$$\frac{dy(t)}{dt} + 5y(t) = \frac{d^2x(t)}{dt^2} + 2\frac{dx(t)}{dt} - 8x(t)$$

Determine the inverse systems for the above. Are these systems both causal and stable?

Solution:

(a) $\boldsymbol{h(t) = \delta(t) - 3e^{-3t}u(t) + 4e^{-4t}u(t)}$
Taking Laplace transform on both sides we get

$$\begin{aligned} H(s) &= 1 - \frac{3}{s+3} + \frac{4}{s+4} \\ &= \frac{(s^2+8s+12)}{(s+3)(s+4)} \end{aligned}$$

The inverse of the above system is,

$$H^{-1}(s) = \frac{1}{H(s)} = \frac{(s+3)(s+4)}{s^2+8s+12}$$

$$\boxed{H^{-1}(s) = \frac{(s+3)(s+4)}{(s+2)(s+6)}}$$

The poles of H^{-1} are at $s = -2$ and $s = -6$. Hence, the inverse systems is stable. The region of convergence (ROC) (refer Chap. 8) is to the right of right most pole $s = -2$. Hence, it is causal.

$$\boxed{\text{The inverse system is both Causal and Stable.}}$$

(b) $\boldsymbol{\frac{dy(t)}{dt} + 5y(t) = \frac{d^2x(t)}{dt^2} + 2\frac{dx(t)}{dt} - 8x(t)}$
Taking Laplace transform on both sides of the above equation we get,

$$\begin{aligned} (s+5)Y(s) &= (s^2+2s-8)X(s) \\ H(s) = \frac{Y(s)}{X(s)} &= \frac{(s^2+2s-8)}{(s+5)} \\ &= \frac{(s-2)(s+4)}{(s+5)} \end{aligned}$$

The inverse system is,

$$H^{-1}(s) = \frac{1}{H(s)} = \frac{(s+5)}{(s-2)(s+4)}$$

$$\boxed{H^{-1}(s) = \frac{(s+5)}{(s-2)(s+4)}}$$

The poles of the inverse systems are at $s = 2$ and $s = -4$. The pole at $s = 2$ will make the system unstable if the system is causal. For the system to be stable the ROC should form a strip between $s = 2$ and $s = -4$ in which case it includes the $j\omega$ axis. In this case, the system has to be non-causal.

$$\boxed{\text{The system is not both Causal and Stable.}}$$

■ Example 2.9

Determine whether the given system is memoryless, time invariant, linear, causal and stable. Justify your answers.

$$y(t) = (\cos 3t)\,x(t)$$

(Anna University, December, 2006)

Solution:

$$\begin{aligned} y(0) &= x(0) \\ y(1) &= \cos 3x(1) \\ y(-1) &= \cos 1x(-1) \end{aligned}$$

1. The output depends only on the present input. Hence, the systems is memoryless (static). Since the output does not depend on the future input, it is **causal**.
2. The output due to the delayed input is,

$$y(t, t_0) = \cos 3t\, x(t - t_0)$$

The delayed output is obtained by substituting $t = (t - t_0)$ in the given equation

$$\begin{aligned} y(t - t_0) &= \cos 3(t - t_0)x(t - t_0) \\ y(t - t_0) &\neq y(t, t_0) \end{aligned}$$

The system is therefore time varying.

3. To test the linearity of the system, consider the given equation

$$\begin{aligned} y(t) &= (\cos 3t)x(t) \\ y_1(t) &= (\cos 3t)x_1(t) \\ y_2(t) &= (\cos 3t)x_2(t) \end{aligned}$$

The sum of the weighted output is

$$y_3(t) = a_1y_1(t) + a_2y_2(t) = \cos 3t[a_1x_1(t) + a_2x_2(t)]$$

The output due to the weighted sum of input is,

$$y_4(t) = (\cos 3t)[a_1x_1(t) + a_2x_2(t)]$$
$$y_3(t) = y_4(t)$$

The system is Linear.

4.

$$|y(t)| = \cos 3t|x(t)|$$

If $x(t)$ is bounded $|y(t)|$ is also bounded. **Hence, the system is stable.**

The system is,

(a) Static, (b) Time Variant, (c) Linear, (d) Causal and (e) Stable.

■ Example 2.10

Verify whether the system given by

$$y(t) = x(t^2)$$

is causal, instantaneous, linear and shift invariant.

(*Anna University, May, 2006*)

Solution:

1.

$$y(t) = x(t^2)$$
$$y(2) = x(4)$$

See Example 2.5b. The output depends on the future input. Hence, the system is **not causal.**

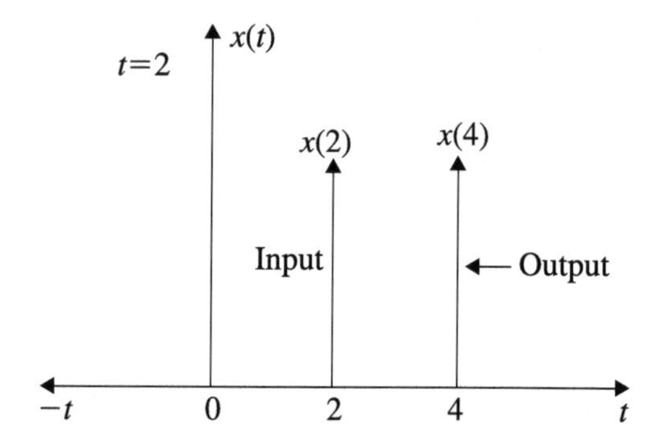

2. Since the output depends on the present, and future inputs, it requires memory. It is, therefore, **not instantaneous.**
3. The response due to the delayed input is,

$$y(t, t_0) = x[(t^2 - t_0)]$$

The delayed output is obtained by putting $t = t - t_0$ in the given equation

$$y(t - t_0) = x[(t - t_0)^2]$$
$$y(t, t_0) \neq y(t - t_0)$$

Hence, the system is shift variant.

4.

$$y(t) = x(t^2)$$
$$y_1(t) = x_1(t^2)$$
$$y_2(t) = x_2(t^2)$$

The sum of the weighted output is,

$$\begin{aligned} y_3(t) &= a_1y_1(t) + a_2y_2(t) \\ &= a_1x_1(t^2) + a_2x_2(t^2) \end{aligned}$$

The output due to the weighted sum of the input is,

$$\begin{aligned} y_4(t) &= f[a_1x_1(t) + a_2x_2(t)] \\ &= a_1x_1(t^2) + a_2x_2(t^2) \\ y_3(t) &= y_4(t) \end{aligned}$$

The system is linear.

The system is

(a) Non-causal, (b) Not Instantaneous, (c) Shift Variant and (d) Linear.

■ Example 2.11

Determine whether the system described by the following equation is static, linear, time variant and causal.

$$y(t) = E_v[x(t)]$$

Solution:

1. $\boldsymbol{y(t) = E_v[x(t)]}$

$$y(t) = E_v[x(t)] = \frac{1}{2}[x(t) + x(-t)]$$

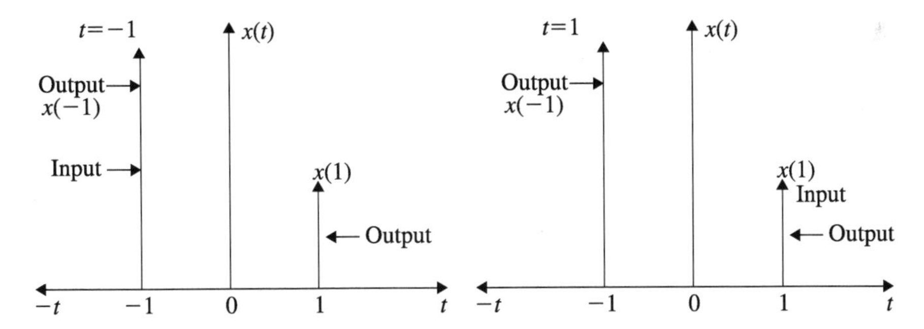

$$\text{For } t = -1, \quad y(-1) = \frac{1}{2}[x(-1) + x(1)]$$

$$\text{For } t = 1, \quad y(1) = \frac{1}{2}[x(1) + x(-1)]$$

For $t = -1$, the output depends on the present value of $x(-1)$ and also the past value of $x(1)$. For $t = 1$, the output depends on the present value of $x(1)$ and future value of $x(-1)$. Hence, the system is **non-causal**. Since $x(1)$ requires memory, **the system is dynamic.**

2. $$y(t) = \frac{1}{2}[x(t) + x(-t)]$$

   The output due to the delayed input is,

$$y(t, t_0) = \frac{1}{2}[x(t - t_0) + x(-t - t_0)]$$

The delayed output is obtained by putting $t = t - t_0$

$$y(t - t_0) = \frac{1}{2}[x(t - t_0) + x(-t + t_0)]$$
$$y(t, t_0) \neq y(t - t_0)$$

Hence, the system is time variant.

3.

$$y(t) = \frac{1}{2}[x(t) + x(-t)]$$
$$y_1(t) = \frac{1}{2}[x_1(t) + x_1(-t)]$$
$$y_2(t) = \frac{1}{2}[x_2(t) + x_2(-t)]$$

The weighted sum of the output is,

$$\begin{aligned} y_3(t) &= a_1 y_1(t) + a_2 y_2(t) \\ &= \frac{1}{2}[a_1 x_1(t) + a_1 x_1(-t) + a_2 x_2(t) + a_2 x_2(-t)] \end{aligned}$$

The output due to the weighted sum of the input is,

$$\begin{aligned} y_4(t) &= f[a_1 x_1(t) + a_2 x_2(t)] \\ &= \frac{1}{2}[a_1\{x_1(t) + x_1(-t)\} + a_2\{x_2(t) + x_2(-t)\}] \\ y_3(t) &= y_4(t) \end{aligned}$$

The system is linear.

The system is

(a) Dynamic, (b) Non-causal, (c) Time Variant and (d) Linear.

■ Example 2.12

Determine whether the following system is static, time invariant, linear and causal.

$$3\frac{dy(t)}{dt} + 5t\,y(t) = x(t)$$

Solution:

1. The system is described by differential equation. Hence, it is dynamic.

2. In the given differential equation, the coefficient of $y(t)$ is $5t$ which is a function of time t. Hence, the system is time varying.
3. The differential equations of the input a_1x_1 and a_2x_2 are written as follows:

$$3\frac{d}{dt}[a_1y_1(t)] + 5t\,a_1y_1(t) = a_1x_1(t)$$
$$3\frac{d}{dt}[a_2y_2(t)] + 5t\,a_2y_2(t) = a_2x_2(t)$$

Adding the above two equations we get

$$3\frac{d}{dt}[a_1y_1(t) + a_2y_2(t)] + 5t[a_1y_1(t) + a_2y_2(t)] = a_1x_1(t) + a_2x_2(t)$$
$$3\frac{d}{dt}y_3(t) + 5t\,y_3(t) = a_1x_1(t) + a_2x_2(t)$$

where

$$y_3(t) = a_1y_1(t) + a_2y_2(t)$$

The differential equation for the weighted sum of input is written as,

$$3\frac{d}{dt}[a_1y_1(t) + a_2y_2(t)] + 5t[a_1y_1(t) + a_2y_2(t)] = a_1x_1(t) + a_2x_2(t)$$
$$3\frac{d}{dt}y_4(t) + 5t\,y_4(t) = a_1x_1(t) + a_2x_2(t)$$

where

$$y_4(t) = a_1y_1(t) + a_2y_2(t)$$
$$y_3(t) = y_4(t)$$

Further when the input $x(t) = 0$, the output $y(t) = 0$. **Hence, the system is linear.**

4. From the given differential equation it is obvious that $y(t)$ depends on the present input only.

$$y(1) = x(1)$$
$$y(2) = x(2)$$

Hence, the system is causal.

The system is

(a) Dynamic, (b) Time Varying, (c) Linear and (d) Causal.

■ Example 2.13

Check whether the system having the input-output relation

$$y(t) = \int_{-\infty}^{t} x(\tau)d\tau$$

is linear and time invariant.

(*Anna University, April, 2004*)

Solution:

1. $\boldsymbol{y(t) = \int_{-\infty}^{t} x(\tau)d\tau}$

$$a_1y_1(t) = \int_{-\infty}^{t} a_1x_1(\tau)d\tau$$
$$a_2y_2(t) = \int_{-\infty}^{t} a_2x_2(\tau)d\tau$$

The weighted sum of the output is,

$$y_3(t) = a_1y_1(t) + a_2y_2(t)$$
$$= \int_{-\infty}^{t} a_1x_1(\tau)d\tau + a_2 \int_{-\infty}^{t} a_2x_2(\tau)d\tau$$

The output due to the weighted sum of input is,

$$y_4(t) = \int_{-\infty}^{t} [a_1x_1(\tau) + a_2x_2(\tau)]d\tau$$
$$y_3(t) = y_4(t)$$

The system is linear.

2. The output due to the delayed input is,

$$y(t, t_0) = \int_{-\infty}^{t} x(\tau - t_0)d\tau$$

The delayed output due to the input is,

$$y(t - t_0) = \int_{-\infty}^{t} x(\tau - t_0)d\tau$$
$$y(t, t_0) = y(t - t_0)$$

The system is time invariant.

The system is both

(a) Linear and (b) Time Invariant.

■ Example 2.14

A certain system is described by the following input-output equation

$$y(t) = x(t+1) + x(t^2)$$

Determine whether the system is static, causal, time invariant, linear and stable.

Solution:

1. $\boldsymbol{y(t) = x(t+1) + x(t^2)}$

$$y(0) = x(1) + x(0)$$

The output component $x(0)$ depends on the present input $x(0)$ and the output component $x(1)$ depends on the future input $x(0)$. To store the future input it requires memory and hence, it is **dynamic system**. Since the output depends on future input it is non-causal.

2. If the input is delayed by t_0, the output is,

$$y(t, t_0) = x(t - t_0 + 1) + x(t^2 - t_0)$$

The delayed output due to the input is obtained by putting $t = t - t_0$.

$$y(t - t_0) = x(t - t_0 + 1) + x(t - t_0)^2$$
$$y(t, t_0) \neq y(t - t_0)$$

The system is time variant.

3. The weighted sum of the output due to input is,

$$\begin{aligned} a_1 y_1(t) &= a_1[x_1(t+1) + x_1(t^2)] \\ a_2 y_2(t) &= a_2[x_2(t+1) + x_2(t^2)] \\ y_3(t) &= a_1 y_1(t) + a_2 y_2(t) \\ &= a_1[x_1(t+1) + x_1(t^2)] + a_2[x_2(t+1) + x_2(t^2)] \end{aligned}$$

The output due to the weighted sum of input is,

$$y_4(t) = a_1[x_1(t+1) + x_1(t^2)] + a_2[x_2(t+1) + x_2(t^2)]$$
$$y_3(t) = y_4(t)$$

The system is linear.

4. For the system, if input $x(t)$ is bounded, then the output $y(t)$ is also bounded. Hence, the system is stable.

The system is

(a) Dynamic, (b) Non-causal, (c) Time Variant, (d) Linear and (e) Stable.

■ Example 2.15

The input-output relationship of a certain system is given by the following equation:

$$y(t) = x(t-7) - x(2-t)$$

Determine whether the above system is linear and causal.

Solution:

1. $\boldsymbol{y(t) = x(t-7) - x(2-t)}$

$$y(t) = x(t-7) - x(2-t)$$

The weighted sum of the output due to the input is given as

$$\begin{aligned} y_3(t) &= a_1 y_1(t) + a_2 y_2(t) \\ a_1 y_1(t) &= a_1[x_1(t-7) - x_1(2-t)] \\ a_2 y_2(t) &= a_2[x_2(t-7) - x_2(2-t)] \\ y_3(t) &= a_1[x_1(t-7) - x_1(2-t)] + a_2[x_2(t-7) - x_2(2-t)] \end{aligned}$$

The output due to the weighted sum of input is,

$$\begin{aligned} y_4(t) &= a_1[x_1(t-7) - x_1(2-t)] + a_2[x_2(t-7) - x_2(2-t)] \\ y_3(t) &= y_4(t) \end{aligned}$$

Further if the input $x(t) = 0$, the output $y(t)$ is also zero. **The system is linear.**

2.

$$y(t) = x(t-7) - x(2-t)$$

For $t = 0$,

$$y(0) = x(-7) - x(2)$$

The output $x(-7)$ depends on the past input $x(0)$ and the output $-x(2)$ depends on the future input $x(0)$. **Hence, it is non-causal.**

The system is

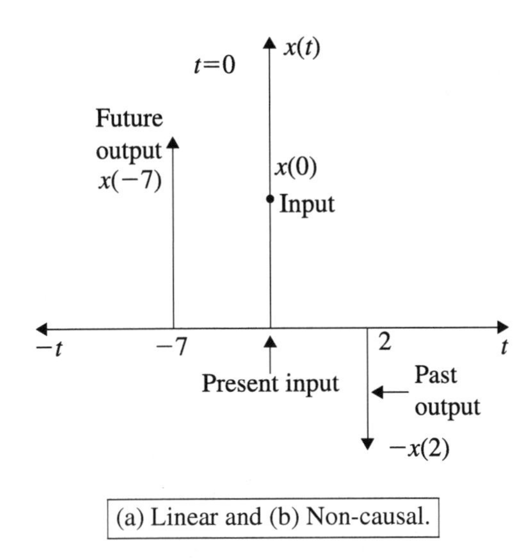

(a) Linear and (b) Non-causal.

2.5 Discrete Time System

The block diagram of a discrete time system is shown in Fig. 2.8. $x[n]$ is the excitation (input) signal and $y[n]$ is the repones (output) signal of the DT system. H represents the functional relationship between the input and output which is described by difference equation. The input-output signals appear at discrete interval of time n where $n = 0, 1, 2\ldots$ which is an integer. n can also take negative value of an integer.

2.6 Properties of Discrete Time System

Like CT systems, DT systems also possesses similar properties which are given below:

1. Linear and non-linear
2. Time varying and time invariant
3. Causal and non-causal
4. Stable and unstable
5. Static (instantaneous) and dynamic (system without and with memory)
6. Invertibility and inverse.

2.6.1 Linear and Non-linear Systems

A linear discrete time system obeys the property of superposition. As discussed for CT system, the superposition property is composed of homogeneity and additivity. Let $x_1[n]$ excitation produce $y_1[n]$ response and $x_2[n]$ produce $y_2[n]$ response. According to additivity property of superposition theorem, if both $x_1[n]$ and $x_2[n]$ are applied simultaneously, then

$$x_1[n] + x_2[n] = y_1[n] + y_2[n]$$

Let $a_1x_1[n]$ and $a_2x_2[n]$ be the inputs. According to the homogeneity (scaling) property, when these signals are separately applied,

$$\begin{aligned} a_1x_1[n] &= a_1y_1[n] \\ a_2x_2[n] &= a_2y_2[n] \end{aligned}$$

If $a_1x_1[n] + a_2x_2[n]$ are simultaneously applied the output is obtained by applying superposition theorem as,

$$a_1x_1[n] + a_2x_2[n] = a_1y_1[n] + a_2y_2[n]$$

In the above equation, $a_1x_1[n] + a_2x_2[n]$ is called the weighted sum of input and $a_1y_1[n] + a_2y_2[n]$ is called the weighted sum of the output. Therefore, the following procedure is followed to test the linearity of a DT system.

1. Express

$$\begin{aligned} y_1[n] &= f(x_1[n]) \\ y_2[n] &= f(x_2[n]) \end{aligned}$$

2. Find the weighted sum of the output as

$$y_3[n] = a_1y_1[n] + a_2y_2[n]$$

3. Find the output $y_4[n]$ due to the weighted sum of input as

$$y_4[n] = f(a_1x_1[n] + a_2x_2[n])$$

4. If $y_3[n] = y_4[n]$, then given DT system is linear. Otherwise it is non-linear.

The following examples illustrate the method of testing a DT system for its linearity.

■ Example 2.16

Test whether the following DT systems are linear or not:

$$\begin{aligned}
&\text{(a)} \quad y[n] = x^2[n] \\
&\text{(b)} \quad y[n] = x[4n+1] \\
&\text{(c)} \quad y[n] = x[n] + \frac{1}{x[n+1]} \\
&\text{(d)} \quad y[n] = x[n^2] \\
&\text{(e)} \quad y[n] = x[n] + nx[n+1]
\end{aligned}$$

Solution:

(a) $\boldsymbol{y[n] = x^2[n]}$

$$y_1[n] = x_1^2[n]$$
$$y_2[n] = x_2^2[n]$$

1. The weighted sum of the output $y_3[n]$ is,

$$\begin{aligned}
y_3[n] &= a_1 y_1[n] + a_2 y_2[n] \\
&= a_1 x_1^2[n] + a_2 x_2^2[n]
\end{aligned}$$

2. The output $y_4[n]$ due to the weighted sum of the input is,

$$\begin{aligned}
y_4[n] &= [a_1 x_1[n] + a_2 x_2[n]]^2 \\
&= a_1^2 x_1^2[n] + a_2^2 x_2^2[n] + 2a_1 a_2 x_1[n] x_2[n]
\end{aligned}$$

3.

$$y_3[n] \neq y_4[n]$$

$$\boxed{\text{The system is Non-linear.}}$$

(b) $\boldsymbol{y[n] = x[4n+1]}$

$$a_1 y_1[n] = a_1 x_1[4n+1]$$
$$a_2 y_2[n] = a_2 x_2[4n+1]$$
$$y_3[n] = a_1 y_1[n] + a_2 y_2[n]$$

1. The weighted sum of the output is,

$$\begin{aligned} y_3[n] &= a_1 y_1[n] + a_2 y_2[n] \\ &= a_1 x_1[4n+1] + a_2 x_2[4n+1] \end{aligned}$$

2. The output due to the weighted sum of the input is,

$$y_4[n] = a_1 x_1[4n+1] + a_2 x_2[4n+1]$$

3.

$$y_3[n] = y_4[n]$$

The system is Linear.

(c) $\mathbf{y[n] = x[n] + \frac{1}{x(n+1)}}$

$$a_1 y_1[n] = a_1 \left[x_1[n] + \frac{1}{x_1(n+1)} \right]$$
$$a_2 y_2[n] = a_2 \left[x_2[n] + \frac{1}{x_2(n+1)} \right]$$

1. The weighted sum of the output $y_3[n]$ is,

$$\begin{aligned} y_3[n] &= a_1 y_1[n] + a_2 y_2[n] \\ &= a_1 \left[x_1[n] + \frac{1}{x_1(n+1)} \right] + a_2 \left[x_2[n] + \frac{1}{x_2(n+1)} \right] \end{aligned}$$

2. The output due to the weighted sum of the input is,

$$\begin{aligned} y_4[n] &= f[a_1 x_1[n] + a_2 x_2[n]] \\ &= a_1 x_1[n] + a_2 x_2[n] + \left[\frac{1}{a_1 x_1[n+1] + a_2 x_2[n+1]} \right] \end{aligned}$$

3.

$$y_3[n] \neq y_4[n]$$

4. Further if $x[n] = 0$, the output $y[n]$ is not zero and it is ∞.

The system is Non-linear.

(d) $\boldsymbol{y[n] = x[n^2]}$

$$\begin{aligned} a_1y_1[n] &= a_1x_1[n^2] \\ a_2y_2[n] &= a_2x_2[n^2] \end{aligned}$$

1. The weighted sum of the output $y_3[n]$ is,

$$\begin{aligned} y_3[n] &= a_1y_1[n] + a_2y_2[n] \\ &= a_1x_1[n^2] + a_2x_2[n^2] \end{aligned}$$

2. The output $y_4[n]$ due to the weighted sum of input is,

$$y_4[n] = a_1x_1[n^2] + a_2x_2[n^2]$$

3.

$$y_3[n] = y_4[n]$$

The system is Linear.

(e) $\boldsymbol{y[n] = x[n] + nx[n+1]}$

$$\begin{aligned} a_1y_1[n] &= a_1[x_1[n] + nx_1[n+1]] \\ a_2y_2[n] &= a_2[x_2[n] + nx_2[n+1]] \end{aligned}$$

1. The weighted sum of the output is,

$$\begin{aligned} y_3[n] &= a_1y_1[n] + a_2y_2[n] \\ &= a_1[x_1[n] + nx_1[n+1]] + a_2[x_2[n] + nx_2[n+1]] \end{aligned}$$

2. The output due to the weighted sum of the input is,

$$y_4[n] = a_1x_1[n] + a_2x_2[n] + a_1nx_1[n+1] + a_2nx_2[n+1]$$

3.

$$y_3[n] = y_4[n]$$

The system is Linear.

2.6.2 *Time Invariant and Time Varying DT Systems*

Consider the discrete time system represented in block diagram of Fig. 2.9a. If the input is $x[n]$ then the output is $y[n]$. If the input is time delayed by n_0, which becomes $x[n - n_0]$, the output becomes $y[n - n_0]$. The signal representation and the delayed signals are shown in Fig. 2.9b and c respectively. Such systems are called time invariant.

If an arbitrary excitation $x[n]$ of a system causes a response $y[n]$ and the delayed excitation $x[n - n_0]$ where **n_0 is any arbitrary integer causes $y[n - n_0]$ then the system is said to be time invariant**.

Procedure to Check Time Invariancy of DT Systems

1. For the delayed input $x[n - n_0]$ find the output $y[n, n_0]$.
2. Obtain the delayed output $y[n - n_0]$ by substituting $n = n - n_0$ in $y[n]$.
3. If $y[n, n_0] = y[n - n_0]$, the system is time invariant. Otherwise the system is time varying.

The following examples illustrate the method of testing the time invariancy of DT systems.

■ Example 2.17

Determine whether the following systems are time invariant or not:

$$
\begin{aligned}
&(a) \qquad y[n] = nx[n] \\
&(b) \qquad y[n] = x[2n] \\
&(c) \qquad y[n] = x[-n] \\
&(d) \qquad y[n] = \sin(x[n]) \\
&(e) \qquad y[n] = x[n]x[n-1]
\end{aligned}
$$

Solution:

(a) $\boldsymbol{y[n] = nx[n]}$

1. The output for the delayed input $x[n - n_0]$ is obtained by delaying the input $x[n]$ as $x[n - n_0]$. Thus

$$y[n, n_0] = nx[n - n_0]$$

2. The delayed output for the input $x[n]$ is obtained by substituting $n = n - n_0$.

$$y[n - n_0] = (n - n_0)x[n - n_0]$$

3.

$$y[n, n_0] \neq y[n - n_0]$$

The system is Time Variant.

(b) $\boldsymbol{y[n] = x[2n]}$

The output due to delayed input is

$$y[n, n_0] = x[2n - n_0]$$

The delayed output is,

$$\begin{aligned} y[n - n_0] &= x[2(n - n_0)] \\ &= x[2n - 2n_0] \\ y[n, n_0] &\neq y[n - n_0] \end{aligned}$$

The system is Time Varying.

(c) $\boldsymbol{y[n] = x[-n]}$

The output due to delayed input is

$$y[n, n_0] = x[-n - n_0]$$

The delayed output is,

$$\begin{aligned} y[n - n_0] &= x[-(n - n_0)] \\ &= x[-n + n_0] \\ y[n, n_0] &\neq y[n - n_0] \end{aligned}$$

The system is Time Varying.

(d) $\boldsymbol{y[n] = \sin(x[n])}$

The output due to delayed input is

$$y[n, n_0] = \sin(x[n - n_0])$$

The delayed output is,

$$y[n-n_0] = \sin(x[n-n_0])$$
$$y[n,n_0] = y[n-n_0]$$

The system is Time Invariant.

(e) $\boldsymbol{y[n] = x[n]x[n-1]}$

The output due to delayed input is

$$y[n,n_0] = x[n-n_0]x[n-n_0-1]$$

The delayed output is,

$$y[n-n_0] = x[n-n_0]x[n-n_0-1]$$
$$y[n,n_0] = y[n-n_0]$$

The system is Time Invariant.

2.6.3 Causal and Non-causal DT Systems

A discrete time system is said to be causal if the response of the system depends on the present or the past inputs applied. The systems is non-causal if the output depends on the future input.

The following examples illustrate the method of identifying causal and non-causal systems.

■ Example 2.18

Determine whether the following systems are causal or not:

(a) $y[n] = x[n-1]$

(b) $y[n] = x[n] + x[n-1]$

(c) $y[n-1] = x[n]$

(d) $y[n] = \sin(x[n])$

(e) $y[n] = \sum_{k=-\infty}^{n+4} x(k)$

(f) $y[n] = \sum_{k=0}^{-3} x(k)$

Solution:

(a) $\boldsymbol{y[n] = x[n-1]}$

$$y[0] = x[-1]$$
$$y[1] = x[0]$$

$x[n-1]$ is the past input for the output $y[n]$. The output depends on the past value of $x[n]$. Hence

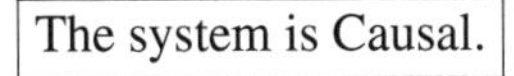

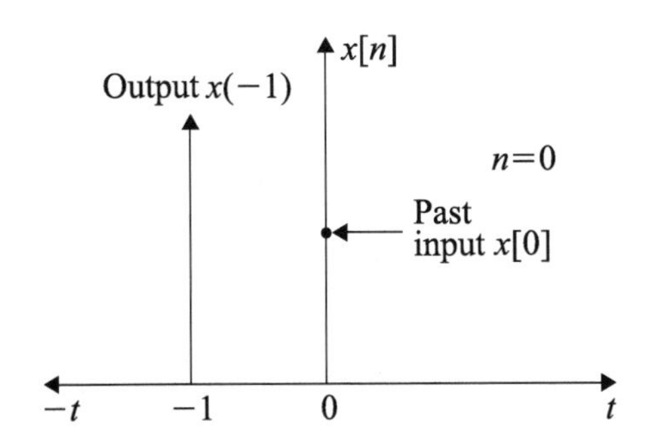

(b) $\boldsymbol{y[n] = x[n] + x[n-1]}$

$$\text{For } n = 0, \quad y[0] = x[0] + x[-1]$$
$$\text{For } n = 1, \quad y[1] = x[1] + x[0]$$

Here $x[n]$ is present value and $x[n-1]$ is past value. The output depends on the present and past inputs. Hence

The system is Causal.

(c) $\boldsymbol{y[n-1] = x[n]}$

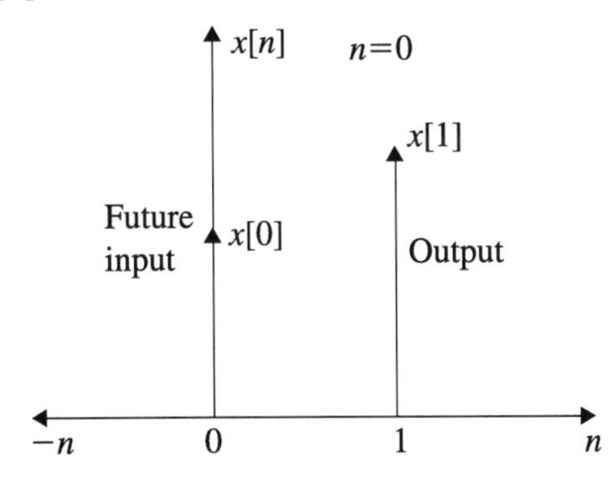

Put $n = n + 1$

$$y[n] = x[n+1]$$
$$y[0] = x[1]$$

The output depends on the future inputs. Hence

The system is Non-causal.

(d) $\boldsymbol{y[n] = \sin x[n]}$

$$y[0] = \sin x[0]$$
$$y[-1] = \sin x[-1]$$

The output depends on the present input only. Hence

The system is Causal.

(e) $\boldsymbol{y[n] = \sum_{k=-\infty}^{n+4} x[k]}$

$$y[0] = \sum_{-\infty}^{4} x[k]$$
$$= x[-\infty]+x[-\infty+1]+\cdots+x[-1]+x[0]+x[1]+x[2]+x[3]+x[4]$$

$$x[-\infty] + x[-\infty+1], \ldots, x[-1] = \text{Future output for past input}$$
$$x[0] = \text{Present output for present input}$$
$$x[1],\ x[2],\ x[3] \text{ and } x[4] = \text{Past output for future input}$$

The output depends on past, present and future inputs. Hence

The system is Non-causal.

(f) $\boldsymbol{y[n] = \sum_{k=0}^{n-3} x[k]}$

$$y[0] = \sum_{k=0}^{-3} x[k]$$
$$= x[0] + x[-1] + x[-2] + x[-3]$$
$$x[0] = \text{Present output for present input}$$
$$x[-1], x[-2], x[-3] = \text{Future outputs for past input}$$

The output depends on the present and past inputs. Hence

The system is Causal.

2.6.4 *Stable and Unstable Systems*

A discrete time system is said to be stable if for any bounded input, it produces a bounded output. This implies that the impulse response

$$y[n] = \sum_{-\infty}^{\infty} |h[n]| < \infty$$

is absolutely summable.
For a bounded input,

$$|x[n]| \leq M_x < \infty$$

the output

$$|y[n]| \leq M_y < \infty$$

From the above two conditions, it can be obtained

$$y[n] = \sum_{-\infty}^{\infty} |h[n]| < \infty$$

The following examples illustrate the above procedure.

■ Example 2.19

Check whether the DT systems described by the following equations are stable or not.

(a) $y[n] = \sin x[n]$

(b) $y[n] = \sum_{k=0}^{n+1} x[k]$

(c) $y[n] = e^{x[n]}$

(d) $h[n] = 3^n u[n+3]$

(e) $y[n] = x[-n-3]$

(f) $y[n] = x[n-1] + x[n] + x[n+1]$

(g) $h[n] = e^{-|n|}$

(h) $h[n] = n\,u[n]$

(i) $h[n] = 3^n u[n-3]$

(j) $h[n] = 2^n u[-n-1]$

Solution:

(a) $\boldsymbol{y[n] = \sin x[n]}$ If $x[n]$ is bounded, then $\sin x[n]$ is also bounded and so $y[n]$ is also bounded

$$\boxed{\text{The system is Stable.}}$$

(b) $\boldsymbol{y[n] = \sum_{k=0}^{n+1} x[k]}$
Here as $n \to \infty$, $y[n] \to \infty$ and the output is unbounded. For bounded input n should be a finite number.
In that case $y[n]$ is bounded and the system is stable.

$$\boxed{\text{The system is Stable.}} \quad \text{for } n = \text{finite}$$

$$\boxed{\text{The system is Unstable.}} \quad \text{for } n = \infty$$

(c) $\boldsymbol{y[n] = e^{x[n]}}$
For $|x[n]|$ bounded, $e^{|x[n]|}$ is bounded and the system is stable.

$$\boxed{\text{The system is Stable.}}$$

(d) $\boldsymbol{h[n] = 3^n u[n+3]}$

$$\begin{aligned}|y[n]| &= \sum_{n=-3}^{\infty} 3^n \\ &= (3)^{-3} + (3)^{-2} + (3)^{-1} + (3)^0 + (3)^1 + \cdots + (3)^{\infty} \\ &= \infty\end{aligned}$$

The output is unbounded.

$$\boxed{\text{The system is Unstable.}}$$

(e) $\boldsymbol{y[n] = x[-n-3]}$ If $x[n]$ is bounded, $x[-n]$ is also bounded, $x[-n-1]$ is bounded and $y[n]$ is bounded.

$$\boxed{\text{The system is Stable.}}$$

(f) $\boldsymbol{y[n] = \delta[n-1] + \delta[n] + \delta[n+1]}$

$$
\begin{aligned}
y[0] &= \delta[-1] + \delta[0] + \delta[1] = 0 + 1 + 0 = 1 \\
y[1] &= \delta[0] + \delta[1] + \delta[2] = 1 + 0 + 0 = 1 \\
y[-1] &= \delta[-2] + \delta[-1] + \delta[0] = 0 + 0 + 1 = 1 \\
y[-2] &= \delta[1] + \delta[2] + \delta[3] = 0 + 0 + 0 = 0 \\
y[2] &= \delta[1] + \delta[2] + \delta[3] = 0 + 0 + 0 = 0 \\
y[n] &= \sum_{-\infty}^{\infty} |h[k]| = 1 + 1 + 1 = 3 < \infty
\end{aligned}
$$

The system is Stable.

(g) $\boldsymbol{h[n] = e^{-|n|}}$

$$
\begin{aligned}
y[n] &= \sum_{-\infty}^{\infty} e^{|n|} = \sum_{-\infty}^{-1} e^{n} + \sum_{0}^{\infty} e^{-n} = \sum_{1}^{\infty} e^{-n} + \sum_{0}^{\infty} e^{-n} \\
&= e^{-1} + e^{-2} + \cdots + 1 + e^{-1} + e^{-2} + \cdots \\
&= e^{-1}[1 + e^{-1} + e^{-2} + \cdots] + 1 + e^{-1} + e^{-2} + \cdots \\
&= e^{-1} \frac{1}{[1 - e^{-1}]} + \frac{1}{[1 - e^{-1}]} \\
&= \frac{e^{-1}}{(1 - e^{-1})} + \frac{1}{(1 - e^{-1})} \\
&= \frac{[1 + e^{-1}]}{[1 - e^{-1}]} = 2.164 < \infty
\end{aligned}
$$

The system is Stable.

(h) $\boldsymbol{h[n] = n\,u[n]}$

$$
y[n] = \sum_{0}^{\infty} n = 1 + 2 + \cdots + \infty = \infty
$$

The system is Unstable.

(i) $\boldsymbol{h[n] = 3^{n} u[n-3]}$

$$
y[n] = \sum_{3}^{\infty} 3^{n} = 3^{3} + 3^{4} + \cdots + \infty = \infty
$$

The system is Unstable.

(j) $\boldsymbol{h[n] = 2^n u[-n-1]}$

$$y[n] = \sum_{-\infty}^{-1} 2^n = \sum_{1}^{\infty} \left(\frac{1}{2}\right)^n$$

$$= \frac{1}{2} + \left(\frac{1}{2}\right)^2 + \cdots$$

$$= \frac{1}{2}\left[1 + \frac{1}{2} + \left(\frac{1}{2}\right)^2 + \cdots\right]$$

$$= \frac{1}{2}\left[\frac{1}{1-\frac{1}{2}}\right] = 1 < \infty$$

The system is Stable.

2.6.5 *Static and Dynamic Systems*

A discrete time system is said to be static (memoryless or instantaneous) if the output response depends on the present value only and not on the past and future values of excitation. Discrete systems described by difference equations require memory and hence they are dynamic systems.

The following examples illustrate the method identifying static and dynamic discrete systems.

■ Example 2.20

Identify whether the following systems are static or dynamic:

(a) $y[n] = x[3n]$

(b) $y[n] = \sin(x[n])$

(c) $y[n-1] + y[n] = x[n]$

(d) $y[n] = \text{sgn}|x[n]|$

Solution:

(a) $\boldsymbol{y[n] = x[3n]}$

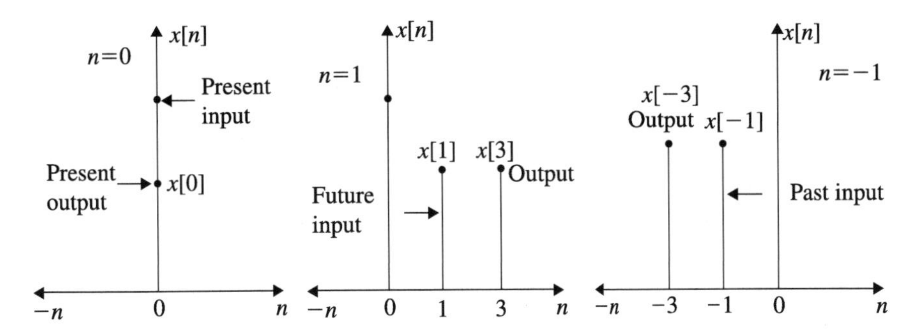

$$\text{For } n = 0, \quad y[0] = x[0]$$
$$\text{For } n = 1, \quad y[1] = x[3]$$
$$\text{For } n = -1, \quad y[-1] = x[-3]$$

The outputs $y[0] = x[0]$, $y[1] = x[3]$ and $y[-1] = x[-3]$ depend upon the present input, future input and past input respectively.

The system is Dynamic.

(b) $\mathbf{y[n] = \sin(x[n])}$

$$y[0] = \sin(x[0])$$
$$y[1] = \sin(x[1])$$

The output depends on the present input at all time. Hence

The system is Static.

(c) $\mathbf{y[n-1] + y[n] = x[n]}$
The system is described by first order difference equation which require memory. Hence

The system is Dynamic.

(d) $\mathbf{y[n] = \mathrm{sgn}|x[n]|}$

$$\begin{aligned} \mathrm{sgn}|x[n]| &= 1 \qquad \text{for } n > 0 \\ &= -1 \qquad \text{for } n < 0 \end{aligned}$$
$$y[1] = x[1] = 1$$
$$y[-1] = x[-1] = -1$$

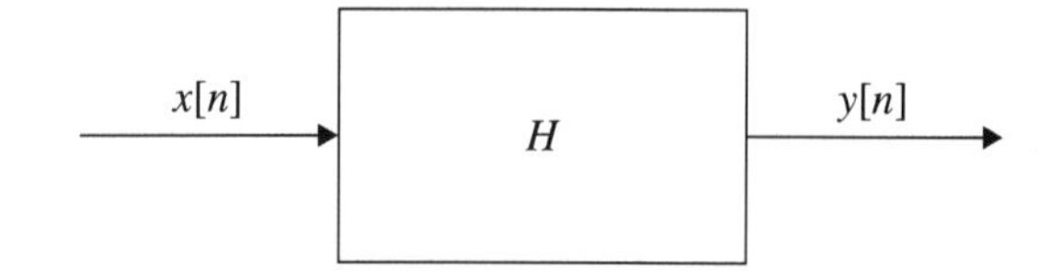

Fig. 2.10 Block diagram representation of discrete time system

The output depends on the present value of the input. Hence

$$\boxed{\text{The system is Static.}}$$

2.6.6 Invertible and Inverse Discrete Time Systems

A discrete time system is said to be invertible if distinct input leads to distinct output. If a system is invertible then an inverse system exists.

Consider the system shown in Fig. 2.10. The input $x[n]$ produces the output $y[n]$. This system is in cascade with its inverse system. The output of this system is nothing but the difference of the two successive inputs $y[n] - y[n-1]$. This is the input to the original system. Thus, by connecting an inverse system in cascade with the original system, the excitation signal $x[n]$ is re-established provided the original system is invertible. The concept of invertibility is very widely used in communications.

■ Example 2.21

Determine whether the following systems are static, causal, time invariant, linear and stable.

$$\begin{array}{ll}
\text{(a)} & y[n] = x[4n+1] \\
\text{(b)} & y[n] = x[n] + n\,x[n+1] \\
\text{(c)} & y[n] = x[n]u[n] \\
\text{(d)} & y[n] = \log_{10} x[n] \\
\text{(e)} & y[n] = x^2[n] \\
\text{(f)} & y[n] = x[n]\cos[\omega n]
\end{array}$$

(*Anna University, 2007 and 2013*)

Solution:

(a) $\boldsymbol{y[n] = x[4n+1]}$

1.

$$\text{For } n = 0, \quad y[0] = x[1]$$

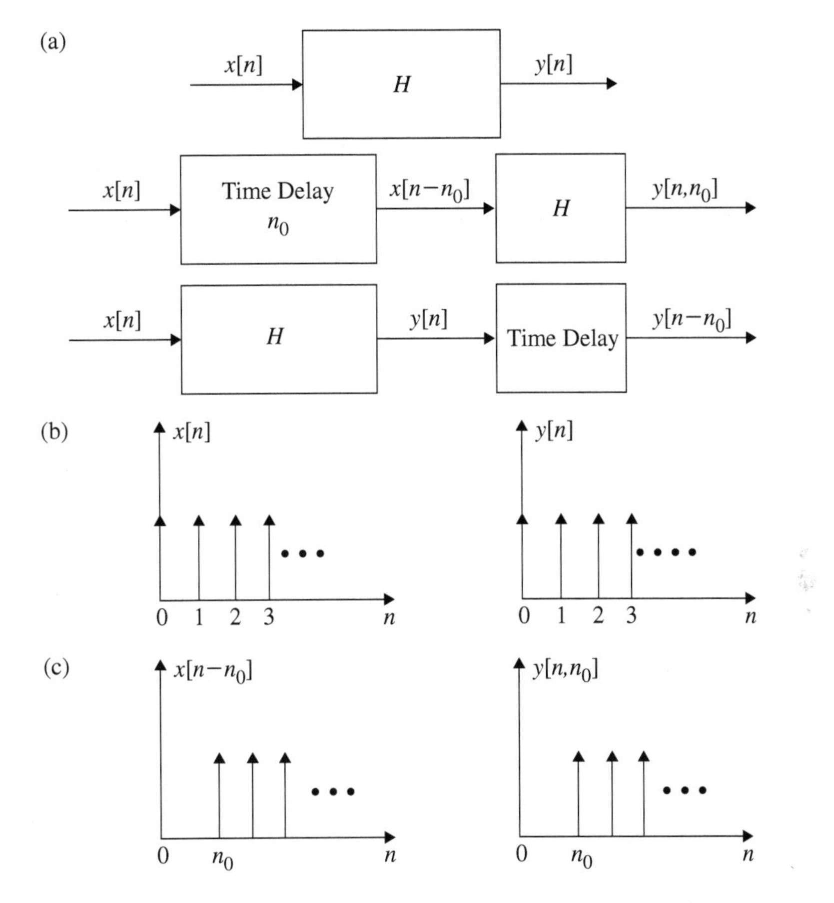

Fig. 2.11 Block diagram and signal representation to illustrate time invariancy of DT system

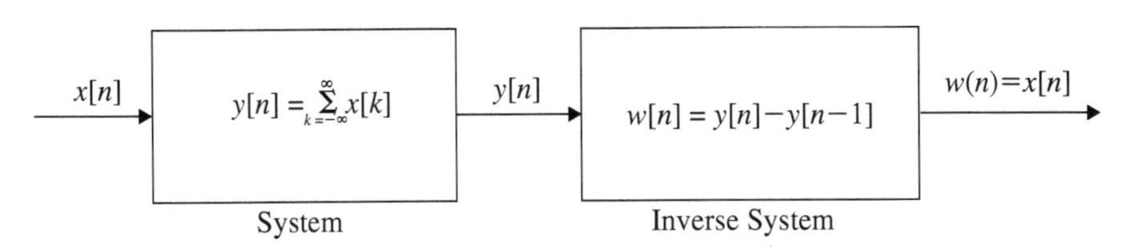

Fig. 2.12 Inverse discrete time system

The output $x[1]$ depends on future input, $[x[0]]$.

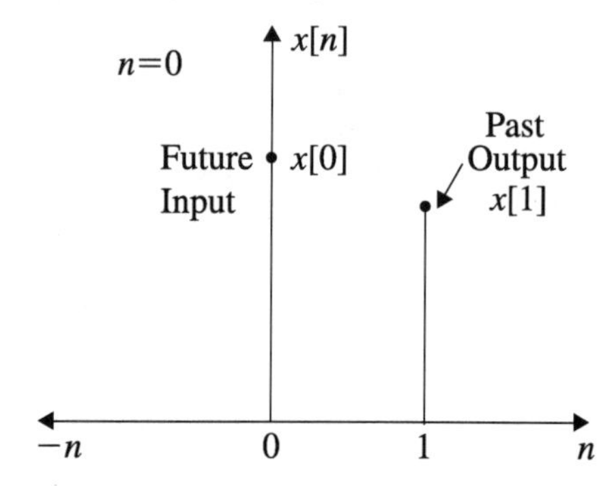

Hence

The system is Dynamic and Non-causal.

2. The output due to the delayed input is,

$$y[n, n_0] = x[4n - n_0 + 1]$$

The delayed output due to the input is,

$$\begin{aligned} y[n - n_0] &= x[4(n - n_0) + 1] \\ &= x[4n - 4n_0 + 1] \\ y[n, n_0] &\neq y[n - n_0] \end{aligned}$$

The system is Time Variant.

3.

$$\begin{aligned} a_1 y_1[n] &= a_1 x_1[4n + 1] \\ a_2 y_2[n] &= a_2 x_2[4n + 1] \end{aligned}$$

The sum of weighted output due to the input is,

$$\begin{aligned} y_3[n] &= a_1 y_1[n] + a_2 y_2[n] \\ &= a_1 x_1[4n + 1] + a_2 x_2[4n + 1] \end{aligned}$$

The output due to the weighted sum of input is,

$$\begin{aligned} y_4[n] &= a_1 x_1[4n + 1] + a_2 x_2[4n + 1] \\ y_3[n] &= y_4[n] \end{aligned}$$

The system is Linear.

4. The input is time shifted and time compressed signal. As long as the input is bounded the output is also bounded.

The system is Stable.

The system is

(1) Dynamic, (2) Non-causal, (3) Time Variant, (4) Linear and (5) Stable.

(b) $\mathbf{y[n] = x[n] + n\,x[n+1]}$

1. For $n = 1$,

$$y[1] = x[1] + 1 \times x[2]$$

The outputs $x[1]$ and $x[2]$ depend upon present and future inputs respectively.

The system is Dynamic and Non-causal.

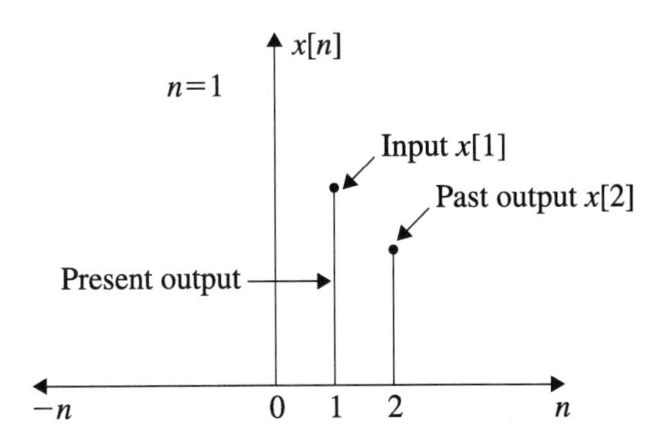

2. The output due to the delayed input is,

$$y[n, n_0] = x[n - n_0] + nx[n - n_0 + 1]$$

The delayed output due to the input is,

$$y[n - n_0] = x[n - n_0] + (n - n_0)x[n - n_0 + 1]$$
$$y[n, n_0] \neq y[n - n_0]$$

The system is Time Variant.

3. The weighted sum of the output due to the input is,

$$\begin{aligned} y_3[n] &= a_1y_1[n] + a_2y_2[n] \\ &= a_1x_1[n] + a_1nx_1[n+1] + a_2x_2[n] + a_2nx_2[n+1] \end{aligned}$$

The output due to the weighted sum of the input is,

$$\begin{aligned} y_4[n] &= a_1\{x_1[n] + nx_1[n+1]\} + a_2\{x_2[n] + nx_2[n+1]\} \\ y_3[n] &= y_4[n] \end{aligned}$$

The system is Linear.

4. As long as $x[n]$ is bounded, $y[n]$ is also bounded provided n is finite.

The system is Stable.

Otherwise the system is unstable. The system is

(1) Dynamic, (2) Non-causal, (3) Time Variant, (4) Linear and (5) Stable.

(c) $\boldsymbol{y[n] = x[n]u[n]}$

1.

$$\begin{aligned} y[0] &= x[0]u[0] \\ y[1] &= x[1]u[1] \end{aligned}$$

The output depends on present input only.

The system is Static and Causal.

2. For a causal signal $n \geq 0$. The weighted sum of the output due to input is,

$$\begin{aligned} y_3[n] &= a_1y_1[n] + a_2y_2[n] \\ &= \{a_1x_1[n] + a_2x_2[n]\}u[n] \end{aligned}$$

The output due to the weighted sum of input is,

$$\begin{aligned} y_4[n] &= \{a_1x_1[n] + a_2x_2[n]\}u[n] \\ y_3[n] &= y_4[n] \end{aligned}$$

The system is Linear.

3. The output due to the delayed input is,

$$y[n, n_0] = x_1[n - n_0]u[n]$$

The delayed output due to the input is,

$$y[n - n_0] = x_1[n - n_0]u[n - n_0]$$
$$y[n, n_0] \neq y[n - n_0]$$

The system is Time Variant.

4. As long as $x[n]$ is bounded, $y[n]$ is also bounded.

The system is Stable.

The system is

(1) Static, (2) Causal, (3) Linear, (4) Time Variant and (5) Stable.

(d) $\mathbf{y[n] = \log_{10} x[n]}$

1.

$$y[0] = \log_{10} x[0]$$
$$y[1] = \log_{10} x[1]$$
$$y[-1] = \log_{10} x[-1]$$

The output depends on present input only.

The system is Static and Causal.

2. The weighted sum of the output due to input is,

$$y_3[n] = a_1 y_1[n] + a_2 y_2[n]$$
$$= a_1 \log_{10} x_1[n] + a_2 \log_{10} x_2[n]$$

The output due to the weighted sum of input is,

$$y_4[n] = \log_{10}(a_1 x_1[n] + a_2 x_2[n])$$
$$y_3[n] \neq y_4[n]$$

The system is Non-linear.

3. The output due to the delayed input is,

$$y[n, n_0] = \log_{10} x[n - n_0]$$

The delayed output due to input is,

$$y[n - n_0] = \log_{10} x[n - n_0]$$
$$y[n, n_0] = y[n - n_0]$$

The system is Time Invariant.

4. As long as $x[n]$ is bounded, $\log_{10} x[n]$ is bounded and $y[n]$ is also bounded.

The system is Stable.

The system is

(1) Static, (2) Causal, (3) Non-linear, (4) Time invariant and (5) Stable.

(e) $\boldsymbol{y[n] = x^2[n]}$

1.
$$y[0] = x^2[0]$$
$$y[1] = x^2[1]$$

The output depends on present input only.

The system is Static and Causal.

2. The weighted sum of the output due to input is,

$$\begin{aligned} y_3[n] &= a_1y_1[n] + a_2y_2[n] \\ &= a_1x_1^2[n] + a_2x_2^2[n] \end{aligned}$$

The output due to weighted sum of input is,

$$\begin{aligned} y_4[n] &= \{a_1x_1[n] + a_2x_2[n]\}^2 \\ &= a_1^2x_1^2[n] + a_2^2x_2^2[n] + 2a_1a_2x_1[n]x_2[n] \\ y_3[n] &\neq y_4[n] \end{aligned}$$

The system is Non-linear.

3. The output due to the delayed input is,

$$y[n, n_0] = x^2[n - n_0]$$

The delayed output due to the input is,

$$y[n - n_0] = x^2[n - n_0]$$
$$y[n, n_0] = y[n - n_0]$$

The system is Time Invariant.

4. If $x[n]$ is bounded, $x^2[n]$ is bounded and $y[n]$ is also bounded.

The system is Stable.

The system is

(1) Static, (2) Causal, (3) Non-linear, (4) Time invariant and (5) Stable.

(f) $\mathbf{y[n] = x[n] \cos[\omega n]}$ For all values of n, the output $y[n]$ depends on the present input $x[n]$ only. Hence the system is static and causal (memoriless). The weighted sum of the output is,

$$\begin{aligned} y_3[n] &= a_1 y_1[n] + a_2 y_2[n] \\ &= a_1 x_1[n] \cos[\omega n] + a_2 x_2[n] \cos[\omega n] \end{aligned}$$

The output due to the weighted sum of the input is,

$$y_4[n] = [a_1 x_1[n] + a_2 x_2[n]] \cos[\omega n]$$
$$y_3[n] = y_4[n]$$

The output is zero when the input is zero. Hence the system is linear. The output due to the delayed input is

$$y[n, n_0] = x[n - n_0] \cos[\omega n]$$

The delayed output is given by

$$y[n - n_0] = x[n - n_0] \cos[\omega (n - n_0)]$$
$$y[n, n_0] \neq y[n - n_0]$$

The system is time varying. As long as $x[n]$ is bounded, $y[n]$ is also bounded. Since $\cos[\omega n]$ varies from -1 to $+1$. Hence the system is BIBO stable. The system is,

$$\boxed{\text{Static, Causal, Linear, Time varying and Stable.}}$$

Summary

1. The system is broadly classified as continuous and discrete time system.
2. The CT and DT systems are further classified based on the property of causality, linearity, time invariancy, invertibility, memory and stability.
3. The definitions of the above properties are given which are same for both CT and DT systems. Illustrative examples are given to explain these properties.

Exercises

I. Short Answer Type Questions

1. **What are the properties of systems?** Systems are generally classified as continuous and discrete time systems. Further classifications of these systems are done based on their properties which include (a) linear and non-linear, (b) time invariant and time variant, (c) static and dynamic, (d) causal and non-causal, (e) stable and unstable and (f) Invertible and non-invertible.
2. **Define system. What is linear system?** A system is defined as the interconnection of objects with a definite relationship between objects and attributes. A system is said to be linear if the weighted sum of several inputs produce weighted sum of outputs. In other words, the system should satisfy the homogeneity and additivity of super position theorem if it is to be linear. Otherwise it is a non-linear system. For a linear system if the input is zero the output should also be zero.
3. **What is time invariant and time varying system?** A system is said to be time invariant if the output due to the delayed input is same as the delayed output due to the input. If the continuous time system is described by the differential equation its coefficients should be time independent for the system to be time invariant. In the case of discrete time system, the coefficients of the difference equation describing the system should be time independent (constant) for the system to be time invariants. If the above conditions are not satisfied the system (CT as well as DT) is said to be time variant.

4. **What are static and dynamic systems?** If the output of the system depends only on the present input, the system is said to be static or instantaneous. If the output of the system depends on the past and future input, the system is not static and it is called dynamic system. Static system does not require memory and so it is called memoryless system. Dynamic system requires memory and hence, it is called system with memory. System which are described by differential and difference equations are dynamic systems.
5. **What are causal and non-causal systems?** If the system output depends on present and on past inputs, it is called causal system. If the system output depends on future input it is called non-causal system.
6. **What are stable and unstable systems?** If the input is bounded and output is also bounded, the system is called BIBO stable system. If the input is bounded and the output is unbounded, the system is unstable. System whose impulse response curve has finite area is also called stable systems.
7. **What are invertible and non-invertible systems?** A system is said to be invertible if the distinct inputs give distinct outputs.
8. **State the condition for a discrete time LTI system to be causal and stable.** (*Anna University, 2008*)
   A discrete time LTI system is said to be causal and stable if the poles of the transfer function all lie in the left half s-plane and the Region of Convergence (ROC) is to the right of the right most pole.
9. **What is the overall impulse response of $h_1(t)$ and $h_2(t)$ when they are in (a) series (b) parallel?** (*Anna University, 2005*)
   (a) The overall impulse repones when $h_1(t)$ and $h_2(t)$ are in series is given by
$$h(t) = h_1(t) * h_2(t)$$
   (b) If $h_1(t)$ and $h_2(t)$ are connected in parallel,
$$h(t) = (h_1(t) + h_2(t)) * x(t)$$
10. **Check whether the system having the input-output relation**
$$y(t) = \int_{-\infty}^{t} x(\tau)d\tau$$
   is linear and time invariant. (*Anna University, 2004*)
   The system is linear. (*See* Example 2.2(f)) By differentiating the above equation we get
$$\frac{dy(t)}{dt} = x(t)$$

The coefficient of the differential equation is time independent and is constant. Hence, it is a time invariant system.

11. **Check whether the system classified by**

$$y(y) = e^{x(t)}$$

is time invariant or not? *(Anna University, 2007)*

$$y(t, t_0) = e^{x(t-t_0)}$$
$$y(t - t_0) = e^{x(t-t_0)}$$
$$y(t, t_0) = y(t - t_0)$$

The system is time invariant.

12. **Determine whether the system described by the following input-output relationship is linear and causal?**

$$y(t) = x(-t)$$

(Anna University, 2007)

The system is linear and non-causal.

13. **Is the system $y(t) = \cos t\, x(t - 5)$ time invariant?**

$$y(t, t_0) = \cos tx(t - t_0 - 5)$$
$$y(t - t_0) = \cos(t - t_0)x(t - t_0 - 5)$$
$$y(t, t_0) \neq y(t - t_0)$$

The system is not time invariant.

14. **A certain LTID time system has the following impulse response.**

$$h[n] = \left(\frac{1}{3}\right)^n u[n - 2]$$

Is the system both causal and stable?

The response depends on the past input $u[n - 2]$ and hence, it is causal.

$$h[n] = \sum_{n=2}^{\infty} \left(\frac{1}{3}\right)^n$$
$$= \frac{1}{6} < \infty$$

The system is stable. The system is both causal and stable.

II. Long Answer Type Questions

The systems given below have input $x(t)$ or $x[n]$ and output $y(t)$ or $y[n]$. Determine whether each of them is (a) Static, (b) Casual, (c) Time Invariant, (d) Linear and (e) Stable.

1.
$$y(t) = \frac{d}{dt}[e^{-2t}x(t)]$$

(a) The system response requires memory. Hence, it is dynamic.
(b) The output depends on the present input only. Hence, it is causal.
(c) The output due to the delayed input is not the same as the delayed output. Hence, it is time variant.
(d) The weighted sum of the output is the same as output due to weighted sum of the input. Hence, the system is linear.
(e) Since $\frac{d}{dt}(e^{-2t}x(t))$ is bounded $y(t)$ is also bounded and hence, the system is stable.

2.
$$y(t) = x(t) + 10x(t-5) \qquad t \geq 0$$

(a) The output response depends on present and past inputs. Hence, it is dynamic.
(b) The output does not depend on the future input. Hence, it is causal.
(c) The output due to the delayed input is same as the delayed output. Hence, the system is time invariant.
(d) The weighted sum of the output is the same as output due to the weighted sum of the input. Hence, it is linear.
(e) As long as the input $x(t)$ is bounded, $x(t-5)$ is also bounded. Hence, $y(t)$ is bounded. The system is stable.

3.
$$y(t) = x(10t)$$

(a) The system response depends on present, past and future inputs. Hence, it is dynamic.
(b) Since the output depends on the future input, it is non-causal.
(c) The output due to the delayed input is not the same as the delayed output. Hence, the system is time variant.
(d) The weighted sum of the output is the same as output due to the weighted sum of the input. Hence, it is linear.
(e) If the input is bounded, the output is also bounded. The system is stable.

4.
$$y(t) = x\left(\frac{t}{10}\right)$$

The output depends on present, past and future inputs.

(a) The system is dynamic.
(b) The system is non-causal.
(c) The output due to the delayed input is not the same as the delayed output. The system is time variant.
(d) The weighted sum of the output will be the same as output due to the weighted sum of the input. The system is linear.
(e) If the input $x(\frac{t}{10})$ is bounded, the output is also bounded. The system is stable.

5.

$$\boldsymbol{y(t) = \frac{d}{dt}x(t-4)}$$

(a) The system requires memory and so it is dynamic.
(b) The output depends on past inputs. Hence, it is causal.
(c) The output due to the delayed input is same as the delayed output. The system is time invariant.
(d) The weighted sum of the output is the same as output due to the weighted sum of the input. The system is linear.
(e) If the input is bounded, the output is also bounded. The system is stable.

6.

$$\boldsymbol{y[n] = x[5n]}$$

(a) The system response depends on present, past and future inputs. Hence, it is dynamic.
(b) Non-causal.
(c) The output due to the delayed input is not same as the delayed output. Hence, it is time variant.
(d) The weighted sum of the output is same as output due to the weighted sum of the input. The system is linear.
(e) If the input $x[5n]$ is bounded, the output $y[n]$ is also bounded. The system is stable.

7.

$$\boldsymbol{y[n+2] + 3y[n+1] + 4y[n] = x[n]}$$

(a) The system is dynamic.
(b) The system is causal.
(c) The system is time invariant.
(d) The system is linear.
(e) The system is stable.

8.

$$\boldsymbol{y[n] = 5x[3^n]}$$

(a) The system is dynamic.
(b) The system is non-causal.
(c) The system is time invariant.
(d) The system is linear.
(e) The system is stable if n is finite.

9.

$$\mathbf{y[n] = \sin(2\pi x[n]) + x[n+1]}$$

(a) The system is dynamic.
(b) The system is non-causal.
(c) The system is time invariant.
(d) The system is non-linear.
(e) The system is stable for n being finite.

Chapter 3
Time Domain Analysis of Continuous and Discrete Time Systems

Learning Objectives

- ♦ To find the time response of an LTIC system by using convolution integral.
- ♦ To find the convolution of two time signals. Both analytical and graphical methods are used.
- ♦ To derive the properties of convolution of CT signals.
- ♦ To get the step response from impulse response and vice versa.
- ♦ To represent discrete time signals in terms of impulses.
- ♦ To establish the properties of convolution sum.
- ♦ To find the convolution of DT signals.
- ♦ To obtain step response, causality and stability of DT system from impulse response.

3.1 Introduction

A system as stated earlier performs a function. It operates on something and produces something else. Thus, when a system is excited by the input, it produces a response (output). Like signals, systems are also classified as Continuous Time (CT) and Discrete Time (DT) systems. If the input to the system is continuous and the output produced is also continuous, the system is called continuous time system. Such systems are described by differential equations. On the other hand if the input to the system is discrete in nature in the form of impulses, and if the output produced is also in the form of impulses such systems are called discrete time systems. These systems are represented in Fig. 3.1a and b, respectively.

S. Palani, *Signals and Systems*,
https://doi.org/10.1007/978-3-030-75742-7_3

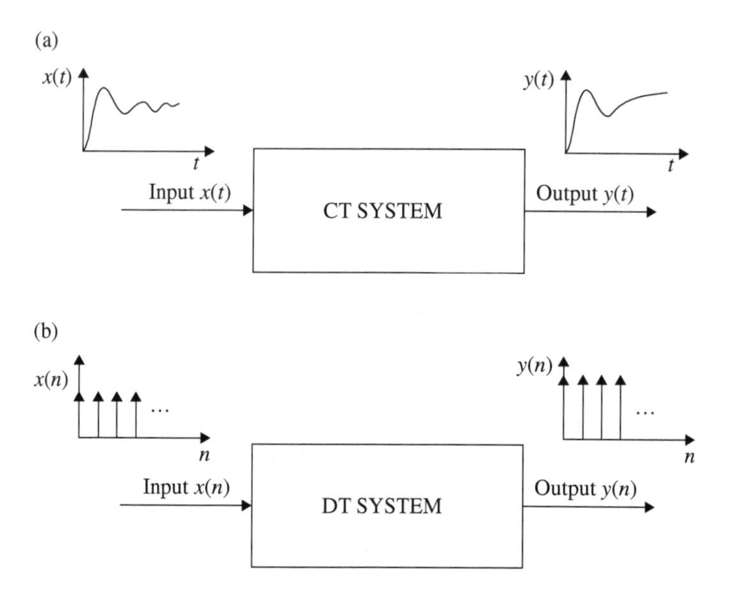

Fig. 3.1 Representation of CT and DT systems

3.2 Time Response of Continuous Time System

A linear time invariant continuous time system is described by the following differential equation:

$$\frac{d^n y(t)}{dt^n} + a_1 \frac{d^{n-1} y(t)}{dt^{n-1}} + \cdots + a_{n-1} \frac{dy(t)}{dt} + a_n y(t)$$
$$= b_{n-m} \frac{d^m x(t)}{dt^m} + b_{n-m-1} \frac{d^{m-1} x(t)}{dt^{m-1}} + \cdots + b_{n-1} \frac{dx(t)}{dt} + b_n x(t) \quad (3.1)$$

The coefficients a_i and b_i are constants. It should be noted that $m \leq n$ for the system to be Bounded Input-Bounded Output (BIBO) stable and to reduce the noise. The total response consists of two parts.

They are

- Zero input response (response due to initial conditions).
- Zero state response (response due to input alone).

Zero input response is the response of the system when the input $x(t) = 0$. The zero input response is obtained due to the initial conditions alone. The zero state response of the system is obtained when all the initial conditions are zero and only the input $x(t)$ alone is applied. The total response is the sum of zero input response and zero state response. The solution of Eq. (3.1) for total response is obtained by the following methods:

- By the application of classical method which gives the complete solution in terms of particular and homogeneous solutions.
- By the application of transform (Laplace and Fourier) techniques.
- By the method of convolution integral.

The transform techniques are very powerful compared to classical method to find the solution of $y(t)$ of Eq. (3.1). These methods are discussed in detail in Chaps. 7 and 8. The solution obtained from convolution integral is discussed in this chapter.

3.3 The Unit Impulse Response

Consider the LTIC time system shown in Fig. 3.2. Let the system be causal. Let the input $x(t) = \delta(t)$, an impulse which is characterized as given below

$$\begin{aligned} \delta(t) &= 1 \quad t = 0 \\ &= 0 \quad t \neq 0 \end{aligned} \tag{3.2}$$

The output response is now denoted by $h(t)$ which is called the impulse response of the system. This is illustrated in Fig. 3.2.

3.4 Unit Impulse Response and the Convolution Integral

Let $x(t)$ be any arbitrary input as shown in Fig. 3.3a. Figure 3.3b shows the input as a sum of narrow rectangular pulses. Consider the rectangular pulse shown in Fig. 3.3b in the shaded area. The width of the pulse is $\Delta\tau = 3\Delta\tau - 2\Delta\tau$. The area of this pulse is $x(2\Delta\tau)\Delta\tau$. As $\Delta\tau \to 0$, the pulse becomes a delta function of strength $x(2\Delta\tau)\Delta\tau$. The above delta function is represented as

$$x(2\Delta\tau)\Delta(\tau)\delta(t - 2\Delta\tau)$$

The function $x(t)$ is continuous and is the sum of the impulses occurring at $t = 0,\ \Delta\tau,\ 2\Delta\tau, \ldots, n\Delta\tau$. This can be expressed as

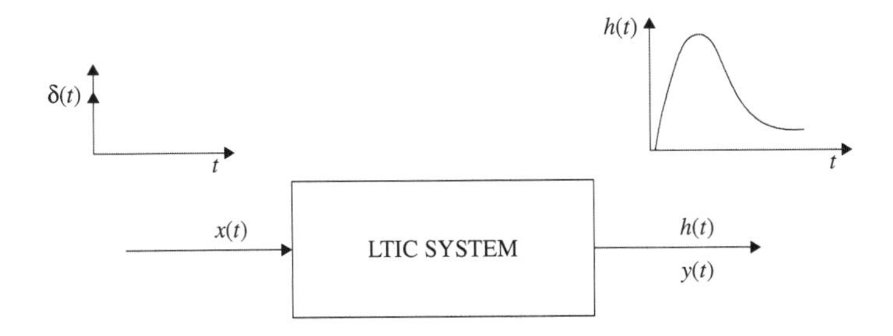

Fig. 3.2 Impulse response of LTIC system

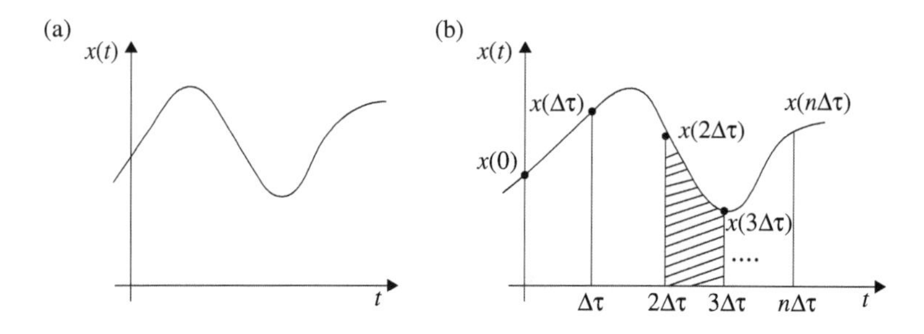

Fig. 3.3 Representation of any arbitrary input signal

Table 3.1 Output due to delta function

S.No	Input $x(t)$	Output $y(t)$
1.	$x(t) = \delta(t)$	$y(t) = h(t)$
2.	$x(t) = \delta(t - n\Delta\tau)$	$y(t) = h(t - n\Delta\tau)$
3.	$x(t) = [x(n\Delta\tau)\Delta\tau][\delta(t - n\Delta\tau)]$	$y(t) = [x(n\Delta\tau)\Delta\tau][h(t - n\Delta\tau)]$
4.	$x(t) = \underset{\Delta\tau\to 0}{Lt} \sum_{n=-\infty}^{\infty} [x(n\Delta\tau)][\delta(t - n\Delta\tau)\Delta\tau]$	$y(t) = \underset{\Delta\tau\to 0}{Lt} \sum_{n=-\infty}^{\infty} [x(n\Delta\tau)][h(t - n\Delta\tau)\Delta\tau]$

$$x(t) = \underset{\Delta\tau\to 0}{Lt} \sum_{-\infty}^{\infty} x(n\Delta\tau)(\Delta\tau)\delta(t - n\Delta\tau) \tag{3.3}$$

Let $h(t)$ be the unit impulse response of a linear time invariant continuous system. This is the output due to the input $\delta(t)$. The response of the system for the delta function at $t = n\Delta\tau$ is therefore

$$y(t) = x(n\Delta\tau)(\Delta\tau)h(t - n\Delta\tau) \tag{3.4}$$

The input and the corresponding output pairs are shown in Table 3.1 and represented in Fig. 3.4.

For the series of impulses, the response is the summation of their respective impulse responses. This is expressed mathematically as

$$y(t) = \underset{\Delta\tau\to 0}{Lt} \sum_{n=-\infty}^{\infty} x(n\Delta\tau)h(t - n\Delta\tau)\Delta\tau \tag{3.5}$$

When the limit $\Delta\tau \to 0$, the summation becomes integration and the integration is expressed as

$$y(t) = \int_{-\infty}^{\infty} x(\tau)h(t - \tau)d\tau \tag{3.5a}$$

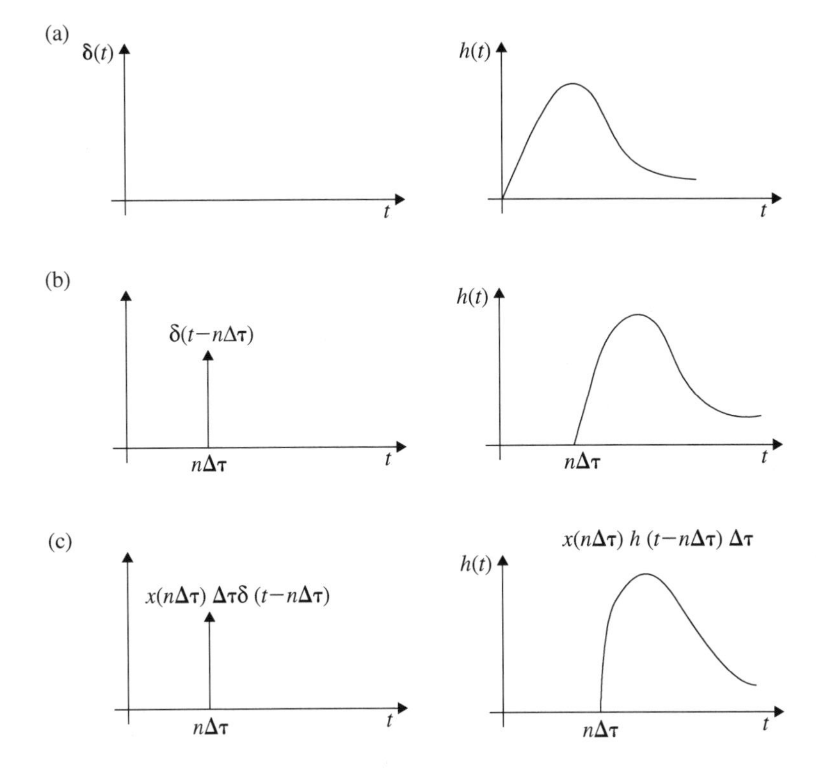

Fig. 3.4 The response of the system for delta function

The above integration is called convolution integral. The convolution operation of the two time functions $x(t)$ and $h(t)$ is symbolically denoted by

$$y(t) = x(t) * h(t) \tag{3.5b}$$

Thus, using convolution integral Eq. 3.5a, one can get output response $y(t)$ if the input $x(t)$ and the impulse response $h(t)$ are known. Equation 3.5a indeed gives the zero state response of the system.

3.5 Step by Step Procedure to Solve Convolution

The following steps are followed to determine the output response using convolution:

Step 1. Let $x(t)$ be the input signal and $h(t)$ the impulse response. Substitute $t = \tau$ where τ is an independent dummy variable and represent $x(\tau)$ and $h(\tau)$.

Step 2. Represent $x(\tau)$ in figure, invert $h(\tau)$ as $h(-\tau)$ and represent it in figure. This is called folding of $h(\tau)$. Shift the inverted $h(-\tau)$ along the τ axis and obtain $h(t-\tau)$ by giving very long negative shift.

Step 3. Multiply the two signals $x(\tau)$ and $h(t-\tau)$ and integrate over the overlapping interval. For this $x(\tau)$ is fixed and $h(t-\tau)$ is moved toward the right so that $x(\tau)$ and $h(t-\tau)$ overlap.

Step 4. Whenever either $x(\tau)$ or $h(t-\tau)$ changes, the new time shift occurs. Identify the end of the current interval and the beginning of the new interval. The output response $y(t)$ is calculated using Step 3.

Step 5. Steps 3 and 4 are repeated for all intervals.

Thus, the output response $y(t)$ can be determined analytically using convolution integral or by graphical method described above. The following are the properties of convolution and are discussed below

3.6 Properties of Convolution

Some important properties of convolution integral include

- The commutative property;
- The distributive property;
- The associative property;
- The shift property;
- The width property and
- The convolution with unit impulse.

The above properties are discussed below.

3.6.1 *The Commutative Property*

According to this property, if $y(t) = x(t) * h(t)$, then $y(t)$ can also be expressed as $y(t) = h(t) * x(t)$.

Proof According to Eq. (3.5), the output $y(t)$ is written in terms of convolution integral as

$$y(t) = \int_{-\infty}^{\infty} x(\tau)h(t-\tau)d\tau$$

Put $t - \tau = p$ and $d\tau = -dp$.

For $\tau = \infty; p = -\infty$; and $\tau = -\infty; p = \infty$

$$\begin{aligned} y(t) &= -\int_{+\infty}^{-\infty} x(t-p)h(p)dp \\ &= \int_{-\infty}^{\infty} x(t-p)h(p)dp \\ &= \int_{-\infty}^{\infty} h(p)x(t-p)dp \\ y(t) &= h(t) * x(t) \end{aligned}$$

$$\boxed{x(t) * h(t) = h(t) * x(t)} \tag{3.6}$$

3.6.2 The Distributive Property

Consider the two systems with impulse responses $h_1(t)$ and $h_2(t)$ connected in parallel as shown in Fig. 3.5a. The reduced block diagram is shown in Fig. 3.5b. According to the distributive property

$$x(t) * h_1(t) + x(t) * h_2(t) = x(t) * (h_1(t) + h_2(t))$$

Proof From Fig. 3.5b, the following equation is written

$$\begin{aligned} y(t) &= x(t) * (h_1(t) + h_2(t)) \\ &= x(t) * h_1(t) + x(t) * h_2(t) \end{aligned}$$

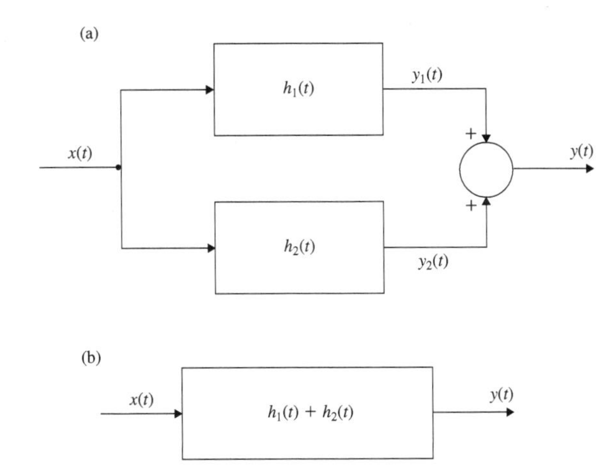

Fig. 3.5 The distributive property of convolution

For Fig. 3.5a, the following equations are written

$$y_1(t) = \int_{-\infty}^{\infty} x(\tau)h_1(t-\tau)d\tau = x(t) * h_1(t)$$

$$y_2(t) = \int_{-\infty}^{\infty} x(\tau)h_2(t-\tau)d\tau = x(t) * h_2(t)$$

$$\begin{aligned} y(t) &= y_1(t) + y_2(t) \\ &= x(t) * h_1(t) + x(t) * h_2(t) \end{aligned}$$

From Fig. 3.5b, the following equation is written

$$\begin{aligned} y(t) &= \int_{-\infty}^{\infty} x(\tau)[h_1(t-\tau) + h_2(t-\tau)]d\tau \\ &= x(t) * (h_1(t) + h_2(t)) \end{aligned}$$

Thus, the distributive property is proved as

$$\boxed{x(t) * h_1(t) + x(t) * h_2(t) = x(t) * [h_1(t) + h_2(t)]} \tag{3.7}$$

3.6.3 The Associative Property

According to this property

$$[x(t) * h_1(t)] * h_2(t) = x(t) * [h_1(t) * h_2(t)]$$

Proof Consider the two systems connected in cascade with their impulse responses $h_1(t)$ and $h_2(t)$ as shown in Fig. 3.6. The excitation signal is $x(t)$, and the output response is $y(t)$. From Fig. 3.6, the following equation is written. The output of the first system is

$$y_1(t) = \int_{\tau=-\infty}^{\infty} x(\tau)h_1(t-\tau)d\tau$$

The output of the second system is

$$y(t) = y_1(t) * h_2(t) = \int_{m=-\infty}^{\infty} y_1(m)h_2(t-m)dm$$

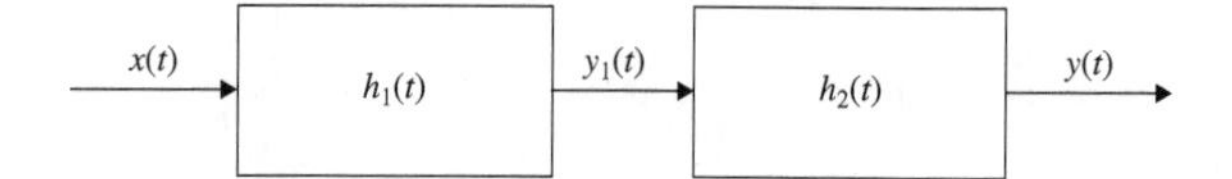

Fig. 3.6 The associative property of convolution

$$= \int_{m=-\infty}^{\infty} \left[\int_{\tau=-\infty}^{\infty} x(\tau)h_1(m-\tau)d\tau \right] h_2(t-m)dm$$

Put $m - \tau = q$ and $dm = dq$

$$y(t) = \int_{\tau=-\infty}^{\infty} x(\tau) \left[\int_{q=-\infty}^{\infty} h_1(q)h_2[(t-(q+\tau)]dq \right] d\tau$$
$$= \int_{\tau=-\infty}^{\infty} x(\tau) \left[\int_{q=-\infty}^{\infty} h_1(q)h_2[(t-\tau)-q]dq \right] d\tau$$

But

$$\int_{q=-\infty}^{\infty} h_1(q)h_2[(t-\tau)-q]dq = h_1(t) * h_2(t) = h(t)$$

Therefore,

$$y(t) = \int_{\tau=-\infty}^{\infty} x(\tau)h(t-\tau)d\tau$$
$$= x(t) * h(t)$$
$$y(t) = x(t) * [h_1(t) * h_2(t)]$$

$$\boxed{[x(t) * h_1(t)] * h_2(t) = x(t) * [h_1(t) * h_2(t)]} \tag{3.8}$$

3.6.4 The Shift Property

According to the shift property

$$x(t) * h(t-T) = y(t-T)$$

Proof Let $x(t)$ and $h(t)$ be the input and the impulse response functions, respectively. Let $h(t)$ be shifted by T as $h(t-T)$. Then, the convolution of these signals is given by

$$x(t) * h(t) = \int_{-\infty}^{\infty} x(\tau)h(t-\tau)d\tau$$

$$x(t) * h(t-T) = \int_{-\infty}^{\infty} x(\tau)h(t-\tau-T)d\tau$$
$$= \int_{-\infty}^{\infty} x(\tau)h[(t-T)-\tau]d\tau$$
$$= y(t-T)$$

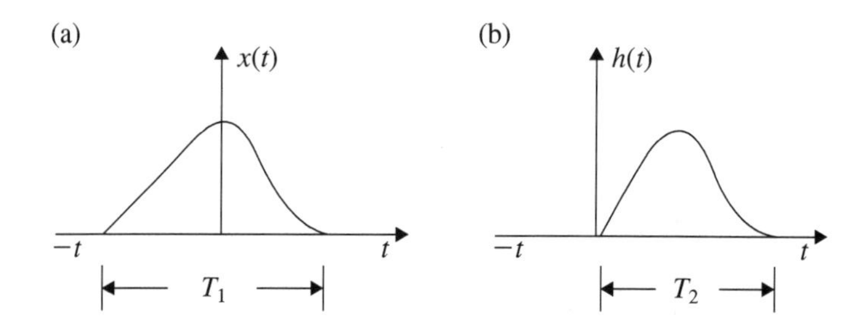

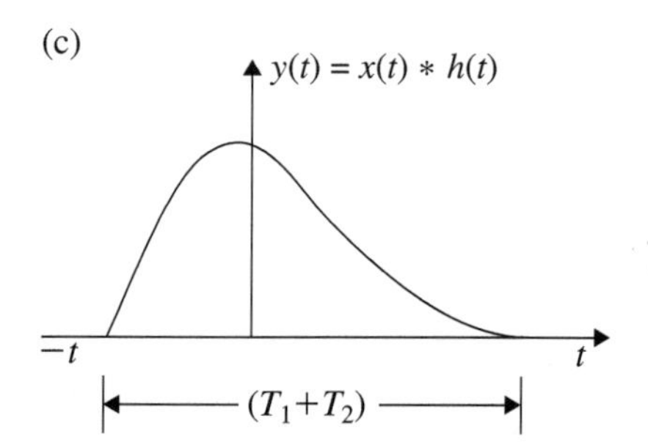

Fig. 3.7 Width property of convolution

Thus,

$$\boxed{x(t) * h(t - T) = y(t - T)} \tag{3.9}$$

3.6.5 *The Width Property*

Let T_1 be the width of $x(t)$ and T_2, the width of $h(t)$. The width of $y(t) = x(t) * h(t)$ is $T_1 + T_2$.

Proof Figure 3.7a shows $x(t)$ of T_1 width and Fig. 3.7b shows $h(t)$ which has a width of T_2. In convolution $h(t - \tau)$ is put in the extreme left and moved toward the right keeping $x(\tau)$ fixed. The leading edge of $h(t - \tau)$ overlaps with the left most edge of $x(\tau)$ and passes through a width of $x(\tau)$ which is T_1. Thus, the duration of overlap is $(T_2 + T_1)$.

3.7 Analytical Method of Convolution Operation

The following are the basic steps involved in convolution integral equation:

1. For the input signal $x(t)$, express it as $x(\tau)$.
2. Express impulse response function $h(t)$ as $h(t - \tau)$ by substituting $t = (t - \tau)$.

3. Integrate $\int_{\tau=-\infty}^{\infty} x(\tau)h(t-\tau)d\tau$ to get $y(t)$.
4. The limit of integration depends on the time limit of $x(t)$. For causal signals $x(t)$ and $h(t)$, the lower limit of convolution integral is zero and the upper limit is t.
5. Find $y(t) = \int_{\tau=-\infty}^{\infty} x(\tau)h(t-\tau)d\tau$.

The following examples, illustrate this.

■ Example 3.1

The impulse response of a certain system is $h(t) = e^{-5t}u(t)$. Find the output response of the system for the input $x(t) = e^{-2t}u(t)$.

Solution:

1.

$$\begin{aligned} x(t) &= e^{-2t} \\ x(\tau) &= e^{-2\tau} \\ h(t) &= e^{-5t} \\ h(\tau) &= e^{-5\tau} \\ h(-\tau) &= e^{5\tau} \\ h(t-\tau) &= e^{-5(t-\tau)} \end{aligned}$$

2. Both $x(t)$ and $h(t)$ are causal. Hence, the limit of convolution integral is from 0 to t. Thus, $y(t)$ is expressed as

$$y(t) = \int_0^t x(\tau)h(t-\tau)d\tau$$

$$\begin{aligned} y(t) &= \int_0^t e^{-2\tau} e^{-5(t-\tau)} d\tau \\ &= e^{-5t} \int_0^t e^{3\tau} d\tau \\ &\quad [e^{-5t} \text{ is a constant when the integration is done for } \tau] \\ &= \frac{e^{-5t}}{3} \left[e^{3\tau} \right]_0^t \\ &= \frac{e^{-5t}}{3} [e^{3t} - 1]u(t) \end{aligned}$$

$$\boxed{y(t) = \frac{1}{3}[e^{-2t} - e^{-5t}]u(t)}$$

The graphical representation of convolution operation is shown in Fig. 3.8.

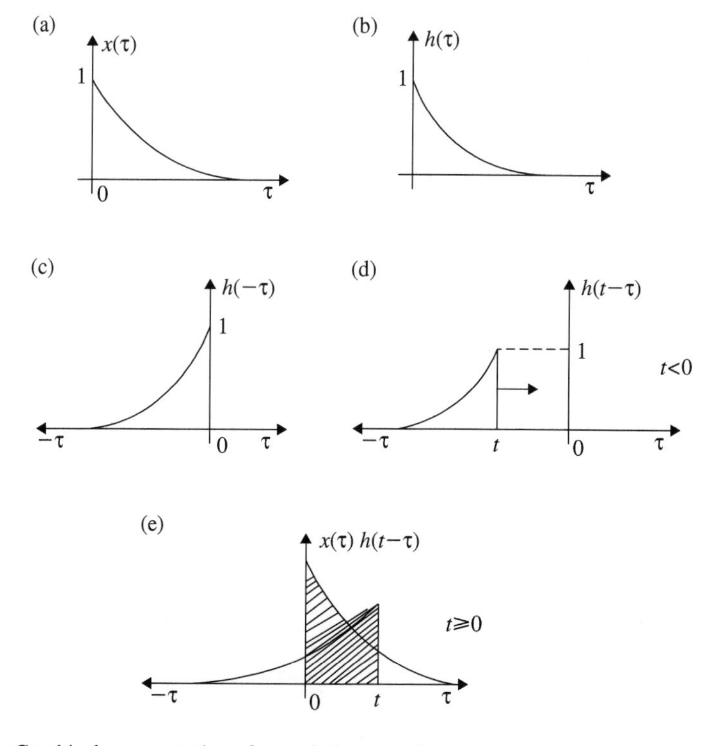

Fig. 3.8 Graphical representation of convolution operation

In Fig. 3.8, $x(\tau)$ is fixed and $h(t-\tau)$ is shifted toward the right so that there is overlapping between $x(\tau)$ and $h(t-\tau)$.

■ Example 3.2

Find $y(t)$ if $x(t) = u(t)$ and $h(t) = u(t)$.

Solution:

1. $$x(\tau) = 1$$

2. $$h(\tau) = 1$$
$$h(-\tau) = 1$$
$$h(t-\tau) = 1$$

3. Both $x(t)$ and $h(t)$ are causal. Hence, the limit of integration is from 0 to t. Thus, $y(t)$ is obtained from

$$y(t) = \int_0^t d\tau = [\tau]_0^t$$

$$\boxed{y(t) = tu(t)}$$

■ Example 3.3

Find $y(t)$ if $x(t) = e^{at}u(t)$ and $h(t) = u(t)$.

Solution:

1.

$$x(t) = e^{at}u(t)$$
$$x(\tau) = e^{a\tau}$$

2.

$$h(t) = u(t)$$
$$h(\tau) = 1$$
$$h(t-\tau) = 1$$

3. Both $x(t)$ and $h(t)$ are causal. Hence, the limit of integration is from 0 to t. Thus, $y(t)$ is obtained from the following integral.

$$y(t) = \int_0^t e^{a\tau} d\tau$$
$$= \frac{1}{a}\left[e^{a\tau}\right]_0^t$$

$$\boxed{y(t) = \frac{1}{a}[e^{at} - 1]u(t)}$$

■ Example 3.4

Find $y(t)$ if $x(t) = e^{a_1 t}u(t)$ and $h(t) = e^{a_2 t}u(t)$.

Solution:

1.

$$x(t) = e^{a_1 t}$$
$$x(\tau) = e^{a_1 \tau}$$

2.

$$h(t) = e^{a_2 t}$$
$$h(-\tau) = e^{-a_2 \tau}$$
$$h(t-\tau) = e^{a_2 (t-\tau)}$$

3. Both $x(t)$ and $h(t)$ are causal signals. Hence, the limit of integration is from 0 to t. Thus, $y(t)$ is obtained from the following integral.

$$\begin{aligned} y(t) &= \int_0^t e^{a_1 \tau} e^{a_2 (t-\tau)} d\tau \\ &= e^{a_2 t} \int_0^t e^{(a_1 - a_2)\tau} d\tau \\ &= \frac{e^{a_2 t}}{(a_1 - a_2)} \left[e^{(a_1 - a_2)\tau} \right]_0^t \\ &= \frac{e^{a_2 t}}{(a_1 - a_2)} [e^{(a_1 - a_2)t} - 1] \end{aligned}$$

$$\boxed{y(t) = \frac{[e^{a_1 t} - e^{a_2 t}]}{(a_1 - a_2)} u(t)} \quad a_1 \neq a_2$$

■ Example 3.5

Find $y(t)$ if $x(t) = e^{at}u(t)$ and $h(t) = e^{at}u(t)$.

Solution:

1.

$$x(t) = e^{at}$$
$$x(\tau) = e^{a\tau}$$

2.

$$h(t) = e^{at}$$
$$h(\tau) = e^{a\tau}$$
$$h(t-\tau) = e^{a(t-\tau)}$$

3. Both $x(t)$ and $h(t)$ are causal signals. Hence, the limit of integration is from 0 to t. Thus, $y(t)$ is obtained from

$$y(t) = \int_0^t e^{a\tau} e^{a(t-\tau)} d\tau$$
$$= e^{at} \int_0^t e^0 d\tau$$
$$= e^{at} \Big[\tau \Big]_0^t$$

$$\boxed{y(t) = te^{at}u(t)}$$

■ Example 3.6

Find $y(t)$ if $x(t) = e^{-3t}u(t)$ and $h(t) = (2 - e^{-2t})u(t)$.

Solution:

1. It can be shown that

$$\int_{-\infty}^{\infty} x(\tau)h(t-\tau)d\tau = \int_{-\infty}^{\infty} x(t-\tau)h(\tau)d\tau$$

The above property is used for convenience.

2.

$$h(t) = (2 - e^{-2t})$$
$$h(\tau) = (2 - e^{-2\tau})$$

3.

$$x(t) = e^{-3t}$$
$$x(\tau) = e^{-3\tau}$$
$$x(t-\tau) = e^{-3(t-\tau)}$$

4. $y(t)$ is obtained by taking the limit of integration from 0 to t since $x(t)$ and $h(t)$ are causal signals

$$y(t) = \int_0^t x(t-\tau)h(\tau)d\tau$$
$$= \int_0^t e^{-3(t-\tau)}(2 - e^{-2\tau})d\tau$$
$$= e^{-3t} \int_0^t (2e^{3\tau} - e^{\tau})d\tau$$
$$= e^{-3t} \left[\frac{2}{3}e^{3\tau} - e^{\tau} \right]_0^t$$
$$= e^{-3t} \left[\frac{2}{3}e^{3t} - \frac{2}{3} - e^t + 1 \right]$$

$$\boxed{y(t) = \left[\frac{2}{3} + \frac{1}{3}e^{-3t} - e^{-2t}\right]u(t)}$$

■ Example 3.7

Convolve the signal

$$x(t) = e^{-2t}u(t) \quad \text{with} \quad h(t) = u(t).$$

(Anna University, April, 2005)

Solution:

1.

$$x(t) = e^{-2t}$$
$$x(\tau) = e^{-2\tau}$$

2.

$$h(t) = u(t)$$
$$h(\tau) = u(\tau)$$
$$h(t-\tau) = u(t-\tau) = 1$$

3. Both $x(t)$ and $h(t)$ are causal signals. Hence,

$$\begin{aligned} y(t) &= \int_0^t x(\tau)h(t-\tau)d\tau \\ &= \int_0^t e^{-2\tau}d\tau \\ &= -\frac{1}{2}\left[e^{-2\tau}\right]_0^t \end{aligned}$$

$$\boxed{y(t) = \frac{1}{2}[1 - e^{-2t}]u(t)}$$

■ Example 3.8

Find $y(t)$ if $x(t) = t$ and $h(t) = u(t)$.

Solution:

1.

$$x(t) = t$$
$$x(\tau) = \tau$$

2.

$$h(t) = u(t)$$
$$h(t-\tau) = u(t-\tau) = 1$$

3. Both $x(t)$ and $h(t)$ are causal. Hence

$$\begin{aligned} y(t) &= \int_0^t x(\tau)h(t-\tau)d\tau \\ &= \int_0^t \tau d\tau \\ &= \frac{1}{2}\left[\tau^2\right]_0^t \end{aligned}$$

$$\boxed{y(t) = \frac{1}{2}t^2u(t)}$$

■ Example 3.9

Find $y(t)$ if $x(t) = \sin at\, u(t)$ and $h(t) = u(t)$.

(*Anna University, December, 2007*)

Solution:

1.

$$x(t) = \sin at\, u(t)$$
$$x(\tau) = \sin a\tau$$

2.

$$h(t) = u(t)$$
$$h(t-\tau) = u(t-\tau) = 1$$

3. Both $x(t)$ and $h(t)$ are causal signals. Hence, $y(t)$ is determined from the following integral:

$$\begin{aligned} y(t) &= \int_0^t x(\tau)h(t-\tau)d\tau \\ &= \int_0^t \sin a\tau d\tau \\ &= \frac{1}{a}\Big[-\cos a\tau\Big]_0^t \end{aligned}$$

$$\boxed{y(t) = \frac{1}{a}[1 - \cos at]u(t)}$$

3.7.1 Convolution Operation of Non-causal Signals

For non-causal signals, the limit of convolution integral depends on the time limit of $x(t)$. These limits should be carefully evaluated and the convolution integral is solved. The following examples illustrate the method of solving convolution integral when the signal is non-causal.

Example 3.10

Solve for $y(t)$ if $h(t) = e^{a_1 t}u(t)$ and $x(t) = e^{a_2 t}u(-t)$.

Solution:

1.

$$x(t) = e^{a_2 t}u(-t)$$
$$x(\tau) = e^{a_2 \tau}u(-\tau)$$

2. The fixed signal is shown in Fig. 3.9a.

$$h(t) = e^{a_1 t}u(t)$$
$$h(\tau) = e^{a_1 \tau}u(\tau)$$
$$h(t-\tau) = e^{a_1 (t-\tau)}u(t-\tau)$$

   $h(\tau)$, $h(-\tau)$, and $h(t-\tau)$ for $t < 0$ and $t > 0$ are shown in Fig. 3.9b, c, d and e, respectively.
3. The overlapping of $h(t-\tau)$ with $x(\tau)$ is shown in Fig. 3.9f for $t < 0$ and in Fig. 3.9g for $t > 0$. The overlapping occurs for $-\infty < \tau < t$ and for $-\infty < \tau < 0$, respectively.
4. For $t < 0$, the convolution integral and the corresponding output $y_1(t)$ are given below

$$\begin{aligned} y_1(t) &= \int_{-\infty}^{t} x(\tau)h(t-\tau)d\tau \\ &= \int_{-\infty}^{t} e^{a_2\tau}e^{a_1(t-\tau)}d\tau \\ &= e^{a_1 t}\int_{-\infty}^{t} e^{(a_2-a_1)\tau}d\tau \\ &= \frac{e^{a_1 t}}{(a_2-a_1)}\left[e^{(a_2-a_1)\tau}\right]_{-\infty}^{t} \end{aligned}$$

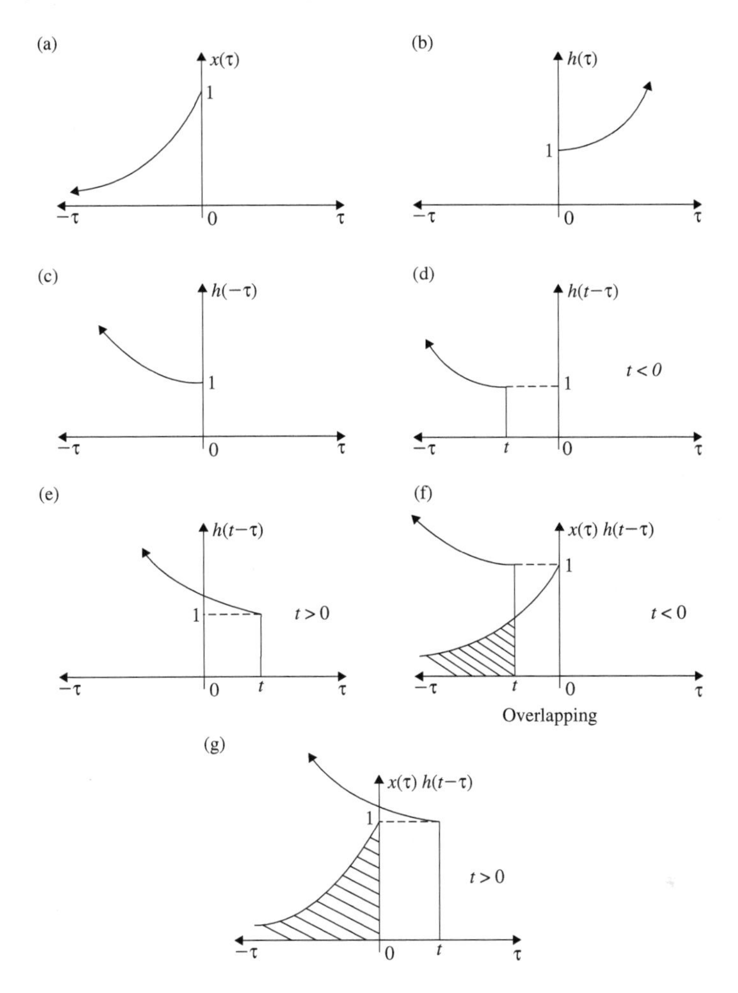

Fig. 3.9 Graphical representation of convolution

$$= \frac{e^{a_1 t}}{(a_2 - a_1)}\left[e^{(a_2-a_1)t}\right]$$

iff $a_2 > a_1$ for convergence.

$$y_1(t) = \frac{e^{a_2 t}}{(a_2 - a_1)}u(-t)$$

5. For $t > 0$, the overlapping occurs for $-\infty < \tau < 0$. The convolution integral and the corresponding output $y_2(t)$ are given by

$$
\begin{aligned}
y_2(t) &= \int_{-\infty}^{0} x(\tau)h(t-\tau)d\tau \\
&= \int_{-\infty}^{0} e^{a_2\tau} e^{a_1(t-\tau)} d\tau \\
&= e^{a_1 t} \int_{-\infty}^{0} e^{(a_2-a_1)\tau} d\tau \\
&= \frac{e^{a_1 t}}{(a_2-a_1)} \left[e^{(a_2-a_1)\tau} \right]_{-\infty}^{0} \\
y_2(t) &= \frac{e^{a_1 t}}{(a_2-a_1)} u(t)
\end{aligned}
$$

6. The output response $y(t)$ is the sum of $y_1(t)$ and $y_2(t)$. Thus,

$$y(t) = y_1(t) + y_2(t)$$

$$\boxed{y(t) = \frac{[e^{a_2 t}u(-t) + e^{a_1 t}u(t)]}{(a_2-a_1)}} \qquad a_2 > a_1$$

■ Example 3.11

Find $y(t)$ if $x(t) = e^{3t}u(-t)$ and $h(t) = u(t-2)$ using convolution integral.

Solution:

1. $x(\tau)$ and $h(\tau)$ are represented in Fig. 3.10a and b, respectively.
2. Fig. 3.10b is folded to get $h(-\tau)$ and then time shifted to the extreme left and is shown in Fig. 3.10c.
3. Figure 3.10a and c is combined in Fig. 3.10d which represents $x(\tau)h(t-\tau)$. When $h(t-\tau)$ is moved toward right it overlaps with $x(\tau)$ for the time interval $-\infty < \tau < t-2$. Hence, the lower limit of integration is $-\infty$ and the upper limit of convolution integral is $(t-2)$. Let $y_1(t)$ be the output now.

$$
\begin{aligned}
y_1(t) &= \int_{-\infty}^{(t-2)} x(\tau)h(t-\tau)d\tau \\
&= \int_{-\infty}^{(t-2)} e^{3\tau} d\tau = \frac{1}{3}\left[e^{3\tau}\right]_{-\infty}^{t-2} \\
&= \frac{1}{3}\left[e^{3(t-2)}\right]u(-t+2) \qquad t < 2
\end{aligned}
$$

4. When $h(t-\tau)$ is moved further toward the right, the right edge of $h(t-\tau)$ slides past the right edge of $x(\tau)$ for $\tau > 0$. The lower limit of the convolution integral is $-\infty$ and the upper limit is zero since for $\tau > 0$, there is no overlapping between $x(\tau)$ and $h(t-\tau)$. Let $y_2(t)$ be the output now.

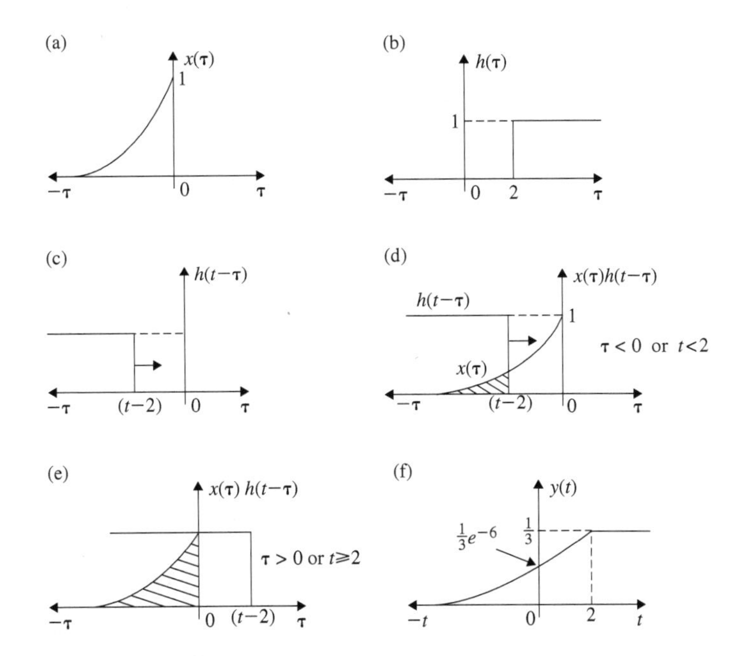

Fig. 3.10 Convolution of $e^{3t}u(-t)$ and $u(t-2)$

$$y_2(t) = \int_{-\infty}^{0} e^{3\tau} d\tau = \left[\frac{1}{3}e^{3\tau}\right]_{-\infty}^{0}$$

$$= \frac{1}{3}u(t-2) \qquad t \geq 2$$

The total output response

$$y(t) = y_1(t) + y_2(t)$$

$$\boxed{y(t) = \frac{1}{3}[e^{3(t-2)}u(-t+2) + u(t-2)]}$$

■ Example 3.12

Find the convolution of the following signals

$$x(t) = e^{-2t}u(t) \text{ and } h(t) = u(-t)$$

and plot the output response with respect to t.

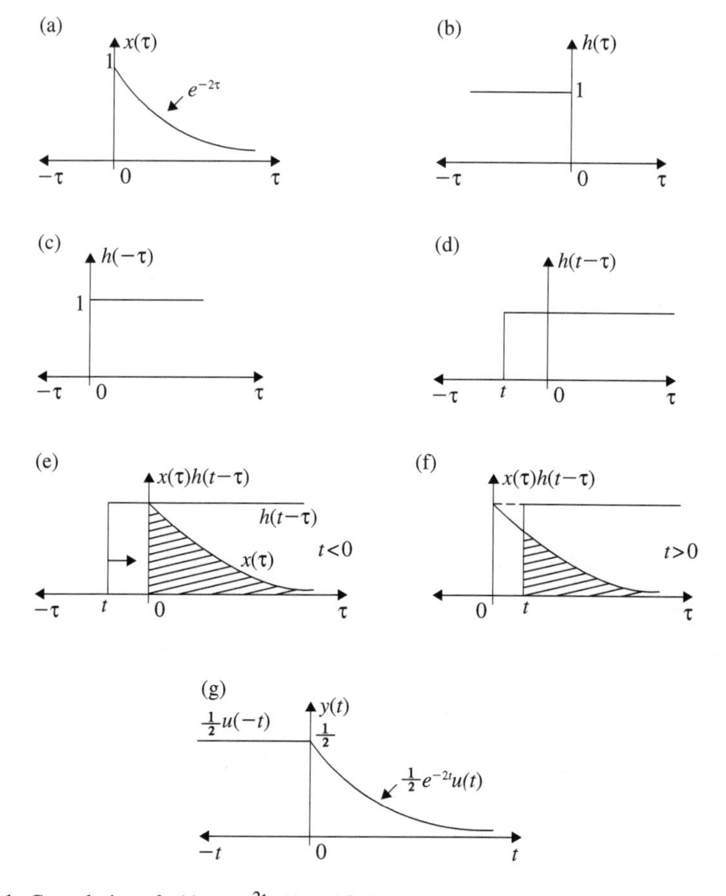

Fig. 3.11 Convolution of $x(t) = e^{-2t}u(t)$ and $h(t) = u(-t)$ and the output response curve

Solution:

1. $x(\tau), h(\tau), h(t-\tau)$ are shown in Fig. 3.11a, b and c, respectively.
2. Figure 3.11d shows $h(t-\tau)$ where $h(t-\tau)$ is left shifted to the extreme.
3. Figure 3.11e shows $x(\tau)$ and $h(t-\tau)$. It is observed that for $-\infty < \tau < t$, there is no overlapping and hence $y(t) = 0$. For $t < 0$ there is overlapping and $y(t)$ is found as follows.
4.

$$\begin{aligned} y(t) &= \int_0^\infty e^{-2\tau} d\tau \\ &= -\frac{1}{2}\Big[e^{-2\tau}\Big]_0^\infty = \frac{1}{2}u(-t) \end{aligned}$$

5. For $t > 0$, $x(\tau)h(t-\tau)$ is shown in Fig. 3.11f.

$$y(t) = \int_t^\infty e^{-2\tau} d\tau = \frac{1}{2} e^{-2t} \qquad t > 0$$

$$\boxed{y(t) = \frac{1}{2}[u(-t) + e^{-2t} u(t)]}$$

6. The plot of $y(t)$ with respect to t is sketched and shown in Fig. 3.11g.

■ Example 3.13

Find the convolution of the following signals and find the output response

$$x(t) = u(t-2)$$
$$h(t) = u(t+2)$$

Solution:

1. $x(t)$, $h(t)$, $x(\tau)$ and $h(-\tau)$ and $h(t-\tau)$ are shown in Fig. 3.12a, b, c, d and e, respectively.
2. $h(t-\tau)$ when shifted to the extreme left is shown in Fig. 3.12f and with $x(\tau)$.
3. $h(t-\tau)$ when moved toward the right, it does not overlap for $-\infty < \tau < 2$. However, it overlaps for $2 \le \tau < (t+2)$. Hence, the limits of the convolution integration is from $\tau = 2$ to $\tau = t+2$. The overlapping area is shown in Fig. 3.12g.
4. $y(t)$ is obtained from the following convolution integral by putting $x(\tau) = 1$ and $h(t-\tau) = 1$

$$y(t) = \int_2^{t+2} x(\tau) h(t-\tau) d\tau$$
$$= \int_2^{t+2} d\tau = [\tau]_2^{t+2}$$

$$\boxed{y(t) = t\,u(t)}$$

5. The plot of $y(t)$ with respect to time t is shown in Fig. 3.12h which is a straight line with unit slope.

■ Example 3.14

Find the convolution of the signal shown in Fig. 3.13a and b by graphical method.

(Anna University, December, 2006)

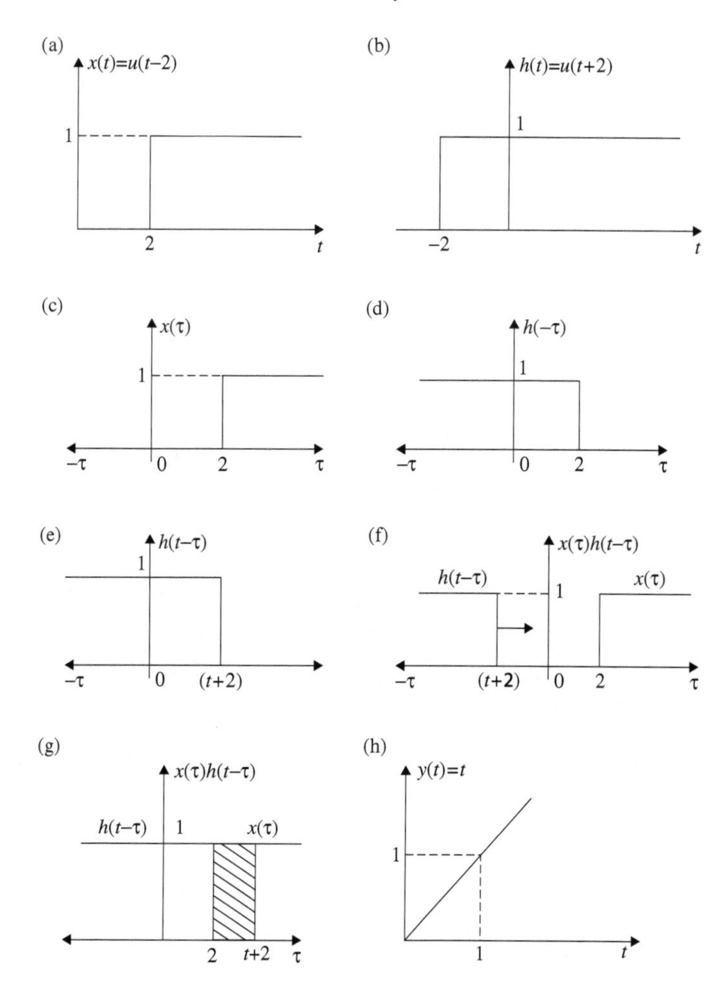

Fig. 3.12 Convolution of $x(t) = u(t-2)$ and $h(t) = u(t+2)$ and the output response curve

Solution:

1. The given rectangular or gate signals $x(t)$ and $h(t)$ are shown in Fig. 3.13a and b, respectively.
2. $h(-\tau)$ is obtained by putting $t = \tau$ in $h(t)$ and then by folding (inversion). This is shown in Fig. 3.13c.
3. $h(t-\tau)$ is obtained by adding t with $-\tau$. $h(t-\tau)$ is shifted to the extreme left so that $x(\tau)$ and $h(t-\tau)$ do not have any overlapping. This is shown in Fig. 3.13d.
4. Now $h(t-\tau)$ is moved toward right so that it overlaps with $x(\tau)$. The right edge CD of $h(t-\tau)$ when slides past the left edge EF of $x(\tau)$, overlapping starts. This is shown in Fig. 3.13e. For one sliding at a time the time interval is $-2 < t < 0$ as seen from Fig. 3.13e. Further, from the overlapping area, the lower limit of

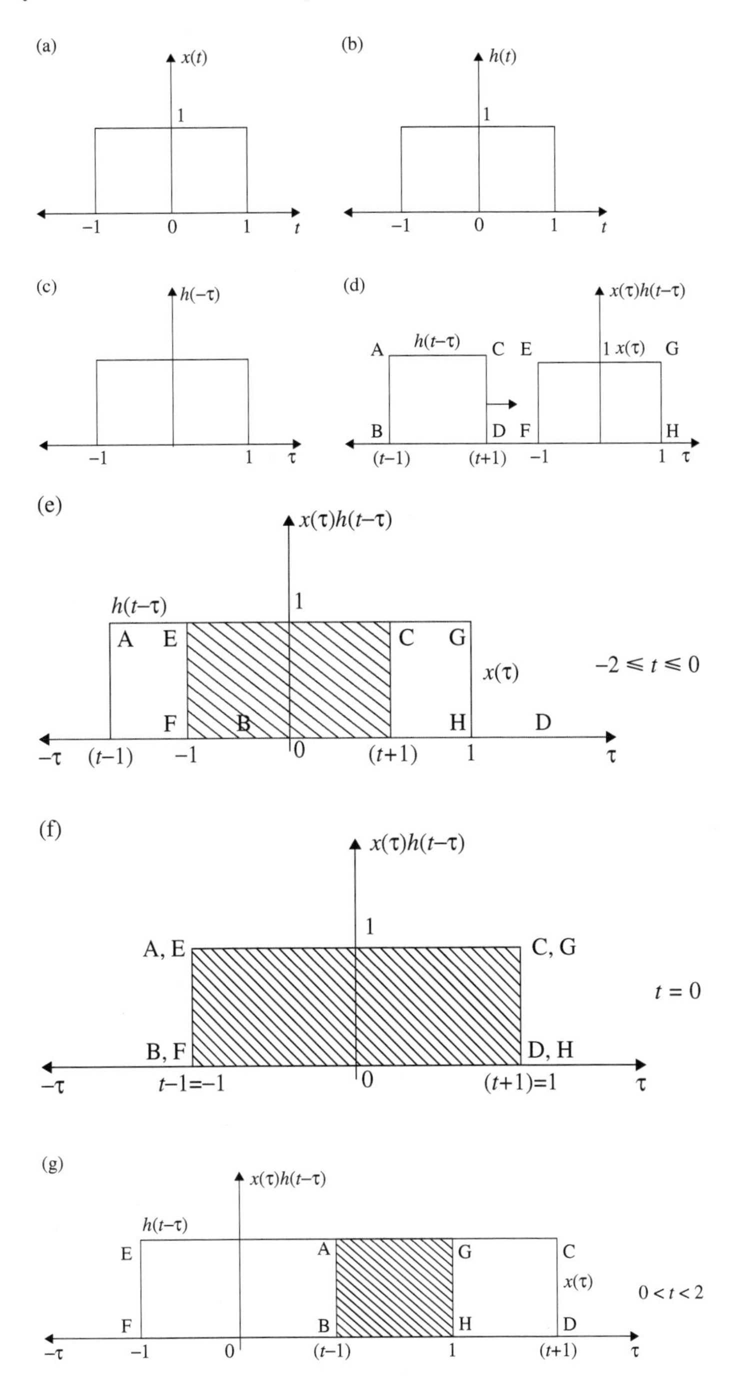

Fig. 3.13 Convolution of two gate signals

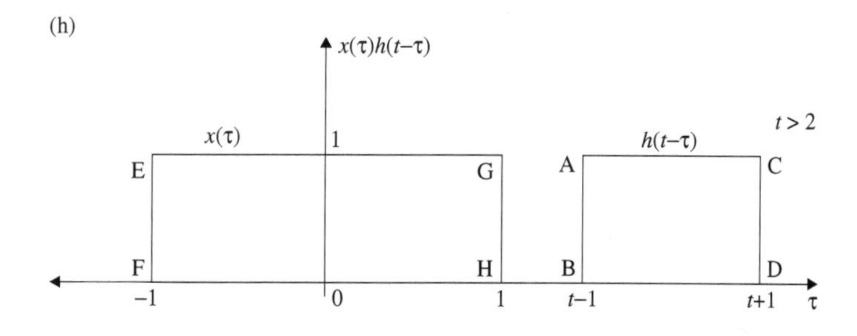

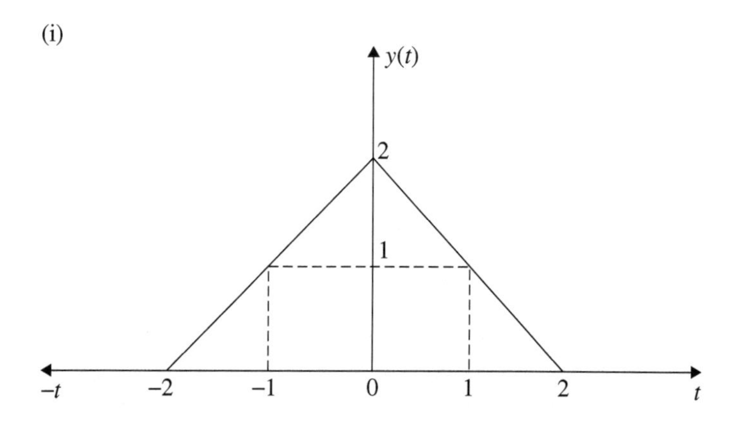

Fig. 3.13 (continued)

integration is -1 and the upper limit is $(t + 1)$. The following integral is solved.

$$\begin{aligned} y(t) &= \int_{-1}^{t+1} x(\tau)h(t-\tau)d\tau \\ &= \int_{-1}^{t+1} d\tau \\ &= \Big[\tau\Big]_{-1}^{t+1} \end{aligned}$$

$$\boxed{y(t) = (t+2)} \quad -2 < t < 0 \tag{a}$$

5. Now $h(t - \tau)$ is further moved toward the right. When $t = 0$, AB edge coincides with EF edge and CD edge coincides with GH edge simultaneously. This is shown in Fig. 3.13f.

   From Fig. 3.13f, the overlapping occurs during the interval $-1 < \tau < 1$. Hence, the lower limit of the convolution integral is -1 and the upper limit is 1.

$$y(t) = \int_{-1}^{1} x(\tau)h(t-\tau)d\tau$$

$$= \int_{-1}^{1} d\tau$$

$$= \Big[\tau\Big]_{-1}^{1}$$

$$\boxed{y(t) = 2} \qquad t = 0 \tag{b}$$

6. $h(t-\tau)$ is further shifted toward the right. Now the right edge of $x(\tau)$ which is CD slides past the right edge of $h(t-\tau)$ which is GH. The overlapping area is shown in Fig. 3.13g. This occurs during the time interval $0 < t < 2$. The lower limit of the integral is $(t-1)$ and the upper limit is 1. Hence,

$$y(t) = \int_{t-1}^{1} x(\tau)h(t-\tau)$$

$$= \int_{t-1}^{1} d\tau$$

$$= \Big[\tau\Big]_{t-1}^{1}$$

$$\boxed{y(t) = (2-t)} \qquad 0 < t < 2 \tag{c}$$

7. Consider $h(t-\tau)$ for $t > 2$. Now $h(t-\tau)$ and $x(\tau)$ do not overlap. Consequently $y(t) = 0$. This is shown in Fig. 3.13h.
8. Now equations (a), (b) and (c) are used to find $y(t)$ in the respective time interval.

$$\begin{aligned} y(t) &= (t+2) \qquad && -2 < t < 0 \\ &= 2 && t = 0 \\ &= (2-t) && 0 < t < 2 \\ &= 0 && t > 2 \end{aligned}$$

t	-2	-1	0	1	2
$y(t)$	0	1	2	1	0

The time response graph is shown in Fig. 3.13i. Here, it should be noted that for equation (a) for $t = 0$, $y(t) = 2$. From equation (b) for $t = 0$, $y(t) = 2$.

9. The result can be easily obtained by using convolution property of Laplace transform (LT) which is discussed in Chap. 8. The output is obtained as

$$\boxed{y(t) = (t+2)u(t+2) + (t-2)u(t-2) - 2tu(t)}$$

■ Example 3.15

Find the convolution of the two signals given below in Fig. 3.14a and b.

(Anna University, December, 2005)

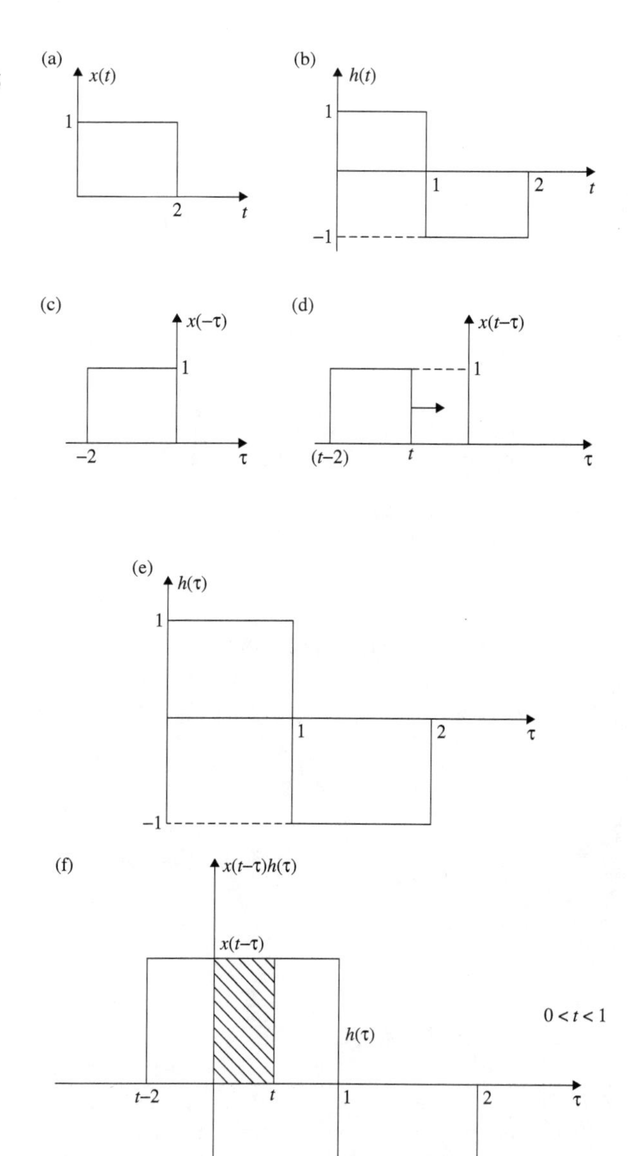

Fig. 3.14 Convolution of two signals of Example 3.15

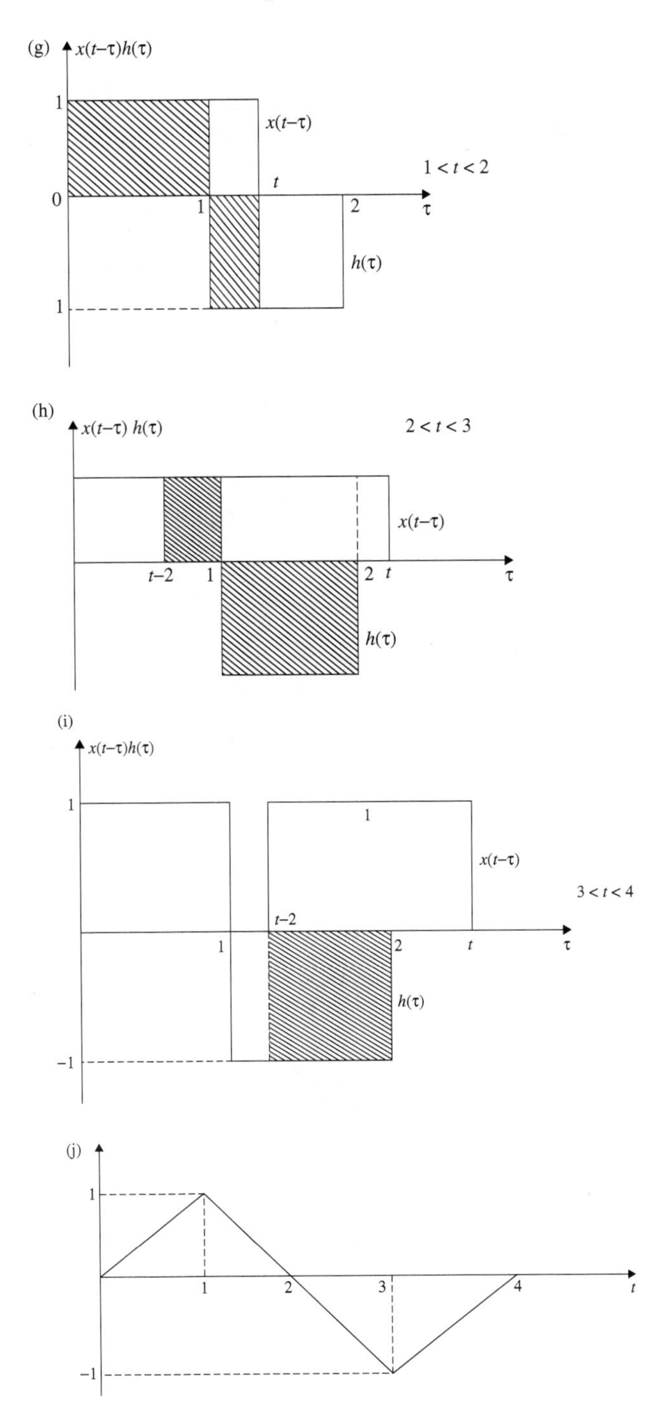

Fig. 3.14 (continued)

Solution:

1. Figure 3.14a and b represents $x(t)$ and $h(t)$. $x(-\tau)$ and $x(t-\tau)$ are shown in Fig. 3.14c and d, respectively. Fig. 3.14e represents $h(\tau)$.
2. In the present case, $h(\tau)$ is kept fixed and $x(t-\tau)$ is moved toward right so that it convolves with $h(\tau)$.
3. The first overlapping of $x(t-\tau)$ and $h(\tau)$ is shown in Fig. 3.14f. This occurs for the time interval $0 < t < 1$. Here, the right edge of $x(t-\tau)$ slides past the left edge of $h(\tau)$.

   The shaded area gives the overlapping. The time limit is from $t = 0$ to t. Accordingly, the integration is carried out as follows:

$$\begin{aligned} y(t) &= \int_0^t x(t-\tau)h(\tau)d\tau \\ &= \int_0^t 1 \times 1 d\tau \\ &= \Big[\tau\Big]_0^t \end{aligned}$$

$$\boxed{y(t) = t} \qquad 0 < t < 1 \tag{a}$$

4. $x(t-\tau)$ is shifted to the right so that the leading edge crosses past of middle edge of $h(\tau)$. The time interval is $1 < t < 2$ before the next overlapping. This is represented in Fig. 3.14g.

   For Fig. 3.14g, the following convolution integral is written

$$\begin{aligned} y(t) &= \int_0^t x(t-\tau)h(\tau)d\tau \\ &= \int_0^1 d\tau - \int_1^t d\tau \\ &= \Big[\tau\Big]_0^1 - \Big[\tau\Big]_1^t \\ &= 1 - t + 1 \end{aligned}$$

$$\boxed{y(t) = 2 - t} \qquad 1 < t < 2 \tag{b}$$

5. $x(t-\tau)$ is moved further toward the right so that its leading edge crosses $\tau = 2$ but the trailing edge is behind $\tau = 1$. This time duration is $2 < t < 3$. The convolution graph is shown in Fig. 3.14h.

   The convolution integral for Fig. 3.14i is written as

$$
\begin{aligned}
y(t) &= \int_{t-2}^{2} x(t-\tau)h(\tau)d\tau \\
&= \int_{t-2}^{1} d\tau - \int_{1}^{2} d\tau \\
&= \Big[\tau\Big]_{t-2}^{1} - \Big[\tau\Big]_{1}^{2} \\
&= 1 - t + 2 - 2 + 1
\end{aligned}
$$

$$\boxed{y(t) = (2-t)} \qquad 2 < t < 3 \tag{c}$$

6. Now $x(t-\tau)$ is moved to the right such that the trailing edge is past $\tau = 1$ but less than $\tau = 2$, the time duration for this convolution is $3 < t < 4$. The convolution graph is shown in Fig. 3.14i.
   For Fig. 3.14i, the following convolution integral is written with the lower limit as $(t-2)$ and the upper limit 2.

$$
\begin{aligned}
y(t) &= \int_{t-2}^{2} x(t-\tau)h(\tau)d\tau \\
&= -\int_{t-2}^{2} d\tau \\
&= -\Big[\tau\Big]_{t-2}^{2} \\
&= t - 2 - 2
\end{aligned}
$$

$$\boxed{y(t) = (t-4)} \qquad 3 < t < 4 \tag{d}$$

7. For $t > 4$, there is no overlapping between $x(t-\tau)$ and $h(\tau)$ and hence $y(t) = 0$.
8. Equations (a), (b), (c) and (d) are used to find $y(t)$ for the respective time duration and are shown in the table below.

t	0	0.5	1	1.5	2	2.5	3	3.5	4
$y(t)$	0	0.5	1	0.5	0	−0.5	−1	−0.5	0

9. The response curve $y(t)$ is plotted and its shown in Fig. 3.14j.
10. Using convolution theorem of LT, the output response can be easily obtained as

$$\boxed{y(t) = t - 2(t-1)u(t-1) + 2(t-3)u(t-3) - (t-4)u(t-4)}$$

The plot of $y(t)$ using the above equation will give the same response as shown in Fig. 3.14j.

Example 3.16

$$x(t) = u(t) - u(t-4)$$
$$h(t) = u(t) - u(t-1)$$

(a) Find $y(t)$ using convolution. Use graphical method.
(b) Verify the width property of convolution.
(c) What is $y(t)$ by LT using convolution theorem?

Solution:

(a)

1. $x(t) = u(t) - u(t-4)$ is represented in Fig. 3.15a.
2. $h(t) = u(t) - u(t-1)$ is represented in Fig. 3.15b.
3. Figure 3.15c represents $x(\tau)$ which is kept fixed.
4. Figure 3.15d represents $h(-\tau)$.
5. By giving very long left shift $h(t-\tau)$ is represented in Fig. 3.15.
6. $x(\tau)h(t-\tau)$ is represented in Fig. 3.15f. Now $h(t-\tau)$ is moved toward the right so that there is overlapping between $x(\tau)$ and $h(t-\tau)$.
7. The overlapping of $x(\tau)$ and $h(t-\tau)$ is shown in shaded area. The movement of $h(t-\tau)$ toward right is such that the right edge crosses past the left edge of $x(\tau)$. From Fig. 3.15g, the limit of integration of the shaded area is from 0 to t. The time interval is $0 < t < 1$ so that the right and left edges of $h(t-\tau)$ do not simultaneously cross over the left edge of $x(\tau)$. The following convolution integral is now solved.

$$y(t) = \int_0^t x(\tau)h(t-\tau)d\tau$$
$$= \int_0^t d\tau$$

$$y(t) = \Big[\tau\Big]_0^t$$

$$\boxed{y(t) = t} \qquad 0 < t < 1 \qquad \text{(a)}$$

8. The left edge of $h(t-\tau)$ now crosses past the left edge of $x(\tau)$. To make the right edge of $h(t-\tau)$ not to cross the right edge of $x(\tau)$, the time duration should be $1 < t < 4$. The overlapping portion of the convolution is shown in Fig. 3.15h. From the overlapping area, the limit of integration is from $(t-1)$ to t. The following equation is solved for $y(t)$.

$$
\begin{aligned}
y(t) &= \int_{t-1}^{t} x(\tau)h(t-\tau)d\tau \\
&= \int_{t-1}^{t} d\tau \\
&= \Big[\tau\Big]_{t-1}^{t}
\end{aligned}
$$

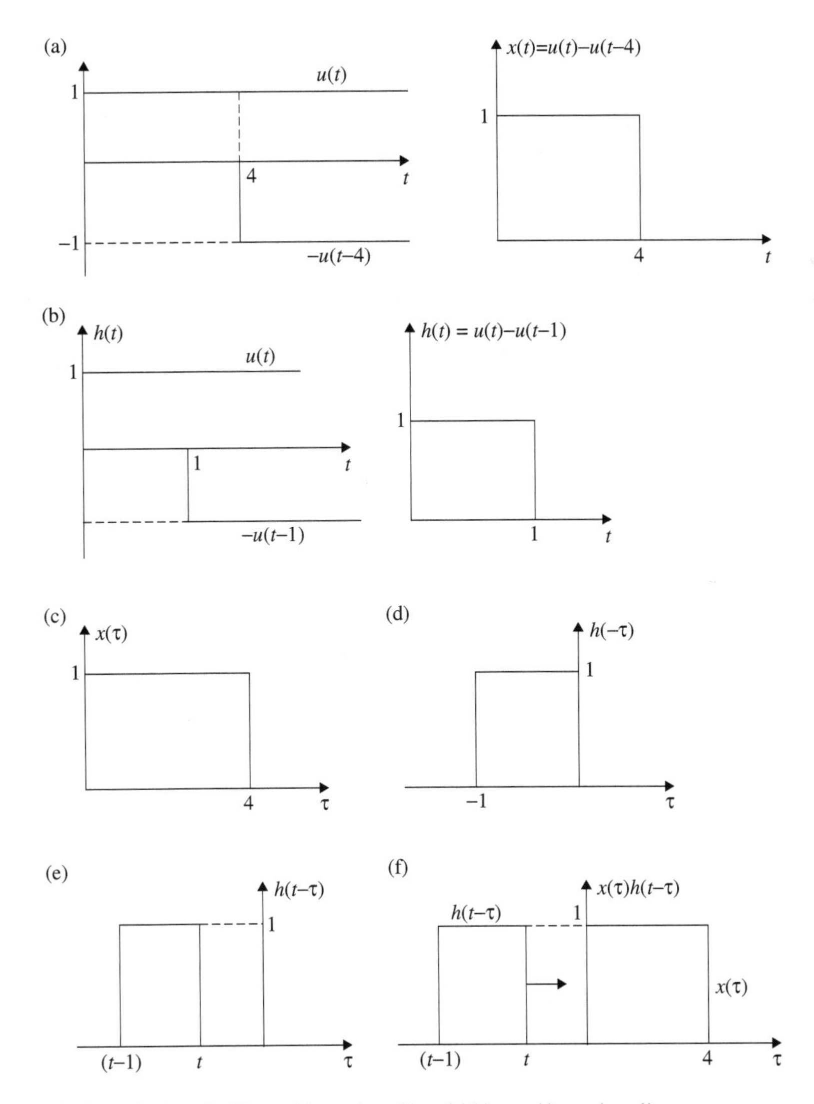

Fig. 3.15 Convolution of $x(t) = u(t) - u(t-4)$ and $h(t) = u(t) - u(t-1)$

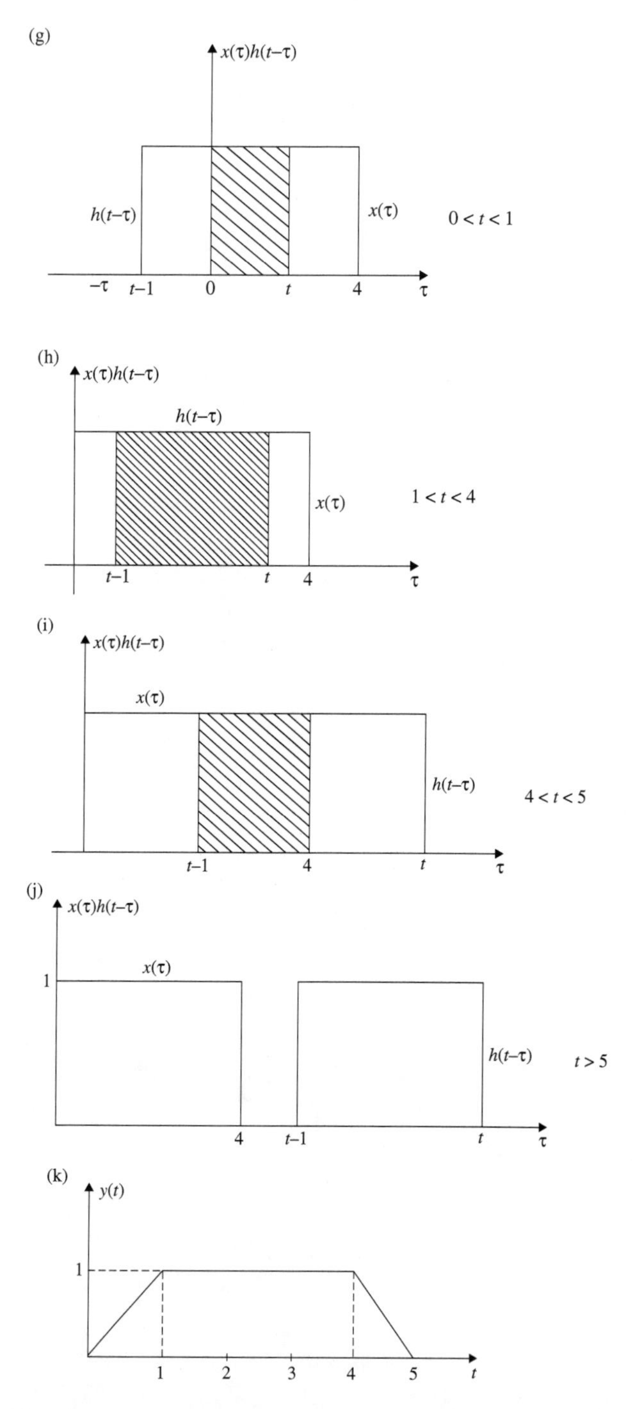

Fig. 3.15 (continued)

$$\boxed{y(t) = 1} \qquad 1 < t < 4 \qquad \text{(b)}$$

9. The right edge of $h(t-\tau)$ crosses past the right edge of $x(\tau)$. The time interval should be $4 < t < 5$. Now the left edge of $h(t-\tau)$ does not cross past the right edge of $x(\tau)$. From the shaded area, the lower limit of integration is $(t-1)$ and the upper limit is 4. Now the following equation is written for $y(t)$.

$$y(t) = \int_{t-1}^{4} x(\tau)h(t-\tau)d\tau$$
$$= \int_{t-1}^{4} d\tau = [\tau]_{t-1}^{4}$$

$$\boxed{y(t) = (5-t)} \qquad 4 < t < 5 \qquad \text{(c)}$$

10. For $t > 5$, there is no overlapping between $x(\tau)$ and $h(t-\tau)$ and hence $y(t) = 0$.
11. Using equations (a), (b) and (c), $y(t)$ is found for the appropriate time interval and tabulated below

t	0	1	2	3	4	5
$y(t)$	0	1	1	1	1	0

12. The response curve $y(t)$ is shown in Fig. 3.14k.

(b) 1. The width of $x(t)$, $T_1 = 4$;
2. The width of $h(t)$, $T_2 = 1$;
3. The width of $y(t) = T_1 + T_2 = 4 + 1 = 5$;
This is verified from Fig. 3.15k.

(c) Convolution by LT method.

1. From Fig. 3.15a

$$X(s) = \frac{1}{s}[1 - e^{-4s}]$$

2. From Fig. 3.15b

$$H(s) = \frac{1}{s}[1 - e^{-s}]$$

3.

$$Y(s) = X(s)H(s)$$
$$= \frac{1}{s^2}[1 - e^{-s}][1 - e^{-4s}]$$
$$= \frac{1}{s^2}[1 - e^{-s} - e^{-4s} + e^{-5s}]$$

$$y(t) = t - (t-1)u(t-1) - (t-4)u(t-4) + (t-5)u(t-5)$$

$$\begin{aligned} \text{For } t = 0, &\quad y(0) = 0 \\ t = 1, &\quad y(1) = 1 \\ t = 2, &\quad y(2) = 2 - 1 = 1 \\ t = 3, &\quad y(3) = 3 - 2 = 1 \\ t = 4, &\quad y(4) = 4 - 3 - 0 = 1 \\ t = 5, &\quad y(5) = 5 - 4 - 1 + 0 = 0 \end{aligned}$$

The same result as in (a) is obtained without any laborious task.

■ Example 3.17

Find the convolution of the signals shown in Fig. 3.16a and b. Verify the result using convolution theorem of LT.

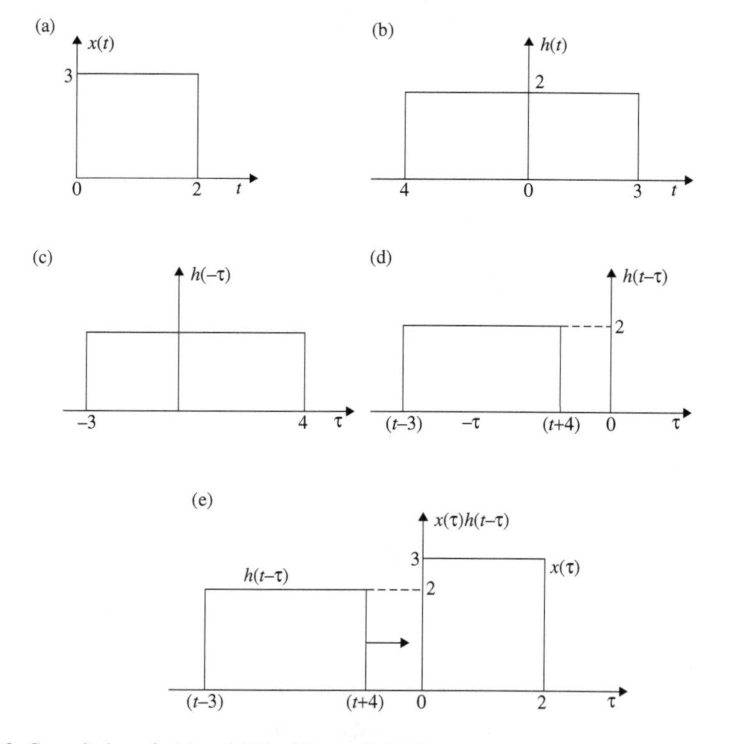

Fig. 3.16 Convolution of $x(t)$ and $h(t)$ of Example 3.17

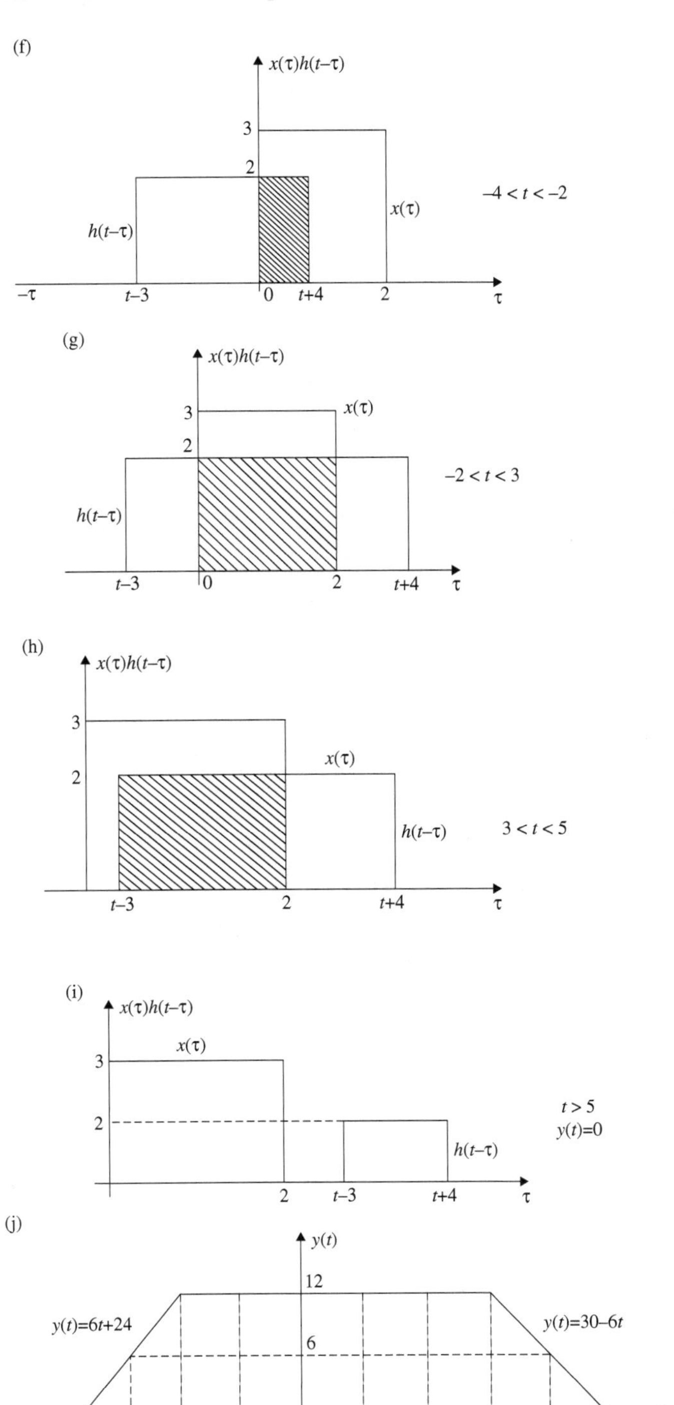

Fig. 3.16 (continued)

Solution:

1. The $h(t)$ is expressed as $h(\tau)$ and folded. The folded $h(-\tau)$ is shown in Fig. 3.16c.
2. $h(t-\tau)$ is obtained by adding t to 4 and -3 in Fig. 3.16c and shifted to the left most and is shown in Fig. 3.16d.
3. $x(\tau)$ and $h(t-\tau)$ are represented in Fig. 3.16e.
4. $h(t-\tau)$ is shifted toward right. The overlapping of $h(t-\tau)$ with $x(\tau)$ is shown in Fig. 3.16f. The time interval is $-4 < t < -2$ so that change occurs one at a time. The limit of integration is form 0 to $(t+4)$. Thus, $y(t)$ is found as given below.

$$\begin{aligned} y(t) &= \int_0^{t+4} x(\tau)h(\tau)d\tau \\ &= \int_0^{t+4} 3 \times 2d\tau \\ &= \Big[6\tau\Big]_0^{t+4} \end{aligned}$$

$$\boxed{y(t) = (6t + 24)} \qquad -4 < t < -2 \qquad \text{(a)}$$

5. $h(t-\tau)$ is further shifted toward right. The right edge of $x(\tau)$ is crossed first. The overlapping area is shown in Fig. 3.16g. The time duration is $-2 < t < 3$ so that only one change occurs at a time. The limit of integration is from 0 to 2. $y(t)$ is obtained as given below.

$$\begin{aligned} y(t) &= \int_0^{2} x(\tau)h(t-\tau)d\tau \\ &= \int_0^{2} 3 \times 2d\tau \\ &= \Big[6\tau\Big]_0^{2} \end{aligned}$$

$$\boxed{y(t) = 12} \qquad -2 < t < 3 \qquad \text{(b)}$$

6. When $h(t-\tau)$ is moved further to the right, its left edge crosses the left edge of $x(\tau)$. The overlapping area is shown in Fig. 3.16h. The time duration is $3 < t < 5$. The limit of integration is from $(t-3)$ to 2. $y(t)$ is obtained as given below

$$
\begin{aligned}
y(t) &= \int_{t-3}^{2} x(\tau)h(t-\tau)d\tau \\
&= \int_{t-3}^{2} 3 \times 2d\tau \\
&= \Big[6\tau\Big]_{t-3}^{2} \\
y(t) &= 6[2 - t + 3]
\end{aligned}
$$

$$\boxed{y(t) = (30 - 6t)} \qquad 3 < t < 5 \tag{c}$$

7. If $h(t-\tau)$ is further moved to the right, for $t > 5$, there is no overlapping between $x(\tau)$ and $h(t-\tau)$ and therefore $y(t) = 0$. This is shown in Fig. 3.16i.
8. For various time intervals, $y(t)$ is found and tabulated as given below

$$
\begin{aligned}
y(t) &= (6t + 24) && -4 < t < -2 \\
y(t) &= 12 && -2 < t < 3 \\
y(t) &= (30 - 6t) && 3 < t < 5 \\
y(t) &= 0 && t > 5
\end{aligned}
$$

t	−4	−3	−2	−1	0	1	2	3	4	5
$y(t)$	0	6	12	12	12	12	12	12	6	0

9. Figure 3.16j shows the output response curve $y(t)$. For the given $x(t)$, the width is $T_1 = 2$ and for $h(t)$ the width is $T_2 = 7$. Thus, the total width of $y(t)$ is $T = T_1 + T_2 = 9$. This is verified from Fig. 3.16j.

$y(t)$ from convolution property of LT.

1. The Laplace transform of $x(t)$ is

$$
\begin{aligned}
X(s) &= \int_{0}^{2} 3e^{-st}dt = \frac{-3}{s}\Big[e^{-st}\Big]_{0}^{2} \\
&= \frac{3}{s}[1 - e^{-2s}]
\end{aligned}
$$

2. The Laplace transform of $h(t)$ is

$$
\begin{aligned}
H(s) &= \int_{-4}^{3} 2e^{-st}dt \\
&= \frac{-2}{s}\Big[e^{-st}\Big]_{-4}^{3} \\
&= \frac{2}{s}[e^{4s} - e^{-3s}]
\end{aligned}
$$

$$\begin{aligned} Y(s) &= X(s)H(s) \\ &= \frac{6}{s^2}[1 - e^{-2s}][e^{4s} - e^{-3s}] \\ &= \frac{6}{s^2}[e^{4s} - e^{2s} - e^{-3s} + e^{-5s}] \end{aligned}$$

$$\begin{aligned} y(t) &= 6[(t+4)u(t+4) - (t+2)u(t+2) - (t-3)u(t-3) \\ &\quad +(t-5)u(t-5))] \\ y(-4) &= 6[0] = 0 \\ y(-3) &= 6 \\ y(-2) &= 6(2) = 12 \\ y(3) &= 6(7 - 5 + 0) = 12 \\ y(4) &= 6 \\ y(5) &= 6(9 - 7 - 2) = 0 \end{aligned}$$

The same response $y(t)$ is obtained here also with ease.

Example 3.18

Find the output response $y(t)$ for the signals shown in Fig. 3.17a and b. Plot the response curve $y(t)$.

Solution:

1. The triangular wave $x(t)$ and the rectangular wave $h(t)$ are shown in Fig. 3.17a and b, respectively.
2. The slope of the triangular wave is $\frac{1}{5}$. Hence,

$$x(t) = \frac{1}{5}t \qquad 0 \le t \le 5$$

3. When $h(t-\tau)$ is moved toward right, it overlaps with $x(\tau)$ for $-2 < t < 2$. The limit of integration is 0 to $t+2$ as seen from the overlapping area of Fig. 3.17e. The output response during the above period is

$$\begin{aligned} y(t) &= \int_0^{t+2} \frac{1}{5}\tau d\tau \\ &= \frac{1}{10}[\tau^2]_0^{t+2} \end{aligned}$$

$$\boxed{y(t) = \frac{1}{10}[(t+2)^2]} \qquad -2 < t < 2 \tag{a}$$

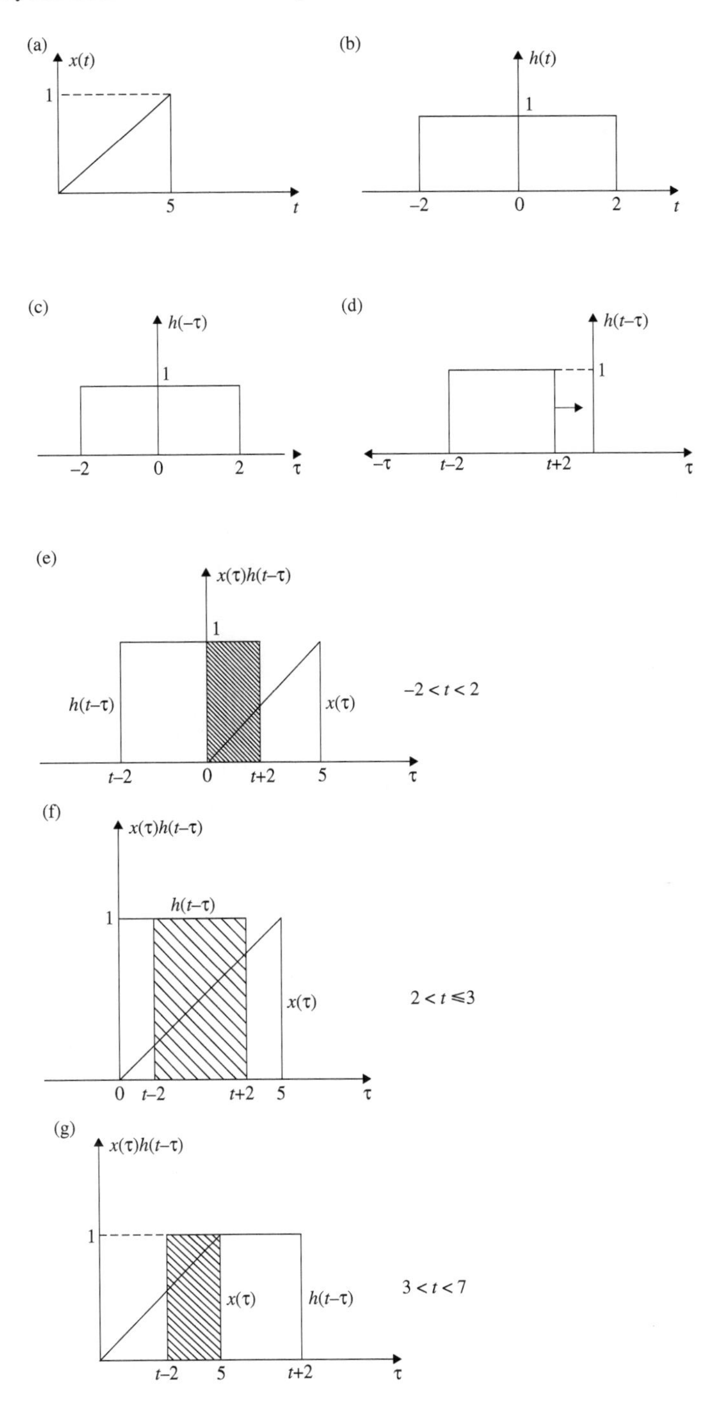

Fig. 3.17 Convolution of a triangular and a rectangular waves

(h)

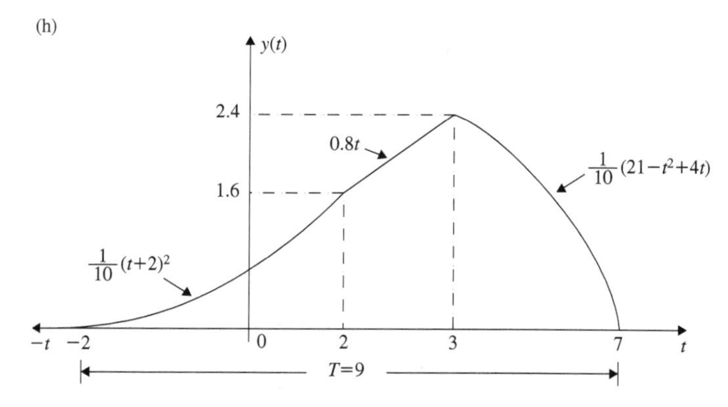

Fig. 3.17 (continued)

4. When the left edge of $h(t-\tau)$ crosses past the left edge of $x(\tau)$, there is overlapping during the time interval $2 < t < 3$ with the limits of integration from $(t-2)$ to $(t+2)$. This is shown in Fig. 3.17f. During this period, the output response is obtained as given below.

$$\begin{aligned} y(t) &= \int_{t-2}^{t+2} \frac{1}{5}\tau d\tau \\ &= \frac{1}{10}[\tau^2]_{t-2}^{t+2} \\ &= \frac{1}{10}[(t+2)^2 - (t-2)^2] \end{aligned}$$

$$\boxed{y(t) = 0.8t} \qquad 2 < t < 3 \tag{b}$$

5. When $h(t-\tau)$ is moved further toward right, the right edge of $h(t-\tau)$ crosses past the right edge of $x(\tau)$. The overlapping occurs during the time interval $3 < t < 7$. The limit of integration as seen from Fig. 3.17g is from $(t-2)$ to 5. During this period, the output response is obtained from the following integral.

$$\begin{aligned} y(t) &= \int_{t-2}^{5} \frac{1}{5}\tau d\tau \\ &= \frac{1}{10}\left[\tau^2\right]_{t-2}^{5} \\ &= \frac{1}{10}[25 - (t-2)^2] \end{aligned}$$

$$\boxed{y(t) = \frac{1}{10}[21 - t^2 + 4t]} \qquad 3 < t < 7 \tag{c}$$

6. Further movement of $h(t-\tau)$ toward the right beyond $\tau = 5$ does not overlap with $x(\tau)$ and hence $y = 0$ for $t > 7$.
7. The expressions for the output response $y(t)$ for different time intervals are given below

$$\begin{aligned}
y(t) &= 0 && t \leq -2 \\
y(t) &= \frac{1}{10}(t+2)^2 && -2 \leq t < 2 \\
y(t) &= 0.8t && 2 < t < 3 \\
y(t) &= \frac{1}{10}(21 - t^2 + 4t) && 3 < t < 7 \\
y(t) &= 0 && t > 7
\end{aligned}$$

The above expressions are used to plot $y(t)$ and the responses curve is shown in Fig. 3.17h.
8. The width property of convolution is checked as follows. From Fig. 3.17a and b, we get $T_1 = 5$, $T_2 = 4$, respectively. Thus, we get $T = T_1 + T_2 = 5 + 4 = 9$. From Fig. 3.17h, the width of $y(t)$ is found as $T = 9$.

■ Example 3.19

Consider the following signals:

$$x(t) = e^{-2|t|}$$
$$h(t) = u(t)$$

Using convolution, find $y(t)$.

Solution:

1. $x(t)$ and $h(t)$ are represented in Fig. 3.18a and b, respectively.
2. $h(t-\tau)$ and $x(\tau)$ are represented in Fig. 3.18c.
3. For $-\infty \leq t \leq 0$, the overlapping is shown in Fig. 3.18c. The signals overlap for the limits of integration from $-\infty$ to t. Hence,

$$\begin{aligned}
y(t) &= \int_{-\infty}^{t} e^{2\tau} d\tau \qquad t < 0 \\
&= \frac{1}{2}\left[e^{2\tau}\right]_{-\infty}^{t}
\end{aligned}$$

$$\boxed{y(t) = \frac{1}{2}e^{2t}} \qquad t < 0 \tag{a}$$

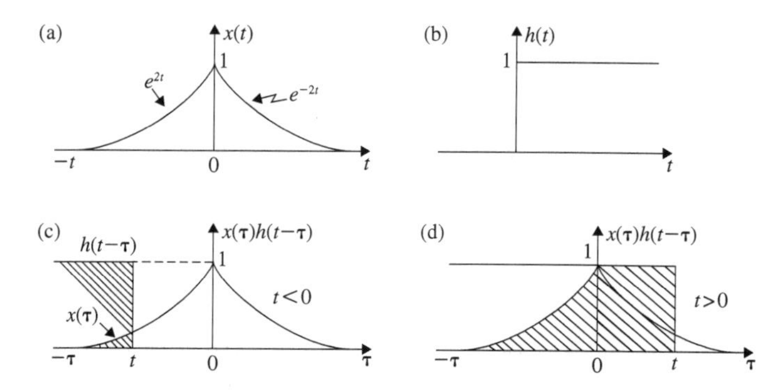

Fig. 3.18 Convolution of $x(t) = e^{-2|t|}$ and $h(t) = u(t)$

4. Consider the overlapping for $t > 0$. The limit of integration is from $-\infty$ to 0 and from 0 to t. The expression for the output response is

$$
\begin{aligned}
y(t) &= \int_{-\infty}^{0} e^{2\tau} d\tau + \int_{0}^{t} e^{-2\tau} d\tau \\
&= \frac{1}{2}\left[e^{2\tau}\right]_{-\infty}^{0} - \frac{1}{2}\left[e^{-2\tau}\right]_{0}^{t}
\end{aligned}
$$

$$
\boxed{y(t) = 1 - \frac{1}{2}e^{-2t}} \qquad t > 0 \qquad \text{(b)}
$$

5. The total response is the sum of (a) and (b).

$$
\boxed{y(t) = \left(1 - \frac{1}{2}e^{-2t}\right)u(t) + \frac{1}{2}e^{2t}u(-t)}
$$

■ Example 3.20

Consider the following signals:

$$
\begin{aligned}
x(t) &= 2t && 0 < t < 1 \\
&= (3 - t) && 1 < t < 3 \\
&= 0 && \text{elsewhere.} \\
h(t) &= u(t) - u(t - 2)
\end{aligned}
$$

Find $y(t)$ by convolving $x(t)$ and $h(t)$.

Solution:

1. The given equation for $x(t)$ is a triangle which is shown in Fig. 3.19a. The equation for $h(t)$ represents a pulse which is shown in Fig. 3.19b. $h(t-\tau)$ is shown in Fig. 3.19c. Figure 3.19d represents the convolution of $x(\tau)$ and $h(t-\tau)$. The overlapping that takes place when $h(t-\tau)$ is moved toward the right at various stages is shown in figures to follow.
2. Consider Fig. 3.19e.

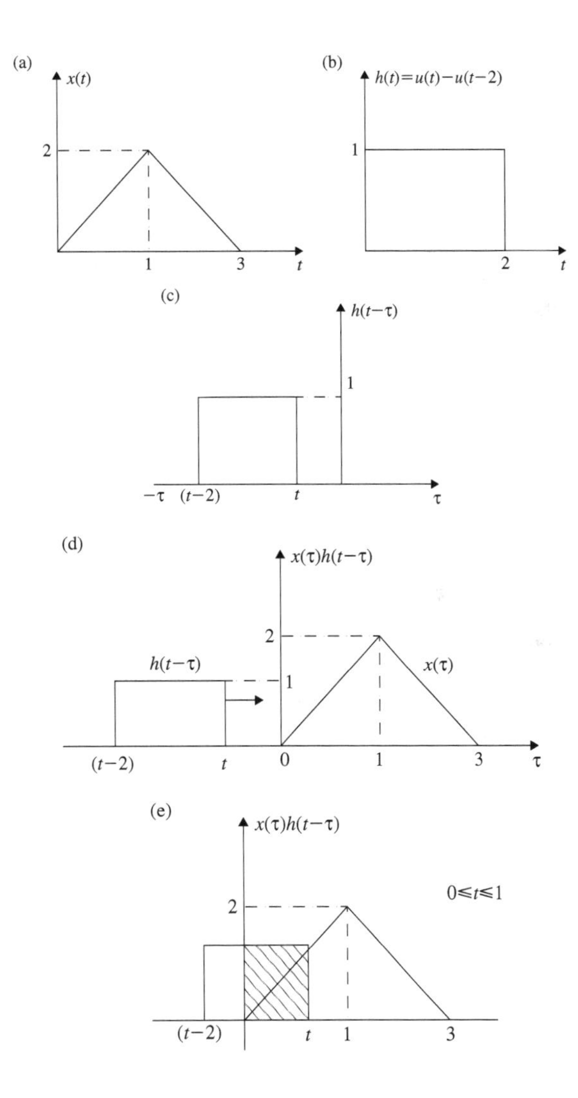

Fig. 3.19 Convolution of a triangle and pulse

Fig. 3.19 (continued)

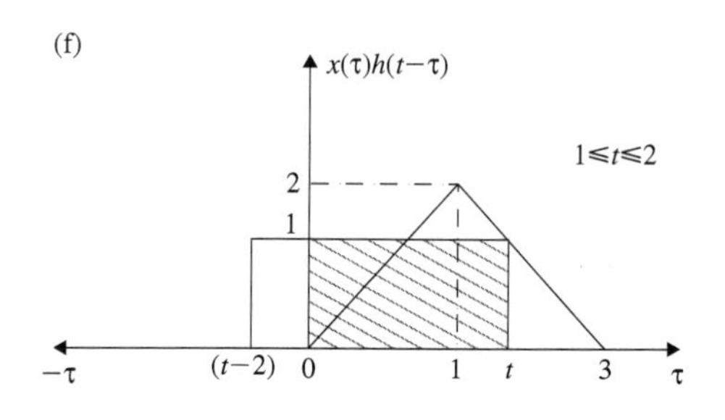

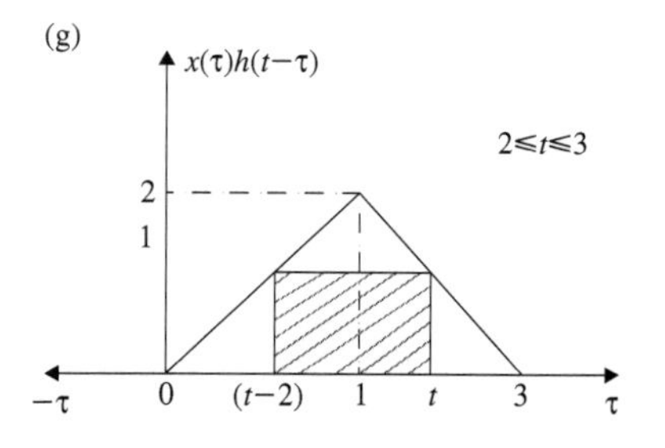

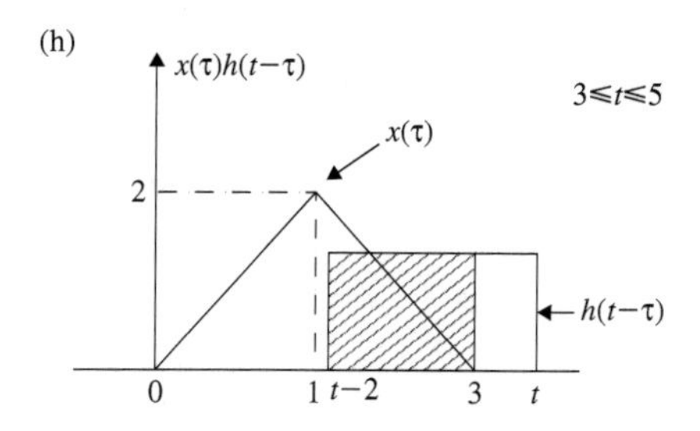

$$
\begin{aligned}
y(t) &= \int_0^t 2\tau d\tau \\
&= \left[\tau^2\right]_0^t
\end{aligned}
$$

$$
\boxed{y(t) = t^2} \qquad 0 \le t \le 1
$$

3. Consider Fig. 3.19f.

$$
\begin{aligned}
y(t) &= \int_0^1 2\tau d\tau + \int_1^t (3-\tau) d\tau \\
&= [\tau^2]_0^1 + \left[3\tau - \frac{\tau^2}{2}\right]_1^t \\
&= \left[1 + 3t - \frac{t^2}{2} - 3 + \frac{1}{2}\right]
\end{aligned}
$$

$$\boxed{y(t) = \left(3t - \frac{t^2}{2} - \frac{3}{2}\right)} \qquad 1 \le t \le 2$$

4. Consider Fig. 3.19g.

$$\begin{aligned} y(t) &= \int_{t-2}^{1} 2\tau d\tau + \int_{1}^{t} (3-\tau)d\tau \\ &= \left[\tau^2\right]_{t-2}^{1} + \left[3\tau - \frac{\tau^2}{2}\right]_{1}^{t} \\ &= \left[1 - t^2 - 4 + 4t + 3t - \frac{t^2}{2} - 3 + \frac{1}{2}\right] \end{aligned}$$

$$\boxed{y(t) = \left(-\frac{3}{2}t^2 + 7t - \frac{11}{2}\right)} \qquad 2 \le t \le 3$$

5. From Fig. 3.19h, the following equation is written for $y(t)$.

$$\begin{aligned} y(t) &= \int_{t-2}^{3} (3-\tau)d\tau \\ &= \left[3\tau - \frac{\tau^2}{2}\right]_{t-2}^{3} \\ &= 9 - \frac{9}{2} - 3t + 6 + \frac{(t-2)^2}{2} \end{aligned}$$

$$\boxed{y(t) = \frac{t^2}{2} - 5t + \frac{25}{2}} \qquad 3 \le t \le 5$$

6. For $t > 5$, there is no overlapping between $x(\tau)$ and $h(t-\tau)$ and hence $y(t) = 0$. The value of $y(t)$ for different time intervals are listed below

$$\begin{aligned} &(a)\ \ y(t) = t^2 && 0 \le t \le 1 \\ &(b)\ \ y(t) = \left(3t - \frac{t^2}{2} - \frac{3}{2}\right) && 1 \le t \le 2 \\ &(c)\ \ y(t) = \left(-\frac{3}{2}t^2 + 7t - \frac{11}{2}\right) && 2 \le t \le 3 \\ &(d)\ \ y(t) = \left(\frac{t^2}{2} - 5t + \frac{25}{2}\right) && 3 \le t \le 5 \\ &(e)\ \ y(t) = 0 && t > 5 \end{aligned}$$

■ Example 3.21

Find the output of a LTIC system with the impulse responses $h(t) = \delta(t-3)$ and $x(t) = (\cos 4t + \cos 7t)$.

(*Anna University, 2004*)

Solution: According to shifting property of convolution

$$\begin{aligned} y(t) &= x(t) * \delta(t - t_0) \\ &= x(t - t_0) \end{aligned}$$

Applying the above property, we get

$$\boxed{y(t) = \cos 4(t-3) + \cos 7(t-3)}$$

■ Example 3.22

The impulse response of an LTIC system is shown in Fig. 3.20a. The input $x(t) = \delta(t) - \delta(t-1.5)$. Find the response $y(t)$ of the system.

(*Anna University, 2004*)

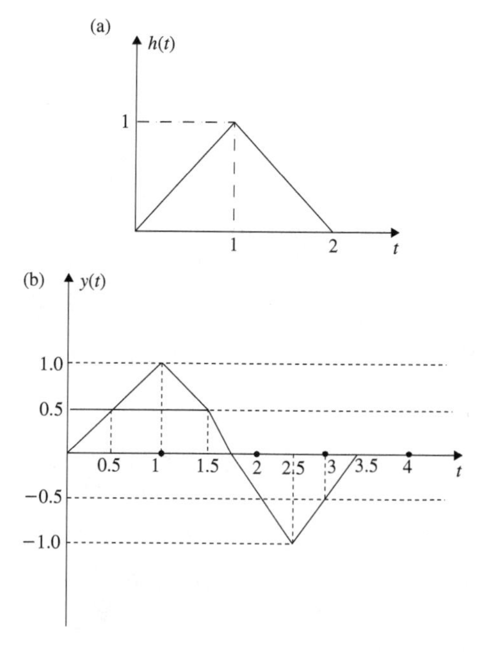

Fig. 3.20 Output response curve

Solution: The given triangular wave $h(t)$ shown in Fig. 3.20a is split up as two ramp signals.

$$h(t) = t[u(t) - u(t-1)] + (2-t)[u(t-1) - u(t-2)]$$

(See also Example 3.88).

$$\begin{aligned} y(t) &= h(t) * x(t) \\ &= h(t) * [\delta(t) - \delta(t-1.5)] \\ &= h(t) * \delta(t) - h(t) * \delta(t-1.5) \\ &= h(t) - h(t) * \delta(t-1.5) \\ h(t) * \delta(t-1.5) &= h(t-1.5) \end{aligned}$$

$$\begin{aligned} y(t) &= t[u(t) - u(t-1)] + (2-t)[u(t-1) - u(t-2)] \\ &\quad -(t-1.5)[u(t-1.5) - u(t-2.5)] \\ &\quad -[2-(t-1.5)][u(t-2.5) - u(t-3.5)] \\ &= tu(t) + (2-2t)u(t-1) - (t-1.5) \\ &\quad \times u(t-1.5) - (2-t)u(t-2) + (2t-5) \\ &\quad \times u(t-2.5) - (t-3.5)u(t-3.5) \end{aligned}$$

The following table is prepared from the above equation.

t	0	1	1.5	2	2.5	3	$t \geq 3.5$
$y(t)$	0	1	0.5	−0.5	−1	−0.5	0

The output response curve $y(t)$ is plotted and is shown in Fig. 3.20b.

■ Example 3.23

The system shown in Fig. 3.21 is formed in connecting two systems in cascade. The impulse response of the systems are given by $h_1(t) = e^{-2t}u(t)$ and $h_2(t) = 2e^{-t}u(t)$. Find the overall impulse response of the system.

(Anna University, June, 2007)

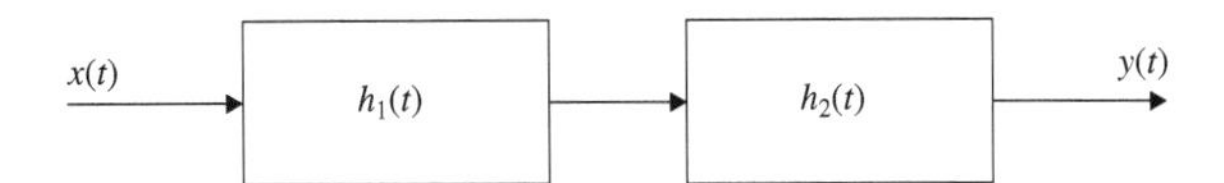

Fig. 3.21 Two systems connected in cascade

Solution: Both $h_1(t)$ and $h_2(t)$ are causal systems. The limits of integration is from 0 to t

$$h_1(\tau) = e^{-2\tau}u(\tau)$$
$$h_2(\tau) = 2e^{-\tau}u(\tau)$$
$$h_2(t-\tau) = 2e^{-(t-\tau)}u(t-\tau)$$

The impulse response of the combined system is given by

$$h(t) = h_1(t) * h_2(t)$$
$$h(t) = \int_0^t 2e^{-2\tau}e^{-(t-\tau)}d\tau$$
$$= 2e^{-t}\int_0^t [e^{-\tau}]d\tau$$
$$= -2e^{-t}\Big[e^{-\tau}\Big]_0^t$$
$$h(t) = -2e^{-t}[e^{-t}-1]$$

$$\boxed{h(t) = 2[e^{-t}-e^{-2t}]u(t)}$$

■ Example 3.24

Show that

$$x(t) * \delta(t-t_0) = x(t-t_0)$$

(*Anna University, June, 2007*)

Solution:

$$x(t) * \delta(t-t_0) = \int_{-\infty}^{\infty} \delta(\tau-t_0)x(t-\tau)d\tau$$
$$= x(t-\tau)|_{\tau=t_0}$$

$$\boxed{x(t) * \delta(t-t_0) = x(t-t_0)}$$

■ Example 3.25

Prove that

$$x(t) * u(t) = \int_{-\infty}^{t} x(\tau)d\tau$$

(*Anna University, June, 2007*)

Solution:

$$x(t) * u(t) = \int_{-\infty}^{t} x(\tau)u(t-\tau)d\tau$$

Since

$$\begin{aligned} u(t-\tau) &= 1 \qquad \tau < t \\ &= 0 \qquad \tau > t \end{aligned}$$

The above equation can be written as

$$\boxed{x(t) * u(t) = \int_{-\infty}^{t} x(\tau)d\tau}$$

■ Example 3.26

The signals $x(t)$ and $y(t)$ shown in Fig. 3.22a and b are, respectively, input and output of an LTIC system. Sketch the response to the following inputs:

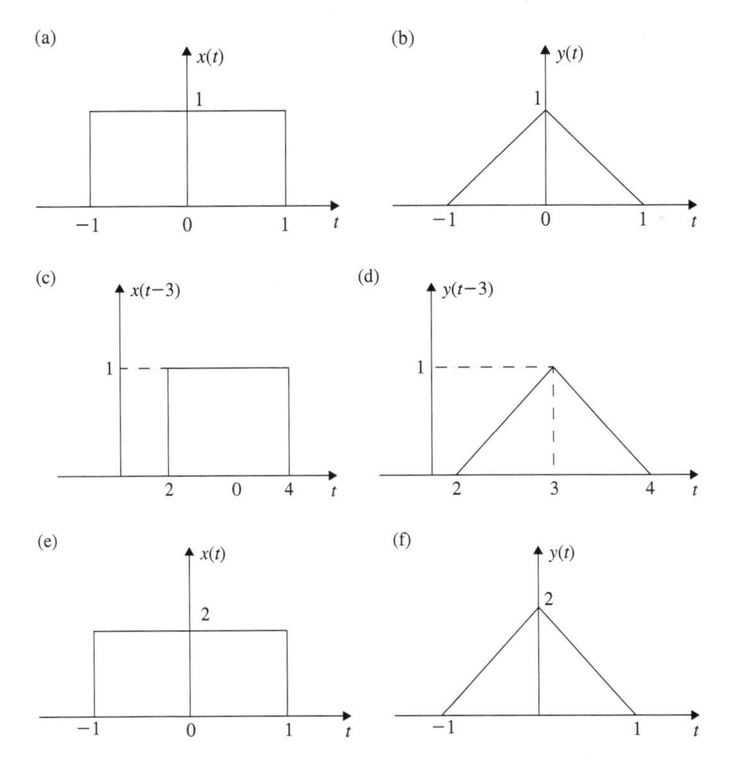

Fig. 3.22 Two signals $x(t)$ and $y(t)$

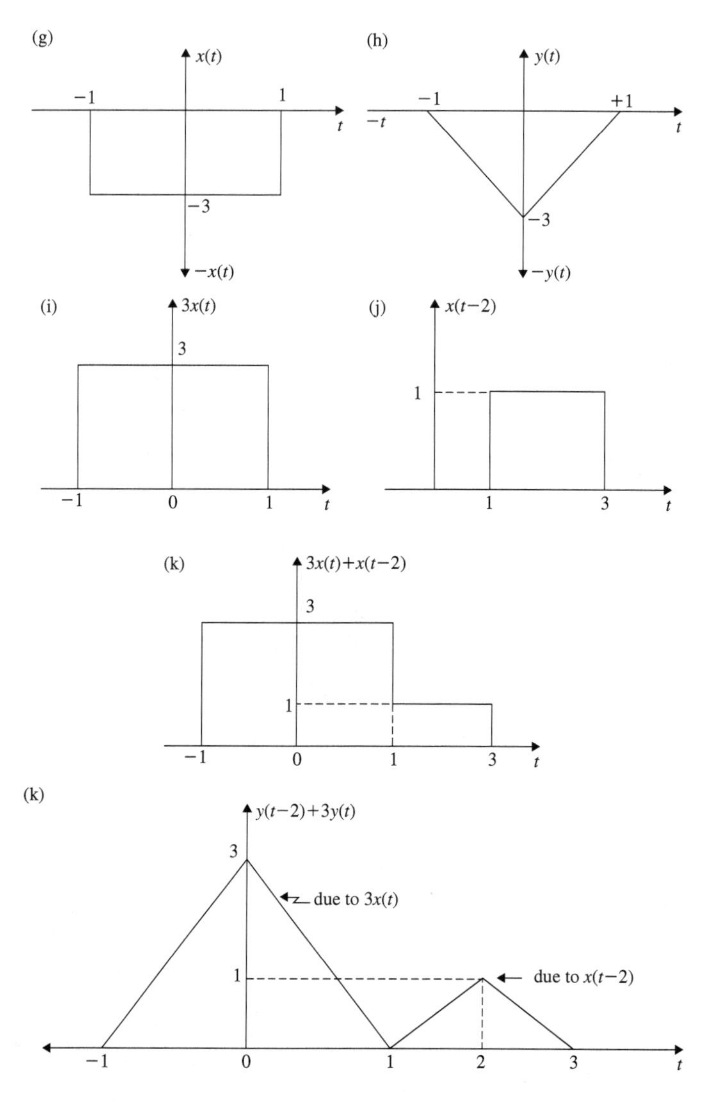

Fig. 3.22 (continued)

1. $x(t) = x(t-3)$
2. $x(t) = 2x(t)$
3. $x(t) = -3x(t)$
4. $x(t) = x(t-2) + 3x(t)$

(*Anna University, May, 2007*)

Solution:

1. $x(t-3)$ is shown in Fig. 3.22c. It is time shifted by 3. The amplitude remains at 1. Hence, the output with the same amplitude is to be time shifted by 3 to the right. This is shown in Fig. 3.22d.
2. The input $x(t) = 2x(t)$.The amplitude is multiplied by a factor 2 without anytime shift. The input and the output are shown in Fig. 3.22e and f, respectively.
3. $x(t) = -3x(t)$ is shown in Fig. 3.22g. Here, the amplitude is multiplied by a factor -3 without anytime advance or delay. Therefore, the output should also be multiplied by the factor -3. This is shown in Fig. 3.22h.
4. The input $x(t) = x(t-2) + 3x(t)$. The input $x(t)$ which is amplified by the factor 3 is shown as $3x(t)$ for $-1 < t < 1$ in Fig. 3.22i. The time delayed input $x(t)$ which is $x(t-2)$ is shown in Fig. 3.22j for the time interval $1 < t < 3$. The output due to input $3x(t)$ for $-1 < t < 1$ and the output due to input $x(t-2)$ for $1 < t < 3$ is shown in Fig. 3.22k.

3.8 Causality of an Linear Time Invariant Continuous Time System

An LTIC system is said to be causal iff the output $y(t)$ depends only on the present and past value of input. Consider the output response of a certain system whose impulse response is $h(t)$ and the input is $x(t)$. Using convolution integral, the output response is obtained from the following equation.

$$y(t) = \int_{-\infty}^{\infty} h(\tau)x(t-\tau)d\tau \tag{3.10}$$

For a causal system $h(\tau) = 0$ for $\tau < 0$. Now Eq. (3.10) is written as

$$y(t) = \int_{0}^{\infty} h(\tau)x(t-\tau)d\tau \tag{3.11}$$

If a causal signal is applied to a non-causal system

$$\begin{aligned} y(t) &= \int_{-\infty}^{\infty} x(\tau)h(t-\tau)d\tau \\ &= \int_{-\infty}^{t} h(\tau)x(t-\tau)d\tau \end{aligned} \tag{3.12}$$

If a causal signal is applied to a causal system

$$\begin{aligned} y(t) &= \int_{0}^{t} x(\tau)h(t-\tau)d\tau \\ &= \int_{0}^{t} h(\tau)x(t-\tau)d\tau \end{aligned} \tag{3.13}$$

3.9 Stability of a Linear Time Invariant System

A linear time invariant continuous time system is said to be Bounded Input Bounded Output (BIBO) stable if every bounded input applied to the input terminal results in a bounded output. Such a stability is called external stability. On the other hand if a system which is in equilibrium state and when a small disturbance is given, the system comes back to the equilibrium state then the system is said to be internally stable. If every bounded input produces bounded output, the system is said to be BIBO stable. On the other hand, if even one bounded input produces unbounded output, the system is said to be BIBO unstable. The BIBO stability can also be expressed in terms of impulse response of the system. Consider the following convolution:

$$\begin{aligned} y(t) &= h(t) * x(t) \\ &= \int_{-\infty}^{\infty} h(\tau)x(t-\tau)d\tau \\ |y(t)| &= \left| \int_{-\infty}^{\infty} h(\tau)x(t-\tau)d\tau \right| \\ &\leq \int_{-\infty}^{\infty} |h(\tau)||x(t-\tau)|d\tau \end{aligned}$$

If $x(t)$ is bounded, then $|x(t-\tau)| < A < \infty$ and

$$|y(t)| \leq A \int_{-\infty}^{\infty} |h(\tau)|d\tau \tag{3.14}$$

For the output also to be bounded which will make the system stable

$$\boxed{\int_{-\infty}^{\infty} |h(\tau)|d\tau < \infty} \tag{3.15}$$

Equation (3.15) is nothing but the area under the impulse response curve. This is the necessary and sufficient condition for the system to be stable. The stable and unstable response curves of an LTIC time system are shown in Fig. 3.23a and b, respectively.

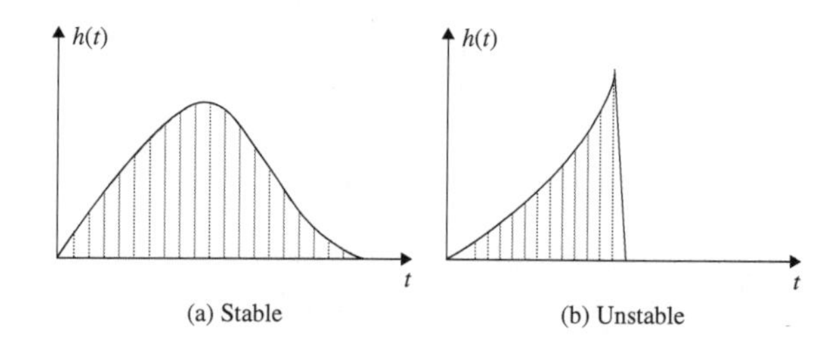

Fig. 3.23 Impulse response curves of stable and unstable systems

■ Example 3.27

The impulse response of a certain system is given by

$$h(t) = e^{-3t}u(t-2)$$

Determine the stability of the system.

Solution:

$$\begin{aligned} y(t) &= \int_{-\infty}^{\infty} |h(t)|dt \\ &= \int_{2}^{\infty} e^{-3t}dt \\ &= -\frac{1}{3}\left[e^{-3t}\right]_2^{\infty} \\ &= \frac{1}{3}e^{-6} \end{aligned}$$

$$\boxed{y(t) = 8.262 \times 10^{-4} < \infty}$$

The output response $y(t) < \infty$ which is finite and hence the system is BIBO stable. The output response curve is shown in Fig. 3.24 which shows the area under the output response curve is finite. Hence, the system is BIBO stable.

■ Example 3.28

The impulse response of an LTIC system is

$$h(t) = e^{-2t} \sin 3t\, u(t)$$

Determine whether the given system is BIBO stable.

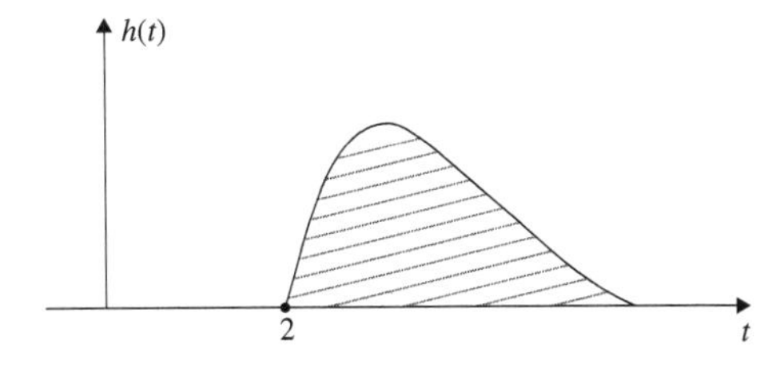

Fig. 3.24 Output response curve of Example 3.27

Solution:

$$h(t) = e^{-2t} \sin 3t\, u(t)$$
$$y(t) = \int_0^{\infty} e^{-2t} \sin 3t dt$$

For a causal system, the lower limit of integration is 0.

$$y(t) = \int_0^{\infty} |e^{-2t} \sin 3t| dt$$

The above integration is solved using the following property of integration

$$y(t) = \int_0^{\infty} e^{-at} \sin bt\, dt = \frac{b}{a^2 + b^2}$$

Here $a = 2$; $b = 3$

$$\therefore \quad y(t) = \frac{3}{4 + 9}$$

$$\boxed{y(t) = \frac{3}{13} < \infty}$$

The system is BIBO stable.

■ Example 3.29

A certain LTIC system has the following impulse response. Determine whether the system is BIBO stable.

$$h(t) = e^{-3t} \cos 2t\, u(t)$$

Solution:

$$h(t) = e^{-3t} \cos 2t\, u(t)$$

This is a causal system and the output is expressed as

$$y(t) = \int_0^{\infty} |e^{-3t} \cos 2t| dt$$

The above integration is solved using the following property

$$\int_0^{\infty} e^{-at} \cos bt\, dt = \frac{a}{a^2 + b^2}$$

Here $a = 3$; $b = 2$

$$\therefore \quad y(t) = \frac{3}{9+4}$$

$$\boxed{y(t) = \frac{3}{13} < \infty}$$

The system is BIBO stable.

Example 3.30

The impulse response of a certain LTIC system is given by

$$h(t) = te^{-2t}u(t)$$

Determine the BIBO stability of the system.

Solution:

$$h(t) = te^{-2t}u(t)$$
$$y(t) = \int_0^\infty |te^{-2t}|dt$$

The above integration is solved using $u - v$ method. Let $u = t,\ du = dt,\ dv = e^{-2t}dt, v = -\frac{1}{2}e^{-2t}$

$$\begin{aligned}
y(t) &= uv - \int_0^\infty vdu \\
&= \left[t\left(-\frac{1}{2}e^{-2t}\right)\right]_0^\infty + \frac{1}{2}\int_0^\infty e^{-2t}dt \\
&= 0 + \frac{1}{2}\int_0^\infty e^{-2t} \\
&= -\frac{1}{4}\left[e^{-2t}\right]_0^\infty
\end{aligned}$$

$$\boxed{y(t) = \frac{1}{4} < \infty}$$

$y(t)$ is finite and hence the system is BIBO stable.

Example 3.31

For a certain system, the impulse response $h(t)$ is given by $h(t) = e^{(-1+j)t}u(t)$. Determine the BIBO stability of the system.

Solution:

$$h(t) = e^{(-1+j)t}u(t)$$

For a causal system

$$\begin{aligned}
y(t) &= \int_0^\infty |h(t)|dt \\
&= \int_0^\infty |e^{(-1+j)t}|dt \\
&= \int_0^\infty |e^{-t}||e^{jt}|dt \\
|e^{jt}| &= |\cos t + j\sin t| = \sqrt{\cos^2 t + \sin^2 t} = 1 \\
y(t) &= \int_0^\infty e^{-t}dt \\
&= \left[-e^{-t}\right]_0^\infty
\end{aligned}$$

$$\boxed{y(t) = 1 < \infty}$$

Since $y(t)$ is finite, the given system is BIBO stable.

■ Example 3.32

Determine the BIBO stability of the system whose impulse response is

$$h(t) = e^{3t}u(-t-3)$$

Solution: The impulse response of the given system is shown in Fig. 3.25. It is non-causal. The lower limit of integration is $-\infty$ and the upper limit is -3. Hence, the output response is obtained from the following equation:

$$\begin{aligned}
y(t) &= \int_{-\infty}^{-3} |e^{3t}|dt \\
&= \frac{1}{3}\left[e^{3t}\right]_{-\infty}^{-3} \\
&= \frac{1}{3}e^{-9}
\end{aligned}$$

$$\boxed{y(t) = 4.11 \times 10^{-5} < \infty}$$

Since $y(t)$ is finite, the system is BIBO stable.

■ Example 3.33

Determine whether the system described by the following response function is BIBO stable.

Solution: From Fig. 3.25a, the area under the impulse response curve is finite. The system is therefore BIBO stable. This can also be proved mathematically as follows:

$$\begin{aligned} y(t) &= \int_{-\infty}^{\infty} |h(t)|dt \\ &= \int_{-\infty}^{0} e^{t}dt + \int_{0}^{\infty} e^{-t}dt \\ &= \left[e^{t}\right]_{-\infty}^{0} + \left[-e^{-t}\right]_{0}^{\infty} \\ &= 1 - 0 - 0 + 1 \end{aligned}$$

$$\boxed{y(t) = 2 < \infty}$$

The system output is finite and hence the system is stable.

■ Example 3.34

The system shown in Fig. 3.26 is formed by connecting two systems in cascade. The impulse response of the systems are given by $h_1(t)$ and $h_2(t)$, respectively. $h_1(t) = e^{-2t}u(t)$ and $h_2(t) = 2e^{-t}u(t)$. Determine if the overall system is BIBO stable.

(Anna University, May, 2007)

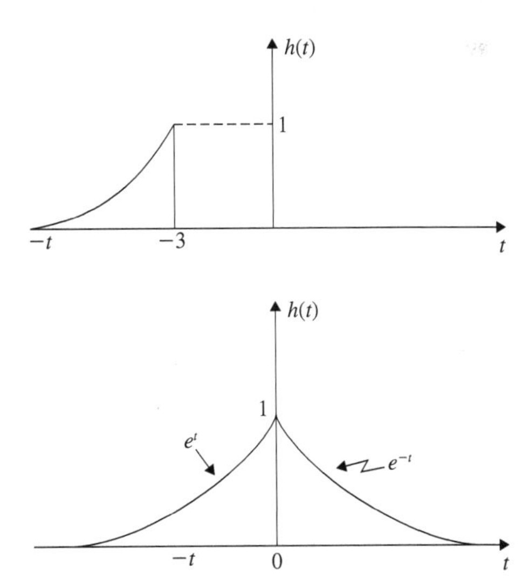

Fig. 3.25 Impulse response of Example 3.32. **a** The Impulse response curve of Example 3.33

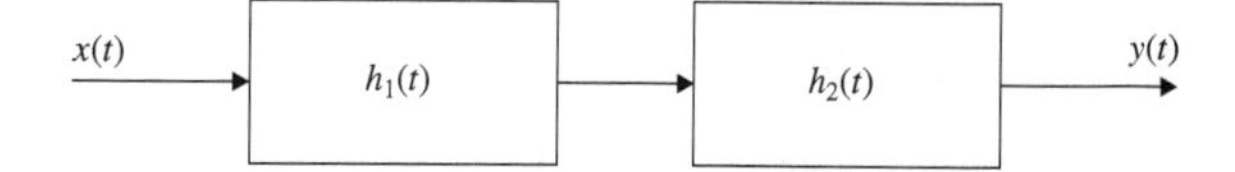

Fig. 3.26 Impulse response of two systems connected in cascade

Solution: The overall impulse response $h(t) = h_1(t) * h_2(t)$ is obtained in Example 3.23 as

$$h(t) = 2[e^{-t} - e^{-2t}]u(t)$$

Since the system is causal, the following equation is obtained to determine the BIBO stability.

$$\begin{aligned} y(t) &= \int_0^{\infty} |h(t)|dt \\ &= 2\int_0^{\infty} [e^{-t} - e^{-2t}]dt \\ &= 2\left[-e^{-t} + \frac{1}{2}e^{-2t}\right]_0^{\infty} \\ &= 2\left[1 - \frac{1}{2}\right] = 1 \\ y(t) &= 1 < \infty \end{aligned}$$

Hence, the system is BIBO stable.

■ Example 3.35

Find which of the following systems with the impulse response given are causal.

$$\begin{aligned} &(a) \quad e^{-at}u(t) \\ &(b) \quad e^{-a|t|} \\ &(c) \quad e^{-at}u(t-1) \\ &(d) \quad e^{+at}u(-t-1) \\ &(e) \quad e^{at}u(-t+1) \\ &(f) \quad e^{at}u(t+1) \\ &(g) \quad e^{at}u(t+1) + e^{-at}u(t-1) \end{aligned}$$

Solution:

(a) $u(t)$ is the present input. Hence, the system is ***causal*** $h(t) = 0$ for $t < 0$.
(b) The impulse response depends on the present and future input. Hence, the system is ***non-causal*** $h(t) \neq 0$ for $t < 0$.

(*c*) The impulse response depends on the past input $u(t-1)$. Hence, the system is ***causal*** $h(t) = 0$ for $t < 0$.
(*d*) The impulse response depends on future input $u(-t-1)$. Hence, the system is ***non-causal*** $h(t) \neq 0$ for $t < 0$.
(*e*) The impulse response depends on future input $u(-t+1)$. Hence, the system is ***non-causal*** $h(t) \neq 0$ for $t < 0$.
(*f*) The impulse response depends on future input $u(t+1)$. Hence, the system is ***non-causal*** $h(t) \neq 0$ for $t < 0$.
(*g*) The impulse response depends on future input $u(t+1)$. Hence, it is ***non-causal*** $h(t) \neq 0$ for $t < 0$.

3.10 Step Response from Impulse Response

The step response of an LTIC system can be obtained from the impulse response by integrating it. Let $s(t)$denote the step response. Consider the system with the impulse response $h(t)$. Let $u(t)$ be the input signal. Using convolution, the following equation is written:

$$\begin{aligned} s(t) &= u(t) * h(t) \\ &= \int_{-\infty}^{t} h(\tau)u(t-\tau)d\tau \end{aligned}$$

$$\boxed{s(t) = \int_{-\infty}^{t} h(\tau)d\tau} \tag{3.16}$$

For the causal system, the lower limit of integration is 0. Hence,

$$s(t) = \int_{0}^{t} h(\tau)d\tau \tag{3.17}$$

Thus, the step response is obtained by integrating the impulse response. The impulse response is obtained by differentiating the step response $s(t)$.

The following examples illustrate the method of obtaining step response from impulse response.

■ Example 3.36

The impulse response of a certain LTIC system is given by

$$h(t) = e^{-4t}u(t) - e^{-2t}u(t)$$

Determine the step response of the system.

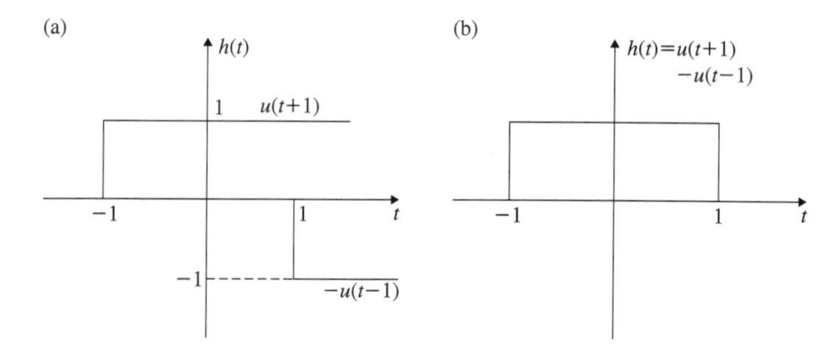

Fig. 3.27 Plot of $h(t) = u(t+1) - u(t-1)$

Solution:

$$h(t) = [e^{-4t} - e^{-2t}]u(t)$$

The step response $s(t)$ is given by

$$\begin{aligned}
s(t) &= \int_0^t h(\tau)d\tau \\
&= \int_0^t [e^{-4\tau} - e^{-2\tau}]d\tau \\
&= \left[-\frac{1}{4}e^{-4\tau} + \frac{1}{2}e^{-2\tau}\right]_0^t \\
&= \left[-\frac{1}{4}e^{-4t} + \frac{1}{4} + \frac{1}{2}e^{-2t} - \frac{1}{2}\right]
\end{aligned}$$

$$\boxed{s(t) = \frac{1}{4}[2e^{-2t} - e^{-4t} - 1]}$$

■ Example 3.37

Find the step response whose impulse response is

$$h(t) = u(t+1) - u(t-1)$$

(*Anna University, May, 2007*)

Solution: The impulse response shown in Fig. 3.27a can be represented as in Fig. 3.27b. From Fig. 3.27b, the step response is obtained as follows:

$$s(t) = \int_{-\infty}^{t} h(\tau)d\tau = \int_{-1}^{1} d\tau = \left[\tau\right]_{-1}^{1}$$

$$\boxed{s(t) = 2}$$

■ Example 3.38

The impulse response of a certain system is given by

$$h(t) = \delta(t) - \delta(t-2)$$

Determine the step response.

Solution:

$$h(t) = \delta(t) - \delta(t-2)$$

The step response is obtained from the following convolution.

$$\begin{aligned} s(t) &= h(t) * u(t) \\ &= [\delta(t) - \delta(t-2)] * u(t) \end{aligned}$$

using convolution property, $\delta(t) * u(t) = u(t)$ and $\delta(t-2) * u(t) = u(t-2)$ we get,

$$\boxed{s(t) = u(t) - u(t-2)}$$

the above result can also be obtained by integrating the impulse function which is a step function.

■ Example 3.39

Find the step response of the system whose impulse response is

$$h(t) = \frac{t^2}{2}u(t)$$

Solution:

$$\begin{aligned} h(t) &= \frac{t^2}{2}u(t) \\ s(t) &= \int_0^t \frac{\tau^2}{2} d\tau \\ &= \frac{1}{6}\left[\tau^3\right]_0^t \end{aligned}$$

$$\boxed{s(t) = \frac{t^3}{6}}$$

■ Example 3.40

Find the step response of the system whose impulse response is

$$h(t) = e^{-2t}u(t+2)$$

Solution:

$$\begin{aligned} h(t) &= e^{-2t}u(t+2) \\ s(t) &= \int_{-2}^{t} e^{-2\tau}d\tau \\ &= -\frac{1}{2}[e^{-2\tau}]_{-2}^{t} \\ &= -\frac{1}{2}[e^{-2t} - e^{4}] \end{aligned}$$

$$\boxed{s(t) = [27.3 - 0.5e^{-2t}]u(t+2)}$$

■ Example 3.41

Find the step response of the system if the impulse response is

$$h(t) = e^{3t}u(t-2)$$

From the results so obtained find the impulse response.

Solution:

$$\begin{aligned} h(t) &= e^{3t}u(t-2) \\ s(t) &= \int_{2}^{t} e^{3\tau}d\tau = \frac{1}{3}\left[e^{3\tau}\right]_{2}^{t} \end{aligned}$$

$$\boxed{s(t) = \frac{1}{3}[e^{3t} - 403.4]u(t-2)}$$

The impulse response $h(t)$ is obtained by differentiating the step response

$$\boxed{h(t) = \frac{ds(t)}{dt} = e^{3t}u(t-2)}$$

Important Points to Remember in Connection with Convolution Integral

1. The convolution operation is expressed as

$$y(t) = \int_{-\infty}^{\infty} x(\tau)h(t-\tau)d\tau$$

The symbolic representation of convolution is

$$y(t) = x(t) * h(t)$$

"$*$" indicates convolution operation.

2. For the causal signal, the convolution integral is given by

$$y(t) = \int_{0}^{t} x(\tau)h(t-\tau)d\tau$$

For the non-causal signal, the convolution is given by

$$y(t) = \int_{-\infty}^{t} x(\tau)h(t-\tau)d\tau$$

3. The commutative property of convolution is

$$x(t) * h(t) = h(t) * x(t)$$
$$\int_{-\infty}^{\infty} x(\tau)h(t-\tau)d\tau = \int_{-\infty}^{\infty} x(t-\tau)h(\tau)d\tau$$

4. The distributive property of convolution is

$$y(t) = x(t) * h(t)$$

where

$$h(t) = (h_1(t) + h_2(t))$$

5. The associative property of convolution is

$$[x(t) * h_1(t)] * h_2(t) = x(t) * [h_1(t) * h_2(t)]$$

6. The shifting property of convolution is

$$x(t-p) * h(t-q) = y(t-p-q)$$

7. The width property of convolution is that the width of

$$x(t) * h(t) = T_1 + T_2$$

where T_1=width of $x(t)$ and T_2=width of $h(t)$.
8. The convolution with an impulse is

$$x(t) * \delta(t) = x(t)$$

9. The convolution with the delayed impulse is

$$x(t) * \delta(t - t_0) = x(t - t_0))$$

10. The convolution with unit step is

$$x(t) * u(t) = \int_0^t x(\tau)d\tau$$

11. The convolution with a delayed step is

$$x(t) * u(t - t_0) = \int_{-\infty}^{t-t_0} x(\tau)d\tau$$

12. **System causality**
For a causal system $h(\tau) = 0$ for $t < 0$.
13. The necessary and sufficient condition for the system to be bounded input bounded output stable is

$$\int_{-\infty}^{\infty} |h(\tau)|d\tau < \infty$$

14. The step response from impulse response is obtained from

$$s(t) = \int_{-\infty}^{t} h(\tau)d\tau$$

The impulse response is obtained by differentiating step response.

3.11 Representation of Discrete Time Signals in Terms of Impulses

As discussed in the previous chapter, the discrete time signal is represented as a sequence of impulses. If these sequences of impulses are expressed mathematically, it will help us develop the characterization of any Linear Time Invariant Discrete (LTID) time system.

Consider the sequence of impulse signals shown in Fig. 3.28a. The time shifted impulse sequences are shown in Fig. 3.28b–e for the sequence interval $-4 \leq n \leq 3$, the impulses are represented as $x(-4)$, $x(-3)$, $x(-2)$, $x(-1)$, $x(0)$, $x(1)$, $x(2)$ and $x(3)$. The signal say at $n = -3$ is mathematically expressed as

$$\begin{aligned} x(-3)\delta(n+3) &= x(-3) \qquad n = -3 \\ &= 0 \qquad\qquad n \neq -3 \end{aligned}$$

$\delta(n+4)$, $\delta(n+3), \ldots, \delta(n-3)$ are unit impulses occurring at $n = -4$, $n = -3$, $\ldots, n = 3$. These sequences of impulses are shown in Fig. 3.28a. These sequences are mathematically expressed as

$$x(n) = x(-4)\delta(n+4) + x(-3)\delta(n+3) + \cdots + x(2)\delta(n-2) + x(3)\delta(n-3)$$

Fig. 3.28 Sequence of impulse signals

(a)

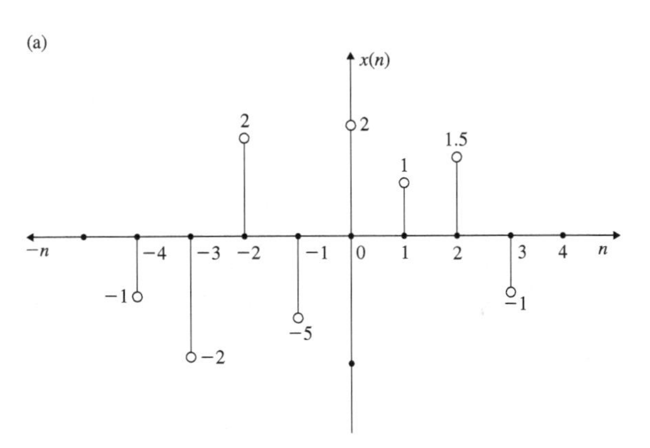

(b)

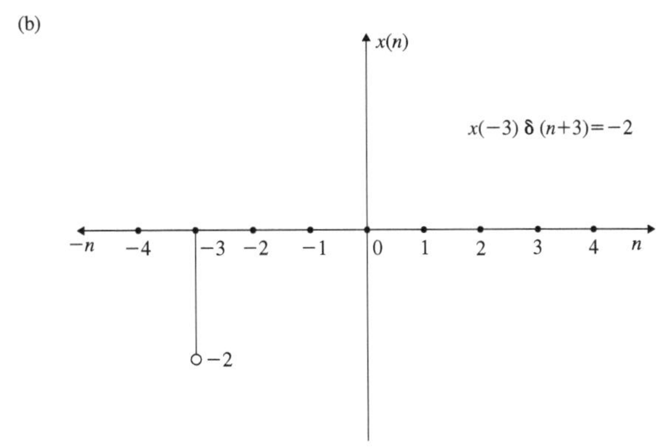

Fig. 3.28 (continued)

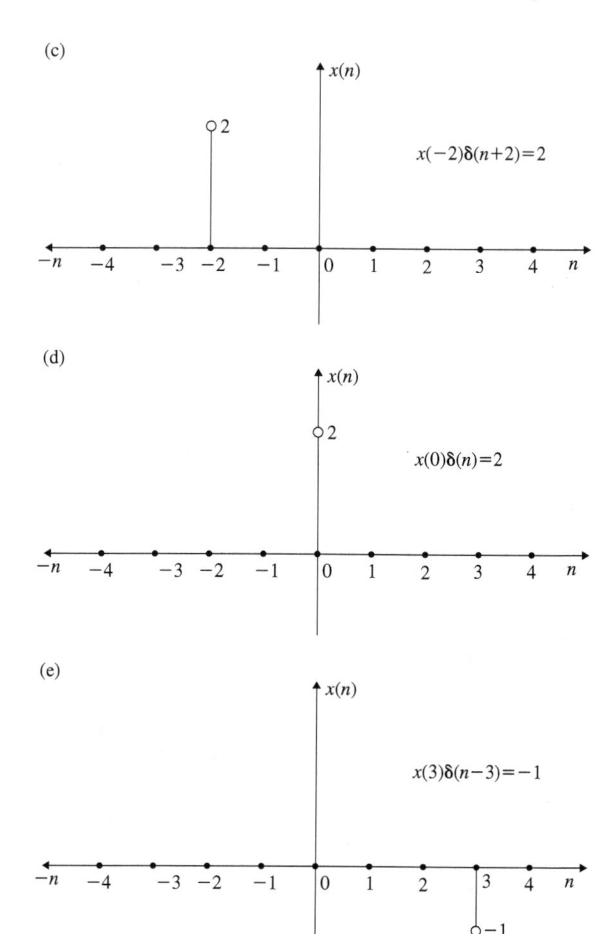

In general, if these sequences occur in the interval $-\infty < n < \infty$, then $x(n)$ is mathematically expressed as

$$\boxed{x(n) = \sum_{k=-\infty}^{\infty} x(k)\delta(n-k)} \tag{3.18}$$

3.12 The Discrete Time Unit Impulse Response

Let $\delta(n-k)$ be unit impulse which is shifted by k. It has value 1 corresponding to value k. Let $h_k(n)$ denote the response of the linear time invariant discrete time system. By superposition theorem, the output response $y(n)$ to the input $x(n)$ is obtained as the weighted linear combination of these responses. Thus, the impulse

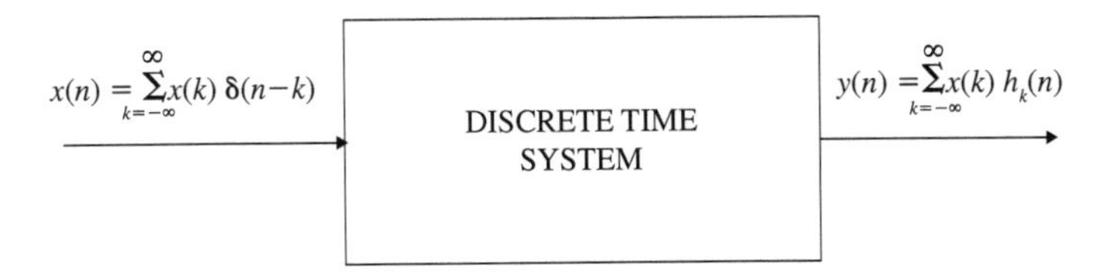

Fig. 3.29 Input-output of DT system

response $y(n)$ can be expressed as

$$y(n) = \sum_{k=-\infty}^{\infty} x(k)h_k(n) \tag{3.19}$$

Thus, if we know the response of a linear system to the set of shifted unit impulses, it is possible to determine the response to any arbitrary input sequence. The block diagram of DT system is shown in Fig. 3.29.

3.13 The Convolution Sum

Similar to linear time invariant continuous time system where the convolution integral was made use of in determining the output response of the system, in discrete time system **the convolution sum or superposition sum** is used to determine the output sequence. For any arbitrary $x(n)$, knowing the impulse response sequence $h_k(n)$, the output response $y(n)$ is obtained as explained below.

Let $h_{-1}(n)$, $h_0(n)$, $h_1(n)$ be the impulse responses to the input $x(n)\delta(n+1)$, $x(n)\delta(n)$ and $x(n)\delta(n-1)$. By applying superposition theorem, it is possible to get $y(n)$ which is nothing but the linear combination of the responses due to the individual shifted impulses. Consider the signals $x(n)$, $h_{-1}(n)$, $h_0(n)$ and $h_1(n)$ represented in Fig. 3.30a, b, c and d, respectively. The signal $x(-1)\delta(n+1)$ is obtained from $x(n)$ at $n=-1$. Similarly, $x(2)\delta(n)$ is obtained from $x(n)$ at $n=0$. Similarly, $x(1)\delta(n-1)$, $x(2)\delta(n-2)$ can be obtained. The product of $x(-1)\delta(n+1)h_{-1}(n)$ gives $y(-1)$. Similarly $x(0)\delta(n)h_0(n)$ gives $y(0)$ and $x(1)\delta(n-1)h_1(n)$ gives $y(1)$ evaluated at $n=1$. These signals are represented in Fig. 3.31.

From Fig. 3.31a–k the values of $y(n)$ for $-\infty < n < \infty$ are calculated from $x(n)h(n)$ by summing up all these values for a given n. Thus,

$$n=-1, \qquad y(-1) = -3 + 2 - 1 = -2$$

$$n=0, \qquad y(0) = 4 = 4$$

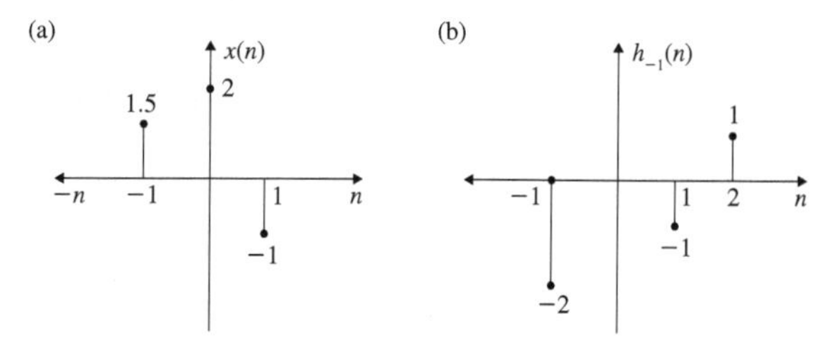

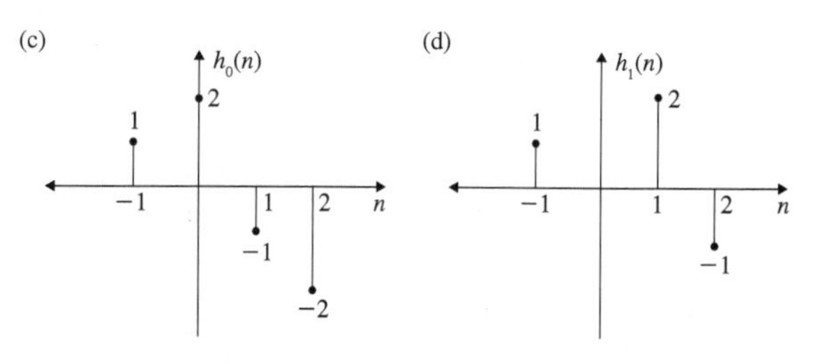

Fig. 3.30 Input and impulse response sequences

$$n = 1, \qquad y(1) = -1.5 - 2 - 2 = -5.5$$

$$n = 2, \qquad y(2) = 1.5 - 4 + 1 = -1.5$$

The sketch of $y(n)$ is shown in Fig. 3.31k. Thus, the response $y(n)$ at any instant n is nothing but the superposition of the input at every point of n. Since $\delta(n-k)$ is the time shifted version of $\delta(n)$, the response $h_k(n)$ is the time shifted version of $h_0(n)$ which can be represented as

$$h_k(n) = h_0(n-k)$$

Now, the equation for the convolution sum can be written as

$$y(n) = \sum_{k=-\infty}^{\infty} x(k)h_0(n-k)$$

For convenience if we drop the subscript on $h_0(n-k)$, the above equation is written as

$$\boxed{y(n) = \sum_{k=-\infty}^{\infty} x(k)h(n-k)} \tag{3.20}$$

The above equation is referred to as the convolution sum or superposition sum. $h(n)$ is the impulse response of the LTID time system for the input $\delta(n)$. The convolution sum is symbolically represented as

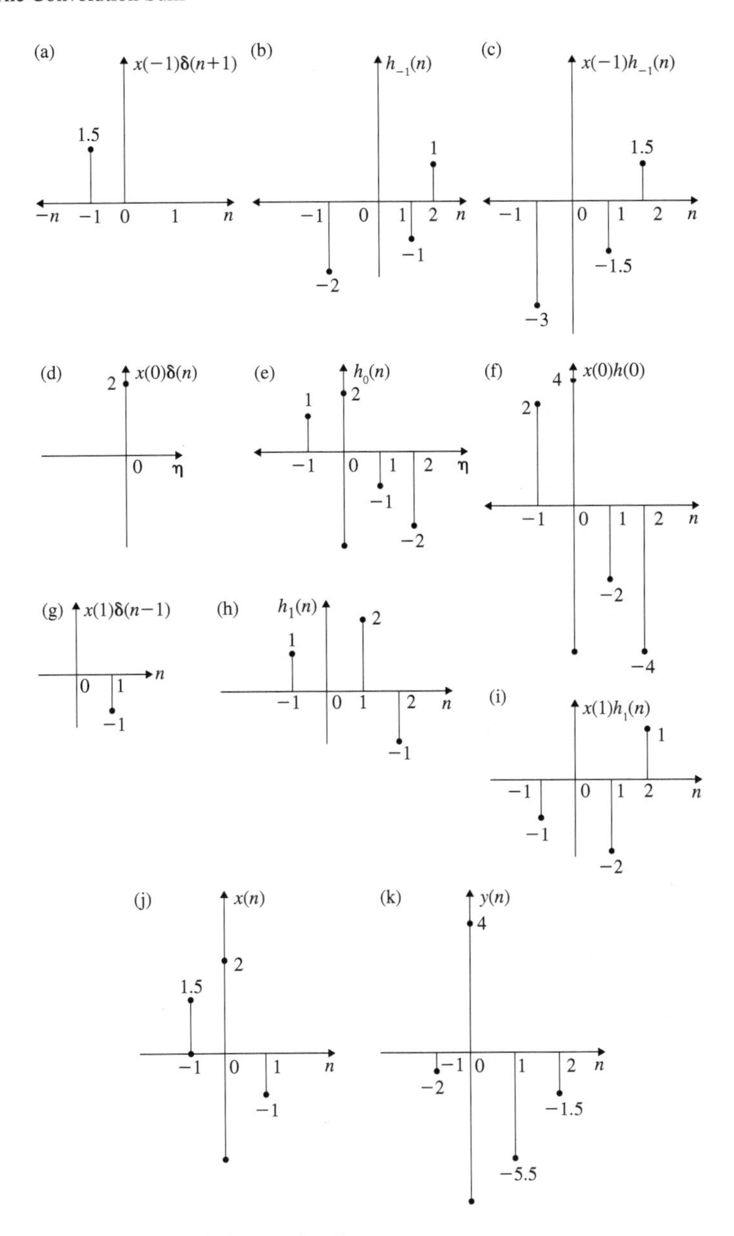

Fig. 3.31 Convolution sum of discrete signals

$$y(n) = x(n) * h(n) \tag{3.21}$$

There are different methods available to get the solution of $y(n)$ by the convolution method, and they are described and illustrated below with examples. The properties of convolution which will be useful for solving convolution sum are discussed first as follows.

3.14 Properties of Convolution Sum

3.14.1 Distributive Property

Two linear time invariant discrete time systems connected in parallel are shown in Fig. 3.32a with their impulse responses $h_1(n)$ and $h_2(n)$. According to convolution sum

$$\begin{aligned} y_1(n) &= x(n) * h_1(n) \\ y_2(n) &= x(n) * h_2(n) \\ y(n) &= y_1(n) + y_2(n) \\ &= x(n) * h_1(n) + x(n) * h_2(n) \end{aligned}$$

$$\begin{aligned} y(n) &= \sum_{k=-\infty}^{\infty} x(k)h_1(n-k) + \sum_{k=-\infty}^{\infty} x(k)h_2(n-k) \\ &= \sum_{k=-\infty}^{\infty} x(k)[h_1(n-k) + h_2(n-k)] \end{aligned}$$

Substituting $h(n-k) = [h_1(n-k) + h_2(n-k)]$, we get

$$\begin{aligned} y(n) &= \sum_{k=-\infty}^{\infty} x(k)h(n-k) \\ &= x(n) * h(n) \end{aligned}$$

$$\boxed{y(n) = x(n) * [h_1(n) + h_2(n)]} \tag{3.22}$$

3.14.2 Associative Property of Convolution

According to this property $y(n) = x(n) * [h_1(n) * h_2(n)] = [x(n) * h_1(n)] * h_2(n)$.

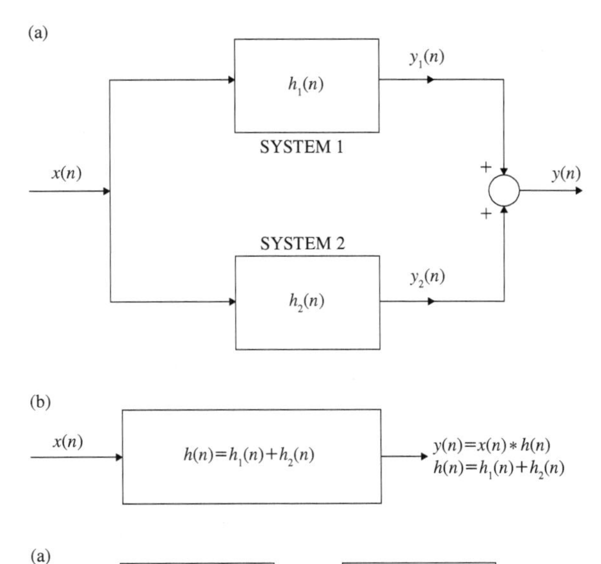

Fig. 3.32 Distributive property of convolution

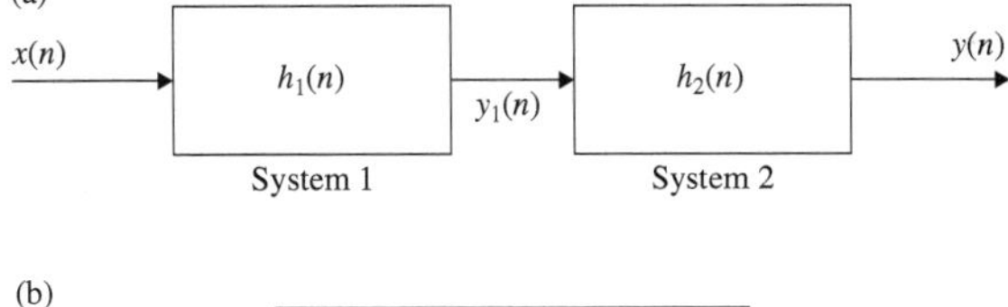

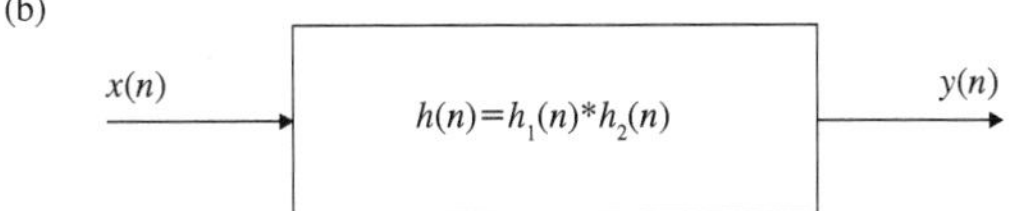

Fig. 3.33 Associative property of convolution

Proof Consider two linear time invariant discrete time systems connected in cascade as shown in Fig. 3.33a with their impulse responses $h_1(n)$ and $h_2(n)$, respectively. From Fig. 3.33a, the following equations are written:

$$
\begin{aligned}
y_1(n) &= x(n) * h_1(n) \\
&= \sum_{k=-\infty}^{\infty} x(k)h_1(n-k) \\
y(n) &= y_1(n) * h_2(n) \\
&= \sum_{k=-\infty}^{\infty} y_1(k)h_2(n-k) \\
&= \sum_{k=-\infty}^{\infty} x(p)h_1(k-p)h_2(n-k)
\end{aligned}
$$

Putting $(k-p)=q$ in the above equation, we get

$$y(n) = \sum_{p=-\infty}^{\infty} x(p) \sum_{q=-\infty}^{\infty} h_1(q)h_2(n-p-q)$$
$$= \sum_{p=-\infty}^{\infty} x(p) \sum_{q=-\infty}^{\infty} h(n-p)$$
$$= \sum_{p=-\infty}^{\infty} x(p)h(n-p)$$

$$\boxed{y(n) = x(n) * h(n)}$$

where

$$h(n) = \sum_{q=-\infty}^{\infty} h_1(q)h_2(n-q)$$

$$\boxed{h(n) = h_1(n) * h_2(n)}$$

Hence,

$$y(n) = x(n) * (h_1(n) * h_2(n))$$

$$y(n) = [x(n) * h_1(n)] * h_2(n) \tag{3.23}$$

$$\boxed{y(n) = x(n) * [h_1(n) * h_2(n)]}$$

3.14.3 Commutative Property of Convolution

According to commutative property

$$h_1(n) * h_2(n) = h_2(n) * h_1(n)$$

Proof Consider the following convolution

$$h_1(n) * h_2(n) = \sum_{k=-\infty}^{\infty} h_1(k)h_2(n-k)$$

Put $n-k=p$ in the above equation.

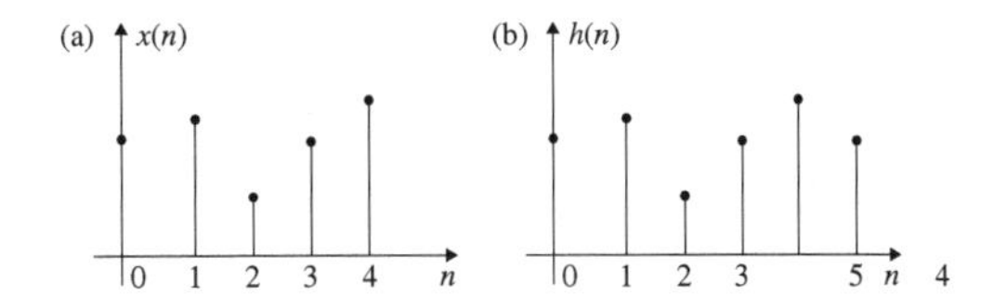

Fig. 3.34 Width property of convolution

$$h_1(n) * h_2(n) = \sum_{p=-\infty}^{\infty} h_1(n-p)h_2(p)$$

$$= \sum_{p=-\infty}^{\infty} h_2(p)h_1(n-p)$$

$$\boxed{h_1(n) * h_2(n) = h_2(n) * h_1(n)} \tag{3.24}$$

3.14.4 *Shifting Property of Convolution*

Let $x(n)$ and $h(n)$ be two sequences. The output response $y(n)$ is expressed as

$$y(n) = x(n) * h(n)$$

If the sequences $x(n)$ and $h(n)$ are shifted by p and q, respectively, as $x(n-p)$ and $h(n-q)$ then

$$\boxed{x(n-p) * h(n-q) = y(n-p-q)} \tag{3.25}$$

3.14.5 *The Width Property of Convolution*

Consider the sequences $x(n)$ and $h(n)$ which are shown in Fig. 3.34a and b, respectively. Let the width of $x(n)$ be T_1 and that of $h(n)$ be T_2. Then, the width of $y(n) = x(n) * h(n)$ is $T = T_1 + T_2$. For the sequences represented in Fig. 3.34, $T_1 = 4$ and $T_2 = 5$. The width of $y(n)$ should be $T = 4 + 5 = 9$.

3.14.6 *Convolution with an Impulse*

Let $x(n)$ be the discrete time signal. When this signal convolves with an impulse $\delta(n)$ then

$$x(n) * \delta(n) = x(n)$$

Proof

$$x(n) * \delta(n) = \sum_{k=-\infty}^{\infty} x(k)\delta(n-k)$$

$$\delta(n-k) = 1 \quad \text{for } n = k$$
$$= 0 \quad \text{for } n \neq k$$

$$x(n) * \delta(n) = \sum_{k=n} x(k)$$

$$\boxed{x(n) * \delta(n) = x(n)} \tag{3.26}$$

3.14.7 Convolution with Delayed Impulse

If the sequence $x(n)$ convolves with $\delta(n - n_0)$ then

$$x(n) * \delta(n - n_0) = x(n - n_0)$$

Proof

$$x(n) * \delta(n - n_0) = \sum_{k=-\infty}^{\infty} x(k)\delta(n - k - n_0)$$

$$\delta(n - k - n_0) = 1 \quad \text{if } k = n - n_0$$
$$= 0 \quad \text{if } k \neq n - n_0$$

$$\therefore \quad x(n) * \delta(n - n_0) = \sum_{k=n-n_0} x(k)\delta((n - n_0) - k)$$

$$\boxed{x(n) * \delta(n - n_0) = x(n - n_0)} \tag{3.27}$$

3.14.8 Convolution with Unit Step

$$x(n) * u(n) = \sum_{k=-\infty}^{\infty} x(k)$$

Proof

$$x(n) * u(n) = \sum_{k=-\infty}^{\infty} x(k)u(n-k)$$

$$u(n-k) = 1 \quad \text{for all } k$$

$$\therefore \quad \boxed{x(n) * u(n) = \sum_{k=-\infty}^{\infty} x(k)} \tag{3.28}$$

3.14.9 Convolution with Delayed Step

If $x(n)$ convolves with delayed step $u(n - n_0)$, then

$$x(n) * u(n - n_0) = \sum_{k=-\infty}^{n-n_0} x(k)$$

Proof

$$x(n) * u(n - n_0) = \sum_{k=-\infty}^{\infty} x(k)u(n - k - n_0)$$

$$\begin{aligned} u(n - n_0) &= 1 \quad \text{for } n \geq n_0 \\ &= 0 \quad \text{for } n < n_0 \end{aligned}$$

$$\therefore \quad \boxed{x(n) * u(n - n_0) = \sum_{k=-\infty}^{n-n_0} x(k)} \tag{3.29}$$

3.14.10 System Causality from Convolution

A linear time invariant discrete time system is said to be causal iff the impulse response does not exist for $n < 0$.

Proof The output response can be written in terms of impulse response and input signal as follows:

$$\begin{aligned} y(n) &= \sum_{k=-\infty}^{\infty} h(k)x(n - k) \\ &= \sum_{k=-\infty}^{1} h(k)x(n - k) + \sum_{k=0}^{n} h(k)x(n - k) \end{aligned}$$

For $k < 0$

$$y(n) = \sum_{k=-\infty}^{1} h(k)x(n-k)$$

The response depends on future inputs $x(n+k), x(n+k-1), \ldots, x(n+1)$. For a causal system, the output response should not depend on future input. Hence,

$$\sum_{k=-\infty}^{1} h(k)x(n-k) = 0 \quad \text{or} \quad h(k) = 0 \quad \text{for} \quad k < 0$$

Changing k as n,

$$\boxed{h(n) = 0} \qquad \text{for } n < 0 \tag{3.30}$$

3.14.11 BIBO Stability from Convolution

A linear time invariant discrete time system is said to be BIBO stable iff its impulse response is absolutely summable. It is mathematically expressed as

$$y(n) = \sum_{k=-\infty}^{\infty} |h(n)| < \infty$$

Proof

$$\begin{aligned} y(n) &= \sum_{k=-\infty}^{\infty} x(k)h(n-k) \\ &= \sum_{k=-\infty}^{\infty} h(k)x(n-k) \\ |y(n)| &= \left| \sum_{k=-\infty}^{\infty} h(k)x(n-k) \right| \end{aligned}$$

The magnitude of the sum of the terms is always less than or equal to the sum of the magnitude. Thus,

$$\left| \sum_{k=-\infty}^{\infty} h(k)x(n-k) \right| \leq \sum_{k=-\infty}^{\infty} |h(k)||x(n-k)|$$

For a bounded input, $|x(n-k)| < \infty$. Let it be M_x

$$|y(n)| \leq M_x \sum_{k=-\infty}^{\infty} |h(k)|$$

For the bounded output $|y(n)| < \infty$. Let it be M_y

$$\sum_{k=-\infty}^{\infty} |h(k)| = M_y < \infty$$
$$y(n) < M_x M_y < \infty$$

Changing $k = n$ in the impulse response summation, we get

$$\sum_{n=-\infty}^{\infty} |h(n)| < \infty \tag{3.31}$$

In other words, $\sum_{n=-\infty}^{\infty} |h(n)|$ is summable and finite for the discrete time system to be BIBO stable.

3.14.12 Step Response in Terms of Impulse Response of a LTDT System

The step response $s(n)$ is obtained from the impulse response as

$$s(n) = \sum_{k=0}^{n} h(k)$$

Proof Let

1. $y(n) = s(n)$ for step input.
2. $h(n) =$ Impulse response
3. $x(n) = u(n)$ =step input sequence

The convolution of $h(n)$ and $x(n)$ is expressed as

$$s(n) = \sum_{k=-\infty}^{\infty} h(k)u(n-k)$$
$$u(n-k) = 1 \quad n > k$$
$$= 0 \quad n < k$$
$$s(n) = \sum_{k=0}^{n} h(k) \times 1$$

$$\boxed{s(n) = \sum_{k=0}^{n} h(k)}$$

The properties of LTID system convolution and the connected results are summarized below.

Important Points to Remember in Connection with Convolution Sum

1. The convolution sum is mathematically expressed as

$$y(n) = x(n) * h(n)$$
$$y(n) = \sum_{k=-\infty}^{\infty} x(k)h(n-k)$$

"$*$" denotes convolution operation.

2. The commutative property of convolution is

$$x(n) * h(n) = h(n) * x(n)$$

3. The distributive property of convolution is

$$y(n) = x(n) * h(n)$$

where

$$h(n) = h_1(n) + h_2(n)$$

4. The associative property of convolution is

$$[x(n) * h_1(n)] * h_2(n) = x(n) * [h_1(n) * h_2(n)]$$

5. The shifting property of convolution is

$$x(n-p) * h(n-q) = y(n-p-q)$$

6. The width property of convolution is

$$[x(n) * h(n)] = T_1 + T_2$$

where T_1=width of $x(n)$ sequence and T_2=width of $h(n)$ sequence.

7. The convolution with an impulse is

$$x(n) * \delta(n) = x(n)$$

8. The convolution with delayed impulse is

$$x(n) * \delta(n - n_0) = x(n - n_0)$$

9. Convolution with unit step is

$$x(n) * u(n) = \sum_{k=-\infty}^{\infty} x(k)$$

10. Convolution with a delayed step is

$$x(n) * u(n - n_0) = \sum_{k=-\infty}^{\infty} x(k - n - n_0)$$

11. System causality in terms of impulse response is

$$h(n) = 0 \qquad \text{for } n < 0$$

12. The necessary and sufficient BIBO stability condition is

$$\sum_{n=-\infty}^{\infty} |h(n)| < \infty$$

In other words $|h(n)|$ should be absolutely summable.

13. Step response $s(n)$ is obtained from impulse response using the following mathematical expression:

$$s(n) = \sum_{k=0}^{n} h(k)$$

3.15 Response Using Convolution Sum

If the impulse response $h(n)$ is known, the output response $y(n)$ can be obtained for any input sequence $x(n)$ by using the following methods:

- Analytical methods and
- Graphical method

In analytical method, $y(n)$ is obtained by

— Using mathematical expression for the convolution sum
— Multiplication method
— Tabulation method.
— Matrix method.

3.15.1 Analytical Method Using Convolution Sum

■ Example 3.42

The impulse response $h(n)$ of a certain LTID system is given by $h(n) = a^n u(n)$ where $0 < a < 1$. The system is excited by $x(n) = u(n)$, a step sequence. Find $y(n)$ using convolution sum.

(*Anna University, December, 2006*)

Solution: The impulse response and input sequences are represented in Fig. 3.35a and b, respectively. They are causal sequences. Hence, the convolution sum given in Eq. (3.20) is written as follows:

$$x(n) = u(n)$$
$$x(k) = u(k) = 1$$
$$h(n) = a^n u(n)$$

$$h(n-k) = a^{n-k} u(n-k)$$
$$u(k) = 1$$
$$u(n-k) = 1$$
$$y(n) = \sum_{k=-0}^{n} x(k)h(n-k)$$
$$= \sum_{k=-0}^{n} (a^{n-k})$$
$$= a^n \sum_{k=-0}^{n} \left(\frac{1}{a}\right)^k$$

The above expression is simplified using the finite summation formula

$$\boxed{\sum_{k=-0}^{n} A^k = \frac{(1-A^{n+1})}{(1-A)}}$$

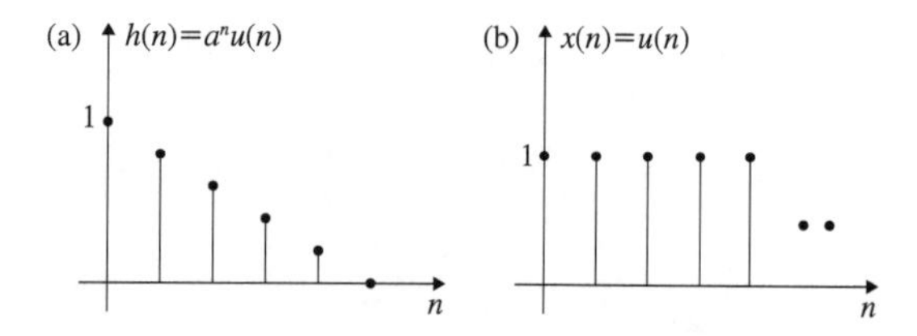

Fig. 3.35 Representation of $h(n)$ and $x(n)$ sequences

Therefore,

$$y(n) = \frac{a^n\left(1 - \frac{1}{a^{n+1}}\right)}{\left(1 - \frac{1}{a}\right)}$$

$$\boxed{y(n) = \frac{(1 - a^{n+1})}{(1 - a)}u(n)}$$

Example 3.43

$$x(n) = u(n)$$
$$h(n) = u(n)$$

Find

$$y(n) = x(n) * h(n)$$

Solution: The excitation sequence $x(n)$ and the impulse response sequence $h(n)$ are shown in Fig. 3.36a and b, respectively. Since these two signals are causal, the convolution sum can be written as follows:

$$y(n) = \sum_{k=0}^{n} (1)^k (1)^{n-k}$$
$$= (1)^n \sum_{k=0}^{n} (1) = \sum_{k=0}^{n} 1$$

According to finite summation formula

$$\boxed{\sum_{k=0}^{n} 1 = (1 + n)}$$

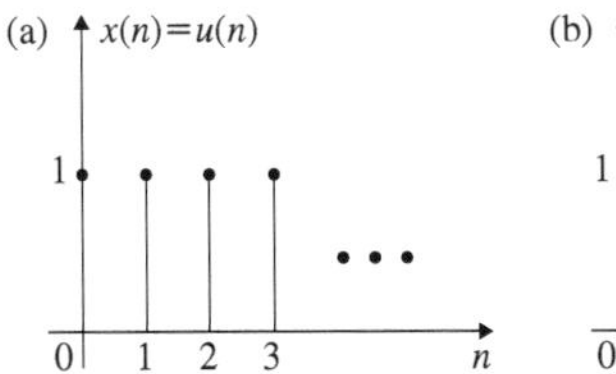

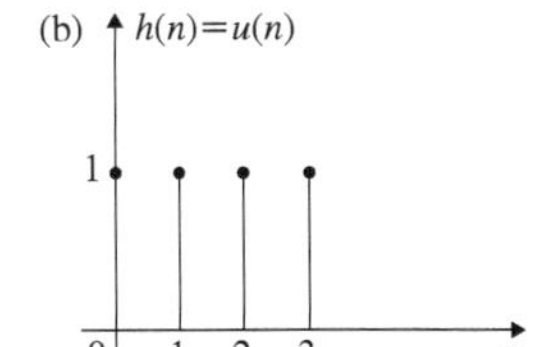

Fig. 3.36 Representation of $x(n) = u(n)$ and $h(n) = u(n)$

Therefore

$$\boxed{y(n) = (1 + n)u(n)}$$

■ Example 3.44

$$x(n) = u(n)$$
$$h(n) = (0.6)^n u(n)$$

Find

$$y(n) = x(n) * h(n)$$

Solution:

$$\begin{aligned}
x(n) &= 1^n \\
x(k) &= 1^k \\
h(n) &= (0.6)^n u(n) \\
h(n-k) &= (0.6)^{n-k} u(n-k) \\
y(n) &= \sum_{k=-0}^{n} 1^k (0.6)^{n-k} \\
&= (0.6)^n \sum_{k=0}^{n} 1^k (0.6)^{-k} \\
&= (0.6)^n \sum_{k=0}^{n} \left(\frac{1}{0.6}\right)^k \\
&= (0.6)^n \frac{\left[1 - \frac{1}{(0.6)^{n+1}}\right]}{\left[1 - \frac{1}{0.6}\right]} \\
&\quad \text{[Using Finite Summation Formula]} \\
&= \frac{[(0.6)^{n+1} - 1]}{(0.6) - 1}
\end{aligned}$$

$$\boxed{y(n) = 2.5[1 - (0.6)^{n+1}]u(n)}$$

■ Example 3.45

$$x(n) = a^n u(n)$$
$$h(n) = b^n u(n)$$

Find

$$y(n) = x(n) * h(n)$$

Solution: $x(n)$ and $h(n)$ are causal and hence the range of convolution sum is $0 < k < n$.

$$\begin{aligned}
x(n) &= a^n u(n) \\
x(k) &= a^k u(k) \\
h(n) &= b^n u(n) \\
h(n-k) &= b^{n-k} u(n-k) \\
y(n) &= \sum_{k=0}^{n} x(k) h(n-k) \\
&= b^n \sum_{k=0}^{n} \left(\frac{a}{b}\right)^k \\
&= b^n \frac{\left[1 - (\frac{a}{b})^{n+1}\right]}{(1 - \frac{a}{b})}
\end{aligned}$$

[Using Finite Summation Formula]

$$\boxed{y(n) = \frac{(b^{n+1} - a^{n+1})}{(b-a)} u(n)}$$

■ Example 3.46

$$x(n) = \left(\frac{1}{5}\right)^n u(n)$$

$$h(n) = 3^n u(n)$$

Find

$$y(n) = x(n) * h(n)$$

Solution:

$$x(n) = \left(\frac{1}{5}\right)^n u(n)$$

$$x(k) = \left(\frac{1}{5}\right)^k u(k)$$

$$h(n) = 3^n u(n)$$
$$h(n-k) = 3^{n-k} u(n-k)$$

For causal $x(n) * h(n)$

$$y(n) = \sum_{k=0}^{n} x(k)h(n-k)$$
$$= \sum_{k=0}^{n} \left(\frac{1}{5}\right)^k 3^{n-k}$$
$$= 3^n \sum_{k=0}^{n} \left(\frac{1}{15}\right)^k$$

Using finite summation formula, we get

$$y(n) = 3^n \frac{\left[1 - \frac{1}{15^{n+1}}\right]}{\left(1 - \frac{1}{15}\right)}$$
$$= 3^n \frac{[15^{n+1} - 1]}{(14)} \frac{15}{15^{n+1}}$$

$$\boxed{y(n) = \frac{1}{14(5)^n}[15^{n+1} - 1]u(n)}$$

■ Example 3.47

$$x(n) = (0.6)^n u(n)$$
$$h(n) = (0.2)^n u(n)$$

Find

$$y(n) = x(n) * h(n)$$

Solution:

$$x(n) = (0.6)^n u(n)$$
$$x(n-k) = (0.6)^{n-k} u(n-k)$$
$$h(n) = (0.2)^n u(n)$$
$$h(k) = (0.2)^k u(k)$$

Both $x(n)$ and $h(n)$ are causal signals. Hence, the summation of convolution is $0 < k < n$. In this problem, we use commutative property of convolution.

$$x(n) * h(n) = h(k) * x(n)$$

$$y(n) = \sum_{k=0}^{n} h(k)x(n-k)$$

$$= \sum_{k=0}^{n} (0.2)^k (0.6)^{n-k}$$

$$= (0.6)^n \sum_{k=0}^{n} \left(\frac{1}{3}\right)^k$$

Using finite summation formula, we get

$$y(n) = (0.6)^n \frac{\left[1 - \frac{1}{3^{n+1}}\right]}{\left(1 - \frac{1}{3}\right)}$$

$$y(n) = (0.6)^n \frac{\left[3^{n+1} - 1\right]}{2} \frac{3}{3^{n+1}}$$

$$\boxed{y(n) = 0.5(0.2)^n[3^{n+1} - 1]u(n)}$$

■ Example 3.48

Two discrete time systems with impulse responses $h_1(n)$ and $h_2(n)$ are connected in cascade as shown in Fig. 3.37. Determine the unit sample response of the interconnected system.

Solution:

$$y(n) = x(n) * h(n)$$

where

$$h(n) = h_1(n) * h_2(n)$$

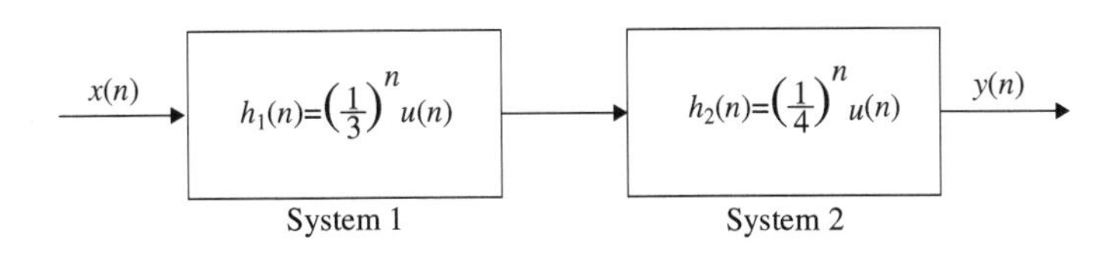

Fig. 3.37 Impulse response of two systems connected in cascade

For a sample

$$\begin{aligned}x(n) &= \delta(n)\\ x(n) * h(n) &= \delta(n) * h(n)\\ &= h(n)\\ &= h_1(n) * h_2(n)\end{aligned}$$

Therefore,

$$\begin{aligned}y(n) &= h_1(n) * h_2(n)\\ h_1(k) &= \left(\frac{1}{3}\right)^k u(k)\\ h_1(n-k) &= \left(\frac{1}{4}\right)^{n-k} u(n-k)\\ y(n) &= \sum_{k=0}^{n}\left(\frac{1}{3}\right)^k\left(\frac{1}{4}\right)^{n-k}\\ &= \left(\frac{1}{4}\right)^n \sum_{k=0}^{n}\left(\frac{4}{3}\right)^k\\ &= \left(\frac{1}{4}\right)^n \frac{\left[1-\left(\frac{4}{3}\right)^{n+1}\right]}{\left(1-\frac{4}{3}\right)}\end{aligned}$$

$$\boxed{y(n) = \left(\frac{1}{4}\right)^n 3\left[\left(\frac{4}{3}\right)^{n+1} - 1\right]u(n)}$$

■ Example 3.49

Determine the convolution of the signals

$$\begin{aligned}x(n) &= \cos \pi n\, u(n)\\ h(n) &= \left(\frac{1}{2}\right)^n u(n)\end{aligned}$$

(*Anna University, May, 2007*)

Solution:

$$\begin{aligned}x(n) &= \cos \pi n\, u(n)\\ &= (-1)^n u(n)\\ x(k) &= (-1)^k u(k)\end{aligned}$$

$$h(n) = \left(\frac{1}{2}\right)^n u(n)$$
$$h(n-k) = \left(\frac{1}{2}\right)^{n-k} u(n-k)$$

Both $x(n)$ and $h(n)$ are casual. Hence, the convolution sum takes the range of $0 < k < n$.

$$\begin{aligned}
y(n) &= \sum_{k=0}^{n} x(k)h(n-k) \\
&= \sum_{k=0}^{n} (-1)^k \left(\frac{1}{2}\right)^{n-k} \\
&= \left(\frac{1}{2}\right)^n \sum_{k=0}^{n} [(-1)(2)]^k \\
&= \left(\frac{1}{2}\right)^n \sum_{k=0}^{n} (-2)^k \\
&= \left(\frac{1}{2}\right)^n \frac{\left[1-(-2)^{n+1}\right]}{[1-(-2)]} \\
&= \frac{1}{3}\left(\frac{1}{2}\right)^n [1-(-2)^{n+1}]
\end{aligned}$$

$$\boxed{y(n) = \frac{1}{3}\left(\frac{1}{2}\right)^n [1+2(-2)^n]u(n)}$$

■ Example 3.50

$$x(n) = (0.2)^n u(n)$$
$$h(n) = (0.2)^{-n} u(-n)$$

Find

$$y(n) = x(n) * h(n)$$

Solution: The signal $x(n)$ is a casual signal and $h(n)$ is an anti-causal signal. $x(k)$ is shown in Fig. 3.38a and $h(k)$ is shown in Fig. 3.38b, respectively. $h(-k)$ is shown in Fig. 3.38c and $h(n-k)$ shifted to the extreme left for $n < 0$ and to the right for $n > 0$. They are shown in Fig. 3.38d and e, respectively.

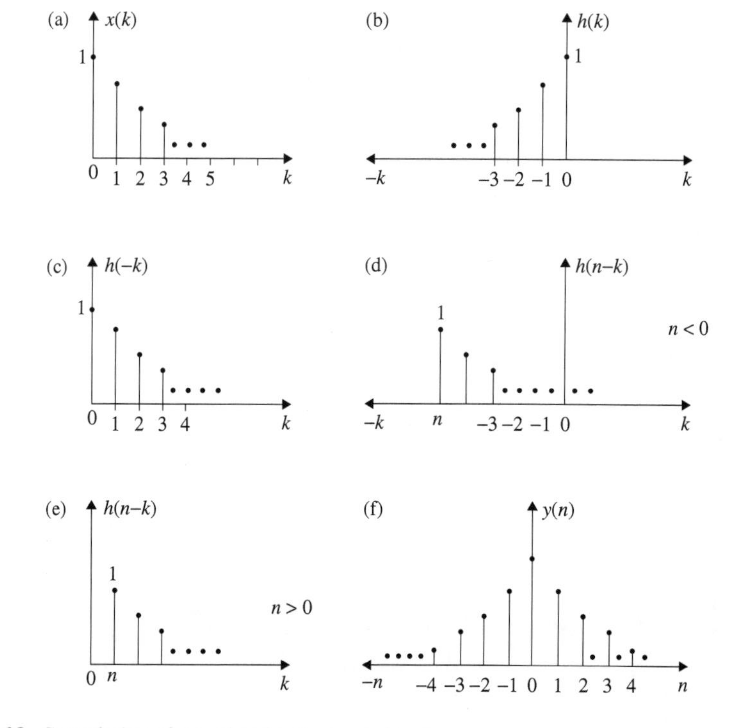

Fig. 3.38 Convolution of casual and anti-causal signals

$$
\begin{aligned}
y(n) &= x(n) * h(n) \\
&= \sum_{k=-\infty}^{\infty} x(k)h(n-k) \\
&= \sum_{k=-\infty}^{\infty} (0.2)^k u(k)(0.2)^{-(n-k)} u(-(n-k)) \\
&= (0.2)^{-n} \sum_{k=-\infty}^{\infty} (0.2)^{2k} u(k)u(k-n)
\end{aligned}
$$

When $h(n-k)$ is moved toward the right, it overlaps with $x(k)$ for $n < 0$ and for $n > 0$. For $n < 0$

$$
u(k)u(k-n) = \begin{cases} 1 & k > 0 \\ 0 & \text{otherwise} \end{cases}
$$

Now the limits of summation of $y(n)$ is $0 < k < \infty$. Therefore,

$$y(n) = (0.2)^{-n} \sum_{k=0}^{\infty} [(0.2)^2]^k$$
$$= (0.2)^{-n} \left[\frac{1}{1 - (0.2)^2} \right]$$
$$y(n) = \frac{(0.2)^{-n}}{0.96} \qquad n < 0$$

For $n > 0$

$$u(k)u(k-n) = \begin{cases} 1 & \text{for } k \geq n \\ 0 & \text{otherwise} \end{cases}$$

The limits of the summation of $y(n)$ is $n \leq k \leq \infty$

$$y(n) = \sum_{k=n}^{\infty} (0.2)^k u(k)(0.2)^{-(n-k)} u(-n+k)$$
$$= (0.2)^{-n} \sum_{k=n}^{\infty} [(0.2)^2]^k$$

Using the finite summation formula

$$\sum_{k=n}^{\infty} a^k = \frac{a^n}{1-a}$$

we get

$$y(n) = (0.2)^{-n} \left[\frac{(0.2)^{2n}}{(1 - 0.2^2)} \right]$$
$$= \frac{(0.2)^n}{0.96} \qquad n \geq 0.$$

In general

$$\boxed{y(n) = \frac{(0.2)^{|n|}}{0.96}} \qquad \text{for all } n$$

The response $y(n)$ is plotted and is shown in Fig. 3.38f.

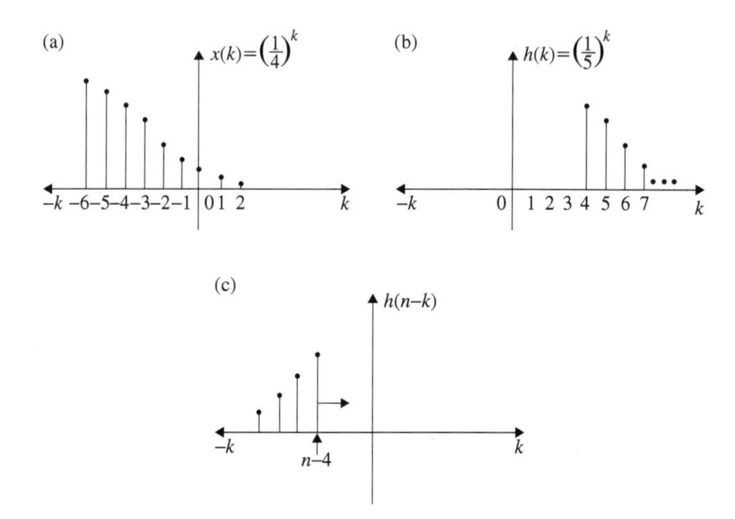

Fig. 3.39 Representation of signals for Example 3.51

■ Example 3.51

$$x(n) = \begin{cases} 0 & n < -6 \\ \left(\frac{1}{4}\right)^n & n \geq -6 \end{cases}$$

$$h(n) = \begin{cases} 0 & n < 4 \\ \left(\frac{1}{5}\right)^n & n \geq 4 \end{cases}$$

Find

$$y(n) = x(n) * h(n)$$

Solution: Figure 3.39a, b and c show $x(k)$, $h(k)$ and $h(n-k)$, respectively.

$$y(n) = x(n) * h(n) = \sum_{k=-6}^{n-4} x(k)h(n-k)$$

From Fig. 3.39a, the following equation is written.

$$x(k) = \begin{cases} 0 & k < -6 \\ \left(\frac{1}{4}\right)^k & k \geq -6 \end{cases}$$

From Fig. 3.39b, the following equation is written.

$$h(n-k) = \begin{cases} 0 & (n-k) < 4 \text{ or } k > n-4 \\ \left(\frac{1}{5}\right)^k & (n-k) > 4 \text{ or } k < n-4 \end{cases}$$

When $h(n-k)$ is moved toward the right, $h(n-k)$ overlaps with $x(k)$ at $n \geq -2$

$$y(n) = \begin{cases} 0 & n < -2 \\ \sum_{k=-6}^{n-4} \left(\frac{1}{4}\right)^k \left(\frac{1}{5}\right)^{n-k} & n \geq -2 \end{cases}$$

using the following summation formula

$$\sum_{k=m}^{n} (a)^k = \frac{a^{n+1} - a^m}{(a-1)}$$

we get

$$\begin{aligned} y(n) &= \sum_{k=-6}^{n-4} \left(\frac{1}{4}\right)^k \left(\frac{1}{5}\right)^{n-k} = \left(\frac{1}{5}\right)^n \sum_{k=-6}^{n-4} \left(\frac{5}{4}\right)^k \\ &= \left(\frac{1}{5}\right)^n \frac{\left[\left(\frac{5}{4}\right)^{n-3} - \left(\frac{5}{4}\right)^{-6}\right]}{\left(\frac{5}{4} - 1\right)} \\ &= 4\left(\frac{1}{5}\right)^n \left[\left(\frac{5}{4}\right)^{n-3} - \left(\frac{5}{4}\right)^{-6}\right] \\ &= 4\left(\frac{1}{5}\right)^n \left[\left(\frac{5}{4}\right)^{n+3} \left(\frac{4}{5}\right)^6 - \left(\frac{4}{5}\right)^6\right] \\ &= 4^7 \left(\frac{1}{5}\right)^{n+6} \left[\left(\frac{5}{4}\right)^{n+3} - 1\right] \end{aligned}$$

■ Example 3.52

A linear time invariant discrete time system has the following impulse response.

$$h(n) = \left(\frac{1}{2}\right)^n u(n)$$

The system is excited by the signal $x(n)$=$u(n)$ determine the output of the system at $n = -6$, $n = 6$ and $n = 12$.

Solution: $x(k)$, $h(k)$ and $h(n-k)$ are shown in Fig. 3.40a, b and c, respectively. At $n = -6$, $h(n-k) < 0$ and it does not overlap with $x(k)$ and hence $y(n) = 0$. For $n > 0$, the following equation is written:

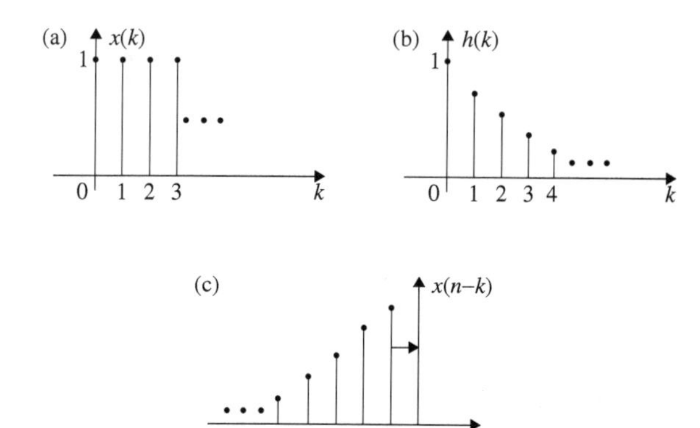

Fig. 3.40 Representation of signals for Example 3.51

$$
\begin{aligned}
y(n) &= x(n) * h(n) \\
&= \sum_{k=0}^{n} x(n)h(k) \\
&= \sum_{k=0}^{n} \left(\frac{1}{2}\right)^{n-k} \\
&= \left(\frac{1}{2}\right)^{n} \sum_{k=0}^{n} (2)^k \\
&= \left(\frac{1}{2}\right)^{n} \frac{1-2^{n+1}}{(1-2)} \\
&= \left(\frac{1}{2}\right)^{n} (2^{n+1}-1)u(n)
\end{aligned}
$$

Substituting $n = 6$

$$y(6) = \left(\frac{1}{2}\right)^{6} [2^7 - 1]$$

$$\boxed{y(6) = \frac{127}{64}}$$

Substituting $n = 12$

$$y(12) = \left(\frac{1}{2}\right)^{12} [2^{13} - 1]$$

$$\boxed{y(12) = \frac{8191}{4096}}$$

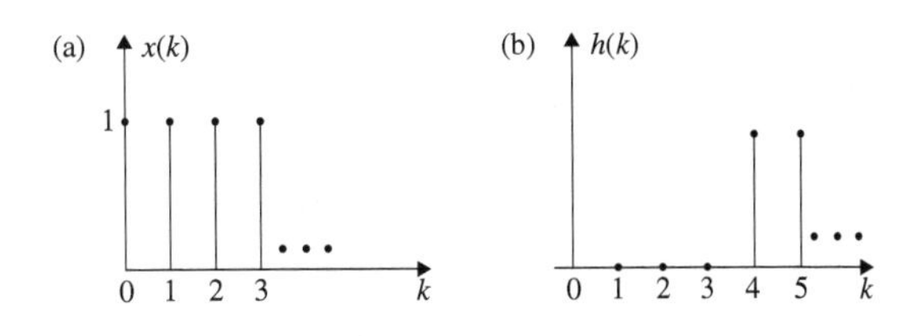

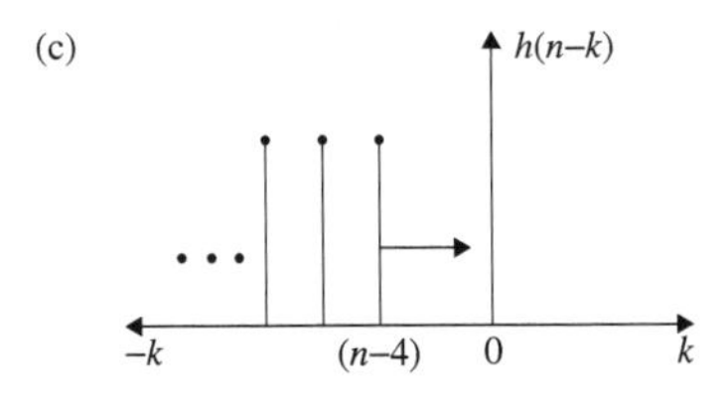

Fig. 3.41 Signals of Example 3.53

$$\boxed{y(-6) = 0}$$

■ Example 3.53

$$x(n) = u(n)$$
$$h(n) = u(n-4)$$

Find

$$y(n) = x(n) * h(n)$$

Solution: The signals $x(k)$, $h(k)$ and $h(n-k)$ are shown in Fig. 3.41a, b and c, respectively. Figure 3.41c is moved toward right so that it overlaps with $x(k)$. Overlapping occurs for $n-4 \leq k \leq \infty$. Therefore, the limit of convolution sum is $k=0$ to $k=(n-4)$

$$y(n) = \sum_{k=0}^{n-4} x(k)h(n-k) \qquad n \geq 4$$
$$= \sum_{k=0}^{n-4} 1 = (n-3)$$

$$\boxed{y(n) = (n-3)} \qquad n \geq 4$$

■ Example 3.54

A linear time invariant system has the following impulse response:

$$h(n) = [u(n) - u(n-6)]$$

The system is excited by

$$x(n) = [u(n-1) - u(n-5)]$$

Determine the output of the system.

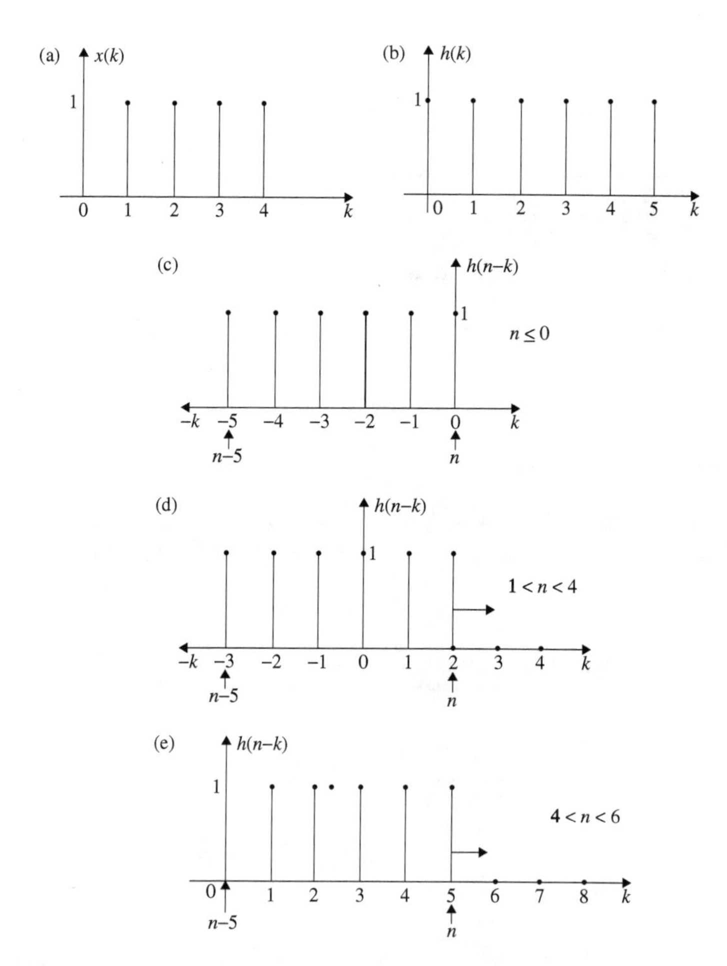

Fig. 3.42 Convolution sum for Example 3.54. Response plot of Example 3.52

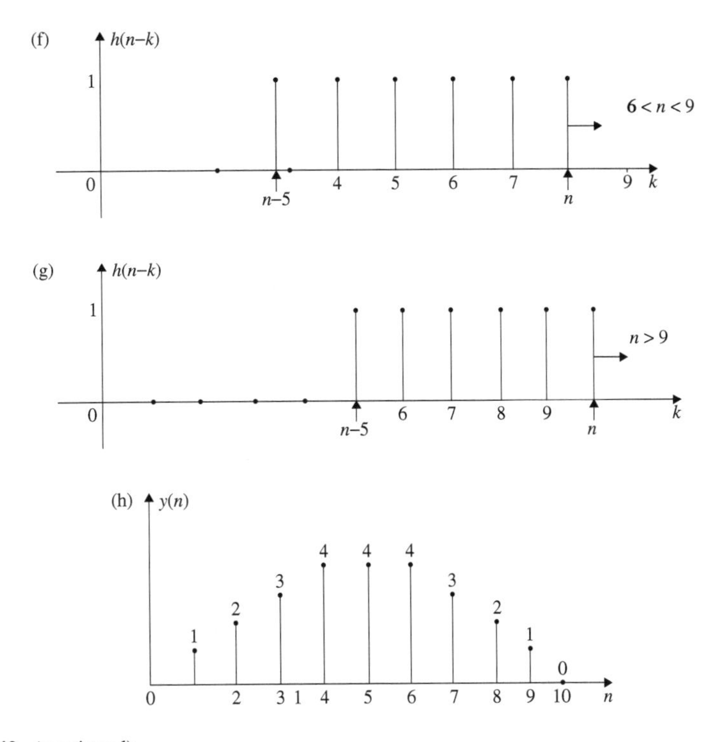

Fig. 3.42 (continued)

Solution:

Method 1

1. $x(k)$, $h(k)$ and $h(n-k)$ are represented in Fig. 3.42a, b and c, respectively. For $n = 0$, there is no overlapping and hence $y(n) = 0$.
2. **Overlapping interval 1:**
   $h(n-k)$ when moved toward right overlaps with $x(k)$ when the leading edge (right edge) of $h(n-k)$ crosses past the left edge $x(k)$. This happens when $n > 1$. This overlapping does not change until $n < 4$. Hence, the overlapping interval is $1 \leq n \leq 4$ The summation interval is from $k = 1$ to $k = n$.

$$y_1(n) = \sum_{k=1}^{n} 1$$

The finite summation formula used is

$$\sum_{k=N_1}^{N_2} 1 = N_2 - N_1 + 1$$

Therefore,

$$y_1(n) = n - 1 + 1 = n \qquad 1 \leq n \leq 4s$$

This is shown in Fig. 3.42d.

3. **Overlapping interval 2:**
   For $n > 4$, the right edge of $h(n-k)$ crosses past the right edge of $x(k)$. This change of overlapping occurs during the time interval $4 < n < 6$. The output during this interval is $y_2(n)$

$$\begin{aligned} y_2(n) &= \sum_{k=1}^{4} 1 \qquad 4 \leq n \leq 6 \\ &= 4 - 1 + 1 \\ &= 4 \end{aligned}$$

   This is shown in Fig. 3.42e.

4. **Overlapping interval 3:**
   When the left edge of $h(n-k)$ crosses past the left edge of $x(k)$, there is overlapping and it continues until $n = 4$. The overlapping interval is therefore $6 \leq n < 9$. The output is $y_3(n)$. The limits of summation is from $k = n - 5$ to $k = 4$

$$\begin{aligned} y_3(n) &= \sum_{k=n-5}^{4} 1 \\ &= -n + 5 + 4 + 1 \\ &= 10 - n \qquad 6 < n < 9 \end{aligned}$$

   This is shown in Fig. 3.42f.

5. For $n > 9$, there is no overlapping and hence $y(n) = 0$. This is shown in Fig. 3.42g. Hence, the output response $y_1(n)$, $y_2(n)$ and $y_3(n)$ are summed up with their respective time interval to get the total response $y(n)$.
6.

$$\begin{aligned} y(n) &= n \qquad 1 < n < 4 \\ y(n) &= 4 \qquad 4 < n < 6 \\ y(n) &= 10 - n \qquad 6 < n < 9 \end{aligned}$$

The values of $y(n)$ for $1 \leq n \leq 10$ is shown in the following table. $y(n) = \{0, 1, 2, 3, 4, 4, 4, 3, 2, 1\}$

n	0	1	2	3	4	5	6	7	8	9	10
$y(n)$	0	1	2	3	4	4	4	3	2	1	0

7. The plot of $y(n)$ with respect to n is shown in Fig. 3.42h.
8. To check the width property of convolution. The width property can be easily checked. The number of elements of $h(n)$ is $T_1 = 4$. The number of elements of $h(n)$ is $T_2 = 6$. The width property of $y(n)$ is $T = T_1 + T_2 - 1 = 6 + 4 - 1 = 9$. From Fig. 3.42h, it is seen that $y(n)$ has 9 elements.

Method 2

1. $x(k)$ and $h(k)$ can be expressed by the following sequences

$$x(n) = \delta(n-1) + \delta(n-2) + \delta(n-3) + \delta(n-4)$$
$$h(n) = \delta(n) + \delta(n-1) + \delta(n-2) + \delta(n-3) + \delta(n-4) + \delta(n-5)$$

2.

$$\begin{aligned} y(n) &= h(n) * x(n) \\ &= h(n) * [\delta(n-1) + \delta(n-2) + \delta(n-3) + \delta(n-4)] \\ &= h(n) * \delta(n-1) + h(n) * \delta(n-2) + h(n) * \delta(n-3) + h(n) * \delta(n-4) \\ &= y_1(n) + y_2(n) + y_3(n) + y_4(n) \end{aligned}$$

where

$$\begin{aligned} y_1(n) &= h(n) * \delta(n-1) \\ y_2(n) &= h(n) * \delta(n-2) \\ y_3(n) &= h(n) * \delta(n-3) \\ y_4(n) &= h(n) * \delta(n-4) \end{aligned}$$

By using the property $h(n) * \delta(n-1) = h(n-1)$, we get

$$\begin{aligned} y_1(n) &= \delta(n-1) + \delta(n-2) + \delta(n-3) + \delta(n-4) + \delta(n-5) + \delta(n-6) \\ y_2(n) &= h(n) * \delta(n-2) \\ &= \delta(n-2) + \delta(n-3) + \delta(n-4) + \delta(n-5) + \delta(n-6) + \delta(n-7) \\ y_3(n) &= h(n) * \delta(n-3) \\ &= \delta(n-3) + \delta(n-4) + \delta(n-5) + \delta(n-6) + \delta(n-7) + \delta(n-8) \\ y_4(n) &= \delta(n-4) + \delta(n-5) + \delta(n-6) + \delta(n-7) + \delta(n-8) + \delta(n-9) \\ y(n) &= y_1(n) + y_2(n) + y_3(n) + y_4(n) \end{aligned}$$

$$\boxed{\begin{aligned} y(n) = \delta(n-1) + 2\delta(n-2) + 3\delta(n-3) + 4\delta(n-4) + 4\delta(n-5) \\ +4\delta(n-6) + 3\delta(n-6) + 2\delta(n-7) + \delta(n-9) \end{aligned}}$$

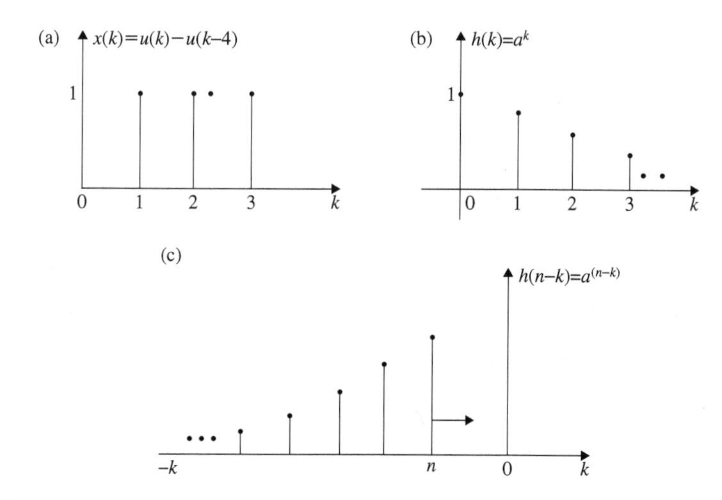

Fig. 3.43 Representation of signals for Example 3.53

$$\boxed{y(n) = \{0,\ 1,\ 2,\ 3,\ 4,\ 4,\ 4,\ 3,\ 2,\ 1\}}$$

The same result is analytically obtained in a simpler way as illustrated above.

■ Example 3.55

$$x(n) = u(n) - u(n-4)$$
$$h(n) = a^n u(n) \qquad 0 < a < 1$$

Find

$$y(n) = x(n) * h(n)$$

Solution:

1. **Time interval $0 < n < 3$:**
   $x(k) = u(k) - u(k-4)$ is shown in Fig. 3.43a. $h(k)$ is represented in Fig. 3.43b and $h(n-k)$ is shown in Fig. 3.43c.
2. Overlapping between $x(k)$ and $h(n-k)$ does not occur for $k < 0$ or $n < 0$. Therefore $y(n) = 0$ for $n < 0$.
3. When $h(n-k)$ is moved toward right, the right edge of $h(n-k)$ overlaps with the left edge of $x(k)$. The overlapping occurs for the time interval $0 \leq n \leq 3$. The summation of limits are therefore from 0 to n. Therefore,

$$y(n) = \sum_{k=0}^{n} a^{n-k} 1$$
$$= a^n \sum_{k=0}^{n} \left(\frac{1}{a}\right)^k$$
$$= a^n \frac{\left[1 - \frac{1}{a^{n+1}}\right]}{(1 - \frac{1}{a})}$$
$$= \frac{(a^{n+1} - 1)}{(a - 1)}$$

$$\boxed{y(n) = \frac{(1 - a^{n+1})}{(1 - a)}} \qquad 0 < n < 3$$

4. **Time interval $n > 3$:**

When $h(n - k)$ moves further toward the right for $n > 3$, the overlapping continues and the limit of the convolution sum is $0 \leq k \leq 3$

$$y(n) = \sum_{k=0}^{3} x(k)h(n - k)$$
$$= \sum_{k=0}^{3} a^{n-k}$$
$$= a^n \sum_{k=0}^{3} \left(\frac{1}{a}\right)^k$$
$$= a^n \frac{\left[1 - \frac{1}{a^4}\right]}{(1 - \frac{1}{a})}$$

$$\boxed{y(n) = a^{n-3} \frac{[1 - a^4]}{(1 - a)}} \qquad n \geq 3$$

■ Example 3.56

$$x(n) = [u(n) - u(n - 6)]$$
$$h(n) = a^n[u(n) - u(n - 3)] \quad \text{where} \quad 0 < a < 1$$

Find (Fig. 3.44)

$$y(n) = x(n) * h(n)$$

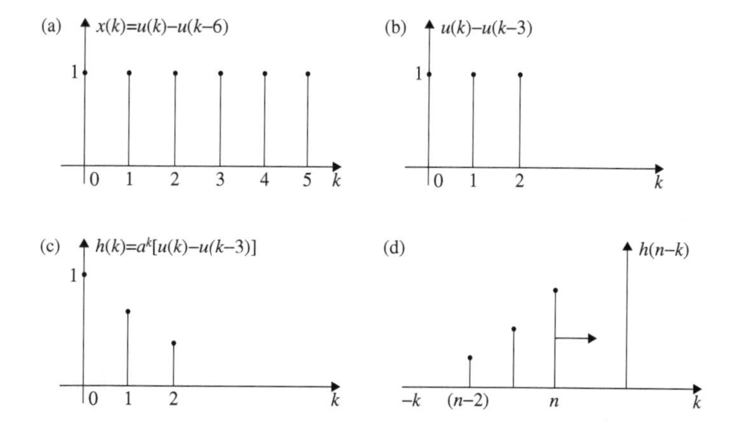

Fig. 3.44 Representation of signals for Example 3.56

Solution:

1. For $n < 0$, $h(n-k)$ and $x(k)$ do not overlap. Hence,

$$y(n) = 0 \text{ for } n < 0$$

2. When $h(n-k)$ is moved toward right $h(n-k)$ overlaps with $x(k)$ for the time interval $0 \leq n \leq 2$. Here, the limits of the convolution sum is $0 \leq k \leq 5$

$$x(k) = \begin{cases} 1 & 0 \leq k \leq 5 \\ 0 & \text{otherwise} \end{cases}$$

$$h(n-k) = \begin{cases} a^{(n-k)} & k \leq n \\ 0 & \text{otherwise} \end{cases}$$

$$\begin{aligned} y(n) &= \sum_{k=0}^{n} a^{n-k} \qquad 0 \leq n \leq 2 \\ &= a^n \left[\sum_{k=0}^{n} \left(\frac{1}{a} \right)^k \right] \\ &= a^n \frac{\left[1 - \frac{1}{a^{n+1}} \right]}{\left(1 - \frac{1}{a} \right)} \end{aligned}$$

$$\boxed{y(n) = \frac{(1 - a^{n+1})}{(1 - a)}} \qquad 0 \leq n \leq 2$$

3. For $n \geq 2$, the left edge of $h(n-k)$ slides over the left edge of $x(k)$. But the right edge of $h(n-k)$ is within the right edge of $x(k)$. For this the time duration is $2 \leq n \leq 5$. Here, the limit of the convolution sum is $(n-2) \leq k \leq n$

$$y(k) = \sum_{k=n-2}^{n} a^{n-k}$$

Put $p = k - n + 2$

$$= \sum_{p=0}^{2} a^{(2-p)}$$

$$= a^2 \left[\sum_{p=0}^{2} \left(\frac{1}{a} \right)^p \right]$$

$$= a^2 \frac{\left[1 - \frac{1}{a^3}\right]}{\left(1 - \frac{1}{a}\right)}$$

$$\boxed{y(n) = \frac{(1-a^3)}{(1-a)}} \qquad 2 \leq n \leq 5$$

4. For $n > 5$, the right edge of $h(n-k)$ slides past the right edge of $x(k)$. The left edge of $h(n-k)$ is within the right edge of $x(k)$ if $n < 7$. Hence, the time interval is $5 \leq n \leq 7$. The limits of the convolution sum is $(n-2) \leq k \leq 5$.

$$y(n) = \sum_{k=n-2}^{k=5} a^{n-k} \qquad 5 \leq n \leq 7$$

using the following summation formula

$$\sum_{k=m}^{n} (a)^k = \frac{a^{n+1} - a^m}{(a-1)}$$

we get

$$y(n) = \sum_{k=n-2}^{k=5} a^{n-k}$$

$$= a^n \sum_{k=n-2}^{k=5} \left(\frac{1}{a} \right)^k$$

$$= \frac{a^n\left[\left(\frac{1}{a}\right)^6 - \left(\frac{1}{a}\right)^{n-2}\right]}{\left(\frac{1}{a} - 1\right)}$$
$$= \frac{a^{n+1}\left[\left(\frac{1}{a}\right)^6 - \left(\frac{1}{a}\right)^n a^2\right]}{(1-a)}$$
$$= \frac{a^{n-5}\left[1 - a^{8-n}\right]}{(1-a)} \qquad 5 \le n \le 7$$

5. For $n > 7$, the left edge of $h(n-k)$ leaves the right edge of $x(k)$ and there is no overlapping between $x(k)$ and $h(n-k)$. Consequently $y(n) = 0$ for $n > 7$.

■ Example 3.57

The impulse response of a certain LTID system is given by

$$h(n) = u(n+1) - u(n-4)$$

The system is excited by the following signal.

$$x(n) = u(n) - 2u(n-2) + u(n-4)$$

Analytically derive an expression for $y(n) = h(n) * x(n)$ and plot the same.

Solution:

1. $u(n)$, $-2u(n-2)$ and $u(n-4)$ are shown in Fig. 3.45a, b and c, respectively. From these figure $x(n) = u(n) - 2u(n-2) + u(n-4)$ is obtained and represented in Fig. 3.45d.
2. $h(n) = u(n+1) - u(n-4)$ is represented in Fig. 3.45e. From this figure, we get $h(n) = \{1, \underset{\uparrow}{1}, 1, 1, 1\}$.
3.

$$\begin{aligned}
y(n) &= h(n) * x(n) \\
x(n) &= \delta(n) + \delta(n-1) - \delta(n-2) - \delta(n-3) \\
y(n) &= h(n) * \delta(n) + h(n) * \delta(n-1) - h(n) * \delta(n-2) - h(n) * \delta(n-3) \\
&= h(n) + h(n-1) - h(n-2) - h(n-3) \\
&= y_1(n) + y_2(n) + y_3(n) + y_4(n)
\end{aligned}$$

where

$$\begin{aligned}
y_1(n) &= h(n) = \delta(n+1) + \delta(n) + \delta(n-1) + \delta(n-2) + \delta(n-3) \\
y_2(n) &= h(n-1) = \delta(n) + \delta(n-1) + \delta(n-2) + \delta(n-3) + \delta(n-4)
\end{aligned}$$

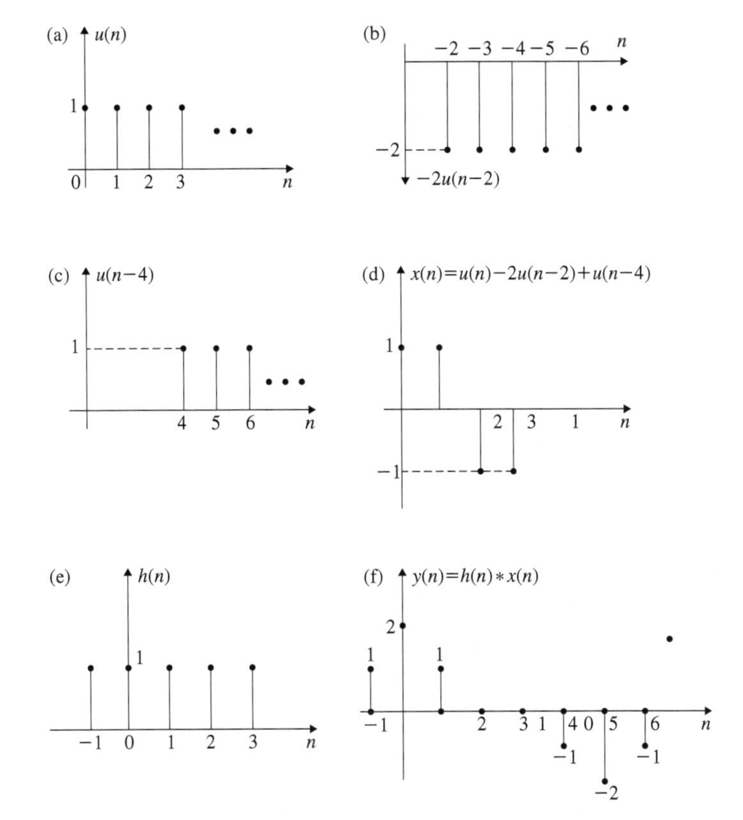

Fig. 3.45 Signal representation of Example 3.57

$$y_3(n) = -h(n-2) = -\delta(n-1) - \delta(n-2) - \delta(n-3) - \delta(n-4) - \delta(n-5)$$
$$y_4(n) = -h(n-3) = -\delta(n-2) - \delta(n-3) - \delta(n-4) - \delta(n-5) - \delta(n-6)$$

$$\boxed{y(n) = \delta(n+1) + 2\delta(n) + \delta(n-1) - \delta(n-4) - 2\delta(n-5) - \delta(n-6)}$$

$$y(n) = \{1, \underset{\uparrow}{2}, 1, 0, 0, -1, -2, -1\}$$

The output response $y(n)$ is represented in Fig. 3.45f.

■ Example 3.58

$$x(n) = 2^n u(-n-2)$$
$$h(n) = u(n-1)$$

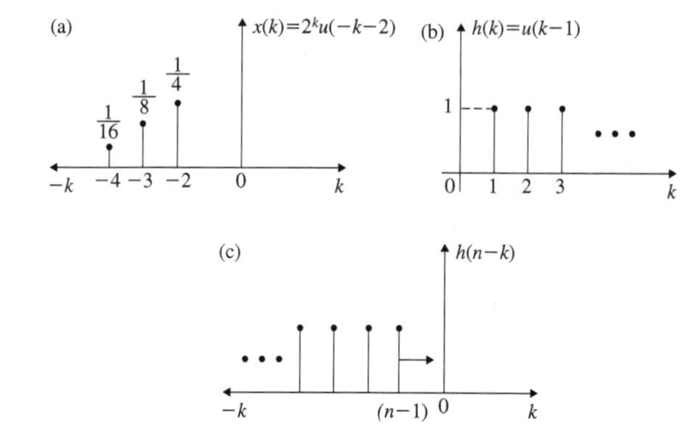

Fig. 3.46 Signal representation of Example 3.58

Find

$$y(n) = x(n) * h(n)$$

by analytical method (Fig. 3.46).

Solution:

$$\begin{aligned} x(k) &= 0 && \text{for } k > -2 \\ &= 2^k && \text{for } k \leq -2 \\ h(n-k) &= 0 && \text{for } k > n-1 \\ &= 1 && \text{for } k \leq n-1 \end{aligned}$$

1. **Time interval $-\infty < n < -1$.**
   When $h(n-k)$ is moved from extreme left toward right, the right edge of $h(n-k)$ starts overlapping with the left edge of $x(k)$ during the interval $-\infty < k < n-1$. Hence, the summation interval is from $-\infty$ to $(n-1)$.

$$\begin{aligned} y(n) &= \sum_{k=-\infty}^{n-1} 2^k \qquad -\infty < n < -1 \\ &= 2^{n-1} + 2^{n-2} + 2^{n-3} + \cdots \end{aligned}$$

$$\begin{aligned} y(n) &= 2^{n-1}\left[1 + \frac{1}{2} + \frac{1}{2^2} + \frac{1}{2^3} + \cdots\right] \\ &= 2^{n-1}\left[\frac{1}{1-\frac{1}{2}}\right] \end{aligned}$$

$$\boxed{y(n) = 2^n} \qquad -\infty < n < -1$$

2. Time interval $n > -1$.
For $n \geq -1$, the right edge of $h(n-k)$ slides past the right edge of $x(k)$ where there is transition. The limits of convolution sum is therefore $-\infty < k < -2$.

$$y(n) = \sum_{k=-\infty}^{-2} 2^k \qquad n > -1$$
$$= \sum_{2}^{\infty} \left(\frac{1}{2}\right)^k$$

Using the summation formula

$$\sum_{k=n}^{\infty} (a)^k = \frac{a^n}{(1-a)}$$

we get

$$y(n) = \left(\frac{1}{2}\right)^2 \frac{1}{1-\left(\frac{1}{2}\right)}$$
$$= 0.5 \qquad n > -1$$

$$\boxed{y(n) = 2^n} \qquad n < -1$$

$$\boxed{y(n) = 0.5} \qquad n > -1$$

■ Example 3.59

What is the response of an LTID system with impulse response $h(n) = \delta(n) + 2\delta(n-1)$ for the input $x(n) = \{1, 2, 3\}$?

(*Anna University, April, 2005*)

Solution:

$$h(n) = \delta(n) + 2\delta(n-1)$$
$$x(n) = \{1, 2, 3\}$$
$$= \delta(n) + 2\delta(n-1) + 3\delta(n-2)$$

The output response $y(n)$ is obtained by the following convolution

$$h(n) = h(n) * x(n)$$
$$= h(n) * \delta(n) + h(n) * 2\delta(n-1) + h(n) * 3\delta(n-2)$$
$$= y_1(n) + y_2(n) + y_3(n)$$

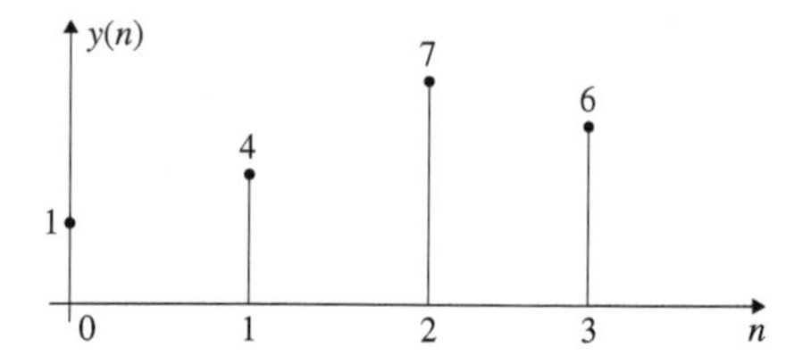

Fig. 3.47 Response of Example 3.59

where

$$
\begin{aligned}
y_1(n) &= h(n) * \delta(n) = h(n) \\
y_2(n) &= h(n) * 2\delta(n-1) = 2h(n-1) \\
y_3(n) &= h(n) * 3\delta(n-2) = 3h(n-2) \\
y_1(n) &= h(n) = \delta(n) + 2\delta(n-1) \\
y_2(n) &= 2h(n-1) = 2\delta(n-1) + 4\delta(n-2) \\
y_3(n) &= 3h(n-2) = 3\delta(n-2) + 6\delta(n-3)
\end{aligned}
$$

$$\therefore \quad \boxed{y(n) = \delta(n) + 4\delta(n-1) + 7\delta(n-2) + 6\delta(n-3)}$$

The response is plotted and is shown in Fig. 3.47.

■ Example 3.60

Find the overall impulse response of the system shown in Fig. 3.48a if

$$
\begin{aligned}
h_1(n) &= \left(\frac{1}{3}\right)^n u(n) \\
h_2(n) &= \left(\frac{1}{2}\right)^n u(n) \\
h_3(n) &= \left(\frac{1}{5}\right)^n u(n)
\end{aligned}
$$

(*Anna University, April, 2004*)

Solution:

1. The block diagram of the system of Example 3.60 is shown in Fig. 3.48a. Its equivalence by block diagram reduction technique is shown in Fig. 3.48b.
2. From Fig. 3.48b, it is seen that the blocks h_1 and h_3 are in cascade and their convolution is $h_1 * h_3$. Similarly, $h_2(n)$ and $h_3(n)$ are in cascade and their convolution is $h_2(n) * h_3(n)$. The block diagram of the above step is shown in Fig. 3.48c.

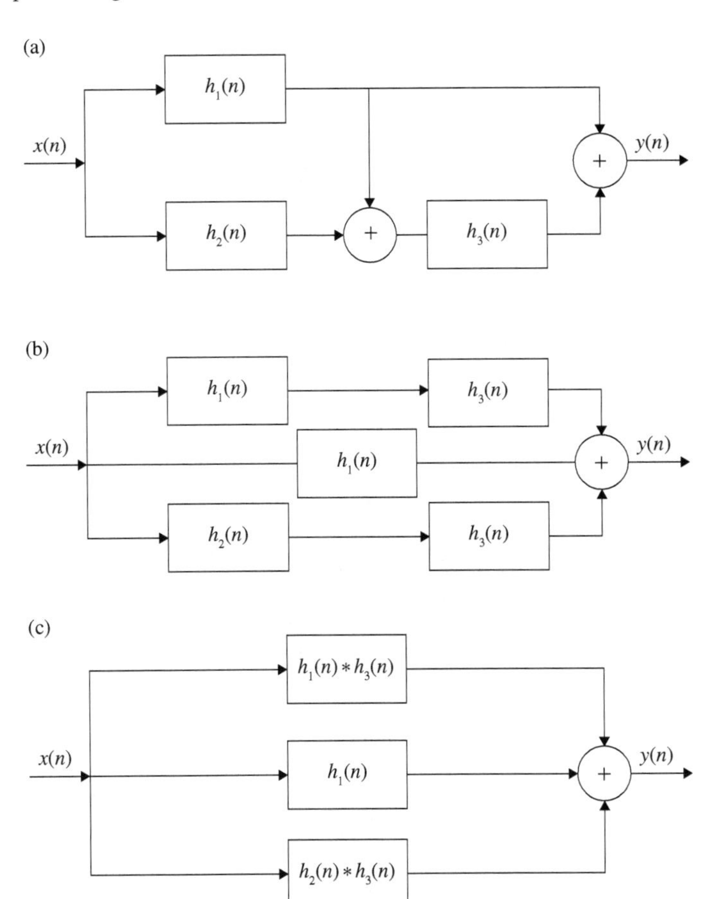

Fig. 3.48 Block diagram representation and its equivalence Example 3.60. **c** Block diagram of Step 2

3. From Fig. 3.48c, the following equations are obtained

$$y(n) = [h_1(n) * h_3(n) + h_1(n) + h_2(n) * h_3(n)] * x(n)$$
$$x(n) = \delta(n)$$

Therefore,

$$\begin{aligned} y(n) &= [h_1(n) * h_3(n) + h_1(n) + h_2(n) * h_3(n)] * \delta(n) \\ &= h_1(n) * h_3(n) + h_1(n) + h_2(n) * h_3(n) \\ &= y_1(n) + y_2(n) + y_3(n) \\ y_1(n) &= h_1(n) * h_3(n) \end{aligned}$$

Since the impulse response of all the blocks are causal, the limits of convolution sum is $0 \le k \le n$.

$$y_1(n) = \sum_{k=0}^{n} \left(\frac{1}{3}\right)^{n-k} \left(\frac{1}{5}\right)^{k}$$

$$= \left(\frac{1}{3}\right)^{n} \sum_{k=0}^{n} \left(\frac{3}{5}\right)^{k}$$

$$= \left(\frac{1}{3}\right)^{n} \frac{\left[1 - \left(\frac{3}{5}\right)^{n+1}\right]}{1 - \frac{3}{5}}$$

$$\boxed{y_1(n) = \frac{5}{2}\left(\frac{1}{3}\right)^{n} \frac{[5^{n+1} - 3^{n+1}]}{5^{(n+1)}} u(n)}$$

$$\boxed{y_2(n) = h_1(n) = \left(\frac{1}{3}\right)^{n} u(n)}$$

$$y_3(n) = h_2(n) * h_3(n)$$

$$= \sum_{k=0}^{n} \left(\frac{1}{2}\right)^{n-k} \left(\frac{1}{5}\right)^{k}$$

$$= \left(\frac{1}{2}\right)^{n} \sum_{k=0}^{n} \left(\frac{2}{5}\right)^{k}$$

$$= \left(\frac{1}{2}\right)^{n} \frac{\left[1 - \left(\frac{2}{5}\right)^{n+1}\right]}{1 - \frac{2}{5}}$$

$$\boxed{y_3(n) = \frac{5}{3}\left(\frac{1}{2}\right)^{n} \frac{[5^{n+1} - 2^{n+1}]}{5^{n+1}} u(n)}$$

$$y(n) = y_1(n) + y_2(n) + y_3(n)$$

$$\boxed{y(n) = \left\{\frac{5}{2}\left(\frac{1}{3}\right)^{n} \frac{[5^{n+1} - 3^{n+1}]}{5^{n+1}} + \left(\frac{1}{3}\right)^{n} + \frac{5}{3}\left(\frac{1}{2}\right)^{n} \frac{[5^{n+1} - 2^{n+1}]}{5^{n+1}}\right\} u(n)}$$

3.15.2 *Convolution Sum of Two Sequences by Multiplication Method*

Let

$$x(n) = \{x_1, x_2, \underset{\uparrow}{x_3}, x_4\}$$

$$h(n) = \{h_1, h_2, \underset{\uparrow}{h_3}\}$$

The convolution of these two sequences $x(n) * h(n)$ is obtained as given below

1. Write down $x(n)$ and $h(n)$ one below the other

$$\begin{array}{cccc} x_1 & x_2 & x_3 & x_4 \\ & h_1 & h_2 & h_3 \end{array}$$

2. Carry out the multiplication of the first row by the second row as given below

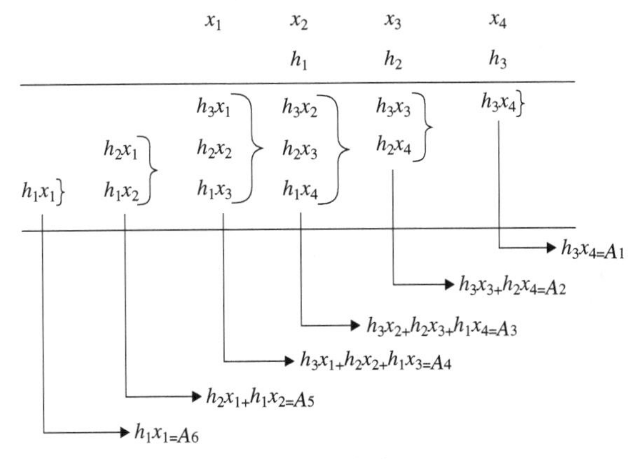

3. Arrange the sequences so obtained as given below

$$y(n) = \{A_6, A_5, A_4, A_3, \underset{\uparrow}{A_2}, A_1\}$$

4. Let N_1 = width of the $x(n)$ sequence to the left and N_2 = width of the $h(n)$ sequence to the left.
   The width of the output sequence $y(n)$ to the left is $N = N_1 + N_2$. In the example illustrated above $N_1 = 2$ and $N_2 = 2$. In the $y(n)$ sequence, the arrow corresponding to $n = 0$ is marked such that to the left of the arrow $N = N_1 + N_2 = 4$ is marked which is at A_2. From $y(n)$ sequence $y(0) = A_2$; $y(1) = A_1$; $y(-1) = A_3$; $y(-2) = A_4$; $y(-3) = A_5$ and $y(-4) = A_6$.
   The width property can also be easily checked. For $x(n)$ the width $T_1 = 4$. For

$h(n)$, the width $T_2 = 3$. The width of $y(n) = T = T_1 + T_2 - 1 = 4 + 3 - 1 = 6$.

It is to be noted here the above method follows the theory explained in Example 3.54, **Method 2 and also in Example** 3.59.

The following examples illustrate the above method.

Example 3.61

What is the response of an LTID system with impulse response $h(n) = \delta(n) + 2\delta(n-1)$ for the input $x(n) = \{1, 2, 3\}$.

Solution:

$$\begin{aligned} h(n) &= \delta(n) + 2\delta(n-1) \\ &= \{\underset{\uparrow}{1},\ 2\} \\ x(n) &= \{\underset{\uparrow}{1},\ 2,\ 3\} \end{aligned}$$

where $N_1 = 0,\ N_2 = 0,\ N = N_1 + N_2 = 0$ and $T_1 = 2,\ T_2 = 3,\ T = 3 + 2 - 1 = 4$

		1	2
	1	2	3
		3	6
	2	4	
1	2		
1	4	7	6

$$\boxed{y(n) = \{\underset{\uparrow}{1},\ 4,\ 7,\ 6\}}$$

The same result as in Example 3.59 is obtained.

Example 3.62

Find the convolution of $x(n) = \{1, 2, 3, 4, 5\}$ with $h(n) = \{1, 2, 3, 3, 2, 1\}$.

(*Anna University, May, 2005*)

Solution: When no arrow is marked in $x(n)$ or $h(n)$, the signals are to be taken as causal and therefore $N_1 = 0, N_2 = 0$ and $N = N_1 + N_2 = 0$; $T_1 = 6,\ T_2 = 5,\ T = 6 + 5 - 1 = 10$.

The following multiplication is done.

				1	2	3	3	2	1
					1	2	3	4	5
				5	10	15	15	10	5
			4	8	12	12	8	4	
		3	6	9	9	6	3		
	2	4	6	6	4	2			
1	2	3	3	2	1				
1	4	10	19	30	36	35	26	14	5

$$y(n) = \{\underset{\uparrow}{1}, 4, 10, 19, 30, 36, 35, 26, 14, 5\}$$

■ Example 3.63

Find the linear convolution of

$$x(n) = \{\underset{\uparrow}{1}, 2, 3, 4, 5, 6\}$$

and

$$h(n) = \{\underset{\uparrow}{2}, -4, 6, -8\}$$

(*Anna University, April, 2004*)

Solution:

			1	2	3	4	5	6
					2	−4	6	−8
			−8	−16	−24	−32	−40	−48
		6	12	18	24	30	36	
	−4	−8	−12	−16	−20	−24		
2	4	6	8	10	12			
2	0	4	0	−4	−8	−26	−4	−48

here $N_1 = 0$; $N_2 = 0$ and $N = 0$; $T_1 = 6$, $T_2 = 4$ and $T = 6 + 4 - 1 = 9$.

$$y(n) = \{\underset{\uparrow}{2}, 0, 4, 0, -4, -8, -26, -4, -48\}$$

■ Example 3.64

$$x(n) = \{-1/2, \underset{\uparrow}{2}, 1/3, 3/2\}$$

$$h(n) = \{1, -1/2, \underset{\uparrow}{2/3}\}$$

Find $y(n)$ by convolution method.

Solution: Here $N_1 = 1$; $N_2 = 2$ and $N_1 + N_2 = 3$; $T_1 = 4$, $T_2 = 3$ and $T = 4 + 3 - 1 = 6$.

$$\begin{array}{cccccc}
 & & -\frac{1}{2} & 2 & \frac{1}{3} & \frac{3}{2} \\
 & & & 1 & -\frac{1}{2} & \frac{2}{3} \\
\hline
 & & -\frac{1}{3} & \frac{4}{3} & \frac{2}{9} & 1 \\
 & \frac{1}{4} & -1 & -\frac{1}{6} & -\frac{3}{4} & \\
-\frac{1}{2} & 2 & \frac{1}{3} & \frac{3}{2} & & \\
\hline
-\frac{1}{2} & \frac{9}{4} & -1 & \frac{8}{3} & -\frac{19}{36} & 1 \\
\hline
\end{array}$$

$$\boxed{y(n) = \left\{-\frac{1}{2}, \frac{9}{4}, -1, \underset{\uparrow}{\frac{8}{3}}, -\frac{19}{36}, 1\right\}}$$

■ Example 3.65

Find the convolution of the following

$$x(n) = u(n) - 3u(n-2) + 2u(n-4)$$
$$h(n) = u(n+1) - u(n-8)$$

Solution:

1. $x(n)$ and $h(n)$ are represented in Fig. 3.49a and d, respectively.

From Fig. 3.49a, $x(n)$ is written as

$$x(n) = \{\underset{\uparrow}{1}, 1, -2, -2\} \qquad N_1 = 0;\ T_1 = 4$$

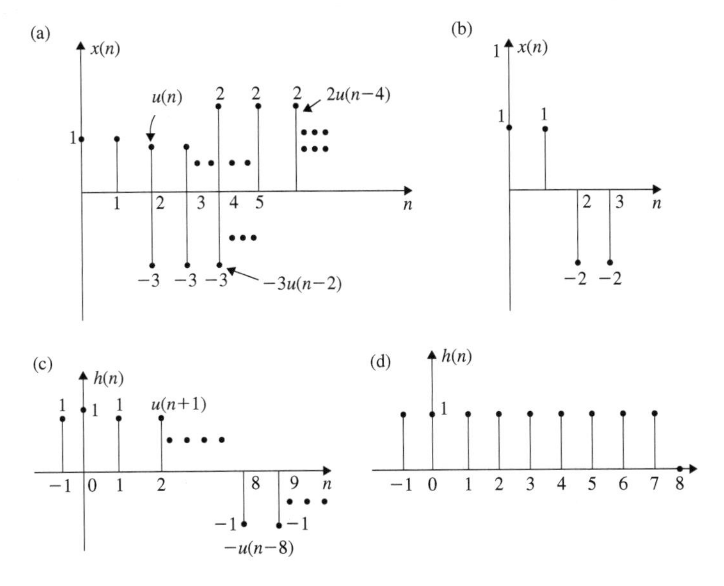

Fig. 3.49 Representation of $x(n)$ and $h(n)$ of Example 3.65

From Fig. 3.49d, $h(n)$ is written as

$$h(n) = \{1, \underset{\uparrow}{1}, 1, 1, 1, 1, 1, 1, 1\} \qquad N_2 = 1;\ T_2 = 9$$

2. The following multiplication is carried out:

			1	1	1	1	1	1	1	1	1
								1	1	−2	−2
			−2	−2	−2	−2	−2	−2	−2	−2	−2
		−2	−2	−2	−2	−2	−2	−2	−2	−2	
	1	1	1	1	1	1	1	1	1		
1	1	1	1	1	1	1	1	1			
1	2	0	−2	−2	−2	−2	−2	−2	−3	−4	−2

$N = N_1 + N_2 = 0 + 1 = 1$ and $T = T_1 + T_2 - 1 = 4 + 9 - 1 = 12$.

$$\boxed{y(n) = \{1, \underset{\uparrow}{2}, 0, -2, -2, -2, -2, -2, -2, -3, -4, -2\}}$$

3.15.3 Convolution Sum by Tabulation Method

The convolution sum of two sequences $x(n)$ and $h(n)$ to obtain $y(n)$ by tabulation method is explained below

1. Let

$$x(n) = \{x_1, x_2, \ldots, x_{(n)}\}$$
$$h(n) = \{h_1, h_2, \ldots, h_{(n)}\}$$

   Mark $x_1, x_2, \ldots, x_n$ in columns and $h_1(n), h_2(n), \ldots, h(n)$ in rows.
2. The h_1 row is completed by multiplying the corresponding column. Thus when h_1 row crosses with x_1 column, the product becomes $x_1 h_1$. Similarly, when h_1 row crosses with x_2 column, the product becomes $h_1 x_2$. Thus, all the elements rowwise are determined and tabulated.
3. Draw the diagonal dotted lines as shown in Table 3.2.
4. By adding the elements in a particular diagonal gives $y(n)$.
5. The value of N and T of the sequence $y(n)$ are determined as explained in the previous examples.
   From Table 3.2, $y(n)$ is obtained and is given below.

From diagonal 1, $y(0) = x_1 h_1$
From diagonal 2, $y(1) = x_1 h_2 + x_2 h_1$
From diagonal 3, $y(2) = x_1 h_3 + x_2 h_2 + x_3 h_1$

From diagonal 4, $y(3) = x_1 h_4 + x_2 h_3 + x_3 h_2 + x_4 h_1$
From diagonal 5, $y(4) = x_1 h_5 + x_2 h_4 + x_3 h_3 + x_4 h_2 + x_5 h_1$
From diagonal 6, $y(5) = x_2 h_5 + x_3 h_4 + x_4 h_3 + x_5 h_2$
From diagonal 7, $y(6) = x_3 h_5 + x_4 h_4 + x_5 h_3$
From diagonal 8, $y(7) = x_4 h_5 + x_5 h_4$
From diagonal 9, $y(8) = x_5 h_5$

$$y(n) = \{y(0), y(1), y(2), y(3), y(4), y(5), y(6), y(7), y(8)\}$$

From the $y(n)$, the width property $T = T_1 + T_2$ can be easily checked. The arrow $\uparrow$ corresponding to $n = 0$ is decided by N_1 and N_2. The following examples illustrate the above method.

Table 3.2 Tabulation method of convolution

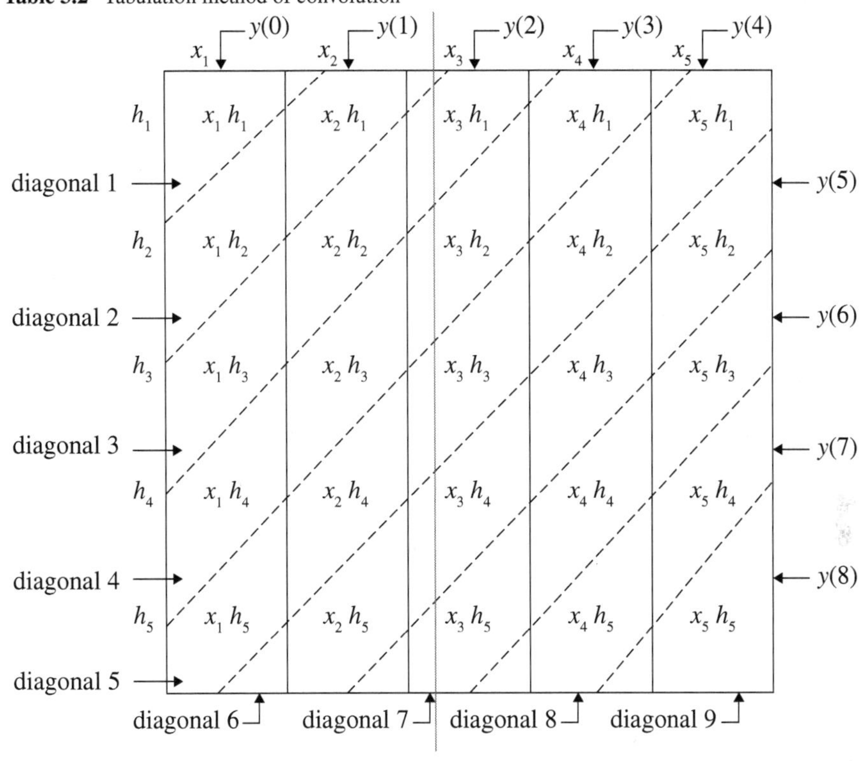

■ Example 3.66

$$x(n) = \{1, 2, 3, 4, 5\}$$
$$h(n) = \{1, 2, 3, 3, 2, 1\}$$

Find

$$y(n) = x(n) * h(n)$$

(*Anna University, April, 2005*)

Solution:

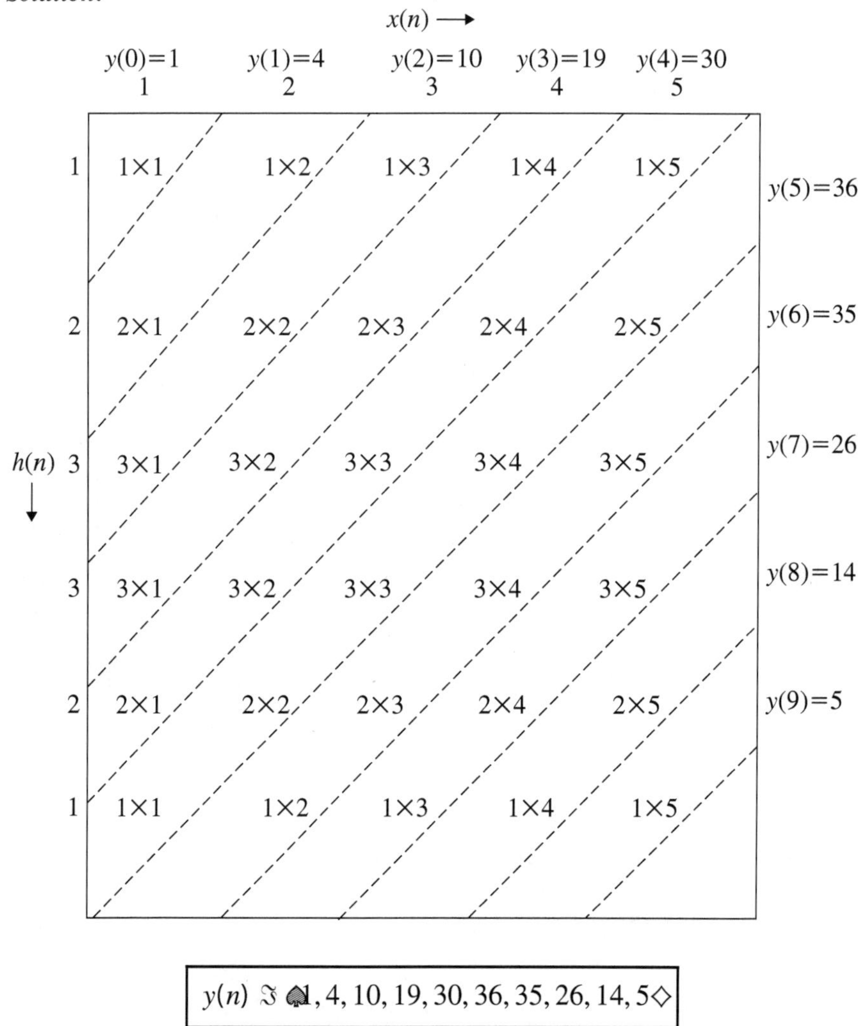

Since $N_1 = 0$ and $N_2 = 0$, $N = N_1 + N_2 = 0$. Therefore, the first diagonal corresponds to $y(0)$. Further, $x(n)$ has $T_1 = 5$ and $h(n)$ has $T_2 = 6$. Hence $y(n)$ has a width of $T = T_1 + T_2 - 1 = 5 + 6 - 1 = 10$. Thus, $y(n)$ starts with $y(0)$ and goes up to $y(9)$. They are calculated from the respective diagonal.

$$y(0) = 1 \times 1 = 1$$
$$y(1) = 2 \times 1 + 1 \times 2 = 4$$
$$y(2) = 3 \times 1 + 2 \times 2 + 1 \times 3 = 10$$
$$y(3) = 3 \times 1 + 3 \times 2 + 2 \times 3 + 1 \times 4 = 19$$
$$y(4) = 2 \times 1 + 3 \times 2 + 3 \times 3 + 2 \times 4 + 1 \times 5 = 30$$

$$y(5) = 1 \times 1 + 2 \times 2 + 3 \times 3 + 3 \times 4 + 2 \times 5 = 36$$
$$y(6) = 1 \times 2 + 2 \times 3 + 3 \times 4 + 3 \times 5 = 35$$
$$y(7) = 1 \times 3 + 2 \times 4 + 3 \times 5 = 26$$
$$y(8) = 1 \times 4 + 2 \times 5 = 14$$
$$y(9) = 1 \times 5 = 5$$

$$\boxed{y(n) = \{1, 4, 10, 19, 30, 36, 35, 26, 14, 5\}}$$

■ Example 3.67

Find the linear convolution of

$$x(n) = \{1,\ 2,\ \underset{\uparrow}{3},\ 4\}$$

$$h(n) = \{2,\ -4,\ 6,\ \underset{\uparrow}{-8}\}$$

Plot the response $y(n)$.

Solution:

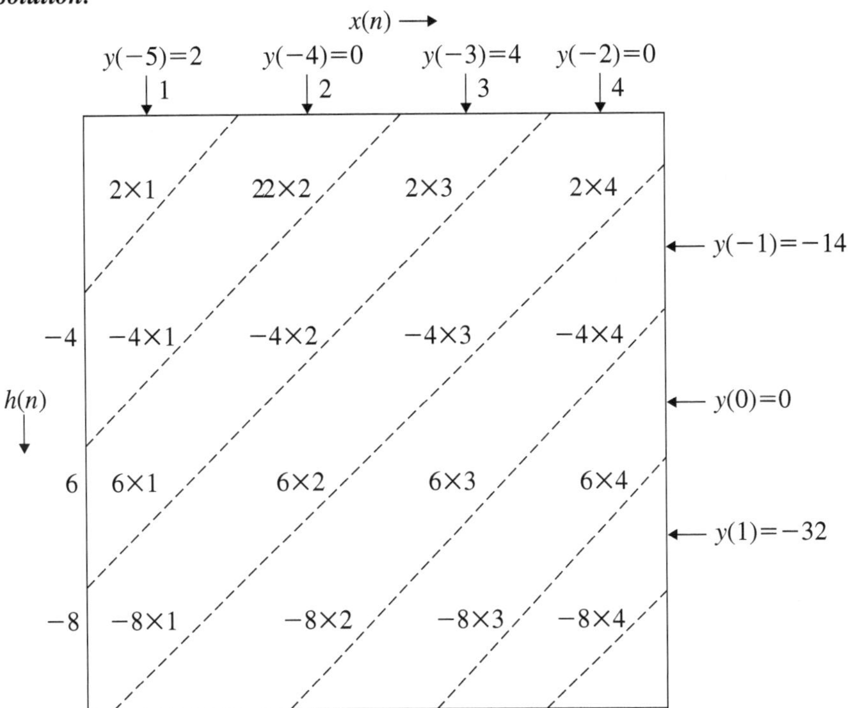

The tabulation of $x(n)$ and $h(n)$ is shown above. $T_1 = 4$ and $T_2 = 4$ and hence $T = T_1 + T_2 - 1 = 4 + 4 - 1 = 7$. There should be 7 diagonals in the table. $N_1 = 2$

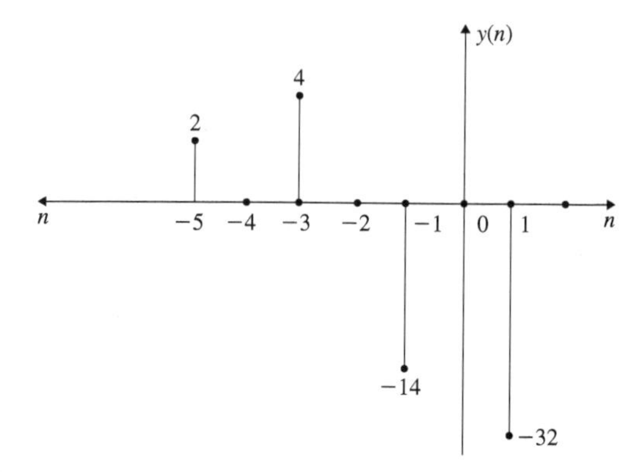

Fig. 3.50 Response plot of Example 3.67

and $N_2 = 3$ and therefore $N = N_1 + N_2 = 2 + 3 = 5$. There should be five elements to the left of $y(0)$ out of the total of seven. Therefore, the first diagonal corresponds to $y(-5)$ and the last diagonal corresponds to $y(1)$. They are calculated as follows:

$$\begin{aligned}
y(-5) &= 2 \times 1 = 2 \\
y(-4) &= -4 \times 1 + 2 \times 2 = 0 \\
y(-3) &= 6 \times 1 - 4 \times 2 + 2 \times 3 = 4 \\
y(-2) &= -8 \times 1 + 6 \times 2 - 4 \times 3 + 2 \times 4 = 0 \\
y(-1) &= -8 \times 2 + 6 \times 3 - 4 \times 4 = -14 \\
y(0) &= -8 \times 3 + 6 \times 4 = 0 \\
y(1) &= -8 \times 4 = -32
\end{aligned}$$

$$\boxed{y(n) = \{2, 0, 4, 0, -14, \underset{\uparrow}{0}, -32\}}$$

$y(n)$ is plotted as shown in Fig. 3.50.

3.15.4 Convolution Sum of Two Sequences by Matrix Method

By this method, the data sequences are represented as a matrix. Let T_1 be the length of the signal $x(n)$ and T_2 be the length of $h(n)$. The X matrix is formed with a dimension of $(T_1 + T_2 - 1) \times T_2$ and a H matrix is formed with a dimension of $T_2 \times 1$. The Y matrix is obtained as

$$Y = XH$$

The output response $y(n)$ is obtained by putting the arrow mark ($\uparrow$) in the appropriate place by determining the range of N as explained in previous cases.

The formation of X matrix is as follows. The first column of X is formed from the sequence of $x(n)$. The other elements of X are formed as explained below. H is a column matrix which is nothing but the transpose of $h(n)$. Consider $x(n) = \{x_1, x_2, x_3, x_4\}$ and $h(n) = \{h_1, h_2, h_3\}$

$$X = \begin{pmatrix} x_1 & 0 & 0 & 0 \\ x_2 & x_1 & 0 & 0 \\ x_3 & x_2 & x_1 & 0 \\ x_4 & x_3 & x_2 & x_1 \\ 0 & x_4 & x_3 & x_2 \\ 0 & 0 & x_4 & x_3 \\ 0 & 0 & 0 & x_4 \end{pmatrix}_{7\times 4} \quad ; \quad H = \begin{pmatrix} h_1 \\ h_2 \\ h_3 \\ 0 \end{pmatrix}_{4\times 1}$$

The elements of the matrix X are completed by diagonalizing with x_1, then x_2, x_3, x_4 and so on. The Y matrix is obtained by multiplying the matrices X and H. To satisfy the multiplication property of matrix, the last row of H is put as 0.

$$Y = XH$$

$$Y = \begin{bmatrix} x_1 & 0 & 0 & 0 \\ x_2 & x_1 & 0 & 0 \\ x_3 & x_2 & x_1 & 0 \\ x_4 & x_3 & x_2 & x_1 \\ 0 & x_4 & x_3 & x_2 \\ 0 & 0 & x_4 & x_3 \\ 0 & 0 & 0 & x_4 \end{bmatrix} \begin{bmatrix} h_1 \\ h_2 \\ h_3 \\ 0 \end{bmatrix}$$

$$Y = \begin{bmatrix} y(0) \\ y(1) \\ y(2) \\ y(3) \\ y(4) \\ y(5) \end{bmatrix} = \begin{bmatrix} x_1h_1 \\ x_2h_1 + x_1h_2 \\ x_3h_1 + x_2h_2 + x_1h_3 \\ x_4h_1 + x_3h_2 + x_2h_3 \\ x_4h_2 + x_3h_3 \\ x_4h_3 \end{bmatrix}$$

The following examples illustrate the above method.

■ Example 3.68

$$x(n) = \{1,\ 2,\ 3,\ 4,\ 5,\ 6\}$$
$$h(n) = \{2,\ -4,\ 6,\ -8\}$$

Find

$$y(n) = x(n) * h(n)$$

by matrix method.

Solution:

$$X = \begin{pmatrix} 1 & 0 & 0 & 0 & 0 & 0 \\ 2 & 1 & 0 & 0 & 0 & 0 \\ 3 & 2 & 1 & 0 & 0 & 0 \\ 4 & 3 & 2 & 1 & 0 & 0 \\ 5 & 4 & 3 & 2 & 1 & 0 \\ 6 & 5 & 4 & 3 & 2 & 1 \\ 0 & 6 & 5 & 4 & 3 & 2 \\ 0 & 0 & 6 & 5 & 4 & 3 \\ 0 & 0 & 0 & 6 & 5 & 4 \\ 0 & 0 & 0 & 0 & 6 & 5 \\ 0 & 0 & 0 & 0 & 0 & 6 \end{pmatrix}_{11\times 6} \quad ; \quad H = \begin{pmatrix} 2 \\ -4 \\ 6 \\ -8 \\ 0 \\ 0 \end{pmatrix}_{6\times 1}$$

$$Y = XH$$

$$Y = \begin{bmatrix} 1\times 2 = 2 \\ 2\times 2 - 1\times 4 = 0 \\ 3\times 2 - 2\times 4 + 1\times 6 = 4 \\ 4\times 2 - 3\times 4 + 2\times 6 - 8 = 0 \\ 5\times 2 - 4\times 4 + 3\times 6 - 2\times 8 = -4 \\ 6\times 2 - 5\times 4 + 4\times 6 - 3\times 8 = -8 \\ -6\times 4 + 5\times 6 - 4\times 8 = -26 \\ 6\times 6 - 5\times 8 = -4 \\ -6\times 8 = -48 \end{bmatrix}$$

For $x(n), N_1 = 0$ and for $h(n), N_2 = 0$. For $y(n), N = 0$ therefore in the above matrix, the first row corresponds to $y(0)$.

$$\boxed{y(n) = \{\underset{\uparrow N=0}{2},\ 0,\ 4,\ 0,\ -4,\ -8,\ -26,\ -4,\ -48\}}$$

■ Example 3.69

Find the convolution sum for the following sequences

$$x(n) = \{1,\ 1,\ 0,\ \underset{\uparrow}{1},\ 1\}$$

$$h(n) = \{1,\ \underset{\uparrow}{-2},\ -3,\ 4\}$$

Solution:

$$X = \begin{pmatrix} 1\,0\,0\,0\,0 \\ 1\,1\,0\,0\,0 \\ 0\,1\,1\,0\,0 \\ 1\,0\,1\,1\,0 \\ 1\,1\,0\,1\,1 \\ 0\,1\,1\,0\,1 \\ 0\,0\,1\,1\,0 \\ 0\,0\,0\,1\,1 \\ 0\,0\,0\,0\,1 \end{pmatrix}_{9\times 5} \quad ; \quad H = \begin{pmatrix} 1 \\ -2 \\ -3 \\ 4 \\ 0 \end{pmatrix}_{5\times 1}$$

$$Y = XH$$

$$Y = \begin{bmatrix} 1 \\ 1-2 \\ -2-3 \\ 1-3+4 \\ 1-2+4 \\ -2-3 \\ -3+4 \\ 4 \\ 0 \end{bmatrix} = \begin{bmatrix} 1 \\ -1 \\ -5 \\ 2 \\ 3 \\ -5 \\ 1 \\ 4 \end{bmatrix}$$

$N_1 = 3$; $N_2 = 1$ and hence $N = 4$; $T_1 = 5$; $T_2 = 4$ and hence $T = 8$.

$$\boxed{\begin{array}{c} \longleftarrow ----T = 8---- \longrightarrow \\ y(n) = \{1,\ -1,\ -5,\ 2,\ 3,\ -5,\ 1,\ -4\} \\ \longleftarrow N = 4 \longrightarrow\uparrow \end{array}}$$

3.16 Convolution Sum by Graphical Method

The procedure for the determination of convolution sum by graphical method is similar to the convolution integral of continuous time system. The following steps are followed.

1. Represent $x(n)$ versus n in a graph. Replace n by k and $x(k)$ versus k is represented in a graph. One can straightaway plot $x(k)$ versus k.
2. Similar to Step 1, plot $h(k)$ versus k.
3. By folding $h(k)$, obtain $h(-k)$.

4. By adding "n" to $h(-k)$, obtain $h(n-k)$. Shift $h(n-k)$ to the extreme left. Start moving $h(n-k)$ toward the right so that $x(k)$ and $h(n-k)$ overlap each other. It is to be noted that $x(k)$ should be kept fixed and $h(n-k)$ alone should be moved by one sample at an instant. Calculate $y(n)$ at that instant.
5. The procedure is repeated at other instants and at each time $y(n)$ is calculated. When there is no overlapping, the movement of $h(n-k)$ is stopped and here $y(n) = 0$.

The following examples illustrate this graphical procedure.

Example 3.70

Compute the convolution of the two sequences $x(n)$ and $h(n)$ shown in Fig. 3.51a and b, respectively, and plot $y(n)$ versus n. Use graphical method.

(Anna University, December, 2006)

Solution:

1. Figure 3.51a shows the sequence of $x(n)$ and Fig. 3.51b the sequence of $h(n)$
2. n is replaced by k and $x(k)$ and $h(k)$ are represented in Fig. 3.51c and d, respectively.
3. $h(-k)$ is obtained from $h(k)$ by folding and is shown in Fig. 3.51e.
4. n is added in $h(-k)$ and $h(n-k)$ is obtained. $h(n-k)$ is moved to the extreme left so that there is no overlapping between $x(k)$ and $h(n-k)$ initially. This is shown in Fig. 3.51f.
5. $x(k)$ is fixed and $h(n-k)$ is moved toward right so that it overlaps with $x(k)$. At $n = -1$, the first overlapping occurs and $y(-1) = 1 \times 2 = 2$. This is shown in Fig. 3.51c and g.
6. When $h(n-k)$ is moved toward right by one more sample (now $n = 0$), the plot of $x(k)$ and $h(n-k)$ is shown in Fig. 3.51h and $y(n) = 1 \times 1 + 2 \times 2 = 5$.
7. For $n = 1$, the plot of $x(k)$ and $h(n-k)$ is shown in Fig. 3.51i and the overlapping is shown in dotted line. $y(1) = 1 \times 2 = 2$.
8. For $n = 2$, the plot of $x(k)$ and $h(n-k)$ is shown in Fig. 3.51j. Here, there is no overlapping. Hence $y(2) = 0$.
9. $N_1 = 2$ and $N_2 = 0$. Therefore, $N = N_1 + N_2 = 2$, $T_1 = 3$ and $T_2 = 3$. Therefore $T = T_1 + T_2 - 1 = 5$
10.

$$\boxed{\begin{array}{c} \longleftarrow T = 5 \longrightarrow \\ y(n) = \{0,\ 2,\ 5,\ 2,\ 0\} \\ |N = 2| \ \uparrow \end{array}}$$

The plot of $y(n)$ is shown in Fig. 3.51k.

■ Example 3.71

Find the linear convolution of

$$x(n) = \{1, 2, 3, 4, 5\}$$
$$h(n) = \{1, 2, 3, 3, 2, 1\}$$

Use graphical methods.

(*Anna University, December, 2007*)

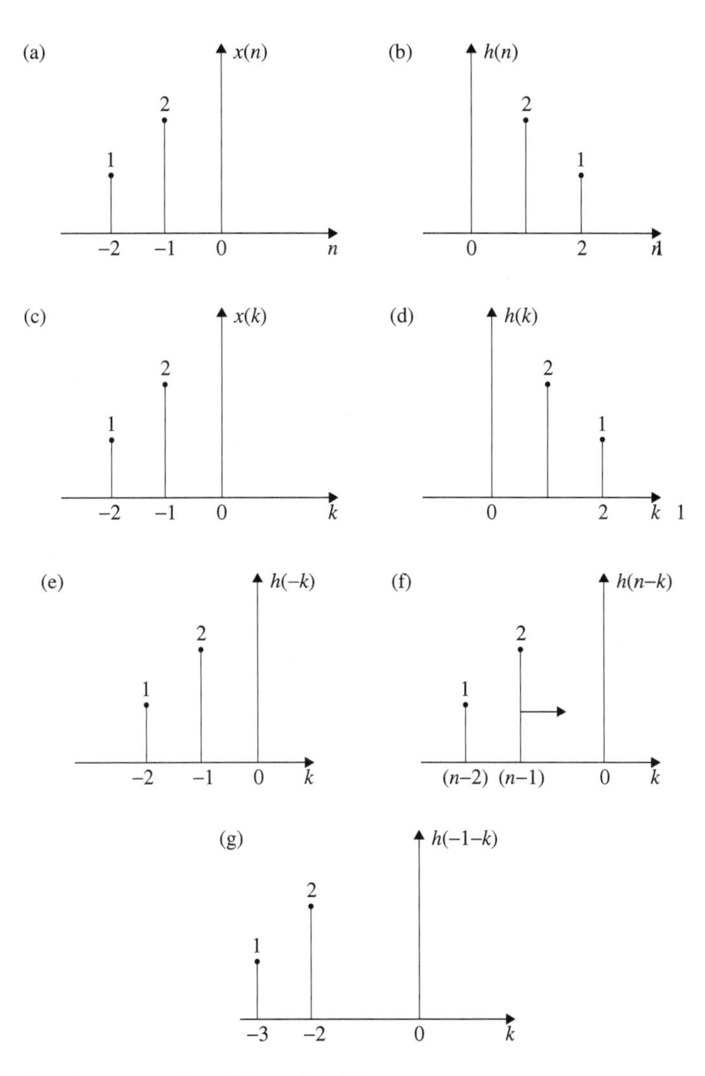

Fig. 3.51 Signal representation of Example 3.57

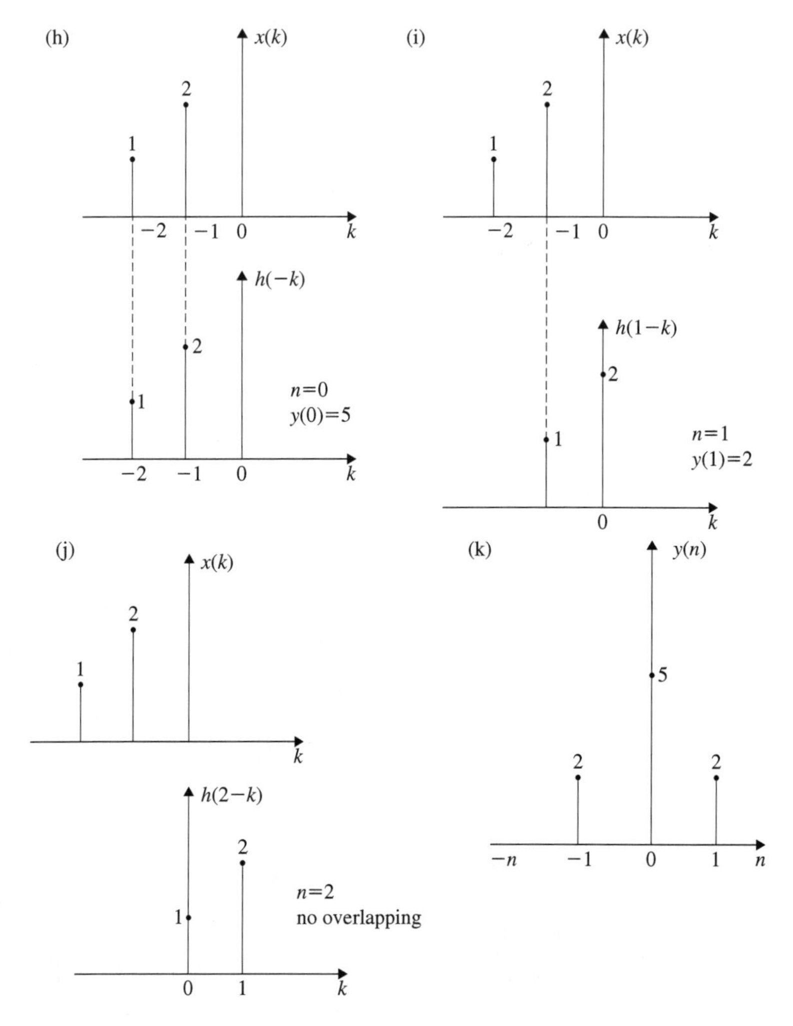

Fig. 3.51 (continued)

Solution:

1. $x(k)$ is shown in Fig. 3.52a and $h(k)$ is shown in Fig. 3.52b. $h(n-k)$ is shown in Fig. 3.52c. $h(n-k)$ is moved toward right so that it overlaps with $x(k)$. This happens at $n = 0$. The sample with strength 1 in $x(k)$ and the sample $h(k)$ with strength 1 at $n = 0$ overlap as in Fig. 3.52d. The product of these two samples is $y(0) = 1 \times 1 = 1$.

2. $h(n-k)$ is moved further toward right by one more sample $n=1$. Now this is shown in Fig. 3.52e. The product of the overlapping samples for $n=1$ is shown by dotted lines) $y(1)=1\times 2+2\times 1=4$.
3. For $n=2$, $x(k)$ and $h(n-k)$ are shown in Fig. 3.52f. 3 samples in each of $x(k)$ and $h(n-k)$ overlap. Hence, $y(2)=1\times 3+2\times 2+3\times 1=10$.
4. For $n=3$, $x(k)$ and $h(n-k)$ are shown in Fig. 3.52g. Here 4 samples overlap. Hence, $y(3)=1\times 3+2\times 3+3\times 2+4\times 1=19$.
5. For $n=4$, $x(k)$ and $h(n-k)$ are shown in Fig. 3.52h $y(4)=1\times 2+2\times 3+3\times 3+4\times 2+5\times 1=30$.
6. For $n=5$, $x(k)$ and $h(n-k)$ are shown in Fig. 3.52i. Here 5 samples overlap. $y(5)=1\times 1+2\times 2+3\times 3+4\times 3+5\times 2=36$.
7. For $n=6$, $x(k)$ and $h(n-k)$ are shown in Fig. 3.52j. Here 4 samples overlap. $y(6)=2\times 1+3\times 2+4\times 3+5\times 3=35$.
8. For $n=7$, $x(k)$ and $h(n-k)$ are shown in Fig. 3.52k. Here 3 samples overlap. Hence, $y(7)=3\times 1+4\times 2+5\times 3=26$.
9. For $n=8$, $x(k)$ and $h(n-k)$ are shown in Fig. 3.52l. Here 2 samples overlap. Hence, $y(8)=4\times 1+5\times 2=14$.
10. For $n=9$, $x(k)$ and $h(n-k)$ are shown in Fig. 3.52m. Here one sample overlaps. $y(9)=5\times 1=5$.
11. For $n=10$, $x(k)$ and $h(n-k)$ does not overlap with any of the samples of $x(k)$. Hence, $y(10)=0$.
12. For $x(n), N_1=0$ and for $h(n), N_2=0$. Hence, $N=N_1+N_2=0$. For $x(n), T_1=5$ and for $h(n), T_2=6$. Hence, width of $y(n)=T_1+T_2-1=5+6-1=10$.

$$\boxed{y(n)=\{\underset{\uparrow}{1},\ 4,\ 10,\ 19,\ 30,\ 36,\ 35,\ 26,\ 14,\ 5\}}$$

Example 3.72

$$x(n)=2u(n+2)-2u(n)-3u(n-1)+3u(n-3)$$
$$h(n)=-u(n+1)+3u(n)-5u(n-1)+3u(n-2)$$

Find

$$y(n)=x(n)*h(n)$$

by graphical convolution.

Solution:

1.

$$\begin{aligned}x(n)&=2u(n+2)-2u(n)-3u(n-1)+3u(n-3)\\&=\{2,\ 2,\ \underset{\uparrow}{0},\ -3,\ -3\}\end{aligned}$$

This is represented in Fig. 3.53(a) with n replaced by k.

$$\begin{aligned} h(n) &= -u(n+1) + 3u(n) - 5u(n-1) + 3u(n-2) \\ &= \{-1, \underset{\uparrow}{2}, -3\} \end{aligned}$$

This is represented in Fig. 3.53b with n replaced by k.

2. $h(n-k)$ is created and put to the extreme left and is shown in Fig. 3.53c.
3. $h(n-k)$ is moved toward right sample by sample. The first overlapping between $x(k)$ and $h(n-k)$ occurs when $n = -3$ which is shown in Fig. 3.53d. The dotted line shows the overlapping with $x(k)$. At $n = -3$, $y(-3)$ is obtained as $y(-3) = 2 \times (-1) = -2$.
4. For $n = -2$, $x(k)$ and $h(-2-k)$ are shown in Fig. 3.53e. The overlaps are shown by dotted lines.
$$y(-2) = 2 \times (-1) + 2 \times 2 = 2$$
5. For $n = -1$, $x(k)$ and $h(-1-k)$ are shown in Fig. 3.53f. The overlaps are shown in dotted lines. $y(-1) = 2 \times (-3) + 2 \times 2 = -2$.
6. For $n = 0$, $x(k)$ and $h(-k)$ are shown in Fig. 3.53g. The overlaps are shown in dotted lines. $y(0) = 2 \times (-3) + 0 + (3 \times 1) = -3$.
7. For $n = 1$, $x(k)$ and $h(1-k)$ are shown in Fig. 3.53h. The overlaps are shown in dotted lines. $y(1) = (-3 \times 2) + (-3 \times -1) = -3$.

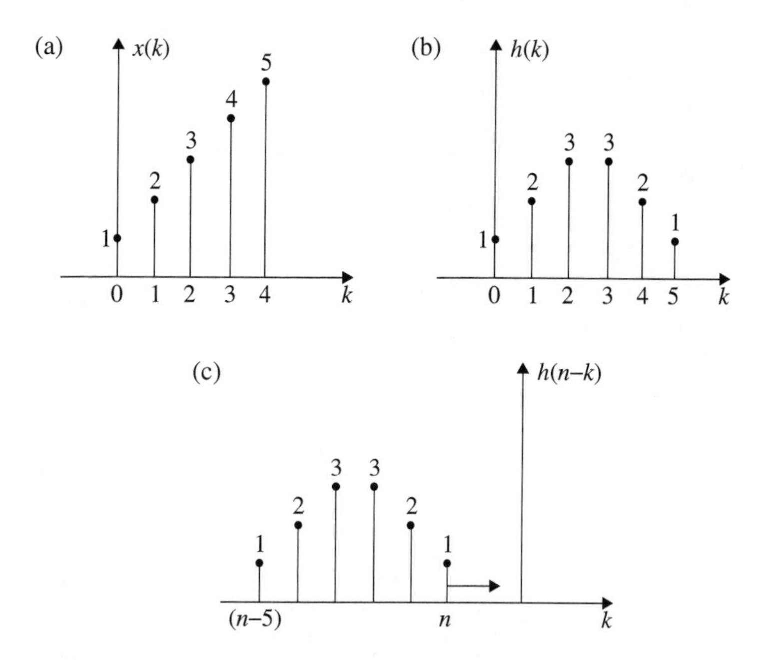

Fig. 3.52 Convolution of Example 3.71

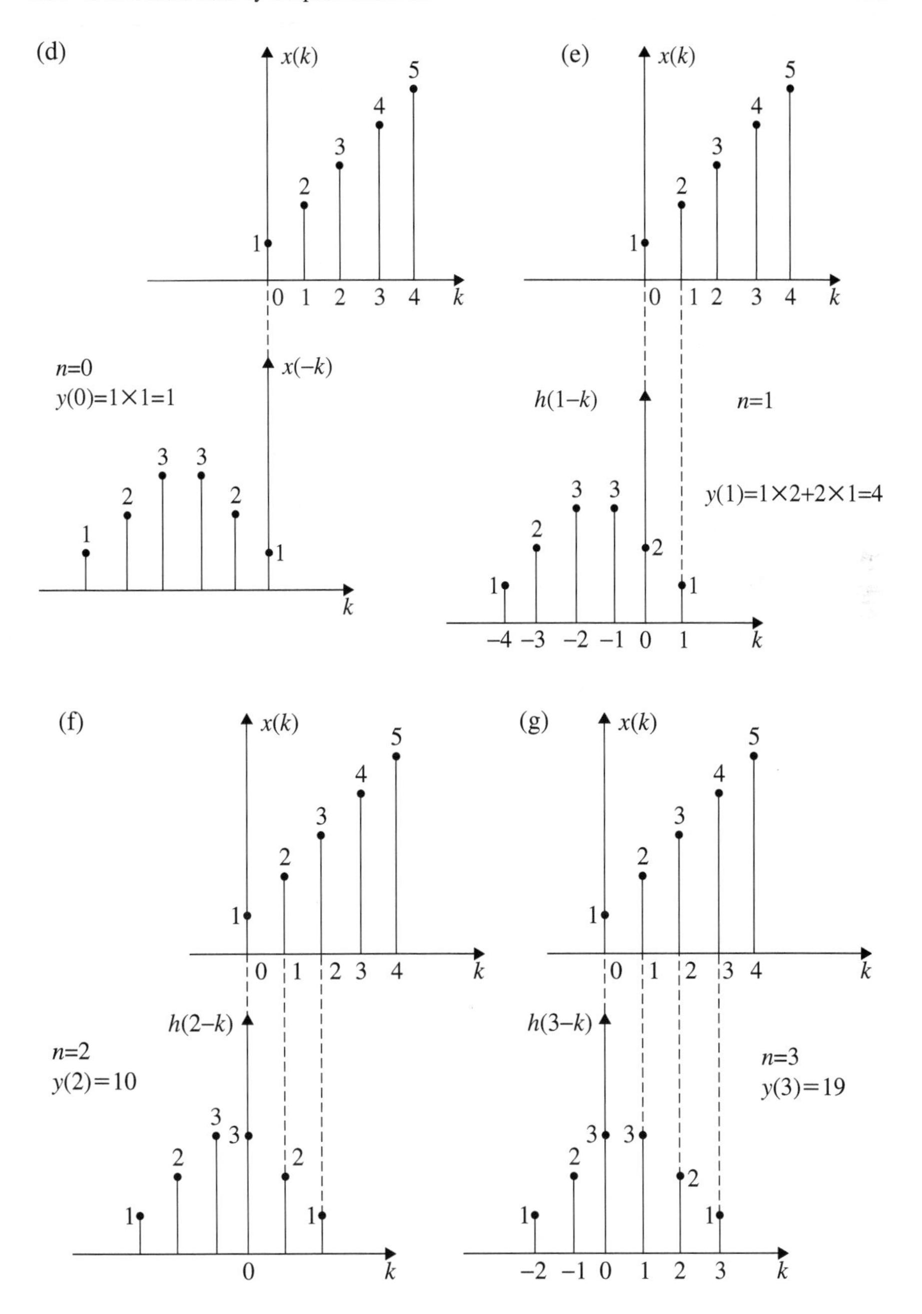

Fig. 3.52 (continued)

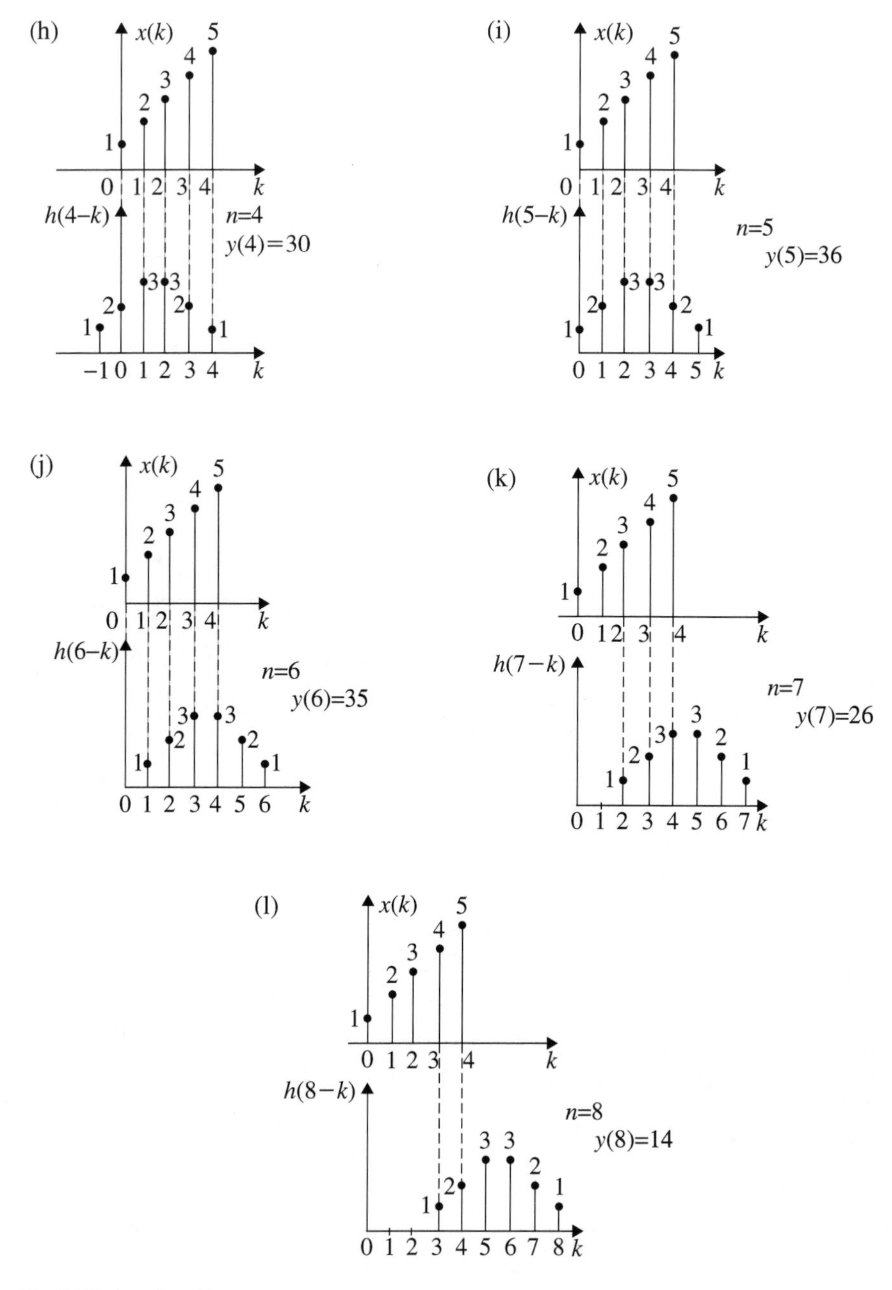

Fig. 3.52 (continued)

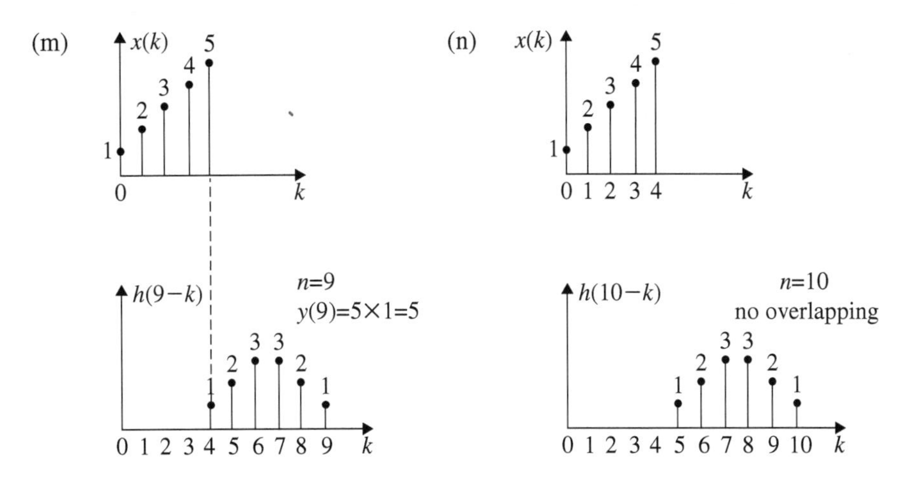

Fig. 3.52 (continued)

8. For $n = 2$, $x(k)$ and $h(2-k)$ are shown in Fig. 3.53i. The overlaps are shown in dotted lines. $y(2) = (-3 \times -3) + (-3 \times 2) = 3$.
9. For $n = 3$, $x(k)$ and $h(3-k)$ are shown in Fig. 3.53j. The overlapping is shown in dotted line. $y(3) = (-3 \times -3) = 9$.
10. For $n = 4$, $x(k)$ and $h(4-k)$ are shown in Fig. 3.53k. There is no overlapping and $y(3) = 0$.
11. For the given $x(n), N_1 = 2$ and for $h(n), N_2 = 1$. Hence for $y(n), N = N_1 + N_2 = 2 + 1 = 3$. The width of $x(n)$ is $T_1 = 5$ and that of $h(n)$ is $T_2 = 3$. Hence, the width of $y(n)$ is $T = T_1 + T_2 - 1 = 5 + 3 - 1 = 7$.
12.

$$\boxed{\begin{array}{c} y(n) = \{-2,\ 2,\ -2,\ -3,\ -3,\ 3,\ 9\} \\ \leftarrow N{=}3 \rightarrow \quad \uparrow \\ n{=}0 \\ \longleftarrow ----T = 7---- \longrightarrow \end{array}}$$

■ Example 3.73

Find the convolution of $x(n)$ with $h(n)$

$$x(n) = a^n u(n) \qquad \text{where } 0 < a < 1$$

$$h(n) = \begin{cases} 1 & 0 \le n \le 9 \\ 0 & n \ge 10 \end{cases}$$

(*Anna University, April, 2004*)

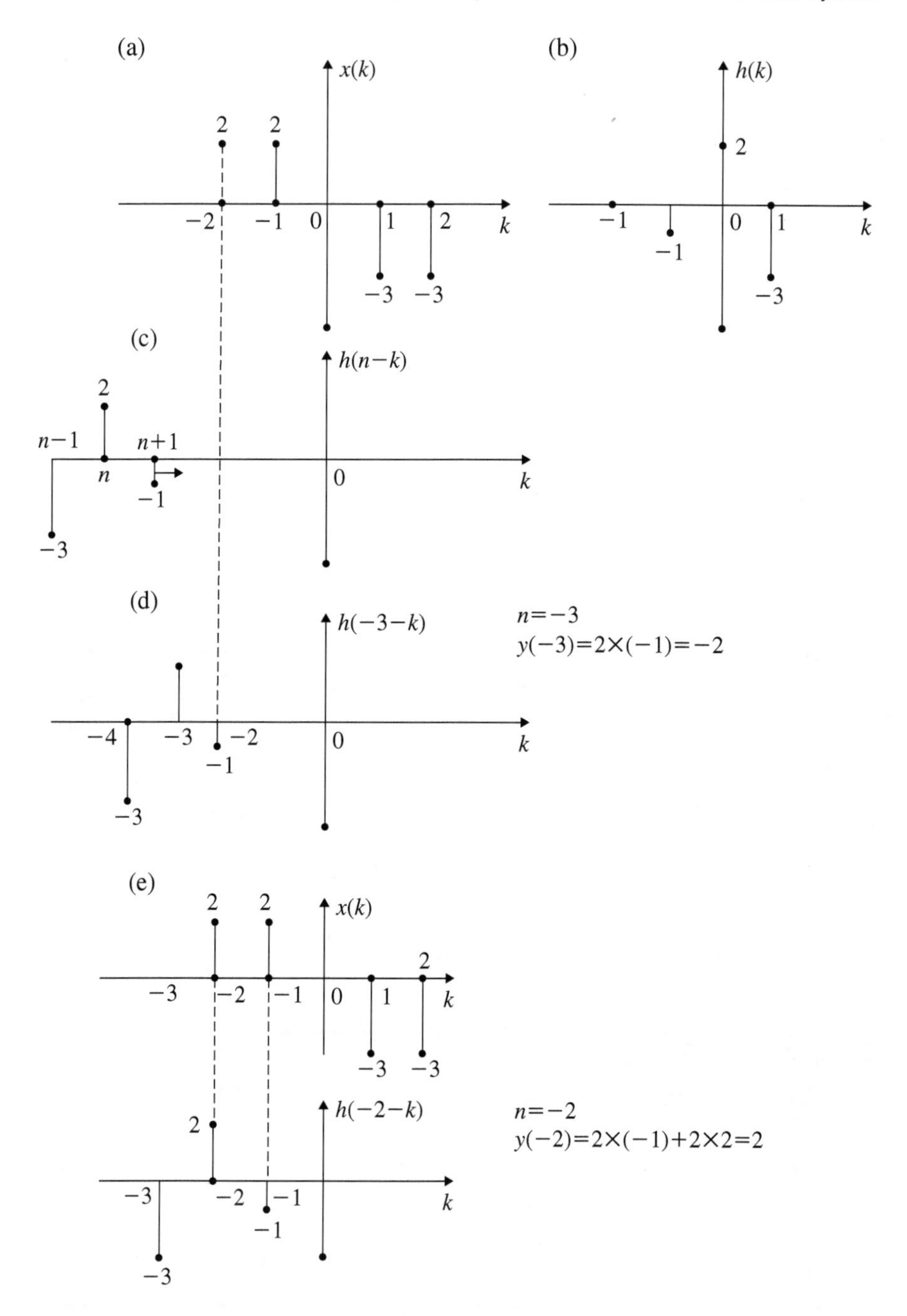

Fig. 3.53 Graphical method of obtaining $y(n)$ of Example 3.72

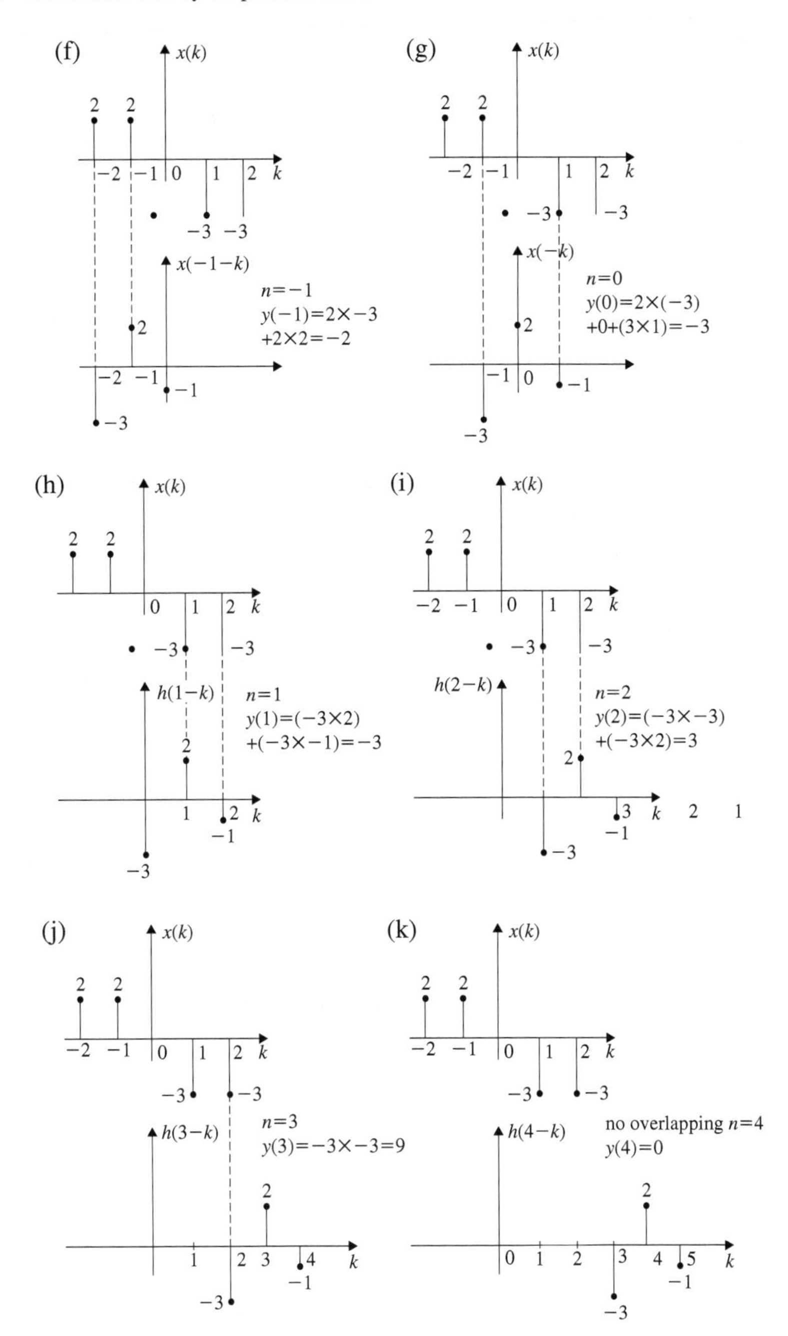

Fig. 3.53 (continued)

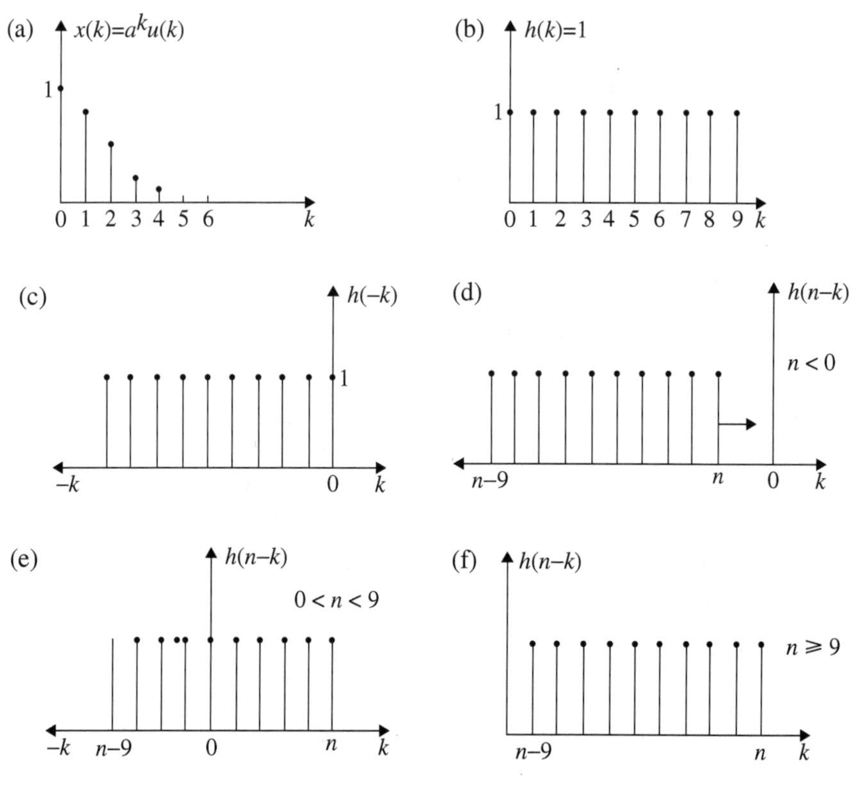

Fig. 3.54 Convolution sum of Example 3.73

Solution:

1. $x(k) = a^k u(k)$ is shown in Fig. 3.54a.
2. $h(k)$, $h(-k)$ are shown in Fig. 3.54b and c, respectively. $h(n-k)$ is shown in Fig. 3.54d.
3. $h(n-k)$ is moved toward right so that it overlaps with $x(k)$. The right edge of $h(n-k)$ slides past the left edge of $x(k)$ for $n = 0$. The overlapping continues without change until the left edge of $h(n-k)$ slides past the right edge of $x(k)$. Hence, the summation limit is $0 \le k \le n$

$$
\begin{aligned}
y(n) &= \sum_{k=0}^{n} x(k)h(n-k) \\
&= \sum_{k=0}^{n} a^k \qquad 0 \le n \le 9
\end{aligned}
$$

Using finite summation formula, we get

$$\boxed{y(n) = \frac{(1-a^{n+1})}{(1-a)}} \qquad 0 \leq n \leq 9$$

4. If $h(n-k)$ is shifted further a change occurs when the left edge of $h(n-k)$ slides past the left edge of $x(k)$ as shown in Fig. 3.54f. The summation interval is from $n-9 \leq k \leq n$

$$y(n) = \sum_{k=n-9}^{k=n} a^k$$

Put $k-n+9 = p$, $k-n = p-9 = 0$ or $p = 9$. The lower limit of summation is $p = 0$ and the upper limit is $p = 9$. Now $a^k = a^{n-9}a$. Substituting the above, we get

$$y(n) = \sum_{p=0}^{p=9} a^{(n-9)} a^p$$
$$= a^{(n-9)} \sum_{p=0}^{p=9} a^p$$

Using the finite summation formula

$$\sum_{p=0}^{p=9} a^p = \frac{(1-a^{10})}{(1-a)}$$

$$y(n) = a^{(n-9)} \frac{(1-a^{10})}{(1-a)} \qquad n > 9$$

$$\boxed{\begin{array}{ll} y(n) = 0 & \text{for } n < 0 \\ y(n) = \dfrac{(1-a^{(n+1)})}{(1-a)} & \text{for } 0 \leq n \leq 9 \\ y(n) = a^{(n-9)} \left(\dfrac{(1-a^{10})}{(1-a)} \right) & n > 9 \end{array}}$$

3.17 Deconvolution

The deconvolution is the process of getting $x(n)$ if $h(n)$ and $y(n)$ are known. As per convolution sum

$$y(n) = \sum_{k=-\infty}^{k=\infty} x(k)h(n-k)$$

If we assume that $x(n)$ and $h(n)$ are one sided finite sequences the above equation is written as

$$y(n) = \sum_{k=-\infty}^{k=n} x(k)h(n-k)$$

$$y(0) = h(0)x(0)$$

$$y(1) = h(1)x(0) + h(0)x(1)$$
$$y(2) = h(2)x(0) + h(1)x(1) + h(0)x(2)$$

Now

$$\boxed{x(0) = \frac{y(0)}{h(0)}}$$

$$y(1) = h(1)x(0) + h(0)x(1)$$

$$\boxed{x(1) = \frac{y(1) - h(1)x(0)}{h(0)}}$$

Knowing $x(0)$, $h(0)$, $h(1)$ and $y(1)$, we can find $x(1)$. From $y(2)$, the following equation is written.

$$x(2) = \frac{y(2) - h(2)x(0) - h(1)x(1)}{h(0)}$$

In general

$$\boxed{x(n) = y(n) - \sum_{k=0}^{k=n-1} x(k)h(n-k)}$$

■ Example 3.74

$$y(n) = \{\underset{\uparrow}{1}, 5, 10, 11, 8, 4, 1\}$$

$$h(n) = \{\underset{\uparrow}{1}, 2, 1\}$$

Find $x(n)$.

(*Anna University, April, 2003*)

Solution: The number of samples in $h(n)$, $y(n)$, and $x(n)$ are $T_1 = 3, T = 7, T_2 = 5$, respectively. The following relationship holds good.

$$\begin{aligned} T &= T_1 + T_2 - 1 \\ T_2 &= T - T_1 + 1 \\ &= 7 - 3 + 1 = 5 \end{aligned}$$

Thus, $x(0), x(1), x(2), x(3)$ and $x(4)$ are to be determined.

For $n = 0$

$$x(0) = \frac{y(0)}{h(0)} = \frac{1}{1} = 1$$

For $n = 1$

$$x(1) = \frac{y(1) - x(0)h(1)}{h(0)} = \frac{5 - 1 \times 2}{1} = 3$$

For $n = 2$

$$\begin{aligned} x(2) &= \frac{y(2) - \sum_{k=0}^{k=1} x(k)h(2-k)}{h(0)} \\ &= \frac{y(2) - x(0)h(2) - x(1)h(1)}{h(0)} \\ &= \frac{10 - 1 \times 1 - 3 \times 2}{1} = 10 - 1 - 6 = 3 \end{aligned}$$

For $n = 3$

$$\begin{aligned} x(3) &= \frac{y(3) - \sum_{k=0}^{k=2} x(k)h(3-k)}{h(0)} \\ &= \frac{y(3) - x(0)h(3) - x(1)h(3) - x(1)h(2) - x(2)h(1)}{h(0)} \\ &= \frac{11 - 1 \times 0 - 3 \times 1 - 3 \times 2}{1} = 2 \end{aligned}$$

For $n = 4$,

$$\begin{aligned} x(4) &= \frac{y(4) - \sum_{k=0}^{k=3} x(k)h(4-k)}{h(0)} \\ &= \frac{y(4) - x(0)h(4) - x(1)h(3) - x(2)h(2) - x(3)h(1)}{h(0)} \\ x(4) &= \frac{8 - 1 \times 0 - 3 \times 0 - 3 \times 1 - 2 \times 2}{1} = 1 \end{aligned}$$

$$\therefore \quad \boxed{x(n) = \{\underset{\uparrow}{1}, 3, 3, 2, 1\}}$$

■ Example 3.75

Find $x(n)$ given $h(n) = \{1, 2, 3, 2\}$ and $y(n) = \{3, 8, 17, 25, 26, 23, 10\}$.

Solution: The number of samples in $y(n)$, $h(n)$, $x(n)$ is $T = 7$, $T_1 = 4$, T_2, respectively.

$$\begin{aligned}
T &= T_1 + T_2 - 1 \\
T_2 &= T - T_1 + 1 \\
&= 7 - 4 + 1 = 4 \\
\therefore \quad x(n) &= \{x(0), x(1), x(2), x(3)\} \\
x(0) &= \frac{y(0)}{h(0)} = \frac{3}{1} = 3 \\
x(1) &= \frac{y(1) - x(0)h(1)}{h(0)} = \frac{8 - 3 \times 2}{1} = 2 \\
x(2) &= \frac{y(2) - \sum_{k=0}^{k=1} x(k)h(2-k)}{h(0)} \\
&= \frac{y(2) - x(0)h(2) - x(1)h(1)}{h(0)} \\
&= \frac{17 - 3 \times 3 - 2 \times 2}{1}
\end{aligned}$$

$$\begin{aligned}
x(2) &= 4 \\
x(3) &= y(3) - \sum_{k=0}^{k=2} x(k)h(3-k) \\
&= \frac{y(3) - x(0)h(3) - x(1)h(2) - x(2)h(1)}{h(0)} \\
&= \frac{25 - 3 \times 2 - 2 \times 3 - 4 \times 2}{1} = 5
\end{aligned}$$

Hence

$$\boxed{x(n) = \{3, 2, 4, 5\}}$$

3.18 Step Response of the System

Equation (3.32) gives the step response of the discrete time system if the impulse response $h(n)$ is known

$$\boxed{s(n) = \sum_{k=0}^{k=n} h(k)} \tag{3.32}$$

The following examples illustrate how to find step response from impulse response.

■ Example 3.76

Find the step response of the system whose impulse response is

$$h(n) = a^n u(n) \qquad \text{where } 0 < a < 1$$

Solution:

$$s(n) = \sum_{k=0}^{n} h(k)$$

$$s(n) = \sum_{k=0}^{n} a^k$$

Using the finite summation formula, we get

$$\boxed{s(n) = \frac{(1 - a^{n+1})}{(1 - a)}}$$

■ Example 3.77

Find the step response whose impulse response is given by

$$h(n) = \delta(n-1) + \delta(n-5)$$

Solution:

$$\begin{aligned} s(n) = h(n) * u(n) &= [\delta(n-1) + \delta(n-5)] * u(n) \\ &= \delta(n-1) * u(n) + \delta(n-5) * u(n) \end{aligned}$$

By using the property $\delta(n-1) * u(n) = u(n-1)$, we get

$$\boxed{s(n) = u(n-1) + u(n-5)}$$

■ Example 3.78

Find the step response of the system whose impulse response is

$$h(n) = \left(\frac{1}{3}\right)^n.$$

Solution:

$$\begin{aligned} s(n) &= \sum_{k=0}^{n} \left(\frac{1}{3}\right)^k \\ &= \frac{1 - \frac{1}{3^{n+1}}}{1 - \frac{1}{3}} = \frac{(3^{n+1} - 1)}{(3 - 1)} \frac{1}{3^n} \end{aligned}$$

$$\boxed{s(n) = \frac{1}{2}\left(\frac{1}{3}\right)^n (3^{n+1} - 1)u(n)}$$

■ Example 3.79

Find the step response if the impulse response $h(n) = u(n)$

Solution:

$$\begin{aligned} s(n) &= \sum_{k=0}^{n} h(k) = \sum_{k=0}^{n} u(k) \\ &= \sum_{k=0}^{n} 1 \end{aligned}$$

$$\boxed{s(n) = (n+1)}$$

3.19 Stability from Impulse Response

From Eq. (3.31), for the discrete time system to be bounded input output stable, $M_y(n) = \sum_{n=-\infty}^{\infty} |h(n)| < \infty$. The following examples illustrate this.

■ Example 3.80

The impulse response of a certain system is

$$h(n) = a^n u(n) \qquad \text{where } 0 < a < 1$$

Find whether the system is BIBO stable.

Solution: $h(n)$ is causal. Hence

$$M_y(n) = \sum_{n=0}^{\infty} a^n = \frac{1}{(1-a)} < \infty$$

Hence, the system is BIBO stable.

■ Example 3.81

$$h(n) = a^n u(-n)$$

Find whether the system is BIBO stable.

Solution:

$$M_y(n) = \sum_{n=-\infty}^{\infty} a^n$$

For $n > 0$, $u(n) = 0$. Hence,

$$\begin{aligned} M_y(n) &= \sum_{n=-\infty}^{0} a^n \\ &= \sum_{n=0}^{\infty} a^{-n} = \sum_{n=0}^{\infty} \left(\frac{1}{a}\right)^n \\ &= \frac{1}{1-\frac{1}{a}} \\ &= \frac{a}{a-1} < \infty \end{aligned}$$

Hence, the system is BIBO stable.

■ Example 3.82

$$h(n) = a^n u(n-1) \quad \text{where} \quad a > 1$$

Find whether the system is BIBO stable.

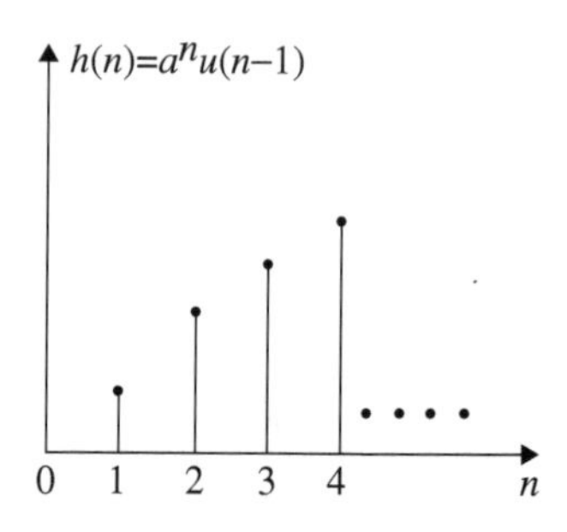

Fig. 3.55 Impulse response plot of Example 3.82

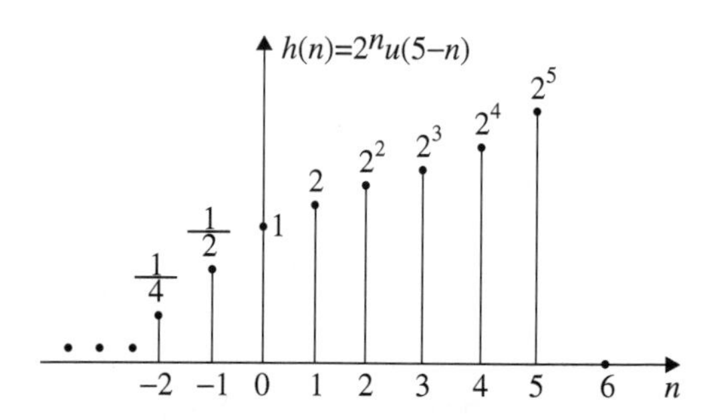

Fig. 3.56 Impulse response plot of Example 3.83

Solution: The impulse response plot of $h(n) = a^n u(n-1)$ is shown in Fig. 3.55, and it is noticed that the sequence is not summable. Hence, the system is unstable.

$$
\begin{aligned}
M_y(n) &= \sum_{n=1}^{\infty} a^n \\
&= a + a^2 + a^3 + \dots
\end{aligned}
$$

$$\boxed{M_y(n) = \infty}$$

The system is BIBO unstable.

■ Example 3.83

$$h(n) = 2^n u(5-n)$$

Find whether the system is BIBO stable

Solution: The impulse response plot of $h(n) = 2^n u(5-n)$ is shown in Fig. 3.56. $u(5-n) = 0$ for $n > 5$. For $-\infty \geq n \leq 5$, $h(n) = 2^n$

$$
\begin{aligned}
M_y(n) &= \sum_{n=-\infty}^{5} 2^n \\
&= \sum_{n=-\infty}^{0} 2^n + \sum_{1}^{5} 2^n \\
&= \sum_{n=0}^{\infty} 2^{-n} + \sum_{1}^{5} 2^n \\
&= \sum_{n=0}^{\infty} \left(\frac{1}{2}\right)^n + \sum_{1}^{5} 2^n \\
&= \frac{1}{1-1/2} + \sum_{0}^{5} 2^n - 1 \\
&= 2 + \frac{(1-2^6)}{1-2} - 1 \\
&= 1 + 2^6 - 1 = 2^6
\end{aligned}
$$

$$
\boxed{M_y(n) = 2^6 < \infty}
$$

The system is BIBO stable.

■ Example 3.84

$$
h(n) = e^{-5|n|}
$$

Find whether the system is BIBO stable (Fig. 3.57).

Solution:

$$
\begin{aligned}
M_y(n) &= \sum_{n=-\infty}^{\infty} e^{-5|n|} \\
&= \sum_{n=-\infty}^{-1} e^{5n} + \sum_{n=0}^{\infty} e^{-5n} \\
&= \sum_{n=1}^{\infty} e^{-5n} + \sum_{n=0}^{\infty} e^{-5n} \\
M_y(n) &= \frac{e^{-5}}{(1-e^{-5})} + \frac{1}{(1-e^{-5})}
\end{aligned}
$$

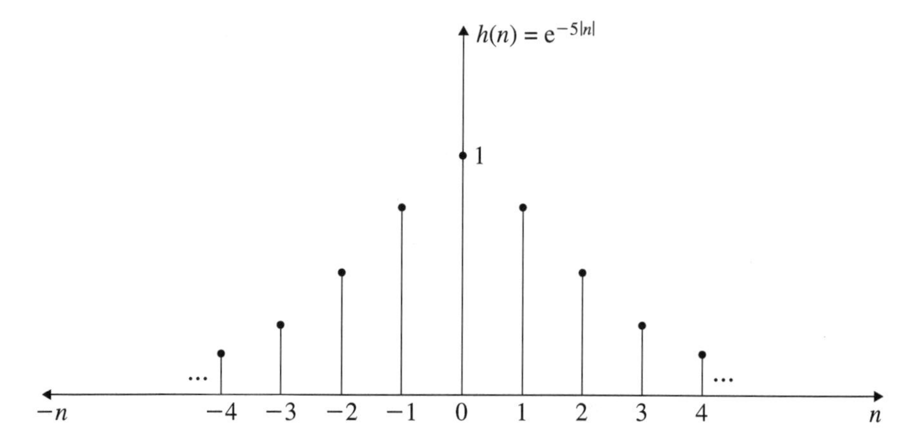

Fig. 3.57 Impulse responses plot of Example 3.84

$$= \frac{(1 + e^{-5})}{(1 - e^{-5})} = 1.0136 < \infty$$

$$\boxed{M_y(n) = 1.0136}$$

$M_y(n)$ is finite and hence the system is stable.

3.20 System Causality

From expression (3.30) the impulse response function $h(n) = 0$ for $n < 0$ for the system to be causal. This is illustrated in the following examples.

■ Example 3.85

$$h(n) = a^n u(-n)$$

Find whether the system is causal.

Solution: The impulse response plot of $h(n)$ is shown in Fig. 3.58. From Fig. 3.58, it is evident that at $n = -1,\ h(-1) = \frac{1}{a}$ and is not zero. Hence, the system is non-causal.

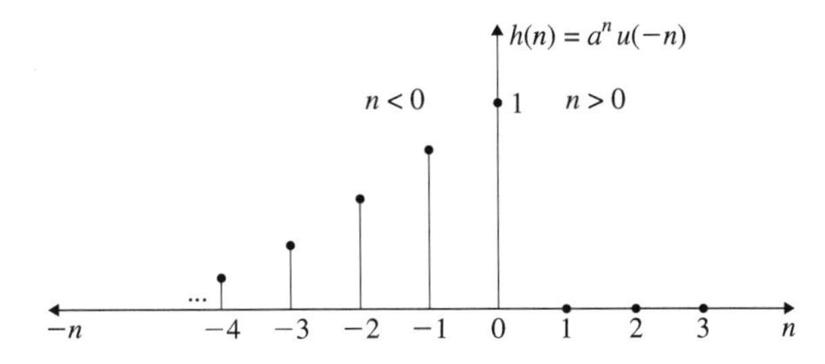

Fig. 3.58 Impulse response of Example 3.85

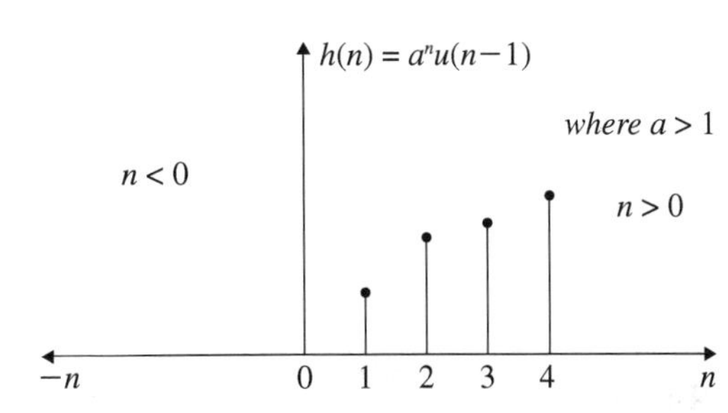

Fig. 3.59 Impulse response of Example 3.86

■ Example 3.86

$$h(n) = a^n u(n-1) \qquad \text{where } a > 1$$

Find whether the system is causal.

Solution: The plot of $h(n) = a^n u(n-1)$ is shown in Fig. 3.59. It is observed that for $n < 0$, $h(n) = 0$. Hence, the system is causal.

■ Example 3.87

$$h(n) = e^{-a|n|}$$

Find whether the system is causal.

Solution: The impulse response plot of $h(n)$ is shown in Fig. 3.60.

$$\begin{aligned} h(n) &= e^{-an} \qquad n \geq 0 \\ &= e^{an} \qquad n < 0 \\ h(n) &\neq 0 \qquad \text{for } n < 0. \end{aligned}$$

Hence, the system is non-causal.

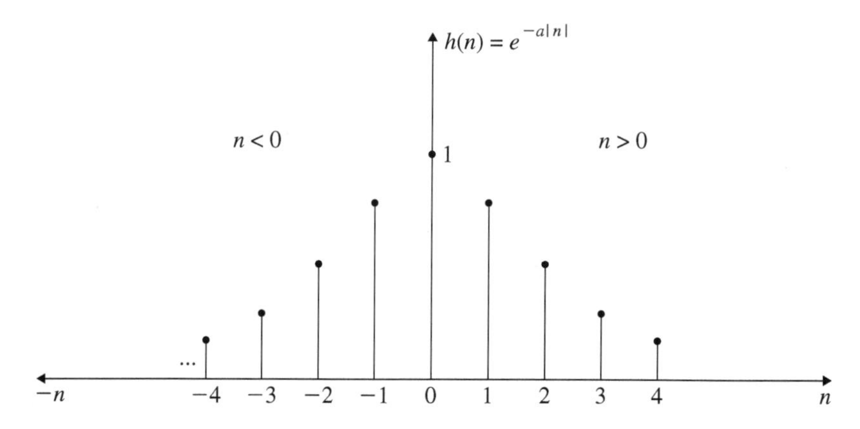

Fig. 3.60 Impulse response of Example 3.87

■ Example 3.88

The impulse response of a certain linear time invariant continuous time system is shown in Fig. 3.61a. The system is excited with the input $x(t)$ which is shown in Fig. 3.61b. Derive expressions for the output response $y(t)$ using convolution method.

Solution:

1. Figure 3.61a and b show $x(t)$ and $h(t)$, respectively. $x(\tau)$ and $h(\tau)$ are obtained by putting $t = \tau$ in $x(t)$ and $h(t)$, respectively. $x(-\tau)$ is obtained by folding $x(\tau)$ and is shown in Fig. 3.61c. $x(t - \tau)$ is obtained by adding t in $x(-\tau)$. $x(t - \tau)$ is shifted to the extreme left and is shown in Fig. 3.61d.
2. $h(\tau)$ is kept fixed and $x(t - \tau)$ is shifted toward right so that there is overlapping between $h(\tau)$ and $x(t - \tau)$.
3. **Time interval $-1 < t < 0$.**
   During the above time interval the right edge of $x(t - \tau)$ slides past the left edge of $h(\tau)$. Here the input is in the form of impulse. The output $y(t)$ is obtained using the property,

$$h(t) * x(t - t_0) = h(t - t_0)$$

   In the above interval (Fig. 3.61e).

$$\begin{aligned} h(t) &= 3t \\ x(t) &= 2\delta(t + 1) \\ y(t) &= h(t) * x(t) \\ &= 3 \times 2(t + 1) \\ &= 6(t + 1) \end{aligned}$$

4. **Time interval $0 < t < 1$ (Fig. 3.61f).**
During the above time interval, the second transition occurs. The transition should be one at a time where there is a change in $h(\tau)$. The right edge of $x(t-\tau)$ slides past $\tau = 1$. Here,

$$\begin{aligned}
h(t) &= (4-t) \\
x(t) &= 2\delta(t+1) \\
y(t) &= h(t) * x(t) \\
&= 2[4-(t+1)] \\
&= 2(3-t)
\end{aligned}$$

5. **Time interval $1 < t < 2$ (Fig. 3.61g).**
During the interval, the left edge of $x(t-\tau)$ slides past the left edge of $h(\tau)$.

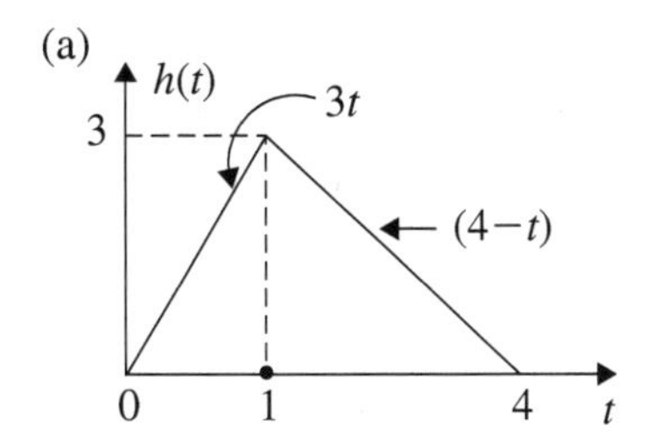

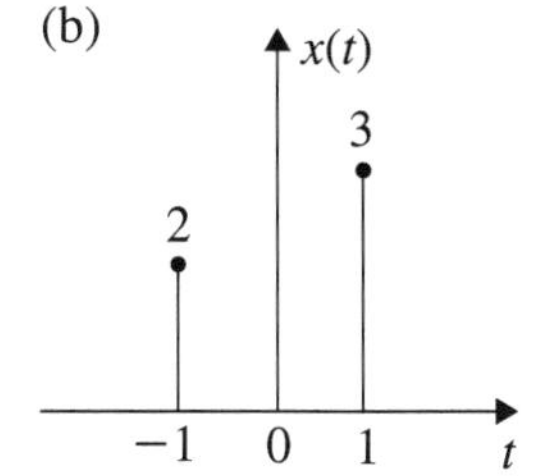

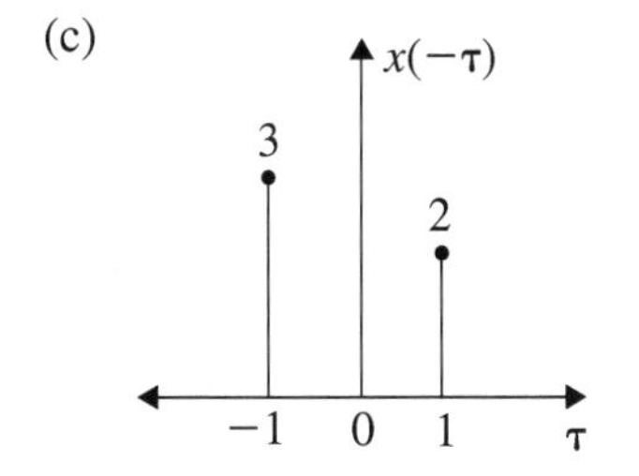

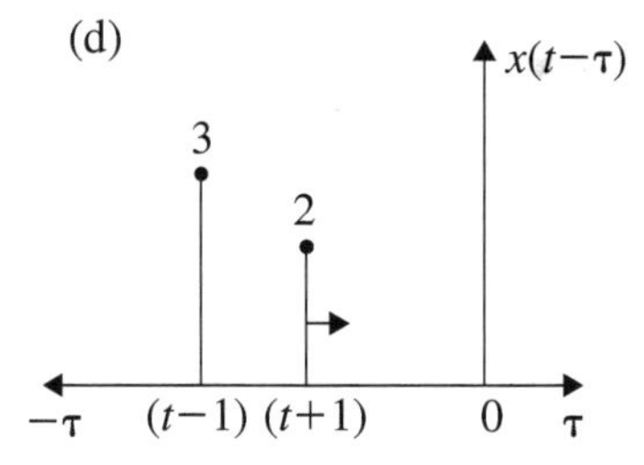

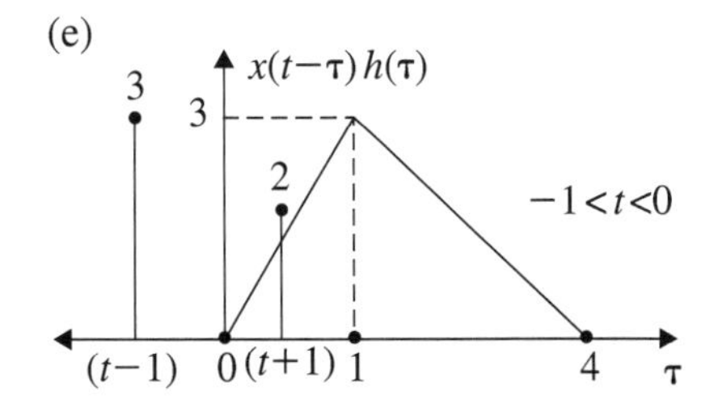

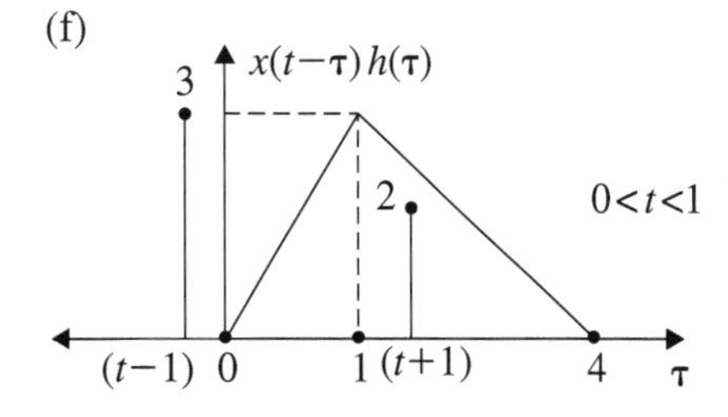

Fig. 3.61 Convolution of two signals of Example 3.88

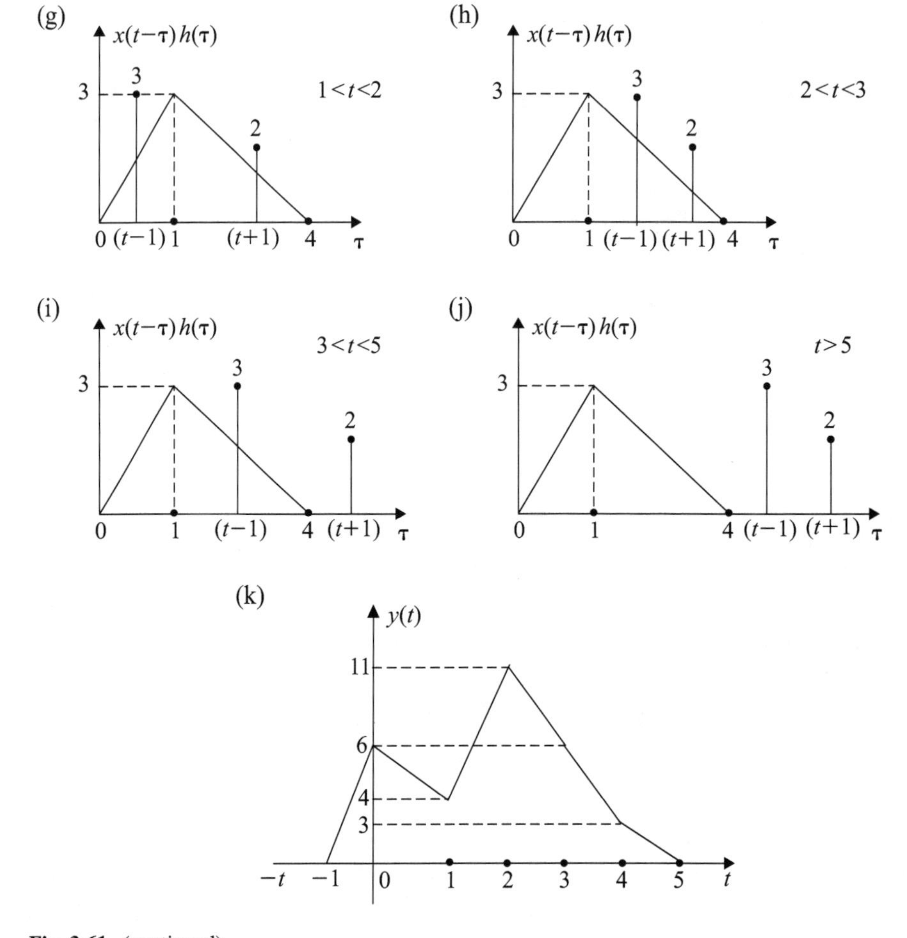

Fig. 3.61 (continued)

$$
\begin{aligned}
y(t) &= h(t) * x(t) \\
&= 3 \times 3(t-1) + 2(4-(t+1)) \\
&= (7t-3)
\end{aligned}
$$

6. **Time interval $2 < t < 3$ (Fig. 3.61h).**
   During the above interval the left edge of $x(t-\tau)$ slides past $h(\tau)$ at $\tau = 1$.

$$
\begin{aligned}
y(t) &= h(t) * x(t) \\
&= 3[4-(t-1)] + 2[4-(t+1)] \\
&= 21 - 5t
\end{aligned}
$$

7. **Time interval $3 < t < 5$ (Fig. 3.61i).**
   In this interval the right edge of $x(t-\tau)$ slides past the right edge of $h(\tau)$.

$$y(t) = h(t) * x(t) = 3[4 - (t-1)] = 15 - 3t$$

8. **Time interval $t > 5$ (Fig. 3.61j).**
   When $t > 5$, $x(t-\tau)$ does not overlap with $h(\tau)$ and hence $y(t) = 0$.

$$\boxed{\begin{aligned} y(t) &= 6(t+1) \\ &= 2(3-t) \\ &= (7t-3) \\ &= (21-5t) \\ &= (15-3t) \\ &= 0 \end{aligned}} \qquad \begin{aligned} -1 < t &< 0 \\ 0 < t &< 1 \\ 1 < t &< 2 \\ 2 < t &< 3 \\ 3 < t &< 5 \\ t &> 5 \end{aligned}$$

The output response is given in Fig. 3.61k.

t	-1	0	1	2	3	4	≥ 5
$y(t)$	0	6	4	11	6	3	0

■ Example 3.89

$$x[n] = 3^n u[-n] \quad \text{and} \quad h[n] = u[n].$$

Find

$$y[n] = x[n] * h[n]$$

Solution:

1. $x[k]$ which is obtained from $x[n]$ is shown in Figs. 3.62a. Similarly, $h[k]$ is sketched in Fig. 3.62b. $h(n-k)$ for $n \leq 0$ and for $n > 0$ are shown in Fig. 3.62c and d, respectively.
2. When $h[n-k]$ is moved toward the right it overlaps with $x[n]$ and the overlapping interval is $-\infty < k \leq n$. The output response $y[n]$ is obtained from the following convolution sum.

$$\begin{aligned} y[n] &= \sum_{k=-\infty}^{n} (3)^k \qquad\qquad n \leq 0 \\ &= \sum_{k=-n}^{\infty} \left(\frac{1}{3}\right)^k \end{aligned}$$

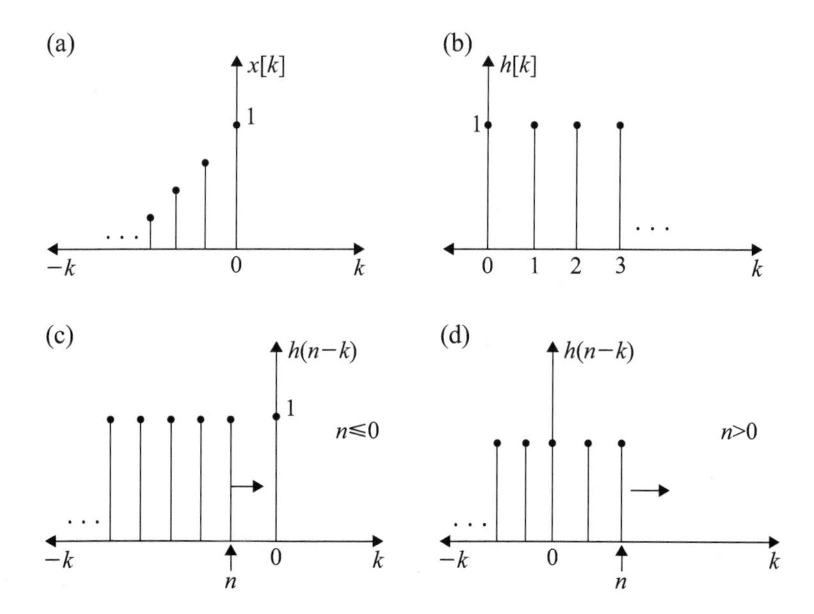

Fig. 3.62 $x[k]$ and $h[k]$ of Example 3.89

Using the following finite summation formula,

$$\sum_{k=n}^{\infty} a^k = \frac{a^n}{(1-a)} \qquad 0 < a < 1$$

we get

$$\begin{aligned} y[n] &= \frac{\left(\frac{1}{3}\right)^{-n}}{1-\frac{1}{3}} \\ &= \frac{1}{2}(3)^{1+n} \qquad n \leq 0 \end{aligned}$$

For the interval $n > 0$,

$$\begin{aligned} y[n] &= \sum_{k=-\infty}^{0} (3)^k = \sum_{k=0}^{\infty} \left(\frac{1}{3}\right)^k = \frac{1}{1-\frac{1}{3}} \\ &= \frac{3}{2} \qquad n > 0 \end{aligned}$$

$$
\boxed{\begin{aligned} y[n] &= \frac{1}{2}(3)^{1+n} \\ &= \frac{3}{2} \end{aligned}} \quad \begin{aligned} &n \leq 0 \\ &n > 0 \end{aligned}
$$

Summary

1. If $h(t)$ is the impulse response of an LTIC system and if $x(t)$ is the input, then the output of the system $y(t)$ is obtained using the convolution integral. The output is given by

$$
y(t) = \int_{-\infty}^{\infty} h(t - \tau)x(\tau)d\tau
$$

2. The convolution integral has commutative, distributive, associative, shift and width properties.
3. Analytical as well as graphical methods are available for solving the convolution integral.
4. If the area under the impulse response curve is finite, then the LTIC system is said to be BIBO stable.
5. From the impulse response of an LTIC system by integrating it, the step response is obtained.
6. For a causal system the impulse response $h(t) = 0$ for $t < 0$.
7. If $h(n)$ is the impulse response of an LTID system and if $x(n)$ is the input, then the output of the system $y(n)$ is obtained using convolution sum. The output is given by

$$
y(n) = \sum_{k=-\infty}^{\infty} x(k)h(n - k)
$$

8. The convolution sum has commutative, distributive, associative, shift and width properties.
9. Analytical as well as graphical methods are available for solving the convolution sum.
10. A discrete time system is said to be causal if the impulse response $h(n) = 0$ for $n < 0$.
11. If the impulse response of a discrete time system is absolutely summable, then the system is said to be BIBO stable.
12. Step response $s(n)$ is obtained from impulse response $h(n)$ using the mathematical expression

$$
s(n) = \sum_{k=0}^{n} h(k)
$$

Exercises

I. Short Answer Type Questions

1. **What is the convolution integral or superposition integral?**
   If $h(t)$ is the impulse response of an LTIC system and if it is excited by the signal $x(t)$, then the output $y(t)$ is expressed as

$$y(t) = \int_{-\infty}^{+\infty} x(\tau)h(t-\tau)d\tau$$

   The above equation is referred to as the convolution or superposition integral. Symbolically, it is written as

$$y(t) = x(t) * h(t)$$

2. **Outline the procedure to evaluate the convolution integral?**
   (a) Graph $x(\tau)$ by substituting $t = \tau$ in $x(t)$ and keep it fixed.
   (b) Obtain $h(\tau)$. By folding get $h(-\tau)$. Graph $h(t-\tau)$.
   (c) Keeping $x(\tau)$ fixed move $h(t-\tau)$ so that it overlaps with $x(\tau)$.
   (d) Multiply $x(\tau)$ and $h(t-\tau)$ and integrate for $-\infty < \tau < \infty$ to obtain $y(t)$.

3. **What are the properties of convolution integral?**
   The properties of convolution integral are

   (a) The commutative property;
   (b) The distributive property;
   (c) The associative property;
   (d) The shift property; and
   (e) The width property.

4. **What is convolution sum?**
   If $h(n)$ is the impulse response of a linear time invariant discrete time system and if $x(n)$ is input, then the output $y(n)$ is obtained from the following equation

$$y(n) = \sum_{k=-\infty}^{\infty} x(k)h(n-k)$$

   The above equation is called convolution sum. Symbolically it is represented as

$$y(n) = x(n) * h(n)$$

5. **How impulse response is related to stability of the system?**
   For an LTIC system if the area under the impulse response curve is finite,

the system is BIBO stable. In other words, $\int_{-\infty}^{\infty} |h(\tau)| d\tau$ should be absolutely integrable. For a LTID time system to be stable the impulse response $\sum_{k=-\infty}^{\infty} h(k)$ should be absolutely summable.

6. **How impulse response is related to causality of the system?**
For an LTIC system to be causal, the impulse response $h(t) = 0$ for $t < 0$. For an LTID system to be causal, the impulse response $h(n) = 0$ for $n < 0$.
7. **How step response is obtained from impulse response of an LTIC and LTID systems?**
The step response of an LTIC system is obtained by integrating its impulse response. The step response of an LTID system is obtained by summing its impulse response (Fig. 3.63).
8. **Evaluate $x(n) * \delta(n)$ and $x(n) * \delta(n - n_0)$?**

$$x(n) * \delta(n) = x(n)$$
$$x(n) * \delta(n - n_0) = x(n - n_0)$$

9. **Evaluate $x(t) * \delta(t)$ and $x(t) * \delta(t - t_0)$?**

$$x(t) * \delta(t) = x(t)$$
$$x(t) * \delta(t - t_0) = x(t - t_0)$$

10. **Evaluate $x(t - t_1) * h(t - t_2)$?**

$$x(t - t_1) * h(t - t_2) = y(t - t_1 - t_2)$$

11. **The response of LTIC system to step input is $y(t) = 1 - e^{-3t}$. Find its impulse response?**

$$h(t) = \frac{dy(t)}{dt} = 3e^{-3t}$$

12. **If $x(n) = \delta(n) - 2\delta(n-1) + 3\delta(n-2) - \delta(n+1)$. Express $x(n-1)$ in terms of sequences?**

$$x(n-1) = \delta(n-1) - 2\delta(n-2) + 3\delta(n-3) - \delta(n)$$

13. **Sketch $x(t) = u(t+3) - u(t-3)$?**
14. **$h(n) = \left(\frac{1}{2}\right)^n u(n)$. Find A so that $h(n) - Ah(n-1) = \delta(n)$?**

$$h(n) - h(n-1) = \left(\frac{1}{2}\right)^n u(n) - A\left(\frac{1}{2}\right)^{n-1} u(n-1)$$

If A = 1/2

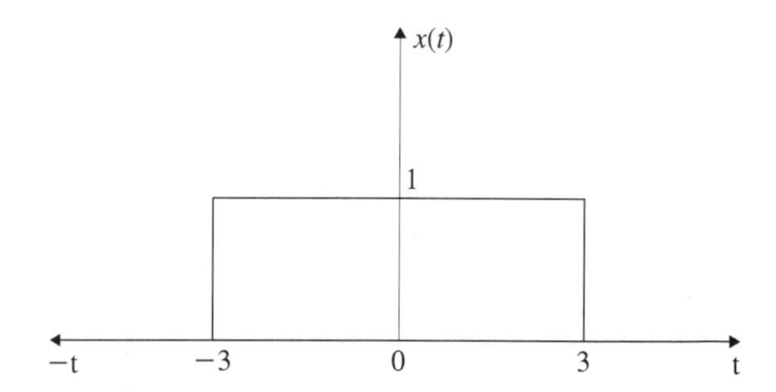

Fig. 3.63 $x(t) = u(t+3) - u(t-3)$.

$$h(n) - h(n-1) = \left(\frac{1}{2}\right)^n u(n) - \left(\frac{1}{2}\right)^n u(n-1) = \delta(n)$$

15. **An LTID system has the following impulse response**

$$\boldsymbol{h(n) = \left(\frac{1}{2}\right)^n u(n+1)}$$

Is the system causal and BIBO stable?
The system is non-zero for $n = -1$. Hence, it is non-causal.

$$\begin{aligned}\sum_{-\infty}^{\infty} h(n) &= \sum_{-1}^{\infty} \left(\frac{1}{2}\right)^n \\ &= \left(\frac{1}{2}\right)^{-1} + \sum_{0}^{\infty} \left(\frac{1}{2}\right)^n \\ &= 2 + \frac{1}{1 - 1/2} = 4 < \infty\end{aligned}$$

The system is BIBO stable.

II. Long Answer Type Questions

1. $\boldsymbol{x(t) = e^{5t}u(-t)}$ **and** $\boldsymbol{h(t) = u(t-10)}$. **Find** $\boldsymbol{y(t) = x(t) * h(t)}$?

$$\begin{aligned}y(t) &= \frac{1}{5}e^{5(t-10)} \qquad t \le 10 \\ &= \frac{1}{5} \qquad t \ge 10\end{aligned}$$

2. $\boldsymbol{x(t) = u(t-2) - u(t-6)}$ **and** $\boldsymbol{h(t) = e^{-5t}u(t)}$. **Find** $\boldsymbol{y(t) = x(t) * h(t)}$?

$$y(t) = \frac{1}{5}[1 - e^{-5(t-2)}] \quad 2 \le t \le 6$$
$$= \frac{1}{5}e^{-5t}[e^{30} - e^{10}] \quad 6 < t \le \infty$$

3. **$x(t) = u(t) - u(t-1)$ and $h(t) = t;\ 0 < t \le 2$. Find $y(t) = x(t) * h(t)$?**

$$y(t) = 0 \quad t < 0$$
$$= \frac{1}{2}t^2 \quad 0 < t < 1$$
$$= \left(t - \frac{1}{2}\right) \quad 1 < t < 2$$
$$= -\frac{1}{2}t^2 + t + \frac{3}{2} \quad 2 < t < 3$$
$$= 0 \quad t > 3$$

4. **$x(t) = u(t-2) - u(t-5)$ and $h(t) = e^{-4t}u(t)$. Find (a) $y(t) = x(t) * h(t)$ and (b) $y(t) = \frac{dx(t)}{dt} * h(t)$?**

(a)
$$y(t) = \frac{1}{4}[1 - e^{-4(t-2)}] \quad 2 \le t \le 6$$
$$= \frac{1}{4}e^{-4t}[e^{24} - e^{8}] \quad 6 \le t \le \infty$$

(b)
$$y(t) = e^{-4(t-2)}u(t-2) - e^{-4(t-5)}u(t-5)$$

5. **Consider the impulse responde $h(t)$ and excitation signal $x(t)$ shown in Fig.** 3.64**a and b, respectively. Using convolution find $y(t) = x(t) * h(t)$?**

$$y(t) = 0 \quad t < -2$$
$$= 6(t+2) \quad -2 \le t \le -1$$
$$= 2(1-2t) \quad -1 \le t \le 0$$

$$y(t) = 2(1-t) \quad 0 \le t \le 1$$
$$= 0 \quad t > 1$$

6. **$x(t)$ and $h(t)$ signals are shown in Fig.** 3.65**a and b, respectively. Find $y(t) = x(t) * h(t)$?**

$$y(t) = 0 \quad t < 0$$
$$= \frac{1}{2}t^2 \quad 0 \le t \le 1$$

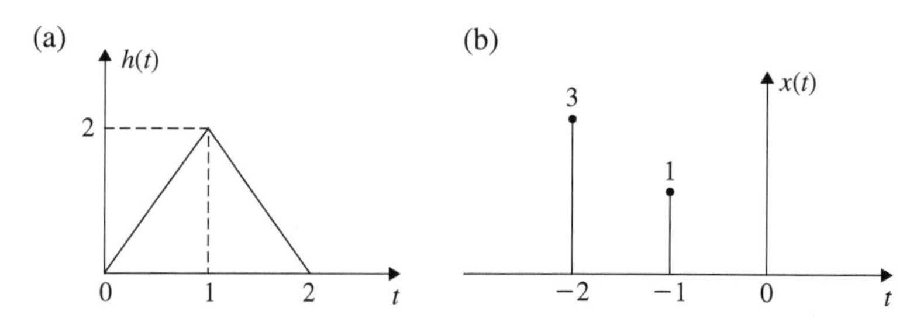

Fig. 3.64 Impulse response and excitation signals of Problem 5

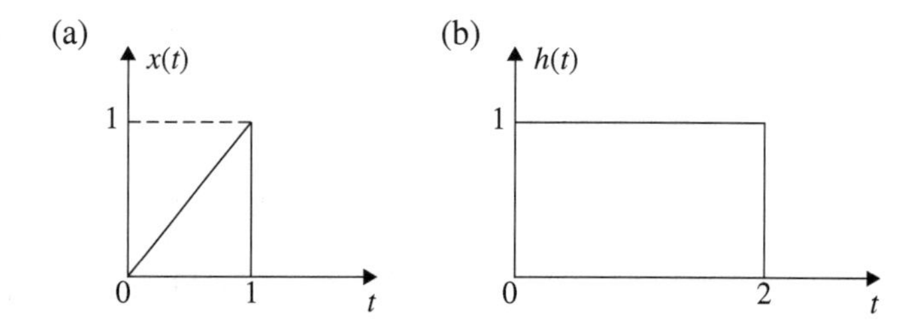

Fig. 3.65 Signals $x(t)$ and $h(t)$ of Problem 6

$$
\begin{aligned}
&= \frac{1}{2} && 1 \leq t \leq 2 \\
&= \left(-\frac{1}{2}t^2 + 2t - \frac{3}{2}\right) && 2 \leq t \leq 3 \\
&= 0 && t > 3
\end{aligned}
$$

7. $\boldsymbol{x(t) = e^{-4t}u(t)}$ **and** $\boldsymbol{h(t) = u(t-3)}$**. Find** $\boldsymbol{y(t) = x(t) * h(t)}$?

$$y(t) = \frac{1}{4}[1 - e^{-4(t-3)}]u(t-3)$$

8. **Find the step response of the system whose impulse response is given as** $\boldsymbol{h(t) = (e^{-5t} - e^{-3t})u(t)}$?

$$s(t) = \left[-\frac{2}{15} - \frac{1}{5}e^{-5t} + \frac{1}{3}e^{-3t}\right]u(t)$$

9. **Determine whether the following LTIC time systems whose impulse response given below are stable. (a)** $\boldsymbol{h(t) = e^{(-3+j5)t}u(t)}$ **and (b)** $\boldsymbol{h(t) = e^{5t}\sin 3tu(t)}$?

$$
\begin{aligned}
&\text{(a)} \qquad y(t) = \frac{1}{3} < \infty \quad \text{B.I.B.O. stable.} \\
&\text{(b)} \qquad y(t) = \frac{3}{34} < \infty \quad \text{B.I.B.O. stable.}
\end{aligned}
$$

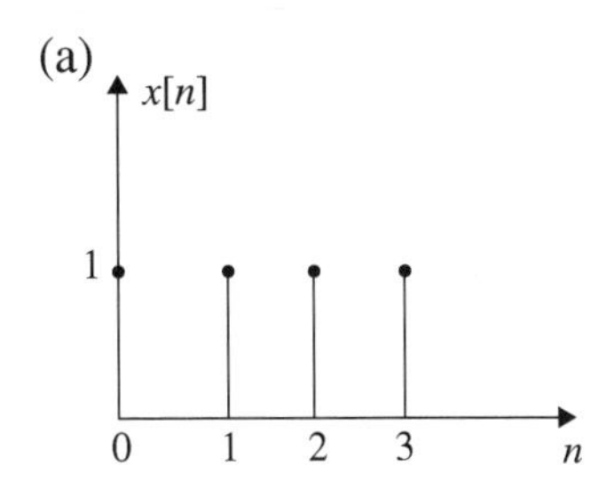

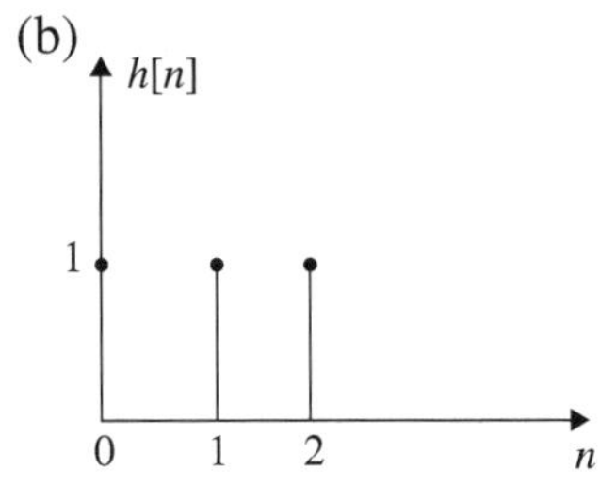

Fig. 3.66 $x[n]$ and $h[n]$ for Problem 13

10. $\boldsymbol{x[n] = u[n-4] - u[n-10]}$ **and** $\boldsymbol{h[n] = u[n-5] - u[n-16]}$. **Find** $\boldsymbol{y[n] = x[n] * h[n]}$?

$$
\begin{aligned}
y[n] &= (n-8) && 9 \le n \le 13 \\
&= 6 && 10 \le n \le 19 \\
&= (25-n) && 20 \le n \le 24 \\
&= 0 && n > 24
\end{aligned}
$$

11. $\boldsymbol{x[n] = 4^n u[-n-2]}$ **and** $\boldsymbol{h[n] = u[n-2]}$. **Find** $\boldsymbol{y[n] = x[n] * h[n]}$?

$$
\begin{aligned}
y[n] &= \frac{1}{3}\left[\frac{1}{4}\right]^{n-1} && n < 0 \\
&= \frac{1}{12} && n > 0
\end{aligned}
$$

12. **Determine whether the following LTID time systems whose impulse response given below are stable. (a)** $\boldsymbol{h[n] = n \sin 2\pi n\, u[n]}$, **(b)** $\boldsymbol{h[n] = 5^n u[-n]}$ **and (c)** $\boldsymbol{h[n] = 2^{-n} u[n-5]}$?

(a) $y[n] = \infty$ B.I.B.O. unstable.

(b) $y[n] = \frac{1}{4} < \infty$ B.I.B.O. stable.

(c) $y[n] = \frac{1}{6} < \infty$ B.I.B.O. stable.

13. $\boldsymbol{x[n]}$**and** $\boldsymbol{y[n]}$ **are shown in Fig.** 3.66a **and b, respectively. Derive expression for** $\boldsymbol{y[n] = x[n] * h[n]}$**and hence obtain the sequence** $\boldsymbol{y[n]}$**by substituting numerical values. Verify the results by multiplication method?**

$$
\begin{aligned}
y[n] &= (n+1) && 0 \le n \le 2 \\
&= (6-n) && 3 \le n \le 5 \\
&= 0 && n > 5
\end{aligned}
$$

$$
y[n] = \{\underset{\uparrow}{1}, 2, 3, 3, 2, 1\}
$$

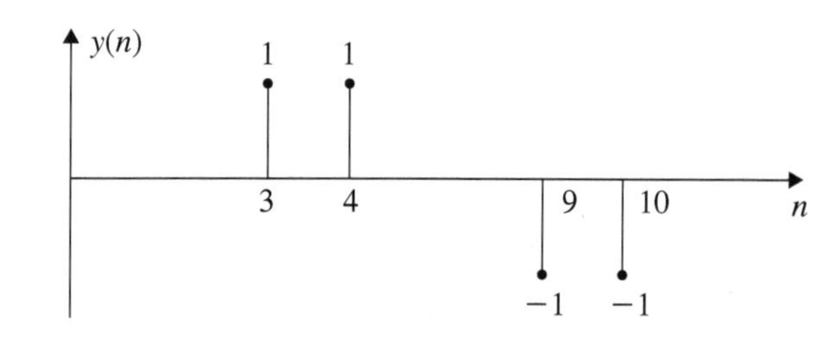

Fig. 3.67 $y[n]$ of Problem 14

The same answer is obtained by multiplication method (Fig. 3.67).

14. $h[n] = u[n] - u[n-6]$ **and** $x[n] = \delta[n-3] - \delta[n-5]$. **Find** $y[n] = x[n] * h[n]$ **and sketch the same?**
15. $x[n] = e^{-2n}u[n]$ **and** $h[n] = 3^{-n}u[n]$. **Find** $y[n] = x[n] * h[n]$?

$$y[n] = \frac{[3e^2 - (3e^2)^{-n}]u[n]}{[3e^2 - 1]}$$

16. $x[n] = (0.5)^n\, u[n]$ **and** $h[n] = (0.9)^n u[n]$. **Find** $y[n] = x[n] * h[n]$?

$$y[n] = 2.5[0.9^{n+1} - 0.5^{n+1}]u[n]$$

Chapter 4
Fourier Series Analysis of Continuous Time Signals

Learning Objectives

- ♦ To represent the periodic continuous time signal by trigonometric Fourier series.
- ♦ To represent the CT signal by polar Fourier series.
- ♦ To determine the exponential Fourier series and Fourier spectra.
- ♦ To establish the properties of Fourier series.
- ♦ To establish Parseval's theorem and Dirichlet conditions.

4.1 Introduction

Sinusoidal input signals are often used to study the response of the system which gives useful information. If a linear time invariant system is excited by a complex sinusoid, the output response is also a complex sinusoid of the same frequency as the input. However, the amplitude of such a sinusoid is different from the input amplitude and also has a phase shift. If the system is excited by the signal which is a weighted superposition of the complex sinusoids, the system output is also a weighted superposition of the system response to each complex sinusoid. Thus, any arbitrary excitation signal $x(t)$ can be expressed as a linear combination of complex sinusoids. The output is obtained by summing up the responses to the individual complex sinusoids using superposition. However, expressing any arbitrary real function as a linear combination of complex sinusoids is a matter of concern. **Baron Jean Baptiste Joseph Fourier (1768–1830)**, a French mathematician represented an arbitrary signal $x(t)$ in the form of a linear combination of complex sinusoids and is called as **Fourier Series**. In a Fourier series, representation of a periodical signal, the higher frequency sines and cosines have frequencies that are integer multiples of the fundamental frequency. These multiples are called **harmonic numbers**. The study of signals using sinusoids has widespread applications in every branch of science

S. Palani, *Signals and Systems*,
https://doi.org/10.1007/978-3-030-75742-7_4

and engineering. This great mathematical poem, which finds wide applications in modern communication, signal processing, antenna design and several other fields, was not shown much enthusiasm by the scientific world during its inception. Fourier could not get the results published for the lack of mathematical rigor. The vehement opposition came from his fellow countrymen and great mathematical wizards Lagrange and Laplace. However, fifteen years later, after several tireless attempts, Fourier successfully published the results in the form of text which is a classic now.

Fourier, born on 21-03-1768, in France, was the son of a tailor. Being orphaned at the age of eight, Fourier was educated in a local military college where he showed his brilliance in mathematics. When the French revolution broke out, many intellectuals decided to leave France to save themselves from the growing barbarism. Fourier escaped the guillotine twice. Napoleon Bonaparte, a soldier scientist captured power in France, after the historical French revolution and stopped prosecution of intellectuals. The French ruler, who himself was a great scientist, appointed Fourier chair of mathematics academy in which he served with distinction when he was just 26 years of age. He was honored as the Baron of the empire by Napoleon in 1809. When Napoleon was exiled by King Louis XVIII, Fourier was identified as a Bonapartist and was treated with all disgrace. Napoleon came back to power within a year of his exile from Elba. However, he was defeated by the English captain Nelson in the battle of Waterloo and the great warrior scientist died in 1821 at St. Helena Island, where he was in exile for the second time. Fourier should have again become an orphan, but with the help of his former student who was now a prefect of Paris. He was appointed as the statistical bureau of the seine and subsequently, in 1827, elected to the powerful position of secretary of the Paris Academy of Science.

While carrying out investigations on the propagation of heat in solid bodies, Fourier was able to establish the Fourier series and Fourier integral. In 1807, when he was 40 years of age, Fourier published his results. He claimed that any arbitrary function can always be expressed as a sum of sinusoids. For the lack of rigor and generality, the judges, including the great French mathematicians Lagrange, Laplace, Legendre, Monge and Lacroix, criticized Fourier's work, but appreciated the novelty and importance of the work. Fourier could not defend the criticisms since the necessary tools were not available with him at that time. However, in the year 1829, Dirichlet proved most of the claims of Fourier by putting a few restrictions (Dirichlet conditions).

Fifteen years after the paper was rejected mainly due to the vehement opposition given by Lagrange and to some extent by Laplace, Fourier published his results in expanded form as a text which has now become a classic in the area of mathematics, science and engineering applications. The great mathematician who laid the foundation of signal representation and analysis died on 16-05-1830, when he was 63 years old.

4.2 Periodic Signal Representation by Fourier Series

A continuous time signal $x(t)$ is said to be periodic if there is a positive non-zero value of T for which

$$x(t+T) = x(t) \qquad \text{for all } t \tag{4.1}$$

The fundamental period T_0 of $x(t)$ is the smallest positive value of T for which Eq. (4.1) is satisfied. $\frac{1}{T_0}$ is called fundamental frequency f_0 and $\omega_0 = \frac{2\pi}{T_0}$ is called fundamental radian frequency. The real sinusoidal signal

$$x(t) = \cos(\omega_0 t + \phi) \tag{4.2}$$

and the complex exponential signal

$$x(t) = e^{j\omega_0 t} \tag{4.3}$$

have been proved in Chap. 1 as periodic signals as Eq. (4.1) is applicable in the above cases. **The prerequisite for the representation of any arbitrary continuous signal $x(t)$ in Fourier series is that it should be periodic. Non-periodic signals cannot be represented by Fourier series but can be represented by Fourier transform which is discussed later.**

4.3 Different Forms of Fourier Series Representation

Any arbitrary real or complex $x(t)$ signal which is periodic with fundamental period T_0 can be expressed as a sum of a sinusoid of period T_0 and its harmonics. They are represented in the following forms of Fourier series:

1. Trigonometric Fourier series.
2. Complex exponential Fourier series.
3. Polar or Harmonic form Fourier series.

The above Fourier series representations are described below with illustrated examples.

4.3.1 Trigonometric Fourier Series

Consider any arbitrary continuous time signal $x(t)$. This arbitrary signal can be split up as sines and cosines of fundamental frequency ω_0 and all of its harmonics are expressed as given below.

$$\boxed{x(t) = a_0 + \sum_{n=1}^{\infty} a_n \cos n\omega_0 t + b_n \sin n\omega_0 t} \tag{4.4}$$

Equation (4.4) is the Fourier series representation of an arbitrary signal $x(t)$ in trigonometric form.

In Eq. (4.4), a_0 corresponds to the zeroth harmonic or DC. The expression for the constant term a_0 and the amplitudes of the harmonic can be derived as

$$\boxed{a_0 = \frac{1}{T_0} \int_{T_0} x(t)\, dt} \tag{4.5}$$

$$\boxed{a_n = \frac{2}{T_0} \int_{T_0} x(t) \cos n\omega_0 t\, dt} \tag{4.6}$$

$$\boxed{b_n = \frac{2}{T_0} \int_{T_0} x(t) \sin n\omega_0 t\, dt} \tag{4.7}$$

In Eqs. (4.5), (4.6) and (4.7)

$$T_0 = \frac{1}{f_0} = \frac{2\pi}{\omega_0}$$

$T_0 =$ Fundamental period of $x(t)$ in seconds;
$f_0 =$ Fundamental frequency in Hz.;
$\omega_0 =$ Radian frequency in rad/second.

For the detailed derivation of the above equations, one may refer to standard textbooks. **Equation (4.4) is valid iff $x(t)$ is periodic.**

To Prove the periodicity of x(t)

The periodicity $x(t)$ is proved if $x(t) = x(t + T_0)$. Substituting $(t + T_0)$ in place of t in Eq. (4.4), the following equation is obtained.

$$\begin{aligned} x(t + T_0) &= a_0 + \sum_{n=1}^{\infty} a_n \cos n\omega_0 (t + T_0) + \sum_{n=1}^{\infty} b_n \sin n\omega_0 (t + T_0) \\ &= a_0 + \sum_{n=1}^{\infty} a_n \cos(n\omega_0 t + n\omega_0 T_0) + \sum_{n=1}^{\infty} b_n \sin(n\omega_0 t + n\omega_0 T_0) \end{aligned}$$

$$x(t+T_0) = a_0 + \sum_{n=1}^{\infty} a_n \cos(n\omega_0 t + 2\pi n) + \sum_{n=1}^{\infty} b_n \sin(n\omega_0 t + 2\pi n)$$
$$= a_0 + \sum_{n=1}^{\infty} a_n \cos(n\omega_0 t) + \sum_{n=1}^{\infty} b_n \sin n\omega_0 t$$

$$\boxed{x(t+T_0) = x(t)} \quad (4.8)$$

Thus, it is established, if $x(t)$ is periodic, at $t = T_0$ every sinusoid starts and repeats the same over the next T_0 seconds and so on. The followings points are to be noted while the coefficients a_0, a_n and b_n are determined. It can be proved that

1. If the periodical signal $x(t)$ is symmetrical with respect to the time axis, then the coefficient $a_0 = 0$.
2. If the periodical signal $x(t)$ represents an even function, only cosine terms in FS exists and therefore $b_n = 0$.
3. If the periodical signal $x(t)$ represents an odd function, only sine terms in FS exists and therefore $a_n = 0$.

4.3.2 *Complex Exponential Fourier Series*

By using Euler's identity, the complex sinusoids can always be expressed in terms of exponentials. Thus, the trigonometric Fourier series of Equation (4.4) can be represented as

$$\boxed{x(t) = \sum_{n=-\infty}^{\infty} D_n e^{j\omega_0 nt}} \quad (4.9)$$

where

$$\boxed{D_n = \frac{1}{T_0} \int_{T_0} x(t) e^{-j\omega_0 nt} dt} \quad (4.10)$$

Equation (4.9) represents exponential Fourier series and D_n is the coefficient of the exponential Fourier series. For detailed derivation of Equation (4.10), one may refer to standard textbooks. It is to be noticed here that Eq. (4.9) is in a compact form and it is much more convenient to handle compared to trigonometric Fourier series. Further, determination of the coefficients D_n using Eq. (4.10) is much easier compared to a_0, a_n and b_n in Eq. (4.4). For these reasons many authors prefer exponential Fourier series representation of signals. The coefficients D_n are related to trigonometric Fourier series coefficients a_n and b_n as

$$\boxed{\begin{aligned} D_0 &= a_0 \\ D_n &= \frac{1}{2}(a_n - jb_n) \\ D_n^* &= \text{conjugate of } D_n \\ &= \frac{1}{2}(a_n + jb_n) \end{aligned}} \tag{4.11}$$

4.3.3 *Polar or Harmonic Form Fourier Series*

The results derived in Sects. 4.31 and 4.32 are applicable if $x(t)$ is real or complex. When $x(t)$ is real, the coefficients of trigonometric Fourier series a_n and b_n are real. In such cases, Eq. (4.4) can be expressed in a compact form as

$$\boxed{x(t) = C_0 + \sum_{n=1}^{\infty} C_n \cos(n\omega_0 t - \theta_n)} \tag{4.12}$$

where C_n and θ_n are related to a_n and b_n as

$$\boxed{\begin{aligned} C_0 &= a_0 \\ C_n &= \sqrt{a_n^2 + b_n^2} \\ \theta_n &= \tan^{-1}\left(\frac{b_n}{a_n}\right) \end{aligned}} \tag{4.13}$$

Equation (4.12) is also called as compact form Fourier series or cosine form Fourier series.

The coefficients of compact form Fourier series and exponential form Fourier series are related as

$$\boxed{\begin{aligned} D_0 &= C_0 \\ |D_n| &= \left|D_n^*\right| = \frac{1}{2}C_n \\ \angle D_n &= \theta_n; \quad \angle D_n^* = -\theta_n \end{aligned}} \tag{4.14}$$

For detailed derivations of Equations (4.13) and (4.14), one may refer to standard textbooks. Table 4.1 gives the different form of Fourier series representation, their coefficients and their equivalence.

Table 4.1 Different forms of fourier series representation

FS Form	Coefficients	Equivalence
1. Trigonometric	$a_0 = \frac{1}{T_0}\int_{T_0} x(t)dt$	$a_0 = C_0 = D_0$
$x(t) = a_0 + \sum_{n=1}^{\infty} a_n \cos n\omega_0 t$	$a_n = \frac{2}{T_0}\int_{T_0} x(t)\cos n\omega_0 t\, dt$	$a_n - jb_n = C_n e^{j\theta n} = 2D_n$
$+b_n \sin n\omega_0 t$	$b_n = \frac{2}{T_0}\int_{T_0} x(t)\sin n\omega_0 t\, dt$	$a_n + jb_n = C_n e^{-j\theta n} = 2D_n^*$
2. Exponential		
$x(t) = \sum_{n=-\infty}^{\infty} D_n e^{jn\omega_0 t}$	$D_n = \frac{1}{T_0}\int_{T_0} x(t)e^{-jn\omega_0 t}dt$	$C_n = 2\lvert D_n\rvert \quad n \geq 1$
3. Polar or compact cosine	$C_0 = a_0$	$\theta_n = \angle D_n$
$x(t) = C_0 + \sum_{n=1}^{\infty} C_n \cos(n^{\omega_0 t} - \theta_n)$	$C_n = \sqrt{a_n^2 + b_n^2}$	
	$\theta_n = \tan^{-1}\left(\frac{b_n}{a_n}\right)$	

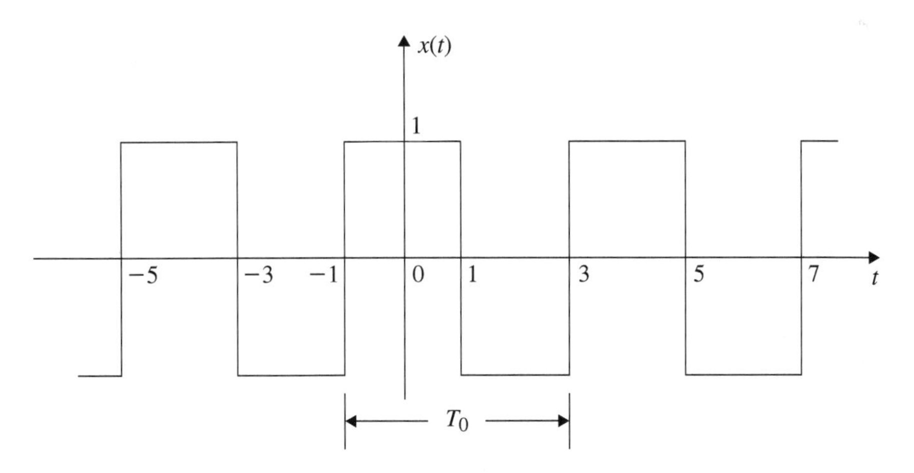

Fig. 4.1 A rectangular wave of Example 4.1

The following examples illustrate the method of determining the Fourier series (FS) in the above three forms.

■ Example 4.1

Find the trigonometric Fourier series for the periodic signal shown in Fig. 4.1.

Solution:

1. From Fig. 4.1, it is evident that the waveform is symmetrical with respect to the time axis t. Hence, $a_0 = 0$.
2. By folding $x(t)$ across the vertical axis, it is observed that $x(t) = x(-t)$ which shows that the function of the signal is even. Hence, $b_n = 0$.

3. From Fig. 4.1, it is easily obtained that the fundamental period $T_0 = 4$ seconds and the fundamental radian frequency $\omega_0 = \frac{2\pi}{T_0} = \frac{\pi}{2}$ radians per second. From Eq. (4.4), the trigonometric Fourier series is written as

$$x(t) = a_0 + \sum_{n=1}^{\infty} [a_n \cos n\omega_0 t + b_n \sin n\omega_0 t]$$

But

$$\begin{aligned} x(t) &= 1 \qquad \text{for} \quad -1 \le t \le 1 \\ &= -1 \qquad \text{for} \quad 1 \le t \le 3 \end{aligned}$$

Substituting $a_0 = 0$ and $b_n = 0$, and $\omega_0 = \frac{\pi}{2}$

$$\begin{aligned} x(t) &= \sum_{n=1}^{\infty} a_n \cos \frac{n\pi}{2} t \\ a_n &= \frac{2}{T_0} \int_{-1}^{3} x(t) \cos\left(\frac{n\pi}{2} t\right) dt \\ &= \frac{1}{2} \left[\int_{-1}^{1} \cos \frac{n\pi}{2} t \, dt + \int_{1}^{3} (-1) \cos \frac{n\pi}{2} t \, dt \right] \\ &= \frac{1}{2} \left[\left\{ \frac{2}{n\pi} \sin \frac{n\pi}{2} t \right\}_{-1}^{1} - \left\{ \frac{2}{n\pi} \sin \frac{n\pi}{2} t \right\}_{1}^{3} \right] \\ &= \frac{1}{n\pi} \left[\sin \frac{n\pi}{2} + \sin \frac{n\pi}{2} + \sin \frac{n\pi}{2} + \sin \frac{n\pi}{2} \right] \\ &= \frac{4}{n\pi} \sin \frac{n\pi}{2} \\ &= 0 \qquad \text{for } n = \text{even} \\ &= \frac{4}{n\pi} \qquad \text{for } n = 1, 5, 9, 13, \ldots \\ &= -\frac{4}{n\pi} \qquad \text{for } n = 3, 7, 11, 15, \ldots \\ x(t) &= \sum_{n=1}^{\infty} a_n \cos \frac{n\pi}{2} t \end{aligned}$$

$$\boxed{x(t) = \frac{4}{\pi} \left[\cos \frac{\pi}{2} t - \frac{1}{3} \cos \frac{3\pi}{2} t + \frac{1}{5} \cos \frac{5\pi}{2} t - \frac{1}{7} \cos \frac{7\pi}{2} t \right]}$$

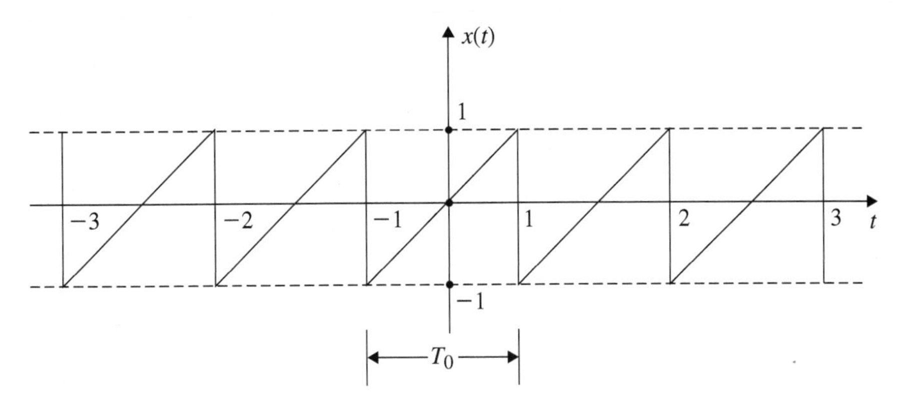

Fig. 4.2 Saw tooth waveform

■ Example 4.2

For the periodic signal shown in Fig. 4.2, determine the trigonometric Fourier series.

Solution:

1. From Fig. 4.2, $T_0 = 2$ seconds and $\omega_0 = \frac{2\pi}{T_0} = \pi$. The signal is symmetrical with respect to time axis and hence $a_0 = 0$. Also, from Fig. 4.2, it is evident that $x(t) = -x(-t)$, and therefore, the signal is an odd signal and $a_n = 0$. The Fourier series for such a signal is, therefore

$$x(t) = \sum_{n=1}^{\infty} b_n \sin n\omega_0 t.$$

2. The coefficient b_n is determined as follows:

$$\begin{aligned} x(t) &= t \qquad -1 \leq t \leq 1 \\ b_n &= \frac{2}{T_0}\int_{-1}^{1} t \sin n\omega_0 t \, dt \\ &= \int_{-1}^{1} t \sin n\pi t \, dt \end{aligned}$$

The above integral is solved using the infinite integral

$$\int u dv = uv - \int v du$$

Let $u = t$, $du = dt$

$$dv = \int \sin n\pi t \, dt; \qquad v = -\frac{1}{n\pi} \cos n\pi t$$

$$b_n = \left[-\frac{t}{n\pi} \cos n\pi t \right]_{-1}^{1} + \frac{1}{n^2\pi^2} \left[\sin n\pi t \right]_{-1}^{1}$$

$$= -\frac{2}{n\pi} \cos n\pi + \frac{1}{n^2\pi^2} [\sin n\pi + \sin n\pi]$$

since $\sin n\pi = 0$,

$$\boxed{b_n = -\frac{2}{n\pi} \cos n\pi}$$

$$x(t) = \sum_{n=1}^{\infty} b_n \sin n\pi t$$

$$\boxed{x(t) = \frac{2}{\pi} \left[\sin \pi t - \frac{1}{2} \sin 2\pi t + \frac{1}{3} \sin 3\pi t + \ldots \right]}$$

■ Example 4.3

Find the trigonometric Fourier series for the signal shown in Fig. 4.3.

(*Anna University, December 2006*)

Solution:

1. From Fig. 4.3, $T_0 = 2\pi$ and $\omega_0 = \frac{2\pi}{T_0} = 1$. The signal is neither odd nor even. Further, it is not symmetrical with respect to the time axis. So the coefficients, a_0, a_n and b_n are to be evaluated.

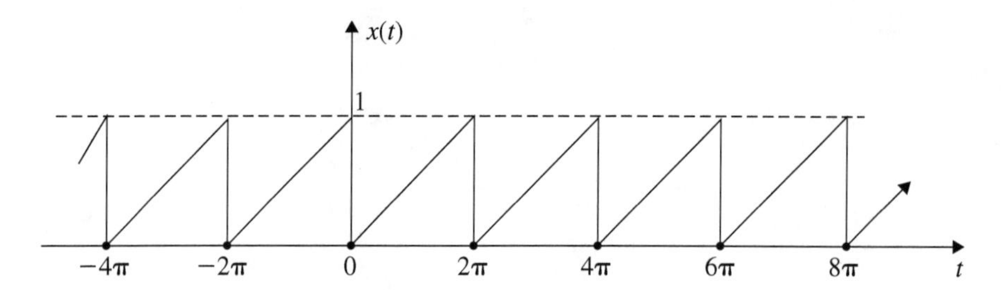

Fig. 4.3 Saw tooth signal of Example 4.3

2.

$$x(t) = \frac{t}{2\pi} \qquad 0 \leq t \leq 2\pi \quad \left(\text{for a ramp signal the slope is } \frac{1}{2\pi}\right)$$

$$a_0 = \frac{1}{T_0}\int_0^{2\pi} \frac{t}{2\pi} dt = \frac{1}{4\pi^2}\left[\frac{t^2}{2}\right]_0^{2\pi}$$

$$\boxed{a_0 = \frac{1}{2}}$$

$$\begin{aligned} a_n &= \frac{2}{T_0}\int_0^{2\pi} \frac{t}{2\pi}\cos nt\, dt \\ &= \frac{1}{2\pi^2}\int_0^{2\pi} t\cos nt\, dt \end{aligned}$$

Let $u = t$; $du = dt$

$$\begin{aligned} dv &= \int \cos nt\, dt; \quad v = \frac{\sin nt}{n} \\ a_n &= uv - \int v du \\ &= \frac{1}{2\pi^2}\left[\frac{t\sin nt}{n} + \frac{\cos nt}{n^2}\right]_0^{2\pi} \\ &= \frac{1}{2\pi^2}[0 + 0 + 1 - 1] \end{aligned}$$

$$\boxed{a_n = 0}$$

(This is due to half wave symmetry).

$$\begin{aligned} b_n &= \frac{2}{T_0}\int_0^{2\pi} \frac{t}{2\pi}\sin nt\, dt \\ &= \frac{1}{2\pi^2}\left[-\frac{t\cos nt}{n} + \frac{\sin nt}{n^2}\right]_0^{2\pi} \qquad \text{[using } u\text{-}v \text{ method]} \\ &= \frac{1}{2\pi^2}\left[-\frac{2\pi}{n}\cos 2\pi n\right] \end{aligned}$$

$$\boxed{b_n = -\frac{1}{n\pi}}$$

$$\boxed{x(t) = \frac{1}{2} - \sum_{n=1}^{\infty} \frac{1}{n\pi} \sin nt}$$

■ Example 4.4

Determine the trigonometric Fourier series representation of a full wave rectified signal.

(*Anna University, April 2005*)

Solution:

1. The full wave rectified signal is shown in Fig. 4.4. Here $T_0 = \pi$ and $\omega_0 = \frac{2\pi}{T_0} = 2$.
2. The signal is not symmetrical with respect to time axis. Therefore, a_0 is calculated as follows:

$$a_0 = \frac{1}{T_0} \int_0^{\pi} x(t) dt$$

where

$$x(t) = \sin t \qquad 0 \leq t \leq \pi$$

$$a_0 = \frac{1}{\pi} \int_0^{\pi} \sin t \, dt$$
$$= \frac{1}{\pi} \Big[-\cos t \Big]_0^{\pi} = \frac{2}{\pi}$$

$$\boxed{a_0 = \frac{2}{\pi}}$$

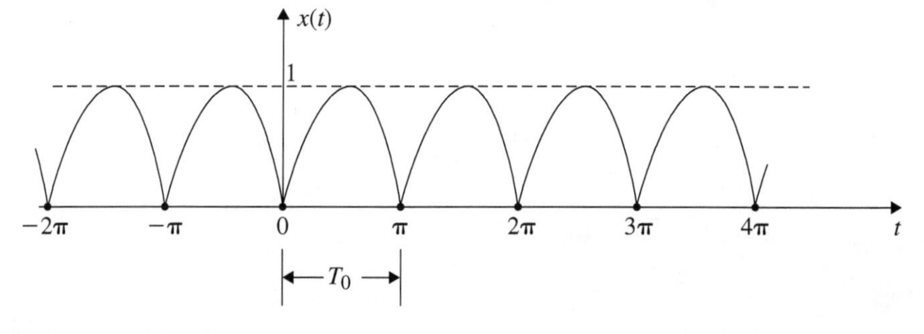

Fig. 4.4 A full wave rectifier

3.

$$x(t) = x(-t)$$

The given signal represents an even function, and therefore

$$\boxed{b_n = 0}$$

4.

$$a_n = \frac{2}{\pi}\int_0^{\pi} \sin t \cos n\omega_0 t \, dt$$
$$= \frac{2}{\pi}\int_0^{\pi} \sin t \cos 2nt \, dt$$

Using the property,

$$\sin A \cos B = \frac{1}{2}\left[\sin(A+B) + \sin(A-B)\right]$$

the above integral is written as

$$a_n = \frac{1}{\pi}\int_0^{\pi} \sin(2n+1)t \, dt + \frac{1}{\pi}\int_0^{\pi} \sin(1-2n)t \, dt$$
$$= \frac{1}{\pi}\left[-\frac{\cos(2n+1)t}{(2n+1)}\right]_0^{\pi} + \frac{1}{\pi}\left[-\frac{\cos(1-2n)t}{(1-2n)}\right]_0^{\pi}$$
$$= \frac{1}{\pi}\left[-\frac{\cos(2n+1)\pi + 1}{(2n+1)}\right] + \frac{1}{\pi}\left[-\frac{\cos(1-2n)\pi + 1}{(1-2n)}\right]$$
$$= \frac{1}{\pi}\left[\frac{1-(-1)^{2n+1}}{(2n+1)} + \frac{1-(-1)^{1-2n}}{(1-2n)}\right]$$
$$= \frac{1}{\pi}\left[\frac{2}{(2n+1)} + \frac{2}{(1-2n)}\right]$$
$$= \frac{2}{\pi}\left[\frac{1-2n+2n+1}{(1-4n^2)}\right]$$

$$\boxed{a_n = \frac{4}{\pi(1-4n^2)}}$$

$$\boxed{x(t) = \frac{2}{\pi} + \frac{4}{\pi}\sum_{n=1}^{\infty}\frac{1}{(1-4n^2)}\cos 2nt}$$

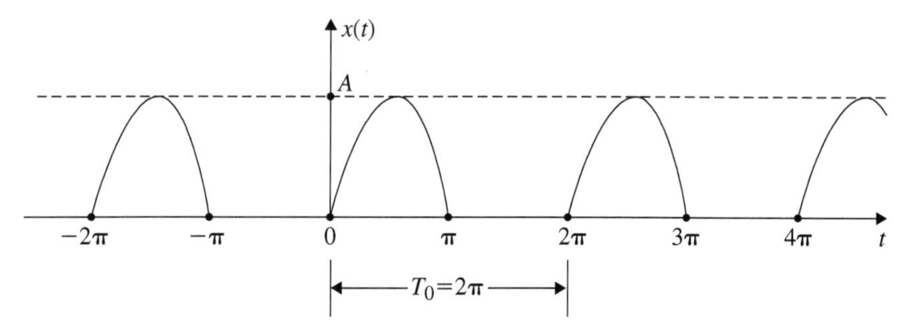

Fig. 4.5 A half wave rectified sine function

■ Example 4.5

Obtain the Fourier series expression of a half wave sine wave.

(Anna University, December 2007)

Solution:

1. $T_0 = 2\pi$ and $\omega_0 = \frac{2\pi}{T_0} = \frac{2\pi}{2\pi} = 1$

$$\begin{aligned} x(t) &= A \sin t \qquad 0 \le t \le \pi \\ &= 0 \qquad \pi \le t \le 2\pi \\ a_0 &= \frac{1}{2\pi} \int_0^{\pi} A \sin t \, dt \\ &= \frac{A}{2\pi} \Big[-\cos t \Big]_0^{\pi} = \frac{A}{\pi} \end{aligned}$$

$$\boxed{a_0 = \frac{A}{\pi}}$$

2.

$$\begin{aligned} a_n &= \frac{2}{2\pi} \int_0^{\pi} A \sin t \cos nt \, dt \\ &= \frac{A}{2\pi} \left[\int_0^{\pi} \sin(1+n)t \, dt + \int_0^{\pi} \sin(1-n)t \, dt \right] \\ &= \frac{A}{2\pi} \left[-\frac{\cos(1+n)t}{(1+n)} - \frac{\cos(1-n)t}{(1-n)} \right]_0^{\pi} \\ &= \frac{A}{2\pi} \left[\frac{1-\cos(1+n)\pi}{(1+n)} + \frac{1-\cos(1-n)\pi}{(1-n)} \right] \end{aligned}$$

$$= \frac{A}{2\pi}\left[\frac{2}{(1+n)} + \frac{2}{(1-n)}\right] = \frac{2A}{\pi(1-n^2)}$$

$$\boxed{a_n = \frac{2A}{\pi(1-n^2)}} \qquad n \neq 1$$

Since for $n = 1$, $a_n = \infty$, a_1 is calculated as follows: For $n = 1$,

$$\begin{aligned}
a_1 &= \frac{1}{2\pi}\int_0^{\pi} A \sin t \cos t\, dt \\
&= \frac{A}{2\pi}\int_0^{\pi} \sin 2t\, dt \\
&= \frac{A}{4\pi}\big[-\cos 2t\big]_0^{\pi} = 0
\end{aligned}$$

$$\boxed{a_1 = 0}$$

3.

$$\begin{aligned}
b_n &= \frac{2}{2\pi}\int_0^{\pi} A \sin t \sin nt\, dt \\
&= \frac{A}{2\pi}\left[\int_0^{\pi} \{\cos(1-n)t - \cos(1+n)t\}\, dt\right] \\
&= \frac{A}{2\pi}\left[\frac{\sin(1-n)t}{(1-n)} - \frac{\sin(1+n)t}{(1+n)}\right]_0^{\pi} \\
&= \frac{A}{2\pi}\left[\frac{\sin(1-n)\pi - \sin 0}{(1-n)} - \frac{\sin(1+n)\pi + \sin 0}{(1+n)}\right]
\end{aligned}$$

$$\boxed{b_n = 0 \quad n \neq 1}$$

For $n = 1$, $b_1 = \infty$ and therefore b_1 is calculated as follows:

$$\begin{aligned}
b_1 &= \frac{2}{2\pi}\int_0^{\pi} A \sin t \sin t\, dt \\
&= \frac{A}{\pi}\int_0^{\pi} A \sin^2 t\, dt \\
&= \frac{A}{2\pi}\left[\int_0^{\pi} (1 - \cos 2t) dt\right] = \frac{A}{2\pi}\left[t - \frac{\sin 2t}{2}\right]_0^{\pi}
\end{aligned}$$

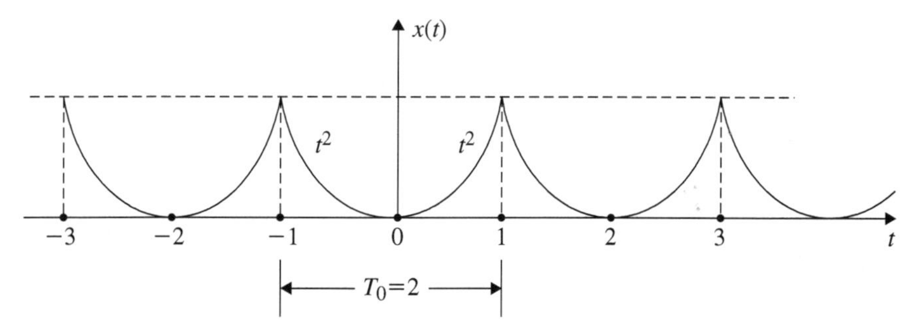

Fig. 4.6 Representation of $x(t) = t^2$

$$\boxed{b_1 = \frac{A}{2}}$$

$$\boxed{x(t) = \frac{A}{\pi} + \frac{A}{2}\sin t + \sum_{n=2}^{\infty} \frac{2A}{\pi(1-n^2)}\cos nt}$$

■ Example 4.6

Determine the Fourier series representation of the signal $x(t) = t^2$ for all values of 't' which exists in the interval $(-1, 1)$.

(Anna University, May, 2007)

Solution:

1. For the given signal $T_0 = 2$ and $\omega_0 = \frac{2\pi}{T_0} = \pi$.

$$a_0 = \frac{1}{2}\int_{-1}^{1} t^2 dt = \frac{1}{2}\left[\frac{t^3}{3}\right]_{-1}^{1} = \frac{1}{3}$$

$$\boxed{a_0 = \frac{1}{3}}$$

2.

$$a_n = \frac{2}{2}\int_{-1}^{1} t^2 \cos n\pi t\, dt$$
$$= \int_{-1}^{1} t^2 \cos n\pi t\, dt$$

Applying $\int udv = uv - \int vdu$ twice for the above equation, we get

$$a_n = \left[t^2 \frac{\sin n\pi t}{n\pi} + \frac{2t}{n^2\pi^2} \cos n\pi t - \frac{2}{n^3\pi^3} \sin n\pi t\right]_{-1}^{1}$$

$$= \left[\frac{\sin n\pi}{n\pi} + \frac{2}{n^2\pi^2} \cos n\pi - \frac{2}{n^3\pi^3} \sin n\pi + \frac{\sin n\pi}{n\pi}\right.$$
$$\left. + \frac{2}{n^2\pi^2} \cos n\pi - \frac{2}{n^3\pi^3} \sin n\pi \right]$$

$$\sin n\pi = 0 \quad \text{for all } n$$

$$a_n = \frac{4}{n^2\pi^2} \cos n\pi$$

$$\boxed{a_n = \frac{4}{n^2\pi^2}(-1)^n}$$

3. From Fig. 4.6, it is evident that $x(t)$ is an even function and therefore $b_n = 0$.
4.

$$x(t) = \frac{1}{3} + \frac{4}{\pi^2} \sum_{n=1}^{\infty} \frac{(-1)^n}{n^2} \cos n\pi t$$

$$\boxed{x(t) = \frac{1}{3} + \frac{4}{\pi^2} \left[- \cos \pi t + \frac{1}{4} \cos 2\pi t - \frac{1}{9} \cos 3\pi t + \ldots\right]}$$

4.4 Properties of Fourier Series

4.4.1 Linearity

Let $x_1(t)$ and $x_2(t)$ be two periodic signals with the same period T_0. Let D_{n1} and D_{n2} be the Fourier series coefficients in complex exponential form. Let $x(t)$ be the composite signal of $x_1(t)$ and $x_2(t)$ which are related as

$$x(t) = Ax_1(t) + Bx_2(t) \tag{4.15}$$

where A and B are constants.

From Eq. (4.10)

$$D_{n1} = \frac{1}{T_0} \int_{T_0} x_1(t) e^{-jn\omega_0 t} dt \tag{4.16}$$

$$D_{n2} = \frac{1}{T_0} \int_{T_0} x_2(t) e^{-jn\omega_0 t} dt \tag{4.17}$$

Let D_n be the Fourier series coefficient of $x(t)$

$$D_n = \frac{1}{T_0} \int_{T_0} x(t) e^{-jn\omega_0 t} dt \tag{4.18}$$

$$= \frac{1}{T_0} \int_{T_0} [Ax_1(t) + Bx_2(t)] e^{-jn\omega_0 t} dt \tag{4.19}$$

$$= \frac{1}{T_0} \int_{T_0} Ax_1(t) e^{-jn\omega_0 t} dt + \frac{1}{T_0} \int_{T_0} Bx_2(t) e^{-jn\omega_0 t} dt \tag{4.20}$$

$$\boxed{D_n = AD_{n1} + BD_{n2}} \tag{4.21}$$

The Fourier series coefficient of the composite signal $x(t)$ is the linear combination of individual signal.

4.4.2 *Time Shifting Property*

According to the time shifting property, if the periodic signal $x(t)$ with fundamental period T_0 is time shifted, the periodicity remains the same and the FS coefficient is multiplied by the factor $e^{-jnw_0 t_0}$.

Proof Let $x(t)$ be time shifted by t_0. Now the time shifted signal is $x(t - t_0)$. The Fourier series coefficient of $x(t)$ is

$$D_n = \frac{1}{T_0} \int_{T_0} x(t) e^{-jn\omega_0 t} dt \tag{4.22}$$

Let D_{n0} be the FS coefficient for the time shifted signal.

$$D_{n0} = \frac{1}{T_0} \int_{T_0} x(t - t_0) e^{-jn\omega_0 t} dt \tag{4.23}$$

Substitute $\tau = (t - t_0)$ in the above equation

$$D_{n0} = \frac{1}{T_0} \int_{T_0} x(\tau) e^{-jn\omega_0(\tau + t_0)} d\tau$$

$$= e^{-jn\omega_0 t_0} \frac{1}{T_0} \int_{T_0} x(\tau) e^{-jn\omega_0 \tau} d\tau \tag{4.24}$$

$$\boxed{D_{n0} = e^{-jn\omega_0 t_0} D_n} \tag{4.25}$$

4.4.3 *Time Reversal Property*

According to the time reversal property, if the signal $x(t)$ is time reversed, the periodicity remains the same with the time reversal in the FS coefficient.

Proof Let $x(t)$ be the signal with period T_0 and the FS coefficient D_n. If $x(t)$ is time reversed, the signal becomes $x(-t)$. Let D_{-n} be the FS coefficient of $x(-t)$.

$$D_n = \frac{1}{T_0} \int_{T_0} x(-t) e^{-jn\omega_0 t} dt \tag{4.26}$$

Let us substitute $\tau = -t$

$$D_n = \frac{1}{T_0} \int_{T_0} x(\tau) e^{-j(-n)\omega_0 \tau} (-d\tau) \tag{4.27}$$

$$= -\frac{1}{T_0} \int_{T_0} x(\tau) e^{-j(-n)\omega_0 \tau} d\tau \tag{4.28}$$

$$\boxed{D_n = -D_{-n}}$$

4.4.4 *Time Scaling Property*

According to time scaling property, if $x(t)$ is periodic with fundamental period T_0, then $x(at)$ where a *is any positive real number, is also periodic but with a fundamental period of $\frac{T_0}{a}$.*

Proof Let D_s be the FS coefficient of $x(at)$.

$$D_s = \frac{1}{T_0} \int_{T_0} x(at) e^{-jn\omega_0 t} dt \tag{4.29}$$

Let $at = \tau$

$$D_s = \frac{1}{aT_0} \int_{T_0} x(\tau) e^{-jn\omega_0 \frac{\tau}{a}} d\tau$$

$$\boxed{D_s = \frac{1}{a} D_{n/a}} \tag{4.30}$$

4.4.5 Multiplication Property

According to multiplication property, if $x_1(t)$ and $x_2(t)$ are the two signals having the periodicity T_0, then the Fourier coefficient of the product of these two signals is given by

$$D_n = \sum_{l=-\infty}^{\infty} A_l B_{n-l}$$

where A_l and B_l are the FS coefficients of $x_1(t)$ and $x_2(t)$ respectively.

Proof Let

$$x(t) = x_1(t) \times x_2(t)$$
$$D_n = \frac{1}{T_0} \int_{T_0} x(t) e^{-jn\omega_0 t} dt$$

$$\begin{aligned} D_n &= \frac{1}{T_0} \int_{T_0} [x_1(t) \times x_2(t)]\, e^{-jn\omega_0 t} dt \\ &= \frac{1}{T_0} \int_{T_0} \left[\sum_{l=-\infty}^{\infty} A_l e^{jl\omega_0 t} \right] x_2(t) e^{-jn\omega_0 t} dt \\ &= \sum_{l=-\infty}^{\infty} A_l \frac{1}{T_0} \int_{T_0} x_2(t) e^{-j(n-l)\omega_0 t} dt \end{aligned}$$

$$\boxed{D_n = \sum_{l=-\infty}^{\infty} A_l B_{n-l}} \tag{4.31}$$

4.4.6 Conjugation Property

According to this property, that the FS coefficients have conjugate symmetric property

Proof

$$D_{-n} = D_n^*$$

$$x(t) = \sum_{n=-\infty}^{\infty} D_n e^{jn\omega_0 t}$$

$$x^*(t) = \left[\sum_{n=-\infty}^{\infty} Dn^{ejn\omega_0 t}\right]^*$$

$$= \sum_{n=-\infty}^{\infty} D_n^* e^{-jn\omega_0 t}$$

Let $l = -n$,

$$x^*(t) = \sum_{l=-\infty}^{\infty} D_{-l}^* e^{jl\omega_0 t} \tag{4.32}$$

Thus during conjugation, FS coefficient becomes conjugate and time reversed.

4.4.7 Differentiation Property

If a periodical signal $x(t)$ is differentiated, the FS coefficient is multiplied by the factor $jn\omega_0$.

Proof

$$x(t) = \sum_{n=-\infty}^{\infty} D_n e^{j\omega_0 nt}$$

$$\frac{dx(t)}{dt} = \sum_{n=-\infty}^{\infty} j\omega_0 n D_n e^{j\omega_0 nt}$$
$$= \sum_{n=-\infty}^{\infty} D_n^1 e^{j\omega_0 nt} \tag{4.33}$$

where $D_n^1 = j\omega_0 n D_n$. Thus, when the signal $x(t)$ is differentiated, its FS coefficient is multiplied by the factor $j\omega_0 n$.

4.4.8 Integration Property

According to the integration property, the FS coefficient of $x(t)$ when $x(t)$ is integrated becomes.

$$\frac{1}{j\omega_0 n} D_n$$

Proof

$$x(t) = \sum_{n=-\infty}^{\infty} D_n e^{jn\omega_0 t}$$

Integrating both sides we get

$$\int_{-\infty}^{t} x(t) = \int_{-\infty}^{t} \sum_{n=-\infty}^{\infty} D_n e^{jn\omega_0 t} dt$$
$$= \sum_{n=-\infty}^{\infty} \frac{D_n e^{jn\omega_0 t}}{j\omega_0 n}$$
$$= \sum_{n=-\infty}^{\infty} D_n^1 e^{jn\omega_0 t} \tag{4.34}$$

where $D_n^1 = \frac{1}{j\omega_0 n} D_n$. Thus, when the signal $x(t)$ is integrated, its FS coefficient is divided by the factor $j\omega_0 n$.

4.4.9 Parseval's Theorem

According to Parseval's theorem, that the total average power in a periodic signal is the sum of the average powers in all its components which is the sum of the squared value of FS coefficients.

Proof The average power in a periodic signal is given by

$$P = \frac{1}{T_0} \int_{T_0} |x(t)|^2 \, dt$$

$$\begin{aligned} P &= \frac{1}{T_0} \int_{T_0} x(t) \, [x(t)]^* \, dt \\ &= \frac{1}{T_0} \int_{T_0} x(t) \left[\sum_{n=-\infty}^{\infty} D_n e^{j\omega_0 nt} \right]^* dt \\ &= \sum_{n=-\infty}^{\infty} D_n^* \frac{1}{T_0} \int_{T_0} x(t) e^{-j\omega_0 nt} dt \\ &= \sum_{n=-\infty}^{\infty} D_n^* D_n \end{aligned}$$

$$P = \sum_{n=-\infty}^{\infty} |D_n|^2 \qquad (4.35a)$$

For a real $x(t)$, $|D_{-n}| = |D_n|$

$$P = D_0^2 + 2 \sum_{n=1}^{\infty} |D_n|^2 \qquad (4.35b)$$

For a trigonometric Fourier series,

$$P = C_0^2 + \frac{1}{2} \sum_{n=1}^{\infty} C_n^2 \qquad (4.35c)$$

■ Example 4.7

Find the Fourier series representation for the signal

$$x(t) = 3 \cos \left(\frac{\pi}{2} t + \frac{\pi}{4} \right)$$

and hence find the power.

(*Anna University, April, 2008*)

Solution:

$$\begin{aligned} x(t) &= 3\cos\left(\frac{\pi}{2}t + \frac{\pi}{4}\right) \\ &= \frac{3}{2}\left[e^{j(\pi/2t+\pi/4)} + e^{-j(\frac{\pi}{2}t+\pi/4)}\right] \\ &= \frac{3}{2}e^{j\pi/4}e^{j(\pi/2)t} + \frac{3}{2}e^{-j\pi/4}e^{-j\pi/2t} \end{aligned}$$

Compare this with complex exponential Fourier series

$$\begin{aligned} x(t) &= \sum_{n=-\infty}^{\infty} D_n e^{jn\omega_0 t} \qquad \text{where } \omega_0 = \frac{\pi}{2} \\ &= \sum_{n=-\infty}^{\infty} D_n e^{jn\frac{\pi}{2}t} \end{aligned}$$

$$x(t) = D_{-1}e^{-j\frac{\pi}{2}t} + D_1 e^{j\frac{\pi}{2}t} \tag{b}$$

$$D_1 = \frac{3}{2}e^{j\frac{\pi}{4}} = \frac{3}{2}\left[\cos\frac{\pi}{4} + j\sin\frac{\pi}{4}\right]$$

Comparing Eqs. (a) and (b), we get

$$\boxed{\begin{aligned} D_1 &= \frac{3}{2\sqrt{2}}(1+j); \quad |D_1| = \frac{3}{2} \\ D_{-1} &= \frac{3}{2\sqrt{2}}(1-j); \quad |D_{-1}| = \frac{3}{2} \end{aligned}}$$

$$P = \sum_{n=-\infty}^{\infty} |D_n|^2 = D_{-1}^2 + D_1^2 = \left(\frac{3}{2}\right)^2 + \left(\frac{3}{2}\right)^2 = \frac{9}{2}.$$

■ Example 4.8

Find the Fourier series of the following signals. Also, find the power using Fourier series coefficients.

(a) $x(t) = 2\cos 3t + 3\sin 2t$

(b) $x(t) = \cos^2 t$

Solution:

(a) $\boldsymbol{x(t) = 2\cos 3t + 3\sin 2t}$

1.

$$\begin{aligned}
\omega_{01} &= 3; \qquad T_{01} = \frac{2\pi}{\omega_{01}} = \frac{2\pi}{3} \\
\omega_{02} &= 2; \qquad T_{02} = \frac{2\pi}{\omega_{02}} = \frac{2\pi}{2} = \pi \\
\frac{T_{01}}{T_{02}} &= \frac{2\pi}{3\pi} = \frac{2}{3} \\
T_0 &= 3T_{01} = 2T_{02} = 2\pi \\
\omega_0 &= \frac{2\pi}{T_0} = \frac{2\pi}{2\pi} = 1
\end{aligned}$$

2. Using Euler's Formula, $x(t)$ can be expressed as

$$\begin{aligned}
x(t) &= (e^{j3t} + e^{-j3t}) + \frac{3}{j2}(e^{j2t} - e^{-j2t}) \\
&= e^{-j3t} + j\frac{3}{2}e^{-j2t} + e^{j3t} - j\frac{3}{2}e^{j2t}
\end{aligned}$$

$x(t)$ can also be expressed in complex exponential form as

$$\begin{aligned}
x(t) &= \sum_{n=-\infty}^{\infty} D_n e^{jn\omega_0 t} \\
&= \sum_{n=-\infty}^{\infty} D_n e^{jnt}
\end{aligned}$$

Equating the two equations for $x(t)$, we get

$$e^{-j3t} + j\frac{3}{2}e^{-j2t} + e^{j3t} - j\frac{3}{2}e^{j2t} = \sum_{n=-\infty}^{\infty} D_n e^{jnt}$$

Putting $n = \pm 3$

$$\boxed{D_3 = 1 \quad \text{and} \quad D_{-3} = 1}$$

Putting $n = \pm 2$

$$\boxed{D_2 = -j\frac{3}{2} \quad \text{and} \quad D_{-2} = j\frac{3}{2}}$$

All other $D_n = 0$.

$$\text{Power } P = |D_{-3}|^2 + |D_{-2}|^2 + |D_3|^2 + |D_2|^2$$
$$= 1^2 + \left(\frac{3}{2}\right)^2 + 1^2 + \left(\frac{3}{2}\right)^2 = \frac{13}{2}.$$

(b) $\boldsymbol{x(t) = \cos^2 t}$

$$x(t) = \cos^2 t$$
$$= \frac{1}{2}[1 + \cos 2t]$$
$$\omega_0 = 2$$
$$x(t) = \frac{1}{2} + \frac{1}{2}\frac{\left[e^{j2t} + e^{-j2t}\right]}{2} = \sum_{n=-\infty}^{\infty} D_n e^{j2nt}$$

For $n = 0$;

$$\boxed{D_0 = \frac{1}{2}}$$

For $n = \pm 1$;

$$\boxed{D_1 = \frac{1}{4} \quad \text{and} \quad D_{-1} = \frac{1}{4}}$$

$$\text{Power } P = D_0^2 + D_{-1}^2 + D_1^2 = \frac{1}{4} + \frac{1}{16} + \frac{1}{16} = \frac{3}{8}.$$

■ Example 4.9

Find the exponential Fourier series for the signal shown in Fig. 4.7.

(Anna University, December 2007)

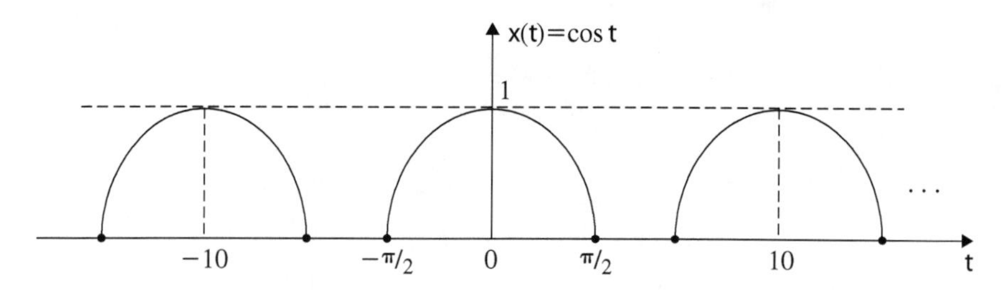

Fig. 4.7 Signal of Example 4.9

Solution:

$$x(t) = \cos t \qquad \frac{-\pi}{2} \le t \le \frac{\pi}{2}$$

$$T_0 = 10$$

$$\omega_0 = \frac{2\pi}{T_0} = 0.2\pi$$

$$D_n = \frac{1}{T_0}\int_0^{T_0} x(t)e^{-j\omega_0 nt}dt$$

$$= \frac{1}{10}\int_{-\pi/2}^{\pi/2} \cos t\, e^{-j\omega_0 nt}dt$$

$$= \frac{1}{20}\int_{-\pi/2}^{\pi/2} (e^{jt} + e^{-jt})e^{-j0.2n\pi t}dt$$

$$= \frac{1}{20}\left[\int_{-\pi/2}^{\pi/2} e^{j(1-.2n\pi)t}dt + \int_{-\pi/2}^{\pi/2} e^{-j(1+.2n\pi)t}dt\right]$$

$$= \frac{1}{20}\left\{\frac{1}{j(1-.2n\pi)}\left[e^{j(1-.2n\pi)t}\right]_{-\pi/2}^{\pi/2} + \frac{1}{j(1+.2n\pi)}\left[e^{-j(1+.2n\pi)t}\right]_{-\pi/2}^{\pi/2}\right\}$$

$$= \frac{1}{20}\left\{\frac{1}{j(1-.2n\pi)}\left[e^{j\frac{\pi}{2}(1-.2n\pi)} + e^{+j\pi/2(1-.2n\pi)}\right]\right.$$
$$\left. -\frac{1}{j(1+.2n\pi)}\left[e^{-j\pi/2(1+.2n\pi)} - e^{j\pi/2(1+.2n\pi)}\right]\right\}$$

$$= \frac{1}{10}\left[\frac{1}{(1-.2n\pi)}\sin\frac{\pi}{2}(1-.2n\pi) + \frac{1}{(1+.2n\pi)}\sin\frac{\pi}{2}(1+.2n\pi)\right]$$

$$= \frac{1}{10(1-.04n^2\pi^2)}\left[(1+.2n\pi)\cos 0.1n\pi^2 + (1-.2n\pi)\cos(0.1n\pi^2)\right]$$

$$\boxed{\begin{aligned} D_n &= \frac{0.2\cos 0.1n\pi^2}{(1-0.04n^2\pi^2)} \\ x(t) &= \sum_{n=-\infty}^{\infty} D_n e^{j0.2\pi nt} \end{aligned}}$$

$$D_0 = 0.2$$
$$D_{-1} = D_{+1} = 0.1818$$
$$D_{-2} = D_{+2} = 0.135$$
$$D_{-3} = D_{+3} = 0.0783$$
$$D_{-4} = D_{+4} = 0.026$$

$$\begin{aligned} x(t) &= \sum_{n=-\infty}^{\infty} D_n e^{j\omega_0 nt} \\ &= D_0 + D_1(e^{-j0.2\pi t} + e^{j0.2\pi t}) + D_2(e^{-j0.4\pi t} + e^{j0.4\pi t}) \\ &\quad + D_3(e^{-j0.6\pi t} + e^{j0.6\pi t}) + D_4(e^{-j0.8\pi t} + e^{j0.8\pi t}) + \cdots \end{aligned}$$

$$\boxed{\begin{aligned} x(t) = [&0.2 + 0.3636 \cos 0.2\pi t + 0.27 \cos 0.4\pi t \\ &+ 0.1566 \cos 0.6\pi t + 0.052 \cos 0.8\pi t + \cdots] \end{aligned}}$$

■ Example 4.10

Consider the waveform shown in Fig. 4.8. Determine the complex exponential Fourier series.

Solution:

1. From Fig. 4.8, $T_0 = 2$ and $\omega_0 = \frac{2\pi}{T_0} = \frac{2\pi}{2} = \pi$.

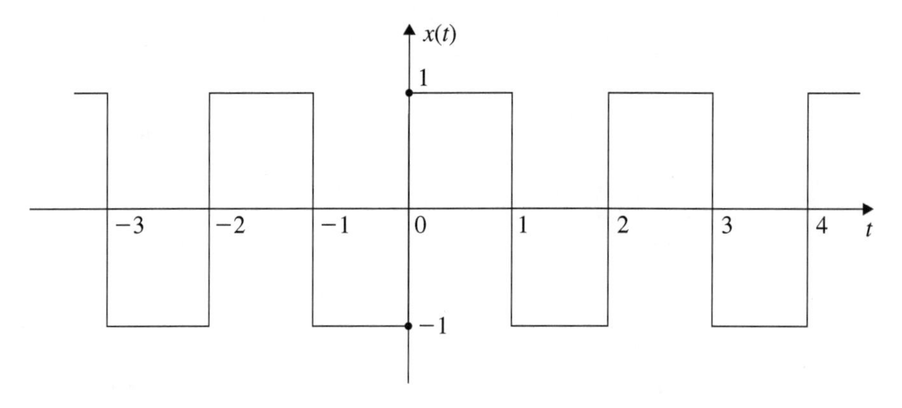

Fig. 4.8 Signal of Example 4.10

2.

$$D_n = \frac{1}{T_0}\int_0^{T_0} x(t)e^{-jn\omega_0 t}dt$$

$$\begin{aligned}
D_n &= \frac{1}{2}\int_0^1 e^{-jn\pi t}dt - \frac{1}{2}\int_1^2 e^{-jn\pi t}dt \\
&= \frac{1}{2}\frac{1}{(-jn\pi)}\left[e^{-jn\pi t}\right]_0^1 - \frac{1}{2(-jn\pi)}\left[e^{-jn\pi t}\right]_1^2 \\
&= \frac{1}{2(-jn\pi)}\left[e^{-jn\pi} - 1 - e^{-jn\pi 2} + e^{-jn\pi}\right] \\
&= \frac{1}{-2n\pi j}\left[2e^{-jn\pi} - 2\right] \qquad \left[\because\ e^{-jn\pi 2} = 1\right] \\
&= \frac{1}{jn\pi}\left[1 - e^{-jn\pi}\right] \\
&= \frac{1}{jn\pi}\left[1 - \cos n\pi\right]
\end{aligned}$$

$$\boxed{D_n = \frac{2}{jn\pi}}$$

where n is an odd number. For even number n, $\cos n\pi = 1$

3.

$$x(t) = \sum_{n=-\infty}^{\infty} D_n e^{jn\pi t}$$

$$\boxed{x(t) = \frac{2}{j\pi}\sum_{m=-\infty}^{\infty}\frac{1}{2m+1}e^{j(2m+1)\pi t}}$$

where m is any integer which will be equivalent to n being odd integer.

$$\begin{aligned}
D_{-1} &= -\frac{2}{j\pi}; \\
x_{-1}(t) &= D_{-1}e^{-j\pi t} \\
D_{+1} &= \frac{2}{j\pi}; \\
x_{+1}(t) &= D_{+1}e^{j\pi t}
\end{aligned}$$

$$\begin{aligned}
x_1(t) &= x_{+1}(t) + x_{-1}(t) \\
&= \frac{2}{j\pi}[e^{j\pi t} - e^{-j\pi t}] \\
&= \frac{4}{\pi}\sin(\pi t) \qquad \text{(Fundamental component)} \\
D_{-3} &= -\left(\frac{2}{j\pi}\right)\frac{1}{3}; \\
x_{-3}(t) &= D_{-3}e^{-j3\pi t} \\
D_{+3} &= \left(\frac{2}{j\pi}\right)\frac{1}{3}; \\
x_{+3}(t) &= D_{+3}e^{j3\pi t} \\
x_3(t) &= x_{+3}(t) + x_{-3}(t) \\
&= \left(\frac{2}{j\pi}\right)\frac{1}{3}[e^{j3\pi t} - e^{-j3\pi t}] \\
&= \left(\frac{4}{\pi}\right)\frac{1}{3}\sin(3\pi t) \qquad \text{(Third harmonic)}
\end{aligned}$$

Similarly, $x_5(t)$ can be obtained as

$$\begin{aligned}
x_5(t) &= \left(\frac{4}{\pi}\right)\frac{1}{5}\sin(5\pi t) \qquad \text{(Fifth harmonics)} \\
x(t) &= [x_1(t) + x_3(t) + x_5(t) + \cdots]
\end{aligned}$$

$$\boxed{x(t) = \frac{4}{j\pi}\left[\sin\pi t + \frac{1}{3}\sin 3\pi t + \frac{1}{5}\sin 5\pi t + \cdots\right]}$$

■ Example 4.11

Let

$$x(t) = \begin{cases} t & 0 \le t \le 1 \\ 2 - t & 1 \le t \le 2 \end{cases}$$

be a periodic signal with fundamental period $T_0 = 2$ and Fourier coefficients a_k.

(a) Determine the value of a_0.
(b) Determine the Fourier series representation of $\frac{dx(t)}{dt}$.
(c) Use the result of part (b) and the differentiation property of FS to help determine the Fourier series coefficients of $x(t)$.

(*Anna University, May 2008*)

Solution:

(a)

$$x(t) = \begin{cases} t & 0 \le t \le 1 \\ 2 - t & 1 \le t \le 2 \end{cases}$$

The above equation represents a triangle in the given time interval and the periodical signal with period $T_0 = 2$ is shown in Fig. 4.9.

$$\omega_0 = \frac{2\pi}{T_0} = \pi$$

The Fourier series coefficient a_0 is determined as follows:

$$a_0 = \frac{1}{T_0}\int_0^{T_0} x(t)dt$$

$$= \frac{1}{2}\int_0^1 t\,dt + \frac{1}{2}\int_1^2 (2-t)dt$$

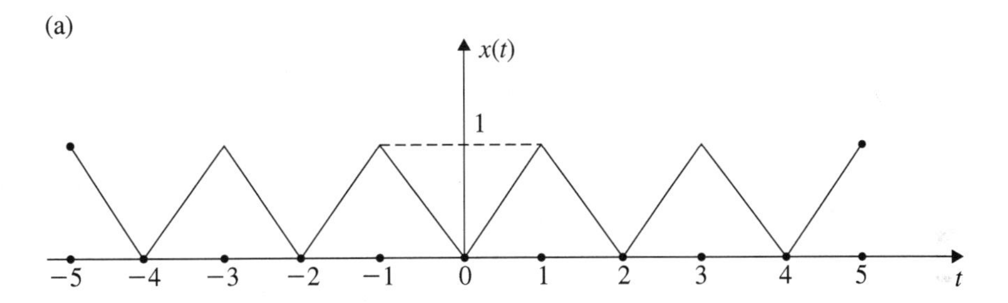

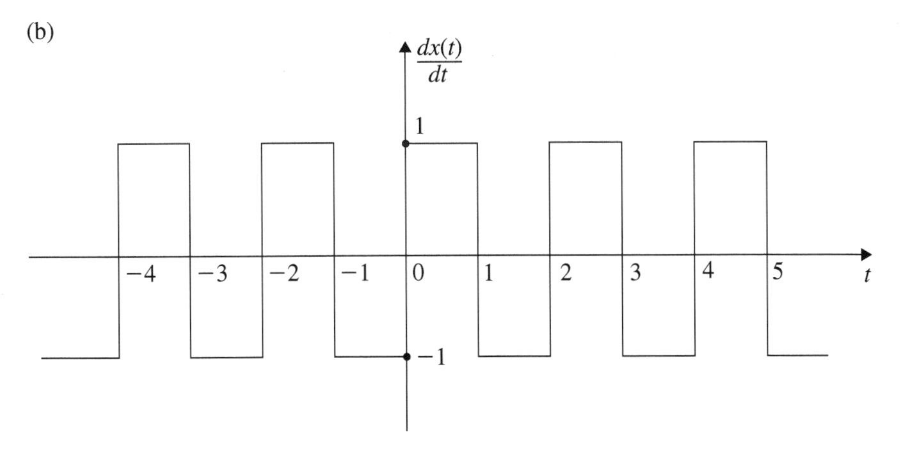

Fig. 4.9 **a** A triangular wave and **b** Derivative of triangular wave

$$= \frac{1}{2}\left[\frac{t^2}{2}\right]_0^1 + \frac{1}{2}\left[2t - \frac{t^2}{2}\right]_1^2$$
$$= \frac{1}{4} + \frac{1}{2}\left[4 - 2 - 2 + \frac{1}{2}\right]$$

$$\boxed{a_0 = \frac{1}{2}}$$

(b) Differentiating the given $x(t)$, we get

$$\frac{dx(t)}{dt} = \begin{cases} 1 & 0 \leq t \leq 1 \\ -1 & 1 \leq t \leq 2 \end{cases}$$

This is the square wave and is shown in Fig. 4.9b. Figures 4.8 and 4.9b are the rectangular waves with the amplitude and periodicity. The exponential FS coefficient of Fig. 4.8 has been determined as

$$D'_n = \frac{2}{jn\pi} \quad \text{where } n \text{ is an odd integer.}$$
$$= \frac{2}{j(2m+1)\pi} \quad \text{where } m \text{ is any integer.}$$

$$\boxed{\frac{dx(t)}{dt} = \dot{x}(t) = \frac{2}{j\pi}\sum_{m=-\infty}^{\infty}\frac{1}{(2m+1)}e^{j(2m+1)\pi t}}$$

(c) $x(t)$ in the Fourier exponential form can be written as follows:

$$x(t) = \sum_{n=-\infty}^{\infty} D_n e^{jn\omega_0 t}$$
$$\frac{dx(t)}{dt} = \sum_{n=-\infty}^{\infty} (jn\omega_0) D_n e^{jn\omega_0 t}$$

From the result derived in part (b)

$$D_n = \frac{D'_n}{jn\omega_0}$$
$$jn\omega_0 D_n = \frac{2}{jn\pi}$$

$$\boxed{D_n = \frac{-2}{n^2\pi^2}}$$

where n is an odd integer.

$$x(t) = D_0 + \sum_{n=-\infty}^{\infty} D_n e^{jn\pi t}$$
$$D_0 = a_0 = \frac{1}{2}$$
$$n = 2m + 1 \quad \text{where } m \text{ is any integer}$$

$$\boxed{x(t) = \frac{1}{2} - \frac{2}{\pi^2} \sum_{m=-\infty}^{\infty} \frac{1}{(2m+1)^2} e^{j(2m+1)\pi t}}$$

Since n is a squared function. $D_{-1} = D_{+1}, D_{-2} = D_{+2}$ *etc*

$$D_{-1} = D_{+1} = -\frac{2}{\pi^2}$$
$$x_1(t) = -\frac{2}{\pi^2}[e^{-j\pi t} + e^{+j\pi t}]$$
$$= -\frac{4}{\pi^2}\cos \pi t \quad \text{(Fundamental component)}$$
$$D_{-3} = D_{+3} = -\left(\frac{2}{\pi^2}\right)\frac{1}{9}$$
$$x_3(t) = -\frac{2}{\pi^2}\frac{1}{9}[e^{-j3\pi t} + e^{+j3\pi t}]$$
$$= -\frac{4}{\pi^2}\left[\frac{1}{9}\cos(3\pi t)\right] \quad \text{(Third harmonic)}$$

Similarly, $x_5(t)$ can be obtained as

$$x_5(t) = -\frac{4}{\pi}\left[\frac{1}{25}\cos(5\pi t)\right]$$

$$\boxed{x(t) = \frac{1}{2} - \frac{4}{\pi^2}\left[\cos(\pi t) + \frac{1}{9}\cos(3\pi t) + \frac{1}{25}\cos(5\pi t) + \cdots\right]}$$

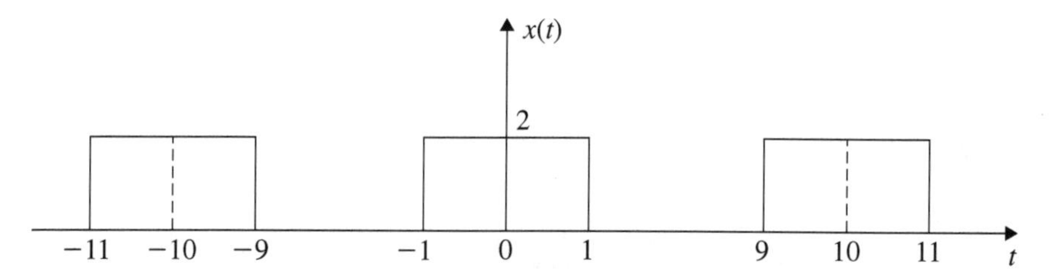

Fig. 4.10 Signal of Example 4.12

■ Example 4.12

For the signal shown in Fig. 4.10. Determine the exponential Fourier series.

Solution:

$$
\begin{aligned}
T_0 &= 10 \\
\omega_0 &= \frac{2\pi}{10} = \frac{\pi}{5} \\
D_n &= \frac{1}{T_0}\int_{-1}^{1} 2e^{-j\omega_0 nt}\,dt \\
&= \frac{2}{10}\int_{-1}^{1} e^{-j\frac{\pi}{5}nt}\,dt \\
&= -\frac{1}{5}\frac{5}{\pi jn}\left[e^{-j\frac{\pi n}{5}t}\right]_{-1}^{1} \\
&= -\frac{1}{j\pi n}\left[e^{-j\frac{\pi n}{5}} - e^{+j\frac{n\pi}{5}}\right]
\end{aligned}
$$

$$
\boxed{D_n = \frac{2}{\pi n}\sin\frac{\pi n}{5} \quad \text{for all } n \text{ but } n \neq 0}
$$

For $n = 0$

$$
\begin{aligned}
D_0 &= \underset{n\to 0}{Lt}\ \frac{2}{5}\frac{\sin\frac{\pi n}{5}}{\frac{\pi n}{5}} \\
D_0 &= \frac{2}{5} = 0.4
\end{aligned}
$$

$$\boxed{x(t) = 0.4 + \frac{2}{\pi} \sum_{n=-\infty}^{\infty} \frac{1}{n} \sin \frac{\pi}{5} n e^{\frac{-j\pi n t}{5}}}$$

$$\begin{aligned}
D_{-1} &= -\frac{2}{\pi} \sin \frac{-\pi}{5} = 0.374 \\
D_{+1} &= \frac{2}{\pi} \sin \frac{\pi}{5} = 0.374 \\
x_1(t) &= D_{-1}(e^{-j0.2\pi t}) + D_{+1}(e^{+j0.2\pi t}) \\
&= 0.374[e^{-j0.2\pi t} + e^{-j0.2\pi t}] \\
&= 0.748 \cos 0.2\pi t \qquad \text{(Fundamental component)} \\
D_{-2} &= -\frac{1}{\pi} \sin \frac{-2\pi}{5} \\
&= 0.303 \\
D_{+2} &= D_{-2} = 0.303 \\
x_2(t) &= 0.303[e^{-j0.4\pi t} + e^{+j0.4\pi t}] \\
&= 0.606 \cos(0.4\pi t) \qquad \text{(Second harmonic)} \\
D_{-3} &= -\frac{2}{3\pi} \sin \frac{-3\pi}{5} \\
&= 0.2 \\
D_{+3} &= D_{-3} = 0.2 \\
x_3(t) &= 0.2[e^{-j0.6\pi t} + e^{+j0.6\pi t}] \\
&= 0.4 \cos(0.6\pi t) \qquad \text{(Third harmonic)} \\
x(t) &= D_0 + x_1(t) + x_2(t) + x_3(t) + \cdots
\end{aligned}$$

$$\boxed{x(t) = 0.4 + [0.748 \cos(0.2\pi t) + 0.606 \cos(0.4\pi t) + 0.4 \cos(0.6\pi t) + \cdots]}$$

Note: For $n = 5, 10, 15, 20, \ldots$

$$D_n = 0$$

■ Example 4.13

Determine the exponential and trigonometric Fourier series of a train of impulse with periodicity $T_0 = 1$. Verify the exponential and trigonometric coefficients relationship.

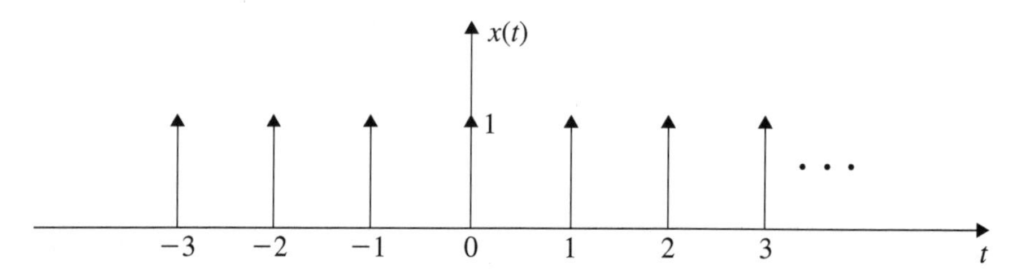

Fig. 4.11 Periodic train of impulses

Solution:

$$T_0 = 1 \quad \text{and} \quad \omega_0 = 2\pi$$

To determine the exponential FS coefficients

$$D_n = \frac{1}{T_0}\int_0^{T_0} \delta(t)\, e^{-jn\omega_0 t} dt = \frac{1}{T_0}\int_{-1/2}^{1/2} \delta(t)\, e^{-j2\pi nt} dt$$

Over this interval, $D_n = \frac{1}{T_0}$

$$\boxed{\begin{aligned} D_n &= \frac{1}{T_0} = 1 \\ D_0 &= 1 \end{aligned}}$$

$$x(t) = \sum_{n=-\infty}^{\infty} D_n e^{jn\omega_0 t}$$

$$\boxed{x(t) = \sum_{n=-\infty}^{\infty} e^{j2\pi nt}}$$

To determine the Trigonometric Fourier series

$$a_0 = \frac{1}{T}\int_0^{T_0} \delta(t)\, dt$$

$$a_0 = \frac{1}{T_0} = 1$$

Since the train of impulses is an even signal $b_n = 0$.

$$a_n = \frac{2}{T_0} \int_0^{T_0} \delta(t) \cos n\omega_0 t \, dt$$
$$= \frac{2}{T_0} = 2$$

$$x(t) = a_0 + \sum_{n=1}^{\infty} a_n \cos n\omega_0 t$$

$$\boxed{x(t) = 1 + \sum_{n=1}^{\infty} 2 \cos 2\pi n t}$$

$$a_0 = D_0 = 1$$
$$D_n = \frac{a_n}{2} = \frac{2}{2} = 1$$

Thus, the relationships between trigonometric and exponential Fourier series coefficients are verified.

■ Example 4.14

For the periodic signal $x(t) = e^{-t}$ with a period $T_0 = 1$ second, find the Fourier series in

(a) Exponential form,
(b) Trigonometric form,
(c) Polar form, and
(d) Verify the relationships of FS coefficients.

Solution:

(a) **Exponential Fourier series**
$x(t)$ is plotted as shown in Fig. 4.12.

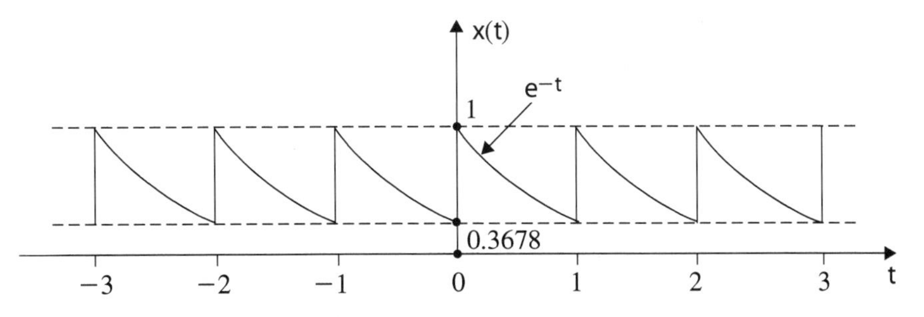

Fig. 4.12 Exponentially decaying periodic signal

$$
\begin{aligned}
T_0 &= 1 \\
\omega_0 &= \frac{2\pi}{T_0} = 2\pi \\
D_n &= \frac{1}{T_0}\int_0^{T_0} x(t)e^{-jn\omega_0 t}dt \\
&= \int_0^1 e^{-t}e^{-jn2\pi t}dt \\
&= \int_0^1 e^{-(1+j2\pi n)t}dt \\
&= -\frac{1}{(1+j2\pi n)}\left[e^{-(1+j2\pi n)t}\right]_0^1 \\
&= \frac{1}{(1+j2\pi n)}\left[1-e^{-(1+j2\pi n)}\right] \\
&= \frac{1}{(1+j2\pi n)}\left[1-e^{-1}\right] \quad \left[\because\ e^{-j2\pi n}=1\right]
\end{aligned}
$$

$$
\boxed{D_n = \frac{0.632}{(1+j2\pi n)}; \quad |D_n| = \frac{0.632}{\sqrt{1+4\pi^2 n^2}}}
$$

$$
\boxed{D_0 = 0.632}
$$

$$
x(t) = \sum_{n=-\infty}^{\infty} D_n e^{jn\omega_0 t}
$$

$$
\boxed{x(t) = 0.632\sum_{n=-\infty}^{\infty}\frac{1}{\sqrt{(1+4\pi^2 n^2)}}e^{j2\pi nt}}
$$

$$
D_n = \frac{0.632}{\sqrt{1+4\pi^2 n^2}}
$$

Since n is a squared function

$$
\begin{aligned}
D_{-n} &= D_{+n} \\
D_1 &= \frac{0.632}{\sqrt{1+4\pi^2}} \\
&= 0.1 \\
x_1(t) &= D_{-1}[e^{j2\pi t}+e^{-j2\pi t}] \\
&= 2\times 0.1\cos 2\pi t
\end{aligned}
$$

$$= 0.2\cos 2\pi t \qquad \text{(Fundamental component)}$$

$$D_{-2} = \frac{0.632}{\sqrt{1+4\pi^2\times 4}}$$

$$= 0.05$$

$$x_2(t) = 0.1\cos 4\pi t \qquad \text{(Second harmonic)}$$

$$D_{-3} = \frac{0.632}{\sqrt{1+4\pi^2\times 9}}$$

$$= 0.0335$$

$$x_3(t) = 0.067\cos(0.6\pi t) \qquad \text{(Third harmonic)}$$

$$x(t) = D_0 + x_1(t) + x_2(t) + x_3(t) + \cdots$$

$$\boxed{x(t) = [0.632 + 0.2\cos 2\pi t + 0.1\cos 4\pi t + 0.067\cos 6\pi t + \cdots]}$$

(b) **Trigonometric Fourier series**

$$a_0 = \frac{1}{T_0}\int_0^{T_0} x(t)dt$$

$$= \int_0^1 e^{-t}dt$$

$$a_0 = -\left[e^{-t}\right]_0^1$$

$$= (1 - e^{-1})$$

$$\boxed{a_0 = 0.632}$$

$$a_n = \frac{2}{T_0}\int_0^{T_0} x(t)\cos\omega_0 nt\,dt$$

$$= 2\int_0^1 e^{-t}\cos 2\pi nt\,dt$$

Using the property

$$\int_a^b e^{at}\cos bt\,dt = \left[\frac{e^{at}(a\cos bt + b\sin bt)}{(a^2+b^2)}\right]_a^b$$

$$a_n = \frac{2}{(1+4\pi^2n^2)}\left[-\cos 2\pi nt(e^{-t}) + e^{-t}2\pi n\sin 2\pi n\right]_0^1$$

$$= \frac{2}{(1+4\pi^2n^2)}\left[e^{-1}\{-\cos 2\pi n + 2\pi n \sin 2\pi n\} + 1\right]$$
$$= \frac{2}{(1+4\pi^2n^2)}[1 - e^{-1}]$$

$$\boxed{a_n = \frac{1.264}{(1+4\pi^2n^2)}}$$

$$b_n = \frac{2}{T_0}\int_0^{To} x(t)\sin \omega_0 nt\, dt$$
$$= 2\int_0^1 e^{-t}\sin 2\pi nt\, dt$$

Using the property

$$\int_a^b e^{at}\sin bt\, dt = \frac{1}{(a^2+b^2)}\left\{e^{at}[a\sin bt - b\cos bt]\right\}_a^b$$

we get,

$$b_n = \frac{2}{\left(1+4\pi^2n^2\right)}\left\{e^{-t}\left[-\sin 2\pi nt - 2\pi n\cos 2\pi nt\right]\right\}_0^1$$
$$= \frac{2}{\left(1+4\pi^2n^2\right)}\left[-e^{-1}(\sin 2\pi n + 2\pi n\cos 2\pi n) + 2\pi n\right]$$

$$b_n = \frac{4\pi n}{\left(1+4\pi^2n^2\right)}(1 - e^{-1})$$

$$\boxed{b_n = \frac{2.53\pi n}{1+4\pi^2n^2}}$$

$$x(t) = a_0 + \sum_{n=1}^{\infty} a_n\cos \omega_0 nt + \sum_{n=1}^{\infty} b_n \sin \omega_0 nt$$

$$\boxed{\begin{aligned} x(t) =& 0.632 + 1.264 \sum_{n=1}^{\infty} \frac{n}{(1+4\pi^2 n^2)} \cos 2\pi nt \\ &+ 2.53\pi \sum_{n=1}^{\infty} \frac{n}{(1+4\pi^2 n^2)} \sin 2\pi nt \end{aligned}}$$

For $n = 1$

$$\begin{aligned} x_1(t) &= 0.312 \cos 2\pi t + 0.196 \sin 2\pi t \\ &= \sqrt{(0.312)^2 + (0.196)^2} \cos\left(2\pi t - \tan^{-1} \frac{0.196}{0.312}\right) \\ &= 0.2 \cos(2\pi t - 0°) \\ &= 0.2 \cos(2\pi t) \qquad \text{(Fundamental component)} \end{aligned}$$

For $n = 2$

$$\begin{aligned} x_2(t) &= 0.159 \cos 4\pi t + 0.1 \sin 4\pi t \\ &= 0.1 \cos(4\pi t) \qquad \text{(First harmonic)} \end{aligned}$$

For $n = 3$

$$\begin{aligned} x_3(t) &= 0.0106 \cos 6\pi t + 0.0666 \sin 6\pi t \\ &= 0.067 \cos 6\pi t \\ x(t) &= a_0 + x_1(t) + x_2(t) + x_3(t) \end{aligned}$$

$$\boxed{x(t) = [0.632 + 0.2 \cos 2\pi t + 0.1 \cos 4\pi t + 0.067 \cos 6\pi t + \cdots]}$$

(c) **Polar form Fourier series (cosine form FS)**

$$\begin{aligned} x(t) &= C_0 + \sum_{n=1}^{\infty} C_n \cos(\omega_0 nt - \theta_n) \\ C_0 &= a_0 \end{aligned}$$

$$\begin{aligned} C_n &= \sqrt{a_n^2 + b_n^2} \\ \theta_n &= \tan^{-1}\left(\frac{b_n}{a_n}\right) \end{aligned}$$

$$\boxed{C_0 = 0.632}$$

$$C_n = \frac{\sqrt{(1.6 + 6.4\pi^2 n^2)}}{(1 + 4\pi^2 n^2)};$$

$$\boxed{C_n = \frac{1.265}{\sqrt{1 + 4\pi^2 n^2}}}$$

$$\theta_n = \tan^{-1} \frac{2.53\pi n}{1.264} = \tan^{-1} 2\pi n = 0$$

$$x(t) = 0.632 + \sum_{n=1}^{\infty} \frac{\sqrt{1.6 + 6.4\pi^2 n^2}}{(1 + 4\pi^2 n^2)} \cos[2\pi nt]$$

$$\boxed{x(t) = 0.632 + \sum_{n=1}^{\infty} \frac{1.265}{\sqrt{(1 + 4\pi^2 n^2)}} \cos 2\pi nt}$$

(d) 1. $a_0 = C_0 = D_0 = 0.632$
2.

$$|D_n| = \frac{C_n}{2} = \frac{1.265}{2\sqrt{1 + 4\pi^2 n^2}} = \frac{0.632}{\sqrt{1 + 4\pi^2 n^2}}$$

3.

$$\begin{aligned} C_n &= \sqrt{a_n^2 + b_n^2} \\ &= \frac{\sqrt{(1.264)^2 + 2.53^2\pi^2 n^2}}{(1 + 4\pi^2 n^2)} \\ &= \frac{\sqrt{1.6(1 + 4\pi^2 n^2)}}{(1 + 4\pi^2 n^2)} \\ &= \frac{1.265}{\sqrt{(1 + 4\pi^2 n^2}} \end{aligned}$$

4.5 Existence of Fourier Series—the Dirichlet Conditions

The continuous Fourier series of the signal $x(t)$, is represented in the following form.

$$x(t) = \sum_{n=-\infty}^{\infty} D_n e^{j2\pi nt} \tag{4.36}$$

where

$$D_n = \frac{1}{T_0} \int_0^{T_0} x(t) e^{-j2\pi nt} dt \tag{4.37}$$

and n represents the harmonic number.

If the integral in Eq. (4.37) diverges, CTFS cannot be found for $x(t)$. If certain constraints are put on $x(t)$, Eq. (4.37) converges and the conditions are called Dirichlet conditions. The Dirichlet Conditions are:

1. The signal $x(t)$ must be absolutely integrable over the time interval $t_0 < t < t_0 + T_0$. The above condition implies that

$$\int_{t_0}^{t_0+T_0} |x(t)|\, dt < \infty \tag{4.38}$$

2. The signal $x(t)$ must have a finite number of maxima and minima in the time interval $t_0 < t < t_0 + T_0$.
3. The signal $x(t)$ must have finite number of discontinuities in the time interval $t_0 < t < t_0 + T_0$.

4.6 Convergence of Continuous Time Fourier Series

The arbitrary signal $x(t)$ can be expressed by FS in Eq. (4.4), if it is periodic. It does not mean that every periodic signal can be expressed by FS. When the series uses a fixed number of terms, then it guarantees convergence. If the energy difference between the signal $x(t)$ and the corresponding finite term series approaches zero, as the number of terms approaches infinity, such a series is said to be convergent in the mean. The Fourier series of $x(t)$ converges in the mean if it has finite energy over one period. This can be expressed as

$$E = \int_{T_0} |x(t)|^2 dt < \infty \tag{4.39}$$

When conditions (4.39) are satisfied, the Fourier series converges in the mean and also guarantees that the Fourier coefficient is finite.

4.7 Fourier Series Spectrum

The plot of Fourier series coefficients with respect to ω is called Fourier series spectrum. In exponential Fourier series and in polar Fourier series, the Fourier series, the FS coefficients D_n and C_n are complex. Thus, these coefficients have magnitude and angle. Thus, the plots of D_n versus n and $\angle D_n$ versus n are called exponential Fourier spectra. Similarly, the plots of $|C_n|$ versus n and $\angle C_n$ versus n are called trigonometric Fourier spectra. The following examples illustrate the above methods.

Example 4.15

For the Example 4.14, plot the exponential Fourier spectra for the periodic signal $x(t)$ shown in Fig. 4.12.

Solution: The exponential Fourier series coefficient of Fig. 4.12 has been derived as

$$D_n = \frac{0.632}{1 + j2\pi n} = \frac{0.632}{\sqrt{1 + 4\pi^2 n^2}} \angle - \tan^{-1} 2\pi n$$

For $n = 0$

$$D_0 = 0.632\angle 0^\circ$$

For $n = \pm 1$,

$$\begin{aligned}
D_1 &= D_{-1} = 0.1\angle \mp 81^\circ \\
D_2 &= D_{-2} = 0.05\angle \mp 85.5^\circ \\
D_3 &= D_{-3} = 3.35 \times 10^{-2}\angle \mp 87^\circ \\
D_4 &= D_{-4} = 2.5 \times 10^{-2}\angle \mp 87.7^\circ \\
D_5 &= D_{-5} = 2 \times 10^{-2}\angle \mp 88.2^\circ \\
D_6 &= D_{-6} = 1.68 \times 10^{-2}\angle \mp 88.5^\circ \\
D_7 &= D_{-7} = 1.44 \times 10^{-2}\angle \mp 88.7^\circ
\end{aligned}$$

The magnitude spectrum of D_n is shown in Fig. 4.13a and the phase spectrum is Fig. 4.13b. **Note:** $\omega = n\omega_0 = 2\pi n$ or $n = \frac{\omega}{2\pi}$ which is a function of frequency.

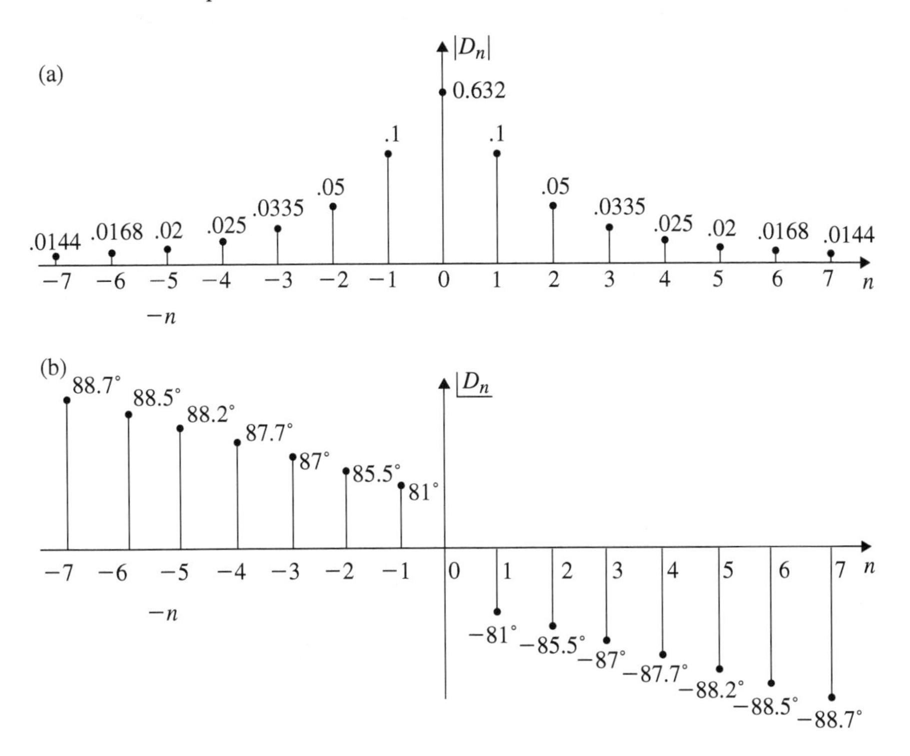

Fig. 4.13 Frequency spectra of Example 4.15. **a** Magnitude spectrum and **b** Phase angle spectrum

■ Example 4.16

Find the trigonometric and exponential series representation of the signal whose mathematical description is given as

$$x(t) = \begin{cases} 1 & 0 \leq t < \frac{T}{2} \\ 0 & \frac{T}{2} \leq t < T \end{cases}$$

$$x(t+T) = x(t)$$

Using the differentiation property of F.S., determine the exponential F.S. coefficient and verify the results.

(*Anna University, 2009 and 2013*)

Solution:

1. **Trigonometric F.S. Representation**
   The mathematical description is sketched as the waveform in Fig. 4.14a and its derivative in Fig. 4.14b. From Fig. 4.14a

$$T_0 = T$$

and

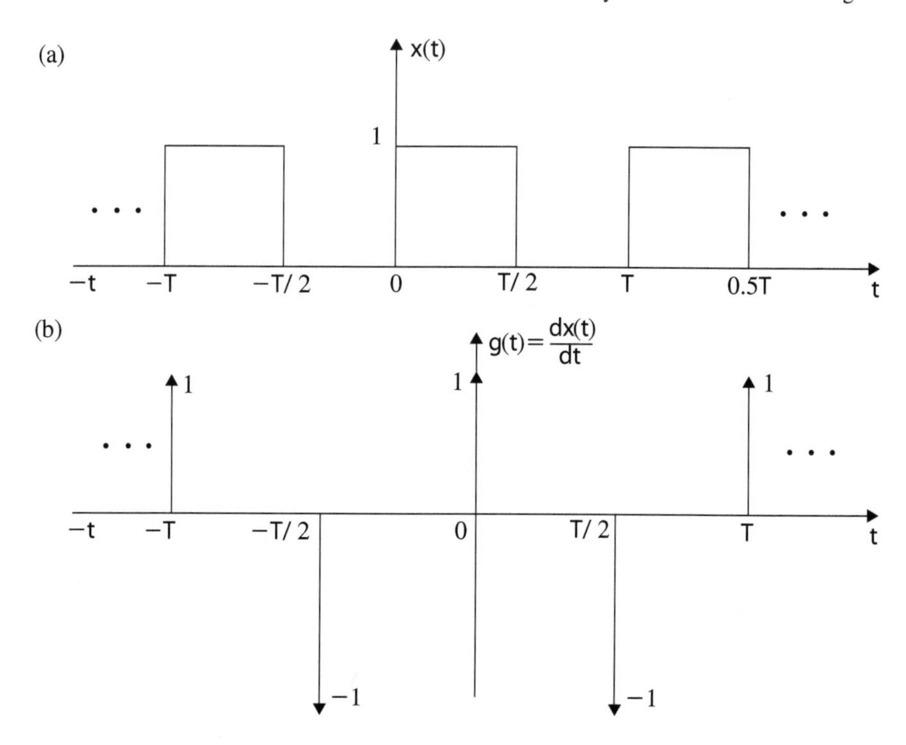

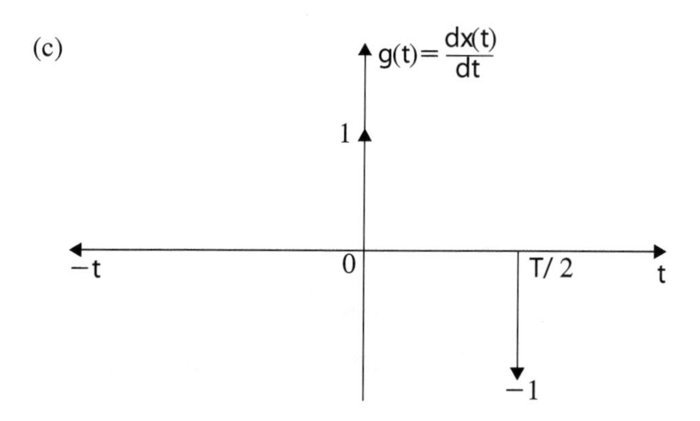

Fig. 4.14 **a**, **b** Representation of $x(t)$ and $dx(t)/dt$ of Example 4.16. **c** Representation of impulses in one period

$$\omega_0 = \frac{2\pi}{T_0} = \frac{2\pi}{T}$$

The trigonometric F.S. coefficients are determined using Eqs. (4.5), (4.6) and (4.7)

$$\begin{aligned} a_0 &= \frac{1}{T_0}\int_{T_0} x(t)dt \\ &= \frac{1}{T_0}\int_0^{T/2} 1dt \end{aligned}$$

$$\begin{aligned} a_0 &= \frac{1}{T}\Big[t\Big]_0^{T/2} \\ &= \frac{1}{2} \\ a_n &= \frac{2}{T_0}\int_{T_0} x(t)\cos\omega_0 ntdt \\ &= \frac{2}{T}\int_0^{T/2} 1\cos\omega_0 ntdt \\ &= \frac{2}{T}\frac{1}{\omega_0 n}[\sin\omega_0 nt]_0^{T/2} \\ &= \frac{2}{T}\frac{T}{2\pi n}\left[\sin\frac{2\pi}{T}nt\right]_0^{T/2} \\ &= \frac{1}{\pi n}\sin\pi n \\ &= 0 \qquad \text{(for all } n\text{)} \\ b_n &= \frac{2}{T_0}\int_{T_0} x(t)\sin\omega_0 ntdt \\ &= \frac{2}{T}\int_0^{T/2} 1\sin\omega_0 ntdt \end{aligned}$$

$$\begin{aligned} b_n &= \frac{2}{T}\frac{1}{\omega_0 n}[-\cos\omega_0 nt]_0^{T/2} \\ &= \frac{2}{T}\frac{T}{2\pi n}\left[-\cos\frac{2\pi}{T}n\frac{T}{2} - 1\right] \\ &= \frac{1}{\pi n}[1-\cos\pi n] \\ &= 0 \qquad \text{(for even values of } n\text{)} \\ &= \frac{2}{\pi n} \qquad \text{(for odd values of } n\text{)} \end{aligned}$$

$$x(t) = a_0 + \sum_{n=1}^{\infty} a_n \cos \omega_0 nt + \sum_{n=1}^{\infty} b_n \sin \omega_0 nt$$
$$= \frac{1}{2} + \frac{2}{\pi} \sum_{n=1}^{\infty} \sin \omega_0 nt \qquad \text{(for odd values of } n\text{)}$$

$$\boxed{x(t) = \left[\frac{1}{2} + \frac{2}{\pi}\left\{\sin\left(\frac{2\pi}{T}t\right) + \frac{1}{3}\sin\left(3\frac{2\pi}{T}t\right) + \frac{1}{5}\sin\left(5\frac{2\pi}{T}t\right) + \cdots\right\}\right]}$$

2. **Exponential F.S. Representation of** $x(t)$
   The exponential F.S. coefficient is given by

$$D_n = \frac{1}{T_0}\int_{T_0} x(t)e^{-jn\omega_0 t}dt$$
$$= \frac{1}{T}\int_0^{T/2} 1e^{-jn\omega_0 t}dt$$
$$= \frac{1}{T(-jn\omega_0)}\left[e^{-jn\omega_0 t}\right]_0^{T/2}$$
$$= -\frac{1}{jT}\frac{T}{2\pi n}\left[e^{-jn(2\pi/T)(T/2)} - 1\right]$$
$$= -\frac{1}{j2\pi n}\left[e^{-j\pi n} - 1\right]$$
$$= -\frac{1}{j2\pi n}\left[\cos \pi n - j\sin \pi n - 1\right]$$
$$= \frac{1}{j2\pi n}\left[1 - \cos \pi n\right] \qquad [\sin \pi n = 0 \text{ for any integer value of } n]$$
$$= 0 \qquad \text{(for even values of } n\text{)}$$
$$= \frac{1}{j\pi n} \qquad \text{(for odd values of } n\text{)}$$

But D_0 is to be calculated from first principle since $D_n|_{n=0} = \infty$.

$$D_n = \frac{1}{T_0}\int_0^{T/2} 1dt$$

$$D_0 = \frac{1}{T}\left[t\right]_0^{T/2}$$
$$= \frac{1}{2}$$

The F.S. is written as

$$x(t) = D_0 + \sum_{n=-\infty}^{\infty} D_n e^{jn\omega_0 t}$$

For $n = \pm 1$

$$\begin{aligned}
x_1(t) &= \frac{1}{j\pi} e^{j(2\pi/T)t} \\
x_{-1}(t) &= -\frac{1}{j\pi} e^{-j(2\pi/T)t} \\
x_1(t) + x_{-1}(t) &= \frac{1}{j\pi}\left[e^{j(2\pi/T)t} - e^{-j(2\pi/T)t}\right] \\
&= \frac{2}{\pi}\sin\frac{2\pi}{T}t
\end{aligned}$$

For $n = \pm 3$

$$\begin{aligned}
x_3(t) &= \frac{1}{j3\pi} e^{+j(2\pi/T)3t} \\
x_{-3}(t) &= -\frac{1}{j3\pi} e^{-j(2\pi/T)3t} \\
x_3(t) + x_{-3}(t) &= \frac{1}{j3\pi}\left[e^{j(2\pi/T)3t} - e^{-j(2\pi/T)3t}\right] \\
&= \frac{2}{3\pi}\sin\left(3\frac{2\pi}{T}\right)t
\end{aligned}$$

Thus, $x(t)$ is written as

$$\boxed{x(t) = \left[\frac{1}{2} + \frac{2}{\pi}\left\{\sin\left(\frac{2\pi}{T}t\right) + \frac{1}{3}\sin 3\left(\frac{2\pi}{T}3t\right) + \cdots\right\}\right]}$$

3. **Exponential F.S. Using Differentiation Technique**
The derivative of $x(t)$ is sketched as a train of pulses in Fig. 4.14b. Let

$$g(t) = \frac{dx(t)}{dt}$$

Let D'_n be the exponential F.S. coefficient of $g(t)$. Consider the pulses in one cycle which is represented in the Fig. 4.14c. From Fig. 4.14c, the F.S. coefficient is written as

$$\begin{aligned}
D'_n &= \frac{1}{T}\left[1 - e^{-j\omega_0 n(T/2)}\right] \\
&= \frac{1}{T}\left[1 - e^{-j(2\pi/T)n(T/2)}\right]
\end{aligned}$$

$$= \frac{1}{T}\left[1 - e^{-j\pi n}\right]$$

$$= \frac{1}{T}\left[1 - \cos \pi n\right] \qquad \text{(which exists for } n \text{ is odd)}$$

$$= 0 \qquad \text{(for } n \text{ is even)}$$

$$D_n' = \frac{2}{T}$$

Using the differentiation property 4.47, we get

$$(j\omega_0 n)D_n = D_n'$$

$$D_n = \frac{D_n'}{j\omega_0 n}$$

$$= \frac{2}{jT(2\pi/T)n}$$

$$= \frac{1}{j\pi n}$$

Thus, identical results are obtained as in part (2) of the problem in few simpler steps. The F.S. of $x(t)$ following end stage of part (2) of the problem we write

$$\boxed{x(t) = \left[\frac{1}{2} + \frac{2}{\pi}\left\{\sin\left(\frac{2\pi}{T}t\right) + \frac{1}{3}\sin 3\left(\frac{2\pi}{T}t\right) + \frac{1}{5}\sin 5\left(\frac{2\pi}{T}t\right) + \cdots\right\}\right]}$$

■ Example 4.17

Consider the triangular wave $x(t)$ shown in Fig. 4.15a.

(a) Find the trigonometric Fourier series using differentiation technique.
(b) Find the exponential F.S. using differentiation technique.

Solution: The equation of the triangle of $x(t)$ can be written as follows.

The slope of the straight line of triangle is,

$$m = -\frac{3}{4}$$

The equation of the straight line of the triangle is,

$$x(t) = -\frac{3}{4}t + c$$

At $t = 0$; $x(t) = 3$ and hence $c = 3$.

$$x(t) = -\frac{3}{4}t + 3 \qquad 0 \le t < 4$$

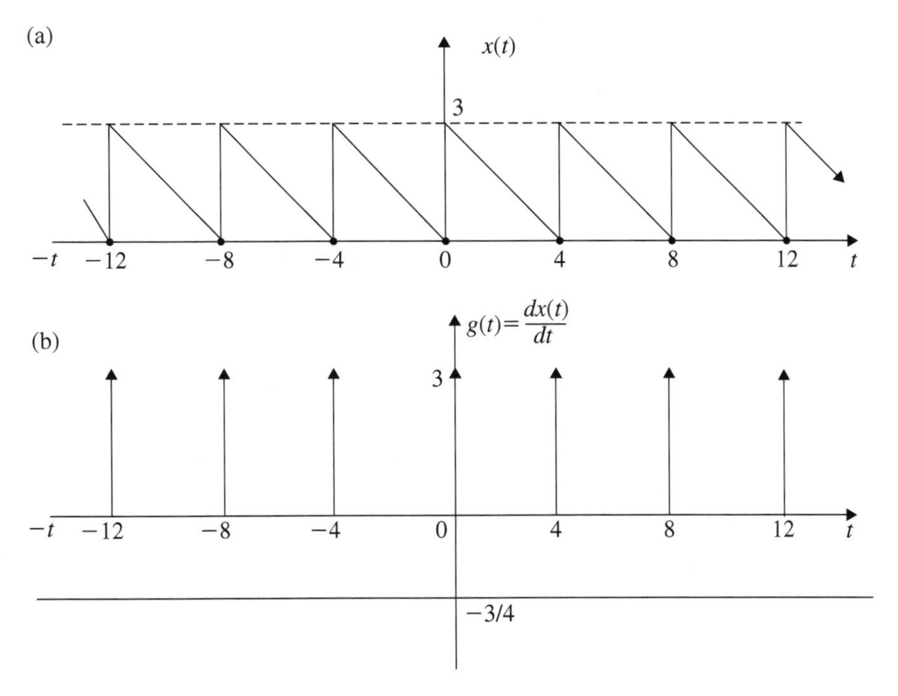

Fig. 4.15 Triangular wave of Example 4.17. **a** $x(t)$ and **b** $dx(t)/dt$

(a) **To Determine F.S. of the Train of Impulses and the DC Term**
Differentiating the above equation with respect to t we get

$$\frac{dx(t)}{dt} = -\frac{3}{4} + 3\delta(t)$$

Considering all the triangles, the above equation is written as

$$\frac{dx(t)}{dt} = -\frac{3}{4} + 3\sum_{n=-\infty}^{\infty} \delta(t - 4n)$$

where n is an integer. The above equation is represented in Fig. 4.15b. Using trigonometric Fourier series, the train of impulses can be represented as follows.

The periodicity of the impulses is given by

$$T_0 = 4 \text{ sec}$$

and

$$\omega_0 = \frac{2\pi}{T_0} = \frac{\pi}{2}$$

$$a_0 = \frac{1}{T_0}\int_0^{T_0} 3\delta(t)dt$$
$$= \frac{3}{4}\int_0^{4} \delta(t)dt$$
$$= \frac{3}{4}$$

Since the train of impulses represents an even function $b_n = 0$ and a_n is determined using Eq. (4.6).

$$a_n = \frac{2}{T_0}\int_0^{T_0} x(t)\cos n\omega_0 t dt$$
$$= \frac{2\times 3}{4}\int_0^{4} \delta(t)\cos\frac{\pi}{2}nt dt \qquad \left[\delta(t)\cos\frac{\pi}{2}nt = \delta(t)\right]$$
$$= \frac{3}{2}\int_0^{4} \delta(t)dt$$
$$= \frac{3}{2}$$

Thus, the F.S. of the train of pulses is written using Eq. (4.4)

$$\sum_{n=-\infty}^{\infty} 3\delta(t-4n) \overset{\text{FS}}{\longleftrightarrow} a_0 + \sum_{n=1}^{\infty} a_n \cos n\omega_0 t + \sum_{n=1}^{\infty} b_n \sin n\omega_0 t$$
$$= \frac{3}{4} + \frac{3}{2}\sum_{n=1}^{\infty}\cos\left(\frac{\pi}{2}nt\right)$$

Taking the DC term into account we get the F.S. of $dx(t)/dt$ as given below

$$\frac{dx(t)}{dt} \overset{\text{FS}}{\longleftrightarrow} \frac{3}{4} + \frac{3}{2}\sum_{n=1}^{\infty}\cos\left(\frac{\pi}{2}nt\right) - \frac{3}{4}$$

$$\frac{dx(t)}{dt} \overset{\text{FS}}{\longleftrightarrow} \frac{3}{2}\sum_{n=1}^{\infty}\cos\left(\frac{\pi}{2}nt\right) \qquad (a)$$

To Find the F.S. of the Train of Triangle $(x(t))$

Using Eq. (4.4), the F.S. of the series of the triangles is obtained as explained below

$$x(t) \overset{\text{FS}}{\longleftrightarrow} a_0 + \sum_{n=1}^{\infty} a_n \cos \omega_0 nt + \sum_{n=1}^{\infty} b_n \sin \omega_0 nt$$

Differentiating the above equation, we get

$$\frac{dx(t)}{dt} \overset{\text{FS}}{\longleftrightarrow} \sum_{n=1}^{\infty} a_n \omega_0 n(-\sin \omega_0 nt) + \sum_{n=1}^{\infty} b_n \omega_0 n(\cos \omega_0 nt) \qquad (b)$$

Comparing Eqs. (a) and (b), we get

$$\begin{aligned}
a_n &= 0 \\
b_n \omega_0 n &= \frac{3}{2} \\
b_n &= \left(\frac{3}{2n}\right)\frac{2}{n} \\
&= \frac{3}{\pi n} \\
a_0 &= \frac{2}{T_0}\int_0^{T_0} x(t)dt \\
&= \frac{1}{4}\int_0^4 \left(3 - \frac{3}{4}t\right) dt \\
&= \frac{1}{4}\left[3t - \frac{3}{8}t^2\right]_0^4 \\
&= \frac{3}{2}
\end{aligned}$$

The F.S. of $x(t)$ is written as

$$x(t) = \frac{3}{2} + \sum_{n=1}^{\infty} \frac{3}{\pi n} \sin \frac{\pi}{2} nt$$

$$\boxed{x(t) = \frac{3}{2} + \frac{3}{\pi}\left\{\sin\left(\frac{\pi}{2}t\right) + \frac{1}{2}\sin 2\left(\frac{\pi}{2}t\right) + \frac{1}{3}\sin 3\left(\frac{\pi}{2}t\right) + \cdots\right\}}$$

Summary of Steps Followed

1. The given signal $x(t)$ is differentiated and $dx(t)/dt$ is represented in graph which is in the form of train of impulses and a DC term.
2. Since the series of impulses represent an even function, the F.S. coefficient $b_n = 0$. Find a_0 and a_n using Eqs. (4.5) and (4.6).
3. With these coefficients express $dx(t)/dt$ in F.S.

4. $x(t)$ is expressed by the analysis equation as

$$x(t) \stackrel{\text{FS}}{\longleftrightarrow} a_0 + \sum_{n=1}^{\infty}(a_n \cos \omega_0 nt + b_n \sin \omega_0 nt)$$

$$\frac{dx(t)}{dt} \stackrel{\text{FS}}{\longleftrightarrow} \sum_{n=1}^{\infty} -a_n\omega_0 n \sin \omega_0 nt + b_n\omega_0 n \cos \omega_0 nt$$

5. Compare the coefficients obtained in step (3) with those obtained in step (4) and get a_n and b_n which are F.S. coefficients of $x(t)$.
6. Find a_0 using Eq. (4.5).
7. Express $x(t)$ in F.S. using Eq. (4.4).

(b) **To Determine the F.S. Using Exponential F.S. The Train of Impulses**
The exponential Fourier series and the F.S. coefficients are obtained using Eqs. (4.9) and (4.10), respectively. The first derivative of the signal $x(t)$ is represented in Fig. 4.15b. The F.S. coefficient for the train of impulses is determined as

$$\begin{aligned} D_n &= \frac{1}{T_0}\int_{T_0} \delta(t)e^{-jn\omega_0 t}dt \qquad [\delta(t)e^{-jn\omega_o t} = 1] \\ &= \frac{3}{4}\int_0^4 \delta(t)dt \\ &= \frac{3}{4} \end{aligned}$$

$$\frac{dx(t)}{dt} \stackrel{\text{FS}}{\longleftrightarrow} -\frac{3}{4} + \frac{3}{4}\sum_{n=-\infty}^{\infty} e^{j\omega_0 nt}$$

$$\frac{dx(t)}{dt} \stackrel{\text{FS}}{\longleftrightarrow} -\frac{3}{4} + \frac{3}{4}\sum_{n=-\infty}^{\infty} e^{j(\pi/2)nt} \qquad (a)$$

Fourier Series for the Triangle

The F.S. for the triangle is determined as follows:

$$D_n = \frac{1}{T_0}\int_{T_0} x(t)e^{-jn\omega_0 t}dt$$

$$x(t) = \sum_{n=-\infty}^{\infty} D_n e^{j\omega_0 nt}$$

$$\frac{dx(t)}{dt} = \sum_{n=-\infty}^{\infty} D_n(jn\omega_0)e^{j\omega_0 nt}$$

$$\frac{dx(t)}{dt} = \sum_{n=-\infty}^{\infty} D_n \left(jn\frac{\pi}{2}\right) e^{j(\pi/2)nt} \tag{b}$$

Comparing Eqs. (a) and (b), we get

$$D_n \left(j\frac{\pi}{2}n\right) = \frac{3}{4}$$

$$\boxed{D_n = -j\frac{3}{2\pi n}}$$

$$x(t) = \sum_{n=-\infty}^{\infty} D_n e^{j\omega_0 nt}$$
$$= -j\frac{3}{2\pi} \sum_{n=-\infty}^{\infty} \frac{1}{n} e^{j(\pi/2)nt}$$

The values of n is substituted from $-\infty$ to $+\infty$ except $n = 0$. At $n = 0$; $D_0 = a_0 = \frac{3}{2}$

$$x_0(t) = \frac{3}{2}$$

For $n = 1$

$$x_1(t) = -j\frac{3}{2\pi} e^{j(\pi/2)t}$$
$$= -j\frac{3}{2\pi}\left(\cos\frac{\pi}{2}t + j\sin\frac{\pi}{2}t\right)$$

For $n = -1$

$$x_{-1}(t) = +j\frac{3}{2\pi}\left(\cos\frac{\pi}{2}t - j\sin\frac{\pi}{2}t\right)$$
$$x_1(t) + x_{-1}(t) = -j\frac{3}{2\pi}\left[2j\sin\frac{\pi}{2}t\right]$$
$$= \frac{3}{\pi}\left[\sin\frac{\pi}{2}t\right]$$

For $n = \pm 2$

$$x_2(t) = -j\frac{3}{2\pi}\frac{1}{2}(e^{j\pi t})$$
$$x_{-2}(t) = +j\frac{3}{2\pi}\frac{1}{2}(e^{-j\pi t})$$
$$x_2(t) + x_{-2}(t) = -j\frac{3}{2\pi}\frac{1}{2}\left[e^{j\pi t} - e^{-j\pi t}\right]$$
$$= \frac{3}{2\pi}\left[\sin \pi t\right]$$
$$= \frac{3}{2\pi}\sin\left(2\frac{\pi}{2}t\right)$$

Similarly for $n = 3$

$$x_3(t) + x_{-3}(t) = \frac{1}{\pi}\sin\left(3\frac{\pi}{2}t\right)$$

In general, $x(t)$ is written as

$$\boxed{x(t) = \frac{3}{2} + \frac{3}{\pi}\left[\sin\left(\frac{\pi}{2}t\right) + \frac{1}{2}\sin\left(2\frac{\pi}{2}t\right) + \frac{1}{3}\sin\left(3\frac{\pi}{2}t\right) + \cdots\right]}$$

The same result using trigonometric F.S. as well as exponential F.S. was obtained.

■ Example 4.18

Consider the waveform shown in Fig. 4.1. Using differentiation and integration properties of F.S., determine the exponential Fourier series coefficient and hence $x(t)$. Verify the results so obtained with that of Example 4.1.

Solution: The first derivative of the signal shown in Fig. 4.1 is shown in Fig. 4.16a which is in the form of train of impulses which is alternatively going positive and negative with amplitude 2. Consider the impulses in one cycle which is shown in Fig. 4.16b. Here,

$$\frac{dx(t)}{dt} = 2[\delta(t+1) - \delta(t-1)]$$

Let D'_n be the F.S. coefficient for these impulses. From Fig. 4.1.

$$T_0 = 4 \text{ sec}$$

and

$$\omega_0 = \frac{2\pi}{T_0} = \frac{\pi}{2} \text{ rad/s}$$

From Fig. 4.16b, D'_n can be written as

$$\begin{aligned} D'_n &= \frac{2}{4}[e^{j\omega_0 n} - e^{-j\omega_0 n}] \\ &= j \sin \omega_0 n \end{aligned}$$

According to integration property, D_n, the complex exponential F.S. coefficient of $x(t)$ can be written as

$$\begin{aligned} D_n &= \frac{D'_n}{j\omega_0 n} \\ &= \frac{j \sin \omega_0 n}{j\omega_0 n} \\ &= \frac{2 \sin(\pi/2)n}{\pi n} \\ &= \frac{2}{\pi n} \sin \frac{\pi n}{2} \qquad \text{(for } n \text{ is odd)} \\ &= 0 \qquad \text{(for } n \text{ is even)} \end{aligned}$$

(a)

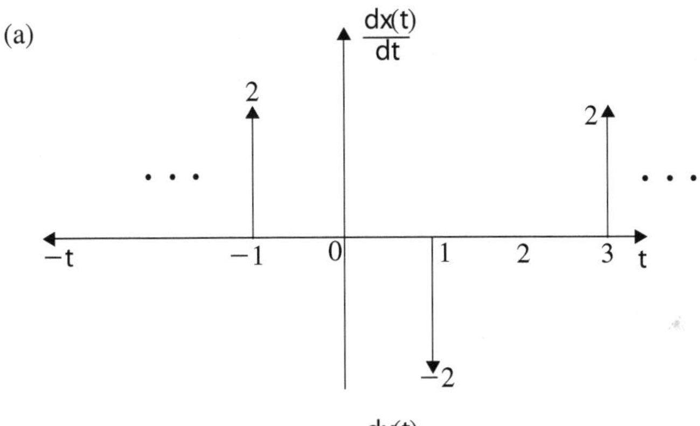

(b)

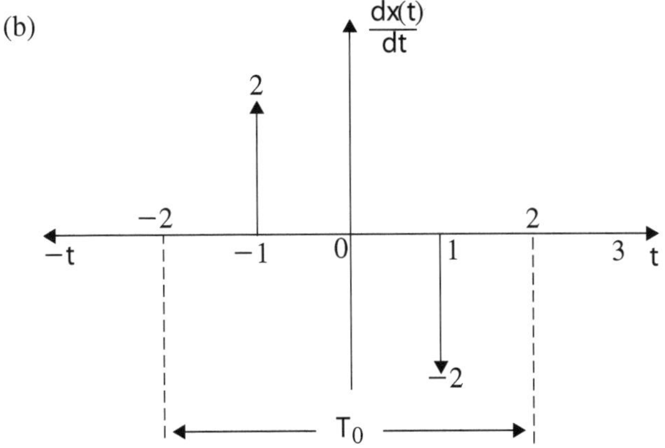

Fig. 4.16 Signal representation of first derivative of signal in Fig. 4.1. **a** $dx(t)/dt$ and **b** $dx(t)/dt$ in one period T_0

Since the waveform $x(t)$ is symmetrical with respect to t, $D_0 = 0$.

Using Eq. (4.9), we write

$$x(t) = \sum_{n=-\infty}^{\infty} D_n e^{j\omega_0 nt}$$

where $n \neq 0$. For $n = \pm 1$

$$\begin{aligned} x_1(t) &= \frac{2}{\pi}[e^{j\omega_0 t} + e^{-j\omega_0 t}] \\ &= \frac{4}{\pi} \cos \omega_0 t \\ &= \frac{4}{\pi} \cos \frac{\pi}{2} t \end{aligned}$$

For $n = -3$

$$\begin{aligned} D_{-3} &= \frac{2}{-\pi \times 3} \sin\left(-\frac{\pi}{2} \times 3\right) \\ &= -\frac{2}{\pi \times 3} \end{aligned}$$

For $n = 3$

$$\begin{aligned} D_{+3} &= \frac{2}{\pi \times 3} \sin\left(\frac{\pi}{2} \times 3\right) \\ &= -\frac{2}{\pi \times 3} \\ x_3(t) &= (D_{-3}e^{-j\omega_0 t} + D_3 e^{j\omega_0 t}) \\ &= -\frac{2}{\pi}\frac{1}{3}[e^{-j(3\pi/2)t} + e^{j(3\pi/2)t}] \end{aligned}$$

For $n = \pm 3$

$$x_3(t) = -\frac{4}{\pi}\frac{1}{3} \cos\left(3\frac{\pi}{2}t\right)$$

Thus, we write

$$\boxed{x(t) = \frac{4}{\pi}\left[\cos\left(\frac{\pi}{2}t\right) - \frac{1}{3}\cos\left(3\frac{\pi}{2}t\right) + \frac{1}{5}\cos 5\frac{\pi}{2}t + \cdots\right]}$$

The same result is obtained in Example 4.1. The number of steps and the mathematical operations involved in the above method are very less.

■ Example 4.19

Consider periodic waveform shown in Fig. 4.17a. Using differentiation and integration properties of F.S. determine the exponential F.S. coefficient and hence F.S. $x(t)$.

Solution: The given signal is represented in Fig. 4.17a. The signal $x(t)$ repeats itself for every period $T_0 = 2$ sec. Hence, the fundamental frequency is

$$\omega_0 = \frac{2\pi}{T_0} = \pi \text{ rad/s.}$$

From Fig. 4.17a, the mathematical description for $x(t)$ is given as

$$x(t) = \begin{cases} 4t & -\frac{1}{2} \leq t < \frac{1}{2} \\ 4(1-t) & \frac{1}{2} \leq t < \frac{3}{2} \end{cases}$$

$$x(t+T) = x(t)$$
$$\frac{dx(t)}{dt} = \begin{cases} 4 & -\frac{1}{2} \leq t < \frac{1}{2} \\ -4 & \frac{1}{2} \leq t < \frac{3}{2} \end{cases}$$
$$x(t+T) = x(t)$$

$dx(t)/dt$ is represented in Fig. 4.17b. When $dx(t)/dt$ is further differentiated at multiplies of $\pm\frac{1}{2}$ the up going and down going impulses occur and in other places, the signal is zero. This is represented in Fig. 4.17c. Let us consider only the impulses within one period $T_0 = 2$ sec. At $t = -1/2$ we have $8\delta(t + 1/2)$ and at $t = 1/2$ we have $8\delta(t - 1/2)$. If we denote D_n' as the exponential F.S. coefficient, then we may write

$$D_n' = \frac{8}{T_0}\left[e^{j(1/2)\omega_0 n} - e^{-j(\frac{1}{2})\omega_0 n}\right]$$

where $T_0 = 2$ sec.

$$D_n' = j8 \sin\frac{1}{2}\omega_0 n$$
$$= j8 \sin\frac{1}{2}\pi n$$

Using integration property of F.S., the exponential F.S. of $x(t)$ is written as

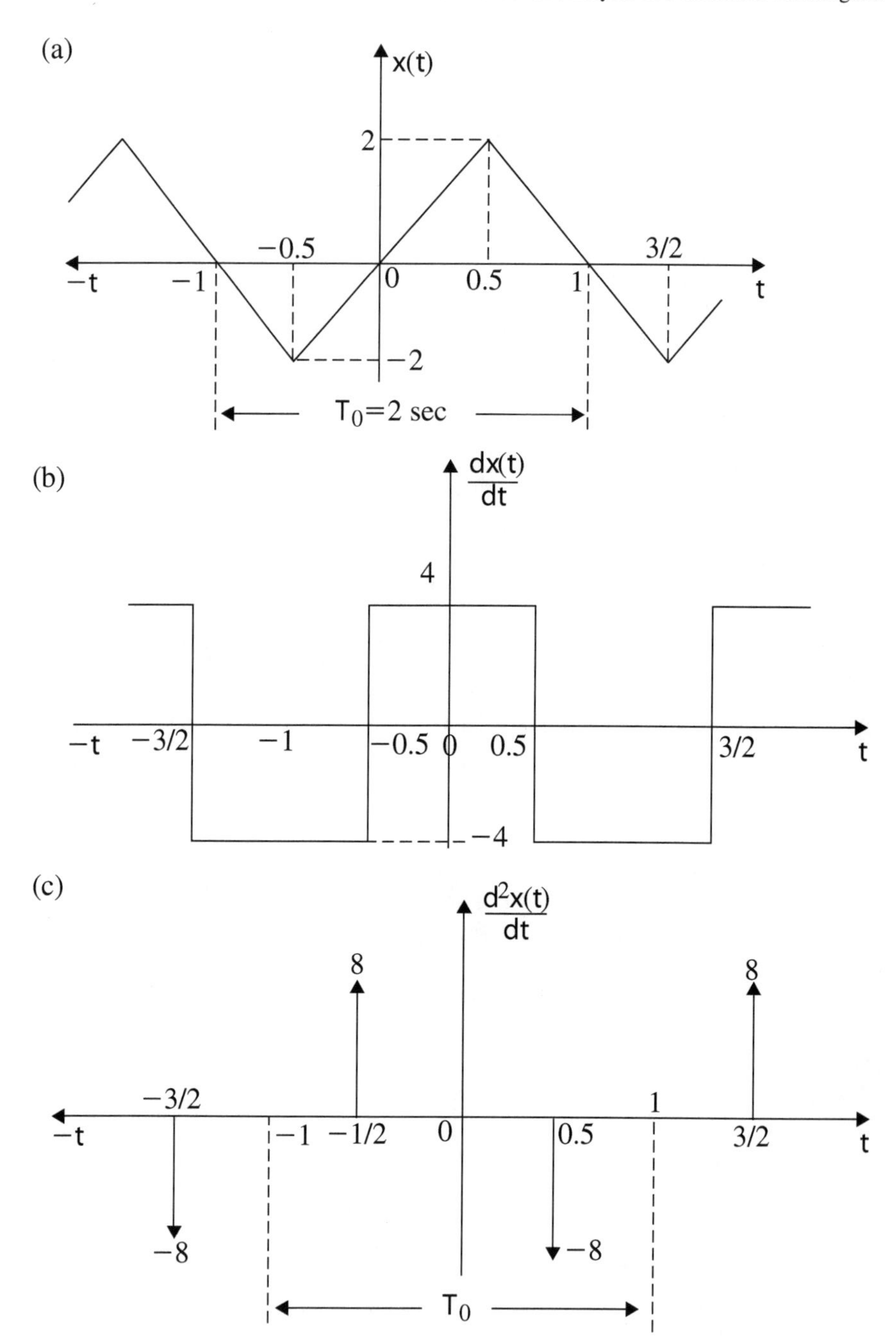

Fig. 4.17 Representation of $x(t)$ of Example 4.19 and its derivatives

$$\begin{aligned} D_n &= \frac{D'_n}{(j\omega_0 n)^2} \\ &= \frac{-j8}{\pi^2 n^2} \sin \frac{\pi}{2} n \qquad \text{(for all odd values of} n) \\ &= 0 \qquad \text{(for even values of} n) \end{aligned}$$

Since $x(t)$ is symmetrical with respect to time axis, $D_0 = 0$. The F.S. of $x(t)$ is written using Eq. (4.9) as

$$\begin{aligned} x(t) &= \sum_{n=-\infty}^{\infty} D_n e^{j\omega_0 nt} \\ &= \frac{-j8}{\pi^2} \sum_{n=-\infty}^{\infty} \frac{1}{n^2} \sin \frac{\pi}{2} n e^{j\pi nt} \end{aligned}$$

For $n = -1$

$$D_{-1} = \frac{-j8}{\pi^2} \sin\left(-1\frac{\pi}{2}\right) = \frac{j8}{\pi^2}$$

For $n = 1$

$$\begin{aligned} D_1 &= \frac{-j8}{\pi^2} \sin\left(\frac{\pi}{2}\right) \\ &= -\frac{j8}{\pi^2} \end{aligned}$$

$$\begin{aligned} x_1(t) &= \left[D_{-1} e^{-j\omega_0 nt} + D_1 e^{j\omega_0 nt}\right] \\ &= -\frac{j8}{\pi^2} \frac{[e^{j\pi t} - e^{-j\pi t}]}{2j} \times 2j \end{aligned}$$

For $n = \pm 1$

$$x_1(t) = \frac{16}{\pi^2} \sin \pi t$$

For $n = \pm 3$

$$x_3(t) = -\frac{16}{\pi^2} \left[\frac{1}{9} \sin 3\pi t\right]$$

For $n = \pm 5$

$$x_5(t) = \frac{16}{\pi^2} \left[\frac{1}{25} \sin 5\pi t\right]$$

Summing up $x_1(t), x_2(t), x_3(t), \ldots$, we get $x(t)$ as

$$\boxed{x(t) = \frac{16}{\pi^2}\left[\sin \pi t - \frac{1}{9}\sin 3\pi t + \frac{1}{25}\sin 5\pi t - \cdots\right]}$$

■ Example 4.20

Consider the waveform whose mathematical description is given by

$$x(t) = t^2 \quad |t| < 1$$
$$x(t+T) = x(t)$$

Sketch the signal $x(t)$. Using differentiation and integration properties, find the exponential F.S. coefficient, and hence express $x(t)$ in Fourier series.

Solution:

$$x(t) = |t^2| \qquad |t| < 1$$

This waveform which is periodic is shown in Fig. 4.18a. The same waveform is given in Example 4.6 and is represented in Fig. 4.6. Example 4.6 was solved using trigonometric F.S. The same problem is solved, using differentiation and integration properties of exponential F.S.

$$x(t) = t^2$$

when differentiated becomes

$$\frac{dx(t)}{dt} = 2t$$

This waveform is shown in Fig. 4.18b. When $dx(t)/dt$ is further differentiated, we get

$$\begin{aligned}\frac{d^2x(t)}{dt^2} &= 2 - 2\delta(t+1) - 2\delta(t-1)\\ &= g_1(t) + g_2(t)\end{aligned}$$

where

$$g_1(t) = 2 \qquad |t| < 1$$

and

$$g_2(t) = 2[\delta(t+1) + \delta(t-1)]$$

$g_1(t)$ and $g_2(t)$ are represented in Fig. 4.18c and d, respectively. When $g_1(t)$ is differentiated

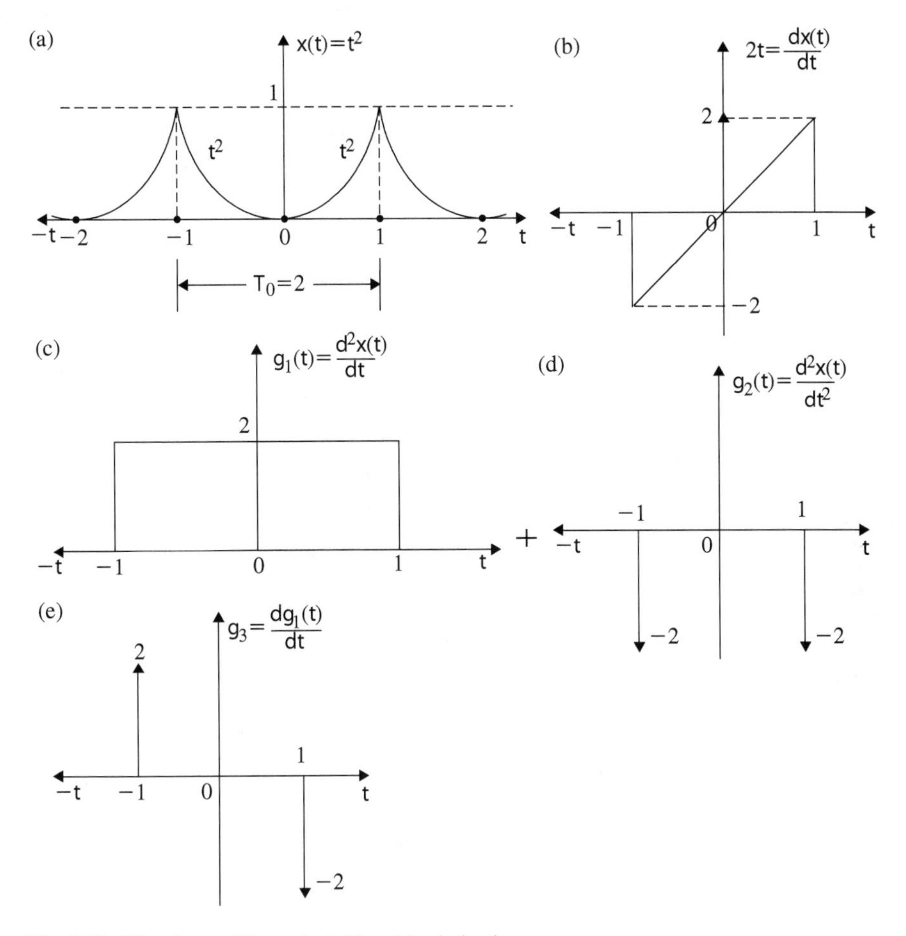

Fig. 4.18 Waveform of Example 4.20 and its derivatives

$$\frac{dg_1(t)}{dt} = 2[\delta(t+1) - \delta(t-1)]$$

is obtained and is sketched in Fig. 4.18e. From these, the exponential F.S. coefficient is obtained as given below

$$\begin{aligned} Dg_2 &= -\frac{2}{T_0}[e^{j\omega_0 n} + e^{-j\omega_0 n}] \\ &= -2\cos\omega_0 n \\ Dg_3 &= \frac{2}{T_0}[e^{j\omega_0 n} - e^{-j\omega_0 n}] \\ &= +j2\sin\omega_0 n \end{aligned}$$

where $T_0 = 2$ sec;

$$\omega_0 = \frac{2\pi}{T_0} = \frac{2\pi}{2} = \pi \text{ rad/s}$$

The exponential F.S. coefficient of $x(t)$ is obtained using integration property as

$$\begin{aligned} D_n &= \frac{Dg_2}{(j\omega_0 n)^2} + \frac{Dg_3}{(j\omega_0 n)^3} \\ &= \frac{+2\cos\pi n}{\pi^2 n^2} + j\frac{2\sin\pi n}{(j\omega_0 n)^3} \end{aligned}$$

Since $\sin\pi n = 0$ for all integer values of n

$$\begin{aligned} D_n &= \frac{+2\cos\pi n}{\pi^2 n^2} \\ &= 2\frac{(-1)^n}{\pi^2 n^2} \end{aligned}$$

where $n \neq 0$

$$\begin{aligned} D_0 &= \frac{1}{T_0}\int_{-1}^{1} x(t)dt \\ &= \frac{1}{2}\int_{-1}^{1} t^2 dt \\ &= \frac{1}{2}\left[\frac{t^3}{3}\right]_{-1}^{1} = \frac{1}{3} \end{aligned}$$

For $n = \pm 1$

$$x(t) = \sum_{n=-\infty}^{\infty} D_n e^{j\pi n t}$$

$$x_1(t) = -\frac{4}{\pi^2}\cos\pi t$$

For $n = \pm 2$

$$x_2(t) = +\frac{4}{\pi^2}\left[\frac{1}{4}\cos 2\pi(t)\right]$$

For $n = \pm 3$

$$x_3(t) = -\frac{4}{\pi^2}\left[\frac{1}{9}\cos 3\pi(t)\right]$$

Using Eq. (4.9) we write

$$\boxed{x(t) = \frac{1}{3} + \frac{4}{\pi^2}\left[-\cos\pi t + \frac{1}{4}\cos 2\pi t - \frac{1}{9}\cos 3\pi t + \cdots\right]}$$

■ Example 4.21

Consider the waveform which has the following mathematical description

$$x(t) = \begin{cases} 2+t & -2 \le t < 1 \\ 1 & |t| < 1 \\ 2-t & 1 \le t < 2 \end{cases}$$
$$x(t+4) = x(t)$$

(a) Sketch the form.
(b) Using differentiation technique find the exponential F.S. coefficient.
(c) Using exponential F.S. coefficient write the F.S. of $x(t)$.

Solution:

(a) For the given mathematical description, the waveform is represented as $x(t)$ in Fig. 4.19a.
(b) The first and second derivatives of $x(t)$ are shown in Fig. 4.19b and c, respectively.
(c) Let D'_n be the exponential F.S. coefficient of the waveform shown in Fig. 4.19c. From this figure, D'_n is written as

$$D'_n = \frac{1}{T_0}\left[e^{j2\omega_0 n} + e^{-j2\omega_0 n}\right] - \frac{1}{T_0}\left[e^{j\omega_0 n} + e^{-j\omega_0 n}\right]$$

where $T_0 = 4$ sec;

$$\omega_0 = \frac{2\pi}{T_0} = \frac{\pi}{2} \text{ rad/s}$$

$$D'_n = \frac{1}{2}\left[\cos\pi n - \cos\frac{\pi}{2}n\right]$$

Using the integration property of F.S.; D_n is calculated as

$$D_n = \frac{D'_n}{(j\omega_0 n)^2}$$

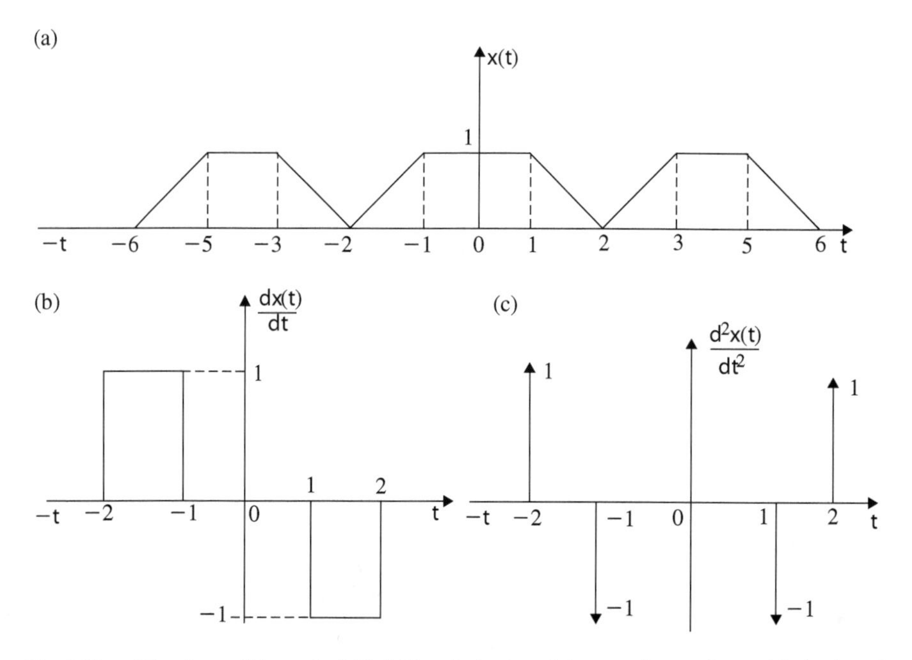

Fig. 4.19 **a** Waveform of Example 4.23. **b** First derivative of the waveform. **c** Second derivative of the waveform

$$\boxed{D'_n = -\frac{2}{\pi^2 n^2}\left[\cos \pi n - \cos \frac{\pi}{2} n\right]}$$

$x(t)$ can be expressed in Fourier series as

$$x(t) = -\frac{2}{\pi^2 n^2} \sum_{n=-\infty}^{\infty} \left[\cos \pi n - \cos \frac{\pi}{2} n\right] e^{j\omega_0 nt}$$

For $n = \pm 1$

$$\begin{aligned} x_1(t) &= -\frac{2}{\pi^2 n^2}\left[\cos \pi - \cos \frac{\pi}{2}\right]\left[e^{-j(\pi/2)nt} + e^{j(\pi/2)nt}\right] \\ &= \frac{4}{\pi^2} \cos\left(\frac{\pi}{2} t\right) \quad \text{(Fundamental component)} \end{aligned}$$

For $n = \pm 2$

$$\begin{aligned} x_2(t) &= -\frac{2}{\pi^2}\left[\frac{\cos 2\pi - \cos \pi}{2}\right]\left[e^{-j\pi t} + e^{j\pi t}\right] \\ &= \frac{4}{\pi^2}\left[-\frac{1}{2}\cos\left(2\frac{\pi}{2} t\right)\right] \quad \text{(Second harmonic)} \end{aligned}$$

For $n = \pm 3$

$$x_3(t) = -\frac{2}{\pi^2 \times 9}\left[\cos 3\pi - \cos\frac{3}{2}n\right]\left[e^{-j3(\pi/2)t} + e^{j3(\pi/2)t}\right]$$

$$= \frac{4}{\pi^2}\left[\frac{1}{9}\cos\left(3\frac{\pi}{2}t\right)\right] \quad \text{(Third harmonic)}$$

$$D_0 = \frac{1}{T_0}\left[\int_{-2}^{1}(t+2)dt + \int_{-1}^{1}(2-t)dt\right]$$

$$= \frac{1}{4}\left\{\left[\frac{t^2}{2} + 2t\right]_{-2}^{-1} + \left[t\right]_{-1}^{1} + \left[2t - \frac{t^2}{2}\right]_{1}^{2}\right\}$$

$$= \frac{1}{4}\left\{\left[\frac{1}{2} - 2 - \frac{4}{2} + 4 + 2 + 4 - \frac{4}{2} - 2 + \frac{1}{2}\right]\right\}$$

$$= \frac{3}{4}$$

The F.S. representation of $x(t)$ is

$$\boxed{x(t) = \frac{3}{4} + \frac{4}{\pi^2}\left[\cos\left(\frac{\pi}{2}t\right) - \frac{1}{2}\cos\left(2\frac{\pi}{2}t\right) + \frac{1}{9}\cos\left(3\frac{\pi}{2}t\right) + \cdots\right]}$$

■ Example 4.22

Consider the periodic signal shown in Fig. 4.20a. By using the differentiation technique, determine the exponential F.S. coefficient, and hence obtain the F.S.

Solution: The given signal is shown in Fig. 4.20a. As seen from this figure $T_0 = 4$ sec and hence

$$\omega_0 = \frac{2\pi}{T_0} = \frac{\pi}{2}\ \text{rad/s}$$

For $x(t)$ the following mathematical description is given using straight line equation for one period.

$$x(t) = \begin{cases} 2t + 2 & -2 \le t < 0 \\ -2t + 2 & 0 \le t < 2 \end{cases}$$

Differentiating the above equations, we get

$$\frac{dx(t)}{dt} = \begin{cases} 2 & -2 \le t < 0 \\ -2 & 0 \le t < 2 \end{cases}$$

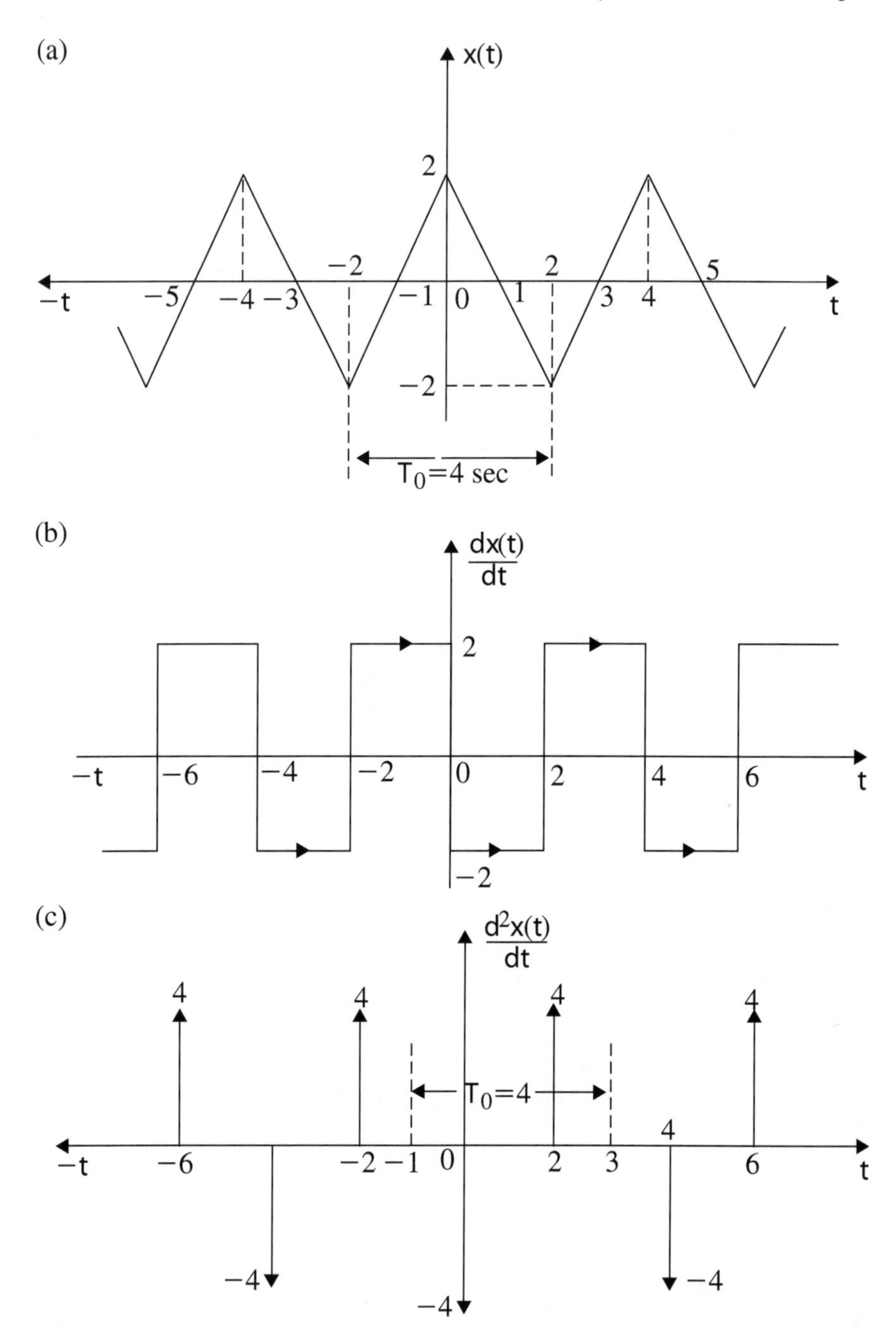

Fig. 4.20 The derivative signals of $x(t)$. **(a)** Triangular waveform of Example 4.22. **b** First derivative and **c** Second derivative

The first derivative of the signal is shown in Fig. 4.20b. Further, differentiating the signal, we get

$$\frac{d^2x(t)}{dt^2} = \begin{cases} 4\delta(t+2) & -2 \leq t < 0 \\ -4\delta(t) & t = 0 \\ 4\delta(t-2) & 0 \leq t < 2 \end{cases}$$

The train of impulses is shown in Fig. 4.20c. Consider the impulses between any one period $T_0 = 4$sec as shown in Fig. 4.20c. The two impulses -4δ and $4\delta(t-2)$ lie within a period. Let D'_n be the exponential F.S. coefficient of these impulses. This can be written as

$$\begin{aligned} D'_n &= \frac{1}{T_0}[-4 + 4e^{-j2\omega_0 n}] \\ &= \frac{4}{4}[-1 + \cos 2\omega_0 n - j \sin 2\omega_0 n] \\ &= [-1 + \cos \pi n - j \sin \pi n] \qquad (\sin \pi n = 0 \text{ for all } n) \\ &= (-1 + \cos \pi n) \\ &= -2 \qquad (\text{for odd values of } n) \\ &= 0 \qquad (\text{for even values of } n) \end{aligned}$$

The exponential series coefficient D_n for the original signal is obtained from

$$\begin{aligned} D_n &= \frac{D'_n}{(j\omega_0 n)^2} \\ &= \frac{8}{\pi^2 n^2} \end{aligned}$$

Using Eq. (4.9), $x(t)$ is expressed in F.S. as

$$x(t) = \sum_{n=-\infty}^{\infty} \frac{8}{\pi^2 n^2} e^{j\omega_0 nt}$$

where n is odd and $D_0 = 0$ since $x(t)$ is symmetrical with respect to the time axis. For $n = \pm 1$

$$\begin{aligned} x_1(t) &= \frac{8}{\pi^2}[e^{j(\pi/2)t} + e^{-j(\pi/2)t}] \\ &= \frac{16}{\pi^2} \cos \frac{\pi}{2} t \end{aligned}$$

For $n = \pm 3$

$$\begin{aligned} x_3(t) &= \frac{8}{\pi^2}\frac{1}{9}[e^{j3(\pi/2)t} + e^{-j3(\pi/2)t}] \\ &= \frac{16}{\pi^2}\left[\frac{1}{9}\cos 3\frac{\pi}{2}t\right] \\ x(t) &= x_1(t) + x_3(t) + \cdots \end{aligned}$$

$$\boxed{x(t) = \frac{16}{\pi^2}\left[\cos\left(\frac{\pi}{2}t\right) + \frac{1}{9}\cos\left(3\frac{\pi}{2}t\right) + \frac{1}{25}\cos\left(5\frac{\pi}{2}t\right) + \cdots\right]}$$

■ Example 4.23

Consider the signal shown in Fig. 4.21. Using the property of the F.S. and the results obtained in Example 4.22 determine $x(t)$. The signal is shown for one period of the periodic signal whose $T_0 = 4$ sec.

Solution: Let $x_f(t)$ be the signal of Example 4.22.

$$x(t) = 2 + x_f(t)$$

The signal shown in Fig. 4.20a is shifted up by a magnitude of 2 to get the signal represented in Fig. 4.21. Hence, a D.C. term of +2 is introduced in the F.S. of Example 4.22. Thus the F.S. of Example 4.23 is written as

$$\boxed{x(t) = 2 + \frac{16}{\pi^2}\left[\cos\frac{\pi}{2}t + \frac{1}{9}\cos 3\frac{\pi}{2}t + \frac{1}{25}\cos 5\frac{\pi}{2}t + \cdots\right]}$$

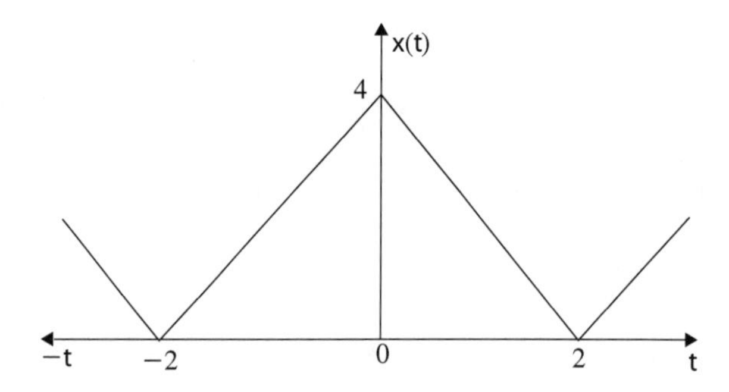

Fig. 4.21 Signal of Example 4.23

■ Example 4.24

Consider the signal shown in Fig. 4.22. Using the property of the F.S. and the results obtained in Example 4.22 determine $x(t)$.

Solution: The signal shown in Fig. 4.22 lags behind the signal shown in Fig. 4.20 by $\pi/2$ radians. Other things remaining the same, the results obtained in Example 4.22 is modified with a phase shift of $\pi/2$ rad. lagging.

$$x(t) = \frac{16}{\pi^2}\left[\cos\left(\frac{\pi}{2}t - \frac{\pi}{2}\right) + \frac{1}{9}\cos\left(3\frac{\pi}{2}t - \frac{\pi}{2}\right) + \frac{1}{25}\cos\left(5\frac{\pi}{2}t - \frac{\pi}{2}\right) + \cdots\right]$$

$$\boxed{x(t) = \frac{16}{\pi^2}\left[\sin\frac{\pi}{2}t + \frac{1}{9}\sin 3\frac{\pi}{2}t + \frac{1}{25}\sin 5\frac{\pi}{2}t + \cdots\right]}$$

■ Example 4.25

Consider the waveform shown in Fig. 4.23. Using the property of the F.S. and the results obtained in Example 4.23 express $x(t)$ of the signal shown in Fig. 4.23 in F.S.

Solution: The signal shown in Fig. 4.23 is same as shown in Fig. 4.20 except $x(t)$ of Fig. 4.20 is time expanded as $x(t/2)$ in Fig. 4.23. The periodicity T_0 is changed from 4 sec. to 8 sec. now and the fundamental frequency as

$$\omega_0 = \frac{\pi}{4}\ \text{rad/s}$$

The time compression or elongation does not bring any change in the F.S. coefficients. However, the fundamental frequency and the harmonics have changes according to

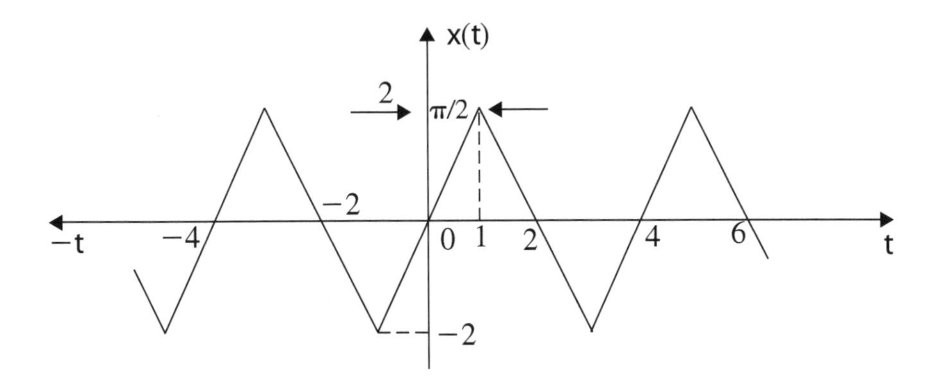

Fig. 4.22 Signal waveform of Example 4.24

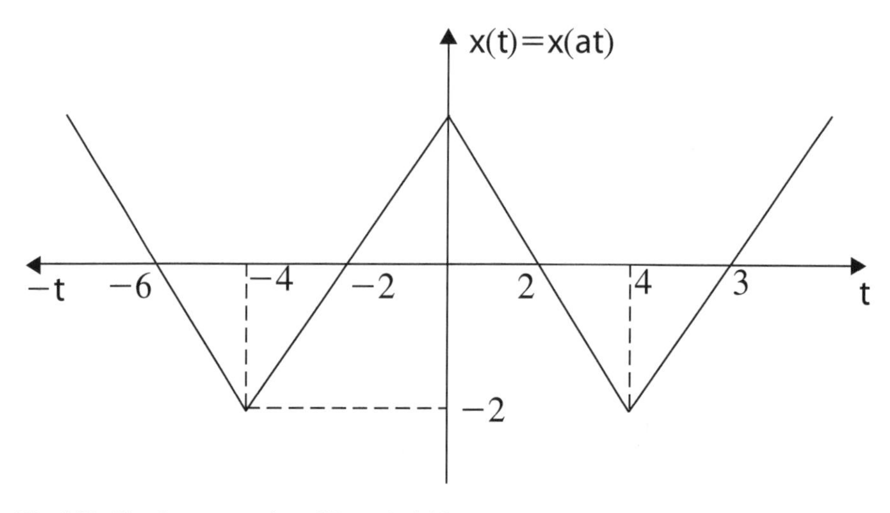

Fig. 4.23 Signal representation of Example 4.25

the changed radian frequency $\omega_0/2$. Thus, the result obtained in Example 4.22 is modified as follows for the Example 4.25.

$$\boxed{x(t) = \frac{16}{\pi^2}\left[\cos\left(\frac{\pi}{4}t\right) + \frac{1}{9}\cos\left(3\frac{\pi}{4}t\right) + \frac{1}{25}\cos\left(5\frac{\pi}{4}t\right) + \cdots\right]}$$

Summary

1. Any arbitrary periodic signal $x(t)$ can be represented in the form a linear combination of complex sinusoids. Such a representation is called Fourier series. The higher frequency sines and cosines have frequencies that are integer multiples of the fundamental frequency.
2. The Fourier series can be represented in any one of the following forms:

   (a) Trigonometry form.
   (b) Complex exponential form.
   (c) Polar or Harmonic or Cosine form.

   The coefficients of the above forms have definite relationships between them.
3. The Fourier series possesses the following properties:

   (a) Linearity,
   (b) Time shifting,
   (c) Time reversal,

   (d) Time scaling,
   (e) Multiplication,
   (f) Conjugation,
   (g) Differentiation and
   (h) Integration.

4. The Parseval's theorem on Fourier series states that the total average power in a periodic signal is the sum of the average powers in all its components which is the sum of the squared value of Fourier series coefficients.
5. Dirichlet showed that if $x(t)$ satisfies certain conditions, the Fourier series of $x(t)$ is guaranteed. These conditions are called Dirichlet conditions.
6. The magnitude and phase angle of Fourier series coefficients plotted versus frequency ω are called Fourier spectra of the signal $x(t)$.
7. The exponential Fourier series is preferred to other types of representations because it is more compact and the mathematical operations involved are less.

Exercises

I. Short Answer Type Questions

1. **What is a Fourier series?**
   Any arbitrary periodic signal $x(t)$ can be expressed as a sum of sinusoids and all its harmonics. Such an infinite series is known as Fourier series.
2. **What are the different forms of representing Fourier series?**
   The different forms of representing Fourier series are:

   (a) Trigonometric Fourier series
   (b) Polar (compact or cosine form) Fourier series
   (c) Exponential form Fourier series

3. **Give mathematical expression for trigonometric Fourier series?**

$$x(t) = a_0 + \sum_{n=1}^{\infty} (a_n \cos n\omega_0 t + b_n \sin n\omega_0 t)$$

   where

$$a_0 = \frac{1}{T_0}\int_{T_0} x(t)dt$$

$$a_n = \frac{2}{T_0}\int_{T_0} x(t)\cos n\omega_0 t dt$$

$$b_n = \frac{2}{T_0}\int_{T_0} x(t)\sin n\omega_0 t dt$$

a_0, a_n, and b_n are called the coefficients of trigonometric Fourier series.

4. **What is the effect of symmetry in trigonometric Fourier series?**
   If $x(t)$ has an odd symmetry, $a_n = 0$. If $x(t)$ has even symmetry $b_n = 0$. If $x(t)$ is symmetrical with respect to the time axis, $a_0 = 0$.
5. **What is half wave symmetry?**
   If the periodic signal $x(t)$ when shifted by half the period remains unchanged except for a sign, the signal is said to be half wave symmetry. Mathematically, it is expressed as

$$x\left(t - \frac{T_0}{2}\right) = -x(t)$$

   For the signal with half wave symmetry, all the even numbered harmonics vanish.
6. **Give the mathematical expression for the cosine Fourier series?**

$$x(t) = C_0 + \sum_{n=1}^{\infty} C_n \cos n(n\omega_0 t - \theta_n)$$

   where

$$C_0 = a_0$$
$$C_n = \sqrt{a_n^2 + b_n^2}$$
$$\theta_n = \tan^{-1}\frac{b_n}{a_n}$$

7. **Give mathematical expression for the exponential Fourier series?**

$$x(t) = \sum_{n=-\infty}^{\infty} D_n e^{j\omega_0 nt}$$

   where

$$D_n = \frac{1}{T_0} \int_{T_0} x(t) e^{-j\omega_0 nt} dt$$

8. **How the coefficients of exponential Fourier series are related to the coefficients of trigonometric and cosine Fourier series?**

$$D_0 = a_0 = C_0$$
$$D_n = \frac{1}{2} [a_n - jb_n]$$
$$|D_n| = \frac{1}{2} C_n$$

9. **Why exponential Fourier series is preferred to represent the Fourier series?**
   The exponential Fourier series is more compact and the system's response to exponential signal is simpler.
10. **What do you understand by Fourier spectrum?**

    The Fourier series expresses a periodic signal $x(t)$ as a sum of sinusoids of fundamental frequency ω_0 and their higher harmonics $2\omega_0, 3\omega_0, \ldots, n\omega_0$. Corresponding to these frequencies, the amplitudes and phases are determined. The plot of these amplitudes versus n which is proportional to $n\omega_0$ is termed as amplitude spectrum. The plot of phase angle θ_n versus n is called phase spectrum.
11. **What do you understand by existence of Fourier series?**
    For the existence of Fourier series, its coefficients should exist. The existence of these coefficients is guaranteed iff $x(t)$ is absolutely integrable. In other words

$$\int_{T_0} |x(t)| \, dt < \infty$$

12. **What do you understand by convergence of Fourier series in the mean?**
    The periodic signal $x(t)$ which has finite energy over one period guarantees the convergence in the mean of its Fourier series. Mathematically, it is expressed as

$$\int_{T_0} |x(t)|^2 \, dt < \infty$$

13. **What are Dirichlet conditions?**
    Fourier at the time of presenting his papers, could not successfully defend,

the existence Fourier series which is infinite. He could not also give convincing reply when there is discontinuities in $x(t)$. The answers to these questions came from the great mathematician Dirichlet in the form of **certain constraints**. These constraints are called Dirichlet conditions and they are:

(a) The function $x(t)$ must be absolutely integrable.
(b) The function $x(t)$ should have finite number discontinuities in one period.
(c) The function $x(t)$ should contain only a finite number of maxima and minima in one period.

14. **What do you understand by Parseval's theorem as applied to Fourier series?**
According to Parseval's theorem, that the power of the periodic signal is equal to the sum of the powers of its Fourier coefficients

$$P = C_0^2 + \frac{1}{2}\sum_{n=1}^{\infty} C_n^2 \qquad \text{(For cosine FS)}$$

$$P = \sum_{n=-\infty}^{\infty} |D_n|^2 \qquad \text{(Exponential FS)}$$

$$P = D_0^2 + 2\sum_{n=1}^{\infty} |D_n|^2 \qquad (x(t) = \text{real})$$

15. **What are differentiating and integrating properties of Fourier series?**
If a periodical signal $x(t)$ is differentiated the Fourier series coefficient gets multiplied by the factor $jn\omega_0$. Suppose D_n is the Fourier series coefficient t of $x(t)$. Then the Fourier series coefficient of $\frac{dx(t)}{dt}$ is $j\omega_0 n D_n$. This is the differentiation property of Fourier series. If the periodic signal $x(t)$ is integrated, then the Fourier series coefficient gets divided by $j\omega_0 n$. If D_n is the coefficient of exponential Fourier series of $x(t)$, then the Fourier series coefficient of $\int_{T_0} x(t)dt$ is $\frac{1}{j\omega_0 n}D_n$. This is called the integration property of Fourier series.

II. Long Answer Type Questions

1. **Determine the Trigonometric and exponential Fourier series representation of the signal $x(t)$ shown in Fig.** 4.24**?**

$$T_0 = T; \qquad \omega_0 = \frac{2\pi}{T_0} = \frac{2\pi}{T}$$

(a) **Trigonometric or quadratic Fourier series.**

$$a_0 = \frac{\tau}{T}$$
$$b_n = 0 \quad \text{since } x(t) \text{ is even}$$
$$a_n = \frac{2}{n\pi}\sin\left(\frac{n\pi\tau}{T}\right)$$
$$x(t) = \frac{\tau}{T} + \frac{2}{\pi}\sum_{n=1}^{\infty}\frac{1}{n}\sin\left(\frac{n\pi\tau}{T}\right)\cos n\frac{2\pi t}{T}$$

(b) **Exponential Fourier series.**

$$D_n = \frac{\tau}{2}\sin c\left(\frac{n\pi\tau}{T}\right)$$
$$x(t) = \sum_{n=1}^{\infty}\frac{\tau}{2}\sin c\left(\frac{n\pi\tau}{T}\right)e^{jn\frac{2\pi t}{T}}$$

2. **Consider the following signal.**

$$x(t) = \cos\left(\frac{1}{3}t + 30^\circ\right) + \sin\left(\frac{2}{5}t + 60^\circ\right)$$

Determine (a) whether the signal is periodic, (b) find the fundamental period and frequency, (c) what harmonics are present in $x(t)$, (d) Determine the coefficients of exponential Fourier series and (e) Determine the power of the signal using Parseval's theorem?

(a) The signal is periodic.
(b) The fundamental period $T_0 = 30\pi$ and the fundamental radian frequency $\omega_0 = \frac{1}{15}$.
(c) Third and sixth harmonics are present.
(d)

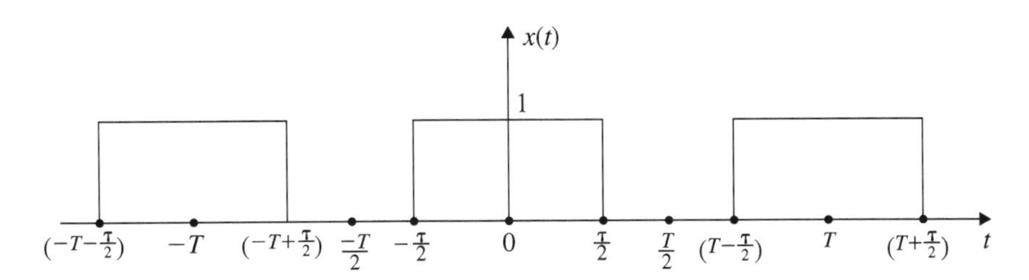

Fig. 4.24 Signal $x(t)$ for Problem 1

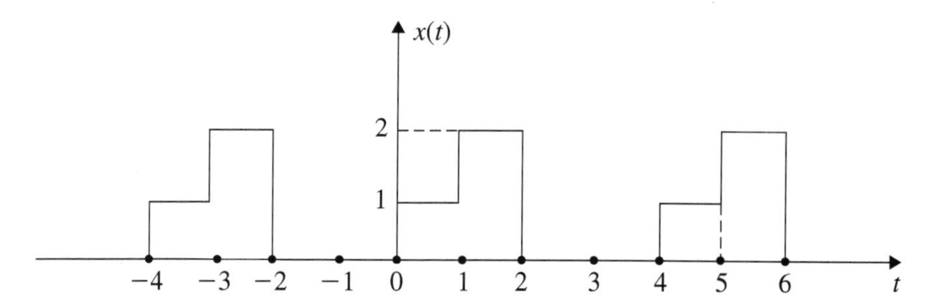

Fig. 4.25 Signal $x(t)$ for Problem 3

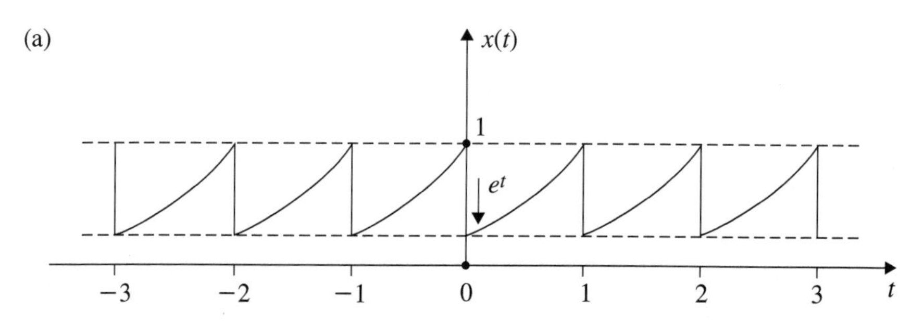

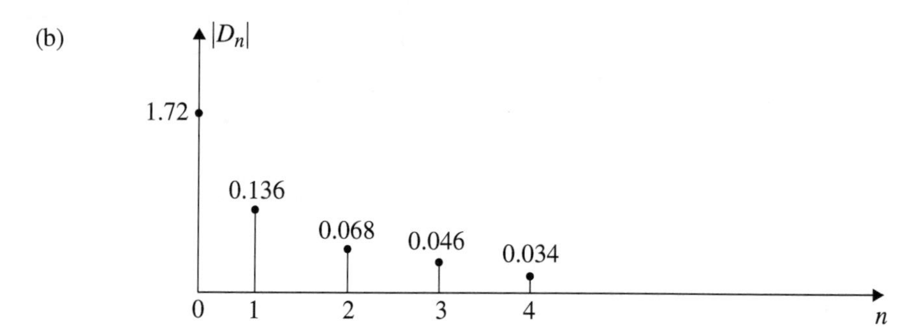

Fig. 4.26 **a** $x(t)$ signal and **b** Amplitude spectrum of D_n

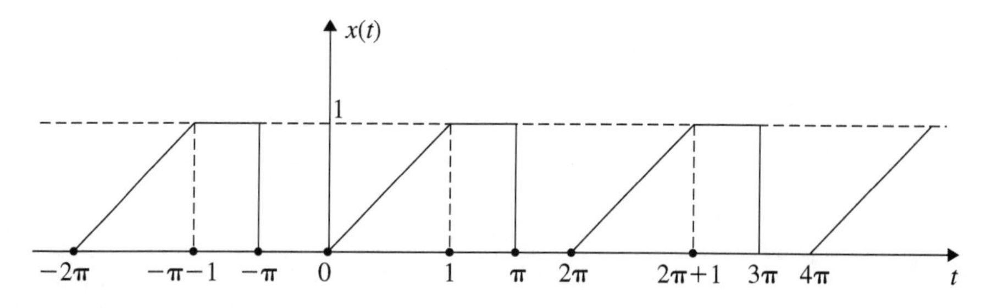

Fig. 4.27 Signal of Problem 5

$$D_3 = \frac{1}{4}[\sqrt{3}+j];\ \ D_{-3} = \frac{1}{4}[\sqrt{3}-j]$$
$$D_6 = \frac{1}{4}[\sqrt{3}-j];\ \ D_{-6} = \frac{1}{4}[\sqrt{3}+j]$$

(e)

$$P = |D_3|^2 + |D_{-3}|^2 + |D_6|^2 + |D_{-6}|^2 = \frac{1}{4} + \frac{1}{4} + \frac{1}{4} + \frac{1}{4} = 1.$$

3. **For the signal shown in Fig.** 4.25, **determine the coefficients of exponential Fourier series?**
4. **Find the exponential Fourier series coefficients for the signal shown in Fig.** 4.26**a and plot its amplitude and phase spectrum?**

$$T_0 = 1;\ \ \omega_0 = 2\pi$$
$$D_n = \frac{1.72}{\sqrt{1+4\pi^2 n^2}}$$
$$\theta_n = 0$$

The amplitude spectrum in shown in Fig. 4.16b.

5. **Consider the signal shown in Fig.** 4.27. **Determine the exponential Fourier series coefficients?**

$$D_0 = \frac{(2\pi - 1)}{4\pi}$$
$$D_n = \frac{1}{2\pi n^2}\left[e^{-jn} - 1\right]$$

Chapter 5
Fourier Series Analysis of Discrete Time Signals

Learning Objectives

- To represent the discrete time signal by exponential Fourier series.
- To determine the exponential Fourier series coefficients.
- To determine the Fourier spectra of discrete time signal.
- To determine the properties of discrete time Fourier series.

5.1 Introduction

In Chap. 4, the continuous time periodic signal $x(t)$ was represented in Fourier series as a sum of sinusoids or exponentials. In this chapter, a similar development is made to represent the periodic discrete time signal $x[n]$ by Fourier series. Even though the Fourier series can be expressed in trigonometric form, because of its compactness and ease of getting the solution, exponential form of Fourier series is preferred and is discussed in this chapter.

5.2 Periodicity of Discrete Time Signal

A periodic discrete time signal $x[n]$ is said to be periodic if it repeats itself after every N_0 samples. Consider the sinusoids $\cos \Omega n$. This is said to be periodic if $\Omega/2\pi$ is a rational number. This can be proved as follows:

$$\begin{aligned} x[n] &= \cos \Omega n \\ x[n+N_0] &= \cos \Omega (n+N_0) \\ &= \cos \Omega n \qquad \text{iff } \Omega N_0 = 2\pi m \end{aligned}$$

S. Palani, *Signals and Systems*,
https://doi.org/10.1007/978-3-030-75742-7_5

where m is an integer. Since N_0 is an integer

$$\frac{\Omega}{2\pi} = \frac{m}{N_0} = \text{a rational number.}$$

A sinusoid $\cos \Omega n$ or exponential $e^{j\Omega n}$ is periodic only if

$$\boxed{\frac{\Omega}{2\pi} = \frac{m}{N_0} \text{ a rational number}} \tag{5.1}$$

The periodicity N_0 is determined by choosing the smallest value of m that will make $m(2\pi/\Omega)$ an integer. The fundamental radian frequency

$$\boxed{\Omega_0 = \frac{2\pi}{N_0} \text{ rad/sample}} \tag{5.2}$$

5.3 DT Signal Representation by Fourier Series

Consider the following exponential Fourier series

$$e^{jon}, e^{\pm j\Omega_0 n}, e^{\pm 2\Omega_0 n}, \ldots$$

The above series will have infinite number of harmonics. Now consider the discrete time exponentials whose frequencies are multiplied by integer multiples of 2π. Thus

$$\begin{aligned} e^{j(\Omega \pm 2\pi m)n} &= e^{j\Omega n} e^{\pm j2\pi mn} \\ &= e^{j\Omega n} \qquad [\because\ e^{\pm j2\pi mn} = 1 \text{ for any integer value } m] \end{aligned} \tag{5.3}$$

Equation (5.3) implies that any kth harmonic is identical to $(k + N_0)$th harmonic. Thus, the first harmonic is identical to $(N_0 + 1)$nd harmonic, the second harmonic is identical to $(N_0 + 2)$nd harmonic and so on. **Thus, if N_0 is the periodicity of $x[n]$ there will be only N_0 independent harmonics and they are repeated in identical manner for every N_0. Unlike continuous time signal which has infinite number of harmonics, DT signal has finite harmonics.**

Now consider the exponentials $e^{jk\Omega_0 n}$ where $k = 0, 1, 2, \ldots, (N_0 - 1)$. The Fourier series for the N_0 harmonics can be expressed as

$$x[n] = \sum_{k=0}^{(N_0-1)} D_k e^{jk\Omega_0 n} \tag{5.4}$$

To determine D_k in Eq. (5.4), multiply both sides of (5.4) by $e^{-jm\Omega_0 n}$

$$x[n]e^{-jm\Omega_0 n} = \sum_{k=0}^{(N_0-1)} D_k e^{jk\Omega_0 n} e^{-jm\Omega_0 n}$$

Summing both sides of the above equation from $n = 0$ to $(N_0 - 1)$, we get

$$\sum_{n=0}^{(N_0-1)} x[n]e^{-jm\Omega_0 n} = \sum_{n=0}^{(N_0-1)} \sum_{k=0}^{(N_0-1)} D_k e^{j(k-m)\Omega_0 n} \tag{5.5}$$

$$= \sum_{k=0}^{(N_0-1)} D_k \left[\sum_{n=0}^{(N_0-1)} e^{j(k-m)\Omega_0 n} \right] \tag{5.6}$$

But

$$\sum_{n=0}^{(N_0-1)} e^{j(k-m)\Omega_0 n} = 0 \qquad \text{for } k \neq m$$

$$= \sum_{n=0}^{(N_0-1)} 1 \qquad \text{for } k = m$$

$$= N_0$$

Equation (5.6) becomes

$$\sum_{n=0}^{(N_0-1)} x[n]e^{-jm\Omega_0 n} = D_k N_0$$

$$\boxed{D_k = \frac{1}{N_0} \sum_{n=0}^{(N_0-1)} x[n]e^{-jm\Omega_0 n}} \tag{5.7}$$

Now the DTFS of $x[n]$ is, therefore, written as by changing $m = k$

$$x[n] = \sum_{k=0}^{(N_0-1)} D_k e^{jk\Omega_0 n} \tag{5.8}$$

$$D_k = \frac{1}{N_0} \sum_{n=0}^{(N_0-1)} x[n]e^{-jk\Omega_0 n} \tag{5.9}$$

Equations (5.8) and (5.9) are called Discrete time Fourier series pairs.

5.4 Fourier Spectra of $x[n]$

In Eq. (5.9), if we substitute for k in the range $0 \leq k < (N_0 - 1)$, we obtain the exponential Fourier series coefficients $D_0, D_1 e^{j\Omega_0 n}, D_2 e^{j2\Omega_0 n}, \ldots, D_{(N_0-1)} e^{j(N_0-1)\Omega_0 n}$. The corresponding frequencies are $0, \Omega_0, 2\Omega_0, \ldots, (N_0 - 1)\Omega_0$, where $\Omega_0 = \frac{2\pi}{N_0} = 2\pi F_0$, where Ω_0 is radian frequency in radian per sample, N_0 is fundamental period and $F_0 = \frac{1}{N_0}$ is the fundamental frequency in cycles per sample. $x[n]$ with periodicity N_0 can be represented by DTFS with sinusoids of fundamental frequency Ω_0 and its higher harmonics as given by Eq. (5.8). The coefficient D_k can be expressed as

$$D_k = |D_k| e^{j\angle D_k} \tag{5.10}$$

The plot of $|D_k|$ versus Ω is known as magnitude spectrum. Similarly, the plot $\angle D_k$ versus Ω is known as phase spectrum.

5.5 Properties of Discrete Time Fourier Series

5.5.1 *Linearity Property*

Let $x_1[n]$ and $x_2[n]$ be two periodic signals with fundamental period N_0. According to linearity property, the linear combinations of these two signals is also periodic with the same fundamental frequency N_0.

Proof Let

$$x[n] = Ax_1[n] + Bx_2[n]$$

$$D_{k1} = \frac{1}{N_0} \sum_{k=0}^{(N_0-1)} x_1[n] e^{-jk\Omega_0 n}$$

$$D_{k2} = \frac{1}{N_0} \sum_{k=0}^{(N_0-1)} x_2[n] e^{-jk\Omega_0 n}$$

$$\begin{aligned} D_k &= \frac{1}{N_0} \sum_{k=0}^{(N_0-1)} x[n] e^{-jk\Omega_0 n} \\ &= \sum_{k=0}^{(N_0-1)} \left\{ \frac{1}{N_0} A x_1[n] + \frac{1}{N_0} B x_2[n] \right\} e^{-jk\Omega_0 n} \\ &= \frac{A}{N_0} \sum_{k=0}^{(N_0-1)} x_1[n] e^{-jk\Omega_0 n} + \frac{B}{N_0} \sum_{k=0}^{(N_0-1)} x_2[n] e^{-jk\Omega_0 n} \end{aligned}$$

$$\boxed{D_k = AD_{k1} + BD_{k2}}$$

$$x[n] = Ax_1[n] + Bx_2[n] \overset{\text{DTFS}}{\longleftrightarrow} AD_{k1} + BD_{k2}$$

5.5.2 Time Shifting Property

If $x[n]$ is time shifted by n_0, then the periodicity of $x[n - n_0]$ is same as $x[n]$.

Proof The Fourier coefficients of $x[n]$ is

$$D_k = \frac{1}{N_0} \sum_{k=0}^{(N_0-1)} x[n]e^{-jk\Omega_0 n}$$

The Fourier coefficients of $x[n - n_0]$ is D_{kn_0}

$$D_{kn_0} = \frac{1}{N_0} \sum_{k=0}^{(N_0-1)} x[n - n_0]e^{-jk\Omega_0 n}$$

Let $(n - n_0) = l$ or $n = (l + n_0)$

$$\begin{aligned} D_{kn_0} &= \frac{1}{N_0} \sum_{k=0}^{(N_0-1)} x[l]e^{-jk\Omega_0(l+n_0)} \\ &= \frac{1}{N_0} e^{-jk\Omega_0 n_0} \sum_{k=0}^{(N_0-1)} x[l]e^{-jk\Omega_0 l} \end{aligned}$$

$$\boxed{D_{kn_0} = e^{-jk\Omega_0 n_0} D_k}$$

$$\boxed{x[n - n_0] \overset{\text{DTFS}}{\longleftrightarrow} e^{-jk\Omega_0 n_0} D_k}$$

5.5.3 Time Reversal Property

If $x[n]$ with fundamental period N_0, is the time reversal, the fundamental period is not changed but the Fourier coefficient changes its sign.

Proof For $x[n]$;

$$D_k = \frac{1}{N_0} \sum_{k=0}^{(N_0-1)} x[n] e^{-jk\Omega_0 n}$$

For $x[-n]$;

$$D_k = \frac{1}{N_0} \sum_{k=0}^{(N_0-1)} x[-n] e^{jk\Omega_0 n}$$

Let us substitute $l = -n$

$$D_k = \frac{1}{N_0} \sum_{k=0}^{(N_0-1)} x[l] e^{jk\Omega_0 l}$$

$$= \frac{1}{N_0} \sum_{k=0}^{(N_0-1)} x[l] e^{-j(-k)\Omega_0 l}$$

$$\boxed{D_k = D_{-k}}$$

5.5.4 Multiplication Property

According to this property, if $x_1[n]$ and $x_2[n]$ are two DT signals with Fourier series coefficients as D_l and D_q then the Fourier series coefficient of $Z[n] = x_1[n] \times x_2[n]$ is

$$D_k = \sum_{q=0}^{N_0-1} D_l D_{k-q}$$

Proof

$$D_l = \frac{1}{N_0} \sum_{n=0}^{(N_0-1)} x_1[n] e^{-jl\Omega_0 n}$$

$$D_q = \frac{1}{N_0} \sum_{n=0}^{(N_0-1)} x_2[n] e^{-jq\Omega_0 n}$$

$$D_k = \frac{1}{N_0} \sum_{n=0}^{(N_0-1)} x_1[n] \times x_2[n] e^{-jk\Omega_0 n}$$

$$= \frac{1}{N_0} \sum_{n=0}^{(N_0-1)} x_1[n] \left\{ \sum_{q=0}^{(N_0-1)} D_q e^{jq\Omega_0 n} \right\} e^{-jk\Omega_0 n}$$

$$= \frac{1}{N_0} \sum_{q=0}^{(N_0-1)} D_q \sum_{n=0}^{(N_0-1)} x_1[n] e^{jq\Omega_0 n} e^{-jk\Omega_0 n}$$

$$= \sum_{q=0}^{(N_0-1)} D_q \frac{1}{N_0} \sum_{n=0}^{(N_0-1)} x_1[n] e^{-j(-q)\Omega_0 n} e^{-jk\Omega_0 n}$$

$$= \sum_{q=0}^{(N_0-1)} D_q \frac{1}{N_0} \sum_{n=0}^{(N_0-1)} x_1[n] e^{-j(k-q)\Omega_0 n}$$

$$\boxed{D_k = \sum_{q=0}^{(N_0-1)} D_q D_{k-q}}$$

5.5.5 Conjugation Property

According to this property, the discrete time Fourier coefficient of $x^*[n]$ conjugate and time reversal of that of $x[n]$.

Proof

$$x^*[n] = \sum_{k=0}^{(N_0-1)} [D_k e^{jk\Omega_0 n}]^*$$

$$= \sum_{k=0}^{(N_0-1)} D_k^* e^{-jk\Omega_0 n}$$

Let $l = -k$

$$x^*[n] = \sum_{l=0}^{(N_0-1)} D_l^* e^{jl\Omega_0 n}$$

$$\boxed{D_k \overset{\text{DTFS}}{\longleftrightarrow} D_{-k}^*}$$

5.5.6 Difference Property

According to this property, the Fourier series of the first difference is given by

$$x[n] - x[n-1] \overset{\text{DTFS}}{\longleftrightarrow} [1 - e^{-jk\Omega_0}]D_k$$

Proof

$$x[n] = \sum_{k=0}^{(N_0-1)} D_k e^{jk\Omega_0 n}$$

$$x[n-1] = \sum_{k=0}^{(N_0-1)} D_k e^{jk\Omega_0 (n-1)}$$

$$x[n] - x[n-1] = \sum_{k=0}^{(N_0-1)} D_k e^{jk\Omega_0 n} - \sum_{k=0}^{(N_0-1)} D_k e^{jk\Omega_0 (n-1)}$$

$$= \sum_{k=0}^{(N_0-1)} D_k e^{jk\Omega_0 n}(1 - e^{-jk\Omega_0})$$

$$\boxed{x[n] - x[n-1] \overset{\text{DTFS}}{\longleftrightarrow} (1 - e^{-jk\Omega_0})D_k}$$

5.5.7 Parseval's Theorem

According to Parseval's theorem, the total average power of a discrete periodic signal $x[n]$ equals the sum of the average powers in individual harmonic components which is expressed as the sum of the squared value of the Fourier series coefficients D_k (Table 5.1).

Proof The average power of discrete periodic signal is given by

$$P = \frac{1}{N_0} \sum_{n=0}^{(N_0-1)} |x[n]|^2$$

$$= \frac{1}{N_0} \sum_{n=0}^{(N_0-1)} |x[n]x[n]|^*$$

$$= \frac{1}{N_0} \sum_{n=0}^{(N_0-1)} x[n] \left\{ \sum_{k=0}^{(N_0-1)} D_k e^{jk\Omega_0 n} \right\}^*$$

Table 5.1 Properties of Discrete time Fourier series

Property	Periodic signal $\boldsymbol{x[n]}$	Fourier series coefficients
Linearity	$Ax_1[n] + Bx_2[n]$	$AD_{k1} + BD_{k2}$
Time shifting	$x[n - n_0]$	$e^{-jk\Omega_0 n_0} D_k$
Time reversal	$x[-n]$	D_{-k}
Multiplication	$x_1[n] \times x_2[n]$	$\sum_{q=<N_0>}^{(N_0-1)} D_q D_{k-q}$
Conjugation	$x^*[n]$	D^*_{-k}
First difference	$x[n] - x[n-1]$	$(1 - e^{-jk\Omega_0 n}) D_k$
Parseval's theorem	$P = \frac{1}{N_0} \sum_{n=<N_0>} \lvert x[n]\rvert^2$	$P = \sum_{k=<N_0>} \lvert D_k\rvert^2$

$$= \sum_{k=0}^{(N_0-1)} D_k^* \frac{1}{N_0} \sum_{n=0}^{(N_0-1)} x[n] e^{-jk\Omega_0 n}$$

$$= \sum_{k=0}^{(N_0-1)} D_k^* D_k$$

$$\boxed{P = \sum_{k=0}^{(N_0-1)} |D_k|^2}$$

■ Example 5.1

Find the discrete time Fourier series of

$$x[n] = \sin 0.2\pi n$$

Sketch the amplitude and phase spectra.

Solution:

$$x[n] = \sin 0.2\pi n$$

$$\Omega_0 = 0.2\pi$$

$$N_0 = \frac{2\pi m}{\Omega_0} = \frac{2\pi}{0.2\pi} = 10\,\text{m}$$

$$N_0 = 10 \qquad \text{for } m = 1$$

$$D_k = \frac{1}{N_0} \sum_{n=0}^{(N_0-1)} x[n] e^{-j\Omega_0 kn}$$

choosing $-5 \leq k < 5$ we get

$$\begin{aligned}
D_k &= \frac{1}{10}\sum_{n=-5}^{4} \sin 0.2\pi n e^{-j0.2\pi kn} \\
&= \frac{1}{20j}\sum_{n=-5}^{4} [e^{j.2\pi n} - e^{-.2\pi n}]e^{-j.2\pi kn} \\
&= \frac{1}{20j}\left[\sum_{n=-5}^{4} e^{j.2\pi n(1-k)} - \sum_{n=-5}^{4} e^{-j.2\pi n(1+k)}\right]
\end{aligned}$$

In the above equation, the summation on the right hand side is zero for all values of k except for $k = 1$ and $k = -1$. For these values, these summations will be equal to $N_0 = 10$. Hence

$$\begin{aligned}
D_1 &= \frac{1}{20j}[N_0] = \frac{1}{2j} = 0.5\angle -90^\circ \\
D_{-1} &= -\frac{1}{20j}[N_0] = -\frac{1}{2j} = 0.5\angle 90^\circ
\end{aligned}$$

$$\begin{aligned}
x[n] &= D_{-1}e^{-j\Omega_0 n} + D_1 e^{j\Omega_0 n} \\
&= \frac{1}{2j}[-e^{-j0.2\pi n} + e^{j0.2\pi n}]
\end{aligned}$$

The plot of D_k and $\angle D_k$ are shown in Figs. 5.1a, b, respectively.

Alternative Method

$$\begin{aligned}
x[n] &= \sin 0.2\pi n \\
\Omega_0 &= 0.2\pi \\
N_0 &= \frac{2\pi}{0.2\pi} = 10 \\
\sin 0.2\pi n &= \frac{1}{2j}[e^{j.2\pi n} - e^{-j.2\pi n}]
\end{aligned}$$

Also

$$\sin 0.2\pi n = \frac{1}{2j}[e^{j\Omega_0 n} - e^{-j\Omega_0}]$$

Comparing the above two equations we get

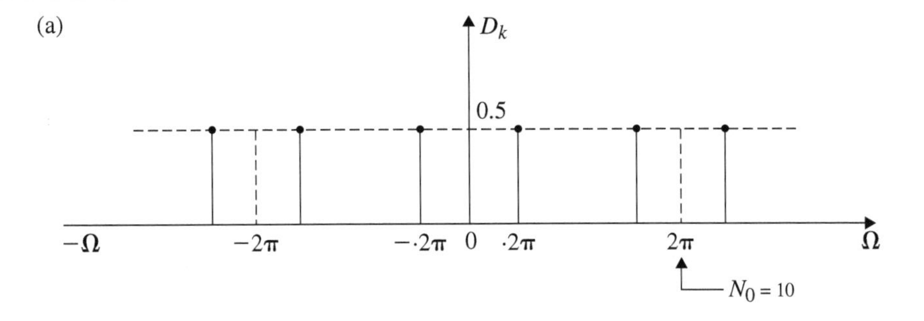

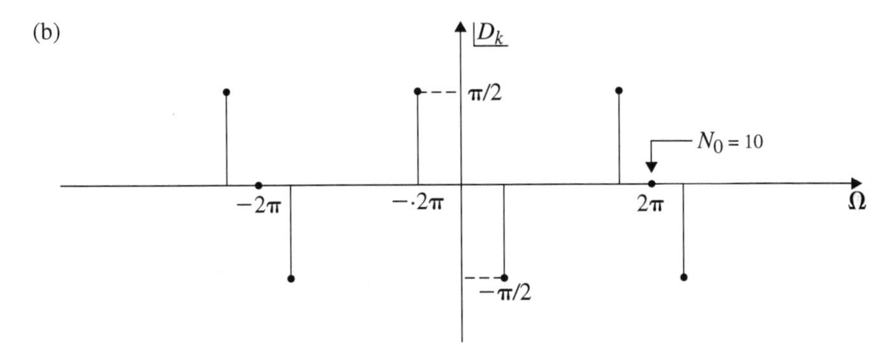

Fig. 5.1 Discrete Fourier spectra of D_k of Example 5.1

$$\boxed{D_1 = \frac{1}{2j}}$$

$$\boxed{D_{-1} = D_{-1+10} = D_9 = -\frac{1}{2j}}$$

$$\boxed{\sin(0.2\pi n) = \frac{1}{2j}[e^{j.2\pi n} - e^{j1.8\pi n}]}$$

■ Example 5.2

Find the discrete Fourier series representation of

$$x[n] = \cos 0.2\pi n$$

Solution:

$$\begin{aligned} x[n] &= \cos 0.2\pi n \\ \Omega_0 &= 0.2\pi \\ N_0 &= \frac{2\pi}{\Omega_0} = 10 \\ \cos 0.2\pi n &= \frac{1}{2}[e^{j.2\pi n} + e^{-j.2\pi n}] \end{aligned}$$

Also

$$\cos 0.2\pi n = \frac{1}{2}[e^{j\Omega_0 n} + e^{-j\Omega_0 n}]$$

Comparing the above two equations of $x[n]$ we get

$$\begin{aligned} D_1 &= \frac{1}{2} \\ D_{-1} &= D_{-1+10} = D_9 = \frac{1}{2} \\ \cos[0.2\pi n] &= \frac{1}{2}[e^{j\Omega_0 n} + e^{j9\Omega_0 n}] \end{aligned}$$

$$\boxed{\cos[0.2\pi n] = \frac{1}{2}[e^{j.2\pi n} + e^{j1.8\pi n}]}$$

Note:

$$\begin{aligned} e^{j1.8\pi n} &= e^{j.2\pi n} e^{-j.2\pi n} \\ &= e^{-j.2\pi n} \qquad [\because\ e^{j2\pi n} = 1] \end{aligned}$$

■ Example 5.3

Find the Fourier series of

$$x[n] = \cos \frac{\pi}{5} n + \sin \frac{\pi}{6} n$$

Solution:

$$\begin{aligned} x[n] &= \cos \frac{\pi}{5} n + \sin \frac{\pi}{6} n \\ \Omega_{01} &= \frac{\pi}{5} \end{aligned}$$

$$\begin{aligned}
N_{01} &= \frac{2\pi}{(\pi/5)} = 10 \\
\Omega_{02} &= \frac{\pi}{6} \\
N_{02} &= \frac{2\pi}{(\pi/6)} = 12 \\
N_0 &= \text{LCM of 10 and 12} \\
&= 60 \\
\Omega_0 &= \frac{2\pi}{N_0} = \frac{2\pi}{60} \\
&= \frac{1}{30}\pi \\
x[n] &= \frac{1}{2}\left[e^{j\frac{\pi}{5}n} + e^{-j\frac{\pi}{5}n}\right] + \frac{1}{2j}\left[e^{j\frac{\pi}{6}n} - e^{-j\frac{\pi}{6}n}\right]
\end{aligned}$$

Expressing the above series in terms of Ω_0, we get

$$\begin{aligned}
x[n] &= \frac{1}{2}\left[e^{j6\Omega_0 n} + e^{-6\Omega_0 n}\right] + \frac{1}{2j}\left[e^{j5\Omega_0 n} - e^{-j5\Omega_0 n}\right] \\
D_6 &= \frac{1}{2}; \qquad D_{-6} = D_{-6+60} = D_{54} = \frac{1}{2} \\
D_5 &= \frac{1}{2j}; \qquad D_{-5} = D_{-5+60} = D_{55} = -\frac{1}{2j}
\end{aligned}$$

$$\boxed{x[n] = \left[\frac{1}{2}e^{j6\Omega_0 n} + \frac{1}{2}e^{j54\Omega_0 n} + \frac{1}{2j}e^{j5\Omega_0 n} - \frac{1}{2j}e^{j55\Omega_0 n}\right]}$$

where $\Omega_0 = \frac{\pi}{30}$

$$\boxed{\begin{aligned}
D_6 &= \frac{1}{2}; \qquad D_{54} = \frac{1}{2} \\
D_5 &= \frac{1}{2j}; \qquad D_{55} = -\frac{1}{2j}
\end{aligned}}$$

Other values of $D_k = 0$.

■ Example 5.4

Find the discrete Fourier series coefficients and the Fourier series for the function

$$x[n] = \sin^2\left(\frac{\pi}{6}n\right)$$

Solution:

$$\left[\sin\frac{\pi}{6}n\right]^2 = \left[\frac{1}{2j}(e^{j\frac{\pi}{6}n} - e^{-j\frac{\pi}{6}n})\right]^2$$

$$= -\frac{1}{4}[e^{j\frac{\pi}{3}n} + e^{-j\frac{\pi}{3}n} - 2]$$

$$\sin^2\left(\frac{\pi}{6}n\right) = \left[\frac{1}{2} - \frac{1}{2}\cos\frac{\pi n}{3}\right]$$

$$\Omega_0 = \frac{\pi}{3}$$

$$N_0 = \frac{2\pi}{\pi} \times 3 = 6$$

$$x[n] = \sin^2\frac{\pi}{6}n$$

$$= \frac{1}{2} - \frac{1}{4}[e^{j\frac{\pi}{3}n} + e^{-j\frac{\pi}{3}n}]$$

$$= \frac{1}{2} - \frac{1}{4}[e^{j\Omega_0 n} + e^{-j\Omega_0 n}]$$

$$D_0 = \frac{1}{2}$$

$$D_1 = -\frac{1}{4}$$

$$D_{-1} = D_{-1+6} = D_5 = -\frac{1}{4}$$

$$\boxed{x[n] = \frac{1}{2} - \frac{1}{4}e^{j\Omega_0 n} - \frac{1}{4}e^{j5\Omega_0 n}}$$

where $\Omega_0 = \frac{\pi}{3}$

$$\boxed{\begin{aligned} D_0 &= \frac{1}{2} \\ D_1 &= -\frac{1}{4} \\ D_5 &= -\frac{1}{4} \end{aligned}}$$

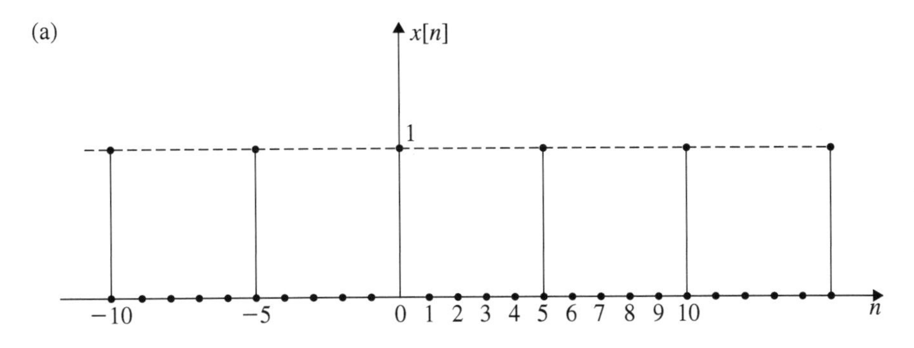

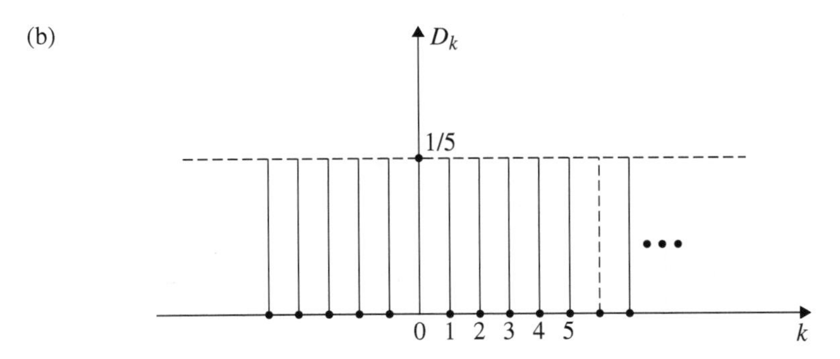

Fig. 5.2 **a** $x[n]$ and **b** Fourier spectra representation

All other Fourier series coefficients are zero.

■ Example 5.5

Find the following sequence find Fourier series coefficients and plot the frequency spectra.

$$x[n] = \sum_{k=-\infty}^{\infty} \delta[n - 5k]$$

Solution: The sequence

$$x[n] = \sum_{k=-\infty}^{\infty} \delta[n - 5k] = \{1, 0, 0, 0, 0\}$$

and $N_0 = 5$. The plot of $x[n]$ is shown in Fig. 5.2a.

$$x[n] = \sum_{k=0}^{N_0-1} D_k e^{jk\Omega_0 n}$$

$$D_k = \frac{1}{N_0} \sum_{n=0}^{N_0-1} x[n] e^{-jk\Omega_0 n}$$

$$N_0 = 5$$

$$\Omega_0 = \frac{2\pi}{N_0} = \frac{2\pi}{5}$$

$$D_k = \frac{1}{5} \sum_{n=0}^{4} x[0] e^{-jk\frac{2\pi}{5}n} \qquad [\because\ x[1], x[2], x[3], x[4] \text{ are all } 0]$$

$$= \frac{1}{5} \quad \text{all}$$

The Fourier coefficients of $x[n]$ are sketched in Fig. 5.2b.

■ Example 5.6

Consider the periodic sequence with period $N_0 = 12$

$$x[n] = \begin{cases} (1)^n & 0 \leq n < 6 \\ 0 & 6 \leq n < 11 \end{cases}$$

Sketch $x[n]$, also sketch the magnitude spectrum of the Fourier coefficient.

Solution:

The given $x[n]$ is sketched as shown in Fig. 5.3a. From Fig. 5.3a, it is evident that $N_0 = 12$, and therefore, $\Omega_0 = \frac{2\pi}{N_0} = \frac{2\pi}{12} = \frac{\pi}{6}$

$$D_k = \frac{1}{N_0} \sum_{n=0}^{5} x[n] e^{-j\Omega_0 k\pi n}$$

$$= \frac{1}{12} \sum_{n=0}^{5} e^{-j\frac{\pi}{6}kn}$$

Applying the summation formula

$$\sum_{n=0}^{N-1} \alpha^n = \frac{(1-\alpha^N)}{(1-\alpha)}$$

the above equation is written as

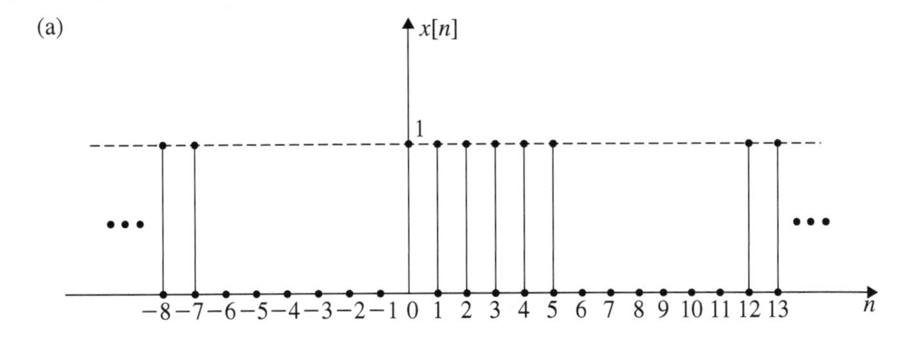

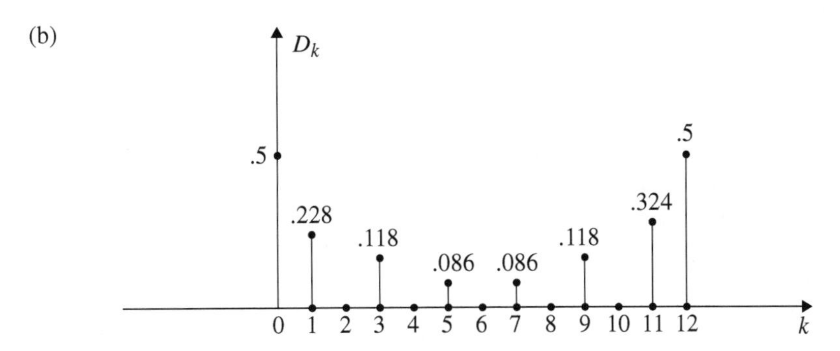

Fig. 5.3 **a** $x[n]$ sequence and **b** D_k magnitude spectra

$$
\begin{aligned}
D_k &= \frac{1}{12}\left[\frac{(1-e^{-j\pi k})}{(1-e^{-j\frac{\pi}{6}k})}\right] \\
&= \frac{1}{12}\left[\frac{e^{-j\frac{k\pi}{2}}[e^{j\frac{k\pi}{2}} - e^{-j\frac{k\pi}{2}}]}{e^{-j\frac{k\pi}{12}}[e^{j\frac{k\pi}{12}} - e^{-j\frac{k\pi}{12}}]}\right]
\end{aligned}
$$

$$
\boxed{D_k = \frac{1}{12} - e^{-jk\frac{5\pi}{12}}\left[\frac{\sin\frac{k\pi}{2}}{\sin\frac{k\pi}{12}}\right]}
$$

where $k = 0, 1, 2, \ldots, 11$.

For $k = 0$,

$$
D_0 = \frac{1}{12}\sum_{n=0}^{5} 1 = \frac{6}{12} = 0.5
$$

The following table is prepared:

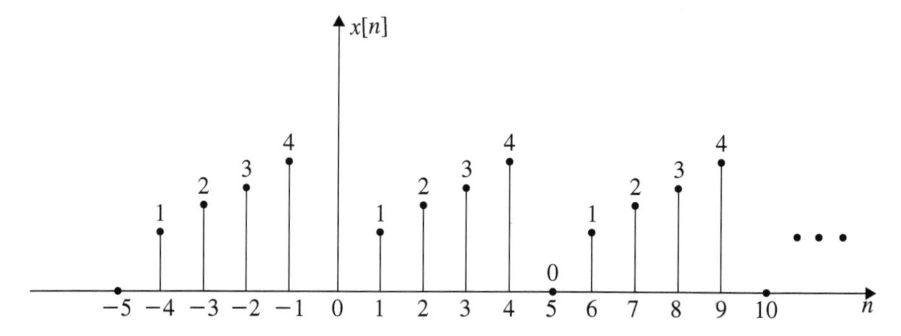

Fig. 5.4 $x[n]$ of Example 5.7

k	0	1	2	3	4	5	6	7	8	9	10	11
$\lvert D_k \rvert$	0.5	0.228	0	0.118	0	0.086	0	0.086	0	0.118	0	0324

the plot of D_k versus k is shown in Fig. 5.3b.

■ Example 5.7

Determine the Fourier series coefficients for the sequence shown in Fig. 5.4.

Solution:

$$N_0 = 5$$

$$\Omega_0 = \frac{2\pi}{N_0} = \frac{2\pi}{5}$$

$$D_k = \frac{1}{N_0}\sum_{n=0}^{(N_0-1)} x[n]e^{-j\Omega_0 kn}$$

$$D_0 = \frac{1}{5}\sum_{n=0}^{4}[0+1+2+3+4]$$

$$\boxed{D_0 = 2}$$

$$\begin{aligned}
D_1 &= \frac{1}{5}\sum_{n=0}^{4} x[n]e^{-j\frac{2\pi}{5}n} \\
&= \frac{1}{5}[0 + e^{-j\frac{2\pi}{5}} + 2e^{-j\frac{4\pi}{5}} + 3e^{-j\frac{6\pi}{5}} + 4e^{-j\frac{8\pi}{5}}] \\
&= \frac{1}{5}[-2.5 + j3.46] = 0.85\angle 125.85^\circ
\end{aligned}$$

$$\boxed{D_1 = 0.85\angle 125.85^\circ}$$

$$\begin{aligned}
D_2 &= \frac{1}{5}\sum_{n=0}^{4} x[n]e^{-j\frac{4\pi}{5}n} \\
&= \frac{1}{5}[0 + e^{-j\frac{4\pi}{5}} + 2e^{-j\frac{8\pi}{5}} + 3e^{-j\frac{12\pi}{5}} + 4e^{-j\frac{16\pi}{5}}] \\
&= \frac{1}{5}[-2.5 + j0.812] = 0.526\angle 162^\circ
\end{aligned}$$

$$\boxed{D_2 = 0.526\angle 162^\circ}$$

$$\begin{aligned}
D_3 &= \frac{1}{5}\sum_{n=0}^{4} x[n]e^{-j\frac{6\pi}{5}n} \\
&= \frac{1}{5}[0 + e^{-j\frac{6\pi}{5}} + 2e^{-j\frac{12\pi}{5}} + 3e^{-j\frac{18\pi}{5}} + 4e^{-j\frac{24\pi}{5}}]
\end{aligned}$$

$$\boxed{D_3 = 0.526\angle 198^\circ}$$

$$\begin{aligned}
D_4 &= \frac{1}{5}\sum_{n=0}^{4} x[n]e^{-j\frac{8\pi}{5}n} \\
&= \frac{1}{5}[0 + e^{-j\frac{8\pi}{5}} + 2e^{-j\frac{16\pi}{5}} + 3e^{-j\frac{24\pi}{5}} + 4e^{-j\frac{32\pi}{5}}]
\end{aligned}$$

$$\boxed{D_4 = 0.85\angle 234^\circ}$$

■ Example 5.8

Find the power of the following signals using Parseval's theorem.

$$\text{(a)} \qquad x[n] = \cos\frac{\pi n}{5} + \sin\frac{\pi n}{6}$$

$$\text{(b)} \qquad x[n] = \sin^2\frac{\pi}{6}n$$

Solution:

(a) In Example 5.3, for $x[n] = \cos\frac{\pi n}{5} + \sin\frac{\pi n}{6}$ the Fourier series coefficients have been determined as

$$D_5 = \frac{1}{2j}; \qquad |D_5| = \frac{1}{2}$$

$$D_{-5} = -\frac{1}{2j}; \qquad |D_{-5}| = \frac{1}{2}$$

$$|D_6| = \frac{1}{2}; \qquad |D_{-6}| = \frac{1}{2}$$

$$P = |D_5|^2 + |D_{-5}|^2 + |D_6|^2 + |D_{-6}|^2$$

$$= \frac{1}{4} + \frac{1}{4} + \frac{1}{4} + \frac{1}{4}$$

$$\boxed{P = 1}$$

(b) In Example 5.4, the Fourier series coefficient for $x[n] = \sin^2\frac{\pi n}{6}$ have been determined as $D_0 = \frac{1}{2}$; $|D_1| = -\frac{1}{4}$; $|D_5| = -\frac{1}{4}$

$$\text{Power } P = |D_0|^2 + |D_1|^2 + |D_5|^2$$

$$= \frac{1}{4} + \frac{1}{16} + \frac{1}{16}$$

$$\boxed{P = \frac{3}{8}}$$

■ Example 5.9

For a periodic sampled gate function shown in Fig. 5.5, find the discrete Fourier series.

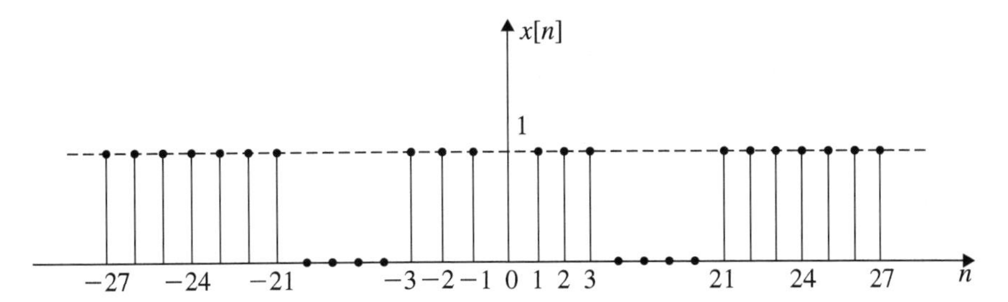

Fig. 5.5 Sampled gate function

Solution:

$$N_0 = 24$$
$$\Omega_0 = \frac{2\pi}{N_0} = \frac{\pi}{12}$$
$$D_k = \frac{1}{N_0} \sum_{n=<N_0>} x[n]e^{-jk\Omega_0 n}$$
$$= \frac{1}{24} \sum_{-3}^{3} e^{-j\frac{k\pi}{12}n}$$

Using the summation formula

$$\sum_{n=m}^{r} x^n = \frac{x^{r+1} - x^m}{(x-1)}$$

we get

$$D_k = \frac{1}{24} \left\{ \frac{e^{-j4\frac{k\pi}{12}} - e^{-j3\frac{k\pi}{12}}}{e^{-j\frac{k\pi}{12}} - 1} \right\}$$
$$= \frac{1}{24} \frac{e^{-\frac{j.5k\pi}{12}}}{e^{-\frac{j.5k\pi}{12}}} \left\{ \frac{e^{-j3.5\frac{k\pi}{12}} - e^{j3.5\frac{k\pi}{12}}}{e^{-j.5\frac{k\pi}{12}} - e^{j.5\frac{k\pi}{12}}} \right\}$$

$$\boxed{D_k = \frac{1}{24} \frac{\sin\left(\frac{3.5\pi k}{12}\right)}{\sin\left(\frac{.5\pi k}{12}\right)}}$$

■ Example 5.10

Find the Fourier series coefficients for the following sequences.

$$\begin{aligned} &(a) \qquad x[n] = 2\cos 2.2\pi n + 4\sin 3.4\pi n \\ &(b) \qquad x[n] = 2\cos 2.2\pi n + 4\sin 3.3\pi n \end{aligned}$$

Solution:

(a) $\boldsymbol{x[n] = 2\cos 2.2\pi n + 4\sin 3.4\pi n}$

$$\begin{aligned} \Omega_{01} &= 2.2\pi \\ N_{01} &= \frac{2\pi}{\Omega_{01}}m = \frac{2\pi m}{2.2\pi} = 10 \qquad \text{for } m = 11 \\ \Omega_{02} &= 3.4\pi \\ N_{02} &= \frac{2\pi}{3.4\pi}m = 10 \qquad \text{for } m = 17 \\ N_0 &= \text{LCM of } N_{01} \text{ and } N_{02} \\ &= 10 \\ \Omega_0 &= \frac{2\pi}{N_0} = 0.2\pi \\ x[n] &= [e^{j2.2\pi n} + e^{-j2.2\pi n}] + \frac{2}{j}[e^{j3.4\pi n} - e^{-j3.4\pi n}] \end{aligned}$$

Expressing the above sequence in terms of Ω_0, we get

$$\begin{aligned} x[n] &= [e^{j11\Omega_0 n} + e^{-j11\Omega_0 n}] + \frac{2}{j}[e^{j17\Omega_0 n} - e^{-j17\Omega_0 n}] \\ D_{11} &= 1; \qquad D_{-11} = D_{-11+10} = D_{-1} = 1 \\ D_{17} &= \frac{2}{j}; \qquad D_{-17+10} = D_{-7} = -\frac{2}{j} \end{aligned}$$

$$\boxed{x[n] = [e^{j11\Omega_0} + e^{-j\Omega_0}] + \frac{1}{2j}[e^{17\Omega_0 n} - e^{-7\Omega_0 n}]}$$

$$\boxed{\begin{aligned} D_{11} &= 1 \\ D_{-1} &= 1 \\ D_{17} &= \frac{2}{j} \\ D_{-7} &= -\frac{2}{j} \end{aligned}}$$

(b) $\boldsymbol{x[n] = 2\cos 2.2\pi n + 4\sin 3.3\pi n}$

$$\begin{aligned}
\Omega_{01} &= 2.2\pi \\
N_{01} &= \frac{2\pi}{2.2\pi}m = 10 \quad (m = 22). \\
\Omega_{02} &= 3.3\pi \\
N_{02} &= \frac{2\pi}{3.3\pi}m = 20 \quad (m = 33). \\
N_0 &= \text{ LCM of } N_{01} \text{ and } N_{02} \\
&= 20 \\
\Omega_0 &= \frac{2\pi}{20} = 0.1\pi \\
x[n] &= [e^{j2.2\pi n} + e^{-j2.2\pi n}] + \frac{1}{2j}[e^{j3.3\pi n} - e^{-j3.3\pi n}]
\end{aligned}$$

Expressing the above sequence in terms of Ω_0, we get

$$\begin{aligned}
x[n] &= [e^{j22\Omega_0 n} + e^{-j22\Omega_0 n}] + \frac{1}{2j}[e^{j33\Omega_0 n} - e^{-j33\Omega_0 n}] \\
D_{22} &= 1; \qquad D_{-22} = D_{-22+20} = D_{-2} = 1 \\
D_{33} &= \frac{1}{2j}; \qquad D_{-33} = D_{-33+20} = D_{-13} = -\frac{1}{2j}
\end{aligned}$$

$$\boxed{x[n] = [e^{j22\Omega_0 n} + e^{-j2\Omega_0 n}] + \frac{1}{2j}[e^{j33\Omega_0 n} - e^{-j13\Omega_0 n}]}$$

$$\boxed{\begin{aligned}
D_{22} &= 1 \\
D_{-2} &= 1 \\
D_{33} &= \frac{1}{2j} \\
D_{-13} &= -\frac{1}{2j}
\end{aligned}}$$

■ Example 5.11

Find the Fourier coefficients and Fourier series of the discrete signal.

$$x[n] = 4\cos 3.3\pi(n-4)$$

Solution:

$$\begin{aligned} \Omega_0 &= 3.3\pi \\ N_0 &= \frac{2\pi}{\Omega_0} m = \frac{2\pi m}{2.2\pi} \times 33 \qquad m = 33 \\ &= 20 \end{aligned}$$

Let

$$\begin{aligned} x_1[n] &= 4\cos 3.3\pi n \\ &= 2[e^{j3.3\pi n} + e^{-j3.3\pi n}] \\ &= [e^{j\Omega_0 \pi n} + e^{-j\Omega_0 \pi n}] \end{aligned}$$

Comparing the above equations, we get

$$\begin{aligned} D_1 &= 2 \\ D_{-1} &= D_{-1+20} = D_{19} = 2 \\ x_1[n] &= 2[e^{j\Omega_0 n} + e^{19\Omega_0 n}] \end{aligned}$$

According to time shifting theory, the Fourier coefficients of time shifted signal is

$$\begin{aligned} D_k &= e^{-jk\Omega_0 n_0} D_k^1 \quad \text{where } n_0 = 4 \\ &= e^{-jk13.2\pi} D_k^1 \end{aligned}$$

Thus, the modified Fourier series coefficients are

$$\begin{aligned} D_1 &= 2e^{-13.2\pi} \\ D_2 &= 2e^{-19\times 13.2\pi} \end{aligned}$$

$$\boxed{x[n] = 2[e^{j\Omega_0 (n-4)} + e^{j19\Omega_0 (n-4)}]}$$

■ Example 5.12

Find the discrete time Fourier series, and also find the Fourier coefficients for the signal $x[n]$ shown in Fig. 5.6.

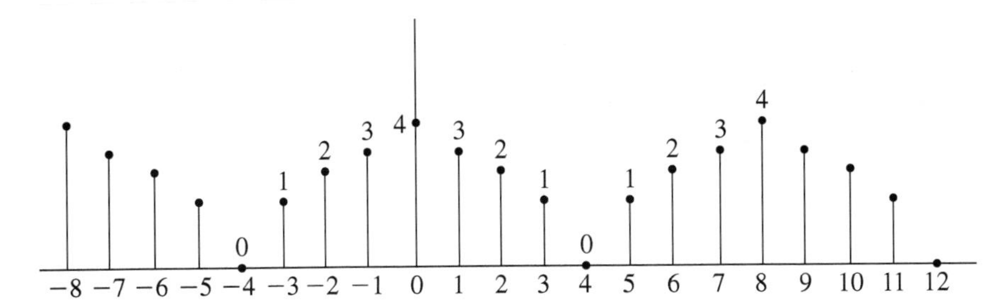

Fig. 5.6 $x[n]$ of Example 5.12

Solution:

$$N_0 = 8$$
$$\Omega_0 = \frac{2\pi}{N_0} = \frac{\pi}{4}$$
$$D_k = \frac{1}{N_0} \sum_{n=-(N_0/2)}^{(N_0/2)} x[n]e^{-j\Omega_0 kn}$$
$$= \frac{1}{8} \sum_{n=-4}^{3} x[n]e^{-j\frac{\pi}{4}kn}$$

where $k = 0, 1, 2, 3, 4, 5, 6$ and 7.

For $k = 0$,

$$D_0 = \frac{1}{8}[0 + 1 + 2 + 3 + 4 + 3 + 2 + 1]$$

$$\boxed{D_0 = 2}$$

For $k = 1$,

$$D_1 = \frac{1}{8} \sum_{n=-4}^{3} x[n]e^{-j\frac{\pi}{4}n}$$
$$= \frac{1}{8}\left[0 + e^{j\frac{3\pi}{4}} + 2e^{j\frac{\pi}{2}} + 3e^{j\frac{\pi}{4}} + 4 + 3e^{-j\frac{\pi}{4}} + 2e^{-j\frac{\pi}{2}} + e^{j\frac{3\pi}{4}}\right]$$
$$= \frac{1}{4}\left[\cos\frac{3\pi}{4} + 2\cos\frac{\pi}{2} + 3\cos\frac{\pi}{4} + 2\right]$$

$$\boxed{D_1 = 0.8535}$$

For $k = 2$,

$$\begin{aligned}
D_2 &= \frac{1}{8}\sum_{n=-4}^{3} x[n]e^{-j\frac{\pi}{2}n} \\
&= \frac{1}{8}\left[0 + e^{j\frac{3\pi}{2}} + 2e^{j\pi} + 3e^{j\frac{\pi}{2}} + 4 + 3e^{-j\frac{\pi}{2}} + 2e^{-j\pi} + e^{-j\frac{3\pi}{2}}\right] \\
&= \frac{1}{4}\left[\cos\frac{3\pi}{2} + 2\cos\pi + 3\cos\frac{\pi}{2} + 2\right]
\end{aligned}$$

$$\boxed{D_2 = 0}$$

For $k = 3$

$$\begin{aligned}
D_3 &= \frac{1}{8}\sum_{n=-4}^{3} x[n]e^{-j\frac{3\pi}{4}n} \\
&= \frac{1}{8}\left[0 + e^{j\frac{9\pi}{4}} + 2e^{j\frac{3\pi}{2}} + 3e^{j\frac{3\pi}{4}} + 4 + 3e^{-j\frac{3\pi}{4}} + 2e^{-j\frac{3\pi}{2}} + e^{-j\frac{9\pi}{4}}\right] \\
&= \frac{1}{4}\left[\cos\frac{9\pi}{4} + 2\cos\frac{3\pi}{2} + 3\cos\frac{3\pi}{4} + 2\right]
\end{aligned}$$

$$\boxed{D_3 = 0.146}$$

For $k = 4$

$$\begin{aligned}
D_4 &= \frac{1}{8}\sum_{n=-4}^{3} x[n]e^{-j\pi n} \\
&= \frac{1}{8}\left[0 + e^{j3\pi} + 2e^{j2\pi} + 3e^{j\pi} + 4 + e^{-j3\pi} + 2e^{-j2\pi} + 3e^{-j\pi}\right] \\
&= \frac{1}{4}\left[\cos 3\pi + 2\cos 2\pi + 3\cos\pi + 2\right]
\end{aligned}$$

$$\boxed{D_4 = 0}$$

For $k = 5$

$$
\begin{aligned}
D_5 &= \frac{1}{8}\sum_{n=-4}^{3} x[n]e^{-j\frac{5\pi}{4}n} \\
&= \frac{1}{8}\left[0 + e^{j\frac{15\pi}{4}} + 2e^{j\frac{10\pi}{4}} + 3e^{\frac{5\pi}{4}} + 4 + e^{-j\frac{15\pi}{4}} + 2e^{-j\frac{10\pi}{4}} + 3e^{-j\frac{5\pi}{4}}\right] \\
&= \frac{1}{4}\left[\cos\frac{15\pi}{4} + 2\cos\frac{10\pi}{4} + 3\cos\frac{5\pi}{4} + 2\right]
\end{aligned}
$$

$$\boxed{D_5 = 0.146}$$

For $k = 6$

$$
\begin{aligned}
D_6 &= \frac{1}{8}\sum_{n=-4}^{3} x[n]e^{-j1.5\pi n} \\
&= \frac{1}{8}\left[0 + e^{j4.5\pi} + 2e^{j3\pi} + 3e^{j1.5\pi} + 4 + e^{-j4.5\pi} + 2e^{-j3\pi} + 3e^{-j1.5\pi}\right] \\
&= \frac{1}{4}\left[\cos 4.5\pi + 2\cos 3\pi + 3\cos 1.5\pi + 2\right]
\end{aligned}
$$

$$\boxed{D_6 = 0}$$

For $k = 7$

$$
\begin{aligned}
D_7 &= \frac{1}{8}\sum_{n=-4}^{3} x[n]e^{-j1.75\pi n} \\
&= \frac{1}{8}\left[0 + e^{j3\times 1.75\pi} + 2e^{j3.5\pi} + 3e^{j1.75\pi} + 4 + e^{-j3\times 1.75\pi} + 2e^{-j3.5\pi} + 3e^{-j1.75\pi}\right] \\
&= \frac{1}{4}\left[\cos 5.25\pi + 2\cos 3.5\pi + 3\cos 1.75\pi + 2\right]
\end{aligned}
$$

$$\boxed{D_7 = 0.8535}$$

Summary

1. Any arbitrary periodic discrete time signal $x[n]$ with fundamental period N_0 and fundamental radian frequency Ω_0 is expressed in Fourier series as

$$x[n] = \sum_{k=0}^{(N_0-1)} D_k e^{jk\Omega_0 n}$$

where D_k is called exponential Fourier series coefficient.

2. The Fourier series coefficient D_k is determined from

$$D_k = \frac{1}{N_0} \sum_{n=0}^{(N_0-1)} x[n]e^{-jk\Omega_0 n}$$

3. The expression for $x[n]$ and D_k are called Fourier series pair.
4. For discrete time Fourier signal $x[n]$, the Fourier series coefficients are finite which are repeated for every fundamental period N_0. This is contrary to continuous time signal $x(t)$ which has infinite number of harmonics.
5. Fourier series possesses, linearity, time shifting, time reversal, multiplication, conjugation and first difference properties.
6. Using Parseval's theorem, the average power of discrete time signal $x[n]$ can be determined by summing up the squared values of discrete Fourier series coefficients over one period using the following formula.

$$P = \sum_{k=0}^{(N_0-1)} |D_k|^2$$

Chapter 6
Fourier Transform Analysis of Continuous Time Signals

Learning Objectives

- To define the Fourier transform for continuous time signal which is aperiodic.
- To derive the properties of Fourier transform and demonstrate with examples.
- To find the magnitude and phase angle spectrum of Fourier transform.
- To solve the differential equation by partial fraction method using Fourier transform (FT).

6.1 Introduction

In Chap. 4, periodic signals were represented as a sum of **everlasting sinusoids or exponentials**. The Fourier series method of analysis of such periodic signals is indeed a very powerful tool. However, FS fails when applied to aperiodic signals. To overcome this major limitation, an aperiodic signal $x(t)$ is expressed as a continuous sum (integral) of **everlasting exponentials**. Such a representation is called Fourier integral which is basically a Fourier series with fundamental frequency tending to zero. By such representation, the aperiodic signal $x(t)$ in the time domain is transformed to $X(j\omega)$ in the frequency domain. The transformations from $x(t)$ to $X(j\omega)$ and from $X(j\omega)$ to $x(t)$ are called Fourier transform and inverse Fourier transform, respectively. They are also called Fourier transform pairs.

S. Palani, *Signals and Systems*,
https://doi.org/10.1007/978-3-030-75742-7_6

6.2 Representation of Aperiodic Signal by Fourier Integral—The Fourier Transform

If an aperiodic signal is viewed as a periodic signal with an infinite period, then it can be represented by Fourier series. In such a situation, as the period increases, the fundamental frequency decreases, and the frequency components become closer. Now the Fourier series sum becomes an integral.

Consider the periodic signal $x(t)$ defined as follows:

$$x(t) = \begin{cases} 1, & |t| < T_1 \\ 0, & T_1 < |t| < \frac{T}{2} \end{cases}$$

The above signal is represented as a periodic square wave in Fig. 6.1. The exponential Fourier series coefficients D_n can be determined as

$$D_n = \frac{2\sin(n\omega_0 T_1)}{(n\omega_0 T)} \tag{6.1}$$

where $\omega_0 = \frac{2\pi}{T}$. The Fourier series coefficients TD_n are obtained as

$$TD_n = \frac{2\sin(n\omega_0 T_1)}{(n\omega_0)} \tag{6.2}$$

For a fixed value of T_1, the plot of TD_n represents a sinc function. Equation (6.2) is plotted for $2\omega_0$, $4\omega_0$ and $8\omega_0$, and they are represented in Fig. 6.2a, b and c, respectively.

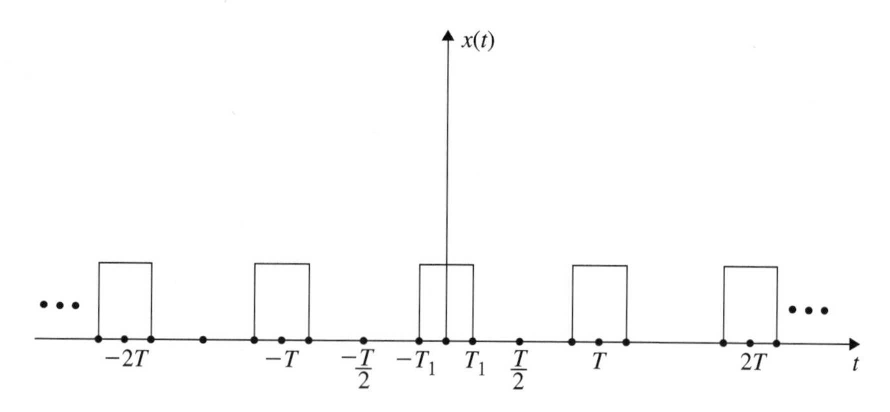

Fig. 6.1 A continuous time periodic square wave

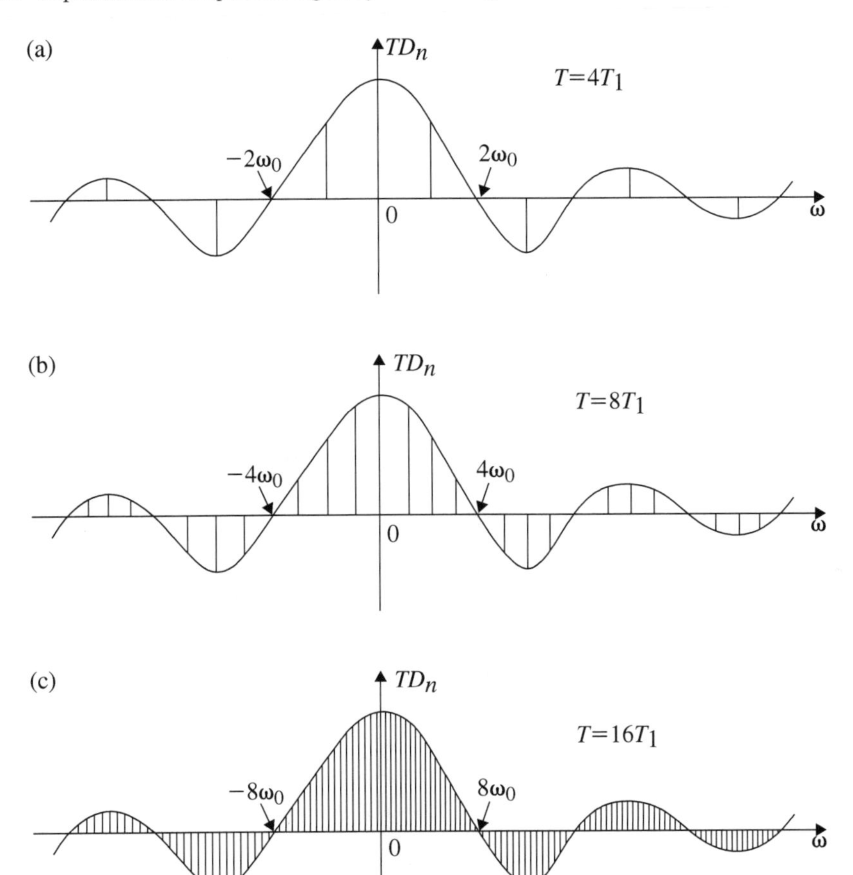

Fig. 6.2 Fourier series coefficients for different values of T

From Fig. 6.2, it is evident that as T increases (the fundamental frequency $\omega_0 = \frac{2\pi}{T}$ decreases), the samples of TD_n become closer and closer. As T becomes very large, the original periodic square wave becomes a rectangular pulse. As $T \to \infty$, TD_n becomes continuous.

Let $\bar{x}(t)$ be a non-periodic square wave as represented in Fig. 6.3.

$$\bar{x}(t) = 0 \qquad |t| > T_1$$

The periodic signal $x(t)$ formed by repeating $\bar{x}(t)$ with fundamental period T is shown in Fig. 6.1. If $T \to \infty$

$$\underset{T\to\infty}{Lt}\ x(t) = \bar{x}(t)$$

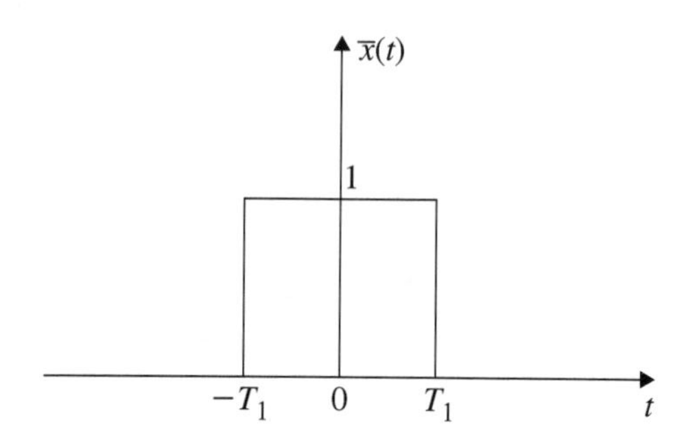

Fig. 6.3 A continuous time aperiodic square wave

The Fourier series coefficients of a periodic signal are written as (Fig. 6.1)

$$D_n = \frac{1}{T}\int_{-T/2}^{T/2} x(t)e^{-jn\omega_0 t}\,dt \tag{6.3}$$

The periodic signal $x(t)$ can be expressed in Fourier series as

$$x(t) = \sum_{n=-\infty}^{\infty} D_n\, e^{jn\omega_0 t} \tag{6.4}$$

$$Tx(t) = \sum_{n=-\infty}^{\infty} TD_n\, e^{jn\omega_0 t} \tag{6.5}$$

Let

$$\begin{aligned} X(n\omega_0) &= TD_n \\ &= \int_{-T/2}^{T/2} x(t)e^{-jn\omega_0 t}\,dt \\ x(t) &= \frac{1}{T}\sum_{n=-\infty}^{\infty} TD_n\, e^{jn\omega_0 t} \\ &= \frac{1}{2\pi}\sum_{n=-\infty}^{\infty} X(n\omega_0)e^{jn\omega_0 t}\omega_0 \end{aligned} \tag{6.6}$$

As $T \to \infty$, $\omega_0 = \frac{2\pi}{T} \to 0$ and $n\omega_0 = \omega$ which is continuous. Further, the summation in Eq. (6.6) becomes an integration. Thus, Eq. (6.6) is written as

$$\boxed{X(j\omega) = \int_{-\infty}^{\infty} x(t)e^{-j\omega t}\,dt} \quad \text{for all } \omega \tag{6.7}$$

$$x(t) = \frac{1}{2\pi}\int_{-\infty}^{\infty} X(j\omega)e^{j\omega t}d\omega \quad \text{for all } t \tag{6.8}$$

Equation (6.7) is called analysis equation while Eq. (6.8) is called synthesis equation. Equations (6.7) and (6.8) are called Fourier transform pair. Equation (6.7) transforms the time function $x(t)$ to frequency function $X(j\omega)$ and so it is called Fourier transform. Equation (6.8) converts the frequency function to time function and hence it is called inverse Fourier transform. These transformations are also denoted as given below.

$$\begin{aligned} X(j\omega) &= F[x(t)] \\ x(t) &\overset{\text{FT}}{\longleftrightarrow} X(j\omega) \\ x(t) &= F^{-1}[X(j\omega)] \\ X(j\omega) &\overset{\text{IFT}}{\longleftrightarrow} x(t) \end{aligned} \tag{6.9}$$

Note: The time function $x(t)$ is always denoted by a lower case letter and the frequency function $X(j\omega)$ by a capital letter. Further, when $x(t)$ is Fourier transformed, it becomes complex and so it is denoted as $X(j\omega)$. In some literature, $X(j\omega)$ is also represented simply as $X(\omega)$.

6.3 Convergence of Fourier Transforms—The Dirichlet Conditions

As in the case of continuous time periodic signals, the following conditions (Dirichlet Conditions) are sufficient for the convergence of $X(j\omega)$.

1. $x(t)$ is absolutely integrable or square integrable. That is

$$\int_{-\infty}^{\infty} |x(t)|\, dt < \infty$$

$$\int_{-\infty}^{\infty} |x(t)|^2\, dt < \infty$$

2. $x(t)$ should have finite number of maxima and minima within any finite interval.
3. $x(t)$ has a finite number of discontinuities within any finite interval.

However, signals which do not satisfy these conditions can have Fourier transforms if impulse functions are included in the transform.

6.4 Fourier Spectra

The Fourier transform of $X(j\omega)$ of $x(t)$ is, in general, complex and can be expressed as

$$X(j\omega) = |X(j\omega)| \ \underline{|X(j\omega)}$$

The plot of $|X(j\omega)|$ versus ω is called magnitude spectrum of $X(j\omega)$. The plot of $\underline{|X(j\omega)}$ versus ω is called phase spectrum. The amplitude (magnitude) and phase spectra are together called Fourier spectrum which is nothing but frequency response of $X(j\omega)$ for the frequency range $-\infty < \omega < \infty$.

6.5 Connection Between the Fourier Transform and Laplace Transform

By definitions,

$$X(j\omega) = \int_{-\infty}^{\infty} x(t)e^{-j\omega t}\, dt \tag{6.10}$$

and the Laplace transform is given by

$$X(s) = \int_{-\infty}^{\infty} x(t)e^{-st}\, dt \tag{6.11}$$

From Eqs. (6.10) and (6.11), it is observed that the Fourier transform is a special case of the Laplace transform in which $s = j\omega$. Substituting $s = \sigma + j\omega$ in Eq. (6.11), we get

$$\begin{aligned} X(\sigma + j\omega) &= \int_{-\infty}^{\infty} x(t)e^{-(\sigma+j\omega)t}\, dt \\ &= \int_{-\infty}^{\infty} [x(t)e^{-\sigma t}]e^{-j\omega t}\, dt \\ &= F[x(t)e^{-\sigma t}] \end{aligned}$$

Thus, the bilateral Laplace transform of $x(t)$ is nothing but the Fourier transform of $x(t)e^{-\sigma t}$. This implies that if the $j\omega$ axis is the ROC of the Laplace transformation, it is the Fourier transform.

Note: The statement that Fourier transform can be obtained from Laplace transform by replacing s by $j\omega$ is true only if $x(t)$ is absolutely integrable. If $x(t)$ is not absolutely integrable, the above statement is erroneous.

The following examples illustrate the method of finding Fourier transform of non-periodic signals.

Example 6.1

Find the Fourier transform of the following time functions and sketch their Fourier spectra (amplitude and phase).

(a) $x(t) = \delta(t)$

(b) $x(t) = \text{sgn}\,(t)$

(c) $x(t) = 1 \quad \text{for all } t$

(d) $x(t) = u(t)$ and $x(t) = u(-t)$

(e) $x(t) = e^{-at}u(t); \quad a > 0$

(f) $x(t) = e^{-|a|t}; \quad a > 0$

(g) $x(t) = e^{at}u(t); \quad a > 0$

$x(t) = e^{at}u(-t)$

Solution:

(a) $\boldsymbol{x(t) = \delta(t)}$

$$
\begin{aligned}
X(j\omega) &= \int_{-\infty}^{\infty} \delta(t)e^{-j\omega t}\,dt \\
&= \int_{-\infty}^{\infty} \delta(t)\,dt \qquad [\delta(t) = 0 \quad \text{for } t \neq 0 \\
&= 1 \qquad\qquad\qquad\qquad = 1 \quad \text{for } t = 0]
\end{aligned}
$$

$$
\boxed{\delta(t) \xleftrightarrow{\text{FT}} 1}
$$

Fourier Spectra of $\delta(t)$

$\delta(\omega) = 1$ which is independent of frequency. Hence, the amplitude spectrum is constant at all ω and the phase spectrum is zero at all ω. $\delta(t)$ and its Fourier spectra are shown in Fig. 6.4a, b and c, respectively.

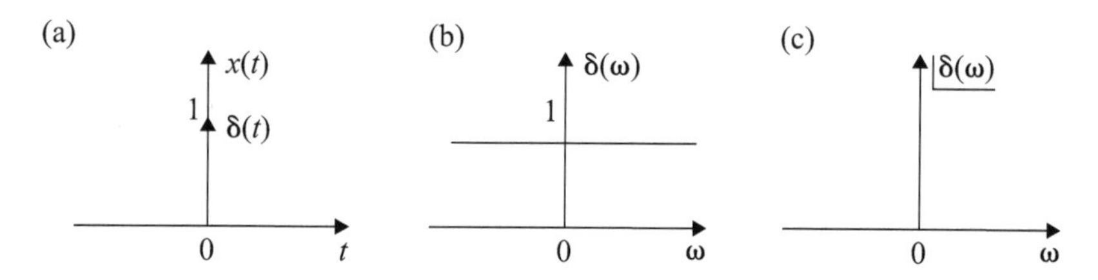

Fig. 6.4 Representation of $\delta(t)$ and its spectra

(b) $\boldsymbol{x(t) = \mathrm{sgn}(t)}$

$$\mathrm{sgn}(t) = \begin{cases} 1 & t > 0 \\ 0 & t = 0 \\ -1 & t < 0 \end{cases}$$

$$F[\mathrm{sgn}(t)] = \int_{-\infty}^{\infty} x(t)e^{-j\omega t}\, dt$$

$$= -\int_{-\infty}^{0} e^{-j\omega t}\, dt + \int_{0}^{\infty} e^{-j\omega t}\, dt$$

For the first integral in the right side of the above equations when the lower limit $-\infty$ is applied, it becomes indeterminate and is not integrable. The problem can be solved by the use of a **TRICK**. $x(t)$ is multiplied by $e^{-a|t|}$ and the limiting value of $a \to 0$ is considered.

$$F[e^{-a|t|}\mathrm{sgn}(t)] = \int_{-\infty}^{0} -e^{at}e^{-j\omega t}\, dt + \int_{0}^{\infty} e^{-at}e^{-j\omega t}\, dt$$

$$F[e^{-a|t|}\mathrm{sgn}(t)] = \int_{-\infty}^{0} -e^{(a-j\omega)t}\, dt + \int_{0}^{\infty} e^{-(a+j\omega)t}\, dt$$

$$F[e^{-a|t|}\mathrm{sgn}(t)] = \underset{a\to 0}{Lt}\left[\frac{-1}{a-j\omega}\left\{e^{(a-j\omega)t}\right\}_{-\infty}^{0} - \frac{1}{(a+j\omega)}\left\{e^{-(a+j\omega)t}\right\}_{0}^{\infty}\right]$$

$$= \underset{a\to 0}{Lt}\left[\frac{-1}{(a-j\omega)} + \frac{1}{a+j\omega}\right] = \frac{1}{j\omega} + \frac{1}{j\omega} = \frac{2}{j\omega}$$

$$\boxed{\mathrm{sgn}(t) \overset{\mathrm{FT}}{\longleftrightarrow} \frac{2}{j\omega}}$$

The same result is derived by a simpler method at a later stage.

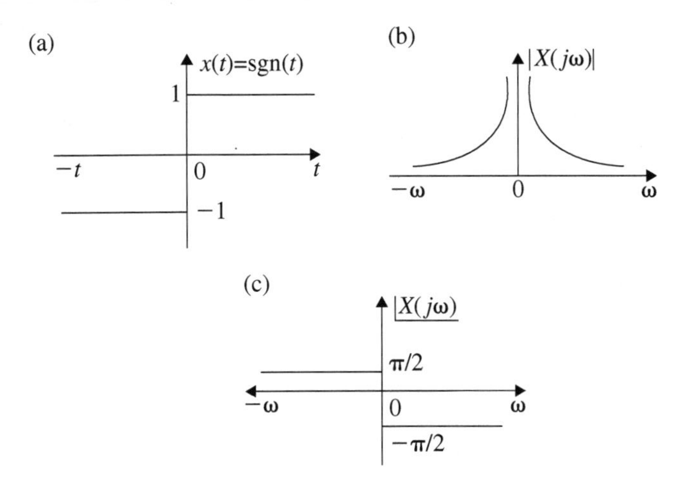

Fig. 6.5 Representation of sgn(t) and its spectra

Fourier Spectra of sgn(t)

$$X(j\omega) = \frac{2}{j\omega} = \begin{cases} \frac{2}{\omega}\angle -90^\circ & \omega \geq 0 \\ \frac{2}{\omega}\angle 90^\circ & \omega < 0 \end{cases}$$

$x(t) = \text{sgn}(t)$, $|X(j\omega)| = \frac{2}{\omega}$ and $\underline{|X(j\omega)}$ are represented in Fig. 6.5a, b and c, respectively.

(c) $\boldsymbol{x(t) = 1;}$ **for all** $\boldsymbol{t}$

$$F^{-1}[\delta(\omega)] = \frac{1}{2\pi}\int_{-\infty}^{\infty} \delta(\omega)e^{j\omega t}\,d\omega$$

Since $\delta(\omega)e^{j\omega t} = \delta(\omega)$,

$$\begin{aligned} F^{-1}[\delta(\omega)] &= \frac{1}{2\pi}\int_{-\infty}^{\infty} \delta(\omega)\,d\omega \\ &= \frac{1}{2\pi} \qquad \text{since } \delta(\omega) = \begin{cases} 1 & \omega = 0 \\ 0 & \text{otherwise} \end{cases} \end{aligned}$$

$$\frac{1}{2\pi} \overset{\text{FT}}{\longleftrightarrow} = \delta(\omega)$$

$$1 \overset{\text{FT}}{\longleftrightarrow} = 2\pi\delta(\omega)$$

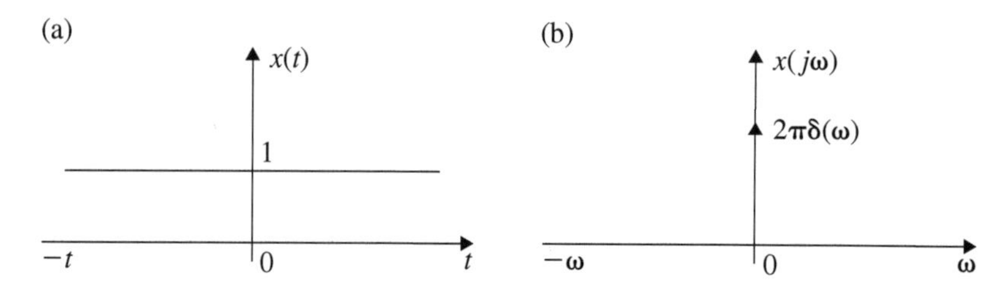

Fig. 6.6 Representation of $x(t) = 1$ and its FT

The above result shows that a constant signal $x(t) = 1$ for all t, when Fourier transformed becomes an impulse $2\pi\,\delta(\omega)$. $x(t)$ and $X(j\omega)$ are represented in Fig. 6.6a and b, respectively.

(d) $\boldsymbol{x(t) = u(t)}$ **and** $\boldsymbol{x(t) = u(-t)}$

$$x(t) = \begin{cases} 0 & t < 0 \\ 1 & t \geq 0 \end{cases}$$

Finding the FT of unit step $u(t)$ by direct integration yields an indeterminate value as is evident from the following equation because it has a jump discontinuity at $t = 0$.

$$\begin{aligned} X(j\omega) &= \int_0^{\infty} e^{-j\omega t}\, dt \\ &= -\frac{1}{j\omega}\left[e^{-j\omega t}\right]_0^{\infty} \end{aligned}$$

When the upper limit ∞ is applied, the integral does not converge. So the problem is approached by considering $u(t)$ as

$$u(t) = \frac{1}{2} + \frac{1}{2}\mathrm{sgn}(t)$$

Figure 6.7 represents $\frac{1}{2}\mathrm{sgn}(t)$, $\frac{1}{2}$ and $u(t)$. From Fig. 6.7, $u(t) = \frac{1}{2} + \frac{1}{2}\mathrm{sgn}(t)$

$$\begin{aligned} F[u(t)] &= F\left[\frac{1}{2}\right] + \frac{1}{2}F\,\mathrm{sgn}(t) \\ F\left[\frac{1}{2}\right] &= \pi\,\delta(\omega) \qquad \text{[From Example 6.1(c)]} \\ F\left[\frac{1}{2}\mathrm{sgn}(t)\right] &= \frac{1}{j\omega} \qquad \text{[From Example 6.1(b)]} \end{aligned}$$

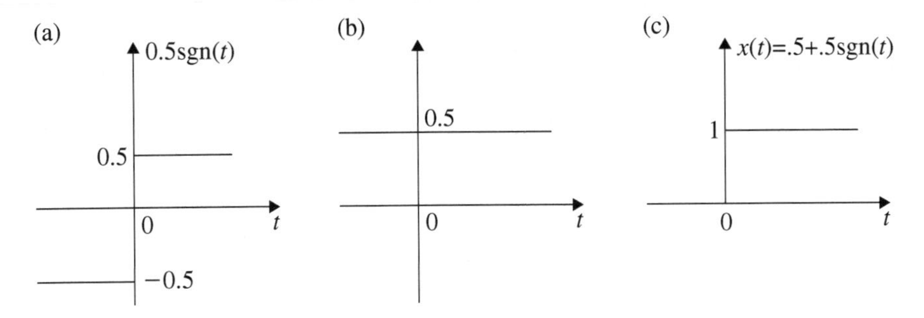

Fig. 6.7 Representation of $u(t)$ in terms of signum function

$$\boxed{F[u(t)] = \pi\,\delta(\omega) + \frac{1}{j\omega}}$$

The same result is obtained in a simpler way which is presented at a later stage. From FT property, which is explained later,

$$F[x(-t)] = X(-j\omega)$$

$$\boxed{F[u(-t)] = \pi\,\delta(\omega) - \frac{1}{j\omega}}$$

(e) $\boldsymbol{x(t) = e^{-at}u(t);\ a > 0}$

$$\begin{aligned} X(j\omega) &= \int_0^\infty e^{-at}e^{-j\omega t}\,dt \\ &= \int_0^\infty e^{-(a+j\omega)t}\,dt \\ &= -\frac{1}{(a+j\omega)}\left[e^{-(a+j\omega)t}\right]_0^\infty \end{aligned}$$

$$\boxed{X(j\omega) = \frac{1}{(a+j\omega)}}$$

$$\begin{aligned} |X(j\omega)| &= \frac{1}{\sqrt{a^2+\omega^2}} \\ \underline{|X(j\omega)} &= -\tan^{-1}\frac{\omega}{a} \end{aligned}$$

The signal $x(t)$, the amplitude spectrum $|X(j\omega)|$ and phase spectrum $\underline{|X(j\omega)}$ are shown in Fig. 6.8a, b and c, respectively.

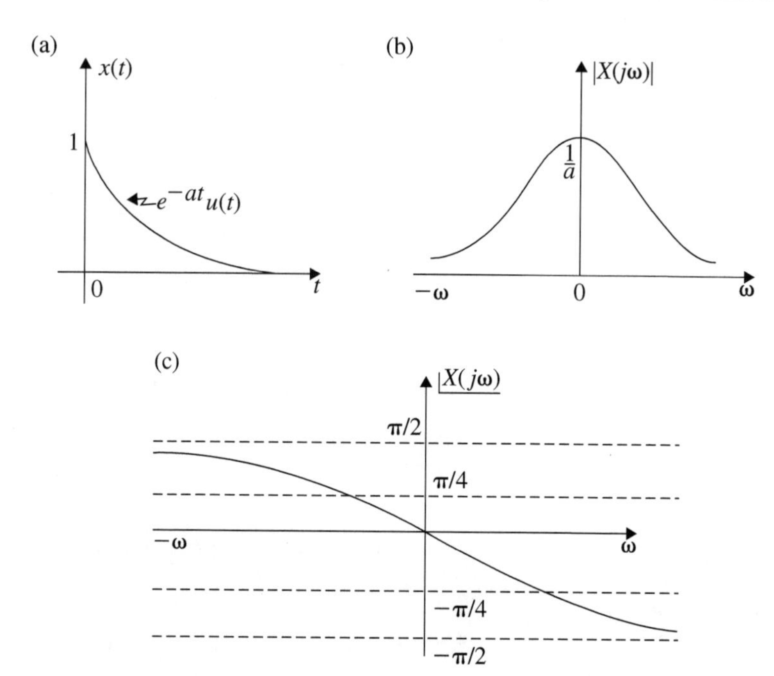

Fig. 6.8 Representation of $x(t) = e^{-at}u(t)$ and its FT spectra

(f) $\boldsymbol{x(t) = e^{-a|t|}; a > 0}$

$$\begin{aligned}
X(j\omega) &= \int_{-\infty}^{\infty} x(t)e^{-j\omega t}\,dt \\
&= \int_{-\infty}^{0} e^{at}e^{-j\omega t}\,dt + \int_{0}^{\infty} e^{-at}e^{-j\omega t}\,dt \\
&= \int_{-\infty}^{0} e^{(a-j\omega)t}\,dt + \int_{0}^{\infty} e^{-(a+j\omega)t}\,dt \\
X(j\omega) &= \frac{1}{(a-j\omega)}\left[e^{(a-j\omega)t}\right]_{-\infty}^{0} - \frac{1}{(a+j\omega)}\left[e^{-(a+j\omega)t}\right]_{0}^{\infty} \\
&= \frac{1}{(a-j\omega)} + \frac{1}{(a+j\omega)}
\end{aligned}$$

$$\boxed{X(j\omega) = \frac{2a}{a^2+\omega^2}}$$

$$\left[e^{-a|t|}\right] \xleftrightarrow{\text{FT}} \frac{2a}{a^2+\omega^2}$$

Fourier Spectra

$$|X(j\omega)| = \frac{2a}{a^2 + \omega^2}$$
$$\underline{|X(j\omega)} = 0$$

The Fourier phase spectrum is zero at all frequencies. The representation of $x(t)$ and its Fourier amplitude spectrum are shown in Fig. 6.9a and b, respectively.

(g) $\boldsymbol{x(t) = e^{at}u(t);\ a > 0}$

$$X(j\omega) = \int_0^\infty e^{at}e^{-j\omega t}\,dt$$
$$= \int_0^\infty e^{(a-j\omega)t}\,dt$$
$$= \frac{1}{(a-j\omega)}\left[e^{(a-j\omega)t}\right]_0^\infty$$

If the upper limit is applied to the above integral, the Fourier integral does not converge. **Hence, FT does not exist for $\boldsymbol{x(t) = e^{at}u(t)}$.**

$$x(t) = e^{at}u(-t) \qquad a > 0$$
$$x(-t) = e^{-at}u(t)$$

From Example 6.1(e), it is derived as

$$F[e^{-at}u(t)] = \frac{1}{(a+j\omega)}$$
$$F[x(-t)] = X(-j\omega)$$

$$\boxed{F[e^{at}u(-t)] = \frac{1}{a-j\omega}}$$

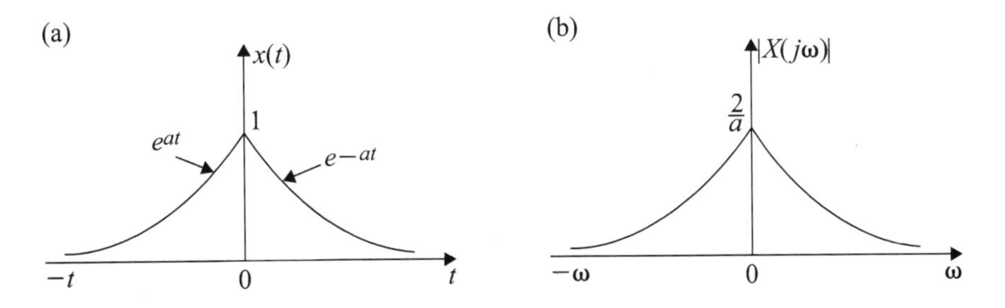

Fig. 6.9 Representation of $e^{-a|t|}$ and its amplitude spectrum

The above result can be derived from the first principle as explained below.

$$F[e^{at}u(-t)] = \int_{-\infty}^{0} e^{at}e^{-j\omega t}\,dt$$
$$= \int_{-\infty}^{0} e^{(a-j\omega)t}\,dt$$
$$= \frac{1}{(a-j\omega)}\left[e^{(a-j\omega)t}\right]_{-\infty}^{0}$$

$$\boxed{F[e^{at}u(-t)] = \frac{1}{(a-j\omega)}}$$

■ Example 6.2

Consider the rectangular pulse shown in Fig. 6.10a which is the gate function. Find the FT and sketch the Fourier spectra.

(*Anna University, April, 2004*)

Solution: **Method 1:**

$$x(t) = 1 \qquad |t| \leq T$$
$$X(j\omega) = \int_{-T}^{T} 1e^{-j\omega t}\,dt$$
$$= \frac{-1}{j\omega}\left[e^{-j\omega t}\right]_{-T}^{T}$$
$$= \frac{\left[e^{j\omega T} - e^{-j\omega T}\right]}{j\omega}$$
$$= \frac{2T\sin\omega T}{\omega T} = 2T\text{sinc}\,\omega T$$

$$\boxed{X(j\omega) = 2T\text{sinc}\,\omega T}$$

Method 2: From Fig. 6.10b, the FT is obtained as

$$F\left[\frac{dx(t)}{dt}\right] = [e^{jT\omega} - e^{-jT\omega}] \qquad \text{(a)}$$

Using the integration property of FT, we get

$$F[x(t)] = \frac{1}{j\omega}[e^{jT\omega} - e^{-jT\omega}] + \pi\dot{X}(0)\delta(\omega)$$

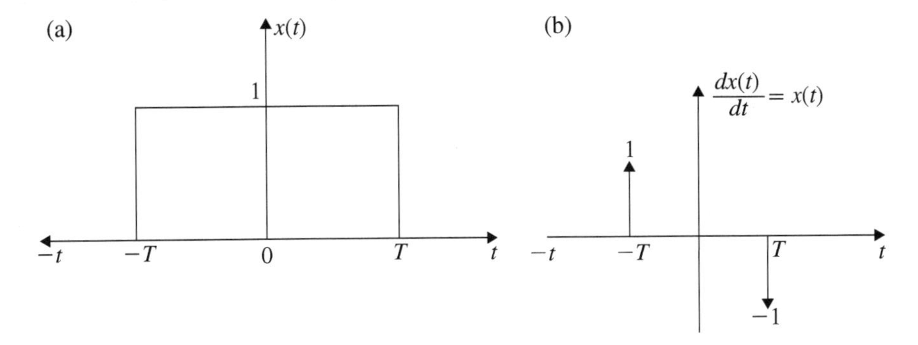

Fig. 6.10 **a** Representation of gate function. **b** Differentiated gate function

where $\dot{x} = dx/dt$. Putting $\omega = 0$ in equation (a), $\dot{X}(0) = 0$

$$\begin{aligned} F\left[x(t)\right] &= \frac{2}{\omega}\frac{\left[e^{jT\omega} - e^{-jT\omega}\right]}{2j} \\ &= \frac{2}{\omega}\sin \omega T \\ &= 2T\frac{\sin \omega T}{\omega T} \end{aligned}$$

$$\boxed{X(j\omega) = 2T\text{sinc}\omega T}$$

Frequency Spectra of Gate Function
Amplitude Spectrum
At $\omega = 0$,

$$|X(j\omega)| = \frac{2\sin \omega T}{\omega T} = \frac{2\sin 0}{0} = 2$$

At $\omega = \pm\frac{n\pi}{T}$,

$$|X(j\omega)| = 0, \qquad \text{where } n = 1, 2, 3, \ldots$$

Phase Spectrum

$$\begin{aligned} &\text{For } 0 < \omega < \frac{\pi}{2}, \qquad \underline{|X(j\omega)} = 0 \\ &\text{For } \frac{\pi}{T} < \omega < \frac{2\pi}{T}, \qquad \underline{|X(j\omega)} = \pi \end{aligned}$$

The amplitude and phase spectra are shown in Fig. 6.11a and b, respectively.
Note: Since $\pi = -\pi$, in Fig. 6.11b, $\underline{|X(j\omega)}$ **is marked as π.**

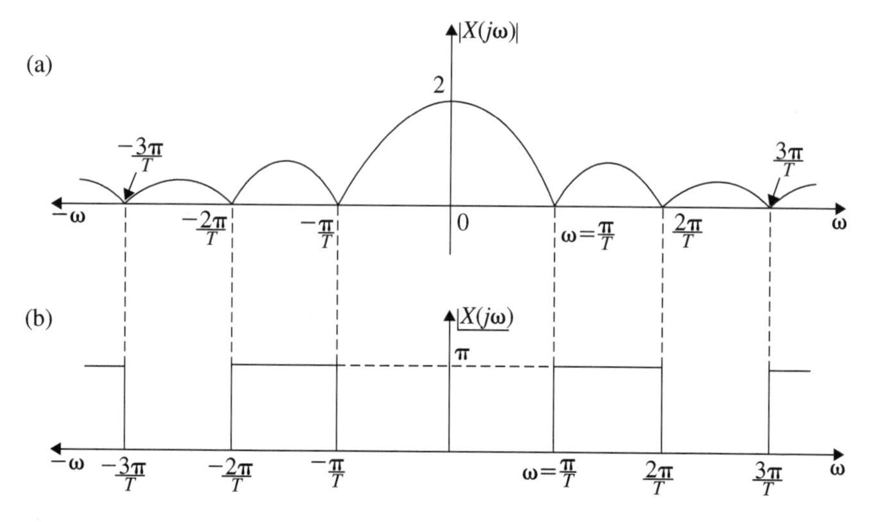

Fig. 6.11 Fourier spectra of gate function

■ Example 6.3

For the following signal $x(t)$, find the FT and FT spectra

$$x(t) = \begin{cases} e^{-at} & t > 0 \\ |1| & t = 0 \\ -e^{+at} & t < 0 \end{cases}$$

Solution: The signal $x(t)$ is sketched as shown in Fig. 6.12.

$$\begin{aligned} X(j\omega) &= \int_{-\infty}^{\infty} x(t)e^{-j\omega t}\,dt \\ &= \int_{-\infty}^{0} -e^{at}e^{-j\omega t}\,dt + \int_{0^-}^{0^+} 1e^{-j\omega t}\,dt + \int_{0^+}^{\infty} e^{-at}e^{-j\omega t}\,dt \\ &= -\int_{-\infty}^{0} e^{(a-j\omega)t}\,dt + \int_{0^-}^{0^+} e^{-j\omega t}\,dt + \int_{0^+}^{\infty} e^{-(a+j\omega)t}\,dt \end{aligned}$$

$$\begin{aligned} X(j\omega) &= \frac{-1}{(a-j\omega)}\left[e^{(a-j\omega)t}\right]_{-\infty}^{0} + 0 - \frac{1}{(a+j\omega)}\left[e^{-(a+j\omega)t}\right]_{0^+}^{\infty} \\ &= \frac{-1}{(a-j\omega)} + \frac{1}{(a+j\omega)} \end{aligned}$$

$$\boxed{X(j\omega) = \frac{-2j\omega}{(a^2+\omega^2)}}$$

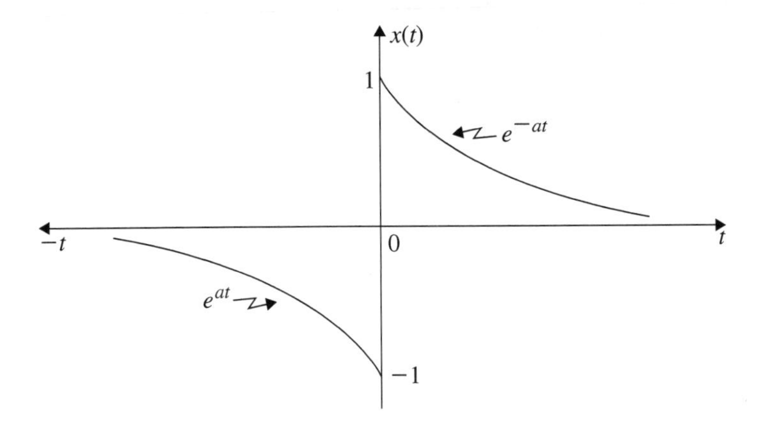

Fig. 6.12 Anti-symmetry exponential decay pulse

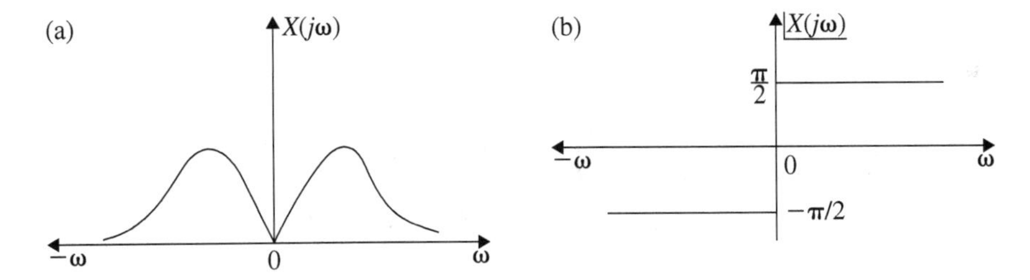

Fig. 6.13 **a** Amplitude spectra and **b** Phase spectra

Fourier Transform Spectra

$$|X(j\omega)| = \frac{2\omega}{(a^2 + \omega^2)}$$

$$\underline{X(j\omega)} = \begin{cases} -\frac{\pi}{2} & \omega > 0 \\ \frac{\pi}{2} & \omega < 0 \end{cases}$$

The frequency spectra for $-\infty < \omega < \infty$ are shown in Fig. 6.13a and b.

■ Example 6.4

Consider the triangular pulse shown in Fig. 6.14. Find the FT and its amplitude spectrum.

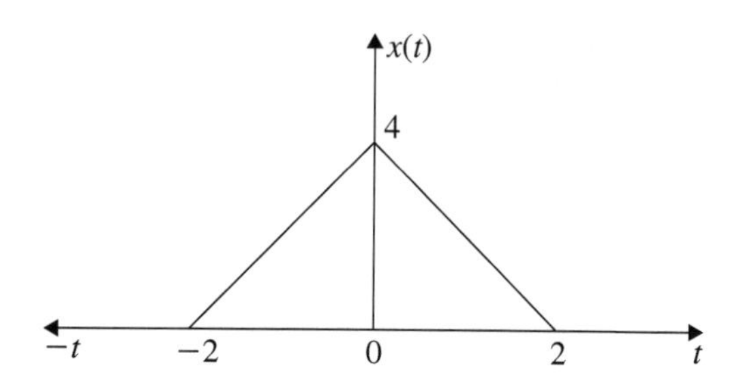

Fig. 6.14 Representation of triangular pulse

Solution:

$$x(t) = 2(|-t| + 2) \qquad |t| \leq 2$$

$$x(t) = \begin{cases} (2t+4) & -2 \leq t \leq 0 \\ (4-2t) & 0 \leq t \leq 2 \end{cases}$$

$$\begin{aligned} X(j\omega) &= \int_{-2}^{0} (2t+4)e^{-j\omega t}\,dt + \int_{0}^{2} (4-2t)e^{-j\omega t}\,dt \\ &= X_1(j\omega) + X_2(j\omega) \\ X_1(j\omega) &= \int_{-2}^{0} (2t+4)e^{-j\omega t}\,dt \end{aligned}$$

Let $u = 2t + 4$; $du = 2\,dt$; $dv = e^{-j\omega t}\,dt$; and $v = -\frac{1}{j\omega}e^{-j\omega t}$

$$\begin{aligned} X_1(j\omega) &= uv - \int v\,du \\ &= \left[(2t+4)\left(\frac{-1}{j\omega}\right)e^{-j\omega t}\right]_{-2}^{0} + \frac{2}{j\omega}\int_{-2}^{0} e^{-j\omega t}\,dt \\ X_1(j\omega) &= \frac{-4}{j\omega} + \frac{2}{\omega^2} - \frac{2}{\omega^2}e^{j2\omega} \\ X_2(j\omega) &= \int_{0}^{2} (4-2t)e^{-j\omega t}\,dt \end{aligned}$$

Let $u = (4 - 2t)$; $du = -2\,dt$; $dv = e^{-j\omega t}\,dt$; and $v = -\frac{1}{j\omega}e^{-j\omega t}$

$$\begin{aligned} X_2(j\omega) &= uv - \int v\,du \\ &= \left[(4-2t)\left(\frac{-1}{j\omega}\right)e^{-j\omega t}\right]_{0}^{2} - \frac{2}{j\omega}\int_{0}^{2} e^{-j\omega t}\,dt \end{aligned}$$

$$
\begin{aligned}
X_2(j\omega) &= \frac{4}{j\omega} - \frac{2}{\omega^2}\left[e^{-j\omega t}\right]_0^2 \\
&= \frac{4}{j\omega} - \frac{2}{\omega^2}\left[e^{-j2\omega} - 1\right] \\
X(j\omega) &= X_1(j\omega) + X_2(j\omega) \\
&= -\frac{4}{j\omega} + \frac{2}{\omega^2} - \frac{2}{\omega^2}e^{j2\omega} + \frac{4}{j\omega} - \frac{2}{\omega^2}e^{-j2\omega} + \frac{2}{\omega^2} \\
&= \frac{4}{\omega^2} - \frac{4}{\omega^2}\cos 2\omega \\
&= \frac{4}{\omega^2}[-\cos 2\omega + 1] \\
&= \frac{8}{\omega^2}\sin^2\omega \\
&= 8\left[\frac{\sin\omega}{\omega}\right]^2
\end{aligned}
$$

$$
\boxed{X(j\omega) = 8\text{sinc}^2\,\omega}
$$

The above problem can be solved using FT property in a few steps which are explained at a later stage.

Fourier Spectra

$$
\begin{aligned}
|X(j\omega)| &= 8\text{sinc}^2\,\omega \\
\underline{|X(j\omega)} &= 0^\circ \quad \text{for all } \omega
\end{aligned}
$$

The magnitude spectra are represented in Fig. 6.15.

Note: The FT of rectangular, triangular and other signals can be easily determined by following the properties of FT which are discussed below.

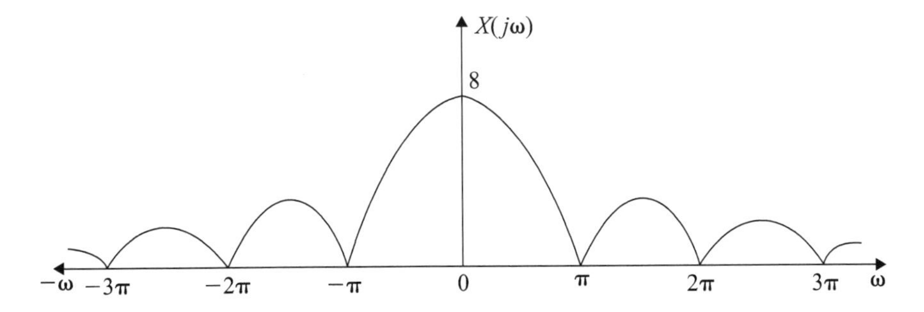

Fig. 6.15 Magnitude spectrum of a triangular wave

6.6 Properties of Fourier Transform

The Fourier transform possesses the following properties and using them, the same results are easily obtained. These properties are

1. Linearity,
2. Time shifting,
3. Conjugation and conjugation symmetry,
4. Differentiation,
5. Integration,
6. Time scaling and time reversal,
7. Frequency shifting,
8. Duality,
9. Time convolution and
10. Parseval's Theorem.

6.6.1 *Linearity*

If

$$x_1(t) \stackrel{\text{FT}}{\longleftrightarrow} X_1(j\omega)$$

$$x_2(t) \stackrel{\text{FT}}{\longleftrightarrow} X_2(j\omega)$$

then

$$[A x_1(t) + B x_2(t)] \stackrel{\text{FT}}{\longleftrightarrow} [A X_1(j\omega) + B X_2(j\omega)]$$

Proof Let $x(t) = A x_1(t) + B x_2(t)$

$$\begin{aligned} X(j\omega) &= \int_{-\infty}^{\infty} x(t) e^{-j\omega t}\, dt \\ &= \int_{-\infty}^{\infty} [A x_1(t) + B x_2(t)] e^{-j\omega t}\, dt \\ &= A \int_{-\infty}^{\infty} x_1(t) e^{-j\omega t} + B \int_{-\infty}^{\infty} x_2(t) e^{-j\omega t}\, dt \end{aligned}$$

$$\boxed{X(j\omega) = A X_1(j\omega) + B X_2(j\omega)} \tag{6.12}$$

6.6.2 Time Shifting

If

$$x(t) \overset{\text{FT}}{\longleftrightarrow} X(j\omega)$$

then

$$x(t-t_0) \overset{\text{FT}}{\longleftrightarrow} e^{-j\omega t_0} X(j\omega)$$

Proof

$$F[x(t-t_0)] = \int_{-\infty}^{\infty} x(t-t_0)e^{-j\omega t}\,dt$$

Let $(t-t_0) = p$ and $dt = dp$

$$\begin{aligned} F[x(t-t_0)] &= \int_{-\infty}^{\infty} x(p)e^{-j\omega(p+t_0)}\,dp \\ &= e^{-j\omega t_0}\int_{-\infty}^{\infty} x(p)e^{-j\omega p}\,dp \end{aligned}$$

$$\boxed{F[x(t-t_0)] = e^{-j\omega t_0} X(j\omega)} \tag{6.13}$$

6.6.3 Conjugation and Conjugation Symmetry

If

$$x(t) \overset{\text{FT}}{\longleftrightarrow} X(j\omega)$$

then

$$x^*(t) \overset{\text{FT}}{\longleftrightarrow} X^*(-j\omega)$$

Proof

$$\begin{aligned} F[x^*(t)] = X^*(j\omega) &= \left[\int_{-\infty}^{\infty} x(t)e^{-j\omega t}\,dt\right]^* \\ &= \int_{-\infty}^{\infty} x^*(t)e^{j\omega t}\,dt \end{aligned}$$

Replacing ω by $(-\omega)$,

$$X^*(-j\omega) = \int_{-\infty}^{\infty} x^*(t)e^{-j\omega t}\, dt$$

$$\boxed{X^*(-j\omega) = X(j\omega)} \qquad \text{if } x(t) \text{ is real } x^*(t) = x(t)$$

Also

$$\boxed{X(-j\omega) = X^*(j\omega)} \tag{6.14}$$

6.6.4 *Differentiation in Time*

If

$$x(t) \overset{\text{FT}}{\longleftrightarrow} X(j\omega)$$

then

$$\frac{dx(t)}{dt} \overset{\text{FT}}{\longleftrightarrow} j\omega X(j\omega)$$

Proof

$$\begin{aligned} F[x(t)] &= \frac{1}{2\pi}\int_{-\infty}^{\infty} X(j\omega)e^{j\omega t}\, d\omega \\ F\left[\frac{dx(t)}{dt}\right] &= \frac{j\omega}{2\pi}\int_{0}^{\infty} X(j\omega)e^{j\omega t}\, d\omega \\ &= j\omega X(j\omega) \end{aligned}$$

$$\boxed{\frac{dx(t)}{dt} \overset{\text{FT}}{\longleftrightarrow} j\omega X(j\omega)} \tag{6.15}$$

In general,

$$F\left[\frac{d^n x(t)}{dt^n}\right] = (j\omega)^n X(j\omega)$$

6.6.5 Differentiation in Frequency

If

$$F[x(t)] = X(j\omega)$$

then

$$F[tx(t)] = j\frac{d}{d\omega}X(j\omega)$$

Proof

$$\begin{aligned} X(j\omega) &= \int_{-\infty}^{\infty} x(t)e^{-j\omega t}\,dt \\ \frac{d}{d\omega}[X(j\omega)] &= \int_{-\infty}^{\infty} -jtx(t)e^{-j\omega t}\,dt \\ &= -jF[tx(t)] \end{aligned}$$

$$\boxed{[tx(t)] \stackrel{\text{FT}}{\longleftrightarrow} j\frac{dX(j\omega)}{d\omega}} \tag{6.16}$$

6.6.6 Time Integration

If

$$F[x(t)] = X(j\omega)$$

then

$$F\left[\int_{-\infty}^{t} x(\tau)\,d\tau\right] = \frac{1}{j\omega}X(j\omega) + \pi X(0)\,\delta(\omega)$$

Proof Let

$$y(t) = \int_{-\infty}^{t} x(\tau)\,d\tau$$

Differentiating the above equation, we get

$$x(t) = \frac{dy(t)}{dt}$$

Using differentiation property, we get

$$X(j\omega) = j\omega Y(j\omega)$$

The differentiation in the time domain corresponds to multiplication by $j\omega$ in the frequency domain.

$$Y(j\omega) = \left(\frac{1}{j\omega}\right) X(j\omega)$$

if the initial condition $X(0) = 0$.

If $X(j\omega) \neq 0$ at $\omega = 0$, then $y(t)$ is not integrable and FT does not exist. However, this problem is overcome by including impulses in the transform. The value at $\omega = 0$ is modified by adding $\pi X(0)$ and the FT is written as

$$\boxed{F\left[\int_{-\infty}^{t} x(\tau)\, d\tau\right] \overset{\text{FT}}{\longleftrightarrow} \frac{1}{j\omega} X(j\omega) + \pi X(0)\, \delta(\omega)} \tag{6.17}$$

6.6.7 Time Scaling

If

$$F[x(t)] = X(j\omega)$$

then

$$F[x(at)] = \frac{1}{|a|} X\left(\frac{j\omega}{a}\right)$$

Proof

$$F[x(at)] = \int_{-\infty}^{\infty} x(at) e^{-j\omega t}\, dt$$

Let $at = p$; and $dt = \frac{1}{a} dp$, $a > 0$

$$F[x(p)] = \frac{1}{a} \int_{-\infty}^{\infty} x(p) e^{-\frac{j\omega p}{a}}\, dp$$

By definition of FT, we get

$$\boxed{F[x(at)] = \frac{1}{a} X\left(j\frac{\omega}{a}\right)}$$

For $a < 0$,

$$F[x(at)] = \frac{-1}{a} X\left(j\frac{\omega}{a}\right)$$

Hence

$$\boxed{F[x(at)] = \frac{1}{|a|} X\left(j\frac{\omega}{a}\right)} \tag{6.18}$$

For time reversal,

$$\boxed{F[x(-t)] = X(-j\omega)} \tag{6.19}$$

6.6.8 Frequency Shifting

If

$$F[x(t)] = X(j\omega)$$

then

$$F[x(t)e^{j\omega_0 t}] = X[j(\omega - \omega_0)]$$

Proof

$$\begin{aligned} F[x(t)e^{j\omega_0 t}] &= \int_{-\infty}^{\infty} x(t)e^{j\omega_0 t} e^{-j\omega t}\, dt \\ &= \int_{-\infty}^{\infty} x(t)e^{-j(\omega - \omega_0)t}\, dt \end{aligned}$$

By definition of FT, we get

$$\boxed{F[x(t)e^{j\omega_0 t}] = X[j(\omega - \omega_0)]} \tag{6.20}$$

6.6.9 Duality

If

$$F[x(t)] = X(j\omega)$$

then

$$F[X(t)] = 2\pi x(j\omega)$$

Proof. From Eq. (6.8), we write

$$\begin{aligned} x(t) &= \frac{1}{2\pi}\int_{-\infty}^{\infty} X(j\omega)e^{j\omega t}\,d\omega \\ x(-t) &= \frac{1}{2\pi}\int_{-\infty}^{\infty} X(j\omega)e^{-j\omega t}\,d\omega \\ 2\pi x(-t) &= \int_{-\infty}^{\infty} X(j\omega)e^{-j\omega t}\,d\omega \end{aligned}$$

By definition of FT, we get

$$2\pi x(-t) = F[X(j\omega)]$$

Changing t to $j\omega$, we get

$$\boxed{2\pi x(j\omega) = F[X(t)]} \tag{6.21}$$

6.6.10 The Convolution

Let

$$\begin{aligned} y(t) &= x(t) * h(t) \\ F[y(t)] &= Y(j\omega) = X(j\omega)H(j\omega) \end{aligned}$$

Proof

$$\begin{aligned} y(t) &= \int_{-\infty}^{\infty} x(\tau)h(t-\tau)\,d\tau \\ F[y(t)] = Y(j\omega) &= \int_{-\infty}^{\infty}\left[\int_{-\infty}^{\infty} x(\tau)h(t-\tau)\,d\tau\right]e^{-j\omega t}\,dt \end{aligned}$$

Interchanging the order of integration, we get

$$Y(j\omega) = \int_{-\infty}^{\infty} x(\tau) \left[\int_{-\infty}^{\infty} h(t-\tau)e^{-j\omega t}\, dt \right] d\tau$$

By time shifting property, the term inside the bracket becomes $e^{-j\omega\tau}H(j\omega)$.

$$\begin{aligned} Y(j\omega) &= \int_{-\infty}^{\infty} x(\tau)e^{-j\omega\tau}H(j\omega)\, d\tau \\ &= H(j\omega)\int_{-\infty}^{\infty} x(\tau)e^{-j\omega\tau}\, d\tau \end{aligned}$$

By definition of FT, we get

$$\boxed{Y(j\omega) = H(j\omega)X(j\omega)} \tag{6.22}$$

6.6.11 *Parseval's Theorem (Relation)*

According to Parseval's theorem, the total energy in a signal is obtained by integrating the energy per unit frequency $\frac{|X(j\omega)|^2}{2\pi}$.

Proof

$$\begin{aligned} E &= \int_{-\infty}^{\infty} |x(t)|^2\, dt \\ &= \int_{-\infty}^{\infty} x(t)x^*(t)\, dt \\ &= \int_{-\infty}^{\infty} x(t)\left[\frac{1}{2\pi}\int_{-\infty}^{\infty} X^*(j\omega)e^{-j\omega t}\, d\omega\right] dt \end{aligned}$$

$$\begin{aligned} E &= \frac{1}{2\pi}\int_{-\infty}^{\infty} X^*(j\omega)\left[\int_{-\infty}^{\infty} x(t)e^{-j\omega t}\, dt\right] d\omega \\ &= \frac{1}{2\pi}\int_{-\infty}^{\infty} X^*(j\omega)X(j\omega)\, d\omega \end{aligned}$$

$$\boxed{E = \frac{1}{2\pi}\int_{-\infty}^{\infty} |X(j\omega)|^2\, d\omega}$$

$$\int_{-\infty}^{\infty} |x(t)|^2 dt = \frac{1}{2\pi}\int_{-\infty}^{\infty} |X(j\omega)|^2 d\omega$$

Table 6.1 Fourier transform properties

Property	Time signal $x(t)$	Fourier transform $X(j\omega)$
1. Linearity	$x(t) = A\,x_1(t) + B\,x_2(t)$	$X(j\omega) = A\,X_1(j\omega) + B\,X_2(j\omega)$
2. Time shifting	$x(t-t_0)$	$e^{-j\omega t_0}X(j\omega)$
3. Conjugation	$x^*(t)$	$X^*(-j\omega)$
4. Differentiation in time	$\frac{d^n x(t)}{dt^n}$	$(j\omega)^n X(j\omega)$
5. Differentiation in frequency	$tx(t)$	$j\frac{d}{d\omega}X(j\omega)$
6. Time integration	$\int_{-\infty}^{t} x(\tau)\,d\tau$	$\frac{1}{j\omega}X(j\omega) + \pi X(0)\delta(\omega)$
7. Time scaling	$x(at)$	$\frac{1}{\lvert a\rvert}X\left(j\frac{\omega}{a}\right)$
8. Time reversal	$x(-t)$	$X(-j\omega)$
9. Frequency shifting	$x(t)e^{j\omega_0 t}$	$X[j(\omega-\omega_0)]$
10. Duality	$X(t)$	$2\pi x(j\omega)$
11. Time convolution	$x(t) * h(t)$	$X(j\omega)H(j\omega)$
12. Parseval's theorem	$E = \int_{-\infty}^{\infty} \lvert x(t)\rvert^2\,dt$	$E = \frac{1}{2\pi}\int_{-\infty}^{\infty} \lvert X(j\omega)\rvert^2\,d\omega$

Table 6.2 Basic Fourier transform pairs

Signal	Fourier transform
1. $\delta(t)$	1
2. $u(t)$	$\frac{1}{j\omega} + \pi\delta(\omega)$
3. $\delta(t-t_0)$	$e^{-j\omega t_0}$
4. $te^{-at}u(t)$	$\frac{1}{(a+j\omega)^2}$
5. $u(-t)$	$\pi\delta(\omega) - \frac{1}{j\omega}$
6. $e^{at}u(-t)$	$\frac{1}{(a-j\omega)}$
7. $e^{-a\lvert t\rvert}$	$\frac{2a}{a^2+\omega^2}$
8. $\cos\omega_0 t$	$\pi[\delta(\omega-\omega_0) + \delta(\omega+\omega_0)]$
9. $\sin\omega_0 t$	$-j\pi[\delta(\omega-\omega_0) - \delta(\omega+\omega_0)]$
10. $\frac{1}{(a^2+t^2)}$	$e^{-a\lvert\omega\rvert}$
11. $\mathrm{sgn}(t)$	$\frac{2}{j\omega}$
12. 1; for all t	$2\pi\,\delta(\omega)$

The above equation is called Parseval's relation. The Fourier transform properties are summarized and given in Table 6.1. The basic Fourier transform pairs are given in Table 6.2.

6.7 Fourier Transform of Periodic Signal

■ Example 6.5

Find the Fourier transform of the following periodic signals:

$$\text{(a)} \quad x(t) = e^{j\omega_0 t}$$
$$\text{(b)} \quad x(t) = e^{-j\omega_0 t}$$
$$\text{(c)} \quad x(t) = \cos \omega_0 t$$
$$\text{(d)} \quad x(t) = \sin \omega_0 t$$

Solution:

(a) $\boldsymbol{x(t) = e^{j\omega_0 t} = 1e^{j\omega_0 t}}$
Let $y(t) = 1$. From Example 6.1c

$$Y(j\omega) = 2\pi\delta(\omega)$$

By using the frequency shifting property, we get

$$\boxed{X(j\omega) = 2\pi\delta(\omega - \omega_0)}$$

(b) $\boldsymbol{x(t) = e^{-j\omega_0 t}}$

$$x(t) = e^{-j\omega_0 t}$$
$$= e^{-j\omega_0 t}1$$

Since $1 \overset{\text{FT}}{\longleftrightarrow} 2\pi\delta(\omega)$, by using the frequency shifting property we get

$$\boxed{X(j\omega) = 2\pi\delta(\omega + \omega_0)}$$

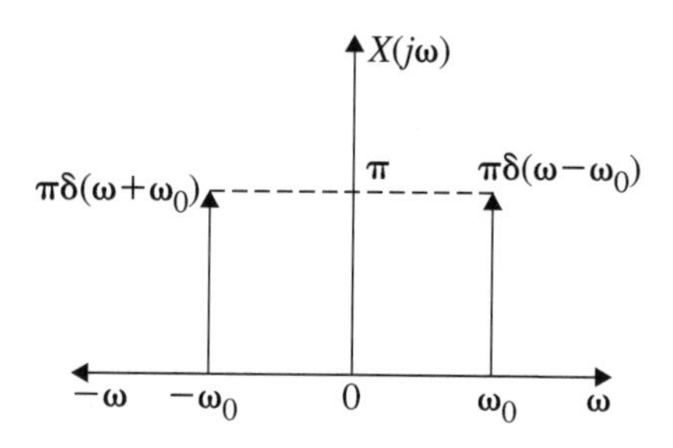

Fig. 6.16 FT of $\cos(\omega_0 t)$

(c) $\boldsymbol{x(t) = \cos(\omega_0 t)}$

$$\begin{aligned} x(t) &= \cos(\omega_0 t) \\ &= \frac{1}{2}\left[e^{j\omega_0 t} + e^{-j\omega_0 t}\right] \end{aligned}$$

Using the results obtained in 6.5(a) and 6.5(b) above, we get

$$\boxed{X(j\omega) = \pi\left[\delta(\omega + \omega_0) + \delta(\omega - \omega_0)\right]}$$

The frequency spectrum is shown in Fig. 6.16.

(d) $\boldsymbol{x(t) = \sin \omega_0 t}$

$$\begin{aligned} x(t) &= \sin \omega_0 t \\ &= \frac{1}{2j}\left[e^{j\omega_0 t} - e^{-j\omega_0 t}\right] \end{aligned}$$

$$\boxed{X(j\omega) = -j\pi\left[\delta(\omega - \omega_0) - \delta(\omega + \omega_0)\right]}$$

The Fourier spectra of $\sin \omega_0 t$ are shown in Fig. 6.17.

6.7.1 Fourier Transform Using Differentiation and Integration Properties

Using differentiation and integration properties, most of the problems encountered in CT system can be easily and quickly solved while determining the Fourier transform. Let $x(t)$ be a signal with Fourier transform $X(j\omega)$. The FT of $dx(t)/dt$ is obtained using Eq. (6.15). Here, $X(j\omega)$ is simply multiplied by $j\omega$. Thus, we get

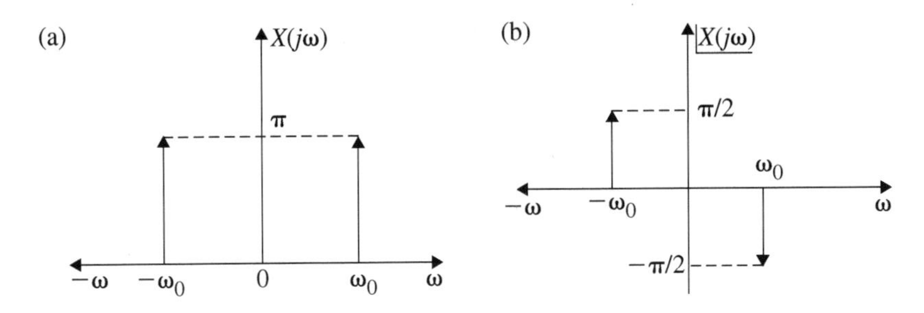

Fig. 6.17 Fourier spectra of sin $\omega_0 t$

$$x(t) \overset{\text{FT}}{\longleftrightarrow} X(j\omega)$$
$$\frac{dx(t)}{dt} \overset{\text{FT}}{\longleftrightarrow} j\omega X(j\omega)$$

Similarly, the FT of $\int_{-\infty}^{t} x(\tau)d\tau$ is obtained by dividing by $j\omega$ which is given in Eq. (6.17).

$$\int_{-\infty}^{t} x(\tau)d\tau \overset{\text{FT}}{\longleftrightarrow} \frac{1}{j\omega}X(j\omega) + \pi X(0)\delta(\omega)$$

where $X(0) = X(\omega)|_{\omega=0}$. $\pi X(0)\delta(\omega)$ accounts for the total area of $x(t)$. If this area is zero then $X(0) = 0$. The use of differentiation and integration properties saves time while solving mathematical equation. The following step-by-step procedure is followed:

1. The signal $x(t)$ is sketched.
2. $x(t)$ is differentiated and $dx(t)/dt$ is sketched.
3. The differentiation procedure is repeated until the FT could be easily obtained just by observation. Usually, the differentiation process is continued until the signal appears in the form of impulse and time shifted impulses. The FT of these impulses can be easily obtained.
4. Obtain $G(0)$ which is nothing but $X(0)$ when ω is substituted in the FT of the last derivative of $x(t)$. Thus, the DC average value $\pi X(0)\delta(\omega)$ that results from integration is added. To this value, $G(j\omega)/(j\omega)^n$ is added to get FT of $x(t)$. In other words,

$$X(j\omega) = \frac{G(j\omega)}{(j\omega)^n} + \pi X(0)\delta(\omega)$$

where

$$G(j\omega) = \text{FT of } n\text{th derivative of } x(t)$$
$$X(0) = G(j\omega)|_{\omega=0}$$

The use of differentiation and integration properties is illustrated in a few examples.

■ Example 6.6

Find the CTFT for the following signal $x(t)$:

$$x(t) = \begin{cases} 2t + 4 & -2 \leq t < 2 \\ 8 & 2 \leq t < \infty \\ 0 & \text{otherwise} \end{cases}$$

Solution: $x(t), dx(t)/dt$ and $d^2x(t)/dt^2$ are represented in Fig. 6.18a to c, respectively. From Fig. 6.18c, we get the following CTFT for $d^2x(t)/dt^2 = g_2(t)$.

$$\begin{aligned} G_2(j\omega) &= [2e^{j2\omega} - 2e^{-j2\omega}] \\ &= 2\frac{[e^{j2\omega} - e^{-j2\omega}]2j}{2j} \\ &= j4 \sin 2\omega \\ G_2(0) &= 0 \end{aligned}$$

$G_1(j\omega)$ is obtained by dividing $G_2(j\omega)$ by $j\omega$ and adding the DC term.

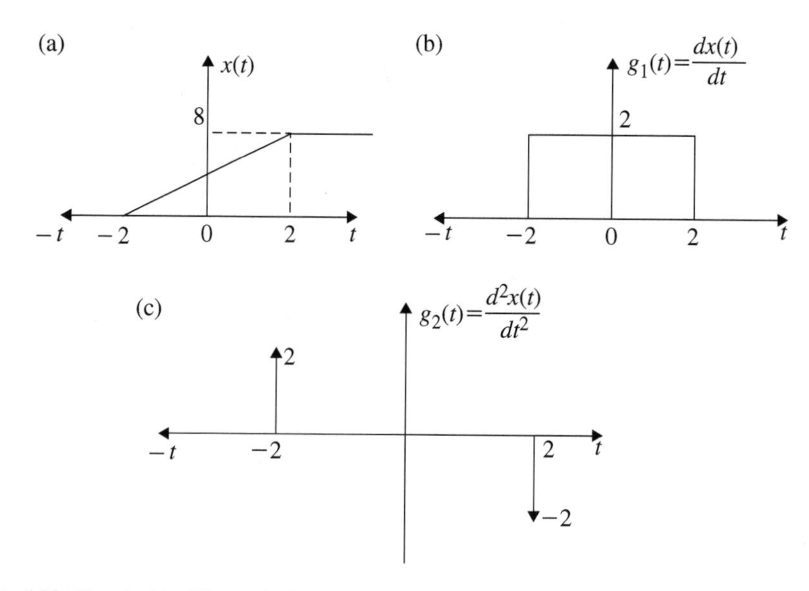

Fig. 6.18 Signal $x(t)$ of Example 6.6

$$
\begin{aligned}
G_1(j\omega) &= \frac{G_2(j\omega)}{j\omega} + \pi\delta(\omega)G_2(0) \\
&= \frac{j4\sin 2\omega}{j\omega} \\
&= 8\left(\frac{\sin 2\omega}{2\omega}\right) \\
G_1(0) &= 8 \\
G_1(j\omega) &= 8\left(\frac{\sin 2\omega}{2\omega}\right) \quad \left[\left.\frac{\sin 2\omega}{2\omega}\right|_{\omega=0} = 1\right] \\
X(j\omega) &= \frac{G_1(j\omega)}{j\omega} + \pi\delta(\omega)G_1(0)
\end{aligned}
$$

$$
\boxed{X(j\omega) = \frac{4\sin 2\omega}{j\omega^2} + 8\pi\delta(\omega)}
$$

■ Example 6.7

Find the FT of the step function $u(t)$ using the integration property of FT.

Solution:

The step function is shown in Fig. 6.19. The step function $u(t)$ and impulse function $\delta(t)$ are related as

$$
\begin{aligned}
\delta(t) &= \frac{du(t)}{dt} \\
du(t) &= \delta(t)dt
\end{aligned}
$$

Substituting $x(t) = u(t)$ and $\delta(t) = g(t)$, the above equation is written after integrating both sides as

$$
x(t) = \int_{-\infty}^{t} g(\tau)d\tau
$$

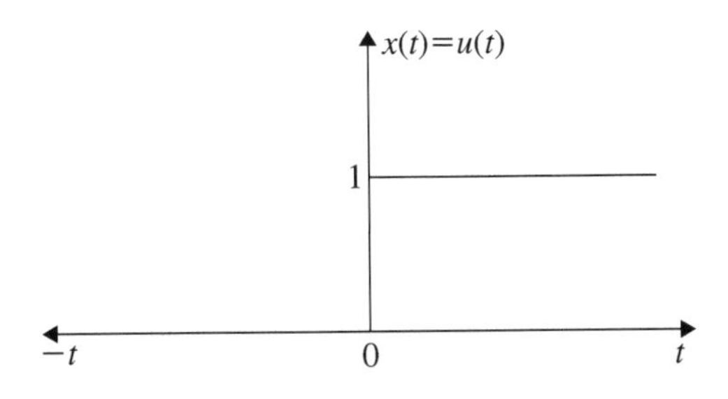

Fig. 6.19 Signal $x(t)$ of Example 6.7

Taking FT on both sides, we get

$$X(j\omega) = \frac{1}{j\omega}G(j\omega) + \pi G(0)\delta(\omega)$$

But

$$g(t) = \delta(t) \stackrel{\text{FT}}{\longleftrightarrow} G(j\omega) = 1$$
$$G(0) = 1$$

Substituting the above in $X(j\omega)$, we get

$$\boxed{X(j\omega) = \frac{1}{j\omega} + \pi\delta(\omega)}$$

The same result is obtained in Example 6.1(d).

■ Example 6.8

For the following signals, determine the FT using FT properties.

$$x(t) = 5\sin 10t$$

(a) $y(t) = x(t-3)$
(b) $y(t) = x(4(t-3))$
(c) $y(t) = x(4t-3)$
(d) $y(t) = x(-3t+4)$

Solution:

(a) $x(t) = 5\sin 10t$

From Example 6.5(d), the FT of $x(t)$ is obtained as

$$X(j\omega) = j5\pi[\delta(\omega+10) - \delta(\omega-10)]$$
$$y(t) = x(t-3)$$

FT of $y(t)$ is obtained using the time shifting property (right shift) as

$$\begin{aligned} Y(j\omega) &= X(j\omega)e^{-j3\omega} \\ &= j5\pi[\delta(\omega+10) - \delta(\omega-10)]e^{-j3\omega} \\ &= j5\pi[\delta(\omega+10)e^{-j3\omega} - \delta(\omega-10)e^{-j3\omega}] \end{aligned}$$

Using the property

$$X(j\omega)\delta(\omega-\omega_0)=X(j\omega_0)\delta(\omega-\omega_0)$$

we get

$$\boxed{Y(j\omega)=j5\pi[\delta(\omega+10)e^{j30}-\delta(\omega-10)e^{-j30}]}$$

(b)

$$\begin{aligned}y(t)&=x(4(t-3))\\&=x(4t-12)\end{aligned}$$

$$x(t)\overset{FT}{\longleftrightarrow}j5\pi[\delta(\omega+10)-\delta(\omega-10)]$$

Using the time shifting property (right shift), we get

$$x(t-12)\overset{FT}{\longleftrightarrow}j5\pi[\delta(\omega+10)-\delta(\omega-10)]e^{-j12\omega}$$

Using the time scaling property,

$$\begin{aligned}x(at)&\overset{FT}{\longleftrightarrow}\frac{1}{a}X\left(j\frac{\omega}{a}\right),\\x(4t-12)&\overset{FT}{\longleftrightarrow}j\frac{5}{4}\pi\left[\delta\left(\frac{\omega}{4}+10\right)-\delta\left(\frac{\omega}{4}-10\right)\right]e^{-j3\omega}\\&\overset{FT}{\longleftrightarrow}j\frac{5}{4}\pi\left[\delta\left(\frac{\omega+40}{4}\right)-\delta\left(\frac{\omega-40}{4}\right)\right]e^{-j3\omega}\end{aligned}$$

Using the property

$$\delta(a\omega)=\frac{1}{a}\delta(\omega)$$

we get

$$x(4t-12)\overset{FT}{\longleftrightarrow}j\frac{5}{4}\pi\left[4\delta\left(\omega+40\right)e^{-j3\omega}-4\delta\left(\omega-40\right)e^{-j3\omega}\right]$$

$$\boxed{Y(j\omega)=j5\pi\left[\delta\left(\omega+40\right)e^{j120}-\delta\left(\omega-40\right)e^{-j120}\right]}$$

(c) $y(t) = x(4t - 3)$

$$x(t) \overset{\text{FT}}{\longleftrightarrow} j5\pi[\delta(\omega + 10) - \delta(\omega - 10)]$$
$$x(t-3) \overset{\text{FT}}{\longleftrightarrow} j5\pi[\delta(\omega + 10) - \delta(\omega - 10)]e^{-j3\omega}$$
$$x(4t-3) \overset{\text{FT}}{\longleftrightarrow} \frac{j5\pi}{4}\left[\delta\left(\frac{\omega}{4} + 10\right) - \delta\left(\frac{\omega}{4} - 10\right)\right]e^{-j(3/4)\omega}$$

$$\boxed{Y(j\omega) = j5\pi\left[\delta\left(\omega + 40\right)e^{j30} - \delta\left(\omega - 40\right)e^{-j30}\right]}$$

(d) $y(t) = x(-3t + 4)$

Using the time reversal property,

$$x(-t) \overset{\text{FT}}{\longleftrightarrow} X(-j\omega)$$
$$x(-t) \overset{\text{FT}}{\longleftrightarrow} j5\pi[\delta(-\omega + 10) - \delta(-\omega - 10)]$$

Using the time shifting property (right shifted), we get

$$x(-t+4) \overset{\text{FT}}{\longleftrightarrow} j5\pi[\delta(-\omega + 10) - \delta(-\omega - 10)]e^{-j4\omega}$$

Using the time scaling property, we get

$$x(-3t+4) \overset{\text{FT}}{\longleftrightarrow} \frac{j5\pi}{3}\left[\delta\left(-\frac{\omega}{3} + 10\right) - \delta\left(\frac{-\omega}{3} - 10\right)\right]e^{-j(4/3)\omega}$$
$$Y(j\omega) = j5\pi\left[\delta\left(-\omega + 30\right)e^{-j(4/3)\omega} - \delta\left(-\omega - 30\right)e^{-j(4/3)\omega}\right]$$

$$\boxed{Y(j\omega) = j5\pi\left[\delta\left(-\omega + 30\right)e^{-j40} - \delta\left(\omega + 30\right)e^{j40}\right]}$$

$\because \delta(-\omega - 30) = \delta(\omega + 30)$

$$Y(j\omega) = j5\pi\left[\delta\left(\omega - 30\right)e^{-j40} - \delta\left(\omega + 30\right)e^{j40}\right]$$

■ Example 6.9

Consider the following CT signal.

$$x(t) = 4\cos 3t$$

Determine the FT of the following signals:

(a) $y(t) = x(2 - t) + x(-2 - t)$
(b) $y(t) = x(3t + 5)$
(c) $y(t) = \frac{d^2}{dt^2}x(t - 2)$

Solution:

(a) $x(t) = 4\cos 3t$
From Example 6.5, the FT of $x(t)$ is obtained as

$$X(j\omega) = 4\pi[\delta(\omega + 3) + \delta(\omega - 3)]$$
$$y(t) = x(2 - t) + x(-2 - t)$$

$$\begin{aligned}
x(2 - t) &\overset{\text{FT}}{\longleftrightarrow} X(-j\omega)e^{-j2\omega} \\
x(-2 - t) &\overset{\text{FT}}{\longleftrightarrow} X(-j\omega)e^{j2\omega} \\
x(2 - t) + x(-t - 2) &\overset{\text{FT}}{\longleftrightarrow} X(-j\omega)[e^{j2\omega} + e^{-j2\omega}] \\
&= X(-j\omega)2\cos 2\omega \\
X(-j\omega) &= 4\pi[\delta(-\omega + 3) + \delta(-\omega - 3)] \\
Y(j\omega) &= X(-j\omega)2\cos 2\omega
\end{aligned}$$

$$\boxed{Y(j\omega) = 8\pi\cos 2\omega\left[\delta\left(\omega + 3\right) + \delta\left(\omega - 3\right)\right]}$$

(b)

$$y(t) = x(3t + 5)$$

$$x(t + 5) \overset{\text{FT}}{\longleftrightarrow} 4\pi[\delta(\omega + 3) + \delta(\omega - 3)]e^{j5\omega}$$

Using the time scaling property of FT, we get

$$\begin{aligned}
x(3t + 5) &\overset{\text{FT}}{\longleftrightarrow} \frac{4}{3}\pi\left[\delta\left(\frac{\omega}{3} + 3\right) + \delta\left(\frac{\omega}{3} - 3\right)\right]e^{j(5/3)\omega} \\
&= 4\pi\left[\delta\left(\omega + 9\right) + \delta\left(\omega - 9\right)\right]e^{j(5/3)\omega} \\
&= 4\pi\left[\delta\left(\omega + 9\right)e^{-j15} + \delta\left(\omega - 9\right)e^{j15}\right]
\end{aligned}$$

$$\boxed{Y(j\omega) = 4\pi\left[\delta\left(\omega + 9\right)e^{-j15} + \delta\left(\omega - 9\right)e^{j15}\right]}$$

(c) $y(t) = \frac{d^2}{dt^2}x(t-2)$

$$\begin{aligned} x(t-2) &\overset{\text{FT}}{\longleftrightarrow} 4\pi[\delta(\omega+3)+\delta(\omega-3)]e^{-j2\omega} \\ &= 4\pi[\delta(\omega+3)e^{j6}+\delta(\omega-3)e^{-j6}] \end{aligned}$$

Using differentiation property

$$\frac{d^2x}{dt^2} \overset{\text{FT}}{\longleftrightarrow} (j\omega)^2X(j\omega)$$

we get

$$\begin{aligned} Y(j\omega) &= 4(j\omega)^2\pi[\delta(\omega+3)e^{j6}+\delta(\omega-3)e^{-j6}] \\ &= 4\pi[-\delta(\omega+3)9e^{j6}-\delta(\omega-3)9e^{-j6}] \end{aligned}$$

$$\boxed{Y(j\omega) = -36\pi\left[\delta(\omega+3)\,e^{j6}+\delta(\omega-3)\,e^{-j6}\right]}$$

■ Example 6.10

A signal has the following FT:

$$X(j\omega) = \frac{\omega^2+j4\omega+2}{-\omega^2+j4\omega+3}$$

Find the FT of $x(-2t+1)$.

(*Anna University, 2011*)

Solution:

$$X(j\omega) = \frac{\omega^2+j4\omega+2}{-\omega^2+j4\omega+3}$$

By using the time reversal property, the FT of $x(-t) = X(-j\omega)$ is obtained as

$$x(-t) \overset{\text{FT}}{\longleftrightarrow} \frac{\omega^2-j4\omega+2}{-\omega^2-j4\omega+3}$$

By using the time shifting (right shift) property, we get

$$x(-t+1) \overset{\text{FT}}{\longleftrightarrow} \left(\frac{\omega^2-j4\omega+2}{-\omega^2-j4\omega+3}\right)e^{-j\omega}$$

By using the time scaling property, we get

$$\boxed{x(-2t+1) \overset{\text{FT}}{\longleftrightarrow} \frac{1}{2}e^{-j\omega/2}\frac{\left[\frac{\omega^2}{4} - j2\omega + 2\right]}{\left[-\frac{\omega^2}{4} - j2\omega + 2\right]}}$$

■ Example 6.11

By using differentiation and integration property of FT, determine the FT of $x(t) = \text{sgn}(t)$.

Solution: $x(t) = \text{sgn}(t)$ is shown in Fig. 6.20a, and its derivative $dx(t)/dt = 2\delta(t)$ is shown in Fig. 6.20.

$$\frac{dx(t)}{dt} = 2\delta(t) \overset{\text{FT}}{\longleftrightarrow} 2$$

Using the integration property, we get

$$\begin{aligned} X(j\omega) &= \frac{1}{j\omega} FT\left[\frac{dx(t)}{dt}\right] + \pi\delta(\omega)G(0) \\ &= \frac{2}{j\omega} \end{aligned}$$

Since the area under the impulse is zero, the initial condition $G(0) = 0$.

$$\boxed{X(j\omega) = \frac{2}{j\omega}}$$

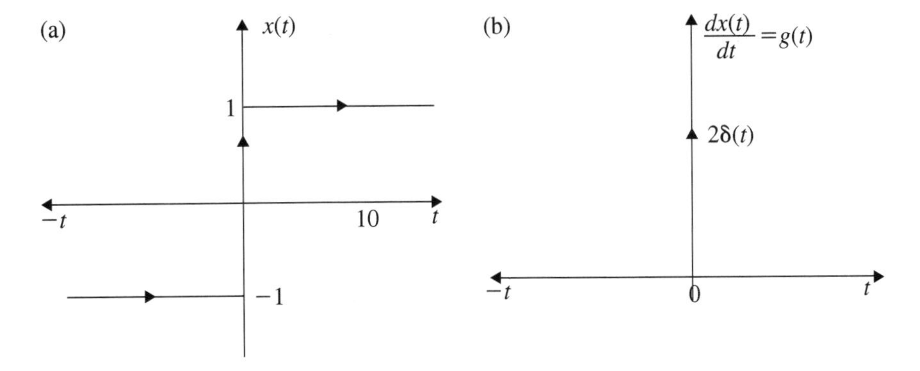

Fig. 6.20 Signals representing sgn(t) and its derivatives

Example 6.12

Consider the signal described by the following signal:

$$x(t) = 1 + 2|t| \quad |t| \leq 2$$

(a) Sketch $x(t)$ and its derivatives.
(b) Using FT integration property, determine $X(j\omega)$.
(c) Determine the odd and even components of $X(j\omega)$.

Solution:

(a)

$$x(t) = 1 + 2|t| \quad |t| \leq 2$$

$$x(t) = \begin{cases} 1 + 2t & 0 \leq t \leq 2 \\ 1 - 2t & -2 \leq t \leq 0 \\ 0 & \text{otherwise} \end{cases}$$

$x(t)$ is sketched as shown in Fig. 6.21a. From Fig. 6.21a, the following equations are written:

$$\frac{dx(t)}{dt} = \begin{cases} 5\delta(t+2) - 2 & -2 \leq t \leq 0 \\ -5\delta(t-2) + 2 & 0 \leq t \leq 2 \end{cases}$$

$dx(t)/dt$ is sketched as the sum of $g_1(t)$ and $g_2(t)$ in Fig. 6.21b and c, respectively. Thus

$$\frac{dx(t)}{dt} = g_1(t) + g_2(t)$$

The derivatives of $g_2(t)$ is sketched as $g_3(t)$ in Fig. 6.21d.

(b) From Fig. 6.21d, the FT of $g_3(t)$ is obtained as

$$\begin{aligned} G_3(j\omega) &= -2[e^{j2\omega} + e^{-j2\omega}] + 4 \\ G_3(0) &= -2(1+1) + 4 \\ &= 0 \\ G_3(j\omega) &= -4\cos 2\omega + 4 \\ &= 4[1 - \cos 2\omega] \\ &= 8\sin^2 \omega \end{aligned}$$

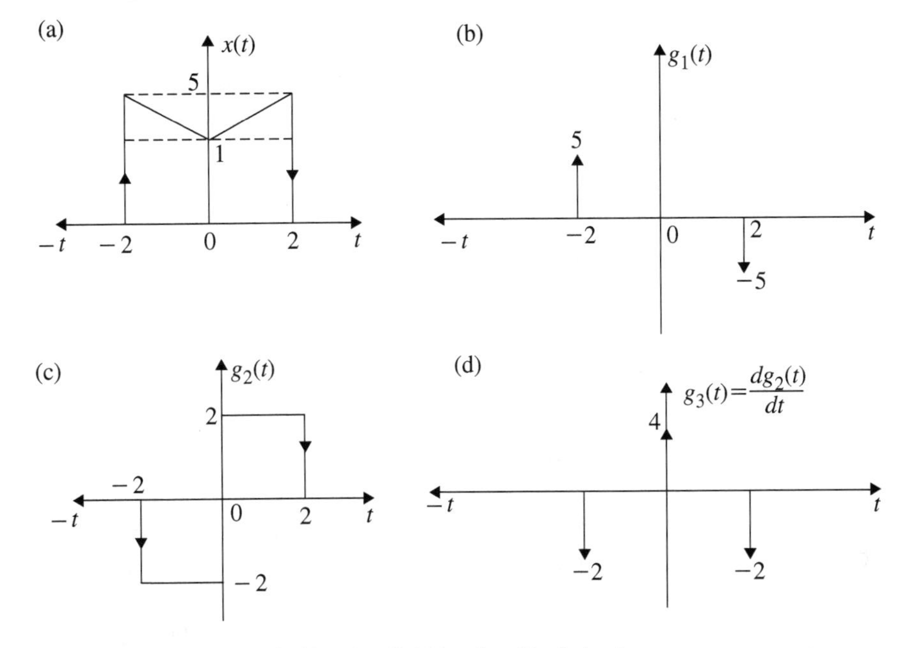

Fig. 6.21 Representation of $x(t) = 1 + 2|t|, |t| \leq 2$ and its derivatives

From Fig. 6.21b

$$\begin{aligned} G_1(j\omega) &= 5(e^{j2\omega} - e^{-j2\omega}) \\ &= j10 \sin 2\omega \\ G_1(0) &= 0 \end{aligned}$$

Using the integration property, we obtain $X(j\omega)$ as

$$X(j\omega) = \frac{G_1(j\omega)}{(j\omega)} + \frac{G_3(j\omega)}{(j\omega)^2} + \pi[G_1(0)\delta(\omega) + G_3(0)\delta(\omega)]$$

$$X(j\omega) = j10\frac{\sin 2\omega}{j\omega} + 8\frac{\sin^2 \omega}{(j\omega)^2}$$

$$\boxed{X(j\omega) = 20\text{sinc}2\omega - 8\text{sinc}^2\omega}$$

$$X(-j\omega) = 20\text{sinc}2\omega - 8\text{sinc}^2\omega$$

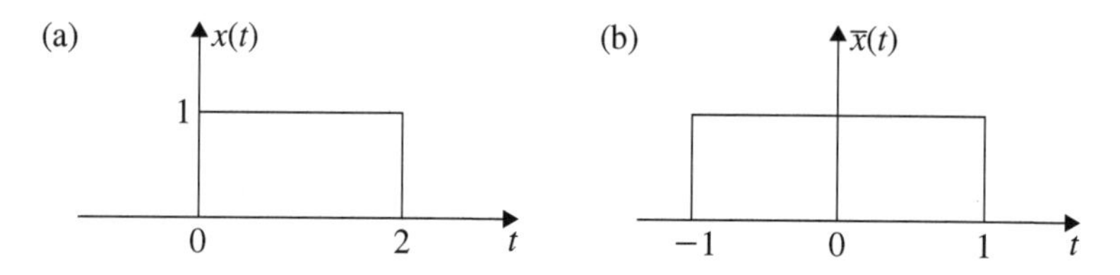

Fig. 6.22 **a** Rectangular time shifted pulse and **b** Rectangular or gate pulse

$$\begin{aligned}X(j\omega)_{\text{ev}} &= \frac{1}{2}\left[X(j\omega)+X(-j\omega)\right]\\ &= \frac{1}{2}[20\text{sinc}\,2\omega - 8\text{sinc}^2\omega + 20\text{sinc}\,2\omega - 8\text{sinc}^2\omega]\end{aligned}$$

$$\boxed{X(j\omega)_{\text{ev}} = -8\text{sinc}^2\omega + 20\text{sinc}\,2\omega}$$

$$\begin{aligned}X(j\omega)_{\text{odd}} &= \frac{1}{2}\left[X(j\omega)+X(-j\omega)\right]\\ &= \frac{1}{2}[20\text{sinc}\,2\omega - 8\text{sinc}^2\omega - 20\text{sinc}\,2\omega + 8\text{sinc}^2\omega]\end{aligned}$$

$$\boxed{X(j\omega)_{\text{odd}} = 0}$$

■ Example 6.13

Consider the signal $x(t)$ shown in Fig. 6.22a. The rectangular pulse $\bar{x}(t)$ is shown in Fig. 6.22b. From $\bar{X}(j\omega)$, determine $X(j\omega)$ using shift property.

Solution: In Example 6.2, the FT of $\bar{x}(t)$ has been derived as

$$\bar{X}(j\omega) = 2\text{sinc}\,\omega$$

From Fig. 6.22,

$$\begin{aligned}x(t) &= \bar{x}(t-1)\\ X(j\omega) &= \bar{X}(j\omega)e^{-j\omega}\end{aligned}$$

Using shift property, the FT of $x(t)$ is obtained as

$$\boxed{X(j\omega) = 2e^{-j\omega}\text{sinc}\,\omega}$$

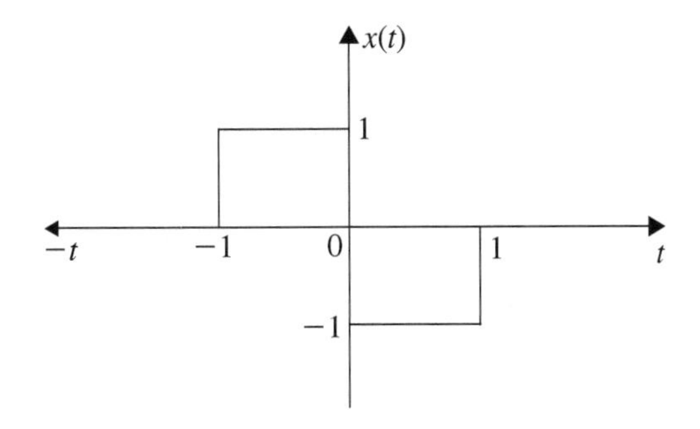

Fig. 6.23 $x(t)$ signal of Example 6.14

■ Example 6.14

Find the Fourier transform of the signal shown in Fig. 6.23 and plot its magnitude spectrum.

(*Anna University, April, 2005*)

Solution: **Method 1:**

$$x(t) = \begin{cases} 1 & -1 \le t \le 0 \\ -1 & 0 \le t \le 1 \end{cases}$$

Using definition of FT, we get

$$\begin{aligned} X(j\omega) &= \int_{-1}^{0} e^{-j\omega t}\, dt - \int_{0}^{1} e^{-j\omega t}\, dt \\ &= \frac{-1}{j\omega}\left\{\left[e^{-j\omega t}\right]_{-1}^{0} - \left[e^{-j\omega t}\right]_{0}^{1}\right\} \\ &= \frac{-1}{j\omega}\left[1 - e^{j\omega} - e^{-j\omega} + 1\right] \end{aligned}$$

$$\boxed{X(j\omega) = \frac{2}{j\omega}[\cos\omega - 1]}$$

Method 2:

Differentiating the signal in Fig. 6.23, $\frac{dx(t)}{dt}$ is obtained and is represented in Fig. 6.24.
Using the time shifting property, FT of Fig. 6.24 is written as follows:

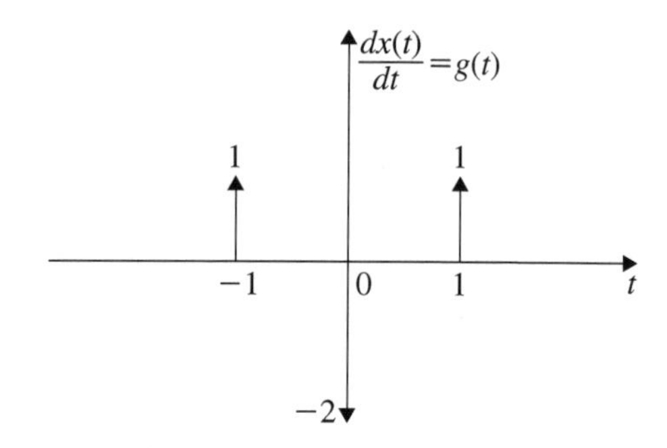

Fig. 6.24 Derivative of the signal represented in Fig. 6.23

$$G(j\omega) = F\left[\frac{dx(t)}{dt}\right] = \left[e^{j\omega} - 2 + e^{-j\omega}\right] = 2[\cos\omega - 1]$$
$$G(0) = 2[1 - 1] = 0$$

Using the time integration property, we get

$$F[x(t)] = X(j\omega) = \frac{G(j\omega)}{j\omega} + \pi G(0)\delta(\omega) = \frac{2}{j\omega}[\cos\omega - 1] + \pi G(0)\delta(\omega)$$

$$\boxed{X(j\omega) = \frac{2}{j\omega}[\cos\omega - 1]}$$

To Plot the Magnitude Spectrum

$$\begin{aligned}
|X(j\omega)| &= \frac{2}{\omega}[\cos\omega - 1] \\
&= \frac{2}{\omega}\left[\cos^2\frac{\omega}{2} - \sin^2\frac{\omega}{2} - 1\right] \\
&= \frac{-4}{\omega}\sin^2\omega/2 \\
&= -\omega\left[\frac{\sin\omega/2}{\frac{\omega}{2}}\right]^2
\end{aligned}$$

$$\boxed{|X(j\omega)| = \left|\omega\,\text{sinc}^2\frac{\omega}{2}\right|}$$

The amplitude spectrum of $X(j\omega)$ is shown in Fig. 6.25.

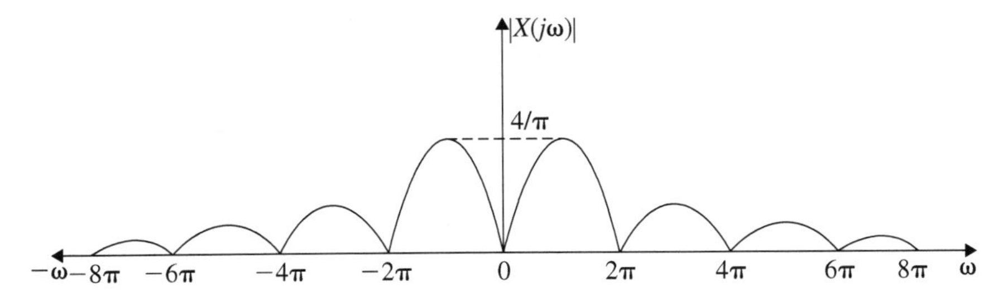

Fig. 6.25 Amplitude spectrum of $\omega\,\mathrm{sinc}^2(\omega/2)$

■ Example 6.15

Using Fourier transform properties, find the Fourier transform of the signal shown in Fig. 6.26a: (a) Time shifting and (b) Differentiation and integration.

(Anna University, December, 2007)

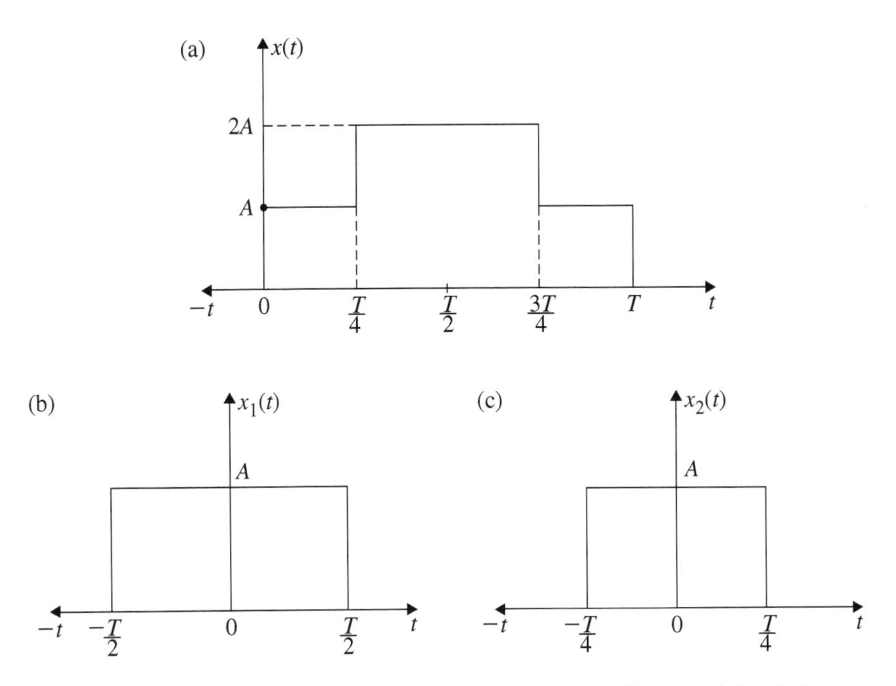

Fig. 6.26 **a**, **b** and **c** Decomposition of signal of Example 6.15. **d** Differentiated signal of $x(t)$

Fig. 6.26 (continued)

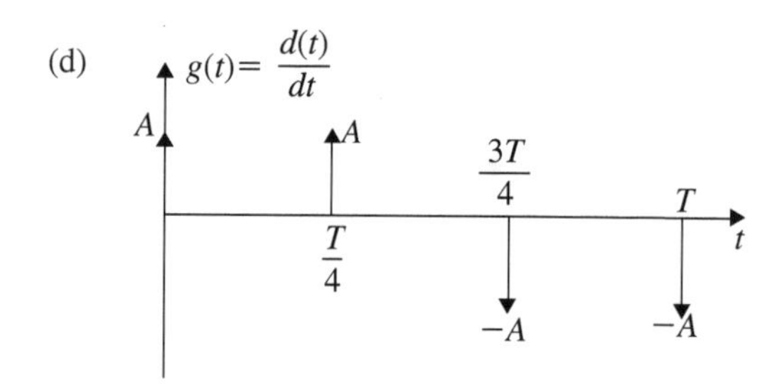

Solution:

Method 1: Time Shifting Property

The given signal $x(t)$ represented in Fig. 6.26a can be decomposed as $x_1(t)$ and $x_2(t)$ and represented in Fig. 6.26b and c, respectively. $x(t)$ can be represented as

$$x(t) = A\left[x_1\left(t - \frac{T}{2}\right) + x_2\left(t - \frac{T}{2}\right)\right]$$

Thus, the FT of $x(t)$ can be obtained using linearity and time shifting. From Example 6.2,

$$X_1(j\omega) = AT\text{sinc}\frac{\omega T}{2}$$
$$X_2(j\omega) = \frac{1}{2}AT\text{sinc}\frac{\omega T}{4}$$
$$X(j\omega) = [X_1(j\omega) + X_2(j\omega)]e^{-j\frac{\omega T}{2}}$$

$$\boxed{X(j\omega) = AT\left[\text{sinc}\frac{\omega T}{2} + \frac{1}{2}\text{sinc}\frac{\omega T}{4}\right]e^{-j\frac{\omega T}{2}}}$$

Method 2: Using Differentiation and Integration Properties

Using differentiation and integration properties of FT, the results obtained using the conventional method can be obtained in a few steps. Differentiating $x(t)$ shown in Fig. 6.26a, $g(t) = dx/dt$ can be obtained which are time shifted impulses shown in Fig. 6.26b. From Fig. 6.26d, the following equation is written as

$$g(t) = \frac{dx}{dt}$$
$$= A\delta(t) + A\delta\left(t - \frac{T}{4}\right) - A\delta\left(t - \frac{3T}{4}\right) - A\delta(t - T)$$

Taking FT on both sides, we get

$$\begin{aligned}
G(j\omega) &= A[1 + e^{-j\omega(T/4)} - e^{-j\omega(3T/4)} - e^{-j\omega T}] \\
G(0) &= A[1 + 1 - 1 - 1] \\
&= 0 \qquad \text{Note: If} x(t) \text{ is finite for } t \to \infty, G(0) = 0 \\
G(j\omega) &= Ae^{-j\omega(T/2)}[(e^{j\omega(T/2)} - e^{-j\omega(T/2)}) + (e^{j\omega(T/4)} - e^{-j\omega(T/4)})] \\
&= 2Aj\left[\sin\frac{\omega T}{2} + \sin\frac{\omega T}{4}\right]e^{-j\omega(T/2)}
\end{aligned}$$

The FT of $x(t)$ is obtained by integrating $G(j\omega)$. Thus

$$\begin{aligned}
X(j\omega) &= \frac{1}{j\omega}G(j\omega) + \pi G(0)\delta(\omega) \\
&= \frac{2A}{\omega}\left[\sin\frac{\omega T}{2} + \sin\frac{\omega T}{4}\right]e^{-j\omega(T/2)} + 0 \\
&= AT\left[\frac{\sin\frac{\omega T}{2}}{\left(\frac{\omega T}{2}\right)} + \frac{1}{2}\frac{\sin\frac{\omega T}{4}}{\left(\frac{\omega T}{4}\right)}\right]e^{-j\omega(T/2)} \\
&= AT\left[\text{sinc}\frac{\omega T}{2} + \frac{1}{2}\text{sinc}\frac{\omega T}{4}\right]e^{-j\omega(T/2)}
\end{aligned}$$

■ Example 6.16

1. Find the Fourier transform $X(j\omega)$ of the signal $x(t)$ represented in Fig. 6.27a using the differentiation property of FT. Verify the same using Fourier integral.
2. Sketch the signal

$$x(t) = 2|t| \qquad |t| \leq 2$$

Sketch $dx(t)/dt$ and using the time integration property, find $X(j\omega)$.

Solution:

1. (a) **FT Using Differentiation Property**

$$\begin{aligned}
x(t) &= 2t \qquad\qquad\qquad\qquad -1 \leq t \leq 1 \\
\frac{dx(t)}{dt} &= 2 - 2\delta(t-1) - 2\delta(t+1) \qquad -1 \leq t \leq 1
\end{aligned}$$

$x(t)$ is represented in Fig. 6.27a and $\frac{dx(t)}{dt}$ is shown in Fig. 6.27b. In Fig. 6.27b, $x_1(t)$ represents the gate function and $x_2(t)$ represents impulse functions. From Fig. 6.27c, the FT of dx_1/dt is obtained as

$$F\left[\frac{dx_1}{dt}\right] = 2[e^{j\omega} - e^{-j\omega}]$$

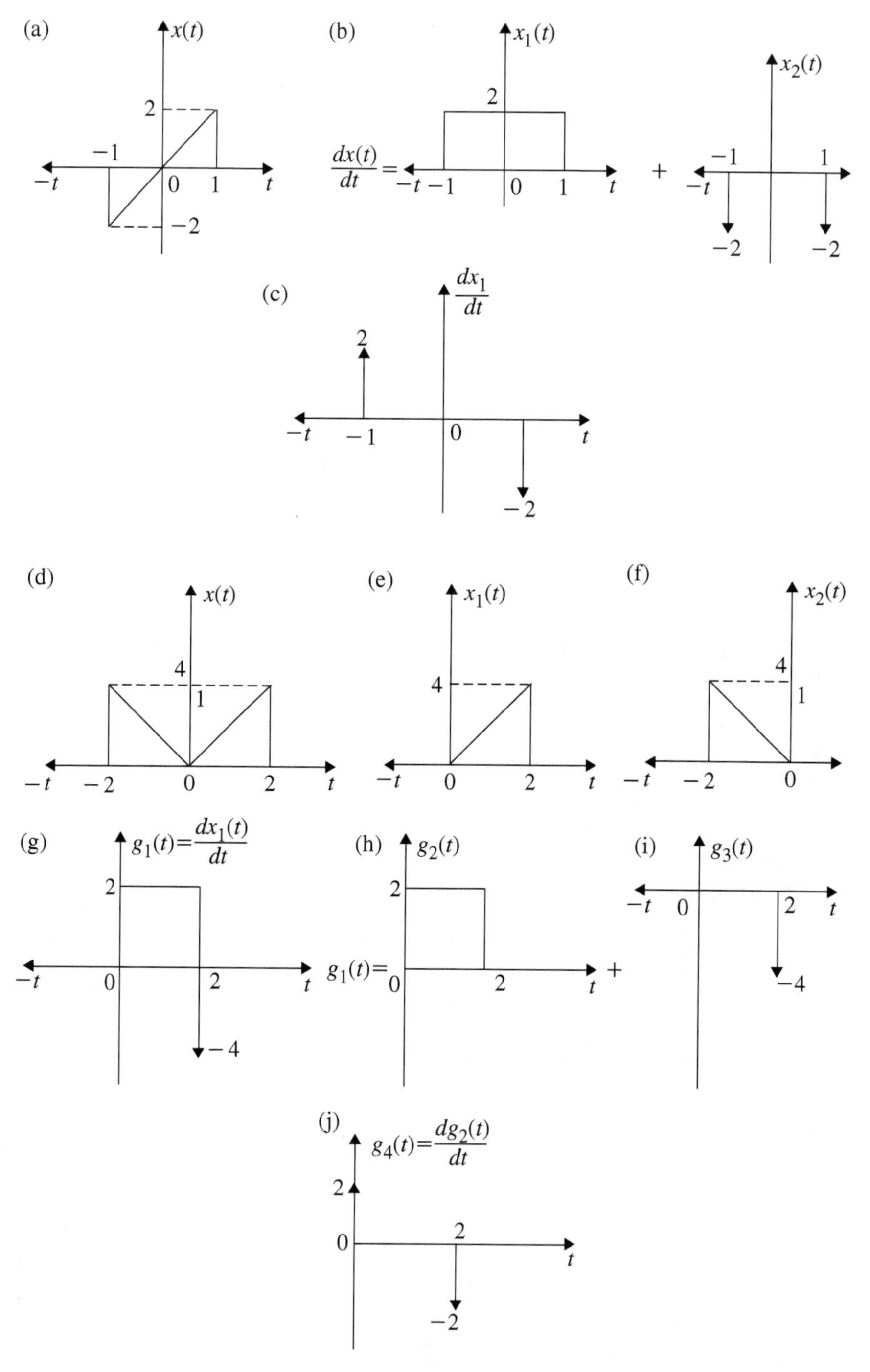

Fig. 6.27 **a** Representation of $x(t)$ and $\frac{dx(t)}{dt}$. **b** Representation of $x(t) = 2|t| \quad |t| \leq 2$

By using integration properties of FT, we get

$$\begin{aligned} X_1(j\omega) &= \frac{4}{j\omega}\frac{[e^{j\omega} - e^{-j\omega}]}{2} \\ &= 4\frac{\sin\omega}{\omega} \\ &= 4\text{sinc}\omega \end{aligned}$$

The FT of $x_2(t)$ is obtained as

$$\begin{aligned} X_2(j\omega) &= -2\left(e^{j\omega} + e^{-j\omega}\right) \\ &= -4\frac{[e^{j\omega} + e^{-j\omega}]}{2} \\ &= -4\cos\omega \\ FT\left[\frac{dx(t)}{dt}\right] &= X_1(j\omega) + X_2(j\omega) \\ &= 4\frac{\sin\omega}{\omega} - 4\cos\omega \\ X_1(0) + X_2(0) &= 4 - 4 \\ &= 0 \end{aligned}$$

By using integration properties of FT, we get

$$\begin{aligned} X(j\omega) &= \frac{1}{j\omega}[X_1(j\omega) + X_2(j\omega)] + \pi[X_1(0) + X_2(0)]\delta(\omega) \\ &= \frac{4}{j\omega}[\text{sinc}\omega - \cos\omega] + 0 \end{aligned}$$

$$\boxed{X(j\omega) = \frac{4}{j\omega}[\text{sinc}\,\omega - \cos\omega]}$$

The above result can be obtained using the Fourier integral as explained below.

(b) **FT Using Fourier Integral**

$$\begin{aligned} x(t) &= 2t \\ X(j\omega) &= \int_{-1}^{1} 2te^{-j\omega t}\,dt \end{aligned}$$

Let $u = 2t$; $du = 2dt$ and $dv = \int e^{-j\omega t}\,dt$; $v = \frac{-1}{j\omega}e^{-j\omega t}$

$$
\begin{aligned}
X(j\omega) &= uv - \int v\,du \\
&= \left[\frac{-2t}{j\omega}e^{-j\omega t}\right]_{-1}^{1} + \frac{2}{j\omega}\int_{-1}^{1} e^{-j\omega t}\,dt \\
&= \left[\frac{-2t}{j\omega}e^{-j\omega t} + \frac{2}{\omega^2}e^{-j\omega t}\right]_{-1}^{1} \\
&= 2\left[\frac{-e^{-j\omega}}{j\omega} + \frac{1}{\omega^2}e^{-j\omega} - \frac{1}{j\omega}e^{j\omega} - \frac{1}{\omega^2}e^{j\omega}\right] \\
&= 2\left[-\frac{1}{j\omega}\left(e^{j\omega} + e^{-j\omega}\right) - \frac{1}{\omega^2}\left(e^{j\omega} - e^{-j\omega}\right)\right] \\
&= 4\left[-\frac{1}{j\omega}\cos\omega + \frac{1}{j\omega}\frac{\sin\omega}{\omega}\right]
\end{aligned}
$$

$$
\boxed{X(j\omega) = \frac{4}{j\omega}[\text{sinc}\,\omega - \cos\omega]}
$$

2. Sketch the signal $x(t) = 2|t| \quad |t| \leq 2$.

   Using differentiation, integration and time reversal properties, find the FT of $X(j\omega)$

$$
x(t) = \begin{cases} 2|t| & |t| \leq 2 \\ 2t & 0 \leq t \leq 2 \\ -2t & -2 \leq t \leq 0 \\ 0 & \text{otherwise} \end{cases}
$$

   $x(t)$ is shown in Fig. 6.27d. From Fig. 6.27, the following equations are written:

$$
x(t) = x_1(t) + x_2(t)
$$

   Let

$$
\begin{aligned}
g_1(t) &= \frac{dx_1(t)}{dt} \\
g_1(t) &= g_2(t) + g_3(t) \\
g_4(t) &= \frac{d}{dt}g_2(t)
\end{aligned}
$$

   FT of $x_1(t)$ is obtained as explained below using the integration property of FT

$$G_4(j\omega) \xleftrightarrow{\text{FT}} 2[1 - e^{-j2\omega}]$$

$$\begin{aligned} G_4(j\omega) &= 2e^{-j\omega}[e^{j\omega} - e^{-j\omega}] \\ &= j4e^{-j\omega} \sin \omega \\ G_4(0) &= 0 \end{aligned}$$

Using the time integration properties of FT, we get

$$\begin{aligned} G_2(j\omega) &= \frac{G_4(j\omega)}{j\omega} + \pi G_4(0)\delta(\omega) \\ &= \frac{j_4}{j\omega} e^{-j\omega} \sin \omega \\ &= 4e^{-j\omega} \frac{\sin \omega}{\omega} \\ G_3(j\omega) &= -4e^{-j2\omega} \\ G_1(j\omega) &= G_2(j\omega) + G_3(j\omega) \\ &= 4e^{-j\omega} \frac{\sin \omega}{\omega} - 4e^{-j2\omega} \end{aligned}$$

$$\begin{aligned} G_1(0) &= 4 - 4 \\ G_1(0) &= 0 \qquad \left[\underset{\omega=0}{Lt} \frac{\sin \omega}{\omega} = 1 \right] \\ g_1(t) &= \frac{dx_1(t)}{dt} \end{aligned}$$

Using the time integration property of FT, we get

$$\begin{aligned} X_1(j\omega) &= \frac{1}{j\omega} G_1(j\omega) + G_1(0)\pi\delta(\omega) \\ &= \frac{4e^{-j\omega}}{j\omega^2} \sin \omega - \frac{4e^{-j2\omega}}{j\omega} \end{aligned}$$

Using the time reversal property of FT, we get $[x_1(-t) = x_2(t)]$

$$\begin{aligned} X_2(j\omega) &= X_1(-j\omega) \\ &= \frac{4e^{j\omega}}{j\omega^2}\sin(-\omega) + \frac{4e^{j2\omega}}{j\omega} \end{aligned}$$

Using the linearity property of FT, we get (Fig. 6.27a)

$$\begin{aligned} X(j\omega) &= X_1(j\omega) + X_2(j\omega) \\ &= \frac{4e^{-j\omega}}{j\omega^2}\sin\omega - \frac{4e^{-j2\omega}}{j\omega} - \frac{4e^{j\omega}\sin\omega}{j\omega^2} + \frac{4e^{j2\omega}}{j\omega} \\ &= \frac{4}{j\omega}\frac{\sin\omega}{\omega}\left[e^{-j\omega} - e^{j\omega}\right] + \frac{4}{j\omega}\left[e^{j2\omega} - e^{-j2\omega}\right] \\ &= -8\text{sinc}\omega\frac{\sin\omega}{\omega} + 8\frac{\sin 2\omega}{\omega} \end{aligned}$$

$$\boxed{X(j\omega) = -8\text{sinc}^2\omega + 16\text{sinc}2\omega}$$

■ Example 6.17

Find the Fourier transform of the impulse train shown in Fig. 6.28.

Solution: For Fig. 6.28a,

$$x(t) = \sum_{n=-\infty}^{\infty} \delta(t - nT)$$

where T is the periodicity. The Fourier series coefficients are determined as

$$\begin{aligned} D_n &= \frac{1}{T}\int_{-T/2}^{T/2} \delta(t)e^{-jn\omega_0 t}\,dt \\ &= \frac{1}{T}\int_{-T/2}^{T/2} \delta(t)e^{-0}\,dt \\ &= \frac{1}{T}\int_{-T/2}^{T/2} \delta(t)dt \\ &= \frac{1}{T} \end{aligned}$$

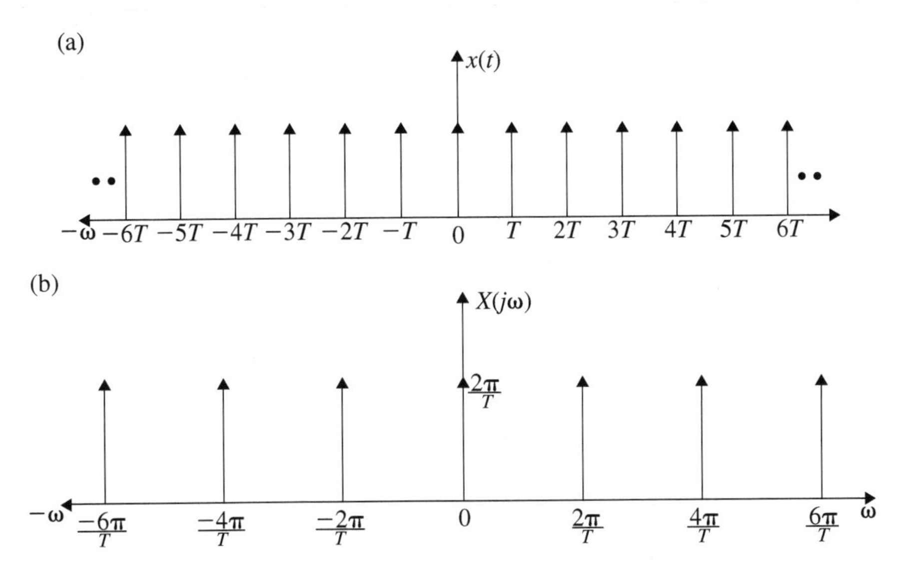

Fig. 6.28 **a** Impulse train and **b** FT of Impulse train

For a periodic signal

$$x(t) = \sum_{n=-\infty}^{\infty} D_n e^{j\omega_0 nt}$$

where

$$\omega_0 = \frac{2\pi}{T}$$

and

$$X(j\omega) = 2\pi \sum_{n=-\infty}^{\infty} D_n \delta(\omega - n\omega_0)$$

$$X(j\omega) = \frac{2\pi}{T} \sum_{n=-\infty}^{\infty} \delta\left(\omega - \frac{2\pi n}{T}\right)$$

The above expression is represented in Fig. 6.28b.

■ Example 6.18

For the triangular wave shown in Fig. 6.29a, find the Fourier transform using differentiation and integration properties.

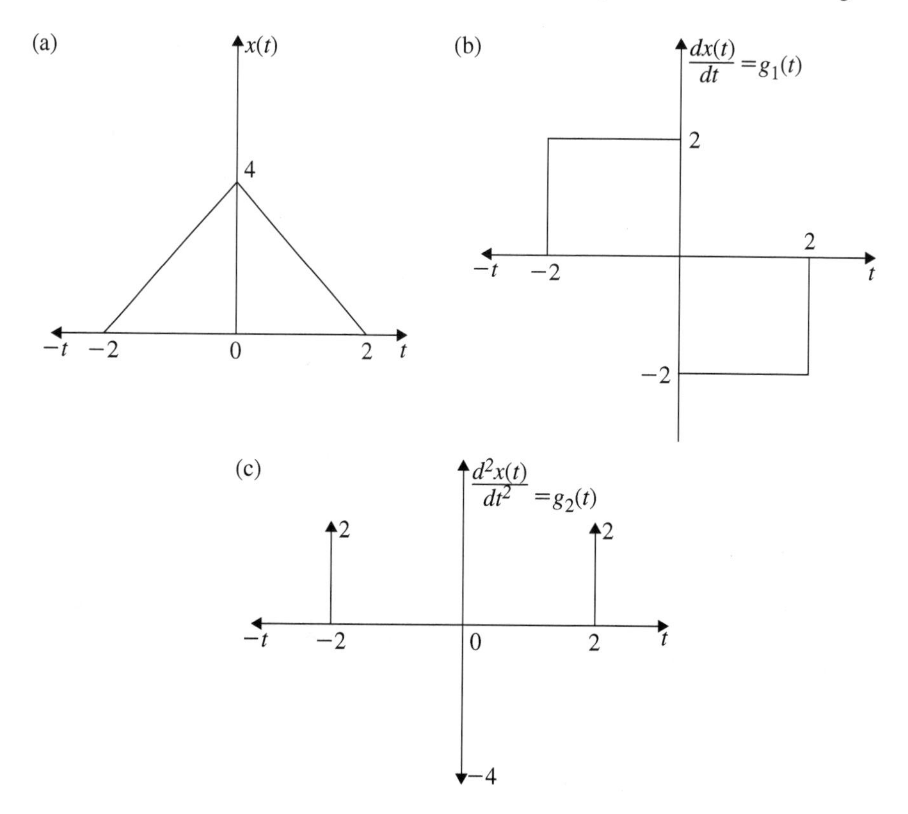

Fig. 6.29 **a** Triangular wave; **b** First derivative and **c** Second derivative of Example 6.18

Solution: The triangular signal $x(t)$ is represented in Fig. 6.29a. It is mathematically expressed as

$$x(t) = \begin{cases} 2t + 4 & -2 \leq t < 0 \\ 4 - 2t & 0 \leq t \leq 2 \end{cases}$$

$$\frac{dx(t)}{dt} = \begin{cases} 2 & -2 \leq t < 0 \\ -2 & 0 \leq t \leq 2 \end{cases}$$

$\left.\frac{dx(t)}{dt}\right|_{t=0}$ varies from $+2$ to -2. $\frac{dx(t)}{dt}$ is represented in Fig. 6.29b.

$$\frac{d^2x(t)}{dt^2} = \begin{cases} 2\delta(t + 2) & t = -2 \\ -4 & t = 0 \\ 2\,\delta(t - 2) & t = 2 \end{cases}$$

$\frac{d^2x(t)}{dt^2} = g_2(t)$ is shown in Fig. 6.29c. From Fig. 6.29c, using linearity and time shifting properties of FT, we get

$$\begin{aligned} F\left[\frac{d^2x(t)}{dt^2}\right] &= G_2(j\omega) = 2e^{j2\omega} - 4 + 2e^{-j2\omega} \\ &= 4[\cos 2\omega - 1] \\ G_2(j\omega) &= -8\sin^2\omega \\ G_2(0) &= 0 \end{aligned}$$

$X(j\omega)$ is obtained by dividing $G_1(j\omega)$ by $(j\omega)^2$ and adding initial condition

$$\begin{aligned} X(j\omega) &= \frac{G_2(j\omega)}{(j\omega)^2} + \pi G_2(0)\delta(\omega) \\ &= \frac{-8}{(j\omega)^2}\sin^2\omega \\ &= 8\left[\frac{\sin\omega}{\omega}\right]^2 \end{aligned}$$

$$\boxed{X(j\omega) = 8\text{sinc}^2\,\omega}$$

The same result is obtained in Example 6.4 which is obtained directly using Fourier integral.

Example 6.19

Consider the signal described below.

$$x(t) = \begin{cases} (t+2) & -2 \le t \le 2 \\ 4 & t \ge 2 \\ 0 & t \le -2 \end{cases}$$

Sketch the signal $x(t)$. Determine $X(j\omega)$ using differentiation and integration properties. Also determine even and odd components of $X(j\omega)$.

Solution:

$$x(t) = \begin{cases} (t+2) & -2 \le t \le 2 \\ 4 & t \ge 2 \\ 0 & t \le -2 \end{cases}$$

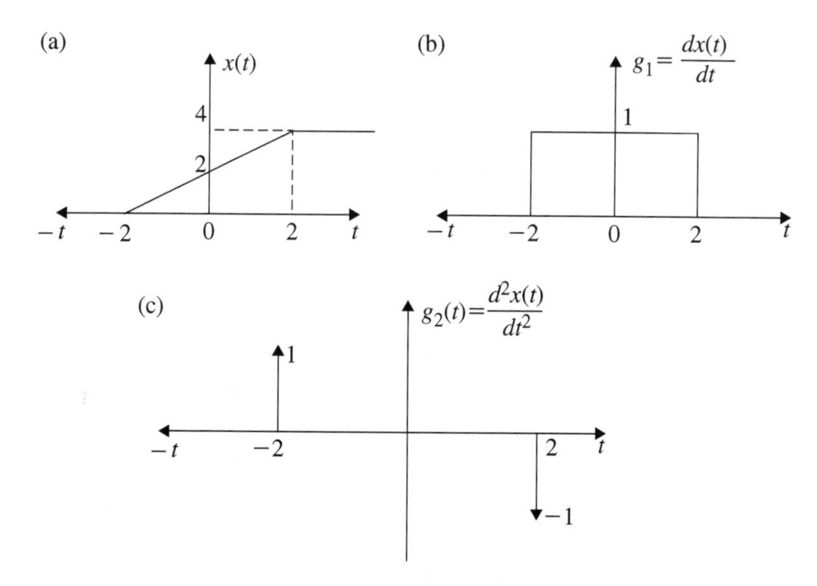

Fig. 6.30 Representation of $x(t)$ and its derivatives of Example 6.19

The signal $x(t)$ corresponding to the above equation is shown in Fig. 6.30a. The signal corresponding to $dx(t)/dt$ is shown in Fig. 6.30b and $d^2x(t)/dt^2$ in Fig. 6.30c.

From Fig. 6.30a,

$$\begin{aligned}
g_2(t) &= \frac{d^2x(t)}{dt^2} \\
&= \delta(t+2) - \delta(t-2) \\
F\left[\frac{d^2x}{dt^2}\right] &= e^{j2\omega} - e^{-j2\omega} \\
G_2(j\omega) &= 2j\frac{[e^{j2\omega} - e^{-j2\omega}]}{2j} \\
&= 2j \sin 2\omega \\
G_2(0) &= 0
\end{aligned}$$

Using the integration property of FT, $G_1(j\omega)$ is obtained as

$$\begin{aligned}
G_1(j\omega) &= \frac{1}{j\omega}G_2(j\omega) + \pi G_2(0)\delta(\omega) \\
G_1(j\omega) &= \frac{2j}{j\omega} \sin 2\omega + \pi G_2(0)\delta(\omega) \\
&= 4\frac{\sin 2\omega}{2\omega} \\
G_1(0) &= 4
\end{aligned}$$

Again using the integration property of FT, $X(j\omega)$ is obtained as

$$X(j\omega) = \frac{G_1(j\omega)}{j\omega} + \pi G_1(0)\delta(\omega)$$

$$X(j\omega) = \frac{4 \sin 2\omega}{j\omega(2\omega)} + 4\pi\delta(\omega)$$

$$\boxed{X(j\omega) = \frac{2 \sin 2\omega}{j\omega^2} + 4\pi\delta(\omega)}$$

The real part of $X(j\omega)$ corresponds to the even component while the imaginary part corresponds to the odd component. Thus

$$\begin{aligned} X_{\text{ev}}(j\omega) &= 4\pi\delta(\omega) \\ X_{\text{odd}}(j\omega) &= \frac{2 \sin 2\omega}{j\omega^2} \end{aligned}$$

Also, we may obtain this as follows:

$$\begin{aligned} X(-j\omega) &= -2\frac{\sin 2\omega}{j\omega^2} + 4\pi\delta(\omega) \\ X_{\text{ev}}(j\omega) &= \frac{1}{2}[X(j\omega) + X(-j\omega)] \\ &= \frac{1}{2}\left[\frac{2 \sin 2\omega}{j\omega^2} + 4\pi\delta(\omega) - \frac{2 \sin 2\omega}{j\omega^2} + 4\pi\delta(\omega)\right] = 4\pi\delta(\omega) \\ X_{\text{odd}}(j\omega) &= \frac{1}{2}\left[\frac{2 \sin 2\omega}{j\omega^2} + 4\pi\delta(\omega) + \frac{2 \sin 2\omega}{j\omega^2} - 4\pi\delta(\omega)\right] \\ &= \frac{2 \sin 2\omega}{j\omega^2} \end{aligned}$$

Example 6.20

Consider the following signal:

$$x(t) = \begin{cases} t & -2 \le t \le 2 \\ 2 & t \ge 2 \\ -2 & t \le -2 \end{cases}$$

Sketch the signal. Determine $X(j\omega)$.

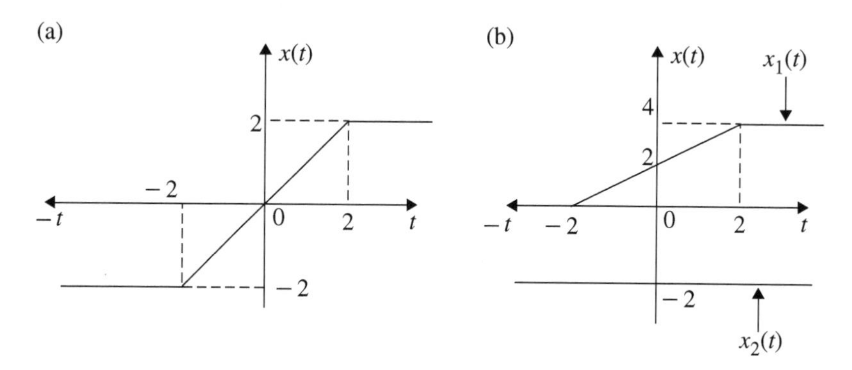

Fig. 6.31 Representation of $x(t)$ of Example 6.20

Solution:

$$x(t) = \begin{cases} t & -2 \leq t \leq 2 \\ 2 & t \geq 2 \\ -2 & t \leq -2 \end{cases}$$

The signal is sketched as shown in Fig. 6.31a. $x(t)$ of Fig. 6.31a can be split up as $x_1(t)$ and $x_2(t)$ as shown in Fig. 6.31b.

$$x(t) = x_1(t) + x_2(t)$$
$$X(j\omega) = X_1(j\omega) + X_2(j\omega)$$

$X_1(j\omega)$ is nothing but the FT of the signal shown in Fig. 6.30a which is written as

$$\begin{aligned} X_1(j\omega) &= \frac{2\sin 2\omega}{j\omega^2} + 4\pi\delta(\omega) \\ x_2(t) &= -2 \\ X_2(j\omega) &= -2 \times 2\pi\delta(\omega) \\ &= -4\pi\delta(\omega) \end{aligned}$$

Therefore

$$\begin{aligned} X(j\omega) &= \frac{2\sin 2\omega}{j\omega^2} + 4\pi\delta(\omega) - 4\pi\delta(\omega) \\ &= \frac{2\sin 2\omega}{j\omega^2} \end{aligned}$$

It can be easily verified that $x(t)$ shown in Fig. 6.31a is the odd component of $x(t)$ shown in Fig. 6.30a. $-x_2(t)$ is the even component of $x(t)$ of Fig. 6.30a.

■ Example 6.21

Find the Fourier transform of

$$x(t) = \frac{2a}{a^2 + t^2}$$

using the duality property of FT.

Solution:

Method 1

From Example 6.1(f), the FT of $x(t) = e^{-a|t|}$ is obtained as

$$x(t) = e^{-a|t|} \stackrel{\text{FT}}{\longleftrightarrow} \frac{2a}{a^2 + \omega^2}$$

By the application of inverse Fourier transform, we get

$$e^{-a|t|} = \frac{1}{2\pi} \int_{-\infty}^{\infty} \frac{2a}{a^2 + \omega^2} e^{j\omega t}\, d\omega$$

$$2\pi e^{-a|t|} = \int_{-\infty}^{\infty} \frac{2a}{a^2 + \omega^2} e^{j\omega t}\, d\omega$$

Replacing t by $-t$ in the above equation, we get

$$2\pi e^{-a|t|} = \int_{-\infty}^{\infty} \frac{2a}{a^2 + \omega^2} e^{-j\omega t}\, d\omega$$

Interchanging t and ω in the above equation, we get

$$2\pi e^{-a|\omega|} = \int_{-\infty}^{\infty} \frac{2a}{(a^2 + t^2)} e^{-j\omega t}\, dt$$

The right-hand side of the above equation is nothing but the FT of $\frac{2a}{a^2+t^2}$.

$$2\pi e^{-a|\omega|} = F\left[\frac{2a}{(a^2 + t^2)}\right]$$

$$\boxed{\left[\frac{2a}{(a^2 + t^2)}\right] \stackrel{\text{FT}}{\longleftrightarrow} 2\pi e^{-a|\omega|}}$$

Method 2: The duality property of $X(t) = 2\pi x(-\omega)$. From Example 6.1(f), the FT of $e^{-|t|}$ is obtained as

$$e^{-a|t|} \overset{\text{FT}}{\longleftrightarrow} \frac{2a}{a^2+\omega^2}$$

$$X(t) = \frac{2a}{a^2+t^2}$$

$$x(-\omega) = e^{-a|\omega|}$$

$$X(t) \overset{\text{FT}}{\longleftrightarrow} 2\pi x(-\omega)$$

$$\boxed{\frac{2a}{a^2+t^2} \overset{\text{FT}}{\longleftrightarrow} 2\pi e^{-a|\omega|}}$$

Example 6.22

For the Fourier transforms shown in Fig. 6.32a, b and c, find the energy of the signals using Parseval's theorem

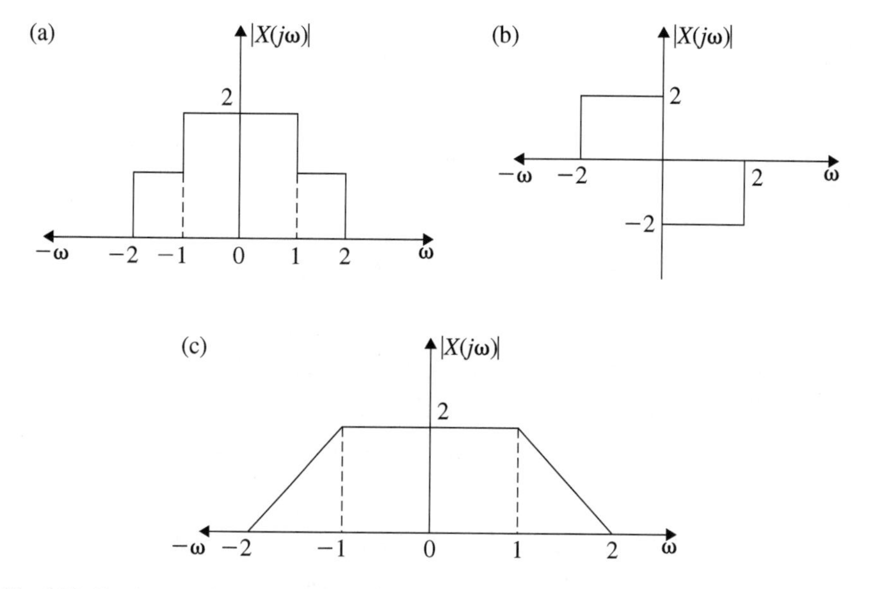

Fig. 6.32 Fourier transformed signal of Example 6.22

Solution:

(a)

$$E = \frac{1}{2\pi}\int_{-\infty}^{\infty} |X(j\omega)|^2\, d\omega$$

$$E = \frac{1}{2\pi}\left\{\int_{-2}^{-1} 1^2\, d\omega + \int_{-1}^{1} 2^2\, d\omega + \int_{1}^{2} 1^2\, d\omega\right\}$$

$$= \frac{1}{2\pi}\left\{\Big[\omega\Big]_{-2}^{-1} + 4\Big[\omega\Big]_{-1}^{1} + \Big[\omega\Big]_{1}^{2}\right\}$$

$$= \frac{1}{2\pi}\{-1+2+4+4+2-1\}$$

$$\boxed{E = \frac{5}{\pi}}$$

(b)

$$E = \frac{1}{2\pi}\left\{\int_{-2}^{0} 2^2\, d\omega + \int_{0}^{2} (-2)^2\, d\omega\right\}$$

$$= \frac{1}{2\pi}\left\{4\Big[\omega\Big]_{-2}^{0} + 4\Big[\omega\Big]_{0}^{2}\right\}$$

$$\boxed{E = \frac{8}{\pi}}$$

(c)

$$|X(j\omega)| = \begin{cases} 2\omega + 4 & -2 \le \omega \le -1 \\ 2 & -1 \le \omega \le 1 \\ (4 - 2\omega) & 1 \le \omega < 2 \end{cases}$$

$$E = \frac{1}{2\pi}\left\{\int_{-2}^{-1}(2\omega+4)^2\,d\omega + \int_{-1}^{1}(2)^2\,d\omega + \int_{1}^{2}(4-2\omega)^2\,d\omega\right\}$$

$$= \frac{1}{2\pi}\left\{\int_{-2}^{-1}(4\omega^2+16\omega+16)\,d\omega + 4\int_{-1}^{1}d\omega + \int_{1}^{2}(4\omega^2-16\omega+16)\,d\omega\right\}$$

$$= \frac{1}{2\pi}\left\{\left[\frac{4}{3}\omega^3+8\omega^2+16\omega\right]_{-2}^{-1} + 4\Big[\omega\Big]_{-1}^{1} + \left[\frac{4}{3}\omega^3-8\omega^2+16\omega\right]_{1}^{2}\right\}$$

$$= \frac{1}{2\pi}\left\{\left[-\frac{4}{3}+8-16+\frac{32}{3}-32+32\right]+[4+4]\right.$$

$$\left.+\left[\frac{32}{3}-32+32-\frac{4}{3}+8-16\right]\right\}$$

$$\boxed{E = \frac{16}{3\pi}}$$

■ Example 6.23

Find the Fourier transform of the following continuous time functions by applying Fourier transform properties or otherwise.

1. $x(t) = \delta(t-2)$
2. $x(t) = \delta(t-1) - \delta(t+1)$
3. $x(t) = \delta(t+2) + \delta(t-2)$
4. $x(t) = u(t+2) - u(t-2)$
5. $x(t) = [u(-t-3) + u(t-3)]$
6. $x(t) = e^{-3t}u(t-1)$
7. $x(t) = te^{-at}u(t)$
8. $x(t) = e^{-a(t-2)}u(t-2)$
9. $x(t) = e^{-a|t-2|}$
10. $x(t) = \cos(\omega_0 t + \phi)$
11. $x(t) = \sin(\omega_0 t + \phi)$
12. $x(t) = \sin\left(2\pi t + \frac{\pi}{4}\right)$
13. $x(t) = \cos\left(3\pi t + \frac{\pi}{8}\right) + 1$
14. $x(t) = \cos\left(6\pi t - \frac{\pi}{8}\right)$

15. $x(t) = x(4t - 8)$

16. $x(t) = \dfrac{d^2}{dt^2}x(t-2)$

17. $x(t) = x(2-t) + x(-2-t)$

18. $x(t) = \text{rect}\left(\dfrac{t+2}{4}\right)$

19. $x(t) = \text{tri}\left(\dfrac{t-4}{10}\right)$

20. $x(t) = \dfrac{d}{dt}\left[5\,\text{rect}\dfrac{t}{8}\right]$

21. $x(t) = \delta(t+2) + 5\delta(t+1) + \delta(t-1) + 5\delta(t-2)$

22. $x(t) = \begin{cases} e^{j6|t|} & |t| \leq \pi \\ 0 & \text{elsewhere} \end{cases}$

23. $x(t) = \begin{cases} 0 & |t| > 1 \\ \dfrac{(t+1)}{2} & -1 \leq t \leq 1 \end{cases}$

24. $x(t) = \begin{cases} t & 0 \leq t < 1 \\ 0 & \text{elsewhere} \end{cases}$

25. $x(t) = \begin{cases} t & 0 \leq t < 1 \\ 1 & 1 \leq t \leq 2 \\ 0 & \text{elsewhere} \end{cases}$

26. $x(t) = \begin{cases} 1 & |t| < 1 \\ 2 - |t| & 1 < |t| < 2 \\ 0 & \text{elsewhere} \end{cases}$

Solution:

1. $\boldsymbol{x(t) = \delta(t-2)}$

The impulse is time shifted by $t_0 = 2$.

$$\begin{aligned} F[\delta(t-2)] &= e^{-j\omega t_0}\,F[\delta(t)] \\ &= e^{-j2\omega} \end{aligned}$$

$$\boxed{F[\delta(t-2)] = e^{-j2\omega}}$$

2. $\boldsymbol{x(t) = \delta(t-1) - \delta(t+1)}$

$$\begin{aligned} F[\delta(t-1)] &= e^{-j\omega} \\ F[\delta(t+1)] &= e^{j\omega} \\ F[\delta(t-1) + \delta(t+1)] &= e^{-j\omega} - e^{j\omega} \\ &= -2j \sin \omega \end{aligned}$$

$$\boxed{F[\delta(t-1) - \delta(t+1)] = -2j \sin \omega}$$

3. $\boldsymbol{x(t) = \delta(t+2) + \delta(t-2)}$

$$\begin{aligned} F[\delta(t+2)] &= e^{j2\omega} \\ F[\delta(t-2)] &= e^{-j2\omega} \\ F[\delta(t+2) + \delta(t-2)] &= e^{j2\omega} + e^{-j2\omega} \\ &= 2 \cos 2\omega \end{aligned}$$

$$\boxed{X(j\omega) = 2 \cos 2\omega}$$

4. $\boldsymbol{x(t) = u(t+2) - u(t-2)}$

$$\begin{aligned} F[u(t+2)] &= \frac{1}{j\omega} e^{j2\omega} + \pi \delta(\omega) e^{j2\omega} \\ F[u(t-2)] &= \frac{1}{j\omega} e^{-j2\omega} + \pi \delta(\omega) e^{-j2\omega} \\ F[u(t+2) - u(t-2)] &= \frac{1}{j\omega} \left[e^{j2\omega} - e^{-j2\omega} \right] + \pi \delta(\omega) - \pi \delta(\omega) \\ &= \frac{2}{\omega} \sin 2\omega \end{aligned}$$

$$\boxed{X(j\omega) = 4 \text{sinc}\, 2\omega}$$

5. $\boldsymbol{x(t) = [u(-t-3) + u(t-3)]}$. **What is** $\boldsymbol{X(j\omega)}$?

$x(t)$ and $\dfrac{dx(t)}{dt}$ are shown in Fig. 6.33a and b, respectively.

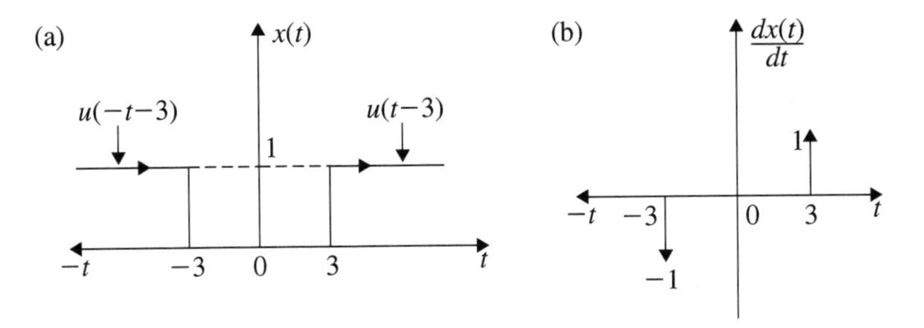

Fig. 6.33 Signal $x(t)$ of Example 6.21.5

From Fig. 6.31b,

$$F\left[\frac{dx(t)}{dt}\right] = e^{-j3\omega} - e^{+j3\omega}$$
$$= -2j\frac{\left[e^{j3\omega} - e^{-j3\omega}\right]}{2j}$$

$$\boxed{F\left[\frac{dx(t)}{dt}\right] = -2j\sin 3\,\omega}$$

Let

$$\frac{dx(t)}{dt} \overset{\text{FT}}{\longleftrightarrow} G(j\omega)$$

Using the integration property of FT, we get

$$X(j\omega) = \frac{1}{j\omega}G(j\omega) + \pi G(0)\delta(\omega)$$

where

$$G(j\omega) = -2j\sin 3\omega$$
$$G(0) = 0$$
$$X(j\omega) = \frac{1}{j\omega}(-1)2j\sin 3\omega$$
$$= -6\left(\frac{\sin 3\omega}{3\omega}\right)$$

$$\boxed{X(j\omega) = -6\text{sinc}\,3\omega}$$

6. $\boldsymbol{x(t) = e^{-3t}u(t-1)}$

Method 1

$$F\left[e^{-3t}u(t)\right] = \frac{1}{(3+j\omega)}$$

Using the time shifting property, we get

$$F\left[e^{-3(t-1)}u(t-1)\right] = \frac{e^{-j\omega}}{(3+j\omega)}$$
$$e^{3}F\left[e^{-3t}u(t-1)\right] = \frac{e^{-j\omega}}{(3+j\omega)}$$

$$\boxed{F\left[e^{-3t}u(t-1)\right] = \frac{e^{-(j\omega+3)}}{(3+j\omega)}}$$

Method 2

Using FT definition, from Fig. 6.34, we get

$$F[x(t)] = \int_{1}^{\infty} e^{-3t}e^{-j\omega t}\,dt$$
$$= \int_{1}^{\infty} e^{-(3+j\omega)t}\,dt$$

$$F[x(t)] = \frac{-1}{(3+j\omega)}\left[e^{-(3+j\omega)t}\right]_{1}^{\infty}$$

$$\boxed{F[x(t)] = \frac{e^{-(3+j\omega)}}{j\omega+3}}$$

7. $\boldsymbol{x(t) = te^{-at}u(t)}$

$$F[e^{-at}u(t)] = \frac{1}{(a+j\omega)}$$

Using the FT property of differentiation in frequency, we get

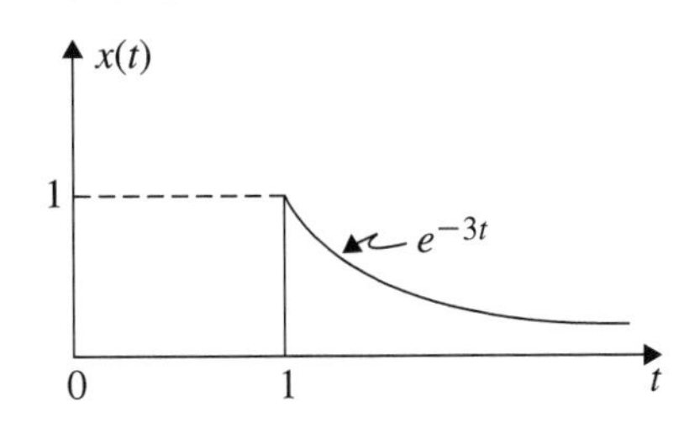

Fig. 6.34 Representation of $x(t) = e^{-3t}u(t-1)$

$$F\left[te^{-at}u(t)\right] = j\frac{d}{d\omega}\left[\frac{1}{(a+j\omega)}\right]$$

$$= \frac{j(-j)}{(a+j\omega)^2}$$

$$\boxed{X(j\omega) = \frac{1}{(a+j\omega)^2}}$$

8. $\boldsymbol{x(t) = e^{-a(t-2)}u(t-2)}$

Method 1

$$x(t) \overset{\text{FT}}{\longleftrightarrow} X(j\omega)$$

$$x(t-t_0) \overset{\text{FT}}{\longleftrightarrow} X(j\omega)e^{-j\omega t_0}$$

$$\boxed{F\left[e^{-a(t-2)}u(t-2)\right] = \frac{1}{(a+j\omega)}e^{-j2\omega}}$$

Method 2

Using the definition of FT, we get

$$X(j\omega) = \int_2^{\infty} e^{-a(t-2)}e^{-j\omega t}\,dt$$

$$= e^{2a}\int_2^{\infty} e^{-(a+j\omega)t}\,dt$$

$$= \frac{-e^{2a}}{(a+j\omega)}\left[e^{-(a+j\omega)t}\right]_2^{\infty}$$

$$= \frac{+e^{2a}}{(a+j\omega)}e^{-(a+j\omega)2}$$

$$\boxed{X(j\omega) = \frac{e^{-j2\omega}}{(a + j\omega)}}$$

9. $\boldsymbol{x(t) = e^{-a|t-2|}}$

$$x(t) = \begin{cases} e^{-a(t-2)} & 0 \leq t \leq \infty \\ e^{a(t+2)} & -\infty \leq t < 0 \end{cases}$$

Let $|t - 2| = \tau$

$$x(\tau) = e^{-a|\tau|}$$

From Example 6.1(e),

$$F\left[e^{-a|\tau|}\right] = \frac{2a}{a^2 + \omega^2}$$

Using the time shifting property,

$$\boxed{F\left[e^{-a|t-2|}\right] = \frac{2a}{a^2 + \omega^2} e^{-j2\omega}}$$

10. $\boldsymbol{x(t) = \cos(\omega_0 t + \phi)}$

$$\begin{aligned} \cos(\omega_0 t + \phi) &= \frac{1}{2}\left[e^{j(\omega_0 t + \phi)} + e^{-j(\omega_0 t + \phi)}\right] \\ &= \frac{1}{2}\left[e^{j\phi} e^{j\omega_0 t} + e^{-j\phi} e^{-j\omega_0 t}\right] \end{aligned}$$

By frequency shifting property,

$$\begin{aligned} F\left[e^{j\omega_0 t}\right] &= 2\pi\delta(\omega - \omega_0) \\ F\left[e^{-j\omega_0 t}\right] &= 2\pi\delta(\omega + \omega_0) \\ F[x(t)] = X(j\omega) &= \frac{2\pi}{2}\left[e^{j\phi}\delta(\omega - \omega_0) + e^{-j\phi}\delta(\omega + \omega_0)\right] \end{aligned}$$

$$\boxed{X(j\omega) = \pi\left[e^{j\phi}\delta(\omega - \omega_0) + e^{-j\phi}\delta(\omega + \omega_0)\right]}$$

11. $\boldsymbol{x(t) = \sin(\omega_0 t + \phi)}$

$$\sin(\omega_0 t + \phi) = \frac{1}{2j}\left[e^{+j(\omega_0 t+\phi)} - e^{-j(\omega_0 t+\phi)}\right]$$
$$= \frac{1}{2j}\left[e^{j\phi}e^{j\omega_0 t} - e^{-j\phi}e^{-j\omega_0 t}\right]$$
$$F[x(t)] = X(j\omega) = \frac{2\pi}{2j}\left[e^{j\phi}\delta(\omega-\omega_0) - e^{-j\phi}\delta(\omega+\omega_0)\right]$$

$$\boxed{X(j\omega) = -j\pi\left[e^{j\phi}\delta(\omega-\omega_0) - e^{-j\phi}\delta(\omega+\omega_0)\right]}$$

12. $\boldsymbol{x(t) = \sin\left(2\pi t + \frac{\pi}{4}\right)}$. (*Anna University, December, 2006*)

Let $\omega_0 = 2\pi$ and $\phi = \frac{\pi}{4}$

$$F[x(t)] = -j\pi\left[e^{j\phi}\delta(\omega-\omega_0) - e^{-j\phi}\delta(\omega+\omega_0)\right]$$

From Example 6.23.11, we write

$$\boxed{X(j\omega) = -j\pi\left[e^{\frac{j\pi}{4}}\delta(\omega-2\pi) - e^{-\frac{j\pi}{4}}\delta(\omega+2\pi)\right]}$$

13. $\boldsymbol{x(t) = \cos\left(3\pi t + \frac{\pi}{8}\right) + 1}$

$$F[\cos(\omega t + \phi)] = \pi\left[e^{j\phi}\delta(\omega-\omega_0) + e^{-j\phi}\delta(\omega+\omega_0)\right]$$

Let $\omega_0 = 3\pi$ and $\phi = \frac{\pi}{8}$. From Example 6.23.10, we write

$$F\left[\cos 3\pi t + \frac{\pi}{8}\right] = \pi\left[e^{j\frac{\pi}{8}}\delta(\omega-3\pi) + e^{-j\frac{\pi}{8}}\delta(\omega+3\pi)\right]$$
$$F[1] = 2\pi\delta(\omega)$$

$$\boxed{\left[\cos\left(3\pi t + \frac{\pi}{8}\right) + 1\right] \stackrel{\text{FT}}{\longleftrightarrow} \pi\left[e^{j\frac{\pi}{8}}\delta(\omega-3\pi) + e^{-j\frac{\pi}{8}}\delta(\omega+3\pi) + 2\delta(\omega)\right]}$$

14. $\boldsymbol{x(t) = \cos\left(6\pi t - \frac{\pi}{8}\right)}$

Let $\omega_0 = 6\pi$ and $\phi = \frac{-\pi}{8}$

$$F[\cos \omega_0 t + \phi] = \pi \left[e^{j\phi} \delta(\omega - \omega_0) + e^{-j\phi} \delta(\omega + \omega_0) \right]$$

$$\boxed{F\left[\cos\left(6\pi t - \frac{\pi}{8}\right)\right] = \pi \left[e^{-j\frac{\pi}{8}} \delta(\omega - 6\pi) + e^{j\frac{\pi}{8}} \delta(\omega + 6\pi)\right]}$$

15. $\boldsymbol{x(t) = x(4t - 8)}$

$$x(t) \overset{\text{FT}}{\longleftrightarrow} X(j\omega)$$

Using the time shifting property of FT, we write

$$x(t - 8) \overset{\text{FT}}{\longleftrightarrow} e^{-j8\omega} X(j\omega)$$

Using the time scaling property, we write

$$x(4t - 8) \overset{\text{FT}}{\longleftrightarrow} \frac{1}{4} e^{-j2\omega} X\left(j\frac{\omega}{4}\right)$$

$$\boxed{F[x(4t - 8)] = \frac{1}{4} X\left(\frac{j\omega}{4}\right) e^{-j2\omega}}$$

16. $\boldsymbol{x(t) = \frac{d^2}{dt^2} x(t - 2)}$

Using differentiation property, we write

$$F\left[\frac{d^2 x(t)}{dt^2}\right] = -\omega^2 X(j\omega)$$

For the time delay t_0,

$$F[x(t - t_0)] = e^{-j\omega t_0} X(j\omega)$$

Here $t_0 = 2$.

$$\boxed{F\left[\frac{d^2}{dt^2} x(t - 2)\right] = -\omega^2 e^{-j2\omega} X(j\omega)}$$

17. $\boldsymbol{x(t) = x(2 - t) + x(-2 - t)}$

$$x(t) = x_1(t) + x_2(t)$$

where

$$x_1(t) = x(2-t)$$
$$x_2(t) = x(-2-t)$$
$$F[x(-t)] = X(-j\omega)$$

Using the time shifting property of FT, we get

$$X_1(j\omega) = F[x(2-t)] = e^{-j2\omega}X(-j\omega)$$
$$X_2(j\omega) = F[x(-2-t)] = e^{j2\omega}X(-j\omega)$$
$$X(j\omega) = X_1(j\omega) + X_2(j\omega)$$
$$= X(-j\omega)\left[e^{-j2\omega} + e^{j2\omega}\right]$$

$$\boxed{X(j\omega) = 2X(-j\omega)\cos 2\,\omega}$$

18. $\boldsymbol{x(t) = \mathbf{rect}\left(\frac{t+2}{4}\right)}$

$$x(t) = \mathrm{rect}\left(\frac{t}{4} + 0.5\right)$$

From Example 6.2, the following equation is written (Fig. 6.35):

$$\mathrm{rect}\left(\frac{t}{4}\right) \overset{\mathrm{FT}}{\longleftrightarrow} \frac{2}{\omega}\sin 2\omega$$

Using the time shifting property, we get

$$\mathrm{rect}\left(\frac{t}{4} + 0.5\right) \overset{\mathrm{FT}}{\longleftrightarrow} \frac{2}{\omega}\sin 2\omega\, e^{+0.5j\omega}$$

$$\boxed{X(j\omega) = \frac{2}{\omega}\sin 2\omega\, e^{j0.5\omega}}$$

19. $\boldsymbol{x(t) = \mathbf{tri}\left(\frac{t-4}{10}\right)}$

$$\mathrm{tri}\left(\frac{t-4}{10}\right) = \mathrm{tri}\left(\frac{t}{10} - 0.4\right)$$

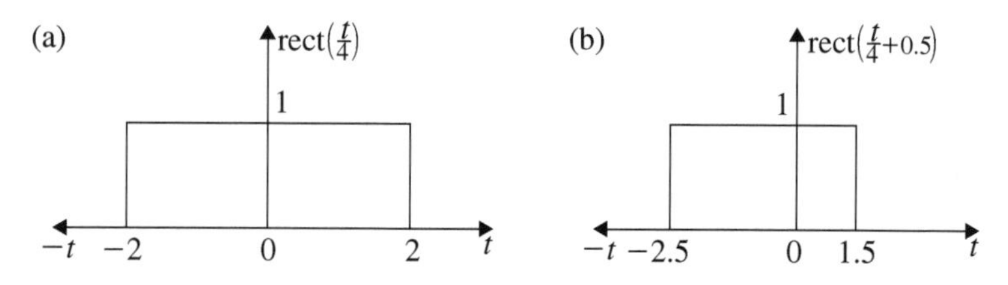

Fig. 6.35 $x(t) = \text{rect}\left(\frac{t}{4}\right)$

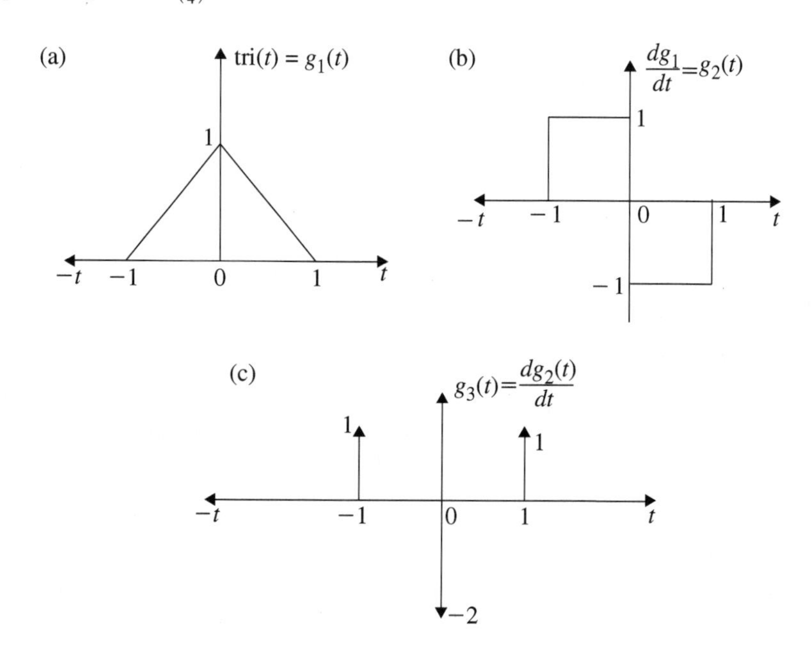

Fig. 6.36 tri(t) and its derivatives

The signal tri(t) is plotted as $g_1(t)$ and is shown in Fig. 6.36a. Its first derivative is plotted as $g_2(t)$ and the second derivative as $dg_2(t)/dt$ which are shown in Fig. 6.36b and c, respectively. From Fig. 6.36c, the FT is obtained as

$$
\begin{aligned}
\frac{dg_3(t)}{dt} &\overset{\text{FT}}{\longleftrightarrow} (e^{j\omega} + e^{-j\omega} - 2) \\
G_3(j\omega) &= 2[\cos\omega - 1] \\
&= -4\sin^2\frac{\omega}{2} \\
G_3(0) &= 0
\end{aligned}
$$

$G_2(\omega)$ is obtained by using the integrating property

$$G_2(j\omega) = -\frac{4}{j\omega}\sin^2\frac{\omega}{2} + \pi G_3(0)\delta(\omega)$$
$$= -\frac{4}{j\omega}\sin^2\frac{\omega}{2}$$
$$G_2(0) = \frac{-\omega}{j}\left(\frac{\sin\frac{\omega}{2}}{\frac{\omega}{2}}\right)^2 \quad \left(\frac{\sin\frac{\omega}{2}}{\frac{\omega}{2}}\bigg|_{\omega=0} = 1\right)$$
$$= 0$$

$G_1(\omega)$ is obtained using the integrating property

$$G_1(j\omega) = \frac{-4}{(j\omega)(j\omega)}\sin^2\frac{\omega}{2} + \pi G_2(0)\delta(\omega)$$
$$= \frac{4}{\omega^2}\sin^2\frac{\omega}{2}$$
$$= \mathrm{sinc}^2\left(\frac{\omega}{2}\right)$$

Using the time shifting property of FT, we get

$$\mathrm{tri}(t-0.4) \overset{\mathrm{FT}}{\longleftrightarrow} G_1(j\omega)e^{-j0.4\omega}$$
$$= \mathrm{sinc}^2\left(\frac{\omega}{2}\right)e^{-j0.4\omega}$$

Using the time scaling property of FT, we get

$$\mathrm{tri}\left(\frac{t}{10} - 0.4\right) \overset{\mathrm{FT}}{\longleftrightarrow} 10\mathrm{sinc}^2 5\omega e^{-j4\omega}$$

$$\boxed{X(j\omega) = 10\mathrm{sinc}^2 5\omega e^{-j4\omega}}$$

20. $\boldsymbol{x(t) = \frac{d}{dt}\left[5\,\mathrm{rect}\left(\frac{t}{8}\right)\right]}$. **What is the FT of** $\boldsymbol{y(t) = \int x(t)dt}$**?**

Figure 6.37a represents 5 rect(t). The time expanded 5 rect $\left(\frac{t}{8}\right)$ is shown in Fig. 6.37b and its derivative is shown in Fig. 6.37c. From Fig. 6.37c,

$$X(j\omega) = 5e^{j4\omega} - 5e^{-j4\omega}$$

$$\boxed{X(j\omega) = j10\sin 4\,\omega}$$

$$X(0) = 5 - 5 = 0$$

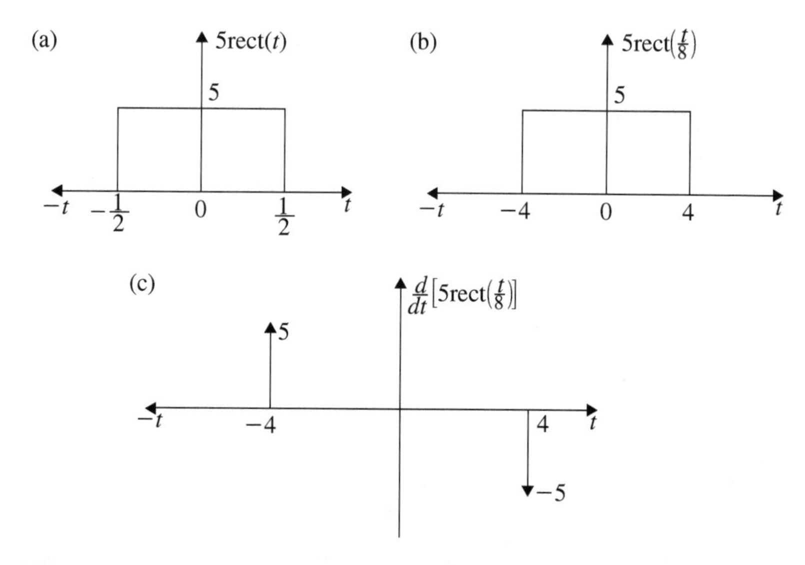

Fig. 6.37 Representation of rectangular wave and its derivatives

Using the integration property, we get

$$
\begin{aligned}
Y(j\omega) &= \frac{1}{j\omega}X(j\omega) + \pi X(0)\delta(\omega) \\
&= \frac{j10}{j\omega}\sin 4\omega
\end{aligned}
$$

$$\boxed{Y(j\omega) = 40\text{sinc}\,4\omega}$$

21. $\boldsymbol{x(t) = \delta(t+2) + 5\delta(t+1) + \delta(t-1) + 5\delta(t-2)}$

The given $x(t)$ is represented in Fig. 6.38. By applying the time shifting property to each impulse, we get

$$\boxed{X(j\omega) = e^{j2\omega} + 5e^{j\omega} + e^{-j\omega} + 5e^{-j2\omega}}$$

22. $\boldsymbol{x(t) = \begin{cases} e^{j6t} & |t| \le \pi \\ 0 & \text{elsewhere} \end{cases}}$

The above signal is represented as a product of a rectangular pulse of width 2π and a complex sinusoid e^{j6t}.

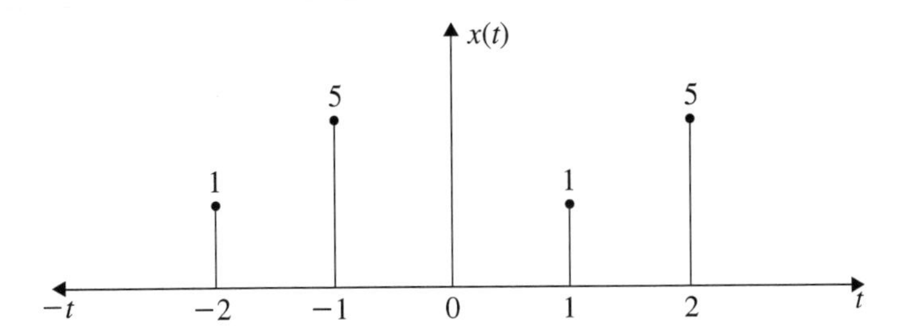

Fig. 6.38 Discrete time signal

Fig. 6.39 Representation of rectangular pulse

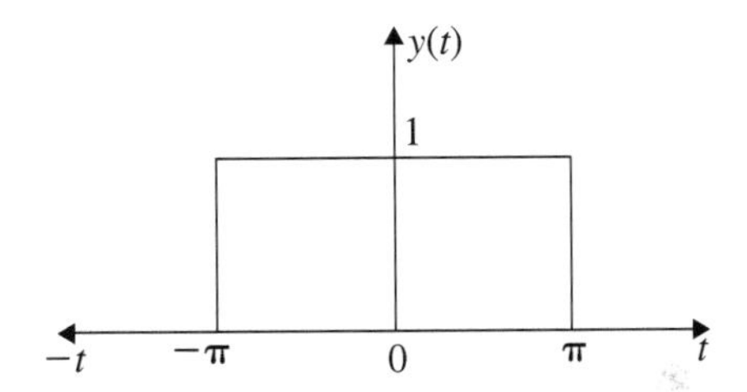

$$x(t) = \begin{cases} 1\, e^{j6t} & |t| \leq \pi \\ 0 & \text{otherwise} \end{cases}$$

For $-\pi \leq t \leq \pi$, the rectangular pulse $y(t)$ is shown in Fig. 6.39.

The FT of the rectangular pulse shown in Fig. 6.35 in Example 6.23.18 is

$$Y(j\omega) = \frac{2}{\omega} \sin \omega\pi$$

Using the frequency shifting property

$$y(t)e^{j6t} \overset{\text{FT}}{\longleftrightarrow} = Y(j(\omega - 6))$$
$$X(j\omega) = F\left[y(t)e^{j6t}\right]$$

$$\boxed{X(j\omega) = \frac{2\sin((\omega - 6)\pi)}{(\omega - 6)}}$$

23. $\boldsymbol{x(t) = \begin{cases} 0 & |t| > 1 \\ \dfrac{(t+1)}{2} & -1 \le t \le 1 \end{cases}}$

Figure 6.40a gives $x(t)$ and Fig. 6.40b gives $\frac{dx(t)}{dt}$. From Fig. 6.40b, signals $g_2(t)$ and $g_3(t)$ are separated and are shown in Fig. 6.40c and d, respectively.

$$
\begin{aligned}
g_1(t) &= g_2(t) + g_3(t) \\
g_4(t) &= \frac{dg_2(t)}{dt} \\
G_4(j\omega) &= 0.5(e^{j\omega} - e^{-j\omega}) \\
&= 2j0.5\frac{(e^{j\omega} - e^{-j\omega})}{2j} \\
&= j\sin\omega \\
G_4(0) &= 0
\end{aligned}
$$

Using the integration property, we get

$$
\begin{aligned}
G_2(j\omega) &= \frac{1}{j\omega}G_4(j\omega) + \pi G_4(0)\delta(\omega) \\
&= \frac{j\sin\omega}{j\omega} = \frac{\sin\omega}{\omega}
\end{aligned}
$$

From Fig. 6.40d, we get

$$
\begin{aligned}
G_3(j\omega) &= -e^{-j\omega} \\
G_1(j\omega) &= G_2(j\omega) + G_3(j\omega) \\
&= \frac{\sin\omega}{\omega} - e^{-j\omega} \\
G_1(0) &= 1 - 1 = 0
\end{aligned}
$$

By using the integration property, we get

$$
\begin{aligned}
X(j\omega) &= \frac{1}{j\omega}G_1(j\omega) + \pi G_1(0)\delta(\omega) \\
&= \frac{1}{j\omega}\left(\frac{\sin\omega}{\omega} - e^{-j\omega}\right)
\end{aligned}
$$

$$
\boxed{X(j\omega) = \frac{1}{j\omega}\left[\frac{\sin\omega}{\omega} - e^{-j\omega}\right]}
$$

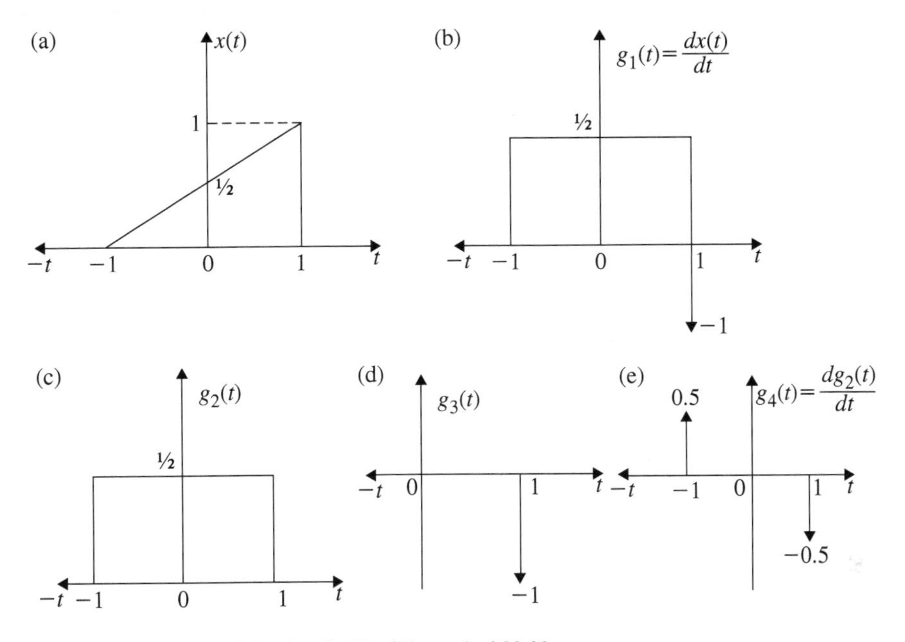

Fig. 6.40 Derivatives of the signal $x(t)$ of Example 6.23.23

24. $\boldsymbol{x(t) = \begin{cases} t & 0 \leq t < 1 \\ 0 & \text{otherwise} \end{cases}}$

$x(t) = t$; $0 \leq t \leq 1$ is shown in Fig. 6.41a; $\frac{dx(t)}{dt}$ is shown in Fig. 6.41b. The Fourier transform of the time shifted rectangle is $2\frac{\sin(\omega/2)}{\omega}e^{-j\omega/2}$ (see Example 6.2) and that of the negative impulse is $-e^{-j\omega}$.

$$G_1(j\omega) = F\left[\frac{dx(t)}{dt}\right]$$

$$G_1(j\omega) = \left[2\frac{\sin(\omega/2)}{\omega}e^{-\frac{j\omega}{2}} - e^{-j\omega}\right] \quad \text{(see Example 6.2)}$$

$$G_1(0) = 1 - 1 = 0$$

Using the integration property of FT,

$$F[x(t)] = \frac{1}{j\omega}F\left[\frac{dx(t)}{dt}\right] + \pi G_1(0)\delta(\omega)$$

$$\boxed{X(j\omega) = \frac{1}{j\omega}\left[\frac{2\sin(\omega/2)}{\omega}e^{-\frac{j\omega}{2}} - e^{-j\omega}\right]}$$

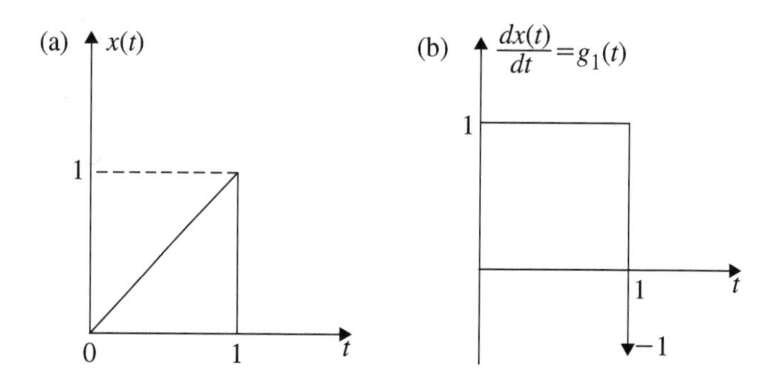

Fig. 6.41 Signal representation of Example 6.23.24 and its derivatives

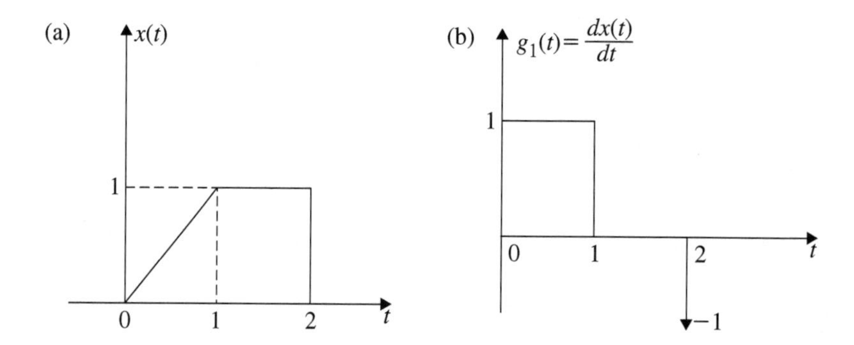

Fig. 6.42 Signal representation of Example 6.23.25 and its derivatives

25. $\boldsymbol{x(t) = \begin{cases} t & 0 \le t < 1 \\ 1 & 1 \le t \le 2 \\ 0 & \text{elsewhere} \end{cases}}$

The signal $x(t)$ shown in Fig. 6.42a when differentiated takes the shape as shown in Fig. 6.42b. For the time shifted square pulse, the FT is (see example 6.2)

$$X_1(j\omega) = \frac{2\sin\frac{\omega}{2}}{\omega}e^{-\frac{j\omega}{2}}$$

For the negative impulse, the FT is

$$\begin{aligned} X_2(j\omega) &= -e^{-j2\omega} \\ G_1(j\omega) &= X_1(j\omega) + X_2(j\omega) \\ &= \left[\frac{2}{\omega}\sin\frac{\omega}{2}e^{-\frac{j\omega}{2}} - e^{-j2\omega}\right] \\ G_1(0) &= 1 - 1 = 0 \end{aligned}$$

The Fourier transform of the given signal is obtained using the integration property.

$$X(j\omega) = \frac{1}{j\omega}G_1(j\omega) + \pi G_1(0)\delta(\omega)$$

$$\boxed{X(j\omega) = \frac{1}{j\omega}\left[\frac{2}{\omega}\sin\frac{\omega}{2}e^{-\frac{j\omega}{2}} - e^{-j2\omega}\right]}$$

26. $x(t) = \begin{cases} 1 & |t| < 1 \\ 2 - |t| & 1 < |t| < 2 \\ 0 & \text{elsewhere} \end{cases}$

The given signal $x(t)$ is represented in Fig. 6.43a. The first and second derivatives are shown in Fig. 6.43b and c, respectively. From Fig. 6.43c, the FT of the impulses are obtained using the time shifting property.

$$\begin{aligned} G(j\omega) &= F\left[\frac{d^2x(t)}{dt^2}\right] \\ &= \left[e^{j2\omega} - (e^{j\omega} + e^{-j\omega}) + e^{-j2\omega}\right] \\ &= 2\left[\cos 2\omega - \cos\omega\right] \\ G(0) &= 1 - 1 - 1 + 1 = 0 \end{aligned}$$

Using the integration property of FT, we get

$$F\left[x(t)\right] = \frac{2}{(j\omega)^2}F\left[\frac{d^2x(t)}{dt^2}\right] + \pi G(0)\delta(\omega)$$

$$\boxed{X(j\omega) = \frac{2}{\omega^2}\left[\cos\omega - \cos 2\omega\right]}$$

Example 6.24

Find the magnitude spectrum of FT and plot it where

$$H(j\omega) = \frac{(1 + 2e^{-j\omega})}{(1 + \frac{1}{2}e^{-j\omega})}$$

(Anna University, April, 2004)

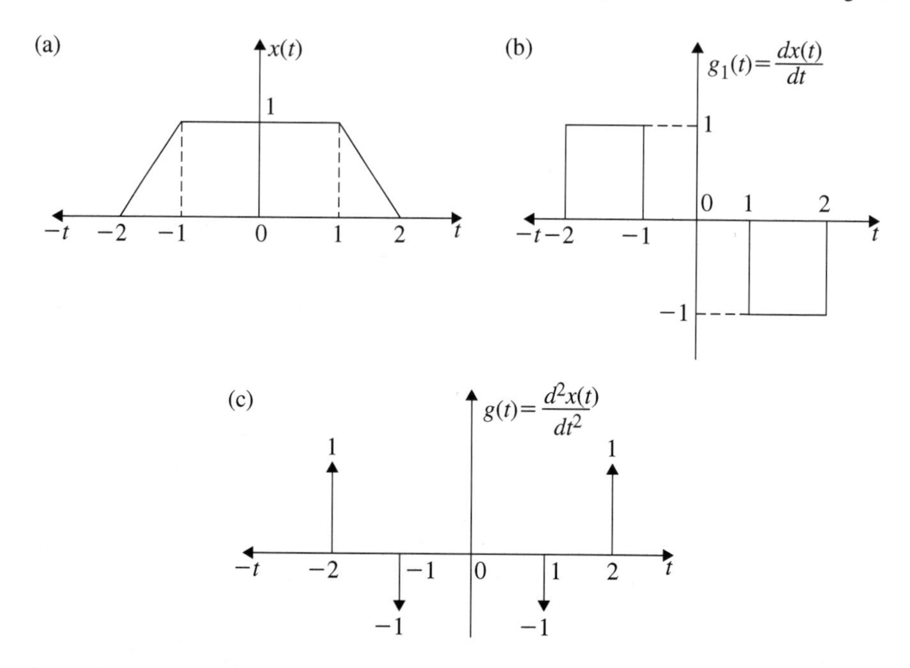

Fig. 6.43 Signal representation of Example 6.23.26 and its derivatives

Solution:

$$
\begin{aligned}
H(j\omega) &= \frac{(1+2e^{-j\omega})}{(1+\frac{1}{2}e^{-j\omega})} \\
&= \frac{(1+2\cos\omega)-j2\sin\omega}{(1+\frac{1}{2}\cos\omega)-\frac{j}{2}\sin\omega} \\
|H(j\omega)| &= \frac{\sqrt{(1+2\cos\omega)^2+4\sin^2\omega}}{\sqrt{(1+\frac{1}{2}\cos\omega)^2+\frac{1}{4}\sin^2\omega}} \\
&= \frac{\sqrt{1+4\cos^2\omega+4\cos\omega+4\sin^2\omega}}{\sqrt{1+\frac{1}{4}\cos^2\omega+\cos\omega+\frac{1}{4}\sin^2\omega}} \\
&= \frac{\sqrt{5+4\cos\omega}}{\sqrt{\frac{5}{4}+\cos\omega}} = 2
\end{aligned}
$$

$$
\boxed{|H(j\omega)| = 2}
$$

$|H(j\omega)|$ is independent of frequency and is shown in Fig. 6.42.

■ Example 6.25

Using the properties of continuous time Fourier transform, determine the time domain signal $x(t)$.

If the frequency domain signal is described as given below.

$$X(j\omega) = j\frac{d}{d\omega}\left[\frac{e^{j2\omega}}{(1+\frac{j\omega}{3})}\right]$$

(*Anna University, December, 2007*)

Solution: From the inspection of $X(j\omega)$, the given problem can be solved using differentiation in frequency, time shifting and scaling in the proper order.

First, the time scaling property is applied. Let

$$X_1(j\omega) = \frac{1}{1+j\omega}$$

$$x_1(t) = e^{-t}u(t)$$

$$F\left[x_1[3t]\right] = 3e^{-3t}u(3t)$$

$$F\left[3e^{-3t}u(3t)\right] = \frac{1}{\left[1+\frac{j\omega}{3}\right]}$$

$$F^{-1}\left[\frac{1}{\left(1+\frac{j\omega}{3}\right)}\right] = 3e^{-3t}u(t) \quad [\because u(t) = u(3t)]$$

According to the time shifting property,

$$e^{j2\omega}Y(j\omega) = y(t+2)$$

$$F^{-1}\left[\frac{e^{j2\omega}}{\left(1+\frac{j\omega}{3}\right)}\right] = 3e^{-3(t+2)}u(t+2)$$

According to differentiating property,

$$j\frac{d}{d\omega}X(j\omega) = tx(t)$$

Applying the above property, we have

$$F^{-1}\left[j\frac{d}{d\omega}\frac{e^{j2\omega}}{\left(1+\frac{j\omega}{3}\right)}\right] = 3te^{-3(t+2)}u(t+2)$$

$$\therefore\ X(j\omega) = \frac{jd}{d\omega}\left[\frac{e^{j2\omega}}{\left(1+\frac{j\omega}{3}\right)}\right]$$

$$\boxed{x(t) = 3te^{-3(t+2)}u(t+2)}$$

Example 6.26

Find the inverse Fourier transform of the following functions:

1. $$X(j\omega) = \delta(\omega - \omega_0)$$
2. $$X(j\omega) = \frac{j\omega}{(2+j\omega)^2}$$
3. $$X(j\omega) = \begin{cases} 1 & |\omega| < 2 \\ 0 & \text{elsewhere} \end{cases}$$
4. $$X(j\omega) = \frac{6}{(\omega^2+9)}$$
5. $$X(j\omega) = \frac{(j\omega+2)}{\left[(j\omega)^2+4j\omega+3\right]}$$
6. $$X(j\omega) = \frac{(j\omega+1)}{\left[(j\omega+2)^2(j\omega+3)\right]}$$

Solution:

1. $\boldsymbol{X(j\omega) = \delta(\omega - \omega_0)}$

The IFT of $\delta(\omega) = \frac{1}{2\pi}$. $\delta(\omega)$ is frequency-shifted by ω_0.

$$F^{-1}\left[X(j\omega)\right] = e^{j\omega_0 t}\frac{1}{2\pi}$$

$$\boxed{F^{-1}\left[\delta(\omega - \omega_0)\right] = \frac{1}{2\pi}e^{j\omega_0 t}}$$

Fig. 6.44 Magnitude spectrum of $H(j\omega)$

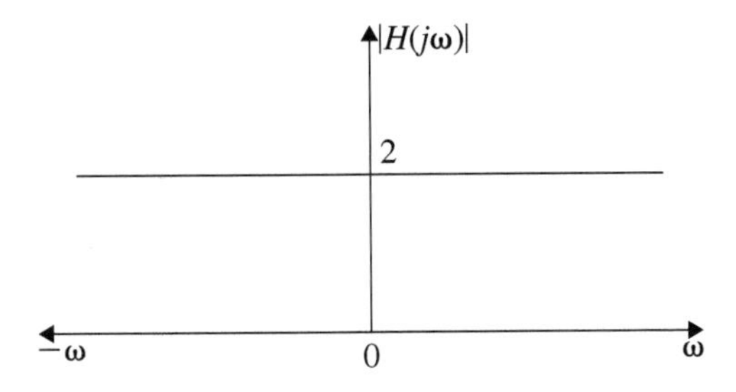

The above result can also be got from the first principle of inverse Fourier transform

$$F^{-1}\left[\delta(\omega-\omega_0)\right] = \frac{1}{2\pi}\int_{-\infty}^{\infty}\delta(\omega-\omega_0)e^{j\omega t}d\omega$$

Using the sampling property of the impulse function which exists only at $\omega = \omega_0$, we get (Fig. 6.44)

$$\boxed{F^{-1}\left[\delta(\omega-\omega_0)\right] = \frac{1}{2\pi}e^{j\omega_0 t}}$$

2. $\boldsymbol{X(j\omega) = \dfrac{j\omega}{(2+j\omega)^2}}$

$$F\left[e^{-2t}\right] = \frac{1}{(2+j\omega)}$$

By applying

$$F\left[te^{-2t}\right] = \frac{d}{d\omega}\frac{1}{(2+j\omega)}$$

(Applying frequency differentiation)

$$F\left[te^{-2t}\right] = \frac{1}{(2+j\omega)^2}$$

$$\therefore \quad F^{-1}\left[\frac{1}{(2+j\omega)^2}\right] = te^{-2t}$$

By applying time differentiation, namely

$$\frac{dx(t)}{dt} = j\omega X(j\omega)$$

Fig. 6.45 Representation of $X(j\omega)$

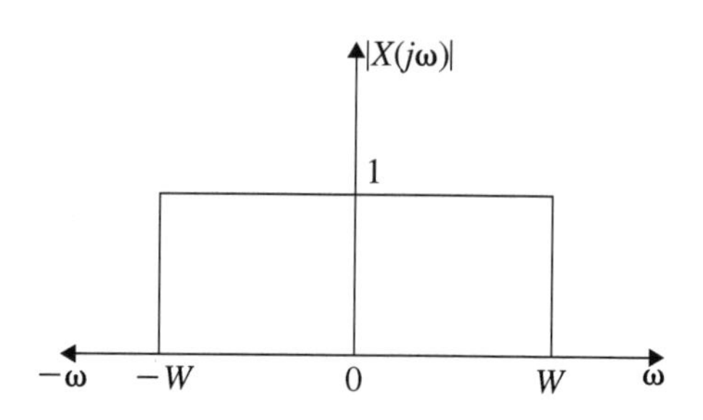

$$\boxed{F^{-1}\left[\frac{j\omega}{(2+j\omega^2)}\right] = \frac{d}{dt}\left(te^{-2t}\right)}$$

3. $X(j\omega) = \begin{cases} 1 & |\omega| < W \\ 0 & \text{otherwise} \end{cases}$

(*Anna University, December, 2013*)

The frequency spectrum of the above function is shown in Fig. 6.45.

Using the definition of inverse FT, we get

$$\begin{aligned} x(t) &= \frac{1}{2\pi}\int_{-W}^{W} X(j\omega)e^{j\omega t}d\omega \\ &= \frac{1}{2\pi jt}\left[1e^{j\omega t}\right]_{-W}^{W} \\ &= \frac{1}{2\pi jt}\left[e^{jWt} - e^{-jWt}\right] \\ &= \frac{1}{\pi t}\sin Wt \end{aligned}$$

$$\boxed{x(t) = \frac{W}{\pi}\text{sinc}\,Wt}$$

4. $X(j\omega) = \frac{6}{(\omega^2+9)}$

$$\begin{aligned} X(j\omega) &= \frac{-6}{(j\omega+3)(j\omega-3)} \\ &= \frac{A_1}{j\omega+3} + \frac{A_2}{j\omega-3} \\ -6 &= A_1(j\omega-3) + A_2(j\omega+3) \end{aligned}$$

Let $j\omega = -3$

$$A_1 = 1$$

Let $j\omega = 3$

$$A_2 = -1$$
$$X(j\omega) = \frac{1}{j\omega + 3} - \frac{1}{j\omega - 3}$$
$$x(t) = F^{-1}\left[X(j\omega)\right] = e^{-3t}u(t) + e^{3t}u(-t)$$

$$\boxed{x(t) = e^{-3t}u(t) + e^{3t}u(-t)}$$

5. $\boldsymbol{X(j\omega) = \frac{(j\omega+2)}{[(j\omega)^2+4j\omega+3]}}$

$$X(j\omega) = \frac{(j\omega + 2)}{(j\omega + 1)(j\omega + 3)}$$
$$= \frac{A_1}{(j\omega + 1)} + \frac{A_2}{(j\omega + 3)}$$
$$(j\omega + 2) = A_1(j\omega + 3) + A_2(j\omega + 1)$$

Let $j\omega = -1$,

$$1 = 2A_1$$
$$A_1 = \frac{1}{2}$$

Let $j\omega = -3$, $A_2 = \frac{1}{2}$

$$X(j\omega) = \frac{1}{2}\left[\frac{1}{j\omega + 1} + \frac{1}{j\omega + 3}\right]$$

$$\boxed{x(t) = \frac{1}{2}\left[e^{-t} + e^{-3t}\right]u(t)}$$

6. $\boldsymbol{X(j\omega) = \frac{(j\omega+1)}{(j\omega+2)^2(j\omega+3)}}$

$$X(j\omega) = \frac{A_1}{(j\omega+2)^2} + \frac{A_2}{(j\omega+2)} + \frac{A_3}{(j\omega+3)}$$
$$(j\omega+1) = A_1(j\omega+3) + A_2(j\omega+2)(j\omega+3) + A_3(j\omega+2)^2$$

Let $j\omega = -2$;

$$-1 = A_1$$

Let $j\omega = -3$;

$$-2 = A_3$$
$$(j\omega+1) = A_1(j\omega+3) + A_2\left[(j\omega)^2 + 5j\omega + 6\right] + A_3\left[(j\omega)^2 + 4j\omega + 4\right]$$

Compare the coefficients of $j\omega$ on both sides,

$$\begin{aligned} 1 &= A_1 + 5A_2 + 4A_3 \\ &= -1 + 5A_2 - 8 \\ A_2 &= 2 \\ X(j\omega) &= \frac{-1}{(j\omega+2)^2} + \frac{2}{(j\omega+2)} - \frac{2}{(j\omega+3)} \\ x(t) &= F^{-1}[x(j\omega)] \end{aligned}$$

$$\boxed{x(t) = \left[-te^{-2t} + 2e^{-2t} - 2e^{-3t}\right]u(t)}$$

■ Example 6.27

Consider a causal LTI system with frequency response,

$$H(j\omega) = \frac{1}{j\omega+3}$$

For a particular input $x(t)$, this system is to produce the output

$$y(t) = e^{-3t}u(t) - e^{-4t}u(t)$$

Determine $x(t)$.

(*Anna University, April, 2008*)

Solution:

$$y(t) = e^{-3t}u(t) - e^{-4t}u(t)$$

$$Y(j\omega) = \frac{1}{(j\omega+3)} - \frac{1}{(j\omega+4)}$$

$$= \frac{1}{(j\omega+3)(j\omega+4)}$$

$$H(j\omega) = \frac{Y(j\omega)}{X(j\omega)}$$

$$X(j\omega) = \frac{Y(j\omega)}{H(j\omega)}$$

$$= \frac{(j\omega+3)}{(j\omega+3)(j\omega+4)}$$

$$= \frac{1}{(j\omega+4)}$$

$$x(t) = F^{-1}X(j\omega) = e^{-4t}u(t)$$

$$\boxed{x(t) = e^{-4t}u(t)}$$

■ Example 6.28

Find the Fourier transform of the following signals using convolution theorem.

1. $x(t) = e^{-2t}u(t) * e^{-5t}u(t)$
2. $x(t) = \frac{d}{dt}\left[e^{-2t}u(t) * e^{-5t}u(t)\right]$
3. $x(t) = \left[e^{-2t}u(t) * e^{-5t}u(t-5)\right]$

Determine $x(t)$ in all the above cases.

Solution:

1. $\boldsymbol{x(t) = e^{-2t}u(t) * e^{-5t}u(t)}$

$$X(j\omega) = F\left[e^{-2t}u(t)\right]F\left[e^{-5t}u(t)\right]$$

$$F\left[e^{-2t}u(t)\right] = \frac{1}{(j\omega+2)}$$

$$F\left[e^{-5t}u(t)\right] = \frac{1}{(j\omega+5)}$$

$$\boxed{X(j\omega) = \frac{1}{(j\omega+2)(j\omega+5)}}$$

$$X(j\omega) = \frac{1}{3}\left[\frac{1}{j\omega+2} - \frac{1}{(j\omega+5)}\right]$$
$$x(t) = F^{-1}[X(j\omega)] = \frac{1}{3}[e^{-2t}u(t) - e^{-5t}u(t)]$$

$$\boxed{x(t) = \frac{1}{3}\left[e^{-2t} - e^{-5t}\right]u(t)}$$

2. $\boldsymbol{x(t) = \frac{d}{dt}\left[e^{-2t}u(t) * e^{-5t}u(t)\right]}$

Let

$$x_1(t) = e^{-2t}u(t) * e^{-5t}u(t)$$
$$X_1(j\omega) = \frac{1}{(j\omega+2)(j\omega+5)}$$

Using the time differentiation property of FT, we get

$$x(t) = \frac{d\,x_1(t)}{dt}$$
$$X(j\omega) = j\omega X_1(j\omega)$$

$$\boxed{X(j\omega) = \frac{j\omega}{(j\omega+2)(j\omega+5)}}$$

Putting into partial fraction, we get

$$X(j\omega) = \frac{A_1}{j\omega+2} + \frac{A_2}{j\omega+5}$$
$$j\omega = A_1(j\omega+5) + A_2(j\omega+2)$$

Let $j\omega = -2$;

$$A_1 = -\frac{2}{3}$$

Let $j\omega = -5$;

$$\begin{aligned} A_2 &= \frac{5}{3} \\ X(j\omega) &= \frac{1}{3}\left[-\frac{2}{j\omega+2} + \frac{5}{j\omega+5}\right] \\ x(t) &= F^{-1}\left[X(j\omega)\right] = \frac{1}{3}\left[-2e^{-2t} + 5e^{-5t}\right]u(t) \end{aligned}$$

$$\boxed{x(t) = \frac{1}{3}\left[-2e^{-2t} + 5e^{-5t}\right]u(t)}$$

3. $\boldsymbol{x(t) = e^{-2t}u(t) * e^{-5t}u(t-5)}$

$$\begin{aligned} x(t) &= x_1(t) * x_2(t) \\ X(j\omega) &= X_1(j\omega)X_2(j\omega) \\ X_1(j\omega) &= \frac{1}{(j\omega+2)} \\ x_2(t) &= e^{-5t}u(t-5) \\ &= e^{-25}e^{-5(t-5)}u(t-5) \\ X_2(j\omega) &= e^{-25}\frac{1}{(j\omega+5)} \end{aligned}$$

$$\boxed{X(j\omega) = e^{-25}\left[\frac{1}{(j\omega+2)(j\omega+5)}\right]}$$

$$X(j\omega) = \frac{1}{3}e^{-25}\left[\frac{1}{j\omega+2} - \frac{1}{j\omega+5}\right]$$

$$\boxed{x(t) = \frac{e^{-25}}{3}\left[e^{-2t} - e^{-5t}\right]u(t)}$$

■ Example 6.29

Consider the following signals $x_1(t)$ and $x_2(t)$. Find

$$y(t) = x_1(t) * x_2(t)$$

1.

$$x_1(t) = e^{-2t}u(t)$$
$$x_2(t) = e^{3t}u(-t)$$

2.

$$x_1(t) = e^{2t}u(-t)$$
$$x_2(t) = e^{4t}u(-t)$$

Solution:

1. $\boldsymbol{x_1(t) = e^{-2t}u(t)}$ **and** $\boldsymbol{x_2(t) = e^{3t}u(-t)}$

$$X_1(j\omega) = \frac{1}{(j\omega + 2)}$$
$$X_2(j\omega) = -\frac{1}{(j\omega - 3)}$$

$$\begin{aligned}
x_1(t) * x_2(t) &= X_1(j\omega)X_2(j\omega)\\
Y(j\omega) &= \frac{1}{(j\omega+2)}\frac{(-1)}{(j\omega-3)}\\
Y(j\omega) &= \frac{A_1}{(j\omega+2)} + \frac{A_2}{(j\omega-3)}\\
&= \frac{1}{5}\left[\frac{1}{j\omega+2} - \frac{1}{j\omega-3}\right]\\
y(t) = F^{-1}[Y(j\omega)] &= \frac{1}{5}\left[e^{-2t}u(t) + e^{3t}u(-t)\right]
\end{aligned}$$

$$\boxed{y(t) = \frac{1}{5}\left[e^{-2t}u(t) + e^{3t}u(-t)\right]}$$

2. $\boldsymbol{x_1(t) = e^{2t}u(-t)}$ **and** $\boldsymbol{x_2(t) = e^{4t}u(-t)}$

$$X_1(j\omega) = \frac{-1}{(j\omega - 2)}$$

$$X_2(j\omega) = \frac{-1}{(j\omega - 4)}$$

$$x_1(t) * x_2(t) = X_1(j\omega)X_2(j\omega)$$

$$Y(j\omega) = \frac{1}{(j\omega - 2)(j\omega - 4)}$$

$$= \frac{A_1}{(j\omega - 2)} + \frac{A_2}{(j\omega - 4)}$$

$$= \frac{1}{2}\left[\frac{-1}{(j\omega - 2)} + \frac{1}{(j\omega - 4)}\right]$$

$$y(t) = F^{-1}[Y(j\omega)] = \frac{1}{2}\left[e^{2t} - e^{4t}\right]u(-t)$$

$$\boxed{y(t) = \frac{1}{2}\left[e^{2t} - e^{4t}\right]u(-t)}$$

■ Example 6.30

Find the Fourier transform of the following functions:

1. $x(t) = e^{j\omega_0 t}\, u(t)$
2. $x(t) = \cos\omega_0 t\, u(t)$
3. $x(t) = \sin\omega_0 t\, u(t)$
4. $x(t) = e^{-at}\cos\omega_0 t\, u(t); \quad a > 0$
5. $x(t) = e^{-at}\sin\omega_0 t\, u(t); \quad a > 0$
6. $x(t) = [u(t + 2) - ut - 2]\cos 3t$
7. $x(t) = e^{-2|t|}\cos 5t$
8. $x(t) = e^{-3|t|}\sin 2t$

Solution:

1. $\boldsymbol{x(t) = e^{j\omega_0 t}u(t)}$

$$F\,[u(t)] = \frac{1}{j\omega} + \pi\delta(\omega)$$

By using the frequency shifting property, the FT of $x(t)$ is obtained.

$$\boxed{F\left[e^{j\omega_0 t}u(t)\right] = \frac{1}{j(\omega - \omega_0)} + \pi\delta(\omega - \omega_0)}$$

2. $\boldsymbol{x(t) = \cos\omega_0 t u(t)}$

$$\cos\omega_0 t = \frac{1}{2}\left[e^{j\omega_0 t} + e^{-j\omega_0 t}\right]$$

$$\cos\omega_0 t u(t) = \frac{1}{2}\left[e^{j\omega_0 t}u(t) + e^{-j\omega_0 t}u(t)\right]$$

By using the frequency shifting property, $F[x(t)]$ is obtained.

$$X(j\omega) = F[\cos\omega_0 t u(t)]$$

$$= \frac{1}{2}\left[\frac{1}{j(\omega - \omega_0)} + \pi\delta(\omega - \omega_0) + \frac{1}{j(\omega + \omega_0)} + \pi\delta(\omega + \omega_0)\right]$$

$$X(j\omega) = \frac{1}{2}\left[\frac{2\omega}{j(\omega - \omega_0^2)} + \pi\delta(\omega - \omega_0) + \pi\delta(\omega + \omega_0)\right]$$

$$\boxed{X(j\omega) = \frac{j\omega}{(\omega_0^2 - \omega^2)} + \frac{1}{2}\pi\delta(\omega - \omega_0) + \frac{1}{2}\pi\delta(\omega + \omega_0)}$$

3. $\boldsymbol{x(t) = \sin\omega_0 t u(t)}$

$$\sin\omega_0 t = \frac{1}{2j}\left[e^{j\omega_0 t} - e^{-j\omega_0 t}\right]$$

$$\sin\omega_0 t u(t) = \frac{1}{2j}\left[e^{j\omega_0 t}u(t) - e^{-j\omega_0 t}u(t)\right]$$

By using the frequency shifting property, $F[x(t)]$ is obtained.

$$F[x(t)] = \frac{1}{2j}\left[\frac{1}{j(\omega - \omega_0)} + \pi\delta(\omega - \omega_0) - \frac{1}{j(\omega + \omega_0)} - \pi\delta(\omega + \omega_0)\right]$$

$$\boxed{X(j\omega) = \left[\frac{\omega_0}{\omega_0^2 - \omega^2} + \frac{\pi}{2j}\delta(\omega - \omega_0) - \frac{\pi}{2j}\delta(\omega + \omega_0)\right]}$$

4. $\boldsymbol{x(t) = e^{-at}\cos\omega_0 t u(t)}$

$$\cos\omega_0 t = \frac{1}{2}\left[e^{j\omega_0 t} + e^{-j\omega_0 t}\right]$$

$$\begin{aligned}
X(j\omega) &= \int_0^\infty e^{-at}\cos\omega_0 t e^{-j\omega t}\,dt \\
&= \frac{1}{2}\int_0^\infty e^{-at}e^{j\omega_0 t}e^{-j\omega t}\,dt + \frac{1}{2}\int_0^\infty e^{-at}e^{-j\omega_0 t}e^{-j\omega t}\,dt \\
&= \frac{1}{2}\int_0^\infty e^{-(a-j\omega_0+j\omega)t}\,dt + \frac{1}{2}\int_0^\infty e^{-(a+j\omega_0+j\omega)t}\,dt \\
&= \frac{1}{2}\left[\frac{-1}{(a-j\omega_0+j\omega)}e^{-(a-j\omega_0+j\omega)t} - \frac{e^{-(a+j\omega_0+j\omega)t}}{(a+j\omega_0+j\omega)}\right]_0^\infty \\
&= \frac{1}{2}\left[\frac{1}{(a+j\omega)-j\omega_0} + \frac{1}{(a+j\omega)+j\omega_0}\right] \\
&= \frac{1}{2}\frac{[a+j\omega+j\omega_0+a+j\omega-j\omega_0]}{(a+j\omega)^2+\omega_0^2}
\end{aligned}$$

$$\boxed{X(j\omega) = \frac{(a+j\omega)}{(a+j\omega)^2+\omega_0^2}}$$

Note: The property used to solve this problem is called the "Modulation" property which states that

$$\boxed{\boldsymbol{x(t)\cos\omega_0 t \stackrel{\text{FT}}{\longleftrightarrow} \frac{1}{2}[X(\omega-\omega_0)+X(\omega+\omega_0)]}}$$

where $x(t)$ is the modulating signal and $\cos\omega_0 t$ is the carrier signal.

5. $\boldsymbol{x(t) = e^{-at}\sin\omega_0 t u(t)}$

$$\sin\omega_0 t = \frac{1}{2j}\left[e^{j\omega_0 t} - e^{-j\omega_0 t}\right]$$

$$\begin{aligned}
X(j\omega) &= \frac{1}{2j}\int_0^\infty e^{-at}e^{j\omega_0 t}e^{-j\omega t}\,dt - \frac{1}{2j}\int_0^\infty e^{-at}e^{-j\omega_0 t}e^{-j\omega t}\,dt \\
&= \frac{1}{2j}\int_0^\infty e^{-(a-j\omega_0+j\omega)}\,dt - \frac{1}{2j}\int_0^\infty e^{-(a+j\omega_0+j\omega)t}\,dt \\
&= \frac{1}{2j}\left[\frac{-e^{-(a-j\omega_0+j\omega)t}}{(a-j\omega_0+j\omega)} + \frac{e^{-(a+j\omega_0+j\omega)t}}{(a+j\omega_0+j\omega)}\right]_0^\infty \\
&= \frac{1}{2j}\left[\frac{1}{(a+j\omega)-j\omega_0} - \frac{1}{(a+j\omega)+j\omega_0}\right] \\
&= \frac{1}{2j}\left[\frac{a+j\omega+j\omega_0-a-j\omega+j\omega_0}{(a+j\omega)^2+\omega_0^2}\right]
\end{aligned}$$

$$\boxed{X(j\omega) = \frac{\omega_0}{[(a + j\omega)^2 + \omega_0^2]}}$$

6. $\boldsymbol{x(t) = [u(t+2) - u(t-2)] \cos 3t}$

In Fig. 6.46, $[u(t+2) - u(t-2)]$ is represented as

$$x_1(t) = u(t+2) - u(t-2) = 1; \quad |t| < 2$$

$$\begin{aligned}
X_1(j\omega) &= \int_{-2}^{2} e^{-j\omega t}\, dt \\
&= -\frac{1}{j\omega}\left[e^{-j\omega t}\right]_{-2}^{2} \\
&= -\frac{1}{j\omega}\left[e^{-j2\omega} - e^{j2\omega}\right] \\
&= \frac{2}{\omega}\frac{\left[e^{j2\omega} - e^{-j2\omega}\right]}{2j}
\end{aligned}$$

$$\begin{aligned}
X_1(j\omega) &= \frac{2}{\omega}\sin 2\,\omega \\
\cos 3t &= \frac{e^{j3t} + e^{-j3t}}{2} \\
F[x(t)\cos\omega_0 t] &= \frac{1}{2}[X(\omega - \omega_0) + X(\omega + \omega_0)]
\end{aligned}$$

Using the above frequency shifting FT property, we get

$$F[\{u(t+2) - u(t-2)\}\cos 3t] = \frac{[\sin 2(\omega - 3)]}{(\omega - 3)} + \frac{[\sin 2(\omega + 3)]}{(\omega + 3)}$$

$$\boxed{X(j\omega) = \left[\frac{\sin 2(\omega - 3)}{(\omega - 3)} + \frac{\sin 2(\omega + 3)}{(\omega + 3)}\right]}$$

7. $\boldsymbol{x(t) = e^{-2|t|} \cos 5t}$

$$F[e^{-2|t|}] = \frac{4}{\omega^2 + 4} \quad [\textit{see} \text{ Example 6.1(f)}]$$

$$F[x(t)\cos\omega_0 t] = \frac{1}{2}[X(\omega - \omega_0) + X(\omega + \omega_0)]$$

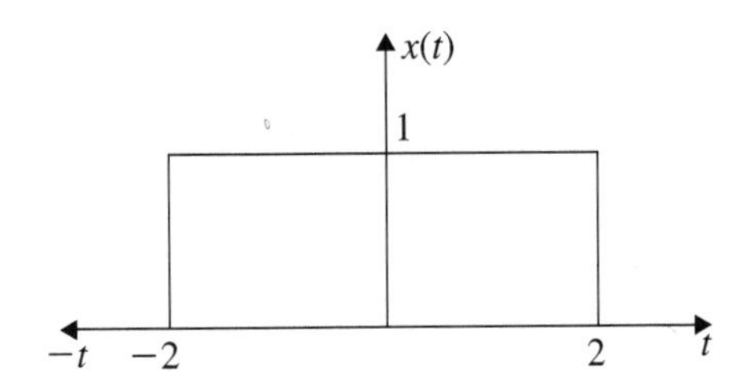

Fig. 6.46 $x(t) = [u(t+2) - u(t-2)]$

In the given problem, $\omega_0 = 5$

$$\boxed{X(j\omega) = \frac{2}{[(\omega-5)^2+4]} + \frac{2}{[(\omega+5)^2+4]}}$$

8. $\boldsymbol{x(t) = e^{-3|t|} \sin 2t}$

From the result obtained in Example 6.1(f), we write

$$F\left[e^{-3|t|}\right] = \frac{6}{(9+\omega^2)}$$

As per the modulation property,

$$x(t)\sin\omega_0 t \overset{\text{FT}}{\longleftrightarrow} \frac{1}{2j}[X(\omega-\omega_0) - X(\omega+\omega_0)]$$

$$F\left[e^{-3|t|}\sin 2t\right] \overset{\text{FT}}{\longleftrightarrow} \frac{1}{2j}\left[\frac{6}{[9+(\omega-2)^2]} - \frac{1}{[9+(\omega+2)^2]}\right]$$

$$\boxed{x(j\omega) = \frac{-j24}{[9+(\omega-2)^2][9+(\omega+2)^2]}}$$

Example 6.31

Consider the following differential equation. Determine the frequency response.

$$\frac{d^2y(t)}{dt^2} + 5\frac{dy(t)}{dt} + 6y(t) = \frac{dx(t)}{dt} + 4x(t)$$

Solution:
Taking FT on both sides of the above differential equation, we get the following algebraic equation. In the above equation, $(j\omega)^2 = \frac{d^2}{dt^2}$; $(j\omega) = \frac{d}{dt}$ are substituted in the Fourier integral.

$$(j\omega)^2 Y(j\omega) + 5(j\omega)Y(j\omega) + 6Y(j\omega) = [(j\omega) + 4]X(j\omega)$$

$$\frac{Y(j\omega)}{X(j\omega)} = H(j\omega) = \frac{(j\omega + 4)}{[(j\omega)^2 + 5j\omega + 6]}$$

$$\boxed{H(j\omega) = \frac{(j\omega + 4)}{(j\omega + 2)(j\omega + 3)}}$$

$$|H(j\omega)| = \frac{\sqrt{(\omega^2 + 16)}}{\sqrt{(\omega^2 + 4)(\omega^2 + 9)}}$$

$$\angle H(j\omega) = \tan^{-1}\frac{\omega}{4} - \tan^{-1}\frac{\omega}{2} - \tan^{-1}\frac{\omega}{3}$$

$H(j\omega)$ is the ratio of the Fourier transform of the output variable to the Fourier transform of the input variable. It is called "**Sinusoidal Transfer Function**".

To draw the frequency response plot (magnitude and phase plot), for $-\infty \leq \omega \leq \infty$, different values for ω are substituted in $H(j\omega)$ and $\angle H(j\omega)$ and the following table is prepared.

ω	0	± 1	± 2	± 4	± 6	$\pm\infty$
$\|H(j\omega)\|$	0.667	0.58	0.439	0.29	0.03	0
$\angle H(j\omega)$	0°	$\mp 31°$	$\mp 52°$	$\mp 72°$	$\mp 79°$	$\mp 90°$

From the above table, the frequency response magnitude plot is sketched as shown in Fig. 6.47a and the phase plot as in Fig. 6.47b.

■ Example 6.32

A certain continuous linear time invariant system is described by the following differential equation:

$$\frac{dy(t)}{dt} + 5y(t) = x(t)$$

Determine $y(t)$, using FT for the following input signals:

(a) $x(t) = e^{-2t}u(t)$
(b) $x(t) = 10u(t)$

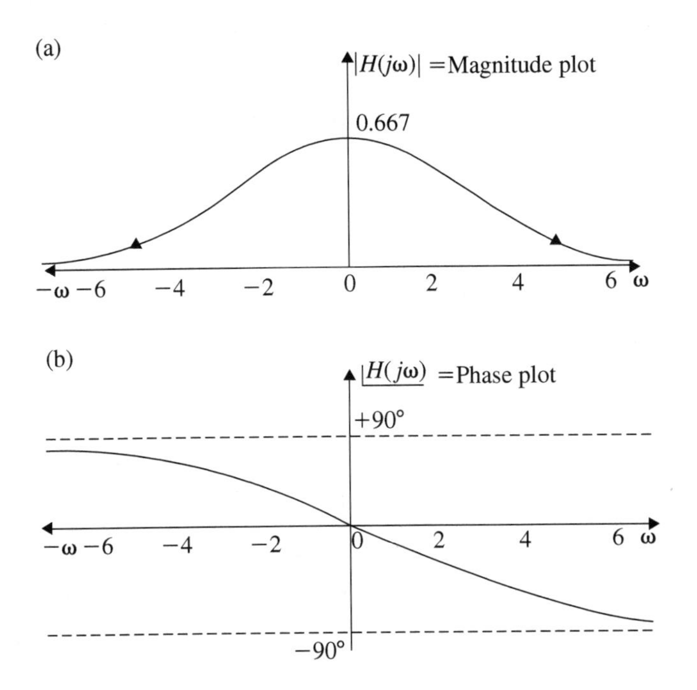

Fig. 6.47 Frequency response plot of Example 6.31. **a** Magnitude plot and **b** Phase plot

(c) $x(t) = \delta(t)$

Solution:

(a) $\boldsymbol{x(t) = e^{-2t}u(t)}$

Taking FT on both sides, we get

$$(j\omega + 5)Y(j\omega) = X(j\omega)$$

$$F[e^{-2t}u(t)] = \frac{1}{(j\omega + 2)}$$

$$Y(j\omega) = \frac{1}{(j\omega + 2)(j\omega + 5)}$$

$$= \frac{1}{3}\left[\frac{1}{j\omega + 2} - \frac{1}{j\omega + 5}\right]$$

$$y(t) = F^{-1}[Y(j\omega)] = \frac{1}{3}\left[e^{-2t} - e^{-5t}\right]u(t)$$

$$\boxed{y(t) = \frac{1}{3}\left[e^{-2t} - e^{-5t}\right]u(t)}$$

(b) $\boldsymbol{x(t) = 10u(t)}$

$$X(j\omega) = F[10u(t)] = \left[10\pi\delta(\omega) + \frac{10}{j\omega}\right]$$

$$Y(j\omega) = \frac{X(j\omega)}{(j\omega+5)}$$

$$= \left[\pi\delta(\omega) + \frac{1}{j\omega}\right]\frac{10}{(j\omega+5)}$$

$$= \frac{10\pi\delta(\omega)}{(j\omega+5)} + \frac{10}{j\omega(j\omega+5)}$$

$$= \frac{10\pi\delta(\omega)}{(j\omega+5)} + \frac{2}{j\omega} - \frac{2}{(j\omega+5)}$$

Applying the property $X(j\omega)\delta(\omega) = X(0)\delta(\omega)$ in the above equation, we get

$$Y(j\omega) = \frac{10}{5}\pi\delta(\omega) + \frac{2}{j\omega} - \frac{2}{(j\omega+5)}$$

$$= 2\left[\pi\delta(\omega) + \frac{1}{j\omega}\right] - \frac{2}{(j\omega+5)}$$

$$y(t) = F^{-1}Y(j\omega) = 2\left[u(t) - e^{-5t}u(t)\right]$$

$$\boxed{y(t) = 2\left[1 - e^{-5t}\right]u(t)}$$

Note:

$$F^{-1}\left[\pi\delta(\omega) + \frac{1}{j\omega}\right] = u(t).$$

The above response is called "Step Response" because the input $u(t)$ is a step signal.

(c) $\boldsymbol{x(t) = \delta(t)}$

$$X(j\omega) = 1$$

$$Y(j\omega) = \frac{1}{j\omega+5}$$

$$y(t) = F^{-1}[Y(j\omega)] = e^{-5t}u(t)$$

$$\boxed{y(t) = e^{-5t}u(t)}$$

The above response is called "Impulse Response of the System" because the input $\delta(t)$ is an impulse.

■ Example 6.33

Consider an LTI system with the differential equation.

$$\frac{d^2y(t)}{dt^2} + 4\frac{dy(t)}{dt} + 3y(t) = \frac{dx(t)}{dt} + 2x(t)$$

Find the frequency response and impulse response.

(Anna University, December, 2006)

Solution:
Taking FT on both sides of the above equation, we get

$$[(j\omega)^2 + 4j\omega + 3]Y(j\omega) = (j\omega + 2)X(j\omega)$$

$$H(j\omega) = \frac{Y(j\omega)}{X(j\omega)} = \frac{(j\omega + 2)}{[(j\omega)^2 + 4j\omega + 3]}$$

$$= \frac{(j\omega + 2)}{(j\omega + 1)(j\omega + 3)}$$

$$\boxed{|H(j\omega)| = \frac{\sqrt{(\omega^2 + 4)}}{\sqrt{(\omega^2 + 1)(\omega^2 + 9)}}}$$

$$\boxed{\underline{|H(j\omega)} = \tan^{-1}\frac{\omega}{2} - \tan^{-1}\omega - tan^{-1}\frac{\omega}{3}}$$

The above expressions give the magnitude and phase of the frequency response.

To draw the magnitude and phase of the frequency response plot, different values for ω are substituted in $H(j\omega)$ and $\angle H(j\omega)$ and the following table is prepared.

ω	0	± 2	± 4	± 6	$\pm\infty$
$\lvert H(j\omega)\rvert$	0.667	0.35	0.216	0.155	0
$\angle H(j\omega)$	0°	$\mp 15.3^\circ$	$\mp 65.7^\circ$	$\mp 72.4^\circ$	$\mp 90^\circ$

From the above table, the frequency response magnitude plot is sketched as shown in Fig. 6.48a and the phase plot as in Fig. 6.48b.

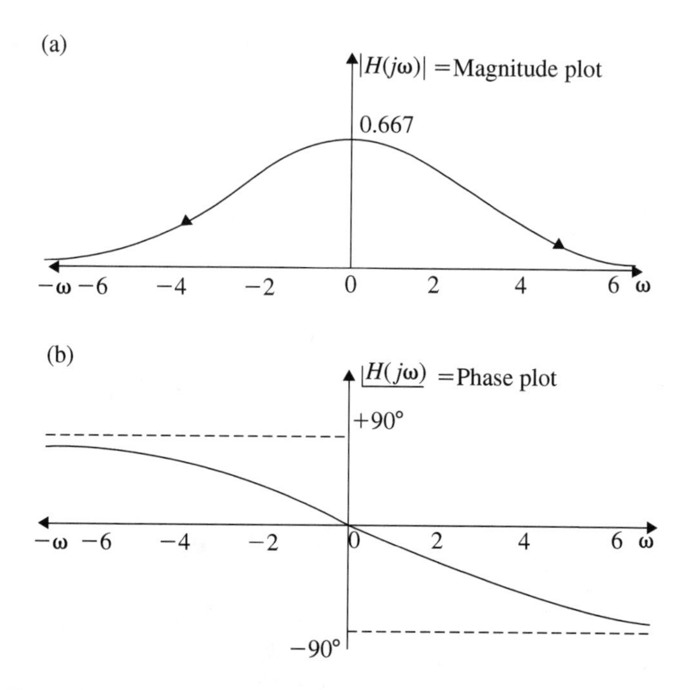

Fig. 6.48 Frequency response plot of Example 6.33. **a** Magnitude plot and **b** Phase plot

To find the impulse response,

$$
\begin{aligned}
x(t) &= \delta(t) \\
F[x(t)] &= F[\delta(t)] \\
&= 1 \\
Y(j\omega) &= \frac{(j\omega+2)}{(j\omega+1)(j\omega+3)} \\
&= \frac{A_1}{(j\omega+1)} + \frac{A_2}{(j\omega+3)} \\
(j\omega+2) &= A_1(j\omega+3) + A_2(j\omega+1)
\end{aligned}
$$

Let $j\omega = -1$;

$$1 = 2A_1 \quad \text{or} \quad A_1 = \frac{1}{2}$$

Let $j\omega = -3$;

$$-1 = -2A_2 \quad \text{or} \quad A_2 = \frac{1}{2}$$

$$Y(j\omega) = \frac{1}{2}\left[\frac{1}{(j\omega+1)} + \frac{1}{(j\omega+3)}\right]$$

Taking inverse FT, we get

$$y(t) = F^{-1}[Y(j\omega)] = \frac{1}{2}\left[e^{-t} + e^{-3t}\right] u(t)$$

$$\boxed{y(t) = \frac{1}{2}\left[e^{-t} + e^{-3t}\right] u(t)}$$

■ Example 6.34

An LTI continuous time system is described by the following differential equation:

$$\frac{d^2y(t)}{dt^2} + 2\frac{dy(t)}{dt} + 2y(t) = x(t)$$

Determine the impulse response of the system using FT and inverse FT.

Solution:
Taking FT on both sides, we get the following equation:

$$\left[(j\omega)^2 + 2j\omega + 2\right] Y(j\omega) = X(j\omega)$$

For an impulse input $x(t) = \delta(t)$,

$$\begin{aligned}
X(j\omega) &= 1 \\
Y(j\omega) &= \frac{1}{(j\omega)^2 + 2j\omega + 2} \\
&= \frac{1}{(j\omega+1+j)(j\omega+1-j)} \\
&= \frac{A_1}{(j\omega+1+j)} + \frac{A_2}{(j\omega+1-j)} \\
1 &= A_1(j\omega+1-j) + A_2(j\omega+1+j)
\end{aligned}$$

Let $j\omega = -(1+j)$

$$1 = A_1(-1-j-1-j)$$
$$A_1 = \frac{-1}{2j}; \qquad A_2 = A_1^* = \frac{1}{2j}$$

$$Y(j\omega) = \frac{1}{2j}\left[\frac{-1}{j\omega + (1+j)} + \frac{1}{j\omega + (1-j)}\right]$$

Taking inverse FT, we get

$$y(t) = \frac{1}{2j}\left[-e^{-(1+j)t} + e^{-(1-j)t}\right]$$
$$= e^{-t}\left[\frac{e^{jt} - e^{-jt}}{2j}\right]$$

$$\boxed{y(t) = e^{-t}\sin t}$$

■ Example 6.35

Find the unit step response of the circuit shown in Fig. 6.49. Use the Fourier transform method.

(*Anna University, December, 2007*)

Solution: For the circuit shown in Fig. 6.49, the following equation is written:

$$L\frac{di(t)}{dt} + Ri(t) = x(t)$$
$$5\frac{di(t)}{dt} + 10i(t) = x(t)$$

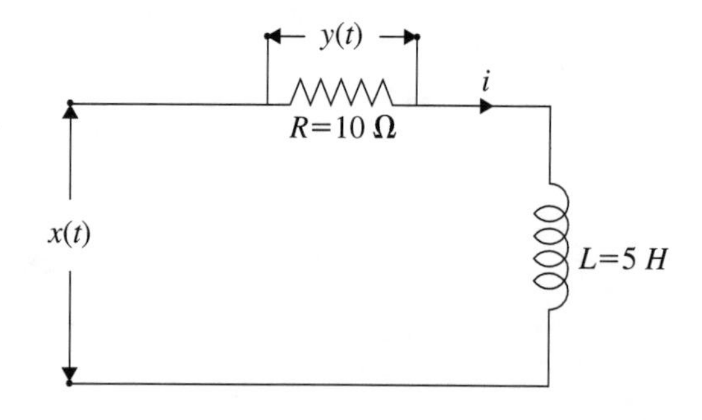

Fig. 6.49 Time response of R-L circuit

Taking FT on both sides, we get

$$[5j\omega + 10]I(j\omega) = X(j\omega)$$
$$I(j\omega) = \frac{0.2X(j\omega)}{(j\omega + 2)}$$

For a step input,

$$X(j\omega) = \pi\delta(\omega) + \frac{1}{j\omega}$$

$$I(j\omega) = \frac{0.2\pi\delta(\omega)}{(j\omega + 2)} + \frac{0.2}{j\omega(j\omega + 2)}$$

Applying the property $X(j\omega)\delta(\omega) = X(0)\,\delta(\omega)$, the above equation is written as

$$\begin{aligned} I(j\omega) &= 0.1\pi\delta(\omega) + 0.1\left[\frac{1}{j\omega} - \frac{1}{j\omega + 2}\right] \\ &= 0.1\left[\pi\delta(\omega) + \frac{1}{j\omega}\right] - \frac{0.1}{j\omega + 2} \\ i(t) &= 0.1\left[u(t) - e^{-2t}u(t)\right] \\ y(t) &= i(t)R \end{aligned}$$

$$\boxed{y(t) = \left[1 - e^{-2t}\right] u(t)}$$

Summary

1. Periodic signals are represented by Fourier series as a sum of complex sinusoids or exponentials. However, FS is not applicable to aperiodic signals. Fourier transform gives spectral representation to aperiodic signal. Thus, FT is applicable to periodic and non-periodic signals as well as to transform time domain signal $x(t)$ to frequency domain signal $X(j\omega)$. Here the frequency domain representation is continuous.
2. It is possible to transform time domain specifications to frequency domain specifications and *vice versa*. The former is called Fourier transform and the latter is called inverse Fourier transform which are denoted as $F[x(t)]$ and $F^{-1}[X(j\omega)]$, respectively, and they are called Fourier transform pair.

3. Fourier transform does not exist for some useful signals. For example, for $x(t) = e^{at}u(t)$ FT does not converge.
4. Fourier and Laplace were contemporaries and great mathematicians who were encouraged by the French ruler Napoleon Bonaparte. Laplace, by introducing an exponential decay in the everlasting exponential, made many functions converge while FT failed in these cases. Further, the Laplace transform is more powerful especially in getting the solution of differential equations compared to FT.
5. FT is a special case of LT which is obtained in many cases by replacing s by $j\omega$. But this is not always true. For example, in the case of a step signal, this is not applicable.
6. Fourier transform has many useful properties. By applying these properties, one can easily get the FT pair of even complex signals. They are powerful tools for manipulating signals in time and frequency domains.

Exercises

I. Short Answer Type Questions

1. **What do you understand by Fourier transform pair? What is called analysis equation and what is called synthesis equation?**
When the time function $x(t)$ is transformed to frequency function $X(j\omega)$, the function $x(t)$ is said to be Fourier transformed. When the frequency function $X(j\omega)$ is transformed to $x(t)$, then the function $X(j\omega)$ is said to be inverse Fourier transformed. These transformations are, respectively, defined as follows:

$$X(j\omega) = \int_{-\infty}^{\infty} x(t)e^{-j\omega t}dt$$

$$x(t) = \frac{1}{2\pi}\int_{-\infty}^{\infty} X(j\omega)e^{j\omega t}d\omega$$

The above two equations are called FT pair. The first equation is called the analysis equation while the second equation is called the synthesis equation.
2. **How is the Fourier transform different from Fourier series?**
Fourier series is applicable to periodic signals. Fourier transform is applicable to periodic and aperiodic signals as well.
3. **How is FT developed from Fourier series?**
When the aperiodic signal is considered as a periodic signal with its fundamental period tending to infinity, the fundamental frequency decreases and the higher harmonics become closer. The frequency components form

a continuum, and the Fourier series sum becomes a Fourier integral which is defined as the Fourier transform.

4. **How is Parseval's Energy theorem defined for the frequency domain signal?**
According to Parseval's theorem (French mathematician of the late eighteenth and early nineteenth centuries), the energy of the frequency domain is defined as

$$E = \frac{1}{2\pi}\int_{-\infty}^{\infty} |X(j\omega)|^2\, d\omega$$

5. **What is the connection between the Fourier transform and the Laplace transform?**
The connection between the Fourier transform and the Laplace transform is that the Fourier transform is the Laplace transform with $s = j\omega$. The Laplace transform of $x(t) = e^{-at}u(t)$ is $X(s) = \frac{1}{(s+a)}$ and its Fourier transform is $X(j\omega) = \frac{1}{(j\omega+a)}$. However, this is not generally true of signals which are not absolutely integrable. The Laplace transform of a step signal is $X(s) = \frac{1}{s}$. The Fourier transform of the step signal is $X(j\omega) = \pi\delta(\omega) + \frac{1}{j\omega}$ and not simply $\frac{1}{j\omega}$.
6. **What do you understand by frequency response?**
If $y(t)$ is the output, $x(t)$ the input and $h(t)$ is the impulse response, then they are related as

$$y(t) = x(t) * h(t)$$

By using tne convolution property, we get

$$Y(j\omega) = X(j\omega)H(j\omega)$$
$$\frac{Y(j\omega)}{X(j\omega)} = H(j\omega)$$

The function $H(j\omega)$ is called the frequency response.
7. **What is the condition required for the convergence of the Fourier transform?**
If the signal $x(t)$ has finite energy or if it is square integrable such that

$$\int_{-\infty}^{\infty} |x(t)|^2\, dt < \infty$$

then the Fourier transform $X(j\omega)$ converges.

8. **What is the Fourier transform of**

$$x(t) = \frac{d^2}{dt^2}x(t+1)$$

$$F[x(t)] = (j\omega)^2 e^{j\omega} X(j\omega)$$

9. **What is the FT of $x(t) = [\delta(t+5) - \delta(t-5)]$?**

$$X(j\omega) = 2j \sin 5\omega$$

10. **Find the FT of $x(t) = 2[u(t+6) - u(t-6)]$?**

$$X(j\omega) = \frac{4}{\omega} \sin 6\omega = 24\text{sinc}6\omega$$

11. **What is the CTFT of a DC signal of amplitude 1?**

$$1 \overset{\text{FT}}{\longleftrightarrow} 2\pi\delta(\omega)$$

12. **What is the CTFT of $x(t) = u(t)$?**

$$u(t) \overset{\text{FT}}{\longleftrightarrow} \frac{1}{j\omega} + \pi\delta(\omega)$$

13. **What is integration property of CTFT?**

$$\int_{-\infty}^{t} x(d) \overset{\text{FT}}{\longleftrightarrow} \frac{1}{j\omega}X(j\omega) + \pi X(0)\delta(\omega)$$

14. **What is the differentiation property of CTFT?**

$$\frac{dx(t)}{dt} \overset{\text{FT}}{\longleftrightarrow} j\omega X(j\omega)$$

15. **If $X(j\omega)$ is the CTFT of $x(t)$, what is the CTFT of $x^*(-t)$?**

$$x^*(-t) \overset{\text{FT}}{\longleftrightarrow} X^*(-j\omega)$$

16. **State Parseval's energy intensity.**

$$E = \int_{-\infty}^{\infty} |x(t)|^2 dt = \frac{1}{2\pi}\int_{-\infty}^{\infty} |X(j\omega)|^2 d\omega$$

The energy content of the CT signal $x(t)$ can be determined either by integrating $|x(t)|^2$ over all time t or by integrating $|X(j\omega)|^2$ over all frequencies ω.

17. **What are Dirichlet's conditions for the CTFT?**
Dirichlet's conditions for the existence of CTFT are as follows:

(a) The signal $x(t)$ should be absolutely integrable, that is,

$$\int_{-\infty}^{\infty} |x(t)| dt < \infty$$

(b) $x(t)$ should have finite number of maxima and minima over a finite interval of time.
(c) $x(t)$ should have finite number of discontinuities in the finite time interval.

18. **What is the CTFT of $x(t) = e^{at}u(-t)$?**

$$e^{at}u(-t) \overset{FT}{\longleftrightarrow} \frac{-1}{(j\omega - a)}$$

where a or real part of $a > 0$.

19. **Is $x(t) = u(t)$ absolutely integrable using analysis equation of CTFT?**

$$X(j\omega) = \int_{0}^{\infty} 1e^{-j\omega t} dt$$
$$= -\frac{1}{j\omega}[e^{-j\omega t}]_0^{\infty}$$

The integral becomes indeterminate when the upper limit is applied.

20. **If**

$$X(j\omega) = \frac{(j\omega + 1)}{(j\omega + 3)}$$

find $x(t)$.

$$x(t) = \delta(t) - 2e^{-3t}u(t)$$

21. **If $X(j\omega) = \delta(\omega - 2)$, find $x(t)$.**

$$x(t) = \frac{1}{2\pi}e^{j2t}$$

22. **If $X(j\omega) = \pi\delta(\omega + 2\pi)$, find $x(t)$.**

$$x(t) = \frac{1}{2}e^{-j2\pi t}$$

23. **$y(t) = x(2-t) + x(-t-2)$. Find $Y(j\omega)$.**

$$Y(j\omega) = 2X(-j\omega)\cos 2\omega$$

24. **$y(t) = x(2-t) - x(-t-2)$. Find $Y(j\omega)$.**

$$Y(j\omega) = -2j\sin 2\omega X(-j\omega)$$

25. **$y(t) = x(t+2) - x(t-2)$. Find $Y(j\omega)$.**

$$Y(j\omega) = 2j\sin 2\omega X(j\omega)$$

26.

$$y(t) = \frac{d^2x(t+2)}{dt^2}$$

Find $Y(j\omega)$.

$$Y(j\omega) = e^{2j\omega}(-\omega^2)X(j\omega)$$

27. **$y(t) = x(5t+10)$. Find $Y(j\omega)$.**

$$Y(j\omega) = \frac{1}{5}e^{2j\omega}X\left(\frac{j\omega}{5}\right)$$

II. Long Answer Type Questions

1. **Consider the following continuous time signal.**

$$x(t) = e^{-5|t|}$$

Find the FT. Hence determine the FT of $tx(t)$.

$$X(j\omega) = \frac{10}{(25+\omega^2)}$$

$$F\left[te^{-5|t|}\right] = \frac{-j20\omega}{(25+\omega^2)^2}$$

2. **For the signal $X(j\omega)$ shown in Fig. 6.50, determine $x(t)$.**

$$x(t) = 5\frac{\sin 5t}{\pi t}$$

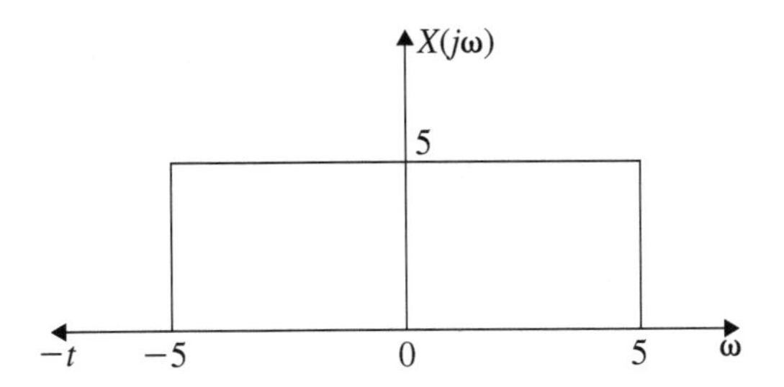

Fig. 6.50 Frequency spectrum of $x(t)$ of question 2

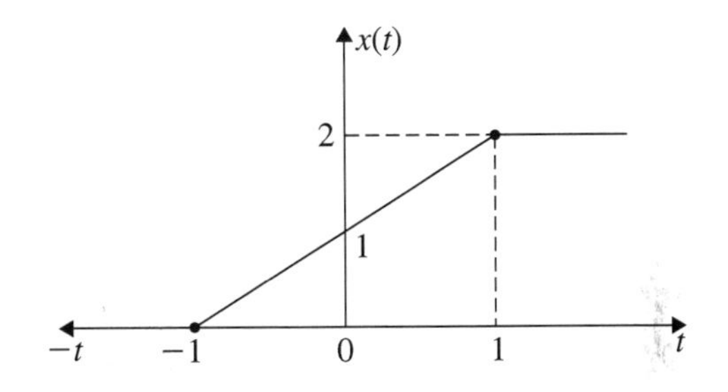

Fig. 6.51 Representation of the signal $x(t)$ for question 3

3. **Consider the signal shown in Fig. 6.51. Find $X(j\omega)$. What is the FT for $x(t-1)$?**

$$X(j\omega) = \frac{2\sin\omega}{j\omega^2} + 2\pi\delta(\omega)$$

$$F[x(t-1)] = X(j\omega)e^{-j\omega}$$

$$= \frac{2\sin\omega}{j\omega^2}e^{-j\omega} + 2\pi\delta(\omega)$$

4. **Using Parseval's theorem, evaluate energy in the frequency domain.**

$$x(t) = e^{-4|t|}$$

$$p = \frac{1}{4}$$

5.

$$\boldsymbol{x(t) = e^{-2t}u(t)}$$

and

$$\boldsymbol{h(t) = e^{-4t}u(t)}$$

$$\boldsymbol{y(t) = x(t) * h(t)}$$

Using the time convolution property, find $Y(j\omega)$ and $y(t)$.

$$Y(j\omega) = \frac{1}{(j\omega + 2)(j\omega + 4)}$$

$$y(t) = \frac{1}{2}\left[e^{-2t} - e^{-4t}\right]u(t)$$

6.

$$x(t) = e^{-2t}u(t)$$
$$h(t) = e^{-2t}u(t)$$
$$y(t) = x(t) * h(t)$$

Find $Y(j\omega)$ and hence $y(t)$.

$$Y(j\omega) = \frac{1}{(j\omega + 2)^2}$$

$$y(t) = te^{-2t}u(t)$$

7. **A certain LTIC system is described by the following differential equation (Fig.** 6.52):

$$\frac{dy(t)}{dt} + 2y(t) = x(t)$$

Determine the Frequency response and the Impulse response.

$$H(j\omega) = \frac{1}{(j\omega + 2)}$$

$$h(t) = e^{-2t}u(t)$$

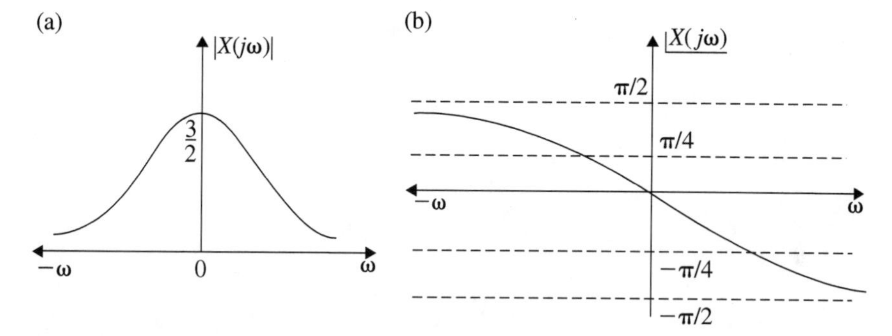

Fig. 6.52 Frequency spectra of Eq. 11

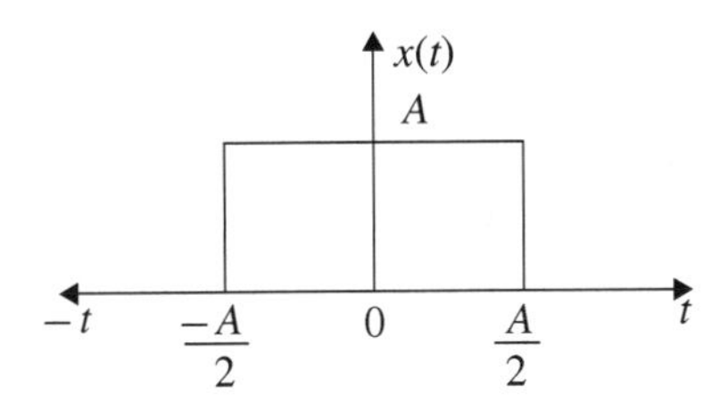

Fig. 6.53 Square wave of amplitude A

8. **Consider the following differential equation (Fig. 6.53):**

$$\frac{d^2y(t)}{dt^2}+8\frac{dy(t)}{dt}+15y(t)=\frac{dx(t)}{dt}+4x(t)$$

   (a) **Find the frequency response.**
   (b) **Find the impulse response.**
   (c) **Find the response $y(t)$ due to the input $x(t)=e^{-3t}u(t)$.**

(a) $$H(j\omega)=\frac{(j\omega+4)}{(j\omega+3)(j\omega+5)}$$

(b) $$h(t)=\frac{1}{2}\left[e^{-3t}+e^{-5t}\right]u(t)$$

(c) $$y(t)=\frac{1}{4}\left[2te^{-3t}+e^{-3t}-e^{-5t}\right]u(t)$$

9. **Determine the impulse response $h(t)$ of the system given by the differential equation**

$$\frac{d^2y(t)}{dt^2}+3\frac{dy(t)}{dt}+2y(t)=x(t)$$

   Assume all initial conditions to be zero.
   (*Anna University, 2013*)

$$y(t)=(e^{-t}-e^{-2t})u(t)$$

10. **The system produces output $y(t)=e^{-t}u(t)$ for the input $x(t)=e^{-2t}u(t)$. Determine**

   (i) **Frequency response.**
   (ii) **Magnitude and phase of the response.**
   (iii) **The impulse response.**

(*Anna University, 2013*)

(i) $$H(j\omega) = \frac{(j\omega + 2)}{(j\omega + 1)}$$

(ii) $$|H(j\omega)| = \sqrt{\frac{(\omega^2 + 4)}{(\omega^2 + 1)}}$$

$$\underline{|H(j\omega)} = \tan^{-1}\frac{\omega}{2} - \tan^{-1}\omega$$

(iii) $$h(t) = \delta(t) + e^{-t}u(t)$$

11. **Obtain the frequency response of a CT-LTI system defined by**

$$\frac{dy(t)}{dt} + 2y(t) = 3x(t)$$

and hence draw the magnitude and phase spectra.
(*Anna University, 2009*)

$$H(j\omega) = \frac{3}{j\omega + 2}$$

$$|H(j\omega)| = \frac{3}{\sqrt{\omega^2 + 4}}$$

$$\underline{|H(j\omega)} = -\tan^{-1}\frac{\omega}{2}$$

12. **Find the CTFT of**

$$x(t) = \frac{1}{2}\left[y(t+1) + y\left(t + \frac{1}{2}\right) + y\left(t - \frac{1}{2}\right) + y(t-1)\right]$$

(*Anna University, 2011*)

$$X(j\omega) = \left[\cos\omega + \cos\frac{\omega}{2}\right]Y(j\omega)$$

13. **Find the CTFT of a square wave of amplitude "A".**

$$X(j\omega) = \frac{2A}{\omega}\sin\frac{A}{2}\omega$$

14. **The input signal $x(t) = e^{-2t}u(t)$ is applied to a relaxed LTI system. The response of the system is**

$$y(t) = \frac{2}{3}[e^{-t} + e^{-2t} - e^{-3t}]u(t)$$

Find the system function using CTFT.

$$H(j\omega) = \frac{2}{3}\frac{(-\omega^2 + j6\omega + 7)}{(j\omega + 1)(j\omega + 3)}$$

Chapter 7
Fourier Transform Analysis of Discrete Time Signals and Systems—DTFT, DFT and FFT

Learning Objectives

- ⧫ To represent aperiodic discrete time signal by Fourier integral.
- ⧫ To define Fourier transform for discrete time signal.
- ⧫ To derive the conditions for the existence of Discrete Time Fourier Transform (DTFT).
- ⧫ To find DTFT for typical discrete time signals.
- ⧫ To establish the properties of the DTFT.
- ⧫ To solve the difference equations using DTFT.
- ⧫ To define DFT and IDFT.
- ⧫ To determine the properties of DFT.
- ⧫ To find the circular convolution using circle method.
- ⧫ To establish the fundamental principle of FFT algorithm.

7.1 Introduction

In Chap. 4, we represented continuous time periodic signals as a sum of everlasting exponentials by Fourier series. Similarly, in Chap. 5, the discrete time periodic signals was represented by discrete time Fourier series using a parallel development of continuous time system. The Fourier series representations of CT and DT signals in these chapters are however applicable only if the signal is periodic. If the signal is non-periodic, then applying a limiting process the aperiodic continuous time signal was expressed as a continuous sum of everlasting exponential or sinusoids and this method was termed as Fourier transform of continuous time signal which was discussed in Chap. 6. On similar line, the discrete time periodic signal represented as a sum of everlasting exponential by Fourier series in Chap. 5, by applying limiting process to aperiodic signal $x[n]$, it can be expressed as a sum of everlasting exponentials. The spectrum $X[\Omega]$ so obtained is called Discrete Time Fourier Transform (DTFT). If the

S. Palani, *Signals and Systems*,
https://doi.org/10.1007/978-3-030-75742-7_7

spectrum obtained by DTFT is sampled for one period of the Fourier transform at a finite number of frequency points, such a transformation is called Discrete Fourier Transform (DFT) which is a very powerful computational tool for the evaluation of FT. Some special algorithms are developed for the easy implementation of DFT which result in saving of considerable computation time. Such algorithms are called Fast Fourier Transform (FFT). The detailed study of DTFT, DFT and FFT are discussed in this chapter with sufficient illustrated examples.

7.2 Representation of Discrete Time Aperiodic Signals

Consider the aperiodic signal $x[n]$ shown in Fig. 7.1a. The periodic signal $x_{N_0}[n]$ is constructed by repeating the signal $x[n]$ every N_0 units as shown in Fig. 7.1b. The period N_0 is chosen large enough to avoid overlapping. If we put $N_0 \longrightarrow \infty$, the signal repeats after an infinite interval and therefore

$$\underset{N_0 \to \infty}{Lt} x_{N_0}[n] = x[n]$$

From Eq. (5.4), for a discrete signal, the FS can be written as

$$x[n] = \sum_{k=[N_0]} D_k e^{jk\Omega_0 n}$$

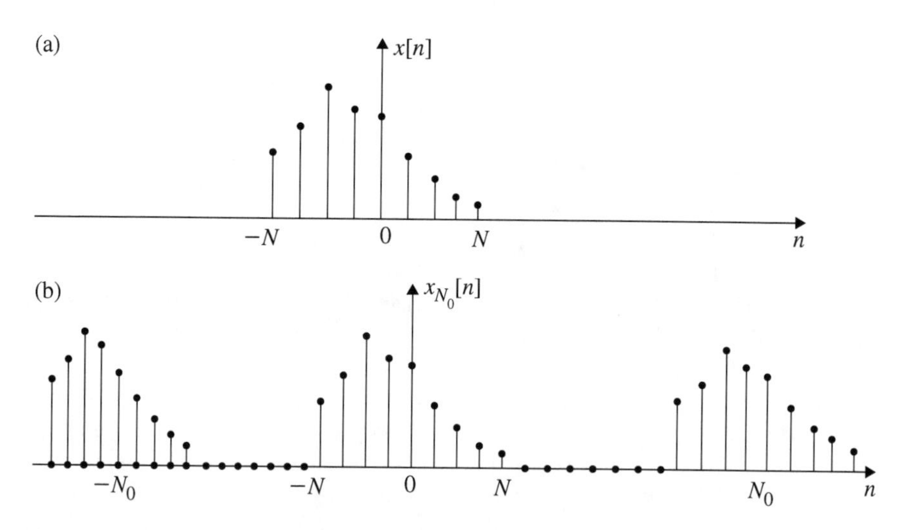

Fig. 7.1 Extension of aperiodic signal to periodic signal

$$x_{N_0}[n] = \sum_{k=[N_0]} D_k e^{jk\Omega_0 n} \tag{7.1}$$

where $\Omega_0 = \frac{2\pi}{N_0}$ and

$$D_k = \frac{1}{N_0} \sum_{n=-\infty}^{\infty} x[n] e^{-jk\Omega_0 n} \tag{7.2}$$

With Ω as continuous function let us define

$$X(\Omega) = \sum_{n=-\infty}^{\infty} x[n] e^{-j\Omega n} \tag{7.3}$$

Substituting Eq. (7.3) in Eq. (7.2), we get

$$D_k = \frac{1}{N_0} X(k\Omega_0) \tag{7.4}$$

Equation (7.4) shows that the Fourier coefficients D_k are $\frac{1}{N_0}$ times the samples of $X(\Omega)$. As $N_0 \to \infty$, the fundamental frequency $\Omega_0 \to 0$ and $D_k \to 0$ and the spectrum becomes continuous. Now consider Eq. (7.3)

$$X(\Omega) = \sum_{n=-\infty}^{\infty} x[n] e^{-jk\Omega_0 n} \tag{7.5}$$

Equation (7.1) can be expressed using Eq. (7.2) as

$$\begin{aligned} x_{N_0}[n] &= \frac{1}{N_0} \sum_{k=[N_0]} X(k\Omega_0) e^{jk\Omega_0 n} \\ &= \sum_{k=[N_0]} X(k\Omega_0) e^{jk\Omega_0 n} \left(\frac{\Omega_0}{2\pi}\right) \end{aligned} \tag{7.6}$$

As $N_0 \to \infty$, $\Omega_0 \to 0$ and $x_{N_0}[n] \to x[n]$

$$x[n] = \underset{\Omega_0 \to 0}{Lt} \sum_{k=N_0} \left[X(k\Omega_0) \frac{\Omega_0}{2\pi} \right] e^{jk\Omega_0 n} \tag{7.7}$$

Since Ω_0 is small, it can be replaced by Ω. Thus,

$$\Delta\Omega = \frac{2\pi}{N_0} \tag{7.8}$$

$$x[n] = \underset{\Delta\Omega_0 \to 0}{Lt} \frac{1}{2\pi} \sum_{k=N_0} X(k\Delta\Omega) e^{jk\Delta\Omega n} \Delta\Omega \tag{7.9}$$

$k = N_0$ implies $N_0 \Delta\Omega = 2\pi$. Hence, Eq. (7.9) becomes the integral

$$\boxed{x[n] = \frac{1}{2\pi} \int_{2\pi} X(\Omega) e^{jn\Omega} d\Omega} \tag{7.10}$$

The spectrum $X(\Omega)$ is given by

$$\boxed{X(\Omega) = \sum_{n=-\infty}^{\infty} x[n] e^{-j\Omega n}} \tag{7.11}$$

Equation (7.10) is called the Fourier integral and $X(\Omega)$ is called the Discrete Time Fourier Transform (DTFT). They are called DFTF pair. Symbolically, they are represented as

$$\begin{aligned} x[n] &= \text{IDTFT}\{X(\Omega)\} \\ X(\Omega) &= \text{DTFT}\{x[n]\} \end{aligned} \tag{7.12}$$

Or

$$x[n] \overset{\text{DTFT}}{\longleftrightarrow} X(\Omega)$$

The Fourier transform $X(\Omega)$ is nothing but the description of $x[n]$ in the frequency domain. From Eq. (7.11), it is proved that the spectrum of a discrete time signal is periodic with fundamental period N_0.

As in the case of continuous time signal, the sufficient condition for the convergence of $X(\Omega)$ is that $x[n]$ is either absolutely summable. That is

$$\sum_{n=-\infty}^{\infty} |x[n]| < \infty$$

or the sequence has finite energy, that is

$$\sum_{n=-\infty}^{\infty} |x[n]|^2 < \infty. \tag{7.13}$$

7.3 Connection Between the Fourier Transform and the z-Transform

From Eq. (7.11)

$$X(\Omega) = \sum_{n=-\infty}^{\infty} x[n]e^{-j\Omega n} \tag{7.14}$$

The z-transform of $x[n]$ is given by

$$X[z] = \sum_{n=-\infty}^{\infty} x[n]z^{-n} \tag{7.15}$$

From Eqs. (7.14) and (7.15), we see that if the ROC of $X[z]$ contains unit circle, then $X(\Omega)$ equals $X[z]$ evaluated on the unit circle. That is

$$X(\Omega) = X[z]\Big|_{z=e^{j\Omega}} \tag{7.16}$$

Note: Using Eq. (7.16), one can obtain Fourier transform by substituting $z = e^{j\Omega}$ provided $x[n]$ is summable. If $x[n]$ is not summable, as in the case of $u[n]$, one cannot obtain $X(\Omega)$ form $X[z]$.

When Fourier transform and z-transform are connected, $X(\Omega)$ is denoted by $X(e^{j\Omega})$ or $X(e^{j\omega})$.

■ Example 7.1

Find the FT of the following DT signals:

1. $x[n] = \delta[n]$
2. $x[n] = a^n u[n]$
3. $x[n] = -a^n u[-n-1]$
4. $x[n] = u[n]$
5. $x[n] = (a)^{|n|} \quad |a| < 1$

Solution

1. $\boldsymbol{x[n] = \delta[n]}$

$$F\{\delta[n]\} = \sum_{-\infty}^{\infty} \delta[n]e^{-j\Omega n}$$

$$\delta[n] = \begin{cases} 1 & n = 0 \\ 0 & n \neq 0 \end{cases}$$

$$\boxed{F\{\delta[n]\} = 1}$$

2. $\boldsymbol{x[n] = a^n u[n]}$

$$X[\Omega] = \sum_{n=-\infty}^{\infty} a^n u[n] e^{-j\Omega n}$$

$$u[n] = \begin{cases} 1 & n \geq 0 \\ 0 & n < 0 \end{cases} \quad \text{since } x[n] \text{ is causal}$$

$$X[\Omega] = \sum_{n=0}^{\infty} a^n u e^{-j\Omega n}$$

$$= \sum_{n=0}^{\infty} [a e^{-j\Omega}]^n$$

By using the summation formula we get

$$\boxed{X[\Omega] = \frac{1}{(1 - a e^{-j\Omega})}} \quad |a| < 1$$

3. $\boldsymbol{x[n] = -a^n u[-n-1]}$

$$X[\Omega] = \sum_{n=-\infty}^{-1} -a^n u[-n-1] e^{-j\Omega n}$$

$$= \sum_{n=-\infty}^{-1} -a^n e^{-j\Omega n}$$

$$= \sum_{n=1}^{\infty} -a^{-n} e^{j\Omega n}$$

$$= -\sum_{n=1}^{\infty} (a^{-1} e^{j\Omega})^n$$

$$= -\{a^{-1} e^{j\Omega} + (a^{-1} e^{j\Omega})^2 + (a^{-1} e^{j\Omega})^3 + \cdots\}$$

$$= -a^{-1} e^{j\Omega} \left[1 + (a^{-1} e^{j\Omega}) + (a^{-1} e^{j\Omega})^2 + \cdots\right]$$

$$= -\frac{a^{-1} e^{j\Omega}}{(1 - a^{-1} e^{j\Omega})}$$

$$\boxed{X(\Omega) = \frac{1}{(1 - ae^{-j\Omega})}} \qquad |a| > 1$$

4. $\boldsymbol{x[n] = u[n]}$
   Let $x[n] = u[n] \longleftrightarrow X(\Omega)$

$$X(\Omega) = \sum_{n=0}^{\infty} u[n]e^{-j\Omega n}$$

$$\begin{aligned} X(\Omega) &= \sum_{n=0}^{\infty} (e^{-j\Omega})^n \\ &\neq \frac{1}{(1 - e^{-j\Omega})} \end{aligned}$$

which is obtained using summation formula because $|e^{j\Omega_0}| = 1$ and $X(\Omega)$ is not summable. The following procedure is followed to evaluate the FT of a causal step sequence.

$$\begin{aligned} \delta[n] &= u[n] - u[n-1] \\ 1 &= [1 - e^{-j\Omega}]X(\Omega) \end{aligned}$$

For $\Omega = 0,\ (1 - e^{-j\Omega}) = 0$. Therefore, $X(\Omega)$ must be written in the following form:

$$X(\Omega) = C\delta(\Omega) + \frac{1}{(1 - e^{-j\Omega})}$$

where C is any constant. The step sequence $u[n]$ can be expressed in terms of odd and even components as

$$\begin{aligned} x_e[n] &= \frac{1}{2} + \frac{1}{2}\delta[n] \\ x_0[n] &= x[n] - x_e[n] \\ &= x[n] - \frac{1}{2} - \frac{1}{2}\delta[n] \\ F[x_0[n]] &= C\delta(\Omega) + \frac{1}{(1 - e^{-j\Omega})} - \pi\delta(\Omega) - \frac{1}{2} \end{aligned}$$

From the property of FT, the FT of an odd sequence must be imaginary. To satisfy the condition, $C = \pi$

$$F[x_0[n]] = \frac{1}{(1 - e^{-j\Omega})} - \frac{1}{2}$$
$$F[x_e[n]] = \pi\delta[\Omega] + \frac{1}{2}$$
$$F[x[n]] = F[x_e[n]] + F[x_0[n]]$$
$$= \pi\delta(\Omega) + \frac{1}{2} + \frac{1}{(1 - e^{-j\Omega})} - \frac{1}{2}$$

$$\boxed{u[n] \longleftrightarrow \pi\delta(\Omega) + \frac{1}{(1 - e^{-j\Omega})}}$$

5. $\boldsymbol{x[n] = (a)^{|n|} \qquad |a| < 1}$

$$X(\Omega) = \sum_{n=-\infty}^{\infty} (a)^{|n|} e^{-j\Omega n}$$
$$= \sum_{n=-\infty}^{1} a^{-n} e^{-j\Omega n} + \sum_{n=0}^{\infty} a^n e^{-j\Omega n}$$
$$= X_1(\Omega) + X_2(\Omega)$$
$$X_2(\Omega) = \frac{1}{(1 - ae^{-j\Omega})} \qquad a < 1$$
$$X_1(\Omega) = \sum_{n=1}^{\infty} (ae^{j\Omega})^n$$
$$= ae^{j\Omega} + (ae^{j\Omega})^2 + (ae^{j\Omega})^3 + \cdots$$
$$= ae^{j\Omega}[1 + ae^{j\Omega} + (ae^{j\Omega})^2 + \cdots]$$
$$= \frac{ae^{j\Omega}}{(1 - ae^{j\Omega})} \qquad a < 1$$
$$X(\Omega) = \frac{ae^{j\Omega}}{(1 - ae^{j\Omega})} + \frac{1}{(1 - ae^{-j\Omega})}$$

$$\boxed{X(\Omega) = \frac{(1 - a^2)}{(1 - 2a\cos\Omega + a^2)}}$$

7.4 Properties of Discrete Time Fourier Transform

The properties of DTFT are very similar to those of CTFT and these properties are very useful to determine the Fourier transform and inverse Fourier transforms very quickly. They are discussed below with proofs.

7.4.1 *Linearity*

If

$$x_1[n] \stackrel{\text{DTFT}}{\longleftrightarrow} X_1(\Omega) \quad \text{and} \quad x_2[n] \stackrel{\text{DTFT}}{\longleftrightarrow} X_2(\Omega)$$

then

$$Ax_1[n] + Bx_2[n] \stackrel{\text{DTFT}}{\longleftrightarrow} AX_1(\Omega) + BX_2(\Omega)$$

Proof

$$X_1(\Omega) = \sum_{n=-\infty}^{\infty} x_1[n]e^{-j\Omega n}$$

$$\begin{aligned} X_2(\Omega) &= \sum_{n=-\infty}^{\infty} x_2[n]e^{-j\Omega n} \\ \{Ax_1[n] + Bx_2[n]\} &\stackrel{\text{DTFT}}{\longleftrightarrow} \sum_{n=-\infty}^{\infty} Ax_1[n]e^{-j\Omega n} + \sum_{n=-\infty}^{\infty} Bx_2[n]e^{-j\Omega n} \\ &= AX_1(\Omega) + BX_2(\Omega) \end{aligned}$$

$$\boxed{\{Ax_1[n] + Bx_2[n]\} \stackrel{\text{DTFT}}{\longleftrightarrow} AX_1(\Omega) + BX_2(\Omega)}$$

7.4.2 *Time Shifting Property*

If

$$x[n] \stackrel{\text{DTFT}}{\longleftrightarrow} X(\Omega)$$

then

$$x[n-n_0] \stackrel{\text{DTFT}}{\longleftrightarrow} X(\Omega)e^{-jn_0\Omega}$$

Proof

$$\begin{aligned}
x[n] &\stackrel{\text{DTFT}}{\longleftrightarrow} X(\Omega) \\
X(\Omega) &= \sum_{n=-\infty}^{\infty} x[n]e^{-j\Omega n} \\
x[n-n_0] &\stackrel{\text{DTFT}}{\longleftrightarrow} \sum_{n=-\infty}^{\infty} x[n-n_0]e^{-j\Omega n}
\end{aligned}$$

Let $(n-n_0)=m$

$$\begin{aligned}
x[n-n_0] &= \sum_{m=-\infty}^{\infty} x[m]e^{-j\Omega(m+n_0)} \\
&= \sum_{m=-\infty}^{\infty} e^{-j\Omega n_0}x[m]e^{-j\Omega m} \\
&= e^{-j\Omega n_0}X(\Omega)
\end{aligned}$$

$$\boxed{x[n-n_0] \stackrel{\text{DTFT}}{\longleftrightarrow} e^{-j\Omega n_0}X(\Omega)}$$

7.4.3 Frequency Shifting

If

$$x[n] \stackrel{\text{DTFT}}{\longleftrightarrow} X(\Omega)$$

then

$$e^{j\Omega_0 n}x[n] \stackrel{\text{DTFT}}{\longleftrightarrow} X(\Omega-\Omega_0)$$

Proof

$$X(\Omega) = \sum_{n=-\infty}^{\infty} x[n]e^{-j\Omega n}$$

$$X(\Omega - \Omega_0) = \sum_{n=-\infty}^{\infty} x[n]e^{-j(\Omega-\Omega_0)n}$$
$$= e^{j\Omega_0 n} \sum_{n=-\infty}^{\infty} x[n]e^{-j\Omega n}$$

The right-hand side of the above equation is the DTFT of $x[n]e^{j\Omega_0 n}$. Therefore,

$$\boxed{e^{j\Omega_0 n}x[n] \overset{\text{DTFT}}{\longleftrightarrow} X(\Omega - \Omega_0)}$$

7.4.4 Time Reversal

If

$$x[n] \overset{\text{DTFT}}{\longleftrightarrow} X(\Omega)$$

then

$$x[-n] \overset{\text{DTFT}}{\longleftrightarrow} X(-\Omega)$$

Proof

$$X(\Omega) = \sum_{n=-\infty}^{\infty} x[n]e^{-j\Omega n}$$

For the time reversal signal $x[-n]$

$$X(\Omega) = \sum_{n=-\infty}^{\infty} x[-n]e^{-j\Omega n}$$

Let $-n = m$

$$X(\Omega) = \sum_{m=-\infty}^{\infty} x[m]e^{-j(-\Omega)m}$$
$$= X(-\Omega)$$

$$\boxed{x[-n] \overset{\text{DTFT}}{\longleftrightarrow} X(-\Omega)}$$

Folding in the time domain corresponds to folding in the frequency domain.

7.4.5 Time Scaling

If

$$x[n] \overset{\text{DTFT}}{\longleftrightarrow} X(\Omega)$$

then

$$x[an] \overset{\text{DTFT}}{\longleftrightarrow} X\left(\frac{\Omega}{a}\right)$$

Proof

$$X(\Omega) = \sum_{n=-\infty}^{\infty} x[an]e^{-j\Omega n}$$

Let $an = p$

$$X(\Omega) = \sum_{p=-\infty}^{\infty} x[p]e^{-j\frac{\Omega}{a}p}$$
$$= X\left(\frac{\Omega}{a}\right)$$

$$\boxed{x[an] \overset{\text{DTFT}}{\longleftrightarrow} X\left(\frac{\Omega}{a}\right)}$$

7.4.6 Multiplication by n

If

$$x[n] \overset{\text{DTFT}}{\longleftrightarrow} X(\Omega)$$

then

$$nx[n] \overset{\text{DTFT}}{\longleftrightarrow} j\frac{dX(\Omega)}{d\Omega}$$

Proof

$$X(\Omega) = \sum_{n=-\infty}^{\infty} x[n]e^{-j\Omega n}$$

Differentiating both sides with respect to Ω

$$\frac{dX(\Omega)}{d\Omega} = \sum_{n=-\infty}^{\infty} -jnx[n]e^{-j\Omega n}$$

$$j\frac{dX(\Omega)}{d\Omega} = \sum_{n=-\infty}^{\infty} \{nx[n]\}e^{-j\Omega n}$$

$$\boxed{\{nx[n]\} \overset{\text{DTFT}}{\longleftrightarrow} j\frac{dX(\Omega)}{d\Omega}}$$

7.4.7 Conjugation

If

$$x[n] \overset{\text{DTFT}}{\longleftrightarrow} X(\Omega)$$

then

$$x^*[n] \overset{\text{DTFT}}{\longleftrightarrow} X^*(-\Omega)$$

Proof

$$X(\Omega) = \sum_{n=-\infty}^{\infty} x[n]e^{-j\Omega n}$$

$$\begin{aligned}X^*(\Omega) &= \left[\sum_{n=-\infty}^{\infty} x[n]e^{-j\Omega n}\right]^* \\ &= \sum_{n=-\infty}^{\infty} x^*[n]e^{j\Omega n} \\ &= \sum_{n=-\infty}^{\infty} x^*[n]e^{-j(-\Omega)n} \\ &= X^*(-\Omega)\end{aligned}$$

$$\boxed{x^*[n] \overset{\text{DTFT}}{\longleftrightarrow} X^*(-\Omega)}$$

7.4.8 Time Convolution

If

$$x_1[n] \overset{\text{DTFT}}{\longleftrightarrow} X_1(\Omega)$$
$$x_2[n] \overset{\text{DTFT}}{\longleftrightarrow} X_2(\Omega)$$

then

$$x_1[n] * x_2[n] \overset{\text{DTFT}}{\longleftrightarrow} X_1(\Omega)X_2(\Omega)$$

Proof When the two signals $x_1[n]$ and $x_2[n]$ are convolved,

$$y[n] = x_1[n] * x_2[n]$$
$$Y(\Omega) = \sum_{m=-\infty}^{\infty} x_1[m] \sum_{n=-\infty}^{\infty} x_2[n-m]e^{-j\Omega n}$$

Let $p = n - m$

$$\begin{aligned} Y(\Omega) &= \sum_{m=-\infty}^{\infty} x_1[m] \sum_{p=-\infty}^{\infty} x_2[p]e^{-j\Omega(p+m)} \\ &= \sum_{m=-\infty}^{\infty} x_1[m]e^{-j\Omega m} \sum_{p=-\infty}^{\infty} x_2[p]e^{-j\Omega p} \\ &= X_1(\Omega)X_2(\Omega) \end{aligned}$$

$$\boxed{x_1[n] * x_2[n] \overset{\text{DTFT}}{\longleftrightarrow} X_1(\Omega)X_2(\Omega)}$$

7.4.9 Parseval's Theorem

$$\sum_{n=-\infty}^{\infty} |x[n]|^2 \overset{\text{DTFT}}{\longleftrightarrow} \frac{1}{2\pi} \int_{2\pi} |X(\Omega)|^2 d\Omega$$

The above relation states that the average power in a DT periodical signal is equal to the squared magnitude of $X(\Omega)$.

Proof

$$X^*(\Omega) = \sum_{n=-\infty}^{\infty} x^*[n]e^{j\Omega n}$$

Now

$$\begin{aligned}
\sum_{n=-\infty}^{\infty} |x[n]|^2 &= \sum_{n=-\infty}^{\infty} x^*[n]x[n] \\
&= \sum_{n=-\infty}^{\infty} x^*[n]\left[\frac{1}{2\pi}\int_{2\pi} X(\Omega)e^{j\Omega n}d\Omega\right] \\
&= \frac{1}{2\pi}\int_{2\pi} X(\Omega)\left[\sum_{n=-\infty}^{\infty} x^*[n]e^{j\Omega n}\right]d\Omega \\
&= \frac{1}{2\pi}\int_{2\pi} X(\Omega)X^*(\Omega)d\Omega \\
&= \frac{1}{2\pi}\int_{2\pi} |X(\Omega)|^2 d\Omega
\end{aligned}$$

7.4.10 Modulation Property

$$x[n]\cos(\Omega_c n + \theta) \stackrel{\text{DTFT}}{\longleftrightarrow} \frac{1}{2}\{X(\Omega - \Omega_c)e^{j\theta} + X(\Omega + \Omega_c)e^{-j\theta}\}$$

Proof From frequency shifting property, the following equation is written

$$x[n]e^{j\Omega_c n} \stackrel{\text{DTFT}}{\longleftrightarrow} X(\Omega - \Omega_c)$$

Multiplying both sides by $e^{j\theta}$, we get

$$x[n]e^{j(\Omega_c n+\theta)} \stackrel{\text{DTFT}}{\longleftrightarrow} X(\Omega - \Omega_c)e^{j\theta}$$

The above equations is generalized as

$$\boxed{x[n]\cos(\Omega_c n + \theta) \stackrel{\text{DTFT}}{\longleftrightarrow} \frac{1}{2}\{X(\Omega - \Omega_c)e^{j\theta} + X(\Omega + \Omega_c)e^{-j\theta}\}}$$

The properties of DTFT are given in Table 7.1.

Table 7.1 Properties of DTFT

Operation	$x[n]$	$X(\Omega)$
1. Linearity	$Ax_1[n]+Bx_2[n]$	$AX_1(\Omega)+BX_2(\Omega)$
2. Time shifting	$x[n-n_0]$	$X(\Omega)e^{-jn_0}$
3. Frequency shifting	$e^{j\Omega_0 n}x[n]$	$X(\Omega-\Omega_0)$
4. Time reversal	$x[-n]$	$X(-\Omega)$
5. Time scaling	$x[an]$	$X\left(\frac{\Omega}{a}\right)$
6. Multiplication by n	$nx[n]$	$j\frac{dX(\Omega)}{d\Omega}$
7. Conjugation	$x^*[n]$	$X^*(\Omega)$
8. Time convolution	$x_1[n]*x_2[n]$	$X_1(\Omega)X_2(\Omega)$
9. Parseval's theorem	$\sum_{n=-\infty}^{\infty}\lvert x[n]\rvert^2$	$\frac{1}{2\pi}\int_{2\pi}\lvert X(\Omega)\rvert^2 d\Omega$
10. Modulation	$x[n]\cos(\Omega_c n+\theta)$	$\frac{1}{2}\{X(\Omega-\Omega_c)e^{j\theta}+X(\Omega+\Omega_c)e^{-j\theta}\}$

■ Example 7.2

Find the discrete time Fourier transform of the following sequences:

1. $x[n]=e^{j\Omega_0 n}$
2. $x[n]=1 \quad \text{all } n$
3. $x[n]=\cos\Omega_0 n \quad |\Omega_0|\leq\pi$
4. $x[n]=u[n]-u[n-N]$
5. $x[n]=\begin{cases}1 & |n|\leq N\\ 0 & |n|>N\end{cases}$
6. $x[n]=a^{-n}u[-n] \quad |a|>1$
7. $x[n]=10\left(\frac{1}{6}\right)^{-n}u[-n]$
8. $x[n]=10\left(\frac{1}{6}\right)^{n}u[n]$
9. $x[n]=n\left(\frac{1}{2}\right)^{n}u[n]$
10. $x[n]=\{2,\ -1,\ 2,\ -2\}$
11. $x[n]=\begin{cases}1 & 0\leq n\leq 5\\ 0 & \text{otherwise}\end{cases}$
12. $y[n]=\left(\frac{1}{2}\right)^{n}u[n]*\left(\frac{1}{3}\right)^{n}h[n]$

13. $x[n] = (n+1)a^n u[n]$
14. $x[n] = u[n-1] - u[n-4]$
15. $x[n] = \left(\frac{1}{3}\right)^{n-1} u[n-1]$
16. $x[n] = \left(\frac{1}{4}\right)^{|n-1|} u[n-1]$
17. $x[n] = \delta[n-2] + \delta[n+2]$
18. $x[n] = \delta[n+2] - \delta[n-2]$
19. $x[n] = \sin\left(\frac{\pi}{4}n + \frac{\pi}{3}\right)$
20. $x[n] = \left(\frac{1}{4}\right)^{-n} u[-n-1]$
21. $x[n] = 10 + \cos\left(\frac{\pi}{4}n - \frac{\pi}{5}\right)$
22. $x[n] = x[2-n] + x[-2-n]$
23. $y[n] = (n-1)^2 x[n]$
24. $x[n] = \left(\frac{1}{2}\right)^{n} u[n+1]$

Solution

1. $\boldsymbol{x[n] = e^{j\Omega_0 n}}$
Consider the following DTFT

$$X(\Omega) = 2\pi\delta(\Omega - \Omega_0) \qquad |\Omega|,\ |\Omega_0| \leq \pi$$

The inverse DTFT is obtained from

$$\begin{aligned} x[n] &= \frac{1}{2\pi}\int_{-\pi}^{\pi} X(\Omega)e^{j\Omega n}d\Omega \\ &= \frac{1}{2\pi}\int_{-\pi}^{\pi} 2\pi\delta(\Omega - \Omega_0)e^{j\Omega n}d\Omega \end{aligned}$$

Using the property $\int_{-\infty}^{\infty} \Phi(\Omega)\delta(\Omega - \Omega_0)d\Omega = \Phi(\Omega_0)$, we get

$$\begin{aligned} x[n] &= \frac{1}{2\pi}[2\pi e^{j\Omega_0 n}] \\ &= e^{j\Omega_0 n} \end{aligned}$$

$$\boxed{e^{j\Omega_0 n} \overset{\text{DTFT}}{\longleftrightarrow} 2\pi\delta(\Omega - \Omega_0) \qquad |\Omega|,\ |\Omega_0| \leq \pi}$$

2. $\boldsymbol{x[n] = 1}$ **all** $\boldsymbol{n}$

$$e^{j\Omega_0 n} \overset{\text{DTFT}}{\longleftrightarrow} 2\pi\,\delta(\Omega - \Omega_0)$$

Substitute $\Omega_0 = 0$ and $x[n] = 1$

$$\boxed{1 \overset{\text{DTFT}}{\longleftrightarrow} 2\pi\,\delta(\Omega)}$$

3. $\boldsymbol{x[n] = \cos \Omega_0 n \qquad |\Omega_0| \leq \pi}$

$$\begin{aligned}
\cos \Omega_0 n &= \frac{1}{2}[e^{j\Omega_0 n} + e^{-j\Omega_0 n}] \\
e^{j\Omega_0 n} &\overset{\text{DTFT}}{\longleftrightarrow} 2\pi\,\delta(\Omega - \Omega_0) \\
e^{-j\Omega_0 n} &\overset{\text{DTFT}}{\longleftrightarrow} 2\pi\,\delta(\Omega + \Omega_0)
\end{aligned}$$

$$\boxed{\cos \Omega_0 n \overset{\text{DTFT}}{\longleftrightarrow} \pi\,[\delta(\Omega - \Omega_0) + \delta(\Omega + \Omega_0)] \qquad |\Omega|,\ |\Omega_0| \leq \pi}$$

4. $\boldsymbol{x[n] = u[n] - u[n - N]}$

$$X(\Omega) = \sum_{n=0}^{N-1} x[n]e^{-j\Omega n}$$

Using the summation formula

$$\sum_{n=0}^{N-1} a^n = \frac{(1 - a^N)}{(1 - a)}$$

we get

$$\begin{aligned}
X(\Omega) &= \frac{1 - e^{-j\Omega N}}{1 - e^{-j\Omega}} \\
&= \frac{e^{-j\frac{\Omega N}{2}}\left[e^{j\frac{\Omega N}{2}} - e^{-j\frac{\Omega N}{2}}\right]}{e^{-j\frac{\Omega}{2}}\left[e^{j\frac{\Omega}{2}} - e^{-j\frac{\Omega}{2}}\right]}
\end{aligned}$$

$$\boxed{X(\Omega) = e^{-j\Omega\frac{(N-1)}{2}}\,\frac{\sin\left(\frac{\Omega N}{2}\right)}{\sin\left(\frac{\Omega}{2}\right)}}$$

5. $\boldsymbol{x[n] = \begin{cases} 1 & |n| \le N \\ 0 & |n| > N \end{cases}}$

$$\begin{aligned} X(\Omega) &= \sum_{-N}^{N} e^{-j\Omega n} \\ &= \sum_{-N}^{-1} e^{-j\Omega n} + \sum_{0}^{N} e^{-j\Omega n} \\ &= e^{j\Omega} \left[\frac{(1 - e^{j\Omega N})}{(1 - e^{j\Omega})} \right] + \left[\frac{(1 - e^{-j\Omega (N+1)})}{(1 - e^{-j\Omega})} \right] \end{aligned}$$

$$\begin{aligned} X(\Omega) &= \left[\frac{e^{j\Omega} - e^{j\Omega (N+1)}}{(1 - e^{-j\Omega})} \right] + \left[\frac{1 - e^{j\Omega (N+1)}}{(1 - e^{-j\Omega})} \right] \\ &= \frac{[e^{j\Omega} - 1 - e^{j\Omega (N+1)} + e^{j\Omega N} + 1 - e^{-j\Omega (N+1)} - e^{j\Omega} + e^{-j\Omega N}]}{1 - (e^{j\Omega} + e^{-j\Omega}) + 1} \\ &= \frac{2 \cos \Omega N - 2 \cos \Omega (N + 1)}{2(1 - cos\Omega)} \\ &= \frac{\sin \left(N + \frac{1}{2}\right) \Omega \sin \frac{\Omega}{2}}{\sin^2 \frac{\Omega}{2}} \end{aligned}$$

$$\boxed{X(\Omega) = \frac{\sin \left(N + \frac{1}{2}\right) \Omega}{\sin \left(\frac{\Omega}{2}\right)}}$$

6. $\boldsymbol{x[n] = a^{-n} u[-n] \qquad |a| < 1}$

$$\begin{aligned} X(\Omega) &= \sum_{n=-\infty}^{0} a^{-n} e^{-j\Omega n} \\ &= \sum_{n=0}^{\infty} (a e^{j\Omega})^n \end{aligned}$$

Using summation formula, we get

$$\boxed{X(\Omega) = \frac{1}{(1 - a e^{j\Omega})}}$$

7. $\boldsymbol{x[n] = 10\left(\frac{1}{6}\right)^{-n} u[-n]}$

$$X(\Omega) = \sum_{n=-\infty}^{0} 10\left(\frac{1}{6}\right)^{-n} e^{-j\Omega n}$$
$$= 10\sum_{n=0}^{\infty}\left(\frac{1}{6}e^{j\Omega n}\right)^{n}$$

$$\boxed{X(\Omega) = \frac{10}{\left(1-\frac{1}{6}e^{j\Omega}\right)}}$$

8. $\boldsymbol{x[n] = 10\left(\frac{1}{6}\right)^{-n} u[n]}$

$$X(\Omega) = \sum_{n=0}^{\infty} 10\left(\frac{1}{6}e^{j\Omega}\right)^{-n}$$
$$= \sum_{n=0}^{\infty} 10\left(6e^{-j\Omega}\right)^{n}$$

RHS of the equation is not summable and $x[n]$ does not have DTFT.

9. $\boldsymbol{x[n] = n\left(\frac{1}{2}\right)^{n} u[n]}$ **(Using multiplication property by $\boldsymbol{n}$)**

$$\left(\frac{1}{2}\right)^{n} \overset{\text{DTFT}}{\longleftrightarrow} \frac{1}{1-\left(\frac{1}{2}\right)e^{-j\Omega}}$$
$$= \frac{e^{j\Omega}}{(e^{j\Omega}-0.5)}$$
$$n\left(\frac{1}{2}\right)^{n} u[n] \overset{\text{DTFT}}{\longleftrightarrow} j\frac{d}{d\Omega}\left[\frac{e^{j\Omega}}{(e^{j\Omega}-0.5)}\right]$$
$$= \frac{j(e^{j\Omega}-0.5)e^{j\Omega}(j) - e^{j\Omega}e^{j\Omega}(j)}{(e^{j\Omega}-0.5)^2}$$
$$= \frac{0.5e^{j\Omega}}{(e^{j\Omega}-0.5)^2}$$

$$\boxed{n\left(\frac{1}{2}\right)^{n} u[n] \overset{\text{DTFT}}{\longleftrightarrow} \frac{0.5e^{j\Omega}}{(e^{j\Omega}-0.5)^2}}$$

10. $\mathbf{x[n] = \{2, -1, 2, -2\}}$

$$\boxed{x(\Omega) = \{2 - e^{-j\Omega} + 2e^{-j2\Omega} - 2e^{-j3\Omega}\}}$$

11. $\mathbf{x[n] = \begin{cases} 1 & 0 \le n \le 5 \\ 0 & \text{otherwise} \end{cases}}$ (*Anna University, April, 2005*).

$$X(\Omega) = \frac{\sin \Omega \left(N + \frac{1}{2}\right)}{\sin \frac{\Omega}{2}}$$

where $N = 5$

$$\boxed{X(\Omega) = \frac{\sin 5.5\Omega}{\sin 0.5\Omega}}$$

12. $\mathbf{y[n] = \left(\frac{1}{2}\right)^n u[n] * \left(\frac{1}{3}\right)^n u[n]}$

$$\begin{aligned} X(\Omega) &= \frac{1}{\left(1 - \frac{1}{2}e^{-j\Omega}\right)} \\ H(\Omega) &= \frac{1}{\left(1 - \frac{1}{3}e^{-j\Omega}\right)} \\ Y(\Omega) &= X(\Omega)H(\Omega) \end{aligned}$$

$$\boxed{Y(\Omega) = \frac{1}{\left(1 - \frac{1}{2}e^{-j\Omega}\right)\left(1 - \frac{1}{3}e^{-j\Omega}\right)}}$$

To find $y[n]$, put $Y(\Omega)$ in partial fraction and take IDTFT.

13. $\mathbf{x[n] = (n + 1)a^n u[n]}$

$$\begin{aligned} x[n] &= na^n u[n] + a^n u[n] \\ na^n u[n] &\overset{\text{DTFT}}{\longleftrightarrow} \frac{ae^{j\Omega}}{(e^{j\Omega} - a)^2} \\ a^n u[n] &\overset{\text{DTFT}}{\longleftrightarrow} \frac{e^{j\Omega}}{(e^{j\Omega} - a)} \\ x[n] &\overset{\text{DTFT}}{\longleftrightarrow} \frac{ae^{j\Omega}}{(e^{j\Omega} - a)^2} + \frac{e^{j\Omega}}{(e^{j\Omega} - a)} \\ &= \frac{e^{j\Omega}(a + e^{j\Omega} - a)}{(e^{j\Omega} - a)^2} \end{aligned}$$

$$= \frac{e^{j2\Omega}}{(e^{j\Omega} - a)^2}$$
$$= \frac{1}{(1 - ae^{-j\Omega})^2}$$

$$\boxed{(n+1)a^n u[n] \overset{\text{DTFT}}{\longleftrightarrow} \frac{1}{(1 - ae^{-j\Omega})^2}}$$

14. $\boldsymbol{x[n] = u[n-1] - u[n-4]}$

$$\boxed{X(\Omega) = e^{-j\Omega} + e^{-j2\Omega} + e^{-j3\Omega}}$$

15. $\boldsymbol{x[n] = \left(\frac{1}{3}\right)^{n-1} u[n-1]}$

$$\left(\frac{1}{3}\right)^n u[n] \overset{\text{DTFT}}{\longleftrightarrow} \frac{1}{1 - \frac{1}{3}e^{-j\Omega}}$$

Using right shift time shifting property, we get

$$\left(\frac{1}{3}\right)^{n-1} u[n-1] \overset{\text{DTFT}}{\longleftrightarrow} \frac{e^{-j\Omega}}{\left(1 - \frac{1}{3}e^{-j\Omega}\right)} = \frac{1}{\left(e^{j\Omega} - \frac{1}{3}\right)}$$

$$\boxed{\left(\frac{1}{3}\right)^{n-1} u[n-1] \overset{\text{DTFT}}{\longleftrightarrow} \frac{1}{\left(e^{j\Omega} - \frac{1}{3}\right)}}$$

16. $\boldsymbol{x[n] = \left(\frac{1}{4}\right)^{|n-1|} u[n-1]}$

From Example 7.1.5,

$$x[n] = (a)^{|n|} \overset{\text{DTFT}}{\longleftrightarrow} \frac{(1 - a^2)}{(1 - 2a\cos\Omega + a^2)}$$

Substitute $|a| = \frac{1}{4}$

$$\left(\frac{1}{4}\right)^{|n|} \overset{\text{DTFT}}{\longleftrightarrow} \frac{(15/16)}{(17/16) - 0.5\cos\Omega}$$
$$= \frac{15}{17 - 8\cos\Omega}$$

Using right shift time shifting property, we get

$$\boxed{\left(\frac{1}{4}\right)^{|n-1|} \overset{\text{DTFT}}{\longleftrightarrow} \frac{15e^{-j\Omega}}{17-8\cos\Omega}}$$

17. $\boldsymbol{x[n] = \delta[n-2] + \delta[n+2]}$

$$\delta[n-2] + \delta[n+2] \overset{\text{DTFT}}{\longleftrightarrow} e^{-j2\Omega} + e^{j2\Omega}$$

$$\boxed{\delta[n-2] + \delta[n+2] \overset{\text{DTFT}}{\longleftrightarrow} 2\cos 2\Omega}$$

18. $\boldsymbol{x[n] = \delta[n+2] + \delta[n-2]}$

$$\begin{aligned}\delta[n+2] - \delta[n-2] &\overset{\text{DTFT}}{\longleftrightarrow} e^{j2\Omega} - e^{-j2\Omega}\\ &= j2\sin 2\Omega\end{aligned}$$

$$\boxed{\delta[n+2] - \delta[n-2] \overset{\text{DTFT}}{\longleftrightarrow} j2\sin 2\Omega}$$

19. $\boldsymbol{x[n] = \sin\left(\frac{\pi}{4}n + \frac{\pi}{3}\right)}$

$$\begin{aligned}\sin\left(\frac{\pi}{4}n + \frac{\pi}{3}\right) &= \frac{e^{j\left(\frac{\pi}{4}n+\frac{\pi}{3}\right)} - e^{-j\left(\frac{\pi}{4}n+\frac{\pi}{3}\right)}}{2j}\\ &= \frac{1}{2j}[e^{j\frac{\pi}{3}}e^{j\frac{\pi}{4}n} - e^{-j\frac{\pi}{3}}e^{-j\frac{\pi}{4}n}]\end{aligned}$$

From Example 7.2.1, it is derived that

$$\begin{aligned}e^{j\Omega_0 n} &\overset{\text{DTFT}}{\longleftrightarrow} 2\pi\delta(\Omega-\Omega_0)\\ \therefore \quad e^{j\frac{\pi}{4}n} &\overset{\text{DTFT}}{\longleftrightarrow} 2\pi\delta\left(\Omega-\frac{\pi}{4}\right)\\ e^{-j\frac{\pi}{4}n} &\overset{\text{DTFT}}{\longleftrightarrow} 2\pi\delta\left(\Omega+\frac{\pi}{4}\right)\end{aligned}$$

$$\sin\left(\frac{\pi}{4}n + \frac{\pi}{3}\right) \overset{\text{DTFT}}{\longleftrightarrow} \frac{2\pi}{2j}\left[e^{j\frac{\pi}{3}}\delta\left(\Omega-\frac{\pi}{4}\right) - e^{-j\frac{\pi}{3}}\delta\left(\Omega+\frac{\pi}{4}\right)\right]$$

$$\boxed{\sin\left(\frac{\pi}{4}n + \frac{\pi}{3}\right) \overset{\text{DTFT}}{\longleftrightarrow} \frac{\pi}{j}\left[e^{j\frac{\pi}{3}}\delta\left(\Omega-\frac{\pi}{4}\right) - e^{-j\frac{\pi}{3}}\delta\left(\Omega+\frac{\pi}{4}\right)\right]}$$

20. $\boldsymbol{x[n] = \left(\frac{1}{4}\right)^{-n} u[-n-1]}$

$$x[n] = \frac{1}{4}\left(\frac{1}{4}\right)^{-n-1} u[-n-1]$$

Using the reversal and left time shift, we get

$$\begin{aligned}\left(\frac{1}{4}\right)^{-n-1} u[-n-1] &\overset{\text{DTFT}}{\longleftrightarrow} \frac{e^{j\Omega}}{1-\frac{1}{4}e^{j\Omega}} \\ &= \frac{1}{\left(e^{-j\Omega}-\frac{1}{4}\right)}\end{aligned}$$

$$\boxed{\left(\frac{1}{4}\right)^{-n} u[-n-1] \overset{\text{DTFT}}{\longleftrightarrow} \frac{1}{4}\frac{1}{\left(e^{-j\Omega}-\frac{1}{4}\right)}}$$

21. $\boldsymbol{x[n] = 10 + \cos\left(\frac{\pi}{4}n - \frac{\pi}{5}\right)}$

$$\begin{aligned}\cos\left(\frac{\pi}{4}n - \frac{\pi}{5}\right) &= \frac{1}{2}[e^{-j\frac{\pi}{5}}e^{j\frac{\pi}{4}n} + e^{j\frac{\pi}{5}}e^{-j\frac{\pi}{4}n}] \\ e^{j\frac{\pi}{4}n} &\overset{\text{DTFT}}{\longleftrightarrow} 2\pi\delta\left(\Omega-\frac{\pi}{4}\right) \\ e^{-j\frac{\pi}{4}n} &\overset{\text{DTFT}}{\longleftrightarrow} 2\pi\delta\left(\Omega+\frac{\pi}{4}\right) \\ 10 &\overset{\text{DTFT}}{\longleftrightarrow} 20\pi\delta(\Omega)\end{aligned}$$

$$\boxed{10+\cos\left(\frac{\pi}{4}-\frac{\pi}{5}\right) \overset{\text{DTFT}}{\longleftrightarrow} 20\pi\delta(\Omega) + \pi\left[e^{-j\frac{\pi}{5}}\delta\left(\Omega-\frac{\pi}{4}\right) + e^{j\frac{\pi}{5}n}\delta\left(\Omega+\frac{\pi}{4}\right)\right]}$$

22. $\boldsymbol{x[n] = x[2-n] + x[-2-n]}$

$$\begin{aligned}x[2-n] &\overset{\text{DTFT}}{\longleftrightarrow} X(\Omega)e^{-j2\Omega} \quad \text{(Right shift)} \\ x[-2-n] &\overset{\text{DTFT}}{\longleftrightarrow} X(\Omega)e^{j2\Omega} \quad \text{(Left shift)} \\ \{x[2-n] + x[-2-n]\} &\overset{\text{DTFT}}{\longleftrightarrow} X(\Omega)[e^{-j2\Omega} + e^{j2\Omega}] \\ &= 2X(\Omega)\cos 2\Omega\end{aligned}$$

$$\boxed{\{x[2-n] + x[-2-n]\} \overset{\text{DTFT}}{\longleftrightarrow} 2X(\Omega)\cos 2\Omega}$$

23. $\boldsymbol{y[n] = (n-1)^2 x[n]}$ *(Anna University, April, 2008)*

$$\begin{aligned} y[n] &= (n^2 - 2n + 1)x[n] \\ &= n^2 x[n] - 2nx[n] + x[n] \end{aligned}$$

$$n^2 x[n] \overset{\text{DTFT}}{\longleftrightarrow} (j)^2 \frac{d^2 X(\Omega)}{d\Omega^2}$$

$$nx[n] \overset{\text{DTFT}}{\longleftrightarrow} j\frac{dX(\Omega)}{d\Omega}$$

$$\boxed{Y(\Omega) = -\frac{d^2 X(\Omega)}{d\Omega^2} - 2j\frac{dX(\Omega)}{d\Omega} + X(\Omega)}$$

24. $\boldsymbol{x[n] = \left(\frac{1}{2}\right)^n u[n+1]}$

$$x[n] = 2\left(\frac{1}{2}\right)^{n+1} u[n+1]$$

Making left shift of $(\frac{1}{2})^n$ we get

$$\left(\frac{1}{2}\right)^{n+1} u[n+1] \overset{\text{DTFT}}{\longleftrightarrow} \frac{e^{j\Omega}}{\left(1 - \frac{1}{2}e^{-j\Omega}\right)}$$

$$\boxed{X(\Omega) = \frac{2e^{j\Omega}}{\left(1 - \frac{1}{2}e^{-j\Omega}\right)}}$$

The above result can be obtained from first principle as follows (Figs. 7.2 and 7.3):

$$\begin{aligned} X(\Omega) &= \sum_{n=-1}^{\infty} \left(\frac{1}{2}\right)^n e^{-j\Omega n} \\ &= \left(\frac{1}{2}\right)^{-1} e^{j\Omega} + \sum_{n=0}^{\infty} \left(\frac{1}{2}\right)^n e^{-j\Omega n} \\ &= 2e^{j\Omega} + \sum_{n=0}^{\infty} \left(\frac{1}{2e^{j\Omega}}\right)^n \\ &= 2e^{j\Omega} + \frac{1}{\left(1 - \frac{1}{2}e^{-j\Omega}\right)} \end{aligned}$$

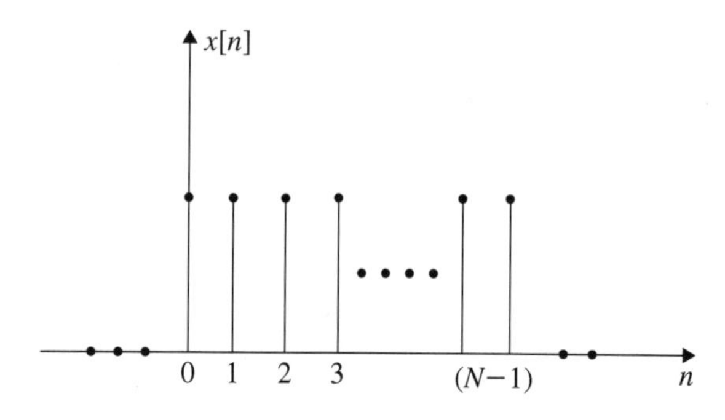

Fig. 7.2 DT sequence of Example 7.2.4

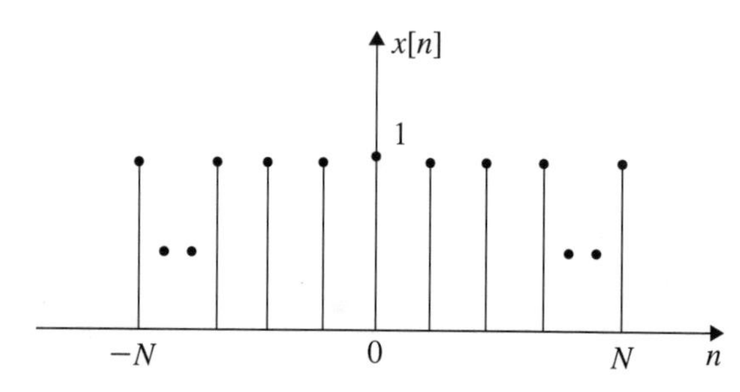

Fig. 7.3 Rectangular pulse $x[n]$ for Example 7.2.5

$$= \frac{2e^{j\Omega} - 1 + 1}{\left(1 - \frac{1}{2}e^{-j\Omega}\right)}$$

$$\boxed{X(\Omega) = \frac{2e^{j\Omega}}{\left(1 - \frac{1}{2}e^{-j\Omega}\right)}}$$

7.5 Inverse Discrete Time Fourier Transform (IDTFT)

If $x[n]$ is given, the discrete time Fourier transform is obtained using Eq. (7.11) which is given below.

$$X(\Omega) = \sum_{n=-\infty}^{\infty} x[n]e^{-j\Omega n} \tag{7.17}$$

If $X(\Omega)$ is given, then the sequence $x[n]$ is obtained from Eq. (7.10) which is given as

Table 7.2 DTFT pair

No.	$x[n]$	$X(\Omega)$
1.	$\delta[n]$	1
2.	$u[n]$	$\pi\delta(\Omega) + \dfrac{1}{(1 - e^{-j\Omega})}$
3.	$a^n u[n]$	$\dfrac{1}{(1 - ae^{-j\Omega})}$ $\quad |a| < 1$
4.	$-a^n u[-n-1]$	$\dfrac{1}{(1 - ae^{-j\Omega})}$ $\quad |a| > 1$
5.	$a^{|n|}$	$\dfrac{(1 - a^2)}{(1 - 2a\cos\Omega + a^2)}$
6.	1	$2\pi\delta(\Omega)$
7.	$\cos\Omega_0 n$	$\pi[\delta(\Omega - \Omega_0) + \delta(\Omega + \Omega_0)]$ $\quad |\Omega|, |\Omega_0| \le \pi$
8.	$\sin\Omega_0 n$	$j\pi[\delta(\Omega + \Omega_0) - \delta(\Omega - \Omega_0)]$ $\quad |\Omega|, |\Omega_0| \le \pi$
9.	$u[n] - u[n-N]$	$e^{-j\Omega\left(\frac{N-1}{2}\right)} \dfrac{\sin(\Omega(N/2))}{\sin(\Omega/2)}$
10.	rect pulse	$\dfrac{\sin\left(\left(N + \frac{1}{2}\right)\Omega\right)}{\sin(\Omega/2)}$

$$x[n] = \frac{1}{2\pi}\int_{2\pi} X(\Omega)e^{j\Omega n}d\Omega \tag{7.18}$$

$x[n]$ is also obtained by putting $X(\Omega)$ by partial fraction and making use of DTFT pair in Table 7.2. The process of getting $x[n]$ from $X(\Omega)$ is called IDTFT. This is illustrated in the following examples.

■ Example 7.3

Find $x[n]$ for the $X(\Omega)$ given below.

1. $$X(\Omega) = 8\pi\delta(\Omega) + 10\pi\delta\left(\Omega - \frac{\pi}{4}\right) + 10\pi\delta\left(\Omega + \frac{\pi}{4}\right)$$

2. $$X(\Omega) = \begin{cases} j & 0 < \Omega \le \pi \\ -j & -\pi < \Omega \le 0 \end{cases}$$

3. $$X(\Omega) = \begin{cases} 2 & 0 \le |\Omega| < \frac{\pi}{3} \\ 0 & \frac{\pi}{3} \le |\Omega| \le \pi \end{cases}$$

   and $\angle X(\Omega) = -\dfrac{4}{3}\Omega$

4. $$X(\Omega) = 6 + 4e^{-j\Omega} + 7e^{-j2\Omega}$$

5. $$x[n] = \left(\frac{1}{2}\right)^n u[n]$$
$$h[n] = \left(\frac{1}{3}\right)^n u[n]$$
$$y[n] = x[n] * h[n]$$

6. $$X(\Omega) = j\Omega$$

Solution

1. $\boldsymbol{X(\Omega) = 8\pi\delta(\Omega) + 10\pi\delta\left(\Omega - \frac{\pi}{4}\right) + 10\pi\delta\left(\Omega + \frac{\pi}{4}\right)}$

$$2\pi\delta(\Omega) \overset{\text{IDTFT}}{\longleftrightarrow} 1$$
$$\pi\delta\left(\Omega - \frac{\pi}{4}\right) + \pi\delta\left(\Omega + \frac{\pi}{4}\right) \overset{\text{DTFT}}{\longleftrightarrow} \cos\frac{\pi}{4}n$$

$$\boxed{x[n] = 4 + 10\cos\frac{\pi}{4}n}$$

2. $\boldsymbol{X(\Omega) = \begin{cases} j & 0 < \Omega \le \pi \\ -j & -\pi < \Omega \le 0 \end{cases}}$

$$\begin{aligned} x[n] &= \frac{1}{2\pi}\left\{\int_{-\pi}^{0} -je^{j\Omega n}d\Omega + \int_{0}^{\pi} je^{j\Omega n}d\Omega\right\} \\ &= \frac{j}{2\pi}\left\{\left[-\frac{1}{jn}e^{j\Omega n}\right]_{-\pi}^{0} + \left[\frac{1}{jn}e^{j\Omega n}\right]_{0}^{\pi}\right\} \\ &= \frac{j}{2\pi}\left\{-\frac{1}{jn} + \frac{1}{jn}e^{-j\pi n} + \frac{1}{jn}e^{j\pi n} - \frac{1}{jn}\right\} \\ &= \frac{1}{2\pi n}\left\{-2 + e^{j\pi n} + e^{-j\pi n}\right\} \\ &= \frac{1}{\pi n}\left\{-1 + \cos\pi n\right\} \end{aligned}$$

$$\boxed{x[n] = -\frac{2}{n\pi}\sin^2\frac{n\pi}{2}}$$

3. $\boldsymbol{X(\Omega) = \begin{cases} 2 & 0 \le |\Omega| < \frac{\pi}{3} \\ 0 & \frac{\pi}{3} \le |\Omega| \le \pi \end{cases}}$
and $\boldsymbol{\angle X(\Omega) = -\frac{4}{3}\Omega}$

$$\begin{aligned}
\angle X(\Omega) &= -\frac{4}{3}\Omega \\
&= e^{-j\frac{4}{3}\Omega} \\
x[n] &= \frac{2}{2\pi}\left\{\int_{-\frac{\pi}{3}}^{0} e^{-j\frac{4}{3}\Omega} e^{j\Omega n} d\Omega + \int_{0}^{\frac{\pi}{3}} e^{-j\frac{4}{3}\Omega} e^{j\Omega n} d\Omega\right\} \\
&= \frac{1}{\pi}\left\{\int_{-\frac{\pi}{3}}^{0} e^{j(n-\frac{4}{3})\Omega} d\Omega + \int_{0}^{\frac{4}{3}} e^{j(n-\frac{4}{3})\Omega} d\Omega\right\} \\
&= \frac{1}{j\pi}\frac{1}{\left(n-\frac{4}{3}\right)}\left[e^{j(n-\frac{4}{3})\Omega}\right]_{-\frac{\pi}{3}}^{0} + \left[e^{j(n-\frac{4}{3})\Omega}\right]_{0}^{\frac{\pi}{3}} \\
&= \frac{1}{j\pi\left(n-\frac{4}{3}\right)}\left[1 - e^{-j(n-\frac{4}{3})\frac{\pi}{3}} - 1 + e^{j(n-\frac{4}{3})\frac{\pi}{3}}\right]
\end{aligned}$$

$$\boxed{x[n] = \frac{2}{\pi\left(n-\frac{4}{3}\right)} \sin\left(n-\frac{4}{3}\right)\frac{\pi}{3}}$$

4. $\boldsymbol{X(\Omega) = 6 + 4e^{-j\Omega} + 7e^{-j2\Omega}}$

$$x[n] = \{6,\ 4,\ 7\}$$

5. $\boldsymbol{x[n] = \left(\frac{1}{2}\right)^n u[n]}$
   $\boldsymbol{h[n] = \left(\frac{1}{3}\right)^n u[n]}$
   $\boldsymbol{y[n] = x[n] * h[n]}$

$$\begin{aligned}
X(\Omega) &= \frac{1}{\left(1-\frac{1}{2}e^{-j\Omega}\right)} \\
H(\Omega) &= \frac{1}{\left(1-\frac{1}{3}e^{-j\Omega}\right)} \\
Y(\Omega) &= \frac{1}{\left(1-\frac{1}{2}e^{-j\Omega}\right)\left(1-\frac{1}{3}e^{-j\Omega}\right)} \\
&= \frac{A_1}{\left(1-\frac{1}{2}e^{-j\Omega}\right)} + \frac{A_2}{\left(1-\frac{1}{3}e^{-j\Omega}\right)} \\
1 &= A_1\left(1-\frac{1}{3}e^{-j\Omega}\right) + A_2\left(1-\frac{1}{2}e^{-j\Omega}\right)
\end{aligned}$$

Substitute $e^{-j\Omega} = 2$; we get $A_1 = 3$ and substitute $e^{-j\Omega} = 3$; we get $A_2 = -2$

$$Y(\Omega) = \frac{3}{\left(1 - \frac{1}{2}e^{-j\Omega}\right)} - \frac{2}{\left(1 - \frac{1}{3}e^{-j\Omega}\right)}$$

Taking inverse Fourier transform, we get

$$\boxed{Y[n] = 3\left(\frac{1}{2}\right)^n u[n] - 2\left(\frac{1}{3}\right)^n u[n]}$$

6. $\boldsymbol{X(\Omega) = j\Omega}$

$$\begin{aligned} x[n] &= \frac{1}{2\pi}\int_{-\pi}^{\pi} j\Omega e^{j\Omega n} d\Omega \\ &= \frac{1}{2\pi}\left[\frac{j\Omega}{jn} e^{j\Omega n}\right]_{-\pi}^{\pi} \\ &= \frac{1}{2\pi n}\left[\pi e^{j\pi n} + \pi e^{-j\pi n}\right] \\ &= \frac{1}{n}\cos \pi n \end{aligned}$$

$$\boxed{x[n] = \frac{(-1)^n}{n} \qquad n \neq 0}$$

7.6 LTI System Characterized by Difference Equation

LTIDT systems are described by linear constant coefficients differential equations of the form

$$\sum_{k=0}^{N} a_k y[n-k] = \sum_{k=0}^{M} b_k x[n-k] \tag{7.19}$$

where $M \leq N$. Taking Fourier transform of both sides of Eq. (7.19) and using the time shifting property, we get

$$\sum_{k=0}^{N} a_k e^{-jk\Omega} Y(\Omega) = \sum_{k=0}^{M} b_k e^{-jk\Omega} X(\Omega) \tag{7.20}$$

$$H(\Omega) = \frac{Y(\Omega)}{X(\Omega)} = \frac{\sum_{k=0}^{M} b_k e^{-jk\Omega}}{\sum_{k=0}^{N} a_k e^{-jk\Omega}} \tag{7.21}$$

In Eq. (7.21) $H(\Omega)$ is called the system transfer function and for any input $X(\Omega)$, the output $y[n]$ can be obtained by taking IDTFT. Further, $H(\Omega)$ gives the frequency responds of DT system which is periodic and is expressed as

$$H(\Omega) = H(\Omega + 2\pi) \tag{7.22}$$

Unlike continuous time system, for a DT system the frequency response is observed for the frequency range $0 \leq \Omega < 2\pi$ or $-\pi \leq \Omega < \pi$.

■ Example 7.4

Consider the system consisting of the cascade of two LTI systems with frequency responses

$$H_1(e^{j\omega}) = \frac{2 - e^{j\omega}}{\left(1 + \frac{1}{2}e^{-j\omega}\right)}$$

and

$$H_2(e^{j\omega}) = \frac{1}{\left(1 - \frac{1}{2}e^{-j\omega} + \frac{1}{4}e^{-j2\omega}\right)}$$

Find the difference equation describing the overall system.

(Anna University, April, 2008)

Solution **Note:** As stated earlier, symbols used here are different from the symbols used in this text book. Their equivalence are

$$\begin{aligned}
\Omega &= \omega \\
H(\Omega) &= H(e^{j\omega}) \\
H(e^{j\omega}) &= H_1(e^{j\omega})H_2(e^{j\omega}) \\
&= \frac{(2 - e^{-j\omega})}{\left(1 + \frac{1}{2}e^{-j\omega}\right)\left(1 - \frac{1}{2}e^{-j\omega} + \frac{1}{4}e^{-j2\omega}\right)} \\
\frac{Y(e^{j\omega})}{X(e^{j\omega})} &= \frac{2 - e^{-j\omega}}{\left(1 + \frac{1}{8}e^{-j3\omega}\right)} \\
\left(1 + \frac{1}{8}e^{-j3\omega}\right)Y(e^{j\omega}) &= (2 - e^{-j\omega})X(e^{j\omega})
\end{aligned}$$

$$\boxed{y[n] + \frac{1}{8}y[n-3] = 2x[n] - x[n-1]}$$

■ Example 7.5

Find the impulse response of the discrete time system described by the difference equation

$$y[n-2]-3y[n-1]+2y[n]=x(n-1)$$

(Anna University, April, 2005)

Solution Taking Fourier transform for the both sides of given difference equation, we get

$$[e^{-j2\Omega}-3e^{-j\Omega}+2]Y(\Omega)=e^{-j\Omega}X(\Omega)$$

$$Y(\Omega)=\frac{e^{-j\Omega}X(\Omega)}{[e^{-j2\Omega}-3e^{-j\Omega}+2]}$$

For an impulse $X(\Omega)=1$

$$[e^{-j2\Omega}-3e^{-j\Omega}+2]Y(\Omega)=(e^{-j\Omega}-1)(e^{-j\Omega}-2)$$

$$Y(\Omega)=\frac{e^{-j\Omega}}{(e^{-j\Omega}-1)(e^{-j\Omega}-2)}$$

$$=\frac{A_1}{(e^{-j\Omega}-1)}+\frac{A_2}{(e^{-j\Omega}-2)}$$

$$e^{-j\Omega}=A_1(e^{-j\Omega}-2)+A_2(e^{-j\Omega}-1)$$

Put $e^{-j\Omega}=1$

$$1=A_1(1-2);\quad A_1=-1$$

Put $e^{-j\Omega}=2$

$$2=A_2(2-1);\quad A_2=2$$

$$Y(\Omega)=\frac{-1}{(e^{-j\Omega}-1)}+\frac{2}{(e^{-j\Omega}-2)}$$

$$=\frac{1}{(1-e^{-j\Omega})}+\frac{1}{\left(1-\frac{1}{2}e^{-j\Omega}\right)}$$

Taking inverse discrete Fourier transform, we get

$$\boxed{y[n] = \left[1 - \left(\frac{1}{2}\right)^n\right] u[n]}$$

■ Example 7.6

Find the DTFT of

$$x[n] = \left(\frac{1}{2}\right)^n u[n]$$

and plot the spectrum.

(Anna University, April, 2005)

Solution

$$\begin{aligned}
X(\Omega) &= \frac{1}{\left(1 - \frac{1}{2}e^{-j\Omega}\right)} \\
&= \frac{1}{\left(1 - \frac{1}{2}\cos\Omega + \frac{1}{2}j\sin\Omega\right)} \\
&= \frac{1\angle -\tan^{-1}\sin\Omega/(1 - \frac{1}{2}\cos\Omega)}{\sqrt{\left(1 - \frac{1}{2}\cos\Omega\right)^2 + \frac{1}{4}\sin^2\Omega}} \\
&= \frac{1\angle -\tan^{-1}\sin\Omega/(1 - \frac{1}{2}\cos\Omega)}{\sqrt{\left(\frac{5}{4} - \cos\Omega\right)}}
\end{aligned}$$

Ω	$-\pi$	$-\frac{\pi}{2}$	0	$\frac{\pi}{2}$	π
$\|X(\Omega)\|$	0.667	0.894	2	0.894	0.667
$\angle X(\Omega)$	0	26.6°	0	−26.6°	0

The frequency spectrum is shown in Fig. 7.4.

■ Example 7.7

Use Fourier transform to find the output of the system whose impulse response $h[n] = \left(\frac{1}{3}\right)^n u[n]$ and the input to the system is $x[n] = \left(\frac{1}{2}\right)^n u[n]$.

(Anna University, May, 2007)

Solution

$$H(\Omega) = \frac{1}{\left(1 - \frac{1}{3}e^{-j\Omega}\right)}$$

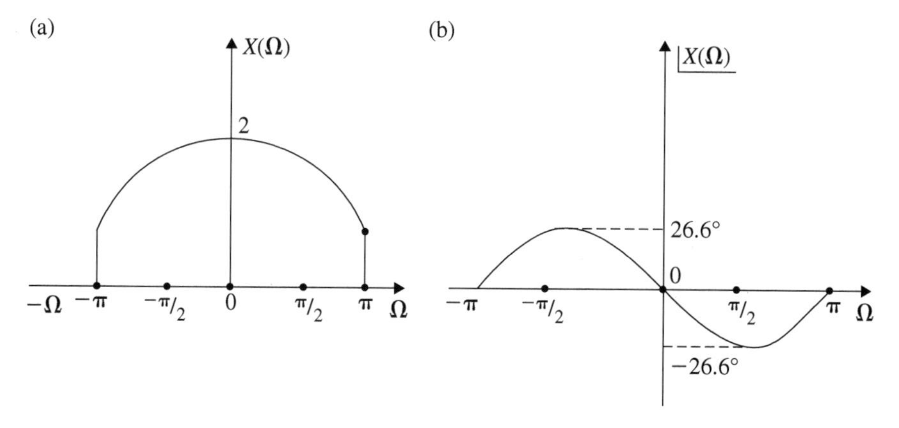

Fig. 7.4 Frequency spectra for Example 7.5. **a** Amplitude spectrum and **b** Phase spectrum

$$\begin{aligned}
X(\Omega) &= \frac{1}{\left(1-\frac{1}{2}e^{-j\Omega}\right)} \\
H(\Omega) &= \frac{Y(\Omega)}{X(\Omega)} \\
Y(\Omega) &= H(\Omega)X(\Omega) \\
&= \frac{1}{\left(1-\frac{1}{3}e^{-j\Omega}\right)\left(1-\frac{1}{2}e^{-j\Omega}\right)} \\
&= \frac{A_1}{\left(1-\frac{1}{3}e^{-j\Omega}\right)} + \frac{A_2}{\left(1-\frac{1}{2}e^{-j\Omega}\right)} \\
1 &= A_1\left(1-\frac{1}{2}e^{-j\Omega}\right) + A_2\left(1-\frac{1}{3}e^{-j\Omega}\right)
\end{aligned}$$

Put $e^{-j\Omega} = 3$

$$1 = A_1\left(-\frac{1}{2}\right); \qquad A_1 = -2$$

Put $e^{-j\Omega} = 2$

$$1 = A_2\left(\frac{1}{3}\right); \qquad A_2 = 3$$

$$Y(\Omega) = \frac{-2}{\left(1-\frac{1}{3}e^{-j\Omega}\right)} + \frac{3}{\left(1-\frac{1}{2}e^{-j\Omega}\right)}$$

$$y[n] = \left[-2\left(\frac{1}{3}\right)^n + 3\left(\frac{1}{2}\right)^n\right]u[n]$$

■ Example 7.8

Given

$$x[n] = \{1, 2, \underset{\uparrow}{0}, 2\} \longrightarrow \text{Input}$$

$$h[n] = \{5, \underset{\uparrow}{A}, 3\} \longrightarrow \text{Impulse response}$$

$$\begin{aligned} y[n] &= x[n] * h[n] \\ &= \{5, 12, 7, \underset{\uparrow}{16}, 4, 6\} \longrightarrow \text{Output} \end{aligned}$$

Find the value of A.

(*Anna University, May, 2007*)

Solution

$$\begin{aligned} x[n] &= \{1, 2, \underset{\uparrow}{0}, 2\} \\ X(\Omega) &= [e^{j2\Omega} + 2e^{j\Omega} + 0 + 2e^{-j\Omega}] \\ h[n] &= \{5, \underset{\uparrow}{A}, 3\} \\ H(\Omega) &= [5e^{j\Omega} + A + 3e^{-j\Omega}] \\ y[n] &= x[n] * h[n] \\ Y(\Omega) &= X(\Omega)H(\Omega) \\ &= [e^{j2\Omega} + 2e^{j\Omega} + 0 + 2e^{-j\Omega}][5e^{j\Omega} + A + 3e^{-j\Omega}] \\ &= 5e^{j3\Omega} + (10 + A)e^{j2\Omega} + (3 + 2A)e^{j\Omega} \\ &\quad +10 + 2Ae^{-j\Omega} + 6e^{-j2\Omega}] \end{aligned} \tag{7.23}$$

Given

$$y[n] = \{5, 12, 7, \underset{\uparrow}{16}, 4, 6\}$$

$$y(\Omega) = 5e^{j3\Omega} + 12e^{j2\Omega} + 7e^{j\Omega} + 16 + 4e^{-j\Omega} + 6e^{-j2\Omega} \tag{7.24}$$

Equate Eqs. (7.23) and (7.24). Equating the coefficients of $e^{j2\Omega}$, we get

$$10 + A = 12; \qquad A = 2$$

Equating the coefficients of $e^{j\Omega}$, we get

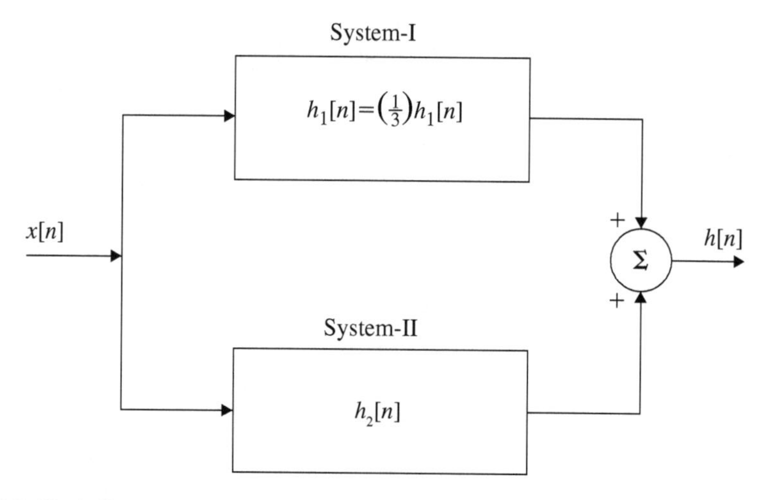

Fig. 7.5 Block diagram representation for Example 7.9

$$3 + 2A = 7; \qquad A = 2$$

Equating the coefficients of $e^{-j\Omega}$, we get

$$2A = 4; \qquad A = 2$$

The value is A is

$$\boxed{A = 2}$$

■ Example 7.9

Two systems connected in parallel are shown in Fig. 7.5. The impulse response of system I is

$$h_1[n] = \left(\frac{1}{3}\right)^n u[n].$$

The impulse response function of the combined system

$$H(\Omega) = \frac{(-18 + 5e^{-j\Omega})}{(6 - 5e^{-j\Omega} + e^{-j2\Omega})}$$

Find the transfer function of system II.

Solution

$$\begin{aligned}H(\Omega) &= \frac{(-18+5e^{-j\Omega})}{(6-5e^{-j\Omega}+e^{-j2\Omega})}\\ &= \frac{\left(-3+\frac{5}{6}e^{-j\Omega}\right)}{\left(1-\frac{5}{6}e^{-j\Omega}+\frac{1}{6}e^{-j2\Omega}\right)}\\ &= \frac{\left(-3+\frac{5}{6}e^{-j\Omega}\right)}{\left(1-\frac{1}{2}e^{-j\Omega}\right)\left(1-\frac{1}{3}e^{-j\Omega}\right)}\end{aligned}$$

$$\begin{aligned}h_1[n] &= \left(\frac{1}{3}\right)^n u[n]\\ H_1(\Omega) &= \frac{1}{\left(1-\frac{1}{3}e^{-j\Omega}\right)}\end{aligned}$$

From Fig. 7.5, $H_2(\Omega)$ can be written as

$$\begin{aligned}H_2(\Omega) &= H(\Omega) - H_1(\Omega)\\ &= \frac{\left(-3+\frac{5}{6}e^{-j\Omega}\right)}{\left(1-\frac{1}{2}e^{-j\Omega}\right)\left(1-\frac{1}{3}e^{-j\Omega}\right)} - \frac{1}{\left(1-\frac{1}{3}e^{-j\Omega}\right)}\\ &= \frac{\left(-4+\frac{4}{3}e^{-j\Omega}\right)}{\left(1-\frac{1}{2}e^{-j\Omega}\right)\left(1-\frac{1}{3}e^{-j\Omega}\right)}\\ &= \frac{-4\left(1-\frac{1}{3}e^{-j\Omega}\right)}{\left(1-\frac{1}{2}e^{-j\Omega}\right)\left(1-\frac{1}{3}e^{-j\Omega}\right)}\\ &= \frac{-4}{\left(1-\frac{1}{2}e^{-j\Omega}\right)}\end{aligned}$$

$$\boxed{h_2[n] = -4\left(\frac{1}{2}\right)^n u[n]}$$

7.7 Discrete Fourier Transform (DFT)

The Fourier transform described above transforms the sequence $x[n]$ to $X(\Omega)$ which is continuous and periodic. The DTFT is defined for sequences with infinite and finite length. A slightly modified transform technique is known as Discrete Fourier Transform (DFT) for finite duration discrete signals. This is a very powerful tool for the analysis and synthesis of discrete signals and systems. The method is ideally suited for use in digital computer or specially designed digital hardware. The DFT is obtained by sampling one period of DTFT only at a finite number of frequency points. It has the following features:

1. The original finite duration signal can be easily recovered from its DFT since there exists one to one correspondence between $x[n]$ and the Fourier transformed discrete signal.
2. For the calculation of the DFT of finite duration sequences, a very efficient and fast techniques known as First Fourier Transform (FFT) has been developed.
3. As far as realization in digital computer is concerned, DFT is the appropriate representation since it is discrete and of finite length in both the time and frequency domains.
4. DFT is closely related to discrete Fourier series, the Fourier transform, convolution, correlation and filtering.

7.7.1 The Discrete Fourier Transform Pairs

Consider the sequence $x[n]$ of length N. The Fourier transform of $x[n]$ is given by

$$X(\Omega) = \sum_{n=-\infty}^{\infty} x[n]e^{-j\Omega n} \tag{7.25}$$

In Eq. (7.25), $X(\Omega)$ is the continuous function of Ω. The range of Ω is fromm $-\pi$ to π or 0 to 2π. Hence calculating $X(\Omega)$ on digital computer or DSP is impossible. It is, therefore, necessary to compute $X(\Omega)$ at discrete values of Ω. When Fourier transform $X(\Omega)$ is calculated at only discrete points k it is called Discrete Fourier Transform (DFT). The DFT is denoted by $X(k)$. For finite discrete points N, Eq. (7.25) is written as

$$X(k) = \sum_{n=0}^{N-1} x[n]e^{-j2\pi kn/N} \tag{7.26}$$

where $k = 0, 1, 2, \ldots, (N-1)$. $X(k)$ is computed at $k = 0, 1, 2, \ldots, (N-1)$ discrete points. $X(k)$ is the sequence of N samples. The sequence $x[n]$ is obtained back fromm

$$X[n] = \frac{1}{N}\sum_{k=0}^{N-1} X(k)e^{j2\pi kn/N} \tag{7.27}$$

Let us define $W_N = e^{-j2\pi/N}$, where W_N is called Twiddle factor. Equations (7.26) and (7.27) are called DFT and IDFT or simply discrete Fourier transform pair. They can be represented in terms of twiddle factor as given below

$$X(k) = \sum_{n=0}^{N-1} x[n] W_N^{kn} \tag{7.28}$$

$$x[n] = \frac{1}{N} \sum_{k=0}^{N-1} X(k) W_N^{-kn} \tag{7.29}$$

Let the sequence $x[n]$ be resentenced as a vector x_N of N samples as

$$x_N = \begin{matrix} n=0 \\ n=1 \\ \vdots \\ n=N-1 \end{matrix} \begin{bmatrix} x(0) \\ x(1) \\ \vdots \\ x(N-1) \end{bmatrix}_{N\times 1} \tag{7.30}$$

and $X(k)$ be represented as a vector X_N of N samples as

$$X_N = \begin{matrix} k=0 \\ k=1 \\ \vdots \\ k=N-1 \end{matrix} \begin{bmatrix} X(0) \\ X(1) \\ \vdots \\ X(N-1) \end{bmatrix}_{N\times 1} \tag{7.31}$$

The twiddle factor W_N^{kn} is represented as a matrix with k rows and n column as

$$W_N = \begin{matrix} k=0 \\ \\ \\ \\ k=N-1 \end{matrix} \begin{bmatrix} W_N^0 & W_N^0 & W_N^0 & \cdots & W_N^0 \\ W_N^0 & W_N^1 & W_N^2 & \cdots & W_N^{N-1} \\ W_N^0 & W_N^2 & W_N^4 & \cdots & W_N^{2(N-1)} \\ \vdots & \vdots & \vdots & \vdots & \vdots \\ W_N^0 & W_N^{N-1} & W_N^{2(N-1)} & \cdots & W_N^{(N-1)(N-1)} \end{bmatrix}_{N\times N} \tag{7.32}$$

Thus, Eqs. (7.28) and (7.29) can be written with matrix form as

$$X_N = [W_N] x_N \tag{7.33}$$

$$x_N = \frac{1}{N} [W_N^*] X_N \tag{7.34}$$

where $W_N^* = W_N^{-kn}$

$$\begin{aligned} W_N &= e^{-j\frac{2\pi}{N}} \\ &= 1\angle -2\pi/N \end{aligned} \tag{7.35}$$

From Eq. (7.35), the magnitude of the twiddle factor is 1 and the phase angle is $-\frac{2\pi}{N}$. It lies on the unit circle in the complex plane from 0 to 2π angle and it gets repeated for every cycle.

7.7.2 Four Point, Six Point and Eight Point Twiddle Factors

As in Eq. (7.35), the magnitude of the twiddle factor is 1 and the angle -2π is equally divided in the interval N. The most commonly used intervals are $N = 4$ and $N = 8$. For $N = 4$, the angle between any $N = 0$ and $N = 1$ is $\frac{\pi}{2}$.

7.7.2.1 Four Point Twiddle Factor

For $N = 4$

$$W_N = \begin{matrix} & n=0 & 1 & 2 & 3 \\ k=0 & W_4^0 & W_4^0 & W_4^0 & W_4^0 \\ 1 & W_4^0 & W_4^1 & W_4^2 & W_4^3 \\ 2 & W_4^0 & W_4^2 & W_4^4 & W_4^6 \\ 3 & W_4^0 & W_4^3 & W_4^6 & W_4^9 \end{matrix} \tag{7.36}$$

Note: $W_4^4 = W_4^0$; $W_4^6 = W_4^2$ and $W_4^9 = W_4^1$. From Eq. (7.35)

$$W_4^1 = 1\angle{-\pi/2}$$

For $N = 4$, the unit circle is divided into four equal segments in the clockwise sequence and labeled as W_4^0, W_4^1, W_4^2 and W_4^3. From Fig. 7.6, the twiddle factor are obtained as

$$W_4^0 = 1; \quad W_4^1 = -j; \quad W_4^2 = -1; \quad W_4^3 = j$$

Equation (7.36) is written as

$$W_N = \begin{bmatrix} 1 & 1 & 1 & 1 \\ 1 & -j & -1 & j \\ 1 & -1 & 1 & -1 \\ 1 & j & -1 & -j \end{bmatrix} \tag{7.37}$$

Equation (7.37) represents the twiddle factor to express DFT of any sequence $x[n]$. Twiddle factors for 6-points DFT and 8-points DFT are derived below.

7.7.2.2 Six Point Twiddle Factor

For $N = 6$, the unit circle is divided into six equal segments and in the clockwise sequence labeled as W_6^0, W_6^1, W_6^2, W_6^3, W_6^4 and W_6^5 noting that $W_6^6 = W_6^0$, $W_6^7 = W_6^1$ and so on. This is shown in Fig. 7.7. Each segment is shifted by $-60°$ on the unit circle. For $N = 6$, W_6 is obtained by multiplying the rows and columns of W_6 and is given below.

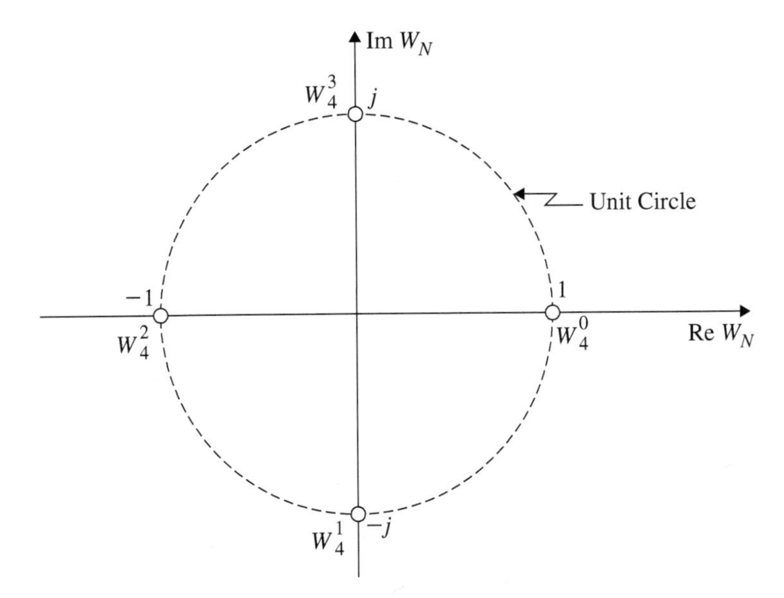

Fig. 7.6 Representation of W_4^{-nk}

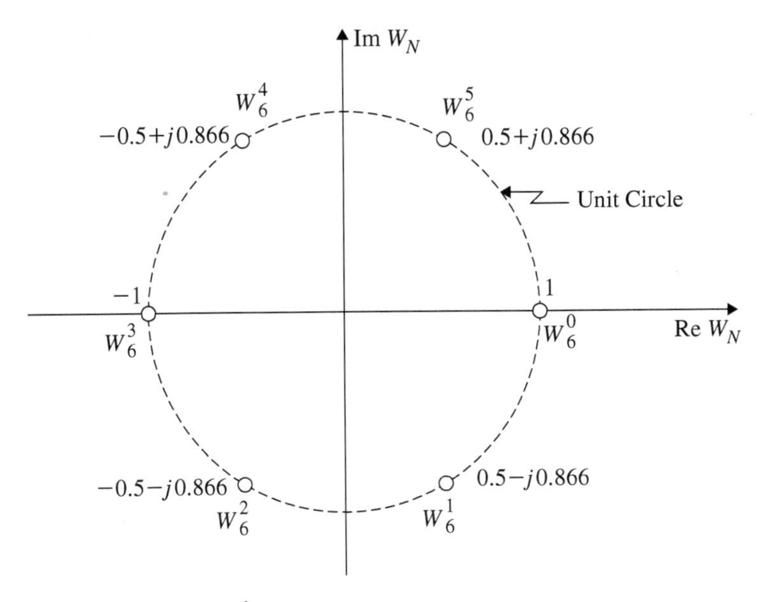

Fig. 7.7 Representation of W_6^{-nk}

$$W_N = \begin{bmatrix} W_6^0 & W_6^0 & W_6^0 & W_6^0 & W_6^0 & W_6^0 \\ W_6^0 & W_6^1 & W_6^2 & W_6^3 & W_6^4 & W_6^5 \\ W_6^0 & W_6^2 & W_6^4 & W_6^6 & W_6^8 & W_6^{10} \\ W_6^0 & W_6^3 & W_6^6 & W_6^9 & W_6^{12} & W_6^{15} \\ W_6^0 & W_6^4 & W_6^8 & W_6^{12} & W_6^{16} & W_6^{20} \\ W_6^0 & W_6^5 & W_6^{10} & W_6^{15} & W_6^{20} & W_6^{25} \end{bmatrix} \tag{7.38}$$

$$W_6^0 = W_6^6 = W_6^{12} = W_6^{18} = W_6^{24} = 1$$
$$W_6^1 = W_6^7 = W_6^{13} = W_6^{19} = W_6^{25} = 1e^{-j\frac{\pi}{3}} = 0.5 - j0.866$$
$$W_6^2 = W_6^8 = W_6^{14} = W_6^{20} = W_6^{26} = 1e^{-j\frac{2\pi}{3}} = -0.5 - j0.866$$
$$W_6^3 = W_6^9 = W_6^{15} = W_6^{21} = W_6^{27} = -1$$
$$W_6^4 = W_6^{10} = W_6^{16} = W_6^{22} = W_6^{28} = 1e^{j\frac{2\pi}{3}} = -0.5 + j0.866$$
$$W_6^5 = W_6^{11} = W_6^{17} = W_6^{23} = W_6^{29} = 1e^{j\frac{\pi}{3}} = 0.5 + j0.866$$

Substituting the values of the elements of the matrix W_6, we get

$$W_6 = \begin{bmatrix} 1 & 1 & 1 & 1 & 1 & 1 \\ 1 & 0.5 - j0.866 & -0.5 - j0.866 & -1 & -0.5 + j0.866 & 0.5 + j0.866 \\ 1 & -0.5 - j0.866 & -0.5 + j0.866 & 1 & -0.5 - j0.866 & -0.5 + j0.866 \\ 1 & -1 & 1 & -1 & 1 & -1 \\ 1 & -0.5 + j0.866 & -0.5 - j0.866 & 1 & -0.5 + j0.866 & -0.5 - j0.866 \\ 1 & 0.5 + j0.866 & -0.5 + j0.866 & -1 & -0.5 - j0.866 & 0.5 - j0.866 \end{bmatrix} \tag{7.39}$$

7.7.2.3 Eight Point Twiddle Factor

$$W_8 = \begin{bmatrix} W_8^0 & W_8^0 & W_8^0 & W_8^0 & W_8^0 & W_8^0 & W_8^0 & W_8^0 \\ W_8^0 & W_8^1 & W_8^2 & W_8^3 & W_8^4 & W_8^5 & W_8^6 & W_8^7 \\ W_8^0 & W_8^2 & W_8^4 & W_8^6 & W_8^8 & W_8^{10} & W_8^{12} & W_8^{14} \\ W_8^0 & W_8^3 & W_8^6 & W_8^9 & W_8^{12} & W_8^{15} & W_8^{18} & W_8^{21} \\ W_8^0 & W_8^4 & W_8^8 & W_8^{12} & W_8^{16} & W_8^{20} & W_8^{24} & W_8^{28} \\ W_8^0 & W_8^5 & W_8^{10} & W_8^{15} & W_8^{20} & W_8^{25} & W_8^{30} & W_8^{35} \\ W_8^0 & W_8^6 & W_8^{12} & W_8^{18} & W_8^{24} & W_8^{30} & W_8^{36} & W_8^{42} \\ W_8^0 & W_8^7 & W_8^{14} & W_8^{21} & W_8^{28} & W_8^{35} & W_8^{42} & W_8^{49} \end{bmatrix} \tag{7.40}$$

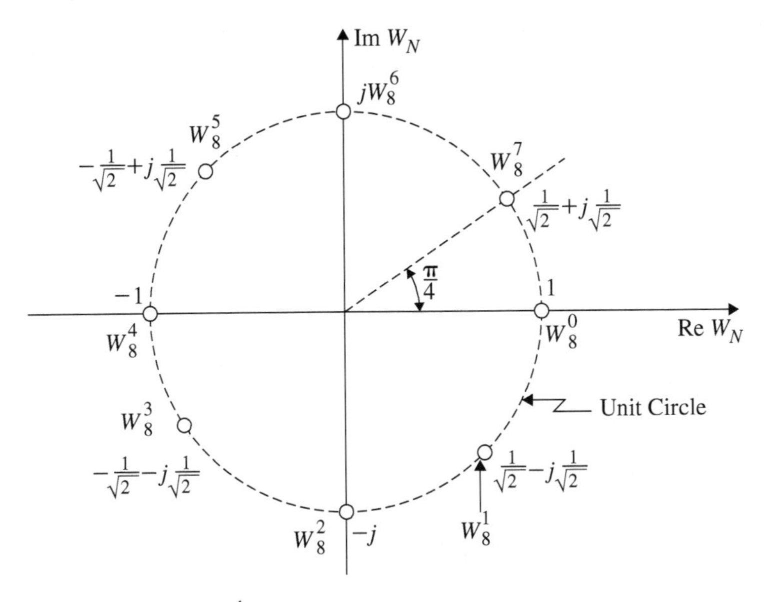

Fig. 7.8 Representation of W_8^{-kn}

$$W_8 =$$
$$\begin{bmatrix} 1 & 1 & 1 & 1 & 1 & 1 & 1 & 1 \\ 1 & \frac{1}{\sqrt{2}}-j\frac{1}{\sqrt{2}} & -j & -\frac{1}{\sqrt{2}}-j\frac{1}{\sqrt{2}} & -1 & -\frac{1}{\sqrt{2}}+j\frac{1}{\sqrt{2}} & j & \frac{1}{\sqrt{2}}+j\frac{1}{\sqrt{2}} \\ 1 & -j & -1 & j & 1 & -j & -1 & j \\ 1 & -\frac{1}{\sqrt{2}}-j\frac{1}{\sqrt{2}} & j & \frac{1}{\sqrt{2}}-j\frac{1}{\sqrt{2}} & -1 & \frac{1}{\sqrt{2}}+j\frac{1}{\sqrt{2}} & -j & -\frac{1}{\sqrt{2}}+j\frac{1}{\sqrt{2}} \\ 1 & -1 & 1 & -1 & 1 & -1 & 1 & -1 \\ 1 & -\frac{1}{\sqrt{2}}+j\frac{1}{\sqrt{2}} & -j & \frac{1}{\sqrt{2}}+j\frac{1}{\sqrt{2}} & -1 & \frac{1}{\sqrt{2}}-j\frac{1}{\sqrt{2}} & j & -\frac{1}{\sqrt{2}}-j\frac{1}{\sqrt{2}} \\ 1 & j & -1 & -j & 1 & j & -1 & -j \\ 1 & \frac{1}{\sqrt{2}}+j\frac{1}{\sqrt{2}} & j & -\frac{1}{\sqrt{2}}+j\frac{1}{\sqrt{2}} & -1 & -\frac{1}{\sqrt{2}}-j\frac{1}{\sqrt{2}} & -j & \frac{1}{\sqrt{2}}-j\frac{1}{\sqrt{2}} \end{bmatrix} \tag{7.41}$$

■ Example 7.10

Compute the DFT of the sequence $x[n] = \{1, j, -1, -j\}$ for $N = 4$ (Fig. 7.8).

(Anna University, November, 2006)

Solution **Method 1**

$$W_4 = \begin{bmatrix} 1 & 1 & 1 & 1 \\ 1 & -j & -1 & j \\ 1 & -1 & 1 & -1 \\ 1 & j & -1 & -j \end{bmatrix}; \quad X_4 = \begin{bmatrix} 1 \\ j \\ -1 \\ -j \end{bmatrix}$$

From Eq. (7.33)

$$\begin{aligned} X_4 &= W_4 x_4 \\ X_4 &= \begin{bmatrix} 1 & 1 & 1 & 1 \\ 1 & -j & -1 & j \\ 1 & -1 & 1 & -1 \\ 1 & j & -1 & -j \end{bmatrix} \begin{bmatrix} 1 \\ j \\ -1 \\ -j \end{bmatrix} \\ &= \begin{bmatrix} 1+j-1-j \\ 1+1+1+1 \\ 1-j-1+j \\ 1-1+1-1 \end{bmatrix} = \begin{bmatrix} 0 \\ 4 \\ 0 \\ 0 \end{bmatrix} \end{aligned}$$

$$\boxed{\begin{aligned} X(0) &= 0 \\ X(1) &= 4 \\ X(2) &= 0 \\ X(3) &= 0 \end{aligned}}$$

Method 2

$$X(k) = \sum_{n=0}^{3} x[n] e^{-j\frac{2\pi kn}{4}}; \quad k = 0, 1, 2, 3, \ldots$$

For $k = 0$

$$\begin{aligned} X(0) &= \sum_{n=0}^{3} x[n] \\ &= x[0] + x[1] + x[2] + x[3] \\ &= 1 + j - 1 - j = 0 \end{aligned}$$

For $k = 1$

$$\begin{aligned}X(1) &= \sum_{n=0}^{3} x[n]e^{-j\frac{\pi n}{2}} \\ &= x[0] + x[1]e^{-j\frac{\pi}{2}} + x[2]e^{-j\pi} + x[3]e^{-j\frac{3\pi}{2}} \\ &= 1 + j(-j) + (-1)(-1) + (-j)(j) \\ &= 1 + 1 + 1 + 1 = 4\end{aligned}$$

For $k = 2$

$$\begin{aligned}X(2) &= \sum_{n=0}^{3} x[n]e^{-j\pi n} \\ &= x[0] + x[1]e^{-j\pi} + x[2]e^{-j2\pi} + x[3]e^{-j3\pi} \\ &= 1 + j(-1) + (-1)(1) + (-j)(-1) \\ &= 1 - j - 1 + j = 0\end{aligned}$$

For $k = 3$

$$\begin{aligned}X(3) &= \sum_{n=0}^{3} x[n]e^{-j\frac{3\pi n}{2}} \\ &= x[0] + x[1]e^{-\frac{3\pi}{2}} + x[2]e^{-j3\pi} + x[3]e^{-j\frac{9\pi}{2}} \\ &= 1 + j(j) + (-1)(-1) + (-j)(-j) \\ &= 1 - 1 + 1 - 1 = 0\end{aligned}$$

$$\boxed{\begin{aligned}X(0) &= 0 \\ X(1) &= 4 \\ X(2) &= 0 \\ X(3) &= 0\end{aligned}}$$

Method 1 is simpler and quicker.

■ Example 7.11

Find 8-point DFT of $x[n] = \{1, -1, 1, -1, 1, -1, 1, -1\}$.

(Anna University, April, 2004)

Solution

$$\begin{aligned}X_N &= W_N x_N \\ &= W_8 x_8\end{aligned} \tag{7.42}$$

W_8 is given in Eq. (7.41)

$$x_8 = \begin{bmatrix} 1 \\ -1 \\ 1 \\ -1 \\ 1 \\ -1 \\ 1 \\ -1 \end{bmatrix}$$

$X(0)$ is obtained by multiplying x_8 with the first row of W_8. Thus,

$$\begin{aligned}
X(0) &= 1-1+1-1+1-1+1-1=0 \\
X(1) &= \text{2nd row of} W_8 \text{to multiply} x_8 \\
&= 1-\frac{1}{\sqrt{2}}+j\frac{1}{\sqrt{2}}-j+\frac{1}{\sqrt{2}}+j\frac{1}{\sqrt{2}}-1+\frac{1}{\sqrt{2}}-j\frac{1}{\sqrt{2}}+j=0 \\
X(2) &= \text{3rd row of} W_8 \text{to multiply} x_8 \\
&= 1+j-1-j+1+j-1-j=0 \\
X(3) &= 1+\frac{1}{\sqrt{2}}+j\frac{1}{\sqrt{2}}+j-\frac{1}{\sqrt{2}}+j\frac{1}{\sqrt{2}}-1-\frac{1}{\sqrt{2}}-j\frac{1}{\sqrt{2}}-j+\frac{1}{\sqrt{2}}-j\frac{1}{\sqrt{2}}=0 \\
X(4) &= \text{5th row of} W_8 \text{to multiply} x_8 \\
&= 1+1+1+1+1+1+1+1=8 \\
X(5) &= \text{6th row of} W_8 \text{to multiply} x_8 \\
&= 1+\frac{1}{\sqrt{2}}-j\frac{1}{\sqrt{2}}-j-\frac{1}{\sqrt{2}}-j\frac{1}{\sqrt{2}}-1-\frac{1}{\sqrt{2}}+j\frac{1}{\sqrt{2}}+j+\frac{1}{\sqrt{2}}+j\frac{1}{\sqrt{2}}=0 \\
X(6) &= 1-j-1+j+1-j-1+j=0 \\
X(7) &= 1-\frac{1}{\sqrt{2}}-j\frac{1}{\sqrt{2}}+j+\frac{1}{\sqrt{2}}-j\frac{1}{\sqrt{2}}-1+\frac{1}{\sqrt{2}}+j\frac{1}{\sqrt{2}}-j-\frac{1}{\sqrt{2}}+j\frac{1}{\sqrt{2}}=0
\end{aligned}$$

$$X_8 = \begin{bmatrix} 0 \\ 0 \\ 0 \\ 0 \\ 8 \\ 0 \\ 0 \\ 0 \end{bmatrix}$$

7.7.3 Zero Padding

In evaluating the DFT, we assumed that the length of the DFT which is N is equal to the length L of the sequence $x[n]$. If $N < L$, time domain aliasing occurs due to under sampling and in the process we could miss out some important details and get misleading information. To avoid this N, the number of samples of $x[n]$ is increased by adding some dummy samples of 0 value. This addition of dummy samples is known as zero padding. The zero padding not only increases the number of samples but also helps in getting a better idea of the frequency spectrum of $X(\Omega)$.

■ Example 7.12

Compute the 4-point DFT of the sequence

$$x[n] = 1 \qquad 0 \leq n < 2$$

Solution For the given sequence $L = 3$ and $N = 4$. By adding a dummy samples of 0 values (zero padding), the given sequence becomes

$$x[n] = \{1,\ 1,\ 1,\ 0\}$$

$$x_N = \begin{bmatrix} 1 \\ 1 \\ 1 \\ 0 \end{bmatrix}$$

W_4 is given in Eq. (7.37).

$$\begin{aligned} X_4 &= W_4 x_4 \\ &= \begin{bmatrix} 1 & 1 & 1 & 1 \\ 1 & -j & -1 & j \\ 1 & -1 & 1 & -1 \\ 1 & j & -1 & -j \end{bmatrix} \begin{bmatrix} 1 \\ 1 \\ 1 \\ 0 \end{bmatrix} \\ X(0) &= [1 + 1 + 1 + 0] = 3 \\ X(1) &= [1 - j - 1 + 0] = -j \\ X(2) &= [1 - 1 + 1 + 0] = 1 \\ X(3) &= [1 + j - 1 + 0] = j \end{aligned}$$

$$X_4 = \begin{bmatrix} 3 \\ -j \\ 1 \\ j \end{bmatrix}$$

■ Example 7.13

Compute the 4-points DFT of the following sequences:

$$\begin{aligned} &(1) \quad x[n] = \{1, 1, 1, 1\} \\ &(2) \quad x[n] = \{1, 1, 0, 0\} \\ &(3) \quad x[n] = \cos \pi n \\ &(4) \quad x[n] = \sin \frac{n\pi}{2} \end{aligned}$$

(*Anna University, April, 2004; November, 2007*)

Solution

(1) $\boldsymbol{x[n] = \{1, 1, 1, 1\}}$

$$x_4 = \begin{bmatrix} 1 \\ 1 \\ 1 \\ 1 \end{bmatrix}$$

$$X_4 = W_4 x_4$$

$$= \begin{bmatrix} 1 & 1 & 1 & 1 \\ 1 & -j & -1 & j \\ 1 & -1 & 1 & -1 \\ 1 & j & -1 & -j \end{bmatrix} \begin{bmatrix} 1 \\ 1 \\ 1 \\ 1 \end{bmatrix}$$

$$X(0) = 1 + 1 + 1 + 1 = 4$$

$$X(1) = 1 - j - 1 + j = 0$$

$$X(2) = 1 - 1 + 1 - 1 = 0$$

$$X(3) = 1 + j - 1 - j = 0$$

$$X_4 = \begin{bmatrix} 4 \\ 0 \\ 0 \\ 0 \end{bmatrix}$$

(2) $\boldsymbol{x[n] = \{1, 1, 0, 0\}}$

$$X_4 = W_4 x_4$$

$$X_4 = \begin{bmatrix} 1 & 1 & 1 & 1 \\ 1 & -j & -1 & j \\ 1 & -1 & 1 & -1 \\ 1 & j & -1 & -j \end{bmatrix} \begin{bmatrix} 1 \\ 1 \\ 0 \\ 0 \end{bmatrix}$$

$$X(0) = 1 + 1 + 0 + 0 = 2$$

$$X(1) = 1 - j + 0 + 0 = (1 - j)$$

$$X(2) = 1 - 1 + 0 + 0 = 0$$
$$X(3) = 1 + j + 0 + 0 = (1 + j)$$
$$X_4 = \begin{bmatrix} 2 \\ 1-j \\ 0 \\ 1+j \end{bmatrix}$$

(3) $\boldsymbol{x[n] = \cos \pi n}$; **where** $\boldsymbol{n = 0, 1, 2, 3, \ldots}$

$$x[n] = \{1,\ -1,\ 1,\ -1\}$$
$$X_4 = W_4 x_4$$
$$= \begin{bmatrix} 1 & 1 & 1 & 1 \\ 1 & -j & -1 & j \\ 1 & -1 & 1 & -1 \\ 1 & j & -1 & -j \end{bmatrix} \begin{bmatrix} 1 \\ -1 \\ 1 \\ -1 \end{bmatrix}$$
$$X(0) = 1 - 1 + 1 - 1 = 0$$
$$X(1) = 1 + j - 1 - j = 0$$
$$X(2) = 1 + 1 + 1 + 1 = 4$$
$$X(3) = 1 - j - 1 + j = 0$$
$$X_4 = \begin{bmatrix} 0 \\ 0 \\ 4 \\ 0 \end{bmatrix}$$

(4) $\boldsymbol{x[n] = \sin \frac{n\pi}{2}}$; **where** $\boldsymbol{n = 0, 1, 2, 3, \ldots}$

$$x[n] = \{0,\ 1,\ 0,\ -1\}$$
$$X_4 = W_4 x_4$$
$$= \begin{bmatrix} 1 & 1 & 1 & 1 \\ 1 & -j & -1 & j \\ 1 & -1 & 1 & -1 \\ 1 & j & -1 & -j \end{bmatrix} \begin{bmatrix} 0 \\ 1 \\ 0 \\ -1 \end{bmatrix}$$
$$X(0) = 0 + 1 + 0 - 1 = 0$$
$$X(1) = 0 - j + 0 - j = -j2$$
$$X(2) = 0 - 1 + 0 + 1 = 0$$
$$X(3) = 0 + j + 0 + j = j2$$

$$X_4 = \begin{bmatrix} 0 \\ -j2 \\ 0 \\ j2 \end{bmatrix}$$

■ Example 7.14

Find the N-point DFT of the following sequences for $0 \leq n \leq N-1$.

$$(1) \quad x[n] = \delta[n]$$
$$(2) \quad x[n] = a^n$$

Solution

(1) $\boldsymbol{x[n] = \delta[n]}$

$$X(k) = \sum_{n=0}^{N-1} x[n] e^{-j\frac{2\pi kn}{N}}$$
$$x[n] = \begin{cases} 1 & n = 0 \\ 0 & n \neq 0 \end{cases}$$

$$\boxed{X(k) = 1}$$

(2) $\boldsymbol{x[n] = a^n}$

$$X(k) = \sum_{n=0}^{N-1} a^n e^{-j\frac{2\pi kn}{N}}$$
$$= \sum_{n=0}^{N-1} \left(ae^{-j\frac{2\pi kn}{N}}\right)^n$$

Using the summation formula

$$\sum_{n=N_1}^{N_2} a^k = \frac{a^{N_1} - a^{N_2+1}}{(1-a)}$$

we get

$$X(k) = \frac{\left(ae^{-j\frac{2\pi k}{N}}\right)^0 - \left(ae^{-j\frac{2\pi k}{N}}\right)^N}{\left(1 - ae^{-j\frac{2\pi k}{N}}\right)}$$

$$\boxed{X(k) = \frac{(1 - a^N)}{\left(1 - ae^{-j\frac{2\pi k}{N}}\right)}} \qquad [e^{-j2\pi k} = 1]$$

■ Example 7.15

Find the IDFT of the following functions with $N = 4$.

$$(1) \quad X(k) = \{1, 0, 1, 0\}$$
$$(2) \quad X(k) = \{6, (-2 + j2), -2, (-2 - j2)\}$$

Solution

(1) $\mathbf{X(k) = \{1, 0, 1, 0\}}$

From Eq. (7.34)

$$x_N = \frac{1}{N}[W_N^*]X_N$$

$$X_N = \begin{bmatrix} 1 \\ 0 \\ 1 \\ 0 \end{bmatrix}$$

$$W_N^* = \begin{bmatrix} 1 & 1 & 1 & 1 \\ 1 & j & -1 & -j \\ 1 & -1 & 1 & -1 \\ 1 & -j & -1 & j \end{bmatrix}$$

For $N = 4$

$$x_N = \frac{1}{4}\begin{bmatrix} 1 & 1 & 1 & 1 \\ 1 & j & -1 & -j \\ 1 & -1 & 1 & -1 \\ 1 & -j & -1 & j \end{bmatrix}\begin{bmatrix} 1 \\ 0 \\ 1 \\ 0 \end{bmatrix}$$

$$x[0] = \frac{1}{4}[1 + 0 + 1 + 0] = 0.5$$

$$x[1] = \frac{1}{4}[1 + 0 - 1 + 0] = 0$$

$$x[2] = \frac{1}{4}[1 + 0 + 1 + 0] = 0.5$$

$$x[3] = \frac{1}{4}[1 + 0 - 1 + 0] = 0$$

$$\boxed{x[n] = \{0.5,\ 0,\ 0.5,\ 0\}}$$

(2) $\mathbf{x[n] = \{6,\ (-2 + j2),\ -2,\ (-2 - j2)\}}$

$$x_N = \frac{1}{4}\begin{bmatrix} 1 & 1 & 1 & 1 \\ 1 & j & -1 & -j \\ 1 & -1 & 1 & -1 \\ 1 & -j & -1 & j \end{bmatrix}\begin{bmatrix} 6 \\ -2+j2 \\ -2 \\ -2-j2 \end{bmatrix}$$

$$x[0] = \frac{1}{4}[6 - 2 + j2 - 2 - 2 - j2] = 0$$

$$x[1] = \frac{1}{4}[6 - j2 - 2 + 2 + j2 - 2] = 1$$

$$x[2] = \frac{1}{4}[6 + 2 - j2 - 2 + 2 + j2] = 2$$

$$x[3] = \frac{1}{4}[6 + j2 + 2 + 2 - j2 + 2] = 3$$

$$\boxed{x[n] = \{0,\ 1,\ 2,\ 3\}}$$

7.8 Properties of DFT

7.8.1 Periodicity

If $x[n]$ is the input sequence and $X(k)$ is the N-point DFT of $x[n]$, then the periodicity of $x[n]$ and $X(k)$ are defined as

$$x[n + N] = x[n] \tag{7.43}$$

$$X(k + N) = X(k) \tag{7.44}$$

7.8.2 Linearity

Let $x_1[n]$ and $x_2[n]$ be two N-point sequences whose DFTs are $X_1(k)$ and $X_2(k)$. Then

$$a_1x_1[n] + a_2x_2[n] \underset{N\text{-points}}{\overset{\text{DFT}}{\longleftrightarrow}} a_1X_1(k) + a_2X_2(k)$$

7.8.3 Complex Conjugate Symmetry

If

$$x[n] \stackrel{\text{DFT}}{\longleftrightarrow} X(k)$$

then

$$x^*[n] \stackrel{\text{DFT}}{\longleftrightarrow} X(N-k) \tag{7.45}$$

Proof

$$\begin{aligned} x^*[n] \stackrel{\text{DFT}}{\longleftrightarrow} &= \sum_{n=0}^{N-1} x^*[n]e^{-j2\pi kn/N} \\ &= \left[\sum_{n=0}^{N-1} x[n]e^{j2\pi kn/N}\right]^* \\ &= \left[\sum_{n=0}^{N-1} x[n]e^{-j\frac{2\pi}{N}(N-k)n}\right]^* \\ x^*[n] \stackrel{\text{DFT}}{\longleftrightarrow} &= X^*(N-k) \end{aligned}$$

7.8.4 Circular Time Shifting

If

$$x[n] \stackrel{\text{DFT}}{\longleftrightarrow} X(k)$$

then

$$x[(n-m)]_N \stackrel{\text{DFT}}{\longleftrightarrow} e^{-j\frac{2\pi km}{N}} X(k) \tag{7.46}$$

7.8.5 Circular Frequency Shifting

If

$$x[n] \stackrel{\text{DFT}}{\longleftrightarrow} X(k)$$

then

$$x[n]e^{j\frac{2\pi nl}{N}} \stackrel{\text{DFT}}{\longleftrightarrow} X(k-l)_N \tag{7.47}$$

7.8.6 Circular Correlation

If

$$x[n] \stackrel{\text{DFT}}{\longleftrightarrow} X(k)$$

and

$$y[n] \stackrel{\text{DFT}}{\longleftrightarrow} Y(k)$$

then

$$r_y \stackrel{\text{DFT}}{\longleftrightarrow} R_{xy}(k) = X(k)Y^*(k)$$

where

$$r_{xy} = \sum_{n=0}^{N-1} x[n]y^*(n-l)_N \tag{7.48}$$

7.8.7 Multiplication of Two DFTs

The multiplication of two DFTs is equal to circular convolution of two sequences in time domain. Let $x_1[n]$ and $x_2[n]$ be finite duration sequences of length N with their DFTs as $X_1(k)$ and $X_2(k)$, respectively. The sequence $x_1[n]$ when circularly convolves with $x_2[n]$ sequence, the circular convolution is represented as $x_1[n] Ⓝ x_2[n]$. The DFT of circular convolution is

$$x_1[n] \,Ⓝ\, x_2[n] \stackrel{\text{DFT}}{\longleftrightarrow} X_1(k)X_2(k) \tag{7.49}$$

7.8.8 Parseval's Theorem

If

$$x[n] \stackrel{\text{DFT}}{\longleftrightarrow} X(k)$$

and

$$y[n] \stackrel{\text{DFT}}{\longleftrightarrow} Y(k)$$

then

$$\sum_{n=0}^{N-1} x[n]y^*[n] = \frac{1}{N}\sum_{k=0}^{N-1} X(k)Y^*(k) \tag{7.50}$$

7.9 Circular Convolution

To determine the circular convolution of any two sequences, $x_1[n]$ and $x_2[n]$ of length N, the following methods are discussed:

1. Circle Method.
2. Matrix multiplication method.
3. DFT-IDFT Method.

7.9.1 *Circular Convolution—Circle Method*

The circular convolution of two sequences is symbolically represented as

$$y[n] = x_1[n] \, \text{Ⓝ} x_2[n] \tag{7.51}$$

The following steps are followed to find $y[n]$:

1. Draw two concentric circles of two different diameters. The data points of $x_1[n]$ are placed on the outer circle in the counter-clockwise direction at equidistance.
2. The data points of $x_2[n]$ are placed on the inner circle in the clockwise direction at equidistance.
3. The first data value of both the sequences should be in alignment.
4. Multiply the corresponding values in both the circles and add them. This corresponds to first data value of the circular convolution.
5. Rotate the inner circle in the counter-clockwise direction by one sample and repeat step 4. This corresponds to the second data value of the circular convolution.
6. Repeat step 5 until one revolution is complete. Each time repeat step 4 to get the data value of the circular convolution.

■ Example 7.16

Consider the following two sequences:

$$x_1[n] = \{2, 1, 4, -3\}$$
$$x_2[n] = \{-1, 2, 3, -2\}$$

Find the circular convolution

$$y[n] = x_1[n] \, Ⓝ x_2[n]$$

Use circle method.

Solution

1. $x_1[n]$ is marked in the outer circle in the anticlockwise direction and $x_2[n]$ is marked in the inner circle in the clockwise direction as shown in Fig. 7.9a.

$$\begin{aligned} y[0] &= x_1[0]x_2[0] + x_1[1]x_2[3] + x_2[2]x_2[2] + x_1[3]x_2[1] \\ &= 2(-1) + 1(-2) + 4(3) + (-3)2 = 2 \end{aligned}$$

2. The outer circle is kept fixed and the inner circle is rotated in the anticlockwise direction by one sample. This is shown in Fig. 7.9b.

$$\begin{aligned} y[1] &= x_1[0]x_2[1] + x_1[1]x_2[0] + x_1[2]x_2[3] + x_1[3]x_2[2] \\ &= 2 \times 2 + 1(-1) + 4(-2) + (-3)3 = -14 \end{aligned}$$

3. Keeping the outer circle of Fig. 7.9b fixed, the inner circle is rotated in the anticlockwise direction by one sample. This is shown in Fig. 7.9c.

$$y[2] = 2 \times 3 + 1 \times 2 + 4(-1) + (-3)(-2) = 10$$

4. The outer circle of Fig. 7.9c is kept fixed and the inner circle is rotated by one sample in the anticlockwise direction by one sample. This is shown in Fig. 7.9d (Fig. 7.10).

$$y[3] = 2(-2) + 1 \times 3 + 4 \times 2 + (-3)(-1) = 10$$

5.

$$\boxed{y[n] = [2, \, -14, \, 10, \, 10]}$$

7.9.2 Circular Convolution-Matrix Multiplication Method

In this method, the circular convolution of two sequences $x_1[n]$ and $x_2[n]$ are obtained by representing these sequences in matrix form as given below.

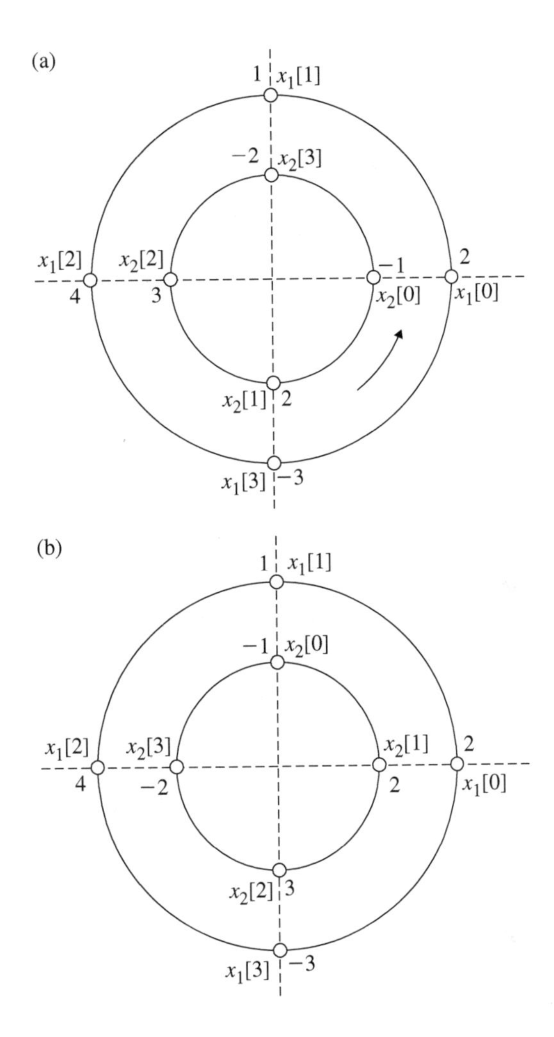

Fig. 7.9 a–d Circular convolution-circle method

$$\begin{bmatrix} x_2[0] & x_2[N-1] & \cdots & x_2[2] & x_2[1] \\ x_2[1] & x_2[0] & \cdots & x_2[3] & x_2[2] \\ x_2[2] & x_2[1] & \cdots & x_2[4] & x_2[3] \\ \vdots & \vdots & \vdots & \vdots & \vdots \\ x_2[N-2] & x_2[N-3] & \cdots & x_2[0] & x_2[N-1] \\ x_2[N-1] & x_2[N-2] & \cdots & x_2[1] & x_2[0] \end{bmatrix} \begin{bmatrix} x_1[0] \\ x_1[1] \\ x_1[2] \\ \vdots \\ x_1[N-2] \\ x_1[N-1] \end{bmatrix} = \begin{bmatrix} y[0] \\ y[1] \\ y[2] \\ \vdots \\ y[N-2] \\ y[N-1] \end{bmatrix} \tag{7.52}$$

Fig. 7.9 (continued)

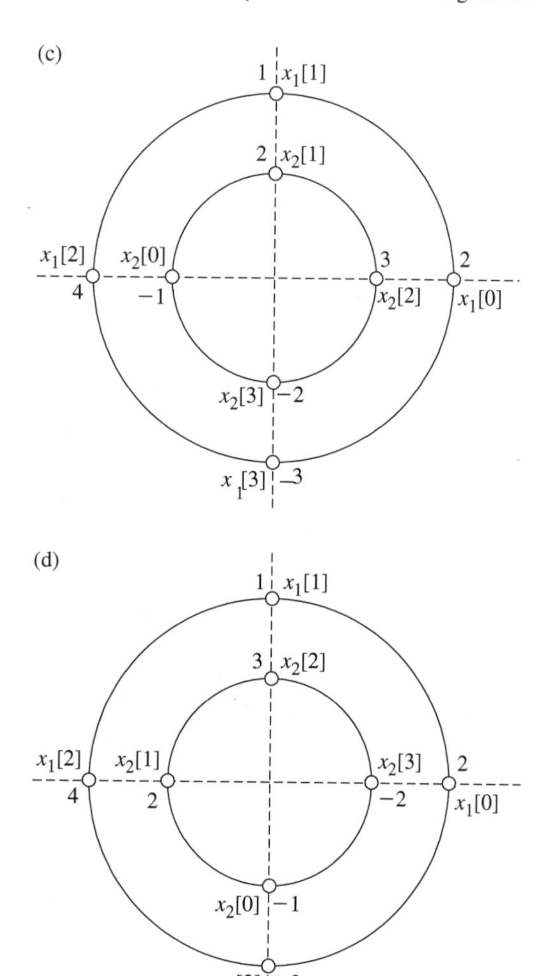

■ Example 7.17

Consider the following two sequences:

$$x_1[n] = \{2,\ 1,\ 4,\ -3\} \quad \text{and} \quad x_2[n] = \{-1,\ 2,\ 3,\ -2\}$$

Find the circular convolution

$$y[n] = x_1[n] \, \circledN \, x_2[n]$$

Use matrix multiplication method.

Solution The following matrices are formed using $x_1[n]$ and $x_2[n]$

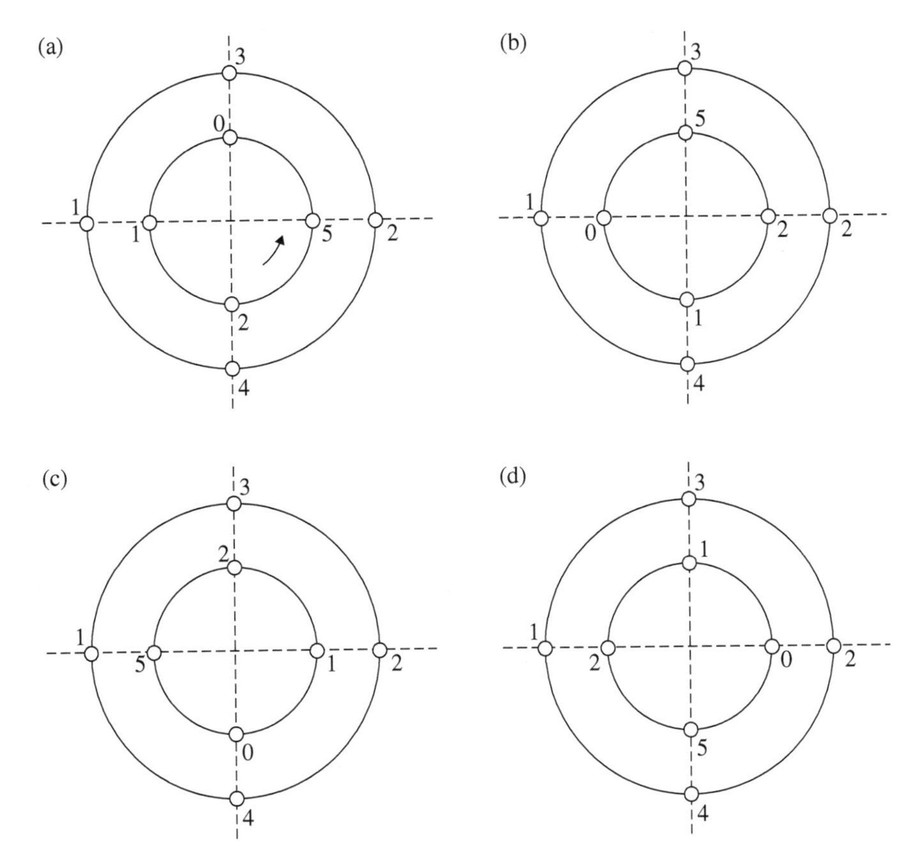

Fig. 7.10 Circular convolution by circle method for Example 7.19

$$\begin{bmatrix} 2 & -3 & 4 & 1 \\ 1 & 2 & -3 & 4 \\ 4 & 1 & 2 & -3 \\ -3 & 4 & 1 & 2 \end{bmatrix} \begin{bmatrix} -1 \\ 2 \\ 3 \\ -2 \end{bmatrix} = \begin{bmatrix} -2-6+12-2=2 \\ -1+4-9-8=-14 \\ -4+2+6+6=10 \\ 3+8+3-4=10 \end{bmatrix} = \begin{bmatrix} 2 \\ -14 \\ 10 \\ 10 \end{bmatrix}$$

$$\boxed{y[n] = \{2, -14, 10, 10\}}$$

7.9.3 *Circular Convolution-DFT-IDFT Method*

According to the DFT property given in Eq. (7.49), the circular convolution is,

$$x_1[n] \, \circledN \, x_2[n] \overset{\text{DFT}}{\longleftrightarrow} X_1(k)X_2(k)$$

For the given $x_1[n]$ and $x_2[n]$, $X_1(k)$ and $X_2(k)$ are found and for this product IDFT is found to get $y(k)$. Use of twiddle factor is to be preferred.

Example 7.18

Consider the following two sequences:

$$x_1[n] = \{2,\ 1,\ 4,\ -3\} \quad \text{and} \quad x_2[n] = \{-1,\ 2,\ 3,\ -2\}$$

Find the circular convolution

$$y[n] = x_1[n] \textcircled{N} x_2[n]$$

Use DFT-IDFT method.

Solution

$$X_4 = W_4 x_4 \qquad \text{where } N = 4$$

$$W_4 = \begin{bmatrix} 1 & 1 & 1 & 1 \\ 1 & -j & -1 & j \\ 1 & -1 & 1 & -1 \\ 1 & j & -1 & -j \end{bmatrix}$$

$$X_1(k) = \begin{bmatrix} 1 & 1 & 1 & 1 \\ 1 & j & -1 & -j \\ 1 & -1 & 1 & -1 \\ 1 & -j & -1 & j \end{bmatrix} \begin{bmatrix} 2 \\ 1 \\ 4 \\ -3 \end{bmatrix} = \begin{bmatrix} 4 \\ -2-j4 \\ 8 \\ -2+j4 \end{bmatrix}$$

$$X_2(k) = \begin{bmatrix} 1 & 1 & 1 & 1 \\ 1 & j & -1 & -j \\ 1 & -1 & 1 & -1 \\ 1 & -j & -1 & j \end{bmatrix} \begin{bmatrix} -1 \\ 2 \\ 3 \\ -2 \end{bmatrix} = \begin{bmatrix} 2 \\ -4-j4 \\ 2 \\ -4+j4 \end{bmatrix}$$

$$\begin{aligned} Y(k) &= X_1(k) X_2(k) \\ &= [4 \quad -2-j4 \quad 8 \quad -2+j4][2 \quad -4-j4 \quad 2 \quad -4+j4] \\ &= \{8 \quad (-8+j24) \quad 16 \quad (-8-j24)\} \\ y[n] &= \frac{1}{4} W_4^* Y_4 \\ &= \frac{1}{4} \begin{bmatrix} 1 & 1 & 1 & 1 \\ 1 & j & -1 & -j \\ 1 & -1 & 1 & -1 \\ 1 & -j & -1 & j \end{bmatrix} \begin{bmatrix} 8 \\ -8+j24 \\ 16 \\ -8-j24 \end{bmatrix} \end{aligned}$$

$$\boxed{y[n] = \{2,\ -14,\ 10,\ 10\}}$$

■ Example 7.19

Consider the following two sequences:

$$x_1[n] = \{2,\ 3,\ 1,\ 4\}$$
$$x_2[n] = \{5,\ 2,\ 1\}$$

Find the circular convolution

$$y[n] = x_1[n] \circledN x_2[n]$$

by the following methods. (1) Circle method; (2) Matrix multiplication method and (3) DFT-IDFT method.

Solution

1. **Circle Method**

$$x_1[n] = \{2,\ 3,\ 1,\ 4\}$$
$$x_2[n] = \{5,\ 2,\ 1\}$$

The length of $x_2[n]$ should be equal to the length of $x_1[n]$ sequence. This is done by zero padding. Thus,

$$x_2[n] = \{5,\ 2,\ 1,\ 0\}$$

The outer circle which is fixed represents $x_1[n]$ in a 4-point DFT. The inner circle which is rotated by one sample at a time represents $x_2[n]$ sequence as shown in Fig. 7.9.

$$y[0] = 5 \times 2 + 3 \times 0 + 1 \times 1 + 4 \times 2 = 19 \quad \text{(From Fig. 7.9a)}$$
$$y[1] = 2 \times 2 + 3 \times 5 + 1 \times 0 + 4 \times 1 = 23 \quad \text{(From Fig. 7.9b)}$$
$$y[2] = 2 \times 2 + 3 \times 2 + 1 \times 5 + 4 \times 0 = 13 \quad \text{(From Fig. 7.9c)}$$
$$y[3] = 2 \times 0 + 3 \times 1 + 1 \times 2 + 4 \times 5 = 25 \quad \text{(From Fig. 7.9d)}$$

$$\boxed{y[n] = \{19\quad 23\quad 13\quad 25\}}$$

2. **Matrix Method:** The following matrices are formed with $x_1[n]$ and $x_2[n]$

$$y[n] = \begin{bmatrix} 2\,4\,1\,3 \\ 3\,2\,4\,1 \\ 1\,3\,2\,4 \\ 4\,1\,3\,2 \end{bmatrix} \begin{bmatrix} 5 \\ 2 \\ 1 \\ 0 \end{bmatrix} = \begin{bmatrix} 10+8+1 \\ 15+4+4 \\ 5+6+2 \\ 20+2+3 \end{bmatrix} = \begin{bmatrix} 19 \\ 23 \\ 13 \\ 25 \end{bmatrix}$$

$$\boxed{y[n] = \{19 \quad 23 \quad 13 \quad 25\}}$$

3. **DFT-IDFT Method**
   For a 4-point DFT

$$W_4 = \begin{bmatrix} 1 & 1 & 1 & 1 \\ 1 & -j & -1 & j \\ 1 & -1 & 1 & -1 \\ 1 & j & -1 & -j \end{bmatrix}$$

$$X_N = W_4 x_n$$

$$X_1(k) = \begin{bmatrix} 1 & 1 & 1 & 1 \\ 1 & j & -1 & j \\ 1 & -1 & 1 & -1 \\ 1 & j & -1 & -j \end{bmatrix} \begin{bmatrix} 2 \\ 3 \\ 1 \\ 4 \end{bmatrix} = \begin{bmatrix} 10 \\ 1+j \\ -4 \\ 1+j \end{bmatrix}$$

$$X_2 = W_4 x_4$$

$$X_2(k) = \begin{bmatrix} 1 & 1 & 1 & 1 \\ 1 & -j & -1 & j \\ 1 & -1 & 1 & -1 \\ 1 & j & -1 & -j \end{bmatrix} \begin{bmatrix} 5 \\ 2 \\ 1 \\ 0 \end{bmatrix} = \begin{bmatrix} 8 \\ 4-j2 \\ 4 \\ 4+j2 \end{bmatrix}$$

$$\begin{aligned} Y(k) &= X_1(k) X_2(k) \\ &= [10 \quad (1+j) \quad -4 \quad (1-j)][8 \quad (4-j2) \quad 4 \quad (4+j2)] \\ &= \{80 \quad (6+j2) \quad -16 \quad (6-j2)\} \end{aligned}$$

$$y[n] = \frac{1}{4} W_4^* Y_4$$

$$W_N^* = \begin{bmatrix} 1 & 1 & 1 & 1 \\ 1 & j & -1 & -j \\ 1 & -1 & 1 & -1 \\ 1 & -j & -1 & j \end{bmatrix}$$

$$y[n] = \frac{1}{4} \begin{bmatrix} 1 & 1 & 1 & 1 \\ 1 & j & -1 & -j \\ 1 & -1 & 1 & -1 \\ 1 & -j & -1 & j \end{bmatrix} \begin{bmatrix} 80 \\ 6+j2 \\ -16 \\ 6-j2 \end{bmatrix} = \begin{bmatrix} 19 \\ 23 \\ 13 \\ 25 \end{bmatrix}$$

$$\boxed{y[n] = \{19,\ 23,\ 13,\ 25\}}$$

7.10 Fast Fourier Transform

For spectral analysis of DT signals, DFT approach is very straight forward. For larger values of N which is greater than 128 points, DFT becomes tedious because of the huge number of mathematical operations required to perform. Several algorithms have been developed to ease the implementation of DFT. The algorithm developed by Cooley and Turkey in 1965 is the most efficient one and its application is discussed here.

Consider Eqs. (7.26) and (7.27) which are given below

$$X(k) = \sum_{n=0}^{N-1} x[n]e^{-j2\pi kn/N}$$

$$x[n] = \frac{1}{N}\sum_{k=0}^{N-1} X(k)W_N^{kn}$$

In direct evaluation of spectral components, the number of complex multiplication and addition required are N^2 and $N(N-1)$ respectively. Such a huge number of mathematical operations limit the bandwidth of digital signal processors. Classical DFT approach does not use the two important properties of twiddle factor, namely symmetry and periodicity properties which are given as

$$W_N^{k+N/2} = -W_N^k \tag{7.53}$$

$$W_N^{k+N} = W_N^k \tag{7.54}$$

Radix-2 FFT algorithm exploits these two properties thereby removing redundant mathematical operations. This results in the required number of complex multiplication for an N-point FFT approximately as $\frac{N}{2}\log_2 N$. However, the results obtained using FFT is exactly the same as that of DFT. The efficiency of the FFT algorithm increases as the number N is increased. For example, if $N = 512$, DFT requires nearly 110 times more multiplications than FFT algorithm.

The basic principle of FFT algorithm is to decompose DFT into successively smaller DFTs. The number of points N must be equal to 2^k where k is some positive integer. The FFT algorithms have been developed in (1) Decimation in time and (2) Decimation in frequency.

7.10.1 FFT Algorithm-Decimation in Time

In Radix-2 FFT each DFT is divided into two smaller DFTs and in Radix-4, each DFT is divided into four smaller DFTs. The N-point DFT is given by the following equation:

$$X(k) = \sum_{n=0}^{N-1} x[n]e^{-j2\pi kn/N}$$
$$= \sum_{n=0}^{N-1} x[n]W_N^{kn} \tag{7.55}$$

The input sequence $x[n]$ is divided into groups of even and odd indexed elements. Hence, Eq. (7.55) can be split up into two parts and represented as given below

$$X(k) = \sum_{\substack{n=0 \\ (n=\text{even})}}^{N-1} x[n]W_N^{kn} + \sum_{\substack{n=0 \\ (n=\text{odd})}}^{N-1} x[n]W_N^{kn} \tag{7.56}$$

Let us introduce the new variable

$$m = \begin{cases} \frac{n}{2} & n = \text{even} \\ \frac{(n-1)}{2} & n = \text{odd} \end{cases} \tag{7.57}$$

Now

$$X(k) = \sum_{m=0}^{\frac{N}{2}-1} x_e[2m]W_N^{2km} + \sum_{m=0}^{N-1} x_o[2m+1]W_N^{(2\,m+1)k} \tag{7.58}$$

In the second summation, taking out the factor W_N^k we get

$$X(k) = \sum_{m=0}^{\frac{N}{2}-1} x_e[2m]W_N^{2km} + W_N^k \sum_{m=0}^{N-1} x_o[2m+1]W_N^{2\,mk} \tag{7.59}$$

But

$$W_N^2 = e^{-j2\pi 2/N} = e^{-j2\pi/(\frac{N}{2})}$$
$$= W_{N/2}$$

Equation (7.59) can therefore be written as

$$X(k) = \sum_{m=0}^{\frac{N}{2}-1} x_e[m]W_{N/2}^{km} + W_N^k \sum_{m=0}^{\frac{N}{2}-1} x_o[m]W_{N/2}^{km} \tag{7.60}$$
$$= X_e(k) \pm W_N^k X_o(k) \tag{7.61}$$

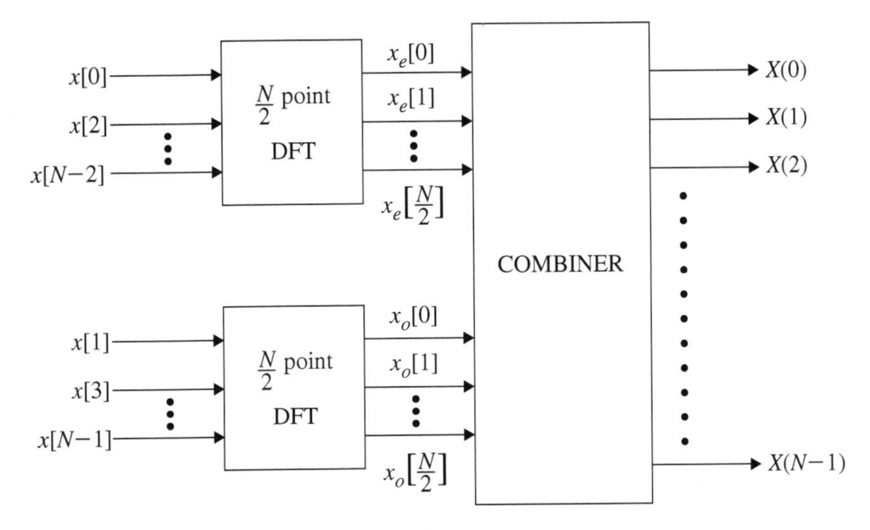

Fig. 7.11 N-point DFT realization using two $\frac{N}{2}$ point DFTs

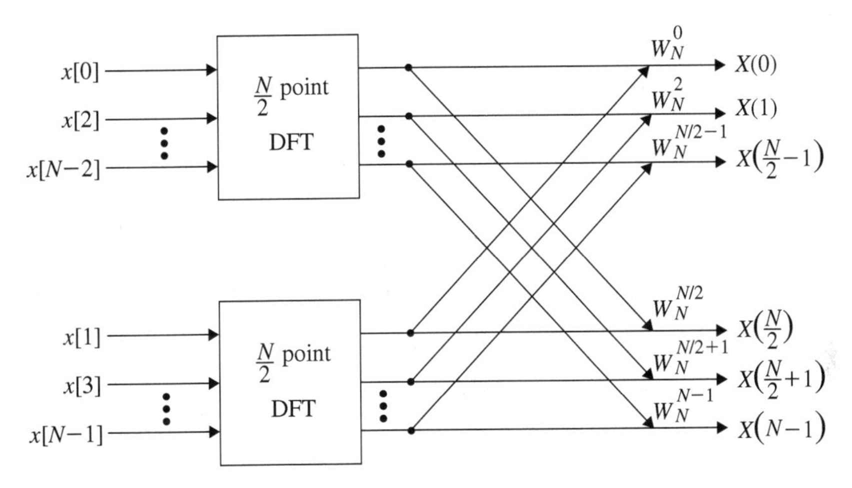

Fig. 7.12 Data flow graph for N-point FFT using $\frac{N}{2}$ points DFTs

Equation (7.61) is represented as shown in Fig. 7.11. Figure 7.12 shows the data flow graph.

In Eq. (7.61) for the first $\frac{N}{2}$ point, transforms are obtained by summing up the weighted outputs of $X_e(k)$ and $X_o(k)$. In view of the symmetry property of twiddle factor $W_N^{k+(N/2)} = -W_N^k$, the remaining $\frac{N}{2}$ transforms are obtained by differencing the weighted outputs.

The $\frac{N}{2}$ point DFT is further divided into two groups so that we get $\frac{N}{4}$-point DFTs and so on until only two points DFTs are used to realize the N-point FFT. This is

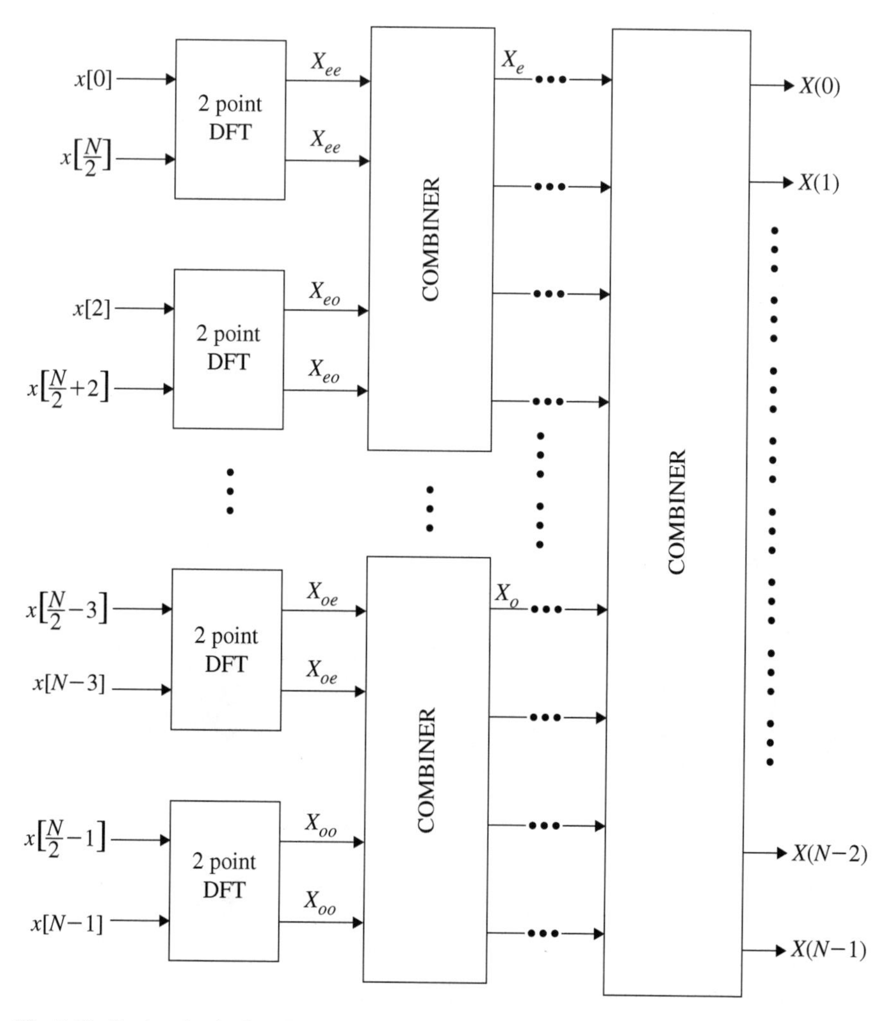

Fig. 7.13 Decimation in time Radix-2 FFT algorithm

the reason to put the constraint that the number of points in FFT are chosen as the power of 2. This is called Radix-2 FFT (Fig. 7.13).

The upper $\frac{N}{2}$ point FFT can be further subdivided into two $\frac{N}{4}$ point DFTs with their odd and even components as

$$X_e(m) = \sum_{n=0}^{\frac{N}{2}-1} x_e[2n]W_{N/2}^{nm} = \sum_{n=0}^{\frac{N}{4}-1} x_{ee}[4n]W_{N/2}^{2nm} \sum_{n=0}^{\frac{N}{4}-1} x_{eo}[4n+2]W_{N/2}^{(2n+1)m} \tag{7.62}$$

Substituting $W_{N/2}^{2nm} = W_{N/2}^{nm}$, Eq. (7.62) is written as

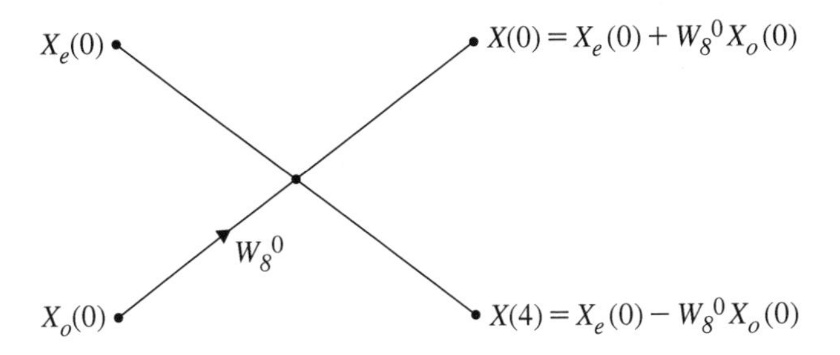

Fig. 7.14 Flow graph for $X(0)$ and $X(4)$ in Eq. (7.65)

$$X_e(m) = \sum_{n=0}^{\frac{N}{4}-1} x_{ee}[4n]W_{N/4}^{nm} + W_{N/2}^{m} \sum_{n=0}^{\frac{N}{4}-1} x_{eo}[4n+2]W_{N/4}^{nm} \tag{7.63}$$

$$X_0(m) = \sum_{n=0}^{\frac{N}{4}-1} x_{oe}[4n+1]W_{N/4}^{nm} + \sum_{n=0}^{\frac{N}{4}-1} x_{oe}[4n+3]W_{N/4}^{nm} \tag{7.64}$$

where $x_{ee}[4n]$ is the DFT for $m = 4n$, $n = 0, 1, \ldots, \frac{N}{4} - 1$.

To illustrate the decimation in time, consider an 8-point FFT. Equation (7.61) becomes

$$X(k) = \begin{cases} X_e(k) + W_8^k X_o(k) & \text{for } 0 \le k \le 3 \\ X_e(k) - W_8^k X_o(k) & \text{for } 4 \le k \le 7 \end{cases}$$

For different values of k, we get

$$\begin{aligned} X(0) &= X_e(0) + W_8^0 X_o(0); \quad X(1) = X_e(1) + W_8^1 X_o(1) \\ X(2) &= X_e(2) + W_8^2 X_o(2); \quad X(3) = X_e(3) + W_8^3 X_o(3) \\ X(4) &= X_e(0) + W_8^0 X_o(0); \quad X(5) = X_e(1) + W_8^1 X_o(1) \\ X(6) &= X_e(2) + W_8^2 X_o(2); \quad X(7) = X_e(8) + W_8^3 X_o(3) \end{aligned} \tag{7.65}$$

From Eq. (7.65), it is obvious that $X(0)$ and $X(4)$ have the same inputs. This is represented in Fig. 7.14.

Using the above representation, it can be extended to 8-point FFT. The 8-point FFT flow graph using 4-point DFTs is shown in Fig. 7.15.

Now the 4-point DFT can be further decomposed into two 2-point DFTs with the odd and even inputs of the respective 4-points DFTs $X_e(k)$ and $X_o(k)$ are obtained using Eqs. (7.63) and (7.64), respectively. Thus,

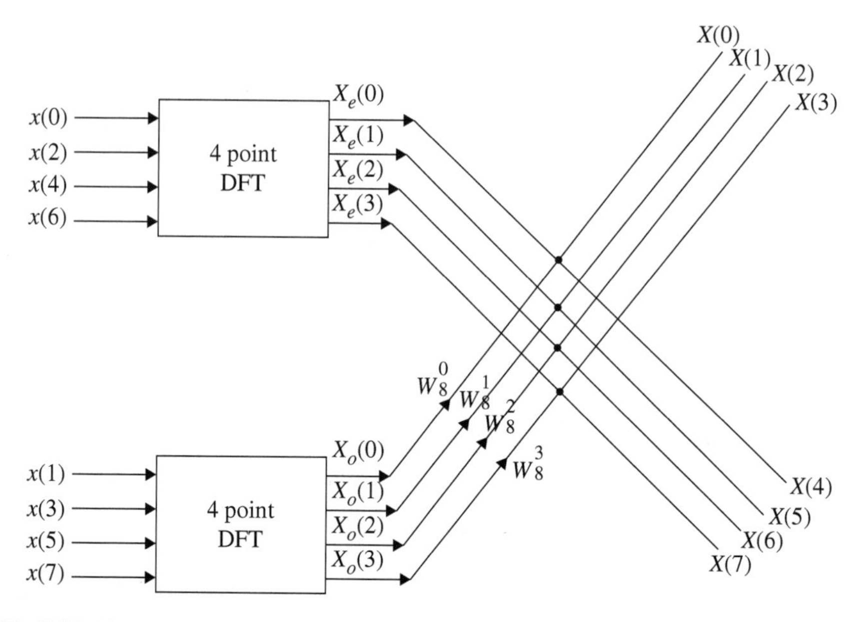

Fig. 7.15 Flow graph for 8-point FFT using 4-point DFTs

$$
\begin{aligned}
X_e(0) &= X_{ee}(0) - W_8^0 X_{eo}(0) \\
X_e(1) &= X_{ee}(1) - W_8^2 X_{eo}(1) \\
X_e(2) &= X_{ee}(0) - W_8^0 X_{eo}(0) \\
X_e(3) &= X_{ee}(1) - W_8^2 X_{eo}(1)
\end{aligned} \tag{7.66}
$$

where X_{ee} is the two points DFT of the even index of $x_e[n]$. X_{0e} is the 2-point DFT of the odd index of $x_e[n]$. Equation (7.66) is represented by the flow chart as shown in Fig. 7.16.

The final 2-point DFT (first stage) involves only addition and subtraction since the twiddle factor present here are W_8^0 and W_8^4 respectively. Therefore

$$
\begin{aligned}
X_{ee}(0) &= x(0) + W_8^0 x(4) = x(0) + x(4) \\
X_{ee}(1) &= x(0) + W_8^4 x(4) = x(0) - x(4)
\end{aligned}
$$

and so on. The signal flow graph for the 8-point FFT is shown in Fig. 7.17.

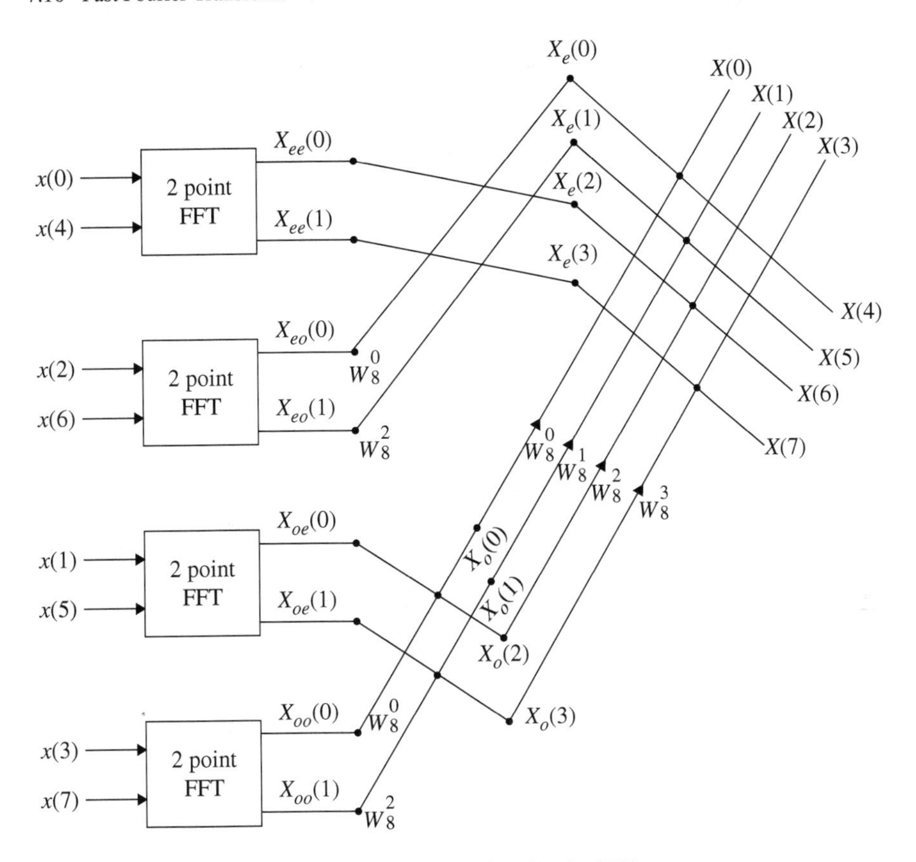

Fig. 7.16 Signal flow graph for 8-point FFT using four 2-point FFTs

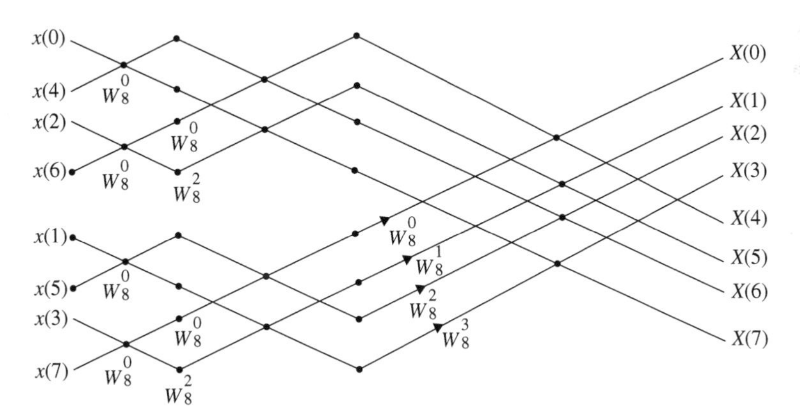

Fig. 7.17 Signal flow chart for 8-point FFT

■ Example 7.20

Compute the 8-point DFT using FFT algorithm for the following sequence

$$x[n] = \{0.5, 0.5, 0.5, 0.5, 0, 0, 0, 0\}$$

(*Anna University, December, 2007*)

Solution **Pass-1**

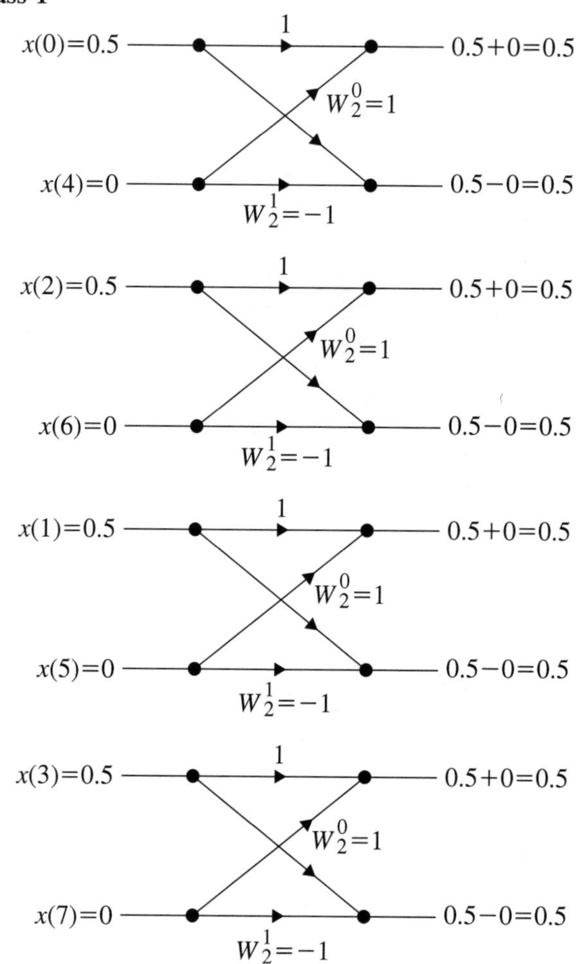

Pass-2

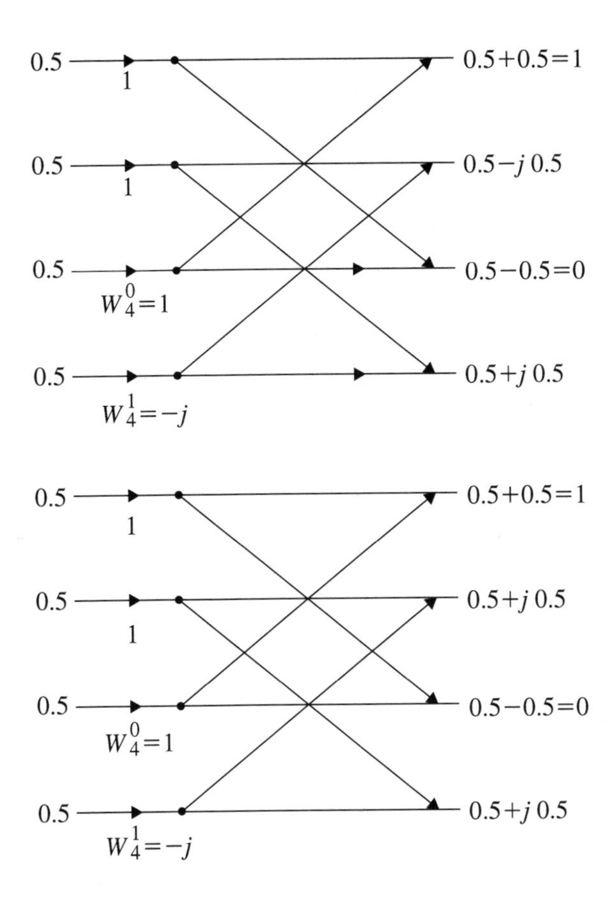
0.5
0.5
0.5
0.5
1
1
$W_4^0=1$
$W_4^1=-j$
$0.5+0.5=1$
$0.5-j\,0.5$
$0.5-0.5=0$
$0.5+j\,0.5$
0.5
0.5
0.5
0.5
1
1
$W_4^0=1$
$W_4^1=-j$
$0.5+0.5=1$
$0.5+j\,0.5$
$0.5-0.5=0$
$0.5+j\,0.5$

Pass-3

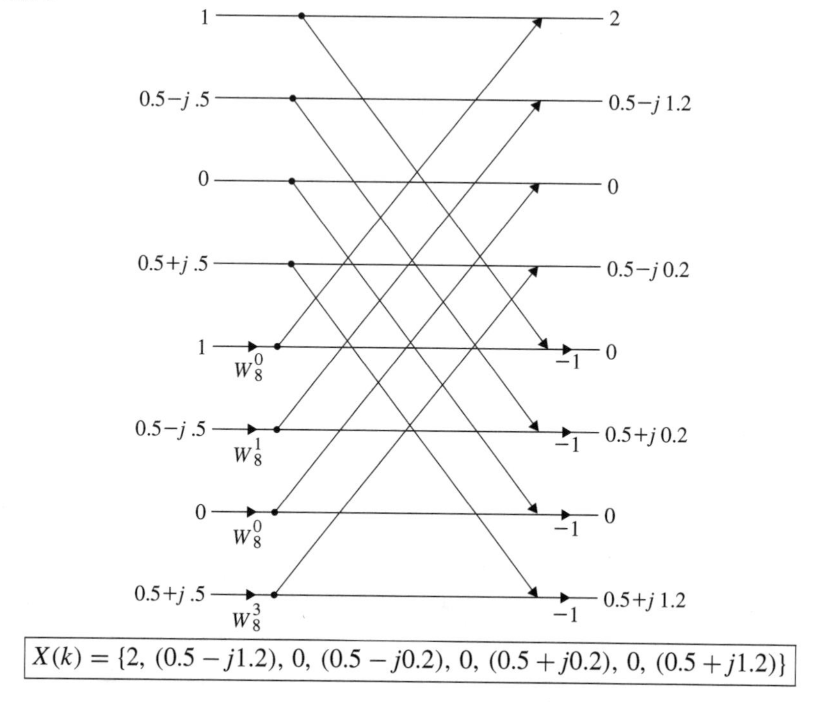

$$X(k) = \{2,\ (0.5 - j1.2),\ 0,\ (0.5 - j0.2),\ 0,\ (0.5 + j0.2),\ 0,\ (0.5 + j1.2)\}$$

7.10.2 FFT Algorithm-Decimation in Frequency

In the decimation in time (DIT) FFT algorithm, the input sequence is divided into even and odd indexed elements. In the decimation in frequency (DIF) FFT algorithm, the input sequence is divided into groups as first half of the sequence and second half of the sequence. The subsequent steps lead to grouping the spectral components into even indexed and odd indexed elements.

The DFT of the sequence $x[n]$ is expressed as

$$X(k) = \sum_{n=0}^{N-1} x[n] W_N^{kn} \tag{7.67}$$

Now divide the input sequence into two groups as

$$\begin{aligned} X(k) &= \sum_{n=0}^{\frac{N}{2}-1} x[n] W_N^{kn} + \sum_{n=\frac{N}{2}}^{N-1} x[n] W_N^{kn} \\ &= \sum_{n=0}^{\frac{N}{2}-1} x[n] W_N^{kn} + \sum_{n=0}^{\frac{N}{2}-1} x\left[n + \frac{N}{2}\right] W_N^{(n+\frac{N}{2})k} \end{aligned} \tag{7.68}$$

$$= \sum_{n=0}^{\frac{N}{2}-1} x[n]W_N^{kn} + W_N^{\frac{N}{2}k} \sum_{n=0}^{\frac{N}{2}-1} x\left[n + \frac{N}{2}\right] W_N^{kn} \tag{7.69}$$

since $W_N^{\frac{N}{2}k} = (-1)^k$, the above equation is written as

$$X(k) = \sum_{n=0}^{\frac{N}{2}-1} x[n]W_N^{kn} + (-1)^k \sum_{n=0}^{\frac{N}{2}-1} x\left[n + \frac{N}{2}\right] W_N^{kn}$$

$$= \sum_{n=0}^{\frac{N}{2}-1} \left[x[n] + (-1)^k x\left[n + \frac{N}{2}\right] W_N^{kn}\right] \tag{7.70}$$

Now dividing $X(k)$ into even and odd indexed elements we get, for even indexed elements

$$X(2k) = \sum_{n=0}^{\frac{N}{2}-1} \left[x[n] + x\left(n + \frac{N}{2}\right)\right] W_N^{2kn} \quad k = 0, 1, \ldots, \left(\frac{N}{2} - 1\right)$$

Since $W_N^{2kn} = W_{\frac{N}{2}}^{kn}$ and $(-1)^k = 1$, the above equation written as

$$X(2k) = \sum_{n=0}^{\frac{N}{2}-1} \left[x[n] + x\left(n + \frac{N}{2}\right)\right] W_N^{kn} \quad k = 0, 1, \ldots, \left(\frac{N}{2} - 1\right) \tag{7.71}$$

Similarly, the odd indexed elements of $X(k)$ are expressed as

$$X(2k+1) = \sum_{n=0}^{\frac{N}{2}-1} \left[x[n] - x\left(n + \frac{N}{2}\right)\right] W_N^{(2k+1)n}$$

$$= \sum_{n=0}^{\frac{N}{2}-1} \left[x[n] - x\left(n + \frac{N}{2}\right) W_N^{n}\right] W_{\frac{N}{2}}^{kn} \tag{7.72}$$

The signal flow graph for DIF-FFT algorithm is shown in Fig. 7.18.

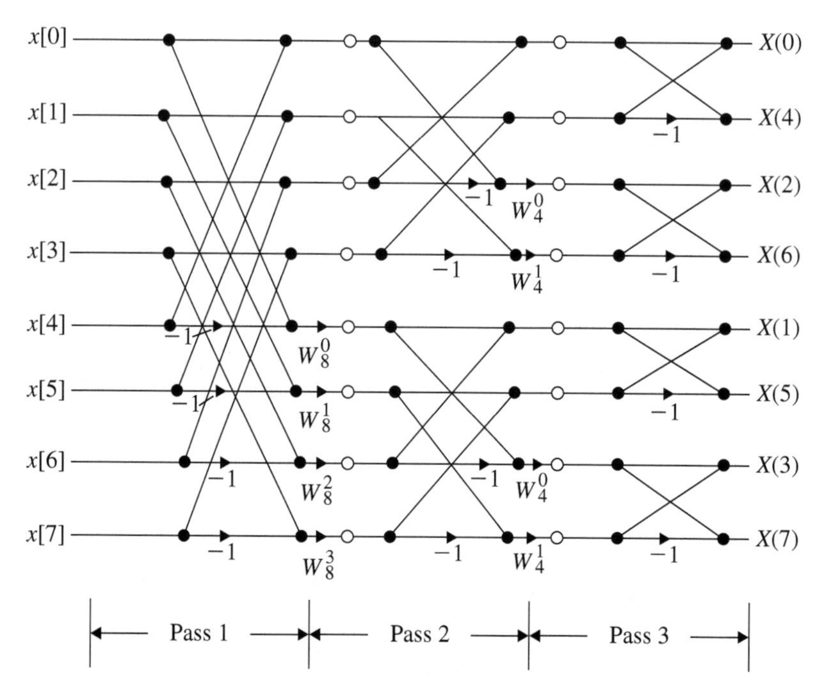

Fig. 7.18 Signal flow graph of 8-point DIF-FFT algorithm. $W_8^0 = 1$; $W_8^1 = \frac{1}{\sqrt{2}} - j\frac{1}{\sqrt{2}}$; $W_8^2 = -j$; $W_8^3 = -\frac{1}{\sqrt{2}}(1+j)$; $W_4^0 = 1$; $W_4^1 = -j$

■ Example 7.21

Compute the 8-point DFT using FFT algorithm in decimation in frequency for the following frequency

$$x[n] = \{0.5, 0.5, 0.5, 0.5, 0, 0, 0, 0\}$$

(Anna University, December, 2007)

Solution

Step 1.

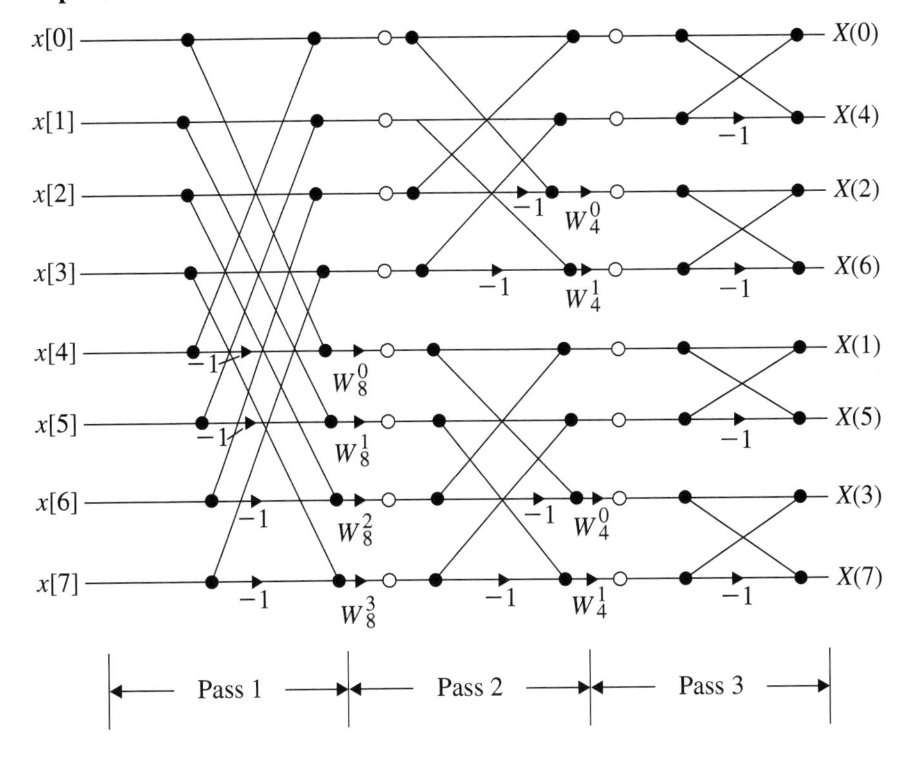

Step 2.

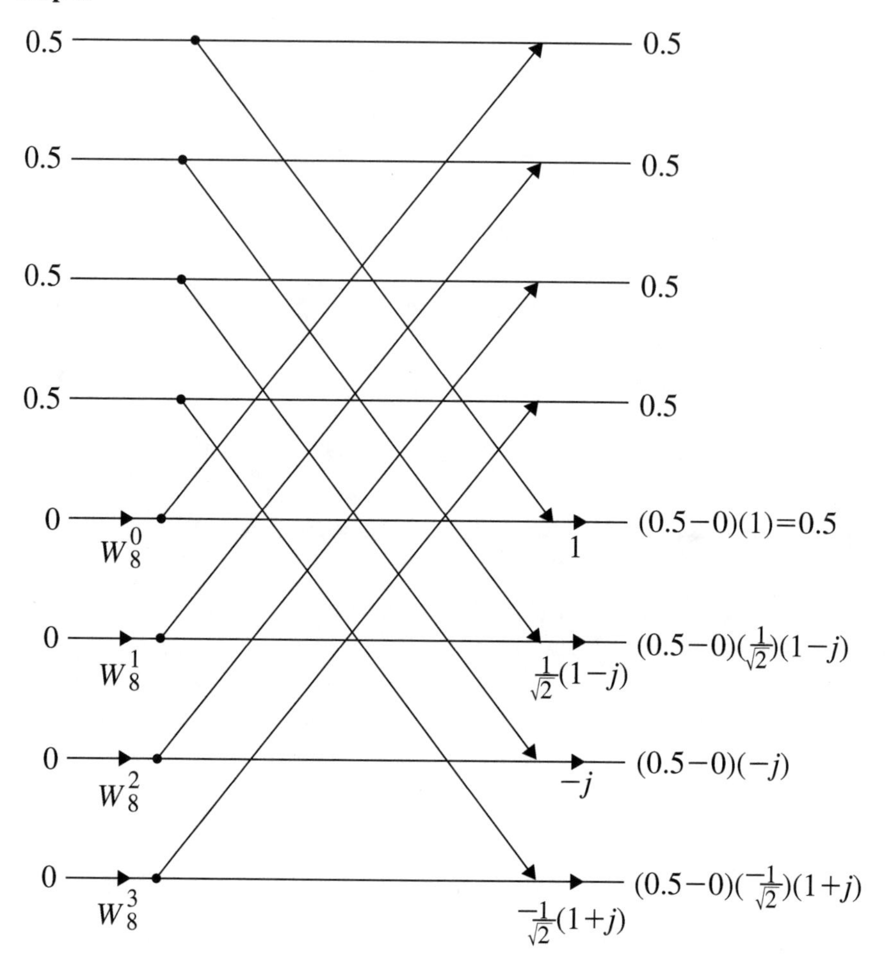

$$X(k) = \{1,\ 1,\ 0,\ 0,\ (0.5 - j0.5),\ -j0.7,\ (0.5 + j0.5),\ -j0.7\}$$

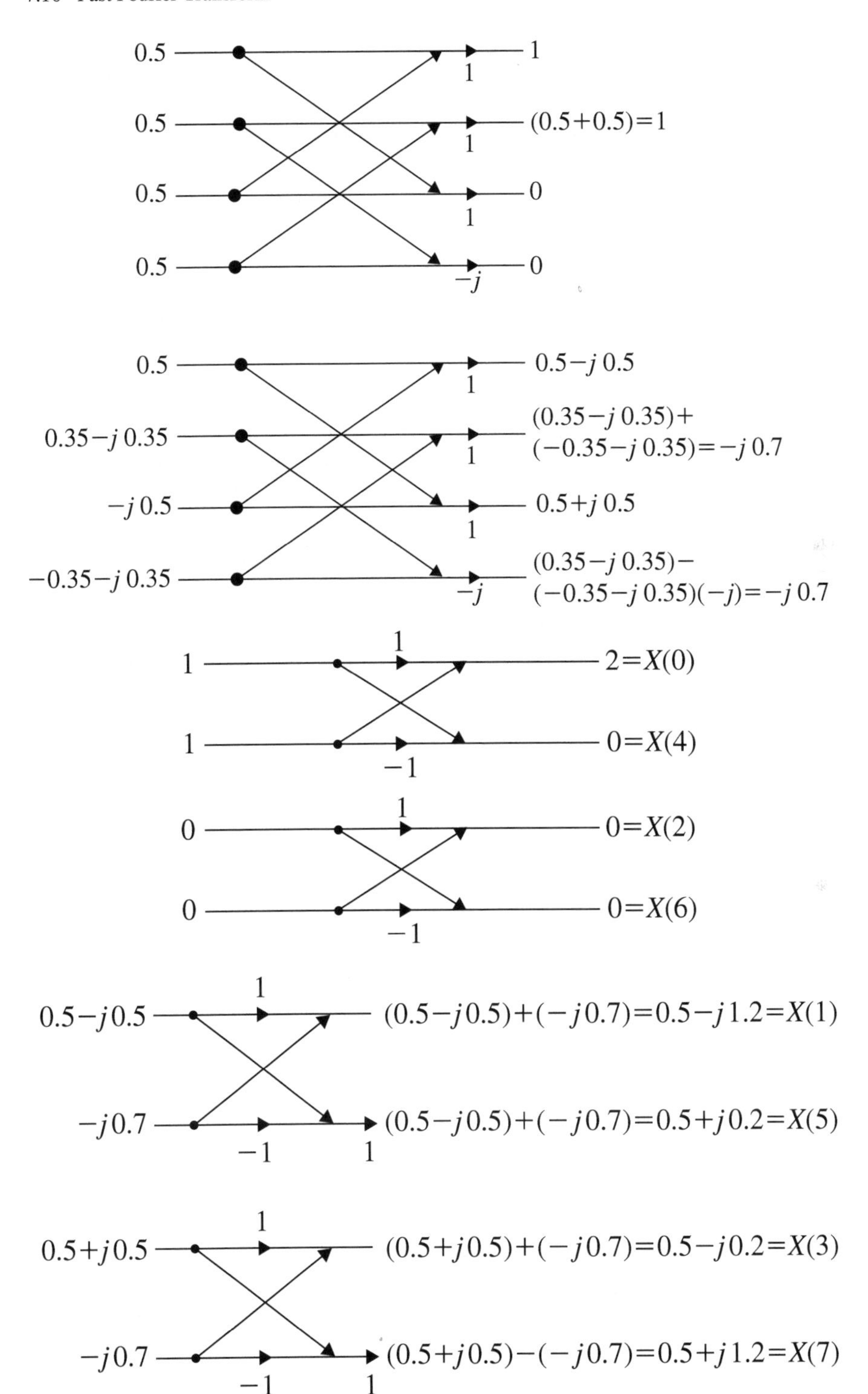

0.5
0.5
0.5
0.5
1
1
1
−j
1
(0.5+0.5)=1
0
0
0.5
0.35−j 0.35
−j 0.5
−0.35−j 0.35
1
1
1
−j
0.5−j 0.5
(0.35−j 0.35)+
(−0.35−j 0.35)=−j 0.7
0.5+j 0.5
(0.35−j 0.35)−
(−0.35−j 0.35)(−j)=−j 0.7
1
1
1
−1
2=X(0)
0=X(4)
0
0
1
−1
0=X(2)
0=X(6)
0.5−j0.5
−j0.7
1
−1
1
(0.5−j0.5)+(−j0.7)=0.5−j1.2=X(1)
(0.5−j0.5)+(−j0.7)=0.5+j0.2=X(5)
0.5+j0.5
−j0.7
1
−1
1
(0.5+j0.5)+(−j0.7)=0.5−j0.2=X(3)
(0.5+j0.5)−(−j0.7)=0.5+j1.2=X(7)

$$\boxed{X(k) = \{2,\ (0.5 - j1.2),\ 0,\ (0.5 - j0.2),\ 0,\ (0.5 + j0.2),\ 0,\ 0.5 + j1.2\}}$$

■ Example 7.22

Find the response of an LTI system with impulse response

$$h[n] = \{1,\ 2\}$$

for the input

$$x[n] = \{1,\ 2,\ 1\}$$

using DIT-FFT algorithm.

(*Anna University, April, 2005*)

Solution

$$h[n] = \{1,\ 2\}$$
$$x[n] = \{1,\ 2,\ 1\}$$

The sequences should be of equal 2^k length. After zero padding, the given sequences are written as

$$x[n] = \{1, 2, 1,\ 0,\} \quad \text{and} \quad h[n] = \{1,\ 2,\ 0,\ 0\}$$

Step 1.

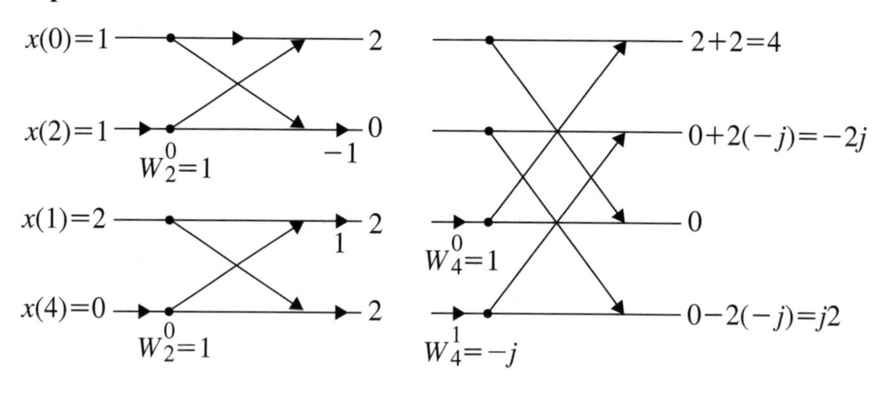

Step 2.

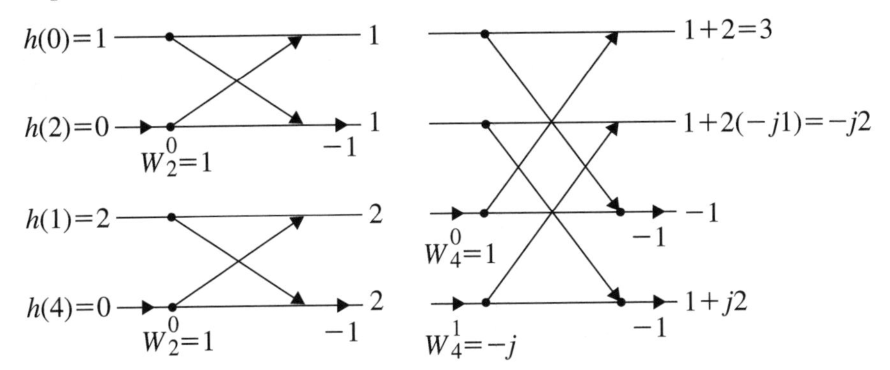

Step 3.

$$y[n] = x[n] * h[n]$$

and

$$Y(k) = X(k)H(k)$$

$$= [4 \quad -j2 \quad 0 \quad j2] \begin{bmatrix} 3 \\ 1-j2 \\ -1 \\ 1+j2 \end{bmatrix}$$

$$Y(k) = \{12,\ (-4-j2),\ 0,\ (-4+j2)\}$$

Step 4. To find $y[n]$

$Y^*(0)=12$ → 12
$Y^*(2)=0$, $W_2^0=1$ → 12
$Y^*(1)=-4+j2$ → $(-4+j2)(-4-j2)=-8$
$Y^*(3)=-4-j2$, $W_2^0=1$ → $(-4+j2)-(-4-j2)=j4$

$$Y^*(k) = \{12,\ (-4+j2),\ 0,\ (-4-j2)\}$$

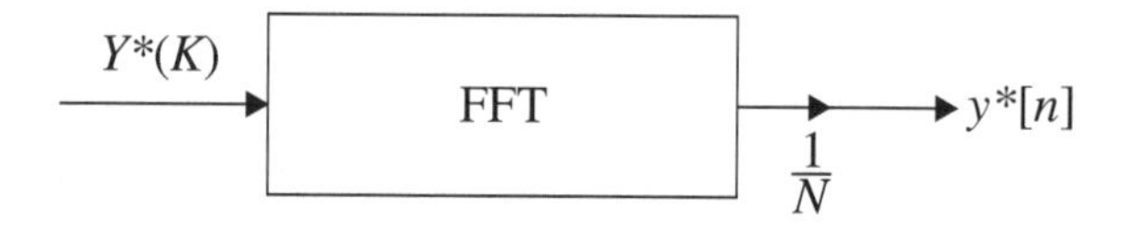

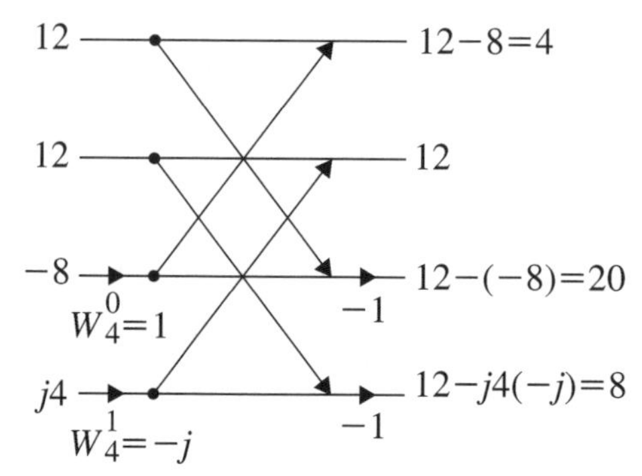

The output sequence $y[n] = \frac{1}{N}[Ny^*[n]]$

$$\boxed{y[n] = \{1, 3, 5, 2\}}$$

Summary

1. The Fourier transform is nothing but the description of $x[n]$ in the frequency domain. The transformed spectrum is periodic.
2. Discrete Time Fourier Transform (DTFT) and Inverse Discrete Time Fourier Transform (IDTFT) are defined.
3. The connection between the DTFT and the z-transform is that if the ROC of z-transform contains unit circle, the DTFT is evaluated on the unit circle. This is equivalent to substituting $z = e^{j\Omega}$ in the z-transform to get DTFT provided $x[n]$ is summable.
4. A certain properties of DTFT have been derived. These properties are useful to determine the DTFT and IDTFT very quickly.
5. Using DTFT and IDTFT, it is possible to solve LTI systems characterized by difference equation.
6. The DTFT is used to transform $x[n]$ to a continuous periodic spectrum. This short of representation is not very much suited for use in digital computer. The DTFT is sampled at finite number of frequency points in one period. This is called Discrete Fourier Transform (DFT) which has many advantages over that of DTFT.
7. Similar to DTFT pair, DFT pair is also defined.
8. The number of computations required to perform DFT is significantly reduced by an algorithm known as Fast Fourier Transform (FFT).

9. FFT using decimation in time and decimation in frequency are explained with numerical examples illustrated.

Exercises

I. Short Answer Type Questions

1. **Define DTFT pair.**
The discrete time Fourier transform is defined as

$$X(\Omega) = \sum_{n=-\infty}^{\infty} x[n]e^{-j\Omega n}$$

The inverse discrete time Fourier transform is defined as

$$x[n] = \frac{1}{2\pi}\int_{2\pi} X(\Omega)e^{jn\Omega}d\Omega$$

2. **What are the necessary and sufficient conditions for the existence of DTFT?**
The necessary and sufficient conditions for the existence of DTFT is that the sequence $x[n]$ should be absolutely summable. That is

$$\sum_{n=-\infty}^{\infty} |x[n]| < \infty$$

Alternatively, if the sequence $x[n]$ has finite energy

$$E = \sum_{n=-\infty}^{\infty} |x[n]|^2 < \infty$$

then DTFT exists.
3. **Define DFT pair.**
The discrete Fourier transform pair is defined as follows. The DFT is given by

$$X(k) = \sum_{n=0}^{N-1} x[n] e^{-j2\pi kn/N}$$

The IDTFT is given by

$$x[n] = \frac{1}{N} \sum_{k=0}^{N-1} X(k) e^{j2\pi kn/N}$$

4. **Is the IDFT of DFT periodic or not? Give your comment.**

$$x[n] = \frac{1}{N} \sum_{k=0}^{N-1} X(\Omega) W_N^{-kn}$$

where $W_N = 1\angle{-2\pi/N}$. W_N lies on the unit circle in the complex plane and gets repeated for every 2π. Hence, IDFT is periodic with period 2π.

5. **What is the DFT of a^n?**

$$a^n \overset{\text{DTFT}}{\longleftrightarrow} \frac{(1 - a^N)}{(1 - ae^{-j\frac{2\pi k}{N}})}$$

6. **What is the connection between the z-transform and the DTFT?**
When the ROC for $X[z]$ includes the unit circle, substituting $z = e^{j\Omega}$ in $X[z]$ gives the DTFT $X(\Omega)$.

7. **Find the system transfer function for the following difference equation?**

$$\mathbf{y[n] - 0.8y[n-1] = x[n]}$$

$$H[\Omega] = \frac{1}{(1 - 0.8e^{-j\Omega})}$$

8. **Find the frequency responses for the following difference equation?**

$$\mathbf{y[n] - 0.1y[n-1] = x[n]}$$

$$H[\Omega] = \frac{1\angle{-\tan^{-1}\frac{0.1\sin\Omega}{(1-0.1\cos\Omega)}}}{\sqrt{(1 - 0.1\cos\Omega)^2 + 0.01\sin^2\Omega}}$$

9. **What is the DTFT of unit step response?**
The DTFT of unit step response is

$$X(\Omega) = \pi\delta(\Omega) + \frac{1}{(1 - e^{-j\Omega})}$$

10. **What is the DTFT of $x[-n]$?**
The DTFT of $x[-n]$ is $X(-\Omega)$
11. **What do you understand by zero padding?**
In evaluating DFT we assume that the length of the DFT which is N which is equal to the length L of the sequence $x[n]$. If $N < L$, time domain aliasing occurs. To avoid this dummy samples are added with value 0 in $X[k]$. This is called zero padding.
12. **Find DTFT and DFT of $\delta[n] = 1$?**

$$\delta[n] \overset{\text{DTFT}}{\longleftrightarrow} 1$$

$$\delta[n] \overset{\text{DFT}}{\longleftrightarrow} 1$$

13. **Find the inverse DTFT of $X[\Omega] = \delta[\Omega]$?**

$$\delta[\Omega] \overset{\text{IDTFT}}{\longleftrightarrow} \frac{1}{2\pi} \qquad -\pi < \Omega < \pi$$

14. **Find $Y(\Omega)$.**

$$x[n] = \left(-\frac{1}{3}\right)^n u[n]$$

$$h[n] = \left(\frac{1}{3}\right)^n u[n]$$

$$y[n] = x[n] * h[n]$$

$$Y(\Omega) = \frac{1}{\left(1 + \frac{1}{3}e^{-j\Omega}\right)\left(1 - \frac{1}{3}e^{-j\Omega}\right)}$$

15. **Find $X(\Omega)$.**

$$x[n] = u[n] - u[n-3]$$

$$X(\Omega) = 1 + e^{-j\Omega} + e^{-j2\Omega}$$

II. Long Answer Type Questions

1. **Find the DTFT of**

$$\mathbf{x[n] = (0.5)^n u[n] - 2^{-n} u[-n-1]}$$

(Anna University, December, 2007)

DTFT for $2^{-n}u[-n-1]$ does not converge and hence does not exist.

2. **Find the DTFT of**

$$\mathbf{x[n] = (0.5)^n u[n] - (.5)^{-n} u[-n-1]}$$

$$X(\Omega) = \frac{(2 - 2.5e^{-j\Omega})}{(1 - 0.5e^{-j\Omega})(1 - 2e^{-j\Omega})}$$

3. **Find the DTFT of**

$$\mathbf{x[n] = \left(\frac{1}{3}\right)^n u[n+2]}$$

$$X(\Omega) = \frac{9e^{j2\Omega}}{\left(1 - \frac{1}{3}e^{-j\Omega}\right)}$$

4. **Find the DTFT of**

$$\mathbf{x[n] = n\left(\frac{1}{4}\right)^n u[n]}$$

$$X(\Omega) = \frac{0.25e^{j\Omega}}{\left(e^{j\Omega} - 0.25\right)^2}$$

5. **Find the DTFT of $x[n]$**

$$\mathbf{x[n] = \begin{cases} 1 & 0 \leq |n| \leq 8 \\ 0 & \text{otherwise} \end{cases}}$$

$$X(\Omega) = \frac{\sin 8.5\Omega}{\sin 0.5\Omega}$$

6. **Find $Y(\Omega)$.**

$$x[n] = \left(\frac{1}{6}\right)^n u[n]$$

$$h[n] = -(3)^n u[-n-1]$$

$$y[n] = x[n] * h[n]$$

$$Y(\Omega) = \frac{1}{\left(1 - \frac{1}{6}e^{-j\Omega}\right)\left(3e^{-j\Omega} - 1\right)}$$

7. **Find $X(\Omega)$.**

$$x[n] = (n+2)\left(\frac{1}{2}\right)^n u[n]$$

$$X(\Omega) = \frac{2(1 + 1.25e^{-j\Omega})}{(1 - 0.5e^{-j\Omega})^2}$$

8. **Find $X(\Omega)$.**

$$x[n] = \left(\frac{1}{3}\right)^{|n-1|} u[n-1]$$

$$X(\Omega) = \frac{8e^{-j\Omega}}{9 - \cos\Omega}$$

9. **Find $X(\Omega)$.**

$$x[n] = (n-2)^2 x[n]$$

$$X(\Omega) = -\frac{d^2X(\Omega)}{d\Omega^2} - 4j\frac{dX(\Omega)}{d\Omega} + 4X(\Omega)$$

10. **Solve the following difference equation.**

$$y[n-2] - \frac{1}{6}y[n-1] - \frac{1}{6}y[n] = x[n]$$

where $x[n] = \left(\frac{1}{4}\right)^n u[n]$

$$y[n] = \left[\frac{6}{5}\left(\frac{1}{2}\right)^n + \frac{8}{35}\left(-\frac{1}{3}\right)^n - \frac{3}{7}\left(\frac{1}{4}\right)^n\right]u[n]$$

11. **Where $\angle X(\Omega) = \Omega$, find $x[n]$.**

$$X(\Omega) = \begin{cases} 5 & 0 \leq |\Omega| < \pi \\ 0 & |\Omega| > \pi \end{cases}$$

$$\begin{aligned} x[n] &= \frac{5}{\pi(n-1)} \sin(n-1)\pi \\ &= 5 \,\text{sinc}\,(n-1)\pi \end{aligned}$$

12. **Find the 4-point DFT of $x[n] = \{0, 1, 2, 3\}$.** (*Anna University, April, 2005*)

$$X(0) = 6; \quad X(1) = -2 + j2; \quad X(2) = -2; \quad X(3) = -2 - j2$$

13. **Find the IDFT of the following functions with $N = 4$**

$$X(k) = \{1, 2, 3, 2\}$$

$$x[n] = \{-2, -0.5, 0, -0.5\}$$

14. **Consider the following two sequences:**

$$x_1[n] = \{1, 2, 2, 1\} \quad \text{and} \quad x_2[n] = \{2, 1, 1, 2\}$$

Find the circular convolution $y[n] = x_1[n] \circledN x_2[n]$

$$y[n] = \{9, 10, 9, 8\}$$

Chapter 8
The Laplace Transform Method for the Analysis of Continuous Time Signals and Systems

Learning Objectives

- To develop a new transform method, the Laplace transform (LT) which is applicable for the analysis of continuous time signals and systems.
- To determine the range of signals to which the LT is applicable.
- To derive the properties of LT.
- To determine the LT of typical Continuous Time (CT) signals.
- To develop inverse LT method and illustrate it with examples.
- To solve differential equations with and without initial conditions using LT and inverse LT and also by classical method.
- To realize the structure of linear time invariant continuous time systems using LT.

8.1 Introduction

The Continuous Time Fourier Transform (CTFT) is a powerful tool for the analysis of CT signals and systems. However, the method has its limitation in that some useful signals do not have CTFT because these signals do not converge. Marquis Pierre Simon de Laplace (1749–1827), the great French mathematician and Astronomer and the contemporary of Fourier (1768–1830), Louis de Lagrange and the French ruler Napoleon, developed a new transform technique which overcame the problem of convergence in CTFT. Laplace, first presented the transform and its applications to solve linear differential equations in a paper published in the year 1779, when he was just 30 years of age. For his excellent contributions to probability theory, astronomy, special functions and celestial mechanics, Laplace was honored by Napoleon, as a policy of honoring and promoting scientists of high caliber, by appointing him as a minister in the French Government. However, Laplace, a born genius, showed more

S. Palani, *Signals and Systems*,
https://doi.org/10.1007/978-3-030-75742-7_8

interest in his research activities and totally neglected the administrative work in the government. It was no surprise that soon Laplace was sacked from the ministerial post by his admirer, Napoleon.

The CTFT expresses signals as linear combinations of complex sinusoids. Some useful signals, when expressed as a combination of complex sinusoids, do not converge and they do not have Fourier transform (FT). However, Laplace made a small modification in his transform technique from time domain to frequency domain by expressing time signals as linear combinations of complex exponential instead of complex sinusoids. LT is more general since complex sinusoids are a special case of complex exponentials. Thus, LT can describe functions that FT cannot describe. Both the FT and LT using mathematical operations, convert the time signal $x(t)$ to frequency function $X(j\omega)$ and $X(s)$, respectively, where $s = \sigma + j\omega$. By introducing σ in LT method, most of the signals become damped waves and convergence becomes possible. However, it is to be noted that there exists a class of signals which do not converge in LT also, and for these signals, LT does not exist.

The LT, even though a very powerful tool in the analysis and design of linear time invariant signals and systems today, did not catch on until nearly a century later. We discuss the development of the LT in the following sections.

8.2 Definition and Derivations of the LT

The time signal $x(t)$ is expressed as a linear combination of complex sinusoids of the form $e^{j\omega t}$ by the FT. Here, $j\omega$ takes only imaginary value of ω which is associated with the frequency f as $\omega = 2\pi f$. Thus, some of the useful time functions such as $x(t) = e^{at}$ do not coverage as per the FT. By changing the complex sinusoid to complex exponential of the form e^{st}, the FT can be generated and is termed as the LT, and is defined as

$$L[x(t)] = X(s) = \int_{-\infty}^{\infty} x(t)e^{-st}dt \tag{8.1}$$

The complex variable s has a real part and an imaginary part and is expressed as

$$s = \sigma + j\omega \tag{8.2}$$

If the real part $\sigma = 0$, then Eq. (8.1) becomes a special case and it becomes the FT. By substituting $s = (\sigma + j\omega)$, Eq. (8.1) can be written as follows:

$$\begin{aligned} X(s) &= \int_{-\infty}^{\infty} x(t)e^{-(\sigma + j\omega)t}dt \\ &= \int_{-\infty}^{\infty} [x(t)e^{-\sigma t}]e^{-j\omega t}dt \end{aligned} \tag{8.3}$$

In Eq. (8.3), the real exponential convergence factor $e^{-\sigma t}$ enables some of the time functions $x(t)$ to converge in the complex s plane. Equation (8.1) is called the two-sided (or bilateral) LT. The signal $x(t)$ is obtained from $X(s)$ by taking inverse LT which is derived as

$$x(t) = L^{-1}X(s) = \frac{1}{2\pi j}\int_{\sigma-j\infty}^{\sigma+j\infty} X(s)e^{st}ds \tag{8.4}$$

Equations (8.1) and (8.4) are called two-sided or bilateral LT pair. The symbol L^{-1} is used when $X(s)$ is inverse Laplace transformed. The following notations are used to represent LT and inverse LT:

$$X(s) = L[x(t)]$$

or

$$\begin{aligned} x(t) &\overset{L}{\longleftrightarrow} X(s) \\ x(t) &= L^{-1}[X(s)] \\ X(s) &= \overset{L^{-1}}{\longleftrightarrow} x(t) \end{aligned} \tag{8.5}$$

It is to be noted that the time function is represented by small case letter and the s function by upper case letter.

8.2.1 LT of Causal and Non-causal Systems

In Eq. (8.1), the transformation of $x(t)$ to $X(s)$ is done for the following conditions:

- $x(t)$ is anti-causal where $t < 0$,
- $x(t)$ is an impulse where $t = 0$,
- $x(t)$ is causal where $t > 0$.

The unilateral LT is a special case of LT and is defined as follows:

$$X(s) = \int_{0}^{\infty} x(t)e^{-st}dt \tag{8.6}$$

It is to be noted here that Eq. (8.6) is valid only for causal signals and systems. For non-causal signals and systems, the limits of integration have to be changed. The following two examples illustrate the method to determine the LT for causal and non-causal signals.

■ Example 8.1

For the following signal, determine the LT

$$x(t) = e^{-at}u(t)$$

Solution: The given signal $x(t)$ is a causal signal. The limit of integration is, therefore, from 0 to ∞. Hence

$$\begin{aligned} X(s) &= \int_0^\infty e^{-at}e^{-st}dt \\ &= \int_0^\infty e^{-(s+a)t}dt \\ &= -\frac{1}{(s+a)}\left[e^{-(s+a)t}\right]_0^\infty = -\frac{1}{(s+a)}\left[e^{-(s+a)\infty} - e^{-(s+a)0}\right] \end{aligned}$$

$$\boxed{X(s) = \frac{1}{(s+a)}}$$

The above integration converges when the upper limit ∞ is applied if $(s+a) > 0$ or $s > -a$. If $(s+a) < 0$, then $e^{(s+a)\infty}$ does not converge. In such a case, LT does not exist.

■ Example 8.2

Consider the following signal

$$x(t) = e^{-at}u(-t)$$

Determine the LT.

Solution: The given signal $x(t)$ is a non-causal signal. Hence, the limit of integration is from $-\infty$ to 0.

$$\begin{aligned} X(s) &= \int_{-\infty}^0 x(t)e^{-st}dt \\ &= \int_{-\infty}^0 e^{-at}e^{-st}dt \\ &= \int_{-\infty}^0 e^{-(s+a)}dt \\ &= \frac{-1}{(s+a)}\left[e^{-(s+a)t}\right]_{-\infty}^0 \end{aligned}$$

$$\boxed{X(s) = \frac{-1}{(s+a)}}$$

The above integration converges when the lower limit $-\infty$ is applied if $(s+a) < 0$ or $s < -a$. The above two examples illustrate that for the same time signal $x(t)$, the LT is also the same with a change of sign. However, the mode of convergence is different which is an important thing to note. This will be discussed in detail in the sections to follow.

8.3 The Existence of LT

Consider the one-sided LT given below.

$$X(s) = \int_0^\infty x(t)e^{-st}dt$$

Substituting $s = \sigma + j\omega$ in the above equation, we get

$$X(s) = \int_0^\infty \left[x(t)e^{-\sigma t}\right] e^{-j\omega t}dt$$

Since $|e^{-j\omega t}| = 1$, the above integral can be written as

$$X(s) = \int_0^\infty \left[x(t)e^{-\sigma t}\right] dt \tag{8.7}$$

The integral in Eq. (8.7) converges if

$$\int_0^\infty \left[x(t)e^{-\sigma t}\right] dt < \infty \tag{8.8}$$

In other words, the LT of (8.7) exists if the integral of Eq. (8.8) is finite for some value of $\sigma > \sigma_0$ or Re(s), which is σ should be greater than σ_0, which is expressed as

$$\boxed{\sigma > \sigma_0}$$

8.4 The Region of Convergence

One of the limitations of CTFT as mentioned earlier is that, some useful functions whether causal or non-causal do not have FT. By making the complex variable s as expressed in Eq. (8.2) and defining LT as in Eq. (8.1), it is possible to overcome

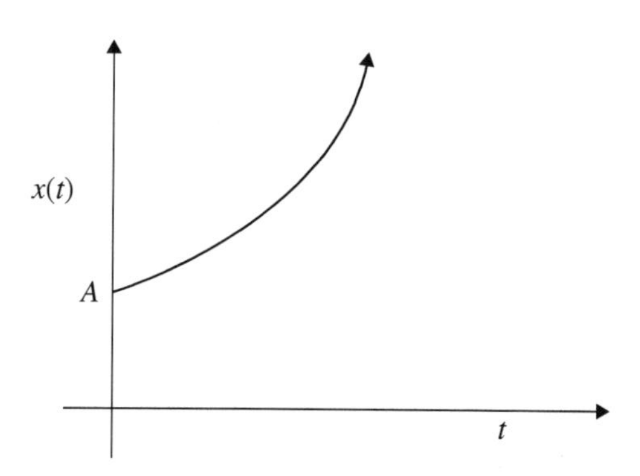

Fig. 8.1 Plot of $x(t) = Ae^{at}u(t)$

this limitation of non-convergence of FT. For example, consider the following causal signal.

$$x(t) = Ae^{at}u(t) \qquad a > 0 \tag{8.9}$$

The plot of Eq. (8.9) as a function of time is shown in Fig. 8.1. From Fig. 8.1, it is evident that $x(t)$ increases without bound as t increases. It can be easily shown that FT does not exist for the above $x(t)$. However, the LT exists for the above $x(t)$ with certain constraints and it is derived as follows. Substituting $x(t) = Ae^{at}$ in (8.1), the following equation is obtained.

$$X(s) = \int_{-\infty}^{\infty} Ae^{at}e^{-st}u(t)dt \tag{8.10}$$

For a causal signal (also called right-sided signal), changing the limit of integration, we get

$$X(s) = \int_{0}^{\infty} Ae^{at}e^{-st}dt \tag{8.11}$$

$$= A\int_{0}^{\infty} e^{-(s-a)t}dt \tag{8.12}$$

$$= \frac{-A}{(s-a)}\left[e^{-(s-a)t}\right]_0^{\infty} \tag{8.13}$$

$$\boxed{X(s) = \frac{A}{(s-a)}} \tag{8.14}$$

Equation (8.13) converges if $(s-a) > 0$. In other words, Re $s > a$. In that case, when the upper limit of $t = \infty$ is applied, $X(s) = 0$ and when the lower limit of $t = 0$ is applied, $X(s)$ is finite. Thus, Eq. (8.13) is simplified and given in Eq. (8.14).

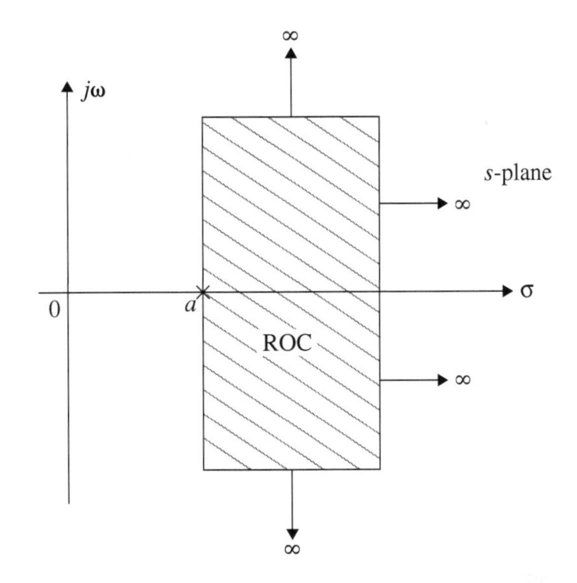

Fig. 8.2 Pole-zero plot and ROC of $X(s) = \dfrac{A}{(s-a)}$

The LT of $x(t)$ of (8.9) exists or Eq. (8.12) converges if $\sigma > a$ in the complex s-plane. This is called the region of convergence.

The region of convergence which is denoted as ROC is, therefore, defined as the set of values of s of the real part of s for which part the integral of Eq. (8.1) **converges.**

The ROC of $x(t)$ in Eq. (8.9) is illustrated in Fig. 8.2. It is to be noted here that $X(s)$ in Eq. (8.14) becomes infinity at $s = a$. Therefore, the points in the s-plane at which the function $X(s)$ becomes infinity are called poles and are marked by a small cross $\times$. Now consider a function $X(s) = (s + a)$. The function $X(s)$ becomes zero at $s = -a$. Therefore, the points in the s-plane at which the function $X(s)$ becomes zero are called zeros and are marked by a small circle O.

Now consider the following non-causal signal or otherwise called left-sided signal shown in Fig. 8.3.

$$x(t) = Ae^{-at}u(-t) \tag{8.15}$$

The LT of the above signal is obtained from

$$\begin{aligned} X(s) &= \int_{-\infty}^{0} x(t)e^{-st}dt \\ &= \int_{-\infty}^{0} Ae^{-at}e^{-st}dt \\ &= \int_{-\infty}^{0} Ae^{-(s+a)t}dt \end{aligned} \tag{8.16}$$

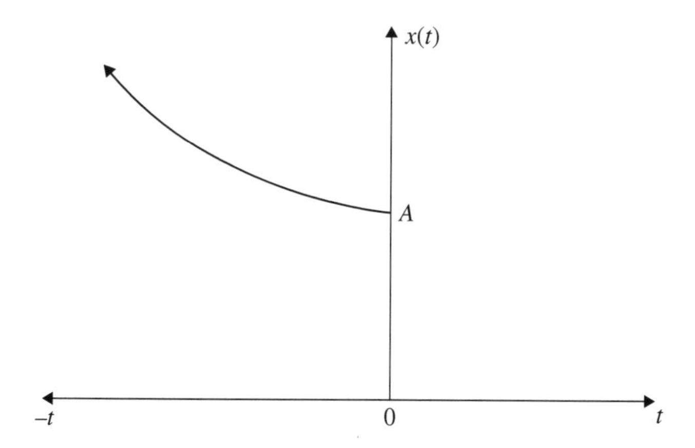

Fig. 8.3 Plot of $x(t) = Ae^{-at}u(-t)$

$$= \frac{-A}{(s+a)}\left[e^{-(s+a)t}\right]_{-\infty}^{0} \tag{8.17}$$

It is evident from Eq. (8.17) that the integral given in Eq. (8.16) will converge if $(s+a) < 0$ when the lower limit of $t = -\infty$ is applied to (8.17). Thus, $X(s)$ is obtained as

$$X(s) = -\frac{A}{(s+a)}$$

The ROC for the left-sided signal is Re $s < -a$. The ROC is shown in Fig. 8.4.

From the above examples illustrated, for the same $X(s)$, different time signals $x(t)$ exist, and therefore, the inverse LT is not unique. Hence, it is necessary to specify the ROC while determining LT and inverse LT. However, for unilateral LT, there exists one-to-one correspondence between the LT pair. For the bilateral or two-sided LT, it is essential to specify the ROC to avoid any ambiguity.

8.4.1 Properties of ROCs for LT

Property 1: The ROC of $X(s)$ consists of parallel strips to the imaginary axis.

Property 2: The ROC of LT does not include any pole of $X(s)$.

Property 3: If $x(t)$ is a finite duration signal, and is absolutely integrable then the ROC of $X(s)$ is the entire s-plane.

Property 4: For the right-sided (causal) signal if the $\text{Re}(s) = \sigma_0$ and is in ROC, then for all the values of s for which $\text{Re}(s) > \sigma_0$ is also in ROC.

Property 5: If $x(t)$ is a left-sided (non-causal) signal and if $\text{Re}(s) = \sigma_0$ is in ROC, then for all the values of s for which $\text{Re}(s) < \sigma_0$ is also in ROC.

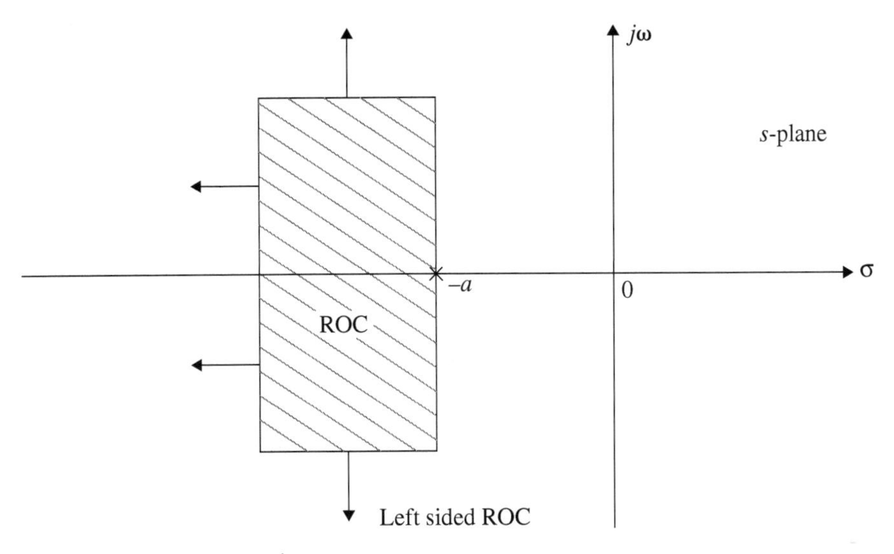

Fig. 8.4 ROC of $X(s) = \frac{A}{(s+a)}$

Property 6: If $x(t)$ is two-sided signal and if $\text{Re}(s) = \sigma_0$ and is in ROC, then the ROC of $X(s)$ will consist of a strip in the s-plane which will include $\text{Re}(s) = \sigma_0$.

The following examples illustrate the above properties of ROC and pole-zero locations of $X(s)$ in the s-plane.

■ Example 8.3

Determine the LT of the following signal. Mark the poles and ROC in the s-plane. $x(t) = Ae^{-at}u(t) + Be^{-bt}u(-t)$ where $a > 0$, $b > 0$ and $|a| > |b|$.

Solution:

1. The given signal $x(t)$ consists of causal and anti-causal signals and can be written as

$$x(t) = x_1(t) + x_2(t)$$

where

$$x_1(t) = Ae^{-at}u(t)$$
$$x_2(t) = Be^{-bt}u(-t)$$

2. $X_1(s)$ is found as follows for a right-sided signal.

$$
\begin{aligned}
X_1(s) &= \int_0^{\infty} Ae^{-at}e^{-st}dt \\
&= A\int_0^{\infty} e^{-(s+a)t}dt \\
&= \frac{-A}{(s+a)}\left[e^{-(s-a)t}\right]_0^{-\infty} \\
&= \frac{A}{(s+a)}
\end{aligned}
$$

The ROC is $\mathrm{Re}(s) > -a$.

3. $X_2(s)$ is found as follows for a left-sided signal.

$$
\begin{aligned}
X_2(s) &= \int_{-\infty}^{0} Be^{-bt}e^{-st}dt \\
&= B\int_{-\infty}^{0} e^{-(s+b)t}dt \\
&= \frac{-B}{(s+b)}\left[e^{-(s+b)t}\right]_{-\infty}^{0} \\
&= -\frac{B}{(s+b)}[1-0] \\
&= \frac{-B}{(s+b)}
\end{aligned}
$$

The ROC is $\mathrm{Re}(s) < -b$.

4.

$$
\begin{aligned}
X(s) &= X_1(s) + X_2(s) \\
&= \frac{A}{(s+a)} - \frac{B}{(s+b)}
\end{aligned}
$$

5. The poles and ROC are marked as shown in Fig. 8.5b. In Fig. 8.5b, $|a| > |b|$. Vertical lines passing through $-a$ and $-b$ are drawn. For $X_1(s)$, the ROC is right sided and for $X_2(s)$ the ROC is left sided. A strip where $-a <\mathrm{Re}\ s < -b$ is drawn and hatched and the ROC is identified.
6. Consider the case where $|b| > |a|$. The poles are located as shown in Fig. 8.5c. Vertical line passing through $-a$ and $-b$ are drawn. For $X_1(s)$, the ROC is right sided and a strip where $\mathrm{Re}(s) > -a$ is drawn and hatched. For $X_2(s)$, the ROC is left sided. A vertical strip to the left of $-b$ is formed and hatched. **It is to be noted that the ROC s of $x_1(t)$ and $x_2(t)$ do not overlap and hence $x(t)$ does not have LT.**

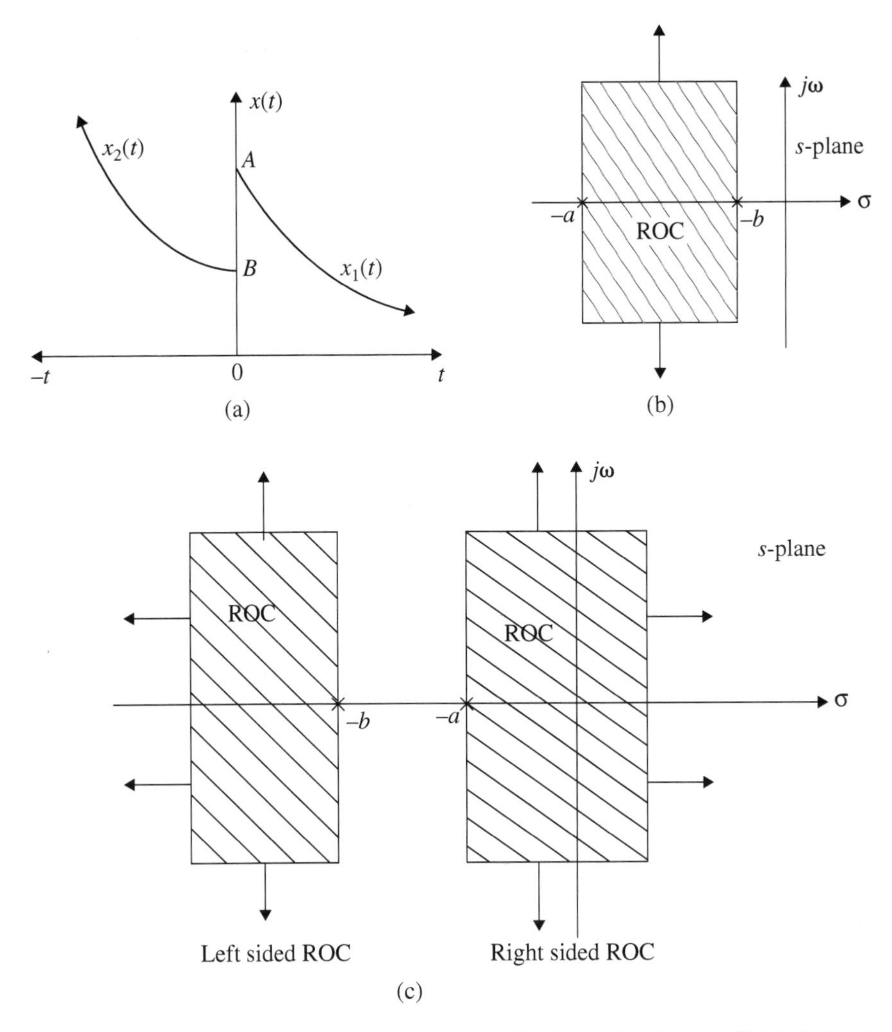

Fig. 8.5 **a** Representation of $x(t)$. **b** ROC and poles of $X(s)$ $|a| > |b|$. **c** Poles and ROC of $X(s)$ for $|b| > |a|$. ROC for right and left sided signals

---

■ Example 8.4

Determine the LT of

$$x(t) = e^{-2t}u(t) + e^{-3t}u(t)$$

and sketch the ROC in the s-plane.

(Anna University, May, 2007)

Solution:

1. $x(t)$ is completely a right-sided signal, and hence the limit of the LT integration is from $t = 0$ to $t = \infty$. Thus, the following equation is written for $X(s)$

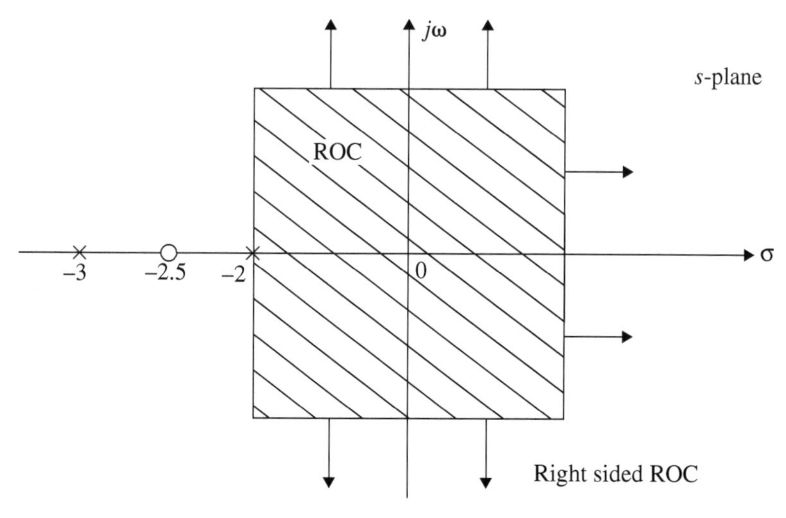

Fig. 8.6 Poles and zeros and ROC of $X(s) = \dfrac{2(s+2.5)}{(s+2)(s+3)}$

$$
\begin{aligned}
X(s) &= \int_0^\infty e^{-2t} e^{-st} dt + \int_0^\infty e^{-3t} e^{-st} dt \\
&= \int_0^\infty e^{-(s+2)t} dt + \int_0^\infty e^{-(s+3)t} dt \\
&= \frac{1}{(s+2)} + \frac{1}{(s+3)} \\
&= \frac{(2s+5)}{(s+2)(s+3)}
\end{aligned}
$$

$$
\boxed{X(s) = \frac{2(s+2.5)}{(s+2)(s+3)}}
$$

2. The poles are at $s = -2$ and $s = -3$ and a zero is at $s = -2.5$ and are marked in Fig. 8.6,
3. For the pole $\frac{1}{s+2}$, the ROC is right sided to the vertical line passing through $\sigma = -2$. For the pole $\frac{1}{s+3}$, the ROC is also right sided passing through $\sigma = -3$. If ROC where $\sigma > -2$ is satisfied then ROC where $\sigma > -3$ is automatically satisfied. Further, no pole of $X(s)$ will be inside the ROC,
4. A strip to the right of $\sigma = -2$ is created and shaded. The strip is enlarged to ∞ in the direction of real and imaginary axis,
5. **Thus, the ROC of a causal signal is to the right of the right most pole of $X(s)$.**

Example 8.5

Determine the LT of

$$x(t) = e^{-2t}u(-t) + e^{-3t}u(-t)$$

Locate the poles and zero of $X(s)$ and also the ROC in the s-plane.

Solution:

1. The given signal is fully a left-sided signal, and hence the limit of LT integration is from $-\infty$ to 0. The LT of $x(t)$ is obtained as follows:

$$X(s) = \int_{-\infty}^{0} e^{-2t}e^{-st}dt + \int_{-\infty}^{0} e^{-3t}e^{-st}dt$$

$$= \int_{-\infty}^{0} e^{-(s+2)t}dt + \int_{-\infty}^{0} e^{-(s+3)t}dt$$

$$= \frac{-1}{(s+2)}\left[e^{-(s+2)t}\right]_{-\infty}^{0} - \frac{1}{(s+3)}\left[e^{-(s+3)t}\right]_{-\infty}^{0}$$

$$X(s) = -\frac{1}{(s+2)} - \frac{1}{(s+3)} \qquad \text{ROC Re } s < -3$$

$$\boxed{X(s) = \frac{-2(s+2.5)}{(s+2)(s+3)}}$$

2. The poles are at $s = -2$ and $s = -3$ and a zero is at $s = -2.5$ and are marked in Fig. 8.7.
3. For the pole $\frac{1}{(s+2)}$, the ROC is left sided to the vertical line passing through $\sigma = -2$. For the pole $\frac{1}{s+3}$, the ROC is also left sided to the vertical line passing through $\sigma = -3$. If ROC, where $\sigma = -3$, is satisfied, then ROC where $\sigma = -2$ is also satisfied. Further, no pole of $X(s)$ will be inside the ROC.
4. A vertical strip to the left of $\sigma = -3$ is created and shaded. The strip is enlarged to ∞ in the direction of real and imaginary axis.
5. **Thus, the ROC of a non-causal signal is to the left of the leftmost pole of $X(s)$.**

Example 8.6

Consider the following signal

$$x(t) = e^{-2t}u(-t) + e^{-3t}u(t)$$

Determine the LT and locate the poles and zeros and the ROC in the s-plane.

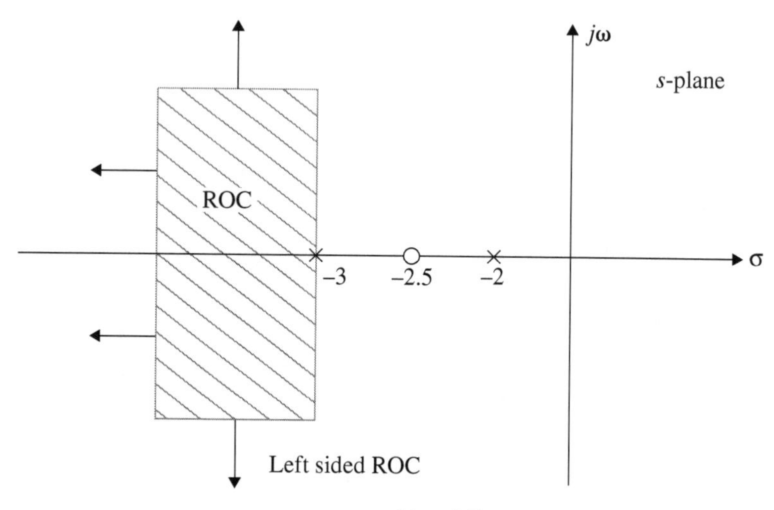

Fig. 8.7 Poles and zeros and ROC of $X(s) = \dfrac{-2(s+2.5)}{(s+2)(s+3)}$

Solution:

1. The given signal is a combination of left- and right-sided. The integration is performed as given below.

$$
\begin{aligned}
X(s) &= \int_{-\infty}^{0} e^{-2t}e^{-st}dt + \int_{0}^{\infty} e^{-3t}e^{-st}dt \\
&= \int_{-\infty}^{0} e^{-(s+2)t}dt + \int_{0}^{\infty} e^{-(s+3)t}dt \\
&= \frac{1}{(s+2)}\left[e^{-(s+2)t}\right]_{-\infty}^{0} - \frac{1}{(s+3)}\left[e^{-(s+3)t}\right]_{0}^{\infty} \\
&= -\frac{1}{(s+2)} + \frac{1}{(s+3)}
\end{aligned}
$$

$$
\boxed{X(s) = \frac{-1}{(s+2)(s+3)}} \quad \text{ROC} - 3 < \text{Re}\, s < -2
$$

2. The pole locations are shown in Fig. 8.8. For the left-sided signal, the ROC is Re $s < -2$ and for the right-sided signal, the ROC is Re $s > -3$. The resultant ROC is a strip in between the vertical lines passing through $\sigma = -2$ and $\sigma = -3$. The strip is shaded as shown in Fig. 8.8. It is enlarged in the vertical direction. The poles are at $s = -2$ and $s = -3$. There is no zero for this function.

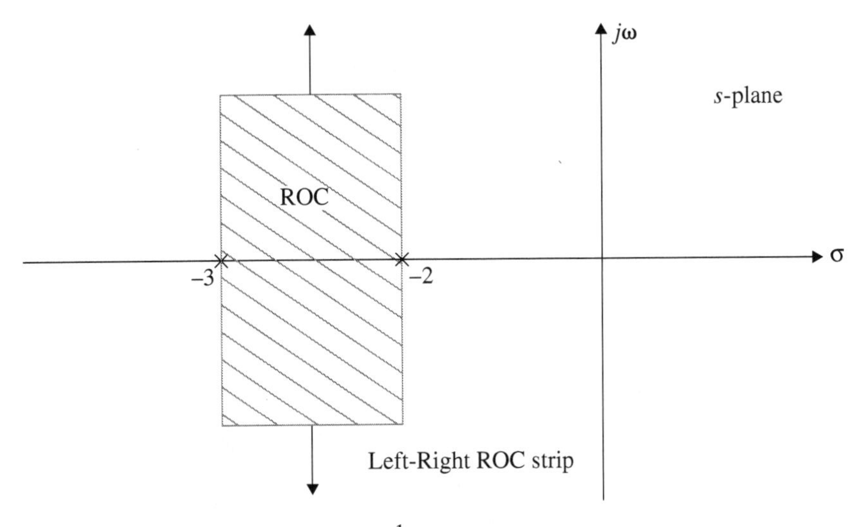

Fig. 8.8 Poles and zeros of $X(s) = \frac{-1}{(s+2)(s+3)}$ and the ROC

■ Example 8.7

Determine the LT and locate the poles and zeros and ROC in the s-plane for the following signal.

$$x(t) = Au(t)$$

Solution:

1. The given signal is right-sided signal. Its LT is obtained as follows:

$$X(s) = \int_0^{\infty} Ae^{-st}\,dt$$
$$= \frac{-A}{s}\left[e^{-st}\right]_0^{\infty}$$

$$\boxed{X(s) = \frac{A}{s}} \qquad \text{ROC Re } s > 0.$$

2. For the given signal, a pole at the origin exists and it is marked in Fig. 8.9b,
3. The LT converges only if $\sigma > 0$. Thus, the ROC is the entire right half of s-plane.

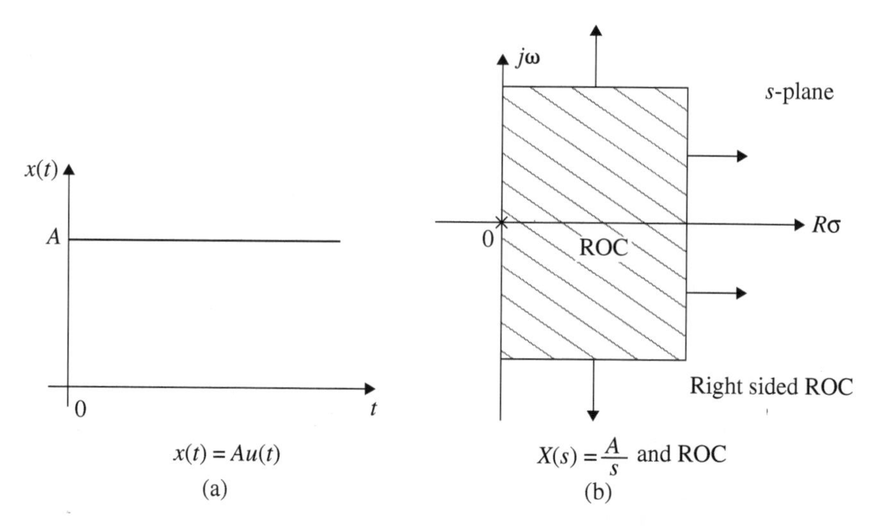

Fig. 8.9 Representation of $x(t)$ and ROC

8.5 The Unilateral Laplace Transform

The unilateral LT is a special case of bilateral LT and is defined as

$$X(s) = \int_{0}^{\infty} x(t)e^{-st}dt \tag{8.18}$$

The unilateral LT has the following features

1. The unilateral LT simplifies the system analysis considerably.
2. The signals are restricted to causal signals.
3. There is one-to-one correspondence between LT and inverse LT.
4. In view of the above advantages, Laplace transform means unilateral LT as defined in Eq. (8.18), unless otherwise it is specifically mentioned that the signal is anti-causal.

Before we go for the determination of LT of some of the commonly used signals, we give below some of the properties of LT which will be useful to determine $X(s)$ from $x(t)$ and *vice versa* in a simplified way.

8.6 Properties of Laplace Transform

8.6.1 Linearity

$$x_1(t) \overset{L}{\longleftrightarrow} X_1(s)$$

$$x_2(t) \overset{L}{\longleftrightarrow} X_2(s)$$

$$[a_1x_1(t) + a_2x_2(t)] \overset{L}{\longleftrightarrow} [a_1X_1(s) + a_2X_2(s)] \tag{8.19}$$

8.6.2 Time Shifting

Let $x(t)$ be time shifted to the right (time delay) by a real constant t_0. The delayed time function is written as $x(t - t_0)$. As per the time shifting property

$$\begin{aligned} x(t) &\overset{L}{\longleftrightarrow} X(s) \\ x(t - t_0) &\overset{L}{\longleftrightarrow} X(s)e^{-st_0} \end{aligned} \tag{8.20}$$

Proof By definition of LT

$$L[x(t - t_0)] = \int_0^{\infty} x(t - t_0)e^{-st}dt \tag{8.21}$$

Let

$$\begin{aligned} t - t_0 &= \lambda \\ dt &= d\lambda \end{aligned}$$

For the integration of Eq. (8.21), the lower and upper limits are determined as follows:

When $t = 0$, $\lambda = -t_0$ and when $t = \infty$, then $\lambda = \infty$. Thus, Eq. (8.21) is written as follows:

$$L[x(t - t_0)] = \int_{-t_0}^{\infty} x(\lambda)e^{-s(\lambda + t_0)}d\lambda \tag{8.22}$$

For a causal signal, $x(t) = 0$ for $t < 0$ and the lower limit of integration is zero. Now Eq. (8.22) is written as follows:

$$L[x(\lambda)] = e^{-st_0}\int_0^\infty x(\lambda)e^{-s\lambda}d\lambda$$
$$= e^{-st_0}X(s)$$

Thus

$$\boxed{x(t-t_0) \overset{L}{\longleftrightarrow} X(s)e^{-st_0} \qquad t_0 > 0} \tag{8.23}$$

8.6.3 *Frequency Shifting*

According to frequency shifting property, if

$$x(t) \overset{L}{\longleftrightarrow} X(s)$$
$$x(t)e^{s_0t} \overset{L}{\longleftrightarrow} X(s-s_0)$$

Proof

$$L[x(t)e^{s_0t}] = \int_0^\infty x(t)e^{s_0t}e^{-st}dt$$

$$L[x(t)e^{s_0t}] = \int_0^\infty x(t)e^{-(s-s_0)t}dt$$
$$= X(s-s_0)$$

$$\boxed{x(t)e^{s_0t} \overset{L}{\longleftrightarrow} X(s-s_0)} \tag{8.24}$$

8.6.4 *Time Scaling*

The time scaling property states that, if

$$x(t) \overset{L}{\longleftrightarrow} X(s)$$
$$x(at) \overset{L}{\longleftrightarrow} \frac{1}{|a|}X\left(\frac{s}{a}\right)$$

Proof

$$L[x(at)] = \int_0^\infty x(at)e^{-st}dt \tag{8.25}$$

Let

$$\lambda = at \quad \text{and} \quad d\lambda = adt$$

For the lower limit of integration of Eq. (8.25), when $t = 0$, $\lambda = 0$ and for the upper limit of integration when $t = \infty$, then $\lambda = \infty$. Hence, Eq. (8.25) is written as follows:

$$\begin{aligned} L[x(at)] &= \int_0^\infty x(\lambda)e^{-\frac{\lambda s}{a}} \frac{1}{a} d\lambda \\ &= \frac{1}{|a|} \int_0^\infty x(\lambda)e^{-\frac{s}{a}\lambda} d\lambda \\ &= \frac{1}{a} X\left(\frac{s}{a}\right) \end{aligned}$$

$$\boxed{x(at) \stackrel{L}{\longleftrightarrow} \frac{1}{a} X\left(\frac{s}{a}\right)} \tag{8.26}$$

8.6.5 Frequency Scaling

According to frequency scaling property, if

$$\begin{aligned} x(t) &\stackrel{L}{\longleftrightarrow} X(s) \\ \frac{1}{a} x\left(\frac{t}{a}\right) &\stackrel{L}{\longleftrightarrow} X(as) \end{aligned}$$

Proof According to time scaling property

$$x(at) \stackrel{L}{\longleftrightarrow} \frac{1}{a} X\left(\frac{s}{a}\right)$$

Let

$$b = \frac{1}{a}$$

$$x\left(\frac{t}{b}\right) \stackrel{L}{\longleftrightarrow} bX(bs)$$

Replacing b by a, we get

$$\boxed{\frac{1}{a}x\left(\frac{t}{a}\right) \overset{L}{\longleftrightarrow} X(as)} \tag{8.27}$$

8.6.6 Time Differentiation

$$\begin{aligned} x(t) &\overset{L}{\longleftrightarrow} X(s) \\ \frac{dx(t)}{dt} &\overset{L}{\longleftrightarrow} sX(s) - x(0^-) \\ \frac{d^2x(t)}{dt^2} &\overset{L}{\longleftrightarrow} s^2X(s) - sx(0^-) - \frac{d}{dt}x(0^-) \end{aligned}$$

Proof

$$X(s) = \int_0^\infty x(t)e^{-st}dt \tag{8.28}$$

The above integral is evaluated by parts using

$$\int udv = uv - \int vdu$$

Let $u = x(t)$ and $dv = e^{-st}dt$; $du = \frac{d}{dt}x(t)dt$ and $v = -\frac{1}{s}e^{-st}$

$$\int_0^\infty x(t)e^{-st}dt = \left[\frac{-1}{s}x(t)e^{-st}\right]_0^\infty - \int_0^\infty -\frac{1}{s}e^{-st}\frac{d}{dt}x(t)dt$$

or

$$X(s) = \frac{1}{s}x(0) + \frac{1}{s}\int_0^\infty e^{-st}\frac{d}{dt}x(t)dt.$$

But

$$L\left[\frac{d}{dt}(x(t))\right] = \int_0^\infty \frac{d}{dt}(x(t))e^{-st}dt$$

$$\boxed{\therefore \quad L\frac{d}{dt}(x(t)) \overset{L}{\longleftrightarrow} sX(s) - x(0^-)} \tag{8.29}$$

The time differentiation twice is proved as follows:

$$\frac{d^2}{dt^2}(x(t)) = \frac{d}{dt}\left(\frac{d}{dt}(x(t))\right)$$

Using the property

$$\frac{d}{dt}(x(t)) \overset{L}{\longleftrightarrow} sX(s) - x(0^-)$$

we get

$$L\left[\frac{d^2(x(t))}{dt^2}\right] = sL\left[\frac{d}{dt}(x(t))\right] - \frac{d}{dt}(x(0^-))\bigg|_{t=0}$$

$$\boxed{\frac{d^2(x(t))}{dt^2} \overset{L}{\longleftrightarrow} s^2X(s) - sx(0^-) - \frac{d}{dt}(x(0^-))}$$

In general

$$\boxed{\begin{array}{c} \dfrac{d^n x(t)}{dt^n} \overset{L}{\longleftrightarrow} s^nX(s) - s^{n-1}x(0^-) - s^{n-2}x(0^-)\cdots x^{n-1}(0^-) \\ \text{OR} \\ \dfrac{d^n x(t)}{dt^n} \overset{L}{\longleftrightarrow} s^nX(s) - \sum_{k=1}^{n} s^n x^{k-1}(0^-) \end{array}} \tag{8.30}$$

8.6.7 Time Integration

The time integration property states that, if

$$x(t) \overset{L}{\longleftrightarrow} X(s)$$

$$\int_0^t x(\tau)d\tau \overset{L}{\longleftrightarrow} \frac{X(s)}{s}$$

Proof We define

$$f(t) = \int_0^t x(\tau)d\tau.$$

Differentiating the above equation, we get

$$\frac{df(t)}{dt} = x(t) \quad \text{and} \quad x(0^-) = 0$$

if

$$f(t) \overset{L}{\longleftrightarrow} F(s)$$

$$X(s) = L\left[\frac{d}{dt}f(t)\right] = sF(s) - f(0^-) = sF(s) \quad \text{if} \quad f(0^-) = 0$$

$$F(s) = \frac{X(s)}{s}$$

$$\boxed{\int_0^t x(\tau)d\tau \overset{L}{\longleftrightarrow} \frac{X(s)}{s}} \tag{8.31}$$

8.6.8 Time Convolution

The time convolution property states that, if

$$\begin{aligned} x_1(t) &\overset{L}{\longleftrightarrow} X_1(s) \\ x_2(t) &\overset{L}{\longleftrightarrow} X_2(s) \\ x_1(t) * x_2(t) &\overset{L}{\longleftrightarrow} X_1(s)X_2(s) \end{aligned} \tag{8.32}$$

Proof

$$\begin{aligned} L[x_1(t) * x_2(t)] &= \int_{-\infty}^{\infty} e^{-st} \left[\int_{-\infty}^{\infty} x_1(\tau)x_2(t-\tau)d\tau \right] dt \\ &= \int_{-\infty}^{\infty} x_1(\tau) \left[\int_{-\infty}^{\infty} e^{-st} x_2(t-\tau)dt \right] d\tau \end{aligned}$$

The inner integral is the LT of $x_2(t - \tau)$ with *a* time delay τ. Substituting

$$\int_{-\infty}^{\infty} e^{-st} x_2(t-\tau)dt = X_2(s)e^{-\tau s}$$

in the above equation, we get

$$\begin{aligned} L[x_1(t) * x_2(t)] &= \int_{-\infty}^{\infty} x_1(\tau)X_2(s)e^{-\tau s}d\tau \\ &= X_2(s) \int_{-\infty}^{\infty} x_1(\tau)e^{-\tau s}d\tau \\ &= X_2(s)X_1(s) \end{aligned}$$

$$\boxed{[x_1(t) * x_2(t)] \overset{L}{\longleftrightarrow} X_1(s)X_2(s)}$$

8.6.9 Complex Frequency Differentiation

According to this property

$$-\,tx(t) \overset{L}{\longleftrightarrow} \frac{d}{ds}(X(s)) \tag{8.33}$$

Proof By definition of LT,

$$X(s) = \int_0^\infty x(t)e^{-st}dt$$

Differentiating both sides with respect to s

$$\begin{aligned}\frac{d}{ds}(X(s)) &= \frac{d}{ds}\int_0^\infty x(t)e^{-st}dt \\ &= -\int_0^\infty tx(t)e^{-st}dt \\ &= -L[tx(t)]\end{aligned}$$

$$\therefore \quad \boxed{-tx(t) \overset{L}{\longleftrightarrow} \frac{d}{ds}(X(s))}$$

8.6.10 Complex Frequency Shifting

According to this property

$$[e^{s_0t}x(t)] \overset{L}{\longleftrightarrow} X(s-s_0) \tag{8.34}$$

$$L[e^{s_0t}x(t)] = \int_0^\infty e^{s_0t}x(t)e^{-st}dt \qquad \text{where } s_0 \text{ is a constant}$$

$$= \int_0^\infty x(t)e^{-(s-s_0t)}dt = X(s-s_0)$$

$$\boxed{[e^{s_0t}x(t)] \overset{L}{\longleftrightarrow} X(s-s_0)}$$

8.6.11 Conjugation Property

According to this property, if $x(t) \overset{L}{\longleftrightarrow} X(s)$, then

$$x^*(t) \overset{L}{\longleftrightarrow} X^*(-s) \tag{8.35}$$

Proof By definition of LT

$$\begin{aligned} L[x^*(t)] &= \int_0^\infty x^*(t)e^{-st}dt \\ &= \int_0^\infty [x(t)e^{-(-s)t}dt]^* \\ &= X^*(-s) \end{aligned}$$

$$\boxed{x^*(t) \overset{L}{\longleftrightarrow} X^*(-s)}$$

8.6.12 Initial Value Theorem

According to this theorem

$$\underset{t\to 0}{Lt}\, x(t) = \underset{s\to\infty}{Lt}\, sX(s) \tag{8.36}$$

Proof

$$L\left[\frac{d}{dt}x(t)\right] = \int_0^\infty \frac{d}{dt}(x(t))e^{-st}dt = sX(s) - x(0)$$

Let $s \to \infty$; then

$$\begin{aligned} \underset{s\to\infty}{Lt} \int_0^\infty \frac{d}{dt}(x(t))e^{-st}dt &= \underset{s\to\infty}{Lt}\, [sX(s) - x(0)] \\ 0 &= \underset{s\to\infty}{Lt}\, [sX(s) - x(0)] \end{aligned}$$

Since $x(0) = \underset{t\to 0}{Lt}\, x(t)$

$$\boxed{\underset{t\to 0}{Lt}\, x(t) = \underset{s\to\infty}{Lt}\, sX(s)}$$

8.6.13 *Final Value Theorem*

According to this theorem

$$\underset{t\to\infty}{Lt}\, x(t) = \underset{s\to 0}{Lt}\, sX(s) \tag{8.37}$$

Proof The LT of $\frac{d}{dt}(x(t))$ could be written as

$$\int_0^\infty \frac{d}{dt}(x(t))e^{-st}dt = [sX(s) - x(0)]$$

Taking $\underset{s\to 0}{Lt}$ on both sides of the above equation, we get

$$\int_0^\infty \frac{d}{dt}(x(t))dt = \underset{s\to 0}{Lt}[sX(s) - x(0)]$$
$$\underset{t\to\infty}{Lt}[x(t) - x(0)] = \underset{s\to 0}{Lt}[sX(s) - x(0)]$$

$$\boxed{\underset{t\to\infty}{Lt}\, x(t) = \underset{s\to 0}{Lt}\, sX(s)}$$

The above theorem is valid if $X(s)$ has no poles in RHP of s-plane Table 8.1 gives the summary of properties of LT.

The following examples illustrate the method of determining LT.

■ Example 8.8

Determine the LT of unit impulse function $\delta(t)$ shown in Fig. 8.10.

Solution: The unit impulse function is represented as

$$\begin{aligned}\delta(t) &= 1 \quad \text{for} \quad t = 0\\ &= 0 \quad \text{otherwise}\\ L[\delta(t)] &= \int_{0^-}^{\infty} \delta(t)e^{-st}dt\\ &= \int_{0^-}^{0^+} e^{-st}dt\\ &= 1\end{aligned}$$

$$\boxed{\delta(t) \stackrel{L}{\longleftrightarrow} 1} \qquad \text{ROC : all } s \tag{8.38}$$

Table 8.1 Summary of properties of LT

S.No	Property	Time function $x(t)$	Frequency function $X(s)$
1.	Linearity	$a_1x_1(t) + a_2x_2(t)$	$a_1X_1(s) + a_2X_2(s)$
2.	Time shifting	$x(t - t_0)$	$X(s)e^{-st_0}$
3.	Frequency shifting	$x(t)e^{at}$	$X(s - a)$
4.	Time scaling	$x(at)$	$\frac{1}{a}X\left(\frac{s}{a}\right)$
5.	Frequency scaling	$\frac{1}{a}x\left(\frac{t}{a}\right)$	$X(as)$
6.	Time differentiation	$\frac{d}{dt}(x(t))$	$sX(s) - x(0^-)$
		$\frac{d^2}{dt^2}(x(t))$	$s^2X(s) - sx(0^-) - \dot{x}(0^-)$
		$\frac{d^n}{dt^n}(x(t))$	$s^nX(s) - \sum_{k=1}^{n} s^n x^{(k-1)}(0^-)$
7.	Time integration	$\int_0^t x(\tau)d\tau$	$\frac{X(s)}{s}$
8.	Time convolution	$x_1(t) * x_2(t)$	$X_1(s)X_2(s)$
9.	Complex frequency differentiation	$-tx(t)$	$\frac{d}{ds}(X(s))$
		$t^nx(t)$	$(-1)^n\frac{d^n}{ds^n}X(s)$
10.	Complex frequency shifting	$e^{-at}x(t)$	$X(s + a)$
11.	Conjugation	$x^*(t)$	$X^*(-s)$
12.	Initial value theorem	$\underset{t\to 0}{Lt}\ x(t)$	$\underset{s\to\infty}{Lt}\ sX(s)$
13.	Final value theorem	$\underset{t\to\infty}{Lt}\ x(t)$	$\underset{s\to 0}{Lt}\ sX(s)$
14.	Shift theorem	$x(t - a)$	$X(s)e^{-as}$

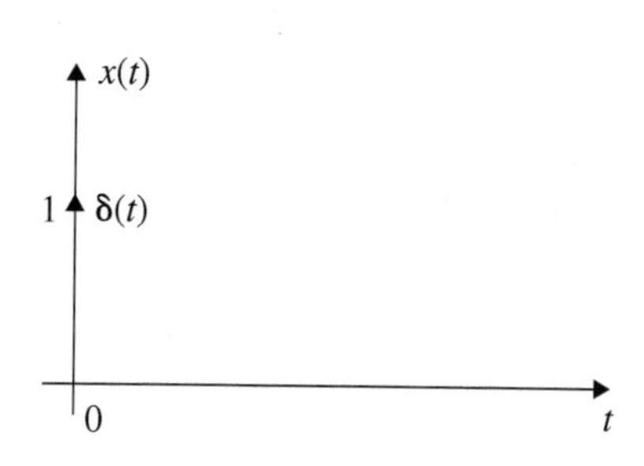

Fig. 8.10 The unit impulse (or delta) function

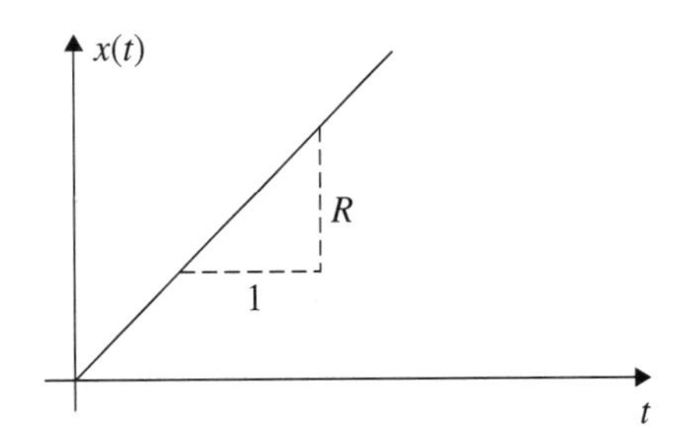

Fig. 8.11 Ramp (or velocity) function

■ Example 8.9

Determine the LT of a ramp function of slope R which is shown in Fig. 8.11.

Solution: The ramp function of slope R is represented in Fig. 8.11 and it is mathematically expressed as

$$x(t) = Rt\,u(t) \quad t \geq 0$$

Taking LT, the following equation is written.

$$L[Rt] = \int_0^{\infty} Rt\,e^{-st}dt$$

The above integration is solved by the well-known integration by parts using the following relationship

$$\int udv = uv - \int vdu$$

Let $u = Rt$ and $du = Rdt$; $dv = e^{-st}dt$ and $v = \int e^{-st}dt = -\frac{e^{-st}}{s}$

$$\therefore \quad L[Rt] = R\left[\frac{te^{-st}}{(-s)}\right]_0^{\infty} - R\int_0^{\infty}\frac{e^{-st}}{(-s)}dt$$

$$= R[0-0] + R\left[\frac{e^{-st}}{-s^2}\right]_0^{\infty}$$

$$= \frac{R}{s^2} \tag{8.39}$$

$$\boxed{L(Rt) \overset{L}{\longleftrightarrow} \frac{R}{s^2}}$$

ROC: The entire right half s-plane (RHP) except for the origin.

The LT of unit ramp ($R = 1$) is

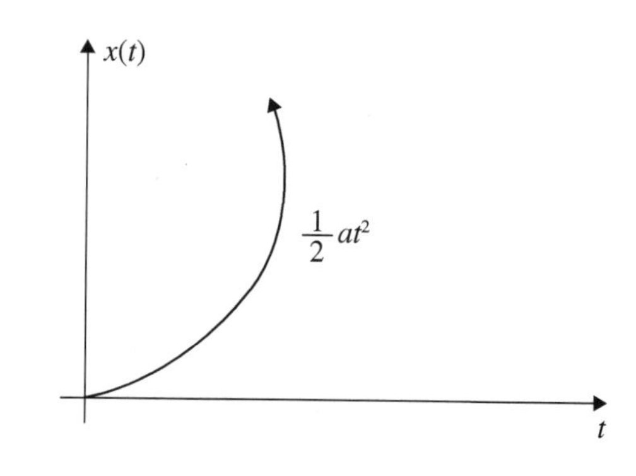

Fig. 8.12 Acceleration function

$$\boxed{L(t) \longleftrightarrow \frac{1}{s^2}}$$

■ Example 8.10

Determine the LT of the acceleration function shown in Fig. 8.12.

Solution: The acceleration function is expressed by the following equation

$$x(t) = \frac{1}{2}at^2u(t) \quad t \geq 0.$$

Taking LT for the above function, we get

$$L\left[\frac{1}{2}at^2\right] = \int_0^\infty \frac{1}{2}at^2e^{-st}dt$$

The above integration is solved using integration by parts as described below

$$u = \tfrac{1}{2}at^2 \quad \text{and} \quad du = at$$
$$dv = \int e^{-st}dt \quad \text{and} \quad v = \frac{e^{-st}}{(-s)}$$

$$L\left[\frac{1}{2}at^2\right] = uv - \int_0^\infty vdu = \left[\frac{1}{2}at^2\frac{e^{-st}}{(-s)}\right]_0^\infty - \int_0^\infty \frac{ate^{-st}}{(-s)}dt$$
$$= 0 + 0 + \frac{a}{s}\int_0^\infty te^{-st}dt.$$

The integration in the right-hand side of the equation is nothing but a ramp signal whose LT is $\frac{1}{s^2}$. Hence

Fig. 8.13 Exponential decay

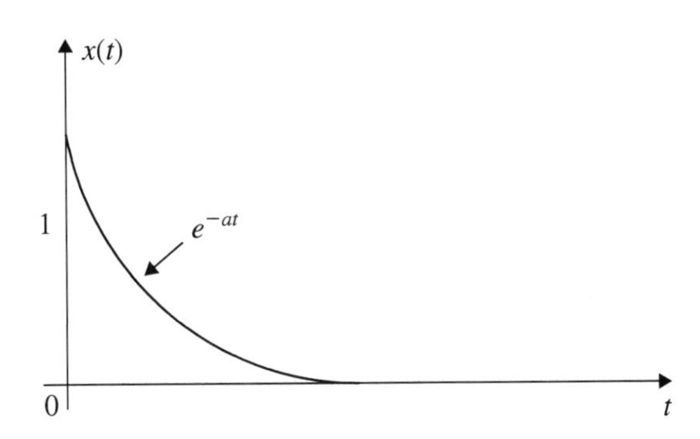

$$\boxed{L\left[\frac{1}{2}at^2\right] = \frac{a}{s^3}} \tag{8.40}$$

The ROC is the entire RHP except for the origin of the s-plane.

■ Example 8.11

Determine the LT of an exponential decay which is shown in Fig. 8.13.

Solution: The exponential decay is represented by

$$x(t) = e^{-at}u(t) \quad t \geq 0.$$

Taking LT for the above function, we get

$$L[e^{-at}u(t)] = \int_0^\infty e^{-at}e^{-st}dt$$
$$= \int_0^\infty e^{-(s+a)t}dt$$

$$L[e^{-at}u(t)] = -\frac{1}{(s+a)}\left[e^{-(s+a)t}\right]_0^\infty$$
$$= \frac{1}{(s+a)} \quad \text{with ROC: Re } s > -a$$

$$\boxed{L[e^{-at}u(t)] = \frac{1}{(s+a)}} \tag{8.41}$$

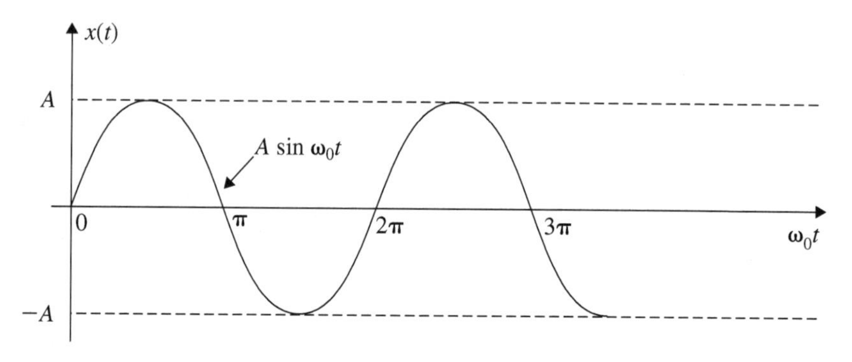

Fig. 8.14 A sine function

■ Example 8.12

Determine the LT of a sine function which is shown in Fig. 8.14.

Solution: A sinusoidal function shown in Fig. 8.14 is mathematically expressed as follows:

$$x(t) = A \sin \omega_0 t \, u(t) \quad t \leq 0$$

The given sinusoidal function is written as follows using Euler's identity.

$$\sin \omega_0 t = \frac{1}{2j}(e^{j\omega_0 t} - e^{-j\omega_0 t})$$
$$L[A \sin \omega_0 t] = \frac{A}{2j}[L(e^{j\omega_0 t}) - Le^{-j\omega_0 t}]$$

From Eq. (8.41), the above equation is written as

$$L[A \sin \omega_0 t] = \frac{A}{2j}\left[\frac{1}{s - j\omega_0} - \frac{1}{s + j\omega_0}\right]$$
$$= \frac{A}{2j}\frac{2j\omega_0}{(s^2 + \omega_0^2)}$$

$$\boxed{L[A \sin \omega_0 t] = \frac{A\omega_0}{(s^2 + \omega_0^2)}} \quad \text{ROC: Re } s > 0. \tag{8.42}$$

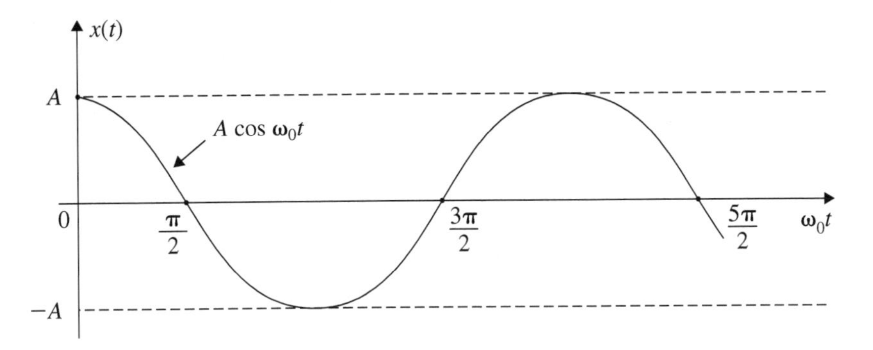

Fig. 8.15 A cosine function

■ Example 8.13

Determine the LT of a cosine function which is shown in Fig. 8.15.

Solution: A cosine function shown in Fig. 8.15 is mathematically expressed as follows:

$$x(t) = A\cos\omega_0 t u(t) \quad t \geq 0.$$

Using Euler's identity, the above equation is written as follows:

$$A\cos\omega_0 t = \frac{A}{2}(e^{j\omega_0 t} + e^{-j\omega_0 t})$$

Taking LT for $x(t)$, the following equation is written

$$L[A\cos\omega_0 t u(t)] = \frac{A}{2}[Le^{j\omega_0 t}u(t) + Le^{-j\omega_0 t}u(t)]$$

Using the results obtained in Eq. (8.41), we get

$$\begin{aligned} L[A\cos\omega_0 t u(t)] &= \frac{A}{2}\left[\frac{1}{(s+j\omega_0)} + \frac{1}{(s-j\omega_0)}\right] \\ &= \frac{As}{(s^2+\omega_0^2)} \end{aligned}$$

$$\boxed{L[A\cos\omega_0 t u(t)] = \frac{As}{(s^2+\omega_0^2)}} \quad \text{ROC: Re}s > 0. \quad (8.43)$$

Example 8.14

Determine the LT of hyperbolic sine function

$$x(t) = \sin h\omega_0 t.$$

Solution:

$$\sin h\omega_0 t = \frac{1}{2}[e^{\omega_0 t} - e^{-\omega_0 t}]$$
$$L[\sin h\omega_0 t] = \frac{1}{2}L[e^{\omega_0 t}] - \frac{1}{2}L[e^{-\omega_0 t}]$$

Using the results obtained in (8.41), we get

$$L[\sin h\omega_0 t] = \frac{1}{2(s - \omega_0)} - \frac{1}{2(s + \omega_0)}$$

$$\boxed{L[\sin h\omega_0 t] = \frac{\omega_0}{s^2 - \omega_0^2}} \quad \text{ROC: Re } s > \omega_0. \tag{8.44}$$

Example 8.15

Determine the Laplace transform of hyperbolic cosine function

$$x(t) = \cos h\omega_0 t.$$

Solution:

$$\cos h\omega_0 t = \frac{1}{2}[e^{\omega_0 t} + e^{-\omega_0 t}]$$

Taking LT on both sides, we get

$$\begin{aligned} L[\cos h\omega_0 t] &= \frac{1}{2}L[e^{\omega_0 t}] + \frac{1}{2}L[e^{-\omega_0 t}] \\ &= \frac{1}{2}\left[\frac{1}{s - \omega_0} + \frac{1}{s + \omega_0}\right] \\ &= \frac{s}{(s^2 - \omega_0^2)} \end{aligned}$$

$$\boxed{L[\cos h\omega_0 t] = \frac{s}{(s^2 - \omega_0^2)}} \quad \text{ROC: Re } s > \omega_0. \quad (8.45)$$

■ Example 8.16

Determine the LT of

$$x(t) = t^n u(t).$$

Solution: Using the definition of LT for the given function, we get

$$L[x(t)] = \int_0^\infty t^n e^{-st} dt$$

Let

$$u = t^n \quad \text{and} \quad du = nt^{n-1} dt$$
$$dv = \int e^{-st} dt \quad \text{and} \quad v = \frac{e^{-st}}{(-s)}$$

Using the property

$$\int u dv = uv - \int v du$$

we get

$$L[t^n] = \left[t^n \frac{e^{-st}}{(-s)} \right]_0^\infty - \int_0^\infty \frac{e^{-st}}{(-s)} nt^{n-1} dt$$
$$= 0 + \frac{n}{s} \int_0^\infty t^{n-1} e^{-st} dt.$$

It can be shown that

$$\int_0^\infty t^{n-1} e^{-st} dt = \frac{(n-1)}{s} \int_0^\infty t^{n-2} e^{-st} dt.$$

Thus, $L[t^n]$ is written as

$$L[t^n] = \frac{n}{s} \frac{(n-1)}{s} \frac{(n-2)}{s} \cdots \frac{2}{s} \frac{1}{s}$$
$$= \frac{n(n-1)(n-2)\ldots 2}{s^n} \frac{1}{s}$$
$$= \frac{\angle n}{s^{n+1}}$$

$$\boxed{L[t^n] = \frac{\angle n}{s^{n+1}}} \quad \text{ROC: Re } s > 0. \tag{8.46}$$

Example 8.17

Using the complex shifting property of LT, determine the LT of

$$x(t) = e^{-at} \sin \omega_0 t.$$

Solution:

$$L[\sin \omega_0 t] = \frac{\omega_0}{(s^2 + \omega_0^2)}$$

From Table 8.2, the complex shifting property is

$$L[e^{-at} x(t)] = X(s + a)$$

Applying the above property, we get

$$\boxed{L[e^{-at} \sin \omega_0 t] = \frac{\omega_0}{(s + a)^2 + \omega_0^2}} \tag{8.47}$$

ROC: Re $s > -a$.

Example 8.18

By applying the complex differentiation property, determine the LT of

$$x(t) = t \sin \omega_0 t.$$

Solution:

$$L[\sin \omega_0 t] = \frac{\omega_0}{(s^2 + \omega_0^2)}$$

According to the complex differentiation property

$$L[-tx(t)] = \frac{d}{ds} X(s)$$

$$\therefore \quad L[\sin \omega_0 t] = \frac{d}{ds} \frac{\omega_0}{(s^2 + \omega_0^2)}$$

$$\boxed{L[t \sin \omega_0 t] = \frac{2\omega_0 s}{(s^2 + \omega_0^2)^2}} \tag{8.48}$$

Fig. 8.16 $x(t) = u(t-3)$

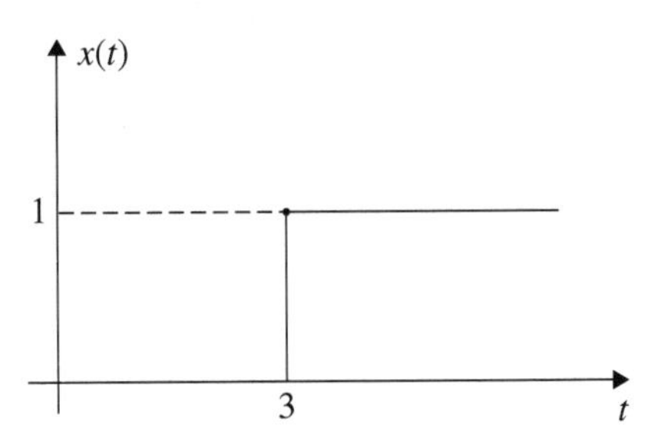

■ Example 8.19

Determine the LT of

$$x(t) = \cos at \sin bt.$$

Solution: The given $x(t)$ is written in the following form

$$x(t) = \frac{1}{2}[\sin(a+b)t - \sin(a-b)t]$$

$$L[\cos at \sin bt] = \frac{1}{2}[L \sin(a+b)t - L \sin(a-b)t]$$

$$\boxed{L[\cos at \sin bt] = \frac{1}{2}\left[\frac{(a+b)}{s^2+(a+b)^2} - \frac{(a-b)}{s^2+(a-b)^2}\right]} \tag{8.49}$$

■ Example 8.20

Consider the following time function $x(t) = u(t-3)$. Determine the LT using shift theorem.

Solution: From Fig. 8.16, for step input, the LT is

$$L[u(t)] = \frac{1}{s}$$

When the signal is shifted by $t = 3$, using time shifting property

$$\boxed{L[u(t-3)] = \frac{1}{s}e^{-3s}}$$

Table 8.2 Laplace transform tables

S.No	$x(t)$	$X(s)$
1	$\delta(t)$	1
2	$u(t)$	$\frac{1}{s}$
3	$tu(t)$	$\frac{1}{s^2}$
4	$t^n u(t)$	$\frac{\angle n}{s^{n+1}}$
5	$e^{at}u(t)$	$\frac{1}{(s-a)}$
6	$e^{-at}u(t)$	$\frac{1}{(s+a)}$
7	$\cos at\, u(t)$	$\frac{s}{(s^2+a^2)}$
8	$\sin at\, u(t)$	$\frac{a}{(s^2+a^2)}$
9	$e^{-bt}\cos at\, u(t)$	$\frac{(s+b)}{(s+b)^2+a^2}$
10	$e^{-bt}\sin at\, u(t)$	$\frac{a}{(s+b)^2+a^2}$
11	$\delta(t-a)$	e^{-as}
12	$u(t-a)$	$\frac{e^{-as}}{s}$
13	$t\sin at\, u(t)$	$\frac{2as}{(s^2+a^2)^2}$
14	$\sin h\, at$	$\frac{a}{(s^2+a^2)}$
15	$\cos h\, at$	$\frac{s}{s^2+a^2}$
16	$\sin(at+\theta)$	$\frac{s\sin\theta+a\cos\theta}{(s^2+a^2)}$
17	$\cos(at+\theta)$	$\frac{s\cos\theta-a\sin\theta}{(s^2+a^2)}$

Table 8.2 gives the LT of some time functions.

■ Example 8.21

Determine the LT for the following time function

$$x(t) = \sin(at+\theta)$$

Solution: The given $x(t)$ can be expanded and written as follows:

$$\begin{aligned} x(t) &= \sin(at+\theta) \\ &= \sin at\cos\theta + \cos at\sin\theta \\ L[\sin(at+\theta)] &= L[\sin at\cos\theta] + L[\cos at\sin\theta] \end{aligned}$$

Substituting $L[\sin at]$ and $L[\cos at]$ from Table 8.2, we get

$$\boxed{L[\sin(at+\theta)] = \frac{a\cos\theta}{(s^2+a^2)} + \frac{s\sin\theta}{(s^2+a^2)}} \tag{8.50}$$

■ Example 8.22

Determine the LT for the following time function

$$x(t) = \cos(at+\theta)$$

Solution: $x(t)$ can be expanded and written as follows:

$$\begin{aligned} x(t) &= \cos at\cos\theta - \sin at\sin\theta \\ L[\cos(at+\theta)] &= \cos\theta L[\cos at] - \sin\theta L[\sin at] \\ &= \frac{s\cos\theta}{(s^2+a^2)} - \frac{a\sin\theta}{(s^2+a^2)} \end{aligned}$$

$$\boxed{L[\cos(at+\theta)] = \frac{(s\cos\theta - a\sin\theta)}{(s^2+a^2)}} \tag{8.51}$$

■ Example 8.23

Determine the LT for the following time function

$$x(t) = \delta(t-2) - \delta(t-5).$$

Solution: The given time function consists of two impulses occurring at $t = 2$ and $t = 5$. By applying shift theorem, we get

$$\begin{aligned} L[\delta(t-2)] &= e^{-2s} \\ L[\delta(t-5)] &= e^{-5s} \end{aligned}$$

$$\boxed{L[\delta(t-2) - \delta(t-5)] = e^{-2s} - e^{-5s}}$$

■ Example 8.24

Determine the LT for the following time function

$$x(t) = u(t-2) - u(t-5).$$

Solution: The given time function $x(t)$ consists of two step functions shifted by $t = 2$ and $t = 5$. By applying shift theorem, we get

$$L[u(t-2)] = \frac{e^{-2s}}{s}$$
$$L[u(t-5)] = \frac{e^{-5s}}{s}$$

$$\therefore \quad \boxed{L[u(t-2) - u(t-5)] = \frac{1}{s}[e^{-2s} - e^{-5s}]}$$

■ Example 8.25

Consider the following function

$$X(s) = \frac{(5s+4)(s+6)}{s(s+2)(3s+1)}$$

Find the initial and final values of $x(t)$.

Solution: The initial value is given by

$$\begin{aligned}\underset{t\to 0}{Lt}\, x(t) = x(0) &= \underset{s\to\infty}{Lt}\, sX(s) \\ &= \underset{s\to\infty}{Lt} \frac{s(5+\frac{4}{s})(1+\frac{6}{s})}{s(1+\frac{2}{s})(3+\frac{1}{s})} \\ &= \frac{5\times 1}{1\times 3} = \frac{5}{3}\end{aligned}$$

$$\boxed{x(0) = \frac{5}{3}}$$

The final value of $x(t)$ is given by

$$\begin{aligned}\underset{t\to\infty}{x(t)} = x(\infty) &= \underset{s\to 0}{Lt}\, sX(s) \\ &= \underset{s\to 0}{Lt} \frac{s(5s+4)(s+6)}{s(s+2)(3s+1)} \\ &= \frac{4\times 6}{2\times 1} = 12\end{aligned}$$

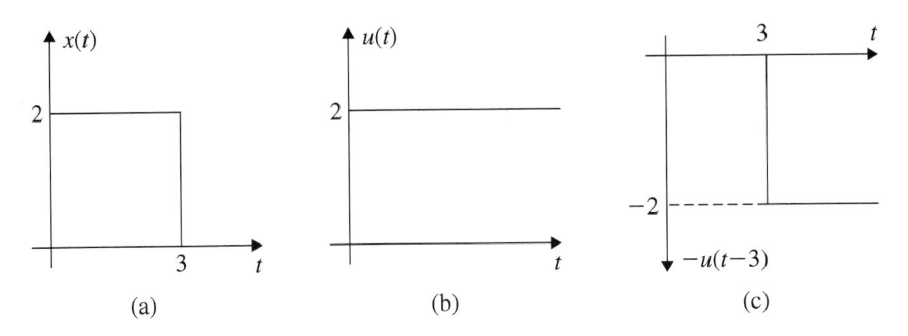

Fig. 8.17 LT of a pulse

$$\boxed{x(\infty) = 12}$$

■ Example 8.26

Consider the pulse shown in Fig. 8.17a. Determine the LT.

Solution:

Method 1: The given signal $x(t)$, which is shown in Fig. 8.17a, could be split up of step signals as shown in Fig. 8.17b, c. Thus, the following equation is written

$$x(t) = u(t) - u(t-3)$$

Taking LT on both sides, we get

$$\begin{aligned} X(s) &= U(s) - U(s)e^{-3s} \\ &= [1 - e^{-3s}]U(s) \end{aligned}$$

$$\text{But } U(s) = \frac{2}{s} \quad \text{(for a step input).}$$

$$\therefore \quad \boxed{X(s) = \frac{2}{s}[1 - e^{-3s}]}$$

Method 2: By definition of LT, the following equation is written for Fig. 8.17a

$$\begin{aligned} X(s) &= \int_0^3 2e^{-st}dt \\ &= \frac{2}{(-s)}\left[e^{-st}\right]_0^3 \end{aligned}$$

Fig. 8.18 A sine wave

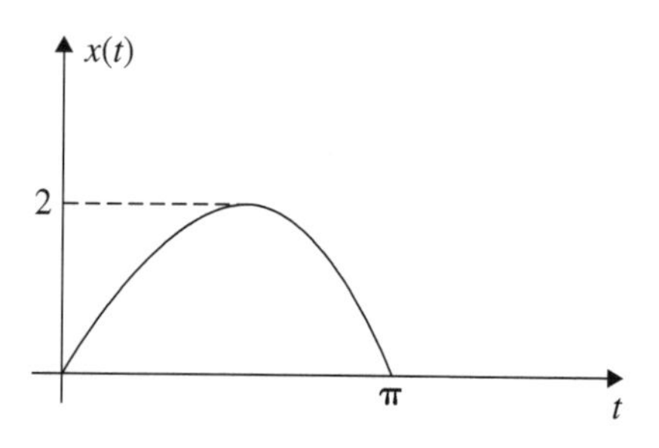

$$\boxed{X(s) = \frac{2}{s}[1 - e^{-3s}]}$$

■ Example 8.27

For the wave form shown in Fig. 8.18, determine the LT.

Solution: For Fig. 8.18, the following equation is written

$$\begin{aligned} x(t) &= 2 \sin t \quad 0 \leq t \leq \pi \\ &= \quad 0 \qquad t > \pi \end{aligned}$$

The LT of the above signal is obtained from the following equation

$$X(s) = \int_0^{\pi} 2 \sin t\, e^{-st} dt$$

Let $u = 2 \sin t$ and $du = 2 \cos t dt$; $dv = e^{-st} dt$ and $v = -\frac{1}{s}e^{-st}$. Applying $\int u dv = uv - \int v du$, we get

$$\begin{aligned} X(s) &= \left[-\frac{2}{s} \sin t e^{-st} \right]_0^{\pi} + \int_0^{\pi} \frac{2}{s} \cos t e^{-st} dt \\ &= 0 + \frac{2}{s} \int_0^{\pi} \cos t e^{-st} dt \end{aligned}$$

Let $u = \cos t$ and $du = -\sin t dt$; $dv = e^{-st} dt$ and $v = -\frac{1}{s}e^{-st}$. Substituting the above in equation for $X(s)$, we get

$$\begin{aligned} X(s) &= \frac{2}{s} \left\{ \left[-\frac{1}{s} \cos t e^{-st} \right]_0^{\pi} - \int_0^{\pi} \frac{1}{s} \sin t e^{-st} dt \right\} \\ &= \frac{2}{s} \left[\left\{ e^{-\pi s} + 1 \right\} \frac{1}{s} - \frac{1}{2s} X(s) \right] \quad \text{since} \int_0^{\pi} \sin t e^{-st} dt = \frac{X(s)}{2} \end{aligned}$$

$$\frac{sX(s)}{2} + \frac{1}{2s}X(s) = \frac{(e^{-\pi s} + 1)}{s}$$
$$\frac{(s^2 + 1)X(s)}{2s} = \frac{(e^{-\pi s} + 1)}{s}$$

$$\boxed{X(s) = \frac{2(e^{-\pi s} + 1)}{(s^2 + 1)}}$$

■ Example 8.28

Determine the LT of the saw tooth wave form shown in Fig. 8.19a.

(Anna University, April, 2005)

Solution: The saw tooth wave form shown in Fig. 8.19 is expressed as

$$x(t) = \frac{3}{2}t \quad 0 \leq t \leq 2$$
$$= 0 \quad \text{otherwise}$$

Taking LT for the time function $x(t)$, we get

$$X(s) = \int_0^2 \frac{3}{2}te^{-st}dt$$

Let

$$u = \frac{3}{2}t \quad \text{and} \quad du = \frac{3}{2}dt$$

$$dv = e^{-st}dt \quad \text{and} \quad v = -\frac{1}{s}e^{-st}$$

Using $\int udv = uv - \int vdu$, we get

$$X(s) = \left[\frac{3}{2}t\left(-\frac{1}{s}\right)e^{-st}\right]_0^2 + \frac{3}{2}\int_0^2 \frac{1}{s}e^{-st}dt$$
$$= \frac{-3}{s}e^{-2s} + \frac{3}{2s^2}\left[-1e^{-st}\right]_0^2$$
$$= \frac{-3}{s}e^{-2s} - \frac{3}{2s^2}e^{-2s} + \frac{3}{2s^2}$$

$$\boxed{X(s) = \frac{3}{2}\frac{1}{s^2} - \left(\frac{3}{s} + \frac{3}{2s^2}\right)e^{-2s}}$$

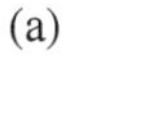

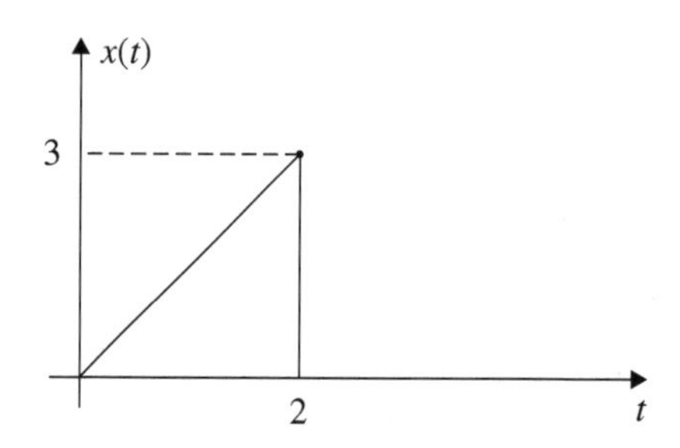

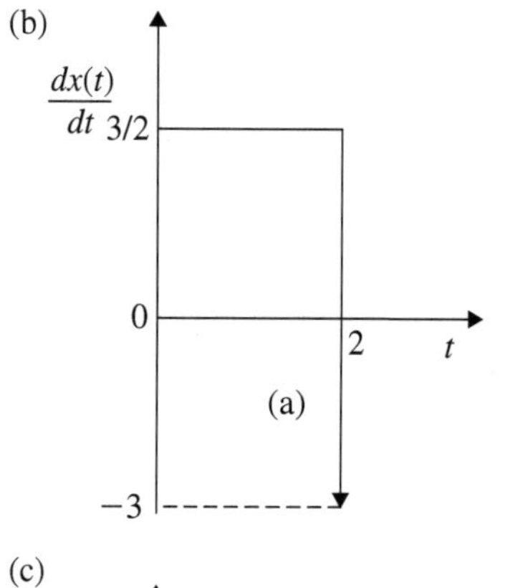

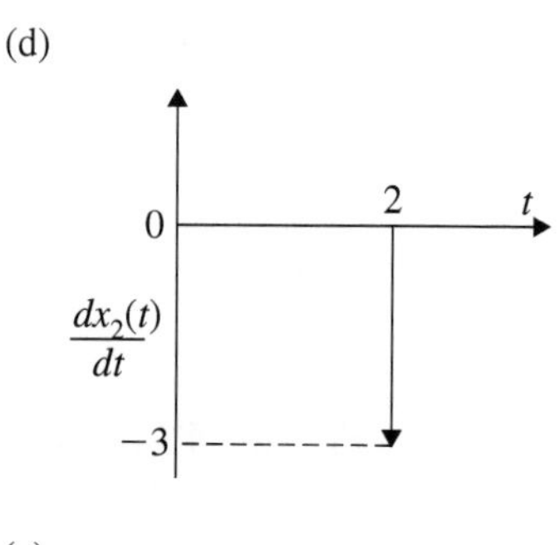

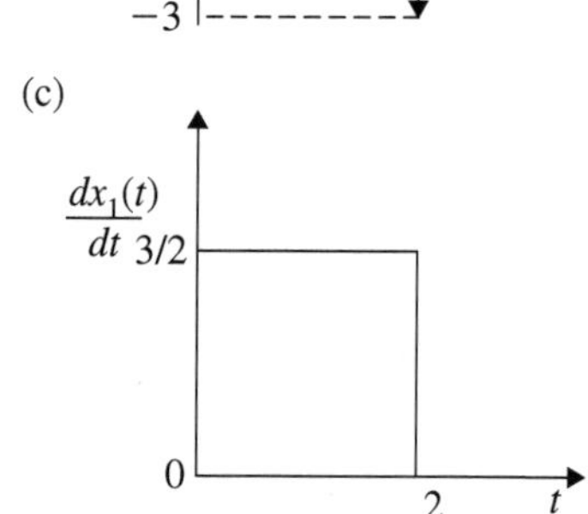

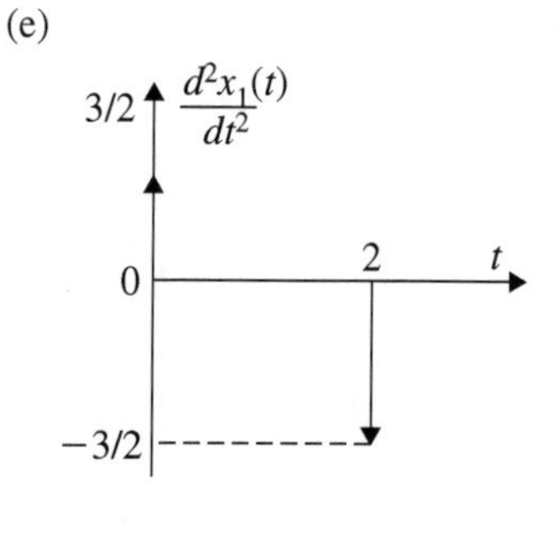

Fig. 8.19 **a** Saw tooth wave form

Method 2:

When $x(t)$ represented in Fig. 8.19a is differentiated we obtain $dx(t)/dt$ and is shown in Fig. 8.19b. Figure 8.19b can be drawn as a sum of the signals shown in Fig. 8.19c, d. Further the signals shown in Fig. 8.19 can be further differentiated and is shown in Fig. 8.19e. From Fig. 8.19d, using integration and time shifting properties of Laplace transform, we get

$$X_2(s) = \frac{-3}{s}e^{-2s}$$

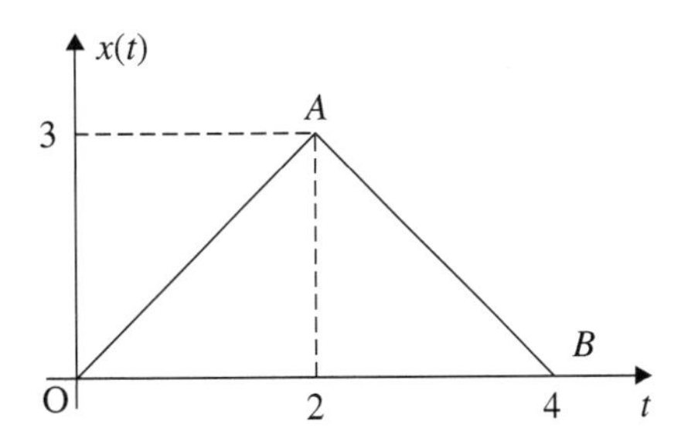

Fig. 8.20 Triangular wave form

From Fig. 8.19e, we get

$$X_1(s) = \frac{1}{s^2}\frac{3}{2}\left[1 - e^{-2s}\right]$$
$$X(s) = X_1(s) + X_2(s)$$

$$\boxed{X(s) = \frac{3}{2s^2}\left[1 - e^{-2s}\right] - \frac{3}{s}e^{-2s}}$$

■ Example 8.29

Consider the triangular wave form shown in Fig. 8.20. Determine the LT.

Solution: For the straight line OA, the slope is $\frac{3}{2}$ and passes through the origin. Hence, the following equation is written

$$x_1(t) = \frac{3}{2}t \quad 0 \leq t \leq 2$$

For the straight line AB, the slope is negative and it is $-\frac{3}{2}$. The following equation is written

$$x_2(t) = -\frac{3}{2}t + C$$

when $t = 2, x_2(t) = 3$. Hence

$$3 = -\frac{3}{2} \times 2 + C$$

or $C = 6$

$$x_2(t) = -\frac{3}{2}t + 6 \quad 2 \leq t \leq 4$$

From Example 8.28, $X_1(s)$ is written as

$$X_1(s) = \frac{3}{2s^2} - \left(\frac{3}{s} + \frac{3}{2s^2}\right)e^{-2s}$$

Now $X_2(s)$ is written as

$$X_2(s) = \int_2^4 \left(6 - \frac{3}{2}t\right) e^{-st} dt$$

Let

$$u = \left(6 - \frac{3}{2}t\right) \quad \text{and} \quad du = -\frac{3}{2}dt$$

$$dv = \int e^{-st} dt \quad \text{and} \quad v = -\frac{1}{s}e^{-st}$$

Using $\int udv = uv - \int vdu$, we get

$$\begin{aligned} X_2(s) &= \left[\left(6 - \frac{3}{2}t\right)\left(-\frac{1}{s}\right)e^{-st}\right]_2^4 - \frac{3}{2s}\int_2^4 e^{-st} dt \\ &= \left[\frac{3}{s}e^{-2s}\right] + \frac{3}{2s^2}\left[e^{-st}\right]_2^4 \\ &= \frac{3}{s}e^{-2s} + \frac{3}{2s^2}e^{-4s} - \frac{3}{2s^2}e^{-2s} \\ X(s) &= X_1(s) + X_2(s) \\ &= \frac{3}{2s^2} - \left(\frac{3}{s} + \frac{3}{2s^2}\right)e^{-2s} + \frac{3}{s}e^{-2s} + \frac{3}{2s^2}e^{-4s} - \frac{3}{2s^2}e^{-2s} \end{aligned}$$

$$\boxed{X(s) = \frac{3}{2s^2} - \left(\frac{3}{s^2}e^{-2s}\right) + \frac{3}{2s^2}e^{-4s}}$$

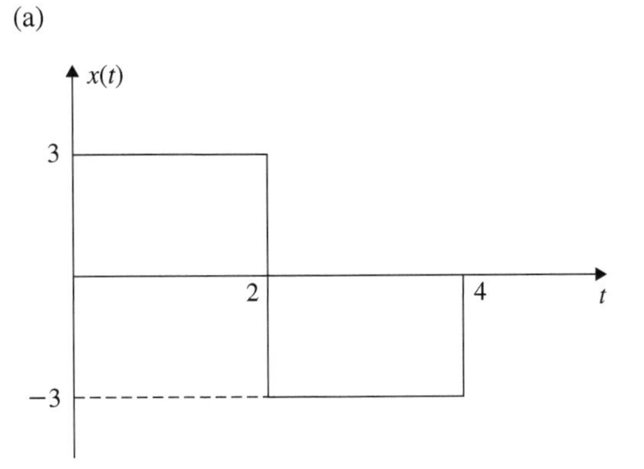

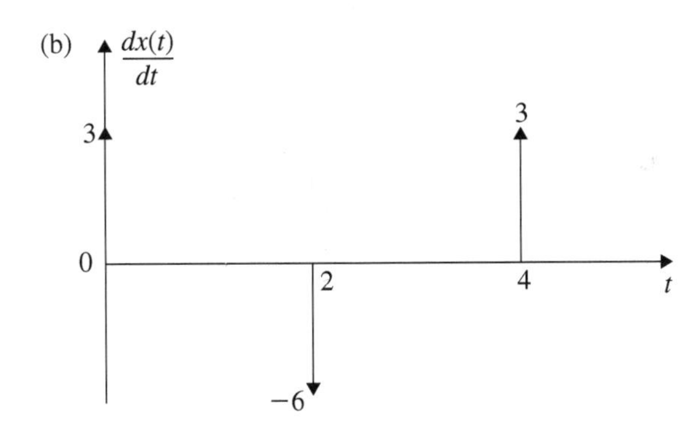

Fig. 8.21 a A rectangular wave

■ Example 8.30

Consider the rectangular wave form shown in Fig. 8.21a. Determine the LT.

Solution: Consider the rectangular wave shown in Fig. 8.21 for the time interval

$$x_1(t) = 3 \quad 0 \le t \le 2.$$

The LT of $x_1(t)$ is found from the equation

$$\begin{aligned} X_1(s) &= \int_0^2 3e^{-st} dt \\ &= \frac{-3}{s}\left[e^{-st}\right]_0^2 \\ &= \frac{3}{s}[1 - e^{-2s}] \end{aligned}$$

Consider rectangular wave

$$x_2(t) = -3 \quad 2 \leq t \leq 4$$

Using shift theorem $X_2(s)$ is obtained as

$$\begin{aligned} X_2(s) &= -X_1(s)e^{-2s} \\ \therefore \quad X(s) &= X_1(s) + X_2(s) \\ &= \frac{3}{s}(1 - e^{-2s}) - \frac{3}{s}(1 - e^{-2s})e^{-2s} \\ &= \frac{3}{s}(1 - e^{-2s})(1 - e^{-2s}) \end{aligned}$$

$$\boxed{X(s) = \frac{3}{s}(1 - e^{-2s})^2}$$

(b)

The differentiated wave form of Fig. 8.21a is shown in Fig. 8.21b. The Laplace transform of the wave form shown in Fig. 8.21b is obtained as

$$L\left[\frac{dx(t)}{dt}\right] = 3 - 6e^{-2s} + 3e^{-4s}$$

using differentiation property of LT, we get

$$X(s) = \frac{3}{s}(1 - 2e^{-2s} + e^{-4s})$$

$$\boxed{X(s) = \frac{3}{s}(1 - e^{-2s})^2}$$

■ Example 8.31

Consider the wave form shown in Fig. 8.22a. Determine the LT.

Solution: The mathematical description of the wave form shown in Fig. 8.22 is written as follows:

$$\begin{aligned} x(t) &= (3t - 6) \quad 2 \leq t \leq 3 \\ &= 3 \qquad\quad 3 \leq t \leq 5 \end{aligned}$$

The LT of $x(t)$ is written as

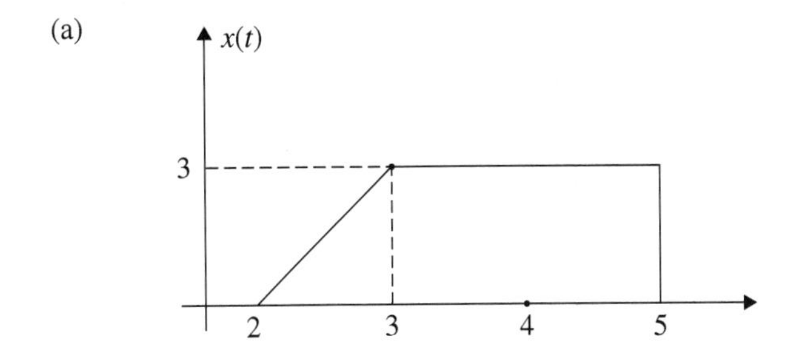

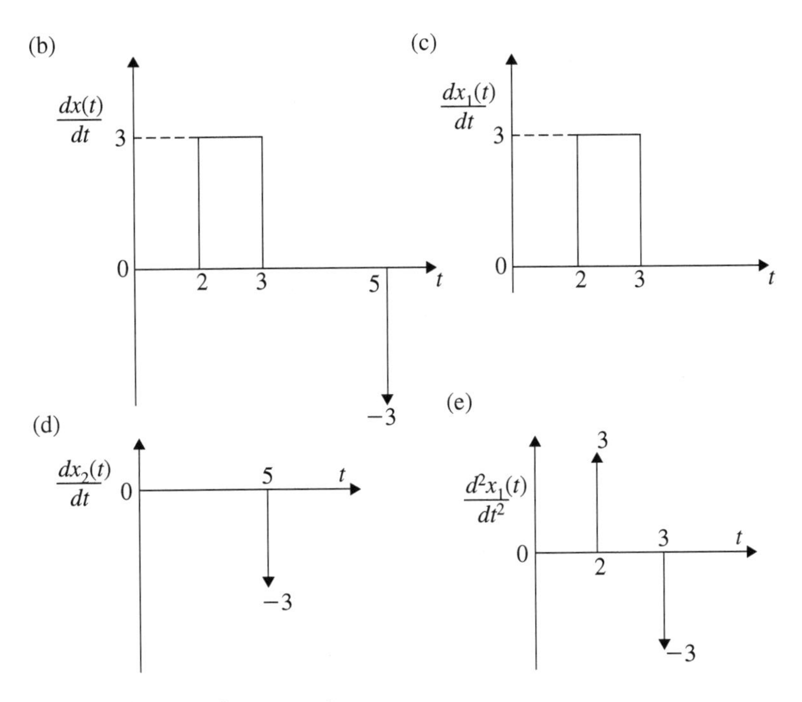

Fig. 8.22 **a** A triangular pulse rectangular wave

$$X(s) = \int_2^3 x_1(t)e^{-st}dt + \int_3^5 x_2(t)e^{-st}dt$$
$$= X_1(s) + X_2(s)$$

where

$$X_2(s) = \frac{3}{(-s)}\left[e^{-st}\right]_3^5 = \frac{3}{s}[e^{-3s} - e^{-5s}]$$

$X_1(s)$ is determined as follows:

For the triangle $x_1(t)$ is written as follows:

$$x_1(t) = 3t + C$$

When $t = 2$, $x_1(t) = 0$

$$0 = 3 \times 2 + C \quad \text{or} \quad C = -6$$
$$x_1(t) = (3t - 6)$$
$$X_1(s) = \int_2^3 (3t - 6)e^{-st}dt$$

Let

$$u = (3t - 6) \quad \text{and} \quad du = 3dt$$
$$dv = \int e^{-st}dt \quad \text{and} \quad v = -\tfrac{1}{s}e^{-st}$$

$$X_1(s) = \left[(3t - 6)\left(-\frac{1}{s}\right)e^{-st}\right]_2^3 - \frac{3}{s^2}\left[e^{-st}\right]_2^3$$
$$= -\frac{3}{s}e^{-3s} - \frac{3}{s^2}(e^{-3s} - e^{-2s})$$
$$X(s) = X_1(s) + X_2(s)$$
$$= -\frac{3}{s}e^{-3s} - \frac{3}{s^2}(e^{-3s} - e^{-2s}) + \frac{3}{s}(e^{-3s} - e^{-5s})$$

$$\boxed{X(s) = -\frac{3}{s}e^{-5s} - \frac{3}{s^2}(e^{-3s} - e^{-2s})}$$

Method (b):

The signal represented in Fig. 8.22a when differentiated is obtained as shown in Fig. 8.22b. The different signal $dx(t)/dt$ is split up into $dx_1(t)/dt$ and $dx_2(t)/dt$ and are represented in Fig. 8.22c and d, respectively. The signal $dx_1(t)/dt$ when further differentiated is shown in Fig. 8.22e. From Fig. 8.22d, using integration property of Laplace transform, we get

$$X_2(s) = \frac{-3}{s}e^{-5s}$$

From Fig. 8.22e, we get

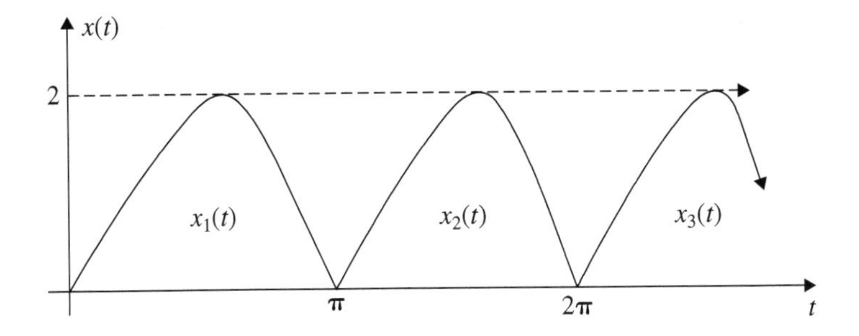

Fig. 8.23 Full wave rectifier

$$
\begin{aligned}
X_1(s) &= \frac{3}{s^2}\left[e^{-2s} - e^{-3s}\right] \\
X(s) &= X_1(s) + X_2(s) \\
&= -\frac{3}{s}e^{-5s} + \frac{3}{s^2}(e^{-2s} - e^{-3s})
\end{aligned}
$$

8.7 Laplace Transform of Periodic Signal

If a signal $x(t)$ is a periodic signal with period T, then the LT of $X(s)$ is given as

$$X(s) = X_1(s)\left[1 + e^{-Ts} + e^{-2Ts} + \ldots\right] = \frac{X_1(s)}{(1 - e^{-Ts})}$$

Here, $x_1(t)$ is the signal which is repeated for every T.

Example 8.32

Consider the output of a full wave rectifier shown in Fig. 8.23. Determine the LT.

Solution: In Example 8.27, $X_1(s)$ is determined as

$$X_1(s) = \frac{2(e^{-\pi s} + 1)}{(s^2 + 1)}$$

If $X(s)$ is the LT of the full wave rectifier

$$X(s) = X_1(s) + X_1(s)e^{-Ts} + X_1(s)e^{-2Ts} + \ldots$$

where $T = \pi$. For $X_1(s)$ see Example 8.27

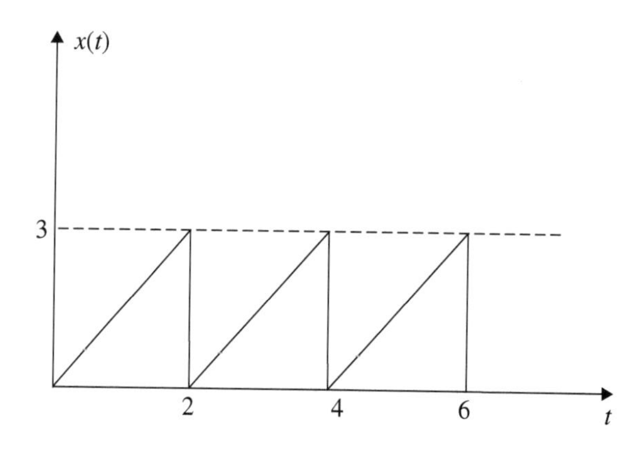

Fig. 8.24 Saw tooth wave

$$
\begin{aligned}
&= X_1(s) + X_1(s)e^{-\pi s} + X_1(s)e^{-2\pi s} + \ldots \\
&= X_1(s)[1 + e^{-\pi s} + e^{-2\pi s} + \ldots] \\
&= \frac{X_1(s)}{(1 - e^{-\pi s})} \\
&= \frac{2(e^{-\pi s} + 1)}{(1 - e^{-\pi s})} \frac{1}{(s^2 + 1)}
\end{aligned}
$$

$$
\boxed{X(s) = \frac{2(1 + e^{-\pi s})}{(1 - e^{-\pi s})(1 + s^2)}}
$$

■ Example 8.33

Consider the saw tooth wave shown in Fig. 8.24. Determine the LT.

Solution: The mathematical description of $x(t)$ for $0 \leq t \leq 2$ is given as $x_1(t)$. In Example 8.28, $X_1(s)$ is determined as

$$
X_1(s) = \frac{3}{2s^2} - \left(\frac{3}{s} + \frac{3}{2s^2}\right) e^{-2s}
$$

from Fig. 8.24

$$
X(s) = X_1(s)[1 + e^{-2s} + e^{-4s} + \ldots] = \frac{X_1(s)}{(1 - e^{-2s})}
$$

$$
\boxed{X(s) = \frac{3}{2(1 - e^{-2s})} \left[\frac{1}{s^2} - \left(\frac{2}{s} + \frac{1}{s^2}\right) e^{-2s}\right]}
$$

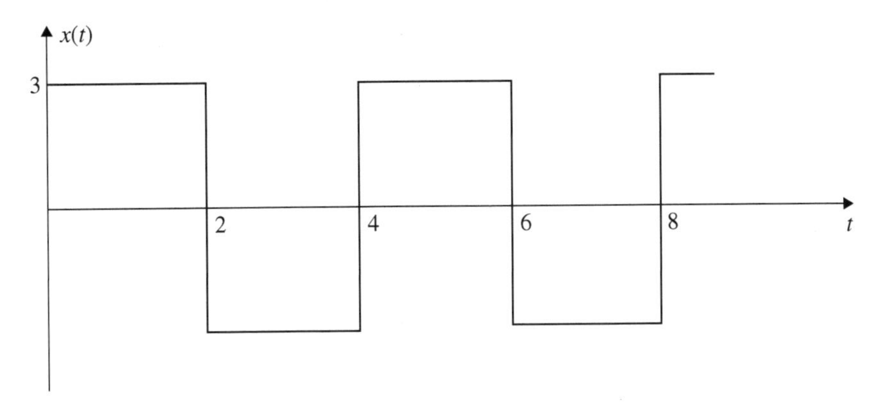

Fig. 8.25 A periodic rectangular wave

■ Example 8.34

Consider the rectangular periodic wave shown in Fig. 8.25. Determine the LT.

Solution: The mathematical description of the periodic wave with period 4 is written as follows:

$$\begin{aligned} x(t) &= 3 \quad 0 \le t \le 2 \\ &= -3 \quad 2 \le t \le 4 \end{aligned}$$

Let $X_1(s)$ be the LT of $x(t)$ for the time $0 \le t \le 4$. $X_1(s)$ in Example 8.30 has been determined as

$$X_1(s) = \frac{3}{s}(1 - e^{-2s})^2$$

$$X(s) = X_1(s)[1 + e^{-4s} + e^{-8s} + \ldots] = \frac{X_1(s)}{(1 - e^{-4s})}$$

$$\boxed{X(s) = \frac{3(1 - e^{-2s})^2}{s(1 - e^{-4s})}}$$

8.8 Inverse Laplace Transform

The time signal $x(t)$ is the Inverse LT of $X(s)$. This is represented by the following mathematical equation.

$$x(t) = \frac{1}{2\pi j} \int_{\sigma - j\infty}^{\sigma + j\infty} X(s) e^{st} ds \tag{8.52}$$

Use of Eq. (8.52) to obtain $x(t)$ from $X(s)$ is really a tedious process. The alternative is to express $X(s)$ in polynomial form both in the numerator and the denominator. Both these polynomials are factorized as

$$X(s) = \frac{(s + z_1)(s + z_2) \ldots (s + z_m)}{(s + p_1)(s + p_2) \cdots (s + p_n)} \tag{8.53}$$

The points in the s-plane at which $X(s) = 0$ are called zeros. Thus, $(s + z_1)$, $(s + z_2), \ldots, (s + z_m)$ are the zeros of $X(s)$ in Eq. (8.53). Similarly, the points in the s-plane at which $X(s) = \infty$ are called poles of $X(s)$ in Eq. (8.53).

The zeros are identified by a small circle O and the poles by a small cross $\times$ in the s-plane. For $m < n$ the degree of the numerator polynomial is less than the degree of the denominator polynomial. Under this condition, $X(s)$ in Eq. (8.53) is written in the following partial fraction form.

$$X(s) = \frac{A_1}{s + p_1} + \frac{A_2}{s + p_2} + \frac{A_3}{s + p_3} + \cdots + \frac{A_n}{s + p_n} \tag{8.54}$$

In Eq. (8.54), A_1, $A_2, \ldots, A_n$ are called the residues and are determined by any convenient method. Once the residues are determined, then by using Table 8.2, one can easily obtain $x(t)$ which is the required inverse LT of $X(s)$.

8.8.1 Graphical Method of Determining the Residues

The residues in Eq. (8.54) are determined by analytical as well as graphical method. The graphical method has the following advantages:

- It is less time-consuming.
- It does not require any graph to be drawn.
- The results are obtained in a compact form very quickly even if the poles and zeros are complex and repeated.

Both analytical and graphical methods are given wherever necessary. The following simple example illustrates both analytical and graphical methods.

■ Example 8.35

$$X(s) = \frac{10(s+2)(s-3)}{s(s+4)(s-5)}$$

Find $x(t)$.

Solution: The given $X(s)$ is expressed in partial fraction form as follows:

$$X(s) = \frac{A_1}{s} + \frac{A_2}{(s+4)} + \frac{A_3}{(s-5)}$$

Method 1. Analytical Method

1. The poles and zeros of $X(s)$ are represented in Fig. 8.26. $X(s)$ is expressed in the following form:

$$X(s) = \frac{A_1(s+4)(s-5) + A_2s(s-5) + A_3s(s+4)}{s(s+4)(s-5)}$$

The numerator polynomial of $X(s)$ should be the same, and therefore, the following equation is written.

$$10(s+2)(s-3) = A_1(s+4)(s-5) + A_2s(s-5) + A_3s(s+4)$$

2. Substitute $s = 0$ in the above equation which will eliminate A_2 and A_3. Thus

$$10(2)(-3) = A_1(4)(-5) + 0 + 0$$
$$A_1 = \frac{60}{20} = 3$$

Substitute $s = -4$ in $X(s)$. This eliminates A_1 and A_3. Thus

$$10(-4+2)(-4-3) = 0 - A_2 4(-4-5) + 0$$
$$10(-2)(-7) = A_2 36$$
$$A_2 = \frac{140}{36} = \frac{35}{9}$$

Substitute $s = 5$ in $X(s)$. This eliminates A_1 and A_2. Thus

$$10(5+2)(5-3) = 0 + 0 + A_3(5)(5+4)$$
$$A_3 = \frac{140}{45} = \frac{28}{9}$$

3. With the values of residues obtained in step 2, $X(s)$ is expressed as follows:

$$X(s) = \frac{3}{s} + \frac{35}{9}\frac{1}{(s+4)} + \frac{28}{9}\frac{1}{(s-5)}$$

4. From the Table 8.2, the inverse Laplace transform is obtained for $\frac{1}{s}$, $\frac{1}{(s+4)}$ and $\frac{1}{s-5}$.
5. To check whether the residues determined are correct, the following procedure is followed

$$X(s) = \frac{10(s+2)(s-3)}{s(s+4)(s-5)} = \frac{3}{s} + \frac{35}{9(s+4)} + \frac{28}{9(s-5)}$$

Choose any value of s, so that, $X(s)$ does not become zero or infinity. Let us choose $s = 1$

$$\frac{10(3)(-2)}{1(5)(-4)} = \frac{3}{1} + \frac{35}{9\times 5} + \frac{28}{9(-4)}$$
$$3 = 3 + \frac{7}{9} - \frac{7}{9} = 3$$
$$\text{LHS} = \text{RHS}.$$

Hence, A_1, A_2 and A_3 determined are correct.

$$\boxed{x(t) = \left(3 + \frac{35}{9}e^{-4t} + \frac{28}{9}e^{5t}\right)u(t)}$$

Method 2. Graphical Method of Determining the Residues

1. According to the graphical method, the residue A at any pole is obtained from

$$A = \frac{\text{Constant term} \times \text{Directed Vector distances drawn from all zeros to the concerned point}}{\text{Directed vector distances drawn from all poles to the concerned point}}$$

2. For the given problem, refer to the pole-zero diagram of Fig. 8.26. From the figure, we obtain A_1 by drawing vectors from poles and zeros of $X(s)$ to $s = 0$.

$$A_1 = \frac{10(2)(-3)}{4(-5)} = 3$$

A_2 is determined by drawing vectors from poles and zeros of $X(s)$ to $s = -4$.

$$A_2 = \frac{10(-2)(-7)}{(-4)(-9)} = \frac{35}{9}$$

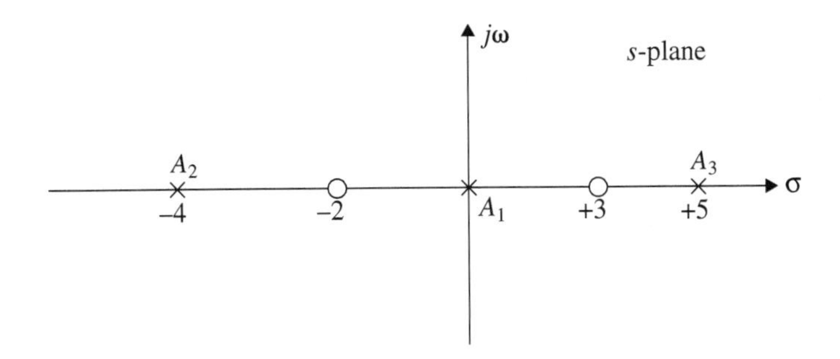

Fig. 8.26 Poles and zeros of $X(s)$ (pole-zero diagram)

A_3 is obtained by drawing vectors from poles and zeros of $X(s)$ to $s = 5$.

$$A_3 = \frac{10(7)(2)}{(5)(9)} = \frac{28}{9}$$

It is to be noted that the directed distances drawn from any pole or zero drawn toward right, a+ve sign is added and for the directed distance drawn toward left, a-ve sign in each case has to be included.

3. It is seen that the residues determined by the analytical method and graphical method are the same. Hence, inverse LT of $X(s)$ is written as

$$\boxed{x(t) = \left(3 + \frac{35}{9}e^{-4t} + \frac{28}{9}e^{5t}\right)u(t)}$$

In the expression for $x(t)$, it is necessary to include $u(t)$ on the right side of the equation. This indicates that the inverse LT is right sided or unilateral. It is also to be noted that the pole-zero diagram of Fig. 8.26 need not be drawn to any scale. The mere location of poles and zeros with the appropriate values is enough.

■ Example 8.36

Find the inverse LT of

$$X(s) = \frac{10e^{-3s}}{(s-2)(s+2)}.$$

Solution: Consider the function

$$X_1(s) = \frac{10}{(s-2)(s+2)}$$

Putting this into partial fraction, we get

$$\begin{aligned}X_1(s) &= \frac{A_1}{(s-2)} + \frac{A_2}{(s+2)} \\ &= \frac{A_1(s+2) + A_2(s-2)}{(s-2)(s+2)} \\ 10 &= A_1(s+2) + A_2(s-2)\end{aligned}$$

Substitute $s = -2$

$$\begin{aligned}10 &= 0 + A_2(-2-2) \\ A_2 &= -2.5\end{aligned}$$

Substitute $s = 2$

$$\begin{aligned}10 &= A_1(2+2) + 0 \\ A_1 &= 2.5 \\ X_1(s) &= 2.5\left[\frac{1}{s-2} - \frac{1}{s+2}\right]\end{aligned}$$

Taking inverse LT we get

$$x_1(t) = 2.5[e^{+2t} - e^{-2t}]u(t)$$

According to time shifting property of LT

$$X(s) = X_1(s)e^{-3s}$$

$$\boxed{x(t) = 2.5[e^{2(t-3)} - e^{-2(t-3)}]u(t-3)}$$

■ Example 8.37

Find the inverse LT of

$$X(s) = \frac{(s+1) + 3e^{-4s}}{(s+2)(s+3)}.$$

Solution: The given function is written in the following form:

$$\begin{aligned}X(s) &= \frac{(s+1)}{(s+2)(s+3)} + \frac{3e^{-4s}}{(s+2)(s+3)} \\ &= X_1(s) + X_2(s) \\ X_1(s) &= \frac{(s+1)}{(s+2)(s+3)}\end{aligned}$$

$$= \frac{A_1}{(s+2)} + \frac{A_2}{(s+3)}$$
$$= \frac{A_1(s+3) + A_2(s+2)}{(s+2)(s+3)}$$
$$(s+1) = A_1(s+3) + A_2(s+2)$$

Put $s = -3$

$$(-3+1) = 0 + A_2(-3+2)$$
$$A_2 = 2$$

Put $s = -2$

$$(-2+1) = A_1(-2+3) + 0$$
$$A_1 = -1$$
$$X_1(s) = -\frac{1}{s+2} + \frac{2}{s+3}$$
$$x_1(t) = (-e^{-2t} + 2e^{-3t})u(t).$$

Now consider $X_2(s)$ without delay as $X_3(s)$

$$X_3(s) = \frac{3}{(s+2)(s+3)}$$
$$= \frac{A_1}{(s+2)} + \frac{A_2}{(s+3)}$$
$$3 = A_1(s+3) + A_2(s+2)$$

Put $s = -2$

$$3 = A_1$$

Put $s = -3$

$$3 = A_2(-3+2)$$

$$A_2 = -3$$
$$X_3(s) = 3\left[\frac{1}{s+2} - \frac{1}{s+3}\right]$$
$$X_2(s) = X_3(s)e^{-4s}$$
$$x_3(t) = 3[e^{-2t} - e^{-3t}]u(t)$$
$$x_2(t) = 3[e^{-2(t-4)} - e^{-3(t-4)}]u(t-4)$$
$$x(t) = x_1(t) + x_2(t)$$

$$\boxed{x(t) = \left[-e^{-2t} + 2e^{-3t}\right]u(t) + 3\left[e^{-2(t-4)} - e^{-3(t-4)}\right]u(t-4)}$$

■ Example 8.38

Find the inverse LT of

$$(1) \quad X(s) = \frac{(s+1)(s+3)}{(s+2)(s+4)}$$

$$(2) \quad X(s) = \frac{(s+1)(s+3)e^{-2s}}{(s+2)(s+4)}$$

Solution:

(1)

$$X(s) = \frac{(s+1)(s+3)}{(s+2)(s+4)}$$

Here, both numerator polynomial and denominator polynomial have the same degree, and therefore, it is an improper function. Now $X(s)$ is written in the polynomial form as given below:

$$X(s) = \frac{(s^2+4s+3)}{(s^2+6s+8)}$$

By synthetic division, we get

$$\begin{array}{r|l} & \quad 1 \\ s^2+6s+8 & \overline{s^2+4s+3} \\ & \underline{s^2+6s+8} \\ & \quad -2s-5 \end{array}$$

$$\therefore \quad X(s) = 1 - \frac{(2s+5)}{(s+2)(s+4)}$$

Now consider

$$X_1(s) = \frac{(2s+5)}{(s+2)(s+4)}$$

$$= \frac{A_1}{(s+2)} + \frac{A_2}{(s+4)}$$

$$(2s+5) = A_1(s+4) + A_2(s+2)$$

Put $s = -4$

$$(-8+5) = 0 + A_2(-4+2)$$
$$A_2 = \frac{3}{2}$$

Put $s = -2$

$$(-4+5) = A_1(-2+4) + 0$$
$$A_1 = \frac{1}{2}$$
$$X_1(s) = \frac{1}{2}\left[\frac{1}{(s+2)} + \frac{3}{(s+4)}\right]$$
$$X(s) = 1 - \frac{1}{2}\left[\frac{1}{(s+2)} + \frac{3}{(s+4)}\right]$$

Taking inverse LT, we get

$$\boxed{x(t) = \delta(t) - \left[0.5e^{-2t} + 1.5e^{-4t}\right]u(t)}$$

(2) Now consider

$$X(s) = \frac{(s+1)(s+3)e^{-2s}}{(s+2)(s+4)}$$

Using the time shifting property the results obtained in the previous example is modified and written as

$$\boxed{x(t) = \delta(t-2) - \left[0.5e^{-2(t-2)} + 1.5e^{-4(t-2)}\right]u(t-2)}$$

■ Example 8.39

Find the inverse LT of the following function.

$$X(s) = \frac{10(s+4)}{s^2(s+2)}$$

Solution: The given function $X(s)$ is written in the partial fraction form as follows:

$$X(s) = \frac{A_1}{s^2} + \frac{A_2}{s} + \frac{A_3}{s+2}$$
$$10(s+4) = A_1(s+2) + A_2 s(s+2) + A_3 s^2$$

Put $s = 0$

$$40 = 2A_1 \quad \text{or} \quad A_1 = 20$$

Put $s = -2$

$$10(-2+4) = 0 + 0 + A_3 4$$
$$A_3 = \frac{20}{4} = 5$$

Comparing the coefficients of s term, we get

$$10 = (A_1 + 2A_2)$$
$$10 = 20 + 2A_2$$
$$A_2 = -5$$
$$X(s) = \frac{20}{s^2} - \frac{5}{s} + \frac{5}{s+2}$$

$$\boxed{x(t) = (20t - 5 + e^{-2t})u(t)}$$

■ Example 8.40

Find the inverse LT of the following function

$$X(s) = \frac{2}{s(s^2 + 2s + 2)}$$

Solution:

Method 1.

$$(s^2 + 2s + 2) = (s + 1 + j)(s + 1 - j)$$

$$X(s) = \frac{A_1}{s} + \frac{A_2}{s+1+j} + \frac{A_3}{s+1-j}$$
$$2 = A_1(s^2 + 2s + 2) + A_2 s(s + 1 - j) + A_3 s(s + 1 + j)$$

Put $s = 0$

$$2 = A_1 2 \quad \text{or} \quad A_1 = 1$$

Put $s = -1 + j$

$$2 = 0 + 0 + A_3(-1 + j)(-1 + j + 1 + j)$$
$$= A_3(-1 + j)2j$$
$$A_3 = \frac{1}{(-1 + j)j}$$

But $(-1+j)$ is expressed in polar form as

$$(-1+j) = \sqrt{2}\angle 135^\circ$$
$$A_3 = \frac{1}{\sqrt{2}\angle 135^\circ + 90^\circ}$$
$$= 0.707\angle{+135^\circ}$$
$$= 0.707e^{+j135^\circ}$$

A_2 is the conjugate of A_3

$$A_2 = 0.707\angle{-135^\circ} = 0.707e^{-j135^\circ}$$
$$X(s) = \frac{1}{s} + 0.707\left[e^{-j135^\circ}\frac{1}{(s+1+j)} + \frac{e^{+j135^\circ}}{s+1-j}\right]$$

Taking inverse LT, we get

$$x(t) = 1 + 0.707[e^{-j135^\circ}e^{-(+1+j)t} + e^{+j135^\circ}e^{-(1-j)t)}]$$
$$= 1 + 1.414e^{-t}\left[\frac{e^{j(135^\circ+t)} + e^{-j(135^\circ+t)}}{2}\right]$$
$$= 1 + 1.414e^{-t}\cos(135^\circ + t)$$
$$= 1 - 1.414e^{-t}\sin(t + 45^\circ)$$

$$\boxed{x(t) = 1 - 1.414e^{-t}\sin\left(t + \frac{\pi}{4}\text{ rad}\right)}$$

Method 2. Graphical Method
From the pole-zero configuration of $X(s)$ shown in Fig. 8.27, we get

$$A_1 = \frac{2}{\sqrt{2}\angle 45^\circ\sqrt{2}\angle{-45^\circ}} = 1$$
$$A_2 = \frac{2}{\sqrt{2}\angle{-135^\circ}2\angle{-90^\circ}}$$
$$= 0.707\angle{-135^\circ}$$
$$A_3 = \text{conjugate of } A_2$$
$$A_3 = 0.707\angle 135^\circ$$

By graphical method, the residues A_1, A_2 and A_3 are obtained with ease. Substituting these values in $X(s)$ and taking inverse LT, the following result is obtained as in Method 1

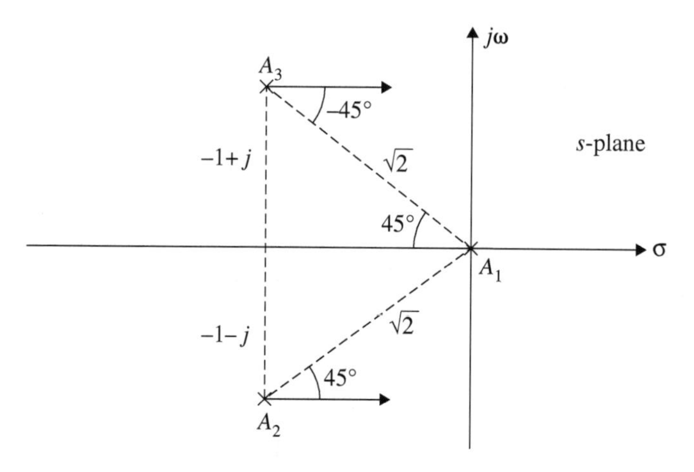

Fig. 8.27 Pole-zero configuration of Example 8.39

$$\boxed{x(t) = 1 - 1.414e^{-t}\sin\left(t + \frac{\pi}{4}\text{ rad}\right)}$$

■ Example 8.41

Find the inverse LT of the following function

$$X(s) = \frac{(3s^2 + 8s + 23)}{(s+3)(s^2 + 2s + 10)}.$$

(*Anna University, April, 2005*)

Solution:

$$s^2 + 2s + 10 = (s + 1 + j3)(s + 1 - j3)$$

The given $X(s)$ is put into partial fraction as follows:

$$X(s) = \frac{A_1}{(s+3)} + \frac{A_2}{(s+1+j3)} + \frac{A_3}{(s+1-j3)}$$

$$(3s^2 + 8s + 23) = A_1(s^2 + 2s + 10) + A_2(s+3)(s+1-j3) + A_3(s+3)(s+1+j3)$$

Let $s = -3$

$$27 - 24 + 23 = A_1(9 - 6 + 10)$$
$$A_1 = \frac{26}{13} = 2$$

Put $s = -1 - j3$

$$3(+1 + j3)^2 - 8(1 + j3) + 23 = A_2(-1 - j3 + 3)(-j6)$$
$$3(-8 + j6) - 8 - 24j + 23 = A_2(j6 - 18 - j18)$$
$$(-24 - 8 + 23) + j18 - 24j = A_2(-18 - j12)$$
$$-9 - j6 = -A_2(18 + j12)$$
$$A_2 = \frac{(3 + j2)}{(6 + j4)} = \frac{3.6\angle 33.7^\circ}{7.2\angle 33.7^\circ}$$
$$= 0.5$$
$$A_3 = \text{conjugate of } A_2$$
$$= 0.5$$
$$X(s) = \left[\frac{2}{s+3} + \frac{0.5}{s+1+j3} + \frac{0.5}{s+1-j3}\right]$$

Taking inverse LT, we get

$$x(t) = 2e^{-3t} + 0.5\left\{e^{-(1+j3)t} + e^{-(1-j3)t}\right\}$$
$$= 2e^{-3t} + e^{-t}\frac{\{e^{-j3t} + e^{j3t}\}}{2}$$

$$\boxed{x(t) = 2e^{-3t} + e^{-t}\cos 3t}$$

■ Example 8.42

Find the inverse LT of

$$X(s) = \frac{3s^2 + 8s + 6}{(s+2)(s^2 + 2s + 1)}.$$

(Anna University, December 2007)

Solution:

$$
\begin{aligned}
(s^2+2s+1) &= (s+1)^2 \\
X(s) &= \frac{(3s^2+8s+6)}{(s+2)(s+1)^2} \\
&= \frac{A_1}{(s+2)} + \frac{A_2}{(s+1)^2} + \frac{A_3}{(s+1)} \\
&= \frac{A_1(s^2+2s+1) + A_2(s+2) + A_3(s+1)(s+2)}{(s+2)(s+1)^2} \\
3s^2+8s+6 &= A_1(s^2+2s+1) + A_2(s+2) + A_3(s+1)(s+2)
\end{aligned}
$$

Put $s=-2$

$$
\begin{aligned}
12-16+6 &= A_1(4-4+1)+0+0 \\
A_1 &= 2 \\
3s^2+8s+6 &= (A_1+A_3)s^2 + (2A_1+A_2+3A_3)s + (A_1+2A_2+2A_3)
\end{aligned}
$$

Comparing the coefficients of s^2, we get

$$
\begin{aligned}
3 &= A_1 + A_3 \\
A_3 &= 3 - A_1 = 3 - 2 \\
A_3 &= 1
\end{aligned}
$$

Comparing the coefficients of s, we get

$$
\begin{aligned}
8 &= 2A_1 + A_2 + 3A_3 \\
&= 4 + A_2 + 3 \\
A_2 &= 1
\end{aligned}
$$

Substituting the values of A_1, A_2 and A_3 in $X(s)$, we get

$$
X(s) = \frac{2}{(s+2)} + \frac{1}{(s+1)^2} + \frac{1}{(s+1)}
$$

Taking inverse LT of $X(s)$, we get

$$
\boxed{x(t) = (2e^{-2t} + te^{-t} + e^{-t})u(t)}
$$

■ Example 8.43

Find the inverse LT of the following function.

$$X(s) = \frac{10s^2}{(s+2)(s^2+4s+5)}$$

Solution:

Method 1.

$$\begin{aligned}
(s^2+4s+5) &= (s+2+j)(s+2-j)\\
X(s) &= \frac{10s^2}{(s+2)(s+2+j)(s+2-j)}\\
&= \frac{A_1}{(s+2)} + \frac{A_2}{(s+2+j)} + \frac{A_3}{(s+2-j)}\\
10s^2 &= A_1(s^2+4s+5) + A_2(s+2)(s+2-j)\\
&\quad + A_3(s+2)(s+2+j)
\end{aligned}$$

Put $s = -2$

$$\begin{aligned}
40 &= A_1(4-8+5)+0+0\\
A_1 &= 40
\end{aligned}$$

Put $s = -2-j$

$$\begin{aligned}
10(-2-j)^2 &= 0 + A_2(-2-j+2)(-2-j+2-j) + 0\\
10(4-1+4j) &= A_2(-j)(-2j)\\
-10(3+4j) &= 2A_2\\
A_2 &= -5(3+4j)\\
&= 25\angle{-126.88^\circ} = 25e^{-j126.88^\circ}
\end{aligned}$$

$A_3 =$ conjugate of A_2

$$\begin{aligned}
A_3 &= 25\angle{+126.88^\circ} = e^{j126.88^\circ}\\
X(s) &= \frac{40}{(s+2)} + \frac{25e^{-j126.88^\circ}}{(s+2+j)} + \frac{25e^{j126.88^\circ}}{(s+2-j)}
\end{aligned}$$

Taking inverse LT, we get

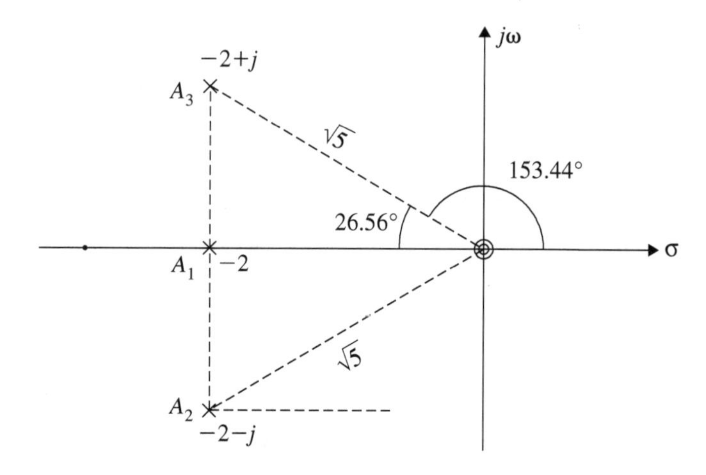

Fig. 8.28 Pole-zero diagram of $X(s)$ of Example 8.43

$$x(t) = 40e^{-2t} + 25\left\{e^{-j126.88^\circ}e^{-(2+j)t} + e^{j126.88^\circ}e^{-(2-j)t}\right\}$$

$$= 40e^{-2t} + e^{-2t}50\frac{\{e^{-j(t+126.88^\circ)} + e^{+j(126.88^\circ+t)}\}}{2}$$

$$x(t) = [40e^{-2t} + 50e^{-2t}\cos(t + 126.88^\circ)]$$

$$\boxed{x(t) = [40 - 50\sin(t + 0.644\,\mathrm{rad})]e^{-2t}u(t)}$$

Method 2. Graphical Method

The pole-zero configuration of $X(s)$ is shown in Fig. 8.28. From Fig. 8.28, the residues A_1, A_2 and A_3 are obtained as follows:

$$A_1 = \frac{10(-2)(-2)}{1\angle 90^\circ\, 1\angle -90^\circ} = 40$$

$$A_3 = \frac{10\sqrt{5}\angle 153.44^\circ\sqrt{5}\angle 153.44^\circ}{1\angle 90^\circ\; 2\angle 90^\circ}$$

$$= 25\angle 126.88^\circ = 25e^{j126.88^\circ}$$

$A_2 =$ conjugate of A_3

$$A_2 = 25\angle -126.88^\circ = 25e^{-j126.88^\circ}$$

The residues A_1, A_2 and A_3 obtained by graphical method are same as obtained by analytical method. Thus, by proceeding as in Method 1, the inverse LT is obtained as

$$\boxed{x(t) = [40 - 50\sin(t + 0.644\,\text{rad})]e^{-2t}u(t)}$$

8.9 Solving Differential Equation

Laplace transform is a very powerful tool in the analysis of linear time invariant dynamic system. It provides

- Solutions to LTI dynamic systems described by linear differential equations by converting the differential equation to algebraic equation.
- For test signals of different kinds, solutions are obtained for the differential equations with and without initial conditions.
- The dynamic systems are represented in terms of a transfer function which is nothing but the ratio of the LT of the output variable to the LT of the input variable.
- The transfer function is made use of to determine the frequency response of the system.
- The transfer function is also made use of to determine the stability of the system using the well-known Routh-Hurwitz criterion and Nyquist stability criterion.
- The structure of the dynamic system is realized using the transfer function.

Now we give below the method of solving differential equation using LT.

8.9.1 Solving Differential Equation without Initial Conditions

1. If $y(t)$ is the output variable and $x(t)$ is the input variable, convert the differential equation to algebraic equation which is obtained by simple multiplication of Laplace complex variable s.
2. These algebraic equations are obtained using the following LT when the initial conditions are zero.

$$L[y(t)] = Y(s)$$
$$L\left[\frac{dy}{dt}\right] = sY(s)$$
$$L\left[\frac{d^2y}{dt}\right] = s^2Y(s)$$

$$L\left[\frac{d^3y}{dt^3}\right] = s^3Y(s)$$

$$\vdots$$

$$L\left[\frac{d^ny}{dt^n}\right] = s^nY(s)$$

Similarly, for the input $x(t)$,, we convert

$$L[x(t)] = X(s)$$

$$L\left[\frac{dx}{dt}\right] = sX(s)$$

$$L\left[\frac{d^2x}{dt^2}\right] = s^2X(s)$$

$$\vdots$$

$$L\left[\frac{d^nx}{dt^n}\right] = s^nX(s)$$

The following examples illustrate the method of solving differential equation using LT when the initial conditions are zero for the input as well as the output.

■ Example 8.44

Consider an LTIC system with the following differential equation with zero initial conditions for the input and output.

$$\frac{d^2y(t)}{dt^2} + 4\frac{dy(t)}{dt} + 3y(t) = \frac{dx(t)}{dt} + 2x(t)$$

Find the impulse response of the system.

(*Anna University, December 2006*)

Solution:

Taking LT on both sides of the given differential equation, we get

$$(s^2 + 4s + 3)Y(s) = (s + 2)X(s)$$

The transfer function is obtained as

$$H(s) = \frac{Y(s)}{X(s)} = \frac{(s + 2)}{(s^2 + 4s + 3)}$$
$$= \frac{(s + 2)}{(s + 1)(s + 3)}$$

$$Y(s) = \frac{(s+2)X(s)}{(s+1)(s+3)}$$

From Table 8.1, for an impulse input $x(t) = \delta(t)$, $X(s) = 1$. Substituting this in the above equation, we get

$$\begin{aligned} Y(s) &= \frac{(s+2)}{(s+1)(s+3)} \\ &= \frac{A_1}{(s+1)} + \frac{A_2}{(s+3)} \\ (s+2) &= A_1(s+3) + A_2(s+1) \end{aligned}$$

Put $s = -1$

$$\begin{aligned} (-1+2) &= A_1(-1+3) + 0 \\ A_1 &= 0.5 \end{aligned}$$

Put $s = -3$

$$\begin{aligned} (-3+2) &= 0 + A_2(-3+1) \\ A_2 &= 0.5 \\ \therefore \quad Y(s) &= \frac{0.5}{(s+1)} + \frac{0.5}{(s+3)} \end{aligned}$$

Taking inverse LT, we get

$$\boxed{y(t) = 0.5\left[e^{-t} + e^{-3t}\right]u(t)}$$

■ Example 8.45

Using LT, find the impulse response of an LTI system described by the following differential equation.

$$\frac{d^2y(t)}{dt^2} - \frac{dy(t)}{dt} - 2y(t) = x(t)$$

Assume zero initial conditions.

(Anna University, April, 2004)

Solution: Taking LT on both sides of the given differential equation, we get

$$\begin{aligned} (s^2 - s - 2)Y(s) &= X(s) \\ Y(s) &= \frac{X(s)}{(s^2 - s - 2)} \end{aligned}$$

$$= \frac{X(s)}{(s+1)(s-2)}$$

For an impulse $X(s) = 1$

$$Y(s) = \frac{1}{(s+1)(s-2)}$$
$$= \frac{A_1}{(s+1)} + \frac{A_2}{(s-2)}$$
$$1 = A_1(s-2) + A_2(s+1)$$

Put $s = -1$

$$A_1 = -\frac{1}{3}$$

Put $s = 2$

$$A_2 = \frac{1}{3}$$
$$Y(s) = \frac{1}{3}\left[\frac{1}{s-2} - \frac{1}{s+1}\right]$$

$$\boxed{y(t) = \frac{1}{3}\left[e^{2t} - e^{-t}\right]u(t)}$$

■ Example 8.46

Consider the LTI system with the following differential equation with zero initial conditions.

$$\frac{d^2y(t)}{dt^2} + 5\frac{dy(t)}{dt} + 6y(t) = x(t)$$

where $x(t) = e^{-4t}u(t)$. Find an expression for $y(t)$ using LT method.

Solution: The given differential equation is written as follows:

$$\frac{d^2y}{dt^2} + 5\frac{dy}{dt} + 6y = e^{-4t}u(t)$$

Taking LT on both sides, we get

$$(s^2+5s+6)Y(s) = \frac{1}{(s+4)}$$

$$Y(s) = \frac{1}{(s+4)(s^2+5s+6)}$$

$$= \frac{1}{(s+3)(s+2)(s+4)}$$

$$= \frac{A_1}{(s+3)} + \frac{A_2}{(s+2)} + \frac{A_3}{(s+4)}$$

$$1 = A_1(s+2)(s+4) + A_2(s+3)(s+4) + A_3(s+3)(s+2)$$

Put $s = -3$

$$1 = A_1(-3+2)(-3+4)$$

$$A_1 = -1$$

Put $s = -2$

$$1 = A_2(-2+3)(-2+4)$$

$$A_2 = \frac{1}{2} = 0.5$$

Put $s = -4$

$$1 = A_3(-4+3)(-4+2)$$

$$A_3 = \frac{1}{2} = 0.5$$

$$Y(s) = \frac{-1}{(s+3)} + \frac{0.5}{(s+2)} + \frac{0.5}{(s+4)}$$

$$\boxed{y(t) = \left(-e^{-3t} + 0.5e^{-2t} + 0.5e^{-4t}\right)u(t)}$$

■ Example 8.47

Consider the following differential equation with zero initial conditions.

$$\frac{d^2y(t)}{dt} + 2\frac{dy(t)}{dt} + 2y(t) = \frac{dx(t)}{dt} + x(t)$$

For $x(t) = u(t)$, a unit step input find the response $y(t)$ of the system.

Solution:

Method 1

Taking LT on both sides of the differential equation, we get

$$(s^2 + 2s + 2)Y(s) = (s + 1)X(s)$$
$$Y(s) = \frac{(s + 1)}{(s^2 + 2s + 2)}X(s)$$

For unit step $X(s) = \frac{1}{s}$. Substituting this in the above equation, we get

$$Y(s) = \frac{(s + 1)}{s(s^2 + 2s + 2)}$$
$$(s^2 + 2s + 2) = (s + 1 + j)(s + 1 - j)$$
$$\therefore \quad Y(s) = \frac{(s + 1)}{s(s + 1 + j)(s + 1 - j)}$$
$$= \frac{A_1}{s} + \frac{A_2}{(s + 1 + j)} + \frac{A_3}{(s + 1 - j)}$$
$$(s + 1) = A_1(s^2 + 2s + 2) + A_2 s(s + 1 - j) + A_3 s(s + 1 + j)$$

Put $s = 0$

$$1 = 2A_1 \quad \text{or} \quad A_1 = 0.5$$

Put $s = -(1 + j)$

$$(-1 - j + 1) = 0 + A_2(-1 - j)(-1 - j + 1 - j) + 0$$
$$-j = A_2(-1 - j)(-2j)$$
$$= A_2(2j - 2) = 2A_2(j - 1)$$
$$A_2 = \frac{0.5j}{1 - j} = \frac{0.5}{\sqrt{2}}\frac{\angle 90^{\circ}}{\angle -45^{\circ}}$$
$$= 0.354\angle 135^{\circ} = 0.354e^{j135^{\circ}}$$
$$A_3 = \text{conjugate of } A_2$$
$$= 0.354e^{-j135^{\circ}}$$
$$Y(s) = \frac{0.5}{s} + \frac{0.354e^{j135^{\circ}}}{s + 1 + j} + \frac{0.354e^{-j135^{\circ}}}{(s + 1 - j)}$$

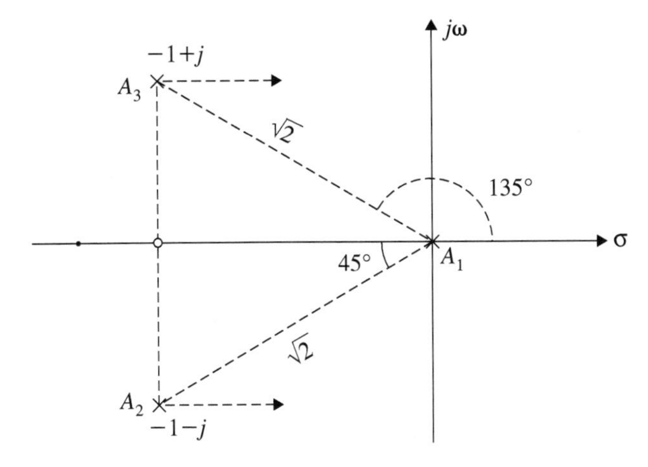

Fig. 8.29 Pole-zero diagram of Example 8.47

$$\begin{aligned}
y(t) &= 0.5 + 0.354e^{j135^\circ}e^{-(1+j)t} + 0.354e^{-j135^\circ}e^{-(1-j)t} \\
&= 0.5 + 0.708e^{-t}\frac{[e^{j(135^\circ - t)} + e^{-j(135^\circ - t)}]}{2} \\
&= 0.5 + 0.708e^{-t}\cos(135^\circ - t) \\
&= 0.5 - 0.708e^{-t}\sin(45^\circ - t)
\end{aligned}$$

$$\boxed{y(t) = \left[0.5 + 0.708e^{-t}\sin\left(t - \frac{\pi}{4}\text{rad}\right)\right]u(t)}$$

Method 2

The pole-zero diagram of $Y(s)$ is shown in Fig. 8.29. The residues A_1, A_2 and A_3 are determined as follows:

$$\begin{aligned}
A_1 &= \frac{1\angle 0^\circ}{\sqrt{2}\angle 45^\circ \sqrt{2}\angle -45^\circ} = 0.5 \\
A_2 &= \frac{1\angle -90^\circ}{\sqrt{2}\angle 225^\circ 2\angle -90^\circ} \\
&= 0.354\angle 135^\circ \\
A_3 &= \text{conjugate of } A_2 \\
&= 0.354\angle -135^\circ
\end{aligned}$$

The residues determined by graphical method is same as determined by analytical method. Therefore, $y(t)$ is written as

$$\boxed{y(t) = \left[0.5 + 0.708e^{-t} \sin\left(t - \frac{\pi}{4} \text{ rad}\right)\right] u(t)}$$

8.9.2 *Solving Differential Equation with the Initial Conditions*

1. When the initial conditions are specified for the given differential equation, they have to be accounted for when LT is taken to convert the differential equation to algebraic equation. Thus

$$L\left[\frac{dy}{dt}\right] = sY(s) - y(0^-)$$
$$L\left[\frac{d^2y}{dt^2}\right] = s^2Y(s) - sy(0^-) - \dot{y}(0^-)$$
$$L\left[\frac{d^3y}{dt^3}\right] = s^3Y(s) - s^2y(0^-) - s\dot{y}(0^-) - \ddot{y}(0^-)$$

   The initial conditions $y(0^-)$, $\dot{y}(0^-)$ and $\ddot{y}(0^-)$, are meant that the system initial conditions are given just before the input is applied to the system.
   The initial condition $y(0^+)$ indicates that the initial condition is given to the system after the input is applied which is not realistic. Unless otherwise mentioned, $y(0^-)$ means $y(0)$ and $y(0)$ is not $y(0^+)$.
2. The zero initial conditions explained in step 1 is applicable to the input also. Thus

$$\frac{dx}{dt} = sX(s) - x(0^-)$$

3. The initial conditions for an input multiplied by $u(t)$ imply that the signals are zero prior to $t = 0$.
4. The solution of the differential equation contains two components. The first component is the response due to the initial conditions only where the input is assumed to be absent. The response is called the zero input response. The second component is the response due to the input alone and the initial conditions here are assumed to be zero. Such response is called zero state response.
5. The total response = zero state response + zero input response.
6. If one is interested to find out the zero initial conditions for verification of the results, only the zero input response has to be considered and not the total response. The total response satisfies the initial conditions at $t = 0^+$.

The following examples illustrate the method of obtaining total response which is due to initial conditions and the input.

■ Example 8.48

A certain system is described by the following differential equation

$$\frac{d^2y(t)}{dt^2}+7\frac{dy(t)}{dt}+12y(t)=x(t)$$

Use LT to determine the response of the system to unit step input applied at $t=0$. Assume the initial conditions are $y(0^-)=-2$ and $\frac{dy(0^-)}{dt}=0$.

(Anna University, May 2007)

Solution: Taking LT on both sides of the given equation, we get

$$s^2Y(s)-sy(0^-)-\dot{y}(0^-)+7Y(s)-7y(0^-)+12Y(s)=X(s)$$

$$(s^2+7s+12)Y(s)+2s+14=\frac{1}{s}$$

$$(s^2+7s+12)Y(s)=-2s-14+\frac{1}{s}$$

$$=\frac{(-2s^2-14s+1)}{s}$$

$$Y(s)=\frac{(-2s^2-14s+1)}{s(s^2+7s+12)}$$

$$=\frac{(-2s^2-14s+1)}{s(s+3)(s+4)}$$

$$=\frac{A_1}{s}+\frac{A_2}{(s+3)}+\frac{A_3}{(s+4)}$$

$$-2s^2-14s+1=A_1(s+3)(s+4)+A_2s(s+4)+A_3s(s+3)$$

Put $s=0$

$$1=A_1(12)$$

$$A_1=\frac{1}{12}$$

Put $s=-3$

$$-18 + 42 + 1 = A_2(-3)(-3+4)$$
$$A_2 = \frac{-25}{3}$$

Put $s = -4$

$$-32 + 56 + 1 = A_3(-4)(-4+3)$$
$$A_3 = \frac{25}{4}$$
$$Y(s) = \frac{1}{12}\frac{1}{s} - \frac{25}{3}\frac{1}{(s+3)} + \frac{25}{4}\frac{1}{(s+4)}$$

The total response is obtained by taking inverse LT

$$\boxed{y(t) = \left[\frac{1}{12} - \frac{25}{3}e^{-3t} + \frac{25}{4}e^{-4t}\right]u(t)}$$

■ Example 8.49

Solve
$$\frac{d^2y(t)}{dt^2} + 4\frac{dy(t)}{dt} + 4y(t) = \frac{dx(t)}{dt} + x(t)$$

if the initial conditions are $y(0^+) = \frac{9}{4}$; $\dot{y}(0^+) = 5$, if the input is $e^{-3t}u(t)$.

(*Anna University, December 2007*)

Solution: Taking LT on both sides of the equation, we get

$$s^2Y(s) - sy(0^+) - \dot{y}(0^+) + 4sY(s) - 4y(0^+) + 4Y(s) = sX(s) + X(s) - x(0^+)$$

If the initial conditions are given at $t = 0^+$ for the output, then the initial conditions must be applied to the input also

$$L\left[\frac{d}{dt}x(t) + x(t)\right] = sX(s) - x(0^+) + X(s)$$
$$x(t) = e^{-3t}$$
$$X(s) = \frac{1}{(s+3)}$$

Since $x(0^+) = \underset{t\to 0}{Lt}\, e^{-3t} = 1$

$$\begin{aligned}(s+1)X(s) - x(0^+) &= \frac{(s+1)}{s+3} - 1 \\ &= \frac{(s+1-s-3)}{(s+3)} \\ &= \frac{-2}{(s+3)}\end{aligned}$$

Alternatively

$$\begin{aligned}x(t) &= e^{-3t} \\ \frac{dx(t)}{dt} &= -3e^{-3t} \\ L\left[\frac{dx(t)}{dt}\right] &= \frac{-3}{s+3} \\ L[x(t)] &= \frac{1}{(s+3)} \\ \therefore \quad L\left[\frac{dx(t)}{dt} + x(t)\right] &= \frac{1}{s+3}[-3+1] \\ &= \frac{-2}{(s+3)}\end{aligned}$$

Substituting $y(0^+)$ and $\dot{y}(0^+)$ in the given equation, we get

$$(s^2+4s+4)Y(s) - \frac{9}{4}s - 5 - 9 = \frac{-2}{(s+3)}$$

$$(s^2+4s+4)Y(s) = \frac{9}{4}s + 14 - \frac{2}{(s+3)}$$

$$Y(s) = \frac{(9s^2+83s+160)}{4(s+3)(s^2+4s+4)}$$

$$= \frac{(9s^2+83s+160)}{4(s+3)(s+2)^2} \qquad \text{(a)}$$

$$= \frac{A_1}{(s+3)} + \frac{A_2}{(s+2)^2} + \frac{A_3}{(s+2)} \qquad \text{(b)}$$

$$\frac{1}{4}[9s^2 + 83s + 160] = A_1(s + 2)^2 + A_2(s + 3) + A_3(s + 2)(s + 3)$$

Put $s = -3$

$$\frac{1}{4}[81 - 249 + 160] = A_1 \quad \text{or} \quad A_1 = -2$$

Put $s = -2$

$$\frac{1}{4}[36 - 166 + 160] = A_2 \quad \text{or} \quad A_2 = 7.5$$

Compare the coefficients of s^2 on both sides

$$\frac{9}{4} = A_1 + A_3 = -2 + A_3 \quad \text{or} \quad A_3 = 4.25$$

$$Y(s) = \frac{-2}{s+3} + \frac{7.5}{(s+2)^2} + \frac{4.25}{(s+2)}$$

Taking inverse LT, we get

$$\boxed{y(t) = -2e^{-3t} + 7.5te^{-2t} + 4.25e^{-2t}} \quad t \geq 0$$

To check whether the residues are correctly determined.

Choose any value of s such that, when substituted in $X(s)$, it does not become zero or infinitive. Find the value of $Y(s)$ in (a) and (b). If both are the same, the residues determined are correct.

For $s = 0$

$$\frac{160}{4 \times 3 \times 4} = \frac{-2}{3} + \frac{7.5}{4} + \frac{4.25}{2}$$
$$\frac{40}{12} = \frac{-8 + 22.5 + 25.5}{12} = \frac{40}{12}$$
$$\text{LHS} = \text{RHS}$$

Hence, the values of A_1, A_2 and A_3 determined are correct.

8.9.3 *Zero Input and Zero State Response*

As described earlier, the response of the system due to the input $x(t)$ with all initial conditions are zero is called zero state response. The response of the system due to the initial conditions with zero input is called zero input response. The total response of the system is the sum of the zero state response and zero input response. This is illustrated in the following example.

Example 8.50

Consider the following differential equation:

$$\frac{d^2y(t)}{dt^2} + 6\frac{dy(t)}{dt} + 8y(t) = \frac{dx(t)}{dt} + 3x(t)$$

$$x(t) = u(t)$$

$$y(0^-) = 1 \quad \text{and} \quad \dot{y}(0^-) = 2$$

Find the zero state, zero input and total response. Verify, from the expression for the response, the initial conditions given.

Solution:

Zero State Response

For zero state response, the initial conditions are assumed to be zero. Under this condition, the output is denoted as $y_i(t)$.

Taking LT on both sides of the given differential equation, we get

$$(s^2 + 6s + 8)Y_i(s) = (s + 3)X(s)$$

Substituting $X(s) = \frac{1}{s}$ and $(s^2 + 6s + 8) = (s + 2)(s + 4)$, we get

$$\begin{aligned} Y_i(s) &= \frac{(s+3)}{s(s+2)(s+4)} \\ &= \frac{A_1}{s} + \frac{A_2}{(s+2)} + \frac{A_3}{(s+4)} \end{aligned}$$

$$(s+3) = A_1(s+2)(s+4) + A_2s(s+4) + A_3s(s+2)$$

Put $s = 0$

$$3 = A_1 8 \quad \text{or} \quad A_1 = \frac{3}{8}$$

Put $s = -2$

$$(-2+3) = A_2(-2)(-2+4)$$

$$A_2 = -\frac{1}{4}$$

Put $s = -4$

$$(-4+3) = A_3(-4)(-4+2)$$
$$A_3 = -\frac{1}{8}$$
$$Y_i(s) = \frac{3}{8}\frac{1}{s} - \frac{1}{4}\frac{1}{(s+2)} - \frac{1}{8}\frac{1}{(s+4)}$$

$$\boxed{y_i(t) = \left(\frac{3}{8} - \frac{1}{4}e^{-2t} - \frac{1}{8}e^{-4t}\right)u(t)}$$

Zero Input Response

Under zero input condition the output is denoted as $y_s(t)$. The given differential equation becomes

$$\frac{d^2y_s(t)}{dt^2} + 6\frac{dy_s(t)}{dt} + 8y_s(t) = 0$$

Taking LT with initial conditions, we get

$$s^2Y_s(s) - sy_s(0^-) - \dot{y}_s(0^-) + 6Y_s(s) - 6y_s(0) + 8Y_s(s) = 0$$

$$(s^2 + 6s + 8)Y_s(s) = (s + 2 + 6) = (s + 8)$$
$$Y_s(s) = \frac{(s+8)}{(s+2)(s+4)}$$
$$= \frac{A_1}{(s+2)} + \frac{A_2}{(s+4)}$$
$$(s+8) = A_1(s+4) + A_2(s+2)$$

Put $s = -2$

$$(-2+8) = A_1(-2+4)$$
$$A_1 = 3$$

Put $s = -4$

$$(-4+8) = A_2(-4+2)$$

$$A_2 = -2$$
$$Y_s(s) = \frac{3}{s+2} - \frac{2}{s+4}$$

$$\boxed{y_s(t) = \left(3e^{-2t} - 2e^{-4t}\right) u(t)}$$

Total Response

The total response is denoted by the letter $y(t)$.

$$\begin{aligned} y(t) &= y_i(t) + y_s(t) \\ &= \left(\frac{3}{8} - \frac{1}{4}e^{-2t} - \frac{1}{8}e^{-4t}\right) u(t) + (3e^{-2t} - 2e^{-4t})u(t) \end{aligned}$$

$$\boxed{y(t) = \left[\frac{3}{8} + \frac{11}{4}e^{-2t} - \frac{17}{8}e^{-4t}\right] u(t)}$$

To verify the initial condition, consider the zero input response $y_s(t)$

$$\begin{aligned} y_s(t) &= 3e^{-2t} - 2e^{-4t} \\ y_s(0) &= y(0) = 3 - 2 = 1 \\ \dot{y}(0) &= \left.\frac{dy_s(t)}{dt}\right|_{t=0} = \left.-6e^{-2t} + 8e^{-4t}\right|_{t=0} = -6 + 8 \\ \dot{y}(0) &= 2 \end{aligned}$$

The given initial conditions are satisfied. On the other hand, consider the expression for the total response

$$\begin{aligned} y(t) &= \frac{3}{8} + \frac{11}{4}e^{-2t} - \frac{17}{8}e^{-4t} \\ y(0) &= \frac{3}{8} + \frac{11}{4} - \frac{17}{8} \\ &= 1 \\ \dot{y}(t) = \frac{dy(t)}{dt} &= -\frac{22}{4}e^{-2t} + \frac{68}{8}e^{-4t} \\ \dot{y}(0) &= \frac{-22}{4} + \frac{68}{8} \\ \dot{y}(0) &= 3 \end{aligned}$$

The result obtained is erroneous. **Therefore, the initial conditions are verified from zero input response and not the total response**.

8.9.4 Natural and Forced Response Using LT

Consider the differential equation of Example 8.50 which is given below

$$\frac{d^2y(t)}{dt^2} + 6\frac{dy(t)}{dt} + 8y(t) = \frac{d}{dx}x(t) + 3x(t)$$

Taking LT on both sides of the above equation with zero conditions, we get

$$(s^2 + 6s + 8)Y(s) = (s + 3)X(s)$$
$$(s + 2)(s + 4)Y(s) = (s + 3)X(s)$$

The transfer function is the ratio of $Y(s)$ to $X(s)$ and is written as

$$\frac{Y(s)}{X(s)} = \frac{(s + 3)}{(s + 2)(s + 4)}$$

In the above equation, $s^2 + 6s + 8 = 0$ is called characteristic equation and $s = -2$ and $s = -4$ are called characteristic roots or Eigen values. In the total response of $y(t)$, corresponding to these Eigen values, the characteristic modes are found. In the above example, the characteristic modes are e^{-2t} and e^{-4t}. In the total response of the system, which is composed of zero input response and zero state response, if we can lump together all the terms corresponding to the characteristic mode, it is called natural response $y_n(t)$. The remaining non-characteristic mode terms are lumped together and the response is called forced response and is denoted by $y_f(t)$. Thus, in Example 8.49, the Eigen values are $s = -2$ and $s = -4$. The characteristic modes are e^{-2t} and e^{-4t}. Thus

$$\boxed{y(t) = \frac{3}{8} + \frac{11}{4}e^{-2t} - \frac{17}{8}e^{-4t}}$$

The natural response

$$\boxed{y_n(t) = \frac{11}{4}e^{-2t} - \frac{17}{8}e^{-4t}}$$

The forced response

$$\boxed{y_f(t) = \frac{3}{8}}$$

■ Example 8.51

Find the forced response of the following differential equation

$$\frac{d^2y(t)}{dt^2} + 6\frac{dy(t)}{dt} + 8y(t) = \frac{dx}{dt} + x(t)$$

where $x(t) = t^2$.

Solution: Taking LT of the given differential equation, we get

$$(s^2 + 6s + 8)Y(s) = (s + 1)X(s)$$
$$(s^2 + 6s + 8) = (s + 2)(s + 4)$$

The Eigen values are $s = -2$ and $s = -4$. The characteristic modes are e^{-2t} and e^{-4t}. The terms involving these characteristic mode will correspond to the natural response of the system. The remaining terms will correspond to forced response of the system. Substituting $X(s) = \frac{2}{s^3}$, we get

$$\begin{aligned} Y(s) &= \frac{2(s+1)}{(s+2)(s+4)s^3} \\ &= \frac{A_1}{(s+2)} + \frac{A_2}{(s+4)} + \frac{A_3}{s^3} + \frac{A_4}{s^2} + \frac{A_5}{s} \\ 2(s+1) &= A_1 s^3(s+4) + A_2 s^3(s+2) + A_3(s+2)(s+4) \\ &\quad + A_4 s(s+2)(s+4) + A_5 s^2(s+2)(s+4) \end{aligned}$$

Put $s = 0$

$$2 = 8A_3 \quad \text{or} \quad A_3 = \frac{1}{4}$$

Compare the coefficients of s

$$\begin{aligned} 2 &= 6A_3 + 8A_4 \\ A_4 &= \frac{1}{16} \end{aligned}$$

Compare the coefficients of s^2

$$\begin{aligned} 0 &= A_3 + 6A_4 + 8A_5 \\ &= \frac{1}{4} + \frac{6}{16} + 8A_5 \\ A_5 &= -\frac{10}{128} = -\frac{5}{64} \end{aligned}$$

The residues A_1 and A_2 are determined as follows: Put $s = -2$

$$\begin{aligned} 2(-2+1) &= A_1(-8)(-2+4) + 0 + 0 + 0 + 0 \\ A_1 &= \frac{1}{8} \end{aligned}$$

Put $s = -4$

$$2(-4+1) = A_2 64(-4+2)$$
$$A_2 = \frac{3}{64}$$

$$Y(s) = \frac{1}{8}\frac{1}{(s+2)} + \frac{3}{64}\frac{1}{s+4} + \frac{1}{4}\frac{1}{s^3} + \frac{1}{16}\frac{1}{s^2} - \frac{5}{64}\frac{1}{s}$$

Taking inverse LT, we get

$$\boxed{y(t) = \underbrace{\frac{1}{8}e^{-2t} + \frac{3}{64}e^{-4t}}_{\text{Natural response}} + \underbrace{\frac{1}{8}t^2 + \frac{1}{16}t - \frac{5}{64}}_{\text{Forced response}}}$$

The natural response which is due to the characteristic modes e^{-2t} and e^{-4t} is given by

$$\boxed{y_n(t) = \frac{1}{8}e^{-2t} + \frac{3}{64}e^{-4t}} \qquad t \geq 0$$

The forced response is the response which does not contain the characteristic mode. Thus

$$\boxed{y_f(t) = \frac{1}{8}t^2 + \frac{1}{16}t - \frac{5}{64}} \qquad t \geq 0$$

8.10 Time Convolution Property of the Laplace Transform

If

$$x_1(t) \stackrel{L}{\longleftrightarrow} X_1(s)$$

and

$$x_2(t) \stackrel{L}{\longleftrightarrow} X_2(s)$$

then

$$x_1(t) * x_2(t) \stackrel{L}{\longleftrightarrow} X_1(s)X_2(s)$$

This property of LT is used to determine

$$y(t) = x_1(t) * x_2(t)$$

The following examples illustrate this.

■ Example 8.52

Using the convolution property of the LT determine $y(t) = x_1(t) * x_2(t)$, where $x_1(t) = e^{-2t}u(t)$ and $x_2(t) = e^{-3t}u(t)$.

Solution:

$$X_1(s) = L[e^{-2t}u(t)] = \frac{1}{(s+2)}$$

$$X_2(s) = L[e^{-3t}u(t)] = \frac{1}{(s+3)}$$

$$\begin{aligned} Y(s) &= X_1(s)X_2(s) \\ &= \frac{1}{(s+2)}\frac{1}{(s+3)} \\ &= \frac{1}{(s+2)} - \frac{1}{(s+3)} \end{aligned}$$

$$\boxed{y(t) = \left[e^{-2t} - e^{-3t}\right]u(t)}$$

■ Example 8.53

Given

$$\begin{aligned} x_1(t) &= e^{-2t}u(t) \\ x_2(t) &= (1 + e^{-3t})u(t) \end{aligned}$$

Determine $y(t) = x_1(t) * x_2(t)$.

Solution:

$$X_1(s) = L[x_1(t)] = \frac{1}{(s+2)}$$

$$X_2(s) = L[x_2(t)] = \left[\frac{1}{s} + \frac{1}{s+3}\right]$$

$$\begin{aligned} Y(s) &= X_1(s)X_2(s) \\ &= \frac{1}{(s+2)}\left[\frac{1}{s} + \frac{1}{s+3}\right] = \frac{(2s+3)}{s(s+2)(s+3)} \\ &= \frac{A_1}{s} + \frac{A_2}{s+2} + \frac{A_3}{s+3} \\ (2s+3) &= A_1(s+3)(s+2) + A_2 s(s+3) + A_3 s(s+2) \end{aligned}$$

Put $s = 0$

$$3 = A_1(2)(3)$$
$$A_1 = \frac{1}{2}$$

Put $s = -2$

$$(-4+3) = A_2(-2)(-2+3)$$
$$A_2 = \frac{1}{2}$$

Put $s = -3$

$$(-6+3) = A_3(-3)(-3+2)$$
$$A_3 = -1$$

$$Y(s) = \frac{1}{2s} + \frac{1}{2(s+2)} - \frac{1}{(s+3)}$$

$$\boxed{y(t) = \left(\frac{1}{2} + \frac{1}{2}e^{-2t} - e^{-3t}\right)u(t)}$$

■ Example 8.54

Find $y(t)$ by Convolution method

$$x_1(t) = e^{a_1 t}u(t)$$
$$x_2(t) = e^{a_2 t}u(-t)$$
$$y(t) = x_1(t) * x_2(t)$$

Solution:

$$X_1(s) = \frac{1}{(s-a_1)}$$
$$X_2(s) = \frac{-1}{(s-a_2)}$$
$$Y(s) = X_1(s)X_2(s)$$
$$= \frac{-1}{(s-a_1)(s-a_2)} = \frac{A_1}{(s-a_1)} + \frac{A_2}{(s-a_2)}$$
$$-1 = A_1(s-a_2) + A_2(s-a_1)$$

Put $s = a_1$

$$-1 = A_1(a_1 - a_2)$$
$$A_1 = \frac{1}{a_2 - a_1}$$

Put $s = a_2$

$$-1 = A_2(a_2 - a_1)$$
$$A_2 = \frac{-1}{(a_2 - a_1)}$$
$$Y(s) = \frac{1}{(a_2 - a_1)}\left[\frac{1}{(s - a_1)} - \frac{1}{(s - a_2)}\right]$$

$$\boxed{y(t) = \frac{1}{(a_2 - a_1)}\left[e^{a_1 t}u(t) + e^{a_2 t}u(-t)\right]}$$

■ Example 8.55

Given

$$x_1(t) = e^{3t}u(-t) \quad \text{and} \quad x_2(t) = u(t - 2)$$

Determine $y(t) = x_1(t) * x_2(t)$.

Solution:

$$X_1(s) = L[x_1(t)] = \frac{-1}{(s - 3)}$$
$$X_2(s) = L[x_2(t)] = \frac{e^{-2s}}{s}$$
$$Y(s) = X_1(s)X_2(s) = \frac{-e^{-2s}}{s(s - 3)}$$
$$= \frac{1}{3}\left[\frac{1}{s} - \frac{1}{s - 3}\right]e^{-2s}$$

Let

$$Y_1(s) = \frac{1}{3}\left[\frac{1}{s} - \frac{1}{s - 3}\right]$$
$$y_1(t) = L^{-1}[Y_1(s)]$$
$$= \frac{1}{3}[u(t) + e^{3t}u(-t)]$$

By using shifting property

$$y(t) = y_1(t-2)$$

$$\boxed{y(t) = \frac{1}{3}\left[u(t-2) + e^{3(t-2)}u(-t+2)\right]}$$

8.11 Network Analysis Using Laplace Transform

An electrical network consists of passive elements like resistors, capacitors and inductors. They are connected in series, parallel and series parallel combinations. The currents through and voltages across these elements are obtained by solving integro-differential equations using LT technique. Alternatively, the elements in the network are transformed from time domain and an algebraic equation is obtained which is expressed in terms of input and output. The commonly used inputs are impulse, step, ramp, sinusoids, exponentials etc. The desired response is expressed as a function of time for the given input. When writing the integro-differential equation for a given network, the initial conditions must be taken into account. The energy storing elements such as inductor and a capacitor have initial conditions. At time $t = 0$, the capacitor is initially charged and has the initial voltage $v_c(0)$. Similarly, at $t = 0$, the current through the inductor is denoted as $i_L(0)$. These initial conditions are expressed $v_c(0^-)$ $v_c(0^+)$ and $i_L(0^-)$ and $i_L(0^+)$. The input is assumed to start at $t = 0$ which is considered as the reference point. The condition just before the input is applied $(t = 0^-)$ is denoted as $v_c(0^-)$ and the condition just after the input is applied $(t = 0^+)$ is denoted as $v_c(0^+)$. In many cases, $v_c(0^-)$ and $v_c(0^+)$ are same but not always. Unless otherwise it is specified, $v_c(0)$ or $i_L(0)$ has to be taken as $v_c(0^-)$ or $i_L(0^-)$ which is more practical.

8.11.1 Mathematical Description of R-L-C- Elements

(a) Resistor

Consider the resistor connected across the voltage source $v_i(t)$. The loop equation for the above circuit is written as follows:

$$\begin{aligned} v_i(t) &= i(t)R \qquad (8.57a) \\ v_R(t) &= i(t)R \end{aligned}$$

(b) Inductor

Consider the inductor connected across the voltage source $v_i(t)$ as shown in Fig. 8.30b. The loop equation for the above circuit is written as follows:

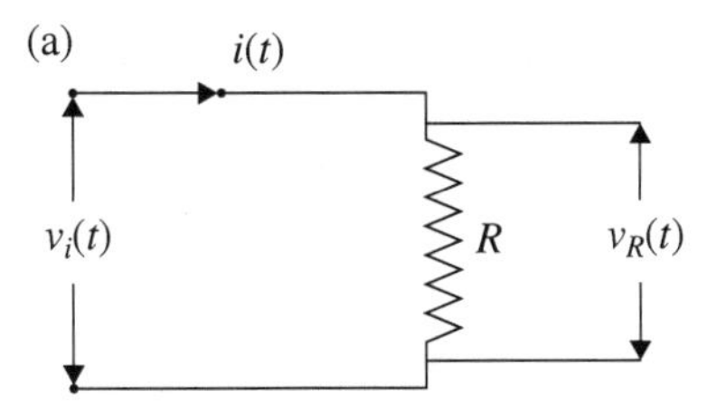

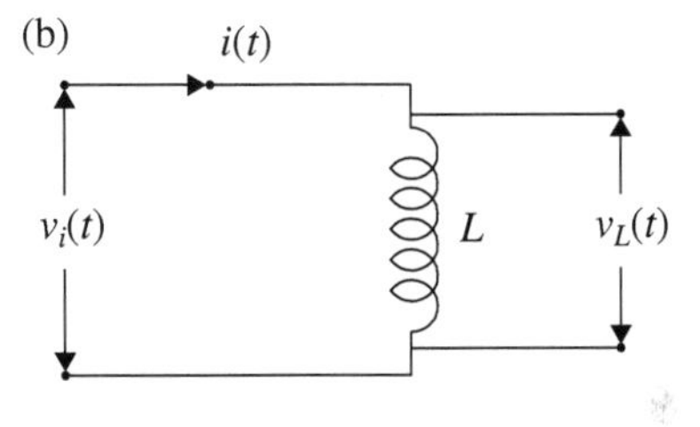

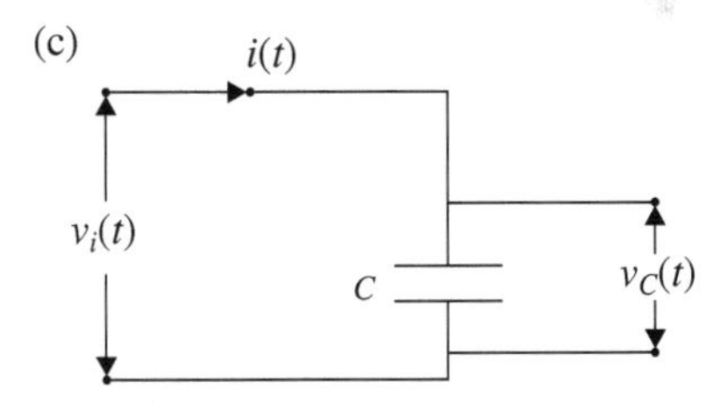

Fig. 8.30 **a** Circuit with a Resistor. **b** Circuit with an Inductor. **c** Circuit with a Capacitor

$$v_i(t) = L\frac{di(t)}{dt}$$
$$v_L(t) = L\frac{di(t)}{dt}$$

Taking LT on both sides of the above equation, with the initial current $i(t) = i(0^-)$, we get

$$V_i(s) = LsI(s) - Li(0^-) \tag{8.57b}$$

(c) Capacitor

For the capacitor circuit shown in Fig. 8.30c, the following equation is written.

$$v_i(t) = \frac{1}{C}\int i(t)dt$$

Taking LT on both sides of the above equation with the capacitor initially charged with $v_c(0^-)$, the following equation is obtained

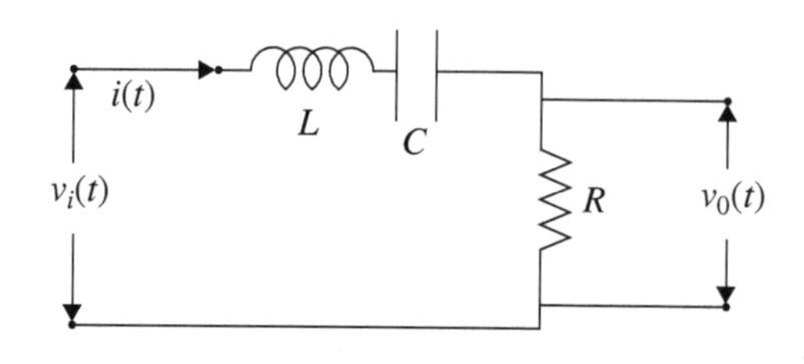

Fig. 8.31 R-L-C series circuit

$$V_i(s) = \frac{1}{Cs}I(s) + v_c(0^-) \tag{8.57c}$$

Equations 8.57a–c are called integro- differential equations. If the initial conditions are zero, these equations can, respectively, be written as

$$\begin{aligned} V_i(s) &= I(s)R \\ V_i(s) &= I(s)Ls \\ V_i(s) &= \frac{1}{Cs}I(s) \end{aligned} \tag{8.58}$$

Equation (8.58) is called algebraic equation. These equations can be written in the frequency domain with the impedance function for the resistor, inductor and capacitor, respectively, as R, Ls and $\frac{1}{Cs}$.

8.11.2 *Transfer Function and Pole-Zero Location*

Consider the R-L-C series circuit shown in Fig. 8.31. $v_i(t)$ is the input, $v_O(t)$ is the output and $i(t)$ is the current flowing through the series circuit. For Fig. 8.31, the following integro- differential equation is written.

$$\begin{aligned} v_i(t) &= L\frac{di(t)}{dt} + \frac{1}{C}\int i(t)dt + Ri(t) \\ v_O(t) &= i(t)R \end{aligned} \tag{8.59}$$

Taking LT on both sides of the above equations, we get the following algebraic equation.

$$\begin{aligned} V_i(s) &= LsI(s) - Li(0^-) + \frac{1}{Cs}I(s) + v_c(0^-) + RI(s) \\ V_O(s) &= RI(s) \end{aligned} \tag{8.60}$$

In Eq. (8.60), if the initial conditions $i(0^-)$ and $v_c(0^-)$ are zero, the following equations could be written.

$$V_i(s) = \left(Ls + \frac{1}{Cs} + R\right) I(s)$$

$$V_0(s) = RI(s)$$

Dividing one by the other, we get

$$\frac{V_0(s)}{V_i(s)} = \frac{R}{Ls + \frac{1}{Cs} + R} \tag{8.61}$$

$$= \frac{RCs}{LCs^2 + RCs + 1}$$

Denoting $\frac{V_0(s)}{V_i(s)} = G(s)$, the above equation can be written in the following form

$$\boxed{G(s) = \frac{RCs}{LCs^2 + RCs + 1}} \tag{8.62}$$

Equation (8.62) is called the transfer function of the given electric circuit.

Transfer function: **Transfer function is therefore defined as the ratio of the LT of the output variable to the LT of the input variable with all the initial conditions being assumed to be zero.**

In Eq. (8.62), if we put $L = 1, C = 1$ and $R = 2.5$, the transfer function is obtained as

$$G(s) = \frac{2.5s}{(s^2 + 2.5s + 1)}$$

$$= \frac{2.5s}{(s + 2)(s + 0.5)} \tag{8.63}$$

The transfer function $G(s)$ becomes zero at $s = 0$.
The points at which the transfer function becomes zero in the s-plane are called zeros and are marked with a circle 0 in the s-plane.

The transfer function $G(s)$ becomes infinity at points $s = -2$ and $s = -0.5$ in the s-plane. These points are called poles of the transfer function and are marked with a small cross $\times$ in the s-plane.

The poles of the transfer function are defined as the points in the s-plane at which the transfer function becomes infinity.

The zeros of the transfer function are obtained by factorizing the numerator polynomial and putting each factor to zero. The poles of a transfer function are obtained by factorizing the denominator polynomial and putting each factor to zero. It is to be noted that the transfer function is not defined if the initial conditions are not zero. The poles and zeros of Eq. (8.63) are shown in Fig. 8.32. The s-plane is a complex plane whose real axis is represented by σ and the imaginary axis by $j\omega$.

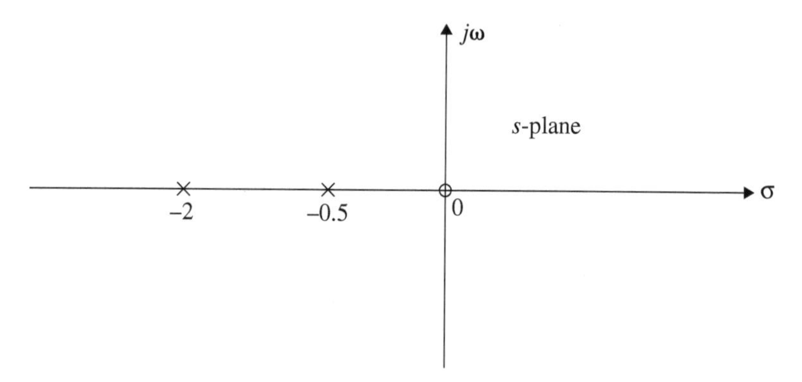

Fig. 8.32 Pole-zero locations of $G(s) = \dfrac{2.5s}{(s+2)(s+0.5)}$

The following examples illustrate electric circuit analysis using LT method.

■ Example 8.56

Consider the R.L.C. series circuit shown in Fig. 8.31 with $L = 1H, C = 1F$ and $R = 2.5$ ohms. Derive an expression for the output voltage $v_0(t)$ if the input is an (a) impulse (b) unit step. Assume zero initial conditions.

Solution: With zero initial conditions, for the circuit shown in Fig. 8.31, the following equation is obtained.

$$L\frac{di(t)}{dt} + \frac{1}{C}\int dt + Ri(t) = v_i(t)$$

$$v_0(t) = i(t)R.$$

Taking LT on both sides and substituting the numerical values for R, L and C, we get

$$\frac{V_0}{V_i}(s) = \frac{2.5s}{(s+2)(s+0.5)}$$

(a) Impulse Response of the System

For an impulse input $V_i(s) = 1$

$$\begin{aligned}
\therefore \qquad V_0(s) &= \frac{2.5s}{(s+2)(s+0.5)} \\
&= \frac{A_1}{(s+2)} + \frac{A_2}{(s+0.5)} \\
2.5s &= A_1(s+0.5) + A_2(s+2)
\end{aligned}$$

Put $s = -2$

$$(2.5)(-2) = A_1(-2+0.5)$$
$$A_1 = \frac{5}{1.5} = \frac{10}{3}$$

Put $s = -0.5$

$$(2.5)(-0.5) = A_2(-0.5+2)$$
$$A_2 = \frac{1.25}{1.5} = \frac{-5}{6}$$
$$\therefore \quad V_0(s) = \frac{10}{3}\frac{1}{(s+2)} - \frac{5}{6}\frac{1}{(s+0.5)}$$

Taking inverse LT, we get

$$\boxed{v_0(t) = \left(\frac{10}{3}e^{-2t} - \frac{5}{6}e^{-0.5t}\right)u(t)}$$

(b) Step Response of the System

$$\frac{V_0}{V_i}(s) = \frac{2.5s}{(s+2)(s+0.5)}$$

For unit step input, $V_i(s) = \frac{1}{s}$

$$\therefore \quad V_0(s) = \frac{2.5s}{s(s+2)(s+0.5)}$$
$$= \frac{2.5}{(s+2)(s+0.5)}$$
$$= \frac{A_1}{(s+2)} + \frac{A_2}{(s+0.5)}$$
$$2.5 = A_1(s+0.5) + A_2(s+2)$$

Put $s = -2$

$$2.5 = A_1(-2+0.5)$$
$$A_1 = -\frac{2.5}{1.5} = \frac{-5}{3}$$

Put $s = -0.5$

$$2.5 = A_2(-0.5 + 2)$$
$$A_2 = \frac{2.5}{1.5} = \frac{5}{3}$$
$$\therefore \quad V_0(s) = \frac{5}{3}\left(-\frac{1}{(s+2)} + \frac{1}{(s+0.5)}\right)$$

$$\boxed{v_0(t) = \frac{5}{3}(-e^{-2t} + e^{-0.5t})u(t)}$$

Note: For an impulse input $V_i(s) = 1$ and for a step input $V_i(s) = \frac{1}{s}$. By integrating the impulse response, one can get the step response. Similarly, by differentiating the step response, the impulse response can be obtained.

In the above example, consider the step response.

$$v_0(t) = \frac{5}{3}(-e^{-2t} + e^{-0.5t})u(t)$$
$$\frac{dv_0(t)}{dt} = \frac{5}{3}(2e^{-2t} - 0.5e^{-0.5t})u(t)$$
$$= \left(\frac{10}{3}e^{-2t} - \frac{5}{6}e^{-0.5t}\right)u(t)$$

The above response is nothing but the impulse response.

■ Example 8.57

Consider the R.L.C circuit shown in Fig. 8.33. The circuit parameters are $R = 3$ ohm; $L = 1H$ and $C = \frac{1}{2}F$. The capacitor C is initially charged with a voltage of $v_c(0^-) = 5$ Volts. The initial current $i(0^-)$ before the input is applied is 2 amps. Find the current in the R-L-C circuit if the input is a unit step. Also, find the voltages across these elements for the above case.

Solution: For the Circuit shown in Fig. 8.33, the loop equations is

$$L\frac{di}{dt} + Ri + \frac{1}{C}\int i(t)dt = x(t)$$

Taking LT on both sides of the equation, we get the following transformation term by term

$$L\left[L\frac{di}{dt}\right] = LsI(s) - Li(0)$$
$$= (sI(s) - 2)$$
$$L[Ri] = RI(s)$$

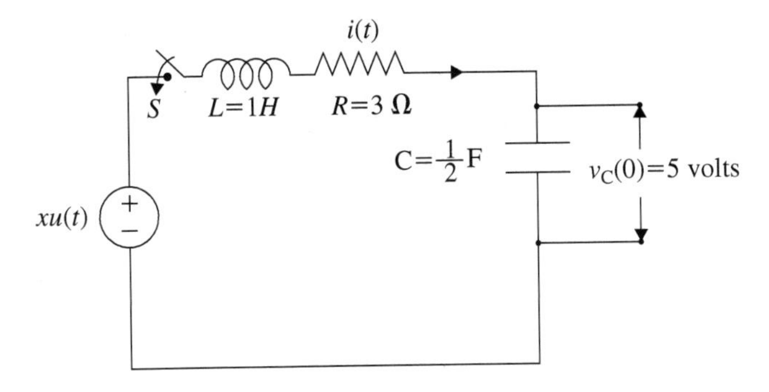

Fig. 8.33 R.L.C. Series Circuit with initial conditions

$$= 3I(s)$$

$$L\left[\frac{1}{C}\int i(t)dt\right] = \frac{1}{Cs}I(s) + \frac{v_c(0^-)}{s}$$
$$= \frac{2I(s)}{s} + \frac{5}{s}$$

Thus, the differential equation after taking LT is written as

$$sI(s) - 2 + 3I(s) + \frac{2I(s)}{s} + \frac{5}{s} = X(s)$$
$$\left[s + 3 + \frac{2}{s}\right]I(s) = 2 - \frac{5}{s} + X(s)$$
$$\frac{(s^2 + 3s + 2)}{s}I(s) = \frac{2s - 5 + sX(s)}{s}$$

Step Response

For step input $X(s) = \frac{1}{s}$

$$I(s) = \frac{(2s - 5) + s\frac{1}{s}}{(s + 1)(s + 2)}$$
$$= \frac{(2s - 4)}{(s + 1)(s + 2)}$$
$$= \frac{A_1}{(s + 1)} + \frac{A_2}{(s + 2)}$$
$$(2s - 4) = A_1(s + 2) + A_2(s + 1)$$

Put $s = -1$

$$(-2-4) = A_1(-1+2)$$
$$A_1 = -6$$

Put $s = -2$

$$(-4-4) = A_2(-2+1)$$
$$A_2 = 8$$
$$I(s) = \frac{-6}{(s+1)} + \frac{8}{s+2}$$

Taking inverse LT we get

$$\boxed{i(t) = (-6e^{-t} + 8e^{-2t})u(t)}$$

The voltage across the resistor is given by

$$\begin{aligned} v_R(t) &= i(t)R \\ &= 3i(t) \end{aligned}$$

$$\boxed{v_R(t) = (-18e^{-t} + 24e^{-2t})u(t)}$$

The voltage across the inductor is given by

$$\begin{aligned} v_L(t) &= L\frac{di(t)}{dt} \\ &= \frac{di(t)}{dt} \end{aligned}$$

$$\boxed{v_L(t) = (6e^{-t} - 16e^{-2t})u(t)}$$

The voltage across the capacitor is given by

$$\begin{aligned} v_c(t) &= \frac{1}{C}\int i(t)dt \\ &= 2\int(-6e^{-t} + 8e^{-2t})dt \\ &= 12e^{-t} - 8e^{-2t} + C. \end{aligned}$$

At $t = 0$, $v_c(0) = 5$

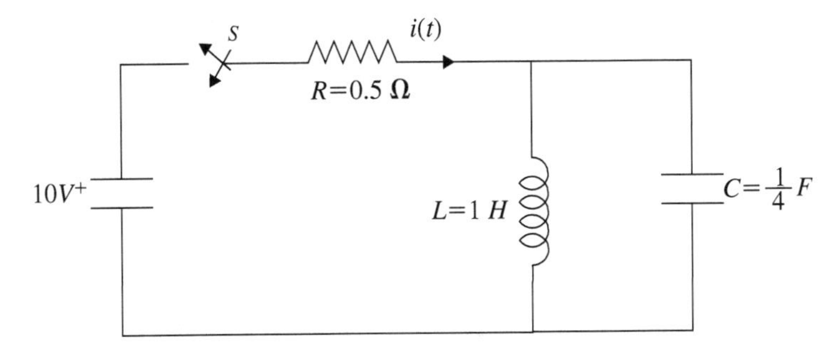

Fig. 8.34 R-L-C- Circuit of Example 8.66

$$5 = 12 - 8 + C \quad \text{or} \quad C = 1$$

$$\boxed{v_c(t) = (12e^{-t} - 8e^{-2t} + 1)u(t)}$$

Example 8.58

Consider the R-L-C circuit shown in Fig. 8.34 with the numerical values given. The initial current through the inductor and the initial voltage across the capacitor at $t = 0^+$ is zero. Derive an expression for the source current as a function of time for $t \geq 0$ when the switch S is closed.

Solution: The expression for $i(t)$ is obtained by writing the algebraic equation rather than the integro-differential equation when the initial conditions are zero.

1. The impedance function for the inductor L is taken as $Z_1(s)$.

$$Z_1(s) = Ls$$
$$= s$$

2. The impedance function for the capacitor C is taken as $Z_2(s)$.

$$Z_2(s) = \frac{1}{Cs} = \frac{4}{s}$$

3. $Z_1(s)$ and $Z_2(s)$ are in parallel. Let $Z_3(s)$ be impedance of the parallel combination of $Z_1(s)$ and $Z_2(s)$. Thus

$$Z_3(s) = \frac{Z_1(s)Z_2(s)}{Z_1(s)+Z_2(s)}$$
$$= \frac{\frac{4}{s}s}{\frac{4}{s}+s}$$
$$= \frac{4s}{s^2+4}$$

4. R and $Z_3(s)$ are in series. Let $Z(s)$ be the impedance of the series combination of R and $Z_3(s)$. Thus

$$Z(s) = R + Z_3(s)$$

$$Z(s) = 0.5 + \frac{4s}{s^2+4}$$
$$Z(s) = \frac{(0.5s^2+4s+2)}{s^2+4}$$

5.

$$I(s) = \frac{V(s)}{Z(s)}$$

For a step input $V(s) = \frac{V}{S} = \frac{10}{s}$

$$I(s) = \frac{10}{s}\frac{(s^2+4)}{(0.5s^2+4s+2)}$$
$$= \frac{20(s^2+4)}{s(s^2+8s+4)}$$

But $(s^2+8s+4) = (s+7.464)(s+0.536)$.

6. Putting $I(s)$ into partial fraction, we get

$$I(s) = \frac{A_1}{s} + \frac{A_2}{(s+7.464)} + \frac{A_3}{(s+0.536)}$$
$$20(s^2+4) = A_1(s^2+8s+4) + A_2s(s+0.536) + A_3s(s+7.464)$$

Put $s = 0$

$$80 = 4A_1$$
$$A_1 = 20$$

Put $s = -7.464$

Fig. 8.35 R.L. series circuit

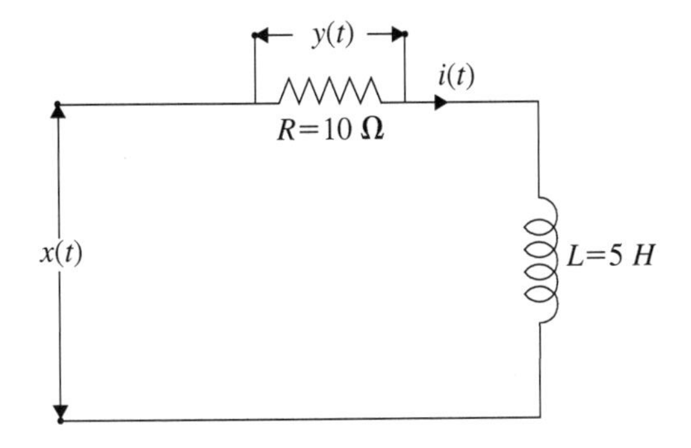

$$(1114.23 + 80) = A_2(-7.464)(-7.464 + 0.536)$$
$$A_2 = 23.1$$

Put $s = -0.536$

$$(5.746 + 80) = A_3(-0.536)(-0.536 + 7.464)$$
$$A_3 = -23.1$$
$$I(s) = \frac{20}{s} + \frac{23.1}{s + 7.464} - \frac{23.1}{s + 0.536}$$

7. Taking inverse LT, we get

$$\boxed{i(t) = \left(20 + 23.1e^{-7.464t} - 23.1e^{-0.536t}\right) u(t)}$$

■ Example 8.59

Find the unit step response of the circuit shown in Fig. 8.35.

(Anna University, December 2007)

Solution:

1. Since the initial condition is zero, the total impedance of the circuit is written as

$$Z(s) = R + Ls$$
$$= 10 + 5s$$

2. The current through the series circuit is

$$I(s) = \frac{X(s)}{Z(s)}$$

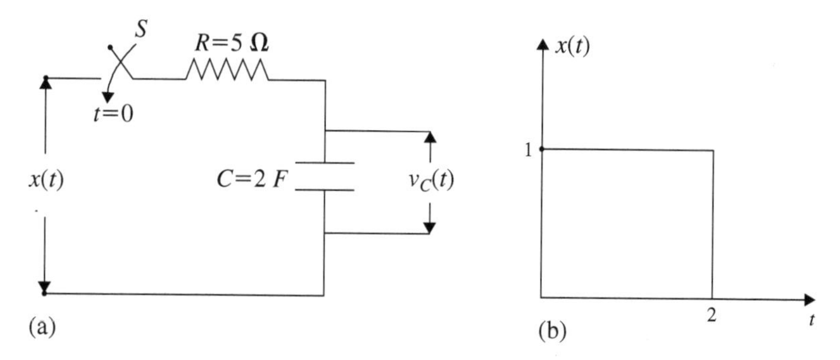

Fig. 8.36 **a** R.C. Circuit and **b** $x(t) = u(t) - u(t-2)$

Since $X(s) = \frac{1}{s}$ for unit step

$$I(s) = \frac{1}{sZ(s)} = \frac{1}{s(10+5s)}$$

3. The output $Y(s) = I(s)R$

$$= \frac{10}{s(10+5s)} = \frac{2}{s(s+2)}$$
$$= \frac{1}{s} - \frac{1}{s+2}$$

4. Taking inverse LT, we get

$$\boxed{i(t) = L^{-1}I(s) = (1 - e^{-2t})u(t)}$$

■ Example 8.60

Consider the R-C-Circuit shown in Fig. 8.36. The input $x(t) = u(t) - u(t-2)$. Derive an expression for the voltage output across the capacitor C as a function of time when the switch S is closed at $t = 0$. Assume zero initial condition.

Solution:

1.

$$x(t) = u(t) - u(t-2)$$
$$X(s) = \left[\frac{1}{s} - \frac{1}{s}e^{-2s}\right]$$

2. Since the initial condition is zero, the impedance of the circuit is written as

$$Z(s) = R + \frac{1}{Cs}$$
$$= 5 + \frac{1}{2s}$$
$$= \frac{(10s+1)}{2s}$$

3. The current in the circuit is

$$I(s) = \frac{X(s)}{Z(s)}$$
$$= \left[\frac{1}{s} - \frac{1}{s}e^{-2s}\right]\frac{2s}{(10s+1)}$$
$$= (1-e^{-2s})\frac{2}{(10s+1)}$$

4. The impedance of the capacitor C is

$$Z_c(s) = \frac{1}{Cs}$$
$$= \frac{1}{2s}$$

5. The output voltage across the capacitor C is given by

$$V_c(s) = I(s)Z_c(s)$$
$$= (1-e^{-2s})\frac{2}{(10s+1)}\frac{1}{2s}$$
$$= \frac{(1-e^{-2s})0.1}{s(s+0.1)}$$
$$= \frac{0.1}{s(s+0.1)} - \frac{0.1e^{-2s}}{s(s+0.1)}$$

6.

$$v_c(t) = L^{-1}V_c(s)$$
$$v_c(t) = L^{-1}\left[\frac{0.1}{s(s+0.1}\right] - L^{-1}\left[\frac{0.1e^{-2s}}{s(s+0.1)}\right]$$

Now consider $\frac{0.1}{s(s+0.1)}$, which can be expressed as $\frac{1}{s} - \frac{1}{s+0.1}$. Thus

$$L^{-1}\left[\frac{1}{s} - \frac{1}{s+0.1}\right] = (1-e^{-0.1t})u(t)$$

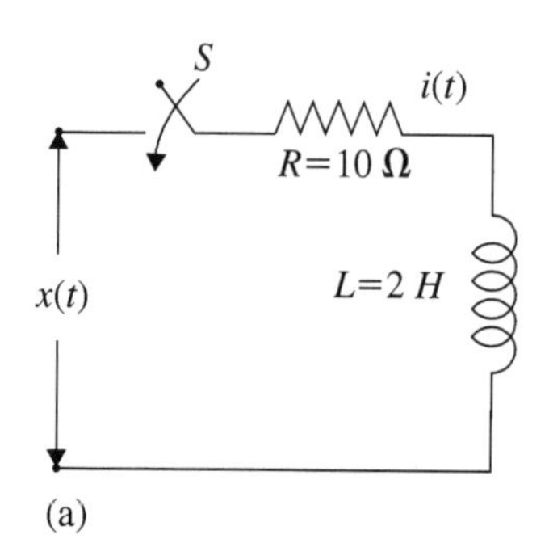

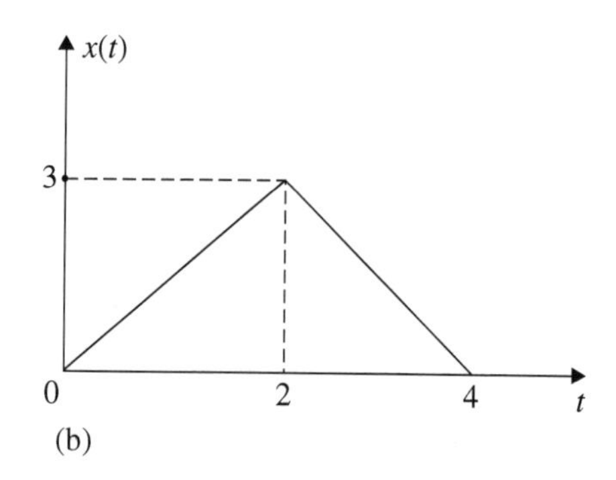

Fig. 8.37 **a** R-L-series circuit; **b** $x(t)$

Using shift theorem, $L^{-1}\left[\frac{0.1e^{-2s}}{s(s+0.1)}\right] = (1 - e^{-0.1(t-2)})u(t-2)$. Thus

$$\boxed{v_c(t) = (1 - e^{-0.1t})u(t) - (1 - e^{-0.1(t-2)})u(t-2)}$$

■ Example 8.61

Consider the R.L. series circuit shown in Fig. 8.37a. At $t = 0$, the switch S is closed. Derive an expression for the current in the series circuit as a function of time. The mathematical description of the input is given by

$$\begin{aligned} x(t) &= \frac{3}{2}t \qquad & 0 \le t \le 2 \\ &= \left(6 - \frac{3}{2}t\right) \qquad & 2 \le t \le 4 \end{aligned}$$

Solution:

1. The mathematical description of $x(t)$ is represented as a triangular wave and is shown in Fig. 8.37b.
2. For Fig. 8.37b, the LT is determined (*see* Example 8.28) as

$$X(s) = \left[\frac{3}{2}\frac{1}{s^2} - \frac{3}{s^2}e^{-2s} + \frac{3}{2s^2}e^{-4s}\right]$$

3. For the circuit shown in Fig. 8.37a, the impedance function is

$$\begin{aligned} Z(s) &= (R + Ls) \\ &= (10 + 2s) \\ &= 2(s + 5) \end{aligned}$$

4. The current through the series circuit is

$$I(s) = \frac{X(s)}{Z(s)}$$
$$= \left[\frac{3}{2}\frac{1}{s^2} - \frac{3}{s^2}e^{-2s} + \frac{3}{2}\frac{1}{s^2}e^{-4s}\right]\frac{1}{2(s+5)}$$

5.

$$i(t) = L^{-1}I(s)$$
$$= L^{-1}\left[\frac{3}{4}\frac{1}{s^2(s+5)} - \frac{3}{2s^2}\frac{e^{-2s}}{(s+5)} + \frac{3}{4s^2}\frac{e^{-4s}}{(s+5)}\right]$$
$$\frac{1}{s^2(s+5)} = \frac{A_1}{s^2} + \frac{A_2}{s} + \frac{A_3}{(s+5)}$$
$$1 = A_1(s+5) + A_2s(s+5) + A_3s^2$$

Put $s = 0$

$$A_1 = \frac{1}{5}$$

Put $s = -5$

$$1 = A_3 25$$
$$A_3 = \frac{1}{25}$$

Compare the coefficients of s^2.

$$A_2 + A_3 = 0$$
$$A_2 = -A_3$$
$$= -\frac{1}{25}$$

6.

$$L^{-1}\left[\frac{3}{4}\frac{1}{s^2(s+5)}\right] = \frac{3}{4}L^{-1}\left[\frac{1}{5s^2} - \frac{1}{25s} + \frac{1}{25(s+5)}\right] \tag{a}$$
$$= \frac{3}{4}\left[\frac{1}{5}t - \frac{1}{25} + \frac{1}{25}e^{-5t}\right]u(t)$$

$$L^{-1}\left[-\frac{3}{2}\frac{e^{-2s}}{s^2(s+5)}\right] = -\frac{3}{2}\left[\frac{1}{5}(t-2) - \frac{1}{25} + \frac{1}{25}e^{-5(t-2)}\right]u(t-2) \tag{b}$$

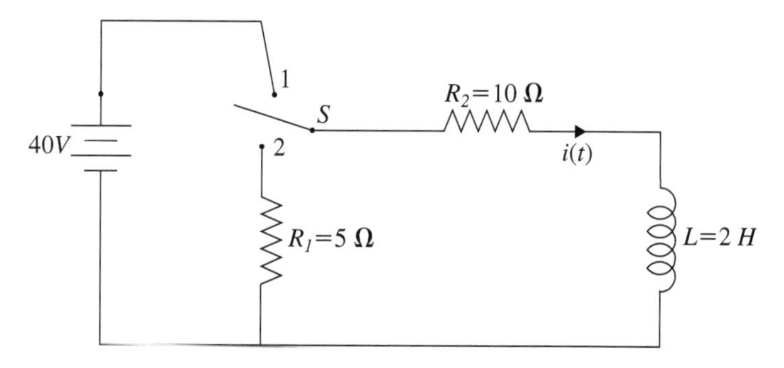

Fig. 8.38 Circuit of Example 8.70

$$L^{-1}\left[\frac{3}{4}\frac{e^{-4s}}{s^2(s+5)}\right] = \frac{3}{4}\left[\frac{1}{5}(t-4) - \frac{1}{25} + \frac{1}{25}e^{-5(t-4)}\right]u(t-4) \qquad \text{(c)}$$

7.

$$\begin{aligned} i(t) &= (a) + (b) + (c) \\ i &= \left[\frac{3}{4}\left\{\frac{1}{5}t - \frac{1}{25} + \frac{1}{25}e^{-5t}\right\}u(t) \right. \\ &\quad - \frac{3}{2}\left\{\frac{1}{5}(t-2) - \frac{1}{25} + \frac{1}{25}e^{-5(t-2)}\right\}u(t-2) \\ &\quad \left. + \frac{3}{4}\left\{\frac{1}{5}(t-4) - \frac{1}{25} + \frac{1}{25}e^{-5(t-4)}\right\}u(t-4)\right] \end{aligned}$$

■ Example 8.62

Consider the circuit shown in Fig. 8.38. Initially, the switch is in position 1. At $t = 0$, the switch is moved to position 2. Find the expression for the current in the inductor L as a function of time t.

Solution:

1. When the switch S is in position 1, the current i passing through the inductor at steady state is decided by the source voltage and resistance R_2.

$$i = \frac{40}{10} = 4 \text{ amp.}$$

2. When the switch S is in position 2, the following dynamic equation for the $R_1 - R_2 - L$ series circuit is written

$$L\frac{di}{dt} + (R_1 + R_2)i = 0$$

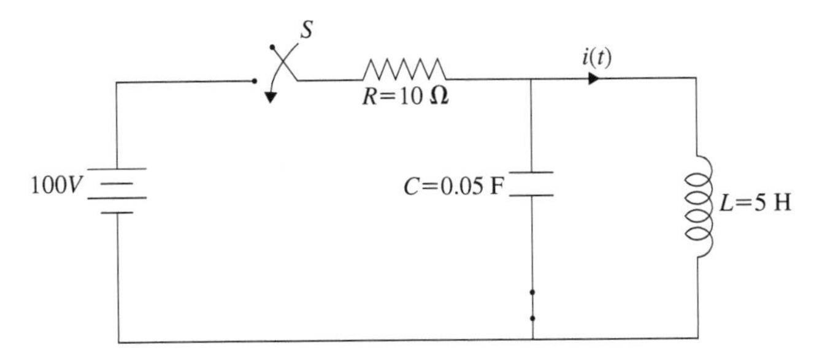

Fig. 8.39 Circuit for Example 8.71

3. Taking LT on both sides of the above equation, we get

$$sLI(s) - Li(0^+) + (R_1 + R_2)I(s) = 0$$

Substituting the numerical values, we get

$$\begin{aligned}(2s + 15)I(s) &= 2 \times 4 = 8\\ I(s) &= \frac{8}{2s + 15}\\ &= \frac{4}{s + 7.5}\end{aligned}$$

4. Taking inverse LT, we get

$$\boxed{i(t) = 4e^{-7.5t}u(t)}$$

■ Example 8.63

Consider the circuit shown in Fig. 8.39. The switch S is initially closed. Derive an expression for the current through the inductor as a function of time when the switch S is suddenly opened at $t = 0$.

Solution:

1. When the switch S is closed, the current is passing through R, and L and C is open circuited. Under this condition the initial current is limited by R only. Thus

$$i(0) = \frac{100}{10} = 10 \text{ Amps.}$$

2. The initial charge across the capacitor is zero because the entire voltage is applied across R only. Therefore, the following Loop equation for L.C. circuit is written when the switch S is open.

$$L\frac{di}{dt} - Li(0) + \frac{1}{C}\int i(t)dt + v_c(0) = 0$$

3. Taking LT and substituting $i(0) = 10$ and $v_c(0) = 0$, we get

$$\left[Ls + \frac{1}{Cs}\right] I(s) = Li(0)$$
$$\left(5s + \frac{1}{0.05s}\right) I(s) = 50$$
$$(5s^2 + 20)I(s) = 50s$$
$$(s^2 + 4)I(s) = 10s$$
$$(s + j2)(s - j2)I(s) = 10s$$
$$I(s) = \frac{A_1}{(s + j2)} + \frac{A_2}{(s - j2)}$$
$$10s = A_1(s - j2) + A_2(s + j2)$$

Put $s = -j2$

$$10(-j2) = 4A_1(-j)$$
$$A_1 = 5$$
$$A_2 = A_1^* = 5$$
$$I(s) = 5\left[\frac{1}{s + j2} + \frac{1}{s - j2}\right]$$

4. Taking inverse LT, we get

$$i(t) = 5[e^{-j2t} + e^{j2t}]$$

$$\boxed{i(t) = 10\cos 2t}$$

■ Example 8.64

Find the transfer function of LTI system described by the differential equation

$$\frac{d^2y(t)}{dt^2} + 3\frac{dy(t)}{dt} + 2y(t) = 2\frac{dx(t)}{dt} - 3x(t)$$

(*Anna University, May, 2008*)

Solution: Taking LT on both sides assuming zero initial conditions, we get

$$(s^2 + 3s + 2)Y(s) = (2s - 3)X(s)$$

The transfer function is $\frac{Y(s)}{X(s)}$

$$\boxed{\frac{Y(s)}{X(s)} = \frac{(2s - 3)}{(s^2 + 3s + 2)}}$$

Example 8.65

Consider an LTI system with input $x(t) = e^{-t}u(t)$ and impulse response $h(t) = e^{-2t}u(t)$

- Determine the LT of $x(t)$ and $h(t)$.
- Using the convolution property, determine the LT $Y(s)$ of the output $y(t)$.
- From the LT of $y(t)$ as obtained in part (2) determine $y(t)$.
- Verify your result in part (2) by explicitly convolving $x(t)$ and $h(t)$.

(Anna University, May 2008)

Solution:

1.

$$x(t) = e^{-t}u(t)$$

From LT table

$$\boxed{X(s) = \frac{1}{(s+1)}} \quad \text{ROC: Re}(s) > -1$$

$$h(t) = e^{-3t}u(t)$$

From LT table

$$\boxed{H(s) = \frac{1}{(s+3)}} \quad \text{ROC: Re}(s) > -3$$

2.

$$Y(s) = X(s)H(s)$$

$$\boxed{Y(s) = \frac{1}{(s+1)(s+3)}}$$

3.

$$
\begin{aligned}
Y(s) &= \frac{1}{(s+1)(s+3)} \\
&= \frac{A_1}{(s+1)} + \frac{A_2}{(s+3)} \\
1 &= A_1(s+3) + A_2(s+1)
\end{aligned}
$$

Put $s = -1$

$$
\begin{aligned}
1 &= A_1(-1+3) \\
A_1 &= \frac{1}{2}
\end{aligned}
$$

Put $s = -3$

$$
\begin{aligned}
1 &= A_2(-3+2) \\
A_2 &= -\frac{1}{2} \\
Y(s) &= \frac{1}{2}\left[\frac{1}{s+1} - \frac{1}{s+3}\right] \\
y(t) &= L^{-1}Y(s) = \frac{1}{2}[e^{-t} - e^{-3t}]u(t)
\end{aligned}
$$

$$
\boxed{y(t) = \frac{1}{2}[e^{-t} - e^{-3t}]u(t)}
$$

4.

$$
\begin{aligned}
x(t) &= e^{-t} \\
x(t-\tau) &= e^{-(t-\tau)} \\
h(\tau) &= e^{-3\tau}
\end{aligned}
$$

Since $x(t)$ and $h(t)$ are causal, the limit of integration varies from 0 to t. Thus

$$
\begin{aligned}
y(t) &= \int_0^t e^{-(t-\tau)} e^{-3\tau} d\tau \\
&= e^{-t} \int_0^t e^{-2\tau} d\tau \\
&= \frac{e^{-t}}{(-2)} \left[e^{-2\tau}\right]_0^t
\end{aligned}
$$

$$\boxed{y(t) = \frac{1}{2}\left[e^{-t} - e^{-3t}\right]u(t)}$$

■ Example 8.66

Determine the impulse response $h(t)$ of the system whose input-output is related by the differential equation; where $x(t)$ is the input, $y(t)$ is the output

$$\frac{d^2y(t)}{dt^2} + 3\frac{dy(t)}{dt} + 2y(t) = x(t)$$

with all initial conditions to be zeros.

(Anna University, April, 2004)

Solution:

1. Taking LT on both sides of the given differential equation, we get

$$(s^2 + 3s + 2)Y(s) = X(s)$$
$$s^2 + 3s + 2 = (s + 1)(s + 2)$$

For an impulse

$$X(s) = 1$$
$$Y(s) = \frac{1}{(s + 1)(s + 2)}$$

2. Putting into partial fraction, we get

$$Y(s) = \frac{A_1}{(s + 1)} + \frac{A_2}{(s + 2)}$$
$$1 = A_1(s + 2) + A_2(s + 1)$$

Put $s = -1$

$$1 = A_1(-1 + 2)$$
$$A_1 = 1$$

Put $s = -2$

$$1 = A_2(-2 + 1)$$
$$A_2 = -1$$
$$Y(s) = \frac{1}{s + 1} - \frac{1}{s + 2}$$

3. Taking inverse LT, we get

$$y(t) = (e^{-t} - e^{-2t})u(t)$$

For impulse input $y(t) = h(t)$

$$\boxed{h(t) = (e^{-t} - e^{-2t})u(t)}$$

Example 8.67

Determine the output response of the system whose impulse response

$$h(t) = e^{-at}u(t)$$

for the step input.

(*Anna University, April, 2004*)

Solution:

1.

$$\begin{aligned} H(s) &= L[h(t)] = L[e^{-at}u(t)] \\ &= \frac{1}{s+a} \\ H(s) &= \frac{Y(s)}{X(s)} \end{aligned}$$

For step input $X(s) = \frac{1}{s}$.

2. Substituting in $H(s)$, we get

$$Y(s) = \frac{1}{s(s+a)}$$

The residues are obtained by intuition

$$Y(s) = \left[\frac{1}{s} - \frac{1}{s+a}\right]\frac{1}{a}$$

3. Taking inverse LT, we get

$$\boxed{y(t) = \frac{1}{a}[1 - e^{-at}]u(t)}$$

Example 8.68

Consider an LTI system whose response to the input $x(t) = (e^{-t} + e^{-3t})u(t)$ is $y(t) = (2e^{-t} - 2e^{-4t})u(t)$. Find the system's impulse response.

(Anna University, December, 2007)

Solution:

1. The LT of $x(t)$ is $X(s)$

$$X(s) = L[e^{-t} + e^{-3t}] = \frac{1}{(s+1)} + \frac{1}{(s+3)} = \frac{2(s+2)}{(s+1)(s+3)}$$

The LT of $y(t)$ is $Y(s)$

$$Y(s) = L[2e^{-t} - 2e^{-4t}] = 2\left[\frac{1}{s+1} - \frac{1}{s+4}\right] = \frac{6}{(s+1)(s+4)}$$

2. The transfer function is

$$H(s) = \frac{Y(s)}{X(s)} = \frac{6}{(s+1)(s+4)} \frac{(s+1)(s+3)}{2(s+2)}$$
$$= \frac{3(s+3)}{(s+2)(s+4)}$$

3. For an impulse $X(s) = 1$. Now, $Y(s)$ can be put into partial fraction as given below.

$$Y(s) = \frac{3(s+3)}{(s+2)(s+4)}$$
$$= \frac{A_1}{(s+2)} + \frac{A_2}{(s+4)}$$
$$3(s+3) = A_1(s+4) + A_2(s+2)$$

Put $s = -2$

$$3(-2+3) = A_1(-2+4)$$
$$A_1 = \frac{3}{2}$$

Put $s = -4$

$$3(-4+3) = A_2(-4+2)$$
$$A_2 = \frac{3}{2}$$
$$Y(s) = \frac{3}{2}\left(\frac{1}{s+2} + \frac{1}{s+4}\right)$$

4. Taking inverse LT of $Y(s)$, we get

$$y(t) = L^{-1}Y(s)$$
$$= \frac{3}{2}L\left(\frac{1}{s+2} + \frac{1}{s+4}\right)$$

$$\boxed{y(t) = \frac{3}{2}(e^{-2t} + e^{-4t})u(t)}$$

■ Example 8.69

Determine the response of the system with impulse response $h(t) = u(t)$ for the input $x(t) = e^{-2t}u(t)$.

(*Anna University, April 2004*)

Solution: **Method 1:**

1. Taking LT for $h(t)$ and $x(t)$, we get

$$H(s) = L(u(t)) = \frac{1}{s}$$
$$X(s) = L[e^{-2t}u(t)] = \frac{1}{(s+2)}$$

2.

$$y(t) = x(t) * h(t)$$
$$Y(s) = X(s)H(s)$$
$$= \frac{1}{s(s+2)}$$

3. Putting into partial fraction and by intuition the residues are obtained. Thus, $Y(s)$ is written as

$$Y(s) = \frac{1}{2}\left(\frac{1}{s} - \frac{1}{s+2}\right)$$

4. Taking Laplace inverse for $Y(s)$, we get $y(t)$

$$y(t) = L^{-1}Y(s)$$
$$= L^{-1}\frac{1}{2}\left(\frac{1}{s} - \frac{1}{s+2}\right)$$

$$\boxed{y(t) = \frac{1}{2}(1 - e^{-2t})}$$

Method 2: $y(t)$ can be Derived by using Convolution Integral

1. Both $h(t)$ and $x(t)$ are causal. Hence, the following convolution integral is written for $y(t)$

$$y(t) = \int_0^t h(\tau)x(t-\tau)d\tau$$
$$= \int_0^t e^{-2(t-\tau)}d\tau$$
$$= e^{-2t}\int_0^t e^{2\tau}d\tau$$
$$= \frac{e^{-2t}}{2}\left[e^{2\tau}\right]_0^t$$
$$= \frac{e^{-2t}}{2}[e^{2t} - 1]$$

$$\boxed{y(t) = \frac{1}{2}[1 - e^{-2t}]}$$

■ Example 8.70

Find the output of an LTI system with impulse response $h(t) = \delta(t-3)$ for the input $x(t) = \cos 4t + \cos 7t$.

(Anna University, April, 2004)

Solution:

$$
\begin{aligned}
h(t) &= \delta(t-3) \\
H(s) &= e^{-3s} \\
X(s) &= L[\cos 4t + \cos 7t] \\
Y(s) &= H(s)X(s) = L[\cos 4t + \cos 7t]e^{-3s}
\end{aligned}
$$

$$\boxed{y(t) = \cos 4(t-3) + \cos 7(t-3)}$$

■ Example 8.71

Find the initial and final values for

$$X(s) = \frac{(s+5)}{(s^2+3s+2)}$$

(Anna University, June, 2007)

Solution:

1. Initial value of $x(0)$. According to initial value theorem

$$
\begin{aligned}
x(0) &= \underset{s\to\infty}{Lt}\ sX(s) \\
&= \underset{s\to\infty}{Lt}\ \frac{s^2+5s}{s^2+3s+2} \\
&= \underset{s\to\infty}{Lt}\ \frac{1+\frac{5}{s}}{1+\frac{3}{s}+\frac{2}{s^2}}
\end{aligned}
$$

$$\boxed{x(0) = 1}$$

2.

$$(s^2+3s+2) = (s+1)(s+2)$$

Here the poles are at $s = -1$ and $s = -2$ and are in LHP. No pole of $X(s)$ is in RHP. Hence, the application of initial value theorem is correct.

3. Final value of $x(\infty)$. According to final value theorem

$$
\begin{aligned}
x(\infty) &= \underset{s\to 0}{Lt}\ sX(s) \\
&= \underset{s\to 0}{Lt}\ \frac{s^2+5s}{s^2+3s+2}
\end{aligned}
$$

$$\boxed{x(\infty) = 0}$$

■ Example 8.72

Find the step response of the system whose impulse response is given as

$$h(t) = u(t+1) - u(t-1)$$

(*Anna University, June, 2007*)

Solution:

1. By taking LT for $h(t)$, using time shifting property, we get

$$\begin{aligned} H(s) &= L[h(t)] \\ &= \frac{1}{s}e^{s} - \frac{1}{s}e^{-s} \\ &= \frac{1}{s}[e^{s} - e^{-s}] \end{aligned}$$

2.
$$H(s) = \frac{Y(s)}{X(s)} = \frac{1}{s}[e^{s} - e^{-s}]$$

For step input $X(s) = \frac{1}{s}$

$$\begin{aligned} Y(s) &= \frac{1}{s^2}[e^{s} - e^{-s}] \\ &= Y_1(s)[e^{s} - e^{-s}] \end{aligned}$$

where

$$Y_1(s) = \frac{1}{s^2}$$

3.
$$\begin{aligned} y_1(t) &= L^{-1}Y_1(s) \\ &= L^{-1}\frac{1}{s^2} \\ &= t \end{aligned}$$

4.
$$y(t) = y_1(t)[u(t+1) - u(t-1)]$$

$$\boxed{y(t) = (t+1)u(t+1) - (t-1)u(t-1)}$$

Example 8.73

Find the response of the system whose impulse response is

$$h(t) = e^{-3t}u(t)$$
$$x(t) = u(t-3) - u(t-5)$$

(Anna University, June 2007)

Solution:

1. The LT of $h(t)$ is

$$H(s) = L[e^{-3t}u(t)]$$
$$= \frac{1}{(s+3)}$$

2. The LT of the input $x(t)$ is

$$X(s) = L[u(t-3) - u(t-5)]$$
$$= \frac{1}{s}[e^{-3s} - e^{-5s}]$$

3.

$$H(s) = \frac{Y(s)}{X(s)}$$
$$Y(s) = H(s)X(s)$$
$$= \frac{1}{s(s+3)}[e^{-3s} - e^{-5s}]$$
$$= Y_1(s)[e^{-3s} - e^{-5s}]$$

where $Y_1(s) = \frac{1}{s(s+3)}$.

4. Now $Y_1(s)$ can be put into partial fraction as

$$Y_1(s) = \frac{1}{3}\left[\frac{1}{s} - \frac{1}{s+3}\right]$$
$$y_1(t) = L^{-1}Y_1(s)$$
$$= \frac{1}{3}[1 - e^{-3t}]$$

5. The response $y(t)$ is obtained from $y_1(t)$ and applying time shifting property

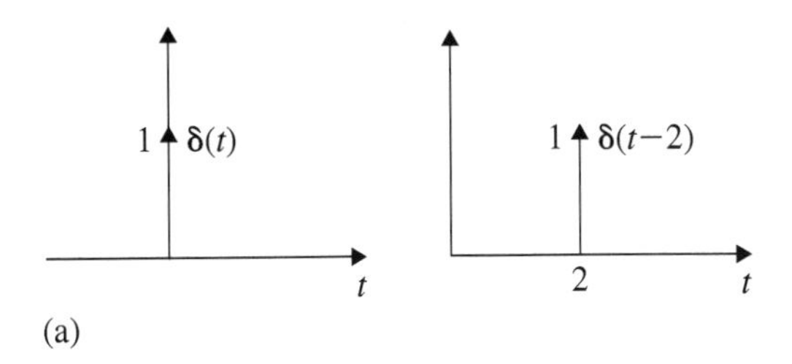

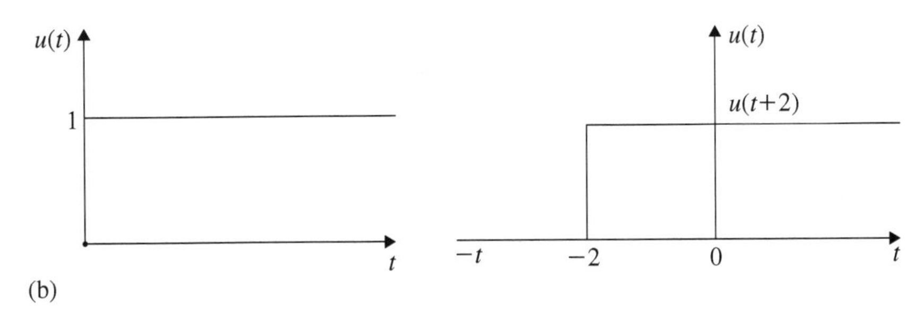

Fig. 8.40 Time shifted unit sample and unit step

$$y(t) = \frac{1}{3}[1 - e^{-3(t-3)}]u(t-3) - \frac{1}{3}[1 - e^{-3(t-5)}]u(t-5)$$

$$\boxed{y(t) = \frac{1}{3}[1 - e^{-3(t-3)}]u(t-3) - \frac{1}{3}[1 - e^{-3(t-5)}]u(t-5)}$$

■ Example 8.74

Draw the wave forms $\delta(t-2)$ and $u(t+2)$.

Solution:

1. The unit sample is shown in Fig. 8.40a. The time delayed signal (right shifted by $t = 2$) is shown by its side.
2. The unit step signal is shown in Fig. 8.40b. The unit step signal is left shifted by $t = -2$ and is shown in the figure shown by its side.

■ Example 8.75

A system has the transfer function

$$H(s) = \frac{(3s-1)}{(s+3)(s-2)}$$

Find the impulse response assuming the system is stable and causal.

(*Anna University, December, 2007*)

Solution:

1.

$$H(s) = \frac{(3s-1)}{(s+3)(s-2)}$$
$$= \frac{A_1}{(s+3)} + \frac{A_2}{(s-2)}$$
$$(3s-1) = A_1(s-2) + A_2(s+3)$$

Put $s = -3$

$$(-9-1) = A_1(-3-2)$$
$$A_1 = 2$$

Put $s = 2$

$$(6-1) = A_2(2+3)$$
$$A_2 = 1$$
$$H(s) = \frac{2}{(s+3)} + \frac{1}{(s-2)}$$

2. The poles of $H(s)$ are at $s = 2$ and $s = -3$. If the system is stable, the pole at $s = 2$ contributes to the left-sided term to the impulse response and the pole at $s = -3$ contributes right-sided term. Thus, we have

$$h(t) = 2e^{-3t}u(t) - e^{2t}u(-t)$$

3. If the system is causal, then both the poles should contribute right-sided term to the impulse response which is obtained as

$$h(t) = [2e^{-3t} + e^{2t}]u(t)$$

   Due to $e^{2t}u(t)$ the system is not stable.
4. Hence, the given system cannot be both stable and causal due to the pole at $s = 2$.

8.12 Connection Between Laplace Transform and Fourier Transform

The bilateral LT of a signal $x(t)$ as defined earlier is written as follows:

$$X(s) = \int_{-\infty}^{\infty} x(t)e^{-st}dt \tag{8.64}$$

Substituting $s = j\omega$ in the above equation, we get

$$X(j\omega) = \int_{-\infty}^{\infty} x(t)e^{-j\omega t}dt \tag{8.65}$$

Thus, the FT is a special case of LT which is obtained by putting $X(s)|_{s=j\omega}$ with the following constraints.

- $x(t)$ is absolutely integrable.
- ROC of $X(s)$ includes the $j\omega$ axis.

Many commonly used signals have $x(t) = 0$ for $t \leq 0$ and ROC of the LT includes the $j\omega$ axis. Under this condition

$$\boxed{X(j\omega) = X(s)|_{s=j\omega}}$$

Consider the following signals

$$x(t) = e^{-2t}u(t)$$
$$X(s) = \frac{1}{(s+2)} \qquad \text{ROC: Re}(s) > -2$$

Put $s = j\omega$

$$X(j\omega) = \frac{1}{j\omega + 2}.$$

Now by FT method, we get

$$X(j\omega) = \int_{0}^{\infty} e^{-2t}e^{-j\omega t}dt$$
$$= \frac{1}{(j\omega + 2)}$$

In the above case, ROC includes the $j\omega$ axis.

Now consider the step function $u(t)$. The LT of a step function is

$$L[u(t)] = \frac{1}{s}.$$

But the FT of $u(t)$ is obtained as

$$F[u(t)] = \pi\delta(\omega) + \frac{1}{j\omega}$$

Thus, the FT of $u(t)$ cannot be obtained from its LT as it is not absolutely integrable.

8.13 Causality of Continuous Time Invariant System

A linear time invariant continuous time system is said to be causal if the impulse response $h(t)$ of the system is zero for $t < 0$. Thus, the system which possesses right-sided impulse response is said to be causal. For this, the ROC of the system transfer function $H(s)$ which is rational, should be in the right half plane and to the right of the rightmost pole.

Consider the following impulse response function

$$\begin{aligned} h(t) &= e^{-2t}u(t) \\ H(s) &= \frac{1}{(s+2)} \qquad \text{ROC: Re}(s) > -2 \end{aligned}$$

The above transfer function is rational because the degree of the denominator polynomial is greater than the degree of the numerator polynomial. The ROC is to the right of the rightmost pole $s = -2$. Hence, the system is causal. The ROC is shown in Fig. 8.41a. Now consider the following impulse response function

$$h(t) = e^{-|t|}$$

The above function can be written as

$$\begin{aligned} h(t) &= e^{-t} \quad t \geq 0 \\ &= e^{t} \quad t \leq 0 \\ H(s) &= \int_{-\infty}^{0} e^{t}e^{-st}dt + \int_{0}^{\infty} e^{-t}e^{-st}dt \\ &= -\frac{1}{(s-1)} + \frac{1}{(s+1)} = \frac{-2}{(s-1)(s+1)} \end{aligned}$$

The transfer function is rational. The ROC is shown in Fig. 8.41b. The rightmost pole is at $s = 1$. The ROC is not to the right of the rightmost pole. Hence, the system is not causal.

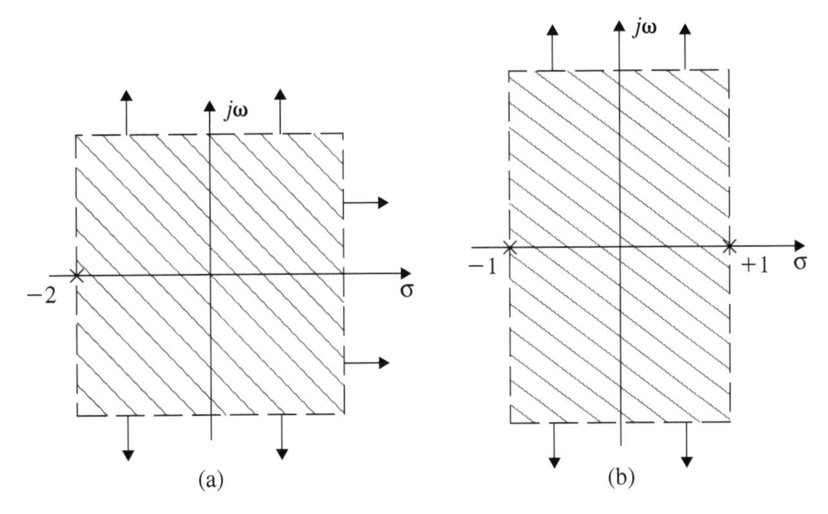

Fig. 8.41 **a** ROC of $h(t) = e^{-2t}$ (causal); **b** ROC of $h(t) = e^{-|t|}$ (non-causal)

8.14 Stability of Linear Time Invariant Continuous System

As already derived a linear time invariant system is said to be stable if the area under the impulse response $h(t)$ curve is finite (absolutely integrable). The impulse response of a causal system is absolutely integrable if the response curve decays exponentially as time increases. Consider the transfer function of an LTIC system.

$$H(s) = \frac{b_n s^n + b_{n-1}s^{n-1} + \cdots + b_0}{a_n s^m + a_{m-1}s^{m-1} + \cdots + a_0}$$

For a rational function $H(s)$, $m > n$. The above transfer function can be written in terms of factors.

$$H(s) = \frac{A_1}{(s+p_1)} + \frac{A_2}{(s+p_2)} + \cdots + \frac{A_n}{(s+p_m)}$$

The impulse response of $H(s)$ is obtained by taking inverse LT.

$$h(t) = L^{-1}H(s) = A_1 e^{-p_1 t} + A_2 e^{-p_2 t} + \cdots + A_m e^{-p_m t}$$

For $h(t)$ to be absolutely integrable, the following conditions are to be satisfied.

- All the poles of $H(s)$ should lie in the left half of the s-plane;
- No repeated pole should be in the imaginary axis. Under these conditions, the system is said to be stable;
- The stability is also assessed by ROC. The ROC of $H(s)$ should include $j\omega$ axis.

Example 8.76

A Certain causal linear time invariant system has the following transfer function. Test whether the system is stable.

$$(a) \quad H(s) = \frac{(s-4)}{(s+2)(s-1)}$$

$$(b) \quad H(s) = \frac{(s-4)}{s^2(s+1)}$$

$$(c) \quad H(s) = \frac{(s-4)}{s(s+1)(s+4)}$$

$$(d) \quad H(s) = \frac{(s-4)}{(s-3)(s+4)} \qquad \text{ROC:} -4 < \text{Re}(s) < 3$$

Solution:

(a) Since the system is causal, the pole $s = 1$, which lies in RHP makes the system unstable.
(b) There are two poles repeated at the origin. The system is unstable.
(c) All the poles are in LHP. The system is stable. It is to be noted that the locations of zeros do not have any influence on the system stability.
(d) This is a non-causal system. ROC strip is enclosing the $j\omega$ axis. Hence, the system is stable.

8.15 The Bilateral Laplace Transform

The unilateral LT is applicable for causal signals and/or systems. However, for non-causal signals and systems, the LT pair is defined as follows:

$$L[x(t)] = X(s) = \int_{-\infty}^{\infty} x(t)e^{-st}dt \tag{8.66}$$

$$L^{-1}[X(s)] = x(t) = \frac{1}{2\pi j}\int_{c-j\infty}^{c+\infty} X(s)e^{st}ds \tag{8.67}$$

It is to be noted here that the unilateral LT pair defined earlier is the special case of bilateral LT.

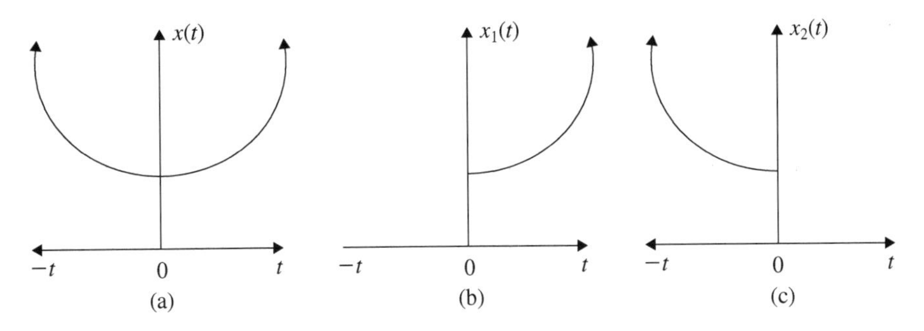

Fig. 8.42 **a** Signal $x(t)$; **b** Causal Signal; **c** Anti-Causal signal

8.15.1 Representation of Causal and Anti-causal Signals

The signal $x(t)$ shown in Fig. 8.42a is a non-causal signal which has two components. $x(t)$ can be split up into two components as $x(t) = x_1(t) + x_2(t)$. The signal $x_1(t)$ is a causal signal (positive time) and is also called as right-sided signal. This is shown in Fig. 8.42b. The signal $x_2(t)$ is called non-causal or anti-causal (negative time) signal. It is also called left-sided signal. $x_2(t)$ is shown in Fig. 8.42c. These signals are given the following mathematical description.

$$x_1(t) = x(t)u(t) \qquad -0 < t < \infty \tag{8.68}$$

$$x_2(t) = x(t)u(-t) \qquad -\infty < t < -0 \tag{8.69}$$

The LT of $x_1(t)$, the causal component is

$$X_1(s) = L[x_1(t)] = \int_{0^-}^{\infty} x_1(t)e^{-st}dt \tag{8.70}$$

The LT of $x_2(t)$, the non-causal component is

$$X_2(s) = L[x_2(t)] = \int_{-\infty}^{0^-} x_2(t)e^{-st}dt \tag{8.71}$$

It is to be noted that if $x(t)$ has any impulse or its derivatives at the origin they should be included in the causal signal $x_1(t)$ and $x_2(t) = 0$ at the origin.

8.15.2 ROC of Bilateral Laplace Transform

Consider the following signal

$$x(t) = e^{-2t}u(t) + e^{3t}u(-t)$$
$$x_1(t) = e^{-2t}u(t)$$
$$X_1(s) = \frac{1}{(s+2)} \qquad \text{ROC: Re } s > -2$$

$$x_2(t) = e^{3t}u(-t)$$
$$\begin{aligned} X_2(s) &= \int_{-\infty}^{-0} e^{3t}e^{-st}dt \\ &= \int_{-\infty}^{-0} e^{-(s-3)t}dt \\ &= -\frac{1}{(s-3)}\left[e^{-(s-3)t}\right]_{-\infty}^{0^-} \\ &= -\frac{1}{(s-3)}[-e^{-(s-3)(-\infty)} + 1] \end{aligned}$$

$e^{-(s-3)(-\infty)}$ converges if $(s-3) < 0$ or $s < 3$. Hence, the ROC of the left sided (anti-causal signal) is to the left of the pole at $s = 3$

$$X_2(s) = -\frac{1}{(s-3)} \qquad \text{ROC: Re } s < 3$$
$$\therefore \quad X(s) = X_1(s) + X_2(s)$$
$$X(s) = \frac{1}{(s+2)} - \frac{1}{s-3} \qquad \text{ROC: } -2 < \text{Re } s < 3$$

Unless the ROC is mentioned, the inverse LT is not unique. In the above case, the ROC is a strip between $-2 < \text{Re } s < 3$ and is shown in Fig. 8.43.

■ Example 8.77

Consider the following function.

$$X(s) = \frac{10}{(s+4)(s-2)}$$

Find $x(t)$ if the ROC is (a) Re $s > 2$; (b) Re $s < -4$; (c) $-4 <$ Re $s > 2$.

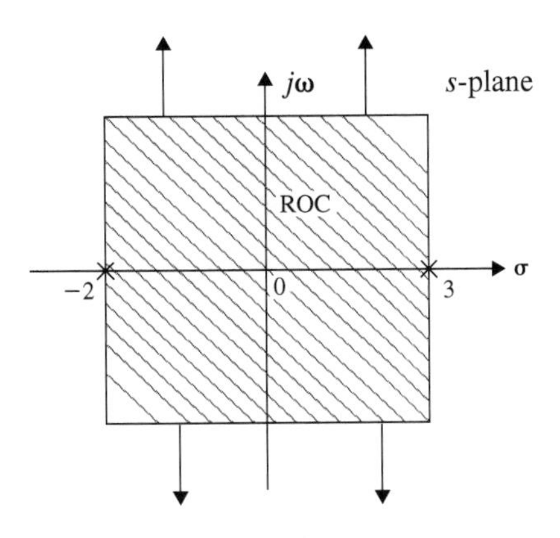

Fig. 8.43 ROC of $x(t)$

Solution:

$$X(s) = \frac{10}{(s+4)(s-2)}$$
$$= \frac{A_1}{(s+4)} + \frac{A_2}{s-2}$$
$$10 = A_1(s-2) + A_2(s+4)$$

Put $s = -4$

$$10 = A_1(-4-2)$$
$$A_1 = -\frac{5}{3}$$

Put $s = 2$

$$10 = A_2(2+4)$$
$$A_2 = \frac{5}{3}$$
$$X(s) = \frac{5}{3}\left[\frac{-1}{s+4} + \frac{1}{s-2}\right]$$

(a) ROC > 2.
Here the ROC is right sided for both the poles at $s = -4$ and $s = 2$. Hence, the system is causal.

$$\boxed{x(t) = \frac{5}{3}[-e^{-4t} + e^{2t}]u(t)}$$

(b) ROC Re $s < -4$ (Fig. 8.44).

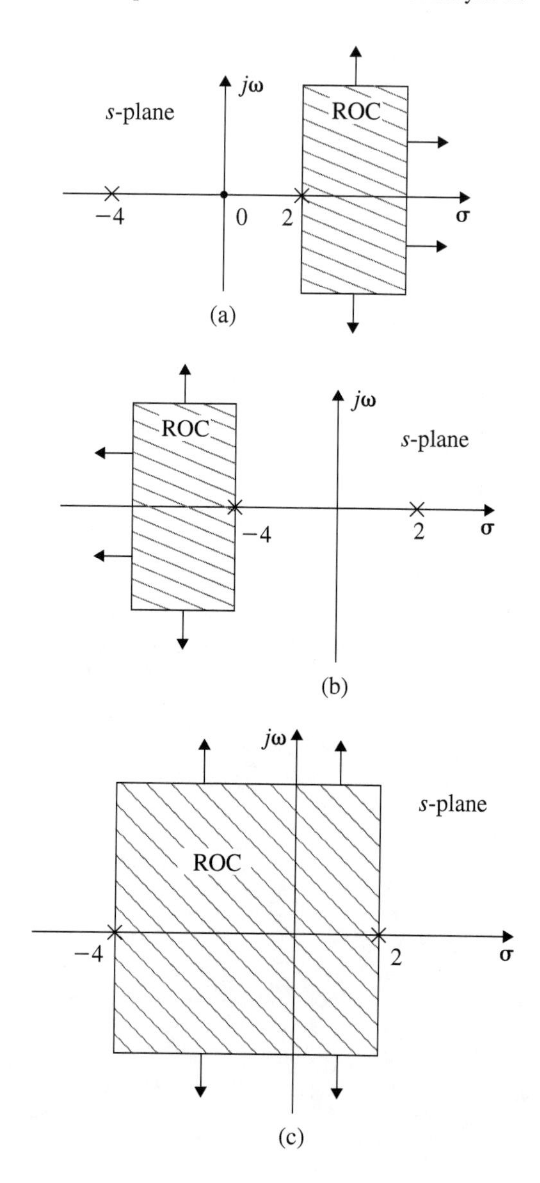

Fig. 8.44 ROCs of $X(s) = \dfrac{10}{(s+4)(s-2)}$

Here the system poles $s = -4$ and $s = 2$ are both left sided since they lie left to the ROC. Both are non-causal.

$$X(s) = \frac{5}{3}\left[\frac{-1}{(s+4)} + \frac{1}{s-2}\right]$$

$$\boxed{x(t) = \frac{5}{3}(e^{-4t} - e^{2t})u(-t)}$$

(c) ROC $-4 < \text{Re } s < 2$.
Here the pole $s = -4$ is to the left of the ROC and it is a right-sided signal. It is, therefore, causal. The pole $s = 2$ is to the right of the ROC, and hence it is a left-sided signal. It is non-causal. Hence

$$\boxed{x(t) = \frac{5}{3}\left[-e^{-4t}u(t) - e^{2t}u(-t)\right]}$$

■ Example 8.78

The impulse response function of a certain system is

$$H(s) = \frac{10}{s-5} \qquad \text{ROC: Re } s < 5$$

The system is excited by $x(t) = e^{-3t}u(t)$. Derive an expression for the output $y(t)$ as a function of time.

Solution:

$$\begin{aligned} H(s) &= \frac{10}{(s-5)} \qquad \text{ROC: Re } s < 5 \\ X(s) &= L^{-1}[e^{-3t}u(t)] \\ &= \frac{1}{(s+3)} \qquad \text{ROC: Re } s > -3 \\ Y(s) &= H(s)X(s) \\ &= \frac{10}{(s-5)(s+3)} \qquad \text{ROC: } -3 < \text{Re } s < 5 \end{aligned}$$

Putting into partial fraction, we get

$$\begin{aligned} Y(s) &= \frac{A_1}{s-5} + \frac{A_2}{s+3} \\ 10 &= A_1(s+3) + A_2(s-5) \end{aligned}$$

Put $s = 5$

$$\begin{aligned} 10 &= A_1(5+3) \\ A_1 &= \frac{5}{4} \end{aligned}$$

Put $s = -3$

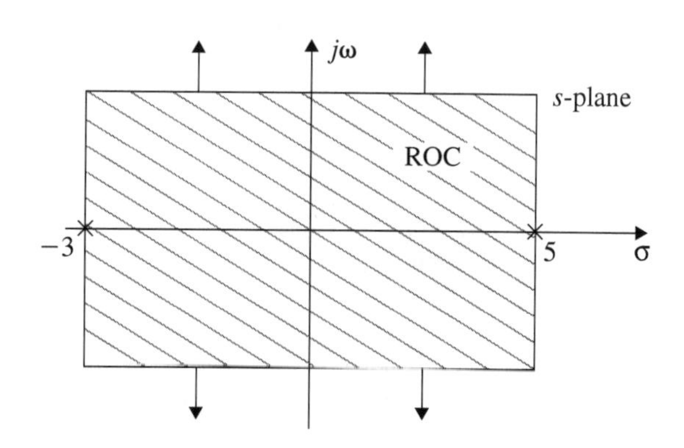

Fig. 8.45 ROC of $Y(s) = \dfrac{10}{(s-5)(s+3)}$

$$10 = A_2(-3-5)$$
$$A_2 = -\frac{5}{4}$$

Hence, $Y(s) = \frac{5}{4}\left(\frac{1}{s-5} - \frac{1}{s+3}\right)$.

The ROC is shown in Fig. 8.45. From the ROC, the pole $\frac{1}{(s-5)}$ is left sided (right to the ROC) and the pole $\frac{1}{(s+3)}$ is right sided (left to the ROC). Hence, $\frac{1}{(s-5)}$ is non-causal and $\frac{1}{(s+3)}$ is causal. $y(t)$ is obtained by taking inverse LT.

$$\boxed{y(t) = \frac{5}{4}\left(-e^{5t}u(-t) - e^{-3t}u(t)\right)}$$

Example 8.79

The impulse response function of a certain system is given by

$$H(s) = \frac{1}{(s+10)} \qquad \text{ROC: Re } s > -10$$

The system is excited by the following input.

$$x(t) = -2e^{-2t}u(-t) - 3e^{-3t}u(t)$$

Derive an expression for the output $y(t)$ as a function of time.

Solution: By taking LT for $x(t)$, we get

$$X(s) = L[-2e^{-2t}u(-t) - 3e^{-3t}u(t)]$$
$$= \frac{2}{(s+2)} - \frac{3}{(s+3)} \qquad \text{ROC: } -3 < \text{Re } s < -2$$

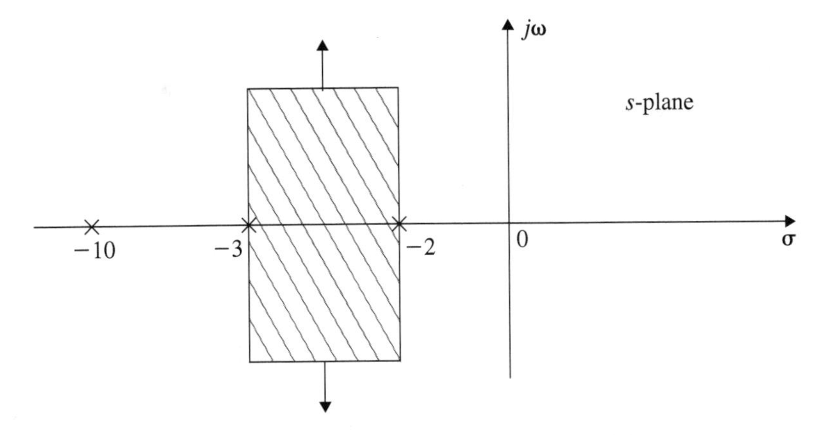

Fig. 8.46 ROC of Example 8.87

$$= \frac{2s + 6 - 3s - 6}{(s+2)(s+3)}$$
$$= \frac{-s}{(s+2)(s+3)}$$

$$H(s) = \frac{1}{(s+10)} \qquad \text{ROC: Re } s > -10$$
$$Y(s) = \frac{-s}{(s+2)(s+3)(s+10)} \qquad \text{ROC: } -3 < \text{Re } s < -2$$

The ROC of $Y(s)$ is shown in Fig. 8.46. The ROC of $H(s)$ is automatically satisfied if ROC Re $s > -3$. Putting $Y(s)$ into partial fraction, we get

$$Y(s) = \frac{A_1}{(s+2)} + \frac{A_2}{(s+3)} + \frac{A_3}{(s+10)}$$
$$-s = A_1(s+3)(s+10) + A_2(s+2)(s+10) + A_3(s+2)(s+3)$$

Put $s = -2$

$$2 = A_1(-2+3)(-2+10)$$
$$A_1 = \frac{1}{4}$$

Put $s = -3$

$$3 = A_2(-3+2)(-3+10)$$
$$A_2 = -\frac{3}{7}$$

Put $s = -10$

$$10 = A_3(-10+2)(-10+3)$$
$$A_3 = \frac{5}{28}$$

Hence

$$Y(s) = \frac{1}{4}\frac{1}{(s+2)} - \frac{3}{7}\frac{1}{(s+3)} + \frac{5}{28}\frac{1}{(s+10)}$$

From Fig. 8.46, it is evident that the pole $\frac{1}{(s+10)}$ of the system and the pole $\frac{1}{(s+3)}$ of the input are right sided (to the left of ROC), and hence causal. On the other hand, the pole $\frac{1}{(s+2)}$ is left sided (right to the ROC), and hence non-causal. Thus, $y(t)$ is obtained by taking inverse LT.

$$\boxed{y(t) = -\frac{1}{4}e^{-2t}u(-t) - \frac{3}{7}e^{-3t}u(t) + \frac{5}{28}e^{-10t}u(t)}$$

Example 8.80

The impulse response of a certain system is given by $h(t) = \delta(t) + e^{-3|t|}$. The system is excited by the following signal $x(t) = e^{-4t}u(t) + e^{-2t}u(-t)$. Find the response of the system $y(t)$.

Solution:

$$\begin{aligned} H(s) &= L[h(t)] \\ &= L(\delta(t)) + L(e^{-3|t|}) \\ &= 1 + L(e^{-3|t|}) \end{aligned}$$

$$\begin{aligned} L(e^{-3|t|}) &= \int_{-\infty}^{0^-} e^{+3t}e^{-st}dt + \int_{0^-}^{\infty} e^{-3t}e^{-st}dt \\ &= -\frac{1}{(s-3)} + \frac{1}{s+3} \end{aligned}$$

$$\begin{aligned} H(s) &= 1 - \frac{1}{(s-3)} + \frac{1}{(s+3)} \qquad \text{ROC: } -3 < \text{Re } s < 3 \\ &= \frac{(s^2-15)}{(s-3)(s+3)} \end{aligned}$$

$$\begin{aligned} X(s) &= L[e^{-4t}u(t) + e^{-2t}u(-t)] \\ &= \frac{1}{(s+4)} - \frac{1}{s+2} \\ &= \frac{-2}{(s+2)(s+4)} \qquad \text{ROC: } -4 < \text{Re } s < -2 \end{aligned}$$

$$\begin{aligned} H(s) &= \frac{Y(s)}{X(s)} \\ Y(s) &= H(s)X(s) \\ &= \frac{(s^2-15)(-2)}{(s-3)(s+3)(s+2)(s+4)} \\ &= \frac{A_1}{s-3} + \frac{A_2}{s+3} + \frac{A_3}{s+2} + \frac{A_4}{(s+4)} \end{aligned}$$

$$\begin{aligned} (15-s^2)2 = {} & A_1(s+3)(s+2)(s+4) + A_2(s-3)(s+2)(s+4) \\ & + A_3(s-3)(s+3)(s+4) + A_4(s-3)(s+3)(s+2) \end{aligned}$$

Put $s = 3$

$$\begin{aligned} 2(15-9) &= A_1(6)(5)(7) \\ A_1 &= \frac{2}{35} \end{aligned}$$

Put $s = -3$

$$\begin{aligned} 2(15-9) &= A_2(-6)(-1)(1) \\ A_2 &= 2 \end{aligned}$$

Put $s = -2$

$$\begin{aligned} 2(15-4) &= A_3(-5)(1)(2) \\ A_3 &= -\frac{11}{5} \end{aligned}$$

Put $s = -4$

$$\begin{aligned} 2(15-16) &= A_4(-7)(-1)(-2) \\ A_4 &= \frac{1}{7} \\ Y(s) &= \frac{2}{35}\frac{1}{s-3} + \frac{2}{s+3} - \frac{11}{5}\frac{1}{(s+2)} + \frac{1}{7}\frac{1}{(s+4)} \end{aligned}$$

The ROC for $Y(s)$ is shown in Fig. 8.47. From Fig. 8.47, the poles $\frac{1}{(s+4)}$ and $\frac{1}{(s+3)}$ are right sided, and hence causal. However, the poles $\frac{1}{(s+2)}$ and $\frac{1}{(s-3)}$ are left sided, and hence non-causal. Taking the ROC into account $y(t)$ is obtained as given below

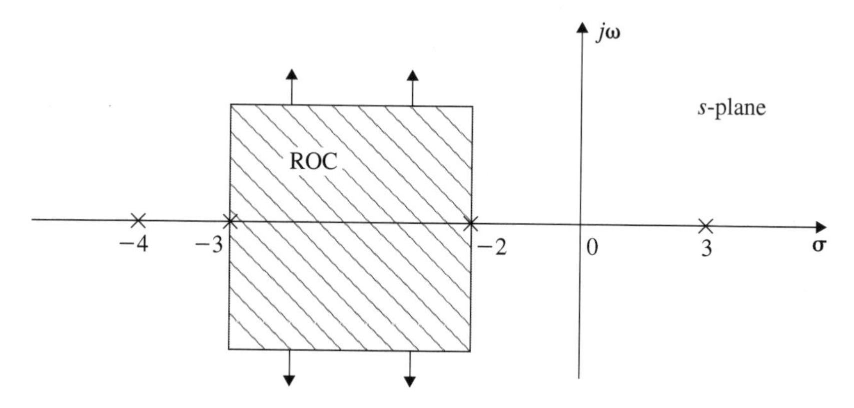

Fig. 8.47 ROC of Example 8.88

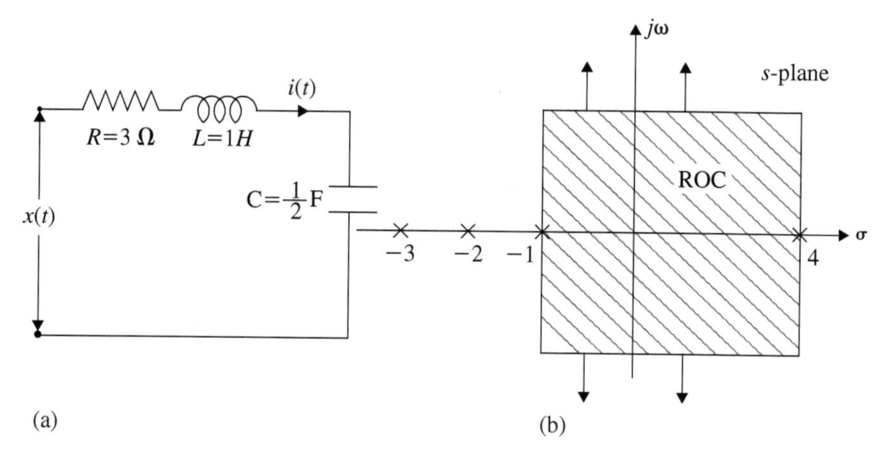

Fig. 8.48 **a** R.L.C Circuit; **b** ROC of Example 8.89

$$y(t) = \left[\left(2e^{-3t} + \frac{1}{7}e^{-4t}\right)u(t) + \left(-\frac{2}{35}e^{3t} + \frac{11}{5}e^{-2t}\right)u(-t)\right]$$

■ Example 8.81

Consider the R.L.C series circuit shown in Fig. 8.48a. The excitation voltage $x(t) = e^{-3t}u(t) + e^{4t}u(-t)$. Derive the expression for the current in the series circuit. Assume zero initial conditions.

Solution:

1. The impedance of the R.L.C circuit is

$$\begin{aligned} Z(s) &= R + Ls + \frac{1}{Cs} \\ &= 3 + s + \frac{2}{s} \\ &= \frac{(s^2+3s+2)}{s} = \frac{(s+1)(s+2)}{s} \end{aligned}$$

2. The excitation voltage

$$\begin{aligned} x(t) &= e^{-3t}u(t) + e^{4t}u(-t) \\ X(s) &= \frac{1}{(s+3)} - \frac{1}{(s-4)} \\ &= \frac{-7}{(s+3)(s-4)} \qquad \text{ROC: } -3 < \text{Re}\, s < 4 \end{aligned}$$

3. The current flowing in the circuit is

$$\begin{aligned} I(s) &= \frac{X(s)}{Z(s)} \\ I(s) &= \frac{-7s}{(s+1)(s+2)(s+3)(s-4)} \end{aligned}$$

The corresponding ROC: $-1 < \text{Re}\, s < 4$. The above ROC satisfies the previous ROC also.

$$\begin{aligned} I(s) &= \frac{A_1}{(s+1)} + \frac{A_2}{(s+2)} + \frac{A_3}{(s+3)} + \frac{A_4}{(s-4)} \\ -7s &= A_1(s+2)(s+3)(s-4) + A_2(s+1)(s+3)(s-4) \\ &\quad + A_3(s+1)(s+2)(s-4) + A_4(s+1)(s+2)(s+3) \end{aligned}$$

Put $s = -1$

$$\begin{aligned} 7 &= A_1(-1+2)(-1+3)(-1-4) \\ A_1 &= -\frac{7}{10} \end{aligned}$$

Put $s = -2$

$$\begin{aligned} 14 &= A_2(-1)(1)(-6) \\ A_2 &= \frac{7}{3} \end{aligned}$$

Put $s = -3$

$$21 = A_3(-2)(-1)(-7)$$
$$A_3 = -\frac{3}{2}$$

Put $s = 4$

$$-28 = A_4(5)(6)(7)$$
$$A_4 = -\frac{2}{15}$$
$$I(s) = \frac{-7}{10}\frac{1}{(s+1)} + \frac{7}{3}\frac{1}{s+2} - \frac{3}{2}\frac{1}{(s+3)} - \frac{2}{15}\frac{1}{(s-4)}$$

The poles $\frac{1}{s+1}$, $\frac{1}{s+2}$ and $\frac{1}{s+3}$ are right sided as seen in ROC of Fig. 8.48b. The pole $\frac{1}{s-4}$ is left sided, and hence non-causal. Taking inverse LT for $I(s)$, we get

$$\boxed{i(t) = \left(\frac{-7}{10}e^{-t} + \frac{7}{3}e^{-2t} - \frac{3}{2}e^{-3t}\right)u(t) + \frac{2}{15}e^{4t}u(-t)}$$

8.16 System Realization

Realization of system transfer function is a synthesis problem and it is realized in many ways and is not unique. Consider the most general form of transfer function which is given below

$$H(s) = \frac{b_0 s^n + b_1 s^{n-1} + b_2 s^{n-2} + \ldots + b_{n-1}s + b_n}{s^n + a_1 s^{n-1} + a_2 s^{n-2} + \ldots + a_{n-1}s + a_n} \tag{8.72}$$

The transfer function $H(s)$ is expressed in terms of numerator polynomial and denominator polynomial. **The denominator polynomial is a monic polynomial (polynomial with the coefficient of s^n is unity).** The above transfer function is realized using integrators or differentiators together with adders and multipliers. For system realization, the differentiators are not used because they enhance the noise level. The symbols used to represent an ideal integrator, adder and multiplier are shown in Fig. 8.49a–c, respectively.

The following methods of realization are described here:

- Direct Form-I
- Direct Form-II
- Cascade Form
- Parallel Form
- Transposed Form

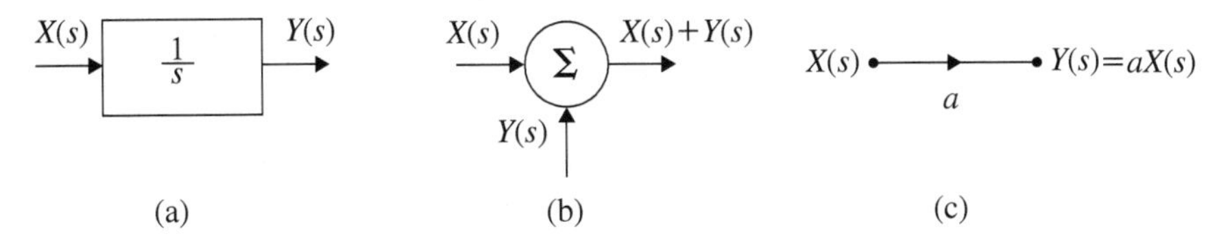

Fig. 8.49 **a** Pure intergrator. **b** Adder. **c** Multiplier. Symbols for integrator, adder and summer

8.16.1 Direct Form-I Realization

Consider the transfer function with third degree polynomial in the numerator and denominator. Thus, Eq. (8.72) is written as follows:

$$H(s) = \frac{b_0 s^3 + b_1 s^2 + b_2 s + b_3}{s^3 + a_1 s^2 + a_2 s + a_3} \tag{8.73}$$

$H(s)$ can be expressed as given below after dividing the numerator and denominator by s^3.

$$H(s) = H_1(s)H_2(s)$$

where

$$\begin{aligned} H_1(s) &= \left(b_0 + \frac{b_1}{s} + \frac{b_2}{s^2} + \frac{b_3}{s^3}\right) \\ H_2(s) &= \frac{1}{\left(1 + \frac{a_1}{s} + \frac{a_2}{s^2} + \frac{a_3}{s^3}\right)} \end{aligned} \tag{8.74}$$

Equation (8.74) can be realized with a transfer functions $H_1(s)$ and $H_2(s)$ in cascade as shown in Fig. 8.50.

From Fig. 8.50, the following equation is written.

$$\begin{aligned} W(s) &= H_1(s)X(s) \\ &= \left[b_0 + \frac{b_1}{s} + \frac{b_2}{s^2} + \frac{b_3}{s^3}\right] X(s) \\ Y(s) &= W(s)H_2(s) \end{aligned} \tag{8.75}$$

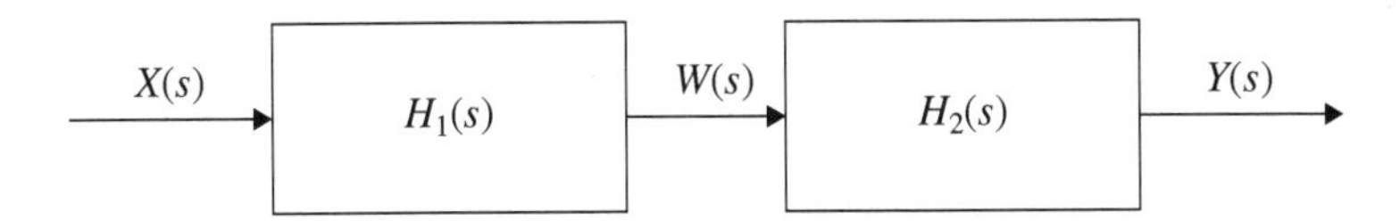

Fig. 8.50 Transfer function in cascade

$$\begin{aligned}
W(s) &= Y(s)\frac{1}{H_2(s)} \\
&= Y(s)\left[1+\frac{a_1}{s}+\frac{a_2}{s^2}+\frac{a_3}{s^3}\right] \\
&= Y(s)+\left(\frac{a_1}{s}+\frac{a_2}{s^2}+\frac{a_3}{s^3}\right)Y(s) \\
Y(s) &= W(s)-\left(\frac{a_1}{s}+\frac{a_2}{s^2}+\frac{a_3}{s^3}\right)Y(s)
\end{aligned} \tag{8.76}$$

Now we realize $H_1(s)$ which is given by Eq. (8.75) and is described below

1. $X(s)$ is multiplied by b_0,
2. $X(s)$ is integrated and multiplied by b_1,
3. $X(s)$ is integrated twice $\left(\frac{1}{s^2}\right)$ and multiplied by b_2,
4. $X(s)$ is integrated thrice $\left(\frac{1}{s^3}\right)$ and multiplied by b_3,
5. Steps 1, 2, 3, and 4 are sent through summers and $W(s)$ is realized. This is shown in Fig. 8.51a.

Realization of $H_2(s)$ is described below. Consider Eq. (8.76).

1. $W(s)$ is taken as reference,
2. $Y(s)$ is integrated once and multiplied by $-a_1$,
3. $Y(s)$ is integrated twice and multiplied by $-a_2$,
4. $Y(s)$ is integrated thrice and multiplied by $-a_3$,
5. Steps 1, 2, 3 and 4 are added through summers to get $Y(s)$ as per Eq. (8.76).

Thus, $H_2(s)$ is realized and is shown in Fig. 8.51b.

Now

$$H(s) = H_1(s)H_2(s)$$

Figure 8.51a, b are connected in cascade at point $W(s)$ where $W(s)$ is the output of $H_1(s)$ and input of $H_2(s)$. The cascaded Direct Form-I, with third degree polynomial of $H(s)$ is shown in Fig. 8.51c, which is the combination of Fig. 8.51a, b.

The above procedure can be generalized to realize the transfer function $H(s)$ of an nth order system. This is shown in Fig. 8.51d.

The following examples illustrate Direct Form-I realization.

Example 8.82

Consider the following transfer function

$$H(s) = \frac{(4s^2+6s+14)}{(2s^2+5s+8)}$$

Realize Direct Form-I structure.

Solution: The denominator polynomial is not a monic polynomial. Therefore, divide the numerator and denominator by a factor 2. The given transfer is written as follows:

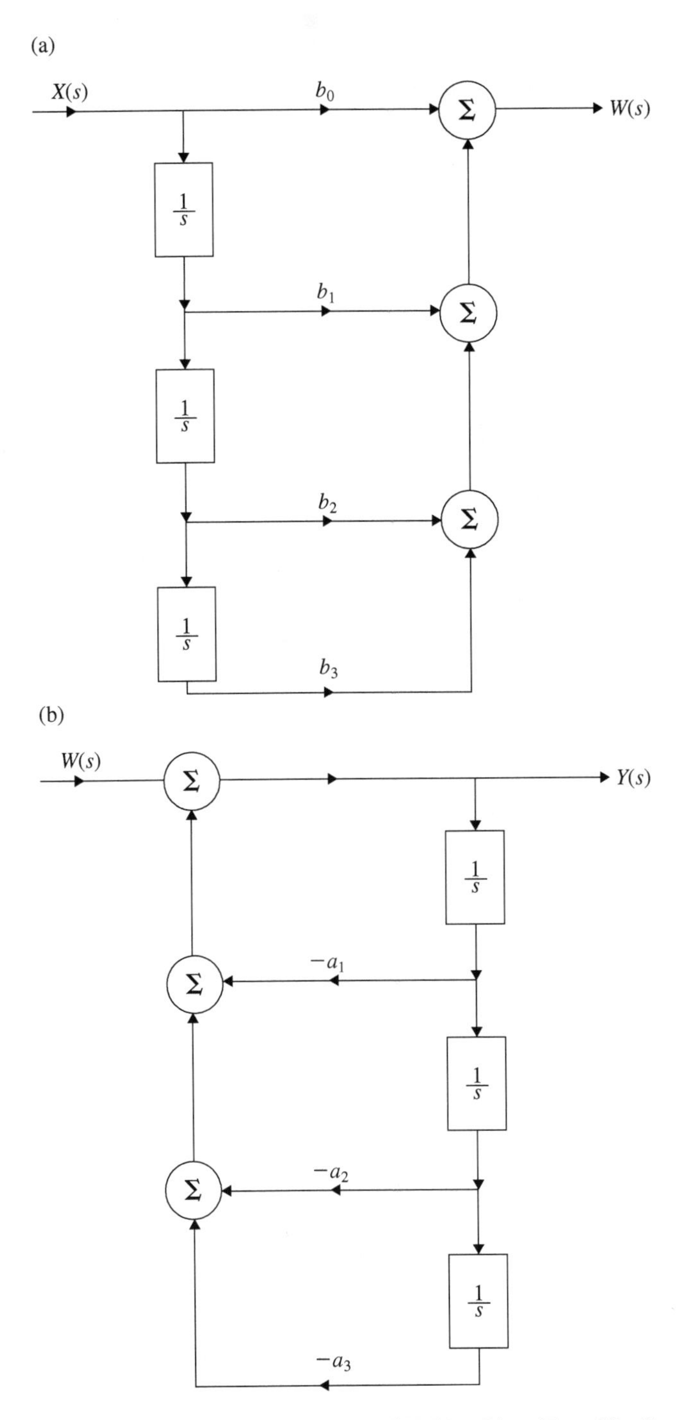

Fig. 8.51 **a** Realization of $H_1(s)$ and **b** Realization of $H_2(s)$. **c** Direct Form-I Realization of third order system. **d** Direct Form-I Realization of *n*th order system

(c)

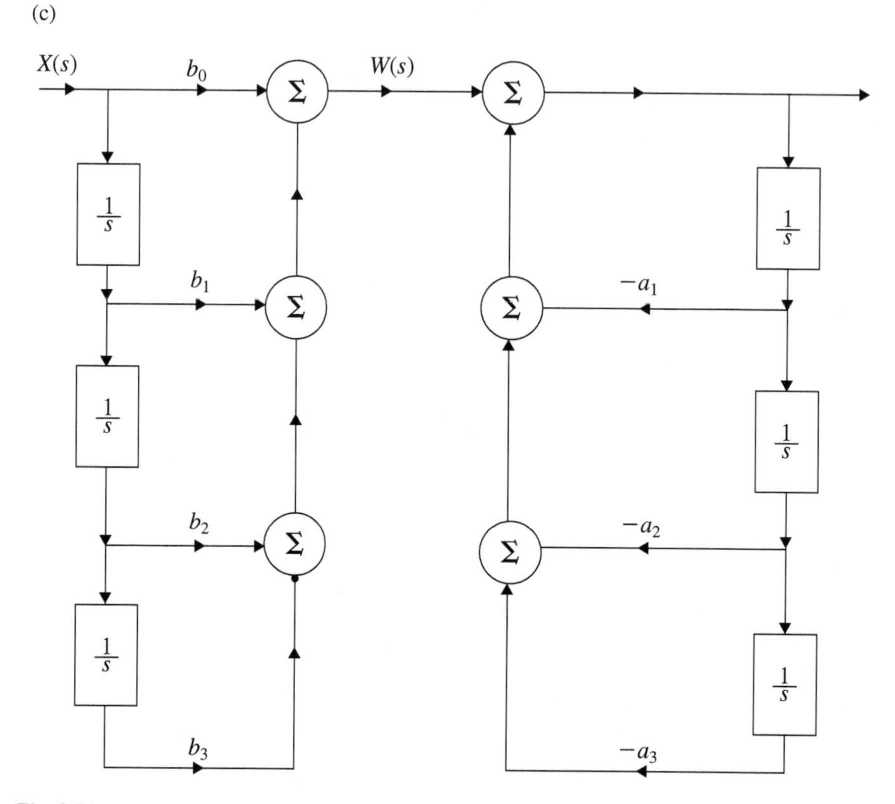

Fig. 8.51 (continued)

$$H(s) = \frac{(2s^2 + 3s + 7)}{s^2 + \frac{5}{2}s + 4}$$

Here $b_0 = 2$; $b_1 = 3$; $b_2 = 7$; $a_1 = \frac{5}{2}$ and $a_2 = 4$.

These values of $H(s)$ are substituted in Fig. 8.51d and is shown in Fig. 8.52. For a second order system, there should be two integrators on each side which amounts to a total of 4 integrators.

■ Example 8.83

Consider the following differential equation

$$\frac{d^2y(t)}{dt^2} + \frac{5dy(t)}{dt} + 4y(t) = \frac{dx(t)}{dt}$$

Realize Direct Form-I structure.

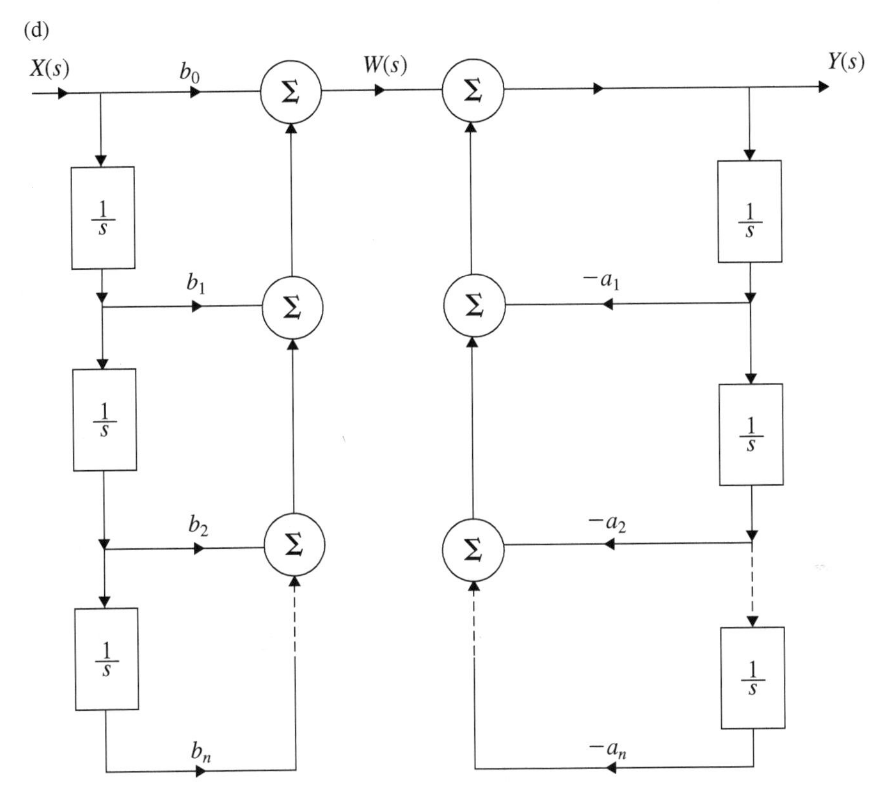

Fig. 8.51 (continued)

Solution: Taking LT on both sides of the given differential equation, we get

$$(s^2 + 5s + 4)Y(s) = sX(s)$$

$$H(s) = \frac{Y(s)}{X(s)} = \frac{s}{(s^2 + 5s + 4)}$$

Here $H_1(s) = s$ and $H_2(s) = \frac{1}{(s^2+5s+4)}$; $b_0 = 0$; $b_1 = 1$ and $b_2 = 0$; $a_1 = 5$ and $a_2 = 4$.

To realize $H_1(s)$ one integrator is needed. To realize $H_2(s)$ two integrators are needed. The Direct Form-I structure is shown in Fig. 8.53.

From the examples illustrated above, it is clear that for a transfer function with numerator polynomial and denominator polynomial of nth degree, Direct Form-I realization requires $2n$ integrators which increases the cost. To avoid this Direct Form-II, structure is suggested and is discussed below.

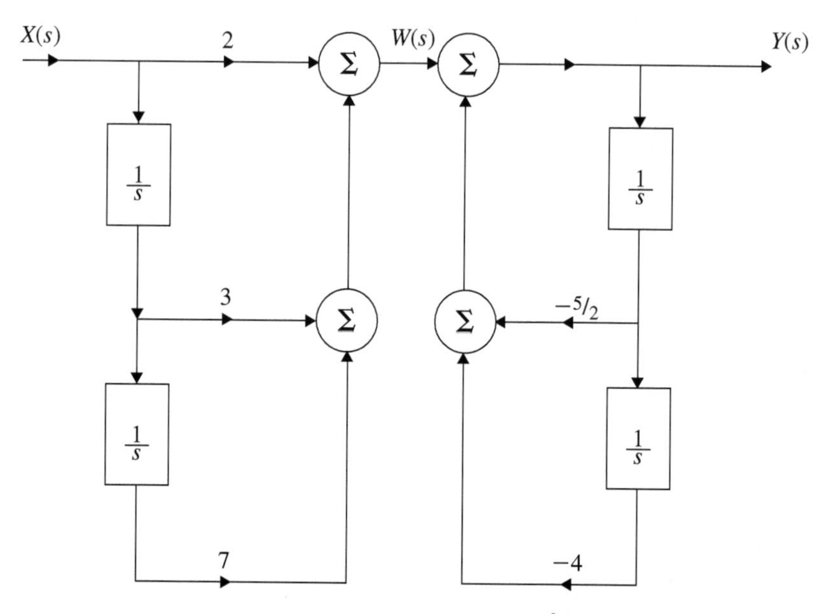

Fig. 8.52 Realization of Direct Form-I structure for $H(s) = \dfrac{(4s^2 + 6s + 14)}{(2s^2 + 5s + 8)}$

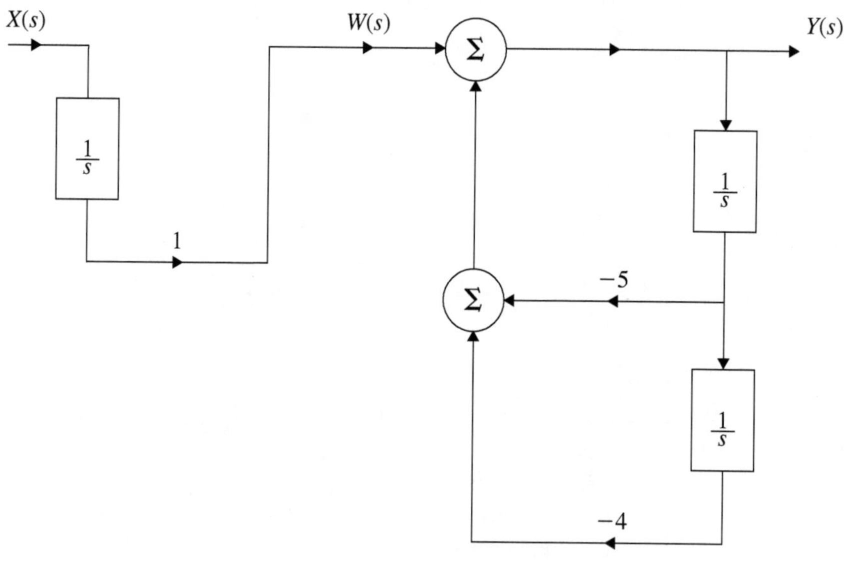

Fig. 8.53 Direct Form-I realization of $H(s) = \dfrac{s}{(s^2 + 5s + 4)}$

8.16.2 Direct Form-II Realization

In Direct Form-I, realization $H_1(s)$, which represents the zeros of $H(s)$, is represented first followed by $H_2(s)$, which represents the poles of $H(s)$. In Director Form-II, the process is reversed. Here $H_2(s)$, which represents the poles is realized first (left half section) followed by the realization of $H_1(s)$, which represents the zeros of $H(s)$. $H_1(s)$ is realized in the right half section. This is shown as a block diagram in Fig. 8.54.

Consider Eq. (8.76). Let

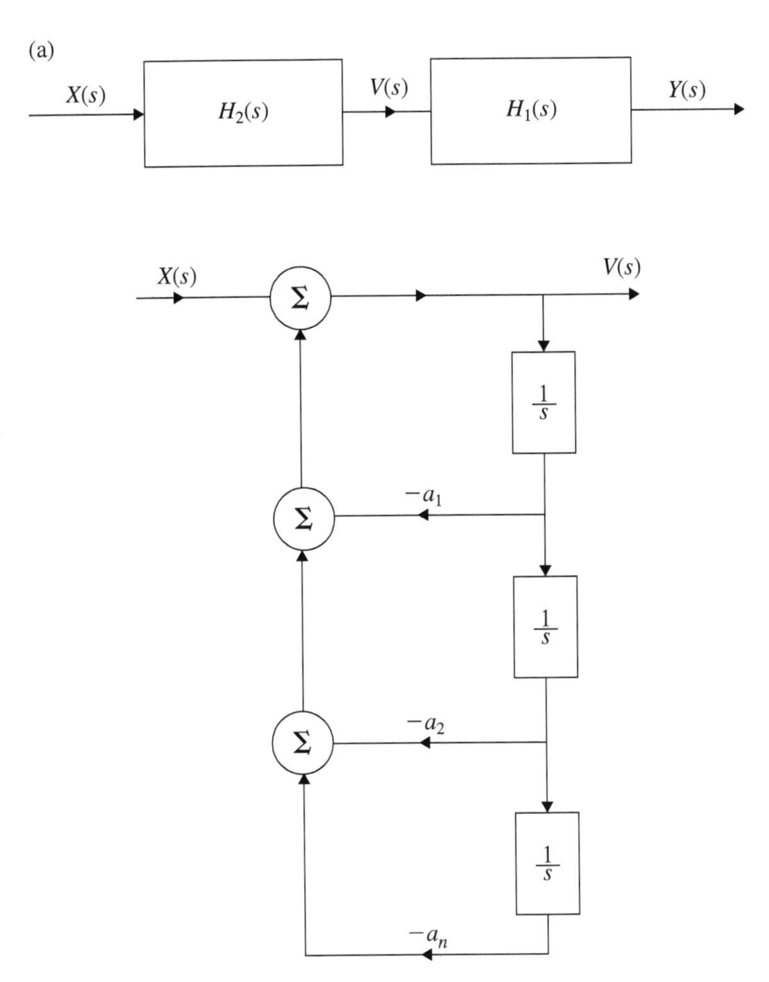

Fig. 8.54 Block diagram representation of Direct Form-II. **a** Realization of $H_2(s) = \dfrac{1}{\left(1 + \dfrac{a_1}{s} + \dfrac{a_2}{s^2} + \ldots + \dfrac{a_n}{s^n}\right)}$. **b** Realization of $H_1(s) = \left(b_0 + \dfrac{b_1}{s} + \ldots + \dfrac{b_n}{s^n}\right)$. **c** Structure realization by Direct Form-II

Fig. 8.54 (continued)

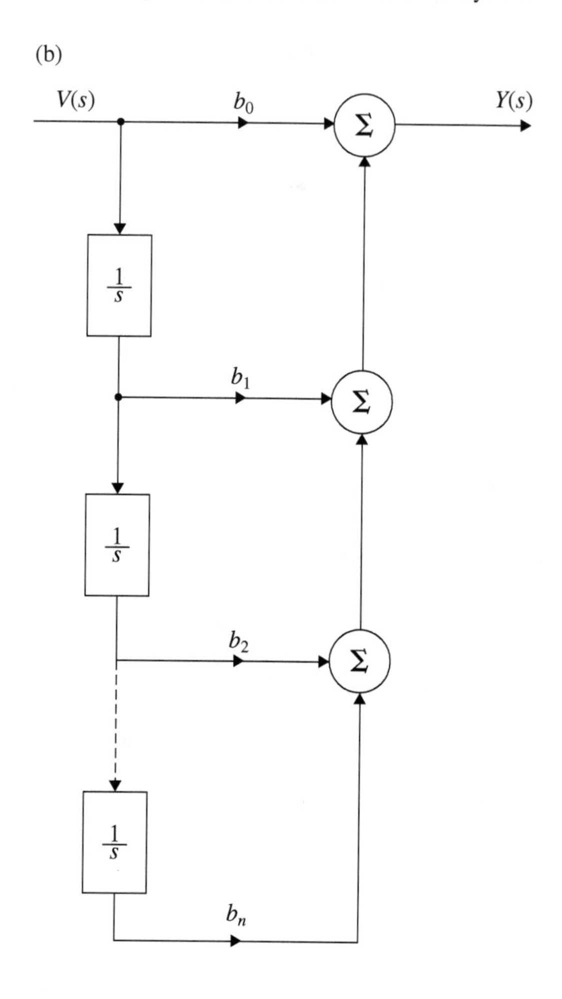

$$H_1(s) = b_0 s^n + b_1 s^{n-1} + \cdots + b_{n-1} s + b_n$$

$$H_2(s) = \frac{1}{s^n + a_1 s^{n-1} + \cdots + a_{n-1} s + a_n}$$

Dividing H_1 and H_2 by s^n we can rewrite $H_1(s)$ and $H_2(s)$ as follows:

$$H_1(s) = b_0 + \frac{b_1}{s} + \frac{b_2}{s^2} + \cdots + \frac{b_n}{s^n}$$

$$H_2(s) = \frac{1}{1 + \frac{a_1}{s} + \frac{a_2}{s^2} + \cdots + \frac{a_n}{s^n}}$$

From Fig. 8.54

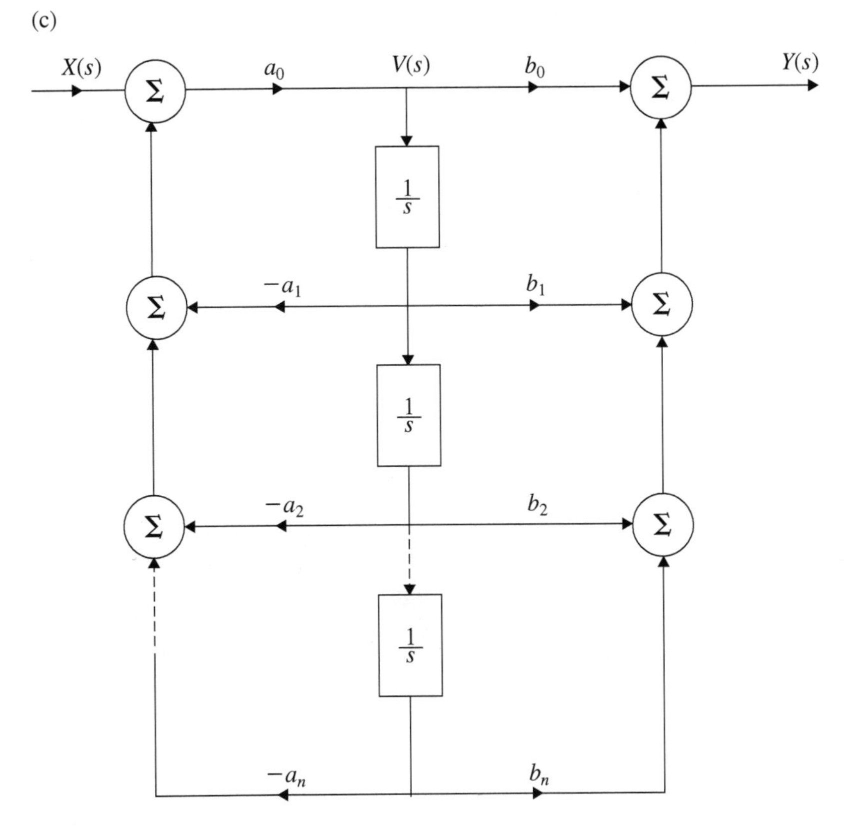

Fig. 8.54 (continued)

$$
\begin{aligned}
V(s) &= H_2(s)X(s) \\
&= \frac{X(s)}{\left[1+\left(\frac{a_1}{s}+\frac{a_2}{s^2}+\ldots+\frac{a_n}{s^n}\right)\right]} \\
& V(s)\left[1+\left(\frac{a_1}{s}+\frac{a_2}{s^2}+\ldots+\frac{a_n}{s^n}\right)\right] = X(s)
\end{aligned}
$$

$$
V(s) = X(s) - \left(\frac{a_1}{s}+\frac{a_2}{s^n}+\ldots+\frac{a_n}{s^n}\right)V(s) \tag{8.77}
$$

Equation (8.77) is realized as shown in Fig. 8.54a.

Now consider

$$
H_1(s) = b_0 + \frac{b_1}{s} + \ldots + \frac{b_n}{s^n}
$$

$$
Y(s) = \left(b_0 + \frac{b_1}{s} + \ldots + \frac{b_n}{s^n}\right)V(s) \tag{8.78}
$$

Equation (8.79) is realized and is shown in Fig. 8.54b.

From Fig. 8.54a and b, it is evidenced that they can be connected in a cascade where $V(s)$ is common. This is shown in Fig. 8.54c.

From Fig. 8.54c, it is clear that, for an nth degree polynomial, either in the numerator or in the denominator, the realization requires only n integrators, whereas in Direct Form-I, structure $2n$ integrators are necessary. For an nth order system, n integrators will suffice for structure realization in Direct Form-II. **Such a realization is said to be canonic realization where the number of integrators required is equal to the order of the system.** Direct Form-I realization is not canonic.

■ Example 8.84

Draw the Direct Form-II realization of the system described by the following differential equation.

$$\frac{d^2y(t)}{dt^2} + 5\frac{dy(t)}{dt} + 4y(t) = \frac{dx(t)}{dt}$$

(*Anna University, May, 2005*)

Solution: Taking LT on both sides of the given differential equation, we get

$$(s^2 + 5s + 4)Y(s) = sX(s)$$

$$H(s) = \frac{Y(s)}{X(s)} = \frac{s}{(s^2 + 5s + 4)}$$

Since the order of the system is two, we require two integrators. In $H(S)$, $b_0 = 0$; $b_1 = 1$; $b_2 = 0$; $a_1 = 5$ and $a_2 = 4$. Direct Form-II structure is shown in Fig. 8.55.

■ Example 8.85

Find the transfer function of LTI system described by the differential equation

$$\frac{d^2y(t)}{dt^2} + 3\frac{dy(t)}{dt} + 2y(t) = 2\frac{dx(t)}{dt} - 3x(t)$$

Also realize the Direct Form-II structure.

Solution: By taking LT on both sides of the given differential equation, we get

$$H(s) = \frac{Y(s)}{X(s)} = \frac{(2s - 3)}{(s^2 + 3s + 2)}$$

Here two integrators are required. For the given $H(s)$, $b_0 = 0$; $b_1 = 2$; $b_2 = -3$ $a_1 = 3$ and $a_2 = 2$.

The Direct Form-II structure is shown in Fig. 8.56.

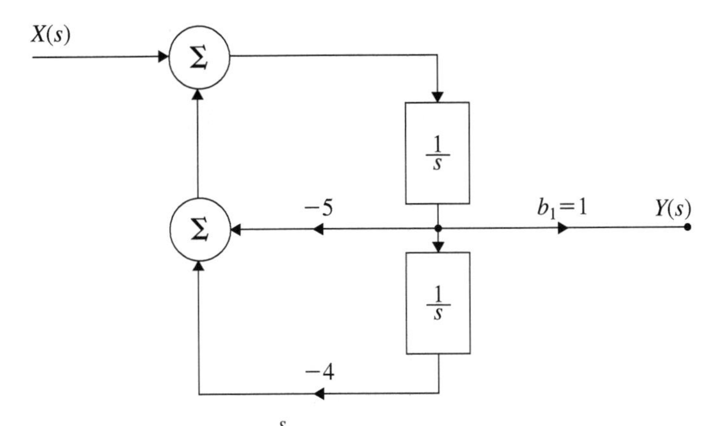

Fig. 8.55 Realization of $H(s) = \dfrac{s}{s^2 + 5s + 4}$ by DF-II

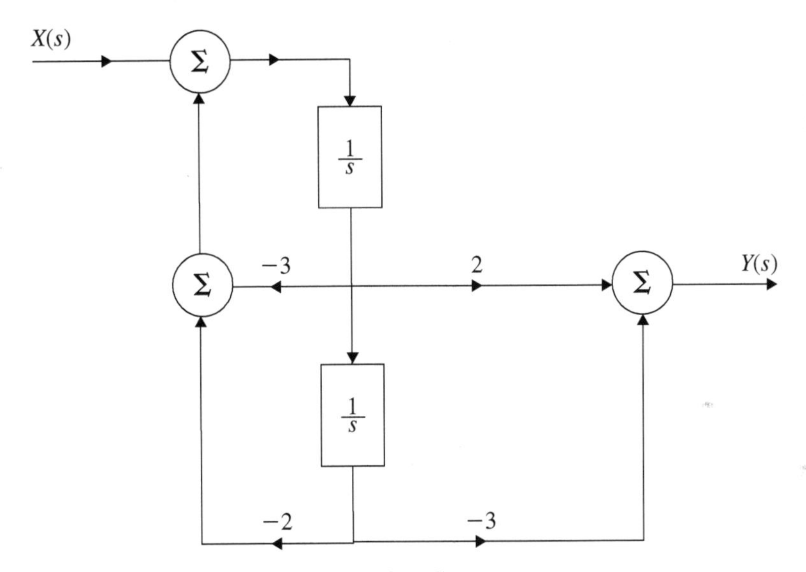

Fig. 8.56 Structure realization of $H(s) = \dfrac{(2s - 3)}{(s^2 + 3s + 2)}$ by DF-II

■ Example 8.86

Realize the following differential equation as a Direct Form-II structure.

$$\frac{d^3y(t)}{dt^3} + 4\frac{d^2y(t)}{dt^2} + 7\frac{dy(t)}{dt} + 8y(t) = 5\frac{d^2x(t)}{dt} + 4\frac{dx(t)}{dt} + 7x(t)$$

(Anna University, June, 2007)

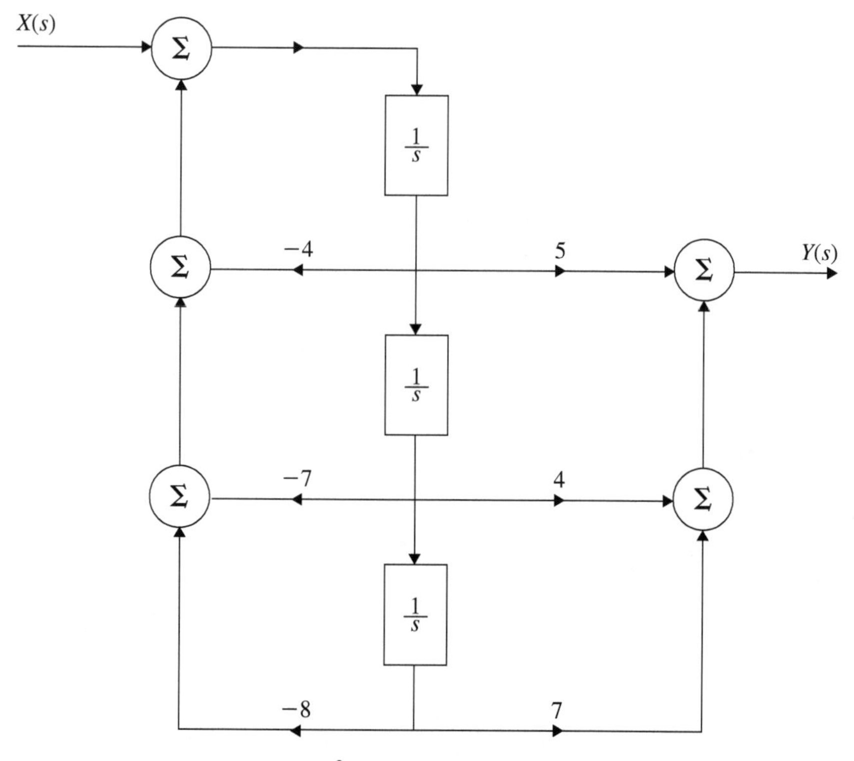

Fig. 8.57 Realization of $H(s) = \dfrac{(5s^2 + 4s + 7)}{s^3 + 4s^2 + 7s + 8}$ by DF-II

Solution: Taking LT on both sides of the given differential equation, we get

$$H(s) = \frac{Y(s)}{X(s)} = \frac{(5s^2 + 4s + 7)}{(s^3 + 4s^2 + 7s + 8)}$$

This is a third order system, and therefore, three integrators are required. For the given $H(s)$, $b_0 = 0$; $b_1 = 5$; $b_2 = 4$; $b_3 = 7$; $a_1 = 4$; $a_2 = 7$ and $a_3 = 8$.

The Direct Form-II structure is shown in Fig. 8.57.

8.16.3 Cascade Form Realization

The nth order transfer function can be expressed as the product of n first order transfer functions. Each transfer function is realized in a Direct Form-II structure and connected in cascade to obtain the realization of total transfer function. Consider the following T.F. $H(s)$ which can be expressed as the product of n first order T.F $H_1(s)$,

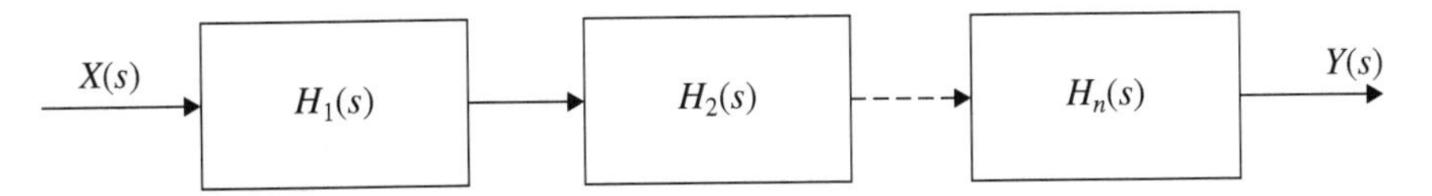

Fig. 8.58 Representation of $H(s)$ in block diagram

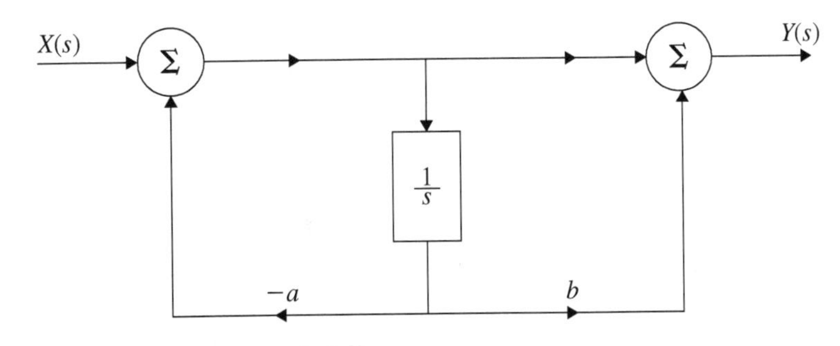

Fig. 8.59 Realization of $H_1(s) = \dfrac{(s+b)}{(s+a)}$

$H_2(s), \ldots, H_n(s)$. Thus, the following equation can be written.

$$H(s) = H_1(s)H_2(s)H_3(s), \ldots, H_n(s) \tag{8.79}$$

Equation (8.79) is represented in Fig. 8.58.

Now, consider $H_1(s)$ in the following form.

$$H_1(s) = \frac{(s+b)}{(s+a)}$$

The above T.F. can be realized by Direct Form-II with a single integrator. This is shown in Fig. 8.59.

On a similar line, $H_2(s), \ldots, H_n(s)$ can be realized and they can be connected in cascade. The following examples illustrate the cascade method of structure realization.

Example 8.87

Consider the following T.F

$$H(s) = \frac{(3s+17)}{(s^3+9s^2+23s+15)}$$

realize the structure in cascade form.

Solution: For the realization cascade structure, the given transfer function should be expressed in terms of poles and zeros. Thus, the denominator polynomial can be

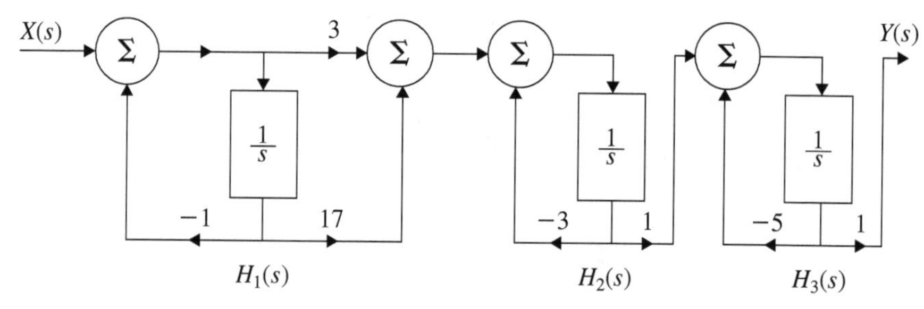

Fig. 8.60 Realization of $H(s) = \dfrac{(3s+17)}{(s^3+9s^2+23s+15)}$ by cascade connection

factorized and put in the following form

$$(s^3 + 9s^2 + 23s + 15) = (s + 1)(s + 3)(s + 5)$$

The given T.F. is written in the following form

$$H(s) = \frac{(3s + 17)}{(s + 1)(s + 3)(s + 5)}$$

Let

$$H_1(s) = \frac{(3s + 17)}{(s + 1)}$$

$$H_2(s) = \frac{1}{(s + 3)}$$

$$H_3(s) = \frac{1}{(s + 5)}$$

$H_1(s)$, $H_2(s)$ and $H_3(s)$ are realized by Direct Form-II as shown in Fig. 8.60. It is to be noted here that in cascade form there are different ways of grouping the factors in the numerator and denominator and hence the realization is not unique.

8.16.4 *Parallel Structure Realization*

In parallel structure realization, the given T.F. $H(s)$ is expressed as a sum of "n" first order equations. The "n" poles are realized by Direct Form-II and they are finally summed up to get the final structure. Here $H(s)$ is expressed in the following form.

$$H(s) = \frac{A_1}{(s + p_1)} + \frac{A_2}{(s + p_2)} + \cdots + \frac{A_n}{(s + p_n)} \tag{8.80}$$

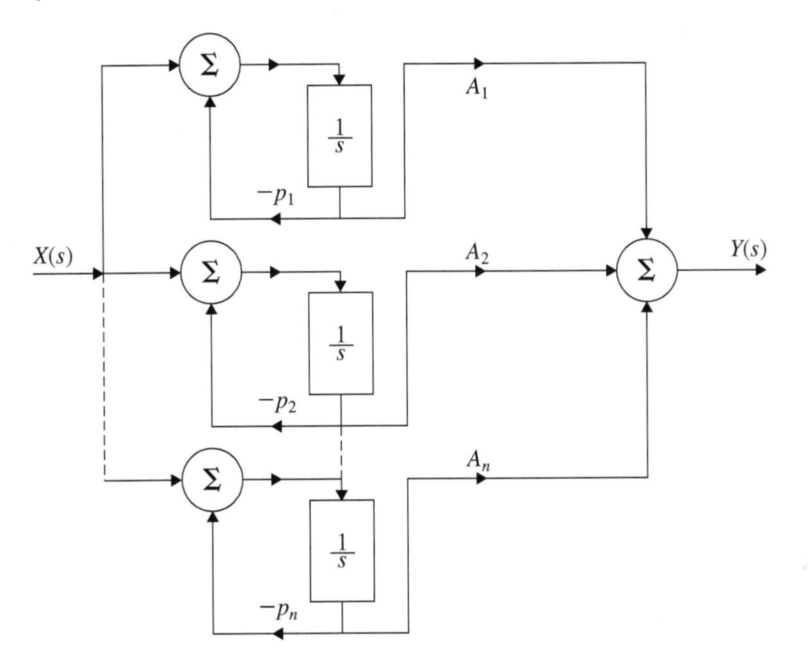

Fig. 8.61 Parallel Realization of $H(s)$

In Eq. (8.79), $A_1, A_2, \ldots, A_n$ are the residues of $H(s)$. $\frac{1}{(s+p_1)}, \frac{1}{(s+p_2)}, \ldots, \frac{1}{(s+p_n)}$ are individually realized by Direct Form-II and summed up. Here the poles may be real, complex and repeated. The Parallel connection realization is represented in Fig. 8.61.

The following examples illustrate the parallel realization.

■ Example 8.88

Consider the following T.F.

$$H(s) = \frac{10(s+1)}{s(s+2)(s+3)}$$

Solution: The given T.F. is written in the following form

$$\begin{aligned} H(s) &= \frac{10(s+1)}{s(s+2)(s+3)} \\ &= \frac{A_1}{s} + \frac{A_2}{(s+2)} + \frac{A_3}{(s+3)} \end{aligned}$$

$$10(s+1) = A_1(s+2)(s+3) + A_2(s+3)s + A_3 s(s+2)$$

Put $s = 0$

$$10 = A_1 6$$
$$A_1 = \frac{5}{3}$$

Put $s = -2$

$$10(-2+1) = A_2(-2)(-2+3)$$
$$A_2 = 5$$

Put $s = -3$

$$10(-3+1) = A_3(-3)(-1)$$
$$A_3 = -\frac{20}{3}$$

$$H(s) = \frac{Y(s)}{X(s)} = \frac{5}{3}\frac{1}{s} + 5\frac{1}{(s+2)} - \frac{20}{3}\frac{1}{(s+3)}$$

The parallel realization is shown in Fig. 8.62.

■ Example 8.89 *Complex Poles*

Consider the following T.F, whose poles are complex.

$$H(s) = \frac{10(s+2)}{(s+1)(s^2+2s+2)}$$

realize the T.F by parallel structure.

Solution:

$$(s^2+2s+2) = (s+1+j)(s+1-j)$$

The given T.F. is written in partial fraction form as given below

$$\begin{aligned} H(s) = \frac{Y(s)}{X(s)} &= \frac{10(s+2)}{(s+1)(s+1+j)(s+1-j)} \\ &= \frac{A_1}{(s+1)} + \frac{A_2}{(s+1+j)} + \frac{A_3}{(s+1-j)} \\ 10(s+2) &= A_1(s^2+2s+2) + A_2(s+1)(s+1-j) \\ &\quad + A_3(s+1)(s+1+j) \end{aligned}$$

Put $s = -1$

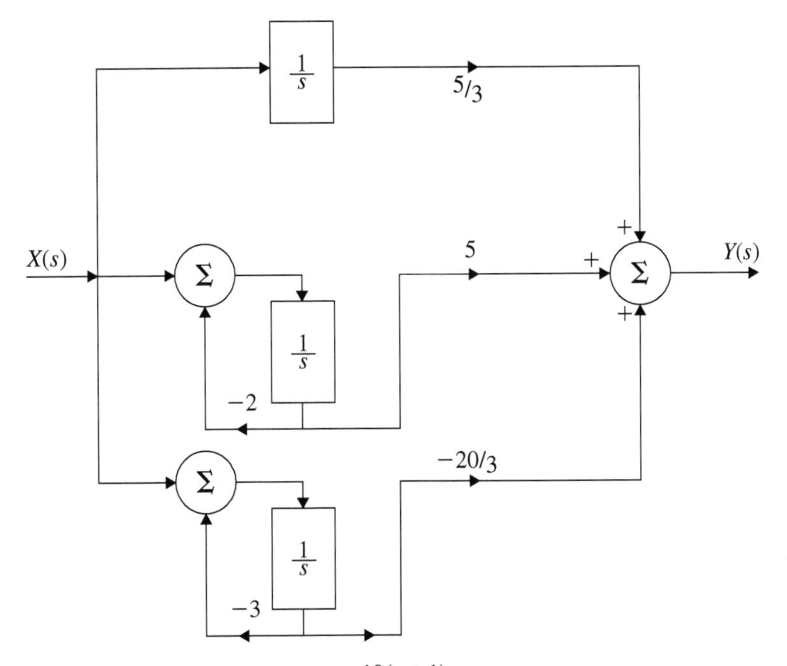

Fig. 8.62 Parallel Realization of $H(s) = \dfrac{10(s+1)}{s(s+2)(s+3)}$

$$10(-1+2) = A_1(1-2+2)$$
$$A_1 = 10$$

Put $s = -(1+j)$

$$10(-1-j+2) = A_2(-1-j+1)(-1-j+1-j)$$
$$10(1-j) = A_2 2$$
$$A_2 = -5+j5$$
$$= 7.07\angle 3\pi/4$$
$$A_3 = \text{conjugateof} A_2$$
$$= 7.07\angle -3\pi/4$$
$$-(+1+j) = \sqrt{2}\angle -3\pi/4$$
$$-(1-j) = \sqrt{2}\angle 3\pi/4$$

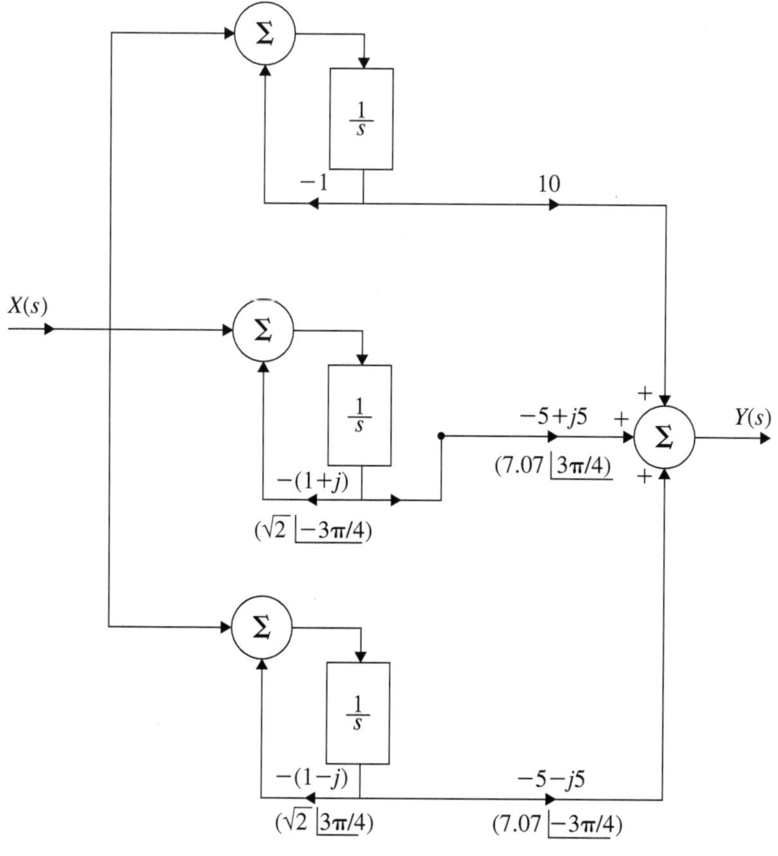

Fig. 8.63 Parallel realization of $H(s) = \dfrac{10(s+2)}{(s+1)(s^2+2s+2)}$

The parallel form of structure realization is shown in Fig. 8.63. The parameters are expressed both in j form as well as polar form.

■ Example 8.90

Repeated Poles

Consider the following T.F, where two poles are repeated.

$$H(s) = \frac{(10s^2 + 27s + 18)}{(s+3)(s+4)^2}$$

By parallel realization construct the structure of $H(s)$.

Solution: The given T.F. $H(s)$ is written in the following form

$$\begin{aligned} H(s) &= \frac{(10s^2 + 27s + 18)}{(s+3)(s+4)^2} \\ &= \frac{A_1}{(s+3)} + \frac{A_2}{(s+4)^2} + \frac{A_3}{(s+4)} \end{aligned}$$

$$10s^2 + 27s + 18 = A_1(s+4)^2 + A_2(s+3) + A_3(s+3)(s+4)$$

Put $s = -3$

$$\begin{aligned} 90 - 27(3) + 18 &= A_1(-3+4)^2 \\ A_1 &= 27 \end{aligned}$$

Put $s = -4$

$$\begin{aligned} 10(16) - 27(4) + 18 &= A_2(-4+3) \\ A_2 &= -70 \end{aligned}$$

Compare the coefficients of s^2 on both sides

$$\begin{aligned} 10 &= A_1 + A_3 = 27 + A_3 \\ A_3 &= -17 \\ H(s) &= \frac{27}{(s+3)} - \frac{70}{(s+4)^2} - \frac{17}{(s+4)} \end{aligned}$$

Let

$$\begin{aligned} H_1(s) &= \frac{27}{(s+3)} \\ H_2(s) &= -\frac{17}{(s+4)} \\ H_3(s) &= -\frac{70}{(s+4)^2} \end{aligned}$$

$H_1(s)$ and $H_2(s)$ are realized by Direct Form-II. The output of $H_2(s)$ is applied as the input of T.F. $-\frac{70}{(s+4)}$ and thus $H_3(s) = -\frac{70}{(s+4)}$ is realized. The parallel realization of the T.F. with repeated roots is shown in Fig. 8.64.

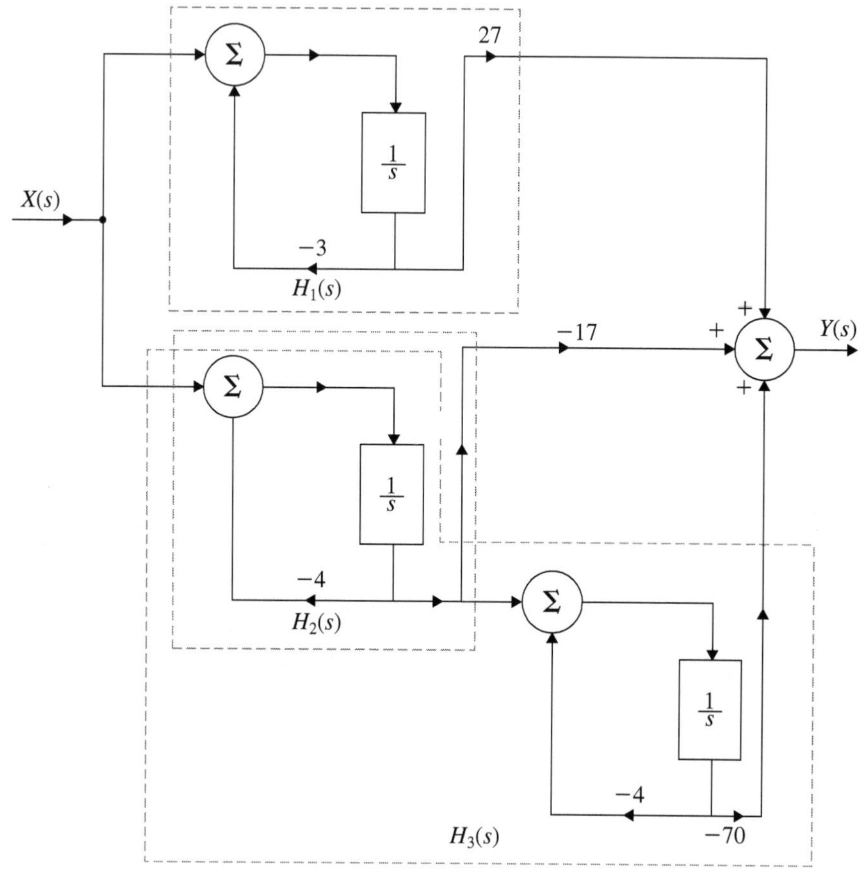

Fig. 8.64 Parallel realization of $H(s) = \dfrac{(10s^2 + 27s + 18)}{(s+3)(s+4)^2}$

8.16.5 *Transposed Realization*

A transposed realization is the realization which has the same transfer function. Here the two realizations are said to be equivalent. The given realization is changed to its transpose by following the steps given below:

- Replace $X(s)$ with the output $Y(s)$ and *vice versa*,
- Reverse all the arrow directions without changing the values of the multiplier,
- Replace pick off nodes by adders and *vice versa*.

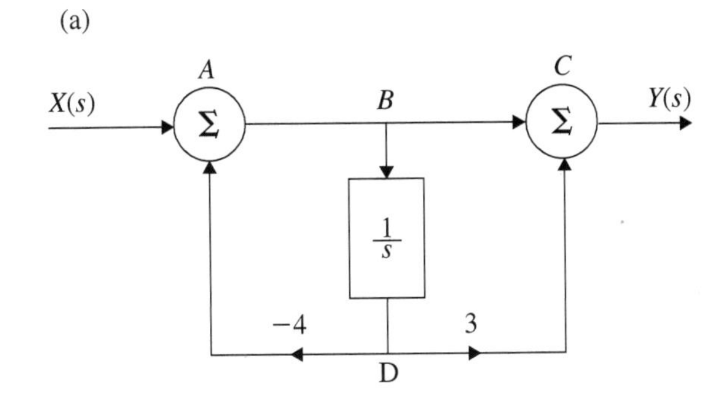

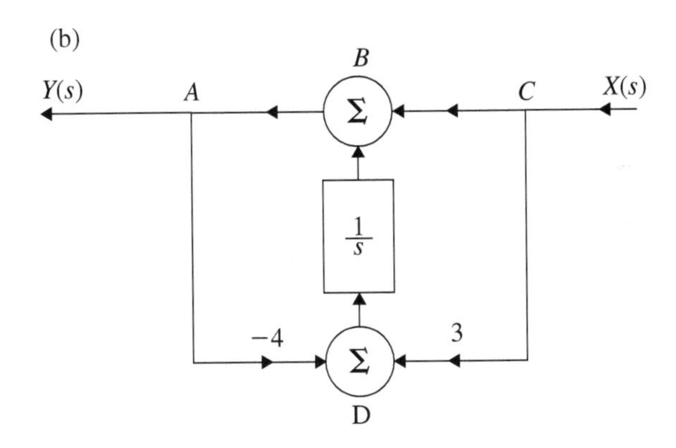

Fig. 8.65 **a** Direct Form-II Realization of $H(s)$. **b** Transposed Realization of $H(s)$

The following examples illustrate the above procedure.

■ Example 8.91

Consider the following T.F

$$H(s) = \frac{(s+3)}{(s+4)}$$

Draw the Direct Form-II realization and its transpose.

Solution:

1. For the given T.F. $H(s)$, the Direct Form-II structure is shown in Fig. 8.65a.
2. To get the transposed realization, $X(s)$ and $Y(s)$ are interchanged.
3. The direction of the arrows are reversed.
4. At points A and C, there are summers. These summers are replaced by take off points. B and D are take off points. These points are replaced by summers.
5. From Fig. 8.65b, the following equations are written.

$$Y(s) = X(s) + 3\frac{1}{s}X(s) - 4\frac{1}{s}Y(s)$$

$$Y(s)\left[1 + \frac{4}{s}\right] = \left[1 + \frac{3}{s}\right]X(s)$$

$$H(s) = \frac{Y(s)}{X(s)} = \frac{(s+3)}{(s+4)}$$

Thus, the transposed realization has the same T.F. as that of the original system.

■ Example 8.92

Realize

$$H(s) = \frac{s(s+2)}{(s+1)(s+3)(s+4)}$$

- in Direct Form-I
- Direct Form-II
- Cascade Form
- Parallel Form
- Transposed Form

(*Anna University, December 2007*)

Solution:

1. **Direct Form-I realization:** The given T.F. $H(s)$ can be written as given below.

$$H(s) = \frac{s(s+2)}{(s+1)(s+4)(s+3)}$$
$$= \frac{s^2 + 2s}{(s^3 + 8s^2 + 19s + 12)}$$

Here $b_0 = 0$; $b_1 = 1$; $b_2 = 2$; $b_3 = 0$; and $a_1 = 8$; $a_2 = 19$; $a_3 = 12$ with the values as given above, the Direct Form-I structure is shown in Fig. 8.66a. It requires five integrators.

2. **Realization by Direct Form-II:**

$$H(s) = \frac{s^2 + 2s}{(s^3 + 8s^2 + 19s + 12)}$$

The order of the system is three and three integrators are required to realize this. $b_1 = 1$; $b_2 = 2$; $a_1 = 8$; $a_2 = 19$; and $a_3 = 12$. With these values, Direct Form-II structure is shown in Fig. 8.66b.

3. **Realization in Cascade Form:** The given T.F. is written in the following form

$$H(s) = \frac{s}{(s+1)}\frac{(s+2)}{(s+3)}\frac{1}{(s+4)}$$

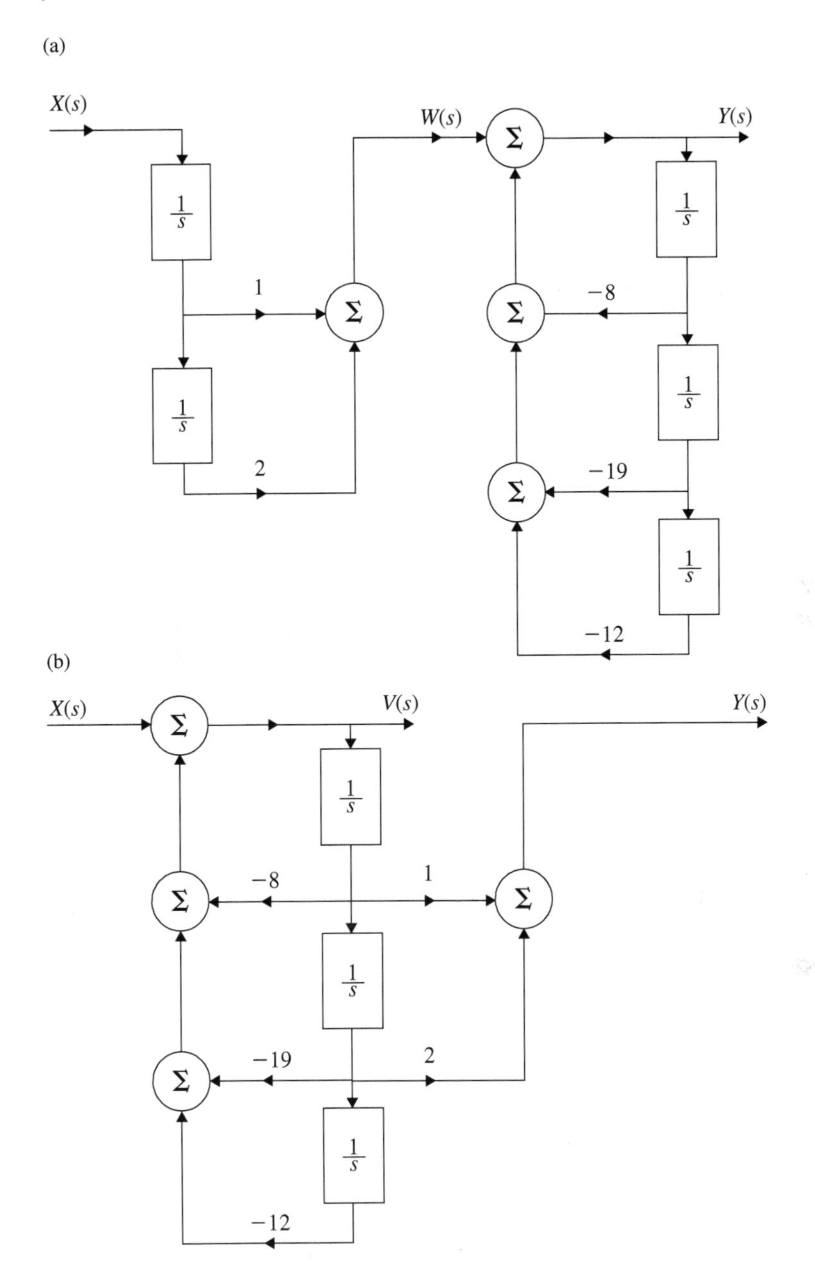

Fig. 8.66 **a** Direct Form-I realization of $H(s) = \dfrac{s(s+2)}{(s+1)(s+4)(s+3)}$. **b** Structure realization by Direct Form-II $H(s) = \dfrac{s(s+2)}{(s+1)(s+3)(s+4)}$. **c** Realization by Cascade Form $H(s) = \dfrac{s(s+2)}{(s+1)(s+3)(s+4)}$. **d** Parallel Form realization $H(s) = \dfrac{s(s+2)}{(s+1)(s+3)(s+4)}$. **e** Transpose Form realization of $H(s) = \dfrac{s(s+2)}{(s+1)(s+3)(s+4)}$

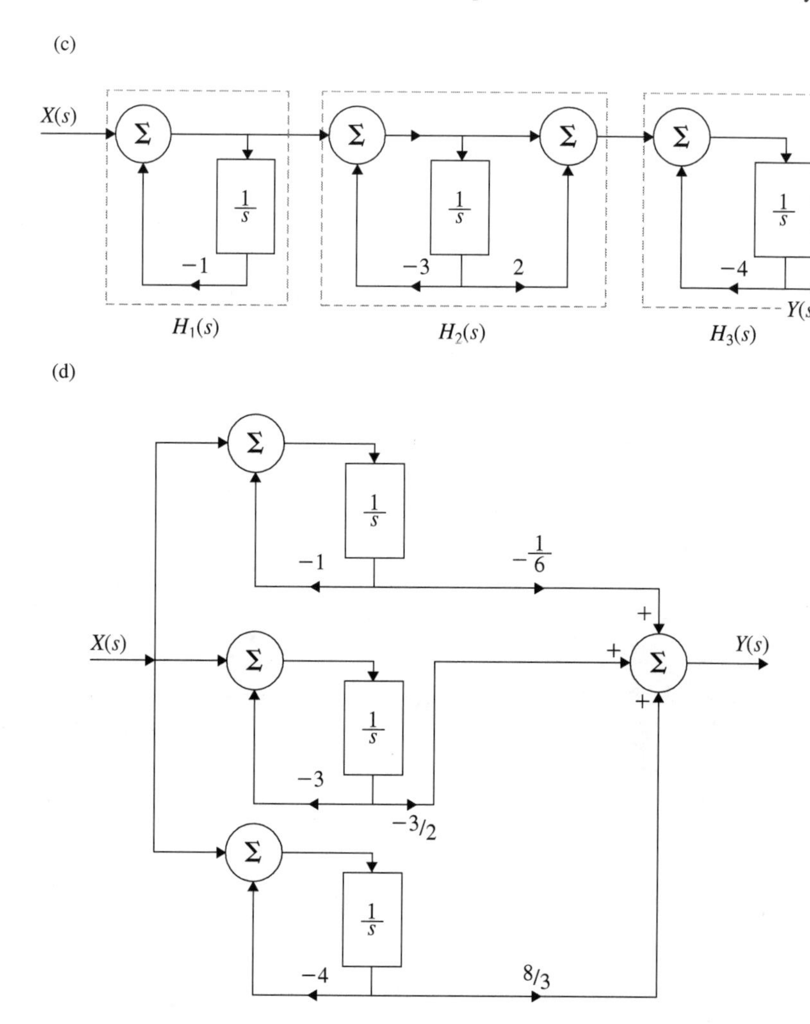

Fig. 8.66 (continued)

Let

$$H_1(s) = \frac{s}{s+1}; \quad H_2(s) = \frac{(s+2)}{(s+3)}; \quad H_3(s) = \frac{1}{(s+4)}$$

$H_1(s)$, $H_2(s)$ and $H_3(s)$ are individually realized by Direct Form-II and they are connected in Cascade. The cascade realized structure is shown in Fig. 8.66c.

4. **Realization in Parallel Form:** For Parallel Form of realization, the T.F. $H(s)$ is expressed in partial fraction form as given below

(e)

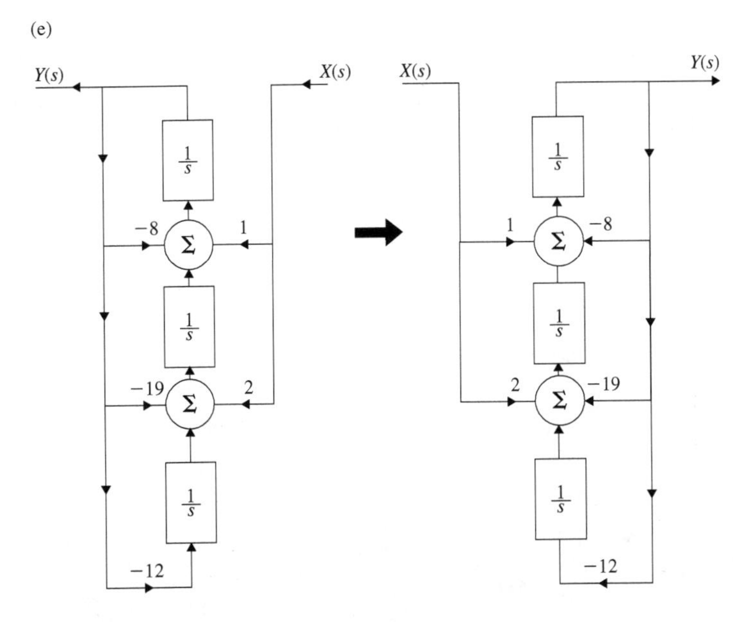

Fig. 8.66 (continued)

$$H(s) = \frac{s(s+2)}{(s+1)(s+3)(s+4)}$$

$$= \frac{A_1}{(s+1)} + \frac{A_2}{(s+3)} + \frac{A_3}{(s+4)}$$

$$s^2 + 2s = A_1(s+3)(s+4) + A_2(s+1)(s+4) + A_3(s+1)(s+3)$$

Put $s = -1$

$$(1-2) = A_1(-1+3)(-1+4)$$

$$A_1 = -\frac{1}{6}$$

Put $s = -3$

$$(9-6) = A_2(-3+1)(-3+4)$$

$$A_2 = -\frac{3}{2}$$

Put $s = -4$

$$(16-8) = A_3(-4+1)(-4+3)$$

$$A_3 = \frac{8}{3}$$

The parallel form realization is shown in Fig. 8.66d.

5. **Transpose Realization:** Consider Fig. 8.66b, where the transfer function is realized by Direct Form-II. By following the rules of transposed realization, Direct Form-II is converted into transposed realization and is shown in Fig. 8.66e.

Summary

1. The LT is a tool to represent any arbitrary signal $x(t)$ in terms of exponential components.
2. The LT is defined as follows:

$$X(s) = \int_{-\infty}^{\infty} x(t)e^{-st}\,dt$$

The Laplace inverse transform which converts $X(s)$ into $x(t)$ is expressed as

$$x(t) = \frac{1}{2\pi j}\int_{\sigma-j\infty}^{\sigma+j\infty} X(s)e^{st}\,ds$$

The above two equations are called LT pair.

3. Fourier transform is a special case of LT. Fourier transform is obtained by substituting $s = j\omega$ in LT in many practical cases even though it is not true always.
4. The LT of a causal signal and system is called unilateral LT. The LT of non-causal signal and system is called bilateral LT.
5. The region in the complex s-plane where the LT converges is called the region of convergence which is written in abbreviated form as ROC. For a causal signal, the ROC exists to the right of the rightmost pole of the transfer function. For a non-causal signal, the ROC exists to the left of the leftmost pole of the transfer function. The ROC will not enclose any pole.
6. The unilateral LT is a special case of bilateral LT. Their properties are discussed in detail.
7. The inverse LT is conveniently obtained using partial fraction method. Analytical, as well as graphical methods, are used to determine the residues in the partial fraction.
8. The integro-differential equation of LTIC system can be converted into algebraic equations using LT and the solution is obtained with the case.
9. By knowing the transfer function using LT, one can easily obtain impulse response and step response. Using LT, it is also possible to get zero state response, zero input response, natural response, forced response and total response of the system.
10. The solutions of differential and integro-differential equations are obtained using LT. The initial conditions are applied for zero input. The differen-

tial equation can also be solved using the classical method. However, in the classical method, the zero initial conditions are applied for the total response. The classical method is restricted to a certain class of input and not applicable to any input. In the classical method, the total response is expressed in terms of natural response and forced response.

11. Using LT, the electrical network which consists of passive elements can be analyzed.
12. Using time convolution property of LT, it is possible to get the system response $y(t)$.
13. Using LT it's possible to obtain the causality and stability of LTIC system.
14. Non-causal signals and/or systems can be analyzed by the bilateral (two-sided) Laplace transform. Here, the ROC is mostly in the form of a strip. Bilateral Laplace transform can also be used for linear system analysis.
15. The transfer function of an nth order system can be realized using integrators, summers and multipliers. The following form of realization which is a synthesis problem have been discussed and illustrated with examples.

    (a) Direct Form-I
    (b) Direct Form-II
    (c) Cascade Form
    (d) Parallel Form
    (e) Transposed Form.

Exercises

I. Short Answer Type Questions

1. **What is Laplace Transform?**
   The representation of a continuous time signal $x(t)$ in terms of complex exponential e^{st} is termed as Laplace transform. Mathematically, it is expressed as

$$L[x(t)] = X(s) = \int_{-\infty}^{\infty} x(t)e^{-st}dt$$

   where s is a complex variable expressed as $s = \sigma + j\omega$. Thus by LT the time function $x(t)$ is expressed as a frequency function.

2. **What do you understand by LT pair?**
   The LT and inverse LT are called Laplace transform pair. Mathematically, they are expressed as

$$X(s) = \int_{-\infty}^{\infty} x(t)e^{-st}dt$$

$$x(t) = \frac{1}{2\pi j}\int_{\sigma-\infty}^{\sigma+\infty} X(s)e^{st}ds$$

3. **What is bilateral Laplace transform?**
The LT to handle non-causal signals and systems is called bilateral LT. It is mathematically expressed as

$$X(s) = \int_{-\infty}^{\infty} x(t)e^{-st}dt$$

4. **What is unilateral Laplace transform?**
The LT to handle causal signals and systems is called unilateral LT. Mathematically, it is expressed as

$$X(s) = \int_{0^-}^{\infty} x(t)e^{-st}dt$$

5. **What do you understand by LT of right-sided and left-sided signals?**
The LT of a causal signal is called the right-sided LT and is mathematically described as

$$X(s) = \int_{0^-}^{\infty} x(t)e^{-st}dt$$

The LT of a non-causal signal is called the left-sided LT and is mathematically expressed as

$$X(s) = \int_{-\infty}^{0^-} x(t)e^{-st}dt$$

6. **What is the connection between LT and FT?**
The FT is a special case of LT which is obtained by putting $X(s)|_{s=j\omega}$ with the constraints that $x(t)$ is absolutely integrable and ROC of $X(s)$ includes the $j\omega$ axis of the s-plane. Thus, the FT $X(j\omega)$ is obtained from LT of $X(s)$ by substituting $s = j\omega$. It is evaluated on the $j\omega$ axis in the s-plane.

7. **What do you understand by Region of convergence?**
The region in the s-plane for which the LT integral

$$X(s) = \int_{-\infty}^{\infty} x(t)e^{-st}dt$$

converges is called the region of convergence which is written in the abbreviated form as ROC.

8. **How do you identify the ROC of a causal signal?**
The ROC of a causal (or right sided) signal is identified in the s-plane in the region to the right of the rightmost pole of the T.F. $H(s)$.

9. **How do you identify the ROC of a non-causal (left sided) signal?**
The ROC of a non-causal signal is identified in the s-plane in the region to the left of the left most pole of the T.F. $H(s)$.

10. **How do you identify the ROC of a bilateral Laplace transform?**
The region to the right of the rightmost pole of the causal signal and the region to the left of the leftmost pole of the non-causal signal are identified as the ROC of bilateral LT. ROC should not include any pole. The ROC is a strip. If ROC does not overlap, LT does not exist.

11. **State any three properties of ROC.**
The three properties of ROC are

(a) The ROC of LT does not include any pole of $X(s)$,
(b) For the right-sided (causal) signal, ROC exists to the right of rightmost pole of $X(s)$,
(c) For the left-sided (non-causal) signal, ROC exists to the left of left most pole of $X(s)$.

12. **Identify the ROCs for the following signals and sketch them in the s-plane?** (Fig. 8.67)

(a) $x(t) = e^{-2t}u(t)$
(b) $x(t) = e^{-3t}u(-t)$
(c) $x(t) = e^{-2t}u(t) + e^{3t}u(-t)$
(d) $x(t) = e^{-2|t|}$
(e) $x(t) = e^{2|t|}$

13. **Sketch the ROC of the following T.F. of a certain causal system and mark the poles and zeros?** (Fig. 8.68)

14. **Sketch the ROC of a non-causal system whose T.F. is given as**

$$H(s) = \frac{(s+2)(s-2)}{s(s+1)(s-3)}$$

Mark the poles and zeros of $H(s)$ (Fig. 8.69).

15. **What are initial and final value theorems?**
Initial value theorem is used to determine the initial value of $x(t)$ (as $t \to 0$) from the LT $X(s)$ which is given below

$$x(0^+) = \underset{s\to\infty}{Lt}\ sX(s)$$

provided $x(t)$ and $\frac{dx(t)}{dt}$ are both Laplace transformable and $X(s)$ is proper.

The final value theorem is used to determine $x(t)$ as t tends to infinity. This can be determined from $X(s)$ using final value theorem as given below.

$$x(\infty) = \underset{s \to 0}{Lt}\, sX(s)$$

provided that $x(t)$ and $\frac{dx(t)}{dt}$ are both Laplace transformable and $sX(s)$ has no poles in the RHP or on the imaginary axis.

16. **Find the initial and final values of $x(t)$ whose LT is given by**

$$X(s) = \frac{(s+5)}{(s^2+3s+2)}$$

(Anna University, June 2007)

Initial Value

$$x(0^+) = \underset{s \to \infty}{Lt} \frac{s(s+5)}{(s^2+3s+2)}$$

$$= \underset{s \to \infty}{Lt} \frac{s^2(1+5/s)}{s^2(1+\frac{3}{s}+\frac{2}{s^2})}$$

$$\boxed{x(0^+) = 1}$$

Final value

$$x(\infty) = \underset{s \to \infty}{Lt} \frac{s(s+5)}{(s^2+3s+2)}$$

$$\boxed{x(\infty) = 0}$$

17. **Define transfer function.**

The transfer function of a linear time invariant continuous system is defined as the ratio of the LT of the output variable to the LT of the input variable with all initial conditions being assumed to be zero. Thus

$$\text{T.F.} \quad H(s) = \frac{\text{Laplace transform of zero state response}}{\text{Laplace transform of input signal}}$$

Transfer function does not exist for non-linear and time-varying systems.

18. **Define poles and zeros of the transfer function.**

The pole of a transfer function is defined as the value of s in the s-plane at which the T.F. becomes infinity. The poles are represented by a small cross $\times$. The poles are the roots of the denominator polynomial of the T.F.

The zero of a transfer function is defined as the value of s in the s-plane at which the T.F. becomes zero. They are represented by a small circle "O" in the s-plane. The zeros are the roots of the numerator polynomial of T.F.

19. **What do you understand by eigenfunction of a system?**
The input for which the system response is also of the same form is called eigenfunction or characteristic function.

20. **What do you understand by causality of a LTIC system?**
A LTIC system with rational T.F. is said to be causal if the impulse response is right sided. For such a system, the ROC is in RHP and to the right of rightmost pole. An ROC to the right of the rightmost pole does not simply guarantee the causality of the system. The ROC should be in RHP also.

21. **What do you understand by stability of a LTIC system?**
The LTIC system is said to be stable if the area under the impulse response $h(t)$ curve is finite. In other words, the impulse response $h(t)$ should be absolutely integrable. In terms of ROC, the T.F of a stable LTIC system includes the $j\omega$ axis of the s-plane.

An LTIC system which is causal is said to be stable if all the poles of the transfer function $H(s)$ lie in the LHP and no repeated poles are at the origin of the s-plane.

22. **What do you understand by impulse response and step response of a system?**
The response of the system for the impulse input is defined as

$$\begin{aligned} \delta(t) &= 1 \qquad t = 0 \\ &= 0 \qquad \text{elsewhere} \end{aligned}$$

is called impulse response of the system. The response of the system for the step input is defined as

$$\begin{aligned} x(t) = u(t) &\qquad t \geq 0 \\ = 0 &\qquad t < 0 \end{aligned}$$

is called step response of the system.

23. **What do you understand by zero state response and zero input response?**
The system response when the system is in zero state (all the initial conditions are zero) is called zero state response. Here, the response is made up of characteristic mode or the Eigen values of the system.

The zero input response of the system is the response due to the initial conditions only. Here the input is made zero. For a LTIC system, the total response is

Total response = zero state response + zero input response

24. **What do you understand by the natural response and forced response of a system?**
The total response of a LTIC system can be expressed in terms of zero input component and zero state component. If we lump together all the characteristic mode terms in the total response, such a response is called the natural response. The remaining part of the total response which consists of non-characteristic mode terms is called the forced response of the system.
25. **Are zero input response and natural response and zero state response and forced response same?**
Zero input response is not the same as the natural response and zero state response is also not the same as the forced response. However, the total response which is the sum of natural response and forced response and also expressed as the sum of zero state response and zero input response will be the same. In few cases, the natural response will be the same as the zero input response and the zero state response is the same as the forced response.
26. **Comment on the solutions of the differential equations obtained by the application of LT and by classical method?**

(a) In the LT method, the initial conditions are applied to zero input response. In the classical method, the total response cannot be represented as zero state response and zero input response. Hence, in the classical method, the zero initial conditions are applied to the total response which begins at $t = 0^+$.
(b) The classical method is restricted to a certain class of inputs, whereas the LT method is applicable to many commonly used signals.

27. **What do you understand by asymptotic stability of an LTIC system?**
An LTIC system is said to be asymptotically stable if all the roots of the T.F., which may be simple or repeated, lie in LHP. Further, there are no repeated roots on the imaginary axis. Under such conditions, the system remains in a particular equilibrium state indefinitely in the absence of an external input.
28. **What do you understand by marginal stability of the system?**
An LTIC system is said to be marginally stable if there are no roots in the RHP and some un-repeated roots are on the imaginary axis.
29. **What do you understand by zero input stability and zero state stability?**
The zero state stability or external stability of the system is obtained when the input is applied with zero initial conditions. The zero input stability or internal stability of the system is obtained by applying initial conditions with no external input.

30. **What do you understand by bounded input and bounded output (BIBO) stability?**
An LTIC system is bounded input bounded output stable if the area under the impulse response curve is finite. Here all the poles of the T.F. lie in LHP. No repeated poles are on the imaginary axis. An asymptotically stable system is BIBO unstable.
31. **Find the transfer function of LTI system described by the differential equation**

$$\frac{d^2y(t)}{dt^2} + 3\frac{dy(t)}{dt} + 2y(t) = 2\frac{dx(t)}{dt} - 3x(t)$$

(*Anna University, May 2008*)

$$H(s) = \frac{Y(s)}{X(s)} = \frac{(2s-3)}{(s^2+3s+2)}$$

32. **Find the LT of $x(t) = e^{-at}u(t)$.** (*Anna University, December 2006*)

$$X(s) = \int_0^{\infty} e^{-(s+a)t}dt$$

$$X(s) = \frac{1}{(s+a)} \qquad \text{ROC: Re } s > -a$$

33. **Given $\frac{dy(t)}{dt} + 6y(t) = x(t)$. Find the T.F.**
(*Anna University, December 2006*)

$$H(s) = \frac{Y(s)}{X(s)} = \frac{1}{(s+6)}$$

34. **Find the LT of $u(t) - u(t-a)$ where $a > 0$.** (*Anna University, December, 2006*)
The LT of $u(t)$ is $\frac{1}{s}$. By using time shifting property of LT

$$L[-u(t-a)] = X(s) = -\frac{1}{s}e^{-as} \qquad \text{ROC: Re } s > 0$$

$$L[u(t) - u(t-a)] = \frac{1}{s}[1 - e^{-as}]$$

35. **Find the LT of $x(t) = +e^{-3t}u(t-10)$?**

$$X(s) = \int_{10}^{\infty} e^{-3t} e^{-st} dt$$
$$= \frac{1}{(s+3)} e^{-10(s+3)} \qquad \text{ROC: Re } s > -3$$

36. **Find the LT of $x(t) = \delta(t-5)$?**

$$X(s) = e^{-5s} \qquad \text{ROC: all } s$$

37. **What is the output of a system whose impulse response $h(t) = e^{-at}$ for a delta input?** *(Anna University, December, 2005)*

$$\frac{Y(s)}{X(s)} = H(s) = \frac{1}{(s+a)} \qquad [X(s) = 1]$$
$$Y(s) = \frac{1}{(s+a)}$$
$$y(t) = e^{-at}u(t) \qquad \text{ROC: } s > -a$$

38. **Find the LT of $x(t) = te^{-at}u(t)$ where $a > 0$?** *(Anna University, May 2005)*

$$L[e^{-at}u(t)] = \frac{1}{(s+a)}$$
$$L[te^{-at}u(t)] = \frac{1}{(s+a)^2}$$

(using frequency differentiation property).

39. **Determine the LT of**

$$\begin{aligned} x(t) &= 2t \qquad 0 \le t \le 1 \\ &= 0 \qquad \text{otherwise.} \end{aligned}$$

(Anna University, May 2005)

$$X(s) = \int_{0}^{1} 2te^{-st} dt$$

Integrating by parts, we get

$$X(s) = \left[\frac{-2t}{s} e^{-st}\right]_0^1 - \frac{2}{s^2}[e^{-st}]_0^1$$
$$= \frac{2}{s^2}[1 - e^{-s}(s+1)]$$

40. **Determine the output response of the system whose impulse response $h(t) = e^{-at}u(t)$ for the step input?** *(Anna University, April 2004)*

$$h(t) = e^{-at}u(t)$$
$$H(s) = \frac{1}{(s+a)}$$
$$\frac{Y(s)}{X(s)} = \frac{1}{(s+a)} \qquad X(s) = \frac{1}{s}$$
$$Y(s) = \frac{1}{a}\left[\frac{1}{s} - \frac{1}{s+a}\right]$$
$$y(t) = \frac{1}{a}[1 - e^{-at}] \qquad \text{ROC: Re } s > 0$$

41. **Find the LT and sketch the pole-zero plot with ROC for $x(t) = (e^{-2t} + e^{-3t})u(t)$** (Fig. 8.70). *(Anna University June 2007)*

$$X(s) = \frac{1}{(s+2)} + \frac{1}{(s+3)}$$
$$= \frac{2(s+2.5)}{(s+2)(s+3)}$$

42. **Find the LT of $x(t) = \delta(t+1) + \delta(t-1)$ and its ROC.**

$$X(s) = e^{s} + e^{-s} \qquad \text{ROC : all } s.$$

43. **Find the LT of $x(t) = u(t+1) + u(t-1)$ and its ROC.**

$$X(s) = \frac{1}{s}[e^{s} + e^{-s}] \qquad \text{ROC: Re } s > 0$$

44. **Using convolution property determine $y(t) = x_1(t) * x_2(t)$ where $x_1(t) = e^{-2t}u(t)$ and $x_2(t) = e^{-3t}u(t)$?**

$$X_1(s) = \frac{1}{(s+2)};$$
$$X_2(s) = \frac{1}{(s+3)}$$
$$Y(s) = X_1(s)X_2(s)$$
$$= \frac{1}{(s+2)(s+3)}$$
$$= \frac{1}{(s+2)} - \frac{1}{(s+3)}$$

$$y(t) = (e^{-2t} - e^{-3t})u(t) \qquad \text{ROC: Re } s > -2$$

45. **Find the zero input response for the following differential equation.**

$$\frac{dy(t)}{dt} + 5y(t) = u(t);$$

$$y(0^-) = 5$$

$$Y(s) = \frac{5}{s+5}$$

$$y(t) = 5e^{-5t}u(t)$$

46. **Find the LT $\frac{d}{dt}[\delta(t)]$.**

$$L\frac{d}{dt}[\delta(t)] = s \qquad \text{ROC: all } s.$$

47. **Find the LT of $x(t) = \delta(2t)$.**

$$X(s) = \frac{1}{2} \qquad \text{ROC: all } s$$

48. **Find the LT of integrated value of $\delta(t)$.**

$$X(s) = \frac{1}{s}$$

49. **Why integrators are preferred to differentiators in structure realization?**
    Use of differentiators in structure realization enhances noise. That is why differentiators are not preferred.
50. **What are the components required in structure realization?**
    The components required in structure realization are:

    (a) Integrators,
    (b) Summers, and
    (c) Multipliers.

51. **Mention the steps to be followed to realize a transposed structure from canonic form structure.**

    (a) Interchange $X(s)$ and $Y(s)$,
    (b) Change the directions of arrows,
    (c) Replace take off points by summers and *vice versa*.

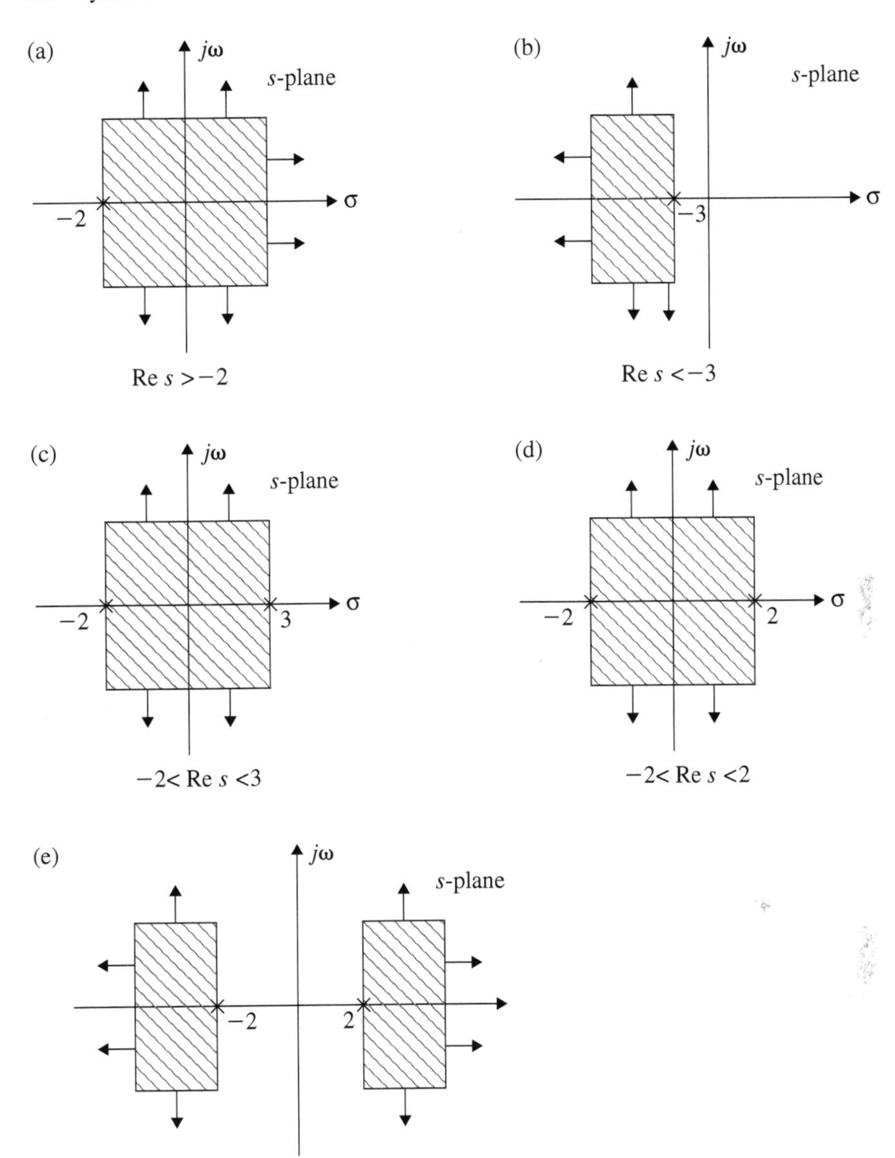

Fig. 8.67 Region of the convergence of different time functions

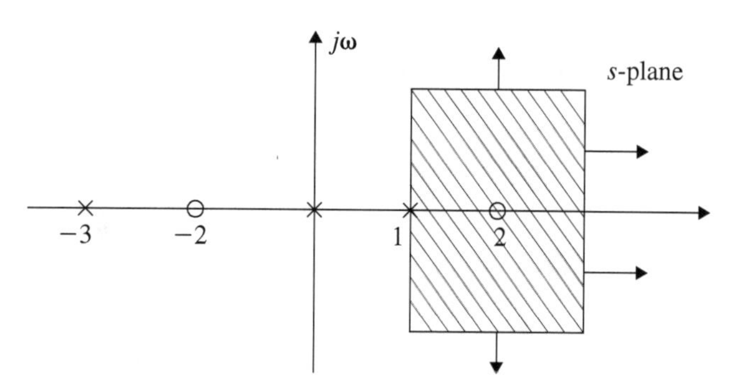

Fig. 8.68 ROC of a causal systematic T.F. $H(s) = \dfrac{10(s-2)(s+2)}{s(s+3)(s-1)}$

Fig. 8.69 ROC of a non-causal system with the T.F

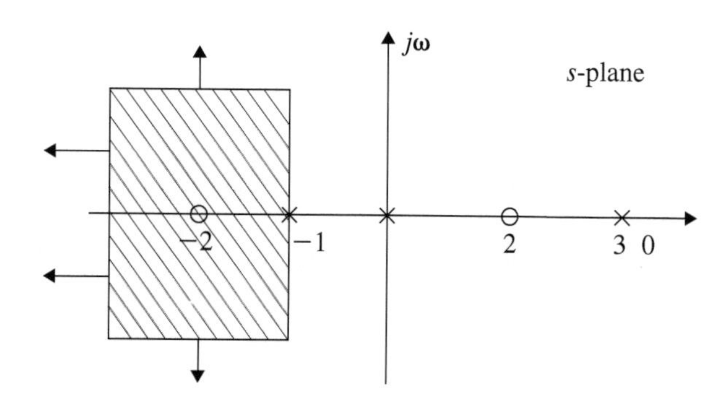

Fig. 8.70 Pole-zero plot and ROC of $X(s) = \dfrac{2(s+2.5)}{(s+2)(s+3)}$

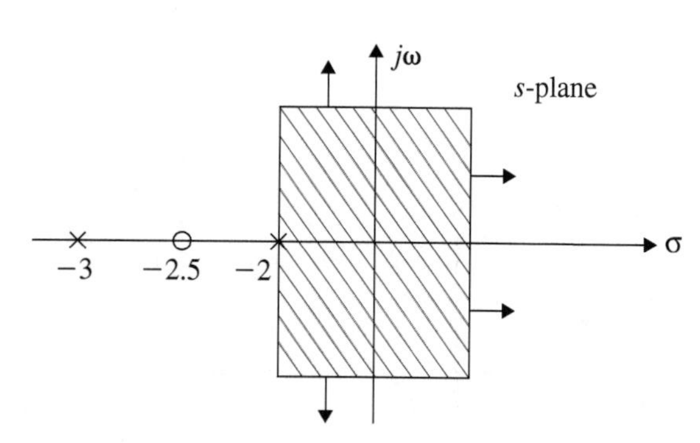

II. Long Answer Type Questions

1. **Find the LT of $x(t) = e^{-2|t|}$ and ROC.**

$$X(s) = \frac{1}{(s+2)} + \frac{1}{s-2} \qquad \text{ROC:} \; -2 < \text{Re}\, s < 2$$

2. **Find the LT of $x(t) = e^{2|t|}$ and ROC.**
   ROC do not overlap and $x(t)$ has no LT $X(s)$.

3. **Find the LT of $x(t) = (e^{2t} + e^{-2t})u(t)$ and the ROC.**

$$X(s) = \frac{1}{(s-2)} + \frac{1}{s+2} \qquad \text{ROC: Re } s > 2$$

4. **Find the LT of $x(t) = (e^{2t} + e^{-2t})u(-t)$ and the ROC.**

$$X(s) = -\left(\frac{1}{(s+2)} + \frac{1}{s-2}\right) \qquad \text{ROC: Re } s < -2$$

5. **Find the LT of $x(t) = (e^{-6t} + e^{-4t})u(t) + (e^{-3t} + e^{-2t})u(-t)$**

$$X(s) = \frac{1}{(s+6)} + \frac{1}{(s+4)} - \left(\frac{1}{(s+3)} + \frac{1}{s+2}\right) \quad \text{ROC: } -4 < \text{Re } s < -3$$

6. **Find the LT of**

$$x(t) = (e^{-6t} + e^{-3t})u(t) + (e^{-4t} + e^{-2t})u(-t)$$

   ROC does not overlap, and hence $x(t)$ has no LT $X(s)$.

7. **Find the LT and ROC of**

$$x(t) = e^{-3t}[u(t) - u(t-4)]$$

$$X(s) = \left[\frac{1}{(s+3)} - \frac{e^{-4(s+3)}}{(s+3)}\right] \qquad \text{ROC: Re} s > -3$$

8. **Find the inverse LT of the following $X(s)$ for all possible combinations of ROC.**

$$X(s) = \frac{4}{(s+1)(s-3)}$$

   (a) $x(t) = (e^{3t} - e^{-t})u(t)$ ROC: Re $s > 3$
   (b) $x(t) = (e^{-t} - e^{3t})u(-t)$ ROC: Re $s < -1$
   (c) $x(t) = (e^{-t}u(t) - e^{3t}u(-t)$ $-1 <$ Re $s < 3$

9. **Find the inverse LT of $X(s)$**

$$X(s) = \frac{8(s+2)}{s(s^2+4s+8)} \qquad \text{ROC: Re } s > -2$$

$$x(t) = 2\left[1 + \sqrt{2}\sin\left(2t - \tfrac{\pi}{4}\right)\right]u(t)$$

10. **Find the inverse LT of**

$$X(s) = \frac{s^2 + 2s + 4)}{(s+2)(s+4)} \qquad \text{ROC: Re } s > -2$$

$x(t) = \delta(t) + \frac{1}{2}[e^{-2t} - e^{-4t}]u(t)$

11. **Find the inverse LT of**

$$X(s) = \frac{(s^2 + 3s + 1)}{(s^2 + 5s + 6)} \qquad \text{ROC: Re } s > -2$$

$x(t) = \delta(t) - (9e^{-2t} - 11e^{-3t})u(t)$

12. **Find the inverse LT of**

$$X(s) = \frac{s^3 + 8s^2 + 21s + 16}{(s^2 + 7s + 12)} \text{ROC: Re } s > -3$$

$x(t) = [\frac{d}{dt}\delta(t) + \delta(t) + 4e^{-4t} - 2e^{-3t}]u(t)$

13. **Find the inverse LT of**

$$X(s) = \frac{10se^{-2s} + 5e^{-4s} + 6}{(s^2 + 13s + 40)} \qquad \text{ROC: Re } s > -5$$

$$x(t) = \left[\frac{80}{3}e^{-8(t-2)} - \frac{50}{3}e^{-5(t-2)}\right]u(t-2)$$
$$+\frac{5}{3}(e^{-5(t-4)} - e^{-8(t-4)})u(t-4) + 2[e^{-5t} - e^{-8t}]u(t)$$

14. **Find the initial and final value of $y(t)$ if its LT $Y(s)$ is given by**

$$Y(s) = \frac{(s^2 + 2s + 5)}{s(s^2 + 4s + 6)}$$

Initial value $y(0) = 1$. Final value $y(\infty) = \frac{5}{6}$

15.

$$x_1(t) = u(t)$$
$$x_2(t) = e^{-2t}u(t)$$

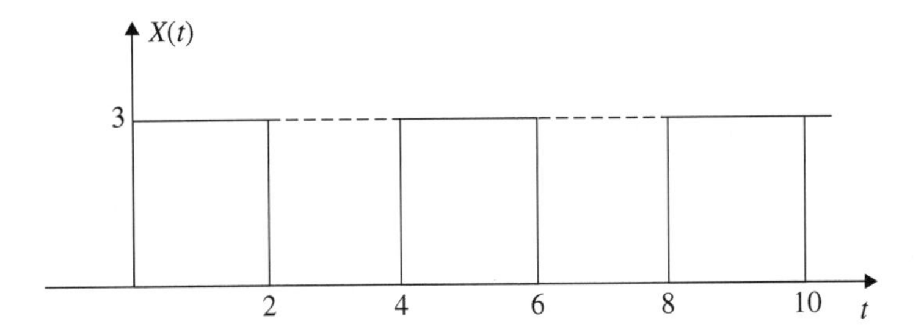

Fig. 8.71 A periodic pulse signal

Using convolution property of LT find $y(t) = x_1(t) * x_2(t)$

$$y(t) = \frac{1}{2}[1 - e^{-2t}]u(t)$$

16. **Consider an LTIC system described by the following differential equation**

$$\frac{d^2y(t)}{dt^2} + \frac{dy(t)}{dt} - 6y(t) = x(t)$$

Determine

(a) the system transfer function.
(b) impulse response of the system if it is causal.
(c) Impulse response of the system if the system is stable.
(d) Impulse response of the system if it is neither causal nor stable.

(a)

$$H(s) = \frac{1}{(s^2 + s - 6)}$$

(b)

$$y(t) = -\frac{1}{5}[e^{-3t} - e^{2t}]u(t) \qquad \text{ROC: Re } s > 2$$

(c)

$$y(t) = \frac{1}{5}[-e^{2t}u(-t) - e^{-3t}u(t)] \qquad \text{ROC: } -3 < \text{Re } s < 2$$

(d)

$$y(t) = \frac{1}{5}[-e^{2t} + e^{-3t}]u(-t) \qquad \text{ROC: Re } s < -3$$

17. **Determine the LT of the periodic signal shown in Fig.** 8.71.
$X(s) = \frac{3}{s}\frac{1}{[1+e^{-2s}]}$

Fig. 8.72 Electrical circuit

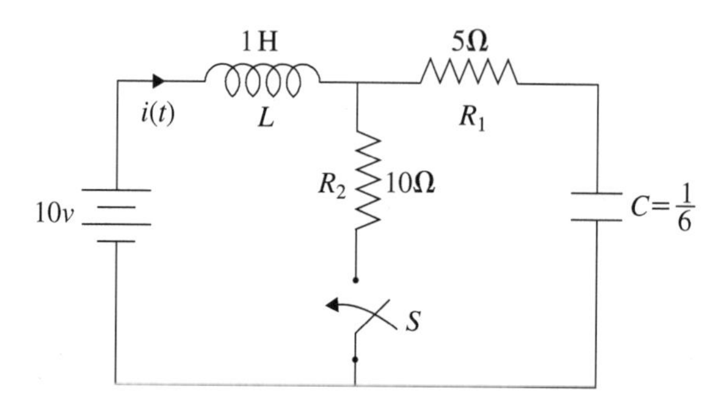

18. **Consider the electrical circuit shown in Fig. 8.72. Initially the switch S is closed. Derive an expression for the current through the inductor as soon as the switch is open.**
$i(t) = [3e^{-3t} - 2e^{-2t}]u(t)$

19. **Find the Laplace inverse of the following $X(s)$**

$$X(s) = \frac{(s+5)}{(s+2)(s+3)^3} \qquad \text{ROC: Re } s > -2$$

$$x(t) = [3e^{-2t} - (t^2 + 3t + 3)e^{-3t}]u(t)$$

20. **Solve the following differential equation**

$$\frac{d^2y(t)}{dt^2} + \frac{dy(t)}{dt} - 2y(t) = \frac{dx(t)}{dt} + x(t)$$

The initial conditions are $y(0^-) = 2$; $\frac{dy(0^-)}{dt} = 1$. The input is

(a) $x(t) = \delta(t)$ an impulse
(b) $x(t) = u(t)$ unit step
(c) $x(t) = e^{-4t}u(t)$ an exponential decay

(a) $x(t) = \left[\frac{2}{3}e^{-2t} + \frac{7}{3}e^{t}\right]u(t)$ ROC: Re $s > 1$

(b) $x(t) = \left[-\frac{1}{2} + \frac{1}{6}e^{-2t} + \frac{7}{3}e^{t}\right]u(t)$ ROC: Re $s > 1$

(c) $y(t) = \left[\frac{1}{2}e^{-2t} - \frac{3}{10}e^{-4t} + \frac{27}{15}e^{t}\right]u(t)$ ROC: Re $s > 1$

21. **The unit step response of a certain LTIC system $y(t) = 10e^{-5t}$. Find (a) the impulse response? (b) the response due to the exponential decay $x(t) = e^{-3t}u(t)$?**

(a) $h(t) = 10\delta(t) - 50e^{-5t}u(t)$ ROC: Re $s > -5$

(b) $y(t) = (25e^{-5t} - 15e^{-3t})u(t)$ ROC: Re $s > -3$

22. **The impulse response of a certain system is $h_1(t) = e^{-3t}u(t)$ and the impulse response of another system is $h_2(t) = e^{-5t}u(t)$. These two systems are connected in cascade. Find (a) the impulse response of the cascade connected system (b). Is the system BIBO stable?**

(a) $h(t) = \frac{1}{2}[e^{-3t} - e^{-5t}]u(t)$ ROC: Re $s > -3$
(b) The system is BIBO stable since the ROC is to the right of rightmost pole at $s = -3$ which includes the $j\omega$ axis.

23. **The impulse response of a certain system is given by $h(t) = e^{-5t}$. The system is excited by $x(t) = e^{-3t}u(t) + e^{-2t}u(-t)$. Determine**

(a) The system transfer function
(b) Output of the system $y(t)$
(c) BIBO stability of the system.

(a) $H(s) = \dfrac{-1}{(s+2)(s+3)(s+5)}$ ROC: $-3 < \text{Re } s < -2$

(b) $y(t) = \left(\frac{1}{2}e^{-3t} - \frac{1}{6}e^{-5t}\right)u(t) + \frac{1}{3}e^{-2t}u(-t)$

(c) The system is not BIBO stable since the ROC does not include the $j\omega$ axis

24. **A certain LTIC system is described by the following differential equation**

$$\frac{d^2y(t)}{dt^2} - \frac{dy(t)}{dt} - 30y(t) = \frac{dx(t)}{dt} + 4x(t)$$

The system is subjected to the following input.

$$x(t) = e^{-3t}u(t)$$

The initial conditions are $y(0^+) = 3$ and $\dot{y}(0^+) = 1$. Derive an expression for the output response as a function of time.

$$y(t) = \left[\frac{35}{22}e^{-5t} + \frac{145}{99}e^{6t} - \frac{1}{18}e^{-3t}\right]u(t) \quad \text{ROC: } -3 < \text{Re } s < 6$$

25. **A certain LTIC system is described by the following differential equation**

$$\frac{d^2y(t)}{dt^2} + 3\frac{dy(t)}{dt} + 2y(t) = \frac{dx(t)}{dt} + 4x(t)$$

where $x(t) = e^{-3t}u(t)$. The initial conditions are $y(0^-) = 2$ and $\dot{y}(0^-) = 1$. Determine

(a) The characteristic polynomial.
(b) The characteristic equation.
(c) The Eigen values.
(d) The zero input response.
(e) The zero state response.
(f) Total response. Use Laplace transform method.

(a) The characteristic polynomial is $F(s) = s^2 + 3s + 2$
(b) The characteristic equation is $\lambda^2 + 3\lambda + 2 = 0$
(c) The Eigen values are $\lambda_1 = -1$ and $\lambda_2 = -2$
(d) Zero input response is $y_s(t) = [5e^{-t} - 3e^{-2t}]u(t)$
(e) Zero state response is

$$y_i(t) = \left[\frac{3}{2}e^{-t} - 2e^{-2t} + \frac{1}{2}e^{-3t}\right]u(t)$$

(f) Total response is $y(t) = y_i(t) + y_s(t)$

$$y(t) = \left[\frac{13}{2}e^{-t} - 5e^{-2t} + \frac{1}{2}e^{-3t}\right]u(t)$$

26. **An LTIC system has the following T.F**

$$H(s) = \frac{(s + 10)}{s^3 + 5s^2 + 3s + 4}$$

Determine the differential equation.

$$\frac{d^3y(t)}{dt^3} + 5\frac{d^2y(t)}{dt^2} + \frac{3dy(t)}{dt} + 4y(t) = \frac{dx(t)}{dt} + 10x(t)$$

27. **An LTIC system is described by the following differential equation**

$$\frac{d^2y(t)}{dt^2} + 4\frac{dy(t)}{dt} + 3y(t) = \frac{dx(t)}{dt} + 4x(t)$$

The system is in the initial state of $y(0^-) = 2$ and $\dot{y}(0^-) = 1$. The system is excited with the input $x(t) = e^{-5t}$. Determine

(a) The natural response of the system.
(b) The forced response of the system.
(c) Total response of the system. Use Laplace transform method.

(a) The natural response of the system is

$$y_n(t) = \left(\frac{31}{8}e^{-t} - \frac{7}{4}e^{-3t}\right)u(t)$$

(b) The forced response of the system is

$$y_f(t) = \left(-\frac{1}{8}e^{-5t}\right)u(t)$$

(c) The total response of the system is

$$y(t) = \left[\frac{31}{8}e^{-t} - \frac{7}{4}e^{-3t} - \frac{1}{8}e^{-5t}\right]u(t)$$

28. **The impulse response of an LTIC system is given by $x(t) = e^{-2t}u(t)$. Is the system causal?**
$X(s) = \frac{1}{s+2}$ and rational ROC: $Re\ s > -2$ which lies in RHP. Hence, the system is causal.
29. **The impulse response of an LTIC system is given by $h(t) = e^{-2|t|}$. Is the system causal.**
$H(s) = \frac{-4}{(s-2)(s+2)}$ which is rational ROC is $-2 < Re\ s < 2$. The ROC is not to the right of the rightmost pole, and hence the system is not causal.

30. **Check the stability of an LTIC system whose impulse response is $h(t) = e^{-2}|t|$**
$H(s) = \frac{-4}{(s-2)(s+2)}$ which is rational. The ROC is $-2 < Re\ s < 2$. This includes the imaginary axis. Hence, the system is stable.

31. **Consider the following transfer function.**

$$X(s) = \frac{1}{(s+2)(s-2)}$$

Identify all possible ROCs and, in each case, find the impulse response, the stability and causality. Also, sketch the ROC.
(1) ROC: Re $s > +2$

$$h(t) = \frac{1}{4}(e^{2t} - e^{-2t})u(t)$$

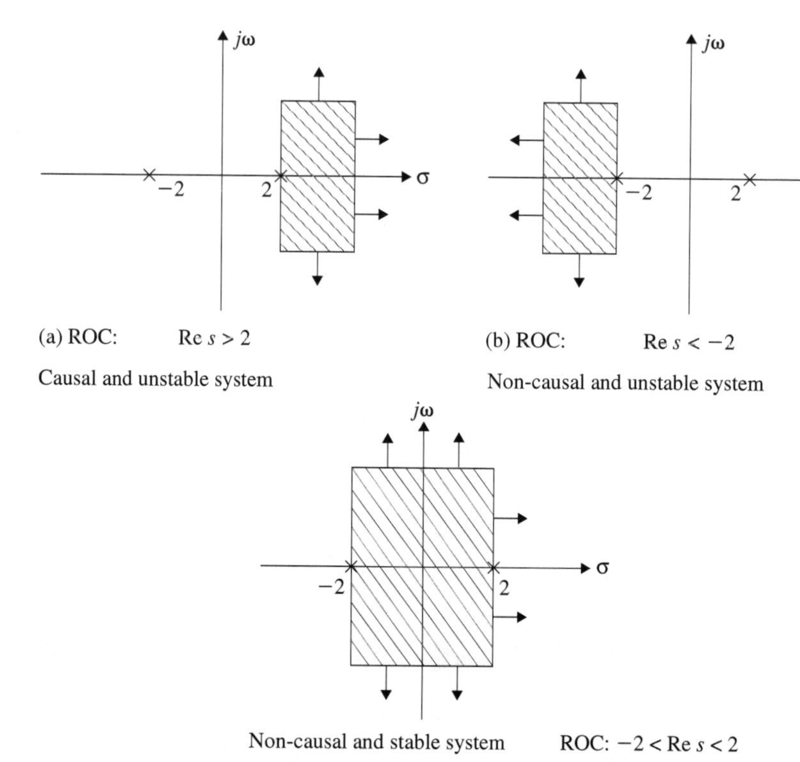

Fig. 8.73 ROC related to causality and stability

ROC does not include $j\omega$ axis. The system is unstable. The system is causal since ROC is right sided and in RHP.

(2) ROC: Re $s < -2$

$$h(t) = \frac{1}{4}[-e^{2t} + e^{-2t}]u(-t)$$

ROC does not include $j\omega$ axis. The system is unstable and non-causal since the ROC is left sided.

(3) ROC: $-2 <$ Re $s < 2$

$$h(t) = \frac{1}{4}[-e^{2t}u(-t) - e^{-2t}u(t)]$$

ROC includes the $j\omega$ axis and the system is stable. The system is non-causal since ROC is a strip (Fig. 8.73).

32. **Find the bilateral LT of**

$$\boldsymbol{x(t) = e^{-10|t|}}$$

(a)

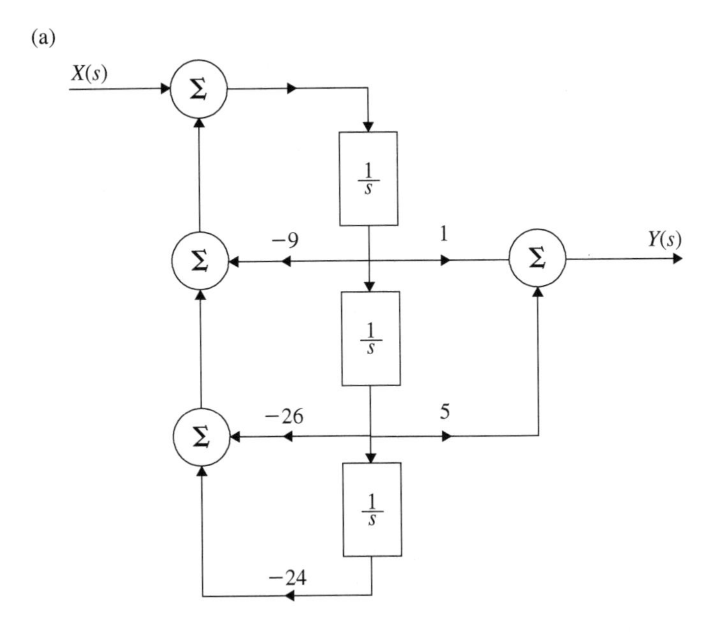

(b)

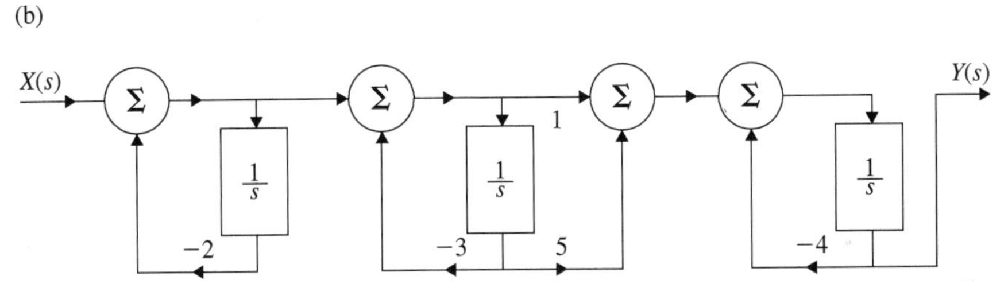

Fig. 8.74 **a** Direct canonic form realization of $H(s) = \frac{s(s+5)}{(s+2)(s+3)(s+4)}$. **b** Cascade form realization of $H(s) = \frac{s(s+5)}{(s+2)(s+3)(s+4)}$

$$X(s) = \frac{-20}{(s^2 - 100)}$$

33. **Find the bilateral LT of**

$$\boldsymbol{x(t) = e^{t}u(t) - e^{3t}u(-t)}$$

$$X(s) = \frac{(2s-4)}{(s-1)(s-3)}$$

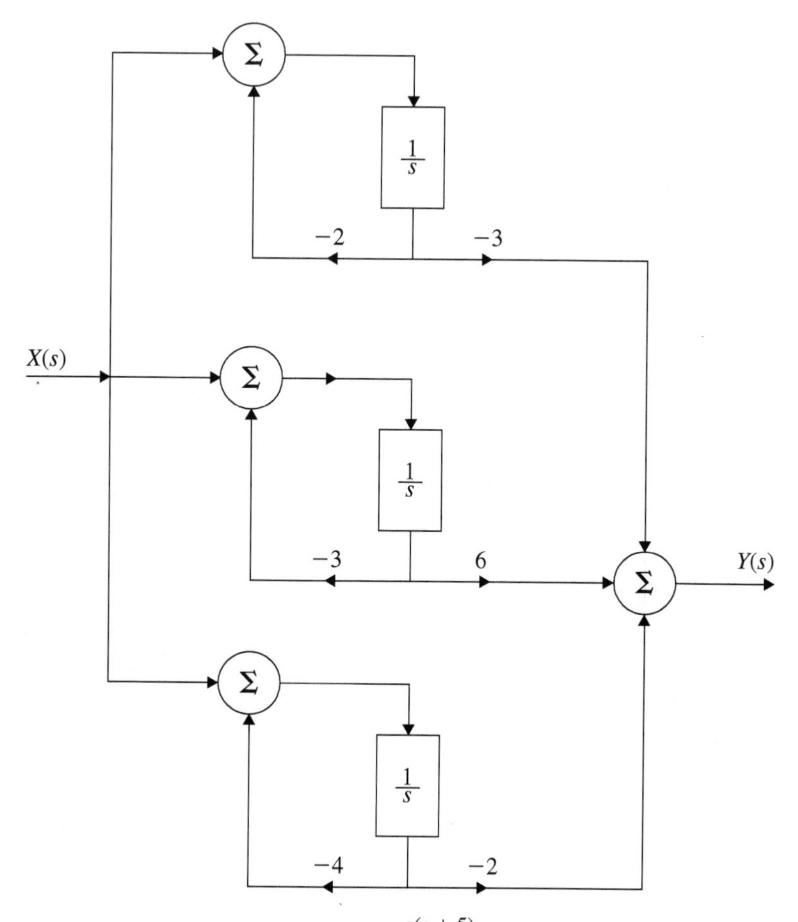

Fig. 8.75 **c** Parallel realization of $H(s) = \dfrac{s(s+5)}{(s+2)(s+3)(s+4)}$

34. **Find the bilateral LT of**

$$\boldsymbol{X(s) = \frac{(s-5)}{(s+2)(s+5)} \quad \text{ROC: } -5 < \text{Re}\, s < -2}$$

$$x(t) = \frac{1}{3}[10e^{-5t}u(t) + 7e^{-2t}u(-t)]$$

35. **Find the inverse bilateral LT of**

$$\boldsymbol{X(s) = \frac{(s+2)}{(s-2)(s-5)} \quad \text{ROC: } 2 < \text{Re}\, s < 5}$$

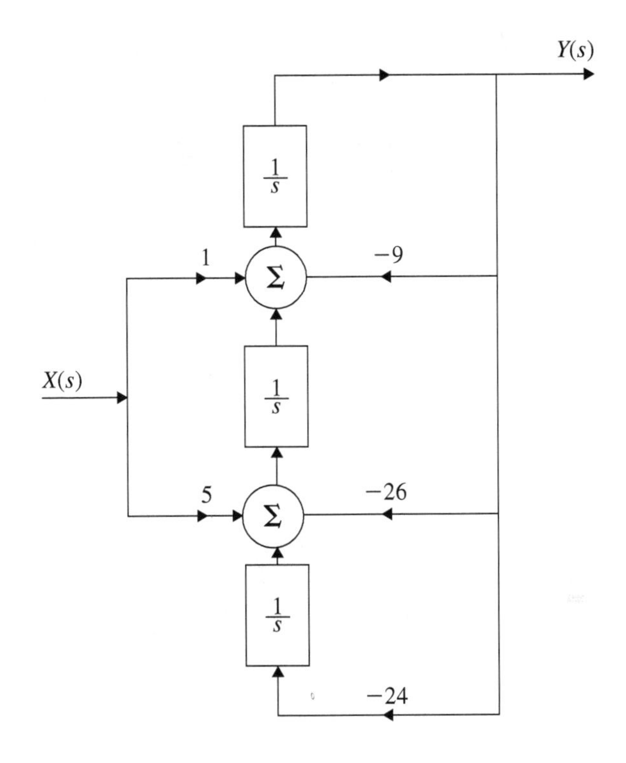

Fig. 8.76 **d** Transpose realization of $H(s) = \dfrac{s(s+5)}{(s+2)(s+3)(s+4)}$

$$x(t) = -\frac{1}{3}[7e^{5t}u(-t) + 4e^{2t}u(t)]$$

36. **Find the inverse bilateral LT of**

$$X(s) = \frac{(s^2 - 2s - 3)}{(s+2)(s+4)(s-6)} \qquad \textbf{ROC: } -2 < \text{Re } s < 6$$

$$x(t) = \left(\frac{-5}{16}e^{-2t} + \frac{21}{20}e^{-4t}\right)u(t) - \frac{21}{80}e^{6t}u(-t)]$$

37. **Realize** (Figs. 8.74, 8.75 and 8.76)

$$H(s) = \frac{s(s+5)}{(s+2)(s+3)(s+4)}$$

by

(a) Direct Canonic Form (Direct Form-II)
(b) Series Form (Cascade)
(c) Parallel Form
(d) Transposed Form

Structure Realization

Chapter 9
The z-Transform Analysis of Discrete Time Signals and Systems

Learning Objectives

- To define the z-transform and the inverse z-transform.
- To find the z-transform and ROC of typical DT signals.
- To find the properties of ROC.
- To find the properties of z-transform.
- To find the inverse z-transform.
- To solve difference equation using the z-transform.
- To establish the relationship between the z-transform, Fourier transform and the Laplace transform.
- To find the causality and stability of DT system.
- To realize the structure of DT system.

9.1 Introduction

The z-transform is the discrete counterpart of the Laplace transform. The Laplace transform converts integro-differential equations into algebraic equations. In the same way, the z-transform converts difference equations of discrete time systems to algebraic equations which simplifies the discrete time system analysis. There are many connections between Laplace and z-transforms except for some minor differences. DTFT represents discrete time signals in terms of complex sinusoids. When this sort of representation is generalized and represented in terms of the complex exponential, it is termed as z-transform. This sort of representation has a broader characterization of the system with signals. Further, the DTFT is applicable only for stable systems whereas z-transform can be applied even to unstable systems which means that z-transform can be used to a larger class of systems and signals. It is to be

S. Palani, *Signals and Systems*,
https://doi.org/10.1007/978-3-030-75742-7_9

noted that many of the properties in DTFT, Laplace transform and z-transform are common except that the Laplace transform deals with continuous time signals and systems.

9.2 The z-Transform

Let z^n be an everlasting exponential. Let $h(n)$ be the impulse response of the discrete time system. The response of a linear, time invariant discrete time system to the everlasting exponential z^n is given as $H(z)z^n$. That is, it is the same exponential within a multiplicative constant. Thus, the system response to the excitation $x[n]$ is the sum of the system's responses to all these exponentials. The tool that is used to represent an arbitrary discrete signal $x[n]$ as a sum of the everlasting exponential of the form z^n is called the z-transform.

Let $x[n] = z^n$ be the input signal applied to an LTI discrete time system whose impulse response is $h[n]$. The system output $y[n]$ is given by

$$\begin{aligned} y[n] &= x[n] * h[n] \\ &= \sum_{k=-\infty}^{\infty} h[k]x[n-k] \end{aligned}$$

Substitute $x[n] = z^n$

$$y[n] = \sum_{k=-\infty}^{\infty} h[k]z^{n-k} = z^n \left[\sum_{k=-\infty}^{\infty} h[k]z^{-k} \right]$$

Define the transfer function

$$H[z] = \sum_{k=-\infty}^{\infty} h[k]z^{-k} \tag{9.1}$$

Equation (9.1) may be written as

$$H[z^n] = H[z]z^n$$

To represent any arbitrary signal as a weighted superposition of the eigenfunction z^n, let us substitute $z = re^{j\Omega}$ into Eq. (9.1)

$$H[re^{j\Omega}] = \sum_{n=-\infty}^{\infty} h[n][re^{j\Omega}]^{-n}$$
$$= \sum_{n=-\infty}^{\infty} \left(h[n]r^{-n}\right) e^{-j\Omega n} \qquad (9.2)$$

Equation (9.2) corresponds to the DTFT of the signal $h[n]r^{-n}$. The inverse of $H[re^{j\Omega}]$, by mathematical manipulation of Eq. (9.2), can be obtained as

$$h[n] = \frac{1}{2\pi j} \oint H(z)z^{n-1}dz \qquad (9.3)$$

More generally, Eqs. (9.2) and (9.3) can be written as

$$X[z] = \sum_{n=-\infty}^{\infty} x[n]z^{-n} \qquad (9.4)$$
$$x[n] = \frac{1}{2\pi j} \oint X(z)z^{n-1}dz \qquad (9.5)$$

The above equations are called z-transform pair. Equation (9.4) is the z-transform of $x[n]$ and Eq. (9.5) is called inverse z-transform. In Eq. (9.4), the range of n is $-\infty < n < \infty$ and hence it is called bilateral z-transform. If $x[n] = 0$ for $n < 0$, Eq. (9.4) can be written as

$$X[z] = \sum_{n=0}^{\infty} x[n]z^{-n} \qquad (9.6)$$

Equation (9.6) is called unilateral or right-sided z-transform. Bilateral z-transform has limited practical applications. Unless otherwise it is specifically mentioned, z-transform means unilateral. z-transform and inverse z-transform are symbolically represented as given below.

$$Z[x[n]] = X[z]$$
$$x[n] \overset{Z}{\longleftrightarrow} X[z]$$
$$z^{-1}[X[z]] = x[n]$$
$$X[z] \overset{Z^{-1}}{\longleftrightarrow} x[n] \qquad (9.7)$$

9.3 Existence of the z-Transform

Consider the unilateral z-transform given by Eq. (9.6)

$$X[z] = \sum_{n=0}^{\infty} x[n]z^{-n}$$
$$= \sum_{n=0}^{\infty} \frac{x[n]}{z^n}$$

For the existence of $X[z]$,

$$|X[z]| \leq \sum_{n=0}^{\infty} \frac{|x[n]|}{|z|^n} < \infty \tag{9.8}$$

If the signal $x[n]$ is expressed in terms of an exponential signal r^n, then if $x[n] \leq r^n$ for some r, then

$$|x[n]| \leq r^n \tag{9.9}$$

Substitute Eq. (9.9) in Eq. (9.8)

$$|X[z]| \leq \sum_{n=0}^{\infty} \left(\frac{r}{z}\right)^n$$
$$= \frac{1}{\left[1 - \frac{r}{|z|}\right]} \quad \text{iff } |z| > r \tag{9.10}$$

From Eq. (9.10), it is evident that the z-transform of $x[n]$ which is $X(z)$ exists for $|z| > r$ and the signal is z-transformable. If the signal $x[n]$ grows faster than the exponential signal r^n for any r_0, Eq. (9.10) is not convergence and $x[n]$ is not z-transformable.

9.4 Connection Between Laplace Transform, z-Transform and Fourier Transform

Consider the Laplace transform of $x(t)$ which is represented below as

$$X(s) = \int_{-\infty}^{\infty} x(t)e^{-st}dt \tag{9.11}$$

When $s = j\omega$, Eq. (9.11) becomes

$$X(j\omega) = \int_{-\infty}^{\infty} x(t)e^{-j\omega t}\,dt \tag{9.12}$$

Equation (9.12) represents the Fourier transform. **The Laplace transform reduces to the Fourier transform on the imaginary axis where $s = j\omega$.** The relationship between these two transforms can also be interpreted as follows. The complex variable s can be written as $(\sigma + j\omega)$. Equation (9.11) is written as

$$\begin{aligned} X(\sigma + j\omega) &= \int_{-\infty}^{\infty} x(t)e^{-(\sigma+j\omega)t}\,dt \\ &= \int_{-\infty}^{\infty} \left[x(t)e^{-\sigma t}\right] e^{-j\omega t}\,dt \end{aligned} \tag{9.13}$$

Equation (9.13) can be recognized as the Fourier transform of $[x(t)e^{-\sigma t}]$. **Thus, the Laplace transform of $x(t)$ is the Fourier transform of $x(t)$ after multiplication by the real exponential $e^{-\sigma t}$ which may be growing or decaying with respect to time.**

The complex variable z can be expressed in polar form as

$$z = re^{j\omega} \tag{9.14}$$

where r is the magnitude of z and ω is the angle of z.

Substitute $z = re^{j\omega}$ in Eq. (9.6)

$$\begin{aligned} X(re^{j\omega}) &= \sum_{n=-\infty}^{\infty} x[n](re^{j\omega})^{-n} \\ &= \sum_{n=-\infty}^{\infty} \{x[n]r^{-n}\}e^{-j\omega n} \\ &= F[x[n]r^{-n}] \end{aligned} \tag{9.15}$$

Thus, $X(re^{j\omega})$ is the Fourier transform of the sequence $x[n]$ which is multiplied by a real exponential r^{-n} which may be growing or decaying with increasing n depending on whether r is greater or less than unity. If $r = 1$, then $|z| = 1$ and the equation becomes

$$X(e^{j\omega}) = \sum_{n=-\infty}^{\infty} x[n]e^{-j\omega n} = F[x[n]]$$

The z-transform reduces to Fourier transform in the complex z-plane on the contour of a circle with a unit radius. The circle which is called the unit circle plays the

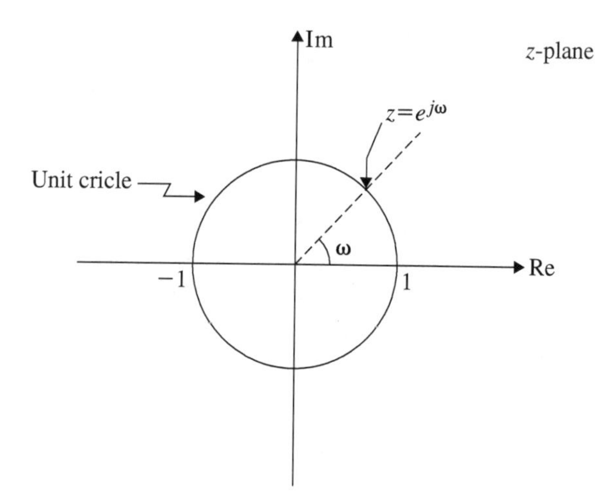

Fig. 9.1 z-transform reduces to FT on the unit circle

role in the z-transform similar to the role of the imaginary axis in the s-plane for the Laplace transform. The unit circle in the z-plane is shown in Fig. 9.1.

9.5 The Region of Convergence (ROC)

In Eq. (9.4) which defines the z-transform $X(z)$, the sum may not coverage for all values of z. The values of z in the complex z-plane for which the sum in the z-transform equation converges is called the region of convergence which is written in abbreviated form as ROC. The concept of ROC is illustrated in the following examples.

■ Example 9.1

Consider the following discrete time signals:

(a) $x[n] = a^n u[n] \quad a < 1$

(b) $x[n] = -a^n u(-n-1) \quad a < 1$

(c) $x[n] = a^n u[n] - b^n u(-n-1) \quad b > a \text{ and } a > b$

Find the z-transform and the ROC in the z-plane.

Solution:

(a) $\boldsymbol{x[n] = a^n u[n]}$

The signal $x[n]$ is shown in Fig. 9.2a which is a right-sided signal.

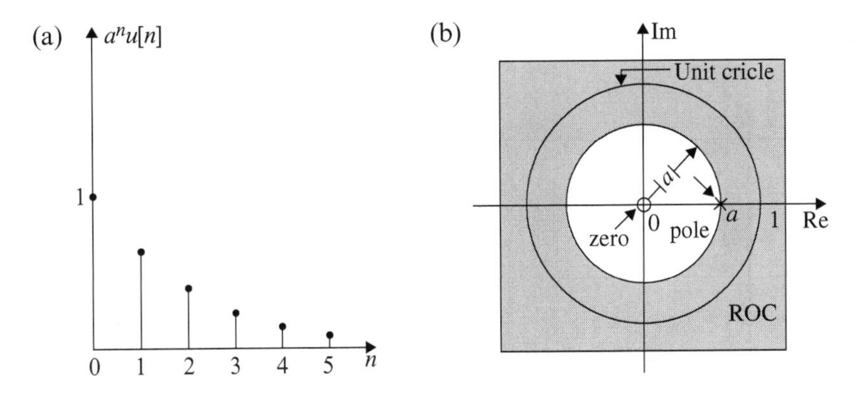

Fig. 9.2 **a** $x[n] = a^n u[n]$ and **b**s ROC: $0 < a < 1$

$$X(z) = \sum_{n=0}^{\infty} a^n u[n] z^{-n}$$

$$= \sum_{n=0}^{\infty} a^n z^{-n} \qquad [\because u[n] = 1 \text{ all } n \geq 0]$$

$$= \sum_{n=0}^{\infty} \left(\frac{a}{z}\right)^n$$

Using the power series, we get

$$X(z) = \frac{1}{\left[1 - \frac{a}{z}\right]}$$

where $\frac{a}{z} < 1$ or $|z| > |a|$.

$$X(z) = \frac{z}{(z-a)} \tag{9.16}$$

$$X(z) = \frac{1}{1 - az^{-1}} \tag{9.17}$$

Fourier transform is represented in the form as shown in Eq. (9.16) to identify poles and zero and system transfer function. Equation (9.17) form is used when inverse z-transform is taken and also for structure realization. z^{-1} is used as time delay operation. z-transform for the causal real exponential converges iff $|z| > |a|$. Thus, the ROC of $X(z)$ is to the exterior of the circle of radius a, which is shown in Fig. 9.2b in shaded area. The ROC includes the unit circle for $|a| < 1$.

(b) $\boldsymbol{x[n] = -a^n[u[-n-1]]}$

The signal $x[n]$ is shown in Fig. 9.3a which is a left-sided signal

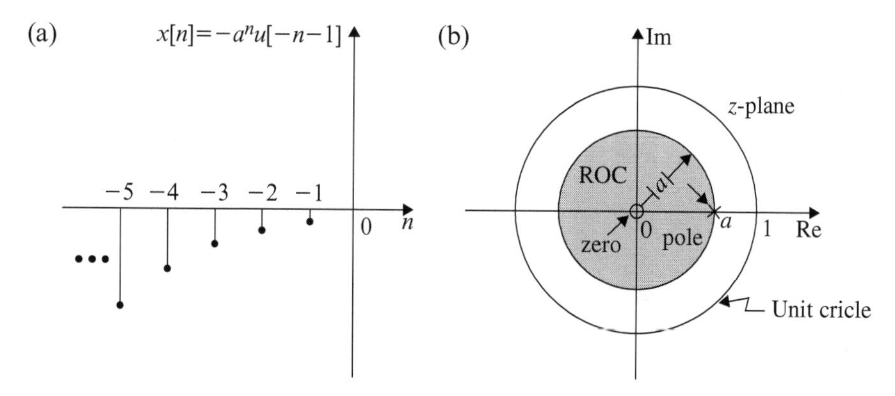

Fig. 9.3 **a** $x[n] = -a^n u[-n-1]$ and **b** ROC: $0 < a < 1$

$$
\begin{aligned}
Z[-a^n u[-n-1]] &= \sum_{n=-\infty}^{-1} -a^n z^{-n} \qquad \because [u(-n-1)] = 1 \text{ for all } -n \\
&= \sum_{n=-\infty}^{-1} -\left[\frac{a}{z}\right]^n = \sum_{n=1}^{\infty} -\left[\frac{z}{a}\right]^n \\
&= -\left[\frac{z}{a} + \frac{z^2}{a^2} + \frac{z^3}{a^3} + \cdots\right] \\
&= 1 - \left[1 + \frac{z}{a} + \left(\frac{z}{a}\right)^2 + \left(\frac{z}{a}\right)^3 + \cdots\right] \\
&= 1 - \frac{1}{1-\frac{z}{a}} \qquad \text{if } \left|\frac{z}{a}\right| < 1
\end{aligned}
$$

$$
\boxed{Z[-a^n u[-n-1]] = \frac{z}{(z-a)}} \quad \text{ROC } |z| < a \quad (9.18)
$$

The z-transforms of $x[n] = a^n u[n]$ which is causal and that of $x[n] = -a^n u[-n-1]$ which is anti-causal are identical. In the former case, the ROC is to the exterior of the circle passing through the outermost pole and in the letter case (anti-causal) the ROC is to the interior of the circle passing through the innermost pole. The ROC is shown in Fig. 9.3b.

(c) $\boldsymbol{x[n] = a^n u[n] - b^n u[-n-1]}$

From the results derived in Example 9.1(a) and (b), we can find the z-transform of $x[n]$ as

$$
X(z) = \frac{z}{(z-a)} + \frac{z}{(z-b)}
$$

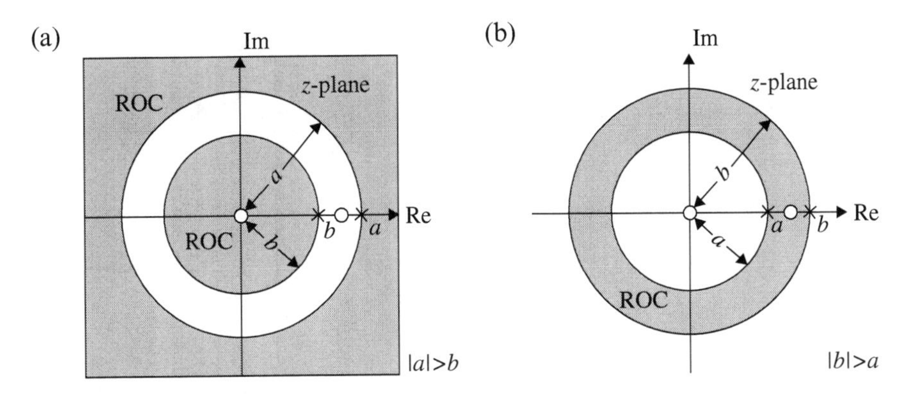

Fig. 9.4 ROC of a two-sided sequence

The right-sided signal $a^n u[n]$ converges if $|z| > a$ and the left-sided signal $-b^n u[-n-1]$ converges if $|z| < b$. The ROC for $|a| > |b|$ and $|a| < |b|$ are shown in Fig. 9.4a and b, respectively. From Fig. 9.4a, it is observed that the two ROCs do not overlap and hence z-transform does not exist for this signal. Now consider Fig. 9.4b; it is observed that the two ROCs overlap and the overlapping area is shaded in the form of a ring. The z-transform exists in the case with ROC as $|a| < |z| < |b|$.

9.6 Properties of the ROC

Assuming that $X(z)$ is the rational function of z, the properties of the ROC are summed up and given below.

1. The ROC is a concentric ring in the z-plane.
2. The ROC does not contain any pole.
3. If $x[n]$ is a finite sequence in a finite interval $N_1 \le n \le N_2$, then the ROC is the entire z-plane except $z = 0$ and $z = \infty$.
4. If $x[n]$ is a right-sided sequence (causal), then the ROC is the exterior of the circle $|z| = r_{\max}$ where $r_{\max}$ is the radius of the outermost pole of $X(z)$.
5. If $x[n]$ is a left-sided sequence (non-causal), then the ROC is the interior of the circle $|z| = r_{\min}$ where $r_{\min}$ is the radius of the innermost pole of $X(z)$.
6. If $x[n]$ is a two-sided sequence, then the ROC is given by $r_1 < |z| < r_2$ where r_1 and r_2 are the magnitudes of the two poles of $X(z)$. Here, ROC is an annular ring between the circle $|z| = r_1$ and $|z| = r_2$ which does not include any poles.

The following examples illustrate the method of finding z-transform $X(z)$ for the discrete time sequence $x[n]$.

■ Example 9.2

Find the z-transform and the ROC for the sequences $x[n]$ given below.

1. $x[n] = \{2,\ -1,\ 0,\ 3,\ 4\}$
   ↑
2. $x[n] = \{1,\ -2,\ 3,\ -1,\ 2\}$
   ↑
3. $x[n] = \{5,\ 3,\ -2,\ 0,\ 4,\ -3\}$
   ↑
4. $x[n] = \delta[n]$
5. $x[n] = u[n]$
6. $x[n] = u[-n]$
7. $x[n] = a^{-n}u[-n]$
8. $x[n] = a^{-n}u[-n-1]$
9. $x[n] = (-a)^n u[-n]$
10. $x[n] = a^{|n|}$ for $|a| < 1$ and $|a| > 1$
11. $x[n] = e^{j\omega_0 n}u[n]$
12. $x[n] = \cos\omega_0 n u[n]$
13. $x[n] = \sin\omega_0 n u[n]$
14. $x[n] = u[n] - u[n-6]$
15. $x[n] = \left[\cos\left(\frac{\pi n}{3} + \frac{\pi}{4}\right)\right]u[n]$

(*Anna University, May, 2007*)

Solution:

1. $\boldsymbol{x[n] = \{2,\ -1,\ 0,\ 3,\ 4\}}$

$$X[z] = \sum_{n=0}^{4} x[n]z^{-n}$$

$$\boxed{X[z] = 2 - z^{-1} + 0 + 3z^{-3} + 4z^{-4}}$$

$X[z]$ will not converge if $|z| = 0$. Hence, ROC is $|z| > 0$.

2. $\boldsymbol{x[n] = \{1,\ -2,\ 3,\ -1,\ 2\}}$
   ↑

$$X[z] = \sum_{n=-4}^{0} x[n]z^{-n}$$

$$\boxed{X[z] = z^4 - 2z^3 + 3z^2 - z + 2}$$

$X[z]$ will not converge if $|z| = \infty$. Hence, ROC is $|z| < \infty$.

3. $\boldsymbol{x[n] = \{5, 3, \underset{\uparrow}{-2}, 0, 4, -3\}}$

$$X[z] = \sum_{n=-2}^{3} x[n]z^{-n}$$

$$\boxed{X[z] = 5z^2 + 3z - 2 + 0 + 4z^{-2} - 3z^{-3}}$$

For $|z| = 0$ and $|z| = \infty$, $X[z]$ is infinity. Hence, ROC is $0 < |z| < \infty$.

4. $\boldsymbol{x[n] = \delta[n]}$

$$X[z] = \sum_{n=-\infty}^{\infty} \delta[n]z^{-n}$$

$$\delta[n] = 1 \quad n = 0$$
$$= 0 \quad n \neq 0$$

$$\boxed{X[z] = 1} \qquad \text{ROC is entire } z\text{-plane}$$

5. $\boldsymbol{x[n] = u[n]}$

$$X[z] = \sum_{n=0}^{\infty} z^{-n}$$
$$= 1 + \frac{1}{z} + \frac{1}{z^2} + \cdots$$
$$= \frac{1}{1 - \frac{1}{z}} \qquad \text{[By using summation formula]}$$

$$\boxed{\begin{aligned} X[z] &= \frac{z}{(z-1)} \\ X[z] &= \frac{1}{(1-z^{-1})} \end{aligned}} \qquad \text{ROC: } |z| > 1 \qquad (9.19)$$

6. $\boldsymbol{x[n] = u[-n]}$

$$\begin{aligned} X[z] &= \sum_{n=-\infty}^{0} z^{-n} \\ &= \sum_{n=0}^{\infty} z^{n} \\ &= 1 + z + z^2 + \cdots \end{aligned}$$

$$\boxed{X[z] = \frac{1}{1-z}} \qquad \text{ROC: } |z| < 1 \qquad (9.20)$$

7. $\boldsymbol{x[n] = a^{-n}u[-n]}$

$$\begin{aligned} X[z] &= \sum_{n=-\infty}^{0} a^{-n}z^{-n} \\ &= \sum_{n=-\infty}^{0} (az)^{-n} \\ &= \sum_{n=0}^{\infty} (az)^{n} \\ &= 1 + (az) + (az)^2 + \cdots \end{aligned}$$

$$\boxed{X[z] = \frac{1}{(1-az)}} \qquad \text{ROC: } |z| < \frac{1}{a} \qquad (9.21)$$

8. $\boldsymbol{x[n] = a^{-n}u[-n-1]}$

$$\begin{aligned} X[z] &= \sum_{n=-\infty}^{-1} a^{-n}z^{-n} \\ &= \sum_{n=-\infty}^{-1} (az)^{-n} \\ &= \sum_{n=1}^{\infty} (az)^{n} \\ &= az + (az)^2 + (az)^3 + \cdots \end{aligned}$$

$$X[z] = az[1 + az + (az)^2 + \cdots]$$
$$= \frac{az}{1 - az}$$

$$\boxed{X[z] = \frac{-z}{\left(z - \frac{1}{a}\right)}} \qquad \text{ROC: } |z| < \frac{1}{a} \tag{9.22}$$

9. $\boldsymbol{x[n] = (-a)^n u[-n]}$

$$X[z] = \sum_{n=-\infty}^{0} (-a)^n z^{-n}$$
$$= \sum_{n=-\infty}^{0} \left(\frac{z}{-a}\right)^{-n}$$
$$= \sum_{n=0}^{\infty} \left(\frac{z}{-a}\right)^{n}$$
$$= 1 + \left(\frac{z}{-a}\right) + \left(\frac{z}{-a}\right)^2 + \left(\frac{z}{-a}\right)^3 + \cdots$$

$$\boxed{X[z] = \frac{a}{(z + a)}} \qquad \text{ROC: } |z| < |a| \tag{9.23}$$

10. $\boldsymbol{x[n] = a^{|n|};\ a < 1}$

$$x[n] = a^n u[n] + a^{-n} u[-n-1]$$
$$Z[a^n u[n]] = \frac{z}{(z - a)} \qquad \text{ROC: } |z| > a$$
$$Z[a^{-n} u[-n-1]] = \frac{-z}{\left(z - \frac{1}{a}\right)} \qquad \text{ROC: } |z| < \frac{1}{a}$$
$$X[z] = \frac{z}{(z - a)} - \frac{z}{\left(z - \frac{1}{a}\right)}$$

$$\boxed{X[z] = \frac{(a^2 - 1)}{a} \frac{z}{(z - a)\left(z - \frac{1}{a}\right)}} \tag{9.24}$$

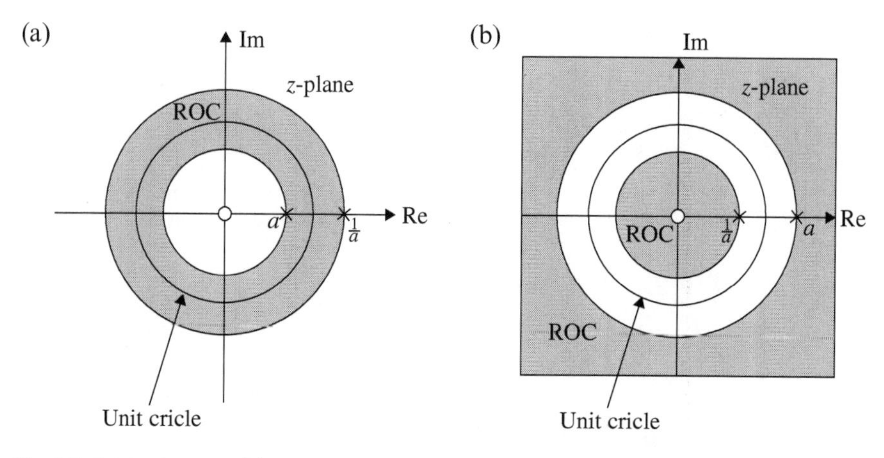

Fig. 9.5 ROC of $x[n] = a^{|n|}$. **a** $a < 1$ and **b** $a > 1$

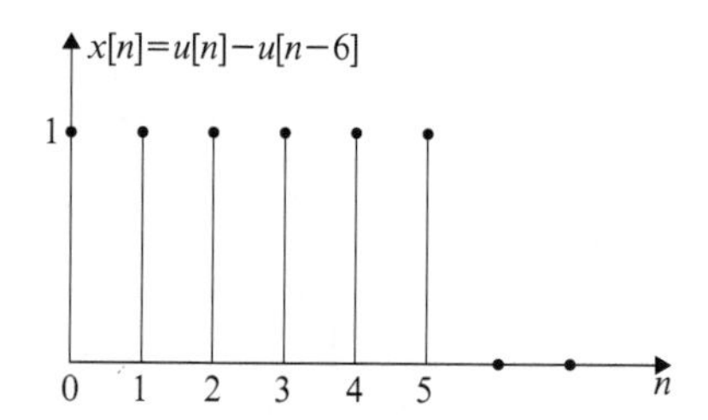

Fig. 9.6 $x[n] = u[n] - u[n-6]$

ROC: $a < |z| < \frac{1}{a}$. The ROC is sketched and shown in Fig. 9.5a for $a < 1$.

$$x[n] = a^{|n|} \qquad a > 1$$

The ROC is sketched and shown in Fig. 9.5b. In Fig. 9.5b, the two ROCs do not overlap and there is no common ROC. Hence, $x[n]$ does not have $X[z]$ (Fig. 9.6).

11. $\boldsymbol{x[n] = e^{j\omega_0 n}u[n]}$

$$\begin{aligned}
X[z] &= \sum_{n=0}^{\infty} e^{j\omega_0 n} z^{-n} \\
&= \sum_{n=0}^{\infty} \left(\frac{e^{j\omega_0}}{z}\right)^n \\
&= \left(\frac{1}{1 - \frac{e^{j\omega_0}}{z}}\right)
\end{aligned}$$

$$X[z] = \frac{z}{(z - e^{j\omega_0})} \qquad \text{ROC: } |z| > |e^{j\omega_0}| \text{ or } |z| > 1 \tag{9.25}$$

12. $\boldsymbol{x[n] = \cos \omega_0 n u[n]}$

$$\begin{aligned}
x[n] &= \frac{1}{2}[e^{j\omega_0 n} + e^{-j\omega_0 n}] \\
Z[e^{j\omega_0 n}] &= \frac{z}{(z - e^{j\omega_0})} \\
Z[e^{-j\omega_0 n}] &= \frac{z}{(z - e^{-j\omega_0})} \\
X[z] &= \frac{1}{2}\left[\frac{z}{(z - e^{j\omega_0})} + \frac{z}{(z - e^{-j\omega_0})}\right] \\
&= \frac{z}{2}\frac{\left[z - e^{-j\omega_0} + z - e^{j\omega_0}\right]}{\left[z^2 - z(e^{-j\omega_0} + e^{j\omega_0}) + 1\right]}
\end{aligned}$$

$$X[z] = \frac{z}{2}\frac{[2z - 2\cos\omega_0]}{\left[z^2 - 2z\cos\omega_0 + 1\right]}$$

$$X[z] = \frac{(1 - z^{-1}\cos\omega_0)}{(1 - z^{-1}2\cos\omega_0 + z^{-2})} \qquad \text{ROC: } |z| > 1 \tag{9.26}$$

13. $\boldsymbol{x[n] = \sin \omega_0 n u[n]}$

$$\begin{aligned}
x[n] &= \frac{1}{2j}[e^{j\omega_0 n} - e^{-j\omega_0 n}] \\
Z[e^{j\omega_0 n}u[n]] &= \frac{z}{(z - e^{j\omega_0})} \\
Z[e^{-j\omega_0 n}u[n]] &= \frac{z}{(z - e^{-j\omega_0})} \\
X[z] &= \frac{z}{2j}\left[\frac{1}{(z - e^{j\omega_0})} - \frac{1}{(z - e^{-j\omega_0})}\right] \\
&= \frac{z}{2j}\frac{\left[z - e^{-j\omega_0} - z + e^{j\omega_0}\right]}{\left[z^2 - 2z\cos\omega_0 + 1\right]} \\
&= \frac{z\sin\omega_0}{(z^2 - 2z\cos\omega_0 + 1)}
\end{aligned}$$

$$X[z] = \frac{z^{-1}\sin\omega_0}{(1 - 2z^{-1}\cos\omega_0 + z^{-2})} \qquad \text{ROC: } |z| > 1 \tag{9.27}$$

14. $\boldsymbol{x[n] = u[n] - u[n-6]}$

$$
\begin{aligned}
x[n] &= \{1,\ 1,\ 1,\ 1,\ 1,\ 1\} \\
X[z] &= 1 + z^{-1} + z^{-2} + z^{-3} + z^{-4} + z^{-5} \\
&= \left[1 + \frac{1}{z} + \frac{1}{z^2} + \frac{1}{z^3} + \frac{1}{z^4} + \frac{1}{z^5}\right]
\end{aligned}
$$

$$
\boxed{X[z] = \frac{[z^5 + z^4 + z^3 + z^2 + z + 1]}{[z^5]} \qquad \text{ROC: all } z \text{ except } z \neq 0} \tag{9.28}
$$

The above result can be represented is a compact form as

$$
\begin{aligned}
X[z] &= \sum_{n=0}^{5} z^{-n} \\
&= \sum_{n=0}^{5} \left(\frac{1}{z}\right)^n
\end{aligned}
$$

The following summation formula is used to simplify this:

$$
\sum_{k=m}^{n} a^k = \frac{a^{n+1} - a^m}{(a-1)}
$$

where $a = \frac{1}{z}$, $k = 0$ and $n = 5$

$$
X[z] = \frac{\left(\frac{1}{z}\right)^6 - \left(\frac{1}{z}\right)^0}{\left(\frac{1}{z} - 1\right)}
$$

$$
\boxed{X[z] = \frac{z}{(z-1)}(1 - z^{-6})}
$$

15. $\boldsymbol{x[n] = \left[\cos\left(\frac{\pi n}{3} + \frac{\pi}{4}\right)\right] u[n]}$

$$
\begin{aligned}
x[n] &= \frac{1}{2}\left[e^{j\left(\frac{\pi n}{3} + \frac{\pi}{4}\right)} + e^{-j\left(\frac{\pi n}{3} + \frac{\pi}{4}\right)}\right] \\
&= \frac{1}{2}\left[e^{j\frac{\pi}{4}} e^{j\frac{\pi n}{3}} + e^{-j\frac{\pi}{4}} e^{-j\frac{\pi}{4}}\right] \\
X[z] &= \frac{1}{2}\left[e^{j\frac{\pi}{4}} \frac{z}{(z - e^{j\frac{\pi}{3}})} + e^{-j\frac{\pi}{4}} \frac{z}{(z - e^{-j\frac{\pi}{3}})}\right]
\end{aligned}
$$

$$X[z] = \frac{z}{2}\frac{\left[ze^{j\frac{\pi}{4}} - e^{-j\frac{\pi}{12}} + ze^{-j\frac{\pi}{4}} - e^{-j\frac{\pi}{12}}\right]}{z^2 - z(e^{j\frac{\pi}{3}} + e^{-j\frac{\pi}{3}}) + 1}$$
$$= \frac{z}{2}\frac{\left[2z\cos\frac{\pi}{4} - 2\cos\frac{\pi}{12}\right]}{\left(z^2 - 2z\cos\frac{\pi}{3} + 1\right)}$$

$$\boxed{X[z] = \frac{z[0.707z - 0.966]}{(z^2 - z + 1)} \qquad \text{ROC: } |z| > 1}$$

9.7 Properties of z-Transform

The transformations of $x(t)$ and $x[n]$ to $X(s)$, and $X(j\omega)$ using Laplace transform and Fourier transform, respectively, as seen from Chaps. 6 and 8 become easier if the properties of these transforms are directly applied. Similarly, if the properties of z-transform are applied directly to $x[n]$, then $X[z]$ can be easily derived. Hence, some of the important properties of z-transform which are applied to signals and systems are derived and the applications illustrated. The following properties are derived:

1. Linearity;
2. Time shifting;
3. Time reversal;
4. Multiplication by n;
5. Multiplication by an exponential;
6. Time expansion;
7. Convolution theorem;
8. Initial value theorem;
9. Final value theorem.

9.7.1 *Linearity*

If

$$x_1[n] \overset{Z}{\longleftrightarrow} X_1[z] \quad \text{and} \quad x_2[n] \overset{Z}{\longleftrightarrow} X_2[z]$$

then

$$\{a_1x_1[n] + a_2x_2[n]\} \overset{Z}{\longleftrightarrow} [a_1X_1[z] + a_2X_2[z]] \tag{9.29}$$

Proof Let

$$x[n] = a_1x_1[n] + a_2x_2[n]$$
$$X[z] = \sum_{n=-\infty}^{\infty} [a_1x_1[n] + a_2x_2[n]]z^{-n}$$
$$= \sum_{n=-\infty}^{\infty} a_1x_1[n]z^{-n} + \sum_{n=-\infty}^{\infty} a_2x_2[n]z^{-n}$$

$$\boxed{X[z] = a_1x_1[z] + a_2x_2[z]}$$

9.7.2 Time Shifting

If

$$x[n] \overset{Z}{\longleftrightarrow} X[z]$$

then

$$x[n-k] \overset{Z}{\longleftrightarrow} z^{-k}X[z]$$

Proof Let

$$Z[x[n-k]] = \sum_{n=-\infty}^{\infty} x[n-k]z^{-n}$$

Substitute $(n-k) = m$

$$Z[x[n-k]] = \sum_{m=-\infty}^{\infty} x[m]z^{-(k+m)}$$
$$= \sum z^{-k}x[m]z^{-m}$$

$$\boxed{Z[x[n-k]] = z^{-k}X[z]} \tag{9.30}$$

9.7.3 Time Reversal

If

$$x[n] \overset{Z}{\longleftrightarrow} X[z] \qquad \text{ROC: } r_1 < |z| < r_2$$

then

$$x[-n] \overset{Z}{\longleftrightarrow} X[z^{-1}] \qquad \text{ROC: } \frac{1}{r_1} < |z| < \frac{1}{r_2}$$

Proof Let

$$Z[x[-n]] = \sum_{n=-\infty}^{\infty} x[-n]z^{-n}$$

Substitute $-n = m$

$$Z[x[-n]] = \sum_{n=\infty}^{-\infty} x[m]z^{m}$$
$$= \sum_{m=-\infty}^{\infty} x[m](z^{-1})^{m}$$

$$\boxed{Z[x[-n]] = X[z^{-1}]} \tag{9.31}$$

Thus, according to time reversal property, folding the signal in the time domain is equivalent to replacing z by z^{-1}. Further, the ROC of $X[z]$ which is $r_1 < |z| < r_2$ becomes $r_1 < |z^{-1}| < r_2$ which is $\frac{1}{r_2} < |z| < \frac{1}{r_1}$.

9.7.4 Multiplication by n

If

$$Z[x[n]] = X[z]$$

then

$$Z[nx[n]] = -z\frac{d}{dz}X[z]$$

Proof Let

$$X[z] = \sum_{n=-\infty}^{\infty} x[n]z^{-n}$$

$$Z[nx[n]] = \sum_{n=-\infty}^{\infty} nx[n]z^{-n}$$

$$= z\sum_{n=-\infty}^{\infty} nx[n]z^{-n-1}$$

$$= z\sum_{n=-\infty}^{\infty} x[n][nz^{-n-1}]$$

$$Z[nx[n]] = z\sum_{n=-\infty}^{\infty} -x[n]\frac{d}{dz}[z^{-n}]$$

$$= -z\frac{d}{dz}\sum_{n=-\infty}^{\infty} x[n]z^{-n}$$

$$\boxed{Z[nx[n]] = -z\frac{d}{dz}X[z]} \tag{9.32}$$

9.7.5 Multiplication by an Exponential

If

$$Z[x[n]] = X[z]$$

then

$$Z[a^n x[n]] = X[a^{-1}z]$$

Proof Let

$$Z[a^n x[n]] = \sum_{n=-\infty}^{\infty} a^n x[n]z^{-n}$$

$$= \sum_{n=-\infty}^{\infty} x[n][a^{-1}z]^{-n}$$

$$\boxed{Z[a^n x[n]] = X[a^{-1}z]} \tag{9.33}$$

ROC: $r_1 < |a^{-1}z| < r_2$ or $ar_1 < |z| < ar_2$. In $X[z]$, z is replaced by $\frac{z}{a}$.

9.7.6 Time Expansion

If

$$Z[x[n]] = X[z]$$

then

$$Z[x_k[n]] = X[z^k]$$

Proof

$$Z[x_k[n]] = \sum_{n=-\infty}^{\infty} x\left[\frac{n}{k}\right] z^{-n}$$

where n is multiple of k. Substitute $\frac{n}{k} = l$

$$\begin{aligned} Z[x_k[n]] &= \sum_{l=-\infty}^{\infty} x[l] z^{-kl} \\ &= \sum_{l=-\infty}^{\infty} x[l][z^k]^{-l} = X[z^k] \end{aligned}$$

$$\boxed{Z[x_k[n]] = X[z^k]} \tag{9.34}$$

9.7.7 Convolution Theorem

If

$$y[n] = x[n] * h[n]$$

then

$$Y[z] = X[z]H[z]$$

Proof

$$\begin{aligned} y[n] &= \sum_{k=-\infty}^{\infty} x[k]h[n-k] \\ Y[Z] &= \sum_{n=-\infty}^{\infty} \left[\sum_{k=-\infty}^{\infty} x[k]h[n-k] \right] z^{-n} \\ &= \sum_{k=-\infty}^{\infty} x[k]z^{-k} \sum_{n=-\infty}^{\infty} h[n-k]z^{-(n-k)} \end{aligned}$$

Substitute $(n-k) = l$

$$Y[z] = \sum_{k=-\infty}^{\infty} x[k]z^{-k} \sum_{l=-\infty}^{\infty} h[l]z^{-l}$$

$$\boxed{Y[z] = X[z]Y[z]} \tag{9.35}$$

9.7.8 Initial Value Theorem

If

$$X[z] = Z[x[n]]$$

where $x[n]$ is causal, then

$$x[0] = \underset{z\to\infty}{Lt}\, X[z]$$

Proof For a causal signal $x[n]$,

$$\begin{aligned} X[z] &= \sum_{n=0}^{\infty} x[n]z^{-n} \\ &= x[0] + x[1]z^{-1} + x[2]z^{-2} + \cdots \end{aligned}$$

Taking $z \to \infty$ on both sides, we get

$$\begin{aligned} \underset{z\to\infty}{Lt}\, X[z] &= \underset{z\to\infty}{Lt}\, [x[0] + x[1]z^{-1} + x[2]z^{-2} + \cdots] \\ &= x[0] \end{aligned}$$

$$\boxed{x[0] = \underset{z\to\infty}{Lt}\, X[z]} \tag{9.36}$$

9.7.9 *Final Value Theorem*

If $Z[x[n]] = X[z]$ where $x[n]$ is a causal signal and the ROC of $X[z]$ has no poles on or outside the unit circle then

$$x[\infty] = \underset{z\to 1}{Lt}\,(z-1)X[z]$$

Proof

$$Z[x[n+1]] - Z[x[n]] = \underset{k\to\infty}{Lt} \sum_{n=0}^{k} [x[n+1] - x[n]]z^{-n}$$

$$x[\infty] = \underset{k\to\infty}{Lt} \sum_{n=0}^{k} [x[n+1] - x[n]]z^{-n}$$

$$zX[z] - x[0] - X[z] = \underset{k\to\infty}{Lt} \sum_{n=0}^{k} [x[n+1] - x[n]]z^{-n}$$

$$(z-1)X[z] - x[0] = \underset{k\to\infty}{Lt} \sum_{n=0}^{k} [x[n+1] - x[n]]z^{-n}$$

Taking $\underset{z\to\infty}{Lt}$ on both sides, we get

$$\begin{aligned}
&\underset{z\to\infty}{Lt}\,(z-1)X[z] - x[0] \\
&= \underset{k\to\infty}{Lt}\,[x[1] - x[0]] + [x[2] - x[-1]] + [x[3] - x[2]] + \cdots + [x[k+1] - x[k]] \\
&= x[\infty] - x[0]
\end{aligned}$$

$$\boxed{x[\infty] = \underset{z\to 1}{Lt}\,(z-1)X[z]} \tag{9.37}$$

■ Example 9.3

Find the z-transform of the following sequences and also ROC using the properties of z-transform:

1. $x[n] = \delta[n - n_0]$
2. $x[n] = u[n - n_0]$
3. $x[n] = a^{n+1}u[n + 1]$
4. $x[n] = a^{n-1}u[n - 1]$
5. $x[n] = \left(\frac{1}{2}\right)^n u[-n]$

   (*Anna University*, *December*, 2007)
6. $x[n] = u[n - 6] - u[n - 10]$

7. $x[n] = nu[n]$
8. $x[n] = n[u[n] - u[n - 8]]$
9. $x[n] = a^n \cos \omega_0 n u[n]$
10. $x[n] = a^n \sin \omega_0 n u[n]$
11. show that $u[n] * u[n - 1] = nu[n]$
12. $x[n] = n\left(-\frac{1}{4}\right)^n u[n] * \left(\frac{1}{6}\right)^{-n} u[-n]$
13. $x[n] = \left[\left(\frac{1}{2}\right)^n - \left(\frac{1}{4}\right)^n\right] u[n]$

    Find $X[z]$ and plot the poles and zeros. (*Anna University*, *December*, 2007)
14. $x[n] = 1 \quad n \geq 0$

    $= z^n \quad n < 0$

    (*AnnaUniversity, April, 2005*)
15. (a) $x[n] = \left[\left(-\frac{1}{3}\right)^n + 3\left(\frac{1}{6}\right)^n\right] u[n]$

    (b) $x[n] = \left[\left(-\frac{1}{3}\right)^n u[-n] + 3\left(\frac{1}{6}\right)^n\right] u[n]$

    (c) $x[n] = \left[\left(-\frac{1}{3}\right)^n + 3\left(\frac{1}{6}\right)^n\right] u[-n]$
16. (a) $x[n] = \left[\left(\frac{1}{4}\right)^n + \left(\frac{1}{5}\right)^n\right] u[n]$

    (b) $x[n] = \left[\left(\frac{1}{5}\right)^n u[n] + \left(\frac{1}{4}\right)^n u[-n - 1]\right]$

    (c) $x[n] = \left(\frac{1}{4}\right)^n u[n] + \left(\frac{1}{5}\right)^n u[-n - 1]$
17. $x[n] = \delta[n] + \frac{1}{2}\delta(n + 1) + \delta(n - 3)$ (*Anna University, December, 2006*)

18. $x[n] = 4^n \cos\left[\frac{2\pi n}{6} + \frac{\pi}{4}\right] u[-n-1]$. Sketch the pole-zero plot and indicate the ROC. *(AnnaUniversity, April, 2008)*

19. $x[n] = nu[n-1]$ *(Anna University, December, 2006)*

20. $$x[n] = (4)^n \quad n < 0$$
$$= \left(\frac{1}{4}\right)^n \quad n = 0, 2, 4, \ldots$$
$$= \left(\frac{1}{5}\right)^n \quad n = 1, 3, 5, \ldots$$

Solution:

1. $\boldsymbol{x[n] = \delta[n - n_0]}$

$$\delta[n] \stackrel{Z}{\longleftrightarrow} 1 \qquad \text{ROC: } |z| > 0$$

By applying the time shifting property, we get

$$Z[\delta[n - n_0]] = z^{-n_0} \tag{9.38}$$

ROC: all z excluding $|z| = 0$.

2. $\boldsymbol{x[n] = u[n - n_0]}$

$$u[n] \stackrel{Z}{\longleftrightarrow} \frac{z}{(z-1)}$$

By applying the time shifting (right shifted) property, we get

$$Z[u[n - n_0]] = \frac{z^{-n_0} z}{(z-1)} = \frac{z^{-(n_0-1)}}{(z-1)}$$

$$\boxed{X[z] = \frac{z^{-(n_0-1)}}{(z-1)} \qquad \text{ROC: } 1 < |z| < \infty} \tag{9.39}$$

3. $\boldsymbol{x[n] = a^{n+1} u[n+1]}$

$$a^n u[n] \stackrel{Z}{\longleftrightarrow} \frac{z}{(z-a)}$$

By applying the time shifting (left shifted) property, we get

$$Z[a^{n+1} u[n+1]] = z \frac{z}{(z-a)}$$

$$\boxed{X[z] = \frac{z^2}{(z-a)} \qquad \text{ROC: } |a| < |z| < \infty} \tag{9.40}$$

4. $\boldsymbol{x[n] = a^{n-1}u[n-1]}$

$$a^n u[n] \stackrel{Z}{\longleftrightarrow} \frac{z}{(z-a)}$$

Applying the time shifting (right shifted) property, we get

$$Z[a^{n-1}u[n-1]] = \frac{z^{-1}z}{(z-a)}$$

$$\boxed{X[z] = \frac{1}{(z-a)} \qquad \text{ROC: } a < |z| < \infty} \tag{9.41}$$

5. $\boldsymbol{x[n] = \left(\frac{1}{2}\right)^n u[-n]}$

$$u[-n] \stackrel{Z}{\longleftrightarrow} \frac{1}{(1-z)}$$

$$x[n] = \left(\frac{1}{2}\right)^n u[-n]$$

By using the multiplication property (replacing z by $(\frac{1}{2})^{-1}z$), we get

$$\boxed{X[z] = \frac{1}{(1-2z)}} \qquad \text{ROC: } |z| < \frac{1}{2}$$

6. $\boldsymbol{x[n] = u[n-6] - u[n-10]}$
The signal is represented in Fig. 9.7.

$$\begin{aligned} X[z] &= z^{-6} + z^{-7} + z^{-8} + z^{-9} \\ &= \frac{1}{z^6} + \frac{1}{z^7} + \frac{1}{z^8} + \frac{1}{z^9} \end{aligned}$$

$$\boxed{X[z] = \frac{z^8 + z^2 + z + 1}{z^9}}$$

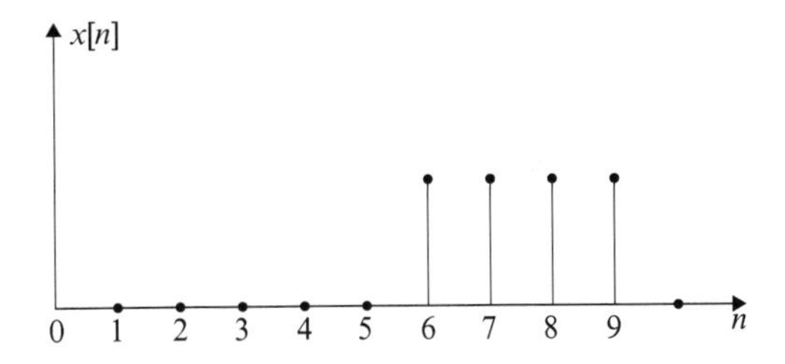

Fig. 9.7 $x[n] = u[n-6] - u[n-10]$

ROC: all z except $z \neq 0$. The above result can be simplified using the summation formula as

$$X[z] = \sum_{n=6}^{9} \left(\frac{1}{z}\right)^n$$

$$= \frac{\left(\frac{1}{z}\right)^{10} - \left(\frac{1}{z}\right)^6}{\left(\frac{1}{z} - 1\right)}$$

$$\boxed{X[z] = \frac{z}{(z-1)}[z^{-6} - z^{-10}]}$$

The above result can be obtained by the time shifting property of unit step sequence.

$$Z[u[n-6]] = \frac{z}{(z-1)}z^{-6}$$

$$Z[u[n-10]] = \frac{z}{(z-1)}z^{-10}$$

$$X[z] = \frac{z}{(z-1)}[z^{-6} - z^{-10}]$$

$$\boxed{X[z] = \frac{(z^{-5} - z^{-9})}{(z-1)}}$$

7. $\boldsymbol{x[n] = nu[n]}$

$$Z[u[n]] = \frac{z}{(z-1)}$$

Applying the differentiation property in z,

$$Z[nu[n]] = -z\frac{dX[z]}{dz}$$
$$Z[nu[n]] = -z\frac{d}{dz}\left[\frac{z}{(z-1)}\right]$$

$$\boxed{X[z] = \frac{z}{(z-1)^2}} \tag{9.42}$$

8. **$x[n] = n[u[n] - u[n-8]]$**
   By using the shift theorem, we get

$$Z[u[n] - u[n-8]] = \frac{z}{(z-1)}[1 - z^{-8}]$$
$$= \frac{(z - z^{-7})}{(z-1)}$$
$$Z[n[u[n] - u[n-8]]] = -z\frac{d}{dz}\frac{[z - z^{-7}]}{z-1}$$
$$X[z] = -z\frac{[(z-1)(1+7z^{-8}) - (z - z^{-7})]}{(z-1)^2}$$
$$X[z] = \frac{(-8z^{-6} + 7z^{-7} + z)}{(z-1)^2}$$

$$\boxed{X[z] = \frac{[z^8 - 8z + 7]}{z^7(z-1)^2}}$$

9. **$x[n] = a^n \cos \omega_0 n u[n]$**
   For Example 9.2.12, we get

$$Z[\cos \omega_0 n u[n]] = \frac{[1 - z^{-1}\cos \omega_0]}{[1 - 2\cos \omega_0 z^{-1} + z^{-2}]}$$

To apply the multiplication property, replace z by $|\frac{z}{a}|$ or $z^{-1} = |\frac{z}{a}|^{-1} = az^{-1}$

$$\therefore\ Z[a^n \cos \omega_0 n u[n]] = \frac{[1 - az^{-1}\cos \omega_0]}{[1 - 2a\cos \omega_0 z^{-1} + a^2 z^{-2}]}$$

$$\boxed{X[z] = \frac{[1 - az^{-1}\cos \omega_0]}{[1 - 2a\cos \omega_0 z^{-1} + a^2 z^{-2}]}} \tag{9.43}$$

10. $\boldsymbol{x[n] = a^n \sin \omega_0 n u[n]}$
For Example 9.2.13, we get

$$Z[\sin \omega_0 n u[n]] = \frac{z^{-1} \sin \omega_0}{[1 - 2\cos \omega_0 z^{-1} + z^{-2}]}$$

To apply the multiplication property, as in the previous example, replace z^{-1} by a z^{-1} and $z^{-2} = a^2 z^{-2}$

$$Z[a^n \sin \omega_0 n u[n]] = \frac{a z^{-1} \sin \omega_0}{[1 - 2a \cos \omega_0 z^{-1} + a^2 z^{-2}]}$$

$$\boxed{X[z] = \frac{[a z^{-1} \sin \omega_0]}{[1 - 2a \cos \omega_0 z^{-1} + a^2 z^{-2}]}} \tag{9.44}$$

11. **Show that** $\boldsymbol{u[n] * u[n-1] = n u[n]}$

$$\begin{aligned}
Z[u[n]] &= \frac{z}{(z-1)} \\
Z[u[n-1]] &= \frac{1}{(z-1)} \\
Z[u[n] * u[n-1]] &= Z[u[n]]Z[u[n-1]] \\
&= \frac{z}{(z-1)} \frac{1}{(z-1)} \\
&= \frac{z}{(z-1)^2}
\end{aligned}$$

Multiplying by Z^{-1} both sides, we get

$$u[n] * u[n-1] = Z^{-1}\left[\frac{z}{(z-1)^2}\right]$$

$$\boxed{u[n] * u[n-1] = n[u[n]]}$$

12. $\boldsymbol{x[n] = n\left(-\frac{1}{4}\right)^n u[n] * \left(-\frac{1}{6}\right)^{-n} u[-n]}$

$$x_1[n] = \left(-\frac{1}{4}\right)^n u[n] \overset{Z}{\longleftrightarrow} \frac{z}{\left(z + \frac{1}{4}\right)}$$

$$n\left[\left(-\frac{1}{4}\right)^n u[n]\right] \overset{Z}{\longleftrightarrow} -z \frac{d}{dz} \frac{z}{\left(z + \frac{1}{4}\right)}$$

$$= -z\frac{\left[z + \frac{1}{4} - z\right]}{\left(z + \frac{1}{4}\right)^2}$$

$$= \frac{\left[-\frac{z}{4}\right]}{\left(z + \frac{1}{4}\right)^2} \qquad \text{ROC: } |z| > \frac{1}{4}$$

$$x_2[n] = \left(\frac{1}{6}\right)^n u[n] \stackrel{Z}{\longleftrightarrow} \frac{z}{\left(z - \frac{1}{6}\right)} \qquad \text{ROC: } |z| > \frac{1}{6}$$

If the time reversal property is used, z is to be replaced by z^{-1}

$$\left(\frac{1}{6}\right)^{-n} u[-n] \stackrel{Z}{\longleftrightarrow} \frac{z^{-1}}{\left(z^{-1} - \frac{1}{6}\right)}$$

$$X_1[z] = -\frac{6}{z-6} \qquad \text{ROC: } |z| < 6$$

$$X[z] = X_1[z]X_2[z]$$

$$= \frac{\frac{z}{4}6}{\left(z + \frac{1}{4}\right)^2 (z-6)}$$

$$\boxed{X[z] = \frac{1.5z}{\left(z + \frac{1}{4}\right)(z-6)}} \qquad \text{ROC: } \frac{1}{4} < |z| < 6$$

13. $\boldsymbol{x[n] = \left[\left(\frac{1}{2}\right)^n - \left(\frac{1}{4}\right)^n\right] u[n]}$
Find $X[z]$ and plot the poles and zeros. *(Anna University, December, 2007)*

$$x_1[n] = \left(\frac{1}{2}\right)^n u[n] \stackrel{Z}{\longleftrightarrow} \frac{z}{\left(z - \frac{1}{2}\right)}$$

$$x_2[n] = \left(\frac{1}{4}\right)^n u[n] \stackrel{Z}{\longleftrightarrow} \frac{z}{\left(z - \frac{1}{4}\right)}$$

$$x[n] = x_1[n] - x_2[n]$$

$$X[z] = X_1[z] - X_2[z]$$

$$= \frac{z}{(z-0.5)} - \frac{z}{(z-0.25)}$$

$$\boxed{X[z] = \frac{z0.25}{(z-0.5)(z-0.25)}}$$

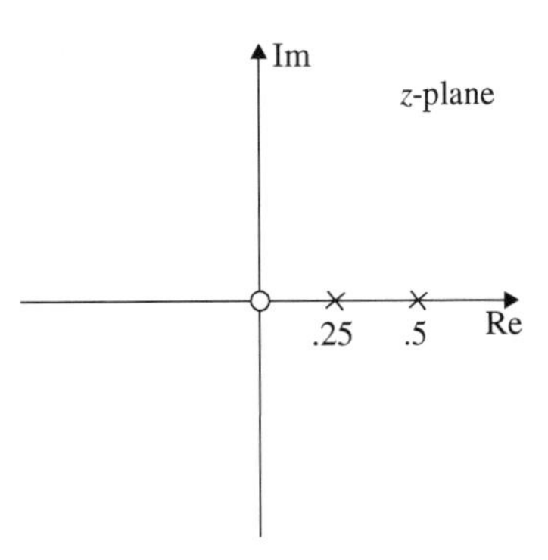

Fig. 9.8 Pole-zero plot

The pole-zero plot is shown in Fig. 9.8.

14.

$$\begin{aligned} \boldsymbol{x[n]} &= \mathbf{1} \qquad \boldsymbol{n \geq 0} \\ &= \mathbf{3^n} \qquad \boldsymbol{n < 0} \end{aligned}$$

(*Anna University, April, 2005*)

$$\begin{aligned} x[n] &= u[n] + 3^n u[-n-1] \\ &= x_1[n] + x_2[n] \\ X_1[z] &= \frac{z}{(z-1)} \qquad \text{ROC: } |z| > 1 \\ x_2[n] &= (3)^n u[-n-1] \end{aligned}$$

Using time reversal and multiplication properties, we get

$$\begin{aligned} X_2[z] &= -\frac{z}{(z-3)} \qquad \text{ROC: } |z| < 3 \\ X[z] &= X_1(z) + X_2(z) \\ &= \frac{z}{(z-1)} - \frac{z}{(z-3)} \end{aligned}$$

$$\boxed{X[z] = \frac{-2z}{(z-1)(z-3)}} \qquad \text{ROC: } 1 < z < 3$$

15.

(a) $\boldsymbol{x[n] = \left(-\frac{1}{3}\right)^n u[n] + 3\left(\frac{1}{6}\right)^n u[n]}$

(b) $\boldsymbol{x[n] = \left[\left(-\frac{1}{3}\right)^n u[-n] + 3\left(\frac{1}{6}\right)^n\right] u[n]}$

(c) $\boldsymbol{x[n] = \left[\left(-\frac{1}{3}\right)^n + 3\left(\frac{1}{6}\right)^n\right] u[-n]}$

(a)

$$x[n] = \left(-\frac{1}{3}\right)^n + 3\left(\frac{1}{6}\right)^n u[n]$$
$$= x_1[n] + x_2[n]$$
$$X_1[z] = \frac{z}{\left(z + \frac{1}{3}\right)} \qquad \text{ROC: } |z| > -\frac{1}{3}$$
$$X_2[z] = 3\frac{z}{\left(z - \frac{1}{6}\right)} \qquad \text{ROC: } |z| > \frac{1}{6}$$

$$X[z] = X_1[z] + X_2[z]$$
$$= z\left[\frac{1}{\left(z + \frac{1}{3}\right)} + \frac{3}{\left(z - \frac{1}{6}\right)}\right] \qquad \text{ROC: } |z| > \frac{1}{6}$$

(b)

$$x[n] = \left[\left(-\frac{1}{3}\right)^n u[-n] + 3\left(\frac{1}{6}\right)^n\right] u[n]$$
$$= x_1[n] + x_2[n]$$
$$x_1[n] = \left(-\frac{1}{3}\right)^n u[-n]$$

Applying the properties of time reversal and multiplication, we get

$$X_1[z] = \frac{1}{(1+3z)} \qquad \textit{See } \text{Example 9.3.12 ROC: } |z| < \frac{1}{3}$$
$$x_2[n] = 3\left(\frac{1}{6}\right)^n u[n]$$
$$X_2[z] = \frac{3z}{\left(z - \frac{1}{6}\right)} \qquad \text{ROC: } |z| > \frac{1}{6}$$
$$X[z] = X_1[z] + X_2[z]$$

$$\boxed{X[z] = \left[\frac{1}{(1+3z)} + \frac{3z}{\left(z - \frac{1}{6}\right)}\right]} \qquad \text{ROC: } \frac{1}{6} < |z| < \frac{1}{3}$$

(c)

$$x[n] = \left[\left(-\frac{1}{3}\right)^n + 3\left(\frac{1}{6}\right)^n\right] u[-n]$$
$$= x_1[n] + x_2[n]$$
$$X_1[z] = \frac{1}{(1+3z)} \qquad \text{ROC: } |z| < \frac{1}{3}$$

The derivation is given in Example 9.3.15(b),

$$x_2[n] = 3\left(\frac{1}{6}\right)^n u[-n]$$
$$u[-n] \overset{Z}{\longleftrightarrow} \frac{1}{(1-z)}$$

Applying the multiplication property, we get

$$Z\left(\frac{1}{6}\right)^n u[-n] \overset{Z}{\longleftrightarrow} \frac{1}{(1-6z)} \qquad \text{ROC: } |z| > \frac{1}{6}$$

$$\boxed{X[z] = \left[\frac{1}{(1+3z)} + \frac{3}{(1-6z)}\right]} \quad \text{ROC: } \frac{1}{6} < |z| < \frac{1}{3}$$

16. (a)

$$\boldsymbol{x[n] = \left[\left(\frac{1}{4}\right)^n + \left(\frac{1}{5}\right)^n\right] u[n]}$$

Applying the results of Eq. (9.16), we get

$$\boxed{X[z] = \left[\frac{z}{\left(z-\frac{1}{4}\right)} + \frac{z}{\left(z-\frac{1}{5}\right)}\right]} \qquad \text{ROC: } |z| > \frac{1}{4}$$

(b)

$$\boldsymbol{x[n] = \left[\left(\frac{1}{5}\right)^n u[n] + \left(\frac{1}{4}\right)^n u[-n-1]\right]}$$

$$\left(\frac{1}{5}\right)^n u[n] \overset{Z}{\longleftrightarrow} \frac{z}{\left(z-\frac{1}{5}\right)} \qquad \text{ROC: } |z| > \frac{1}{5}$$

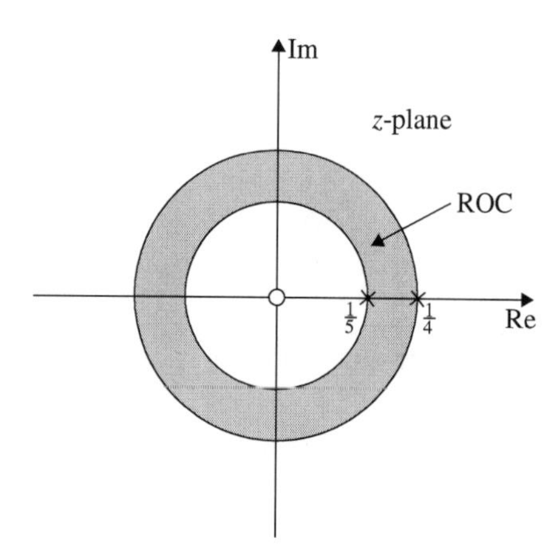

Fig. 9.9 $X[z]$ and its ROC of Example 16(b)

$$\left(\frac{1}{4}\right)^n u[-n-1] \overset{Z}{\longleftrightarrow} \frac{-z}{\left(z-\frac{1}{4}\right)} \qquad \text{ROC: } |z| < \frac{1}{4}$$

$$\left(\frac{1}{5}\right)^n u[n] + \left(\frac{1}{4}\right)^n u[-n-1] \overset{Z}{\longleftrightarrow} \frac{z}{\left(z-\frac{1}{5}\right)} - \frac{z}{\left(z-\frac{1}{4}\right)}$$

$$\boxed{X[z] = \frac{-\frac{z}{20}}{\left(z-\frac{1}{5}\right)\left(z-\frac{1}{4}\right)}} \qquad \text{ROC: } \frac{1}{5} < |z| < \frac{1}{4}$$

The poles and zero and the ROC are marked in Fig. 9.9.

(c)

$$\boldsymbol{x[n] = \left[\left(\frac{1}{4}\right)^n u[n] + \left(\frac{1}{5}\right)^n\right] u[-n-1]}$$

$$\left(\frac{1}{4}\right)^n u[n] \overset{Z}{\longleftrightarrow} \frac{z}{\left(z-\frac{1}{4}\right)} \qquad \text{ROC: } |z| > \frac{1}{4}$$

$$\left(\frac{1}{5}\right)^n u[-n-1] \overset{Z}{\longleftrightarrow} \frac{-z}{\left(z-\frac{1}{5}\right)} \qquad \text{ROC: } |z| < \frac{1}{5}$$

The ROCs of the above two equations are shown in Fig. 9.10, and it is seen that they do not overlap and thus the given $x[n]$ does not have $X[z]$.

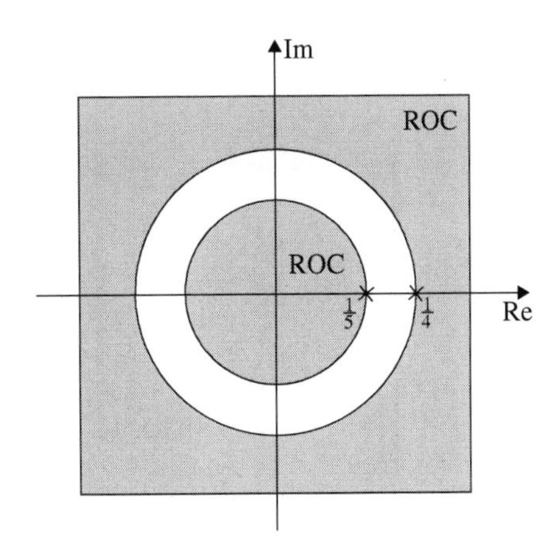

Fig. 9.10 ROC of Example 9.16(c)

17. $\boldsymbol{x[n] = \delta[n] + \frac{1}{2}\delta(n+1) + \delta(n-3)}$

$$\delta[n] \overset{Z}{\longleftrightarrow} 1$$
$$\frac{1}{2}\delta[n+1] \overset{Z}{\longleftrightarrow} \frac{1}{2}z$$
$$\delta[n-3] \overset{Z}{\longleftrightarrow} z^{-3}$$

$$\boxed{X[z] = 1 + \frac{1}{2}z + z^{-3}}$$

18. $\boldsymbol{x[n] = 4^n \cos\left[\frac{2\pi n}{6} + \frac{\pi}{4}\right] u[-n-1]}$

$$\cos\left(\frac{2\pi n}{6}\right) = \cos\frac{\pi n}{3}$$
$$\cos\left(\frac{2\pi n}{6} + \frac{\pi}{4}\right) = \frac{e^{j\left(\frac{\pi}{4}+\frac{\pi n}{3}\right)} + e^{-j\left(\frac{\pi}{4}+\frac{\pi n}{3}\right)}}{2}$$
$$= \frac{1}{2}e^{j\left(\frac{\pi}{4}\right)}e^{j\left(\frac{\pi n}{3}\right)} + \frac{1}{2}e^{-j\left(\frac{\pi}{4}\right)}e^{-j\left(\frac{\pi n}{3}\right)}$$
$$4^n \cos\left(\frac{2\pi n}{6} + \frac{\pi}{n}\right) = \frac{1}{2}e^{j\left(\frac{\pi}{4}\right)}\left(4e^{j\frac{\pi}{3}}\right)^n + \frac{1}{2}e^{-j\frac{\pi}{4}}\left(4e^{-j\frac{\pi}{3}}\right)^n$$

From Eq. (9.18),

$$\left(4e^{j\frac{\pi}{3}}\right)^n u[-n-1] \overset{Z}{\longleftrightarrow} \frac{-z}{\left(z-4e^{j\frac{\pi}{3}}\right)} \qquad \text{ROC: } |z| < 4$$
$$\left(4e^{-j\frac{\pi}{3}}\right)^n u[-n-1] \overset{Z}{\longleftrightarrow} \frac{-z}{\left(z-4e^{-j\frac{\pi}{3}}\right)} \qquad \text{ROC: } |z| < 4$$

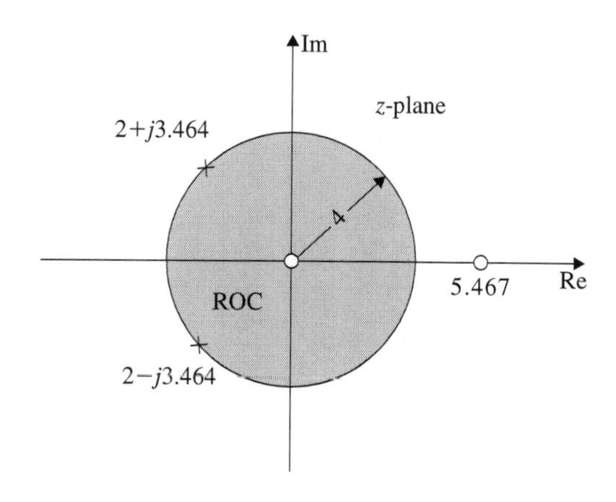

Fig. 9.11 Poles and zeros and ROC of Example 9.3.18

$$\begin{aligned}
X[z] &= -\frac{1}{2}z\left[\frac{e^{j\frac{\pi}{4}}}{(z-4e^{j\frac{\pi}{3}})} + \frac{e^{-j\frac{\pi}{4}}}{(z-4e^{-j\frac{\pi}{3}})}\right] \\
&= -\frac{1}{2}z\left[\frac{ze^{j\frac{\pi}{4}} - 4e^{-j\frac{\pi}{12}} - 4e^{j\frac{\pi}{12}} + ze^{-j\frac{\pi}{4}}}{z^2 - z4(e^{j\frac{\pi}{3}} + e^{-j\frac{\pi}{3}}) + 16}\right] \\
&= \frac{-\frac{1}{2}z[\sqrt{2}z - 7.73]}{(z^2 - 4z + 16)}
\end{aligned}$$

$$\boxed{X[z] = \frac{-0.707z[z-5.467]}{(z-2+j3.464)(z-2-j3.464)}} \quad \text{ROC: } |z| < 4$$

The pole-zero diagram is shown in Fig. 9.11. The ROC is the interior of the circle.

19. $\boldsymbol{x[n] = nu[n-1]}$

Method 1

$$u[n-1] \overset{Z}{\longleftrightarrow} \frac{1}{(z-1)}$$

Using the differential property, we get

$$nu[n-1] \overset{Z}{\longleftrightarrow} -z\frac{d}{dz}\frac{1}{(z-1)}$$

$$\boxed{z[nu[n-1]] = \frac{z}{(z-1)^2}}$$

Method 2

$$nu[n-1] = (n-1)u[n-1] + u[n-1]$$

$$(n-1)u[n-1] \overset{Z}{\longleftrightarrow} \frac{zz^{-1}}{(z-1)^2} = \frac{1}{(z-1)^2}$$

$$u[n-1] \overset{Z}{\longleftrightarrow} \frac{1}{(z-1)}$$

$$nu[n-1] \overset{Z}{\longleftrightarrow} \frac{1}{(z-1)^2} + \frac{1}{(z-1)}$$

$$\boxed{Z[nu[n-1]] = \frac{z}{(z-1)^2}}$$

20.

$$\boldsymbol{x[n] = \begin{cases} (4)^n & n < 0 \\ \left(\frac{1}{4}\right)^n & n = 0, 2, 4, \ldots \\ \left(\frac{1}{5}\right)^n & n = 1, 3, 5, \ldots \end{cases}}$$

$$\begin{aligned} X[z] &= \sum_{n=-\infty}^{-1} (4)^n z^{-n} + \sum_{n=0}^{\infty} \left(\frac{1}{4}\right)^n z^{-n} + \sum_{n=0}^{\infty} \left(\frac{1}{5}\right)^n z^{-n} \\ &= X_1[z] + X_2[z] + X_3[z] \\ X_1[z] &= \sum_{n=-\infty}^{-1} (4)^n z^{-n} \\ &= \sum_{n=-\infty}^{-1} \left(\frac{z}{4}\right)^{-n} \\ &= \sum_{n=1}^{\infty} \left(\frac{z}{4}\right)^{n} \\ &= \frac{z}{4} + \left(\frac{z}{4}\right)^2 + \cdots \\ &= \frac{z}{4}\left[1 + \frac{z}{4} + \left(\frac{z}{4}\right)^2 + \cdots\right] \\ &= \frac{z}{4} \frac{1}{\left(1 - \frac{z}{4}\right)} \\ &= \frac{-z}{(z-4)} \qquad \text{ROC: } |z| < 4 \end{aligned}$$

$$X_2[z] = \sum_{n=0}^{\infty} \left(\frac{1}{4}\right)^n z^{-n}$$
$$= \sum_{n=0}^{\infty} (4z)^{-n}$$
$$= \sum_{p=0}^{\infty} (4z)^{-2p} \quad \text{where } n = 2p \text{ and } p = 0, 1, 2, \ldots$$

$$X_2[z] = \sum_{p=0}^{\infty} \left(16z^2\right)^{-p}$$
$$= \frac{1}{\left(1 - \frac{1}{16z^2}\right)}$$
$$= \frac{z^2}{\left(z^2 - \frac{1}{16}\right)} \qquad \text{ROC: } |z| > \frac{1}{4}$$
$$X_3[z] = \sum_{n=0}^{\infty} \left(\frac{1}{5}\right)^n z^{-n}$$
$$= \sum_{n=0}^{\infty} (5z)^{-n}$$
$$= \sum_{q=0}^{\infty} (5z)^{-(2q+1)} \quad \text{where } n = 2q + 1$$
$$= \frac{1}{5z} \sum_{q=0}^{\infty} \left(25z^2\right)^{-q}$$
$$= \frac{1}{5z} \frac{1}{\left(1 - \frac{1}{25z^2}\right)}$$
$$= \frac{z/5}{\left(z^2 - \frac{1}{25}\right)} \qquad \text{ROC: } |z| > \frac{1}{5}$$

$$\boxed{X[z] = \left[-\frac{z}{(z-4)} + \frac{z^2}{\left(z^2 - \frac{1}{16}\right)} + \frac{z/5}{\left(z^2 - \frac{1}{25}\right)}\right]}$$

ROC: $\frac{1}{4} < |z| < 4$.

■ Example 9.4

Find the initial and final values of the following functions:

$$\text{(a)} \qquad X[z] = \frac{z}{(4z^2 - 5z - 1)} \qquad \text{ROC: } |z| > 1$$

$$\text{(b)} \qquad x[z] = \frac{10z(z - 0.4)}{(z - 0.5)(z - 0.3)} \qquad \text{ROC: } |z| > 0.5$$

Solution:

(a) $\mathbf{X[z] = \frac{z}{(4z^2-5z-1)}}$

Initial Value

$$x[0] = \underset{z\to\infty}{Lt}\ X[z]$$

$$= \underset{z\to\infty}{Lt}\ \frac{z}{z^2\left(4 - \frac{5}{z} - \frac{1}{z^2}\right)}$$

$$= \underset{z\to\infty}{Lt}\ \frac{1}{z\left(4 - \frac{5}{z} - \frac{1}{z^2}\right)}$$

$$\boxed{x[0] = 0}$$

Final Value

$$x[\infty] = \underset{z\to 1}{Lt}\ \frac{(z-1)}{z}$$

Provided all the poles are inside, the unit circle and possibly one pole on the unit circle.

$$(4z^2 - 5z + 1) = 4(z-1)\left(z - \frac{1}{4}\right)$$

$$X[z] = \frac{z}{4(z-1)\left(z - \frac{1}{4}\right)}$$

The poles $(z - 1)$ are on the unit circle and $z = \frac{1}{4}$ within the unit circle. $X[z]$ is valid to apply the final value theorem.

$$x[\infty] = \underset{z\to 1}{Lt}\ \frac{(z-1)}{z}\ \frac{z}{4(z-1)\left(z - \frac{1}{4}\right)}$$

$$\boxed{x[\infty] = \frac{1}{3}}$$

(b) $X[z] = \frac{10z(z-0.4)}{(z-0.5)(z-0.3)}$

$$x[0] = \underset{z\to\infty}{Lt} \frac{10z^2\left(1 - \frac{0.4}{z}\right)}{z^2\left(1 - \frac{0.5}{z}\right)\left(1 - \frac{0.3}{z}\right)}$$

$$\boxed{x[0] = 10}$$

To find the final value $x[\infty]$, the poles of $X[z]$ are all inside the unit circle and hence it is valid to apply the final value theorem.

$$x[\infty] = \underset{z\to 1}{Lt} \frac{10z(z-1)(z-0.4)}{z(z-0.5)(z-0.3)}$$

$$\boxed{x[\infty] = 0}$$

■ Example 9.5

$$X[z] = \frac{\left[1 - \frac{1}{4}z^{-2}\right]}{\left[1 + \frac{1}{4}z^{-2}\right]\left[1 + \frac{5}{4}z^{-1} + \frac{3}{8}z^{-2}\right]}$$

How many different regions of convergence could correspond to $X[z]$?

(*Anna University, May, 2008*)

Solution:

$$X[z] = \frac{z^2\left[z^2 - \frac{1}{4}\right]}{\left(z^2 + \frac{1}{4}\right)\left(z^2 + \frac{5}{4}z + \frac{3}{8}\right)}$$

$$= \frac{z^2\left(z + \frac{1}{2}\right)\left(z - \frac{1}{2}\right)}{\left(z - \frac{j}{2}\right)\left(z + \frac{j}{2}\right)\left(z + \frac{3}{4}\right)\left(z + \frac{1}{2}\right)}$$

$$X[z] = \frac{z^2\left[z - \frac{1}{2}\right]}{\left(z - \frac{j}{2}\right)\left(z + \frac{j}{2}\right)\left(z + \frac{3}{4}\right)}$$

The poles and zeros are located in Fig. 9.12.

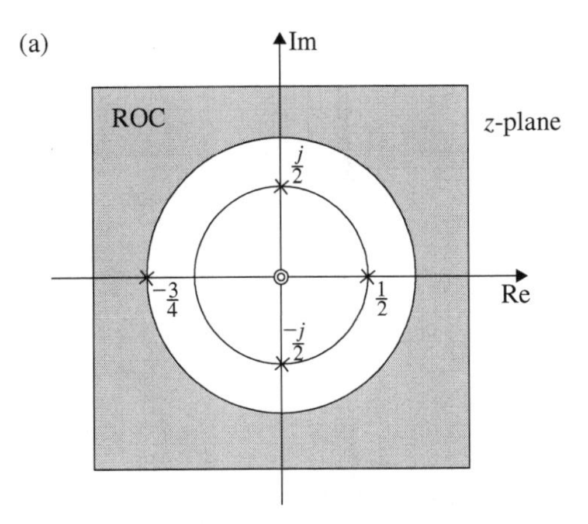

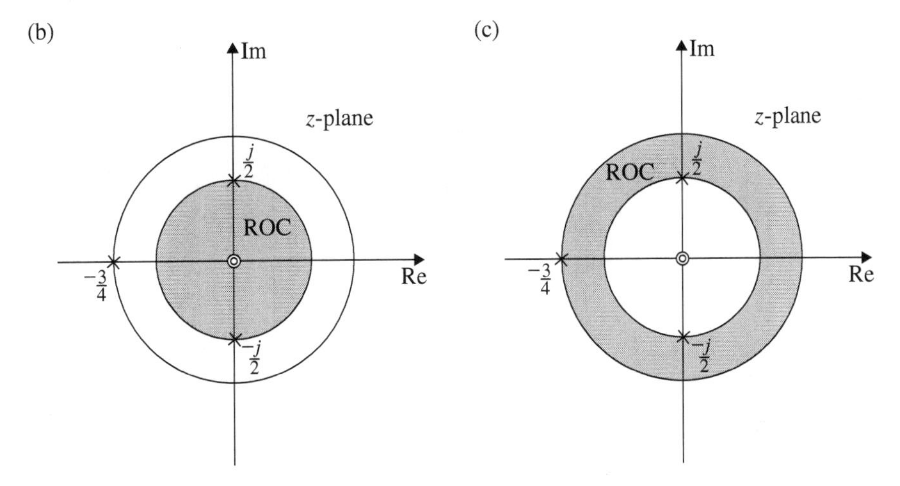

Fig. 9.12 Pole-zero diagram and ROC of $X[z]$

From Fig. 9.12, the circle passing through $|z| = \frac{3}{4}$ and $|z| = \frac{1}{2}$ are drawn. $X[z]$ exists from the following ROCs.

1. $|z| > \frac{3}{4}$. ROC is the exterior of the outermost pole $z = -\frac{3}{4}$. The system is causal and $X[z]$ exits (Fig. 9.12a).
2. $|z| < \frac{1}{2}$. ROC is the interior of the innermost pole $\pm\frac{j}{2}$. The system is anti-causal and $X[z]$ exits (Fig. 9.12b).
3. $\frac{1}{2} < |z| < \frac{3}{4}$. The ROC is a ring between the two circles of radius $r_1 = \frac{3}{4}$ and $r_1 = \frac{1}{2}$. Here $X[z]$ exits. The system is both causal and anti-causal (Fig. 9.12c).

The unilateral z-transform pairs are given in Table 9.1. The properties of z-transform are given in Table 9.2.

Table 9.1 Unilateral z-transform pairs

No.	**x[n]**	**X[z]**
1	$\delta[n]$	1
2	$u[n]$	$\frac{z}{(z-1)}$
3	$nu[n]$	$\frac{z}{(z-1)^2}$
4	$n^2u[n]$	$\frac{z(z+1)}{(z-1)^3}$
5	$a^nu[n]$	$\frac{z}{(z-a)}$
6	$a^{n-1}u[n-1]$	$\frac{1}{(z-a)}$
7	$na^nu[n]$	$\frac{az}{(z-a)^2}$
8	$\cos\omega_0 nu[n]$	$\frac{1-\cos\omega_0 z^{-1}}{1-2\cos\omega_0 z^{-1}+z^{-2}}$
9	$\sin\omega_0 nu[n]$	$\frac{z^{-1}\sin\omega_0}{1-2\cos\omega_0 z^{-1}+z^{-2}}$
10	$a^n\cos\omega_0 nu[n]$	$\frac{1-az^{-1}\cos\omega_0}{1-2a\cos\omega_0 z^{-1}+a^2z^{-2}}$
11	$a^n\sin\omega_0 nu[n]$	$\frac{az^{-1}\sin\omega_0}{1-2a\cos\omega_0 z^{-1}+a^2z^{-2}}$

Table 9.2 z-transform properties (operations)

Operation	**x[n]**	**X[z]**
Linearity	$a_1x_1[n]+a_2x_2[n]$	$a_1X_1[z]+a_2X_2[z]$
Multiplication by a^n	$a^nx[n]u[n]$	$X\left[\frac{z}{a}\right]$
Multiplication by n	$nx[n]u[n]$	$-z\frac{d}{dz}X[z]$
Time shifting	$x[n-n_0]$	$z^{-n_0}X[z]$
Multiplication by $e^{j\omega_0 n}$	$e^{j\omega_0 n}x[n]$	$X[e^{-j\omega_0}z]$
Time reversal	$x[-n]$	$X\left[\frac{1}{z}\right]$
Accumulation	$\sum_{k=-\infty}^{n}x[n]$	$\frac{z}{(z-1)}X[z]$
Convolution	$x_1[n]*x_2[n]$	$X_1[z]X_2[z]$
Initial value	$x[0]$	$\underset{z\to\infty}{Lt}\ X[z]$
Final value	$x[\infty]$	$\underset{z\to 1}{Lt}\ \frac{(z-1)}{z}X[z]$ poles of $(z-1)X[z]$ are inside the unit circle
Right shifting	$x[n-m]u[n-m]$	$\frac{1}{z^m}X[z]$
	$x[n-m]u[n]$	$\frac{1}{z^m}X[z]+\frac{1}{z^m}\sum_{n=1}^{m}x(-m)z^n$
	$x[n-1]u[n]$	$\frac{1}{z}X[z]+x[-1]$
	$x[n-2]u[n]$	$\frac{1}{z^2}X[z]+\frac{1}{z}x[-1]+x[-2]$
Left shifting	$x[n+m]u[n]$	$z^mX[z]-z^m\sum_{n=0}^{m-1}x[n]z^{-n}$
	$x[n+1]u[n]$	$zX[z]-zx(0)$
	$x[n+2]u[n]$	$z^2X[z]-z^2x[0]-zx[1]$

9.8 Inverse z-Transform

If $X[z]$ is given, then the sequence $x[n]$ is determined. This is called inverse z-transform. As in the Laplace transform, in inverse z-transform also, the integration in the complex z-plane using Eq. (9.5) is avoided since it is tedious. Instead, the following methods are used. They are

1. Partial fraction method;
2. Power series expansion;
3. Residue method.

Of these, the partial fraction method is very easy to apply as was done in determining inverse Laplace transform.

9.8.1 Partial Fraction Method

If $X[z]$ is a rational function of z, then it can be expressed as follows:

$$X[z] = \frac{N[z]}{D[z]} = \frac{K(z - z_1)(z - z_2)\ldots(z - z_m)}{(z - p_1)(z - p_2)\ldots(z - p_n)} \tag{9.45}$$

where $n \geq m$ and all the poles are simple.

$$\begin{aligned}\frac{X[z]}{z} &= \frac{K(z - z_1)(z - z_2)\ldots(z - z_m)}{z(z - p_1)(z - p_2)\ldots(z - p_n)} \\ &= \frac{A_0}{z} + \frac{A_1}{z - p_1} + \frac{A_2}{z - p_2} + \cdots + \frac{A_n}{z - p_n}\end{aligned}$$

where

$$\begin{aligned}A_0 &= X[z]|_{z=0} \\ A_1 &= (z - p_1)\frac{X[z]}{z}\bigg|_{z=p_1} \\ X[z] &= A_0 + A_1\frac{z}{z - p_1} + \cdots + \frac{A_n z}{z - p_n}\end{aligned} \tag{9.46}$$

Using z-transform pair table, $x[n]$ can be determined. The following examples illustrate the above method. For repeated poles, the z-transform pairs given in Table 9.3 may be referred to.

Table 9.3 z-transform pairs of repeated poles

X[z]	**x[n] ROC:** $\|z\| > \|a\|$
1. $\dfrac{z}{z-a}$	$a^n u[n]$
2. $\dfrac{z}{(z-a)^2}$	$na^{n-1}u[n]$
3. $\dfrac{z}{(z-a)^3}$	$\dfrac{n(n-1)a^{n-2}}{\angle 2}u[n]$
4. $\dfrac{z}{(z-a)^k}$	$\dfrac{n(n-1)(n-2)\ldots(n-(k-2))a^{n-k+1}}{\angle(k-1)}u[n]$

■ Example 9.6

Find the inverse z-transform of

$$X[z] = \frac{1-\frac{1}{3}z^{-1}}{(1-z^{-1})(1+2z^{-1})} \qquad \text{ROC: } |z| > 2$$

(*Anna University, April, 2004*)

Solution:

$$X[z] = \frac{1-\frac{1}{3}z^{-1}}{(1-z^{-1})(1+2z^{-1})}$$

$$= \frac{z\left(z-\frac{1}{3}\right)}{(z-1)(z+2)}$$

$$\frac{X[z]}{z} = \frac{A_1}{(z-1)} + \frac{A_2}{(z+2)}$$

$$\left(z-\frac{1}{3}\right) = A_1(z+2) + A_2(z-1)$$

Substitute $z = 1$

$$A_1 = \frac{2}{9}$$

Substitute $z = -2$

$$A_2 = \frac{7}{9}$$

$$X[z] = \frac{1}{9}\left[\frac{2z}{z-1} + \frac{7z}{z+2}\right]$$

$$\boxed{x[n] = \frac{1}{9}\left[2(1)^n + 7(-2)^n\right]u[n]}$$

■ Example 9.7

Find the inverse z-transform of

$$X[z] = \frac{1}{1024}\left[\frac{1024 - z^{-10}}{1 - \frac{1}{2}z^{-1}}\right] \qquad \text{ROC: } |z| > 0$$

(*Anna University, April, 2008*)

Solution:

$$X[z] = \frac{1}{1024}\left[\frac{1024 - z^{-10}}{1 - \frac{1}{2}z^{-1}}\right]$$

$$= \frac{z}{\left(z - \frac{1}{2}\right)} - \frac{z}{\left(z - \frac{1}{2}\right)}\frac{z^{-10}}{1024}$$

Taking inverse z-transform, we get

$$x[n] = \left(\frac{1}{2}\right)^n u[n] - \frac{1}{1024}\left(\frac{1}{2}\right)^{n-10} u[n-10]$$

$$= \left(\frac{1}{2}\right)^n u[n] - \frac{1}{1024}\left(\frac{1}{2}\right)^n \left(\frac{1}{2}\right)^{-10} u[n-10]$$

$$= \left(\frac{1}{2}\right)^n u[n] - \frac{1}{1024}\left(\frac{1}{2}\right)^n 1024u[n-10]$$

$$= \left(\frac{1}{2}\right)^n u[n] - \left(\frac{1}{2}\right)^n u[n-10]$$

$$x[n] = \left(\frac{1}{2}\right)^n - 0 \qquad 0 \le n \le 9$$

$$= \left(\frac{1}{2}\right)^n - \left(\frac{1}{2}\right)^n$$

$$= 0 \qquad n \ge 10$$

$$\boxed{\begin{aligned} x[n] &= \left(\frac{1}{2}\right)^n & 0 \le n \le 9 \\ &= 0 & \text{otherwise} \end{aligned}}$$

Example 9.8

Find the inverse z-transform of

$$X[z] = \frac{z^2}{(1 - az)(z - a)}$$

(Anna University, December, 2007)

Solution:

$$X[z] = \frac{z^2}{(1 - az)(z - a)}$$

$$\frac{X[z]}{z} = \frac{-z}{a\left[z - \frac{1}{a}\right][z - a]}$$

$$\frac{X[z]}{z} = \frac{A_1}{\left(z - \frac{1}{a}\right)} + \frac{A_2}{(z - a)}$$

$$-\frac{z}{a} = A_1(z - a) + A_2\left(z - \frac{1}{a}\right)$$

Substitute $z = \frac{1}{a}$

$$-\frac{1}{a^2} = A_1\left(\frac{1}{a} - a\right)$$

$$A_1 = \frac{-1}{a(1 - a^2)}$$

Substitute $z = a$

$$-1 = A_2\left(a - \frac{1}{a}\right)$$

$$A_2 = \frac{a}{(1 - a^2)}$$

$$X[z] = \frac{1}{(1 - a^2)}\left[\frac{-1}{a}\frac{z}{\left(z - \frac{1}{a}\right)} + \frac{az}{(z - a)}\right]$$

For $a > 1$, the ROC is shown in Fig. 9.13a. For $a < 1$, the ROC is shown in Fig. 9.13b.
For $a > 1$, the ROC is exterior of the outermost pole. Hence, the function is casual.

$$x[n] = \frac{1}{(1 - a^2)}\left[\frac{-1}{a}\frac{1}{(a)^n} + a(a)^n\right]u[n]$$

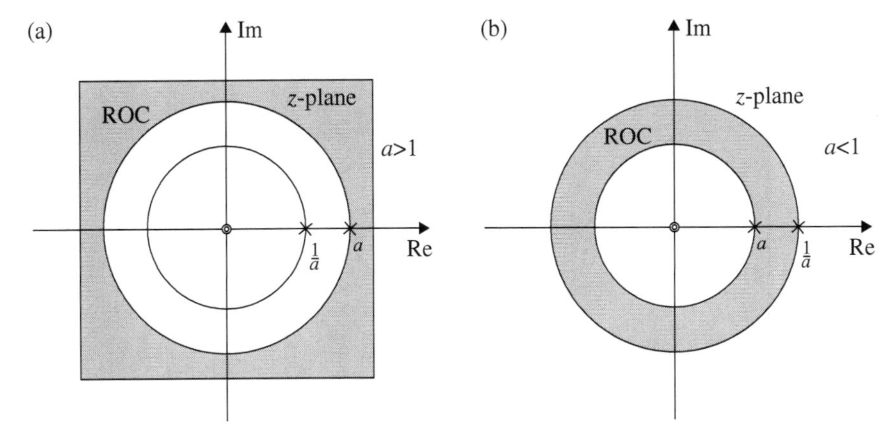

Fig. 9.13 ROC of Example 9.8

$$\boxed{x[n] = \frac{1}{(1-a^2)}\left[-\left(\frac{1}{a}\right)^{n+1} + (a)^{n+1}\right]u[n]}$$

For $a < 1$, the ROC is $a < |z| < \frac{1}{a}$ and it is a concentric strip. The pole at $|z| = \frac{1}{a}$ is anti-causal and $z = a$ is causal.

$$\boxed{x[n] = \frac{1}{(1-a^2)}\left[\left(\frac{1}{a}\right)^{n+1} u[-n-1] + (a)^{n+1}u[n]\right]}$$

■ Example 9.9

$$X[z] = \frac{(7z-23)}{(z-3)(z-4)}$$

Find $x[n]$. ROC: $|z| > 4$.

Solution:

Method 1:

Dividing both sides by z, we get

$$\begin{aligned}\frac{X[z]}{z} &= \frac{(7z-23)}{z(z-3)(z-4)}\\ &= \frac{A_1}{z} + \frac{A_2}{(z-3)} + \frac{A_3}{(z-4)}\\ (7z-23) &= A_1(z-3)(z-4) + A_2 z(z-4) + A_3 z(z-3)\end{aligned}$$

Substitute $z = 0$

$$-23 = 12A_1; \qquad A_1 = -\frac{23}{12}$$

Substitute $z = 3$

$$-2 = A_2(3)(-1); \qquad A_2 = \frac{2}{3}$$

Substitute $z = 4$

$$5 = 4A_3; \qquad A_3 = \frac{5}{4}$$

$$X[z] = -\frac{23}{12} + \frac{2}{3}\frac{z}{(z-3)} + \frac{5}{4}\frac{z}{(z-4)}$$

$$\boxed{X[n] = \left[-\frac{23}{12}\delta[n] + \frac{2}{3}(3)^n + \frac{5}{4}(4)^n\right]u[n]}$$

Method 2:

$$\begin{aligned} X[z] &= \frac{(7z-23)}{(z-3)(z-4)} \\ &= \frac{A_1}{(z-3)} + \frac{A_2}{(z-4)} \\ 7z - 23 &= A_1(z-4) + A_2(z-3) \end{aligned}$$

Substitute $z = 3$

$$-2 = -A_1; \qquad A_1 = 2$$

Substitute $z = 4$

$$\begin{aligned} 5 &= A_2 \\ X[z] &= \frac{2}{(z-3)} + \frac{5}{(z-4)} \\ \frac{2}{(z-3)} &\overset{Z^{-1}}{\longleftrightarrow} 2(3)^{n-1}u[n-1] \\ \frac{5}{(z-4)} &\overset{Z^{-1}}{\longleftrightarrow} 5(4)^{n-1}u[n-1] \end{aligned}$$

$$\boxed{x[n] = [2(3)^{n-1} + 5(4)^{n-1}]u[n-1]}$$

The results of the above two methods are the same even though they are expressed in different forms.

Example 9.10

$$X[z] = \frac{10z}{(z+2)(z+4)^2} \qquad \text{ROC: } |z| > 4$$

Find $x[n]$ using the partial fraction method.

Solution: **This is the case with poles repeated twice**

$$\begin{aligned}
X[z] &= \frac{10z}{(z+2)(z+4)^2} \\
\frac{X[z]}{z} &= \frac{10}{(z+2)(z+4)^2} \\
&= \frac{A_1}{(z+2)} + \frac{A_2}{(z+4)} + \frac{A_3}{(z+4)^2}
\end{aligned}$$

$$10 = A_1(z+4)^2 + A_2(z+2)(z+4) + A_3(z+2)$$

Substitute $z = -2$

$$10 = 4A_1; \qquad A_1 = \frac{5}{2}$$

Substitute $z = -4$

$$10 = -2A_3; \qquad A_3 = -5$$

Compare the coefficients of free terms

$$\begin{aligned}
10 &= 16A_1 + 8A_2 + 2A_3 \\
&= 16\frac{5}{2} + 8A_2 - 10 \\
A_2 &= -\frac{5}{2}
\end{aligned}$$

$$X[z] = \frac{5}{2}\frac{z}{(z+2)} - \frac{5}{2}\frac{z}{(z+4)} - \frac{5}{(z+4)^2}$$

$$\boxed{x[n] = \left[\frac{5}{2}(-2)^n - \frac{5}{2}(-4)^n - 5n(-4)^n\right]u[n]}$$

Example 9.11

$$X[z] = \frac{z(z^2 + z - 30)}{(z - 2)(z - 4)^3} \qquad \text{ROC: } |z| > 4$$

Find $x[n]$ using the partial fraction method.

Solution: **This is the case with poles repeated thrice**

$$X[z] = \frac{z(z^2 + z - 30)}{(z - 2)(z - 4)^3}$$

$$\frac{X[z]}{z} = \frac{(z^2 + z - 30)}{(z - 2)(z - 4)^3} = \frac{(z - 5)(z + 6)}{(z - 2)(z - 4)^3}$$

$$= \frac{A_1}{(z - 2)} + \frac{A_2}{(z - 4)^3} + \frac{A_3}{(z - 4)^2} + \frac{A_4}{(z - 4)}$$

$$(z^2 + z - 30) = A_1(z - 4)^3 + A_2(z - 2) + A_3(z - 2)(z - 4) + A_4(z - 2)(z - 4)^2$$

Substitute $z = 2$

$$(-3)(8) = -8A_1; \qquad A_1 = 3$$

Substitute $z = 4$

$$(-1)(10) = 2A_2; \qquad A_2 = -5$$

$$(z^2 + z - 30) = 3(z^3 - 12z^2 + 48z - 64) - 5(z - 2) + A_3(z^2 - 6z + 8)$$
$$+A_4(z^3 - 10z^2 + 32z - 32)$$

Compare the coefficients of z^2

$$1 = -36 + A_3 - 10A_4$$
$$A_3 - 10A_4 = 37$$

Compare the coefficients of z

$$1 = 144 - 5 - 6A_3 + 32A_4$$
$$6A_3 - 32A_4 = 138$$

Solving the above equation, we get

$$A_3 = 7; \qquad A_4 = -3$$

$$X[z] = \frac{3z}{(z-2)} - \frac{5z}{(z-4)^3} + \frac{7z}{(z-4)^2} - \frac{3z}{(z-4)}$$

$$\frac{z}{(z-4)^3} \overset{Z^{-1}}{\longleftrightarrow} \frac{n(n-1)}{\angle 2}(4)^{n-2}u[n] = \frac{n(n-1)}{32}(4)^n u[n]$$

$$\frac{z}{(z-4)^2} \overset{Z^{-1}}{\longleftrightarrow} n(4)^{n-1}u[n] = \frac{1}{4}n(4)^n u[n]$$

$$\boxed{x[n] = \left[3(2)^n + \left\{-\frac{5}{32}n(n-1) + \frac{7}{4}n - 3\right\}(4)^n\right]u[n]}$$

The values of A_1, A_2 and A_3 determined are checked for their correctness as follows:

$$\frac{X[z]}{z} = \frac{(z-5)(z+6)}{(z-2)(z-4)^3}$$

Substitute $z = 0$

$$\left.\frac{X[z]}{z}\right|_{z=0} = \frac{(-5)(6)}{(-2)(-4)^3} = -\frac{15}{64}$$

$$\frac{X[z]}{z} = \frac{3}{z-2} - \frac{5}{(z-4)^3} + \frac{7}{(z-4)^2} - \frac{3}{(z-4)}$$

Substitute $z = 0$

$$\left.\frac{X[z]}{z}\right|_{z=0} = -\frac{3}{2} + \frac{5}{64} + \frac{7}{16} + \frac{3}{4} = -\frac{15}{64}$$

Hence the values of A_1, A_2, A_3 and A_4 are found to be correct.

■ Example 9.12

$$X[z] = \frac{z(z+10)}{(z-1)(z^2-8z+20)}$$

Find $x[n]$ using the partial fraction method.

Solution: This is the case with complex poles

$$X[z] = \frac{z(z+10)}{(z-1)(z^2-8z+20)}$$

$$\frac{X[z]}{z} = \frac{(z+10)}{(z-1)(z-4+j2)(z-4-j2)}$$

$$= \frac{A_1}{(z-1)} + \frac{A_2}{(z-4+j2)} + \frac{A_3}{(z-4-j2)}$$

$$(z+10) = A_1(z^2-8z+20) + A_2(z-1)(z-4-j2) + A_3(z-1)(z-4+j2)$$

Substitute $z = 1$

$$11 = A_1(13); \qquad A_1 = \frac{11}{13}$$

Substitute $z = 4 + j2$

$$\begin{aligned}
(14 + j2) &= A_3(4 + j2 - 4 + j2)(4 + j2 - 1) \\
A_3 &= \frac{(14 + j2)}{j4(3 + j2)} = \frac{14.142\angle 8.13^\circ}{4\sqrt{13}\angle 123.69^\circ} \\
&= 0.98\angle - 115.56^\circ = 0.98e^{-j115.56^\circ} \\
A_2 &= \text{conjugate of } A_3 \\
&= 0.98\angle 115.56^\circ = 0.98e^{j115.56^\circ} \\
X[z] &= \frac{11}{13}\frac{z}{(z - 1)} + \frac{0.98e^{j115.56^\circ}}{(z - 4 + j2)} + \frac{0.98e^{-j115.56^\circ}}{(z - 4 - j2)}
\end{aligned}$$

Using z-transform pair, we get the following inverse z-transform:

$$x[n] = \frac{11}{13}u[n] + [0.98e^{j115.56^\circ}(4 - j2)^n + 0.98e^{-j115.56^\circ}(4 + j2)^n]u[n]$$

$115.56^\circ = 2$ radians

$$\begin{aligned}
(4 + j2)^n &= (4.47)^n e^{j0.4636n} \\
(4 - j2)^n &= (4.47)^n e^{-j0.4636n}
\end{aligned}$$

$$\begin{aligned}
x[n] &= \frac{11}{13}u[n] + [0.98e^{j2}e^{-j0.4636n}(4.47)^n + 0.98(4.47)^n e^{-j2}e^{j0.4636n}]u[n] \\
&= \frac{11}{13}u[n] + 0.98 * (4.47)^n[e^{j(2-.4636n)} + e^{-j(2-.4636n)}]u[n]
\end{aligned}$$

$$\boxed{x[n] = \left[\frac{11}{13} + 1.96(4.47)^n \cos(2 - 0.4636n)\right] u[n]}$$

■ Example 9.13

$$X[z] = \frac{(5z^3 - 29z^2 + 8z + 60)}{(z^2 - 7z + 10)}$$

Find $x[n]$ by partial fraction method.

Solution: This is the case with irrational system function. The solution of $x[n]$ will have forward and backward shifts. Dividing the numerator polynomial by the denominator polynomial, we get

$$\begin{array}{r} 5z+6 \\ z^2-7z+10\overline{)5z^3-29z^2+8z+60} \\ \underline{5z^3-35z^2+50z\quad\;} \\ 6z^2-42z+60 \\ \underline{6z^2-42z+60} \end{array}$$

$$\begin{aligned}(z^2-7z+10) &= (z-2)(z-5) \\ X[z] &= (5z+6)+\frac{1}{(z-2)(z-5)} \\ &= X_1[z]+X_2[z]\end{aligned}$$

where

$$\begin{aligned}X_1[z] &= (5z+6) \\ X_2[z] &= \frac{1}{(z-2)(z-5)} \\ \frac{X_2[z]}{z} &= \frac{1}{z(z-2)(z-5)} \\ &= \frac{A_1}{z}+\frac{A_2}{z-2}+\frac{A_3}{z-5} \\ 1 &= A_1(z-2)(z-5)+A_2z(z-5)+A_3z(z-2)\end{aligned}$$

Substitute $z=0$

$$A_1=\frac{1}{10}$$

Substitute $z=2$

$$1=A_2(2)(-3);\qquad A_2=-\frac{1}{6}$$

Substitute $z=5$

$$1=A_3(5)(3);\qquad A_3=\frac{1}{15}$$

$$X[z] = 5z + 6 + \frac{1}{10} - \frac{1}{6}\frac{z}{(z-2)} + \frac{1}{15}\frac{z}{(z-5)}$$

$$\boxed{x[n] = \left[\delta(n+1) + 6.1\delta[n] - \frac{1}{6}(2)^n + \frac{1}{15}(5)^n\right]u[n]}$$

■ Example 9.14

Find the inverse z-transform of

$$X[z] = \frac{(5 + z^{-2} + 4z^{-3})}{(z^2 + 7z + 10)}$$

Solution:

$$\begin{aligned} X[z] &= \frac{(5 + z^{-2} + 4z^{-3})}{(z^2 + 7z + 10)} \\ &= \frac{(5 + z^{-2} + 4z^{-3})}{z}\frac{z}{(z+2)(z+5)} \\ &= [5z^{-1} + z^{-3} + 4z^{-4}]\frac{z}{(z+2)(z+5)} \end{aligned}$$

$$\frac{z}{(z+2)(z+5)} = \frac{1}{3}\left[\frac{z}{z+2} - \frac{z}{z+5}\right]$$

$$\frac{z}{(z+2)(z+5)} \overset{Z^{-1}}{\longleftrightarrow} \frac{1}{3}\left[(-2)^n - (-5)^n\right]u[n]$$

Now

$$x[n] = [5z^{-1} + z^{-3} + 4z^{-4}]\frac{1}{3}[(-2)^n - (-5)^n]u[n]$$

Using the time shifting property, we get

$$\boxed{\begin{aligned} x[n] = {} & \frac{5}{3}\left[(-2)^{n-1} - (-5)^{n-1}\right]u[n-1] + \frac{1}{3}\left[(-2)^{n-3} - (-5)^{n-3}\right]u[n-3] \\ & + \frac{4}{3}\left[(-2)^{n-4} - (-5)^{n-4}\right]u[n-4] \end{aligned}}$$

Example 9.15

Find the inverse z-transform for the following system functions:

(a) $X[z] = \dfrac{4}{(z-5)}$ ROC: $|z| < 5$

(b) $X[z] = z(1 - z^{-1})(1 + 2z^{-1})$ ROC: $0 < |z| < \infty$

Solution:

(a) $\mathbf{X[z] = \frac{4}{(z-5)}}$

$$X[z] = \frac{4}{(z-5)} = 4z^{-1}\frac{z}{z-5}$$
$$x[n] = 4z^{-1}[(5)^n]u[n]]$$

$$\boxed{x[n] = 4(5)^{n-1}u[n-1]}$$

(b) $\mathbf{X[z] = z(1 - z^{-1})(1 + 2z^{-1})}$

$$X[z] = z(1 - z^{-1})(1 + 2z^{-1})$$
$$X[z] = z[1 + 2z^{-1} - z^{-1} - 2z^{-2})$$
$$= [z + 1 - 2z^{-1}]$$

$$\boxed{x[n] = \{1, \underset{\uparrow}{1}, -2\}}$$

9.8.2 *Inverse z-Transform Using Power Series Expansion*

The z-transform Eq. (9.4)

$$X[z] = \sum_{n=-\infty}^{\infty} x[n]z^{-n}$$

can be expressed in power series form and the coefficients of $z^{|n|}$ give the values of the sequence. Equation (9.4) can be express as

$$X[z] = \cdots + x[-3]z^3 + x[-2]z^2 + x[-1]z + x[0] + x[1]z^{-1} + x[2]z^{-2} + x[3]z^{-3} + \cdots \tag{9.47}$$

Equation (9.47) does not give closed form. However, if $X[z]$ is not in a simpler form other than the polynomial in z^{-1}, using the power series method $x[n]$ is easily obtained. If $X[z]$ is rational, the power series is obtained by long division. The following examples illustrate the above method.

Example 9.16

Using power series expansion, find the inverse z-transform of the following $X[z]$:

(a) $X[z] = \dfrac{4z}{(z^2 - 3z + 2)}$ ROC: $|z| > 2$

(b) $X[z] = \dfrac{4z}{(z^2 - 3z + 2)}$ ROC: $|z| < 1$

(c) $X[z] = \dfrac{1}{(1 - az^{-1})}$ ROC: $|z| > |a|$ and ROC: $|z| < |a|$

(Anna University, December, 2006)

Solution:

(a) $\mathbf{X[z] = \frac{4z}{(z^2-3z+2)}}$; **ROC: $|z| > 2$**

$$X[z] = \frac{4z}{(z^2 - 3z + 2)}$$
$$= \frac{4z}{(z-1)(z-2)}$$

For ROC: $|z| > 2$, $x[n]$ is a right-sided sequence where $n \geq 0$. Hence, the long division is done in such a way that $X[z]$ is expressed in the power of z^{-1}.

$$\begin{array}{r l}
 & 4z^{-1} + 12z^{-2} + 28z^{-3} + \cdots \\
z^2 - 3z + 2 \,\big) & 4z \\
 & \underline{4z - 12 + 8z^{-1}} \\
 & 12 - 8z^{-1} \\
 & \underline{12 - 36z^{-1} + 24z^{-2}} \\
 & 28z^{-1} - 24z^{-2} \\
 & \underline{28z^{-1} - 84z^{-2} + 56z^{-3}}
\end{array}$$

$$X[z] = 4z^{-1} + 12z^{-2} + 28z^{-3} + \cdots$$

$$\boxed{x[n] = \{\underset{\uparrow}{0},\ 4,\ 12,\ 28, \ldots\}}$$

(b) $\mathbf{X[z] = \frac{4z}{(z^2-3z+2)}}$; **ROC: $|z| < 1$**
For ROC: $|z| < 1$, $x[n]$ sequence is negative where $n \leq 0$. The long division is done in such a way that $X[z]$ is expressed in the power of z.

$$
\begin{array}{r}
2z + 3z^2 + \frac{7}{2}z^3 \\
2 - 3z + z^2 \overline{)\,4z \qquad\qquad\quad} \\
\underline{4z - 6z^2 + 2z^3} \\
6z^2 - 2z^3 \\
\underline{6z^2 - 9z^3 + 3z^4} \\
7z^3 - 3z^4 \\
\underline{7z^3 - \frac{21}{2}z^4 + \frac{7}{2}z^5}
\end{array}
$$

$$X[z] = 2z + 3z^2 + \frac{7}{2}z^3 + \cdots$$

$$\boxed{x[n] = \left\{\cdots \frac{7}{2},\ 3,\ 2,\ \underset{\uparrow}{0}\right\}}$$

(c) $\mathbf{X[z] = \frac{1}{(1-az^{-1})}}$; **ROC: $|z| > |a|$**

$$X[z] = \frac{z}{(z-a)}$$

The ROC: $|z| > a$, and it is exterior of the circle of radius $|a|$. Hence, $x[n]$ is a right-sided sequence where $n \geq 0$. The long division is done such that $X[z]$ is expressed in terms of the power of z^{-1} as shown below.

$$
\begin{array}{l}
\qquad\quad 1 + az^{-1} + a^2z^{-2} + a^3a^{-3} + \cdots \\
z - a \overline{)\,z \qquad\qquad\qquad\qquad\qquad\qquad} \\
\qquad\quad \underline{z - a} \\
\qquad\qquad\quad a \\
\qquad\qquad\quad \underline{a - a^2z^{-1}} \\
\qquad\qquad\qquad\quad a^2z^{-1} \\
\qquad\qquad\qquad\quad \underline{a^2z^{-1} - a^3z^{-2}} \\
\qquad\qquad\qquad\qquad\qquad a^3z^{-2} \\
\qquad\qquad\qquad\qquad\qquad \underline{a^3z^{-2} - a^4z^{-3}}
\end{array}
$$

$$X[z] = 1 + az^{-1} + a^2z^{-2} + a^3z^{-3} + \cdots$$

$$\boxed{\begin{array}{l} x[n] = \{\underset{\uparrow}{1},\ a,\ a^2,\ a^3, \ldots\} \\ x[n] = a^n u[n] \end{array}}$$

For ROC: $|z| < |a|$, $x[n]$ sequence is left-sided

$$\begin{array}{r} -a^{-1}z - a^{-2}z^2 - a^{-3}z^3 \cdots \\ -a + z\overline{)\,z \qquad\qquad\qquad\qquad\qquad} \\ \underline{z - a^{-1}z^2} \\ a^{-1}z^2 \\ \underline{a^{-1}z^2 - a^{-2}z^3} \\ a^{-2}z^3 \\ \underline{a^{-2}z^3 - a^{-3}z^4} \end{array}$$

$$X[z] = -a^{-1}z - a^{-2}z^2 - a^{-3}z^3 + \cdots$$

$$\boxed{\begin{array}{l} x[n] = \left\{\cdots,\ \dfrac{1}{a^3},\ -\dfrac{1}{a^2},\ -\dfrac{1}{a},\ \underset{\uparrow}{0}\right\} \\ x[n] = -a^n u[-n-1] \end{array}}$$

■ Example 9.17

Determine the inverse z-transform of

$$X[z] = \log(1 - 2z), \qquad |z| < \frac{1}{2}$$

by using the power series

$$\log(1 - x) = -\sum_{n=1}^{\infty} \frac{x^n}{n}, \qquad |x| < 1$$

and by first differentiating $X[z]$ and then using this to recover $x[n]$.

(*Anna University, December, 2007*)

Solution:

(a) **Using Power Series**

$$X[z] = \log(1 - 2z)$$
$$= -\sum_{n=1}^{\infty} \frac{1}{n}(2z)^n$$

Replace $n = -n$

$$X[z] = \sum_{n=-1}^{-\infty} \frac{1}{n}(2z)^{-n}$$
$$= \sum_{n=-1}^{-\infty} \left(\frac{1}{2}\right)^n \frac{1}{n} z^{-n}$$

By z-transform definition, it is a left-sided signal

$$X[n] = \frac{1}{n}\left(\frac{1}{2}\right)^n u(-n-1) \qquad n \le -1$$
$$= 0 \qquad n \ge 0$$

(b) **Using the Differentiation Property**

$$X[z] = \log(1 - 2z)$$
$$\frac{d}{dz}X[z] = \frac{-2}{(1-2z)}$$

Multiplying both sides by $-z$, we get

$$-z\frac{d}{dz}X[z] = \frac{2z}{(1-2z)}$$
$$= \frac{-z}{z-\frac{1}{2}}$$

$$-z\frac{d}{dz}X[z] \overset{Z^{-1}}{\longleftrightarrow} nx[n]$$
$$\frac{-z}{z-\frac{1}{2}} \overset{Z^{-1}}{\longleftrightarrow} \left(\frac{1}{2}\right)^n u(-n-1) \qquad \text{ROC: } |z| < \frac{1}{2}$$
$$nx[n] = \left(\frac{1}{2}\right)^n u(-n-1)$$

$$\boxed{x[n] = \left(\frac{1}{2}\right)^n \frac{1}{n} u(-n-1)}$$

■ Example 9.18

Find the inverse z-transform of

$$\begin{array}{lll}\text{(a)} & X[z] = \log(1 + az^{-1}) & |z| > |a| \\ \text{(b)} & X[z] = \log(1 - az^{-1}) & |z| > |a|\end{array}$$

(*Madras University, October, 1998*)

Solution:

(a) The power series expansion for $\log(1 + x)$ is

$$\begin{aligned}\log(1 + x) &= \sum_{n=1}^{\infty} \frac{(-1)^{n+1}}{n} x^n \qquad \text{for } x < 1 \\ \log(1 + az^{-1}) &= \sum_{n=1}^{\infty} \frac{(-1)^{n+1}(az^{-1})^n}{n} \qquad |az^{-1}| < 1 \text{ or } |z| > |a| \\ &= \sum_{n=1}^{\infty} \frac{(-1)^{n+1} a^n z^{-n}}{n}\end{aligned}$$

Since the summation is from $n = 1$, using the time shifting property we get

$$\boxed{x[n] = \frac{(-1)^{n+1} a^n}{n} u[n-1]}$$

(b) The power series expansion for $\log(1 - x)$ is

$$\begin{aligned}\log(1 - x) &= -\sum_{n=1}^{\infty} \frac{1}{n} x^n \qquad |x| < 1 \\ \log(1 - az^{-1}) &= -\sum_{n=1}^{\infty} \frac{1}{n} (az^{-1})^n\end{aligned}$$

$$\log(1 + az^{-1}) = -\sum_{n=1}^{\infty} \frac{a^n}{n} z^{-n}$$

$$\boxed{x[n] = -\frac{a^n}{n} u[n-1]}$$

9.8.3 *Inverse z-Transform Using Contour Integration or the Method of Residue*

The inverse z-transform can be obtained from Eq. (9.2) which is given by

$$x[n] = \frac{1}{2\pi j} \oint_c X[z] z^{n-1} dz \tag{9.48}$$

The above integral can be evaluated by summing up all the residues of the poles which are inside the circle c of Eq. (9.48) which can be expressed as

$$\begin{aligned} x[n] &= \sum \text{(Residues of } X[z]z^{-n} \text{ at the poles inside } (c)) \\ &= \sum_i (z - z_i) X[z] z^{-n-1} \Big|_{z=z_i} \end{aligned} \tag{9.49}$$

For multiple poles of order k, and $z = \alpha$, the residue is written as

$$\text{Residue} = \frac{1}{\angle(k-1)} \underset{z \to \alpha}{Lt} \left\{ \frac{d^{k-1}}{dz^{k-1}} (z - \alpha)^k X[z] z^{n-1} \right\} \tag{9.50}$$

■ Example 9.19

Find the inverse z-transform of the following $X[z]$ using the Residue method:

(a) $X[z] = \dfrac{(1 + z^{-1})}{(1 + 8z^{-1} + 15z^{-2})}$ $\quad |z| > 5$

(b) $X[z] = \dfrac{z^{-1}}{(1 - 10z^{-1} + 24z^{-2})}$ $\quad 4 < |z| < 6$

(c) $X[z] = \dfrac{z}{\left(z - \frac{1}{2}\right)^2}$

Solution:

(a) $\mathbf{X[z] = \frac{(1+z^{-1})}{(1+8z^{-1}+15z^{-2})}; \ |z| > 5}$

$$X[z] = \frac{z(z+1)}{(z^2 + 8z + 15)}$$

$$\begin{aligned} X[z] &= \frac{z(z+1)}{(z+3)(z+5)} \\ x[n] &= \sum \text{Residue of } \frac{z(z+1)}{(z+3)(z+5)} z^{n-1} \end{aligned}$$

$$= \text{Residue of } (z+3)\frac{z(z+1)}{(z+3)(z+5)}z^{n-1}\bigg|_{z=-3} + \text{Residue of } (z+5)\frac{z(z+1)z^{n-1}}{(z+3)(z+5)}\bigg|_{z=-5}$$

$$\boxed{x[n] = -(-3)^n + 2(-5)^n}$$

(b) $\boldsymbol{X[z] = \frac{z^{-1}}{(1-10z^{-1}+24z^{-2})}}$; **4 < |z| < 6**

$$X[z] = \frac{z}{(z^2-10z+24)} = \frac{z}{(z-4)(z-6)}$$

For $n \geq 0$,

$$\begin{aligned} x[n] &= \text{Residue of } X[z]z^{n-1}\Big|_{z=4} \\ &= (z-4)\frac{z(z^{n-1})}{(z-4)(z-6)}\bigg|_{z=4} \\ &= -\frac{1}{2}(4)^n u[n] \end{aligned}$$

For $n < 0$,

$$\begin{aligned} x[n] &= -\left[(z-6)\frac{zz^{n-1}}{(z-4)(z-6)}\right]_{z=6} \\ &= -\frac{1}{2}(6)^n u(-n-1) \end{aligned}$$

$$\boxed{x[n] = -\frac{1}{2}[(4)^n u[n] + (6)^n u(-n-1)]}$$

(c) $\boldsymbol{X[z] = \frac{z}{(z-\frac{1}{2})^2}}$

$$\begin{aligned} x[n] &= \frac{d}{dz}\left[\left(z-\frac{1}{2}\right)^2 \frac{zz^{n-1}}{\left(z-\frac{1}{2}\right)}\right]_{z=\frac{1}{2}} \\ &= \frac{d}{dz}z^n\Big|_{z=1/2} = nz^{n-1}\Big|_{z=1/2} \end{aligned}$$

$$\boxed{x[n] = 2n\left(\frac{1}{2}\right)^n u[n]}$$

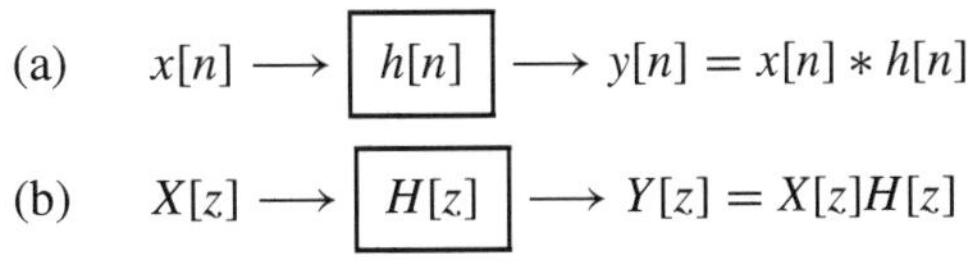

Fig. 9.14 System impulse response and system function

(a) $x[n] \longrightarrow \boxed{h[n]} \longrightarrow y[n] = x[n] * h[n]$

(b) $X[z] \longrightarrow \boxed{H[z]} \longrightarrow Y[z] = X[z]H[z]$

9.9 The System Function of DT Systems

Let

1. $x[n]$ = Input of the system;
2. $y[n]$ = Output of the system;
3. $h[n]$ = Impulse response of the system.

The output $y[n]$ can be expressed as the convolution of $x[n]$ with $h[n]$ as

$$y[n] = x[n] * h[n] \tag{9.51}$$

By applying the convolution property of z-transform, we obtain

$$Y[z] = X[z]H[z] \tag{9.52}$$

where $Y[z]$, $X[z]$ and $H[z]$ are the z-transforms of $y[n]$, $x[n]$ and $h[n]$, respectively. Equation (9.52) can be expressed as

$$H[z] = \frac{Y[z]}{X[z]} \tag{9.53}$$

In Eq. (9.53), $H[z]$ is referred to as the system function or the transfer function. System function is defined as the ratio of the z-transforms of the output $y[n]$ and the input $x[n]$. The system function completely depends on the system characteristic. Equations (9.51) and (9.52) are illustrated in Fig. 9.14a and b, respectively.

9.10 Causality of DT Systems

A linear time invariant discrete time system is said to be causal if the impulse response $h[n] = 0$ for $n < 0$ and it is therefore right-sided. The ROC of such a system $H[z]$ is the exterior of a circle. If $H[z]$ is rational, then the system is said to be causal if the ROC lies exterior of the circle passing through the outermost pole and includes infinity area. A DT system which is linear time invariant with its system function $H[z]$ rational is said to be causal iff the ROC is the exterior of a circle which passes through the outermost pole of $H[z]$. Further, the degree of the numerator polynomial of $H[z]$ should be less than or equal to the degree of the denominator polynomial.

9.11 Stability of DT System

As we discussed in Chap. 2, an LTI discrete time system is said to be BIBO stable if the impulse response $h[n]$ is summable. This is expressed as

$$\sum_{n=-\infty}^{\infty} |h[n]| < \infty \tag{9.54}$$

The corresponding requirement on $H[z]$ is that the ROC of $H[z]$ contains the unit circle. By definition of z-transform,

$$H[z] = \sum_{n=-\infty}^{\infty} h[n]z^{-n}$$

Let $z = e^{j\Omega}$

$$\begin{aligned} |z| &= |e^{j\Omega}| \\ &= 1 \\ |H[e^{j\Omega}]| &= \left| \sum_{n=-\infty}^{\infty} h[n]e^{-j\Omega n} \right| \\ &\leq \sum_{n=-\infty}^{\infty} \left| h[n]e^{-j\Omega n} \right| \\ &= \sum_{n=-\infty}^{\infty} |h[n]| < \infty \end{aligned} \tag{9.55}$$

From Eq. (9.55), we see that the stability condition given by Eq. (9.54) is satisfied if $z = e^{j\Omega}$. Thus, this implies that $H[z]$ must contain unit circle $|z| = 1$.

An LTI system is stable iff the ROC of its system function $H[z]$ contains the unit circle $|z| = 1$.

9.12 Causality and Stability of DT System

For a causal system whose $H[z]$ is rational, the ROC is outside the outermost pole. For the BIBO stability, the ROC should include the unit circle $|z| = 1$. For the system to be causal and stable, the above requirements are satisfied if all the poles are within the unit circle in the z-plane.

An LTID system with the system function $H[z]$ is said to be both causal and stable iff all the poles of $H[z]$ lie inside the unit circle.

The above characteristics of LTI discrete time systems are illustrated in Fig. 9.15 for a causal system.

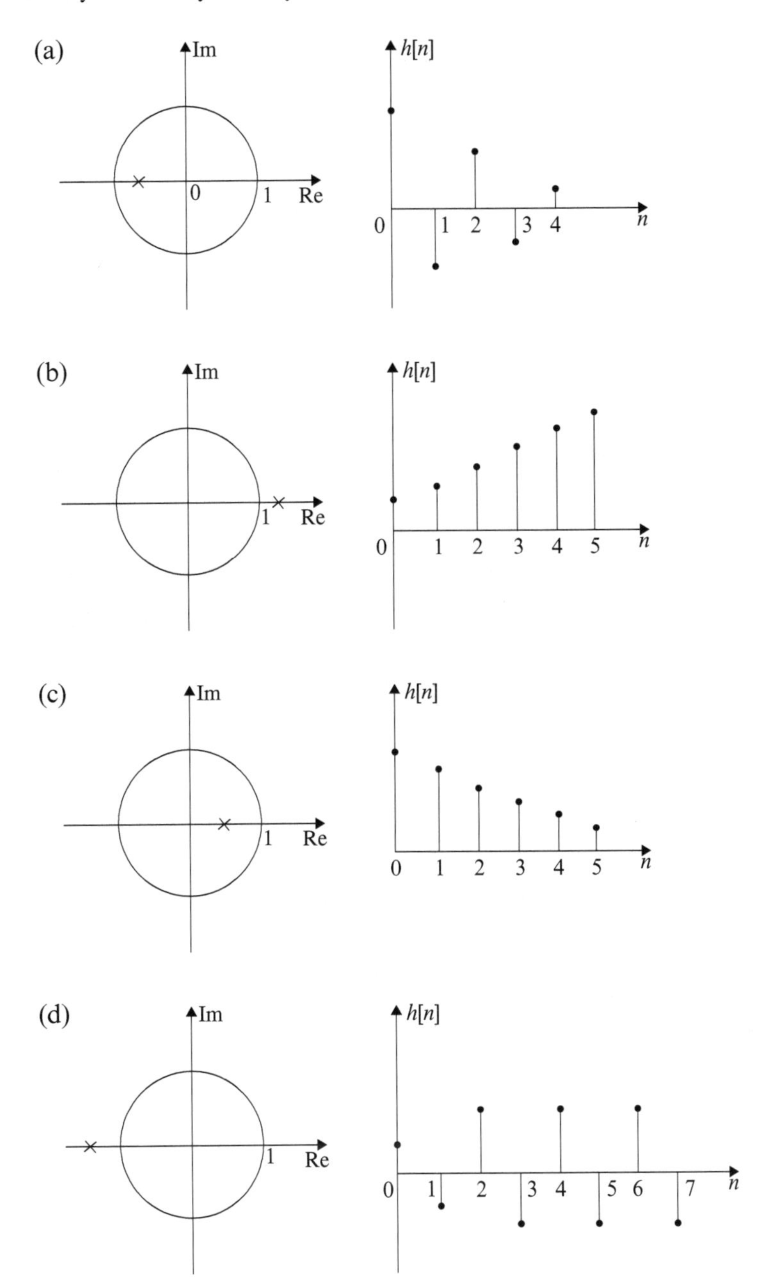

Fig. 9.15 Pole location and impulse response of a causal system

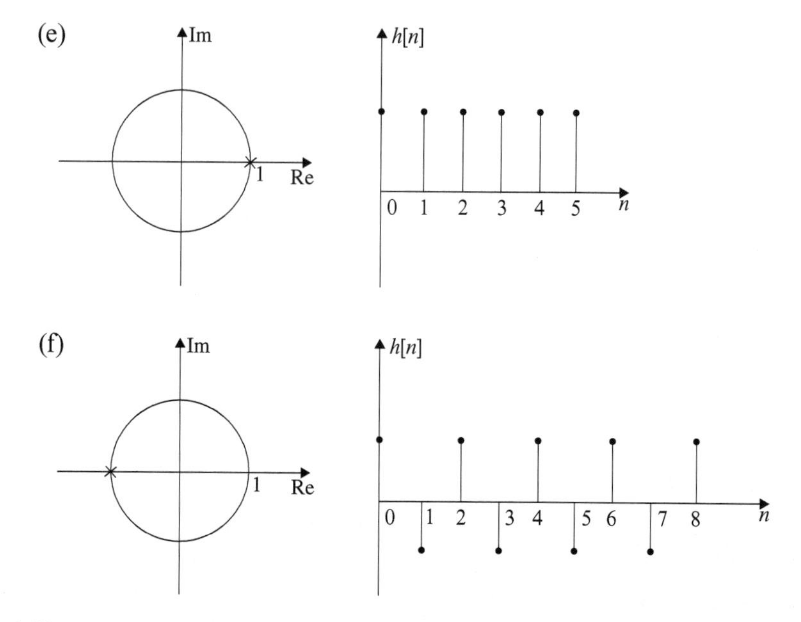

Fig. 9.15 (continued)

■ Example 9.20

The input to the causal LTI system is

$$x[n] = u[-n-1] + \left(\frac{1}{2}\right)^n u[n]$$

The z-transform of the output of the system is

$$Y[z] = \frac{-\frac{1}{2}z^{-1}}{\left(1 - \frac{1}{2}z^{-1}\right)(1 + z^{-1})}$$

Determine $H[z]$, the z-transform of the impulse response and also the output $y[n]$.

(*Anna University, December, 2007*)

Solution:

$$\begin{aligned} X[z] &= -\frac{z}{(z-1)} + \frac{z}{(z-0.5)} \\ &= \frac{-0.5z}{(z-1)(z-0.5)} \end{aligned}$$

$$
\begin{aligned}
Y[z] &= \frac{-\frac{1}{2}z^{-1}}{\left(1-\frac{1}{2}z^{-1}\right)(1+z^{-1})} \\
&= \frac{-\frac{1}{2}z}{(z-0.5)(z+1)} \\
H[z] &= \frac{Y[z]}{X[z]} \\
&= \frac{(-0.5)z(z-1)(z-0.5)}{(z-0.5)(z+1)(-0.5)z} \\
&= \frac{(z-1)}{(z+1)} \\
\frac{H[z]}{z} &= \frac{(z-1)}{z(z+1)} \\
&= \frac{A_1}{z} + \frac{A_2}{z+1} \\
(z-1) &= A_1(z+1) + A_2 z
\end{aligned}
$$

Substitute $z = 0$

$$-1 = A_1$$

Substitute $z = -1$

$$-2 = -A_2; \qquad A_2 = 2$$

$$H[z] = -1 + \frac{2z}{(z+1)}$$

$$\boxed{h[n] = -\delta[n] + (-1)^n 2u[n]}$$

$$
\begin{aligned}
Y[z] &= \frac{-\frac{1}{2}z}{(z-0.5)(z+1)} \\
\frac{Y[z]}{z} &= \frac{-\frac{1}{2}}{(z-0.5)(z+1)} \\
&= \frac{A_1}{z-0.5} + \frac{A_2}{z+1} \\
-\frac{1}{2} &= A_1(z+1) + A_2(z-0.5)
\end{aligned}
$$

Substitute $z = 0.5$

$$-\frac{1}{2} = \frac{3}{2}A_1; \qquad A_1 = -\frac{1}{3}$$

Substitute $z = -1$

$$-\frac{1}{2} = -\frac{3}{2}A_2; \qquad A_2 = \frac{1}{3}$$

$$Y[z] = \frac{1}{3}\left[-\frac{1}{(z-0.5)} + \frac{1}{(z+1)}\right]$$

$$\boxed{y[n] = \frac{1}{3}\left[-\left(\frac{1}{2}\right)^n + (-1)^n\right]u[n]}$$

■ Example 9.21

A certain LTI system is described by the following system function:

$$H[z] = \frac{\left(z+\frac{1}{2}\right)}{(z-1)\left(z-\frac{1}{2}\right)}$$

Find the system response to the input $x[n] = 4^{-(n+2)}u[n]$.

Solution:

$$\begin{aligned}
x[n] &= 4^{-(n+2)}u[n] \\
&= \frac{1}{16}(4)^{-n}u[n] \\
X[z] &= \frac{1}{16}\frac{z}{\left(z-\frac{1}{4}\right)} \\
Y[z] &= H[z]X[z] \\
&= \frac{1\left(z+\frac{1}{2}\right)z}{16(z-1)\left(z-\frac{1}{2}\right)\left(z-\frac{1}{4}\right)} \\
\frac{Y[z]}{z} &= \frac{\left(z+\frac{1}{2}\right)}{16(z-1)\left(z-\frac{1}{2}\right)\left(z-\frac{1}{4}\right)} \\
&= \frac{A_1}{z-1} + \frac{A_2}{\left(z-\frac{1}{2}\right)} + \frac{A_3}{\left(z-\frac{1}{4}\right)}
\end{aligned}$$

$$\frac{1}{16}\left(z+\frac{1}{2}\right) = A_1\left(z-\frac{1}{2}\right)\left(z-\frac{1}{4}\right) + A_2(z-1)\left(z-\frac{1}{4}\right) + A_3(z-1)\left(z-\frac{1}{2}\right)$$

Substitute $z = 1$

$$\left(\frac{1}{16}\right)\left(\frac{3}{2}\right) = \left(\frac{1}{2}\right)\left(\frac{3}{4}\right)A_1; \quad A_1 = \frac{1}{4}$$

Substitute $z = \frac{1}{2}$

$$\frac{1}{16} = -\left(\frac{1}{2}\right)\left(\frac{1}{4}\right)A_2; \quad A_2 = -\frac{1}{2}$$

Substitute $z = \frac{1}{4}$

$$\frac{1}{16}\frac{3}{4} = -\left(\frac{3}{4}\right)\left(-\frac{1}{4}\right)A_3; \quad A_3 = \frac{1}{4}$$

$$Y[z] = \left[\frac{1}{4}\frac{z}{(z-1)} - \frac{1}{2}\frac{z}{\left(z-\frac{1}{2}\right)} + \frac{1}{4}\frac{z}{\left(z-\frac{1}{4}\right)}\right]$$

$$\boxed{y[n] = \left[\frac{1}{4}(1)^n - \frac{1}{2}\left(\frac{1}{2}\right)^n + \frac{1}{4}\left(\frac{1}{4}\right)^n\right]u[n]}$$

■ Example 9.22

Given

$$x[n] = \{2, -3, 1\}$$
$$h[n] = \{1, 2, -1\}$$

Find $y[n]$ using z-transform.

Solution:

$$\begin{aligned} X[z] &= (2 - 3z^{-1} + z^{-2}) \\ H[z] &= 1 + 2z^{-1} - z^{-2} \\ Y[z] &= X[z]H[z] \\ &= [2 - 3z^{-1} + z^{-2}][1 + 2z^{-1} - z^{-2}] \\ &= 2 + z^{-1} - 7z^{-2} + 5z^{-3} - z^{-4} \end{aligned}$$

$$\boxed{y[n] = \{2, 1, -7, 5, -1\}}$$

■ Example 9.23

Given

$$x[n] = u[n]$$
$$y[n] = (2)^n u[n]$$

Find the system function and the impulse response.

Solution:

$$x[n] = u[n]$$
$$X[z] = \frac{z}{(z-1)} \qquad |z| > 1$$
$$y[n] = (2)^n u[n]$$
$$Y[z] = \frac{z}{(z-2)} \qquad |z| > 2$$
$$H[z] = \frac{Y[z]}{X[z]} = \frac{(z-1)}{(z-2)} \qquad |z| > 2$$
$$\frac{H[z]}{z} = \frac{(z-1)}{z(z-2)}$$
$$= \frac{A_1}{z} + \frac{A_2}{(z-2)}$$
$$z - 1 = A_1(z-2) + A_2 z$$

Substitute $z = 0$

$$-1 = A_1(-2); \qquad A_1 = \frac{1}{2}$$

Substitute $z = 2$

$$1 = 2A_2; \qquad A_2 = \frac{1}{2}$$

$$H[z] = \frac{1}{2}\left[1 + \frac{z}{(z-2)}\right]$$

$$\boxed{y[n] = \frac{1}{2}\left[\delta(n) + (2)^n u[n]\right]}$$

■ Example 9.24

Given

$$y[n] = \left(\frac{1}{4}\right)^n u[n]$$

$$x[n] = \left(\frac{1}{2}\right)^n u[-n-1]$$

Find the system function and hence the system impulse response.

Solution:

$$y[n] = \left(\frac{1}{4}\right)^n u[n]$$

$$Y[z] = \frac{z}{\left(z - \frac{1}{4}\right)} \qquad |z| > \frac{1}{4}$$

$$x[n] = \left(\frac{1}{2}\right)^n u[-n-1]$$

$$X[z] = -\frac{z}{\left(z - \frac{1}{2}\right)} \qquad |z| < \frac{1}{2}$$

$$H[z] = \frac{Y[z]}{X[z]}$$

$$\boxed{H[z] = \frac{-\left(z - \frac{1}{2}\right)}{\left(z - \frac{1}{4}\right)}}$$

$$\frac{H[z]}{z} = \frac{-\left(z - \frac{1}{2}\right)}{z\left(z - \frac{1}{4}\right)}$$

$$= \frac{A_1}{z} + \frac{A_2}{z\left(z - \frac{1}{4}\right)}$$

$$\frac{1}{2} - z = A_1\left(z - \frac{1}{4}\right) + A_2 z$$

Substitute $z = 0$

$$\frac{1}{2} = -\frac{1}{4}A_1; \qquad A_1 = -2$$

Substitute $z = \frac{1}{4}$

$$\frac{1}{2} = \frac{1}{4}A_2; \qquad A_2 = 2$$

$$H[z] = 2\left[-1 + \frac{z}{\left(z - \frac{1}{4}\right)}\right]$$

$$\boxed{h[n] = 2\left[-\delta[n] + \left(\frac{1}{4}\right)^n\right]u[n]}$$

■ Example 9.25

Consider the following system functions:

$$\text{(a)} \qquad H[z] = \frac{(1 + 4z^{-1} + z^{-2})}{(2z^{-1} + 5z^{-2} + z^{-3})}$$

$$\text{(b)} \qquad H[z] = \frac{(z-1)(z+2)}{\left(z - \frac{1}{2}\right)\left(z - \frac{3}{4}\right)} \qquad \text{ROC: } |z| > \frac{3}{4}$$

$$\text{(c)} \qquad H[z] = \frac{(z-1)(z+2)}{\left(z - \frac{1}{2}\right)\left(z - \frac{3}{4}\right)} \qquad \text{ROC: } |z| < \frac{1}{2}$$

Determine whether these systems are causal or not.

Solution:

(a) $\boldsymbol{H[z] = \frac{(1+4z^{-1}+z^{-2})}{(2z^{-1}+5z^{-2}+z^{-3})}}$

$$H[z] = \frac{(z^3 + 4z^2 + z)}{(2z^2 + 5z + 1)}$$

$H[z]$ is irrational since the degree of the numerator polynomial is greater than the denominator polynomial.

$$\boxed{\text{The System is Non-causal.}}$$

(b) $\boldsymbol{H[z] = \frac{(z-1)(z+2)}{(z-\frac{1}{2})(z-\frac{3}{4})}}$; **ROC:** $\boldsymbol{|z| > \frac{3}{4}}$
The ROC is the exterior of the circle passing through the outermost pole of $H[z]$. Hence $h[n]$, the impulse response, is right-sided.

$$\boxed{\text{The System is Causal.}}$$

(c) $\boldsymbol{H[z] = \frac{(z-1)(z+2)}{(z-\frac{1}{2})(z-\frac{3}{4})}}$ **ROC :** $|z| < \frac{1}{2}$

The ROC is the interior of the circle passing through the innermost pole of $H[z]$. Hence $h[n]$, the impulse response, is left-sided.

The System is Non-causal.

■ Example 9.26

Consider the following system function:

$$H[z] = \frac{\left(2 - \frac{13}{4}z^{-1}\right)}{\left(1 - \frac{1}{4}z^{-1}\right)\left(1 - 3z^{-1}\right)}$$

Determine the causality and stability of the system for the following cases:

(a) ROC: $|z| > 3$;
(b) ROC: $|z| < \frac{1}{4}$;
(c) ROC: $\frac{1}{4} < |z| < 3$.

Solution:

$$H[z] = \frac{\left(2 - \frac{13}{4}z^{-1}\right)}{\left(1 - \frac{1}{4}z^{-1}\right)\left(1 - 3z^{-1}\right)} = \frac{z\left(2z - \frac{13}{4}\right)}{\left(z - \frac{1}{4}\right)(z - 3)}$$

(a) **ROC** : $|z| > 3$
The ROC is the exterior of the circle passing through the outermost pole of $H[z]$ which is rational (the denominator and numerator polynomials have the same order). The impulse response $h[n]$ is a right-sided sequence. Hence $H[z]$ is causal. The ROC does not contain a unit circle. Hence $h[n]$ is not summable. The system is unstable. Refer to Fig. 9.16a.

The System is Causal and Unstable.

(b) **ROC** : $|z| < \frac{1}{4}$
The ROC is the interior of the circle passing through the innermost pole of $H[z]$. The impulse response is a left-sided sequence. $H[z]$ is therefore non-causal. The ROC does not include the unit circle. The $h[n]$ is a growing exponential negative sequence. The system is unstable. Refer to Fig. 9.16b.

The System is Non-causal and Unstable.

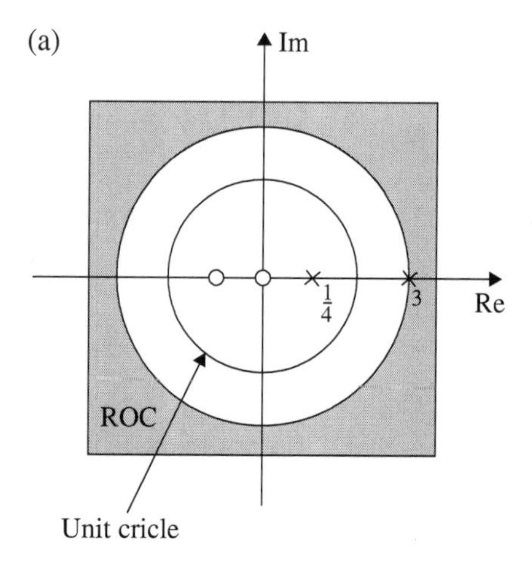

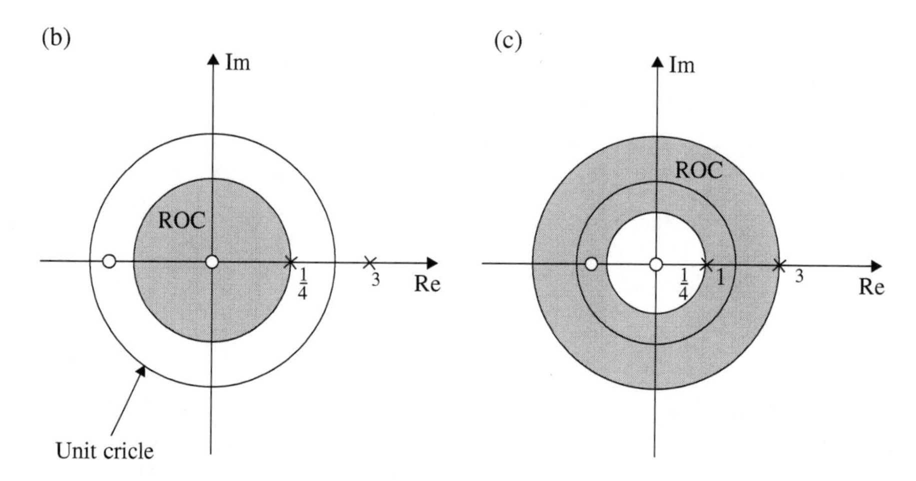

Fig. 9.16 **a** Causal and unstable system, **b** Non-causal and unstable system and **c** Non-causal and stable system

(c) **ROC** : $\frac{1}{4} < |z| < 3$
The ROC is to the left of the outermost pole and to the right of the innermost pole. Hence, $h[n]$ will have right- and left-sided sequences, which is non-causal. The ROC includes a unit circle, which means that the right- and left-side sequences of $h[n]$ will exponentially decay and the system is stable. Refer to Fig. 9.16c.

$$\boxed{\text{The System is Non-causal and Stable.}}$$

The system cannot be both Causal and Stable.

■ Example 9.27

Consider the following system function:

$$H[z] = \frac{z}{\left(z - \frac{1}{4}\right)\left(z + \frac{1}{4}\right)\left(z - \frac{1}{2}\right)}$$

For different possible ROCs, determine the causality, stability and the impulse response of the system.

Solution:

$$H[z] = \frac{z}{\left(z - \frac{1}{4}\right)\left(z + \frac{1}{4}\right)\left(z - \frac{1}{2}\right)}$$

The possible ROCs for $H[z]$ to exist are (a) ROC: $|z| > \frac{1}{2}$, (b) ROC: $|z| < \frac{1}{4}$ and (c) ROC: $\frac{1}{4} < |z| < \frac{1}{2}$.

$$\begin{aligned}\frac{H[z]}{z} &= \frac{1}{\left(z - \frac{1}{4}\right)\left(z + \frac{1}{4}\right)\left(z - \frac{1}{2}\right)} \\ &= \frac{A_1}{\left(z - \frac{1}{4}\right)} + \frac{A_2}{\left(z + \frac{1}{4}\right)} + \frac{A_3}{\left(z - \frac{1}{2}\right)} \\ 1 &= A_1\left(z + \frac{1}{4}\right)\left(z - \frac{1}{2}\right) + A_2\left(z - \frac{1}{4}\right)\left(z - \frac{1}{2}\right) + A_3\left(z - \frac{1}{4}\right)\left(z + \frac{1}{4}\right)\end{aligned}$$

Substitute $z = \frac{1}{4}$

$$1 = A_1\left(\frac{1}{4} + \frac{1}{4}\right)\left(\frac{1}{4} - \frac{1}{2}\right); \quad A_1 = -8$$

Substitute $z = -\frac{1}{4}$

$$1 = A_2\left(-\frac{1}{4} - \frac{1}{4}\right)\left(-\frac{1}{4} - \frac{1}{2}\right); \quad A_2 = \frac{8}{3}$$

Substitute $z = \frac{1}{2}$

$$1 = A_3\left(\frac{1}{2} - \frac{1}{4}\right)\left(\frac{1}{2} + \frac{1}{4}\right); \quad A_3 = \frac{16}{3}$$

$$H[z] = -\frac{8z}{\left(z - \frac{1}{4}\right)} + \frac{8}{3}\frac{z}{\left(z + \frac{1}{4}\right)} + \frac{16}{3}\frac{z}{\left(z - \frac{1}{2}\right)}$$

(a) **ROC** : $|z| > \frac{1}{2}$
The pole-zero diagram and the ROC are shown in Fig. 9.17a. From Fig. 9.17a, the ROC is the exterior of the outermost pole $z = \frac{1}{2}$. Further, ROC includes a unit circle. Thus, $h[n]$ is a right-sided sequence and hence $H[z]$ is causal. Since ROC includes the unit circle and all the poles are within the unit circle, the system is stable.

Now,

$$H[z] = -\frac{8z}{\left(z - \frac{1}{4}\right)} + \frac{8}{3}\frac{z}{\left(z + \frac{1}{4}\right)} + \frac{16}{3}\frac{z}{\left(z - \frac{1}{2}\right)}$$

$$\boxed{h[n] = \left[-8\left(\frac{1}{4}\right)^n + \frac{8}{3}\left(-\frac{1}{4}\right)^n + \frac{16}{3}\left(\frac{1}{2}\right)^n\right]u[n]}$$

The System is Causal and Stable.

(b) **ROC** : $|z| < \frac{1}{4}$
For ROC: $|z| < \frac{1}{4}$, the pole-zero diagram is shown in Fig. 9.17b. The ROC is interior of the circle passing through the innermost pole. Hence the system is non-causal. The condition that the ROC does not include a unit circle implies that the system is unstable. The sequence $h[n]$ is left-sided. This is obtained as follows:

$$H[z] = -\frac{8z}{\left(z - \frac{1}{4}\right)} + \frac{8}{3}\frac{z}{\left(z + \frac{1}{4}\right)} + \frac{16}{3}\frac{z}{\left(z - \frac{1}{2}\right)}$$

$$h[n] = \left[8\left(\frac{1}{4}\right)^n - \frac{8}{3}\left(-\frac{1}{4}\right)^n - \frac{16}{3}\left(\frac{1}{2}\right)^n\right]u[-n-1]$$

The left-sided sequence $u[-n-1]$ will exponentially increase for $n < 0$ and makes the system unstable.

The System is Non-causal and Unstable.

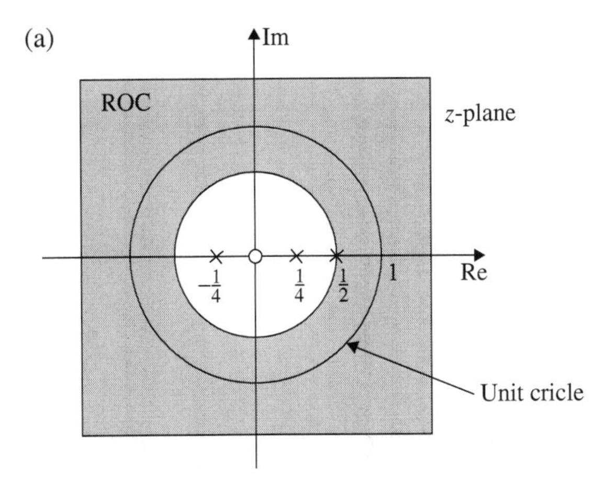

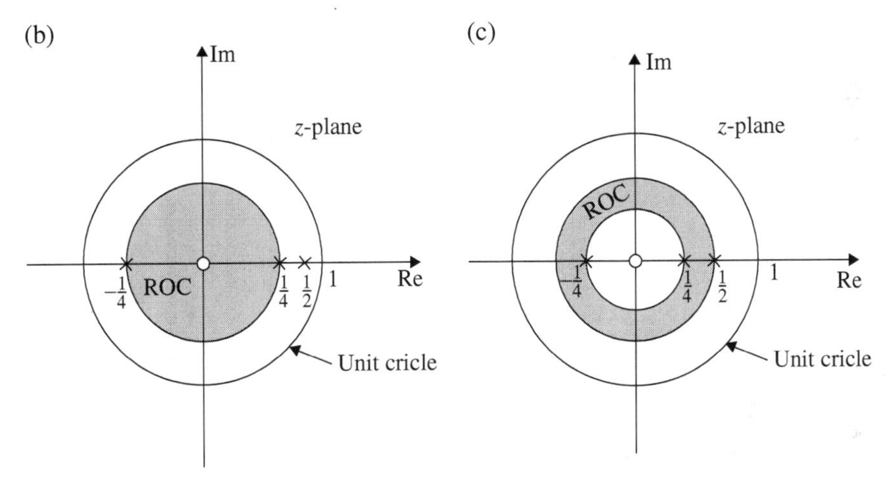

Fig. 9.17 **a** Pole-zero diagram and ROC: $|z| > \frac{1}{2}$ of Example 9.27, **b** Pole-zero diagram and ROC: $|z| < \frac{1}{4}$ and **c** Pole-zero diagram and ROC: $\frac{1}{4} < |z| < \frac{1}{2}$

(c) **ROC** : $\frac{1}{4} < |z| < \frac{1}{2}$
The pole-zero diagram and ROC of $H[z]$ are shown in Fig. 9.17c. The ROC is a concentric ring for $\frac{1}{4} < |z| < \frac{1}{2}$. The $h[n]$ sequences die to the poles at $z = \frac{1}{4}$ and $z = -\frac{1}{4}$ are right-sided and the sequence due to the pole $z = \frac{1}{2}$ is left-sided. Hence the system is non-causal. The ROC does not include the unit circle and hence the system is unstable. The impulse response is obtained as follows.

$$H[z] = -\frac{8z}{\left(z - \frac{1}{4}\right)} + \frac{8}{3}\frac{z}{\left(z + \frac{1}{4}\right)} + \frac{16}{3}\frac{z}{\left(z - \frac{1}{2}\right)}$$

$$\boxed{h[n] = \left[-8\left(\frac{1}{4}\right)^n + \frac{8}{3}\left(-\frac{1}{4}\right)^n\right]u[n] - \frac{16}{3}\left(\frac{1}{2}\right)^n u[-n-1]}$$

The term $-\frac{16}{3}(1/2)^n u[-n-1]$ for $n<0$ yields exponentially increasing sequence.

$$\boxed{\text{The System is Non-causal and Unstable.}}$$

9.13 z-Transform Solution of Linear Difference Equations

As in the case of Laplace transform with a differential equation, to get the solution in time domain z-transform is used to solve the difference equation to get the output sequence as a function of n. By using the time shift property of z-transform, the difference equation is converted into an algebraic equation, taking into account the initial conditions. By taking z-inverse transform, the time domain solution is obtained.

9.13.1 *Right Shift (Delay)*

If

$$x[n]u[n] \overset{Z}{\longleftrightarrow} X[z]$$

then

$$\begin{aligned} x[n-1]u[n-1] &\overset{Z}{\longleftrightarrow} \frac{1}{z}X[z] \\ x[n-1]u[n] &\overset{Z}{\longleftrightarrow} \frac{1}{z}X[z] + x[-1] \\ x[n-2]u[n] &\overset{Z}{\longleftrightarrow} \frac{1}{z^2}X[z] + \frac{1}{z}x[-1] + x[-2] \end{aligned}$$

In general,

$$\boxed{x[n-m]u[n] \overset{Z}{\longleftrightarrow} z^{-m}X[z] + z^{-m}\sum_{n=1}^{m} x[-n]z^n} \tag{9.56}$$

9.13.2 *Left Shift (Advance)*

If

$$x[n]u[n] \overset{Z}{\longleftrightarrow} X[z]$$

$$x[n+1]u[n] \overset{Z}{\longleftrightarrow} zX[z] - zx[0]$$

$$x[n+2]u[n] \overset{Z}{\longleftrightarrow} z^2X[z] - z^2x[0] - zx[1]$$

In general,

$$\boxed{x[n+m]u[n] \overset{Z}{\longleftrightarrow} z^m x[z] - z^m \sum_{n=0}^{m-1} x[n]z^{-n}} \tag{9.57}$$

Equations (9.56) and (9.57) are used to convert difference equations with initial conditions to algebraic equations in z. With the application of Eq. (9.56), the delay shift is more common. The following examples illustrate the above procedure.

■ Example 9.28

Consider the following linear constant coefficient difference equation:

$$y[n] - \frac{3}{4}y[n-1] + \frac{1}{8}y[n-2] = 2x[n-1]$$

Determine $y[n]$ when $x[n] = \delta[n]$ and $y[n] = 0,\ n < 0$.

(Anna University, May and December, 2007)

Solution: If $y[n] = 0,\ n = 0$ implies the initial conditions are zero. Taking z-transform on both sides of the given equation, we get

$$\left[1 - \frac{3}{4}z^{-1} + \frac{1}{8}z^{-2}\right] Y[z] = 2z^{-1}X[z]$$

For $\delta[n],\ X[z] = 1$,

$$Y[z] = \frac{2z^{-1}}{1 - \frac{3}{4}z^{-1} + \frac{1}{8}z^{-2}}$$

$$= \frac{2z}{z^2 - \frac{3}{4}z + \frac{1}{8}}$$

$$\frac{Y[z]}{z} = \frac{2}{\left(z - \frac{1}{2}\right)\left(z - \frac{1}{4}\right)}$$

$$= \frac{A_1}{\left(z - \frac{1}{2}\right)} + \frac{A_2}{\left(z - \frac{1}{4}\right)}$$

$$2 = A_1\left(z - \frac{1}{4}\right) + A_2\left(z - \frac{1}{2}\right)$$

Substitute $z = \frac{1}{2}$

$$2 = A_1\frac{1}{4}; \qquad A_1 = 8$$

Substitute $z = \frac{1}{4}$

$$2 = A_2\left(-\frac{1}{4}\right); \qquad A_2 = -8$$

$$Y[z] = 8\left[\frac{z}{\left(z - \frac{1}{2}\right)} - \frac{z}{\left(z - \frac{1}{4}\right)}\right]$$

$$\boxed{y[n] = 8\left[\left(\frac{1}{2}\right)^n - \left(\frac{1}{4}\right)^n\right]u[n] \qquad \text{ROC: } |z| > \frac{1}{2}}$$

■ Example 9.29

$$y[n+2] + 1.1y[n+1] + 0.3y[n] = x[n+1] + x[n]$$

where $x[n] = (-4)^{-n}u[n]$. Find $y[n]$ if the initial conditions are zero.

Solution: Taking z-transform using the left shift property, we get

$$[z^2 + 1.1z + 0.3]Y[z] = [z+1]X[z]$$
$$x[n] = (-4)^{-n}u[n]$$

$$X[z] = \frac{z}{\left(z + \frac{1}{4}\right)}$$

$$Y[z] = \frac{z(z+1)}{\left(z + \frac{1}{4}\right)(z^2 + 1.1z + 0.3)}$$

$$= \frac{z(z+1)}{\left(z + \frac{1}{4}\right)(z+0.5)(z+0.6)}$$

$$\frac{Y[z]}{z} = \frac{(z+1)}{\left(z + \frac{1}{4}\right)(z+0.5)(z+0.6)}$$

$$= \frac{A_1}{\left(z+\frac{1}{4}\right)} + \frac{A_2}{(z+0.5)} + \frac{A_3}{(z+0.6)}$$

$$(z+1) = A_1\,(z+0.5)\,(z+0.6) + A_2\left(z+\frac{1}{4}\right)(z+0.6)$$
$$+A_3\left(z+\frac{1}{4}\right)(z+0.5)$$

Substitute $z = -\frac{1}{4}$

$$\left(-\frac{1}{4}+1\right) = A_1\left(-\frac{1}{4}+0.5\right)\left(-\frac{1}{4}+0.6\right); \qquad A_1 = 8.57$$

Substitute $z = -0.5$

$$(-0.5+1) = A_2\left(-0.5+\frac{1}{4}\right)(-0.5+0.6); \qquad A_2 = -20$$

Substitute $z = -0.6$

$$(-0.6+1) = A_3\left(-0.6+\frac{1}{4}\right)(-0.6+0.5); \qquad A_3 = 11.43$$

$$Y[z] = \frac{8.57z}{\left(z+\frac{1}{4}\right)} - \frac{20z}{(z+0.5)} + \frac{11.43}{(z+0.6)}$$

$$\boxed{y[n] = \left[8.57\left(-\frac{1}{4}\right)^n - 20\,(-0.5)^n + 11.43(-0.6)^n\right]u[n]}$$

■ Example 9.30

A causal LTI system is described by the difference equation

$$y[n] = y[n-1] + y[n-2] + x[n-1]$$

Find (a) System function for this system and (b) Unit impulse response of the system.

(*Anna University, April, 2008*)

Solution: Taking z-transform on both sides of the equation and making use of right shift property, we get

$$[1 - z^{-1} - z^{-2}]Y[z] = z^{-1}X[z]$$

(a)

$$H[z] = \frac{Y[z]}{X[z]}$$

$$\boxed{H[z] = \frac{z^{-1}}{(1 - z^{-1} - z^{-2})}}$$

(b)

$$\begin{aligned}
H[z] &= \frac{z}{(z^{-2} - z - 1)} \\
\frac{H[z]}{z} &= \frac{1}{(z - 1.618)(z + 0.618)} \\
&= \frac{A_1}{(z - 1.618)} + \frac{A_2}{(z + 1.618)} \\
1 &= A_1(z + 0.618) + A_2(z - 1.618)
\end{aligned}$$

Substitute $z = 1.618$

$$1 = A_1\,(1.618 + 0.618)\,; \qquad A_1 = 0.447$$

Substitute $z = -0.618$

$$1 = A_2\,(-0.618 - 1.618)\,; \qquad A_2 = -0.447$$

$$H[z] = 0.447\left[\frac{z}{z - 1.618} - \frac{z}{z + 0.618}\right]$$

$$\boxed{h[n] = 0.447\left[(1.618)^n - (-0.618)^n\right]u[n]}$$

■ Example 9.31

Find the impulse response of the discrete time system described by the difference equation

$$y[n-2] - 3y[n-1] + 2y[n] = x[n-1]$$

(Anna University, April, 2005)

Solution:

$$[z^{-2} - 3z^{-1} + 2]Y[z] = z^{-1}X[z]$$

$$H[z] = \frac{Y[z]}{X[z]}$$

$$= \frac{z^{-1}}{(z^{-2} - 3z^{-1} + 2)}$$

$$= \frac{z}{(2z^2 - 3z + 1)}$$

$$\frac{H[z]}{z} = \frac{0.5}{(z-1)(z-0.5)}$$

$$= \frac{1}{(z-1)} - \frac{1}{(z-0.5)}$$

$$H[z] = \frac{z}{z-1} - \frac{z}{z-0.5}$$

$$\boxed{h[n] = \left[(1)^n - \left(\frac{1}{2}\right)^n\right]u[n]}$$

■ Example 9.32

Determine the impulse response and frequency response of the system described by the difference equation

$$y[n] - \left(\frac{1}{6}\right)y[n-1] - \frac{1}{6}y[n-2] = x[n-1]$$

(Anna University, May, 2007)

Solution:

To obtain Impulse Response

$$\left[1 - \frac{1}{6}z^{-1} - \frac{1}{6}z^{-2}\right]Y[z] = z^{-1}X[z]$$

$$H[z] = \frac{Y[z]}{X[z]}$$

$$= \frac{z}{\left(z^2 - \frac{1}{6}z - \frac{1}{6}\right)}$$

$$= \frac{z}{\left(z - \frac{1}{2}\right)\left(z + \frac{1}{3}\right)}$$

$$\frac{H[z]}{z} = \frac{1}{\left(z - \frac{1}{2}\right)\left(z + \frac{1}{3}\right)}$$

$$= \frac{6}{5}\left[\frac{1}{\left(z-\frac{1}{2}\right)} - \frac{1}{\left(z+\frac{1}{3}\right)}\right]$$

$$H[z] = \frac{6}{5}\left[\frac{z}{\left(z-\frac{1}{2}\right)} - \frac{1}{\left(z+\frac{1}{3}\right)}\right]$$

$$\boxed{h[n] = \frac{6}{5}\left[\left(\frac{1}{2}\right)^n - \left(\frac{1}{3}\right)^n\right]u[n]}$$

To obtain Frequency Response

Substitute $z = e^{j\omega}$ in $H[z]$

$$H[e^{j\omega}] = \frac{e^{j\omega}}{\left(e^{j\omega}-\frac{1}{2}\right)\left(e^{j\omega}+\frac{1}{3}\right)}$$

This can be expressed in terms of amplitude and phase as follows:

$$H[e^{j\omega}] = \frac{e^{j\omega}}{\left(\cos\omega + j\sin\omega - \frac{1}{2}\right)\left(\cos\omega + j\sin\omega + \frac{1}{3}\right)}$$

since $|e^{j\omega}| = 1$

$$\boxed{|H(e^{j\omega})| = \frac{1}{\left[\left\{\left(\cos\omega - \frac{1}{2}\right)^2 + \sin^2\omega\right\}\left\{\left(\cos\omega + \frac{1}{3}\right)^2 + \sin^2\omega\right\}\right]^{\frac{1}{2}}}}$$

Since $\angle e^{j\omega} = \omega$,

$$\boxed{\angle H(e^{j\omega}) = \omega - \tan^{-1}\frac{\sin\omega}{\left(\cos\omega - \frac{1}{2}\right)} - \tan^{-1}\frac{\sin\omega}{\left(\cos\omega + \frac{1}{3}\right)}}$$

Example 9.33

A causal system is represented by the difference equation

$$y[n] + \frac{1}{4}y[n-1] = x[n] + \frac{1}{2}x[n-1]$$

Use z-transform to determine the

(1) System function;
(2) Unit sample response of the system;

(3) Frequency response of the system.

Solution:

(1)

$$\left[1+\frac{1}{4}z^{-1}\right]Y[z]=\left[1+\frac{1}{2}z^{-1}\right]X[z]$$

$$H[z]=\frac{Y[z]}{X[z]}$$

$$\boxed{H[z]=\frac{\left[1+\frac{1}{2}z^{-1}\right]}{\left[1+\frac{1}{4}z^{-1}\right]}}$$

(2)

$$H[z]=\frac{\left(z+\frac{1}{2}\right)}{\left(z+\frac{1}{4}\right)}$$

$$\frac{H[z]}{z}=\frac{\left(z+\frac{1}{2}\right)}{z\left(z+\frac{1}{4}\right)}=\frac{A_1}{z}+\frac{A_2}{\left(z+\frac{1}{4}\right)}$$

$$\left(z+\frac{1}{2}\right)=A_1\left(z+\frac{1}{4}\right)+A_2 z$$

Substitute $z=0$

$$A_1=2$$

Substitute $z=-\frac{1}{4}$

$$\left[-\frac{1}{4}+\frac{1}{2}\right]=A_2\left(-\frac{1}{4}\right);\qquad A_2=-1$$

$$H[z]=2-\frac{z}{\left(z+\frac{1}{4}\right)}$$

$$\boxed{h[n]=2\delta[n]-\left(\frac{1}{4}\right)^n u[n]}$$

(3)

$$H[z] = \frac{\left(z + \frac{1}{2}\right)}{\left(z + \frac{1}{4}\right)}$$

$$H[e^{j\omega}] = \frac{\left(e^{j\omega} + \frac{1}{2}\right)}{\left(e^{j\omega} + \frac{1}{4}\right)} = \frac{\left(\cos\omega + \frac{1}{2}\right) + j\sin\omega}{\left(\cos\omega + \frac{1}{4}\right) + j\sin\omega}$$

$$\boxed{|H(e^{j\omega})| = \frac{\left[\left(\cos\omega + \frac{1}{2}\right)^2 + \sin^2\omega\right]^{1/2}}{\left[\left(\cos\omega + \frac{1}{4}\right)^2 + \sin\omega\right]^{1/2}}}$$

$$\boxed{\angle H(j\omega) = \tan^{-1}\frac{\sin\omega}{\left(\cos\omega + \frac{1}{2}\right)} - \tan^{-1}\frac{\sin\omega}{\left(\cos\omega + \frac{1}{4}\right)}}$$

Example 9.34

Find the output of the system whose input-output is related by the difference equation

$$y[n] - \frac{5}{6}y[n-1] + \frac{1}{6}y[n-2] = x[n] - \frac{1}{2}x[n-1]$$

for the step input. Assume initial conditions to be zero.

Solution:

$$\left[1 - \frac{5}{6}z^{-1} + \frac{1}{6}z^{-2}\right]Y[z] = \left[1 - \frac{1}{2}z^{-1}\right]X[z]$$

$$Y[z] = \frac{\left[1 - \frac{1}{2}z^{-1}\right]}{\left[1 - \frac{5}{6}z^{-1} + \frac{1}{6}z^{-2}\right]}X[z]$$

For unit step input, $X[z] = \frac{z}{z-1}$

$$Y[z] = \frac{z^2\left[z - \frac{1}{2}\right]}{(z-1)\left(z^2 - \frac{5}{6}z + \frac{1}{6}\right)}$$

$$\frac{Y[z]}{z} = \frac{z\left[z - \frac{1}{2}\right]}{(z-1)\left(z - \frac{1}{2}\right)\left(z - \frac{1}{3}\right)}$$

$$= \frac{z}{(z-1)\left(z - \frac{1}{3}\right)}$$

$$= \frac{A_1}{(z-1)} + \frac{A_2}{\left(z-\frac{1}{3}\right)}$$

$$z = A_1\left(z - \frac{1}{3}\right) + A_2(z-1)$$

Substrate $z = 1$

$$1 = A_1\left(1 - \frac{1}{3}\right); \qquad A_1 = \frac{3}{2}$$

Substrate $z = \frac{1}{3}$

$$\frac{1}{3} = A_2\left(\frac{1}{3} - 1\right); \qquad A_2 = -\frac{1}{2}$$

$$Y[z] = \frac{3}{2}\frac{z}{(z-1)} - \frac{1}{2}\frac{z}{\left(z-\frac{1}{3}\right)}$$

$$\boxed{y[n] = \left[\frac{3}{2}(1)^n - \frac{1}{2}\left(\frac{1}{3}\right)^n\right]u[n]}$$

Example 9.35

Find the output response of the discrete time system described by the following difference equation:

$$y[n] - \frac{3}{4}y[n-1] + \frac{1}{8}y[n-2] = x[n]$$

The initial conditions are $y[-1] = 0$ and $y = [-2] = 1$. The input $x[n] = \left(\frac{1}{5}\right)^n u[n]$.

Solution:
Taking z-transform on both sides of the above equation, we get

$$Y[z] - \frac{3}{4}[z^{-1}Y[z] + y[-1]] + \frac{1}{8}[z^{-2}Y[z]$$
$$+z^{-1}y[-1] + y[-2]] = X[z]$$
$$\left[1 - \frac{3}{4}z^{-1} + \frac{1}{8}z^{-2}\right]Y[z] = -\frac{1}{8} + \frac{z}{\left(z-\frac{1}{5}\right)}$$
$$\frac{\left[z^2 - \frac{3}{4}z + \frac{1}{8}\right]}{z^2}Y[z] = -\frac{1}{8} + \frac{z}{\left(z-\frac{1}{5}\right)}$$

$$\begin{aligned}
\frac{Y[z]}{z} &= -\frac{z}{8\left(z-\frac{1}{4}\right)\left(z-\frac{1}{2}\right)} + \frac{z^2}{\left(z-\frac{1}{5}\right)\left(z-\frac{1}{4}\right)\left(z-\frac{1}{2}\right)} \\
&= Y_1[z] + Y_2[z] \\
Y_1[z] &= -\frac{z}{8\left(z-\frac{1}{4}\right)\left(z-\frac{1}{2}\right)} \\
&= -\frac{A_1}{\left(z-\frac{1}{4}\right)} + \frac{A_2}{\left(z-\frac{1}{2}\right)} \\
-\frac{z}{8} &= A_1\left(z-\frac{1}{2}\right) + A_2\left(z-\frac{1}{4}\right)
\end{aligned}$$

Substitute $z = \frac{1}{4}$

$$-\frac{1}{4}\frac{1}{8} = -A_1\frac{1}{4}; \qquad A_1 = \frac{1}{8}$$

Substrate $z = \frac{1}{2}$

$$-\frac{1}{2}\frac{1}{8} = A_2\frac{1}{4}; \qquad A_2 = -\frac{1}{4}$$

$$\begin{aligned}
Y_1[z] &= \frac{1}{8\left(z-\frac{1}{4}\right)} - \frac{1}{4\left(z-\frac{1}{2}\right)} \\
Y_2[z] &= \frac{z^2}{\left(z-\frac{1}{5}\right)\left(z-\frac{1}{4}\right)\left(z-\frac{1}{2}\right)} \\
z^2 &= A_1\left(z-\frac{1}{4}\right)\left(z-\frac{1}{2}\right) + A_2\left(z-\frac{1}{5}\right)\left(z-\frac{1}{2}\right) + A_3\left(z-\frac{1}{5}\right)\left(z-\frac{1}{4}\right)
\end{aligned}$$

Substitute $z = \frac{1}{5}$

$$\frac{1}{25} = A_1\left(\frac{1}{5}-\frac{1}{4}\right)\left(\frac{1}{5}-\frac{1}{2}\right); \qquad A_1 = \frac{8}{3}$$

Substitute $z = \frac{1}{4}$

$$\frac{1}{16} = A_2\left(\frac{1}{4}-\frac{1}{5}\right)\left(\frac{1}{4}-\frac{1}{2}\right); \qquad A_2 = -5$$

Substitute $z = \frac{1}{2}$

$$\frac{1}{4} = A_3\left(\frac{1}{2}-\frac{1}{5}\right)\left(\frac{1}{2}-\frac{1}{4}\right); \qquad A_3 = \frac{10}{3}$$

$$Y_2[z] = \frac{8}{3}\frac{1}{\left(z-\frac{1}{5}\right)} - \frac{5}{\left(z-\frac{1}{4}\right)} + \frac{10}{3}\frac{1}{\left(z-\frac{1}{2}\right)}$$

$$Y[z] = \frac{z}{8\left(z-\frac{1}{4}\right)} - \frac{z}{4\left(z-\frac{1}{2}\right)} + \frac{8}{3}\frac{z}{\left(z-\frac{1}{5}\right)} - \frac{5z}{\left(z-\frac{1}{4}\right)} + \frac{10}{3}\frac{z}{\left(z-\frac{1}{2}\right)}$$

$$= -\frac{39}{8}\frac{z}{\left(z-\frac{1}{4}\right)} + \frac{37}{12}\frac{z}{\left(z-\frac{1}{2}\right)} + \frac{8}{3}\frac{z}{\left(z-\frac{1}{5}\right)}$$

$$\boxed{y[n] = \left[-\frac{39}{8}\left(\frac{1}{4}\right)^n + \frac{37}{12}\left(\frac{1}{2}\right)^n + \frac{8}{3}\left(\frac{1}{5}\right)^n\right]u[n]}$$

■ Example 9.36

Consider the following difference equation:

$$y[n] + 2y[n-1] + 2y[n-2] = x[n]$$

The initial conditions are $y[-1] = 0$ and $y = [-2] = 2$. Find the step response of the system.

Solution: Taking z-transform on both sides of the above equation, we get

$$X[z] = Y[z] + 2[z^{-1}Y[z] + y[-1]] + 2[z^{-2}Y[z] + z^{-1}y[-1] + y[-2]]$$

$$-4 + X[z] = [1 + 2z^{-1} + 2z^{-2}]Y[z]$$

$$-4 + X[z] = \frac{(z^2+2z+2)}{z^2}Y[z]$$

For step input $X[z] = \frac{z}{(z-1)}$,

$$z^2 + 2z + 2 = (z+1+j)(z+1-j)$$

$$\frac{(z+1+j)(z+1-j)}{z^2}Y[z] = -4 + \frac{z}{z-1}$$

$$= \frac{(4-3z)}{z-1}$$

$$\frac{Y[z]}{z} = \frac{z(4-3z)}{(z-1)(z+1+j)(z+1-j)}$$

$$= \frac{A_1}{(z-1)} + \frac{A_2}{(z+1+j)} + \frac{A_3}{(z+1-j)}$$

$$z[4-3z] = A_1(z^2+2z+2) + A_2(z-1)(z+1-j)$$
$$+A_3(z-1)(z+1+j)$$

Substitute $z = 1$

$$1 = A_1 5; \qquad A_1 = \frac{1}{5}$$

Substitute $z = -1 + j$

$$(-1+j)(4-3+j3) = A_3(-1+j-1)(-1+1+j+j)$$
$$(-1+j)(1+j3) = A_3(-2+j)j2$$
$$\sqrt{2}\angle 135^\circ \sqrt{10}\angle 71.56^\circ = A_3\sqrt{5}\angle 153.43^\circ \sqrt{2}\angle 90^\circ$$
$$A_3 = \frac{\sqrt{2}\angle 135^\circ \sqrt{10}\angle 71.56^\circ}{\sqrt{5}\angle 153.43^\circ \sqrt{2}\angle 90^\circ}$$
$$= 1\angle -36.87^\circ = 1e^{-j0.643}$$

$$A_2 = \text{conjugate of } A_3$$
$$= 1e^{j0.643}$$

The exponentials of A_1 and A_2 are expressed in radians using $57.3^\circ = 1$ radian.

$$Y[z] = \frac{1}{5}\frac{z}{(z-1)} + \frac{e^{j0.643}z}{z+1+j} + \frac{e^{-j0.643}z}{z+1-j}$$
$$Y[z] = \frac{1}{5}\frac{z}{(z-1)} + \frac{e^{j0.643}z}{(z+\sqrt{2}e^{j\frac{\pi}{4}})} + \frac{e^{-j0.643}z}{(z+\sqrt{2}e^{-j\frac{\pi}{2}})}$$

Taking inverse z-transform, we get

$$y[n] = \frac{1}{5} + e^{j0.643}(-\sqrt{2}e^{j\frac{\pi}{4}})^n + e^{-j0.643}(-\sqrt{2}e^{-j\frac{\pi}{4}})^n$$
$$= \frac{1}{5} + (-\sqrt{2})^n \left[e^{j(0.643+\frac{\pi}{4}n)} + e^{-j(0.643+\frac{\pi}{4}n)}\right]$$

$$\boxed{y[n] = \left[\frac{1}{5} + 2(-\sqrt{2})^n \cos\left(\frac{\pi}{4}n + 0.643\right)\right]u[n]}$$

■ Example 9.37

Solve the following difference equation:

$$y[n] + 6y[n-1] + 8y[n-2] = 5x[n-1] + x[n-2]$$

The initial conditions are $y[-1] = 1$ and $y[-2] = 2$. The input $x[n] = u[n]$.

Solution: Taking z-transform on both sides, we get

$$1 + 6(z^{-1}Y[z] + y[-1]) + 8(z^2 Y[z] + z^{-1}y[-1] + y[-2]) = [5z^{-1} + z^{-2}]X[z]$$

For a causal signal $u[n]$, $x[-2]$, $x[-1]$ are zero.

$$[1 + 6z^{-1} + 8z^{-2}]Y[z] + (6 + 8z^{-1} + 16) = [5z^{-1} + z^{-2}]\frac{z}{(z-1)}$$

$$\frac{(z+2)(z+4)}{z^2}Y[z] = -(22 + 8z^{-1}) + (5z^{-1} + z^{-2})\frac{z}{(z-1)}$$

$$= \frac{(-22z^2 + 19z + 9)}{z(z-1)}$$

$$\frac{Y[z]}{z} = \frac{(-22z^2 + 19z + 9)}{(z-1)(z+2)(z+4)}$$

$$= \frac{A_1}{(z-1)} + \frac{A_2}{(z+2)} + \frac{A_3}{(z+4)}$$

$$-22z^2 + 19z + 9 = A_1(z+2)(z+4) + A_2(z-1)(z+4) + A_3(z-1)(z+2)$$

Substitute $z = 1$

$$-22 + 19 + 9 = A_1(3)(5); \qquad A_1 = 0.4$$

Substitute $z = -2$

$$-88 - 38 + 9 = A_2(-3)(2); \qquad A_2 = 19.5$$

Substitute $z = -4$

$$-352 - 76 + 9 = A_3(-5)(-2); \qquad A_3 = -41.9$$

$$Y[z] = \frac{0.4z}{(z-1)} + 19.5\frac{z}{(z+2)} - 41.9\frac{z}{(z+4)}$$

$$\boxed{y[n] = \left[0.4 + 19.5(-2)^n - 41.9(-4)^n\right] u[n]}$$

■ Example 9.38

Find the response of the LTID system described by the following difference equation:

$$y[n+2] + y[n+1] + 0.24y[n] = x[n+1] + 2x[n]$$

where $x[n] = (\frac{1}{2})^n u[n]$ and all the initial conditions are zero.

Solution: When the initial conditions are zero,

$$y[n+2] \overset{Z}{\longleftrightarrow} z^2 Y[z]$$
$$y[n+1] \overset{Z}{\longleftrightarrow} zY[z]$$
$$x[n+1] \overset{Z}{\longleftrightarrow} zX[z]$$
$$\left(\frac{1}{2}\right)^2 u[n] \overset{Z}{\longleftrightarrow} \frac{z}{(z-0.5)}$$

The given difference equation can be written in the following form after taking z-transform on both sides:

$$\begin{aligned}
[z^2 + z + 0.24]Y[z] &= [z+2]\frac{z}{(z-0.5)} \\
(z^2 + z + 0.24) &= (z+0.6)(z+0.4) \\
\frac{Y[z]}{z} &= \frac{(z+2)}{(z-0.5)(z+0.6)(z+0.4)} \\
&= \frac{A_1}{(z-0.5)} + \frac{A_2}{(z+0.6)} + \frac{A_3}{(z+0.4)} \\
(z+2) &= A_1(z+0.6)(z+0.4) + A_2(z-0.5)(z+0.4) \\
&\quad + A_3(z-0.5)(z+0.6)
\end{aligned}$$

Substitute $z = 0.5$

$$2.5 = A_1(1.1)(0.9); \qquad A_1 = 2.525$$

Substitute $z = -0.6$

$$1.4 = A_2(-1.1)(-0.2); \qquad A_2 = 6.36$$

Substitute $z = -0.4$

$$1.6 = A_3(-0.9)(-0.2); \qquad A_3 = -8.89$$

$$Y[z] = 2.525\frac{z}{(z-0.5)} + 6.36\frac{z}{(z+0.6)} - 8.89\frac{z}{(z+0.4)}$$

$$\boxed{y[n] = \left[2.525(0.5)^n + 6.36(-0.6)^n - 8.89(-0.4)^n\right]u[n]}$$

■ Example 9.39

Consider the following difference equation:

$$y[n+2] - 5y[n+1] + 6y[n] = x[n+1] + 4x[n]$$

The auxiliary conditions are $y[0] = 1$ and $y[1] = 2$ and the input $x[n] = u[n]$. Solve for $y[n]$.

Solution:

$$\begin{aligned}
y[n+2] &\overset{Z}{\longleftrightarrow} z^2Y[z] - z^2y(0) - zy(1) \\
&= z^2Y[z] - z^2 - 2z \\
y[n+1] &\overset{Z}{\longleftrightarrow} zY[z] - zy[0] \\
&= zY[z] - z \\
x[n+1] &\overset{Z}{\longleftrightarrow} zX[z] - zx[0] \\
&= zX[z] - z
\end{aligned}$$

Taking z-transform on both sides of the above equation and substituting $X[z] = \frac{z}{(z-1)}$, we get

$$\begin{aligned}
[z^2 - 5z + 6]Y[z] &= z^2 + 2z - 5z + (z+4)\frac{z}{(z-1)} - z \\
(z-2)(z-3)Y[z] &= \frac{z(z-4)(z-1) + z(z+4)}{(z-1)} \\
\frac{Y[z]}{z} &= \frac{(z^2 - 4z + 8)}{(z-1)(z-2)(z-3)} \\
&= \frac{A_1}{(z-1)} + \frac{A_2}{(z-2)} + \frac{A_3}{(z-3)}
\end{aligned}$$

$$(z^{-2} - 4z + 8) = A_1(z-2)(z-3) + A_2(z-1)(z-3) + A_3(z-1)(z-2)$$

Substitute $z = 1$

$$1 - 4 + 8 = A_1(-1)(-2); \qquad A_1 = 2.5$$

Substitute $z = 2$

$$4 - 8 + 8 = A_2(-1); \qquad A_2 = -4$$

Substitute $z = 3$

$$9 - 12 + 8 = A_3(2)(1); \qquad A_3 = 2.5$$

$$Y[z] = 2.5\frac{z}{(z-1)} - 4\frac{z}{(z-2)} + 2.5\frac{z}{(z-3)}$$

$$\boxed{y[n] = \left[2.5 - 4(2)^n + 2.5(3)^n\right]u[n]}$$

Example 9.40

Solve the following difference equation:

$$y[n+2] - 9y[n+1] + 20y[n] = 4x[n+1] + 2x[n]$$

The input $x[n] = (\frac{1}{2})^n u[n]$. The initial conditions are $y[-1] = 2$ and $y[-2] = 1$.

Solution: The given difference equation is in advanced operator form which requires the knowledge of $y[1]$ and $y[2]$. Therefore, the given equation is converted in delay operator form as described below and the given initial conditions are applied. Replacing n with $(n-2)$, the given difference equation is converted as

$$y[n] - 9y[n-1] + 20y[n-2] = 4x[n-1] + 2x[n-2]$$

Since the input is causal, $x[-1] = x[-2] = 0$. Taking z-transform on both sides of the above equation, we get

$$\begin{aligned}
&Y[z] - 9[z^{-1}Y[z] + y[-1]] + 20[z^{-2}Y[z] + z^{-1}y[-1] + y[-2]] \\
&= 4[z^{-1}X[z] + x[-1] + 2[z^{-2}X[z] + z^{-1}x[-1] + z^{-2}x[-2]]] \\
&= [4z^{-1} + 2z^{-2}]X[z] \\
&= [1 - 9z^{-1} + 20z^{-2}]Y[z] - 18 + 40z^{-1} + 20 = (4z^{-1} + 2z^{-2})X[z] \\
&\frac{[z^2 - 9z + 20]}{z^2}Y[z] = -(2 + 40z^{-1}) + (4z^{-1} + 2z^{-2})X[z]
\end{aligned}$$

Substitute $(z^2 - 9z + 20) = (z - 4)(z - 5)$ and $X[z] = \frac{z}{(z-0.5)}$

$$\begin{aligned}\frac{Y[z]}{z} &= \frac{(-2z^2 - 35z + 22)}{(z - 0.5)(z - 4)(z - 5)} \\ &= \frac{A_1}{(z - 0.5)} + \frac{A_2}{(z - 4)} + \frac{A_3}{(z - 5)} \\ (-2z^{-2} - 35z + 22) &= A_1(z - 4)(z - 5) + A_2(z - 0.5)(z - 5) \\ &\quad + A_3(z - 0.5)(z - 4)\end{aligned}$$

Substitute $z = 0.5$

$$-0.5 - 17.5 + 22 = A_1(-3.5)(-4.5); \qquad A_1 = 0.254$$

Substitute $z = 4$

$$-32 - 140 + 22 = A_2(3.5)(-1); \qquad A_2 = 42.86$$

Substitute $z = 5$

$$-50 - 175 + 22 = A_3(4.5); \qquad A_3 = -45.1$$

$$Y[z] = \frac{0.254z}{(z - 0.5)} + \frac{42.86z}{(z - 4)} - \frac{45.1z}{(z - 5)}$$

$$\boxed{y[n] = \left[0.254(0.5)^n + 42.86(4)^n - 45.1(5)^n\right] u[n]}$$

9.14 Zero Input and Zero State Response

The total solution of the difference equation is separated into zero input and zero state components. The response due to the initial conditions alone (in the absence of the input) is called zero input response. The response due to the input alone (assuming that the initial conditions are zero) is called a zero state response. The total response is the sum of zero input response and zero state response. This is illustrated in the following examples.

■ Example 9.41

$$y[n] + 5y[n-1] + 6y[n-2] = x[n-1] + 2x[n]$$

where $x[n] = u[n]$. The initial conditions are $y[-1] = 1$ and $y[-2] = 0$. Find (a) Zero input response, (b) Zero state response and (c) Total response.

Solution:

(a) **Zero Input Response**

$$\begin{aligned}
y[n] &\overset{Z}{\longleftrightarrow} Y[z] \\
y[n-1] &\overset{Z}{\longleftrightarrow} z^{-1}Y[z] + y[-1] \\
y[n-2] &\overset{Z}{\longleftrightarrow} z^{-2}Y[z] + z^{-1}y[-1] + y[-2]
\end{aligned}$$

Assuming the input is zero, taking z-transform on both sides of the given equation, we get

$$\begin{aligned}
Y[z] + 5(z^{-1}Y[z] + y[-1]) + 6(z^{-2}Y[z] & \\
+z^{-1}y[-1] + y[-2]) &= 0 \\
(1 + 5z^{-1} + 6z^{-2})Y[z] + 5 + 6z^{-1} &= 0 \\
\frac{(z+2)(z+3)}{z^2}Y[z] &= -\frac{(5z+6)}{z} \\
\frac{Y[z]}{z} &= -\frac{(5z+6)}{(z+2)(z+3)} \\
&= \frac{A_1}{(z+2)} + \frac{A_2}{(z+3)} \\
-(5z+6) &= A_1(z+3) + A_2(z+2)
\end{aligned}$$

Substitute $z = -2$

$$10 - 6 = A_1; \quad A_1 = 4$$

Substitute $z = -3$

$$15 - 6 = A_2(-1); \quad A_2 = -9$$

$$Y[z] = \frac{4z}{(z+2)} - \frac{9z}{(z+3)}$$

$$\boxed{y[n] = \left[4(-2)^n - 9(-3)^n\right] u[n]}$$

The initial condition can be easily checked as explained below. Substitute $n = -1$

$$\begin{aligned} y[-1] &= 4\frac{1}{(-2)} - 9\left(\frac{1}{-3}\right) \\ &= -2 + 3 = 1 \end{aligned}$$

Substitute $n = -2$

$$\begin{aligned} y[-2] &= 4\frac{1}{(-2)^2} - 9\frac{1}{(-3)^2} \\ &= 1 - 1 = 0 \end{aligned}$$

(b) **Zero State Response**

Assuming the zero initial conditions and noting $x[-1] = 0$, we get

$$\begin{aligned} [1 + 5z^{-1} + 6z^{-2}]Y[z] &= z^{-1}X[z] - x[-1] + 2X[z] \\ \frac{[z^2 + 5z + 6]}{z^2}Y[z] &= [z^{-1} + 2]X[z] \\ &= \frac{(2z+1)}{z}\frac{z}{(z-1)} \\ \frac{Y[z]}{z} &= \frac{z(2z+1)}{(z-1)(z+2)(z+3)} \\ &= \frac{A_1}{(z-1)} + \frac{A_2}{(z+2)} + \frac{A_3}{(z+3)} \\ 2z^2 + 3 &= A_1(z+2)(z+3) + A_2(z-1)(z+3) \\ &\quad + A_3(z-1)(z+2) \end{aligned}$$

Substitute $z = 1$

$$2 + 1 = A_1(3)(4); \qquad A_1 = \frac{1}{4}$$

Substitute $z = -2$

$$8 - 2 = A_2(-3); \qquad A_2 = -2$$

Substitute $z = -3$

$$18 - 3 = A_3(-4)(-1); \qquad A_3 = \frac{15}{4}$$

$$Y[z] = \frac{1}{4}\frac{z}{(z-1)} - \frac{2z}{(z+2)} + \frac{15}{4}\frac{z}{(z+3)}$$

$$\boxed{y[n] = \left[\frac{1}{4} - 2(-2)^n + \frac{15}{4}(-3)^n\right]u[n]}$$

(c) **Total Response**

$$\text{Total response} = \text{Zero input response} + \text{Zero state response}$$

$$y[n] = 4(-2)^n - 9(-3)^n + \frac{1}{4} - 2(-2)^n + \frac{15}{4}(-3)^n$$

$$\boxed{y[n] = \left[\frac{1}{4} + 2(-2)^n - \frac{21}{5}(-3)^n\right]u[n]}$$

9.15 Natural and Forced Responses

In the total response, the response due to the characteristic modes are called forced response. The terms which do not include characteristic modes (eigenvalues) are called natural response. In Example 9.41, the eigenvalues are $\lambda_1 = -2$ and $\lambda_2 = -3$. In the total response, $y[n] = \frac{1}{4}u[n]$ is free from characteristic modes. Hence, it is the natural response. The rest of the terms belong to the forced response. This is illustrated in the following example.

Example 9.42

Consider the following difference equation:

$$y[n+2] - 6y[n+1] + 8y[n] = x[n]$$

where $x[n] = (\frac{1}{4})^n u[n]$. The initial conditions are $y[0] = 1$ and $y[1] = 2$. Find (a) Zero state response, (b) Zero input response, (c) Natural response, (d) Forced response and (e) Total response.

Solution:

(a) Taking z-transform on both sides, we get

$$X[z] = z^2Y[z] - z^2y[0] - zy[1] - 6\{zY[z] - zy[0]\} + 8Y[z]$$
$$X[z] = [z^2 - 6z + 8]Y[z] - z^2 - 2z + 6z$$

Substituting $X[z] = \frac{z}{(z-0.25)}$ and $z^2 - 6z + 8 = (z-2)(z-4)$, we get

$$(z-2)(z-4)Y[z] = z^2 - 4z + \frac{z}{(z-0.25)}$$

$z = 2$ and $z = 4$ are the eigenvalues. If the initial conditions are zero, we get

$$\begin{aligned}\frac{Y[z]}{z} &= \frac{1}{(z-2)(z-4)(z-0.25)} \\ &= \frac{A_1}{(z-2)} + \frac{A_2}{(z-4)} + \frac{A_3}{(z-0.25)} \\ 1 &= A_1(z-4)(z-0.25) + A_2(z-2)(z-0.25) + A_3(z-2)(z-4)\end{aligned}$$

Substitute $z = 2$

$$1 = A_1(-2)(1.75); \qquad A_1 = -\frac{2}{7}$$

Substitute $z = 4$

$$1 = A_2(2)(3.75); \qquad A_2 = \frac{2}{15}$$

Substitute $z = 0.25$

$$1 = A_3(-1.75)(-3.75); \qquad A_3 = \frac{16}{105}$$

Let $y_{0s}[n]$ denote zero state response and $y_{0i}[n]$ denote zero input response.

$$Y_{0s}[z] = -\frac{2}{7}\frac{z}{(z-2)} + \frac{2}{15}\frac{z}{(z-4)} + \frac{16}{105}\frac{z}{(z-0.25)}$$

$$\boxed{y_{0s}[n] = \left[-\frac{2}{7}(2)^n + \frac{2}{15}(4)^n + \frac{16}{105}(0.25)^n\right]u[n]}$$

(b) If we assume the input is zero, $X[z] = 0$

$$\begin{aligned}\frac{Y_{0i}[z]}{z} &= \frac{(z-4)}{(z-2)(z-4)} \\ Y_{0i}[z] &= \frac{z}{(z-2)}\end{aligned}$$

$$\boxed{y_{0i}[n] = (2)^n u[n]}$$

(c) The total response $y[n]$ is given by

$$\begin{aligned} y[n] &= y_{0s}[n] + y_{0i}[n] \\ &= \left[-\frac{2}{7}(2)^n + \frac{2}{15}(4)^n + \frac{16}{105}(0.25)^n + (2)^n \right] u[n] \\ &= \Bigg[\underbrace{\frac{5}{7}(2)^n + \frac{2}{15}(4)^n}_{\text{Natural response}} + \underbrace{\frac{16}{105}(0.25)^n}_{\text{Forced response}} \Bigg] u[n] \end{aligned}$$

Let us denote $y_n[n]$ and $y_f[n]$ as the natural and forced responses, respectively. The natural response is the response which is due to the characteristic roots $z = 2$ and $z = 4$. The remaining portion of $y[n]$ is the forced response.

$$\boxed{y_f[n] = \frac{16}{105}(0.25)^n u[n]}$$

(d) The natural response is

$$\boxed{y_n[n] = \left[\frac{5}{7}(2)^n + \frac{2}{15}(4)^n \right] u[n]}$$

(e) The total response is

$$\boxed{y[n] = \left[\frac{5}{7}(2)^n + \frac{2}{15}(4)^n + \frac{16}{105}(0.25)^n \right] u[n]}$$

9.16 Difference Equation from System Function

Let the system function $H[z]$ be expressed as

$$\frac{Y[z]}{X[z]} = H[z] = \frac{b_0 z^N + b_1 z^{N-1} + \cdots + b_{N-1} z + b_N}{z^N + a_1 z^{N-1} + \cdots + a_{N-1} z + a_N}$$

Cross-multiplying and operating z on $Y[z]$ and $X[z]$, we get

$$\begin{aligned} &y[n+N] + a_1 y[n+N-1] + \cdots + a_{N-1} y[n+1] + a_N y[n] \\ &= b_0 x[n+N] + b_1 x[n+N-1] + \cdots + b_{N-1} x[n+1] \\ &\quad + b_N x[n] \end{aligned} \tag{9.58}$$

A similar procedure has to be followed if the system frequency response $H(e^{j\omega})$ is given. Here $e^{j\omega}$ has to be treated as z. The following examples demonstrate the above methods.

Example 9.43

For the system functions given below, determine the difference equation

(a) $H[z] = \dfrac{(1 - z^{-1})}{\left(1 - \frac{1}{2}z^{-1} + \frac{1}{4}z^{-2}\right)}$

(b) $H[z] = \dfrac{(z - 1)}{(z + 1)(z - 2)}$

(c) $H[z] = \dfrac{1}{\left(1 - \frac{1}{4}z^{-1}\right)}$

(Anna University, December, 2006)

(d) Consider the system consisting of the cascade of two LTI systems with frequency responses

$$H_1(e^{j\omega}) = \frac{2 - e^{j\omega}}{\left(1 + \frac{1}{2}e^{-j\omega}\right)}$$

$$H_2(e^{j\omega}) = \frac{1}{\left(1 - \frac{1}{2}e^{-j\omega} + \frac{1}{4}e^{-j2\omega}\right)}$$

Find the difference equation describing the overall system.

(Anna University, April, 2008)

(e) Write a difference equation that characterizes a system whose frequency response is

$$H(e^{j\omega}) = \frac{\left(1 - \frac{1}{2}e^{-j\omega} + e^{-3j\omega}\right)}{\left(1 + \frac{1}{2}e^{-j\omega} + \frac{3}{4}e^{-2j\omega}\right)}$$

(Anna University, May, 2007)

Solution:

(a) $\boldsymbol{H[z] = \frac{(1-z^{-1})}{(1-\frac{1}{2}z^{-1}+\frac{1}{4}z^{-2})}}$

$$\frac{Y[z]}{X[z]} = \frac{\left(1 - z^{-1}\right)}{\left(1 - \frac{1}{2}z^{-1} + \frac{1}{4}z^{-2}\right)}$$

$$Y[z] - \frac{1}{2}z^{-1}Y[z] + \frac{1}{4}z^{-2}Y[z] = X[z] - z^{-1}X[z]$$

$$\boxed{y[n] - \frac{1}{2}y[n-1] + \frac{1}{4}y[n-2] = x[n] - x[n-1]}$$

(b) $\boldsymbol{H[z] = \frac{(z-1)}{(z+1)(z-2)}}$

$$\begin{aligned}\frac{Y[z]}{X[z]} &= \frac{(z-1)}{(z+1)(z-2)} \\ &= \frac{(z-1)}{(z^2-z-2)} \\ z^2Y[z] - zY[z] - 2Y[z] &= zX[z] - X[z]\end{aligned}$$

$$\boxed{y[n+2] - y[n+1] - 2y[n] = x[n+1] - x[n]}$$

(c) $\boldsymbol{H[z] = \frac{Y[z]}{X[z]} = \frac{1}{(1-\frac{1}{4}z^{-1})}}$

$$\left[1 - \frac{1}{4}z^{-1}\right]Y[z] = X[z]$$

$$\boxed{y[n] - \frac{1}{4}y[n-1] = x[n]}$$

(d) $\boldsymbol{H_1(e^{j\omega}) = \frac{2-e^{j\omega}}{(1+\frac{1}{2}e^{-j\omega})}}$ **and** $\boldsymbol{H_2(e^{j\omega}) = \frac{1}{(1-\frac{1}{2}e^{-j\omega}+\frac{1}{4}e^{-j2\omega})}}$

$$\begin{aligned}H_1H_2(e^{j\omega}) &= \frac{Y(j\omega)}{X(j\omega)} \\ &= \frac{(2-e^{-j\omega})}{\left(1+\frac{1}{2}e^{-j\omega}\right)\left(1-\frac{1}{2}e^{-j\omega}+\frac{1}{4}e^{-j2\omega}\right)} \\ &= \frac{(2-e^{-j\omega})}{\left(1-\frac{1}{2}e^{-j\omega}+\frac{1}{4}e^{-j2\omega}+\frac{1}{2}e^{-j\omega}-\frac{1}{4}e^{-j2\omega}+\frac{1}{8}e^{-j3\omega}\right)} \\ Y[e^{j\omega}]\left[1+\frac{1}{8}e^{-j3\omega}\right] &= [2-e^{-j\omega}]X[e^{j\omega}]\end{aligned}$$

$$\boxed{y[n] + \frac{1}{8}y[n-3] = 2x[n] - x[n-1]}$$

(e) $\boldsymbol{\frac{Y[e^{j\omega}]}{X[e^{j\omega}]} = H(e^{j\omega}) = \frac{(1-e^{-j\omega}+e^{-3j\omega})}{(1+\frac{1}{2}e^{-j\omega}+\frac{3}{4}e^{-2j\omega})}}$

$$\left[1+\frac{1}{2}e^{-j\omega}+\frac{3}{4}e^{-2j\omega}\right]Y[e^{j\omega}]=\left[1-e^{-j\omega}+e^{-3j\omega}\right]X[e^{j\omega}]$$

$$\boxed{y[n]+\frac{1}{2}y[n-1]+\frac{3}{4}y[n-2]=x[n]-x[n-1]+x[n-3]}$$

■ Example 9.44

Obtain the difference equation for the block diagram shown in Fig. 9.18.

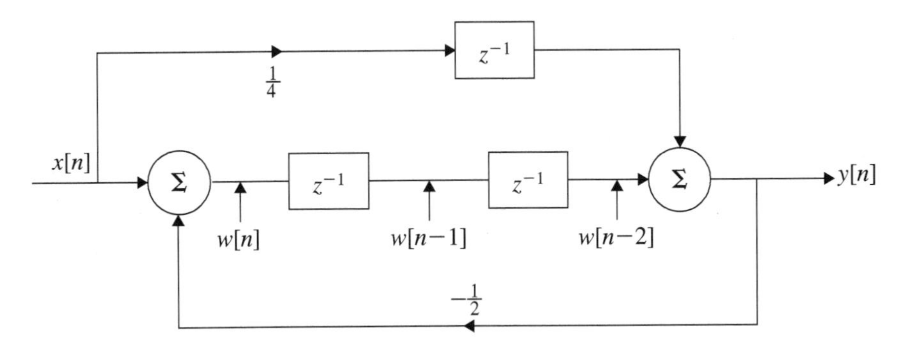

Fig. 9.18 Block diagram of Example 9.44. **a** Block form-I structure

Solution: From Fig. 9.18, the following equations are written:

$$w[n]=x[n]-\frac{1}{2}y[n]$$

Replace n by $(n-2)$

$$\begin{aligned}w[n-2]&=x[n-2]-\frac{1}{2}y[n-2]\\ y[n]&=\frac{1}{4}x[n-1]+w[n-2]\\ &=\frac{1}{4}x[n-1]+x[n-2]-\frac{1}{2}y[n-2]\end{aligned}$$

$$\boxed{y[n]+\frac{1}{2}y[n-2]=\frac{1}{4}x[n-1]+x[n-2]}$$

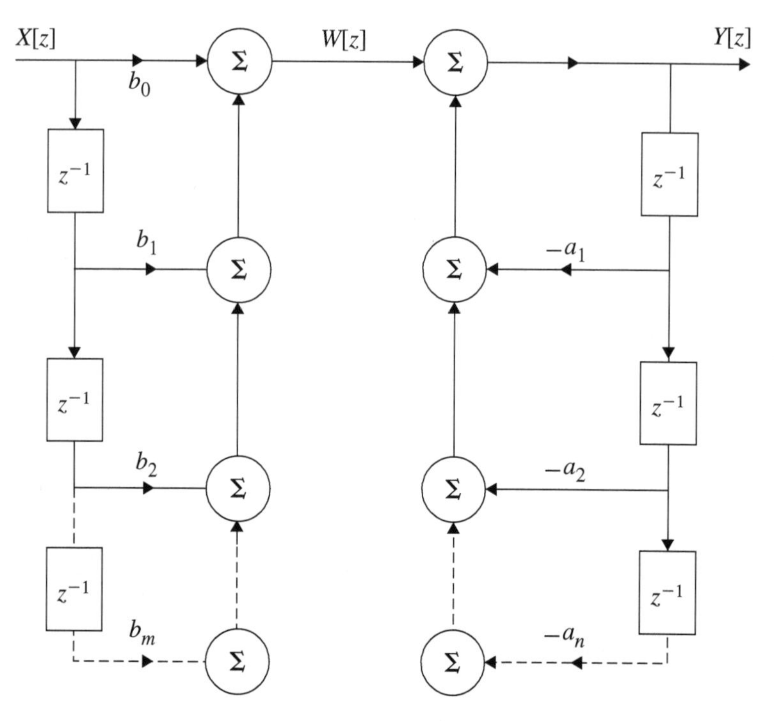

Fig. 9.18 (continued)

9.17 Discrete Time System Realization

Like the continuous time system, realization of discrete time system from the system function $H[z]$ is done in an identical fashion. The only difference is that the integrator $\frac{1}{s}$ in the LTIC system is replaced by the time delay z^{-1} in LTID. The basic operations such as an adder, scalar multiplier and pick-off points remain the same. The following methods or realization are described below:

1. Direct Form-I;
2. Direct Form-II;
3. Cascade Form;
4. Parallel Form;
5. Transposed Form.

9.17.1 Direct Form-I Realization

Consider the system function represented in the following form:

$$H[z] = \frac{Y[z]}{X[z]} = \frac{\sum_{k=0}^{m} b_k z^{-k}}{1 + \sum_{k=1}^{n} a_k z^{-k}} \tag{9.59}$$

$$= \frac{(b_0 + b_1 z^{-1} + b_2 z^{-2} + \cdots + b_m z^{-m})}{(1 + a_1 z^{-1} + a_2 z^{-2} + \cdots + a_n z^{-n})} \tag{9.60}$$

$$\frac{Y[z]}{X[z]} = \frac{Y[z]}{W[z]} \frac{W[z]}{X[z]}$$

$$Y[z] + a_1 z^{-1} Y[z] + a_2 z^{-2} Y[z] + \cdots + a_n z^{-n} Y[z]$$
$$= b_0 X[z] + b_1 z^{-1} X[z] + b_2 z^{-2} X[z] + \cdots + b_m z^{-m} X[z] \tag{9.61}$$

Let $W[z] = b_0 X[z] + b_1 z^{-1} X[z] + \cdots + b_m z^{-m} X[z]$

$$\frac{Y[z]}{W[z]} = \frac{1}{(1 + a_1 z^{-1} + a_2 z^{-2} + \cdots + a_n z^{-n})}$$
$$W[z] = Y[z](1 + a_1 z^{-1} + a_2 z^{-2} + \cdots + a_n z^{-n})$$
$$Y[z] = W[z] - a_1 z^{-1} Y[z] - a_2 z^{-2} Y[z] - \cdots - a_n z^{-n} Y[z] \tag{9.62}$$

Equations (9.61) and (9.62) are represented in Fig. 9.18a.

■ Example 9.45

Realize in direct form-I structure given that

$$y[n] - \frac{5}{6} y[n-1] + \frac{1}{6} y[n-2] = x[n] + 2x[n-1]$$

(Anna University, December, 2007)

Solution:
Taking z-transform on both sides, we get

$$\left[1 - \frac{5}{6} z^{-1} + \frac{1}{6} z^{-2}\right] Y[z] = [1 + 2z^{-1}] X[z]$$

$$H[z] = \frac{Y[z]}{X[z]} = \frac{(1 + 2z^{-1})}{\left(1 - \frac{5}{6} z^{-1} + \frac{1}{6} z^{-2}\right)}$$

The coefficients of numerator and denominator polynomials are identified as

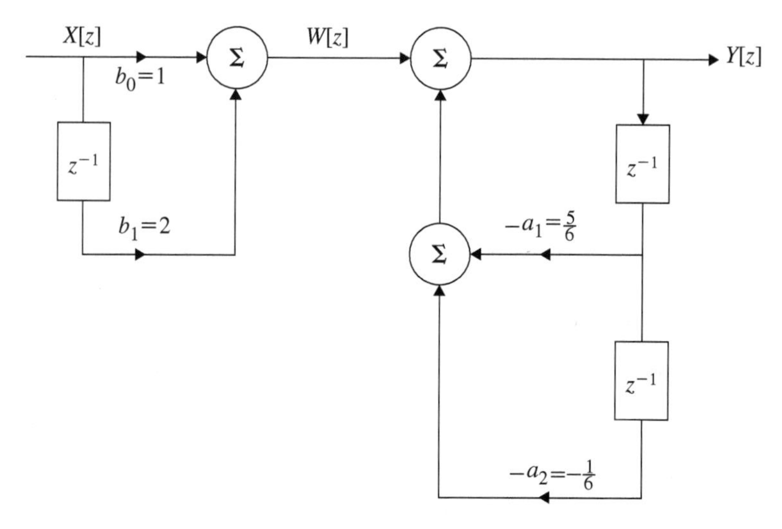

Fig. 9.19 Direct form-I realization

$$b_0 = 1; \quad b_1 = 2; \quad a_1 = -\frac{5}{6}; \quad a_2 = \frac{1}{6}$$

The direct form-I representation is shown in Fig. 9.19.

9.17.2 Direct Form-II Realization

Consider the system function given in Eq. (9.60)

$$\frac{Y[z]}{X[z]} = H[z] = \frac{(b_0 + b_1 z^{-1} + b_2 z^{-2} + \cdots + b_m z^{-m})}{(1 + a_1 z^{-1} + a_2 z^{-2} + \cdots + a_n z^{-n})}$$

Let $H[z] = H_1[z]H_2[z]$ where

$$\begin{aligned} H_1[z] &= (b_0 + b_1 z^{-1} + b_2 z^{-2} + \cdots + b_m z^{-m}) \\ H_2[z] &= \frac{1}{(1 + a_1 z^{-1} + a_2 z^{-2} + \cdots + a_n z^{-n})} \\ Y[z] &= H_1[z]W[z] \end{aligned} \tag{9.63}$$

where

$$W[z] = H_2[z]X[z] \tag{9.64}$$

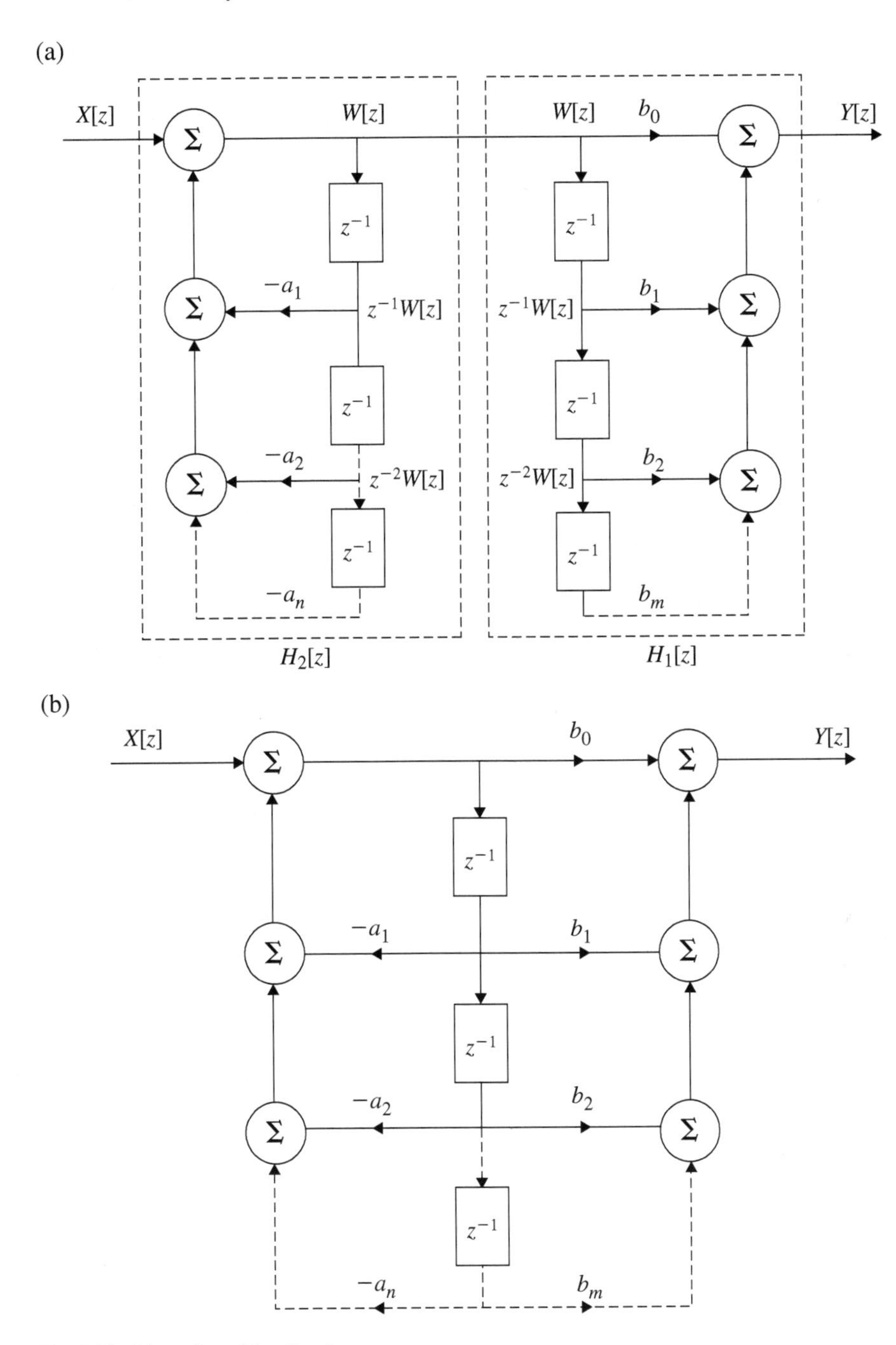

Fig. 9.20 Direct form-II realization

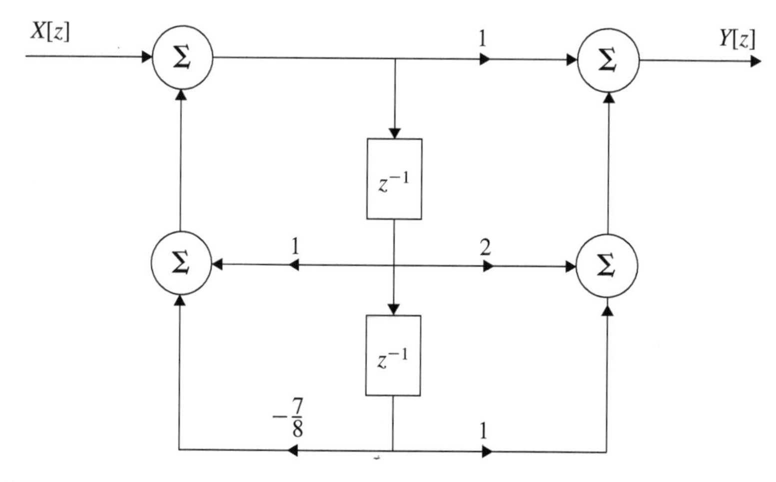

Fig. 9.21 Direct form-II realization for Example 9.46

Equations (9.63) and (9.64) are represented in Fig. 9.20a, where $H_1[z]$ and $H_2[z]$ are in cascade. Collapsing the two sets of z^{-1} blocks, $H[z]$ is obtained and is represented in Fig. 9.20b, as test blocks generate identical quantities. Thus, the time delay blocks necessary to realize the system function are reduced by a factor 2. Direct form-II realization is demonstrated in the following example.

■ Example 9.46

The system function of a discrete time system is

$$H[z] = \frac{(1+z^{-1})^2}{\left(1 - z^{-1} + \frac{7}{8}z^{-2}\right)}$$

Realize the system by direct form-II.

Solution:

$$H[z] = \frac{(1+2z^{-1}+z^{-2})}{\left(1 - z^{-1} + \frac{7}{8}z^{-2}\right)}$$

where $b_0 = 1$, $b_1 = 2$, $b_2 = 1$, $a_1 = -1$ and $a_2 = \frac{7}{8}$. The direct form-II realization is shown in Fig. 9.21.

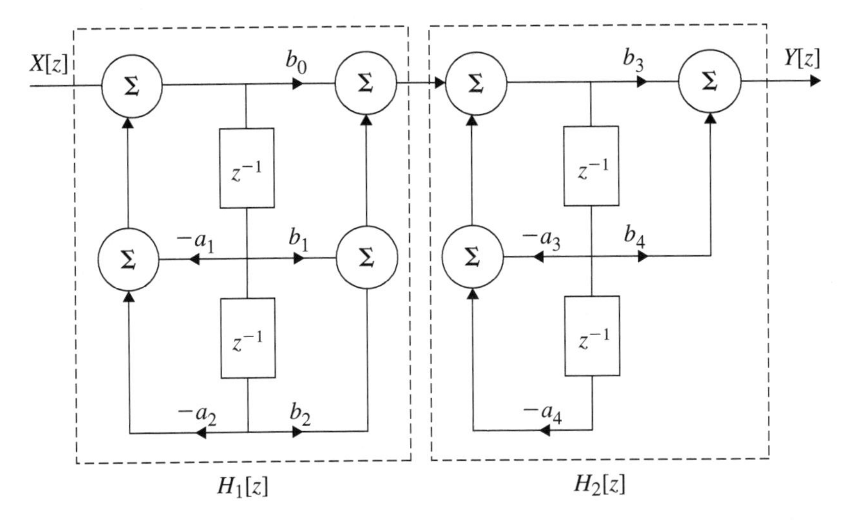

Fig. 9.22 System realization by cascade connection

9.17.3 *Cascade Form Realization*

In cascade realization, the system function is expressed as a product of several subsystems and each subsystem is realized in direct form-II.

Consider the following system function:

$$H[z] = \frac{(b_0 + b_1 z^{-1} + b_2 z^{-2})(b_3 + b_4 z^{-1})}{(1 + a_1 z^{-1} + a_2 z^{-2})(1 + a_3 z^{-1} + a_4 z^{-2})}$$

The above function is expressed as

$$H[z] = H_1[z]H_2[z]$$

where

$$H_1[z] = \frac{(b_0 + b_1 z^{-1} + b_2 z^{-2})}{(1 + a_1 z^{-1} + a_2 z^{-2})}$$

$$H_2[z] = \frac{(b_3 + b_4 z^{-1})}{(1 + a_3 z^{-1} + a_4 z^{-2})}$$

$H_1[z]$ and $H_2[z]$ are realized separately in direct form-II and they are connected in cascade as shown in Fig. 9.22.

The following example illustrates the cascade realization.

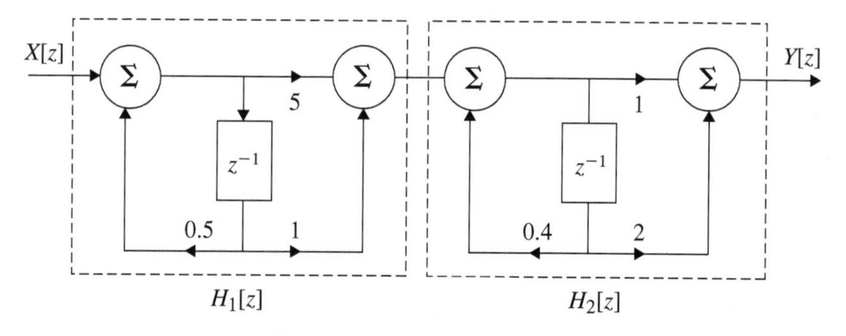

Fig. 9.23 System realization in cascade for Example 9.47

■ Example 9.47

$$H[z] = \frac{(5z+1)(z+2)}{(z-0.5)(z-0.4)}$$

Realize the system function in cascade form.

Solution:

$$H[z] = \frac{(5z+1)(z+2)}{(z-0.5)(z-0.4)} = \frac{(5+z^{-1})(1+2z^{-1})}{(1-0.5z^{-1})(1+0.4z^{-1})}$$

Let

$$H_1[z] = \frac{(5+z^{-1})}{1-0.5z^{-1}}$$

$$H_2[z] = \frac{(1+2z^{-1})}{(1-0.4z^{-1})}$$

$H_1[z]$ and $H_2[z]$ are realized in direct form-II and they are connected in cascade as shown in Fig. 9.23.

9.17.4 Parallel Form Realization

In parallel form realization, the system function $H[z]$ is put into partial fraction. Each term is individually realized and $Y[z]$ is obtained by adding them through a summer. This is illustrated by the following examples.

■ Example 9.48

Obtain the parallel form realization of the following system functions:

$$\text{(a)} \qquad H[z] = \frac{(z^2 + 4z + 10)}{(z+2)(z+4)}$$

$$\text{(b)} \qquad H[z] = \frac{(z+5)}{(z+4)^2}$$

Solution:

(a) $\boldsymbol{H[z] = \frac{(z^2+4z+10)}{(z+2)(z+4)}}$

$$\frac{H[z]}{z} = \frac{(z^2 + 4z + 10)}{z(z+2)(z+4)}$$

$$= \frac{A_1}{z} + \frac{A_2}{(z+2)} + \frac{A_3}{z+4}$$

$$z^2 + 4z + 10 = A_1(z+2)(z+4) + A_2 z(z+4) + A_3 z(z+2)$$

Substitute $z = 0$

$$10 = A_1 8; \qquad A_1 = \frac{5}{4}$$

Substitute $z = -2$

$$4 - 8 + 10 = A_2(-2)(2); \qquad A_2 = -\frac{3}{2}$$

Substitute $z = -4$

$$16 - 16 + 10 = A_3(-4)(-2); \qquad A_3 = \frac{5}{4}$$

$$H[z] = \frac{Y[z]}{X[z]} = \frac{5}{4} - \frac{3}{2}\frac{z}{(z+2)} + \frac{5}{4}\frac{z}{(z+4)}$$

$$Y[z] = \frac{5}{4}X[z] - \frac{3}{2}\frac{X[z]}{(1+2z^{-1})} + \frac{5}{4}\frac{X[z]}{(1+4z^{-1})}$$

The parallel realization is shown in Fig. 9.24.

(b) $\boldsymbol{H[z] = \frac{(z+5)}{(z+4)^2}}$

$$\frac{H[z]}{z} = \frac{(z+5)}{z(z+4)^2}$$
$$= \frac{A_1}{z} + \frac{A_2}{(z+4)} + \frac{A_3}{(z+4)^2}$$
$$(z+5) = A_1(z+4)^2 + A_2 z(z+4) + A_3 z$$

Substitute $z = 0$

$$5 = A_1 16; \qquad A_1 = \frac{5}{16}$$

Substitute $z = -4$

$$1 = A_3(-4); \qquad A_3 = -\frac{1}{4}$$

Compare the coefficient of z

$$1 = 8A_1 + 4A_2 + A_3; \qquad A_2 = -\frac{5}{16}$$

$$H[z] = \frac{5}{16} - \frac{5}{16}\frac{z}{(z+4)} - \frac{1}{4}\frac{z}{(z+4)^2}$$
$$\frac{Y[z]}{X[z]} = \frac{5}{16} - \frac{5}{16}\frac{1}{(1+4z^{-1})} - \frac{1}{4}\frac{z^{-1}}{(1+4z^{-1})^2}$$
$$Y[z] = \frac{5}{16}X[z] - \frac{5}{16}\frac{X[z]}{(1+4z^{-1})} - \frac{1}{4}\frac{z^{-1}X[z]}{(1+4z^{-1})^2}$$

The above equation is represented in Fig. 9.25.

9.17.5 The Transposed Form Realization

A transposed realization is the realization which has the same system function. Here, the realization is said to be equivalent. The transpose realization is exactly the same as for continuous time system. The given realization is changed to its transpose by following the steps given below.

1. Replace the input $X[z]$ with the output $Y[z]$ and *vice versa*.
2. Reverse all the arrow directions without changing the values of the multiplier.
3. Replace pick-off nodes with adders and *vice versa*.

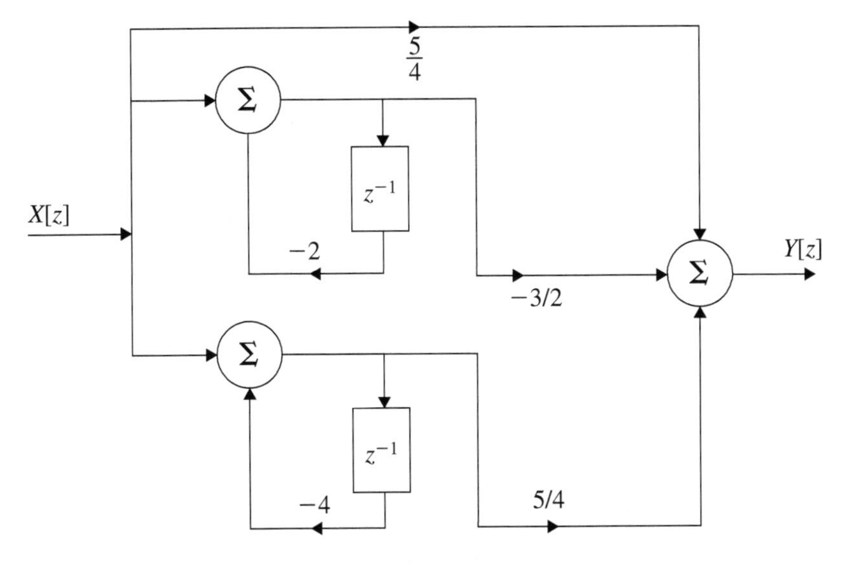

Fig. 9.24 Parallel realization for Example 9.48(a)

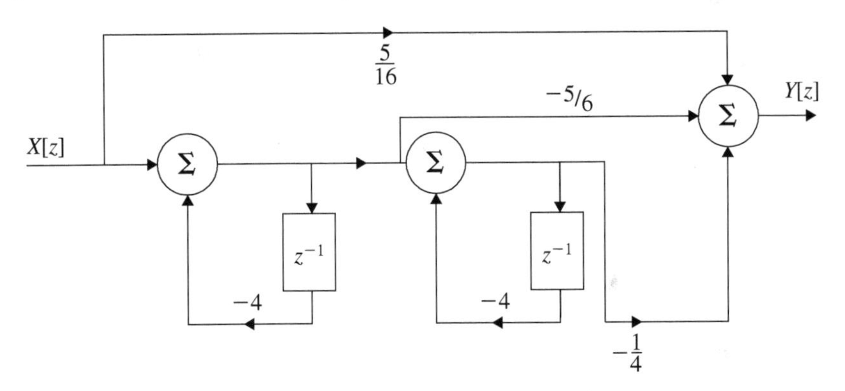

Fig. 9.25 Parallel realization for Example 9.48(b)

The following examples illustrate the above procedure.

■ Example 9.49

Find the transposed direct form-II realization for the system described by the following difference equation:

$$y[n] = \frac{3}{4}y[n-1] - \frac{3}{4}y[n-2] + x[n] - \frac{1}{3}x[n-1]$$

(Anna University, December, 2007)

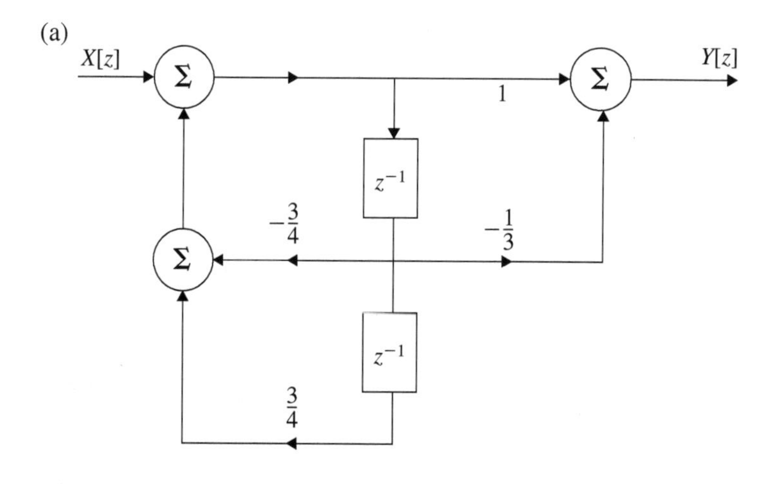

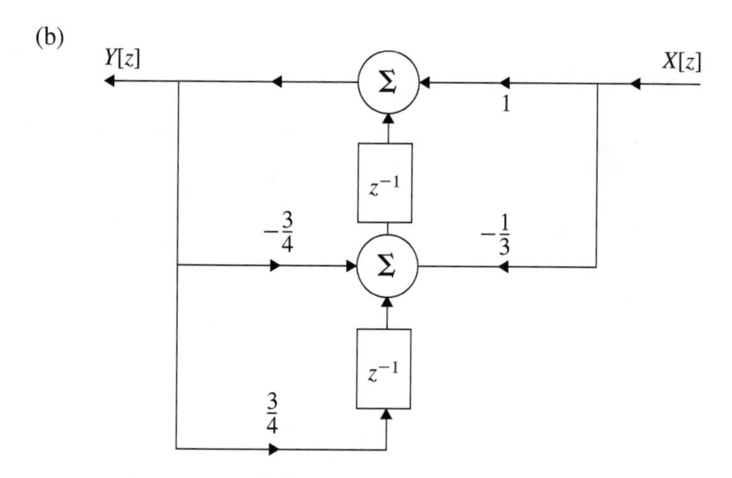

Fig. 9.26 **a** Direct form-II and **b** Transposed form

Solution: Taking z-transform on both sides of the above equation, we get

$$\left[1 - \frac{3}{4}z^{-1} + \frac{3}{4}z^{-2}\right] Y[z] = \left[1 - \frac{1}{3}z^{-1}\right] X[z]$$

$$H[z] = \frac{Y[z]}{X[z]} = \frac{\left[1 - \frac{1}{3}z^{-1}\right]}{\left[1 - \frac{3}{4}z^{-1} + \frac{3}{4}z^{-2}\right]}$$

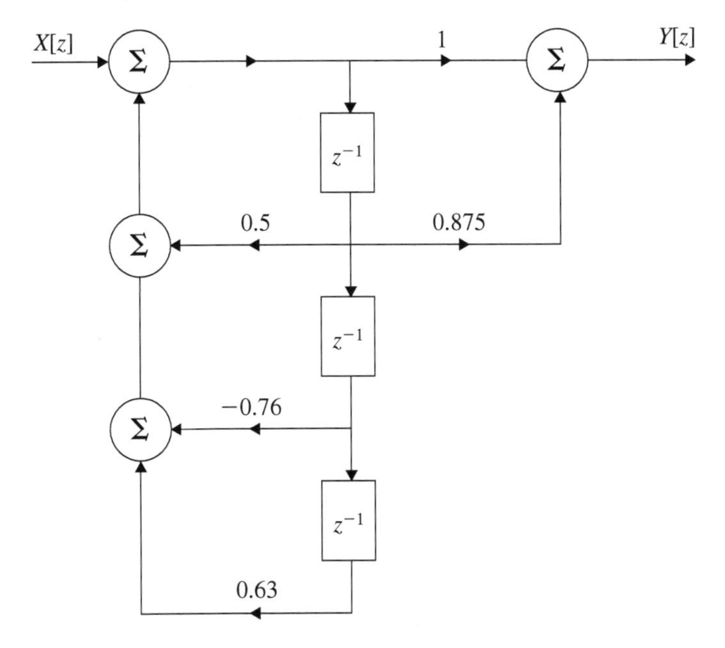

Fig. 9.27 **a** Direct form-II structure and **b** Parallel form realization of first- and second-order systems

To realize direct form-II structure, we have $b_0 = 1$, $b_1 = -\frac{1}{3}$, $b_2 = 0$, $a_1 = -\frac{3}{4}$ and $a_2 = \frac{3}{4}$. The direct form-II structure is shown in Fig. 9.26b.

■ Example 9.50

Consider the causal linear shift invariant filter with system function

$$H[z] = \frac{(1 + 0.875z^{-1})}{(1 + 0.2z^{-1} + 0.9z^{-2})(1 - 0.7z^{-1})}$$

Draw the following realization structure of the system.

(a) Direct form-II;
(b) A parallel form connections of first- and second-order systems realized in direct form-II.

(Anna University, December, 2007)

Solution:

(a) $\boldsymbol{H[z] = \frac{(1+0.875z^{-1})}{(1+0.2z^{-1}+0.9z^{-2})(1-0.7z^{-1})}}$

The denominator is expressed in the following polynomial form:

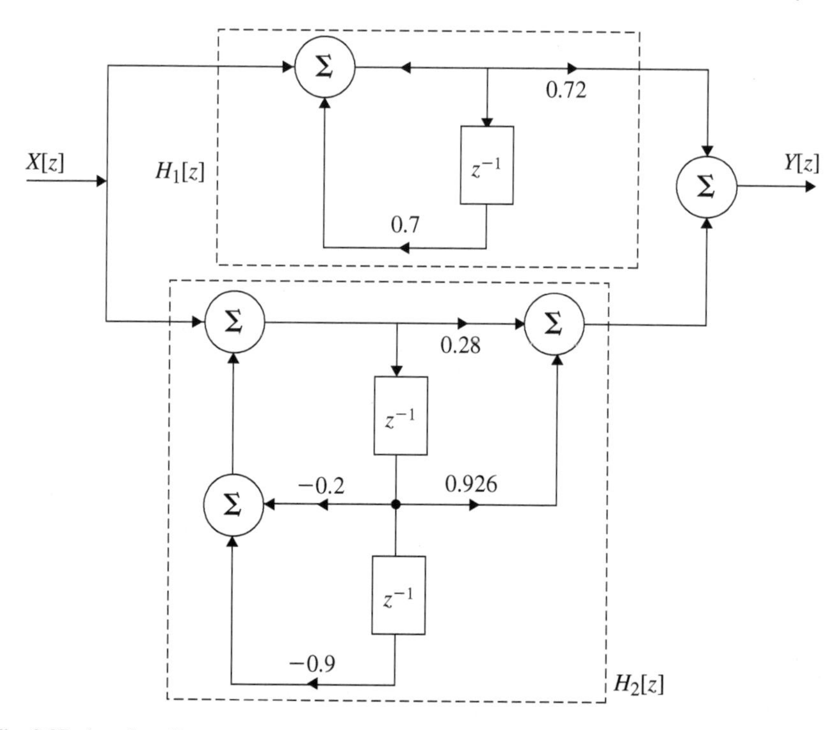

Fig. 9.27 (continued)

$$(1+0.2z^{-1}+0.9z^{-2})(1-0.7z^{-1}) = (1-0.5z^{-1}+0.76z^{-2}-0.63z^{-3})$$

$$H[z] = \frac{(1+0.875z^{-1})}{(1-0.5z^{-1}+0.76z^{-2}-0.63z^{-3})}$$

where $b_0 = 1$, $b_1 = 0.875$, $b_2 = 0$, $b_3 = 0$, $a_1 = -0.5$, $a_2 = 0.76$ and $a_3 = -0.63$. The direct form-II structure is shown in Fig. 9.27a.

(b) **Parallel form Realization**

$$H[z] = \frac{(1+0.875z^{-1})}{(1-0.2z^{-1}+0.9z^{-2})(1-0.7z^{-1})}$$

Let us substitute $z^{-1} = x$

$$H[x] = \frac{(1+0.875x)}{(1+0.2x+0.9x^2)(1-0.7x)}$$

$$= \frac{A_1}{(1-0.7x)} + \frac{(A_2+A_3x)}{(1+0.2x+0.9x^2)}$$

$$(1+0.875x) = A_1(1+0.2x+0.9x^2) + A_2(1-0.7x) + A_3x(1-0.7x)$$

Substitute $x = \frac{1}{0.7}$

$$\left(1 + \frac{0.875}{0.7}\right) = A_1(1 + 0.2857 + 1.836)$$
$$A_1 = 0.72$$

Comparing the coefficients of constant terms, we get

$$A_1 + A_2 = 1$$
$$A_2 = 0.28$$

Comparing the coefficients of x^2, we get

$$0.9A_1 - 0.7A_3 = 0$$
$$A_3 = 0.926$$

$$H[x] = \frac{0.72}{(1 - 0.7x)} + \frac{(0.28 + 0.926x)}{(1 + 0.2x + 0.9x^2)}$$
$$H[z] = \frac{0.72}{(1 - 0.7z^{-1})} + \frac{(0.28 + 0.926z^{-1})}{(1 + 0.2z^{-1} + 0.9z^{-2})}$$
$$= H_1[z] + H_2[z]$$

The parallel form connections of first- and second-order systems realized are shown in Fig. 9.27b.

■ Example 9.51

Realize direct form-I, direct form-II, cascade and parallel realization of the discrete time system having system function as

$$H[z] = \frac{2[z + 2]}{z(z - 0.1)(z + 0.5)(z + 0.4)}$$

(Anna University, April, 2004)

Solution:

(a) **Direct Form-I Realization**

$$H[z] = \frac{(2z + 4)}{z^4 + 0.8z^3 + 0.11z^2 + 0.02z + 0}$$
$$= \frac{2z^{-3} + 4z^{-4}}{1 + 0.8z^{-1} + 0.11z^{-2} + 0.02z^{-3} + 0z^{-4}}$$

From the above equation, the coefficients of the polynomials are obtained as $b_0 = 0$, $b_1 = 0$, $b_2 = 0$, $b_3 = 2$, $b_4 = 4$, $a_1 = 0.8$, $a_2 = 0.11$, $a_3 = 0.02$ and $a_4 = 0$. The direct form-I structure is shown in Fig. 9.28a.

(b) **Direct Form-II Realization**
The direct form-II realization is shown in Fig. 9.28b.

(c) **Cascade Form-II Realization**

$$\begin{aligned} H[z] &= \frac{2}{z}\frac{(z+2)}{(z-0.1)}\frac{1}{(z+0.5)}\frac{1}{(z+0.4)} \\ &= H_1[z]H_2[z]H_3[z]H_4[z] \end{aligned}$$

where

$$\begin{aligned} H_1[z] &= 2z^{-1} \longrightarrow b_0 = 0;\ b_1 = 2 \\ H_2[z] &= \frac{(1+2z^{-1})}{(1-0.1z^{-1})} \longrightarrow b_0 = 1;\ b_1 = 2;\ a_1 = -0.1 \\ H_3[z] &= \frac{z^{-1}}{(1+0.5z^{-1})} \longrightarrow b_0 = 0;\ b_1 = 1;\ a_1 = 0.5 \\ H_4[z] &= \frac{z^{-1}}{(1+0.4z^{-1})} \longrightarrow b_0 = 0;\ b_1 = 1;\ a_1 = 0.4 \end{aligned}$$

(d) **Parallel Form Realization**

$$\begin{aligned} H[z] &= \frac{2(z+2)}{z(z-0.1)(z+0.5)(z+0.4)} \\ &= \frac{A_1}{z} + \frac{A_2}{(z-0.1)} + \frac{A_3}{(z+0.5)} + \frac{A_4}{(z+0.4)} \\ 2z+4 &= A_1(z-0.1)(z+0.5)(z+0.4) + A_2z(z+0.5)(z+0.4) \\ &\quad + A_3z(z-0.1)(z+0.4) + A_4z(z-0.1)(z+0.5) \end{aligned}$$

Substitute $z = 0$

$$4 = A_1(-0.1)(0.5)(0.4); \qquad A_1 = -200$$

Substitute $z = 0.1$

$$(0.2+4) = A_2(.1)(.6)(.5); \qquad A_2 = 140$$

Substitute $z = -0.4$

$$-1+4 = A_3(-0.5)(-0.6)(-0.1); \qquad A_3 = -100$$

Substitute $z = -0.4$

$$-0.8 + 4 = A_4(-0.4)(-0.5)(0.1); \qquad A_4 = 160$$

$$H_1[z] = -\frac{200}{z} = -200z^{-1}$$

$$H_2[z] = \frac{140}{(z-0.1)} = \frac{140z^{-1}}{(1-0.1z^{-1})} \longrightarrow b_0 = 0;\ b_1 = 140;\ a_1 = -0.1$$

$$H_3[z] = -\frac{100}{(z+0.5)} = -\frac{100z^{-1}}{(1+0.5z^{-1})} \longrightarrow b_0 = 0;\ b_1 = -100;\ a_1 = 0.5$$

$$H_4[z] = \frac{160}{(z+0.4)} = \frac{160z^{-1}}{(1+0.4z^{-1})} \longrightarrow b_0 = 0;\ b_1 = 160;\ a_1 = 0.4$$

The parallel form realization is shown in Fig. 9.28d.

■ Example 9.52

The system function of a discrete time system is

$$H[z] = \frac{(1+z^{-1})^4}{\left(1 - z^{-1} + \frac{7}{8}z^{-2}\right)\left(1 + 2z^{-1} + \frac{3}{4}z^{-2}\right)}$$

Realize this system using a cascade of a second-order system in direct form-II. (Anna University, December, 2007).

Solution:

$$H[z] = \frac{(1+z^{-1})^4}{\left(1 - z^{-1} + \frac{7}{8}z^{-2}\right)\left(1 + 2z^{-1} + \frac{3}{4}z^{-2}\right)}$$

$$(1+z^{-1})^4 = (1+z^{-1})^2(1+z^{-1})^2$$

$$= (1+2z^{-1}+z^{-2})(1+2z^{-1}+z^{-2})$$

$$H[z] = \frac{(1+2z^{-1}+z^{-2})}{\left(1 - z^{-1} + \frac{7}{8}z^{-2}\right)} \frac{(1+2z^{-1}+z^{-2})}{\left(1 + 2z^{-1} + \frac{3}{4}z^{-2}\right)}$$

$$= H_1[z]H_2[z]$$

For $H_1[z]$,

$$b_0 = 1;\ b_1 = 2;\ b_2 = 1;\ a_1 = -1;\ a_2 = \frac{7}{8}$$

For $H_2[z]$,

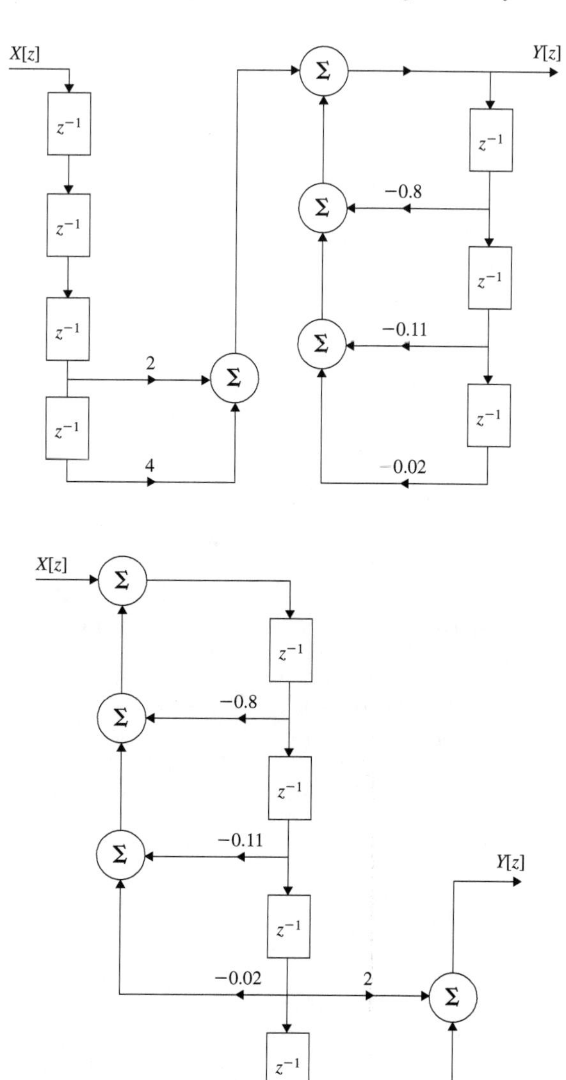

Fig. 9.28 **a** Direct form-I realization, **b** Direct form-II realization, **c** Cascade form realization and **d** Parallel form realization

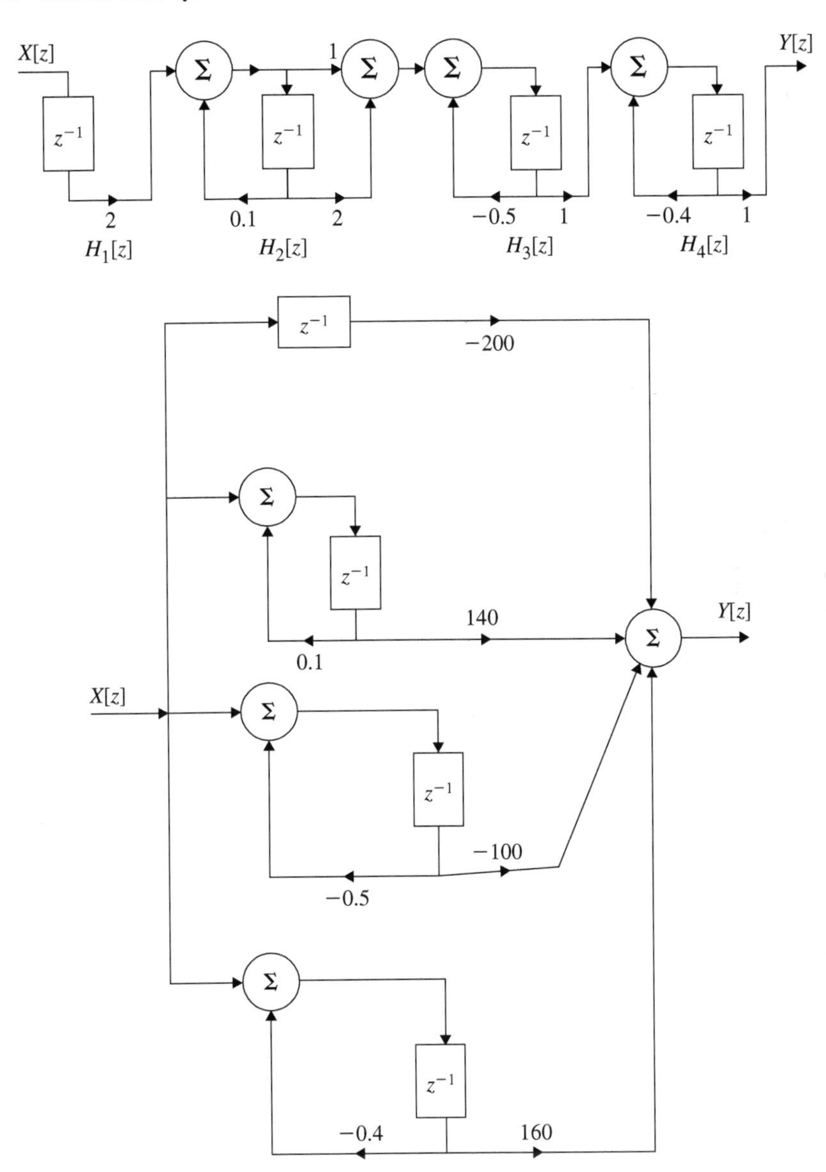

Fig. 9.28 (continued)

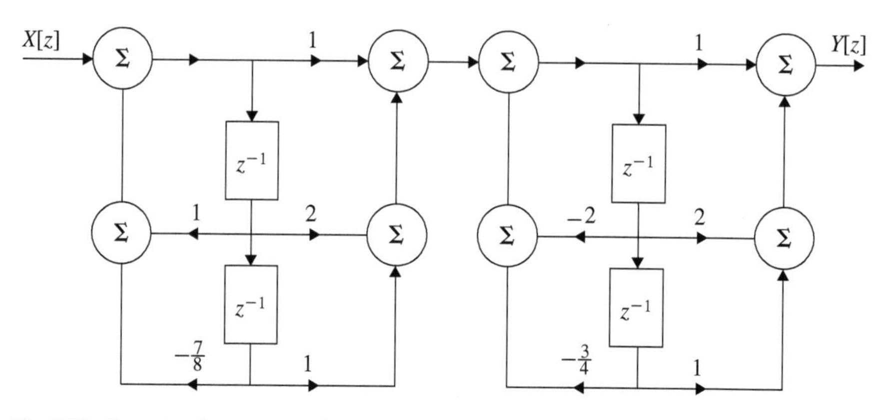

Fig. 9.29 Second-order system realization by cascade connection

$$b_0 = 1;\ b_1 = 2;\ b_2 = 1;\ a_1 = 2;\ a_2 = \frac{3}{4}$$

The cascade system is shown in Fig. 9.29.

■ Example 9.53

The unit system response of an FIR filter is

$$h[n] = a^n\{u[n] - u[n-2]\}$$

Draw the direct form realization of this system.

(Anna University, December, 2007)

Solution:

$$h[n] = a^n\{u[n] - u[n-2]\}$$

$$a^n u[n] \quad \overset{Z}{\longleftrightarrow} \frac{1}{(1-az^{-1})}$$

$$a^n u[n-2] \quad = a^2 a^{n-2} u[n-2]$$

$$a^n u[n-2] \quad \overset{Z}{\longleftrightarrow} a^2 \frac{z^{-2}}{(1-az^{-1})}$$

$$H[z] = \frac{1}{(1-az^{-1})}(1 - a^2 z^{-2})$$

$$b_0 = 1;\ b_1 = 0;\ b_2 = -a^2;\ a_1 = -a;\ a_2 = 0$$

The system realization is shown in Fig. 9.30.

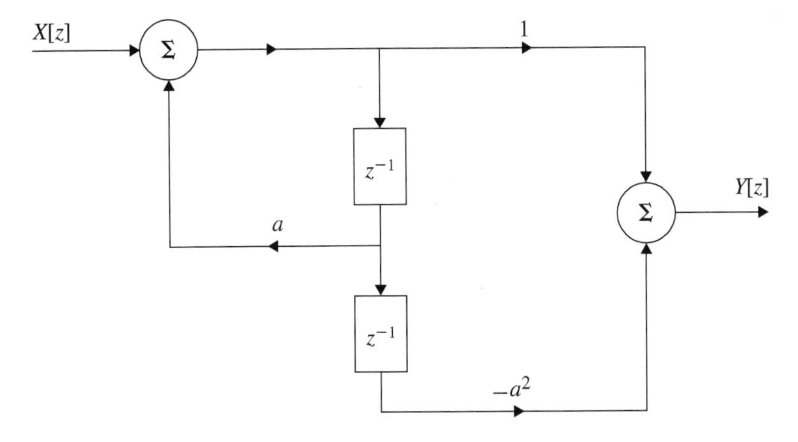

Fig. 9.30 Direct form realization for Example 9.53

■ Example 9.54

Consider a causal LTI system whose input $x[n]$ and output $y[n]$ are related through the block diagram representation shown in Fig. 9.31a. (a) Determine a difference equation relating $y[n]$ and $x[n]$ and (b) Is this system stable?

(Anna University, April, 2008)

Solution:

(a) From Fig. 9.31, $b_0 = 1$, $b_1 = -6$ and $b_2 = 8$. $a_1 = -\frac{2}{3}$ and $a_2 = \frac{1}{9}$. With these coefficients, the following second-order difference equation is written as

$$\boxed{y[n-2] - \frac{2}{3}y[n-1] + \frac{1}{9}y[n] = x[n] - 6x[n-1] + 8x[n-2]}$$

(b) Taking z-transform on both sides of the given differences equation, we get

$$\begin{aligned}H[z] = \frac{Y[z]}{X[z]} &= \frac{(1 - 6z^{-1} + 8z^{-2})}{\left(1 - \frac{2}{3}z^{-1} + \frac{1}{9}z^{-2}\right)} \\ &= \frac{(z^2 - 6z + 8)}{\left(z^2 - \frac{2}{3}z + \frac{1}{9}\right)} \\ &= \frac{(z-2)(z-4)}{\left(z - \frac{1}{3}\right)^2}\end{aligned}$$

For the causal system, ROC is $|z| > \frac{1}{3}$. The ROC includes The unit circle. The system is therefore stable (Fig. 9.31b).

$$\boxed{\text{The System is Stable.}}$$

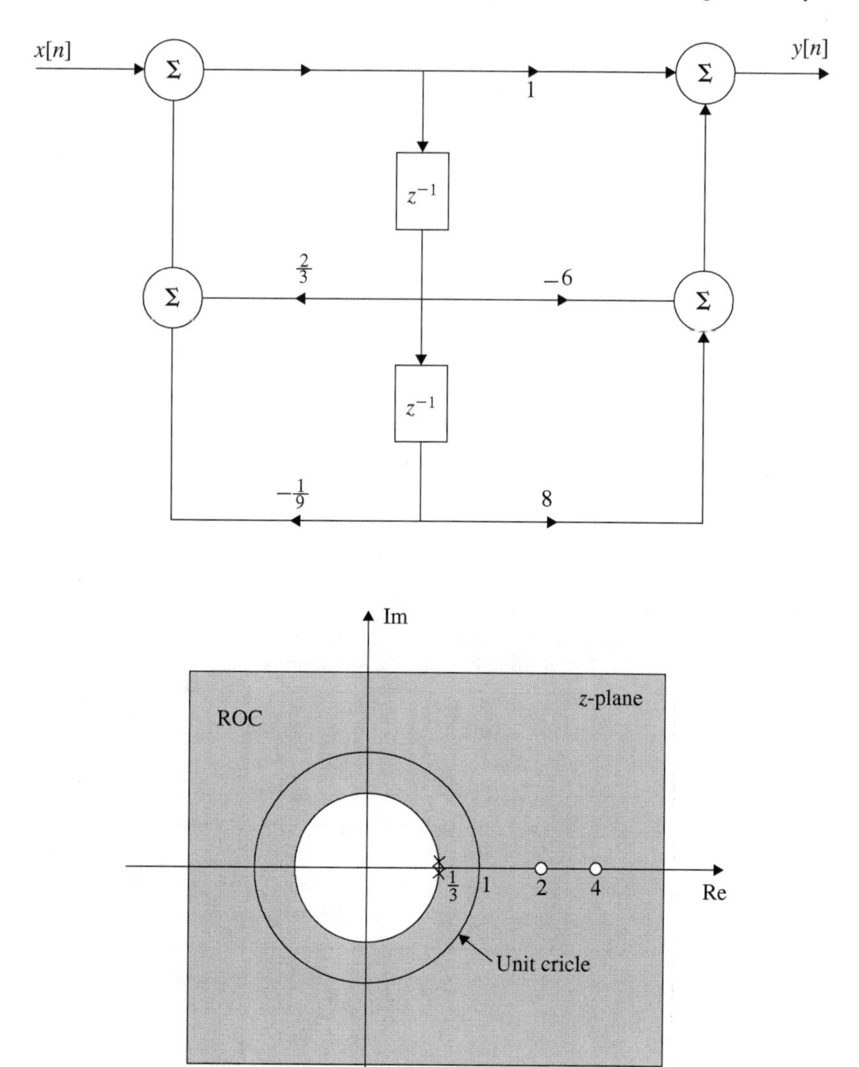

Fig. 9.31 **a** Structure of the system for Example 9.54 and **b** Pole-zero location and ROC for Example 9.54

Summary

1. The z-transform for discrete time signals and systems have been developed. This resembles corresponding treatment of Laplace transform for continuous time system.
2. A definite connection exists between Laplace transform, Fourier transform and z-transform. The Laplace transform reduces to Fourier transform on the imaginary axis in the s-plane. Then z-transform reduces to Fourier transform on the unit circle in the complex z-plane.
3. For the causal signal system (right-sided), the z-transform exists if the ROC is the exterior of the circle which passes through the outermost pole of the system function. For the anti-causal signal and system (left-sided), the z-transform exists if the ROC is the interior of the circle which passes through the innermost pole of the system. For the right- and left-sided signals, the ROC is a ring which does not include any pole of system function.
4. The application of the properties of z-transform very much simplifies the procedure to determine z-transform and inverse z-transform.
5. For an LTID system to be causal, the system function should be rational and the ROC is the exterior of the circle which passes through the outermost pole of the system function $H[z]$.
6. An LTID system is said to be stable if the ROC of the system function $H[z]$ includes the unit circle.
7. An LTID system is said to be causal and stable if all the poles of the system function $H[z]$ lie inside the unit circle in the z-plane.
8. Using the properties of z-transform, LTID systems described by a constant coefficient difference equation can be converted into algebraic equations and easily analyzed. The solution obtained is classified as zero state response, zero input response, natural response and forced response.
9. An LTID system structure is realized using adders, multipliers and unit delay. system is realized in direct form-I, direct form-II, parallel form, cascade form and transposed form.

Exercises

I. Short Answer Type Questions

1. **Define z-transform.**

   The z-transform of a discrete time signal $x[n]$ is defined as

$$X[z] = \sum_{n=-\infty}^{\infty} x[n]z^{-n}$$

   where z is a complex variable.

2. **Define z-transform pair.**
When the discrete time signal $x[n]$ is z-transformed, it is expressed as

$$X[z] = \sum_{n=-\infty}^{\infty} x[n]z^{-n}$$

If we want to recover $x[n]$ from $X[z]$, it is obtained using the following integration:

$$x[n] = \frac{1}{2\pi j} \oint X[z]z^{n-1}dz$$

This equation is called inverse z-transform. The above two equations for z-transform and inverse z-transform are called z-transform pair.
3. **What do you understand by ROC of z-transform?**
The range of values of z for which the function $X[z]$ converges is called region of convergence which is expressed in abbreviated form as ROC.
4. **Mention the properties of ROC.**
    1. The ROC of $X[z]$ is in the form of a ring in the z-plane which is centered about the origin.
    2. The ROC does not include any poles.
    3. For the right-sided sequence $x[n]$, the ROC is the exterior of the outermost pole.
    4. For the left-sided sequence $x[n]$, the ROC is the interior of the innermost pole.
    5. If the sequence $x[n]$ is two-sided, then the ROC consists of a ring in the z-plane.
5. **What is the scaling property of z-transform?**
If

$$x[n] \overset{Z}{\longleftrightarrow} X[z] \qquad \text{ROC: } R$$

then

$$a^n x[n] \longleftrightarrow X\left[\frac{z}{a}\right] \qquad \text{ROC: } aR$$

By using the multiplication property, the z-transform is obtained by replacing z by $\frac{z}{a}$ with ROC R replaced by aR.

6. **What is the convolution property of z-transform?**
If

$$x_1[n] \overset{Z}{\longleftrightarrow} X_1[z] \quad \text{ROC: } R_1$$
$$x_2[n] \overset{Z}{\longleftrightarrow} X_2[z] \quad \text{ROC: } R_2$$

then

$$x_1[n] * x_2[n] \overset{Z}{\longleftrightarrow} X_1[z]X_2[z] \quad \text{ROC: } R_1 \cap R_2$$

7. **What is difference property in the z-transform?**
If

$$x[n] \overset{Z}{\longleftrightarrow} X[z] \quad \text{ROC: } R$$

then

$$nx[n] \overset{Z}{\longleftrightarrow} -z\frac{dx[z]}{dz} \quad \text{ROC: } R$$

8. **What are initial and final value theorems?**
If $x[n] = 0$ for $n < 0$, then

$$x[0] = \underset{z\to\infty}{Lt}\, X[z]$$

is called the initial value theorem. According to the finial value theorem if $X[z]$ is the z-transform $x[n]$ and if all the poles of $X[z]$ are inside the unit circle, then the final value of $x[n] = x[\infty]$ is obtained from

$$x[\infty] = \underset{z\to 1}{Lt}\,(z-1)X[z]$$

9. **What do you understand by the time reversal property of z-transform?**
If

$$x[n] \overset{Z}{\longleftrightarrow} X[z] \quad \text{ROC: } R$$

then

$$x[-n] \overset{Z}{\longleftrightarrow} X\left[\frac{1}{z}\right] \quad \text{ROC: } \frac{1}{R}$$

Thus, the z-transform of the time reversal signal is obtained by replacing z by its reciprocal and also its ROC by its reciprocal.

10. **What do you understand by causality of an LTID system?**
A linear time invariant discrete time system is said to be causal if the ROC of the system function $H[z]$ is the exterior of the circle containing all the poles of $H[z]$.
11. **What do you understand by the stability of an LTID system?**
An LTID system is said to be stable if the ROC of the system function $H[z]$ includes the unit circle in the z-plane.
12. **When is the system said to be both causal and stable?**
An LTID system is said to be both causal and stable if all the poles of the system function $H[z]$ are inside the unit circle in the z-plane.
13. **Define system function.**
System function or transfer function $H[z]$ is defined as the ratio of the z-transform of output sequence $y[n]$ and the input sequence $x[n]$

$$H[z] = \frac{Y[z]}{X[z]}$$

14. **What is the z-transform of $\delta[n-2]$?**

$$\delta[n-2] \overset{Z}{\longleftrightarrow} z^{-2}$$

15. **What is the z-transform of $u[n]$ and $\delta[n]$?**

$$u[n] \overset{Z}{\longleftrightarrow} \frac{z}{(z-1)}$$
$$\delta[n] \overset{Z}{\longleftrightarrow} 1$$

16. **Find the z-transform of $x[n] = u[n] - u[n-5]$.**

$$X[z] = \frac{z}{(z-1)}[1 - z^{-5}]$$

17. **Write the relationship between z-transform and Fourier transform.**
The z-transform reduces to Fourier transform on the unit circle in the complex z-plane.
18. **Write the relationship between z-transform and Laplace transform.**
The Laplace transform and z-transform are related as

$$e^s = z$$
$$X[s] = X[z]\Big|_{z=e^s}$$

19. **What is the inverse z-transform of $X[\frac{z}{a}]$?**

$$X\left[\frac{z}{a}\right] \stackrel{Z^{-1}}{\longleftrightarrow} a^n x[n]$$

20. **Find the system function of the following first-order difference equation $y[n] - 2y[n-1] = x[n] + x[n-1]$?**

$$H[z] = \frac{Y[z]}{X[z]} = \frac{[1+z^{-1}]}{[1-2z^{-1}]} = \frac{[z+1]}{[z-2]}$$

II. Long Answer Type Questions

1. **Find the z-transform of the following sequence.**

$$x[n] = [3^{n-1} - (-3)^{n-1}]u[n]$$

$$X[z] = \frac{2z^2}{3(z^2-9)} \qquad \text{ROC: } |z| > 3$$

2. **Find the z-transform of**

$$x[n] = \sum_{n=0}^{\infty} \frac{1}{3} z^{-n} + \frac{1}{4}(-2)^n z^{-n}$$

$$X[z] = \frac{1}{3}\frac{1}{(1-z^{-1})} + \frac{1}{3}\frac{1}{(1+2z^{-1})} \qquad \text{ROC: } |z| > 2$$

3. **Find the z-transform of**

$$x[n] = \sum_{n=-1}^{\infty} \left(\frac{1}{4}\right)^{n+1} z^{-n}$$

$$X[z] = z + \frac{1}{4}\frac{1}{\left(1-\frac{1}{4}z^{-1}\right)} \qquad \text{ROC: } |z| > \frac{1}{4}$$

4. **Find the z-transform of**

$$x[n] = \sum_{n=1}^{\infty} \left(\frac{1}{4}\right)^{-n+1} z^{-n}$$

$$X[z] = \frac{1}{4} + \frac{4}{(1-4z)} \qquad \text{ROC: } |z| < \frac{1}{4}$$

5. **Find the z-transform of**

$$x[n] = \left(\frac{1}{10}\right)^n u[n-4]$$

$$X[z] = 10^{-4}\frac{z^{-3}}{\left(z - \frac{1}{10}\right)} \qquad \text{ROC: } |z| > \frac{1}{10}$$

6. **Find the z-transform of**

$$\begin{aligned} &\textbf{(a)} \quad x[n] = 1 \quad 0 \le n \le 9 \\ &\qquad\qquad\; = 0 \quad \text{otherwise} \\ &\textbf{(b)} \quad y[n] = x[n] - x[n-1] \end{aligned}$$

$$\begin{aligned} &\text{(a)} \qquad X[z] = \frac{(1-z^{-10})}{(1-z^{-1})} \qquad \text{ROC: } |z| > 0 \\ &\text{(b)} \qquad Y[z] = 1 - z^{-10} \qquad \text{ROC: } |z| > 0 \end{aligned}$$

7. **Find the unilateral z-transform and the ROC for the following sequences:**

$$\begin{aligned} &\textbf{(a)} \quad x[n] = \left(\frac{1}{6}\right)^n u[n+6] \\ &\textbf{(b)} \quad x[n] = 3\delta[n+4] + \delta[n] + (3)^n u[-n] \\ &\textbf{(c)} \quad x[n] = \left(\frac{1}{4}\right)^{|n|} \end{aligned}$$

$$\begin{aligned} &\text{(a)} \qquad X[z] = \frac{1}{(1-6z^{-1})} \qquad \text{ROC: } |z| > 6 \\ &\text{(b)} \qquad X[z] = 4 \qquad \text{ROC: all } z \\ &\text{(c)} \qquad X[z] = \frac{1}{\left(1 - \frac{1}{4}z^{-1}\right)} \qquad \text{ROC: } |z| > \frac{1}{4} \end{aligned}$$

8. **By applying the properties of z-Transform, find the z-transform of the following sequences given $x[n] \overset{Z}{\longleftrightarrow} \frac{z}{(z^2+2)}$:**

$$\begin{aligned}
&\textbf{(a)} \quad y[n] = x[n-3]\\
&\textbf{(b)} \quad y[n] = nx[n]\\
&\textbf{(c)} \quad y[n] = x[n+1] + x[n-1]\\
&\textbf{(d)} \quad x[n] = 2^n x[n]\\
&\textbf{(e)} \quad x[n] = (n-2)x[n-1]\\
&\textbf{(f)} \quad x[n] = x[-n]
\end{aligned}$$

(a) $Y[z] = \dfrac{z^{-2}}{(z^2+2)}$ ROC:$|z| < 2$ (Time shifting property)

(b) $Y[z] = \dfrac{z[z^2-2]}{(z^2+2)^2}$ (Differentiation property)

(c) $Y[z] = \dfrac{(z^2+1)}{(z^2+2)}$ (Time advancing and time delaying)

(d) $Y[z] = \dfrac{2z}{(z^2+8)}$ (Multiplying property)

(e) $Y[z] = \dfrac{-4}{(z^2+2)^2}$ (Time differentiation and time shifting)

(f) $Y[z] = \dfrac{z}{(1+2z^2)}$ (Time reversal)

9. **Find the z-transform of**

$$\begin{aligned}
&\textbf{(a)} \quad x[n] = 2^n u[n-2]\\
&\textbf{(b)} \quad x[n] = \left(\frac{1}{4}\right)^n u[-n]
\end{aligned}$$

(a) $X[z] = \dfrac{4z^{-2}}{(1-2z^{-1})}$

(b) $X[z] = \dfrac{1}{(1-4z)}$

10. **Find the z-transform of**

$$\begin{aligned}
&\textbf{(a)} \quad x[n] = (n-4)u[n-4]\\
&\textbf{(b)} \quad x[n] = u[n] - u[n-4]\\
&\textbf{(c)} \quad x[n] = (n-4)u[n]\\
&\textbf{(d)} \quad x[n] = n[u[n] - u[n-4]]
\end{aligned}$$

(a) $X[z] = \dfrac{z^{-3}}{(z-1)^2}$ ROC: $|z| > 1$

(b) $X[z] = \dfrac{z}{(z-1)}[1 - z^{-4}]$ ROC: $|z| > 1$

(c) $X[z] = \dfrac{(5z - 4z^2)}{(z-1)^2}$ ROC: $|z| > 1$

(d) $X[z] = \dfrac{(z - 3z^{-2} + 2z^{-3})}{(z-1)^2}$ ROC: $|z| > 1$

11. **Find the z-transform of the following sequence:**

$$\begin{aligned} x[n] &= \left(\frac{1}{4}\right)^n u[n] \\ &= \left(-\frac{1}{2}\right)^n u[-n-1] \end{aligned}$$

$$X[z] = \frac{3}{4}\left[\frac{z}{\left(z - \frac{1}{4}\right)\left(z + \frac{1}{2}\right)}\right] \qquad \text{ROC: } \frac{1}{4} < |z| < \frac{1}{2}$$

12. **Using convolution find $y[n]$ given**

$$\begin{aligned} x[n] &= \left(\frac{1}{2}\right)^n u[n] \\ h[n] &= \left(\frac{1}{3}\right)^n u[n] \\ y[n] &= x[n] * h[n] \end{aligned}$$

$$y[n] = \left[3\left(\frac{1}{2}\right)^n - 2\left(\frac{1}{3}\right)^n\right] u[n] \qquad \text{ROC: } |z| < \frac{1}{2}$$

13. **Using partial function, find the inverse z-transform**

$$H[z] = \frac{(1 - z^{-1} + z^{-2})}{(1 - z^{-1})(1 - 2z^{-1})(1 - 4z^{-1})} \qquad \text{ROC: } 2 < |z| < 4$$

$$h[n] = \left[\frac{1}{3} - \frac{3}{2}(2)^n\right] u[n] + \frac{3}{16}(4)^n u[-n-1]$$

14. **Find the inverse z-transform of**

$$H[z] = \frac{4z + 1}{z - \frac{1}{4}}$$

using power series expansion.

$$\text{(a)} \quad \textbf{ROC}: |z| > \frac{1}{4}$$

$$\text{(b)} \quad \textbf{ROC}: |z| < \frac{1}{4}$$

$$\text{(a)} \quad x[n] = \left\{ \underset{\uparrow}{4}, 2, \frac{1}{2}, \frac{1}{8}, \frac{1}{32}, \ldots \right\}$$

$$\text{(b)} \quad x[n] = \{\ldots, 2048, -512, -128, -32, \underset{\uparrow}{-4}\}$$

15. **Consider the algebraic expression for the z-transform of $x[n]$**

$$x[n] = \frac{\left(1 - \frac{1}{4}z^{-2}\right)}{\left(1 + \frac{1}{4}z^{-1}\right)\left(1 - \frac{5}{6}z^{-1} + \frac{1}{6}z^{-2}\right)}$$

How many different ROCs could correspond the $X[z]$?

$$\text{(a)} \quad \text{ROC: } |z| > \frac{1}{2}$$

$$\text{(b)} \quad \text{ROC: } 0 < |z| < \frac{1}{4}$$

$$\text{(c)} \quad \text{ROC: } \frac{1}{3} < |z| < \frac{1}{2}$$

16. **Consider the algebraic expression for the z-transform of $x[n]$**

$$x[n] = \frac{\left(1 + z^{-1} + 4z^{-2}\right)}{\left(1 - \frac{1}{4}z^{-1}\right)\left(1 - \frac{7}{24}z^{-1} + \frac{1}{48}z^{-2}\right)}$$

ROC: $|z| > \frac{1}{4}$. Find whether the system is causal and stable.
$X[z]$ is rational and the poles are at $z = \frac{1}{4}$, $z = \frac{1}{6}$ and $z = \frac{1}{8}$. Since the ROC is the exterior of the outermost pole, the system is causal. The ROC includes the unit circle and the poles are inside the unit circle. The system is stable. Therefore the system is causal and stable.

17. **A system with impulse response $h[n] = 5(3)^n u[n-1]$ produces on output $y[n] = (-4)^n u[n-1]$. Determine the input $x[n]$.**

$$x[n] = \frac{1}{5}\left[(-4)^n u[n] - 3(-4)^{n-1} u[n-1]\right] \qquad \text{ROC: } |z| > 4$$

18. **Consider the following difference equation:**

$$y[n] - y[n-1] - 2y[n-2] = x[n] + 2x[n-1]$$

The initial conditions are $y[-1] = 1$ and $y[-2] = 2$. The input $x[n] = u[n]$. Find (a) Zero input response, (b) Zero state response, (c) Natural response, (d) Forced response and (e) Total response.

(a) $y_{0i}[n] = [(-1)^n + 4(2)^n]u[n]$

(b) $y_{0s}[n] = \left[-\frac{1}{6}(-1)^n + \frac{8}{3}(2)^n - \frac{3}{2}\right]u[n]$

(c) $y_n[n] = -\frac{3}{2}u[n]$

(d) $y_f[n] = \left[\frac{5}{6}(-1)^n + \frac{20}{3}(2)^n\right]u[n]$

(e) $y_{\text{total}}[n] = \left[-\frac{3}{2} + \frac{5}{6}(-1)^n + \frac{20}{3}(2)^n\right]u[n]$

19. **Consider the causal LTID system represented in block diagram shown in Fig.** 9.32. **(a) Determine the difference equation relating the output $y[n]$ and input $x[n]$ and (b) Is the system stable?**

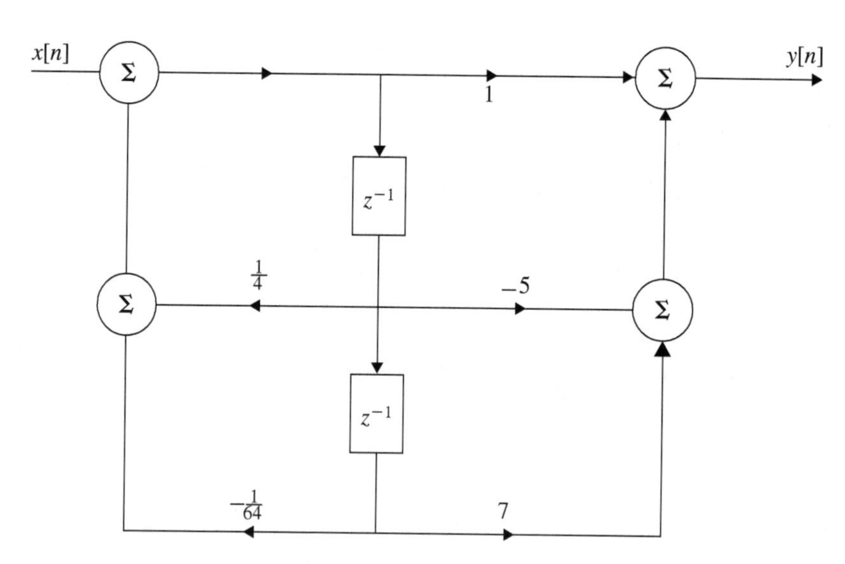

Fig. 9.32 Block diagram of Problem 19

(a)

$$y[n] - \frac{1}{4}y[n-1] + \frac{1}{64}y[n-2] = x[n] - 5x[n-1] + 7x[n-2]$$

(b) The ROC includes the unit circle, and the poles of system function are within unit circle. Hence, the system is stable.

20. **For each of the following difference equations, determine the output response $y[n]$?**

(a) $y[n] - 4y[n-1] = x[n]$ **with** $y[-1] = 2$**and** $x[n] = \left(\frac{1}{3}\right)^n u[n]$

(b) $y[n] + \frac{1}{3}y[n-1] = x[n] - \frac{1}{3}x[n-1]$
with $y[-1] = 1$ **and** $x[n] = u[n]$

(a) $$y[n] = \left[\frac{100}{11}(4)^n - \frac{1}{11}\left(\frac{1}{3}\right)^n\right]u[n]$$

(b) $$y[n] = \left[\frac{3}{2} - \frac{5}{6}\left(-\frac{1}{3}\right)^n\right]u[n]$$

Chapter 10
Sampling

Learning Objectives

- To represent a continuous time signal by a sequence of equally spaced samples.
- To establish the sampling theorem for exact reconstruction of the original signal.
- To reconstruct the original signal by means of low-pass filters.
- To choose the correct sampling rate to avoid aliasing.
- To study the important applications of sampling.

10.1 Introduction

Due to the dramatic development of digital technology in the recent past, continuous time signals are converted into discrete time signals, which are processed by discrete time systems and again converted back to continuous time signals which are applied to continuous time systems. A continuous time signal can be completely represented by and recovered from its values called samples at points equally spaced in time. This process is called sampling. The concept of sampling uses a discrete time system to implement continuous time systems and process continuous time signals. The information present in the sampled continuous time signal is retained in the discrete time signal also. While a sampled continuous time signal is represented by a sequence of impulses, a discrete time signal is represented by a sequence of numbers which carry the sample information as that of the sampled sequence.

In the discussion to follow, the concept of sampling and the process of reconstructing a continuous time signal from its samples are developed. The necessary condition under which a CT signal can be exactly reconstructed from its samples is established through the **Sampling Theorem**. Finally, the consequences that arise when the sampling theorem is not satisfied are discussed.

S. Palani, *Signals and Systems*,
https://doi.org/10.1007/978-3-030-75742-7_10

10.2 The Sampling Process

Figure 10.1a shows the block diagram representation of a continuous signal $x(t)$ being multiplied by a periodic impulse train $\delta_T(t)$ to get the sampled output $g(t)$. The device used for this is called a **sampler**. The sampler is also represented by a switch which opens and closes with periodicity T_s. This is shown in Fig. 10.1b. The continuous time signal $x(t)$ is shown in Fig. 10.1c. The periodic impulse train $\delta_T(t)$ is shown in Fig. 10.1d. The product of $x(t)$ and $\delta_T(t)$ which is the sampled signal $g(t)$ is shown in Fig. 10.1e. Now we develop the sampling theorem as discussed below.

10.3 The Sampling Theorem

The signals $x(t)$, $\delta_T(t)$ and $g(t)$ shown in Fig. 10.1 are connected by the following equation:

$$g(t) = x(t)\delta_T(t) \tag{10.1}$$

where

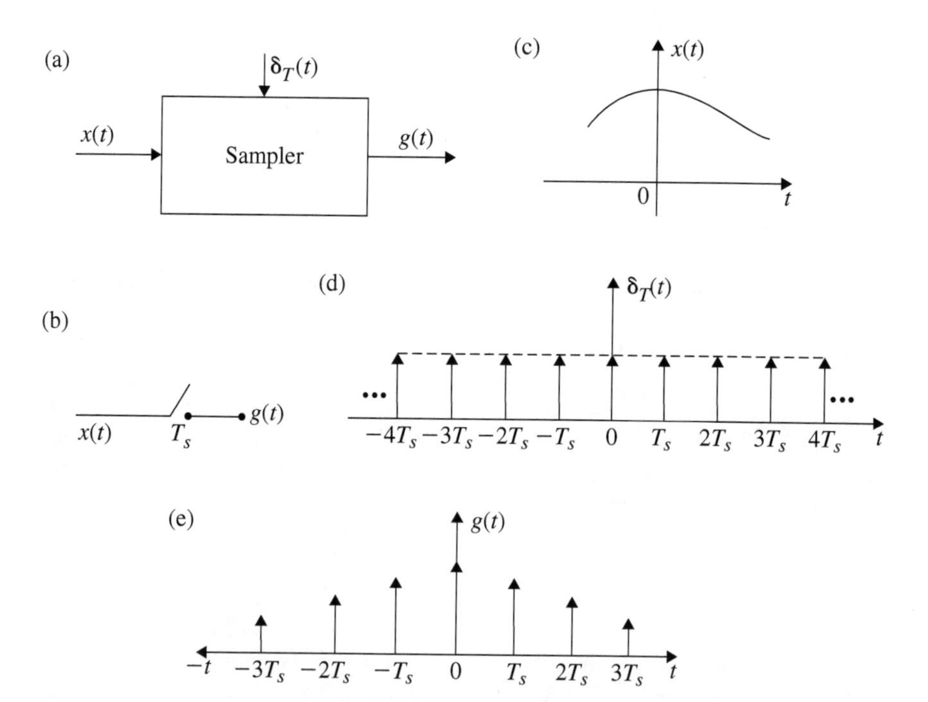

Fig. 10.1 The sampling process representation by schematic diagram and signals

$$\delta_T(t) = \sum_{n=-\infty}^{\infty} \delta(t - nT_s) \tag{10.2}$$

$\delta_{T_s}(t)$ is called a sampling function, T_s the sampling period and $\omega_s = \frac{2\pi}{T_s}$ the fundamental sampling frequency. This is also called radian frequency and is related to cyclic frequency as

$$\omega_s = \frac{2\pi}{T_s} = 2\pi f_s$$

where $f_s = \frac{1}{T_s}$ is the cyclic frequency. The sampled signal $g(t)$ shown in Fig. 10.1e consists of impulses spaced every T_s seconds, which is the sampling interval. $\delta_T(t)$ can be expressed as a trigonometric Fourier series as

$$\delta_T(t) = \frac{1}{T_s}[1 + 2\cos\omega_s t + 2\cos 2\omega_s t + 2\cos 3\omega_s t + \cdots] \tag{10.3}$$

$$\begin{aligned} g(t) &= x(t)\delta_T(t) \\ &= \frac{1}{T_s}[x(t) + 2x(t)\cos\omega_s t + 2x(t)\cos 2\omega_s t + \cdots] \end{aligned} \tag{10.4}$$

From the knowledge of Fourier transform of continuous time signals, the following equations are written:

$$\begin{aligned} x(t) &\overset{\text{FT}}{\longleftrightarrow} X(\omega) \\ 2x(t)\cos\omega_s t &\overset{\text{FT}}{\longleftrightarrow} X(\omega - \omega_s) + X(\omega + \omega_s) \\ 2x(t)\cos 2\omega_s t &\overset{\text{FT}}{\longleftrightarrow} X(\omega - 2\omega_s) + X(\omega + 2\omega_s) \\ g(t) &\overset{\text{FT}}{\longleftrightarrow} G(\omega) \end{aligned} \tag{10.5}$$

Substituting Eq. (10.5) in Eq. (10.4), we get

$$\begin{aligned} G(\omega) = \frac{1}{T_s}[&X(\omega) + X(\omega - \omega_s) + X(\omega + \omega_s) \\ &+ X(\omega - 2\omega_s) + X(\omega - 2\omega_s) + \cdots] \end{aligned} \tag{10.6}$$

$$\boxed{G(\omega) = \frac{1}{T_s}\sum_{n=-\infty}^{\infty} X(\omega - n\omega_s)} \tag{10.7}$$

In Eq. (10.6), the first term in the bracket is $X(\omega)$. The second term is $X(\omega - \omega_s) + X(\omega + \omega_s)$. This represents the spectrum of $X(j\omega)$ shifted to ω_s and $-\omega_s$. Similarly, the third term $X(\omega - 2\omega_s) + X((\omega - 2\omega_s))$ which represents the

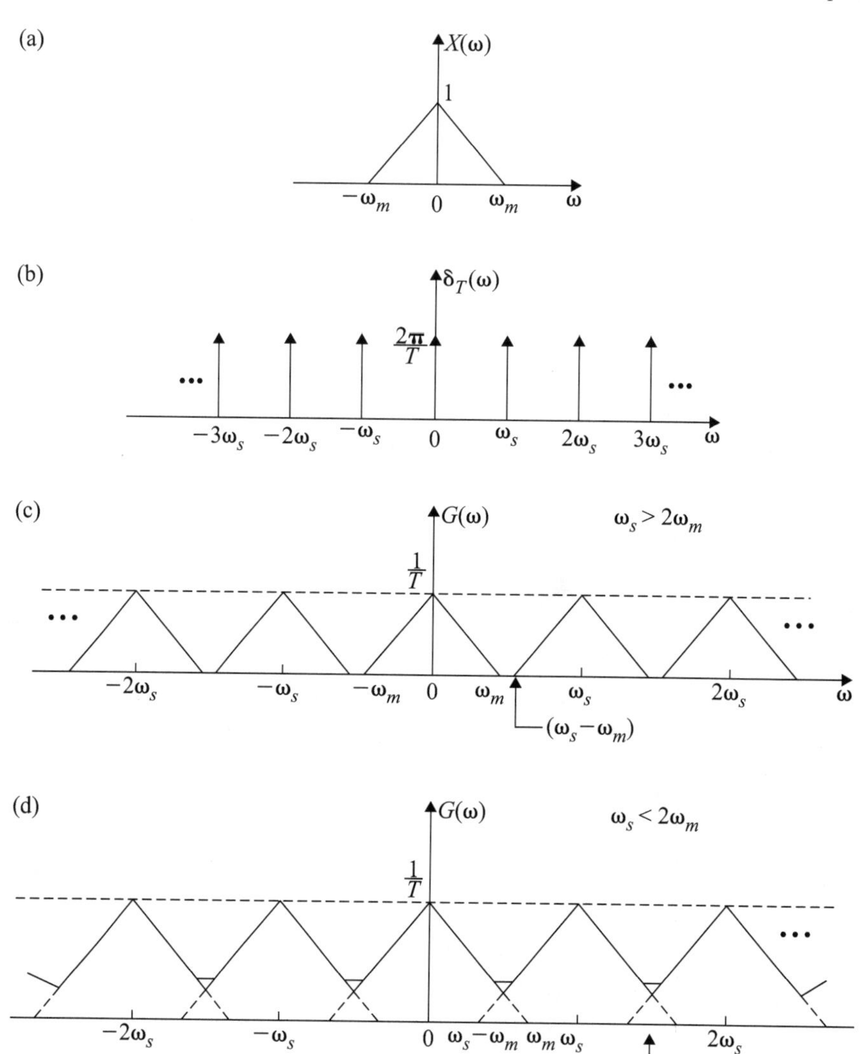

Fig. 10.2 Continuous signal, impulse train sampled signal spectrum

spectrum $X(\omega)$ shifted by $2\omega_s$ and $-2\omega_s$ and so on. The frequency spectrum $X(\omega)$ and $\delta(\omega)$ are represented in Fig. 10.2a and b, respectively. In Fig. 10.2a, ω_m is the maximum frequency content of the continuous time signal. Figure 10.2c represents the frequency spectrum $G(\omega)$ of the continuous sampled signal for $\omega_s > 2\omega_m$. Spectrum of the sampled signal $G(\omega)$ for $\omega_s < 2\omega_m$ is shown in Fig. 10.2d.

To reconstruct the continuous time signal $x(t)$ from sampled signal $g(t)$, we should be able to recover $X(\omega)$ from $G(\omega)$. This recovery is possible of there is no overlap between successive $G(\omega)$. From Fig. 10.2c, this is possible if $\omega_s > 2\omega_m$ or

$$\boxed{f_s > 2f_m} \tag{10.8}$$

The minimum sampling rate $f_s = 2f_m$ is called the **Nyquist rate** of $x(t)$, and the corresponding time interval $T_s = \frac{1}{f_s} = \frac{1}{2f_m}$ is called **Nyquist interval** of $x(t)$. Samples of a signal taken at its Nyquist rate are the **Nyquist samples**.

From Eq. (10.8), the Shannon sampling theorem or simply sampling theorem is stated as follows.

A band-limited signal of finite energy which has no frequency component higher than f_m can be completely described and recovered back if the sampling frequency is twice the highest frequency of the given signal.

The proof of the theorem is given in Eq. (10.7) and Fig. 10.2c. If $f_s < 2f_m$, Eq. (10.7) is represented in Fig. 10.2d, and here overlapping between successive samples occurs and therefore it is not possible to recover $x(t)$ from the frequency spectrum $G(\omega)$ when passed through low-pass filter.

10.4 Signal Recovery

If the condition $\omega_s > 2\omega_m$ is satisfied, $x(t)$ can be recovered exactly from $g(t)$ using an ideal low-pass filter whose characteristic is shown in Fig. 10.3d with a gain T_s and cutoff frequency greater than ω_m and less than $(\omega_s - \omega_m)$.

10.5 Aliasing

Consider the frequency spectrum of sampled signal $g(t)$ which has been obtained by sampling $x(t)$ with a sampling frequency $f_s < 2f_m$. The frequency spectrum of $G(\omega)$ is shown in Fig. 10.2d. When $f_s < 2f_m$, the signal is said to be under-sampled. The spectra located at $G(\omega_m)$, $G(\omega_m - \omega_s)$, $G(\omega_m - 2\omega_s)$, *etc.*, overlap on each other. When the high frequency interferes and appears as low frequency, then the phenomenon is called aliasing.

The effects of aliasing are as follows:

1. Distortion in signal recovery is generated when the high and low frequencies interfere with each other.
2. The data is lost and it cannot be recovered.

Different methods are available to avoid aliasing:

1. To increase the sampling rate f_s so that $f_s > 2f_m$.
2. To put anti-aliasing filter before the signal $x(t)$ is sampled.

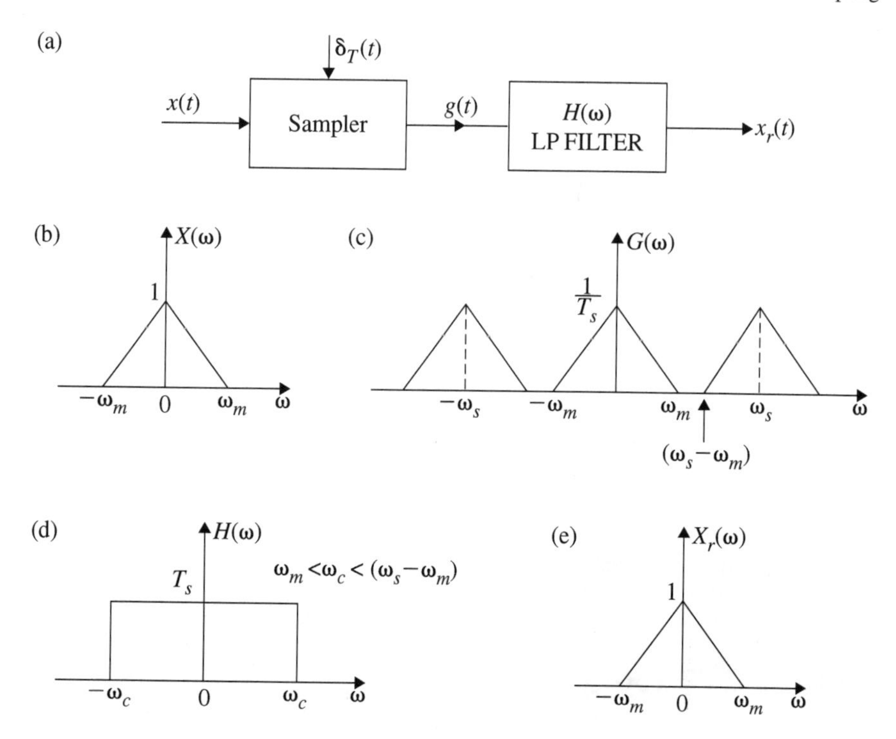

Fig. 10.3 Signal sampling and recovery. **a** Continuous signal for sampling and reconstruction; **b** Frequency spectrum of $x(t)$; **c** Frequency spectrum of sampled signal; (d) Characteristic of a low-pass filter; **e** Spectrum of recovered continuous signal

10.5.1 Sampling Rate ω_s Higher than $2\omega_m$

If the sampling rate $\omega_s > 2\omega_m$, the frequency spectrum of the sampled continuous signal is as shown in Fig. 10.2a and there is no overlapping between the samples and the original signal can be reconstructed without aliasing.

10.5.2 Anti-aliasing Filter

The anti-aliasing filter $H_{aa}(\omega)$ put before the sampler is shown in Fig. 10.4. $x(t)$ is the continuous signal and is passed through the anti-aliasing filter $H_{aa}(\omega)$ which gives the output $\bar{x}(t)$. The signal $\bar{x}(t)$ is passed through the sampler and recovered as $x_r(t)$. The continuous signal $x(t)$ is passed through an anti-aliasing filter whose cutoff frequency is $f_s/2$. All the frequency components of $x(t)$ beyond $f_s/2$ are eliminated before sampling of $x(t)$ is started. The anti-aliasing filter essentially band-limits the signal to $f_s/2$. By this, the components of $x(t)$ beyond $f_s/2$ are lost. However, these suppressed

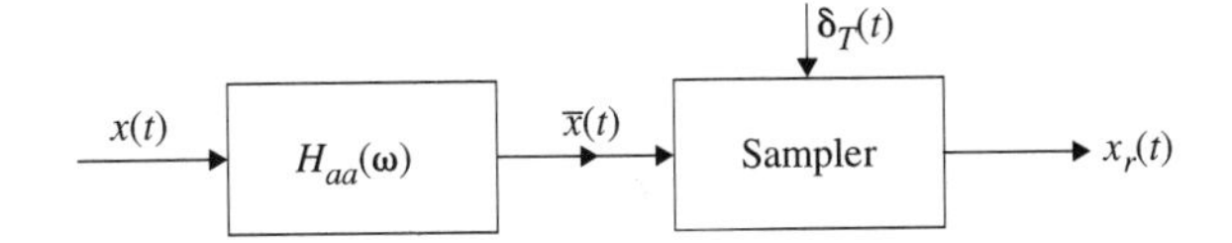

Fig. 10.4 Anti-aliasing filter before the sampler

components cannot corrupt the components of $x(t)$ whose frequency is less than $f_s/2$. Thus, the spectrum below $f_s/2$ remains intact and completely recovered. The noise produced by the aliasing is very much reduced when anti-aliasing filter is used. It also suppresses the entire noise spectrum beyond the frequency $f_s/2$.

10.6 Sampling with Zero-Order Hold

Band-limited signals which are sampled are narrow with large amplitude pulses. These impulses are difficult to generate and transmit. However, a more convenient method is to generate the sampled signal and send it through a zero-order hold. The input-output of a zero-order hold is shown in Fig. 10.5. The Zero-Order Hold (ZOH) samples $x(t)$ at a given time and holds those values until the next instant.

The input of the ZOH is shown in Fig. 10.5b and the output is shown in Fig. 10.5c. The output of the hold circuit is passed through a low-pass filter to recover the continuous time signal. The cascade-connected ZOH with a reconstruction filter is shown in Fig. 10.6.

The transfer function of ZOH is obtained from a unit step function with a time T shifted step being subtracted from that. Thus,

$$\begin{aligned} H_0(s) &= \frac{1}{s}[1 - e^{-sT}] \\ &= \frac{e^{-sT/2}}{s}[e^{sT/2} - e^{-sT/2}] \end{aligned} \tag{10.9}$$

Fig. 10.5 Input-output of zero-order hold

(a)

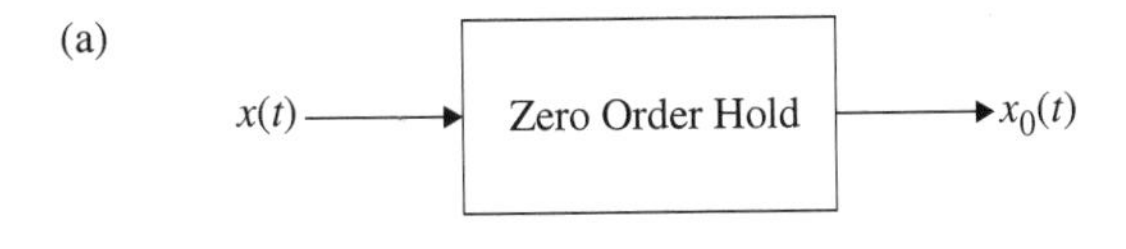

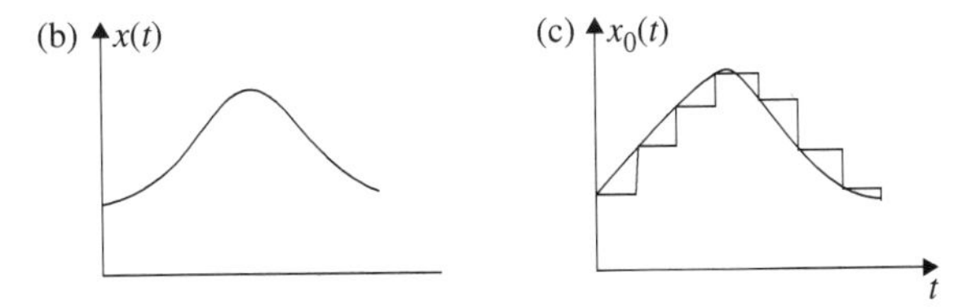

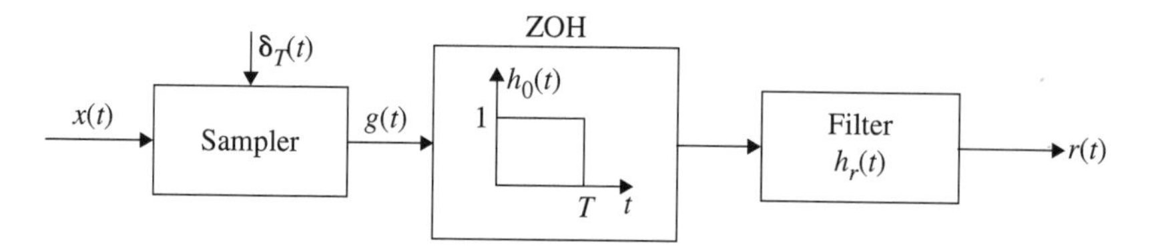

Fig. 10.6 Cascade-connected ZOH and reconstruction filter

The frequency response of the ZOH is therefore

$$\boxed{H_0(\omega) = e^{-j\omega T/2} \frac{\sin \omega T/2}{\omega}} \tag{10.10}$$

If we consider the frequency response of the ideal filter as $H(\omega)$, then

$$H_0(\omega) H_r(\omega) = H(\omega)$$

$$\boxed{H_r(\omega) = \frac{H(\omega)\omega e^{j\omega T/2}}{\sin\left(\frac{\omega T}{2}\right)}} \tag{10.11}$$

10.7 Application of Sampling Theorem

1. The sampling theorem is used in the analysis, processing and transmission of signals.
2. Processing the continuous time signal is equivalent to processing a discrete sequence of numbers, which ultimately leads to the area of digital filtering.
3. In the field of communication, the transmission of continuous time signals reduces to the transmission of a sequence of numbers. While doing so, the amplitude (PAM) of the sample, the width of the sample (PWM) or the position of the sample (PPM) can be varied and transmitted. At the receiver end, the pulse modulated signal is reconstructed and $x(t)$ is received. This process permits simultaneous transmission of several signals on a time-sharing basis which is called Time Division Multiplexing (TDM). By this, we can multiplex several signals in the same channel.
4. The transmission of digital signals is more rugged than that of analog signals because digital signals can withstand channel noise and distortion much better.

■ Example 10.1

Find the Nyquist rate and Nyquist interval for the following signals:

1. $x(t) = \sin 200\pi t$
2. $x(t) = 2 + 3\cos 100\pi t + 2\sin 200\pi t$
3. $x(t) = \dfrac{\sin 100\pi t}{\pi t}$
4. $x(t) = (\sin 200\pi t)^2$
5. $x(t) = \cos 200\pi t \cos 100\pi t$
6. $x(t) = \text{sinc}\, 2000\pi t$

Solution:

1. $\boldsymbol{x(t) = \sin 200\pi t}$

Let $\omega_m = 200\pi$

$$f_m = \frac{\omega_m}{2\pi} = \frac{200\pi}{2\pi} = 100 \text{ Hz}$$
$$\text{Nyquist rate } f_s = 2f_m = 200 \text{ Hz}$$
$$\text{Nyquist width } W = \frac{1}{f_s} = \frac{1}{200} = 5 \text{ m.s.}$$

2. $\boldsymbol{x(t) = 2 + 3\cos 100\pi t + 2\sin 200\pi t}$

Let $x_1(t) = \cos 100\pi t$ and $x_2(t) = \sin 200\pi t$

$$f_{m1} = \frac{100\pi}{2\pi} = 50 \text{ Hz}$$
$$f_{m2} = \frac{200\pi}{2\pi} = 100 \text{ Hz}$$
$$f_{m2} > f_{m1}$$
$$\text{Nyquist rate } f_s = 2f_{m2} = 200 \text{ Hz}$$
$$\text{Nyquist width } W = \frac{1}{f_s} = \frac{1000}{200} = 5 \text{ m.s.}$$

3. $\boldsymbol{x(t) = \dfrac{\sin 100\pi t}{\pi t}}$

Let $\omega_m = 100\pi$

$$f_m = \frac{\omega_m}{2\pi} = \frac{100\pi}{2\pi} = 50 \text{ Hz}$$
$$\text{Nyquist rate } f_s = 2f_m = 2 \times 50 = 100 \text{ Hz}$$
$$\text{Nyquist width } W = \frac{1}{f_s} = \frac{1000}{100} = 10 \text{ m.s.}$$

4. $\boldsymbol{x(t) = (\sin 200\pi t)^2}$

Let

$$\begin{aligned}
x(t) &= \sin^2 200\pi t \\
&= \frac{1}{2}[1 - \cos 400\pi t] \\
\omega_m &= 400\pi \\
f_m &= \frac{400\pi}{2\pi} = 200 \text{ Hz} \\
\text{Nyquist rate } f_s &= 2f_m = 400 \text{ Hz} \\
\text{Nyquist width } W &= \frac{1}{f_s} = \frac{1000}{400} = 2.5 \text{ m.s.}
\end{aligned}$$

5. $\boldsymbol{x(t) = \cos 200\pi t \cos 100\pi t}$

Let

$$\begin{aligned}
x(t) &= \frac{1}{2}[\cos(200 + 100)\pi t + \cos(200 - 100)\pi t] \\
&= \frac{1}{2}[\cos 300\pi t + \cos 100\pi t] \\
x_1(t) &= \cos 300\pi t \\
f_{m1} &= \frac{300\pi}{2\pi} = 150 \text{ Hz} \\
x_2(t) &= \cos 100\pi t \\
f_{m2} &= \frac{100\pi}{2\pi} = 50 \text{ Hz} \\
f_{m1} &> f_{m2} \\
\text{Nyquist rate } f_s &= 2f_{m1} = 2 \times 150 = 300 \text{ Hz} \\
\text{Nyquist width } W &= \frac{1}{f_s} = \frac{1000}{300} = \frac{10}{3} \text{ m.s.}
\end{aligned}$$

6. $\boldsymbol{x(t) = \text{sinc}\, 2000\pi t}$

Let

$$\begin{aligned}
\omega_m &= 2000\pi \\
f_m &= \frac{\omega_m}{2\pi} = \frac{2000\pi}{2\pi} = 1000 \text{ Hz} \\
\text{Nyquist rate } f_s &= 2f_m = 2000 \text{ Hz} \\
\text{Nyquist width } W &= \frac{1}{2000} = 0.5 \text{ m.s.}
\end{aligned}$$

■ Example 10.2

Consider the signal

$$x(t) = \cos 2000\pi t + 10 \sin 10{,}000\pi t + 20 \cos 5000\pi t$$

Determine the (1) Nyquist rate for this signal and (2) If the sampling rate is 5000 samples per sec., then what is the discrete time signal obtained after sampling?

(Anna University, May, 2007)

Solution:

1.

$$\begin{aligned} x_1(t) &= \cos 2000\pi t \\ f_{m1} &= \frac{2000\pi}{2\pi} = 1000 \text{ Hz} \\ x_2(t) &= 10 \sin 10{,}000\pi t \\ f_{m2} &= \frac{10{,}000\pi}{2\pi} = 5000 \text{ Hz} \\ x_3(t) &= 20 \cos 5000\pi t \\ f_{m3} &= \frac{5000\pi}{2\pi} = 2500 \text{ Hz} \\ f_{m2} &> f_{m3} > f_{m1} \end{aligned}$$

Hence, the Nyquist rate

$$f_s = 2f_{m2} = 10{,}000 \text{ Hz}$$

2. The required sampling rate is 5000 samples per second. The maximum frequency of the given signal is 5000 Hz. The sampling rate is not equal to twice the maximum frequency content of the given signal and therefore aliasing occurs. The given signal cannot be recovered. $f_s = 5000$ Hz, $2f_m = 10{,}000$ and $f_s < 2f_m$.

10.8 Sampling of Band-Pass Signals

In the previous sections, we discussed the sampling theorem for low-pass signals. If the signal is band pass, the sampling theorem is stated as follows.

The band-pass signal $x(t)$ whose maximum bandwidth is $2W$ can be completely represented and recovered from the sample if it is sampled at a minimum rate of twice the bandwidth.

Fig. 10.7 Sampling of band-pass signal

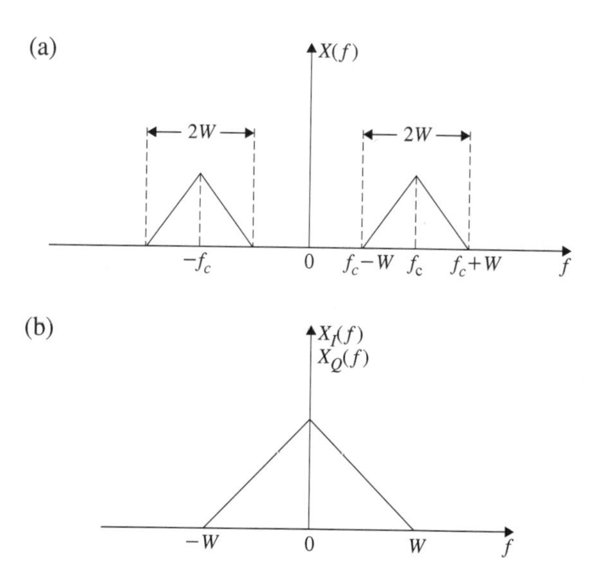

Let $x(t)$ be the band-pass signal whose bandwidth is $2W$ as shown in Fig. 10.7a. The signal can be represented as impulse and quadrature components $x_I(t)$ and $x_Q(t)$. Thus, $x(t)$ which is centered around f_s can be represented as

$$x(t) = x_I(t)\cos 2\pi f_c t - x_Q \sin 2\pi f_c t \tag{10.12}$$

These two components are low pass in nature. Their spectrum is shown in Fig. 10.7b. The in phase and quadrature components are sampled at $f_s = 4W$ rate. The sampled $x_I(nT_s)$ and $x_Q(nT_s)$ are passed through their respective reconstruction filters and $x_I(t)$ and $x_Q(t)$ are reconstructed. The original $x(t)$ is obtained by multiplying $x_I(t)$ by $\cos 2\pi f_c t$ and $x_Q(t)$ by $\sin 2\pi f_c t$. $x(t)$ is obtained using the following relationship:

$$x(t) = x_I(t)\cos 2\pi f_c t + x_Q \sin 2\pi f_e t.$$

Summary

1. The sampling theorem states that a continuous time band-limited signal $x(t)$ can be sampled and reconstructed iff the sampling frequency is greater than twice the maximum frequency of the given signal.
2. If the sampling theorem is not satisfied while sampling $x(t)$, aliasing occurs and it is not possible to reconstruct the original signal. Further, the sampling process creates noise.

3. While the continuous time signal is sampled, if the high frequency of the spectrum interferes with the low frequency spectrum then the phenomenon is called aliasing.
4. The effect of aliasing is the distortion in the recovered signal and also loss of data.
5. To avoid sampling, sampling rate is increased such that the sampling rate (Nyquist rate) $f_s > 2f_m$. Aliasing is also avoided by putting an anti-aliasing filter just before the signal $x(t)$ is sampled.
6. Some of the applications of sampling include signal analysis processing and transmission.
7. The band-pass signal $x(t)$ whose maximum bandwidth is $2W$ can be completely represented and recovered if it is sampled at a minimum rate of twice the bandwidth.

Exercise

I. Short Answer Type Questions

1. **What is sampling theorem?**
   If a continuous time signal $x(t)$ is to be sampled and recovered, then the sampling frequency should be greater than twice the maximum frequency content of the signal. This is called sampling theorem.
2. **What is Nyquist rate and Nyquist interval?**
   The sampling frequency which is greater than twice the maximum frequency content of the signal to be sampled is called the Nyquist rate. The reciprocal of the Nyquist rate is called Nyquist width.
3. **What is aliasing?**
   When the continuous time signal is sampled and if it does not satisfy the sampling theorem, then the high-frequency spectrum of the sampled signal interferes and appears as low-frequency spectrum. This phenomenon is called aliasing.
4. **What are the effects of aliasing?**
   The effects of aliasing are as follows:
   1. Distortion in signal recovery is generated.
   2. The data is lost and the reconstruction of original signal becomes impossible.
5. **What are the methods available to avoid aliasing?**
   Aliasing can be avoided or minimized by increasing the sampling rate which satisfies the sampling theorem. Aliasing can also be minimized by putting anti-aliasing filters just before the signal is sampled.

6. **What is zero-order hold? What is its T.F.?**
Band-limited signals which are to be sampled are passed through the zero-order hold. The ZOH samples the continuous time signal $x(t)$ at a given time and holds that value until the next instant. The T.F. of a ZOH is

$$H(s) = \frac{1}{s}\left[1 - e^{-sT}\right]$$

7. **What are the applications of the sampling theorem?**
The sampling theorem is used in the analysis, processing and transmission of signals.
8. **State the sampling theorem as applied to band-pass signals.**
The band-pass signal $x(t)$ whose maximum bandwidth is $2W$ can be completely represented and recovered from the sample if it is sampled at a minimum rate of twice the bandwidth. This is the sampling theorem as applied to band-pass signal.
9. **Find the Nyquist rate and Nyquist width for the signal given below.**

1. $\quad x(t) = 10\cos 2000\pi t + \sin 3000\pi t + 5\cos 1500\pi t$

2. $\quad x(t) = (10\cos 100\pi t)^2$

3. $\quad x(t) = (10\text{sinc } 2000\pi t)^2$

4. $\quad x(t) = \cos 1000\pi t \cos 2000\pi t$

1. Nyquist rate $f_s = 3000$ Hz.
Nyquist width $W = \frac{1}{3}$ m.s.
2. Nyquist rate $f_s = 200$ Hz.
Nyquist width $W = 5$ m.s.
3. Nyquist rate $f_s = 4000$ Hz.
Nyquist width $W = 0.25$ m.s.
4. Nyquist rate $f_s = 3000$ Hz.
Nyquist width $W = \frac{1}{3}$ m.s.

Appendix
Mathematical Formulae

A.1 Summation Formulae

1. $$\sum_{n=0}^{N-1} \alpha^n = \begin{cases} \frac{1-\alpha^N}{1-\alpha} & \alpha \neq 1 \\ N & \alpha = 1 \end{cases}$$

2. $$\sum_{n=0}^{\infty} \alpha^n = \frac{1}{(1-\alpha)} \qquad |\alpha| < 1$$

3. $$\sum_{n=k}^{\infty} \alpha^n = \frac{\alpha^k}{(1-\alpha)} \qquad |\alpha| < 1$$

4. $$\sum_{n=0}^{\infty} n\alpha^n = \frac{\alpha}{(1-\alpha)^2} \qquad |\alpha| < 1$$

5. $$\sum_{n=0}^{\infty} n^2\alpha^n = \frac{\alpha^2+\alpha}{(1-\alpha)^3} \qquad |\alpha| < 1$$

6. $$\sum_{k=m}^{n} a^k = \frac{a^{n+1}-a^m}{a-1} \qquad a \neq 1$$

7. $$\sum_{k=0}^{n} k = \frac{n(n+1)}{2}$$

8. $$\sum_{k=0}^{n} k^2 = \frac{n(n+1)(2n+1)}{6}$$

9. $$\sum_{k=n_2}^{k=n_1} (1)^k = n_2 - n_1 + 1$$

S. Palani, *Signals and Systems*,
https://doi.org/10.1007/978-3-030-75742-7

A.2 Euler's Formula

1. $e^{\pm j\theta} = \cos\theta \pm j\sin\theta$
2. $\cos\theta = \frac{1}{2}[e^{j\theta} + e^{-j\theta}]$
3. $\sin\theta = \frac{1}{2j}[e^{j\theta} - e^{-j\theta}]$

A.3 Power Series Expansion

1. $e^{\alpha} = \sum_{k=0}^{\alpha} \frac{\alpha^k}{\angle k} = 1 + \alpha + \frac{\alpha^2}{\angle 2} + \frac{\alpha^3}{\angle 3} + \cdots$
2. $(1+\alpha)^n = 1 + n\alpha + \frac{n(n-1)}{\angle 2}\alpha^2 + \cdots + \left(\frac{n}{k}\right)\alpha^k + \cdots$
3. $n(1+\alpha) = \alpha - \frac{1}{2}\alpha^2 + \frac{1}{3}\alpha^3 + \cdots + \frac{(-1)^{k+1}}{k}\alpha^k + \cdots$

A.4 Trigonometric Identities

1. $\sin^2\theta + \cos^2\theta = 1$
2. $\sin^2\theta = \frac{1}{2}(1 - \cos 2\theta)$
3. $\cos^2\theta = \frac{1}{2}(1 + \cos 2\theta)$
4. $\sin 2\theta = 2\sin\theta\cos\theta$
5. $\cos 2\theta = 1 - 2\cos^2\theta$
6. $\cos(a \pm b) = \cos a\cos b \mp \sin a\sin b$
7. $\sin(a \mp b) = \sin a\cos b \pm \cos a\sin b$
8. $\sin a\sin b = \frac{1}{2}[\cos(a-b) - \cos(a+b)]$
9. $\cos a\cos b = \frac{1}{2}[\cos(a-b) + \cos(a+b)]$
10. $\sin a\cos b = \frac{1}{2}[\sin(a-b) + \sin(a+b)]$
11. $\cos a + \cos b = 2\cos\frac{(a+b)}{2}\cos\frac{(a-b)}{2}$

12. $$\sin a + \sin b = 2\sin\frac{(a+b)}{2}\cos\frac{(a-b)}{2}$$

13. $$a\cos\alpha + b\sin\alpha = \sqrt{a^2+b^2}\cos\left(\alpha - \tan^{-1}\frac{b}{a}\right)$$

14. $$\tan(A+B) = \frac{\tan A + \tan B}{1 - \tan A\tan B}$$

15. $$\tan(A-B) = \frac{\tan A - \tan B}{1 + \tan A\tan B}$$

16. $$\tan^{-1} A \pm \tan^{-1} B = \tan^{-1}\frac{A \pm B}{1 \mp AB}$$

A.5 Definite Integrals

1. $$\int_0^\infty \frac{\sin x}{x}dx = \frac{\pi}{2}$$

2. $$\int_0^\infty \left(\frac{\sin x}{x}\right)^2 = \frac{\pi}{2}$$

3. $$\int_0^\infty e^{-ax}\sin bx dx = \frac{b}{(a^2+b^2)}$$

4. $$\int_0^\infty e^{-ax}\cos bx dx = \frac{b}{(a^2+b^2)}$$

A.6 Indefinite Integrals

1. $$\int \frac{dx}{(a+bx)} = \frac{1}{b}\ln[a+bx]$$

2. $$\int x\cos x dx = \cos x + x\sin x$$

3. $$\int e^{ax}\sin x dx = \frac{e^{ax}}{(a^2+1)}(a\sin x - \cos x)$$

4. $$\int x\sin x dx = \sin x - x\cos x$$

5. $$\int x e^{ax} dx = e^{ax}\left[\frac{x}{a} - \frac{1}{a^2}\right]$$

A.7 Derivatives

1. $\int u dv = uv - \int v du$

2. $\frac{d}{dx}(\ln x) = \frac{1}{x}$

3. $\frac{d}{dx}(\log_a x) = \frac{1}{x} \log a$

4. $\frac{d}{dx}(a^x) = a^x \ln a$

References

1. Oppenheim AV, Willsky AS, Young LT (1983) Signals and systems. Prentice Hall, Englewood Cliffs
2. Lindner DK (1999) Introduction to signals and systems. McGraw-Hill International Edition, New York
3. Haykin S, Van Veen B (1998) Signals and systems. Wiley Inc, Hoboken
4. Lathi BP (2005) Linear systems and signals, 2nd edn. Oxford University Press, Oxford
5. Palani S, Kalaiyarasi D (2015) Discrete time systems and digital signal processing, 2nd edn. Ane Books Pvt, Ltd, New Delhi
6. Roberts MJ (2004) Signals and systems. Tata McGraw-Hill Publishing Company Limited, New Delhi
7. Hsu HP (2006) Signals and systems. Schaum's outlines, John McGraw-Hill Company, New Delhi, Reprint
8. Cadzow JA, Landingham HF (1985) Signals and systems. Prentice Hall, Englewood Cliffs
9. Cruz JB, Vankenburg ME (1974) Signals and linear circuits. Houghton Mifflin, Boston
10. Houts RC (1991) Signals analysis in linear systems. Saunders College, New York
11. Jackson LB (1991) Signals and systems and transforms. Addison Wesley, Reading
12. Kamen E (1987) Introduction to signals and systems. Macmillan, New York
13. McGillem CD, Cooper GR (1991) Continuous and discrete signal and system analysis, 3rd edn. Holt, Rinehart and Winston, New York
14. Zeimer RE, Tranter WH, Fannin RD (1989) Signals & systems - continuous and discrete, 2nd edn. Pearson, Macmillan, New York
15. Siebert WM (1986) Signals and systems. MIT Press, Cambridge
16. Taylor FJ (1994) Principles of signals and systems. McGrawHill series in electrical and computer engineering, McGrawHill, New York
17. Soliman S, Srinath M (1990) Continuous and discrete signals and systems. Prentice Hall, New York
18. Bracewell RN (1986) The Fourier transform and its applications, 2nd edn. McGrawHill, New York
19. Doetsch G (1974) Introduction to the theory and applications of the Laplace transformation with a table of Laplace transformations. Springer, New York
20. Jury EI (1982) Theory and applications of the Z-transform method. R.E, Krieger, Malabar
21. Ogata K (1990) Modern control engineering, 2nd edn. Prentice Hall, Englewood Cliffs

S. Palani, *Signals and Systems*,
https://doi.org/10.1007/978-3-030-75742-7

Index

S. Palani, *Signals and Systems*,
https://doi.org/10.1007/978-3-030-75742-7

N

O

P

R

S

T

U

Z

9783030757441